Lecture Notes in Computer Science 5463

Commenced Publication in 1973
Founding and Former Series Editors:
Gerhard Goos, Juris Hartmanis, and Jan van Leeuwen

Xiaofang Zhou Haruo Yokota Ke Deng
Qing Liu (Eds.)

Database Systems for Advanced Applications

14th International Conference, DASFAA 2009
Brisbane, Australia, April 21-23, 2009
Proceedings

Springer

Volume Editors

Xiaofang Zhou
Ke Deng
The University of Queensland
School of Information Technology and Electrical Engineering
Brisbane QLD 4072, Australia
E-mail: {zxf, dengke}@itee.uq.edu.au

Haruo Yokota
Tokyo Institute of Technology
Graduate School of Information Science and Engineering
2-12-1 Oh-Okayama, Meguro-ku, Tokyo 152-8552, Japan
E-mail: yokota@cs.titech.ac.jp

Qing Liu
CSIRO
Castray Esplanade, Hobart, TAS 7000, Australia
E-mail: q.liu@csiro.au

Library of Congress Control Number: Applied for

CR Subject Classification (1998): H.2, H.3, H.4, H.5, J.1

LNCS Sublibrary: SL 3 – Information Systems and Application, incl. Internet/Web and HCI

ISSN 0302-9743
ISBN-10 3-642-00886-0 Springer Berlin Heidelberg New York
ISBN-13 978-3-642-00886-3 Springer Berlin Heidelberg New York

springer.com

Printed in Germany

Typesetting: Camera-ready by author, data conversion by Scientific Publishing Services, Chennai, India
Printed on acid-free paper SPIN: 12645305 06/3180 5 4 3 2 1 0

Message from the DASFAA 2009 Chairs

It is our great pleasure to present this volume of proceedings containing the papers selected for presentation at the 14th International Conference on Database Systems for Advanced Applications (DASFAA 2009), which was held in Brisbane, Australia during April 21–23, 2009.

DASFAA 2009 received 186 submissions. After a thorough review process for each submission by the Program Committee and specialists recommended by Program Committee members, DASFAA 2009 accepted 39 full papers and 22 short papers (the acceptance rates were 21% and 12%, respectively). This volume also includes three invited keynote papers, presented by leading experts in database research and advanced applications at DASFAA 2009: David Lomet (Microsoft Research), Elisa Bertino (Purdue University), Robert Vertessy (Australian Bureau of Meteorology). Other papers in this volume include nine demo papers organized by Marek Kowalkiewicz (SAP Research) and Wei Wang (University of New South Wales), three tutorial abstracts organized by Yanchun Zhang (Victoria University), Alan Fekete (University of Sydney) and Xiaoyong Du (Renmin University of China), and one panel abstract organized by Athman Bouguettaya (CSIRO, Australia) and Jeffrey Yu (Chinese University of Hong Kong).

Six workshops were selected by the Workshop Co-chairs, Lei Chen (Hong Kong University of Science and Technology) and Chengfei Liu (Swinburne University of Technology), and were held in conjunction with DASFAA 2009. They are the Second International Workshop on Managing Data Quality in Collaborative Information Systems (MCIS 2009), First International Workshop on Benchmarking of XML and Semantic Web Applications (BenchmarX 2009), First International Workshop on Data and Process Provenance (WDPP 2009), First International Workshop on Mobile Business Collaboration (MBC 2009), First International Workshop on Privacy-Preserving Data Analysis (PPDA) and the DASFAA 2009 PhD Workshop. The workshop papers are included in a separate volume of proceedings also published by Springer in its *Lecture Notes in Computer Science* series.

The conference received generous financial support from The University of Melbourne, The University of New South Wales, The University of Sydney, The University of Queensland, National ICT Australia (NICTA), the Australian Research Council (ARC) the Research Network in Enterprise Information Infrastructure (EII), the ARC Research Network on Intelligent Sensors, Sensor Networks and Information Processing (ISSNIP), and the ARC Research Network for a Secure Australia. We, the conference organizers, also received extensive help and logistic support from the DASFAA Steering Committee, The University of Queensland, Tokyo Institute of Technology, and the Conference Management Toolkit Support Team at Microsoft.

We are grateful to Shazia Sadiq, Ke Deng, Qing Liu, Gabriel Pui Cheong Fung, Kathleen Williamson, James Bailey, and many other people for their great effort in supporting the conference organization. Special thanks also go to the DASFAA 2009 Regional Chairs: Xiaofeng Meng (Asia), John Roddick (Oceania), Torben Pedersen (Europe), and Jun Yang (America), and Best Paper Award Committee Co-chairs Katsumi Tanaka (University of Kyoto), Kyu-Young Whang (KAIST), and Lizhu Zhou (Tsinghua University).

Finally, we would like to take this opportunity to thank all Program Committee members and external reviewers for their expertise and help in evaluating papers, and to thank all authors who submitted their papers to this conference.

February 2009

Ramamohanarao Kotagiri
Xuemin Lin
Xiaofang Zhou
Haruo Yokota

Organization

General Co-chairs

Ramamohanarao Kotagiri	University of Melbourne, Australia
Xuemin Lin	University of New South Wales, Australia

Program Committee Co-chairs

Xiaofang Zhou	University of Queensland, Australia
Haruo Yokota	Tokyo Institute of Technology, Japan

Tutorial Co-chairs

Yanchun Zhang	Victoria University, Australia
Alan Fekete	University of Sydney, Australia
Xiaoyong Du	Renmin University, China

Workshop Co-chairs

Lei Chen	Hong Kong University of Science and Technology, China
Chengfei Liu	Swinburne University of Technology, Australia

Demo Co-chairs

Wei Wang	University of New South Wales, Australia
Marek Kowalkiewicz	SAP Research, Australia

Panel Co-chairs

Jeffrey Xu Yu	Chinese University of Hong Kong, China
Athman Bouguettaya	CSIRO, Australia

Organizing Committee Chair

Shazia Sadiq	University of Queensland, Australia

Publicity Chair

James Bailey	University of Melbourne, Australia

Publication Co-chairs

Qing Liu	CSIRO, Australia
Ke Deng	University of Queensland, Australia

Web Chair

Gabriel Pui Cheong Fung	University of Queensland, Australia

Regional Chairs

Asia: Xiaofeng Meng	Renmin University, China
Oceania: John Roddick	Flinders University, Australia
Europe: Torben Bach Pedersen	Aalborg University, Denmark
America: Jun Yang	Duke University, USA

Best Paper Committee Co-chairs

Katsumi Tanaka	Kyoto University, Japan
Kyu-Young Whang	Korea Advanced Institute of Science and Technology, Korea
Lizhu Zhou	Tsinghua University, China

DASFAA Awards Committee

Tok Wang Ling (Chair)	National University of Singapore
Jianzhong Li	Harbin Institute of Technology, China
Krithi Ramamritham	IIT Bombay, India
Katsumi Tanaka	Kyoto University, Japan
Kyu-Young Whang	Korea Advanced Institute of Science and Technology, Korea
Jeffrey Xu Yu	Chinese University of Hong Kong, China
Xiaofang Zhou	University of Queensland, Australia

Steering Committee

Kyu-Young Whang (Chair)	Korea Advanced Institute of Science and Technology, Korea
Katsumi Tanaka (Vice Chair)	Kyoto University, Japan
Yoshifumi Masunaga (Advisor)	Ochanomizu University, Japan
Yoshihiko Imai (Treasurer)	Matsushita Electric Industrial Co., Ltd., Japan
Kian Lee Tan (Secretary)	National University of Singapore, Singapore
Ramamohanarao Kotagiri	University of Melbourne, Australia
Yoon Joon Lee	Korea Advanced Institute of Science and Technology, Korea
Qing Li	City University of Hong Kong, China (Hong Kong)
Krithi Ramamritham	Indian Institute of Technology at Bombay, India
Ming-Syan Chen	National Taiwan University, Taiwan
Eui Kyeong Hong	University of Seoul, Korea
Hiroyuki Kitagawa	University of Tsukuba, Japan
Lizhu Zhou	Tsinghua University, China
Jianzhong Li	Harbin Institute of Technology, China

Program Committee

Toshiyuki Amagasa	University of Tsukuba, Japan
Zhou Aoying	East China Normal University, China
Boualem Benatallah	University of New South Wales, Australia
Elisa Bertino	Purdue University, USA
Sourav Bhowmick	Nanyang Technological University, Singapore
Cui Bin	Peking University, China
Sharma Chakravarthy	The University of Texas at Arlington, USA
Yi Chen	Arizona State University, USA
Phoebe Chen	Deakin University, Australia
Zhiming Ding	Chinese Academy of Sciences, China
Gill Dobbie	University of Auckland, New Zealand
Ada Fu	Chinese University of Hong Kong, China
Yu Ge	Northeast University, China
Lukasz Golab	AT&T, USA
Stephane Grumbach	INRIA, France
Wook-Shin Han	Kyung-Pook National University, Korea
Takahiro Hara	Osaka University, Japan

Seung-Won Hwang	POSTECH, Korea
Keun Ho Ryu	Chungbuk National University, Korea
Yoshiharu Ishikawa	Nagoya University, Japan
Mizuho Iwaihara	Kyoto University, Japan
Markus Kirchberg	Institute for Infocomm Research, Singapore
Hiroyuki Kitagawa	University of Tsukuba, Japan
Masaru Kitsuregawa	University of Tokyo, Japan
Sang-Won Lee	Sungkyunkwan University, Korea
Xue Li	University of Queensland, Australia
Qing Li	City University of Hong Kong, China
Cuiping Li	Renmin University of China, China
Guoliang Li	Tsinghua University, China
Jianzhong Li	Harbin Institute of Technology, China
Ling Liu	Georgia Institute of Technology, USA
Chengfei Liu	Swinburne University of Technology, Australia
Li Ma	IBM Research China
Nikos Mamoulis	The University of Hong Kong, China
Jun Miyazaki	Advanced Institute of Science and Technology, Japan
Mukesh Mohania	IBM, India
Atsuyuki Morishima	University of Tsukuba, Japan
Chaoyi Pang	CSIRO, Australia
Jian Pei	Simon Fraser University, Canada
Zhiyong Peng	Wuhan University, China
Wen-Chih Peng	National Chiao Tung University, Taiwan
Krithi Ramamritham	IIT Bombay, India
Uwe Roehm	University of Sydney, Australia
Wasim Sadiq	SAP Research Australia, Australia
Yasushi Sakurai	NTT, Japan
Simonas Šaltenis	Aalborg University, Denmark
Monica Scannapieco	Università degli Studi di Roma, Italy
Jialie Shen	Singapore Management University, Singapore
Timothy Shih	Tamkang University, Taiwan
Divesh Srivastava	AT&T, USA
Jaideep Srivastava	University of Minnesota, USA
Keishi Tajima	Kyoto University, Japan
Kian-Lee Tan	National University of Singapore, Singapore
Egemen Tanin	University of Melbourne, Australia
Heng Tao Shen	University of Queensland, Australia
Anthony Tung	National University of Singapore, Singapore
Agnès Voisard	Fraunhofer ISST and FU Berlin, Germany
Wei Wang	University of North Carolina Chapel Hill, USA
Haixun Wang	IBM, USA
Guoren Wang	Northeast University, China
Jianyong Wang	Tsinghua University, China

Tok Wang Ling	National University of Singapore
Jeffrey Xu Yu	Chinese University of Hong Kong, China
Ke Yi	Hong Kong University of Science and Technology, China
Chee Yong Chan	National University of Singapore, Singapore
Masatoshi Yoshikawa	Kyoto University, Japan
Rui Zhang	University of Melbourne, Australia
Ying Zhang	University of New South Wales, Australia
Lizhu Zhou	Tsinghua University, China

External Reviewers

Mohammed Eunus Ali
Toshiyuki Amagasa
Parvin A. Birjandi
Yasuhito Asano
Ming Gao
Ke Geng
Ming Hua
Gang Li
Dustin Jiang
Bin Jiang
Ning Jin
Akimitsu Kanzaki
Junpei Kawamoto
Hideyuki Kawashima
Woralak Kongdenfha
Mei Ma
Roochi Mishra
Muhammad Asif Naeem
Miyuki Nakano
Wee Siong Ng
Qun Ni
Bo Ning
Shinagawa Norihide
Sarana Nutanong
Shingo Otsuka
Weixiong Rao
Jackie Rong
Martin Stradling
Kun Suo
Aditya Telang
Muhammad Umer
Raymond C. Wong
Hairuo Xie
Lin Hao Xu
Fei Yang
Junjie Yao
Tomoki Yoshihisa
Qi Zhang
Bin Zhou
Pu Zhou
Rui Zhou
Gaoping Zhu
Ke Zhu

Sponsoring Institutions (in alphabetical order)

Australian Research Council (ARC) Research Network in Enterprise Information Infrastructure (EII)
http://www.eii.edu.au/

Australian Research Council (ARC) Research Network on Intelligent Sensors, Sensor Networks and Information Processing (ISSNIP) http://www.issnip.org/

Australia's Information and Communications Technology Centre of Excellence (NICTA) http://nicta.com.au/

The Research Network for a Secure Australia (RNSA)
http://www.secureaustralia.org/

The University of Melbourne, Australia
The University of New South Wales, Australia
The University of Sydney, Australia
The University of Queensland, Australia

Table of Contents

Keynote Papers

Uncertain Data and Ranking

Sensor Networks

Graphs

RFID and Data Streams

Skyline and Rising Stars

Parallel and Distributed Processing

Mining and Analysis

XML Query

Privacy

XML Keyword Search and Ranking

Web and Web Services

XML Data Processing

Multimedia

Advanced Applications

Demonstrations

Tutorials

Panel

Dependability, Abstraction, and Programming

David Lomet

Microsoft Research, Redmond, WA
lomet@microsoft.com

Abstract. In this paper, we look at what is required to produce programs that are dependable. Dependability requires more than just high availability. Rather, a program needs to be "right" as well, solving the problem for which it was designed. This requires a program development infrastructure that can, by means of appropriate abstractions, permit the programmer to focus on his problem, and not be distracted by "systems issues" that arise when high availability is required. We discuss the attributes of good abstractions. We then illustrate this in the programming of dependable systems. Our "abstraction" is a *transparently persistent stateful programming model* for use in the web enterprise setting where exactly-once execution is required. Work on this abstraction is reviewed. The new technical meat of the paper is in (1) describing how to reduce the performance cost of using the abstraction; (2) extending the flexibility of using this abstraction; (3) and showing how to exploit it to achieve dependability.

Keywords: dependability, abstraction, application persistence, availability, scalability, programming model, enterprise applications.

1 Introduction

Systems need to be dependable. But what exactly does that mean? Brian Randell and co-authors [1] defines dependability as *the system property that integrates such attributes as availability, safety, security, survivability, maintainability.* This is more than just high availability (the fraction of the time that the system is operational), with which it is frequently confused. Rather, it encompasses systems that do what they are supposed to (correctness), can be modified to keep them up to date (maintenance), cover all the expected cases, handle exceptions as well as normal cases, protecting data appropriately, etc. It is no accident that dependable systems have been in short supply as achieving them is very difficult.

Tony Hoare argues [14] that *the price of reliability is utter simplicity- and this is a price that major software manufacturers find too high to afford.* Perhaps my view is colored by being employed by a software vendor, but I do not believe the difficulty stems from want of trying. Rather, it is extraordinarily difficult to make systems simple, without drastic reductions in functionality and the very aspects desired, such as availability. The key point is that unless it is easy, natural, and simple, programming for dependability may well compromise it.

This difficulty has led to a very unsatisfactory state. Jim Gray [12] characterized the situation for availability as: (1) *everyone has a serious problem*; (2) *the BEST people publish their stats*; (3) *the others HIDE their stats.* This is not unexpected. It

X. Zhou et al. (Eds.): DASFAA 2009, LNCS 5463, pp. 1–21, 2009.

represents a very basic human reluctance to own up to difficulties, especially when doing so makes companies look worse than their competitors.

This is not simply an abstract problem. It has business impact. Dave Patterson [21] has characterized this as *Service outages are frequent* with *65% of IT managers reporting that their websites were unavailable over a 6-month period.* Further, *outage costs are high with social effects like negative press and loss of customers who click over to a competitor.*

To summarize, dependability is essential and we do not currently achieve it consistently. We all too frequently deploy undependable systems, which result in real and on-going economic costs.

The rest of this paper focuses on abstraction as the key to realizing dependable programs. We first discuss some existing techniques employed to make systems scalable (increasing the hardware supporting the application permits the application to serve a corresponding increase in system load, e.g. number of users served) and highly available (no prolonged down times, high probability of being operational), important aspects of dependability. We argue that these techniques compromise other aspects of dependability, specifically correctness, maintainability, and simplicity. We next discuss the characteristics of good abstractions, and illustrate this with existing examples that have survived the test of time.

The technical meat of this paper is a "new" abstraction, *transparently persistent stateful programming,* which we argue will enable dependable applications to be written more easily. We elaborate on this abstraction, showing how it can be made to work and provide availability and scalability. We then show that unexpected capabilities can be added to this abstraction that increase deployment flexibility and performance, hence increasing its practical utility.

2 The Current Situation

The classical tool in our bag of dependability technology, in the context of enterprise programming where exactly-once execution is required, is transactions. All early enterprise systems used transactions, usually within the context of a TP monitor. So we start by taking a look at how transactions have been used and the implications and problems of using them.

2.1 Using Transactions

Database systems use transactions very successfully to enhance their dependability. TP monitors also use transactions to provide high availability for applications. However, transaction use with applications has some implications for the application and for the context in which the application operates.

System context: TP monitors [8, 13] were developed in the context of "closed systems", where an organization controlled the entire application from user terminal to mainframe computer. This provided a high level of trust among the elements of the system. In particular, it permitted an application with transactional parts to participate in a transaction hosted by a database or transactional queue. Such transactions were distributed, requiring two phase commit, but because of the level of trust, this was acceptable.

Programming model: The key to using transactions for application dependability is to have meaningful application execution state exist ONLY within a transaction. When the transaction commits, the essential state is extracted and committed to a database or queue [7]. This approach is also used in workflows [24]. An application is broken into a series of steps where each step executes a transaction that reads input "state" from a transactional queue, executes business logic, perhaps invoking a backend relational database, and writes output "state" to another transactional queue, followed by the commit of this distributed transaction. These applications have been called "stateless" (now a term that is a bit overloaded) in that no required execution state exists in the application outside of a transactional step.

Web applications: The web setting, because of the trust issue and the uncertain reliability or latency of communication, frequently precludes the use of two phase commit. Thus, as shown in Figure 1, most web application code lives outside any transaction boundary. The key to making such systems available and scalable is again "stateless" applications, the idea being that since there is no essential program execution state that needs to be preserved in the executing application, it can fail without any consequence for availability (though clearly with some performance impact). The term stateless application continues to mean that application execution state outside of transactions is not essential for application availability, but the techniques used for managing what is the state of the application are frequently quite different, involving a mix of posting information to backend databases, including it in cookies that are returned to users, or replicating it elsewhere in the middle tier.

Scalability: Stateless applications can enhance web application scalability as well. Because there is no execution state retained, an application can move to other application servers easily simply by re-directing a request (perhaps with the appropriate cookie) to another application server. This permits new application servers to be easily added to the running system, hence providing scalability.

The above discussion seems to suggest that any debate about the stateless application approach is pretty much over. In the next subsection, we discuss why we do not think that is the case.

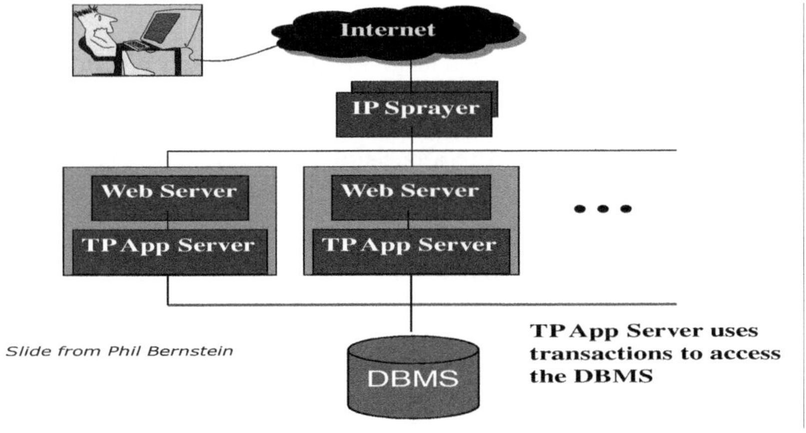

Fig. 1. A Typical e-commerce system deployment

2.2 Problems with Stateless Applications

Stateless applications, and the distributed transactions that are sometimes used to provide durability (and hence availability) are indeed a way to provide highly available applications. But the approach is not without difficulties.

Two phase commit: In a web setting, trust is frequently not present because the level of control necessary for it isn't present. Web host systems frequently have little control over the sea of users running web browsers or other potentially unreliable or even malicious software. Thus they are reluctant to be hostage to the commits of users, or even of other systems that may be involved in an overall business transaction. This has been called the "site autonomy" issue, and it is hard to envision distributed transactions being used with any frequency across organizations.

Latency is also a very big issue in web systems. It is important for performance for transactions to commit promptly. Large latencies introduce corresponding delays in the release of locks, which interferes with concurrency in these systems.

State management: Stateless frameworks force the application programmer to deal explicitly with state. Sometimes this is via cookie management, sometimes via storing information in a transactional queue, sometimes storing state in a database system. The application programmer needs to identify the state that is needed, and manage it explicitly. Then the programmer needs to organize his program into a sequence of steps ("string of beads") where the "state" before and after the execution of a bead is materialized and persisted in some way. Thus, the program needs to be organized to facilitate state management. Because the programmer needs to manage state within his program, the concerns of availability and scalability get mingled with the business logic, making programs more difficult to write and maintain, and sometimes with worse performance as well.

Error handling: With stateless programs, there is no meaningful program execution state outside of the transactions. The assumption is that any execution state that exists can be readily discarded without serious impact on the application. But this poses a problem when handling errors that occur within a transaction that might prevent a transaction from committing.

Since state outside of a transaction might disappear, it is very difficult to guarantee that code will be present that can deal with errors reported by a transaction. In classic transaction processing systems, after a certain number of repeated tries, an error message would eventually be posted to a queue that is serviced MANUALLY. It then becomes people logic, not program logic, which deals with the remaining errors.

3 Abstractions and Applications

Transactions are a great abstraction, but not in the context of reliable, scalable, and highly available applications, as described in section 2.2. We need to take a fresh look at what abstraction to use to make applications robust. To guide us in this process, we first consider the characteristics of good abstractions, so that we have some basis for making judgments about any new proposed abstraction.

3.1 Characteristics of a Good Abstraction

Good abstractions are hard to come by. This can readily be seen by simply observing the small number of abstractions that have made it into wide use—procedures, typed data and objects in programming languages, files, processes, threads in operating systems, and in databases, transactions and the relational model. There are surely other abstractions, but the list is not long. So why are good abstractions so few? It is because there are difficult and sometimes conflicting requirements that good abstractions must satisfy.

Clean and simple semantics: A good abstraction must be simple and easy to understand. But this is surely not sufficient. "Do what I mean" is simple but it is not, of course, implementable.

Good performance and robustness: If performance is bad, then either programmers will not use it at all, or will greatly limit their use of it. If the implementations are not robust as well as performant, it will be impossible to interest enterprise application programmers in using them. A system must perform well and provide reliable service as a basic property.

Good problem match: An abstraction must "match" in some sense the problem for which it is being used. Without this match, the abstraction is irrelevant. The relational model has a tremendous relevance for business data processing, but is surely less relevant for, e.g. managing audio files.

Delegation: When the characteristics above are present, the application programmer can not only exploit the abstraction for his application, but can frequently **delegate** to the system supporting the abstraction some "systems" aspect of his problem. This means that the programmer need no longer dedicate intellectual cycles to this part, permitting him to focus on the application business problem.

Historical examples of abstractions resulting in successful delegation include:

1. **Transactions:** a programmer using transactions delegates concurrency control and recovery to the system. This permits him to construct a program that is to execute within a transaction as if it were the only code executing.
2. **Relational model:** a programmer using the relational model delegates physical database design and query processing/optimization to the system, permitting him to work on a "data independent" conceptual view. This permits him to focus on business relevant issues, not storage/performance relevant issues.

3.2 A New Abstraction for Enterprise Applications

We want to propose a new abstraction to help an application programmer deal with the complex difficulties introduced by the stateless programming model, with its application program requirement to explicitly manage state. Our new abstraction is, in truth, an old abstraction, that has been used by application programmers for years. We call it the stateful programming model, and when applied to enterprise internet applications, it is the *transparently persistent stateful programming* model.

This seems hardly to be an abstraction at all. Application programmers, in simple contexts, naturally tend to write stateful applications. Such programs are easier to write and to understand, as the application programmer can focus on the requirements of the business, rather than the intricacies of managing state so that it will persist even

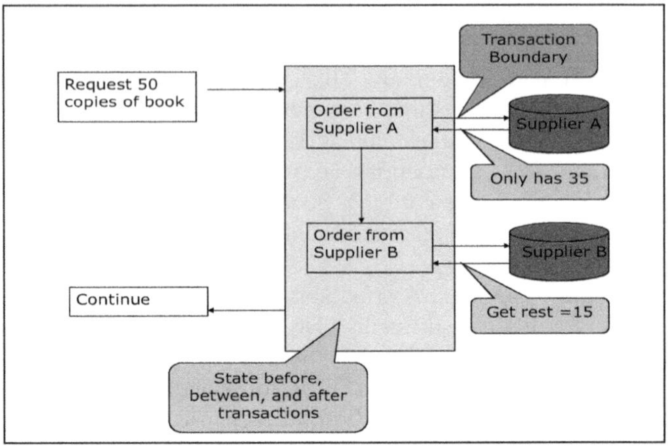

Fig. 2. A stateful application in the web context

if systems crash, messages get lost, etc. Because of this focus on business requirements, such applications are easier to read and maintain, and are more likely to be correct as the program is not cluttered with state management code. This aspect of the program has been delegated to the system infrastructure.

Figure 2 shows a stateful application in a web context. Note that there is important application state between the interactions with suppliers A and B. It is application state that captures the fact that there remain 15 copies of a book that still need to be ordered after supplier A has only been able to commit to providing 35 copies. This information is captured in the *execution state* of the program, without the programmer needing to take any measures to otherwise "manifest" this state. Thus, the application programmer has, in a sense, delegated state management to the system. And this permits the programmer to focus on what the business requires.

3.3 It's Been Tried and Failed

We, of course, want stateful applications to work in a web context, and support mission critical functions. That means such applications must provide "exactly once" execution semantics, i.e. as if, despite a wide range of possible failures, the program has executed without failures and exactly once. We further want our application to be highly available and scalable. To accomplish these things transparently requires that our supporting infrastructure handle state management and state persistence.

The problem for stateful applications, classically, has been that they fail to provide the properties we enumerated above. That is, the application cannot be a made transparently scalable and available with exactly once execution semantics. Consider Figure 2 again. If the system crashes after the application's transaction with Supplier A, the fact that it has arranged for 35 books from Supplier A and needs another 15 books from Supplier B is difficult to reconstruct. The user does not even know yet what the order number used for Supplier A was. So not only is application state lost when a crash happens, which is bad enough, but the state of backend resource managers may now reflect some of the changes induced by committed transactions of the stateful application. Thus, we cannot simply restart the application and let it execute all over again. That strategy works for stateless applications, but not here.

4 Phoenix/App: Transparent Persistence

4.1 How Transparent Stateful Applications Might Work

As indicated in section 2, the stateless application model forces the application programmer to manage state explicitly. Usually the infrastructure supports this by wrapping application steps automatically with a transaction. But transactions are an expensive way to persist state and require explicit state management. So why not use database style recovery for applications: use redo logging with occasional checkpointing of state, all done transparently by the enterprise application support infrastructure. This is an old technology [9, 10], but our Phoenix project [2, 3, 4, 15] applied it to enterprise applications. It was previously applied in the context of scientific applications, not to support exactly once execution, but rather to limit the work lost should systems crash while the application is executing.

The way this works is that the redo log captures the non-deterministic events that a software component is exposed to. To recover the component, it is replayed, and the events on the log are fed to it whenever it reaches a point that in its initial execution experienced a non-deterministic event. Frequently such events are things like receiving a request for service (e.g. an rpc), or perhaps an application read of the state of some other part of the system (e.g. the system clock, etc.).

Unlike scientific applications, with enterprise applications, we need to provide exactly once execution. This requires "pessimistic logging". Logging is pessimistic when there is enough logging (and log forces) so that no non-deterministic event is lost from the log that is needed to provide the exactly once property. Optimistic logging permits later state to be lost, as with scientific applications. Pessimistic logging can require a log force whenever state is revealed to other parts of system (we refer to this as committing state). We focus on improving the performance of pessimistic logging. Many log forces can be eliminated– and our Phoenix project has published some *slick* methods [1].

The result is that Phoenix manages an application's state by capturing its execution state on the redo log. Once the state is successfully captured, (1) it can be used to re-instantiate the application for availability should its system crash; and (2) it can be shipped to other systems for scalability and/or manageability, i.e. the log is shipped to another part of the system. The application programmer need merely focus on writing a correct program to solve his business problem and then to declaratively characterize the attributes he wants of his software components.

4.2 Transparent Robustness

We want to provide robust, dependable applications. If we are to do this, we need a "natural" programming model. Thus, the Phoenix goal has been to provide these enterprise attributes transparently using the stateful programming abstraction. If we can do this, then the application program need not include program logic for robustness. That has been delegated to Phoenix. So how do we do this?

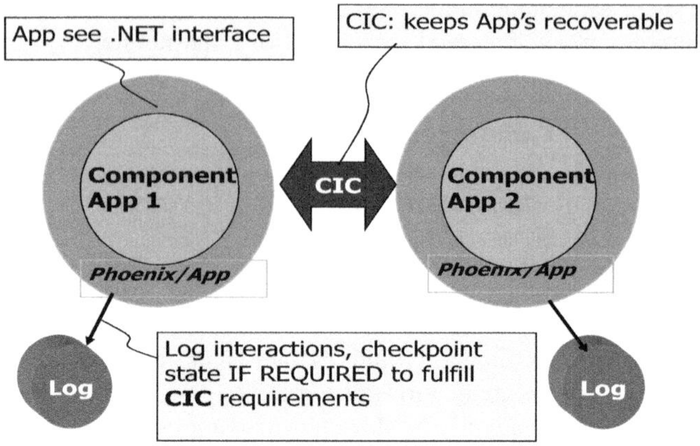

Fig. 3. The Phoenix .NET infrastructure for transparent persistence

Phoenix is implemented using the Microsoft .Net infrastructure [19]. ".Net Remoting" supports the development of component software for web applications. Among its facilities is the notion of a context such that messages (e.g. rpc's) that cross a boundary between two contexts are intercepted by .Net and made available to the Phoenix infrastructure code. This is illustrated in Figure 3. The application program "sees" the standard .Net interfaces for communication between components. When .Net gives Phoenix the intercepted message, Phoenix logs it in order to provide exactly once execution and application availability, scalability, and manageability. These requirements are captured in an interaction contract (CIC) which is shown in Figure 3 and described below.

4.3 Component Types and Interaction Contracts

Phoenix classifies application software components into three types [4] (subsequently, more types were introduced to support various kinds of optimizations). These are:

1. **Persistent** (PCOM): PCOM's represent a part of the application whose persistence, scalability, availability, etc. is being guaranteed by Phoenix. An application executing at an app server, executing business logic and interacting with a backend database system might be represented as a PCOM. A PCOM exploits the transparently persistent stateful programming abstraction provided by Phoenix. Phoenix logs interactions and checkpoints state as required to provide this persistence.
2. **Transactional** (TCOM): Some parts of a web application will execute transactions, e.g. SQL DB sessions. We call the components executing transactions TCOM's. When a PCOM interacts with a TCOM, it must be aware that transactions can either commit or abort. If committed, the post-state of the transaction must be durably remembered, while if aborted, the pre-state of the transaction must be restored.
3. **External** (XCOM): Applications are rarely self-contained. For example, an application frequently is executed in response to end user input or from an input from an autonomous site not controlled by Phoenix. We call these components XCOM's. We need do nothing for the persistence of XCOM's (indeed there is

frequently nothing we can do). But we need to promptly capture via forced logging, the interactions that a PCOM has with an XCOM to minimize the window in which, e.g. a user might need to re-enter data as a result of a system failure.

Figure 4 illustrates the deployment of Phoenix components in support of a web application. We characterize the end user as an XCOM, and a SQL Server database (actually a session with it) as a TCOM, while the mid-tier components execute application logic in a context that ensures its exactly once execution.

Depending upon what component type a PCOM interacts with, Phoenix needs to provide different functionality for exactly once execution. This is captured in what we call interaction contracts [5, 6, 22], of which there is one for each "flavor" of component. Where these interaction contracts are used is also shown also in Figure 4.

1. **Committed interaction contract** (**CIC**): A CIC interaction is between a pair of PCOM's. It must guarantee causality, i.e. that if the receiver PCOM remembers a message, then so must its sender PCOM, regardless of possible failures. We frequently enforce a stronger guarantee, i.e., that the sender guarantees that its state is persistent at the point when it sends a message.
2. **Transactional interaction contract** (**TIC**): A TIC interaction is between a PCOM and a TCOM. The PCOM may or may not remember that it has interacted with a TCOM should the transaction abort. But if the transaction at the TCOM has committed, the PCOM must remember the interactions it had with the TCOM. Transactions permit Phoenix to log less aggressively during the TIC, but require that state persistence be guaranteed at the point of commit.
3. **External interaction contract** (**XIC**): When a PCOM interacts with software not controlled by Phoenix we enforce an external interaction contract that does prompt log forces to reduce the window when failures may be unmasked to the duration of the interaction. Thus, usually failures are not exposed, e.g. to end users. Further, should software external to Phoenix provide messages to PCOM's, or receive messages from PCOM's, prompt logging minimizes the risk of a non-exactly-once execution.

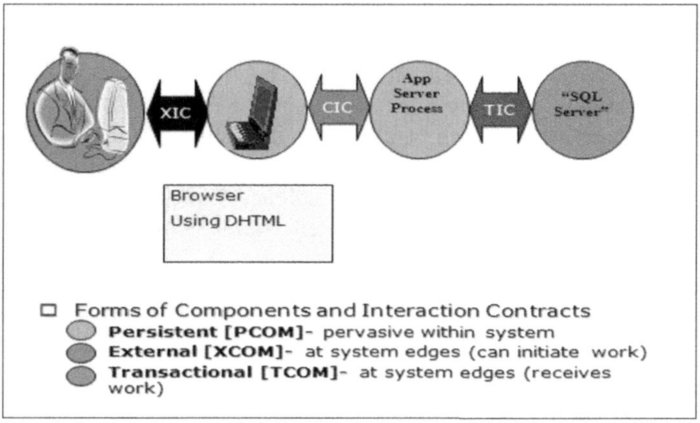

Fig. 4. System schematic of deployed Phoenix components

5 Extending Phoenix

Our Phoenix/App prototype is quite robust. (Indeed, it can be tough to "kill" a Phoenix supported application since the system so promptly brings it back to life.) Nonetheless, there are some limitations that we would like to eliminate or at least ease. In particular, for scalability, we need to move the state of PCOMs from one app server (A) to another (B). This requires that we ship the log of PCOM events from A to B. We might do this regularly by mirrored logging, or periodically by copying the log from place to place. Alternatively, we might do our logging at a backend server where the log might be very widely available to app servers in the middle tier.

While transparent to the application, this incurs costs of the sort that stateless programs incur when managing state. That is, the application must post the state explicitly in a place, e.g. a transactional database, where it can be read by other app servers. It would be nice, for example, if we did not need to have a logging facility in the middle tier. And it would be even nicer if we did not have to always incur the cost of shipping application state (from the log) across the middle tier, especially since the crash frequency of web applications is low. So, *can we have persistent stateful applications without logging?* The answer is "No", but we can remove this requirement from parts of the system, especially parts of the middle tier, while still providing the customary Phoenix robustness attributes, and in particular permitting an application to be re-instantiated anywhere in the middle tier.

5.1 e-Transactions

A PCOM in the middle tier has great flexibility. It can serve multiple client PCOM's, interact with multiple backend TCOMs, read XCOM generated external state, etc. Logging is required at a PCOM to capture the nondeterminism in the order of clients' requests, and to capture the result of reads outside the Phoenix system boundary. The key to "eliminating" logging is to restrict what such PCOMs can do.

There is precursor work called e-transactions [11] that removed the need for logging in the middle tier. A deployment for e-transactions is illustrated in Figure 5. What is it about this deployment that makes middle tier logging unnecessary?

1. All interactions are request/reply and have unique identifiers.
2. Only a single client is served. This means that all mid-tier nondeterminism via servicing client calls can be captured at the client, on the client's log.
3. The single backend server provides idempotence via testable state. Calls from the middle tier to this server can be repeated without compromising exactly once execution.

In this very simple scenario, the client tags its request with a request ID and logs it. The mid-tier, after doing whatever processing is required, forwards a request to the server using the same request ID. If there is a failure anywhere in the system, the request can be replayed based on the client log. The backend server's idempotence ensures that the request is executed at most once. The request is replayed until the client receives a reply to the request. This guarantees that the request is executed at least once. The combination provides exactly once execution.

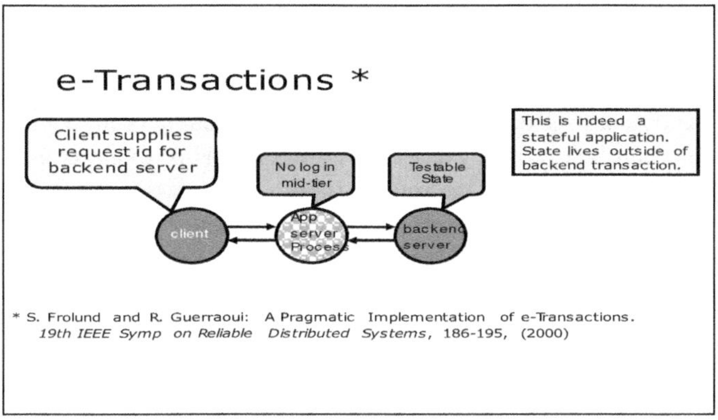

Fig. 5. e-transactions in which the middle tier does no logging, yet a stateful middle tier application component is recoverable

5.2 "Logless" Components

Within the Phoenix framework, we can define a "logless" component type (LLCOM) that can provide application state persistence while doing no logging [16], and can be exploited more flexibly than e-transactions. LLCOM's do not usually require additional logging or log forces from the components with which they interact. Indeed, we can sometimes advantageously reduce some logging. To other components, the LLCOM can usually be treated as if it were a PCOM, though these components may be required to keep messages stable for longer periods.

For LLCOM's to provide persistence without logging, we, as with e-transactions, restrict all interactions to be request/reply. This satisfies a very large class of applications, and indeed is consistent with the way that enterprise applications are typically structured. We need more than this, however, and we describe this below.

Functional Initiation: One source of nondeterminism is how components are named. That nondeterminism must be removed without having to log it. Thus, an LLCOM must have what we call functional initiation, i.e., its creation message fully determines the identity of the LLCOM. This permits a resend of the creation message to produce a component that is logically indistinguishable from earlier incarnations. The initiating component (client) can, in fact, create an LLCOM multiple times such that all instances are logically identical. Indeed, the initiator component might, during replay, create the LLCOM in a different part of the system, e.g. a different application server. The interactions of the LLCOM, regardless of where it is instantiated, are all treated in exactly the same way. During replay, any TCOM or PCOM whose interaction was part of the execution history of the LLCOM responds to the re-instantiated LLCOM in exactly the same way, regardless of where the LLCOM executes.

Determinism for Calls to an LLCOM: A middle tier PCOM's methods might be invoked in a nondeterministic order from an undetermined set of client components. We usually know neither the next method invoked nor the identity of the invoking component. Both these aspects are captured in PCOM's via logging. So, as with e-transactions, we restrict an LLcom to serving only a single client Pcom (i.e., we make it *session-oriented*) and rely on this client Pcom to capture the sequence of method calls.[1] Since our Pcom client has its own log, it can recover itself. Once the client Pcom recovers, it also recovers the sequence of calls to any LLcom with which it interacts, and can, via replay, recover the LLcom. This single client limitation results in an LLcom that can play the role, e.g., of a J2EE session bean [23]. Only now, the "session bean" has persistence across system crashes.

Determinism for Calls from an LLcom: An LLcom's execution must, based on its early execution, identify the next interaction type (send or receive) and with which component. A next message that is a send is identified and re-created via the prior deterministic component replay that leads to the message send. To make receive interactions deterministic requires more. However, a receive message that is a reply to a request is from the recipient of the request message, and the reply is awaited at a deterministic point in the execution. Like e-transactions, what LLcom's need is idempotence from the backend servers for LLcom requests. This is usually required in any case. Replay of the LLcom will retrace the execution path to the first called backend server. That server is required, via idempotence to only execute the request once, and to return the same reply message. That same reply message permits the LLcom to continue deterministic replay to subsequent interactions, etc.

Figure 6 illustrates a deployment of an LLcom in the middle tier. Note that there is only a single client, which is a Pcom, interacting with the LLcom. The client's log will thus enable the calls to the LLcom to be replayed. Further, the only actions permitted of the LLcom at the backend are idempotent requests. Reading of other state is not permitted as it can compromise LLcom deterministic replay. However, unlike e-transactions, our LLcom can interact with multiple backend servers.

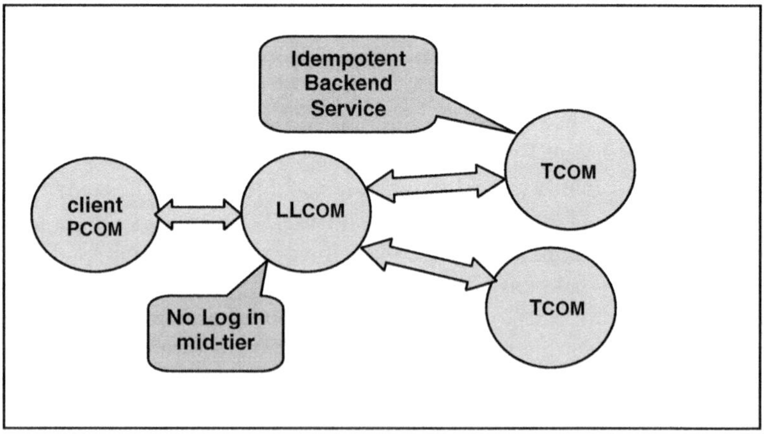

Fig. 6. Logless component in a web deployment

[1] A bit more generality is possible, though it is both harder to define and more difficult to enforce.

6 Middle Tier Reads

Requiring idempotence for servers executing updates has a long history in the context of TP monitors and queued transactions [8, 13, 24]. It is required to provide exactly-once execution in the presence of failures. We do not usually require idempotence for interactions that merely read data without changing it. However, an LLCOM that cannot do such reads is fairly restricted. So, can we permit non-idempotent reads? Superficially, it looks like the answer is "no", but we have identified two situations that we explore next where this is possible [17].

6.1 Exploratory Procedures

Generalized Idempotent Request (GIR): Exploratory reads are reads that precede an invocation of a request to a backend server. These reads permit the middle tier application to specify, e.g., items included in an order, the pickup time for a rental car, the departure time for a subsequent flight, etc. That is, an exploratory read permits an LLCOM to exploit information read outside of idempotent interactions in its dealings with a given backend service. In each case, the read is followed by the sending of an "idempotent" request to a backend service. We need to ensure that the backend server is "idempotent" (note now the quotes) even if exploratory reads are different on replay than during the original execution. Further, we need to prevent exploratory reads from having a further impact, so that the execution path of the LLCOM is returned to the path of its original execution.

"Idempotence" is typically achieved not by remembering an entire request message, but rather by remembering a unique request identifier which is a distinguished argument, perhaps implicit and generated, e.g., by a TP monitor. That technique will surely provide idempotence. Should an identical message arrive at a server, it will detect it as a duplicate and return the correct reply. However, the use of request identifiers to detect duplicate requests enables support for a **generalized idempotence** property. A server supporting generalized idempotent requests (**GIR**'s) permits a resend of a message with a given request identifier to have other arguments in the message that are different. This is exactly what we need for exploratory reads to have an impact on backend requests.

Idempotent requests (IR's) satisfy the property:

$$\mathtt{IR(ID_X,A_X)} = \mathtt{IR(ID_X,A_X)} \circ \mathtt{IR(ID_X,A_X)}$$

where $\mathtt{ID_X}$ denotes the request identifier, $\mathtt{A_X}$ denotes the other arguments of the request, and we use "$\circ$" to denote composition, in this case multiple executions.

Generalized idempotent requests (GIR's), however, satisfy a stronger property:

$$\mathtt{GIR(ID_X,A_1)} = \mathtt{GIR(ID_X,A_X)} \circ \mathtt{GIR(ID_X,A_1)}$$

where $\mathtt{A_1}$ represents the values of the other arguments on the first successful execution of the request. A GIR request ensures that a state change at the backend occurs exactly once, with the first successful execution prevailing for both resulting backend state change and reply.

E-proc: We require every exploratory read to be within an exploratory procedure (**E-proc**). E-proc's always end their execution with a GIR request to the same server, using the same request identifier on all execution paths. Thus it is impossible to exit from the exploratory procedure in any way other than with the execution of a GIR request to the same server using the same request identifier. (More flexibility is possible, e.g. see the next section, but describing and enforcing the required constraints is more difficult.) Since a GIR request can have arguments other than its request ID that differ on replay, and it is guaranteed to return the result of its first successful execution, exploratory reads within the E-proc can determine on first successful E-proc execution, what the GIR request does.

Restricting exploratory reads to E-proc's confines their impact to the arguments for the GIR request. We permit exploratory reads to update local variables, as these have local scope only, but we do not permit non-local variables to be updated from within an E-proc. The GIR influences subsequent LLcom execution via the result that it returns and by how it sets its output parameters. Hence, whenever the E-proc is replayed, its impact on LLcom execution following its return will be dependent only on the first successful execution of its GIR request.

The net effect is that E-proc replay is "faithless", not faithful. But the faithlessness is confined to the internals of the E-proc. At its return, the E-proc provides idempotence. Thus, the LLcom replay is faithful overall, in terms of its effects on backend services and on the LLcom client. An LLcom with an E-Proc for a rental car application is shown in Figure 7. A customer wants a convertible when the deal is good, but otherwise a sedan. The exploratory read permits this choice, with execution guaranteed to be idempotent for the rental request, even when, during replay, the customer's choice within the E-proc is different.

Handling Aborts: Traditional stateless applications make it difficult to handle transaction aborts within an application because there is no *persistent* application code that can respond to the transaction failure. TP monitors usually automatically retry the failed transaction a number of times, which is fine for transient failures, e.g. system

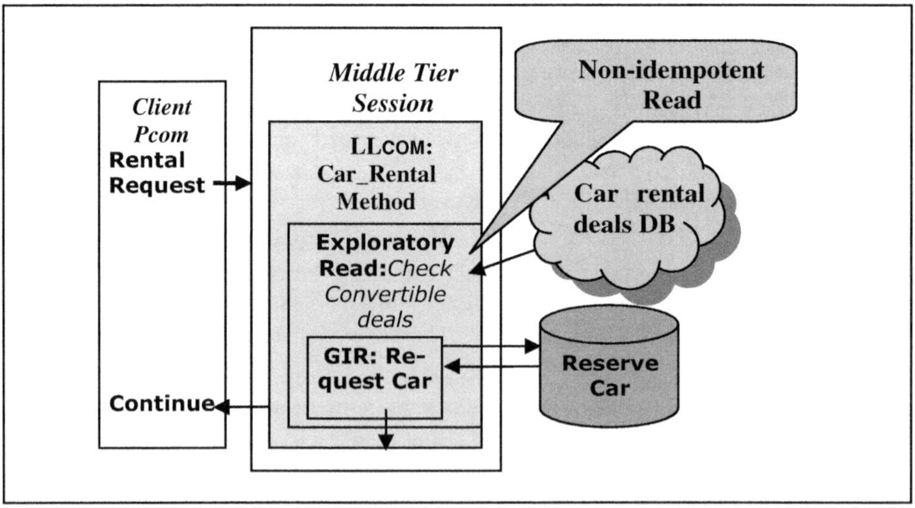

Fig. 7. A GIR request is made to a car rental company, based on the result of an exploratory read about the deals offered by the company for a convertible

crashes, deadlocks, etc.. Hard failures, e.g., application bugs, cannot be dealt with in this manner. For those failures, the TP system usually enqueues an error message on a queue for manual intervention.

A PCOM stateful application can handle transaction failures because it has persistent execution state outside of transactions. It can inspect an error message (outside of a transaction), and provide code to deal with the failure. We want this capability for LLCOMS. However, aborts are not traditionally idempotent. Rather, a backend server "forgets" the aborted transaction, erasing its effects from its state.

Like a read, an abort outcome for a GIR leaves backend state unchanged. By embedding a GIR commit request within an E-proc, we are in a position to respond programmatically to responses reporting that the transaction has aborted. We can then include in the E-proc a loop that repeatedly retries the GIR request until that request commits. Further, we can use additional exploratory reads, and prior error messages, to change the arguments (but not the request identifier) of the GIR request in an effort to commit the transaction. Thus, an E-proc lets us respond programmatically and make multiple attempts to commit the request.

However, we cannot leave the GIR with an abort outcome. The problem here is now the non-idempotence of aborts. Should an LLCOM executing the E-proc abandon its effort to commit the transaction and then fail, upon replay of the E-proc, its GIR request might commit the transaction. Subsequent state of the LLCOM will then diverge from its initial pre-failure execution.

One way out of this "conflicted" situation is to provide a parameter to the GIR that permits the calling LLCOM to request an idempotent abort. Normally, we expect that several executions of the GIR request to be non-idempotent, permitting the request to be retried, perhaps with different values to arguments. But faced with repeated failure, requesting an idempotent abort permits the LLCOM execution to proceed past this request. Once the GIR abort is stable, subsequent logic in the LLCOM could, for example, try other backend services. Idempotent aborts do require the cooperation of backend servers and are not now normally provided. This is unlike GIRs, where the usual implementation of idempotence already relies on request IDs.

6.2 Wrap-up Procedures

Exploiting the Client PCOM Log: We cannot use an E-proc to select which back end server to call. For example, we cannot use an E-proc to explore prices for rental cars and chose the web site of the rental company offering the best rates. To choose a backend server requires this choice to persist across system crashes. So the information from non-idempotent reads that determines our decision needs to be stable, encountered during LLCOM replay, and used to persistently make the same choice. An LLCOM has no log. Further, the "log" at the backend server, which we exploit in E-proc's, requires that we know which server to invoke, which is not very useful here.

With E-proc's, backend logging provides idempotence and E-proc scope forces execution back to its original "trajectory". Here we use client PCOM logging for a similar purpose. This permits added flexibility as logging at the client PCOM does not involve making any additional "persistent" choice (itself needing to be logged) for where the log is since there is only a single client PCOM.

Consider a travel site LLCOM with a read-only method that checks rental car prices and returns that information to our PCOM client. The client PCOM, after interaction with the user, proceeds to make a subsequent call indicating his choice of car and intention to rent. Again, the read-only method is NOT idempotent. Upon replay, the car prices returned might be very different. Despite the lack of idempotence, we can permit read-only LLCOM methods. When the car rates are returned to the client PCOM, the client forces this information to its log prior to initiating the LLCOM call requesting the car. The LLCOM uses the client supplied information to select the rental web site. Replay is deterministic because the client PCOM has made information from the non-idempotent read(s) stable. The client can even avoid repeating the read-only calls to the LLCOM during its replay as it already has the answer it needs on its log. Thus, the first successful read execution "wins", and here we exploit client PCOM logging to guarantee this. Figure 8 shows how client logging, in this case with a read-only method, enables us to choose the rental company at which to make a reservation.

Unlike dealing with an idempotent middle tier component, where log forcing is not required (though it can be useful for efficiency of client recovery), we now need the client PCOM to force the log prior to revealing state via a subsequent interaction since the read-only method execution results in a non-deterministic event that is not captured in any other way. Thus, we need to identify LLCOM methods that need forced logging.

WU-Proc: We can exploit client logging more generally, however we must preclude the LLCOM from making state changes that depend upon unlogged non-idempotent reads. The requirement is that activity preceding wrap-up reads be idempotent. Hence, the wrap-up reads may be preceded by invocations of E-proc's within a **wrap-up procedure** (**WU-proc**). So long as this wrap-up reading of external state does not result in updates to LLCOM state outside of the WU-proc activation, it will have no

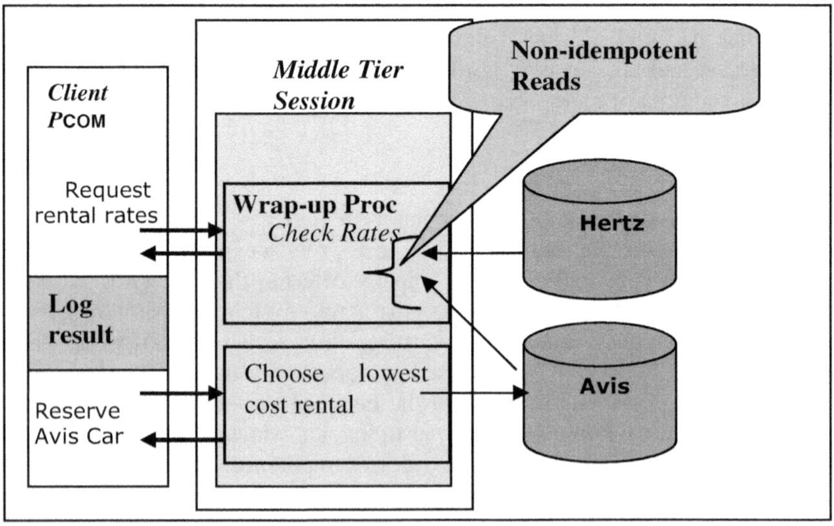

Fig. 8. A request is made to a car rental company, based on the cheapest rates we have read. Logging for this choice is done at the client.

further impact, except via subsequent client requests—and hence after the client has logged the LLcom reply. Thus, with WU-proc's, the client must be aware that it is calling a WU-proc and force log the WU-proc reply. Subsequent client requests become replayable because of the logging at the client Pcom.

This is reminiscent of middle tier components returning a cookie to the client. The client then returns the cookie to the web service so that the web service can restore the appropriate middle tier state for the client. However, here there is more flexibility. The client need not return the reply that it was sent (the "cookie"). Rather, it can do further local execution, perhaps interacting with an end user, before deciding what to do. For example, an end user might see the results of middle tier comparison shopping and make a choice that determines the web site the middle tier visits to consummate a purchase. Logging of the rate information at the client comes before we choose "Hertz" or "Avis". Having logged, we can now select the car rental web site at which to place our reservation. And, of course, the client log needs to be forced before the "reserve" call is made to guarantee deterministic replay.

7 Achieving Availability and Scalability

In this section we provide some perspective on how one might put together an infrastructure that supports multi-tier stateful web applications to achieve dependability using the concepts we have described. This involves not just our logless components, but the surrounding system elements that enable them. We start with these surrounding components and then discuss how to exploit logless middle tier components to provide scalability and availability.

7.1 Enabling Logless Components

The Client Pcom: We require that clients support persistent components (Pcom's) in order for logless middle tier components be enabled. So how is this functionality provided? We could require that this client software be installed on every machine that intends to play the role of client for our enterprise applications. However, that is unnecessary. The EOS system [5] demonstrates a technique for "provisioning" a web browser with Pcom client functionality, while not requiring any modification to the web browser, in this case Internet Explorer (IE). One downloads DHTML from the middle tier to IE that permits IE to act as a Pcom in our multi-level enterprise application. For our purposes, this DHTML needs to include the functional create call that instantiates the logless middle tier session oriented component, the client logging of user input, and the logging of results returned from the middle tier. The DHTML also specifies how to recover a crashed middle tier LLcom. Thus, an enterprise infrastructure in the middle tier can exercise control over the client system by providing the client functionality that it desires or requires. This includes the kind of forced logging that is required for wrap-up reads.

The Backend Server: In order to provide persistence in the presence of system failures, idempotence is required for component interactions. Logging enables replay of the interactions to be avoided most of the time. But it does not cover the case where the failure occurs during the interaction. When application replay once again reaches the interaction, it is uncertain whether the effect intended on the other component (in our case, the backend server) happened or not. Idempotence permits the recovering application to safely resubmit its request. Idempotence is required for the fault tolerance techniques of which we are aware. In TP systems, it is often a transactional queue that captures the request and/or reply to enable idempotence.

In many, if not most, services that provide idempotence, detection of duplicate requests relies on a unique identifier for the request message. The remaining arguments of the message are usually ignored, except in the actual execution of the request. When the request is completed, a reply message is generated, again identified and tied to the incoming request by means of the request identifier only. This style of implementation for idempotence already supports our generalized idempotent requests, though here we require the backend service to provide idempotence in exactly this way when checking for duplicates.

Thus there is frequently nothing new that a backend web service need do to become a GIR request server. The hard part has already been done. Thus, the web service interaction contract that we proposed in [15] can be readily transformed into a generalized form, usually without having to change the implementation. This contract is a unilateral contract by the web service, requiring nothing from the application. However, in its generalized form, it can support exploratory reads by logless persistent session-oriented components.

7.2 Exploiting Logless Components

Scalability: Because there is no log for an LLcom, we can re-instantiate an LLcom at any middle tier site simply by directing its functional initiation call to the site. Should an LLcom take too long to respond to a client request, the client can choose to treat the LLcom and the server upon which it runs as "down". At this point, the client will want to redeploy the component elsewhere. Note, of course, that the original middle tier server might not have crashed, but simply be temporarily slow in responding.

One great advantage of LLcom's is that whether a prior instance has crashed or not, one can initiate a second instance of it safely. This avoids a difficult problem in many other settings, i.e. ensuring that only a single instance is active at a time. The client simply recovers the LLcom at a second server site by replaying its calls from its local log. The new instance repeats the execution of the first instance until it has advanced past the spot where the original instance either crashed or was seriously slowed. From that point on, the secondary instance truly becomes the primary.

So what happens if the original LLcom is, in fact, not dead, and it resumes execution? In this case, it will eventually return a result to the client Pcom. The client Pcom should always be in a position to terminate one or the other of the multiple LLcom instances. An LLcom might support a self-destruct method call that puts it out of its misery. In this case, the client Pcom might invoke that method. Alternatively, the middle tier infrastructure might simply time-out an idle LLcom at some point, which requires nothing from the client.

This failover approach is much simpler than is typically required for stateful system elements. It avoids shipping state information; and it avoids any requirement to ensure that a primary is down before a secondary can take over.

Availability: To a large extent, the persistent component at the client can choose the level of robustness for the middle-tier application. Replication is used for high availability in CORBA [18, 20]. Because there is no requirement to "kill" a primary before a secondary takes over, it is possible to use LLCOM's in a number of replication scenarios. Clients can issue multiple initiation calls, thus replicating a mid-tier LLCOM. By controlling when and how it manages LLCOM instances, a client can achieve varying degrees of availability at varying costs.

1. It may have no standby for its mid-tier session, but rather recover the existing session in place or create and failover to another instantiation should a failure occur.
2. It might create a "warm standby" at another site that has not received any calls following its initiation call. The client would be expected to replay the operational calls, but recovery time would be shortened by avoiding standby creation during recovery.
3. It might create a standby that it keeps current by continuously feeding it the same calls as the primary. Indeed, it can let the two mid-tier logless components race to be the first to execute. If one of them fails, the other can simply continue seamlessly in its execution.

Replicas do not directly communicate with each other, and there is no explicit passing of state between them. Rather, they run independently, except for their interactions with the client PCOM and their visits to the same collection of back end servers—where the first one to execute a service call determines how subsequent execution proceeds for all replicas. No replica needs to know about the existence of any other replica, and indeed, the middle tier application servers need not know anything about replication. They simply host the logless session components.

8 Summary

It has long been recognized that dependability is an essential attribute for enterprise applications. Software developers have known that database transactions are not enough, by themselves, to provide the level of guarantees that these applications require. Techniques for providing persistence in the presence of failures have been around for a long time, as have efforts to improve their performance.

We have argued that *transparently persistent stateful programming* can be highly effective in providing dependability by permitting an application programmer to delegate to application infrastructure the tasks of scalability and availability. This delegation is essential as it is what enables the programmer to focus almost exclusively on the business logic of the application. And it is this focus that results in applications that are simpler, more maintainable, and more likely to be correct.

Our Phoenix/App prototype at Microsoft Research demonstrated the feasibility of the general approach, which uses automatic redo logging to capture the non-deterministic events that permit applications to be replayed to restore their state. This technology was subsequently extended to enable persistent session-oriented middle tier components that themselves do not need to log, greatly simplifying scalability and availability.

In this paper, we have shown how to extend these persistent session-oriented components to enable them to read and respond to system state without requiring that such reads be idempotent. It requires a stylized form of programming, frequently called a "programming model" that includes E-proc's and WU-proc's. However, the resulting model is only mildly restrictive and the payoff is substantial.

The result is application state persistence with exactly once execution semantics despite system crashes that can occur at arbitrary times, including when execution is active within the component or when the component is awaiting a reply from a request. Because no middle tier log is required (i) performance for normal execution of middle tier applications is excellent and (ii) components can be deployed and redeployed trivially to provide availability and scalability. Aside from the stylistic requirements of exploratory and wrap-up reads, persistence is transparent in that the application programmer need not know about the logging being done elsewhere in the system to provide middle tier persistence.

References

1. Avižienis, A., Laprie, J.-C., Randell, B.: Fundamental Concepts of Computer System Dependability. In: IARP/IEEE-RAS Workshop, Seoul (2001)
2. Barga, R., Chen, S., Lomet, D.: Improving Logging and Recovery Performance in Phoenix/App. In: ICDE Conference, Boston, pp. 486–497 (2004)
3. Barga, R., Lomet, D.: Phoenix Project: Fault Tolerant Applications. SIGMOD Record 31(2), 94–100 (2002)
4. Barga, R., Lomet, D., Paparizos, S., Yu, H., Chandrasekaran, S.: Persistent Applications via Automatic Recovery. In: IDEAS Conference, Hong Kong, pp. 258–267 (2003)
5. Barga, R., Lomet, D., Shegalov, G., Weikum, G.: Recovery Guarantees for Internet Applications. ACM Trans. on Internet Technology 4(3), 289–328 (2004)
6. Barga, R., Lomet, D., Weikum, G.: Recovery Guarantees for Multi-tier Applications. In: ICDE Conference, San Jose, pp. 543–554 (2002)
7. Bernstein, P., Hsu, M., Mann, B.: Implementing Recoverable Requests Using Queues. In: SIGMOD Conference, Atlantic City, pp. 112–122 (1990)
8. Bernstein, P.A., Newcomer, E.: Principles of Transaction Processing. Morgan Kaufmann, San Francisco (1996)
9. Borg, A., Baumbach, J., Glazer, S.: A message system supporting fault tolerance. In: Symposium on Operating Systems Principles, pp. 90–99 (1983)
10. Elnozahy, E.N., Alvisi, L., Wang, Y.-M., Johnson, D.B.: A survey of rollback-recovery protocols in message-passing systems. ACM Comput. Surv. 34(3), 375–408 (2002)
11. Frølund, S., Guerraoui, R.: A Pragmatic Implementation of e-Transactions. In: IEEE Symposium on Reliable Distributed Systems, Nürnberg, pp. 186–195 (2000)
12. Gray, J.: Keynote address Internet Reliability. In: 2nd HDCC Workshop, Santa Cruz, CA (May 2001)
13. Gray, J., Reuter, A.: Transaction Processing: Concepts and Techniques. Morgan Kaufmann, San Francisco (1993)
14. Hoare, H.A.R.: As quoted in B. Randell: Turing Memorial Lecture Facing Up to Faults. Comput. J. 43(2), 95–106 (2000)
15. Lomet, D.: Robust Web Services via Interaction Contracts. In: TES Workshop, Toronto, pp. 1–14 (2004)

16. Lomet, D.: Persistent Middle Tier Components without Logging. In: IDEAS Conference, Montreal, pp. 37–46 (2005)
17. Lomet, D.: Faithless Replay. for Persistent Logless Mid-Tier Components. Microsoft Research Technical Report MSR-TR-2008-50 (April 2008)
18. Narasimhan, P., Moser, L., Melliar-Smith, P.M.: Lessons Learned in Building a Fault-Tolerant CORBA System. In: DSN 2002, pp. 39–44 (2002)
19. MSDN Library: Persistence Overview, `http://msdn.microsoft.com/workshop/author/persistence/overview.asp`
20. OMG: CORBA 2000. Fault Tolerant CORBA Spec V1.0, `http://cgi.omg.org/cgi-bin/doc?ptc/00-04-04`
21. Patterson, D.: Recovery Oriented Computing (talk) (September 2001), `http://roc.cs.berkeley.edu`
22. Shegalov, G., Weikum, G.: Formal Verification of Web Service Interaction Contracts. IEEE SCC (2), 525–528 (2008)
23. Sun 2001. Enterprise Java Beans Specification, Vers. 2.0 (2001), `http://java.sun.com/products/ejb/docs.html`
24. Weikum, G., Vossen, G.: Transactional Information Systems. Morgan Kaufmann, San Francisco (2002)

The Challenge of Assuring Data Trustworthiness

Elisa Bertino[1], Chenyun Dai[1], and Murat Kantarcioglu[2]

[1] Computer Science Department and CERIAS, Purdue University
West Lafayette, Indiana, USA
{bertino,daic}@cs.purdue.edu
[2] Computer Science Department, University of Texas at Dallas
Richardson, Texas, USA
muratk@utdallas.edu

Abstract. With the increased need of data sharing among multiple organizations, such as government organizations, financial corporations, medical hospitals and academic institutions, it is critical to ensure that data is trustworthy so that effective decisions can be made based on these data. In this paper, we first discuss motivations and requirement for data trustworthiness. We then present an architectural framework for a comprehensive system for trustworthiness assurance. We then discuss an important issue in our framework, that is, the evaluation of data provenance and survey a trust model for estimating the confidence level of the data and the trust level of data providers. By taking into account confidence about data provenance, we introduce an approach for policy observing query evaluation. We highlight open research issues and research directions throughout the paper.

Keywords: Data Integrity and Quality; Security; Policies.

1 Introduction

Today, more than ever, there is a critical need for organizations to share data within and across the organizations so that analysts and decision makers can analyze and mine the data, and make effective decisions. However, in order for analysts and decision makers to produce accurate analysis and make effective decisions and take actions, data must be trustworthy. Therefore, it is critical that data trustworthiness issues, which also include data quality and provenance, be investigated for organizational data sharing, situation assessment, multi-sensor data integration and numerous other functions to support decision makers and analysts. Indeed, today's demand for data trustworthiness is stronger than ever. As many organizations are increasing their reliance on data for daily operations and critical decision making, data trustworthiness, and integrity in particular, is arguably one of the most critical issues. Without integrity, the usefulness of data becomes diminished as any information extracted from them cannot be trusted with sufficient confidence.

The problem of providing "good data" to users is an inherently difficult problem which often depends on the semantics of the application domain. Also solutions for improving the data, like those found in data quality, may be very expensive and may require access to data sources which may have access restrictions, because of data

X. Zhou et al. (Eds.): DASFAA 2009, LNCS 5463, pp. 22–33, 2009.

sensitivity. Also even when one adopts methodologies to assure that the data is of good quality, errors may still be introduced and low quality data be used; therefore, it is important to assess the damage resulting from the use of such data, to track and contain the spread of errors, and to recover.

The many challenges of assuring data trustworthiness require articulated solutions combining different approaches and techniques. In this paper we discuss some components of such a solution and highlight relevant research challenges.

The rest of this paper is organized as follows. We start by a quick survey of areas that are relevant to the problem of data trustworthiness. We then present a comprehensive framework for policy-driven data trustworthiness, and discuss relevant components of such framework. Finally, we outline a few conclusions.

2 State of the Art

Currently there is no comprehensive approach to the problem of high assurance data trustworthiness. The approach we envision, however, is related to several areas that we discuss in what follows.

Integrity Models. Biba [3] was the first to address the issue of integrity in information systems. His approach is based on a hierarchical lattice of integrity levels, and integrity is defined as a relative measure that is evaluated at the subsystem level. A subsystem is some sets of subjects and objects. An information system is defined to be composed of any number of subsystems. Biba regards integrity threat as that a subsystem attempts to improperly change the behavior of another by supplying false data. A drawback of the Biba approach is that it is not clear how to assign appropriate integrity levels and that there are no criteria for determining them. Clark and Wilson [4] make a clear distinction between military security and commercial security. They then argue that security policies related to integrity, rather than disclosure, are of the highest priority in commercial information systems and that separated mechanisms are required for the enforcement of these policies. The model by Clark and Wilson has two key notions: well-formed transactions and separation of duty. A well-formed transaction is structured such that a subject cannot manipulate data arbitrarily, but only in constrained ways that ensure internal consistency of data. Separation of duty attempts to ensure the external consistency of data objects: the correspondence among data objects of different subparts of a task. This correspondence is ensured by separating all operations into several subparts and requiring that each subpart be executed by a different subject.

Semantic Integrity. Many commercial DBMS enable users to express a variety of conditions, often referred to as *semantic integrity constraints*, that data must satisfy [18]. Such constraints are used mainly for *data consistency and correctness*. As such semantic integrity techniques are unable to deal with the more complex problem of data trustworthiness in that they are not able to determine whether some data correctly reflect the real world and are provided by some reliable and accurate data source.

Data Quality. Data quality is a serious concern for professionals involved with a wide range of information systems, ranging from data warehousing and business intelligence to customer relationship management and supply chain management. One industry study estimated the total cost to the US economy of data quality problems at over US$600 billion per annum [5]. Data quality has been investigated from different perspectives, depending also on the precise meaning assigned to the notion of data quality. Data are of high quality "if they are fit for their intended uses in operations, decision making and planning" [6]. Alternatively, the data are deemed of high quality if they correctly represent the real-world construct to which they refer. There are a number of theoretical frameworks for understanding data quality. One framework seeks to integrate the product perspective (conformance to specifications) and the service perspective (meeting consumers' expectations) [7]. Another framework is based in semiotics to evaluate the quality of the form, meaning and use of the data [8]. One highly theoretical approach analyzes the ontological nature of information systems to define data quality rigorously [9]. In addition to these more theoretical investigation, a considerable amount of research on the data quality has been devoted to investigating and describing various categories of desirable attributes (or dimensions) of data. These lists commonly include accuracy, correctness, currency, completeness and relevance. Nearly 200 such terms have been identified and there is little agreement on their nature (are these concepts, goals or criteria?), their definitions or measures. Tools have also been developed for analyzing and repairing poor quality data, through the use for example of *record linkage techniques* [22].

Reputation Techniques. Reputation systems represent a key technology for securing collaborative applications from misuse by dishonest entities. A reputation system computes reputation scores about the entities in a system, which helps single out those entities that are exhibiting less than desirable behavior. Examples of reputation systems may be found in several application domains; E-commerce websites such as eBay (ebay.com) and Amazon (amazon.com) use their reputation systems to discourage fraudulent activities. The EigenTrust [10] reputation system enables peer-to-peer file sharing systems to filter out peers who provide inauthentic content. The web-based community of Advogato.org uses a reputation system [19] for spam filtering. Reputation techniques can be useful in assessing data sources and data manipulation intermediaries; however their use for such purpose has not been yet investigated.

3 A Comprehensive Approach

Our envisioned approach [2] to data trustworthiness is based on a comprehensive framework composed of four key elements (see Figure 1). The first element is a mechanism for associating *confidence values* with data in the database. A confidence value is a numeric value ranging from 0 to 1 and indicates the trustworthiness of the data. Confidence values can be generated based on various factors, such as the trustworthiness of data providers and the way in which the data has been collected. The second element is the notion of *confidence policy*, indicating which confidence level is required for certain data when used in certain tasks. The third element is the computation of the confidence values of a query results based on the confidence values of each data item and lineage propagation techniques [11]. The fourth element is a set of strategies for incrementing the confidence of query results at query processing time. Such element is a crucial component in that it makes possible to return query results meeting the confidence levels stated by the confidence policies.

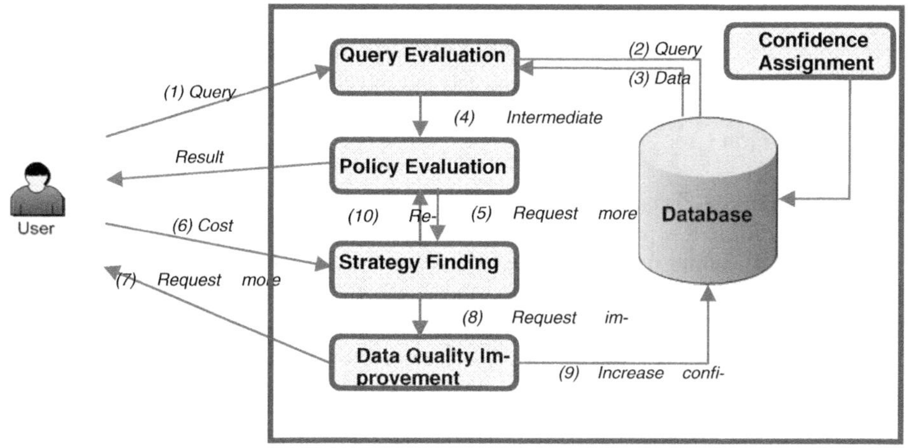

Fig. 1. A framework for assuring data trustworthiness

The notion of confidence policy is a key novel notion of our approach. Such a policy specifies the minimum confidence level that is required for use of a given data item in a certain task by a certain subject. As a complement to the traditional access control mechanism that applies to base tuples in the database before any operation, the confidence policy restricts access to the query results based on the confidence level of the query results. Such an access control mechanism can be viewed as a natural extension to the Role-based Access Control (RBAC) [12] which has been widely adopted in commercial database systems. Therefore, our approach can be easily integrated into existing database systems.

Because some query results will be filtered out by the confidence policy, a user may not receive enough data to make a decision and he/she may want to improve the data quality. To meet the user's need, an approach for dynamically incrementing the data confidence level (the fourth element of our solution) is required. Our envisioned approach will select an optimal strategy which determines which data should be selected and how much the confidence should be increased to satisfy the confidence level stated by the confidence policies.

4 Assigning Confidence Levels to Data

A possible approach to assign confidence levels, also referred to as *confidence scores*, to data is based on the trustworthiness of data provenance [1]. By data provenance we refer to information about the source of the data and the entities, such as agents, applications, users, who have accessed and/or manipulated the data before the data has been entered in the destination database. Though several research efforts have been devoted to data provenance [13, 14, 15, 16], the focus has mainly been on the collection and semantic analysis of provenance information. Little work has been done with respect to the trustworthiness of data provenance. Evaluating the trustworthiness of data provenance requires answering questions like "From which sources did the data originate from? How trustworthy are such data sources? Which entities (applications or users or systems) handled the data? Are these entities trustworthy?" To address these challenges, Dai et al. [1] have proposed a *data provenance trust model* (trust model, for short) which estimates the confidence level of data and the trustworthiness of data sources. In what follows, we first survey such a trust model and then we discuss open research issues.

A Data Provenance Trust Model. The model developed by Dai et al. [1] takes into account three different aspects that may affect data trustworthiness: data similarity, data conflict, and path similarity. Similar data items are considered as support to one another, while conflicting data items compromise the confidence level of one another. Besides data similarity and data conflict, the source of the data is also an important factor for determining the trustworthiness of the data. For example, if several independent sources provide the same data, such data is most likely to be true. We also observe that a data is likely to be true if it is provided by trustworthy data sources, and a data source is trustworthy if most data it provides are true. Due to such inter-dependency between data and data sources, the model is based on an iterative procedure to compute the scores to be assigned to data and data sources. To start the computation, each data source is first assigned an initial trust score which can be obtained by querying available information about data sources. At each iteration, the confidence level of the data is computed based on the combined effects of the aforementioned three aspects, and the trustworthiness of the data source is recomputed by using the confidence levels of the data it provides. When a stable stage is reached, that is, when changes to the scores (of both data and data sources) are negligible, the trust computation process stops.

Table 1. An example data set

RID	SSN	Name	Gender	Age	Location	Date
1	479065188	Tom	Male	38	Los Angeles	3pm 08/18/2007
2	47906518	Tom	Male	38	Los Angeles	3pm 08/18/2007
3	479065188	Bob	Male	38	New York	7pm 08/18/2007
4	4790651887	Tom	Male	38	Pasadena	3pm 08/18/2007

Data similarity in this model refers to the likeness of different items. Similar items are considered as supportive to each other. The challenge here is how to determine whether two items are similar. Consider the example in Table 1. We can observe that the first two items are very similar since they both report the same locations of Tom at the same date. The only difference between these two items is a possible typo error in the person's SSN. In contrast, the third item is different from the first two because it reports a totally different location. In order to determine sets of data items that are very similar likely describe the same real world item, the model employs a clustering algorithm. The clustering process results in a set of item; each such set represents a single real-world data item. For each item the effect of data similarity on its confidence score is determined in terms of the number of items in the same cluster and the size of the cluster. The use of clustering techniques requires developing distance functions to measure the similarities among items and the cost function which the clustering algorithm tries to minimize. The distance functions are usually determined by the type of data being clustered, while the cost function is defined by the specific objective of the clustering problem. Well known approaches can be used for such functions, such as the edit distance for string data types, or hierarchy-based distances for categorical data.

Data conflict refers to inconsistent descriptions or information about the same entity or event. A simple example of a data conflict is that the same person appears at different locations during the same time period. It is obvious that data conflict has a negative impact on the confidence level of items. There are various reasons for data conflicts, such as typos, false data items generated by malicious sources, or misleading knowledge items generated by intermediate parties. Data conflict largely depends on the knowledge domain of the specific application. Therefore, the trust model by Dai et al. [1] allows users to define their own data conflict functions according to their application-dependent requirements. To determine if two items conflict with each other, data users first need to define the exact meaning of conflict, which we call *data consistency rules*. Consider the example in Table 1 again. The attribute value of "SSN" in the first item is the same as that in the third item, but the attribute value of "Name" in the first item is different from that in the third one. This implies a data conflict, since we know that each single SSN should correspond to only one individual. We can further infer that there should be something wrong with either source providers (airports) or intermediate agents (police stations) whichever handled these two items. The data consistency rule we would use in this example is that *if* $r1$(*"SSN"*) = $r2$(*"SSN"*), *then* $r1$(*"Name"*) = $r2$(*"Name"*) (such rule can be simply modeled as functional dependency). If two items cannot satisfy the condition stated by such data consistency rule, these two items are considered conflicting with each other; if two items conflicts, their confidence level will in general be lower. To facilitate automatic conflict detection, a simple language needs to be provided allowing data users and/or domain experts to define data consistency rules; such language can then be implemented by using the trigger and assertion mechanisms provided by DBMS.

Path similarity models how similar are the paths followed by two data items from the sources to the destination. Path similarity is important in that it is used to evaluate the provenance independence of two or more data items. A path in the model by Dai et al. [1] is represented by a list of identifiers; the first element of such list is the data source, whereas the subsequent elements are the identifiers of all the intermediate parties that processed and/or simply retransmitted the data. The similarity of two paths is computed by comparing the lists corresponding to these paths.

Open Research Issues. The above model is still preliminary and requires addressing several research issues which we discuss in what follows.

- *Similarity/dissimilarity of data.* A key component of the model is represented by the factors concerning the similarity/dissimilarity of data. In addition to the techniques mentioned in the above paragraph, like the edit distance, one needs to include modeling techniques able to take into account data semantics. For example, consider the fourth item in Table 1; such item can be considered very similar (or supportive of the first two items) by observing that Pasadena is part of the Los Angeles area. Such an inference requires taking into account knowledge about spatial relationships in the domain of interest. Possible approaches that can be used include semantic web techniques, like ontologies and description logics.

- *Secure data provenance.* An important requirement for a data provenance trust model is that the provenance information be protected from tampering when flowing across the various parties. In particular, we should be able to determine the specific contribution of each party to the provenance information and the type of modification made (insert/delete/update). We may also have constraints on what the intermediate parties processing the data and providing provenance information can see about provenance information from previous parties along the data provisioning chain. An approach to address such problem is based on approaches for controlled and cooperative updates of XML documents in Byzantine and failure-prone distributed systems [20]. One could develop an XML language for encoding provenance information and use such techniques to secure provenance documents.
- *Data validation through privacy-preserving record linkage.* In developing solutions for data quality, the use of record linkage techniques is important. Such techniques allow a party to match, based on similarity functions, its own records with records by another party in order to validate the data. In our context such techniques could be used not only to match the resulting data but also to match the provenance information, which is often a graph structure. Also in our case, we need not only to determine the similarity for the data, but also the dissimilarity for the provenance information. In other words, if two data items are very much similar and their provenance information is very dissimilar, the data item will be assigned a high confidence level. In addition, confidentiality of provenance information is an important requirement because a party may have relevant data but have concerns or restrictions for the data use by another party. Thus application of record linkage technique to our context thus requires addressing the problem of privacy, the extension to graph-structured information, and the development of similarity/dissimilarity functions. Approaches have been proposed for privacy-preserving record linkage [22, 23]. However those approaches have still many limitations, such as the lack of support for graph-structured information.
- *Correlation among data sources.* The relationships among the various data sources could be used to create more detailed models for assigning trust to each data source. For example, if we do not have good prior information about the trustworthiness of a particular data source, we may try to use distributed trust computation approaches such as EigenTrust [10] to compute a trust value for the data source based on the trust relationships among data sources. In addition, even if we observe that the same data is provided by two different sources, if these two sources have a very strong relationship, then it may not be realistic to assume that the data is provided by two independent sources. An approach to address such issue is to develop "source correlation" metrics based on the strength of the relationship among possible data sources. Finally, in some cases, we may need to know "how important is a data sources within our information propagation network?" to reason about possible data conflicts. To address such issue one can apply various social network centrality measures such as degree, betweenness, closeness, and information centralities [21] to assign importance values to the various data sources.

5 Policies

Policies provide a high level mechanism for expressing organizational requirements and policies and simplify administration and management. In our context, a key type of policy is represented by the *confidence policy*, regulating the use of the data according to requirements concerning data confidence levels. Such a policy is motivated by the observation that the required level of data trustworthiness depends on the purpose for which the data have to be used. For example, for tasks which are not critical to an organization, like computing a statistical summary, data with a medium confidence level may be sufficient, whereas when an individual in an organization has to make a critical decision, data with high confidence are required. An interesting example is given by Malin et al. [17] in the context of healthcare applications: for the purpose of generating hypothesis and identifying areas for further research, data about cancer patients' diseases and primary treatment need not be highly accurate, as treatment decisions are not likely to be made on the basis of these results data alone; however, for evaluating the effectiveness of a treatment outside of the controlled environment of a research study, accurate data is desired. In what follows, we first survey a policy model [2] addressing such requirement and then discuss open research issues.

A Confidence Policy Model. A policy in the confidence policy model by Dai et al. [2] specifies the minimum confidence that has to be assured for certain data, depending on the user accessing the data and the purpose the data access. In its essence, a confidence policy contains three components: a *subject specification*, denoting a subject or set of subjects to whom the policy applies; a *purpose specification*, denoting why certain data are accessed; a *confidence level*, denoting the minimum level of confidence that has to be assured by the data covered by the policy when the subject to whom the policy applies requires access to the data for the purpose specified in the policy. In this policy model, subjects to which policies apply are assumed to be roles of a RBAC model, because this access control model is widely used and well understood. However such policy model can be easily extended to the case of attribute-based access control models, as we discuss in the research issues paragraph.

The confidence policies are thus based on the following three sets: *R*, *Pu* and *R+*. *R* is a set of roles used for subject specification; a user is human being and a role represents a job function or job title within the organization that the user belongs to. *Pu* is a set of data usage purposes identified in the system. *R+* denotes non-negative real numbers. The definition of a confidence policy is thus as follows.

[*Confidence Policy*]. Let $r \in R$, $pu \in Pu$, and $\beta \in R+$. A confidence policy is a tuple $< r, pu, \beta >$, specifying that when a user under a role *r* issues a database query *q* for purpose *pu*, the user is allowed to access the results of *q* only if these results have confidence value higher than β.

The confidence policies <Secretary, summary, 0.1>, and <Manager, investment, 0.8> specify, respectively, that a secretary can use data with low confidence value for the purpose a summary reports, whereas a manager must use data with high confidence value when making investment decisions.

Open Research Issues. The development of policies for integrity management requires however addressing several research issues which we discuss in what follows.

- *Expressive power.* A first issue is related to improve the expressivity of the confidence policy model. The simple model outlined in the previous paragraph needs some major extensions. It should be possible to support a more fine-grained specification of confidence requirements concerning data use whereby for a given task and role, one can specify different confidence levels for different categories of data. The model should support the specification of subjects, in terms of subject attributes and profiles other than the subject role. If needed, exceptions to the policies should be supported; a possible approach is to support *strong policies*, admitting no exceptions, and *weak policies*, admitting exceptions. If exceptions are allowed for a policy (set of policies), the policy enforcement mechanism should support the gathering of evidence about the need for exceptions. Such evidence is crucial in order to refine confidence policies.
- *Policy provisioning.* An important issue is related to the confidence policy provisioning. Provisioning refers to assigning confidence policies to specific tasks, users, and data and, because it very much depends from the applications and data semantics, it may be quite difficult. To address such issue, one approach is the use of machine learning techniques.
- *Data validation policies.* Even though confidence policies have a key role in our framework, policies are also required to manage data validation activities, periodically or whenever certain events arise. To address such requirement, one possibility is to design a *data validation policy* (DVP) language that includes the following components: (i) A set of events that trigger the execution of some validation actions. An event can be a data action (read, insert, delete, update) or a user-defined event such as a specific time or a particular situation; for example a source of certain data has been found to be invalid and thus the validation process needs to determine which data, users and application programs may have been affected by the invalid data. Notice that it should also be possible to specify that a validation must be executed before any access is made by a given subject or set of subjects, or even when the data is being accessed (see also next section). (ii) A validation procedure which performs the actual integrity analysis of the data. Such procedure may be complex in that it can involve human users and may result in a number of actions, possibly organized according to a workflow. (iii) A set of actions to be undertaken as consequence of the validation. A large variety of actions are possible, such as blocking all accesses to data, blocking the execution of an application program, invoking some data repair procedures, making changes to the metadata associated with the data. It is important to notice that even though it would be desirable to perform data validation very frequently, the impact on performance may be significant; therefore the availability of a language, like the DVP language, will make easier for the data administrators to fine tune the system according to trade-offs among performance, cost and integrity.

6 Policy Complying Query Evaluation

A critical issue in enforcing confidence policies in the context of query processing is that some of the query results may be filtered out due to confidence policy violation. Therefore a user may not receive enough data to make a decision and he may want to improve the data quality. A possible approach [2] is based on dynamically incrementing the data confidence level. Such approach selects an optimal strategy which determines which data should be selected and how much the confidence should be increased to comply with the confidence level stated by the policies. Such approach assumes that each data item in the database is associated with a cost function that indicates the cost for improving the confidence value of this data item. Such a cost function may be a function on various factors, like time and money. As part of the approach several algorithms have been investigated to determine the increment that has the lowest cost. In what follows we first briefly discuss components of the system proposed by Dai et al. [2] and then we discuss open research issues.

A Policy Complying Query Evaluation System. The query evaluation system (see Figure 1) consists of four main components: *query evaluation*, *policy evaluation*, *strategy finding*, and *data quality improvement*. We elaborate on the data flow within the system. Initially, each base tuple is assigned a confidence value by the *confidence assignment* component (based on the model described in Section 4). A user inputs query information in the form <*Q*, *purpose*, *perc*>, where *Q* is a normal SQL query, *purpose* indicates the purpose for which the data returned by the query will be used, and *perc* indicates percentage of results that the user expects to receive after the confidence policy enforcement. The *query evaluation* component then computes the results of *Q* and the confidence level of each tuple in the result based on the confidence values of base tuples. The intermediate results are sent to the *policy evaluation* component. The *policy evaluation* component selects the confidence policy associated with the role of the user who issued *Q* and checks each tuple in the query result according to the selected confidence policy. Only the results with confidence value higher than the threshold specified in the confidence policy are immediately returned to the user. If less than *perc* of the results satisfy the confidence policy, the *strategy finding* component is invoked to devise an optimal strategy for increasing the confidence values of the base tuples and report the cost of such strategy to the user. If the user agrees about the cost, the *strategy finding* component will inform the *data quality improvement* component to take actions to improve the data quality and then update the database. Finally, new results will be returned to the user.

Open Research Issues. The development of policy complying query evaluation framework requires addressing several research issues which we discuss in what follows.

- *Cost models.* In the above discussion, we assumed that cost models for the quality improvement are available for each tuple. However suitable cost models need to be developed, also depending on the optimization criteria adopted, like time and financial cost.

- *Heuristics for confidence increases.* Algorithms need to be devised for determining suitable base tuples for which the increase in the confidence values can lead to the minimum cost. Because it is likely that finding the optimal solution may be computationally very expensive, heuristics need to be devised.
- *Query based data validation strategies.* Data validation and correction is often an expensive activity and thus needs to be minimized and executed when high-confidence data are actually required. Also because data validation and correction may take some time, approaches must be in place to make sure that the data are corrected by the time they are needed. To address such issue, approaches must be devised that, based on knowing in advance the queries issued by the users and the timeline of these queries, are able to determine which data must be validated and corrected while at the same time minimizing the costs.

7 Conclusions

In this paper we have discussed research directions concerning the problem of providing data that can be trusted to end-users and applications. This is an important problem for which multiple techniques need to be combined in order to achieve good solutions. In addition to approaches and ideas discussed in the paper, many other issues needed to be addressed to achieve high-assurance data trustworthiness. In particular, data need to be protected from attacks carried through unsecure platforms, like the operating system, and unsecure applications, and from insider threats. Initial solutions to some of those data security threats are starting to emerge.

Acknowledgments. The authors have been partially supported by the AFOSR grant FA9550-07-0041 "Systematic Control and Management of Data Integrity, Quality and Provenance for Command and Control Applications".

References

1. Dai, C., Lin, D., Bertino, E., Kantarcioglu, M.: An Approach to Evaluate Data Trustworthiness based on Data Provenance. In: Jonker, W., Petković, M. (eds.) SDM 2008. LNCS, vol. 5159, pp. 82–98. Springer, Heidelberg (2008)
2. Dai, C., Lin, D., Kantarcioglu, M., Bertino, E., Celikel, E., Thuraisingham, B.: Query Processing Techniques for Compliance with Data Confidence Policies. Technical Report, CERIAS (submitted for publication) (2009)
3. Biba, K.J.: Integrity Considerations for Secure Computer Systems. Technical Report TR-3153, Mitre (1977)
4. Clark, D.D., Wilson, D.R.: A Comparison of Commercial and Military Computer Security Policies. In: IEEE Symposium on Security and Privacy, Oakland, CA (1987)
5. Eckerson, W.: Data Warehousing Special Report: Data quality and the bottom line (2002), `http://www.adtmag.com/article.aspx?id=6321`
6. Juran, J.M.: Juran on Leadership for Quality – an Executive Handbook. Free Press, New York (1989)
7. Kahn, B., Strong, D., Wang, R.: Information Quality Benchmarks: Product and Service Performance. In: Communications of the ACM, vol. 45, pp. 184–192. ACM, New York (2002)

8. Price, R., Shanks, G.: A Semiotic Information Quality Framework. In: IFIPInternational Conference on Decision Support Systems: Decision Support in anUncertain and Complex World, Prato, Italy (2004)
9. Wand, Y., Wang, R.Y.: Anchoring Data Quality Dimensions in Ontological Foundations. In: Communications of the ACM, vol. 39, pp. 86–95. ACM, New York (1996)
10. Kamvar, S.D., Schlosser, M.T., Garcia-Molina, H.: The EigenTrust Algorithm for Reputation Management in P2P Networks. In: Twelfth International World Wide Web Conference, pp. 640–651. ACM, New York (2003)
11. Dalvi, N.N., Suciu, D.: Efficient Query Evaluation on Probabilistic Databases. In: Thirtieth International Conference on Very Large Data Bases, pp. 864–875. Morgan Kaufmann, San Fransisco (2004)
12. Ferraiolo, D.F., Sandhu, R.S., Gavrila, S.I., Kuhn, D.R., Chandramouli, R.: Proposed NIST Standard for Role-Based Access Control. In: ACM Transactions on Information and System Security, vol. 4, pp. 224–274. ACM, New York (2001)
13. Simmhan, Y.L., Plale, B., Gannon, D.: A Survey of Data Provenance Techniques. Technical Report, Indiana University, Bloomington (2007)
14. Buneman, P., Khanna, S., Tan, W.C.: Why and where: A characterization of data provenance. In: Van den Bussche, J., Vianu, V. (eds.) ICDT 2001. LNCS, vol. 1973, pp. 316–330. Springer, Heidelberg (2000)
15. Lanter, D.P.: Design of a Lineage-Based Meta-Data Base for GIS. Cartography and Geographic Information Systems 18, 255–261 (1991)
16. Greenwood, M., Goble, C., Stevens, R., Zhao, J., Addis, M., Marvin, D., Moreau, L., Oinn, T.: Provenance of e-Science Experiments – Experiences from Bioinformatics. In: UK OST e-Science All Hands Meeting (2003)
17. Malin, J.L., Keating, N.L.: The Cost-Quality Trade-off: Need for Data Quality Standards for Studies that Impact Clinical Practice and Health Policy. Journal of Clinical Oncology 23, 4581–4584 (2005)
18. Ramakrishnan, R., Gehrke, J.: Database Management Systems. McGraw-Hill, New York (2000)
19. Levien, R.: Attack Resistant Trust Metrics. PhD thesis, University of California, Berkeley, CA, USA (2002)
20. Mella, G., Ferrari, E., Bertino, E., Koglin, Y.: Controlled and cooperative updates of XML documents in byzantine and failure-prone distributed systems. In: ACM Transactions on Information and System Security, vol. 9, pp. 421–460. ACM, New York (2006)
21. Jackson, M.O.: Social and Economics Networks. Princeton University Press, Princeton (2008)
22. Batini, C., Scannapieco, M.: Data Quality: Concepts, Methodologies and Techniques. Springer, Heidelberg (2006)
23. Scannapieco, M., Figotin, I., Bertino, E., Elmagarmid, A.: Privacy Preserving Schema and Data Matching. In: ACM SIGMOD International Conference on Management of Data, pp. 653–664 (2007)
24. Inan, A., Kantarcioglu, M., Bertino, E., Scannapieco, M.: A Hybrid Approach to Private Record Linkage. In: 24th IEEE International Conference on Data Engineering, pp. 496–505 (2008)

Integrating Australia's Water Data

Rob Vertessy

Bureau of Meteorology, Australian Government
r.vertessy@bom.gov.au

Abstract. Australia is facing an unprecedented water scarcity crisis and governments are responding with major policy, regulatory, market and infrastructure interventions. No matter what facet of water reform one is talking about, good intelligence on the country's water resource base and how it is trending is a vital ingredient for good decision making. Though most individual water businesses and agencies have an adequate view of the water resources under their control, gaining a national view has always been a fraught exercise. In recognition of this problem, the Bureau of Meteorology has been tasked with the challenge of integrating water data sets collected across Australia by over 250 organisations. In this paper, we look at how the Bureau is going about this task and at the suite of water information products that will arise from a coherent national water data set.

X. Zhou et al. (Eds.): DASFAA 2009, LNCS 5463, p. 34, 2009.

Probabilistic Inverse Ranking Queries over Uncertain Data

Xiang Lian and Lei Chen

Hong Kong University of Science and Technology
Clear Water Bay, Kowloon,
Hong Kong, China
{xlian,leichen}@cse.ust.hk

Abstract. Query processing in the uncertain database has become increasingly important due to the wide existence of uncertain data in many real applications. Different from handling precise data, the uncertain query processing needs to consider the data uncertainty and answer queries with confidence guarantees. In this paper, we formulate and tackle an important query, namely *probabilistic inverse ranking* (PIR) query, which retrieves possible ranks of a given query object in an uncertain database with confidence above a probability threshold. We present effective pruning methods to reduce the PIR search space, which can be seamlessly integrated into an efficient query procedure. Furthermore, we also tackle the problem of PIR query processing in high dimensional spaces, which reduces high dimensional uncertain data to a lower dimensional space. The proposed reduction technique may shed light on processing high dimensional uncertain data for other query types. Extensive experiments have demonstrated the efficiency and effectiveness of our proposed approaches over both real and synthetic data sets, under various experimental settings.

1 Introduction

Uncertain query processing has become increasingly important due to the wide existence of data uncertainty in many real-world applications such as sensor data monitoring [10], *location-based services* (LBS) [23], object identification [3], moving object search [5], and so on. As an example, sensory data collected from different sites may contain noises [10] because of environmental factors, device failure, or low battery power. In LBS applications [23], for the sake of privacy preserving [24], it is a common practice for a trusted third-party to inject synthetic noises into trajectory data of mobile users. Therefore, in all these scenarios, it is very crucial to answer queries *efficiently* and *effectively* over uncertain data.

In literature [4,12,9,14,20,8,29], the *ranking query* (a.k.a. *top-k* or *ranked query*) has been extensively studied, which is useful in applications like decision making, recommendation raising, and data mining tasks. In particular, in a database $\mathcal{D}$, with respect to any ad-hoc monotonic function (that computes scores of objects), a ranking query retrieves k objects in the database that have the highest scores. In contrast, the *quantile query* [22] returns an object in the database with the k-th rank, where k can be arbitrarily large, for example, 50% of the database size. While the ranking or quantile query

X. Zhou et al. (Eds.): DASFAA 2009, LNCS 5463, pp. 35–50, 2009.

obtains object(s) from the database with k highest ranks or the k-th highest rank, respectively, the *inverse ranking query* over precise data proposed by Li et al. [19] reports the rank of a user-specified query object in the database. Specifically, given a database $\mathcal{D}$, a *monotonic function* $f(\cdot)$[1], and a query point q, an *inverse ranking query* returns the rank of query point q among all the data points $o \in \mathcal{D}$ (or equivalently, the number of points o satisfying $f(o) \geq f(q)$).

Due to the inherent uncertainty in real application data, in this paper, we focus on the inverse ranking query on uncertain data, namely *probabilistic inverse ranking* (PIR) query, which, to our best knowledge, no previous work has studied in the context of uncertain database. In particular, given an uncertain database $\mathcal{D}^U$ and a user-specified query point q, a PIR query computes all the possible ranks of q in $\mathcal{D}^U$ with probability greater than a pre-defined threshold $\alpha \in [0, 1)$.

The PIR query is important in many real applications such as financial or image data analysis, sensor data monitoring, multi-criteria decision making, and business planning, where we need to identify the importance (rank or priority) of a particular object among its peers. For example, in a survey of the laptop market, customers are asked to specify the ranges of attributes for laptop models that they prefer, including the price, weight, size, and so on. Here, each customer's preference can be considered as an uncertain object (with uncertain attribute ranges). Now if a company wants to design a new model, a data analyzer can specify a ranking function (e.g. the summation of price, weight, and size), and issue a PIR query to obtain the rank of this new model (query point) among all the customers' preferences (uncertain objects). Intuitively, the company can make a decision about the attributes of the new model such that it can attract as many customers (having scores of preferences higher than this new model) as possible.

As another example [19], for a newborn baby, we may be interested in his/her health compared with other babies, in terms of height, weight, and so on. In this case, we can infer the baby's health from his/her rank among other babies. Note that, newborn babies in the hospital are confidential. Thus, for the sake of privacy preserving, these data are usually perturbed by adding synthetic noises or generalized by replacing exact values with uncertain intervals, before the public release. Thus, in these situations, we can conduct a PIR query over uncertain data (perturbed or generalized data) in order to obtain all possible ranks that a new baby may have with high confidence.

In this paper, we formulate and tackle the problem of PIR queries in the uncertain database. Note that, previous techniques [19] on inverse ranking query in "certain" database cannot be directly used, since the PIR problem has to deal with data uncertainty during the query processing. In particular, due to the imprecise data attributes, the ranking score of each uncertain object is a variable rather than an exact value. Thus, the inverse ranking query in the context of uncertain database has to be re-defined, considering the confidence (accuracy) of the query results. Motivated by this, we formalize the PIR problem, propose effective pruning methods to reduce the PIR search space, and finally present efficient query procedure to answer PIR queries.

Specifically, we make the following contributions in this paper.

1. We formally define the problem of the *probabilistic inverse ranking* (PIR) query.

[1] Function $f(\cdot)$ is monotonic iff: $\forall i, o.A_i \leq p.A_i \mapsto f(o) \leq f(p)$, where $x.A_i$ is the i-th coordinate of point x.

2. We propose a general framework for answering the PIR query.
3. We provide effective pruning methods to significantly reduce the search space, and seamlessly integrate them into an efficient query procedure.
4. We extend the PIR solutions to the PIR query in high dimensional spaces.

In the sequel, Section 2 formalizes the PIR problem in the uncertain database. Section 3 provides the framework for answering PIR queries, and illustrates the basic idea of pruning methods in order to retrieve PIR results. Section 4 investigates the PIR query processing by applying our pruning methods. Section 5 discusses the PIR query processing in high dimensional space. Section 6 presents the query performance of our proposed approaches. Section 7 reviews previous work on inverse ranking queries over precise data, and uncertain query processing. Finally, Section 8 concludes this paper.

2 Problem Definition

2.1 Uncertainty Model

In an uncertain database $\mathcal{D}^U$, uncertain objects are assumed to definitely belong to the database, and their attributes are imprecise and uncertain [26]. Thus, in literature, they are often modeled by *uncertainty regions* [5]. Each uncertain object o can reside within its uncertainty region, denoted as $UR(o)$, with arbitrary distribution; moreover, object o cannot appear outside $UR(o)$. Formally, assuming the *probability density function* (pdf) of object o's data distribution is $pdf(o_0)$, we have $\int_{o_0 \in UR(o)} pdf(o_0) do_0 = 1$ and $pdf(o_0) = 0$ for $o_0 \notin UR(o)$. Following the convention of uncertain database [6,5,25], uncertain objects in the database are assumed to be independent of each other.

Figure 1(a) illustrates an example of small uncertain database in a 2D space, which contains 5 uncertain objects o_1, o_2, ..., and o_5. In particular, each uncertain object o_i $(1 \leq i \leq 5)$ is represented by an uncertainty region $UR(o_i)$ (i.e. shaded area) of (hyper)rectangular shape [6,25], within which o_i may appear at any location (however, o_i cannot appear outside its own shaded region).

2.2 The PIR Query

After giving the uncertainty model for uncertain objects, we formally define the inverse ranking query over uncertain data as follows.

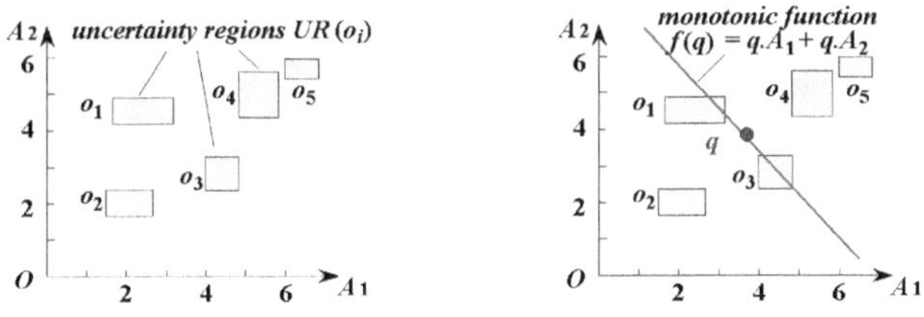

(a) Uncertainty Regions (b) The PIR Query

Fig. 1. Illustration of the PIR Query in the Uncertain Database

Definition 1. (*Probabilistic Inverse Ranking Query, PIR*) *Given an uncertain database $\mathcal{D}^U$ containing N uncertain objects, a query object q, a probability threshold $\alpha \in [0, 1)$, and a monotonic function $f(\cdot)$, a probabilistic inverse ranking (PIR) query computes all the possible ranks, k, of the query object q in the database $\mathcal{D}^U$ each with probability greater than α. That is, the PIR query returns the query object q's rank $k \in [1, N+1]$, if it holds that:*

$$Pr_{k\text{-}PIR}(q, f(\cdot)) = \sum_{\forall T=\{p_1, p_2, \ldots, p_{k-1}\} \subseteq \mathcal{D}^U} \left(\prod_{\forall p_i \in T} Pr\{f(p_i) \geq f(q)\} \cdot \prod_{\forall p_j \in \mathcal{D}^U \setminus T} Pr\{f(p_j) < f(q)\} \right) > \alpha \quad (1)$$

Intuitively, probability $Pr_{k\text{-}PIR}(q, f(\cdot))$ in Inequality (1) calculates the expected probability that query object q has the k-th rank in the database, with respect to monotonic function $f(\cdot)$. More specifically, query object q has rank k, if and only if there exists $(k-1)$ objects $p_1, p_2, \ldots$, and p_{k-1} in set T satisfying $f(p_i) \geq f(q)$, for all $1 \leq i \leq k-1$, and the remaining objects $p_j \in \mathcal{D} \setminus T$ satisfy the condition that $f(p_j) \leq f(q)$. Thus, as given in Inequality (1), $Pr_{k\text{-}PIR}(q, f(\cdot))$ sums up the probabilities that q has rank k (i.e. $\prod_{\forall p_i \in T} Pr\{f(p_i) \geq f(q)\} \cdot \prod_{\forall p_j \in \mathcal{D} \setminus T} Pr\{f(p_j) < f(q)\}$), for all possible object combinations of T. If probability $Pr_{k\text{-}PIR}(q, f(\cdot))$ is above the threshold α, then rank k is considered as the answer to the PIR query; otherwise, we will not report rank k.

Figure 1(b) illustrates a PIR example in the same uncertain database as Figure 1(a). For simplicity, here we assume that a linear ranking function $f(x) = x.A_1 + x.A_2$ is specified by PIR. Note, however, that our proposed approaches in this paper can be easily extended to the case of PIR with *arbitrary* monotonic function $f(\cdot)$. In the figure, we are given a query object q and its score can be captured by the ranking function $f(q)$. Clearly, query object q cannot have rank $k = 1, 2$, or 6, since their corresponding $Pr_{k\text{-}PIR}(q, f(\cdot))$ probabilities are zero. Thus, for $k = 3, 4$, and 5, we can compute their probabilities $Pr_{k\text{-}PIR}(q, f(\cdot))$ in Inequality (1), and output ranks k as the PIR query answers if $Pr_{k\text{-}PIR}(q, f(\cdot)) > \alpha$ holds.

To the best of our knowledge, no previous work has studied the PIR problem over uncertain data. Moreover, the work on inverse ranking query processing [19] over precise data cannot be directly applied, since the proposed methods do not consider the uncertain property of objects. Therefore, the only straightforward method is to sequentially scan uncertain objects in the database, and determine the possible ranks of the query object q with respect to $f(\cdot)$ by checking the PIR probability $Pr_{k\text{-}PIR}(q, f(\cdot))$ according to Definition 1. Finally, we output the final query answer if the PIR probability is above the threshold α. Clearly, this method is inefficient in terms of both CPU time and I/O cost. In order to improve the query performance, below, we present effective pruning methods to reduce the PIR search space.

3 Probabilistic Inverse Ranking Search

3.1 General Framework

In this subsection, we present the general framework for answering PIR queries in the uncertain database. Specifically, the framework consists of four phases, *indexing*, *retrieval*, *pruning*, and *refinement phases*. As mentioned in Section 2.2, the method that sequentially scans the entire database is very inefficient in terms of both CPU time and I/O cost. In order to speed up the PIR query processing, in the indexing phase, we construct an aR-tree index $\mathcal{I}$ [18] over uncertain database $\mathcal{D}^U$. In particular, we assume the uncertainty region of each object has hyperrectanglular shape [25,16,17], and we can insert all the uncertainty regions into an aR-tree, utilizing the standard "insert" operator in aR-tree. In addition, each entry of aR-tree nodes stores an aggregate, which is the number of uncertain objects under this entry. Such aggregates can facilitate our PIR query processing to quickly obtain the number of objects under an entry without accessing leaf nodes.

When a PIR query arrives with a query object q, a monotonic function $f(\cdot)$, and a probabilistic threshold α, we traverse the aR-tree and compute all possible ranks of q with non-zero probability $Pr_{k\text{-}PIR}(q, f(\cdot))$. Moreover, we also obtain a set, R_0, of objects that are involved in the calculation of $Pr_{k\text{-}PIR}(q, f(\cdot))$ for these ranks. The details will be described in Section 3.2. Next, in the pruning phase, we perform the upper-bound pruning to further reduce the PIR search space, which will be discussed in Section 3.3. Finally, for the remaining candidate ranks of q, we check Inequality (1) by computing the actual probability $Pr_{k\text{-}PIR}(q, f(\cdot))$, and return the final PIR answers.

3.2 Reducing the PIR Search Space

In this subsection, we illustrate how to reduce the search space of PIR queries. In particular, as shown in the example of Figure 1(b), no matter where uncertain objects o_4 and o_5 reside within their uncertainty regions, their scores are always higher than that of query object q. Therefore, it is clear that query object q cannot be ranked as the first or second place. Similarly, query object q cannot have the 6-th highest score, since object o_2 always has lower score than query object q. In other words, the probability $Pr_{k\text{-}PIR}(q, f(\cdot))$, for $k = 1, 2$, and 6, is equal to zero ($\leq \alpha$), and ranks 1, 2, and 6 are thus not answers to the PIR query.

Based on the observation above, we can reduce the search space of the PIR query by ignoring those ranks that query object q definitely cannot have. That is, we first find an interval $[k_{min}, k_{max}]$ for k, where k_{min} and k_{max} are the minimum and maximum possible rank values, respectively, such that probability $Pr_{k\text{-}PIR}(q, f(\cdot))$ is not zero.

We can classify uncertain objects in the database into three categories (sets):

- R^+: containing uncertain objects o with scores $f(o)$ always larger than $f(q)$;
- R_0: containing uncertain objects o that might have score $f(o)$ equal to $f(q)$; and
- R^-: containing uncertain objects o with scores $f(o)$ always smaller than $f(q)$.

As in the example of Figure 1(b), $R^+ = \{o_4, o_5\}$, $R_0 = \{o_2, o_3\}$, and $R^- = \{o_1\}$. Note that, from the definition, the three sets have properties that $R^+ \cup R_0 \cup R^- = \mathcal{D}^U$, and any two sets do not intersect with each other.

Among these three sets, R_0 is the only set containing uncertain objects o with scores $f(o)$ whose relationships with $f(q)$ are not definite. Thus, we have the following lemma.

Lemma 1. ($[k_{min}, k_{max}]$) *Let* $k_{min} = |R^+| + 1$ *and* $k_{max} = |R^+| + |R_0| + 1$. *For any rank* k *smaller than* k_{min} *or greater than* k_{max}*, it always holds that* $Pr_{k\text{-}PIR}(q, f(\cdot)) = 0$*, where* $Pr_{k\text{-}PIR}(q, f(\cdot))$ *is given in Inequality (1).*

Proof. We consider the case where $k < k_{min}$. Clearly, there are $(k_{min} - 1)$ (i.e. $|R^+|$) objects p_i in R^+ such that $f(p_i) \geq f(q)$ definitely holds. However, as given in Inequality (1), the cardinality of set T is exactly $(k - 1)$, which is smaller than $(k_{min} - 1)$. Thus, there must exist at least one object $p_l \in R^+$ such that $p_l \in \mathcal{D}^U \backslash T$ and $Pr\{f(p_l) < f(q)\} = 0$. Hence, probability $Pr_{k\text{-}PIR}(q, f(\cdot)) = 0$, for $k < k_{min}$. The case where $k > k_{max}$ is similar and thus omitted. □

From Lemma 1, we find that, instead of paying expensive computation for $Pr_{k\text{-}PIR}(q, f(\cdot))$ where $k \in [1, N+1]$, we can restrict k values to a much smaller interval $[k_{min}, k_{max}]$, where k_{min} and k_{max} are defined in Lemma 1. As a consequence, we can avoid the costly integrations for those rank with zero probabilities.

Specifically, we have the following lemma to rewrite the definition of the PIR query.

Lemma 2. (*Equivalent PIR Query*) *Given an uncertain database $\mathcal{D}^U$, a query object q, a threshold $\alpha \in [0, 1)$, and a monotonic function $f(\cdot)$, a PIR query retrieves all the possible ranks, $k \in [k_{min}, k_{max}]$, of the query object q in $\mathcal{D}^U$ such that:*

$$Pr_{k\text{-}PIR}(q, f(\cdot)) = \sum_{\forall T=\{p_1, p_2, \ldots, p_{k-k_{min}}\} \subseteq R_0} \left(\prod_{\forall p_i \in T} Pr\{f(p_i) \geq f(q)\} \cdot \prod_{\forall p_j \in R_0 \backslash T} Pr\{f(p_j) < f(q)\} \right) > \alpha \quad (2)$$

Proof. Derived from the fact that $\mathcal{D}^U = R^+ \cup R_0 \cup R^-$ and the definitions of sets R^+, R_0, and R^-. □

Lemma 2 indicates that, in order to obtain the probability that q has rank $k \in [k_{min}, k_{max}]$ in the database $\mathcal{D}^U$, we can compute the probability that q has rank $(k - k_{min} + 1)$ in set R_0. The resulting probability can be verified by the condition in Inequality (2).

3.3 Upper-Bound Pruning

In this subsection, we propose an effective pruning method, namely *upper-bound pruning*, to further reduce the search space of PIR queries. In particular, the heuristics of our pruning method is as follows. Although Inequality (2) has significantly reduced the PIR search space (compared with Inequality (1)), the direct computation of probability $Pr_{k\text{-}PIR}(q, f(\cdot))$ is still very costly. Therefore, alternatively, we plan to find a method to calculate its probability upper bound with a lower cost, which is denoted as $UB\_Pr_{k\text{-}PIR}(q, f(\cdot))$. That is, if this upper bound $UB\_Pr_{k\text{-}PIR}(q, f(\cdot))$ is already smaller than threshold $\alpha \in [0, 1)$, then rank k will not be our query result and it can be safely pruned.

Next, we discuss how to obtain the upper bound $UB\_Pr_{k\text{-}PIR}(q, f(\cdot))$. Specifically, we can simplify probability $Pr_{k\text{-}PIR}(q, f(\cdot))$ in Inequality (2) below.

$$Pr_{k\text{-}PIR}(q, f(\cdot)) = S(|R_0|, k - k_{min}) \quad (3)$$

where for objects $p_j \in R_0$, we have:

$$\begin{aligned} S(j, a) &= S(j-1, a) \cdot Pr\{f(p_j) \leq f(q)\} + S(j-1, a-1) \cdot (1 - Pr\{f(p_j) \leq f(q)\}), \\ S(j, 0) &= \prod_{i=1}^{j} Pr\{f(p_i) \leq f(q)\}, \\ S(a, a) &= \prod_{i=1}^{a} Pr\{f(p_i) \geq f(q)\}, \\ S(j, a) &= 0, \text{ for } j < a. \end{aligned} \quad (4)$$

Assume all the objects p_j in set R_0 are in a sequence of p_1, p_2, ..., and $p_{|R_0|}$. Intuitively, $S(j, a)$ is a recursive function which reports the probability that there are a out of j objects having scores higher than q (i.e. $f(q)$) and the remaining $(j - a)$ objects with scores smaller than or equal to $f(q)$.

From Eq. (3), the problem of finding upper bound probability $UB\_Pr_{k\text{-}PIR}(q, f(\cdot))$ is equivalent to that of obtaining the upper bound of $S(|R_0|, k - k_{min})$. In turn, according to recursive functions in Eq. (4), we only need to compute the lower and upper bounds of $S(j, a)$, which can be summarized in the following lemma.

Lemma 3. *We respectively denote the lower and upper bounds of probability $Pr\{f(p_j) \le f(q)\}$ as $LB\_F(p_j)$ and $UB\_F(p_j)$. Moreover, we also denote the lower and upper bounds of $S(j, a)$ as $LB\_S(j, a)$ and $UB\_S(j, a)$, respectively. Then, we have:*

$$LB\_S(j,a) = \begin{cases} (LB\_S(j-1,a) - UB\_S(j-1,a-1)) \cdot LB\_F(p_j) \\ \quad +LB\_S(j-1,a-1), \quad \text{if } LB\_S(j-1,a) > UB\_S(j-1,a-1); \\ (LB\_S(j-1,a) - UB\_S(j-1,a-1)) \cdot UB\_F(p_j) \\ \quad +LB\_S(j-1,a-1), \quad \text{otherwise.} \end{cases} \tag{5}$$

and

$$UB\_S(j,a) = \begin{cases} (UB\_S(j-1,a) - LB\_S(j-1,a-1)) \cdot LB\_F(p_j) \\ \quad +UB\_S(j-1,a-1), \ \text{if } UB\_S(j-1,a) < LB\_S(j-1,a-1); \\ (UB\_S(j-1,a) - LB\_S(j-1,a-1)) \cdot UB\_F(p_j) \\ \quad +UB\_S(j-1,a-1), \ \text{otherwise.} \end{cases} \tag{6}$$

Proof. Since $S(j, a) \in [LB\_S(j, a), UB\_S(j, a)]$ and $Pr\{f(p_j) \le f(q)\} \in [LB\_F(p_j), UB\_F(p_j)]$, we can derive $LB\_S(\cdot, \cdot)$ and $UB\_S(\cdot, \cdot)$ recursively from Eq. (4). Details are omitted due to space limit. □

Therefore, one remaining issue that needs to be addressed is how to obtain the lower and upper bounds of $Pr\{f(p_j) \le f(q)\}$. Our basic idea is to calculate the probability bounds by utilizing some offline pre-computed information for uncertain object p_j. Figure 2(a) illustrates this idea using a 2D example. Assume we pre-process uncertain object p_j by selecting a point s_1 within the uncertainty region $UR(p_j)$. With respect to coordinates of s_1, we obtain 4 quadrants I, II, III, and IV. Note that, since the PIR query can specify arbitrary monotonic function $f(\cdot)$, our pre-computed probability bounds should be able to be correct for any $f(\cdot)$. Clearly, in our example, the probability $Pr\{f(p_j) \le f(s_1)\}$ is always lower bounded by the probability that p_j falls into region III, denoted as $LB\_F(s_1)$, and upper bounded by the probability that p_j falls into regions (II ∪ III ∪ IV), denoted as $UB\_F(s_1)$. Thus, for point s_1, its corresponding pre-computed bounds are $LB\_F(s_1)$ and $UB\_F(s_1)$.

As a result, we have the following lemma to give lower and upper bounds of probability $Pr\{f(p_j) \le f(q)\}$, denoted as $LB\_F(p_j)$ and $UB\_F(p_j)$, respectively.

Lemma 4. *Assume we have l pre-selected points s_1, s_2, ..., and s_l in an uncertain object p_j, where each point s_i has pre-computed probability bounds $[LB\_F(s_i), UB\_F(s_i)]$. Given a monotonic function $f(\cdot)$ and a PIR query point q, without loss of generality, let $f(s_{i+1}) \le f(q) \le f(s_i)$ for $1 \le i < l$. Then, we can set $LB\_F(p_j)$ and $UB\_F(p_j)$ to $LB\_F(s_{i+1})$ and $UB\_F(s_i)$, respectively.*

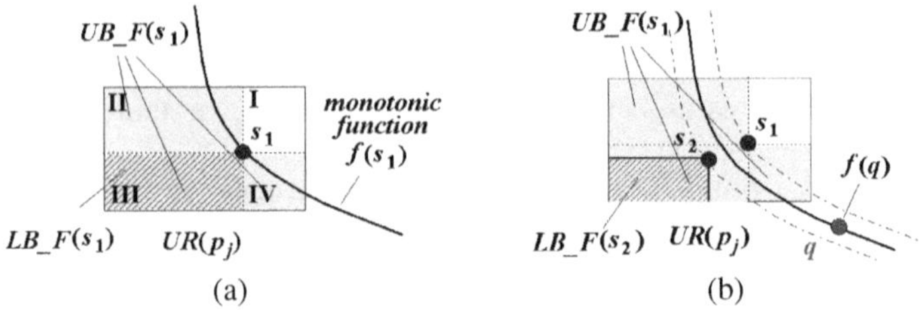

Fig. 2. Illustration of Computing $[LB\_F(p_j), UB\_F(p_j)]$ for any Monotonic Function $f(\cdot)$

Proof. Based on the definition of $LB\_F(s_{i+1})$ and the fact that $f(s_{i+1}) \leq f(q)$, we have $LB\_F(s_{i+1}) \leq Pr\{f(p_j) \leq f(s_{i+1})\} \leq Pr\{f(p_j) \leq f(q)\}$. Thus, $LB\_F(s_{i+1})$ is a lower bound of probability $Pr\{f(p_j) \leq f(q)\}$ (i.e. $LB\_F(p_j)$). The case of $UB\_F(p_j)$ is similar and thus omitted. □

Therefore, by Lemma 4, we can pre-compute probability bounds with respect to pre-selected points in each uncertain object. As an example in Figure 2(b), when we want to compute bounds $LB\_F(p_j)$ and $UB\_F(p_j)$ with respect to a monotonic function $f(\cdot)$, among all pre-selected points, we can choose two points, say s_1 and s_2, that have the closest scores to $f(q)$, say $f(s_2) \leq f(q) \leq f(s_1)$ (note: close scores can give tight bounds). Then, we only need to let $LB\_F(p_j) = LB\_F(s_2)$ and $UB\_F(p_j) = UB\_F(s_1)$. Intuitively, $LB\_F(p_j)$ is given by probability that p_j falls into the region filled with sloped lines (i.e. the rectangle with top-right corner s_2) in Figure 2(b), whereas $UB\_F(p_j)$ is the probability that p_j is in the grey region with respect to s_1.

In Lemma 4, for the l selected points, we need to store l pairs of probability bounds. We found through experiments that the pruning ability can converge with small l value.

4 Query Processing

4.1 Pruning Nodes/Objects

In order to efficiently answer the PIR query, we need to traverse the aR-tree and reduce the search cost by accessing as few nodes/objects as possible. As mentioned in Section 3.2, all the uncertain objects in the database can be partitioned into 3 disjoint sets, R^+, R_0, and R^-, with respect to score of query object (i.e. $f(q)$). In particular, to answer a PIR query, we only need to obtain the cardinality (i.e. aggregate) of sets R^+ and R_0 (i.e. resulting in $[k_{min}, k_{max}]$), as well as the uncertain objects in set R_0. Thus, those objects (or nodes only containing objects) in R^- can be safely pruned.

Therefore, one important issue to enable pruning a node/object is to identify its category. Specifically, each node e (or uncertainty region of uncertain object) in aR-tree is represented by a *minimum bounding rectangle* (MBR) in the form $[e_1^-, e_1^+; e_2^-, e_2^+; ...; e_d^-, e_d^+]$, where $[e_i^-, e_i^+]$ is the interval of e along the i-th dimension. Given a monotonic function $f(\cdot)$, the minimum and maximum scores for points in node e can be achieved (due to monotonic property of $f(\cdot)$) at d-dimensional points $e_{min}\langle e_1^-, e_2^-, ..., e_d^-\rangle$ and $e_{max}\langle e_1^+, e_2^+, ..., e_d^+\rangle$, respectively. Without loss of generality, we let $LB\_score(e) = f(e_{min})$ and $UB\_score(e) = f(e_{max})$. We have the lemma of pruning nodes below.

Lemma 5. *(Pruning Nodes/Objects) Given a monotonic function $f(\cdot)$, a query object q, and a node/object e, if it holds that $UB\_score(e) < f(q)$, then node/object e is in set R^- and it can be safely pruned during PIR query processing; if it holds that $LB\_score(e) > f(q)$, then node e is in set R^+, and we only need to count the aggregate $e.agg$ without accessing its children.*

4.2 Query Procedure

Up to now, we have discussed how to prune nodes/objects during PIR query processing without introducing false dismissals. Now we present the detailed query procedure for answering PIR queries. In particular, our PIR query procedure traverses the aR-tree index in a *best-first* manner. Meanwhile, the query procedure collects the aggregate (i.e. count) of set R^+ and the content of set R_0. Note that, here the number of uncertain objects under the nodes can be provided by aggregates in aR-tree nodes, without the necessity of visiting their children. After we obtain a complete set R_0, the upper-bound pruning can be conducted over R_0, as mentioned in Section 3.3. Finally, the remaining candidates (ranks) can be refined by computing their actual probabilities $Pr_{k\text{-}PIR}(q, f(\cdot))$.

Figure 3 illustrates the pseudo code of the PIR query procedure, namely procedure PIR_Processing. Specifically, we maintain a maximum heap $\mathcal{H}$ in the form $\langle e, key\rangle$, where e is an MBR node and key is defined as the upper bound of score for any point in node e (i.e. $key = UB\_score(e)$ defined in Section 4.1). Intuitively, nodes with high score upper bounds are more likely to contain objects in $R^+ \cup R_0$. Thus, we access

```
Procedure PIR_Processing {
  Input: aR-tree I constructed over D^U, query point q, a monotonic function f(·), and a probability threshold α
  Output: the PIR query result PIR_rlt
  initialize an empty max-heap H accepting entries in the form ⟨e, key⟩
  R_0 = φ; count(R^+) = 0;
  insert (root(I), 0) into heap H
  while H is not empty
     (e, key) = de-heap H
     if e is a leaf node
        for each uncertain object o ∈ e
           obtain lower/upper bound scores LB_score(o) and UB_score(o)
           if UB_score(o) ≥ f(q)               // o is in set R^+ or R_0
              if LB_score(o) ≤ f(q)            // o is in set R_0
                 R_0 = R_0 ∪ {o}
              else                             // o is in R^+
                 count(R^+) = count(R^+) + 1
     else                                      // intermediate node
        for each entry e_i in e
           obtain lower/bound scores LB_score(e_i) and UB_score(e_i)
           if UB_score(e_i) ≥ f(q)             // e_i contains objects in R^+ or R_0
              if LB_score(e_i) ≤ f(q)          // e_i may contain objects in R_0
                 insert ⟨e_i, UB_score(e_i)⟩ into heap H
              else                             // e_i contains objects in R^+
                 count(R^+) = count(R^+) + e_i.agg
  k_min = count(R^+) + 1; k_max = count(R^+) + |R_0| + 1;
  PIR_rlt = Upper-Bound_Pruning (R_0, f(·), q, α, k_max − k_min + 1);
  Refinement (R_0, f(·), q, α, PIR_rlt);
  return PIR_rlt
}
```

Fig. 3. PIR Query Processing

nodes of aR-tree in descending order of the score upper bound (i.e. key), and stop the traversal when all nodes/objects in $R^+ \cup R_0$ are visited/counted.

5 The PIR Query in High Dimensional Spaces

Up to now, we have illustrated how to answer the PIR query over uncertain data through a multidimensional index like R-tree [11]. Note, however, that the query efficiency of multidimensional indexes often rapidly degrades with the increasing dimensionality, which is also known as the "dimensionality curse" problem. That is, in the worse case, the query performance through index is even worse than that of a simple linear scan of the database. Therefore, in this section, we will discuss our PIR query processing problem in high dimensional spaces (i.e. uncertain objects have many attributes).

In order to break the dimensionality curse, previous works usually reduce the dimensionality of data objects and convert them into a lower dimensional points, on which index is constructed and queries (e.g. range or nearest neighbor query) are answered following a filter-and-refine framework. The existing reduction methods, such as SVD [15] and DFT [1], satisfy the property (a.k.a. *lower bounding lemma*) that the distance between any two reduced data points is a lower bound of that between their corresponding points in the original space. Although previous methods can successfully answer queries over precise data points, they are not directly applicable to uncertain scenario.

In the sequel, we propose an effective approach of dimensionality reduction over uncertain data, namely *uncertain dimensionality reduction* (UDR). Specifically, assume we have a d-dimensional uncertain object o with its uncertainty region $UR(o) = \langle o.A_1^-, o.A_1^+; o.A_2^-, o.A_2^+; ...; o.A_d^-, o.A_d^+\rangle$, where $[o.A_i^-, o.A_i^+]$ is the uncertain interval of object o along the i-th dimension. The UDR method first vertically divides dimensions of object o into w portions ($w < d$), each of which contains the same number, $\lceil d/w \rceil$, of dimensions (note: one portion may have less than $\lceil d/w \rceil$ dimensions). Without loss of generality, let the j-th partition ($1 \leq j \leq w$) correspond to $[o.A^-_{j\cdot\lceil d/w\rceil+1}, o.A^+_{j\cdot\lceil d/w\rceil+1}; ...; o.A^-_{(j+1)\cdot\lceil d/w\rceil}, o.A^+_{(j+1)\cdot\lceil d/w\rceil}]$. Next, our UDR approach maps uncertain object o onto a w-dimensional region, $UR(o^{(r)}) = \langle o^{(r)}.A_1^-, o^{(r)}.A_1^+; o^{(r)}.A_2^-, o^{(r)}.A_2^+; ...; o^{(r)}.A_w^-, o^{(r)}.A_w^+\rangle$, where $o^{(r)}.A_j^- = min\{o.A^-_{j\cdot\lceil d/w\rceil+1}, ..., o.A^-_{(j+1)\cdot\lceil d/w\rceil}\}$, and $o^{(r)}.A_j^+ = max\{o.A^+_{j\cdot\lceil d/w\rceil+1}, ..., o.A^+_{(j+1)\cdot\lceil d/w\rceil}\}$. Intuitively, $o^{(r)}.A_j^-$ and $o^{(r)}.A_j^+$ are the minimum and maximum possible values of uncertain intervals for dimensions in the j-th portion. Note that, heuristically, we want $o^{(r)}.A_j^-$ and $o^{(r)}.A_j^+$ are as close as possible such that we can use them to obtain tight score lower/upper bounds during the query processing (as will be discussed later). Therefore, we collect the mean statistics of $(o^{(r)}.A_i^- + o^{(r)}.A_i^+)/2$ for all objects in $\mathcal{D}$ along the i-th dimension ($1 \leq i \leq d$), sort d resulting mean values, and select every $\lceil d/w \rceil$ dimensions corresponding to consecutive mean values as a portion. This way, we can reduce the dimensionality (i.e. d) of each uncertain object to a lower w-dimensional space.

When a PIR query arrives with preference function $f(\cdot)$, we can traverse the index in the reduced space. For each object o we encounter, we can obtain its score upper/lower bounds by using $o^{(r)}.A_j^-$ ($o^{(r)}.A_j^+$) for parameters $o.A^-_{j\cdot\lceil d/w\rceil+1}, ...,$ and $o.A^-_{(j+1)\cdot\lceil d/w\rceil}$ ($o.A^+_{j\cdot\lceil d/w\rceil+1}, ...,$ and $o.A^+_{(j+1)\cdot\lceil d/w\rceil}$) in preference function $f(\cdot)$. The case of computing score upper/lower bounds for nodes is similar. Thus, similar to the query procedure discussed in Section 4.2, we collect the count of set R^+, as well as the content in set R_0, in the reduced space. Then, the upper-bound pruning can be conducted in the original d-dimensional space, and the remaining candidates are refined by computing their actual PIR probabilities.

6 Experimental Evaluation

In this section, we evaluate the performance of the *probabilistic inverse ranking* (PIR) query in the uncertain database. Specifically, we test the efficiency and effectiveness of our approaches on both real and synthetic data sets. For the real data sets, we use 2D geographical data sets[2], denoted as LB and RR, which contain bounding rectangles of 53,145 Long Beach county roads and 128,971 Tiger/Line LA rivers and railways, respectively, where these rectangles are treated as uncertainty regions for object locations.

In order to verify the robustness of our proposed methods, we also synthetically generate uncertain data sets as follows. To produce an uncertain object p, we first randomly generate a center location C_p in a data space $\mathcal{U} = [0, 1000]^d$, where d is the dimensionality of the data set. Then, we pick up a random radius, $r_p \in [r_{min}, r_{max}]$, for uncertainty region of p, where r_{min} and r_{max} are the minimum and maximum possible distances from p to C_p, respectively. As a third step, we randomly obtain a hyperrectangle that is tightly bounded by a hypersphere centered at C_p with radius r_p. Here, the hyperrectangle can be considered as uncertainty region $UR(p)$ of object p. Finally, we obtain 100 samples within $UR(p)$ to represent the data distribution of object p in $UR(p)$. We consider the *center location* C_p of *Uniform* and *Skew* (with skewness 0.8) distributions, denoted as lU and lS, respectively; moreover, we consider *radius* r_p of *Uniform* and *Gaussian* (with mean $\frac{r_{min}+r_{max}}{2}$ and variance $\frac{r_{max}-r_{min}}{5}$) distributions, denoted as rU and rS, respectively. Thus, we can obtain 4 types of data sets, $lUrU$, $lUrG$, $lSrU$, and $lSrG$. Without loss of generality, we assume samples in the uncertainty regions follow uniform distribution. Due to space limit, we only present the experimental results on the data sets mentioned above. Note, however, that, for data sets with other parameters (e.g. mean and variance of Gaussian distribution, the skewness of skew distribution, or sample distributions in uncertainty regions), the trends of the PIR query performance are similar and thus omitted.

For both real and synthetic data sets, we construct a multidimensional R-tree [11] by inserting the uncertainty regions of each uncertain object via standard "insert" function in the R-tree, where the page size is set to 4K. In order to evaluate the query performance on the R-tree index, we also generate 100 query objects following the distribution of center location C_p in synthetic data sets or center of rectangles in real data sets. Moreover, for the monotonic function $f(\cdot)$, we simply test the linear function given by $f(p) = \sum_{i=1}^{d} w_i \cdot p.A_i$, where w_i are weights uniformly distributed within $[0, 1]$, and $p.A_i$ are coordinates of possible object position p.

In our experiments, we evaluate our PIR (as well as high dimensional PIR) query processing, in terms of two measures, *filtering time* and *speed-up ratio*. Specifically, the

[2] The two real data sets are available online at URL: *http://www.rtreeportal.org/*.

filtering time is the time cost of the index traversal, which consists of both CPU time and I/O cost, where we incorporate each page access by penalizing $10ms$ (i.e. 0.01 second) [21]. Furthermore, the speed-up ratio is defined as the time cost of the *linear scan* method, mentioned in Section 2.2, divided by that of our proposed approaches. All the experiments are conducted on Pentium IV 3.4GHz PC with 1G memory, and experimental results are averaged over 100 runs.

6.1 Evaluation of the PIR Query

In the first set of experiments, we first illustrate the query performance of our PIR query processing approach over both real and synthetic data sets. Specifically, Figure 4 illustrates the filtering time of the index traversal on 6 data sets LB, RR, $lUrU$, $lUrG$, $lSrU$, and $lSrG$, where uncertain objects in the last 4 synthetic data sets have the radius range $[r_{min}, r_{max}] = [0, 3]$, probability threshold $\alpha = 0.5$, dimensionality $d = 3$, and data size $N = 30K$. In order to enable the upper-bound pruning, we pre-compute the probability bounds as discussed in Section 3.3 by setting the number of pre-selected points, l, per object to 15. The numbers over columns in the figure are the speed-up ratios of our PIR processing method compared with the linear scan. We can see that the required filtering time of our approach is small (e.g. 0.2-0.7 second), and the speed-up ratio indicates that our method outperforms the linear scan by about two orders of magnitude.

Next, we start to evaluate the robustness of our proposed PIR approach by varying different parameters on synthetic data sets. Due to space limit, we only report α and N.

Figure 5 varies the probability threshold α specified by PIR queries. In particular, we set α to 0.1, 0.3, 0.5, 0.7, and 0.9, where the radius range $[r_{min}, r_{max}] = [0, 3]$, dimensionality $d = 3$, data size $N = 30K$, and parameter $l = 15$. As expected, since the probability threshold α is only used for the upper-bound pruning, that is, checking whether or not the probability upper bound of each PIR candidate is smaller than α, the filtering time of the index traversal remains the same. On the other hand, the speed-up ratio of our approach compared with linear scan slightly increases with the increasing α since fewer candidates need to be refined in the refinement phase.

In the next set of experiments, we perform the scalability test on the size of uncertain data sets. Specifically, we vary the data size N from $10K$ to $100K$, and let the radius range $[r_{min}, r_{max}] = [0, 3]$, probability threshold $\alpha = 0.5$, dimensionality $d = 3$, and parameter $l = 15$. Figure 6 depicts the experimental results over 4 types of data sets, in terms of the filtering time and speed-up ratio. In figures, the filtering time increases

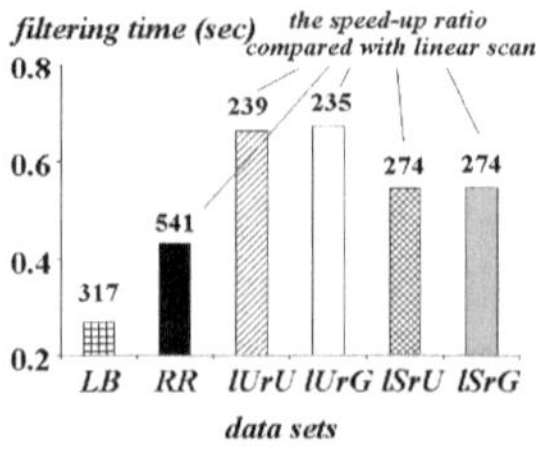

Fig. 4. PIR Performance vs. Real/Synthetic Data Sets

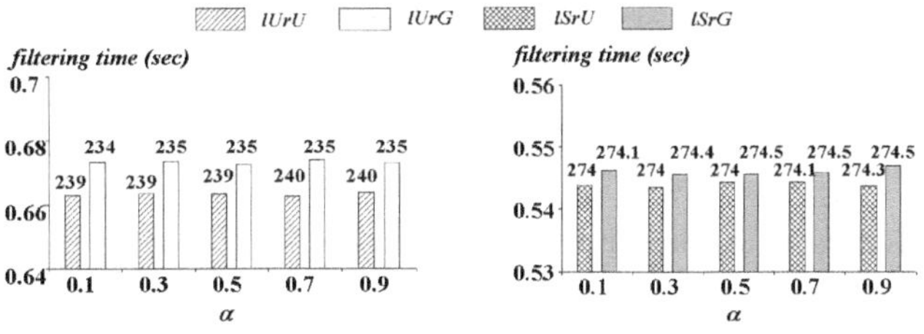

(a) lU (b) lS

Fig. 5. PIR Performance vs. Probability Threshold α (Synthetic Data)

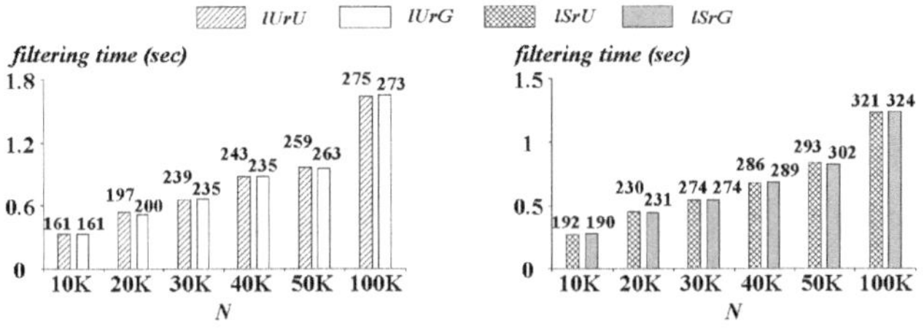

(a) lU (b) lS

Fig. 6. PIR Performance vs. Data Size N (Synthetic Data)

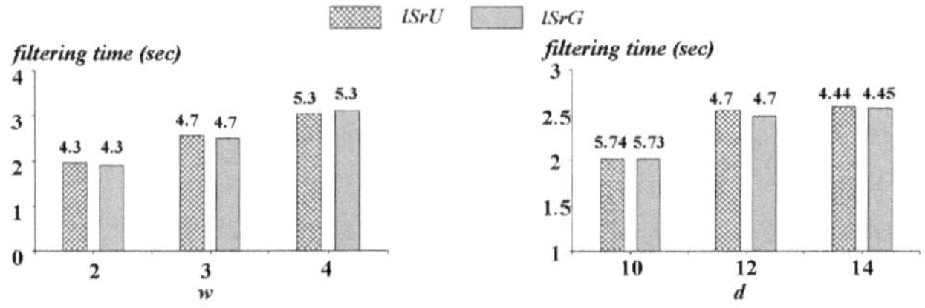

(a) *filtering time* vs. w (b) *filtering time* vs. d

Fig. 7. HD-PIR Performance vs. Parameters w and d (lS Synthetic Data)

smoothly with the increasing data sizes. Moreover, the speed-up ratio also becomes higher, when the data size is larger, which indicates the good scalability and robustness of our PIR query processing approach, compared with the linear scan method.

6.2 Evaluation of PIR Queries in High Dimensional Spaces

In this subsection, we present the experimental results of PIR query processing in high dimensional spaces (denoted as HD-PIR). Due to space limit, we only report results on $lSrU$ and $lSrG$ data sets by varying the reduced dimensionality w via our UDR reduction method, as well as the original dimensionality d. In particular, Figure 7(a) illustrates the filtering time of the HD-PIR query, where $w = 2, 3, 4$, data size $N = 30K$, radius range $[r_{min}, r_{max}] = [0, 3]$, probability threshold $\alpha = 0.5$, dimensionality

$d = 12$, and parameter $l = 15$. When w increases, the filtering time slightly increases, which is mainly due to the increasing I/O cost (as more dimensions take up more space). Furthermore, since large w gives tight bounds of $o^{(r)}.A_j^-$ and $o^{(r)}.A_j^+$ (as mentioned in Section 5), the resulting number of PIR candidates to be refined decreases when w is large. Thus, the speed-up ratio of our approach compared with linear scan increases with the increasing w. Figure 7(b) demonstrates the HD-PIR query performance where $d = 9, 12, 15$, $w = 3$, $N = 30K$, $[r_{min}, r_{max}] = [0, 3]$, $\alpha = 0.5$, and $l = 15$. From the figure, we find that when d becomes large, the filtering time of our HD-PIR query processing smoothly increases, whereas the speed-up ratio slightly decreases. This is reasonable, since more HD-PIR candidates are needed to be retrieved and refined.

7 Related Work

7.1 Inverse Ranking Query

In literature [4,12,9,14,20,8,29], there are many works on the *top-k query*, which is useful in a wide spectrum of applications, including decision making, recommendation raising, and data mining tasks. In particular, the top-k query retrieves k objects in a database that report the highest scores with respect to a monotonic function. The *quantile query* [22] retrieves an object in the database with the k-th highest score, where k is arbitrarily large. Lin et al. [22] studied efficient quantile query processing in the stream environment with precision guarantees. Inversely, Li et al. [19] proposed another important query type, *inverse ranking query*, which computes the rank of a user-specified query point in the "certain" database, which has many practical applications to determine the importance (i.e. rank) of any query object among peers.

Note that, Li et al. [19] studied the inverse ranking query in a database that contains *precise* data points. In contrast, our work focuses on the inverse ranking query over uncertain data (i.e. the PIR query), which is more challenging. Specifically, the inverse ranking query in the uncertain database has to be re-defined, as given in Definition 1, which needs to consider the data uncertainty and provides the confidence of uncertain query processing (i.e. α). Thus, techniques on inverse ranking query processing over precise points cannot be directly used in our problem (since scores of uncertain objects are now variables instead of fixed values).

7.2 Uncertain Query Processing

Uncertain query processing is very important in many applications. For instance, the Orion system [7] can monitor uncertain sensor data; the TRIO system [2] propose working models to capture the data uncertainty on different levels. In the context of uncertain databases, various queries have been proposed, including *range query* [6,5], *nearest neighbor query* [6,5,17], *skyline query* [25], *reverse skyline query* [21], and *similarity join* [16]. In addition, in the probabilistic database, the top-k query [28,27,13] has been studied under *possible worlds* semantics. In this paper, we consider PIR queries, which, to our best knowledge, no previous work has studied in uncertain databases.

8 Conclusions

Uncertain query processing has played an important role in many real-world applications where data are imprecise and uncertain. In this paper, we study an important query type in the context of uncertain database, namely *probabilistic inverse ranking* (PIR) query. Specifically, given a query object q, a PIR query obtains all the possible ranks that q may have in an uncertain database with probability above a threshold. In order to efficiently answer the PIR query, we propose effective pruning methods to reduce the PIR search space, and integrate them into an efficient query procedure. Furthermore, we also investigate the PIR query in high dimensional spaces. Extensive experiments have shown the efficiency and effectiveness of our proposed approaches to answer the PIR query, as well as its variants over real/synthetic data, under various parameter settings.

Acknowledgment

Funding for this work was provided by NSFC/RGC Joint Research Scheme N_HKUST602 /08.

References

1. Agrawal, R., Faloutsos, C., Swami, A.N.: Efficient similarity search in sequence databases. In: Lomet, D.B. (ed.) FODO 1993. LNCS, vol. 730. Springer, Heidelberg (1993)
2. Benjelloun, O., Das Sarma, A., Halevy, A.Y., Widom, J.: ULDBs: Databases with uncertainty and lineage. In: VLDB (2006)
3. Böhm, C., Pryakhin, A., Schubert, M.: The Gauss-tree: efficient object identification in databases of probabilistic feature vectors. In: ICDE (2006)
4. Chang, Y.-C., Bergman, L.D., Castelli, V., Li, C.-S., Lo, M.-L., Smith, J.R.: The Onion technique: indexing for linear optimization queries. In: SIGMOD (2000)
5. Cheng, R., Kalashnikov, D., Prabhakar, S.: Querying imprecise data in moving object environments. In: TKDE (2004)
6. Cheng, R., Kalashnikov, D.V., Prabhakar, S.: Evaluating probabilistic queries over imprecise data. In: SIGMOD (2003)
7. Cheng, R., Singh, S., Prabhakar, S.: U-DBMS: A database system for managing constantly-evolving data. In: VLDB (2005)
8. Das, G., Gunopulos, D., Koudas, N., Tsirogiannis, D.: Answering top-k queries using views. In: VLDB (2006)
9. Fagin, R., Lotem, A., Naor, M.: Optimal aggregation algorithms for middleware. In: PODS (2001)
10. Faradjian, A., Gehrke, J., Bonnet, P.: GADT: A probability space ADT for representing and querying the physical world. In: ICDE (2002)
11. Guttman, A.: R-trees: a dynamic index structure for spatial searching. In: SIGMOD (1984)
12. Hristidis, V., Koudas, N., Papakonstantinou, Y.: PREFER: A system for the efficient execution of multi-parametric ranked queries. In: SIGMOD (2001)
13. Hua, M., Pei, J., Zhang, W., Lin, X.: Ranking queries on uncertain data: A probabilistic threshold approach. In: SIGMOD (2008)
14. Ilyas, I.F., Aref, W.G., Elmagarmid, A.K.: Supporting top-k join queries in relational databases. VLDBJ (2004)

15. Ravi Kanth, K.V., Agrawal, D., Singh, A.: Dimensionality reduction for similarity searching in dynamic databases. In: SIGMOD (1998)
16. Kriegel, H.-P., Kunath, P., Pfeifle, M., Renz, M.: Probabilistic similarity join on uncertain data. In: Li Lee, M., Tan, K.-L., Wuwongse, V. (eds.) DASFAA 2006. LNCS, vol. 3882, pp. 295–309. Springer, Heidelberg (2006)
17. Kriegel, H.-P., Kunath, P., Renz, M.: Probabilistic nearest-neighbor query on uncertain objects. In: Kotagiri, R., Radha Krishna, P., Mohania, M., Nantajeewarawat, E. (eds.) DASFAA 2007. LNCS, vol. 4443, pp. 337–348. Springer, Heidelberg (2007)
18. Lazaridis, I., Mehrotra, S.: Progressive approximate aggregate queries with a multi-resolution tree structure. In: SIGMOD (2001)
19. Li, C.: Enabling data retrieval: By ranking and beyond. Ph.D. Dissertation, University of Illinois at Urbana-Champaign (2007)
20. Li, C., Chang, K.C.-C., Ilyas, I.F., Song, S.: RankSQL: Query algebra and optimization for relational top-k queries. In: SIGMOD (2005)
21. Lian, X., Chen, L.: Monochromatic and bichromatic reverse skyline search over uncertain databases. In: SIGMOD (2008)
22. Lin, X., Xu, J., Zhang, Q., Lu, H., Yu, J.X., Zhou, X., Yuan, Y.: Approximate processing of massive continuous quantile queries over high-speed data streams. In: TKDE (2006)
23. Mokbel, M.F., Chow, C.-Y., Aref, W.G.: The new casper: query processing for location services without compromising privacy. In: VLDB (2006)
24. Papadimitriou, S., Li, F., Kollios, G., Yu, P.S.: Time series compressibility and privacy. In: VLDB (2007)
25. Pei, J., Jiang, B., Lin, X., Yuan, Y.: Probabilistic skylines on uncertain data. In: VLDB (2007)
26. Prabhakar, S., Mayfield, C., Cheng, R., Singh, S., Shah, R., Neville, J., Hambrusch, S.: Database support for probabilistic attributes and tuples. In: ICDE (2008)
27. Re, C., Dalvi, N., Suciu, D.: Efficient top-k query evaluation on probabilistic data. In: ICDE (2007)
28. Soliman, M.A., Ilyas, I.F., Chang, K.C.: Top-k query processing in uncertain databases. In: ICDE (2007)
29. Tao, Y., Hristidis, V., Papadias, D., Papakonstantinou, Y.: Branch-and-bound processing of ranked queries. Inf. Syst. (2007)
30. Theodoridis, Y., Sellis, T.: A model for the prediction of R-tree performance. In: PODS (1996)

Consistent Top-k Queries over Time

Mong Li Lee, Wynne Hsu, Ling Li, and Wee Hyong Tok

School of Computing, National University of Singapore
{leeml,whsu,lil,tokwh}@comp.nus.edu.sg

Abstract. Top-k queries have been well-studied in snapshot databases and data streams. We observe that decision-makers are often interested in a set of objects that exhibit a certain degree of consistent behavior over time. We introduce a new class of queries called *consistent top-k* to retrieve k objects that are always amongst the top at every time point over a specified time interval. Applying top-k methods at each time point leads to large intermediate results and wasted computations. We design two methods, rank-based and bitmap, to address these shortcomings. Experiment results indicate that the proposed methods are efficient and scalable, and consistent top-k queries are practical in real world applications.

1 Introduction

Many real world applications (e.g online stock trading and analysis, traffic management systems, weather monitoring, disease surveillance and performance tracking) rely on a large repository of historical data in order to support users in their decision making. Often, the decisions made are based on the observations at a specific time point. Such observations may not be reliable or durable if we need to have consistent performance over a long time horizon. Indeed, the search for objects that exhibit consistent behavior over a period of time will empower decision-makers to assess, with greater confidence, the potential merits of the objects. We call this class of queries that retrieve objects with some persistent performance over time as *consistent top-k* queries.

Example 1. Stock Portfolio Selection. Investors who select a portfolio of stocks for long-term investment would have greater confidence in stocks that consistently exhibit above industry average in growth in earnings per share and returns on equity. These stocks are more resilient when the stock market is bearish. We can issue a consistent top-k query to return a set of stocks whose growth in earnings per share or return on equity are consistently among the top 20 over a period of time.

Example 2. Targeted Marketing. The ability to identify "high value" customers is valuable to companies who are keen on marketing their new products or services. These customers would have been with the company for some time with regular significant transactions. Marketing efforts directed to this group of customers are likely to be more profitable than those to the general customer base. The consistent top-k query allows these "high value" customers to be retrieved and companies can develop appropriate strategies to further their business goals.

Example 3. Awarding Scholarships. Organizations that provide scholarships have many criteria for selecting suitable candidates. One of the selection criteria often requires the students to have demonstrated consistent performance in their studies. The consistent top-k query can be used to retrieve this group of potential scholarship awardees.

X. Zhou et al. (Eds.): DASFAA 2009, LNCS 5463, pp. 51–65, 2009.

Example 4. Researchers with Sustained Publications. Research communities have started to track the publications of researchers in an effort to recognize their contributions to the respective areas. Recent attempts have applied top-k queries to identify authors with the most number of publications captured in online databases (e.g. dblp). Another possible dimension is to recognize outstanding researchers with sustained contributions. These authors can be identified using a consistent top-k query.

A consistent top-k query over a time series dataset returns a set of time series that are ranked among the top k at *every* time point. The size of the answer set could range from 0 to k. In practice, some applications may require the size of answer set to be precisely k. Figure 1 shows a sample *publication* dataset which records the number of conference and journals papers of researchers in the past 5 years. A consistent top-3 query on this dataset will yield $\{author_2, author_3\}$.

Consistent top-k vs. Top-k and Skyline Queries. Consistent top-k queries are different from top-k and skyline queries. Consistent top-k queries aim to retrieve the set of objects that show some consistent performance over time. But mapping a time series dataset to a multi-dimensional dataset, and using top-k or skyline query methods may not be able to retrieve the desired set of objects. Consider publications for the years 2001 and 2002 in Figure 1. A top-3 query over the period [2001-2002] requires a monotonic function to be defined over the two years. Table 1 lists one such monotonic function, the average publication count of researchers, in [2001-2002]. Based on this monotonic function, a top-3 query will retrieve authors $author_1$, $author_2$ and $author_4$. We observe that $author_1$ and $author_4$ are not consistent in their publications.

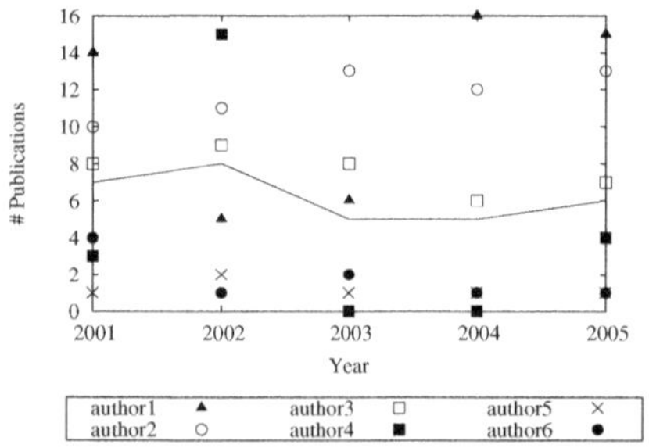

Fig. 1. Example *publication* dataset with $\{author_2, author_3\}$ being consistently in the top 3

Table 1. Average publication count of researchers in 2001 and 2002

id	2001	2002	Average #Publications
author1	14	5	9.5
author2	10	11	10.5
author3	8	9	8.5
author4	3	15	9.0
author5	1	2	1.5
author6	4	1	2.5

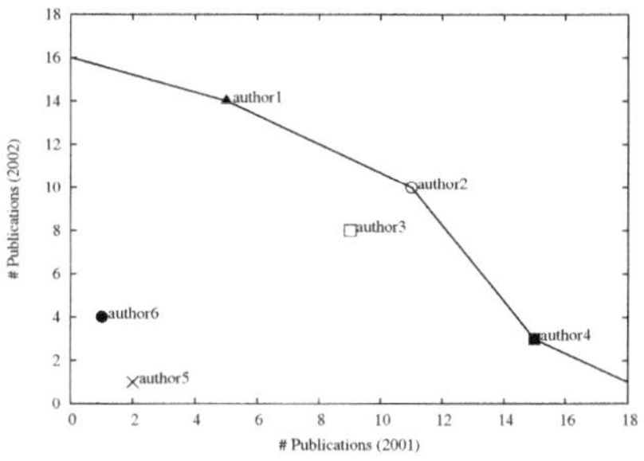

Fig. 2. Mapping 2001 and 2002 publication counts in Figure 1 to a 2-D space

Now let us map the 2001 and 2002 publication counts of the researchers in Figure 1 to a two-dimensional space as shown in Figure 2. The x-axis and y-axis in Figure 2 represent the number of publications of the researchers in 2001 and 2002 respectively. A skyline query retrieves a set of points from a multi-dimensional dataset which are not dominated by any other points [1]. Figure 2 shows the results of a skyline query ($author_1$, $author_2$, $author_4$). Note that $author_1$ and $author_4$ are retrieved although they do not have high publication counts in 2002 and 2001 respectively. Further, $author_3$ who has consistently published 8 or more publications in 2001 and 2002, is not retrieved by the skyline query.

A consistent top-k query can be processed as a set of top-k queries over a continuous time interval. For each time point in the time interval, we obtain the top-k answers and compute their intersections. This naive approach is expensive with many redundant computations. In this paper, we present a rank-based approach and a bitmap method to evaluate consistent top-k queries efficiently. In the rank-based approach, the time series at each time point are ranked; the time series with the highest value at a time point has a rank of 1. The rank of a time series is affected when it intersects with other time series. In Figure 1, $author_1$ is ranked first in 2001 while $author_2$ is ranked second. The rank of $author_1$ drops in 2002 because the time series for $author_1$ intersects with that of $author_2$, $author_3$ and $author_4$ between the two years. Based on this observation, we design an efficient algorithm to construct a compact *RankList* structure from a time series dataset. With this structure, we can quickly answer consistent top-k queries.

The rank-based approach also provides us with insights to design the second bitmap approach. If a time series has a rank of *r*, then it will be in the results for any top-k query, where $k \geq r$. Hence, our bitmap method assumes an application-dependent upper bound K, and utilizes bitmaps to indicate whether a time series is a candidate for a top-k query, $k \leq K$, at each time point. Using the bitwise operation *AND*, we can quickly determine whether a time series is part of the results of a consistent top-k query.

The key contributions of the paper are as follows:

1. We introduce the notion of consistent top-k queries and illustrate its relevance in various applications.
2. We propose a technique that utilizes rank information to answer consistent top-k queries. Algorithms to construct, search and update the *RankList* structure are presented.

3. We also design a bitmap approach to process both consistent top-k efficiently. Algorithms for constructing the *BitMapList*, as well as searching and updating this structure are detailed.
4. We present a suite of comprehensive experiment results to show the efficiency and scalability of the proposed methods. We also demonstrate that consistent top-k queries are able to retrieve interesting results from real world datasets.

To the best of our knowledge, this is the first work to address the problem of finding objects which exhibit some consistent behavior over time.

2 Related Work

Many methods have been developed to process top-k queries. Fagin's algorithm [2] carries out a sorted access by retrieving a list of top matches for each condition in the round-robin fashion until k objects that matches all the conditions are found. The TA algorithm [3] generalizes Fagin's algorithm by computing the score for each object seen during a sorted access. Subsequent work reduces the access cost in the TA algorithm by computing the probability of the total score of an object [4], and exploring bounds for the score for early pruning [5].

Top-k queries can be mapped to multidimensional range queries [6,7]. The key issue is to determine an appropriate *search distance* that would retrieve the k best matches. This distance can be determined from the statistics on the relations [7] or from a multidimensional histogram [6]. The method in [8] computes the search distance by taking into account imprecision in the optimizer's knowledge of data distribution and selectivity estimation while [9] adopts a sampling-based approach.

All the above studies are focused on retrieving top k answers at one time point. The work in [10] designs two algorithms to address the problem of top-k monitoring over sliding windows. The top-k monitoring algorithm re-computes the answer of a query whenever some of the current top-k tuples expire, while the skyband monitoring algorithm partially pre-computes future results by reducing top-k to k-skyband queries. These methods are not applicable to consistent top-k queries because they assume objects in the sliding windows have constant values until they expire whereas each time series may change its value at every time point.

3 Rank-Based Approach to Process Consistent Top-k Queries

A time series s is a sequence of values that change with time. We use $s(t)$ to denote the value of s at time t, $t \in [0, T]$. A time series database TS consists of a set of time series s_i, $1 \leq i \leq N$. Given a time series database TS, an integer k, and a time point t, a top-k query will retrieve k time series with the highest values at t. We use top-$k(TS, k, t)$ to denote the set of top k time series at t.

A consistent top-k query over a time interval $[t_u, t_v]$ retrieves the set of time series $U = \bigcap U_t$ where U_t = top-k(TS, k, t) $\forall\, t \in [t_u, t_v]$. Size of U ranges from 0 to k.

We can leverage the top-k operator in [6] to answer consistent top-k queries. This involves mapping the top-k query at each time point to a range query. The search distance is estimated using any methods in [7,6,8]. The basic framework is as follows:

1. For each time point t in the specified time interval, estimate the search distance $dist$ such that it encompasses at least k tuples with values greater than $dist$. Note that the search distance could vary for the different time points.
2. Use the estimated distances to retrieve the set of top k tuples at each time point, and compute the intersection.

There are two drawbacks to this approach: the intermediate relation to store the top k results at each time point is proportional to k and the number of time points; and many of these computations are wasted. These drawbacks can be overcome if we rank the time series s at each time point t according to their values $s(t)$. This rank, denoted by rank(s,t), is affected by the intersection of s with other time series between the time points $t-1$ and t.

We can construct an inverted list for each time series to store the rank information (see Figure 3). Each entry in the list consists of the rank of the time series at the corresponding time point. Note that it is only necessary to create an entry in the inverted list of a time series when its ranking is affected by an intersection. Further, if an existing time series does not have any value at some time point, then it will be ranked 0 at that time point. We call this structure *RankList*.

A consistent top-k query can be quickly answered with the RankList structure by traversing the list of each time series and searching for entries with rank values greater than k. The result is the set of time series which do not have such entries in their lists. For example, to answer a consistent top-3 query issued over the *publication* dataset in Figure 1, we traverse the list of $author_1$ in Figure 3 and find that the rank in the second entry is greater than 3. Hence, $author_1$ will not be in the answer set. In contrast, there are no entries in the lists of $author_2$ and $author_3$ with rank values greater than 3, and $\{author_2, author_3\}$ are the results of the consistent top-k query. We stop searching a list whenever an entry in the list with rank value greater than k is encountered.

RankList Construction and Search. Algorithm 1 shows the steps to construct the inverted list structure *RankLst* that captures the rank information for each time series in a dataset. The algorithm utilizes two arrays called *PrevRank* and *CurrRank* to determine if the ranking of a time series at the current time point has been affected by some intersection. Lines 3 and 5 initialize each entry in the *PrevRank* and *CurrRank* array to 0. This is because of the possibility of missing values for some time series. If a time series s has a value at time t, CurrRank[s] will be initialized to 1 (lines 7-9). We scan the database once and compare the values of the time series at each time point (lines 10-16). If the ranking of a time series s changes from time point $t-1$ to t, we create an entry in the inverted list of s to record its new rank (lines 17-22).

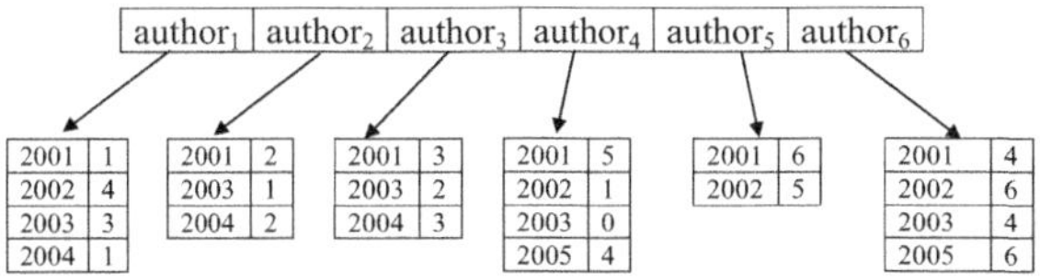

Fig. 3. RankList constructed for the publication dataset in Figure 1

Algorithm 2 takes as input the inverted list structure RankLst for the time series dataset, an integer k, and the start and end time points t_start, t_end. The output is S, a set of time series whose rank is always higher than k over $[t\_start, t\_end]$. For each time series s, we check if its rank is always higher than k in the specified time interval (lines 4-18). The entry with the largest time point that is less than or equal to t_start is located (line 5). If the entry does not exist or there is no value of s or the rank value of the entry is larger than k, then s is removed from S (lines 6-13). Otherwise, we check the ranks of the entries for s until the end time point.

Algorithm 1. Build RankList

```
Input: TS - time series database with
            attributes id, time and value
       T - total number of time points in TS
Output: RankLst - RankList structure for TS
initialize int [] PrevRank to 0;
for each time point t from 0 to T do
    initialize int [] CurrRank to 0;
    let S be the set of tuples with time t;
    for each tuple p ∈ S do
        initialize CurrRank[p.id] to 1;
    end for
    for each pair of tuples p, q ∈ S do
        if p.value < q.value then
            CurrRank[p.id]++;
        else
            CurrRank[q.id]++;
        end if
    end for
    for each time series s in TS do
        if CurrRank[s] != PrevRank[s] then
            Create an entry <t, CurrRank[s]> for time series s in RankLst;
            PrevRank[s] = CurrRank[s];
        end if
    end for
end for
return RankLst;
```

RankList Update. Insertion involves adding new values to an existing time series or adding a new time series into the dataset. This may affect the rankings of existing time series. Hence, we need to compare the new value with the values of existing time series at the same time point. Algorithm 3 takes as input a tuple $<p, t, p(t)>$ to be inserted and checks for the set of existing time series S whose values are smaller than $p(t)$ at time point t (lines 6-17). We obtain the rank of $s \in$ S from the entry which has the largest time point that is less than or equal to t and store it in the variable *PrevRank* (line 7-8), and try to retrieve the entry $<t$, rank$>$ for s. If it exists, we increase the rank by 1, otherwise, we insert a new entry for s at t (lines 9-12). Updating the rank of s at t may affect its rank at time $t+1$. Lines 14-17 check if an entry exists for s at $t+1$. If it does not exist, we create an entry with *PrevRank* at $t+1$ and insert into RankList (lines 15-16). We update the rank for time series p of the new value at t using *CurrRank* (lines 24-28). The algorithm also checks the entry for p at $t+1$ (line 29). If the entry does not exist, we insert an entry with rank 0 for p (lines 30-32). Similarly, the deletion of a value from a time series dataset may affect entries in the RankList structure which needs to be updated accordingly.

Algorithm 2. RankList Search

```
Input: RankLst - RankList structure of TS;
          t_start, t_end - start and end time points;
          integer k;
Output: A - set of time series that are in top k
          over [t_start, t_end];
initialize A to contain all the time series in TS;
for each time series s in A do
    locate the entry <t, rank> for s in the RankLst with the largest time point that is less than or equal to t_start;
    if entry not exist then
        A = A - s;
        CONTINUE;
    end if
    while t ≤ t_end do
        if rank > k or rank = 0 then
            A = A - s;
            break;
        else
            entry = entry.next;
        end if
    end while
end for
return A;
```

Algorithm 3. RankList Insert

```
Input: TS - database with attributes id, time and value
          RankLst - RankList structure of TS;
          <p, t, p(t)> - a tuple to be inserted;
Output: RankLst - updated RankList structure for TS
initialize int CurrRank to 1;
let S be the set of tuples with time t in TS;
for each tuple s ∈ S do
    if p(t) > s.value then
        locate the entry e for s.id in RankLst which has the largest time point that is less than or equal to t;
        let PrevRank = e.rank;
        if e.time = t then
            increment e.rank by 1;
        else
            create an entry <t, PrevRank + 1> for s.id and insert into RankLst;
        end if
        locate the entry e at time t + 1 for s.id in RankLst;
        if entry does not exist then
            create an entry <t + 1, PrevRank> for s.id and insert into RankLst;
        end if
    else
        CurrRank++; /* p(t) < s.value */
    end if
end for
locate the entry e for p in the RankLst which has the largest time point that is less than or equal to t;
if e.rank ≠ CurrRank then
    if e.time = t then
        replace e.rank with CurrRank;
    else
        create an entry <t, CurrRank> for p and insert into RankLst;
    end if
    locate the entry e' at time point t + 1 for p in RankLst;
    if e' does not exist then
        create an entry <t + 1, 0> for p and insert into RankLst;
    end if
end if
return RankLst;
```

4 Bitmap Approach to Process Consistent Top-k Queries

The bitmap approach uses a set of bitmaps to encode if a time series s is in the result set for a top-k query at a time point. The construction of the bitmap assumes an application-dependent upper bound K where K >> k in a top-k query. For each time series s at a time point t, if the rank of s at t is r, then the rth to the Kth bits are set to 1, $r \leq K$.

Figure 4 shows the bitmap that is created for Figure 1 with $K = 6$. In Figure 4, we observe that $author_1$ has the most number of publications at time point *2001*, i.e. $author_1$ is ranked 1. Hence, the bitmap for $author_1$ at 2001 is *111111*. Similarly, for $author_6$ whose rank is 4 at time point *2001*, the fourth-bit from the left and its subsequent bits are all set to 1. This produces a bitmap of *000111*. Note that some of the entries in Figure 4 are marked with a *X* to denote missing values. The bitmap representation can easily handle missing values by mapping *X* to *111111* if the user allows skipping of time points; if not, then the bits for *X* are all set to 0 (i.e. 000000).

	$author_1$	$author_2$	$author_3$	$author_4$	$author_5$	$author_6$
2001	111111	011111	001111	000011	000001	000111
2002	000111	X	X	111111	000011	000001
2003	001111	111111	011111	X	X	000111
2004	111111	011111	001111	X	X	X
2005	X	X	X	0001111	X	000001

Fig. 4. Bitmap Structure

Algorithm 4. Build BitmapList

1: **Input:** TS - time series database with attributes *id*, *time* and *value*;
T - total number of time points in TS;
K - upper bound for top-k queries.
2: **Output:** BitmapLst - Bitmap structure for TS
3: **for** each time point t from 0 to T **do**
4: Let TS_t be the set of time series in TS at t;
5: Sort TS_t in descending order;
6: offset = 0;
7: **for** each time series s in TS_t **do**
8: Create a bitmap b and set to 1 all the bits from 0 to (K-offset);
9: Create an entry <t, b> for time series s in BitmapLst;
10: offset++;
11: **end for**
12: **end for**
13: return BitmapLst;

BitmapList Construction and Search. Algorithm 4 shows the steps to construct the bitmap representation. In Line 5, the time series at time t are sorted in descending order. The bits for each time series are set according to their ranks at each time point (Lines 7 to 11). If a time series has rank r at time t, then we set the rth bit from the left and all its subsequent bits to 1, that is, $(K - r)^{th}$ to the 0th bit from the right.

Algorithm 5 uses the bitmap encodings to find the set of time series that are consistently within the top-k in the specified time interval. In Line 4, we create the bit mask, *topkMask* where the kth bit from the left is set to 1. Lines 6 to 11 iterate through all the time points in the query interval, and perform a bitwise AND between the *topkMask* and the corresponding bitmap entry, $<t, b>$, for a time series, *s*, at a given time point *t*. In Line 12, we check whether *result* is non-zero. If *result* AND with *topkMask* is 1, the time series *s* is added to the answer set for the consistent top-k query.

Algorithm 5. BitmapList Search

```
1:  Input: t_start, t_end - start and end time points;
           integer k;
2:  Output: A - set of time series that are in top k
            over [t_start, t_end];
3:  A = {};
4:  topkMask = (1 << k) ;
5:  for each time series s in TS do
6:      for each time point t between t_start, t_end do
7:          result = topkMask & <t, b> for s ;
8:          if ( result == 0 ) then
9:              break;
10:         end if
11:     end for
12:     if ( (result & topkMask) then
13:         A = A ∪ s;
14:     end if
15: end for
16: return A;
```

Algorithm 6. BitmapList Insert

```
1:  Input: TS - database with attributes id, time and value
    BitmapLst - Bitmap structure of TS;
    <p, t, p(t)> - a tuple to be inserted;
2:  Output: BitmapLst - Updated Bitmap structure for TS
3:  let S be the set of tuples with time t, and value < p(t) in TS
4:  for each tuple s ∈ S do
5:      locate the bitmap b in BitmapLst;
6:      b = b >> 1;
7:  end for
8:  Insert <p, t, p(t)> into TS;
9:  Insert into BitmapLst bitmap for <p, t, p(t)> ;
10: return BitmapLst;
```

BitmapList Updates. The bitmap approach is insertion-friendly. When a new time series needs to be inserted, we only require the bitmap of the affected time series to be right-shifted. Algorithm 6 takes as input a tuple $<p, t, p(t)>$ to be inserted. The algorithm retrieves the set of time series S whose values are smaller than *p(t)* at time point t (line 3). It then retrieves the corresponding bitmap *b* from *BitmapLst* (line 5). *b* is then right-shifted by one bit (line 6).

We use Figure 4 and Table 1 to illustrate the insertion algorithm. Suppose we want to add the time series entry $<author_7, 2001, 80>$ into Table 1. The time series affected at time point 2001 are $\{author_4, author_5, author_6\}$. From Figure 4, we observe that their corresponding bitmaps are {000011, 000001, 000111} respectively. The insertion of this new entry causes their corresponding bitmaps to be right-shifted by 1. The updated bitmaps for $\{author_4, author_5, author_6\}$ are $\{000001, 000000, 000011\}$ respectively. The newly inserted entry is then assigned a bitmap of 000111. Updating the bitmap structure for deletion is similar and omitted due to space constraints.

Table 2. Parameters of dataset generator

Parameter	Range	Default
Number of time series N	[100, 500]	100
Number of time points T	[5000, 20000]	10000
k	[50, 250]	50
Length of query interval L	[5000, 10000]	10000
Percentage of intersection points	[2%, 10%]	5%

5 Performance Study

We present the performance analysis for the rank-based, bitmap, and top-k approaches. The various algorithms including the top-k method in [6] are implemented in Java. All the experiments are carried out on a 2GHz Pentium 4 PC with 1 GB RAM, running WinXP. Each experiment is repeated 5 times, and the average time taken is recorded.

We use synthetic datasets to evaluate the efficiency of the proposed methods, and three real world datasets to demonstrate the effectiveness of consistent top-k queries. The data generator produces time series datasets with attributes *id*, *time* and *value*. Table 2 shows the range of values for the various parameters and their default values. The upper bound K for the *BitmapList* structure is set to 64. The real world datasets comprise of (a) the DBLP publication dataset; (b) a student dataset from our department; and (c) the stock dataset from the UCR Time Series Data Mining Archive.

5.1 Experiments on Synthetic Datasets

We first present the results of experiments to evaluate the sensitivity and scalability of the proposed methods.

Effect of Query Selectivity. The selectivity of consistent top-k queries is determined by the value of k and the length of the query interval.

In the first set of experiments, we vary the value of k between 10 to 60 (inclusive) and set the length of the query interval to be 10000. Figure 5(a) shows the the response time of the three methods in log scale. We observe that the *Rank* and the *Bitmap* methods outperforms the *Top-k* method by a large margin. The latter needs to process all the time series at all the time points in the query interval regardless of the value of k. In contrast, the *Rank* and *Bitmap* methods quickly eliminate those time series which would not be in the consistent top-k. This helps to limit the increase in response time as k increases.

The second set of experiments varies the length of the query interval between 2000 and 10000 (inclusive). The result is shown in Figure 6(a). As expected, the response time for the *Top-k* method increases when we increase the length of the query interval. Figure 6(b) shows that the response time for the *Bitmap* method increases at a slower pace than the *Rank* method, demonstrating the advantages of using bitwise operations.

Effect of Number of Intersection Points. Next, we examine whether the number of intersection points has any effect on the runtime. We fix the number of time series to be 100 and vary the percentage of intersection points among the time series from 2% to 10%. This is achieved by controlling the ranking of the time series from one time

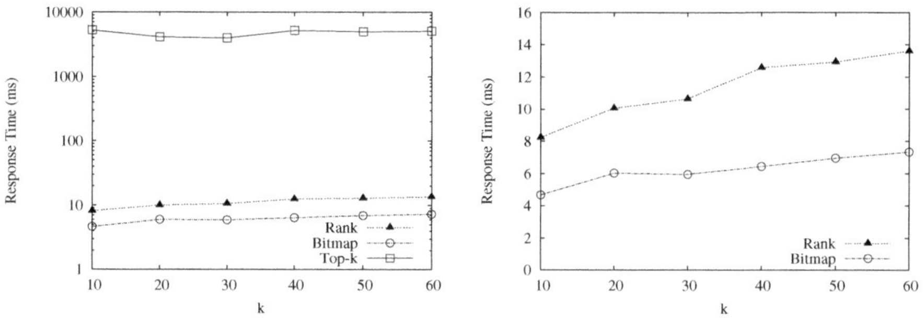

(a) Top-k vs. Rank and Bitmap (in log scale) (b) Rank vs. Bitmap

Fig. 5. Effect of varying k

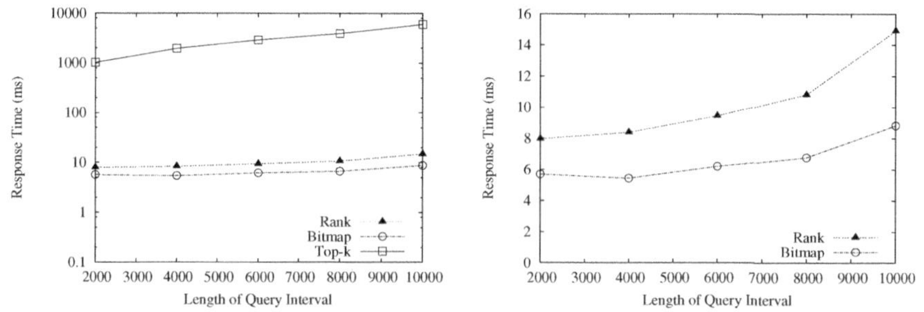

(a) Top-k vs. Rank and Bitmap (in log scale) (b) Rank vs. Bitmap

Fig. 6. Effect of varying query interval

point to another. Figure 7(a) shows the response time of the three methods in log scale. Overall, the *Rank* and *Bitmap* methods outperform the *Top-k* method. A closer look also reveals that increasing the percentage of intersection points has minimal impact on the runtime of both the *Bitmap* and *Rank* methods (see Figure 7(b)).

Scalability. Finally, we analyze the scalability of the three approaches. The size of a time series dataset is determined by the number of time points and the the number of time series in the database. Figures 8(a) and 9(a) show similar trends where the *Top-k* method does not scale well compared to the *Rank* and *Bitmap* methods. Figures 8(b) and 9(b) indicate that the Bitmap approach consistently performs much better than the Rank approach.

5.2 Experiments on Real World Datasets

We demonstrate the usefulness of consistent top-k queries in real world applications.

DBLP Dataset. We extract the number of papers published by researchers in several database conferences (VLDB, SIGMOD, ICDE) from the DBLP dataset (http://dblp.uni-trier.de/xml/) between 1807 to 2007. Each time point consists of a maximum

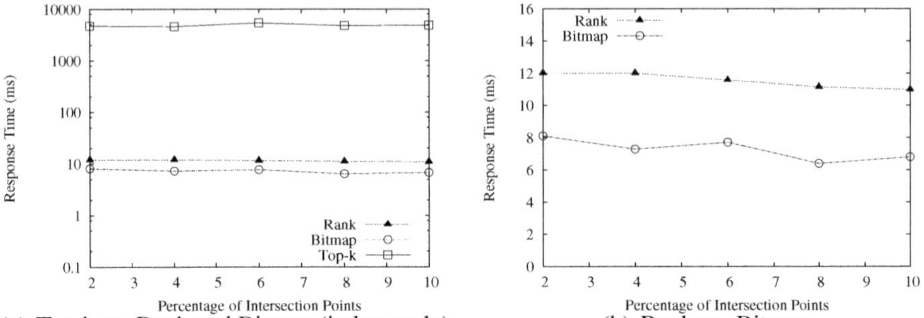

(a) Top-k vs. Rank and Bitmap (in log scale) (b) Rank vs. Bitmap

Fig. 7. Effect of varying number of intersection points

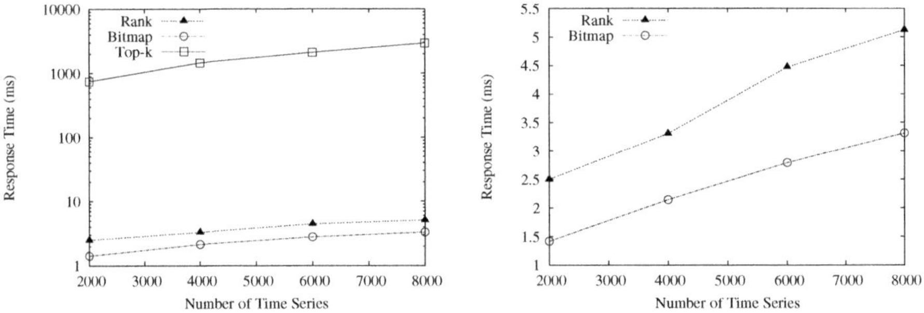

(a) Top-k vs. Rank and Bitmap (in log scale) (b) Rank vs. Bitmap

Fig. 8. Effect of varying number of time series

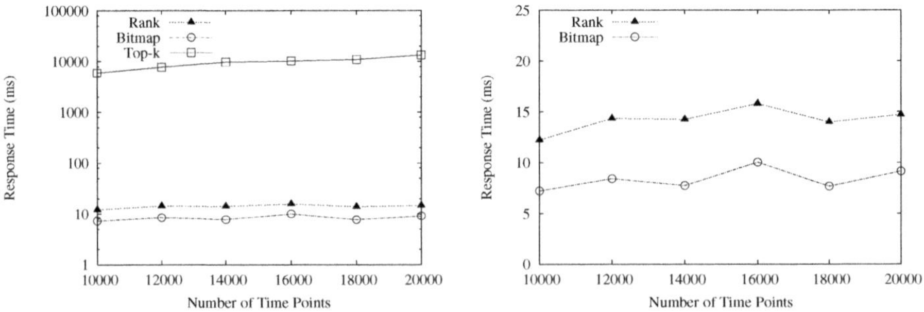

(a) Top-k vs. Rank and Bitmap (in log scale) (b) Rank vs. Bitmap

Fig. 9. Effect of varying number of time points

Table 3. Top-10 database researchers by year

2007		2006		2005	
Author	#Pubs	Author	#Pubs	Author	#Pubs
Philip S. Yu	15	Jiawei Han	11	N. Koudas	9
N. Koudas	11	B.C. Ooi	8	S. Chaudhuri	8
Haixun Wang	11	N. Koudas	8	D. Srivastava	8
Jiawei Han	11	R. Ramakrishnan	8	Kevin Chen	7
D. Srivastava	10	Jun Yang	7	B.C. Ooi	7
H. V. Jagadish	9	H. V. Jagadish	7	C. Faloutsos	6
Kevin Chen	9	G. Weikum	6	D. J. DeWitt	6
AnHai Doan	8	D. Srivastava	6	A. Ailamaki	6
Christoph Koch	8	D. Agrawal	6	K.L. Tan	6
S. Chaudhuri	8	K.L. Tan	6	G. Weikum	6
2004		**2003**		**2002**	
Author	#Pubs	Author	#Pubs	Author	#Pubs
S. Chaudhuri	11	D. Srivastava	16	M. N. Garofalakis	9
B.C. Ooi	8	N. Koudas	9	D. Srivastava	7
Walid G. Aref	7	S. Chaudhuri	9	D. Agrawal	7
Alon Y. Halevy	7	Michael J. Franklin	8	J. F. Naughton	7
J. M. Hellerstein	7	J. F. Naughton	7	G. Weikum	6
Jennifer Widom	6	Stanley B. Zdonik	6	Dimitris Papadias	6
Jiawei Han	6	K.L. Tan	6	C. Mohan	6
D. Srivastava	6	L. V. S. Lakshmanan	6	N. Koudas	6
J. F. Naughton	6	Rakesh Agrawal	6	S. Chaudhuri	6
K.L. Tan	6	B.C. Ooi	6	Rajeev Rastogi	6

Table 4. Consistent Top-10 performance of database researchers

Year	Researchers
2002-2004	D. Srivastava, S. Chaudhuri , J. F. Naughton
2003-2005	D. Srivastava, S. Chaudhuri, B.C. Ooi, K.L. Tan
2004-2006	D. Srivastava, B.C. Ooi, K.L. Tan
2005-2007	D. Srivastava, N. Koudas

3-year period

Year	Researchers
2002-2005	D. Srivastava, S. Chaudhuri
2003-2006	D. Srivastava, B.C. Ooi, K.L. Tan
2004-2007	D. Srivastava

4-year-period

Year	Researchers
2002-2006	D. Srivastava
2003-2007	D. Srivastava

5-year period

of 7263 authors. We use consistent top-k queries where $k = 10$ to analyze the publications of researchers over the period 2002 to 2007.

Table 3 shows the top 10 database researchers with the highest publication count for the years 2002 to 2007. The list of top 10 researchers fluctuates from year to year. Table 4 gives the consistent top-10 researchers over a 3-, 4-, and 5-year period which provide more insights into researchers with sustained publications. For example, there are two researchers (D. Srivastava and N. Koudas) who are consistently within the top-10 for 2005-2007. In addition, D. Srivastava has sustained his performance over a longer time period (as evidence from the 3 to 5 years consistent top-k analysis). Such insights cannot be easily gleaned from the results of top-k queries.

Student Dataset. We obtain a 8-year student dataset from our department. This dataset has 3800 records that capture the students' GPA for each semester. Other attributes in the records include studentID, gender, entrance exam code and year of enrollment. We group the students according to their year of enrollment and issue consistent top-k queries to retrieve the top 20% students that show consistent performance throughout their undergraduate studies. Figure 10 shows the percentage of consistent performers

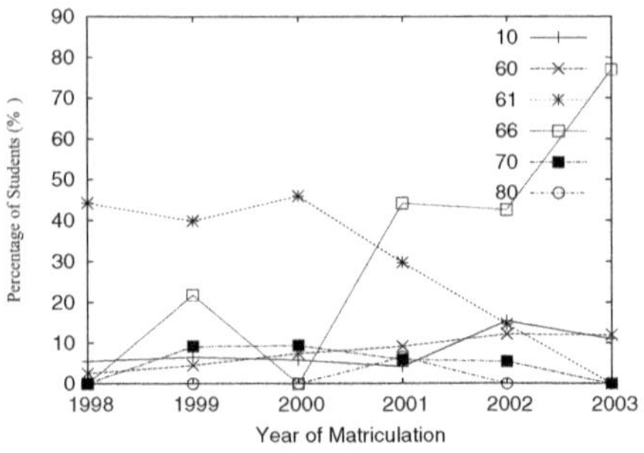

Fig. 10. Top 20% students for each batch group by entrance exam code

Table 5. Percentage gains of stocks retrieved by consistent top-k queries

k	2001	2002	2003	Total
10	8.3%	16.4%	34.8%	18%
20	6.76%	9.8%	17.97%	10.6%
30	7.5%	9.6%	6.75%	8.2%
40	10.9%	11.9%	13.3%	11.8%
50	5.2%	4%	7.2%	5.1%

grouped by entrance exam code (after normalization). The entrance exam code gives an indication of the education background of the students.

We observe that from 1998 to 2000, the majority of the consistent performers (top 20%) are students with an entrance code 61. However, from 2001 onwards, the top consistent performers shifted to students with entrance code 66. An investigation reveals that due to the change in the educational policy, the admission criteria for students with entrance code 61 has been relaxed, leading to a decline in the quality of this group of students. This trend has been confirmed by the data owner. The sudden increase in the quality of students with entrance code of 66 from 2001 onwards is unexpected and has motivated the user to gather more information to explain this phenomena.

Stock Dataset. The stock dataset from the UCR Time Series Data Mining Archive records the daily prices for 408 stocks from 1995 to 2003. We retrieve the opening and closing prices for each stock to determine their gains for each day. Based on the average gains of all the stocks over the period 1997 - 2000, we issue a top-k query to retrieve a set of k stocks, where k varies from 10 to 50. We also issue consistent top-k queries to retrieve a second set of k stocks over the same period. We compare the performance of these two sets of stocks over the period of 2001 - 2003 by computing their gains for each year as well as their total gains over the three years.

Table 5 shows that the set of stocks retrieved by consistent top-k queries consistently attain higher gains than the stocks retrieved by top-k queries. This strengthens our confidence that consistency is important to identify the potential merits of stocks.

6 Conclusion

In this work, we have introduced a new class of consistent top-k queries for time series data. The consistent top-k query retrieves the set of time series that is always within top k for all time points in the specified time interval T. We have examined how consistent top-k queries can be answered using standard top-k methods and proposed two methods to evaluate consistent top-k queries. The first method is a *RankList* structure to capture the rank information of the time series data. The proposed structure can be easily implemented on any relational database system. The second bitmap approach leverages on bitwise operations to answer consistent top-k queries.

The results of extensive experiments on synthetic datasets indicate that both the rank and bitmap approaches are efficient and scalable, and outperform top-k methods by a large margin. Further, the bitmap approach is able to handle consistent top-k queries better than the rank-based approach. We have also demonstrated that consistent top-k queries are useful in three real world applications. In particular, we can easily identify researchers with sustained publications over the years, recognize shifts in student quality as well as select a portfolio of stocks that have good potentials.

References

1. Borzsonyi, S., Kossmann, D., Stocker, K.: The skyline operator. In: IEEE ICDE, pp. 421–430 (2001)
2. Fagin, R.: Fuzzy queries in multimedia database systems. In: PODS, pp. 1–10 (1998)
3. Fagin, R., Lotem, A., Naor, M.: Optimal aggregation algorithms for middleware. In: ACM PODS, pp. 102–113 (2001)
4. Theobald, M., Weikum, G., Schenkel, R.: Top-k query evaluation with probabilistic guarantees. In: VLDB (2004)
5. Bruno, N., Gravano, L., Marian, A.: Evaluating top-k queries over web-accessible databases. In: IEEE ICDE, pp. 369–380 (2002)
6. Bruno, N., Chaudhuri, S., Gravano, L.: Top-k selection queries over relational databases: Mapping strategies and performance evaluation. In: ACM TODS, pp. 153–187 (2002)
7. Chaudhuri, S., Gravano, L.: Evaluating top-k selection queries. In: VLDB, pp. 397–410 (1999)
8. Donjerkovic, D., Ramakrishnan, R.: Probabilistic optimization of top n queries. In: VLDB, pp. 411–422 (1999)
9. Chen, C.-M., Ling, Y.: A sampling-based estimator for top-k selection query. In: IEEE ICDE (2002)
10. Mouratidis, K., Bakiras, S., Papadias, D.: Continuous monitoring of top-k queries over sliding windows. In: ACM SIGMOD, pp. 635–646 (2006)

Effective Fuzzy Keyword Search over Uncertain Data

Xiaoming Song, Guoliang Li, Jianhua Feng, and Lizhu Zhou

Department of Computer Science, Tsinghua University, Beijing 100084, China
songxm07@mails.tsinghua.edu.cn, {liguoliang, fengjh, dcszlz}@tsinghua.edu.cn

Abstract. Nowadays a huge amount of data were automatically generated or extracted from other data sources. The uncertainty and imprecision are intrinsic in those data. In this paper, we study the problem of effective keyword search over uncertain data. We allow approximate matching between input keywords and the strings in the underlying data, even in the presence of minor errors of input keywords. We formalize the problem of fuzzy keyword search over uncertain data. We propose efficient algorithms, effective ranking functions, and early-termination techniques to facilitate fuzzy keyword search over uncertain data. We propose a lattice based fuzzy keyword search method to efficiently identify top-k answers. Extensive experiments on a real dataset show that our proposed algorithm achieves both high result quality and search efficiency.

1 Introduction

Keyword search is a proven and widely accepted mechanism for querying information from various data sources. The database research community has introduced keyword search capability into relational databases [5], XML databases [3], graph databases [1], and heterogeneous data sources [4]. However, existing studies treated the data as precise data, but neglect the fact that the uncertain and imprecise data are inherent in some data sources.

Recently, many algorithms [2,6] have been proposed to study the problem of querying and modeling uncertain data. Most of them use the possible world semantics and use the *generation rule* to manage the conflicts. But in the keyword-search area, users want to get the top-k relevant answers and do not care about the inner relationship. We assign a confidence between 0 and 1 to describe the trustworthy of the data and incorporate it into a universal ranking mechanism to find the most relevant results.

To the best of our knowledge, few existing works has explored the keyword search problem over uncertain data. We in this paper proposed a fuzzy keyword search method. We use edit distance to measure the similarity between a given keyword and a string in the candidates to deal with imprecise data. We propose a lattice based top-k search method to answer the keyword query. We devise a universal ranking mechanism, which takes the textual similarity, edit distance and the confidence into consideration.

X. Zhou et al. (Eds.): DASFAA 2009, LNCS 5463, pp. 66–70, 2009.

2 Gram-Based Fuzzy Keyword Search

2.1 Formulation

Fuzzy Match: Given a query string q and a candidate string s, a similarity function f, a threshold δ, we say string q and string s *fuzzy match* if $f(q, s) \leq \delta$.

Definition 1. *The Fuzzy Keyword Search Problem: Given a collection of documents D, a similarity function f, and a set of query strings $Q = \{k_1, k_2, \cdots, k_n\}$, a threshold δ, retrieve the relevant documents from D, which contain string s_i, such that $f(s_i, k_i) \leq \delta$ for $1 \leq i \leq m$. We use the edit distance to quantify the similarity between two strings.*

Definition 2. *Q-gram: The gram with a fixed length of q is called q-gram. Given a string s and a positive integer q, we extend s to a new string s' by adding $q-1$ copies of a special character in the prefix that do not appear in the string and $q-1$ copies of another special character in the suffix. The set of q-grams of s, denoted by $G(s, q)$ is obtained by sliding a window of length q over the characters of string s'.*

2.2 Finding Similar Strings

In our proposed algorithm, we sort the string ids (denoted by *sid*) on each inverted list in an ascending order. Given a query string s, we first get it's set of *q-grams* $G(s, q)$. Let $N = |G(s, q)|$ denote the number of lists corresponding to the grams from the query string. We scan the N inverted lists one by one. For each string id on each list, We return the string ids that appear at least T times on the lists.

The answer set R returned is a set of candidate string ids. We then use a dynamic programming algorithm to eliminate the false positives in the candidates R, and then get those string ids that have an edit distance smaller than or equals to δ.

2.3 Finding Answers

We use inverted lists to index the documents. For each input keyword, there will be a set of fuzzy matched strings. We first compute the union of document ids for each fuzzy matched strings and get the list of document ids that contain the keyword or its similar words. Then, we compute the intersection of the union lists, which is exactly the set of document ids that contain all the input keywords approximately. There are two basic methods to compute the intersection. The first one is to merge-join the union lists. This method is inefficient if there are large number of elements in each list. The second method is to first select the union list with the smallest size, and then use the elements in the shortest list to probe other lists using binary-search methods or hash-based methods.

3 Ranking

A document with a higher confidence is more trustable and should be ranked higher in the answer list. So in the ranking mechanism, we need to take the confidence together with the textual similarity into consideration.

TF·IDF based methods for ranking relevant documents have been proved to be effective for keyword search in text documents. We employ *TF·IDF* based methods for computing textual similarity. Given an input query $Q = \{k_1, k_2, \cdots, k_n\}$, and a document. We present Formula (1) to compute $Sim_{text}(Q, \mathcal{D})$,

$$Sim_{text}(Q, \mathcal{D}) = \sum_{k_i \in Q \cap \mathcal{D}} \mathcal{W}(k_i, D) \tag{1}$$

where

$$\mathcal{W}(k_i, D) = \frac{ntf * idf}{ntl} \tag{2}$$

Besides, we incorporate the normalized text length (ntl) into our ranking function to normalize text length of a document, as the longer the document the more terms in it.

We consider both the edit distance, the confidence of a document, and the textual similarity. Formula (3) shows the overall similarity between Q and $\mathcal{D}$,

$$Sim(Q, \mathcal{D}) = \frac{1}{(ed(Q, Q') + 1)^2} + \alpha * Conf(D) + \beta * Sim_{text}(Q', \mathcal{D}), \tag{3}$$

where $Conf(D)$ denotes the confidence of the document, Q' is reformulated query from Q' by considering similarity, and $Sim_{text}(Q', \mathcal{D})$ denotes the text similarity between Q' and $\mathcal{D}$. α and β are parameters to differentiate the importance of the three parameters.

4 Lattice Based Approach

A problem of the gram-based approach is that we need to find all the similar strings within a threshold δ. How to choose a proper threshold δ is a no trivial problem. In order to address this problem, we proposed a lattice-based approach to progressively retrieve the answer. We assume that the edit distance between a query string and data string dominates other ranking parameters, such as term frequency, inverse document frequency, and confidence. That is, we prefer the answers with a smaller edit distance. Thus, we can set α and β as very smaller number.

The general idea of lattice based approach is to progressively retrieve the answer set with threshold of the edit distance incrementing from 0 to another threshold named maximum threshold, until we get the top-k answers.

The theoretical maximum edit distance between two strings equals to the sum of the length of the two strings, $max(ed(s_1, s_2)) = |s_1| + |s_2|)$. In practice, we set it to one fifth of the length of the given query string. The maximum threshold is another early-termination condition.

In the same way, if there are many answers for an edit-distance threshold, we assume confidence dominate the textual similarity. Thus we set $\alpha \gg \beta$. We can apply the same lattice idea to address this issue.

5 Experimental Study

We used a real datasets crawled from Internet as our experimental data. The data size is about 780 MB, and there are about 682 K paper items. Each paper was treated as an independent document. In the experiment, as these papers were got from different sources, they have different trustworthy. We synthetically generated a confidence to each paper to simulate the uncertainty of the data.

The experiments were conducted on an Intel(R) Xeon(R) 2.0 GHz computer with 4 GB of RAM running Linux, and the algorithms were implemented in Java.

Figure 1 shows the precision comparison with 1 mistyping. The precisions of the top-1 and top-3 answer sets are higher than that of top-10 answer sets. This result proves that our ranking mechanism is very effective.

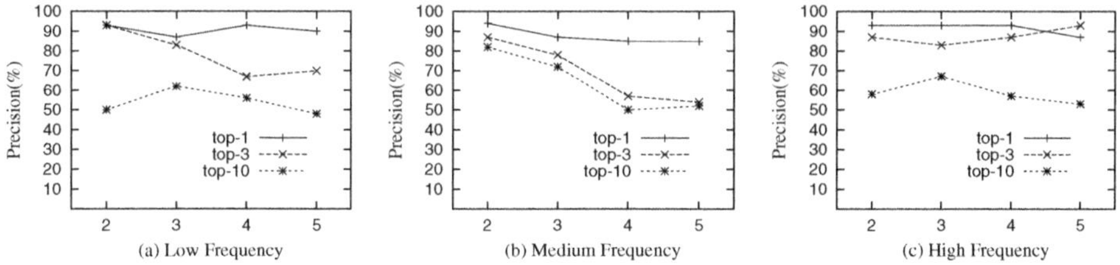

Fig. 1. Precision comparison with one mistyping

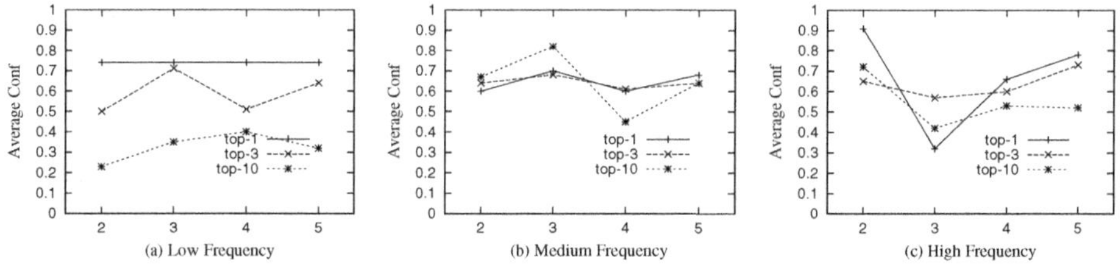

Fig. 2. The Effect of Confidence

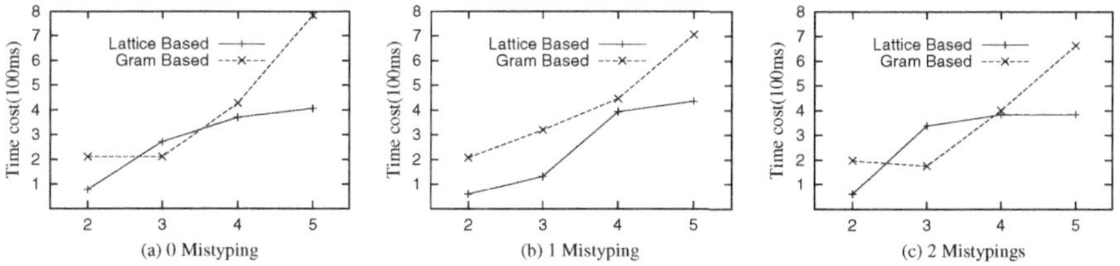

Fig. 3. Efficiency Comparison

Figure 2 shows effect of *confidence* in the ranking mechanism. The document with high confidence tends to appear higher.

Figure 3 shows the time cost of the lattice-based approach is less than that of the gram-based approach.

6 Conclusion

In this paper, we have studied the problem of fuzzy keyword search over uncertain data. We have examined the gram-based method to answer keyword queries approximately. We have proposed ranking functions to rank the answers by considering the edit distance between input keywords and data strings, textual similarity, and confidence. We have developed lattice-based methods and early-termination techniques to improve search efficiency. We have implemented our algorithm, and the experimental results show that our method achieves high search efficiency and result quality.

Acknowledgement

This work is partly supported by the National High Technology Development 863 Program of China under Grant No.2007AA01Z152, the National Grand Fundamental Research 973 Program of China under Grant No.2006CB303103, and 2008 HP Labs Innovation Research Program.

References

1. He, H., Wang, H., Yang, J., Yu, P.S.: Blinks: ranked keyword searches on graphs. In: SIGMOD Conference, pp. 305–316 (2007)
2. Hua, M., Pei, J., Zhang, W., Lin, X.: Efficiently answering probabilistic threshold top-k queries on uncertain data. In: ICDE, pp. 1403–1405 (2008)
3. Li, G., Feng, J., Wang, J., Zhou, L.: Effective keyword search for valuable lcas over xml documents. In: CIKM, pp. 31–40 (2007)
4. Li, G., Ooi, B.C., Feng, J., Wang, J., Zhou, L.: Ease: an effective 3-in-1 keyword search method for unstructured, semi-structured and structured data. In: SIGMOD Conference, pp. 903–914 (2008)
5. Luo, Y., Lin, X., Wang, W., Zhou, X.: Spark: top-k keyword query in relational databases. In: SIGMOD Conference, pp. 115–126 (2007)
6. Soliman, M.A., Ilyas, I.F., Chang, K.C.-C.: Top-k query processing in uncertain databases. In: ICDE, pp. 896–905 (2007)

Adaptive Safe Regions for Continuous Spatial Queries over Moving Objects

Yu-Ling Hsueh[1], Roger Zimmermann[2], and Wei-Shinn Ku[3]

[1] Dept. of Computer Science, University of Southern California, USA
hsueh@usc.edu
[2] Computer Science Department, National University of Singapore, Singapore
rogerz@comp.nus.edu.sg
[3] Dept. of Computer Science and Software Engineering, Auburn University, USA
weishinn@auburn.edu

Abstract. Continuous spatial queries retrieve a set of time-varying objects continuously during a given period of time. However, monitoring moving objects to maintain the correctness of the query results often incurs frequent location updates from these moving objects. To address this problem, existing solutions propose lazy updates, but such techniques generally avoid only a small fraction of all unnecessary location updates because of their basic approach (e.g., safe regions, time or distance thresholds). In this paper, we introduce an *Adaptive Safe Region (ASR)* technique that retrieves an adjustable safe region which is continuously reconciled with the surrounding dynamic queries. In addition, we design a framework that supports multiple query types (e.g., range and c-kNN queries). In this framework, our query re-evaluation algorithms take advantage of $ASRs$ and issue location probes only to the affected data objects. Simulation results confirm that the ASR concept improves scalability and efficiency over existing methods by reducing the number of updates.

1 Introduction

Significant research attention has focused on efficiently processing continuous queries and its extension work that supports location-based services during the recent past. Existing work [2,3,4] has provided significant insight into the cost of the communication overhead by assuming a set of computationally capable moving objects that cache query-aware information (e.g., thresholds or safe regions) and locally determine a mobile-initiated location update. However, the focus of these solutions is mainly on static queries or simple types of queries (e.g., range queries). In this paper, we propose a framework to support multiple types of dynamic, continuous queries. Our goal is to minimize the communication overhead in a highly dynamic environment where both queries and objects change their locations frequently. When a new query enters the system we leverage the trajectory information that it can provide by registering its starting and destination points as a movement segment for continuous monitoring. This assumption can be easily extended to a more realistic scenario which may approximate a curved road segment with several straight-line sub-segments. We propose an *adaptive*

X. Zhou et al. (Eds.): DASFAA 2009, LNCS 5463, pp. 71–76, 2009.

safe region that reconciles the surrounding queries based on their movement trajectories such that the system can avoid unnecessary location probes to the objects in the vicinity (i.e., the ones which overlap with the current query region). Furthermore, our incremental result update mechanisms allow a query to issue location probes only to a minimum area where the query answers are guaranteed to be fulfilled. In particular, to lower the amortized communication cost for c-kNN queries, we obtain extra nearest neighbors (n more NNs) which are buffered and reused later to update the query results. Thus, the number of location updates incurred from the query region expansion due to query movement is reduced.

2 Related Work

Continuous monitoring of queries over moving objects has become an important research topic, because it supports various useful mobile applications. Prabhakar et al. [4] designed two approaches named query indexing and velocity constrained indexing with the concept of *safe regions* to reduce the number of updates. Hu et al. [2] proposed a generic framework to handle continuous queries with safe regions through which the location updates from mobile clients are further reduced. However, these methods only address part of the mobility challenge since they are based on the assumption that queries are static which is not always true in real world applications. A threshold-based algorithm is presented in [3] which aims to minimize the network cost when handling c-kNN queries. To each moving object a threshold is transmitted and when its moving distance exceeds the threshold, the moving object issues an update. However, the system suffers from many downlink message transmissions for refreshing the thresholds of the entire moving object population due to frequent query movements. Cheng et al. [1] proposed a time-based location update mechanism to improve the temporal data inconsistency for the objects relevant to queries. Data objects with significance to the correctness of query results are required to send location updates more frequently. The main drawback of this method is that an object will repeatedly send location updates to the server when it is enclosed by a query region.

In contrast, our proposed techniques aim to reduce the communication cost of dynamic queries over moving objects and also support multiple types of queries. We utilizes adaptive safe regions to reduce the downlink messages of location probes due to query movements. Our *ASR*-based techniques surpass the aforementioned solutions with higher scalability and lower communication cost.

3 The System Overview

The location updates of a query result point (result point for short) and a non-result point (data point for short) are handled with two different mechanisms. An *adaptive safe region* (*ASR*) is computed for each data point. A mobile-initiated voluntary location update is issued when any data point moves out of its safe region. For a result point, to capture its possible movement at the server side, we use a *moving region* (*MR*) whose boundary increases by the maximum moving distance per time unit. The location updates of a result point are requested only when the server sends server-initiated location probes triggered when the moving region of a result point overlaps with some query regions. The details of the *ASR* computation and an efficient mechanism that uses location probes for continuous query updates are described in Sections 3.1 and 3.2, respectively.

3.1 Adaptive Safe Region Computation

We propose a novel approach to retrieve an *ASR*, which is effective in reducing the amortized communication cost in a highly dynamic mobile environment. The key observation lies in the consideration of some important factors (e.g., the velocity or orientation of the query objects) to reconcile the size of the safe regions. The following lemma establishes the *ASR* radius based on the observation.

Lemma 1

$$p_i.ASR.radius = \min(CDist(p_i, q_j) - q_j.QR.radius), \forall q_j \in Q, \text{ where}$$
$$CDist(p_i, q_j) = \begin{cases} \overline{p_i f'} & \text{if } \theta_j \le \frac{\pi}{2} \text{ and } \exists f', \text{ or} \\ \overline{p_i q_j^s} & \text{if } \theta_j > \frac{\pi}{2} \text{ or } \nexists f' \end{cases}$$

As an illustration of Lemma 1 (and to explain the symbol notation), consider Figure 1, where the set of queries $Q = \{q_j, q_k\}$ are visited for retrieving the adaptive safe region (the dashed circle) of the data point p_i. The movement trajectory of q_j (and q_k) is denoted by $\overrightarrow{q_j} = [q_j^s, q_j^e])$, where q_j^s and q_j^e are the starting and ending points, respectively. We measure the Euclidian distance between a query and a data point ($CDist$ in Lemma 1) and then deduct the query range (denoted by $q_j.QR.radius$, where QR represents the query region of q_j). Lemma 1 captures two cases of *CDist*. The first case ($CDist(p_i, q_j)$) computes a distance $\overline{p_i f'} = \overline{q_j^s f}$ in the worst-case scenario where both p_i and q_j move toward each other (under the constraint of the maximum speed). f' represents the border point (on the border of $q_j.QR$ while q_j arrives at f on its movement segment), after which p_i would possibly enter the query region of q_j. f is the closest point to q_j^s on the trajectory of q_j, which satisfies the condition that the distance from p_i to f is equal to $\overline{p_i f'} + \overline{f' f}$, where $\overline{f' f} = q_j.QR.radius = r_j$. Let $\overline{p_i f'} = x$ for short. We can obtain the f and f' points by computing x first, which is considered the safe distance for p_i with respect to q_j. x can be easily computed with the trajectory information of q_j by solving the quadratic equation: $(x + r_j)^2 = h^2 + (\overline{q_j^s m} - x)^2$ (h is the height of triangle $\Delta p_i q_j^s m$). f on $\overrightarrow{q_j}$ exists only when θ_j ($\angle p_i q_j^s q_j^e$) is less or equal to $\frac{\pi}{2}$ and $\overline{p_i q_j^e} - q_j.QR.radius < \overline{q_j^s q_j^e}$ (triangle inequality). If the first case is not satisfied, we consider the second case ($CDist(p_i, q_k)$), which finds the maximum non-overlapping area with $q_j.QR$. Since $\theta > \frac{\pi}{2}$ in the second case, the query range of q_j can never cover p_i due to the opposing movement of q_j. In this example, the safe distance x (with respect to q_j) is smaller than y (with respect to q_k), so x is chosen as the radius of the *ASR* of p_i. In our system, since a c-kNN query can be considered an order-sensitive range query, we use the same principle to compute safe regions for each data object with respect to

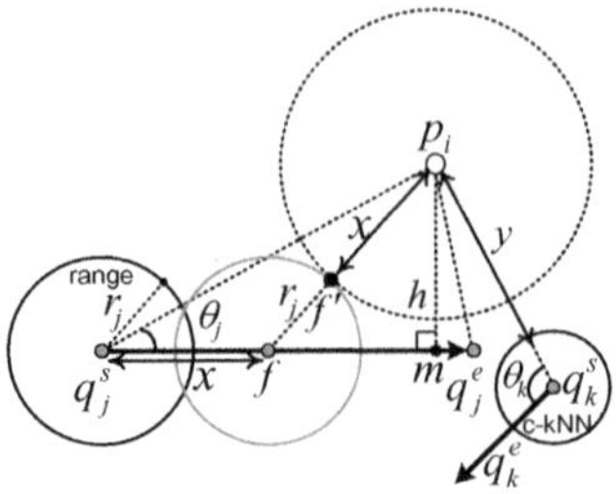

Fig. 1. An adaptive safe region

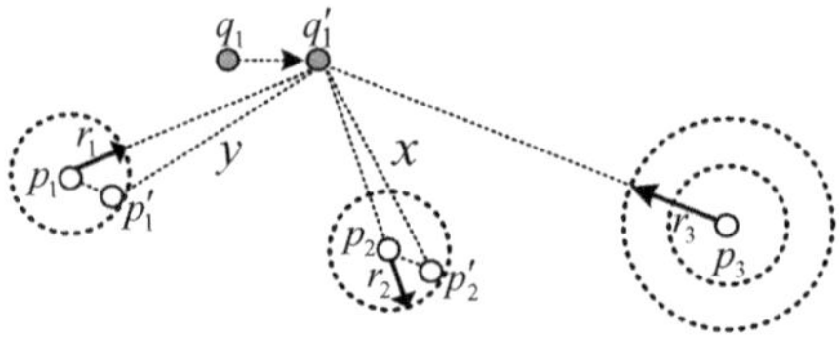

Fig. 2. The order checks of a c-kNN query

range queries and c-kNN queries. In case of a query insertion or query region expansion of a c-kNN query, the *ASRs* of the affected data objects must be reassigned according to current queries to avoid any missing location updates.

3.2 Query Evaluation with Location Probes

We propose an incremental query re-evaluation algorithms for both range and c-kNN queries. While updating the query answers, on-demand server-initiated location probes are issued whenever any location ambiguity exists. To handle a range query, the query processor sends the on-demand location probes to those result points that might move out of the current query regions. A *MR* for each result point is indexed on the grid and the boundary increases at each time step by the maximum moving distance until the result point is probed by the server. Since the number of result points are relatively small, indexing *MRs* does not significantly increase the overall server workload. For a data point, in addition to its adaptive safe region, we also consider the current possible moving boundary to serve as an additional indicator for the server to determine a necessary location probe. To handle a c-kNN query, the cost of updating c-kNN queries is usually higher than updating range queries. In our approach, the strategy to handle such increasing unnecessary location updates incurred from a c-kNN query is that the query processor computes $(k + n)$ NNs for a c-kNN query instead of evaluating exactly k NNs. This approach helps to reduce the number of future query region expansions to retrieve sufficient NNs for the queries. Since a c-kNN query is treated as an order-sensitive range query, we adopt the same principle that is used for a range query to find the new answer set in the current query regions first. A query region is expanded only when there are less than k NNs in the result set. Finally, an order-checking procedure is performed to examine the order of the result points and determine necessary location probes. Let q'_j be the last reported position of the query object q_j. The *OrderCheck* procedure checks each result point p_i (the ith result point sorted by the *mindist to* q'_j), where $i = 1$ to k. A necessary location update of a kNN result point p_i is issued when the following condition is satisfied: $dist(q'_j, p_i) + p_i.MR.radius \geq dist(q'_j, p_{i+1}) - p_{i+1}.MR.r$, where $p_i.MR.radius$ (and $p_{i+1}.MR.radius$) is the maximum moving distance since the last update of p_i. An example is shown in Figure 2. The result set of q'_1 is $\{p_2, p_1, p_3\}$ sorted by the distance between q'_1 and their positions at the server since the last updates. The *OrderCheck* procedure first checks p_2 and

p_1. Since $dist(q_1', p_2) + r_2 > dist(q_1', p_1) - r_1$, the order of p_2 and p_1 might need to be switched. The system needs to probe p_2 and p_1. After the location probes, the order of the NNs becomes $\{p_1', p_2', p_3\}$. Thus, the procedure checks the next pair of p_2' and p_3. Since $dist(q_1', p_2') < dist(q_1', p_3) - r_3$, the location probe of p_3 is not necessary.

4 Experimental Evaluation

We evaluated the performance of the proposed framework that utilizes *ASRs* and compared it with the traditional safe region approach [2,4] and a periodic update approach (*PER*). The periodic technique functions as a baseline algorithm where each object issues a location update (only uplink messages are issued in this approach) every time it moves to a new position. We extended the safe region approach (*SR**) to handle dynamic range and c-kNN queries where the result points are monitored the same way as in *ASR*. We preserve the traditional safe region calculations (maximum non-overlapping area) for the SR* approach. Figure 3(a) shows the communication overhead of *ASR*, *SR** and *PER* with respect to the object cardinality, where all query and object sets move with a mobility rate of 50%. *ASR* outperforms *SR** and *PER*. The difference increases as the number of objects grows. Figure 3(b) shows the impact of the number of queries. Our algorithm achieves about 50% reduction compared with *SR** and 90% reduction compared with *PER*.

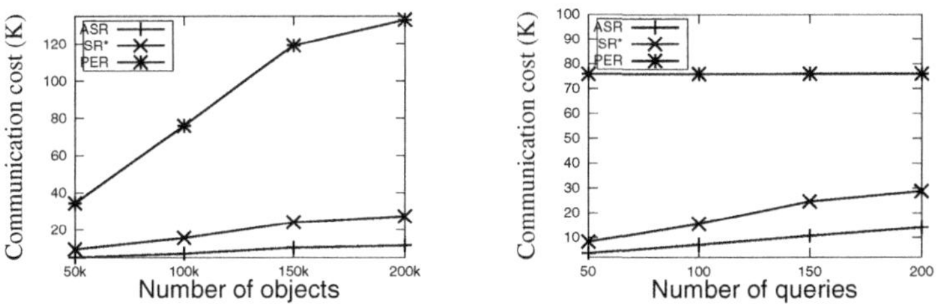

(a) P v.s. communication cost (b) Q v.s. communication cost

Fig. 3. Object and Query Cardinality

5 Conclusions

We have designed an *ASR*-based framework for highly dynamic environments where mobile units may freely change their locations. The novel concept of an adaptive safe region is introduced to provide a mobile object with a reasonable-sized safe region that adapts to the surrounding queries. Hence, the communication overhead resulting from the query movements is greatly reduced. An incremental result update mechanism that checks only the set of affected points to refresh the query answers is presented. Experimental results demonstrate that our approach scales better than existing techniques in terms of the communication cost and the outcome confirms the feasibility of the *ASRs* approach.

References

1. Cheng, R., Lam, K.y., Prabhakar, S., Liang, B.: An Efficient Location Update Mechanism for Continuous Queries Over Moving Objects. Inf. Syst. 32(4), 593–620 (2007)
2. Hu, H., Xu, J., Lee, D.L.: A Generic Framework for Monitoring Continuous Spatial Queries over Moving Objects. In: SIGMOD Conference, pp. 479–490 (2005)
3. Mouratidis, K., Papadias, D., Bakiras, S., Tao, Y.: A Threshold-Based Algorithm for Continuous Monitoring of k Nearest Neighbors. IEEE Trans. Knowl. Data Eng. 17(11), 1451–1464 (2005)
4. Prabhakar, S., Xia, Y., Kalashnikov, D.V., Aref, W.G., Hambrusch, S.E.: Query Indexing and Velocity Constrained Indexing: Scalable Techniques for Continuous Queries on Moving Objects. IEEE Trans. Computers 51(10), 1124–1140 (2002)

Optimization on Data Object Compression and Replication in Wireless Multimedia Sensor Networks

MingJian Tang[1], Jinli Cao[2], Xiaohua Jia[3], and Ke-Yan Liu[4]

[1,2] Department of Computer Science and Computer Engineering, La Trobe University, Melbourne, Australia
mj2tang@students.latrobe.edu.au, j.cao@latrobe.edu.au
[3] Department of Computer Science, City University of Hong Kong, Kowloon, Hong Kong
csjia@cs.cityu.edu.hk
[4] HP Labs China, Beijing, China
keyan.liu@hp.com

Abstract. Wireless Multimedia Sensor Networks (WMSNs) have brought unprecedented potentials for applications requiring ubiquitous access to multimedia contents such as still images. However, new challenges have arisen due to the extra sensor capacity and various requirements of multimedia objects in-network processing. In this paper, we consider a large-scale WMSN comprising multiple storage nodes and many multimedia sensor nodes. In particular, we investigate the Optimal Compression and Replication (OCR) of multimedia data objects. In sharp contrast to earlier research, we integrate both computation and communication energy consumption as a joint optimization problem. We prove that the problem is NP-hard if storage nodes have limited storage capacities. We proposed a solution based on Lagrangian relaxation interwoven with the subgradient method. Extensive simulations are conducted to evaluate the performance of the proposed solution.

Keywords: Data compression and replication, Lagrangian relaxation, NP-hard, subgradient method, Wireless Multimedia Sensor Networks, Optimization.

1 Introduction

The unprecedented focus on Wireless Sensor Networks (WSNs) has made feasible for a wide range of applications [2], such as environmental monitoring, military surveillance, and intelligent information gathering. A typical WSN consists of numerous tiny and battery-powered sensor nodes, which are densely distributed into the physical area feeding the base station with in-situ data. After deployment, replenishments of node batteries become either too expensive or impossible. Additionally, relatively slow advancements in battery technologies make the energy constraint the fundamental challenges in the WSN. As a consequence, power-efficient techniques have been an active research area attracting significant interdisciplinary efforts [5, 19].

The recent emergence of low-energy and low-cost multimedia devices, such as microphones and cameras, has fostered the development of the next generation WSN, known as Wireless Multimedia Sensor Network (WMSN) [1]. WMSNs have brought unprecedented potentials especially for applications requiring ubiquitous access to the

X. Zhou et al. (Eds.): DASFAA 2009, LNCS 5463, pp. 77–91, 2009.

multimedia contents, such as audio and video streams, and still images. The extra sensor capacity and various requirements of multimedia objects in-network processing also pose new and unique challenges. Since most of the multimedia sensors [1] are still operated under limited batter powers, it is imperative to provide effective and efficient collaboration amongst sensor nodes for voluminous multimedia data storage, computations, and transmissions.

In-network storage [7, 16] is a powerful data management scheme. It facilitates energy-efficient sensor data storage and retrieval by allowing sensor nodes to replicate their data at some of the strategically chosen or pre-designated storage nodes. These storage nodes can be augmented general-purpose computers with relatively abundant resources, such as storage and power. They can be deployed proportionally according to different WMSN factors, such as network scale and node density. They play important roles in balancing the deployment cost and the overall functionalities of the WMSN. They can also effectively mitigate the storage constraint of in-network multimedia sensor nodes, and further enhance the availability of important historical data. Furthermore, the approach is also very energy-efficient since user-posed queries now can be answered directly by these storage nodes. In this paper, we consider such an in-network storage infrastructure for managing in-network sensory data.

Unlike the traditional WSN with simple sensing modalities, naively distributing raw data to the storage nodes requires excessive bandwidth and energy consumption for the WMSN. Hence, energy-efficient data compression techniques are utilized in most of the WMSN applications, such as [10] and [13]. Nevertheless, overuse of these techniques can cause detrimental effects on the overall energy efficiency since they make computations comparable in energy consumptions to transmissions. To facilitate the motivations of the study, we show an example as follows.

Motivational example. The example is illustrated in Fig. 1(a), where there are three nodes A, B, and C. Nodes A and B act as two candidates for hosting C's data. Node C is a source node that generates 1 unit-sized raw data every time. The transmission cost from C to A is 10 units for every unit of data object, whereas from C to B it takes 11 units for every unit of data object. Fig. 1(b) captures the different relationships between C's energy consumptions and their respective compression ratios. For example, it costs C, 0.75 of a unit of energy for achieving a 10% compression ratio (i.e., 1 unit-sized raw data becomes 0.9 unit-sized compressed data). Hence, we can see that the optimal replication plan for node C is to distribute its raw data with a 10% compression ratio to storage node A. The incurred total energy consumption is 9.75 (0.9 * 10 + 0.75), which is less than 10 (i.e., transmitting raw data only without any compression). Let us next consider a more common scenario, where node A can only store

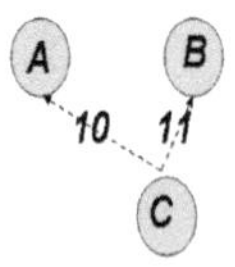

Energy	Compression
0.75	10%
1.9	20%
20	30%

(a) (b)

Fig. 1. A motivational example

another 0.7 unit of data from C, whereas node B can cater any data object size from C. If we still replicate C's data object at A with 30% of compression, the incurred overall energy can be as much as 27 units (0.7 * 10 + 20). However, if we replicate C's data object at B with a 20% compression ratio, the incurred overall energy can be 10.7 (0.8 * 11 + 1.9). The more expensive transmission is leveraged by the judicious usage of compression ratio, which poses a stimulative optimization problem.

In this paper, we initiate the study of the Optimization on data object Compression and Replication in WMSNs (OCR), where we consider the minimization of nodes' overall energy consumption (i.e., computation + transmission). In particular, we consider the OCR while multiple storage nodes (m) and multiple source nodes (n) are presented and each storage node has certain storage capacity.

Contributions. We first present a model for the problem of the OCR, formulated as an optimization problem. In our model, decisions on data compression and data replication are made based on both computation and transmission energy cost. With limited storage capacities at storage nodes, we prove the NP-hardness of OCR. In order to efficiently solve the problem, we relax the problem formulation with Lagrangian dualization, and then decompose the relaxed problem into two sub-problems: Lagrangian problem and Lagrangian dual problem. We propose an efficient sequential search based algorithm for solving the Lagrangian problem, and a subgradient based method for the Lagrangian dual problem. Lastly, we evaluate our model on various performance metrics through extensive simulations. The results show that our solution consistently and significantly outperforms the existing solution.

Overview of the paper. The remainder of the paper proceeds as follows. Section 2 summaries the related work. Section 3 describes the system model along with the problem formulation of the OCR. The formulated problem is also shown to be NP-hard. Section 4 revolves around our Lagrangian relaxation based solution procedure. The proposed approach is eventually evaluated using extensive simulations in Section 5, and section 6 concludes the paper.

2 Related Work

Data replication has been an active research topic in different environment, such as web [3, 4, 6, 9, 12, 13, 14] and WSN [5, 7, 15, 16]. Mainly data replication-related techniques can be used to alleviate server overloading, improve QoS (i.e., response time), and cope with inadequate bandwidth.

In web environment, the placement of web proxies or mirrored web severs has been an active research topic in recent years. In [12], a dynamic programming method was proposed to find the optimal placement of web proxies in a tree. In [14], the problem of placing mirrored web servers to minimize the overall access distance was formulated as the p-median problem, which is NP-complete, and several heuristic algorithms were proposed. In [4], the same problem was investigated yet with the objective of minimizing the maximum distance between any clients to a mirrored server. The authors formulated the problem as the k-center problem (also is NP-complete), and a greedy algorithm was proposed. Some of the research has been done on the replication of data objects at proxies since web proxies are widely used to improve the quality of web data retrievals. The data replication problem in traditional database systems has been extensively studied [3, 6, 8], and it has been proven that optimal replication of data objects in general networks is NP-complete. Though numerous approaches have been emerged for the data replication under the traditional web environment, they are not applicable to WMSN applications due to their web-based assumptions, such as abundant resources on clients and relatively unlimited bandwidth.

In WSN environment, data replication techniques are more energy-oriented. One of the prominent approach is Data-Centric Storage [16], known as DCS. The DCS is a distributed hash table based data dissemination approach for energy-efficient sensor data replications. In [7], the author discusses some of the issues existing in the prevalent DCS mechanisms in WSNs. The focus of the paper is to improve the fault-resilience of the existing DCS schemes. A new method called Resilient Data-Centric Storage (R-DCS) is proposed to achieve both the minimization of the query retrieval traffic and the increase of the data availability. In [11], the authors investigate the hotspot issue that is intrinsic to most of the DCS schemes. A new Dynamic Balanced data-centric Storage (DBAS) scheme, based on a cooperative strategy between the base station and the in-network processing, was proposed to mitigate the hotspot problem. The data storage placement problem is discussed in [15], where the aim is to minimize the total energy cost for data gathering and data queries. In [18], the cache placement problem was first addressed in WSNs with the objective of minimizing the access cost under different constraints (i.e., maximum allowable update costs). For a tree topology, a dynamic programming based approach was proposed to compute the optimal solution. The author further proves that the problem is NP-hard for any general graph topology. Most of these researchers focus on minimizing transmission energy costs since WSN energy consumption is dominated by data communications. On the contrary, we consider a joint optimization problem addressing both computation and communication energy consumptions.

Data compression techniques are of paramount importance to WMSN applications. In [10], different image compression algorithms were proposed for a low-bandwidth wireless camera network. In [13], different video coding techniques were discussed along with their pros and cons. In this paper, we take an optimization approach towards judiciously coordinating computation power spend on data compression.

3 System Model and Problem Formulation

3.1 System Model

In this paper, we consider a system given in Fig. 2, in which a large-scale WMSN is deployed for the continuous surveillance application. The WMSN comprises of four types of nodes: a base station node, storage nodes, relay nodes, and source nodes. The base station is a powerful machine with abundant resources including processing, storage, and energy. It handles user queries with respect to the in-situ information within the monitored area. Storage nodes are the data centers storing all important environmental information (e.g. periodic image captures) that can be retrieved by the base station for answering user queries. They are relatively less powerful than the

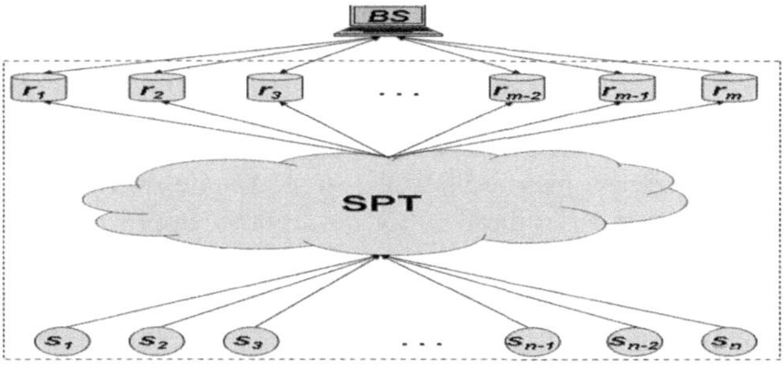

Fig. 2. An auxiliary example of WMSN model

Table 1. Primary notations

Symbol	Meaning
s_i	i-th source node
r_j	j-th storage node
b_j	Storage capacity of r_j
o_i	Multimedia data object generated by s_i
z_i	Size of o_i
o^-_i	Compressed o_i
z^-_i	Size of o^-_i
f_i	Updating frequency of s_i
α_i	Application-dependent compression parameter of s_i
d_{ij}	SPT distance from s_i to r_j
t_i	Transmission energy consumption of s_i
c_i	Computation energy consumption of s_i
x_{ij}	Matrix element indicating the data replication of s_i at r_j

base station yet their power supplies can still be constant. One or more relay nodes can be chosen strategically to form different Shortest Path Trees (SPTs) for connecting storage nodes and source nodes. These SPTs are also responsible for reliable in-network data transmissions. Lastly, source nodes are resource constrained, with limited processing capacity, storage, and power. They are normally equipped with advanced sensory devices for capturing multimedia contents (e.g. still images). Periodically, they compress and send their sensed data to the storage nodes, and their reporting intensities/frequencies are application-dependent.

3.2 Problem Formulation

For the sake of convenience, Table 1 lists some of the primary notations used in the paper. The OCR problem is defined as follows. Suppose there is a set of m storage nodes $R = \{r_j \mid 1 \le j \le m\}$ and a set of n source nodes $S = \{s_i \mid 1 \le i \le n\}$, where $m << n$. We assume that the locations of m storage nodes are known to n source nodes. Therefore, for each source node s_i, it can use one of the storage node r_j for hosting its data replications. Each storage node, r_j, has a limited storage capacity (in bytes), denoted by b_j. The size of b_j depends on the hardware configuration of r_j.

Let o_i be the multimedia data object generated by s_i. We denote the size of o_i by z_i in bytes. We assume a uniform model for z_i, which means it only depends on the node that generates it. It can be justified because source nodes with pre-programmed task loads are sufficient for most of the WMSN surveillance applications.

Since source nodes are extremely power constrained, we consider two types of power consumptions associated with each source node s_i, namely computation power (c_i) and transmission power (t_i). The amount of each computation power, c_i, is directly related to the energy consumption on the data compression. Given an original data object o_i and its size z_i, we denote $o^{\bar{}}_i$ and $z^{\bar{}}_i$ as the compressed data object and its size respectively. The smaller $z^{\bar{}}_i$ becomes, the more computation energy is required. Additionally, for each source node $s_i \in S$, we can know that $z^{\bar{}}_i \leq z_i$ and $z^{\bar{}}_i > 0$. Formally, we define each c_i as below:

$$c_i = \alpha_i \times (z_i - z^{\bar{}}_i) \times f_i, \forall 1 \leq i \leq n, \forall i : \alpha_i > 0, \tag{1}$$

where α_i is an application-dependent parameter indicating the unit energy consumption per byte compressed. f_i is the updating frequency of source node s_i, which can be derived by user queries, and it defines the temporal requirement of the WMSN application.

For calculating each transmission power, t_i, we define an abstract measure d_{ij} signifying the distance between source node s_i and storage node r_j. The value of d_{ij} depends on the underlying routing protocols and the metrics (e.g. bandwidth, energy, and delay) used in the system [20]. Since the focus of this paper is to explore the trade-off between computation powers and transmission powers, we take a more abstract approach for the distance calculation. Based on this premise, we assume that the value d_{ij} can be determined based on the hop-distance of the Shortest Path Tree (SPT) linking s_i and r_j. Eventually, we define each t_i as the distance of data travelling multiplied by the size of the compressed data object during a certain time period:

$$t_i = d_{ij} \times z^{\bar{}}_i \times f_i, \forall 1 \leq i \leq n, \forall 1 \leq j \leq m \tag{2}$$

Equation (1) + (2) is the total power consumption e_i of source node s_i:

$$\begin{aligned} e_i = c_i + t_i = \alpha_i \times (z_i - z^{\bar{}}_i) \times f_i + d_{ij} \times z^{\bar{}}_i \times f_i, \\ 1 \leq i \leq n, \forall i : \alpha_i > 0, 1 \leq j \leq m \end{aligned} \tag{3}$$

We define a matrix $X \in \{0, 1\}^{|S|*|R|}$, and denote each element x_{ij} of X as:

$$x_{ij} = \begin{cases} 1, & \text{if } s_i \text{ chooses } r_j \\ 0, & \text{otherwise} \end{cases} \tag{4}$$

The overall power consumption of n source nodes is:

$$E = \sum_{i=1}^{n} \sum_{j=1}^{m} e_i \times x_{ij} \tag{5}$$

From the above power consumption function, we can see that overall power consumption depends on not only the transmission power but also the computation power. The objective of the paper is to explore the trade-off benefits between these two types of power consumption, so that we can minimize the overall power consumption for n source nodes in the network. The underlying problem is formulated as follows:
Minimize

$$E \tag{6}$$

Subject to

$$\sum_{i=1}^{n}\sum_{j=1}^{m} x_{ij} = n \tag{7}$$

$$\forall j : \sum_{i=1}^{n} x_{ij} \times z^{-}_{i} \leq b_j, 1 \leq j \leq m \tag{8}$$

where equation (6) describes the objective function and equation (7) constrains that each source node can choose one storage node only for replicating its data objects. The total size of the replicas at each storage node r_j, should not exceed its capacity b_j, which is described by equation (8).

3.3 NP-Hardness of OCR

In this section, we prove the NP-hardness of the OCR problem by first proving that a special case of it is NP-hard. We refer to this special case as the Constrained OCR (COCR), where each z^{-}_{i} is assumed to be fixed. To facilitate our analysis, we first introduce the following definition.

Definition 1. *Partition Problem (PP): Given a list of integers* $\{g_1, g_2, \ldots, g_n\}$*, determine whether there is a set of* $I \subseteq \{1, 2, \ldots, n\}$ *such that* $\sum_{i\in I} g_i = \sum_{i\notin I} g_i$.

The PP has been proved to be NP-hard [9]. In the following, we show that the PP is a special case of the COCR.

Theorem 1. *The COCR problem is NP-hard.*

Proof. Suppose there are n compressed multimedia data objects $\{o^{-}_{1}, o^{-}_{2}, \ldots, o^{-}_{n}\}$ produced by n source nodes $\{s_1, s_2, \ldots, s_n\}$ respectively and the sum of the sizes of n objects is $\sum_{i=1}^{n} z^{-}_{i} = 2a$. We consider a special case of the COCR problem that replication of compressed data objects at a storage node is independent from others. Suppose there are m = 2 storage nodes (r_1 and r_2) and their respective storage capacities b_1 and b_2. Consider $b_1 + b_2 = 2a$, and each of r_1 and r_2 is configured with a storage capacity sized a. We first assume all source nodes have the same distance to each of the storage nodes, i.e.:

$$\forall u,v: d_{ur_1} = d_{vr_2}, n > 1, 1 \le u, v \le n \tag{9}$$

We can always design each f_i (updating frequency), z_i (size of the original data object), and z^{-}_i (size of the compressed data object) in such a way to meet the following constraints. For $1 \le u, v \le n$ and $n > 1$:

$$(z_u - z^{-}_u) \times f_u = (z_v - z^{-}_v) \times f_v \tag{10}$$

$$\begin{aligned} & d_{u1} \times z^{-}_u \times f_u = \\ & d_{u2} \times z^{-}_u \times f_u = \\ & d_{v1} \times z^{-}_v \times f_v = \\ & d_{v2} \times z^{-}_v \times f_v \end{aligned} \tag{11}$$

The purpose of these two constraints is to make storage of all the data objects contributing the same to the power consumption. That is, the selection of objects to be replicated at storage nodes can be arbitrary. Since compressed data objects have the same weight to be selected for replication at either r_1 or r_2, the optimal replication of them at r_1 or r_2 with total capacity 2a becomes choosing a subset of objects $I = \{o^{-}_1, o^{-}_2, \ldots, o^{-}_i\}$, $I \subseteq \{o^{-}_1, o^{-}_2, \ldots, o^{-}_n\}$, such that:

$$\sum_{i \in I} z^{-}_i = \sum_{i \notin I} z^{-}_i = a \tag{12}$$

This is exactly the PP, which is NP-hard. Hence, the COCR problem is NP-hard. □

Theorem 2. *The OCR problem is NP-hard.*

Proof. It is obvious that the COCR problem is polynomial time reducible to the OCR problem since it is a special case of the OCR problem. Therefore, the OCR problem is also NP-hard. □

4 Our Solution

The developed OCR problem model is difficult to solve for large-sized problem instances. Largely it is due to the complex storage constraint coupled with the computation power (i.e., z^{-}_i is dependent on the storage constraint of a particular r_j). Therefore, we propose a heuristic algorithm which is based on Lagrangian relaxation of the formulation given in (6) – (8). When the capacity constraint (8) is relaxed, i.e., removed from the constraint set and added to the objective function after being multiplied by a nonnegative parameter λ (also called Lagrange multipliers), we obtain the following Lagrangian (lower bound) problem:

$$
\begin{aligned}
L_{LB}(\lambda) &= \min E + \sum_{j=1}^{m} \lambda_j \left(\sum_{i=1}^{n} x_{ij} \times z^{-}_{i} - b_j\right) \\
&= \min \sum_{i=1}^{n} \sum_{j=1}^{m} x_{ij} \times \left[\alpha_i \times (z_i - z^{-}_{i}) \times f_i + d_{ij} \times z^{-}_{i} \times f_i\right] \\
&\quad + \sum_{j=1}^{m} \sum_{i=1}^{n} \lambda_j \times x_{ij} \times z^{-}_{i} - \sum_{j=1}^{m} \lambda_j \times b_j \\
&= \min \sum_{i=1}^{n} \sum_{j=1}^{m} [\alpha_i \times z_i \times f_i + z^{-}_{i} \times (\alpha_i \times f_i + d_{ij} \times f_i + \lambda_j)] \times x_{ij} \\
&\quad - \sum_{j=1}^{m} \lambda_j \times b_j
\end{aligned}
\tag{13}
$$

s.t. (7)

It can be seen from the above formula that $L_{LB}(\lambda) \leq E$ holds for all values of $\lambda \geq 0$. Hence, our aim is to find the biggest possible value of $L_{LB}(\lambda)$ denoted as L^{*}_{LB}. This problem is called Lagrangian dual problem and is given as follows:

$$
L^{*}_{LB} = \max_{\lambda \geq 0} L_{LB}(\lambda) \tag{14}
$$

With the help of the Lagrangian relaxation we can decompose the primal problem into two separable sub-problems ($L_{LB}(\lambda)$ and L^{*}_{LB}) with dramatically reduced complexity. Firstly, with any given m-vector λ we observe that the $L_{LB}(\lambda)$ problem becomes a multiple-source shortest path problem with a modified cost function as the weight (w) for each edge between a source node and a storage node. Each weight, w_i, essentially embodies each e_i plus certain level of penalties based on the multiplication of the size of compressed data object (z^{-}_{i}) and its λ. For efficiently solving the $L_{LB}(\lambda)$ sub-problem, we devise a Sequential Search-λ algorithm (SS-λ) shown as below:

SS-λ Algorithm

```
START
SET E = 0; w_c = +∞; pos = null; X = [x_ij = 0];
FOR i = 1 TO n
    FOR j = 1 TO m
        Find the minimal w*_i based on replicating o⁻_i at r_j
        IF w*_i ≤ w_c THEN
            SET w_c = w*_i; pos = j;
        END IF
    END FOR
    SET x_ipos = 1; E = E + w_c;
    SET w_c = +∞; pos = null;
END FOR
END
```

The algorithm starts with each source node s_i trying to find its optimal storage node r_i based on its optimal weight w^*_i, which can be calculated based on the formulation given in (13). The inner-loop of the algorithm (j =1 to m) drives this optimality exploration by allowing s_i to determine where to replicate (x_{ij}) and how much to replicate (z^-_i). Line 4 describes a procedure for finding the minimal weight based on replicating the compressed data object at a certain storage node. The complexity of the procedure largely depends on the total number of compression levels supported by the underlying application. In order to facilitate the presentation, we assume that each source node has h (h > 0, $h \in I$) finite levels of compression and can lead to h different sizes of compressed objects $\{o^-_{i[1]}, o^-_{i[2]}, \ldots, o^-_{i[h-1]}, o^-_{i[h]}\}$ with $\{z^-_{i[1]}, z^-_{i[2]}, \ldots, z^-_{i[h-1]}, z^-_{i[h]}\}$ as their respective sizes. Furthermore, given two levels of compression p and q, if $1 \le p < q \le h$, then we say that $z^-_{i[p]} > z^-_{i[q]}$ and intuitively more computation energy is required for achieving a more thorough compression.

Theorem 3. *Given a set of m storage nodes R and a set of n source nodes S each with one data object o_i, the complexity of SS-λ algorithm is O(nh).*

Proof. It is easy to know that the SS-λ algorithm can converge. With the outer for-loop, there are at most n loops. In each outer loop, it takes another O(mh) to find the minimal weight for each source node. Therefore, the whole algorithm ends in the time O(nmh). Given that m << n, so the SS-λ algorithm can be completed in time O(nh). □

For solving the Lagrangian dual problem (L^*_{LB}), we use the subgradient method. It is commonly used as a subordinate heuristic for iteratively improving the tightness of $L_{LB}(\lambda)$ until it converges to the optimal solution. The method starts with a set of initial nonnegative Lagrangian multipliers $\lambda_j[0]$ (for all j, $1 \le j \le m$). A possible choice for the initial Lagrangian multipliers can be all zeros since the Lagrangian multipliers are penalty parameters that reflect the overcrowded status on each storage node.

Adjustments are required for the Lagrangian multipliers so that solutions generated by the SS-λ algorithm can gradually approach the optimal solution. Firstly, for a given vector of λ[k] (e.g. k = 0), we use SS-λ[k] to generate a solution denoted as $L_{LB}(\lambda[k])$, and then we update λ[k] for the next iteration denoted as λ[k+1] until the algorithm converges. Each value of $\lambda_j[k+1]$ (for all j, $1 \le j \le m$) can be derived by:

$$\lambda_j[k+1] = [\lambda_j[k] + \theta[k](\sum_{i=1}^{n} x_{ij}[k] \times z^-_i[k] - b_j)]^+ \quad (15)$$

where $[y]^+ = \max\{0, y\}$ and θ is a prescribed sequence of step sizes. The value of θ is vital to the subgradient method, and according to [17] the convergence is guaranteed when the given θ value is satisfied with the following conditions (regardless the values of initial Lagrangian multipliers):

$$\theta[k] \ge 0, \lim_{k \to \infty} \theta[k] = 0, \sum_{k=1}^{\infty} \theta[k] = \infty \quad (16)$$

In this paper, we set θ[k] = β / (γ + δk), where β, γ, and δ are all constants and greater than zero. In summary, our solution procedure for solving the OCR problem can be described as follows:

Solution procedure
Step 1: Set k = 0 and $\lambda_j[k] = 0$ for all j
Step 2: Solve sub-problem $L_{LB}(\lambda[k])$ with algorithm SS-λ[k]
Step 3: Set k = k + 1 and update $\lambda_j[k]$ for all j based on equation (15)
Step 4: Repeat from step 2 until convergence.

Table 2. Parameters and settings

Parameter	Setting
m: *no. of storage nodes*	{1, 2, 3, 4, 5, 6, 7, 8, 9, 10}
n: *no. of source nodes*	{10, 20, 30, 40, 50, 60, 70, 80, 90, 100}
d_{ij}: *SPT distance*	{5, 6, 7, 8, 9, 10}
z_i: *data object size*	{100, 110, 120, 130, 140}
f_i: *updating frequency*	{1}
LEV: *compression level*	{5, 10, 15, 20, 25, 30, 35, 40}
k: *no. of iterations*	1000

5 Performance Evaluation

We have conducted extensive experimental studies to evaluate the performance of our approach (LR-SUB). For comparison we consider a naive approach [15] that only relies on finding SPTs for source-storage transmissions without utilizing any compression technique. In addition, the naïve algorithm also does not take account of the limited storage constraint. It can serve as a baseline for studying the worthiness of exploring the trade-off between energy consumptions on computation and transmission. We also include an algorithm called SS-0 (i.e., we fix and set all λ values as 0). The SS-0 allows us to obtain the slackest lower bound of the primal problem since the storage constraint is not considered. We have implemented all three algorithms using C#. The total energy consumption (E) is used as the performance metric for evaluating all three algorithms. Table 2 presents parameters and their settings used throughout the performance evaluation. We set the ratio between m and n to 1:10 so that for each generated storage node we generate 10 source nodes accordingly. The value of d_{ij} is randomly chosen from a set of values given in Table 2. Each uncompressed data object size, z_i, is also randomly selected from 100 to 140. We set f_i to a constant since it does not affect exploring the balanced benefits between computation and transmission consumption. In order to attain the convergence of LR-SUB, the total number of iterations (k) is set to be 1000, and $\theta[k]$ is set to be 4 / (2 + 1*k) during the k-th iteration. We also define each storage capacity as follows:

$$\forall j : b_j = \frac{\sum_{i=1}^{n} z_i}{m} \times PER, 1 \le j \le m \tag{17}$$

where PER represents a percentage, and we then can tune the degree of the storage scarcity by varying PER. We allow a set of eight compression levels (LEV) representing the amount of data reduction can be achieved. For instance, an uncompressed data object with a size of 100, if we apply a LEV of 10, then the size of the compressed data object becomes 90. α_i is an application-dependent parameter that determines the energy consumption ratio between computation and transmission. For brevity, the ratio is referred to as the Computation and Transmission energy Ratio (CTR). We adopt two models for CTR: a uniform model and a non-uniform model. For the uniform model, we consider a fixed ratio (e.g. 1:2), which implies a proportional relationship between energy consumptions on compressing one unit of data and sending one unit of data. In order to consider the possible worst case scenario, we also consider a non-uniform model for CTR given as below:

$$CTR = \begin{cases} 1:2, & if \quad LEV \in \{5,10,15,20\} \\ 2:1, & if \quad LEV \in \{25,30,35,40\} \end{cases} \tag{18}$$

For a particular experiment setting, we run it repeatedly for 5 times in order to get the mean values and standard deviations as well.

5.1 Uniform CTR

Fig. 3 shows three different experiments based on the uniform CTR model. We first study effects of varying CTR on the total energy consumption for a fixed network topology (200 source nodes and 20 storage nodes) with fixed storage capacities (PER = 100%). The experimental results are plotted in Fig. 3(a). Though the total energy consumptions derived from LR-SUB gradually increase while increasing CTR values, the LR-SUB can still achieve a higher energy-efficiency than the naïve approach and keeps outperforming the naïve algorithm. The consistency indicates that the joint optimization approach is more energy efficient. On the other hand, the gap between the LR-SUB and the SS-0 almost keeps constant showing that the LR-SUB solutions are always quality-bounded. For the second experiment, we fix CTR as 1:2 and PER as 100% while varying the network topology in terms of various numbers of source nodes and storage nodes. Fig. 3(b) shows the results for this experiment. The LR-SUB provides competitive solutions comparing with the SS-0 algorithm, and it is still more energy-efficient than the naïve algorithm regardless the network topology changes. The last experiment, Fig. 3(c), examines effects of various PERs on the results of LR-SUB. We use different PERs from 60% to 240% while fixing the network topology (200 source nodes and 20 storage nodes) and the CTR (1:2). With extremely limited storage capacity on each storage node (e.g. 60%), it is possible that the LR-SUB incurs more energy than the naïve approach. Mainly the naïve approach does not consider the storage constraint so the solution provided is not even feasible, whereas the LR-SUB can guarantee the feasibility of the derived solution while trying to bind the solution quality. Additionally, the benefits from compression are evened out since large number of source nodes are forced to choose "far away" storage nodes. Nevertheless, the performance of the LR-SUB dramatically improves if there are reasonable amount of storage spaces, and it can even achieve the optimal value (e.g. PER = 240%) as the SS-0.

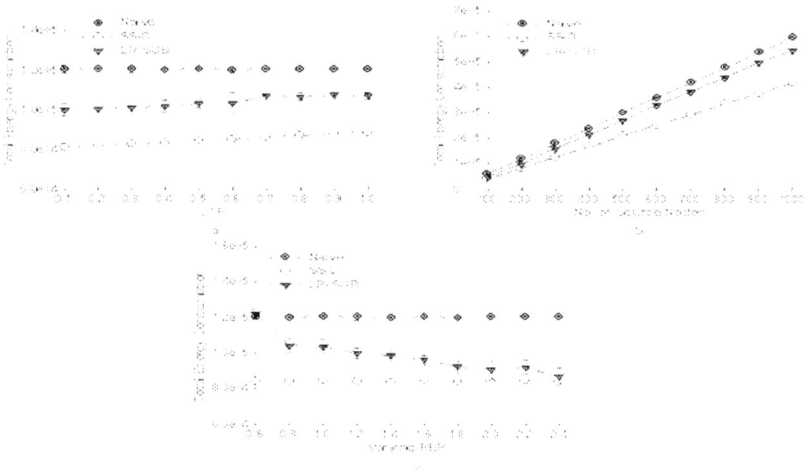

Fig. 3. Uniform CTR model. (a) varying CTR. (b) varying network topology. (c) varying PER.

5.2 Non-uniform CTR

The intuition behind adopting a non-uniform CTR is because whenever a storage node runs out of its storage space, if a CTR is still set to be uniformly increasing, then theoretically the optimal solution can still be achieved (i.e., by infinitely decreasing sizes of stored objects). Instead of being optimistic about the problem, we devised a worst case representative that can be encountered with the LR-SUB algorithm. Based on this CTR model (equation (18)), we first study the effects of various PERs on total energy consumption (Fig. 4(a)). Meanwhile, we fix the network topology with a total of 200 source nodes and 20 storage nodes. With the most stringent storage constraint (e.g. PER = 60%), the solution of LR-SUB has to compromise its quality for its feasibility (i.e., applying much more expensive LEVs to meet the storage requirement). However, we can observe that with gradually relaxing the storage constraint (from 60% to 150%), the LR-SUB also gradually outperforms the naïve approach showing that it

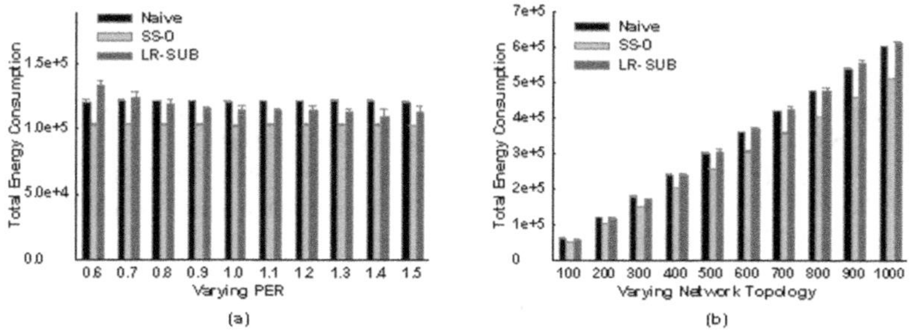

Fig. 4. Non-uniform CTR model. (a) varying PER. (b) varying network topology.

can effectively utilize these additional storage spaces for finding a better solution. Fig. 4(b) shows different results of total energy consumption obtained from varying network topology while fixing PER as 100%. For a fixed PER, the results from LR-SUB tend to deviate gradually from the SS-0 with increases of total number of source nodes. It is because more source nodes can potentially enlarge the shortage of storage spaces. Though the LR-SUB can incur more energy consumption than the naïve algorithm for some cases, it can guarantee feasible solutions every time.

6 Conclusion

In this paper, we have examined an optimization problem on how to leverage data compression for in-network storage based WMSN infrastructures. The objective is to minimize both computation and transmission energy consumption. We have further delved into a NP-hard case where the storage nodes have limited storage capacities. In order to efficiently solve the problem, we proposed a novel solution procedure based on first dualizing the storage capacity constraint (Lagrangian relaxation), and then applied a sequential search based algorithm and a subgradient method to solve the relaxed Lagrangian problem and its dual problem respectively. Extensive simulations have been conducted for evaluating the performance of our solution procedure. Preliminary results have shown that it is worthy of carefully balancing data compression and transmission, and our approach is more energy efficient than the state-of-the-art.

References

1. Akyildiz, F.I., Melodia, T., Chowdhury, K.R.: A survey on wireless multimedia sensor networks. Computer Networks: The International Journal of Computer and Telecommunications Networking 51, 921–960 (2007)
2. Akyildiz, I.F., Su, W., Sankarasubramaniam, Y., Cayirci, E.: A Survey on sensor networks. IEEE Communications Magazine 40(8), 102–114 (2002)
3. Ceri, S., Martella, G., Pelagatti, G.: Optimal File Allocation in a Computer Networks: A Solution Method Based on the Knapsack Problem. Computer Networks 6, 345–357 (1982)
4. Cronin, E., Jamin, S., Jin, C., Kurc, A., Raz, D., Shavitt, Y.: Constrained Mirror Placement on the Internet. IEEE J. Selected Areas in Comm. 20, 1369–1382 (2002)
5. Diao, Y., Ganesan, D., Mathur, G., Shenoy, P.: Rethinking Data Management for Storage-centric Sensor Networks. In: CDIR 2007 (2007)
6. Dowdy, L.W., Foster, D.V.: Comparative Models of the File Assignment Problem. ACM Computing Survey 14, 287–313 (1982)
7. Ghose, A., Grossklags, J., Chuang, J.: Resilient data-centric storage in wireless ad-hoc sensor networks. In: Chen, M.-S., Chrysanthis, P.K., Sloman, M., Zaslavsky, A. (eds.) MDM 2003. LNCS, vol. 2574, pp. 45–62. Springer, Heidelberg (2003)
8. Irani, K.B., Khabbaz, N.G.: A Methodology for the Design of Communication Networks and the Distribution of Data in Distributed Supercomputer Systems. IEEE Trans. Computers 31, 419–434 (1982)
9. Jia, X., Li, D., Cao, J.: On the Optimal Replication of Data Object at Hierarchical and Transparent Web Proxies. IEEE Trans. On Parallel and Distributed Systems 16 (2005)
10. Karlsson, J., Wark, T., Valencia, P., Ung, M., Corke, P.: Demonstration of Image Compression in a Low-Bandwidth Wireless Camera Network. In: Information Processing in Sensor Networks (IPSN 2007), Cambridge, Massachusetts, USA (2007)

11. Lai, Y., Chen, H., Wang, Y.: Dynamic balanced storage in wireless sensor networks. In: The 4th workshop on Data management for sensor networks (DMSN 2007), Vienna, Austria (2007)
12. Li, B., Golin, M.J., Italiano, G.F., Deng, X.: On the Optimal Placement of Web Proxies in the Internet. In: IEEE INFOCOM 1999 (1999)
13. Puri, R., Majumdar, A., Ishwar, P., Ramchandran, K.: Distributed Video Coding in Wireless Sensor Networks. IEEE Signal Processing 94 (2006)
14. Qiu, L., Padmanabhan, V., Voelker, G.: On the Placement of Web Server Replicates. In: IEEE INFOCOM 2001 (2001)
15. Sheng, B., Li, Q., Mao, W.: Data Storage Placement in Sensor Network. In: MobiHoc 2006, Florence, Italy (2006)
16. Shenker, S., Ratnasamy, S., Karp, B., Govindan, R., Estrin, D.: Data-centric storage in sensornets. In: The ACM SIGCOMM Workshop on Hot Topics in Networks (HOTNETS) (2002)
17. Shor, N.Z.: Minimization Methods for Non-differentiable Functions. Springer, New York (1985)
18. Tang, B., Gupta, H.: Cache placement in sensor networks under an update cost constraint. Journal of Discrete Algorithms 5, 422–435 (2007)
19. Tang, M., Cao, J.: Optimization on Distributed User Management in Wireless Sensor Networks. In: ICPADS 2007, Taiwan (2007)
20. Xing, G., Lu, C., Zhang, Y., Huang, Q.: Minimum Power Configuration in Wireless Sensor Networks. In: MobiHoc 2005, Urbana-Champaign, Illinois, USA (2005)

An Effective and Efficient Method for Handling Transmission Failures in Sensor Networks

Heejung Yang and Chin-Wan Chung

Korea Advanced Institute of Science and Technology, Korea
heejung@islab.kaist.ac.kr, chungcw@kaist.edu

Abstract. The suppression scheme is a solution for limited energy constraints in sensor networks. Temporal suppression, spatial suppression and spatio-temporal suppression are proposed to reduce energy consumption by transmitting data only if a certain condition is violated. Among these suppression schemes, spatio-temporal suppression is the most energy efficient than others because it combines the advantages of temporal suppression and spatial suppression. A critical problem of these suppression schemes is the transmission failure because every nonreport is considered as a suppression. This causes the accuracy problem of query results. In this paper, we propose an effective and efficient method for handling transmission failures in the spatio-temporal suppression scheme. In order to detect transmission failures, we devise an energy efficient method using Bloom Filter. We also devise a novel method for recovering failed transmissions which can save energy consumption and recover failed values more accurately. The experimental evaluation shows the effectiveness of our approach. On the average, the energy consumption of our approach is about 39% less than that of a recent approach and the accuracy of the query results of our approach is about 55% more accurate than that of the recent approach in terms of the error reduction.

Keywords: Sensor networks, Spatio-temporal suppression, Transmission failures.

1 Introduction

Sensor networks give us new opportunities for observing and interacting with the physical world. They are composed of a large number of sensor nodes and each sensor node has capabilities of sensing, processing, and communication. These sensor nodes are deployed in environments where they may be hard to access and provide various useful data. Habitat and environmental monitoring are representative applications of sensor networks. For example, in Great Duck Island project[6], sensor nodes were deployed in the nests of the storm petrels and biologists can collect various scientific data to analyze their lifestyle.

While sensor networks enable continuous data collection on unprecedented scales, there are challenges because of the limited battery resources on each sensor node. Since sensor nodes are usually deployed in an unattended manner, it is

X. Zhou et al. (Eds.): DASFAA 2009, LNCS 5463, pp. 92–106, 2009.

not easy to replace their batteries. Therefore reducing energy consumption is a major concern in sensor networks. Batteries of sensor nodes are depleted by sensing, computation, and communication. Among these tasks, communication is the primary source of energy consumption. Thus several techniques were proposed to resolve the limited energy constraint by reducing the communication.

The suppression is one solution to reduce communication using the temporal/spatial correlation of sensor readings. Each sensor node transmits its sensor reading only if the value of the sensor reading violates a certain condition related to the temporal/spatial correlation. In the temporal suppression scheme, sensor nodes do not transmit their readings, if the current reading is similar to the last transmitted reading. The base station assumes that any nonreport values are unchanged from the previously received ones. In the spatial suppression scheme, there are several groups of sensor nodes having similar sensor readings. Each group has one leader node and this leader node reports for its group. Both of these suppression schemes can lower energy consumption by reducing the number of transmissions. And it is possible to combine these two approaches. For example, sensor nodes are grouped by using the spatial correlation, and the leader and member nodes use the temporal suppression. The leader node makes the representative value for the group based on sensor readings received from its member nodes. This approach can greatly reduce the communication using the temporal and spatial correlation of sensor nodes.

However this suppression has a critical weakness. This problem is caused by the fact that every nonreport is considered as a suppression. Sensor networks are prone to transmission failures due to interference, obstacles, and congestions, etc. In the suppression scheme, these transmission failures create ambiguity. A nonreport may either be a suppression or a failure but there is no way to differentiate between them. [10] proposed a framework, BaySail (*BAYesian analysis of Suppression and fAILures*), to deal with transmission failures in the suppression scheme. Each node adds some redundant information on every report which consists of the last r transmission timestamps and direction bits indicating whether each report is increased or decreased compared to the previously reported sensor reading. Using this information, the base station estimates missing readings using the Bayesian inference. Therefore the missing readings are estimated more accurately. However this approach cannot be applied to the spatio-temporal suppression scheme.

The spatio-temporal suppression scheme is to combine the advantages of both spatial and temporal suppression schemes and can reduce more energy consumption. In this suppression scheme, leader nodes make their representative values based on their member nodes' sensor readings, and transmit these representative values to the base station. Therefore transmission failures can be classified into two categories such as failures within each group and failures from each leader node to the base station. BaySail only considered the latter transmission failures. It is difficult to apply BaySail approach on transmission failures within groups because the Bayesian inference is complex and time-consuming while leader nodes have very limited resource constraints compared to the base station. If we do not resolve transmission failures within groups in using the spatio-temporal suppression scheme, the leader node will treat every nonreport as a suppression and make the representative value based on these inaccurate data and then transmit this to the base station. Thus we need a new method to deal with transmission failures more effectively in the spatio-temporal suppression scheme.

In this paper, we propose an effective and efficient method to resolve the transmission failures in the spatio-temporal suppression scheme. Our approach can detect transmission failures more energy efficiently and recover the failed values more accurately. The contributions of this paper are as follows:

- We extend the transmission failure handling for the spatio-temporal suppression scheme.
- We devise an energy efficient method to distinguish a suppression and a transmission failure while considering resource constraints of leader nodes. Some additional information related to the previous transmissions has to be added on every report to distinguish suppression and transmission failure although this increases energy consumption. We propose a method to represent this information compactly and to identify transmission failures effectively.
- We devise an effective method to recover the value of a failed transmission. Since sensor networks have very limited energy constraints, we propose an energy efficient method to notify the transmission failure to the sender while recovering the failed value more accurately.
- We experimentally evaluate our approach against other approaches to deal with transmission failures. Experimental results show the effectiveness of our approach in the spatio-temporal suppression scheme. On the average, the energy consumption of our approach is about 39% less than that of BaySail and the accuracy of the query results of our approach is about 55% more accurate than that of BaySail in terms of the error reduction.

The remainder of this paper is organized as follows. Section 2 reviews related works of reducing energy consumption and handling transmission failures in sensor networks. In Section 3, we describe our proposed approach to resolve transmission failures in the spatio-temporal suppression scheme. The experimental results are shown in Section 4. Finally, in Section 5, we conclude our work.

2 Related Work

The suppression is proposed to reduce energy consumption in sensor networks. In suppression, data is transmitted only if a certain kind of condition is violated. Suppression schemes can be classified into three categories such as temporal suppression, spatial suppression and spatio-temporal suppression.

The temporal suppression scheme uses the temporal correlation of sensor readings. If the current value is different from the previously transmitted value by more than a certain threshold, this value is transmitted. [7] and [4] use temporal suppression. [7] uses bounded filters to suppress stream data. If the current value lies inside a bounded filter, the data source does not transmit this value. In [4], dual Kalman Filter is used to suppress data as much as possible. The server activates a Kalman Filter KF_s and at the same time, a sensor activates a mirror Kalman Filter KF_m. The dual filters KF_s and KF_m predict future data values. Only when the filter at a sensor KF_m fails to predict future data within the precision constraint then the sensor sends updates to KF_s.

The spatial correlation between sensor nodes is used in the spatial suppression scheme. In this scheme, a node suppresses its sensor reading if it is similar to those of its neighboring nodes. There is a group of nodes having similar values and one node is selected to represent the group. [5] proposes the spatial suppression scheme using a small set of representative nodes. These representative nodes constitute a network snapshot and are used to provide approximate answers to user queries.

The spatio-temporal suppression scheme combines the advantages of both of the above schemes. [2] uses replicated dynamic probabilistic models for groups. These groups are made using the disjoint-cliques approach and then build models for each of them. Both the base station and sensor nodes maintain a pair of dynamic probabilistic models of how data evolves and these models are kept synchronized. The base station computes the expected values of sensor readings according to the model and uses it as the answer. Data is transmitted only if the predicted value is not within the error bound. [9] suggests a novel technique called CONCH (*Constraint Chaining*) combining both suppression schemes. Based on the minimum spanning forest covering all sensor nodes, CONCH temporally monitors spatial constraints which are differences in values between neighboring nodes. For each edge in the minimum spanning forest, one node is designated *updater* and the other one is designated *reporter*. *Updater* triggers a report if its value has changed. *Reporter* triggers a report if the difference between its node's value and *reporter*'s value has changed. The set of reports collected at the base station is used to derive all node values. To cope with transmission failures, it uses multiple, different forests over the network. Failure probabilities are integrated into edge weights to get more reliable forests. If transmission failure occurs, reported difference values from each forest are inconsistent. To recover the failed values, the maximum-likelihood approach is used.

All these suppression schemes can reduce energy consumption by suppressing the transmission. But they do not consider transmission failures except CONCH. But the method used in CONCH is only applicable to their approach. In a general suppression scheme, since every nonreport is considered as a suppression, it causes the accuracy problem. Because transmission failures are prone to sensor networks, it is needed to resolve this problem. [10] addresses this problem and proposes a solution in the temporal suppression scheme. It adds some redundant information about the previous transmissions to every report and infers the failed value using the Bayesian inference. However, this approach does not consider the spatio-temporal suppression scheme. In the spatio-temporal suppression scheme, there are two kinds of transmission failures which are failures from a member node to a leader node and failures from a leader node to the base station. Although this approach can be applied to failures from a leader node to the base station, it cannot be used at a leader node due to resource constraints of sensor nodes. Therefore we need a new approach to resolve transmission failures in the spatio-temporal suppression scheme.

3 Proposed Approach

The goal of our approach is to resolve transmission failures more energy efficiently and accurately in the spatio-temporal suppression scheme. We assume that sensor nodes are grouped according to the spatial correlation and each group has a leader node. Each leader node makes a representative value for its group and transmits this to the base station using a temporal suppression policy. In other words, the transmission will occur only when the current representative value for a group has been changed more than a user-specified threshold compared to the previously transmitted one. We consider the cases when the user queries are in the form of an aggregation of sensor readings. Therefore, the representative value of a leader is the aggregation of sensor readings of the group to which the leader belongs. For the general cases where the user queries require individual sensor readings, a leader sends all received readings after possible compression. Therefore, the extension of the proposed method to general cases is straightforward. The representative value for each group is made by aggregating over received values and the previous values for the suppressed values. Any aggregation functions such as MIN, MAX, SUM, and AVG can be used for making the representative value. Member nodes are also using the temporal suppression. We assume that the min value and the max value of sensor readings are known previously. In this section, we describe the details of our approach to handle transmission failures in the spatio-temporal suppression scheme.

3.1 Overall Approach

In the spatio-temporal suppression scheme, there are two kinds of transmission failures, that is, failures from a member node to a leader node and those from a leader node to the base station. As mentioned, since sensor nodes have very limited resource constraints, we cannot apply a complex method to resolve transmission failures from a member node to a leader node. Therefore, we devise a method which does not require much resource to deal with transmission failures within groups.

First, we have to distinguish a suppression and a transmission failure. We use the compressed history of the previous transmissions to distinguish them. This history information consists of timestamps of the previous transmissions and we compress the history information using Bloom Filter. Based on this compressed history information, we can identify transmission failures using the membership test of Bloom Filter.

Table 1. Notation

Notation	Description
$T_{transmitted}$	the set of timestamps of previous transmissions
$T_{succeeded}$	the set of timestamps of succeeded transmissions
T_{failed}	the set of timestamps of failed transmissions
$T_{suppressed}$	the set of timestamps of suppressions
T_{test}	the set of timestamps for transmission failure detection
t_{last}	the timestamp of the last received data
$t_{current}$	the timestamp of the currently received data
t_{prevS}	the earlier timestamp of a pair of adjacent timestamps in $T_{succeeded}$
t_{nextS}	the later timestamp of a pair of adjacent timestamps in $T_{succeeded}$
$t_{duration}$	the difference between t_{prevS} and t_{nextS}, i.e. $t_{duration} = t_{nextS} - t_{prevS}$
v_{prevS}	the sensor reading value at t_{prevS}
B	compressed history information (Bloom Filter)
m	Bloom Filter size
$h_1, h_2, ..., h_k$	k hash functions used in Bloom Filter
B[h_i]	the bit position in Bloom Filter by applying h_i
x	sampling interval specified in the user query
δ	user-specified error threshold
$D_{received}$	data buffer for successfully received data
D_{sent}	data buffer for transmitted data

After detecting transmission failures, a leader node or the base station requests the retransmission for failures to senders. A retransmission request is constituted of two successfully transmitted timestamps having failed transmissions between them and the quantization is applied to reduce the size of this.

When a member node receives the retransmission request, it can identify two successfully transmitted timestamps having transmission failures between them. Thus the node can calculate differences between failed values and the sensor value of the first timestamp in the retransmission request. The quantization is applied to each difference and this data is transmitted.

Finally, the leader node can recover failed values using the received quantized value. We can recover the range of a failed value based on the sensor value of the first timestamp in the retransmission request. We assign the average value of the range to the corresponding failed value.

In the case of transmission failures from a leader node to the base station, we can apply our approach or the Bayesian inference of BaySail because the base station has no resource constraints. But our approach shows better performance than BaySail according to experimental results.

Table. 1 summarizes the notation used in this paper.

3.2 Transmission Failure Detection

In the suppression scheme, if we don't use any specific method to distinguish a suppression and a transmission failure, every nonreport is considered as a suppression. This causes the accuracy problem of query results. Therefore we add timestamps of previous transmissions on every report and use Bloom Filter[1] to compress this history of transmissions. Bloom Filter is a space-efficient probabilistic data structure that is used to test whether an element is a member of a set.

Fig. 1 shows the algorithm of compressed history information. We make m-bit Bloom Filter to represent the set of timestamps of previous transmissions

Algorithm CompressHistoryInfo
Input Previously transmitted timestamp set $T_{transmitted} = \{t_1, t_2, ..., t_n\}$
Output m-bit Bloom Filter B

begin
1. **for** each timestamp t in $T_{transmitted}$
2.   Compute $h_1, h_2, ..., h_k$
3.   Set B[h_1(t)] = B[h_2(t)] = ... = B[h_k(t)] = 1
4. **return** B

end

Fig. 1. Algorithm of Compressed History Information

Algorithm FailureDetection
Input Received Bloom Filter B, t_{last}, $t_{current}$, $D_{received}$
Output The set of failed transmission timestamps T_{failed},
the set of succeeded transmission timestamps $T_{succeeded}$

begin
1. $t_{test} := t_{last}$
2. **while** $t_{test} < t_{current}$
3. t_{test} += x
4.   Insert t_{test} into the set of test timestamps T_{test}
5. **for** each timestamp t in T_{test}
6.   **for** each hash function h_i
7.   Compute h_i(t)
8.   **if** all B[h_i(t)] == 1
9.   **if** data of t is not in the data buffer $D_{received}$ // Transmission Failure
10.   Insert t into the set of failed timestamps T_{failed}
11.   **else** // Transmission Success
12.   Insert t into the set of succeeded timestamps $T_{succeeded}$
13.   **else** // Suppression
14.   Insert t into the set of suppressed timestamps $T_{suppressed}$
15. **return** T_{failed}, $T_{succeeded}$

end

Fig. 2. Algorithm of Transmission Failure Detection

$T_{transmitted}$. It has n timestamps. The number of histories (the number of timestamps in $T_{transmitted}$) affects the accuracy of query results. The consecutive transmission failures cause the history information losses. If the number of histories is large, it consumes more energy but lowers the loss rate of the history information. We vary the number of histories in our experiments and show its effect on the energy consumption and the accuracy. For each timestamp in $T_{transmitted}$, we compute k hash functions $h_1, h_2, ..., h_k$ with range $\{0, ..., m-1\}$ and all bit positions B[$h_1(t)$], B[$h_2(t)$], ..., B[$h_k(t)$] are set to 1 in Bloom Filter (Line (2), Line (3)). This Bloom Filter is added to data when data is transmitted.

When receiving such a report at a leader node or at the base station, it applies the membership test of Bloom Filter to distinguish a suppression and a transmission failure. Fig. 2 shows the algorithm of transmission failure detection.

To detect transmission failures, we make the test timestamp set T_{test} from the timestamp of the last received data t_{last} (Line(1) - Line(4)). Since the sampling interval is specified in the user query, we can know the timestamps of possible transmissions. If the sampling interval is x seconds, we know that the data will be transmitted every x seconds. Therefore, to make the T_{test}, we start t_{last} and add x to the previously generated test timestamp until it reaches the currently received timestamp $t_{current}$. After that, we check a timestamp t in the T_{test}

whether this is in the reported Bloom Filter or not (Line(5) - Line(13)). To check whether t is in the Bloom Filter, we apply k hash functions to t. If all k bits of $h_i(t)$ are set in the reported Bloom Filter and the leader node or the base station has data transmitted at t, the data is successfully transmitted (Line(11)). If all k bits of $h_i(t)$ are set but the data of the corresponding time is not in the leader node or the base station, we know that the transmission is failed at that time (Line(9)). If any $h_i(t)$ is not set in the reported Bloom Filter, the data is suppressed (Line(13)).

Bloom Filter may yield false positives. To minimize the false positive rate, $k = \ln 2 \times (m/\#\ of\ histories)$ hash functions are used [1]. Note that if $m = 10 \times \#\ of\ histories$, the false positive rate is less than 1%[3]. Therefore, we use $m = 10 \times \#\ of\ histories$ bits for Bloom Filter and find the optimal number of hash functions based on that. In our approach, a false positive means that a suppression is identified as a transmission failure. Specifically, a certain timestamp is considered as transmitted in the reported Bloom Filter but actually it is not. Since the data of this timestamp is not in the leader node or the base station, it is considered as a transmission failure. Although this causes unnecessary retransmission of suppressed data, it does not decrease the accuracy of query results but may slightly increase the accuracy.

3.3 Retransmission Request

Using the membership test of Bloom Filter, we can find timestamps of failed transmissions. After detecting transmission failure, we make a retransmission request to the sender. A retransmission request consists of nodeID and two successfully transmitted timestamps, t_{prevS} and t_{nextS}, having failed transmissions between them. Fig. 3 shows the algorithm of retransmission request.

```
Algorithm RetransmissionRequest
Input t_current, T_failed, T_succeeded
Output Retransmission request R

begin
    for i = 0; i < length(T_succeeded) - 1; i++
        t_prevS := T_succeeded[i]
        t_nextS := T_succeeded[i+1]
        if t_nextS − t_prevS > x
            for j = 0; j < length(T_failed) - 1; j++
                t_prevF := T_failed[j] // t_prevF, t_nextF: local variables
                t_nextF := T_failed[j+1]
                if t_prevS < t_prevF && t_prevF < t_nextS
                    if t_nextF < t_nextS
                        j++
                    else
                        t_duration := t_nextS - t_prevS
                        n := t_duration / x
                        n_binary := convert n into binary string
                        R := Request(nodeID, t_prevS, n_binary)
    return R
end
```

Fig. 3. Algorithm of Retransmission Request

We search $T_{succeeded}$ and T_{failed} to find two timestamps having failed transmissions between them (Line(1) - Line(11)). If we find these two timestamps, we make the retransmission request (Line(12) - Line(15)). To reduce the energy consumption, we transmit the first timestamp of success transmission t_{prevS} and the difference $t_{duration} = t_{nextS} - t_{prevS}$ after encoding it. If the sampling interval is x, the range of $t_{duration}$ is $\{0, 1x, 2x, ..., nx\}$ where n is a positive integer. So we represent the $t_{duration}$ using n (Line(12) - Line(14)). We use a small number of bits for representing n more compactly. After making the retransmission request, we transmit this to the corresponding sender.

3.4 Failed Value Retransmission

Whenever the sensor node receives the retransmission request, it retransmits the values of failed transmissions. But a naive retransmission of failed values consumes much energy. Therefore, we use the quantization to reduce energy consumption for the failed value retransmission.

Before applying the quantization, we have to decide how many bits will be used in the quantization. We can decide the number of bits for the quantization before starting the query processing. The number of bits is determined by the user-specified error threshold δ and the range of the input. The difference of the min value v_{min} and the max value v_{max} of sensor readings is the range of the input. By dividing the range of the input by the user-specified error threshold v_{max} - v_{min} / δ, we can get the number of intervals to represent the sensor values. The number of intervals has to satisfy $\#\, of\ intervals \leq 2^{\#\ of\ bits}$. Therefore, we can choose the minimum number of bits satisfying this condition.

```
Algorithm ValueRetransmission
Input Request R(nodeID, t_prevS, n_binary), D_sent
Output Retransmission message retransmissionMSG

begin
    retransmissionMSG := NULL
    Calculate t_nextS from t_prevS and n_binary
    Find transmitted data between t_prevS and t_nextS in D_sent
    num_of_failures := the number of transmitted data between t_prevS and t_nextS in D_sent
    for i = 0; i < num_of_failures; i++
        diff := v_prevS - v_failed // v_failed is the value of the i-th failed transmission
        interval := ⌊|diff|/δ⌋ + 1
        binary := convert interval into binary string
        signIndex := i * num_of_bits // num_of_bits is the number of bits necessary for binary
        if diff < 0
            quantizedValue[signIndex] := 1
            // quantizedValue is an array to store binaries for the transmitted values
        else
            quantizedValue[signIndex] := 0
        for k = 0, nbit = 1; k < length(binary); k++, nbit++ // Quantized Value Setting
            if binary[k] == 1
                quantizedValue[signIndex + nbit] := 1
            else if binary[k] == 0
                quantizedValue[signIndex + nbit] := 0
        retransmissionMSG := (nodeID, t_prevS, quantizedValue)
    return retransmissionMSG
end
```

Fig. 4. Algorithm of Failed Value Retransmission

```
Algorithm Recovery
Input Received retransmissionMSG
Output Recovered values for failed transmissions

begin
1.  if retransmissionMSG != NULL
2.    for i = 0; i < num_of_failures; i++
3.      signIndex := i * num_of_bits
4.      if quantizedValue[signIndex] == 1
5.        sign := -1
6.      else
7.        sign := 1
8.      for j = signIndex + 1, nbit = 1; j < signIndex + num_of_bits; j++, nbit++
9.        if quantizedValue[j] == 1
10.           binary[nbit] := 1
11.     interval := convert quantizedValue into decimal number
12.     rangeL := ((interval - 1) * δ) + v_prevS
13.     rangeH := (interval * δ) + v_prevS
14.     recoveredValue := (rangeL + rangeH) / 2
15.     Insert recoveredValue into corresponding failed data value in D_received
end
```

Fig. 5. Algorithm of Failed Value Recovery

Fig. 4 shows the algorithm of the failed value retransmission. Based on the received retransmission request, the node identify two timestamps of successful transmissions, t_{prevS} and t_{nextS} (Line(1)). From this, the sensor node can find the number of failed transmissions (Line(3)). We calculate the difference between failed value v_{failed} and the value v_{prevS} for each failed transmission (Line(6)). The difference belongs to a certain interval $(L, H]$ where $L = \lfloor |diff|/\delta \rfloor \times \delta$ and $H = (\lfloor |diff|/\delta \rfloor + 1) \times \delta$. We transform $(diff/\delta) + 1$ into the bit representation using the quantization (Line(7) - Line(18)). For example, let the number of bits for quantization be 3. If the user-specified error threshold δ is 5 and the difference $diff$ is 12, this belongs to the interval of $(2 \times 5, 3 \times 5]$. Then the quantized value for 3 is 011 and it is transmitted to the leader node or the base station.

3.5 Failed Value Recovery

Finally, the leader node or the base station can recover the failed values using received quantized data. Fig. 5 shows the algorithm of failed value recovery.

We can recover the failed values based on v_{prevS} because each quantized value represents the interval to which failed value belongs. Let $q_1, q_2, ..., q_n$ be quantized values for failed values. Because q_i is the bit representation of $diff_i/\delta + 1$, we can get the range $(L, H]$ of the difference using q_i (Line(1) - Line(13)). Then we assign the average value of the range to the corresponding failed value (Line(14)). In the above example used in the failed value retransmission, let v_{prevS} be 33. If we receive 011 at the leader node or the base station, the range of difference value is $(10, 15]$. Thus the original failed value belongs to $(33 + 10, 33 + 15]$ and we assign 45.5 as the failed value.

When a failure occurs during the transmission of a retransmission request or a failed value retransmission, the requestor resends the request after a certain waiting time. This is possible because the requestor is expecting the retransmission. The experimental result shows that this type of failures has little effect on the transmission cost and the accuracy.

4 Experimental Evaluation

We perform the experimental analysis to validate our approach using our own simulator. The simulated network consists of one group that is a rectangular grid. We performed experiments for multiple groups, but the pattern of the result for multiple groups was similar to that for a single group. Sensor nodes are placed on grid points and we varied the size of the group as 3 × 3, 4 × 4, and 5 × 5. We assume that the leader node of the group is placed at one hop distance from the base station. The minimum spanning tree is built over all member nodes where each member node can reach to the leader node with the number of hops as small as possible. Sensor readings follow the Gaussian model and are produced at the user-specified sampling rate. Each sensor node generates sensor readings which follow a Gaussian distribution with the mean μ and the standard deviation σ. The ranges of μ and σ are 0 to 100 and 0 to 20 respectively. These two parameters for each sensor node are randomly generated between their ranges. The user-specified sampling rate is 10 seconds and the error threshold is 5.

We compare the performance using the energy consumption and the accuracy of query results. The comparison schemes are as follows:

- **ACK**: The acknowledgement is used to detect transmission failures. If a sensor node does not receive an acknowledgement, the corresponding data is retransmitted.
- **BF**: All member nodes and leader nodes are using our proposed approach utilizing Bloom Filter to resolve transmission failures.
- **BF + BaySail**: Our proposed approach BF is used within groups and BaySail is applied to transmissions from the leader node to the base station.
- **Leader BaySail**: A leader node uses BaySail to resolve transmission failures. This assumes that the leader node has no resource constraints.
- **BS BaySail**: Each node transmits its data to the base station using BaySail. This is the original BaySail proposed in [10]. There is no concept of a group in the spatio-temporal suppression scheme.

We change the failure rate from 10% to 50% and vary the number of history information from 1 to 5. Each experiment is run for 5 times and the results are averaged.

4.1 Energy Consumption

We measure the energy consumption using the amount of transmitted data because the larger the amount of transmission, the more energy is consumed. The basic data sizes used in our experiments are as follows:

Component	Size (bits)
Acknowledgement	40 (B-MAC protocol [8])
NodeID	32
Sensor Reading	32
Timestamp	32

The energy consumption for 5×5 grid is shown in (a), (c), (e), (g), (i) of Fig. 6. H is the number of histories used in BF and BaySail. We do not show the results for 3×3 and 4×4 due to the space limitation. But they also have similar results to those of 5×5. BF consumes less energy than the other schemes in almost all cases. ACK's consumption steeply increases when the failure rate increases. Since it has to retransmit the original data until it is successfully transmitted. Energy consumption of BF also increases when the failure rate increases because the number of retransmission requests and value retransmissions increase in accordance with the increased failure rate. Although the number of retransmission requests and value retransmissions increase when the failure rate is high, BF does not require too much energy due to Bloom Filter and the quantization technique. BaySail has the constant energy consumption when failure rates are varied because it transmits a fixed size of data having history information only once and failed values are inferred at the base station. The number of histories in BF and BaySail increases the size of transmitted data. But the transmitted data size of BF is much less than that of BaySail because BF compresses the history information using Bloom Filter. Specifically, we set the size of Bloom Filter to 10 times larger than the number of histories to reduce the false positive rate less than 1%. But this is very small compared to the history size used in BaySail. In the case that the number of histories is 3, the history size is 96 bits in BaySail while 30 bits in BF. Therefore BF does not increase the size of transmitted data severely when the number of histories increases. Consequently, BF is more energy efficient than other schemes. We compare the energy consumption between BF and the original BaySail (BS BaySail) by calculating the reduction of the transmission cost of BS BaySail by using BF. Let the average of the transmission costs for all numbers of histories and all failure rates for BS BaySail be T(BS BaySail) and that for BF be T(BF). Then (T(BS BaySail)$-T$(BF))/T(BS BaySail) is about 0.39.

4.2 Accuracy

We evaluate the accuracy of query results using the relative error rate. Relative error rate is calculated by $|(\textit{recovered value} - \textit{original value})/\textit{original value}|$, where *recovered value* is the estimated value after applying our approach to handle transmission failures. The query result is an aggregated value of the group and we use AVG as the aggregation function. We assume that ACK can successfully retransmit failed values in the end. (b), (d), (f), (h), (j) of Fig. 6 show the result of the accuracy for each scheme. BF shows better accuracy than

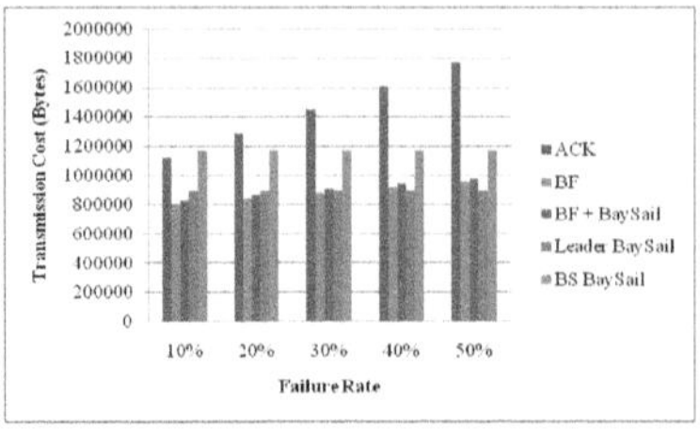

(a) Energy Consumption (H = 1)

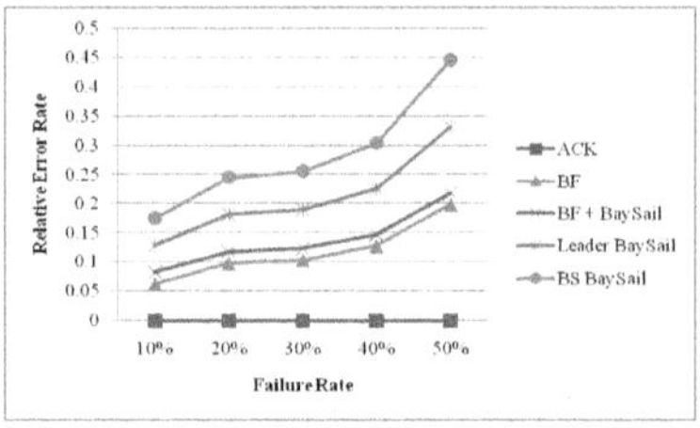

(b) Accuracy (H = 1)

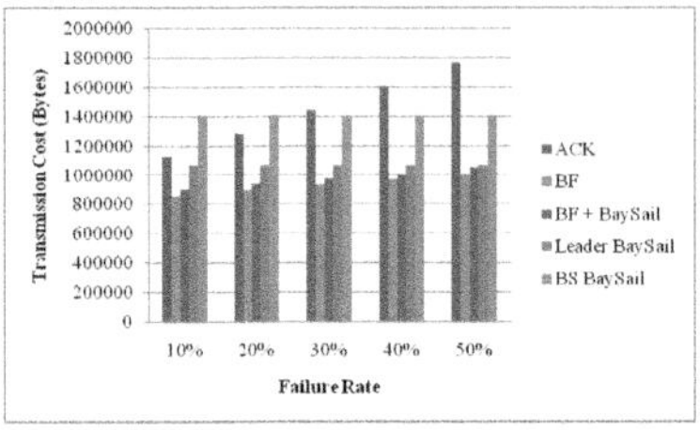

(c) Energy Consumption (H = 2)

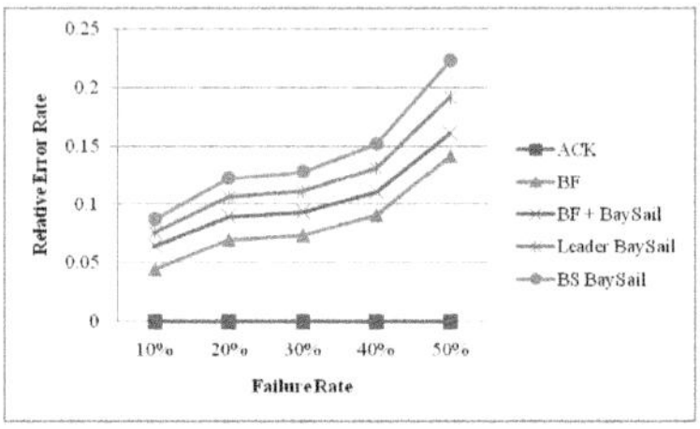

(d) Accuracy (H = 2)

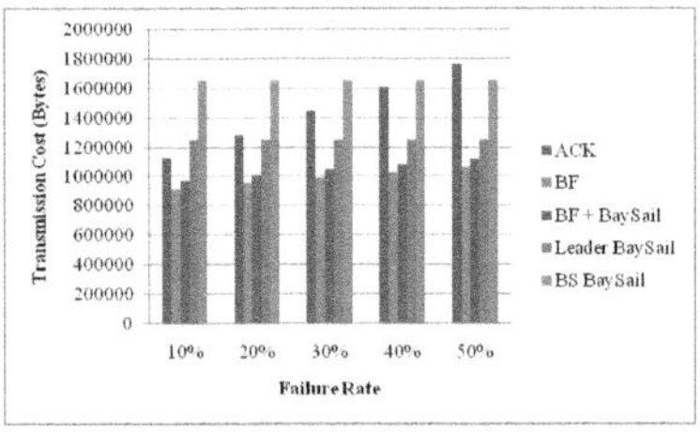

(e) Energy Consumption (H = 3)

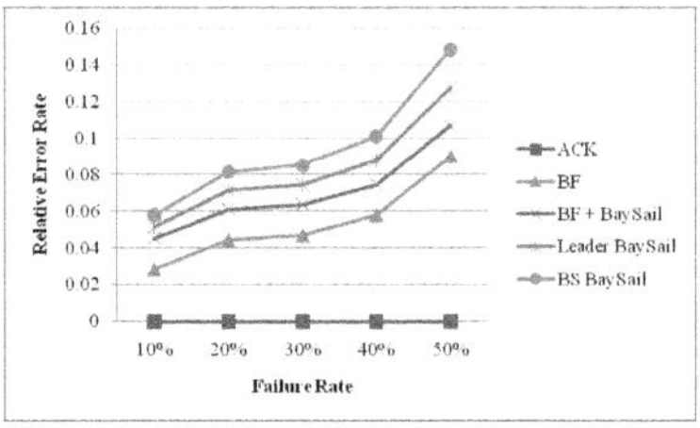

(f) Accuracy (H = 3)

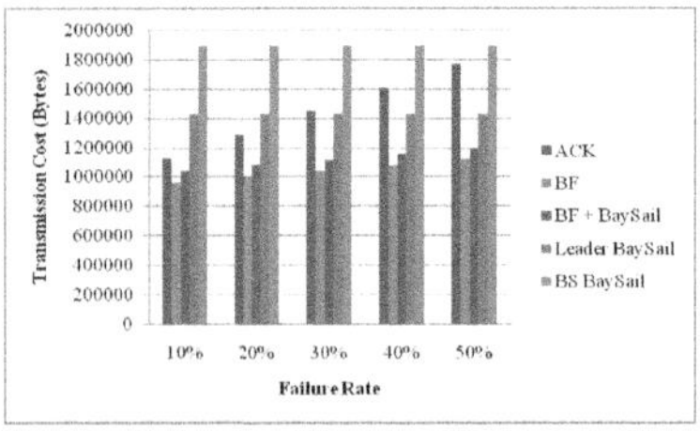

(g) Energy Consumption (H = 4)

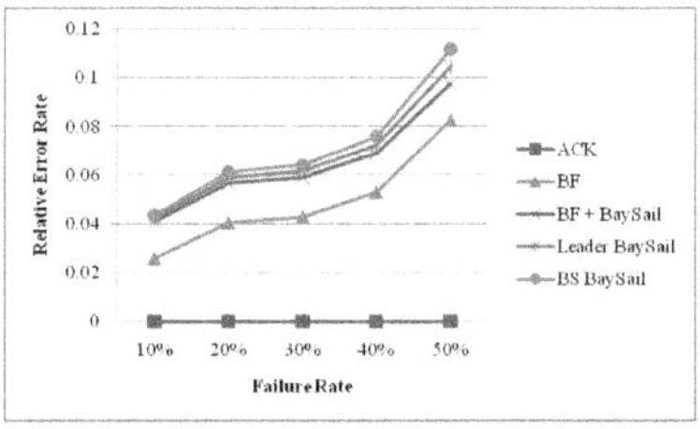

(h) Accuracy (H = 4)

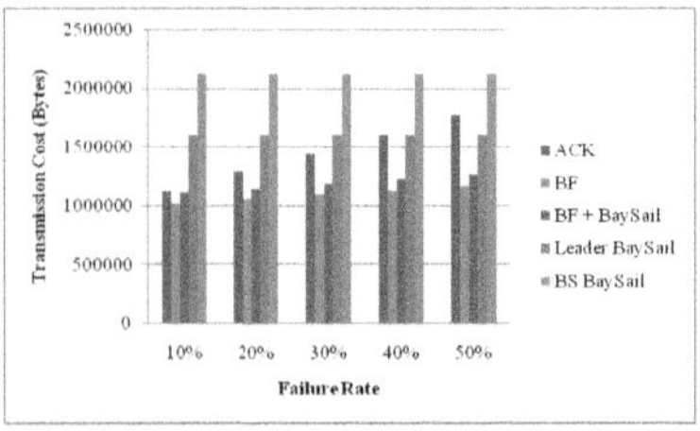

(i) Energy Consumption (H = 5)

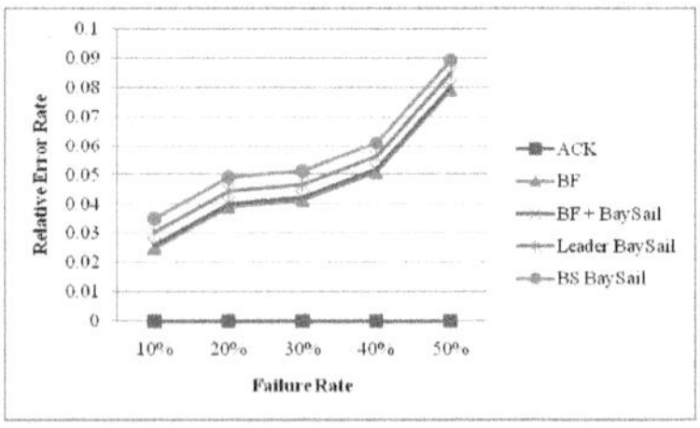

(j) Accuracy (H = 5)

Fig. 6. Results for 5 × 5 grid

all other schemes. As for the energy consumption, we compare the accuracy between BF and the BS BaySail by calculating the reduction of the relative error rate of BS BaySail by using BF. Similarly as above, the reduction by using BF is 55%.

In BF, each failed value is represented as the difference from the previously successfully transmitted data value, and this difference is retransmitted using the quantization. The quantization is used to reduce energy consumption while guaranteeing that the recovered values are not different from the original values more than the user-specified error threshold. Therefore we can recover the failed value more accurately while using less energy than the other schemes. When the transmission failure rate increases, the relative error rate of each scheme also increases. The number of histories also affects the relative error rate. If the transmission failure rate is high, the probability of consecutive transmission failures is also high. Thus the history information could be lost. For example, let the number of histories is 1. If data is transmitted at t_n, t_{n+1}, t_{n+2} but data is successfully transmitted only at t_{n+2}, then the history information about t_n is lost and it is considered as a suppression. Therefore the relative error rate is higher than those for larger numbers of histories.

We set the minimum number of bits for the value quantization to satisfy $\#\ of\ intervals \leq 2^{\#\ of\ bits}$. The number of intervals is calculated by $(v_{max} - v_{min})/\delta$ where v_{max}, v_{min} and δ are the max value, min value for sensor readings, and the user-specified error threshold respectively. If we use a value smaller than δ to calculate the number of intervals, the interval becomes narrowed and the number of bits for the quantization increases. Using this quantization bits and interval, we can get a tighter range for a failed value. This can increase the accuracy while consuming more energy.

5 Conclusion

Sensor networks usually have very limited energy resources. To reduce energy consumption, suppression schemes are proposed. Among these suppression schemes, spatio-temporal suppression can dramatically reduce the energy consumption. But the critical weakness of suppression is the transmission failure because this is considered as a suppression. This causes the accuracy problem in the query result.

We propose an effective and efficient method for handling transmission failures in the spatio-temporal suppression scheme. In the spatio-temporal suppression, transmission failures can occur from the member node to the leader node of a group and from the leader node to the base station. In resolving transmission failures, we have to consider the resource constraints of each sensor node. Therefore, we devise an energy efficient method to distinguish a suppression and a failure using Bloom Filter. History information of previous transmissions is inserted into Bloom Filter and we can effectively identify failures using the membership test of Bloom Filter. After detecting transmission failures, the receiver notifies the transmission failures to the sender, which retransmits these failed values using quantization. This quantization can reduce the size of transmitted data and recover the failed values more accurately. The experimental results show that our approach resolves the transmission failures energy efficiently and accurately.

Acknowledgments. This work was partially supported by Defense Acquisition Program Administration and Agency for Defense Development under the contract.

References

1. Bloom, B.H.: Space/time trade-offs in hash coding with allowable errors. Communications of the ACM 13, 422–426 (1970)
2. Chu, D., Deshpande, A., Hellerstein, J.M., Hong, W.: Approximate data collection in sensor networks using probabilistic models. In: ICDE 2006, p. 48. IEEE Computer Society, Los Alamitos (2006)
3. Fan, L., Cao, P., Almeida, J., Broder, A.Z.: Summary cache: A scalable wide-area web cache sharing protocol. In: IEEE/ACM Transactions on Networking, pp. 254–265 (1998)
4. Jain, A., Chang, E.Y., Wang, Y.-F.: Adaptive stream resource management using kalman filters. In: SIGMOD 2004, pp. 11–22. ACM, New York (2004)
5. Kotidis, Y.: Snapshot queries: Towards data-centric sensor networks. In: ICDE 2005, pp. 131–142. IEEE Computer Society, Los Alamitos (2005)
6. Mainwaring, A., Polastre, J., Szewczyk, R., Culler, D.: Wireless sensor networks for habitat monitoring (2002)
7. Olston, C., Jiang, J., Widom, J.: Adaptive filters for continuous queries over distributed data streams. In: SIGMOD 2003, pp. 563–574. ACM, New York (2003)
8. Polastre, J., Hill, J., Culler, D.: Versatile low power media access for wireless sensor networks. In: SenSys 2004, pp. 95–107. ACM, New York (2004)
9. Silberstein, A., Braynard, R., Yang, J.: Constraint chaining: on energy-efficient continuous monitoring in sensor networks. In: SIGMOD 2006, pp. 157–168. ACM, New York (2006)
10. Silberstein, A., Puggioni, G., Gelfand, A., Munagala, K., Yang, J.: Suppression and failures in sensor networks: a bayesian approach. In: VLDB 2007, pp. 842–853. VLDB Endowment (2007)

Query Allocation in Wireless Sensor Networks with Multiple Base Stations

Shili Xiang[1], Yongluan Zhou[2], Hock-Beng Lim[3], and Kian-Lee Tan[1]

[1] Department of Computer Science, National University of Singapore
[2] Department of Mathematics and Computer Science, University of Southern Denmark
[3] Intelligent Systems Center, Nanyang Technological University

Abstract. To support large-scale sensor network applications, in terms of both network size and query population, an infrastructure with multiple base stations is necessary. In this paper, we optimize the allocation of queries among multiple base stations, in order to minimize the total data communication cost among the sensors. We first examine the query allocation problem in a static context, where a two-phase semi-greedy allocation framework is designed to exploit the sharing among queries. Also, we investigate dynamic environments with frequent query arrivals and terminations and propose adaptive query migration algorithms. Finally, extensive experiments are conducted to compare the proposed techniques with existing works. The experimental results show the effectiveness of our proposed query allocation schemes.

1 Introduction

Wireless sensor networks (WSNs) are widely used in many applications, such as environmental monitoring, smart home environment and road traffic monitoring etc. To ease the deployment of Wireless Sensor Network (WSN) applications, researchers have proposed techniques to treat the sensor network as a database. This approach operates as follows: user interests are expressed as queries and submitted to a powerful base station; the base station disseminates the queries into the sensor network, more specifically, to the sensor nodes that are involved in the queries; the sensor nodes generate, process and transfer sensory data back to the base station, which will then correspondingly return the query result back to the users. However, sensor nodes are quite resource-constrained with limited processing capacity, storage, bandwidth and power.

To better realize the potential of WSNs, several query processing techniques have been specially designed to optimize the processing of each query [16,8,10,1]. For example, for aggregation queries, a tree-like routing structure is often utilized and in-network aggregation is done at intermediate nodes to reduce the amount of data sent to the base station [17,4,6]. To further enable the sensor network to scale well with the number of users and queries, multi-query optimization techniques have been investigated to minimize the energy consumption among sensors. These techniques either exploit the commonality among all running

X. Zhou et al. (Eds.): DASFAA 2009, LNCS 5463, pp. 107–122, 2009.

queries at a particular base station as a whole [13,11], or enable new queries to salvage the routing/aggregation subtree of existing queries [6].

All the above works focused on WSNs with a single base station. However, for a large scale WSN, it is necessary and beneficial to have multiple base stations. Firstly, it provides the WSN with better coverage and scalability. The limited radio range of sensor nodes leads to multi-hop routing, where the nodes nearer to the base station need to relay the messages for other nodes and hence become the bottleneck [4,10]. Using multiple base stations can alleviate this problem. Secondly, multiple base stations provide the WSN with better reliability [5]. The communication among sensor nodes are prone to failures, due to collision, node failure and environmental noise etc. With more base stations in the network, the average number of hops each datum travels is fewer, and correspondingly the reliability of the data transmission is better. Lastly, it extends the life time of the WSN. The sensor nodes nearer to the base stations are likely to have higher load and the energy consumption there is greater than other nodes; with more base stations, the burden of nodes nearer to each base station can be relieved.

Thus, we study how to perform multi-query optimization within a WSN with multiple base stations, to minimize the total communication cost among sensor nodes. We assume that, once the queries are sent out to the WSN, the WSN can exploit the sharing of data communication among queries from the same base station to minimize the communication cost by using the existing approaches [13,11]. Within this context, the allocation of queries to the base stations plays a critical role as it determines how much sharing can be exploited by the WSN.

In our previous work [12], we proposed similarity-aware query allocation algorithms for region-based aggregation queries, namely SDP-K-Cut approach and greedy query insertion algorithm (incremental insertion + heuristic ordering). However, more investigations suggest that both of them do not scale well enough with large number of queries, because the SDP-K-Cut approach is too memory-consuming and time-consuming, while the incremental query insertion is sensitive to the insertion sequence of queries. Moreover, they do not consider the effect of query termination, which would happen frequently in a dynamic environment.

In this paper, we address the limitations identified above to better support large-scale WSNs. For static environment, we design a semi-greedy allocation framework, which comprises a greedy insertion phase and an iterative refinement phase. Furthermore, we propose adaptive migration algorithms to deal with the effect of query termination which was not addressed before. This adaptive scheme also serves as an online refinement of the incremental query insertion algorithm in dynamic environment. An extensive performance study is conducted to compare the performance of the proposed techniques with some simpler schemes as well as previous work. The results show that our techniques can effectively minimize the communication cost of a large-scale WSN.

The rest of this paper is organized as follows. In Section 2, we formulate our query allocation problem and examine existing works. Sections 3 and 4 describe the semi-greedy query allocation framework and adaptive query migration algorithm for static and dynamic environment respectively. In Section 5, we present our experimental results and finally, we conclude the paper in Section 6.

2 Preliminaries

Consider a large scale sensor network that comprises K base stations and hundreds of (say N) sensor nodes. The base stations are powerful machines, with abundant processing, storage, and memory capacity and can be recharged easily. On the other hand, the sensor nodes are resource constrained, with limited processing capacity, storage, bandwidth and power. Thus, we focus on conserving the resources of sensor nodes. More specifically, we focus on minimizing the communication cost among sensor nodes, instead of that among base stations which is comparatively negligible.

To exploit the sharing among queries, one solution is to tightly integrate our query allocation work with a specific multiple query optimization scheme that runs at each base station, such as [13,11]. This means the query allocation scheme has to be aware of the cost models used by the specific multi-query optimization scheme at each base station. In this paper, to be general, we adopt a more flexible approach. To guide query allocation, we exploit the inherent sharing among queries without knowledge of the specific underlying multiple query optimization scheme.

2.1 Problem Statement

Our query allocation problem is defined as follows. Suppose there are K base stations and currently M queries are running in the sensor network of size N. For a query q_i running exclusively at a specific base station b_j, a number of radio messages will be incurred to transmit the sensory data to the base station, and we refer to the number of radio messages as the communication cost, denote as c_{ij}. We further denote the query set allocated to base station b_j as Q_j, and the amount of sharing (redundant requests) among these queries as S_j. Then the objective of the query allocation problem, to minimize the communication cost among sensor nodes in order to answer the queries, can be expressed as:

$$minimize \sum_{j=1}^{K}\Big(\sum_{q_i \in Q_j} c_{ij} - S_j\Big)$$

To determine the optimal allocation, we need to find the optimal balance between minimizing $\sum_{j=1}^{K}\sum_{q_j \in Q_j} c_{ij}$ and maximizing $\sum_{j=1}^{K} S_j$.

2.2 System Model

In this paper, we adopt the following assumptions and simplifications of the system model.

First, we focus on region-based aggregation queries, such as SUM, MAX and AVG. More specifically, they belong to the category of *distributive* and *algebraic* aggregation queries, as defined in [4]. A region-based query specifies a fixed set of nodes that are involved in the query and hence simplifies the cost estimation and sharing estimation. Our method can be generalized to other queries as long as the sharing among queries can be estimated, e.g., by semantics exploration or maintaining a statistical model such as [2]. As it is an independent problem, we do not study it in this paper.

Second, for each base station, a routing infrastructure (a routing tree or a routing Directed Acyclic Graph (DAG)) rooted at the base station is constructed. Each such routing infrastructure only involves those sensor nodes that are necessary to process the queries allocated to the corresponding base station.

Third, in-network aggregation [13,11] with multi-query optimization is performed in the routing infrastructure.

Finally, it is assumed that there is a controller which maintains the query allocation information for all the base stations and optimizes the allocation of queries. Such a controller-based architecture is often used in load management within locally distributed systems which are under centralized administrations [15].

With the above assumptions, we can compute the cost function as follows. c_{ij} is computed by counting the number of sensors involved in processing query q_i (including those in the query region and those used to relay the message to the base station). This is because each sensor node only has to send (or relay) one message (due to in-network aggregation).

To estimate the value of S_j, we keep bitmap m_j of size N maintained for base station b_j, whose default values are zero. If a sensor node x is queried by q_i, where $q_i \in Q_j$, we check the value of $m_j[x]$. If it is 0, we set it to 1. Otherwise, some other queries have already requested data from a sensor node x at base station b_j, and this cost is shared and correspondingly we add 1 to S_j.

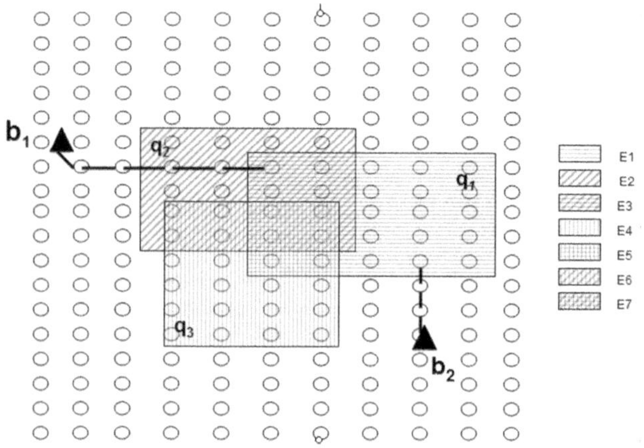

Fig. 1. A scenario with multiple base stations and queries

Note that, if different parts of the network have different reliability in transmission, a weight factor should be assigned to each queried node to represent its reliability during the computation of c_{ij} and S_j.

Figure 1 shows an example of how the query cost c and sharing S are computed in our model. Each of the small circles denotes one sensor, while each rectangular region represents one query and each triangle denotes a base station. q_1 covers 25 sensors and its minimal distance to b_1 is 5, so $c_{11} = 30$. Likewise, we can compute c_{12} to be equal to 28. If both q_1 and q_3 are allocated to b_1, the regions E_5 and E_7 can be shared, hence $S_1 = |E_5| + |E_7|$. It is worth noting that when q_1, q_2 and q_3 are all allocated to b_1, since E_7 has been shared twice, $S_1 = |E_3| + |E_5| + |E_6| + 2 * |E_7|$.

2.3 Existing Works

Query and/or operator allocation have been studied in both traditional distributed database systems [9] and stream processing systems [7,18]. However, these techniques cannot be applied here. Apart from the different application context, our objective is to minimize the communication cost inside the sensor network instead of among the base stations. Moreover, here we endeavor to maximize the sharing of the data collection cost among various queries allocated to the same base station, which is typically not considered in existing works.

In this section, we review methods that allocate queries among base stations in wireless sensor networks. More specifically, we examine the best naive solutions *Nearest* and *Partition*, and then review the similarity-aware query allocation algorithm *SDP-K-Cut* proposed in [12].

Nearest: This strategy minimizes $\sum_{j=1}^{K} \sum_{q_j \in Q_j} c_{ij}$. For each query q_i, we assign it to its nearest base station b_j, where b_j has the smallest distance to the query region of q_i. In this way, since the number of sensors in the query region is fixed, b_j also incurs the smallest allocation cost for q_i. We note that *Nearest* also has the tendency to assign similar queries to the same base station because the region-based aggregation queries whose nearest base station is the same are inherently likely to have overlap with each other; however, it does not explicitly exploit the sharing when making its decision.

Partition: This strategy maximizes $\sum_{j=1}^{K} S_j$. One can simply divide the network into non-overlapping subregions (e.g., according to the voronoi graph for the base stations) and get each base station to take care of the subregion that is closest to it. Each query is then partitioned into sub-queries according to the partitioned subregions, and each sub-query is allocated to its respective base station. Finally, the partially aggregated results for sub-queries are further aggregated through communication among base stations to get the final result of the original aggregation query. In this way, it maximizes the sharing among sub-queries because such sharing can all be automatically captured at each sensor node. However, it sacrifices the benefit of in-network aggregation and introduces considerable relaying overhead for each partitioned query.

SDP-K-Cut: This scheme attempts to minimize $\sum_{j=1}^{K} \sum_{q_j \in Q_j} c_{ij}$ and maximize $\sum_{j=1}^{K} S_j$. It models the problem in an undirected complete graph

G = (V, E, W), where each vertex denotes either a query or a base station, each edge weight denotes either the cost of assigning the particular query to the particular base station, or the negative value of the amount of sharing between two queries. In this way, the problem of assigning each query to the base station that incurs the least communication cost while allocating similar queries to the same base station, is essentially to partition V into K disjoint sets so that the sum of the weight of the edges between the disjoint sets is maximized. Thus, the query allocation problem is approximated as a classical Max-K-Cut problem. The Max-K-Cut problem is then relaxed into a semidefinite programming problem (SDP), and the algorithm in [3] is adapted to solve it.

Given a static set of queries, the SDP-K-Cut approach can generate good query allocation plans. However, its high complexity in terms of time and space makes it impractical to be deployed if the number of queries is huge. Also, upon the insertion of a new query, the SDP-K-Cut algorithm has to recompute from scratch instead of incrementally optimizing the new query set based on the previous status. Therefore, it is computationally too expensive to deploy the scheme in a dynamic context where queries frequently arrive and depart.

3 A Semi-greedy Allocation Framework

In this section, we design a semi-greedy allocation framework for static environment where queries are known apriori. The framework comprises two phases. In the first phase, an allocation plan is generated quickly. The plan essentially determines how the set of queries are to be allocated to the base stations. In the second phase, the generated plan is iteratively refined.

3.1 Greedy Insertion

For the first phase, we adopt the *Greedy Insertion* scheme [12], which comprises incremental insertion and heuristical ordering.

Incremental: this strategy greedily assigns one query at a time to the base station that results in the smallest additional cost for that query. The additional cost ac_{ij} incurred by q_i at b_j (i.e., c_{ij} subtracted by the amount of sharing between q_i and the other queries at b_j) reflects not only the amount of non-sharing region but also the cost of executing q_i at b_j by itself. In this way, *Incremental* is able to efficiently find the best allocation for the new query, which incurs the least communication cost inside the sensor network and does not affect other allocated queries.

Since the order in which queries are assigned may affect the optimality of the query plan, two heuristics have also been proposed to order the queries before they are assigned:

- *Area*: Queries are ordered in descending order of the areas of their query regions.
- *Diff*: Queries are ordered in descending order of the values of function $\text{diff}(q_i) = c_{im} - c_{ij}$, where c_{ij} is the cost of the minimum allocation of

q_i and c_{im} is the cost of the sub-minimum allocation. More formally, $c_{ij} = \min(c_{i1}, c_{i2} ... c_{iK})$, and $c_{im} = \min(c_{i1}, ... c_{ij-1}, c_{ij+1}, ... c_{iK})$.

Greedy Insertion phase essentially benefits to-be-assigned query. However, a query that has already been assigned cannot benefit from the queries that are subsequently allocated if the latter are not allocated to the same base station. Hence, it is sensitive to the sequence of query insertion. The task of phase 2 of the semi-greedy framework is to attempt to relieve the sensitivity.

3.2 Iterative Refinement

Our proposed iterative refinement phase aims to further optimize the query allocation plan generated by the above greedy insertion phase, before the queries are physically disseminated into the sensor network.

The algorithm is shown in Algorithm 1. It runs in multiple iterations. Within each iteration, it tries to refine the current allocation plan (lines 5-11). If a plan better than the current one is found, it will replace the current plan with the best plan in this iteration and continue the refinement process by restarting another iteration (lines 12-14), otherwise it will stop as no more refinement can be found (lines 15-16).

Algorithm 1. Iterative Refinement Algorithm

Input: The initial query allocation plan $QB[0..M-1]$
Output: The refined query allocation plan $QB[0..M-1]$

```
1  SmallestCost ← CompCost ();
2  while True do
3      Count ← 0; Changed ← 0;
4      while Count < M do
5          Qnode ← FindNextQuery (QList); /*Qnode =(qid,bid,costdiff);*/
6          reallocate q_qid from b_QB[qid] to b_bid;
7          TmpCost ← CompCost ();
8          if costdiff > 0 and TmpCost < SmallestCost then
9              SmallestCost ← TmpCost; TempQB[0..M-1] ← QB[0..M-1];
10             Changed ← 1;
11         Count++; Remove q_qid from QList;
12     if Changed == 1 then
13         QB[0..M-1] ← TempQB[0..M-1];
14         Restore QList to contain all queries;
15     else
16         QB[0..M-1] ← TempQB[0..M-1]; Break;
17 return QB[0..M-1];
```

In each iteration, one chance is given for each of the M queries in the whole query list *QList* to reallocate itself, and hence it takes M *rounds*. As shown in line 5, in each *round*, the function *FindNextQuery (QList)* examines all the choices of reallocating a query in the current *QList* to another base station and returns the query q_{qid} whose reallocation results in the largest cost reduction $costdiff$. Note that the largest cost reduction $costdiff$ could be a negative value here. In line 6, we update the bitmaps of the base stations affected by the reallocation of query q_{qid}, and update the current allocation plan by setting QB[qid] to bid. Then, in line 7, we recompute the cost of the current plan QB[0..M-1] and store it in *Tmpcost*. If the current plan has a cost smaller than *SmallestCost*, the cost of the best plan we have visited, then it caches the current plan at *TempQB[0..M-1]* and set *SmallestCost* as the current cost (lines 8-10). Before we continue to start the next *round* to reallocate the next query, we remove the current query q_{qid} from the *QList*(lines 11). Note that if extra gain can be further achieved through reallocating q_{qid} again after the reallocation of other queries, it will be exploited in the next iteration.

It is worthy to note that, the algorithm still continues the refinement (lines 4-11) even if the current cost $TmpCost$ is larger than the one before the refinement. This is to capture the opportunities where performance gain can be achieved through the relocation of *multiple* queries altogether, and allows the algorithm to jump out of the local optima.

4 Adaptive Migration Algorithm

In many of the real sensor applications, new users may issue new requests and existing users' demands may have been satisfied and their running queries terminate as well. This calls for adaptive algorithms that are able to adjust to the dynamic context. We adopt *Incremental* for query insertion since it is able to efficiently find the best allocation for newly inserted query without affecting other running queries. However, *Incremental* on the other hand does not involve any re-allocation of running queries to benefit from the newly inserted query. To deal with this deficiency and also to deal with the effect of termination of queries, existing queries may need to be re-allocated if necessary.

We note that migration during running time incurs overhead for the communication inside the sensor network. Taking the detailed query message dissemination mechanism into consideration [4,13], for a specific query q_i to migrate from base station b_j to b_k, b_j needs to send its query abort message to relevant sensor nodes that are within the query region of q_i, which incurs cost c_{ij}; similarly, b_k needs to send query insert message for q_i at cost c_{ik}. That is, a one time cost of $(c_{ij} + c_{ik})$ will be incurred. If this particular migration improves the cost by Δc at each epoch, it takes at least $(c_{ij} + c_{ik})/\Delta c$ epochs for the migration gain to compensate for the migration cost. Therefore, migration needs to consider the trade-off between the possible gain and the migration overhead.

Below, we present the adaptive query migration techniques, which include the following two parts:

- A migration detection algorithm detects when it is beneficial to perform query migration. It considers the trade-off between migration gain and its overhead mentioned above.
- A migration algorithm selects which queries to be migrated. Basically, it greedily selects the most beneficial migrations.

Migration Detection. To detect when to perform the migration, the controller maintains some information of the current queries at the system. Recall that the controller maintains bitmap m_j (j=0,...,K-1) for base station b_j to denote whether a sensor is involved in the queries from b_j. Here, we extend m_j to be a countmap, which denotes the number of queries from b_j that request data from each sensor. Furthermore, the controller also dynamically keeps a M*K two dimensional matrix $a[\,][\,]$ to record the additional cost of allocating each query to each base station. For instance, $a[i][j]$ keeps the additional cost of allocating q_i to b_j.

To detect whether a query should be migrated, we associate a *pressure* value p_i with each query q_i. In general, a higher p_i value represents a higher benefit to migrate query q_i. In particular, p_i is defined as $a[i][bid] - a[i][j]$, where *bid* is the *id* of the base station that q_i is currently allocated to, and j is the *id* of another base station b_j which satisfies $a[i][j] == MIN(a[i][0], ..., a[i][bid-1], a[i][bid+1], ...a[i][K-1])$. One can note that p_i is essentially the gain (of migrating q_i) that can be achieved at each epoch.

The migration detection algorithm is presented from line 1 to line 10 in Algorithm 2. It considers two factors. First, if the gain of a migration is high, the migration should tend to be performed. Second, if there is too frequent query arrival/termination, where the benefit for migration is not stable, migration should tend to be suppressed to avoid the thrashing effect and migration overhead.

We provide more details here. If there is a gain through migration (p_j is positive), the algorithm accumulates the gain over the epochs (lines 3-5). Otherwise, if p_j is negative and its accumulated value is positive, it means that other query insertion/termination has reduced the additional cost for q_j on the current base station or increased the additional cost for q_j to be reinserted into other base stations, during a short period of time, before q_j triggers migration. This suggests that there is frequent insert/termination of queries in the system to adjust the query allocation by itself, and hence we discourage migration by resetting the accumulated value of p_j to restart the accumulation and increasing parameter f_j to increase the migration threshold (lines 9-10). When the accumulated value has reached the adaptive threshold $f_j * p_j$, which suggests either the extra gain in each epoch p_j is big and/or the dynamics of queries is not frequent, under the assumption that query workload and patterns in the past is similar to that in the future under most cases, we choose to trigger the migration (lines 6-7).

Now we present how to maintain the parameters that are required to implement the above migration detection algorithm. As shown in Algorithm 2, when a new query q_i arrives, we record the additional cost of allocating it to each base station (lines 11 to 14).

Furthermore, the q_i will also affect the optimal allocation decision of other existing queries. First, for another query q_j allocated to the same base station as q_i, if the countmap value for a sensor at (x,y) in their overlapped area is equal to two (line 20), it means that data at (x,y) which was exclusively requested by q_j before is now shared by both q_i and q_j. Hence, with q_i, the additional cost of q_j on its assigned base station b_{bid} decreases, the probability that other base station is better for q_j reduces, and we correspondingly reduce the pressure value p_j. Second, for a query

q_j at another base station that overlaps with q_i, if the overlapped area is exclusively requested by q_i at base station b_{bid} (line 24), the additional cost of q_j on b_{bid} (i.e. $a[j][bid]$) should be decreased. However, p_j may not increase as $a[j][bid]$ may not be the smallest among all the $a[j][\,]$ values. Therefore, we recompute p_j instead of directly increasing p_j (line 23).

Symmetrically, when existing query terminates, the parameters are adjusted in a similar way, as shown in "Upon Termination of q_i" part of Algorithm 2.

Algorithm 2. Migration Detection Algorithm

```
   Migration detection:
1  for each data fetching epoch do
2      for j=0; j<M; j++ do
3          if p[j] > 0 then
4              if f_j == 0 then f_j ← q_j.area/p[j];
5              Accumulate[j] ← Accumulate[j] + p[j];
6              if Accumulate[j] >= p[j] * f_j then
7                  Migrate (); f_j=0; Accumulate[j]=0;
8          else
9              if Accumulate[j] > 0 then
10                 Accumulate[j] ← 0; f_j ++;

   Upon New Arrival of Query q_i:
11 for j =0; j<K; j++ do
12     a[i][j] = c_ij;
13     for all (x,y) in q_i.area do
14         if (x,y) of m_j >= 1 then a[i][j] ← a[i][j]-1;
15 if a[i][bid]==MIN(a[i][0],...a[i][K-1]) then
16 Allocate q_i to b_bid;
17 Update m_bid and QB[i] accordingly; Compute p[i];
18 for all q_j overlaps with q_i do
19     for all (x,y) in q_i.area ∩ q_j.area do
20         if QB[j] == bid AND (x,y)of m_bid == 2 then
21         a[j][bid] ← a[j][bid] − 1; p[j] ← p[j] − 1;
22         if QB[j] ≠ bid AND (x,y)of m_bid == 1 then
23         a[j][bid] ← a[j][bid] − 1; Recompute p[j];

   Upon Termination of Query q_i:
24 Update m_QB[i] accordingly;
25 for all q_j overlaps with q_i do
26     for all (x,y) in q_i.area ∩ q_j.area do
27         if QB[j] == QB[i] AND (x,y)of m_QB[i] == 1 then
           a[j][QB[i]] ← a[j][QB[i]] + 1; p[j] ← p[j] + 1;
28         if QB[j] ≠ QB[i] AND (x,y)of m_QB[i] == 0 then
           a[j][QB[i]] ← a[j][QB[i]] + 1; Recompute p[j];
```

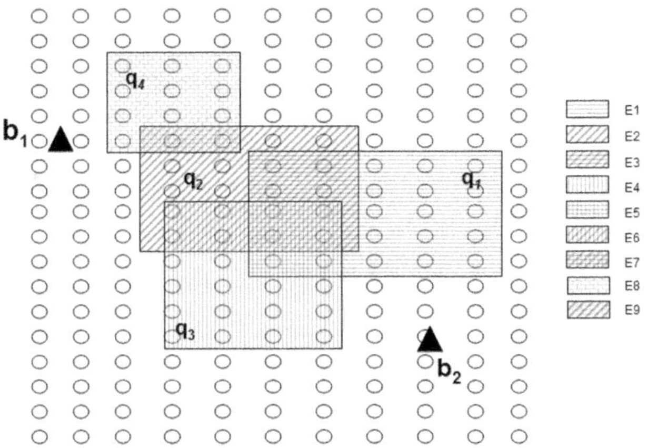

Fig. 2. A Scenario to illustrate migration detection algorithm

Below, we illustrate by example the process of keeping track of information needed for migration detection, such as migration pressure $p[\ ]$ and additional cost matrix $a[\ ][\ ]$. As shown in Figure 2, suppose queries q_1, q_2 and q_3 are allocated to base station b_2. Now:

1. Query q_4 arrives at the system. $a[4][1] = 12 + 1 = 13$, $a[4][2] = 12 + 12 - |E9| = 22$, while $a[4][1] = MIN(a[4][1], a[4][2])$, hence q_4 is assigned to b_1, $QB[4] = 1$ and $p[4] = a[4][1] - a[4][2] = -9$. Since q_4 overlaps with q_2, $a[2][1] = a[2][1] - |E9|$, and $p[2] = p[2] + |E9|$.
2. Query q_2 terminates from the system. q_1 overlaps with q_2, and their previous sharing E3 is now exclusively for q_1, so $a[1][2] = a[1][2] + |E3|$, $p[2] = p[2] + |E3|$; similarly, for q_3, $a[3][2] = a[3][2] + |E6|$, $p[3] = p[3] + |E6|$. For q_4, $a[4][2] = a[4][2] + |E9|$ and $p[2] = p[2] - |E9|$.

Migration Algorithm. Once the above migration detection issues request to perform migration, the migration algorithm shown in Algorithm 3 will be run. In each round, through function *FindNextQuery* as we introduced in Section 3.2,

---

Algorithm 3. Migration Algorithm

Input: The initial query allocation plan $QB[0..M-1]$
Output: The query allocation plan after migration $QB[0..M-1]$

while *True* **do**
  Qnode ← `FindNextQuery` (QList);
  if *costdiff* > *0* **then**
    migrate q_{qid} from $b_{QB[qid]}$ to b_{bid}; `/*Through information update at the coordinator, such as countmap etc.*/`
  else
    Break;

---

we pick the query that will result in the largest cost improvement to migrate. It is worthy to note that the migration here only modifies the query allocation plan by changing the information kept at the controller, such as *countmap* etc, and the intermediate plan is not disseminated into the sensor network. This migration process repeats until no beneficial query migration exists any more, and the final migration plan is disseminated into the network. In this way, local optimum can be achieved.

5 Experimental Study

In the experiments, we assume N sensor nodes are deployed uniformly in a two-dimensional grid square. For every 100 sensor nodes, there is one base station at the center. Each query is expressed as a rectangular region $((x_1, x_2), (y_1, y_2))$, where (x_1, y_1) is randomly selected from any point in the network, and the lengths on the x-axis $(x_2 - x_1)$ and y-axis $(y_2 - y_1)$ satisfy the uniform distribution. We also assume lossless communications among the sensor nodes.

It is worthy to note that our query allocation method is general, and it is not constrained to distribution of sensors/base stations, the region shape of the queries and the assumption of lossless communication in the experiments. Through the cost estimation function, the properties of the network can be captured and the process of query allocation decision is the same.

5.1 Importance of Leveraging Query Sharing

Firstly, we evaluate the importance of leveraging query sharing. We compare the performance of *Nearest* and *Partition* proposed in Section 2.3 with the following two strategies:

Collection: it is a data collection approach, instead of a query-driven approach. Each sensor sends its raw data to its nearest base station.

Random: Each query is randomly allocated to a base station.

Note that even though *Random*, *Nearest* and *Partition* are oblivious of the query similarity during allocation, queries allocated to the same base station

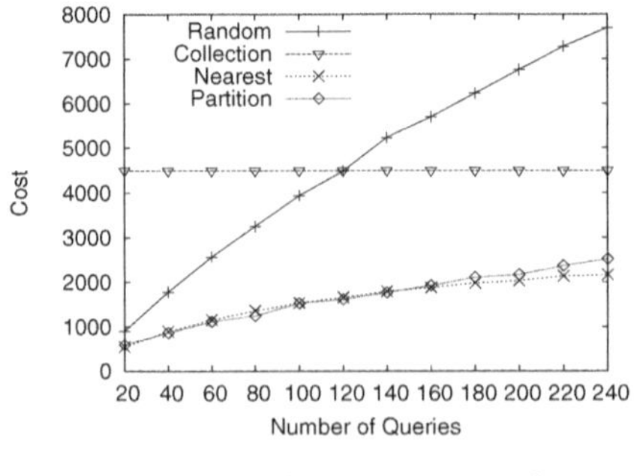

Fig. 3. Communication cost over random queries with average QR=5*5, N=900

may still benefit from data sharing through the multiple query optimization at the base station.

As shown in Figure 3, *Nearest* and *Partition* perform much better than *Random* and *Collection*. *Collection* scales well with the number of queries, but its communication cost is generally much higher than necessary, because it is always fetching all the data from the network to satisfy possible queries and it cannot take advantage of in-network aggregation. *Random* cannot effectively make use of the underlying multiple query optimization scheme, and hence its cost grows linearly with the number of queries, which makes it unattractive for large scale networks. On the other hand, *Nearest* and *Partition* both scale well with the number of queries and incur low communication cost.

5.2 Performance in the Static Context

In this section, we compare the effectiveness of our semi-greedy query allocation framework against the naive *Nearest* and *Partition*, in a static environment, under large-scale scenarios. As for small-scale scenarios with fewer queries and base stations, the results are similar with the ones in [12]. Due to space limit, we do not show them here. Interested readers are referred to [14] for details.

From the experimental results in Figures 4, we observe that neither *Nearest* nor *Partition* always outperforms the other but both strategies perform worse than our schemes. This is expected because both of them excel in one aspect but neglect the other aspect as explained in Section 2.3. Figure 4 also shows the performance of each strategy in our semi-greedy allocation scheme, under various network size, average query region, and number of queries. More specifically, the greedy insertion phase, no matter with area-based sorting or diff-based sorting, have considerable improvement over our shown not-so-bad baselines *Nearest* and *Partition*. The iterative refinement is further shown to effectively reduce the communication cost in the process of refining the initial query allocation plans. As a side note, it is also efficient and converges fast, taking less than 10ms.

5.3 Performance in the Dynamic Context

In this section, we evaluate the effectiveness of the adaptive migration algorithm in the dynamic context. Basically, we evaluate the following two aspects.

Firstly, we evaluate its ability to improve the current query allocation and compare the performance of migration algorithm (Algorithm 3) against SDP-K-Cut. The current query allocation is the allocation by incremental insertion algorithm *Incremental*. Since SDP-K-Cut is only suitable for static scenario, here we only note down the cost of *Incremental* after a specific number of queries have been inserted and running in the network; we take the same point of time as migration point and run our migration Algorithm 3, and see the amount of gain in cost we can achieve; the same set of queries will then be allocated by SDP-K-Cut as a whole and we compare their costs.

Figure 5 shows that migration effectively bridges the gap between *Incremental* and SDP-K-Cut. It is essentially one kind of online refinement strategy. When

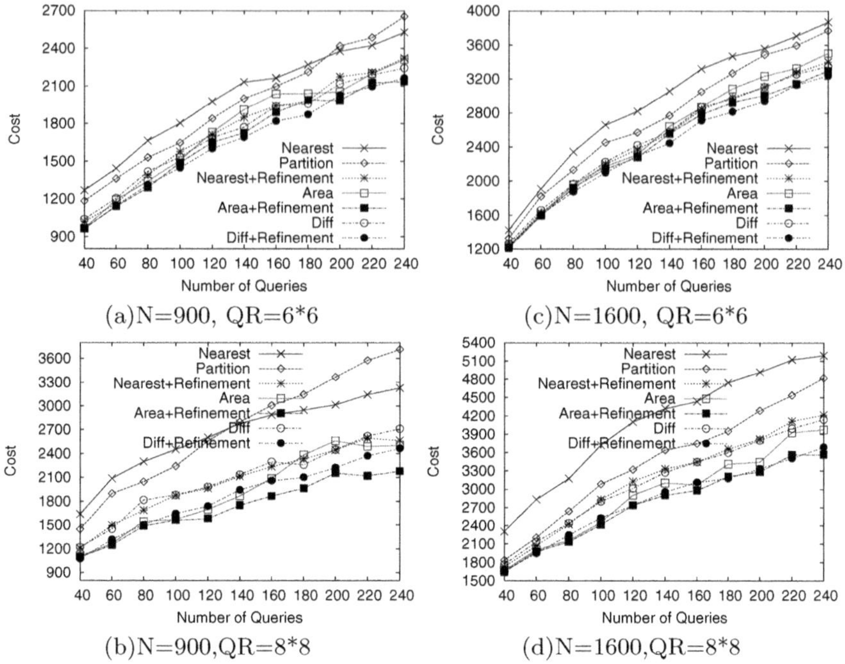

(a)N=900, QR=6*6 (c)N=1600, QR=6*6

(b)N=900,QR=8*8 (d)N=1600,QR=8*8

Fig. 4. Effectiveness of greedy query allocation for random queries in large-scale WSN

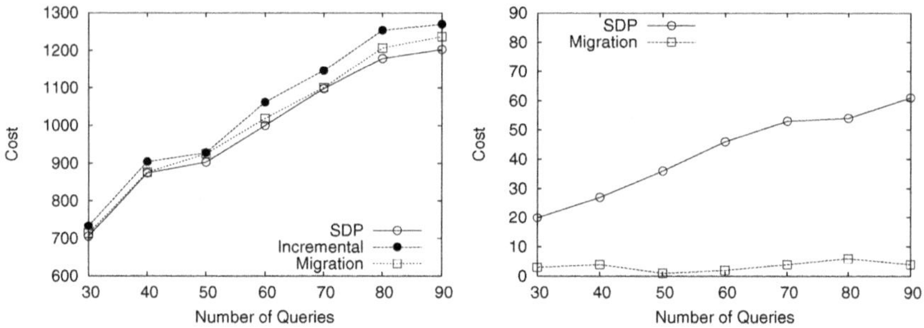

(a)Communication Cost (b)Number of Migrated Queries

Fig. 5. Evaluating migration over Random Queries with N=900, QR=5*5

the number of queries is bigger, there will be more cases that previously inserted query cannot benefit from the newly inserted query, and hence *Incremental* is further from being optimal. Through greedily exploiting the sharing among all the queries, much communication cost inside the network can be reduced. In this experiment, our migration algorithm is able to achieve around 70% of the gain SDP-K-Cut can obtain, at the cost of migrating around 10% of the number of queries that SDP-K-Cut needs to migrate.

Secondly, we examine the query migration detection (Algorithm 2) and migration algorithm (Algorithm 3) as a whole and evaluate its integration with the incremental insertion algorithm. Queries arrive and terminate whenever they want,

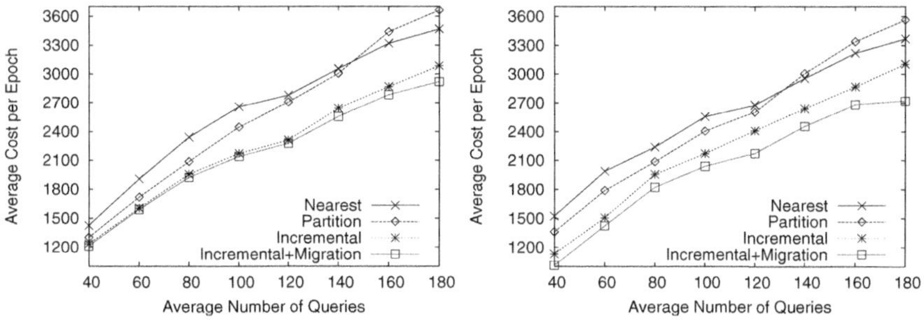

(a)Average Interval=10 (b)Average Interval= 40

Fig. 6. Evaluating adaptive migration in large-scale WSN over random queries with N=900, QR=6*6

but the average frequency of arrival and duration is controlled, and the average number of queries remains the same. After a set of 300 queries terminates, we note down the total communication cost in the network so far and use the cost of communication per epoch as the metric to compare different approaches. As shown in Figure 6, when the average interval of query arrival/termination is low, the performance gain of *Migration+Incremental* over *Incremental* is less than the situation when the interval is higher (e.g., a comparatively steady environment). This is because when there is frequent insertion/termination, migration has less job to do, since incremental insertion itself has high chance to reduce the migration pressure. From the above, we can see that migration algorithm can effectively adapt to the frequency of query arrivals and terminations.

6 Conclusion

In this paper, we have studied the query allocation problem in large scale sensor network, with the objective to minimize the total data communication cost among the sensor nodes. We designed a semi-greedy query allocation algorithm, which enhanced the existing greedy insertion scheme with iterative refinement to enable greater sharing among queries. We also proposed adaptive optimization algorithms to handle the dynamic change of query set.

The experimental results showed that the proposed query allocation schemes outperform the existing methods in both static and dynamic environments without introducing much overhead. The results also suggested that a good allocation scheme should take both the inherent sharing among queries and the power of in-network aggregation into consideration.

References

1. Deligiannakis, A., Kotidis, Y., Roussopoulos, N.: Hierarchical in-network data aggregation with quality guarantees. In: Bertino, E., Christodoulakis, S., Plexousakis, D., Christophides, V., Koubarakis, M., Böhm, K., Ferrari, E. (eds.) EDBT 2004. LNCS, vol. 2992, pp. 658–675. Springer, Heidelberg (2004)

2. Deshpande, A., Guestrin, C., Madden, S., et al.: Model-driven data acquisition in sensor networks. In: Proc. of VLDB (2004)
3. Frieze, A.: Improved approximation algorithms for max k-cut and max bisection. Algorithmica 18(1), 67–81 (1997)
4. Madden, S., Franklin, M.J., Hellerstein, J.M., et al.: TinyDB: An acquisitional query processing system for sensor networks. ACM TODS 30(1) (November 2005)
5. Munteanu, A., Beaver, J., Labrinidis, A., et al.: Multiple query routing trees in sensor networks. In: Proc. of the IASTED International Conference on Databases and Applications (DBA) (2005)
6. Pavan Edara, A.L., Ramamritham, K.: Asynchronous In-Network Prediction: Efficient Aggregation in Sensor Networks. ACM TOSN 4(4), 25 (2008)
7. Pietzuch, P.R., Ledlie, J., Shneidman, J., et al.: Network-aware operator placement for stream-processing systems. In: ICDE (2006)
8. Silberstein, A., Yang, J.: Many-to-many aggregation for sensor networks. In: ICDE (2007)
9. Stonebraker, M., Aoki, P.M., Litwin, W., et al.: Mariposa: A wide-area distributed database system. VLDB Journal 5(1), 48–63 (1996)
10. Tang, X., Xu, J.: Extending network lifetime for precision-constrained data aggregation in wireless sensor networks. In: INFOCOM (2006)
11. Trigoni, N., Yao, Y., Demers, A., Gehrke, J., Rajaraman, R.: Multi-query optimization for sensor networks. In: Prasanna, V.K., Iyengar, S.S., Spirakis, P.G., Welsh, M. (eds.) DCOSS 2005. LNCS, vol. 3560, pp. 307–321. Springer, Heidelberg (2005)
12. Xiang, S., Lim, H.B., Tan, K.L., Zhou, Y.: Similarity-aware query allocation in sensor networks with multiple base stations. In: Proc. of DMSN (2007)
13. Xiang, S., Lim, H.B., Tan, K.L., Zhou, Y.: Two-tier multiple query optimization for sensor networks. In: Proc. of ICDCS (2007)
14. Xiang, S., Zhou, Y., Lim, H.B., et al.: Leveraging Query Sharing in Sensor Networks with Multiple Base Stations, http://www.comp.nus.edu.sg/~xiangshi/QAfull.pdf
15. Xing, Y., Zdonik, S.B., Hwang, J.-H.: Dynamic load distribution in the borealis stream processor. In: ICDE (2005)
16. Yang, X., Lim, H.B., Ozsu, T., Tan, K.L.: In-network execution of monitoring queries in sensor networks. In: Proc. of SIGMOD (2007)
17. Yao, Y., Gehrke, J.: Query processing for sensor networks. In: CIDR (2003)
18. Zhou, Y., Ooi, B.C., Tan, K.-L., Wu, J.: Efficient dynamic operator placement in a locally distributed continuous query system. In: Meersman, R., Tari, Z. (eds.) OTM 2006. LNCS, vol. 4275, pp. 54–71. Springer, Heidelberg (2006)

GraphREL: A Decomposition-Based and Selectivity-Aware Relational Framework for Processing Sub-graph Queries

Sherif Sakr

NICTA and University of New South Wales
Sydney, Australia
Sherif.Sakr@nicta.com.au

Abstract. Graphs are widely used for modelling complicated data such as: chemical compounds, protein interactions, XML documents and multimedia. Retrieving related graphs containing a query graph from a large graph database is a key issue in many graph-based applications such as drug discovery and structural pattern recognition. Relational database management systems (RDBMSs) have repeatedly been shown to be able to efficiently host different types of data which were not formerly anticipated to reside within relational databases such as complex objects and XML data. The key advantages of relational database systems are its well-known maturity and its ability to scale to handle vast amounts of data very efficiently. RDMBSs derive much of their performance from sophisticated optimizer components which makes use of physical properties that are specific to the relational model such as: sortedness, proper join ordering and powerful indexing mechanisms. In this paper, we study the problem of indexing and querying graph databases using the relational infrastructure. We propose a novel, decomposition-based and selectivity-aware SQL translation mechanism of sub-graph search queries. Moreover, we carefully exploit existing database functionality such as partitioned B-trees indexes and influencing the relational query optimizers by selectivity annotations to reduce the access costs of the secondary storage to a minimum. Finally, our experiments utilise an IBM DB2 RDBMS as a concrete example to confirm that relational database systems can be used as an efficient and very scalable processor for sub-graph queries.

1 Introduction

Graphs are among the most complicated and general form of data structures. Recently, they have been widely used to model many complex structured and schemaless data such as XML documents [24], multimedia databases [18], social networks [3] and chemical compounds [16]. Hence, retrieving related graphs containing a query graph from a large graph database is a key performance issue in all of these graph-based applications. It is apparent that the success of any graph database application is directly dependent on the efficiency of the graph indexing and query processing mechanisms. The fundamental sub-graph

X. Zhou et al. (Eds.): DASFAA 2009, LNCS 5463, pp. 123–137, 2009.

search operation on graph databases can be described as follows: given a graph database $D = \{g_1, g_2, ..., g_n\}$ and a graph query q expressed as *find all graphs* g_i *which belongs to the graph database* D *such that* q *is a subgraph of* g_i. Clearly, it is an inefficient and a very time consuming task to perform a sequential scan over the whole graph database D and then to check whether q is a subgraph of each graph database member g_i. Hence, there is a clear necessity to build graph indices in order to improve the performance of processing sub-graph queries.

Relational database management systems (RDBMSs) have repeatedly shown that they are very efficient, scalable and successful in hosting types of data which have formerly not been anticipated to be stored inside relational databases such complex objects [7,21], spatio-temporal data [8] and XML data [6,14]. In addition, RDBMSs have shown its ability to handle vast amounts of data very efficiently using its powerful indexing mechanisms. In this paper we focus on employing the powerful features of the relational infrastructure to implement an efficient mechanism for processing sub-graph search queries.

In principle, XPath-based XML queries [4] are considered to be a simple form of graph queries. Over the last few years, various relational-based indexing methods [5,6,11,23] have been developed to process this type of XML queries. However, these methods are optimized to deal only with tree-structured data and path expressions. Here, we present a purely relational framework to *speed up* the search efficiency in the context of graph queries. In our approach, the graph data set is firstly encoded using an intuitive *Vertex-Edge* relational mapping scheme after which the graph query is translated into a sequence of SQL evaluation steps over the defined storage scheme. An obvious problem in the relational-based evaluation approach of graph queries is the huge cost which may result from the large number of join operations which are required to be performed between the encoding relations. In order to overcome this problem, we exploit an observation from our previous works which is that the size of the intermediate results dramatically affect the overall evaluation performance of SQL scripts [12,19,20]. Hence, we use an effective and efficient pruning strategy to *filter* out as many as possible of the false positives graphs that are guaranteed to not exist in the final results first before passing the candidate result set to an *optional verification* process. Therefore, we keep statistical information about the *less frequently* existing nodes and edges in the graph database in the form of simple Markov Tables [2]. This statistical information is also used to influence the decision of relational query optimizers by selectivity annotations of the translated query predicates to make the right decision regarding selecting the most efficient join order and the cheapest execution plan to get rid of the *non-required* graphs very early out of the intermediate results. Moreover, we carefully exploit the fact that the number of distinct vertices and edges labels are usually far less than the number of vertices and edges respectively. Therefore, we try to achieve the maximum performance improvement for our relation execution plans by utilizing the existing powerful *partitioned B-trees* indexing mechanism of the relational databases [10] to reduce the access costs of the secondary storage to the minimum [13]. In summary, we made the following contributions in this paper:

1) We present a purely relational framework for evaluating directed and labelled sub-graph search queries. In this framework, we encode graph databases using an intuitive *Vertex-Edge* relational schema and translate the sub-graph queries into standard SQL scripts.
2) We describe an effective and very efficient pruning strategy for reducing the size of the intermediate results of our relational execution plans by using a simple form of statistical information in the form of Markov tables.
3) We describe our approach of using summary statistical information about the graph database vertices and edges to consult the relational query optimizers through the use of accurate selectivity annotations for the predicates of our queries. These selectivity annotations help the query optimizers to decide the right join order and choose the most efficient execution plan.
4) We exploit a *carefully tailored* set of the powerful partitioned B-trees relational indexes to reduce the secondary storage access costs of our SQL translation scripts to a minimum.
5) We show the efficiency and the scalability of the performance of our approach through an extensive set of experiments.

The remainder of the paper is organized as follows: We discuss some background knowledge in Section 2. Section 3 describe the different components of our relational framework for processing graph queries including the graph coding method, pruning strategy and the SQL translation mechanisms. We evaluate our method by conducting an extensive set of experiments which are described in Section 4. We discuss the related work in Section 5. Finally, we conclude the paper in Section 6.

2 Preliminaries

2.1 Labelled Graphs

In labelled graphs, vertices and edges represent the entities and the relationships between them respectively. The attributes associated with these entities and relationships are called labels. A graph database D is a collection of member graphs $D = \{g_1, g_2, ..., g_n\}$ where each member graph g_i is denoted as (V, E, L_v, L_e) where V is the set of vertices; $E \subseteq V \times V$ is the set of edges joining two distinct vertices; L_v is the set of vertex labels and L_e is the set of edge labels.

In principal, labelled graphs can be classified according to the direction of their edges into two main classes: 1) *Directed-labelled graphs* such as XML, RDF and traffic networks. 2) *Undirected-labelled graphs* such as social networks and chemical compounds. In this paper, we are mainly focusing on dealing with directed labelled graphs. However, it is straightforward to extend our framework to process other kinds of graphs. Figure 1(a) provides an example of a graph database composed of three directed-labelled graphs $\{g_1, g_2, g_3\}$.

2.2 Subgraph Search Queries

In principal, the subgraph search operation can be simply described as follows: given a graph database $D = \{g_1, g_2, ..., g_n\}$ and a graph query q, it returns the

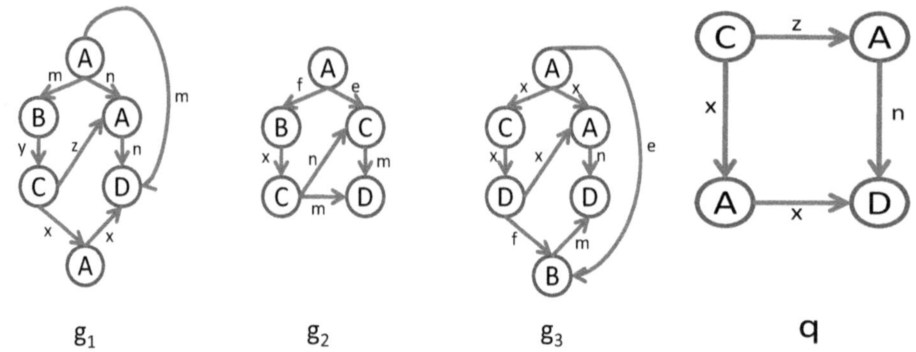

(a) Sample graph database (b) Graph query

Fig. 1. An example graph database and graph query

query answer set $A = \{g_i | q \subseteq g_i, g_i \in D\}$. A graph q is described as a sub-graph of another graph database member g_i if the set of vertices and edges of q form subset of the vertices and edges of g_i. To be more formal, let us assume that we have two graphs $g_1(V_1, E_1, L_{v1}, L_{e1})$ and $g_2(V_2, E_2, L_{v2}, L_{e2})$. g_1 is defined as sub-graph of g_2 , if and only if:

1) For every distinct vertex $x \in V_1$ with a label $vl \in L_{v1}$, there is a distinct vertex $y \in V_2$ with a label $vl \in L_{v2}$.
2) For every distinct edge edge $ab \in E_1$ with a label $el \in L_{e1}$, there is a distinct edge $ab \in E_2$ with a label $el \in L_{e2}$.

Figure 1(b) shows an example of graph query q. Running the example query q over the example graph database D (Figure 1(a)) returns an answer set consists of the graph database member g_1.

3 GraphREL Description

3.1 Graph Encoding

The starting point of our relational framework for processing sub-graph search queries is to find an efficient and suitable encoding for each graph member g_i in the graph database D. Therefore, we propose the *Vertex-Edge* mapping scheme as an efficient, simple and intuitive relational storage scheme for storing our targeting *directed labelled graphs.* In this mapping scheme, each graph database member g_i is assigned a unique identity *graphID*. Each vertex is assigned a sequence number (*vertexID*) inside its graph. Each vertex is represented by one tuple in a single table (*Vertices table*) which stores all vertices of the graph database. Each vertex is identified by the *graphID* for which the vertex belongs to and the *vertex ID*. Additionally, each vertex has an additional attribute to store the vertex label. Similarly, all edges of the graph database are stored in a single table (*Edges table*) where each edge is represented by a single tuple in this table. Each edge tuple describes the graph database member which the edge

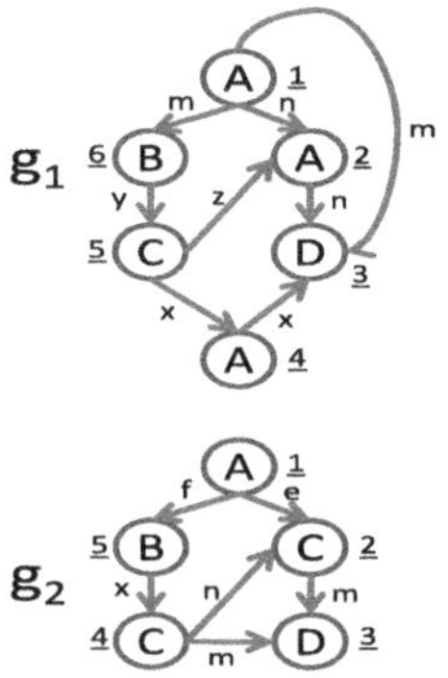

graphID	vertexID	vLabel
1	1	A
1	2	A
1	3	D
1	4	A
1	5	C
1	6	B
2	1	A
2	2	C
2	3	D
2	4	C
2	5	B

Vertices Table

graphID	sVertex	dVertex	eLabel
1	1	2	n
1	1	3	m
1	2	3	n
1	4	3	x
1	5	4	x
1	6	5	y
1	5	2	z
1	1	6	m
2	1	2	e
2	2	3	m
2	4	3	m
2	4	2	n
2	5	4	x
2	1	5	f

Edges Table

Fig. 2. Vertex-Edge relational mapping scheme of graph database

belongs to, the id of the *source vertex* of the edge, the id of the *destination vertex* of the edge and the edge label. Therefore, the relational storage scheme of our *Vertex-Edge* mapping is described as follows:

– $Vertices(graphID, vertexID, vertexLabel)$
– $Edges(graphID, sVertex, dVertex, edgeLabel)$

Figure 2 illustrates an example of the *Vertex-Edge* relational mapping scheme of graph database. Using these mapping scheme, we employ the following SQL-based *filtering-and-verification* mechanism to speed up the search efficiency of sub-graph queries.

– **Filtering phase:** in this phase we use an effective and efficient pruning strategy to filter out as many as possible of the non-required graph members very early. Specifically, in this phase we specify the set of graph database members contain the set of vertices and edges which are describing the sub-graph query. Therefore, the filtering process of a sub-graph query q consists of a set of vertices QV with size equal m and a set of edges QE equal n (see Figure 1(b)) can be achieved using the following SQL translation template:

1 SELECT DISTINCT $V_1.graphID, V_i.vertexID$
2 FROM Vertices as V_1,..., Vertices as V_m, Edges as E_1,..., Edges as E_n
3 WHERE
4 $\forall_{i=2}^{m}(V_1.graphID = V_i.graphID)$
5 AND $\forall_{j=1}^{n}(V_1.graphID = E_j.graphID)$
6 AND $\forall_{i=1}^{m}(V_i.vertexLabel = QV_i.vertexLabel)$
7 AND $\forall_{j=1}^{n}(E_j.edgeLabel = QE_j.edgeLabel)$
8 AND $\forall_{j=1}^{n}(E_j.sVertex = V_f.vertexID$ AND $E_j.dVertex = V_f.vertexID)$;

(TRANSTEMPLATE)

Where each referenced table V_i (Line number 2) represents an instance from the table $Vertices$ and maps the information of one vertex of the set of vertices QV which is belonging to the sub-graph query q. Similarly, each referenced table E_j represents an instance from the table $Edges$ and maps the information of one edge of the set of edges QE which is belonging to the sub-graph query q. f is the mapping function between each vertex of

QV and its associated vertices table instance V_i. Line number 4 of the SQL translation template represents a set of $m-1$ conjunctive conditions to ensure that all queried vertices belong to the same graph. Similarly, Line number 5 of the SQL translation template represents a set of n conjunctive conditions to ensure that all queried edges belong to the same graph of the queried vertices. Lines number 6 and 7 represent the set of conjunctive predicates of the vertex and edge labels respectively. Line number 8 represents the edges connection information between the mapped vertices.

- **Verification phase:** this phase is an *optional* phase. We apply the verification process only if more than one vertex of the set of query vertices QV have the same label. Therefore, in this case we need to verify that each vertex in the set of filtered vertices for each candidate graph database member g_i is distinct. This can be easily achieved using their *vertex ID*. Although the fact that the conditions of the verification process could be injected into the SQL translation template of the filtering phase, we found that it is more efficient to avoid the cost of performing these conditions over each graph database members g_i by delaying their processing (*if required*) in a separate phase after pruning the candidate list.

Clearly, an obvious problem of the SQL translation template of the *filtering* is that it involves a large number of conjunctive SQL predicates $(2m + 4n)$ and join $(m + n)$ *Vertices* and *Edges* tables instances. Hence, although this template can be efficiently used with relatively small sub-graph search queries, most of relational query engines will certainly fail to execute the SQL translation queries of medium size or large sub-graph queries because they are too long and too complex (this does not mean they must consequently be too expensive). In the following subsections we will describe our approach to effectively deal with this problem by carefully and efficiently *decomposing* this complex one step evaluation step into a series of well designed relational evaluation steps.

3.2 Relational Indexes Support for Vertex-Edge Mapping Scheme

Relational database indexes have proven to be very efficient tools to speed up the performance of evaluating the SQL queries. Moreover, the performance of queries evaluation in relational database systems is very sensitive to the defined indexes structures over the data of the source tables. In principal, using relational indexes can accelerate the performance of queries evaluation in several ways. For example, applying highly selective predicates first can limit the data that must be accessed to only a very limited set of rows that satisfy those predicates. Additionally, query evaluations can be achieved using index-only access and save the necessity to access the original data pages by providing all the data needed for the query evaluation. In [15], He and Singh have presented an indexing scheme for processing graph queries which is very similar to R-Tree index structure. However, R-tree indexing technique is not commonly supported by many of the RDBMS systems where B-tree indexing is still the most commonly used technique. Therefore, in our purely relational framework, we use a *standard*, powerful and matured indexing mechanisms to accelerate the processing performance of our SQL evaluation of the sub-graph queries, namely *partitioned B-tree indexes* and *automated relational index advisors.*

Partitioned B-tree Indexes Partitioned B-tree indexes are considered to be a slight variant of the B-tree indexing structure. The main idea of this indexing technique has been represented by Graefe in [10] where he recommended the use of low-selectivity leading columns to maintain the partitions within the associated B-tree. For example, in labelled graphs, it is generally the case that the number of *distinct* vertices and edges labels are far less than the number of vertices and edges respectively. Hence, for example having an index defined in terms of columns $(vertexLabel, graphID)$ can reduce the access cost of subgraph query with only one label to one disk page which is storing a list of *graphID* of all graphs which are including a vertex with the target query label. On the contrary, an index defined in terms of the two columns $(graphID, vertexLabel)$ requires scanning a large number of disk pages to get the same list of targeted graphs. Conceptually, this approach could be considered as a horizontal partitioning of the *Vertices* and *Edges* table using the high selectivity partitioning attributes. Therefore, instead of requiring an execution time which is linear with the number of graph database members (graph database size), having partitioned B-trees indexes of the high-selectivity attributes can achieve fixed execution times which are no longer dependent on the size of the whole graph database [10,13].

Automated Relational Index Advisor Leveraging the advantage of relying on a pure relational relational infrastructure, we are able to use the ready made tools provided by the RDBMSs to propose the candidate indexes that are effective for accelerating our query work loads. In our work, we were able to use the *db2advis* tool provided by the DB2 engine (our hosting experimental engine) to recommend the suitable index structure for our query workload. Through the use of this tool we have been able significantly improve the quality of our designed indexes and to speed up the evaluation of our queries by reducing the number of calls to the database engine. Similar tools are available in most of the widely available commercial RDBMSs.

3.3 Statistical Summaries Support of Vertex-Edge Mapping Scheme

In general, one of the most effective techniques for optimizing the execution times of SQL queries is to select the relational execution based on the accurate selectivity information of the query predicates. For example, the query optimizer may need to estimate the selectivities of the occurrences of two vertices in one subgraph, one of these vertices with label A and the other with label B to choose the more selective vertex to be filtered first. Providing an accurate estimation for the selectivity of the predicates defined in our SQL translation template requires having statistical information that contain information about the structure of the stored graph data. Additionally, these statistics must be small enough to be processed efficiently in the short time available for query optimization and without any disk accesses. Therefore, we construct three Markov tables to store information about the frequency of occurrence of the distinct labels of vertices,

Vertex Label	Frequency
A	100
B	200
C	38
D	4
E	50
L	6
M	10
N	250
O	3
P	40
R	55

Markov Table summary of vertices labels

Edge Label	Frequency
a	40
c	5
e	28
l	54
m	140
n	3
o	20
p	15
x	8
y	60
z	15

Markov Table summary of edges labels

Edge Label Connection	Frequency
ab	3
ac	15
ae	45
ec	14
em	103
la	5
pc	18
px	45
xy	25
xz	2
za	1

Markov Table summary of pair-wise edge connections

Fig. 3. Sample Markov tables summaries of Vertex-Edge mapping

distinct labels of edges and connection between pair of vertices (edges). Figure 3 presents an example of our Markov table summary information. In our context, we are only interested in label and edge information with low frequency. Therefore, it is not necessary and not useful to keep all such frequency information. Hence, we summarize these Markov tables by deleting high-frequency tuples up to certain defined threshold $freq$. The following subsection will explain how we can use these information about the low frequency labels and edges to effectively prune the search space, reduce the size of intermediate results and influence the decision of the relational query optimizers to select the most efficient join order and the cheapest execution plan in our decomposed and selectivity-aware SQL translation of sub-graph queries.

3.4 Decomposition-Based and Selectivity-Aware SQL Translation of Sub-graph Queries

In Section 3, we described our preliminary mechanism for translating sub-graph queries into SQL queries using our *Vertex-Edge mapping*. As discussed previously, the main problem of this *one-step* translation mechanism is that it cannot be used with medium or large sub-graph queries as it generate SQL queries that are too long and too complex. Therefore, we need a *decomposition* mechanism to divide this large and complex SQL translation query into a sequence of intermediate queries (using temporary tables) before evaluating the final results. However, applying this decomposition mechanism blindly may lead to inefficient execution plans with very large, non-required and expensive intermediate results. Therefore, we use the statistical summary information described in Section 3.3 to perform an effective selectivity-aware decomposition process. Specifically, our decomposition-based and selectivity-aware translation mechanism goes through the sequence of following steps:

- **Identifying the pruning points.** The frequency of the labels of vertices and edges in addition to the frequency of edge connection play a crucial role in our decomposition mechanism. Each vertex label, edge label or edge connection with low frequency is considered as a pruning point in our relational evaluation mechanism. Hence, given a query graph q, we first check the structure of q against our summary Markov tables to identify the possible *pruning* points. We refer to the number of the identified pruning points by NPP.

- **Calculating the number of partitions.** As we previously discussed, having a sub-graph query q consists of a m vertices and a set of n edges requires $(2m + 4n)$ conjunctive conditions. Assuming that the relational query engine can evaluate up to number of conjunctive condition equal to NC in one query then the number of partitions (NOP) can be simply computed as follows : $(2m + 4n)/NC$
- **Decomposed SQL translation.** Based on the identified number of pruning points (NPP) and the number of partitions (NOP), our decomposition process can be described as follows:
  - *Blindly Single-Level Decomposition.* if $NPP = 0$ then we blindly decompose the sub-graph query q into the calculated number of partition NOP where each partition is translated using our translation template into an intermediate evaluation step S_i. The final evaluation step FES represents a join operation between the results of all intermediate evaluation steps S_i in addition to the conjunctive condition of the sub-graphs connectors. The unavailability of any information about effective pruning points could lead to the result where the size of some intermediate results may contain a large set of non-required graph members.
  - *Pruned Single-Level Decomposition.* if $NPP >= NOP$ then we distribute the pruning points across the different intermediate NOP partitions. Therefore, we ensure a balanced effective pruning of *all* intermediate results, by getting rid of the non-required graph database member early which consequently results in a highly efficient performance. All intermediate results S_i of all pruned partitions are constructed before the final evaluation step FES joins all these intermediate results in addition to the connecting conditions to constitute the final result.
  - *Pruned Multi-Level Decomposition.*if $NPP < NOP$ then we distribute the pruning points across a first level intermediate results of NOP partitions. This step ensures an effective pruning of a percentage of NPP/NOP % partitions. An *intermediate* collective pruned step IPS is constructed by joining all these pruned first level intermediate results in addition to the connecting conditions between them. Progressively, IPS is used as an entry pruning point for the rest ($NOP - NPP$) non-pruned partitions in a hierarchical multi-level fashion to constitute the final result set. In this situation, the number of *non-pruned* partitions can be reduced if any of them can be connected to one of the pruning points. In other words, each pruning point can be used to prune more than one partition (if possible) to avoid the cost of having any large intermediate results.

Figure 4 illustrates two example of our selectivity-aware decomposition process where the pruning vertices are marked by solid fillings, pruning edges are marked by bold line styles and the connectors between subgraphs are

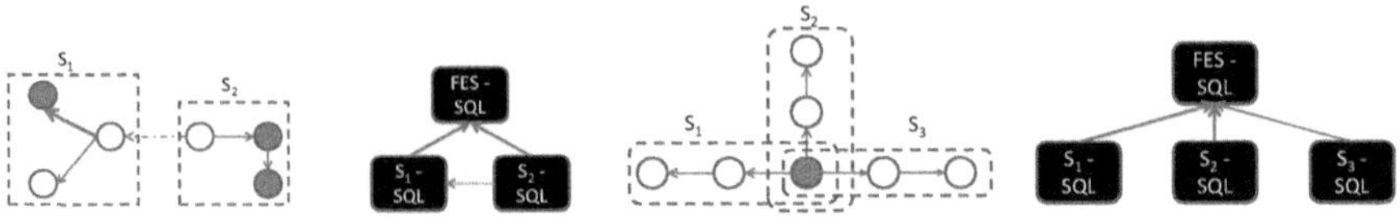

(a) $NPP > NOP$ (b) $NPP < NOP$

Fig. 4. Selectivity-aware decomposition process

marked by dashed edges. Figure 4(a) represents an example where the number of pruning points is greater than the number of partitions. Figure 4(b) represents an example where one pruning vertex is shared by the different partitioned sub-graphs because the number of pruning points is less than the number of partitions.

- **Selectivity-aware Annotations.** In principal, the main goal of RDBMS query optimizers is to find the most efficient execution plan for every given SQL query. For any given SQL query, there are a large number of alternative execution plans. These alternative execution plans may differ significantly in their use of system resources or response time. Sometimes query optimizers are not able to select the most optimal execution plan for the input queries because of the unavailability or the inaccuracy of the required statistical information. To tackle this problem, we use our statistical summary information to give influencing hints for the query optimizers by *injecting* additional selectivity information for the individual query predicates into the SQL translations of the graph queries. These hints enable the query optimizers to decide the optimal join order, utilizing the most useful indexes and select the cheapest execution plan. In our context, we used the following syntax to pass the selectivity information to the DB2 RDBMS query optimizer:

  SELECT *fieldlist* FROM *tablelist*
  WHERE P_i SELECTIVITY S_i

  Where S_i indicates the *selectivity* value for the query predicate P_i. These selectivity values are ranging between 0 and 1. Lower selectivity values (close to 0) will inform the query optimizer that the associated predicates will effectively prune the number of the intermediate result and thus they should be executed first.

4 Performance Evaluation

In this section, we present a performance evaluation of *GraphREL* as a purely relational framework for storing graph data and processing sub-graph queries. We conducted our experiments using the IBM DB2 DBMS running on a PC with 2.8 GHZ Intel Xeon processors, 4 GB of main memory storage and 200 GB of SCSI secondary storage. In principle, our experiments have the following goals:

1) To demonstrate the efficiency of using *partitioned B-tree* indexes and *selectivity injections* to improve the execution times of the relational evaluation of sub-graph queries.

2) To demonstrate the efficiency and scalability of our decomposition-based and selectivity-aware relational framework for processing sub-graph queries.

4.1 Datasets

In our experiments we use two kinds of datasets:

1) The real DBLP dataset which presents the famous database of bibliographic information of computer science journals and conference proceedings [1]. We converted the available XML tree datasets into labelled directed graphs by using edges to represent the relationship between different entities of the datasets such as: the ID/IDREF, cross reference and citation relationships. Five query sets are used, each of which has 1000 queries. These 1000 queries are constructed by randomly selecting 1000 graphs and then extracting a connected m edge subgraph from each graph randomly. Each query set is denoted by its edge size as Q_m.
2) A set of synthetic datasets which is generated by our implemented data generator which is following the same idea proposed by Kuramochi et al. in [17]. The generator allows the user to specify the number of graphs (D), the average number of vertices for each graph (V), the average number of edges for each graph (E), the number of distinct vertices labels (L) and the number of distinct edge labels (M). We generated different datasets with different parameters according to the nature of each experiment. We use the notation $DdEeVvLlMm$ to represent the generation parameters of each data set.

4.2 Experiments

The effect of using partitioned B-tree indexes and selectivity injections. Figures 5(a) and 5(b) indicate the percentage of speed-up improvement on the execution times of the SQL-based relational evaluation sub-graph queries using the partitioned B-tree indexing technique and the selectivity-aware annotations respectively. In these experiments we used an instance of a syntactic database that was generated with the parameters $D200kV15E20L200M400$ and a DBLP instance with a size that is equal to 100 MB. We used query groups with different edge sizes of 4,8,12,16 and 20. The groups with sizes of 4 and 8 are translated into one SQL evaluation step, the queries with sizes of 12 and 16 are *decomposed* into two SQL evaluation steps and the queries with of size 20 are *decomposed* into three SQL evaluation steps. The reported percentage of speed up improvements are computed using the formula: $(1 - \frac{G}{C})$ %. In Figure 5(a) G represents the execution time of the SQL execution plans using our defined set of the partitioned B-tree indexes while C represents the execution time of the SQL execution plans using the traditional B-tree indexes. Similarly, in Figure 5(b) G represents the execution time of the SQL execution plans with the injected selectivity annotations while C represents the execution time of the SQL execution plans without the injected selectivity annotations. The results of both experiments confirm the efficiency of both optimization techniques on both data sets. Clearly, using partitioned B-tree indexes has a higher effect on improving the

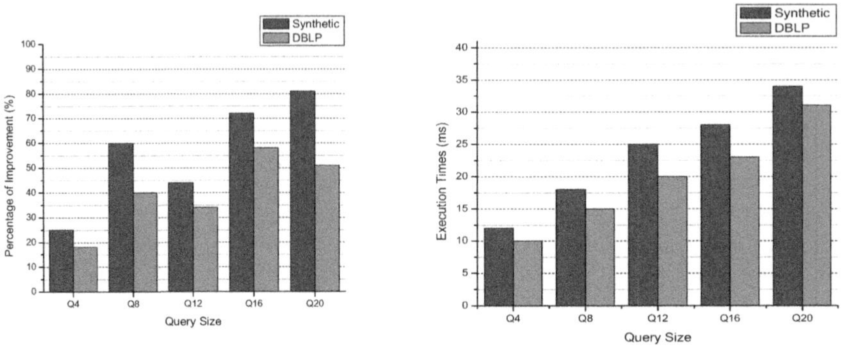

(a) Partitioned B-tree indexes (b) Injection of selectivity annotations

Fig. 5. The *speedup improvement* for the relational evaluation of sub-graph queries using partitioned B-tree indexes and selectivity-aware annotations

execution of the SQL plans because it dramatically reduce the access cost of the secondary storage while selectivity annotations only improve the ability to select the more optimal execution plans. Additionally, the effect on the synthetic database is greater than the effect on the DBLP database because of the higher frequency on the vertex and edge labels and thus reducing the cost of accessing the secondary storage is more effective. The bigger the query size, the more join operations are required to be executed and consequently the higher the effect of both optimization techniques on pruning the cost of accessing the secondary storage and improving the execution times.

Performance and Scalability. One of the main advantages of using a relational database to store and query graph databases is to exploit their well-know scalability feature. To demonstrate the scalability of our approach, we conducted a set of experiments using different database sizes of our datasets and different query sizes. For the DBLP data sets, we used different subsets with sizes of 1,10,50 and 100MB. For the synthetic datasets, we generate four databases with the following parameters: $D2kV10E20L40M50$, $D10kV10E20L40M50$, $D50kV30E40L90M150$ and $D100kV30E40L90M150$. For each dataset, we generated a set of 1000 queries. Figures 6(a) and 6(b) illustrate the average execution times for the SQL translations scripts of the 1000 sub-graph queries.

In these figures, the running time for sub-graph query processing is presented in the Y-axis while the X-axis represents the size of the query graphs. The running time of these experiments include both the *filtering* and *verification* phases. However, on average the running time of the verification phase represents 5% of the total running time and can be considered as have a negligible effect on all queries with small result set. Obviously, the figures show that the execution times of our system performs and scales in a near linear fashion with respect to the graph database and query sizes. This linear increase of the execution time starts to decrease with the very large database sizes (DBLP 100MB and Synthetic $D100kV30E40L90M150$) because of the efficiency of the partitioned B-tree indexing mechanism which

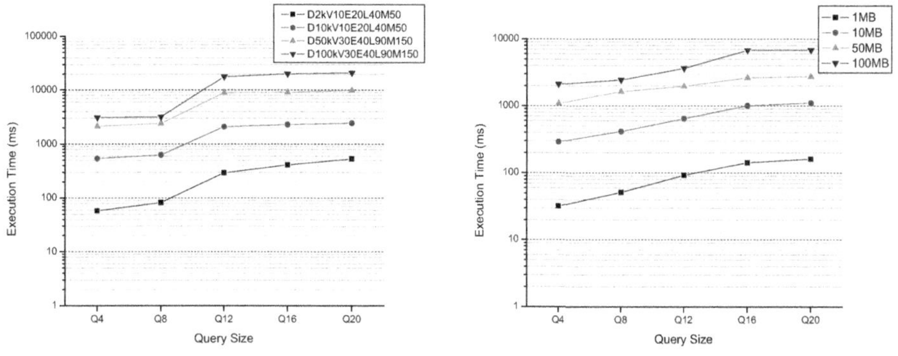

(a) Synthetic Dataset (b) DBLP Dataset

Fig. 6. The scalability of GraphREL

decouples the query evaluation cost from the total database size. . To the best of our knowledge, this work is the first to successful demonstrate a feasible approach of processing large subgraph queries up to 20 vertices and 20 edges over graph databases with such very large sizes up to 100 MB.

5 Related Work

Recently, graph database has attracted a lot of attentions from the database community. In [9], Shasha et al. have presented *GraphGrep* as a path-based approach for processing graph queries. It enumerates all paths through each graph in a database until a maximum length L and records the number of occurrences of each path. An index table is then constructed where each row stands for a path, each column stands for a graph and each entry is the number of occurrences of the path in the graph. The main problem of this approach is that many false positive graphs could be returned in the filtering phase. In addition, enumerating the graphs into a set of paths may cause losing some of their structural features. Some researchers have focused on indexing and querying graph data using data mining techniques such as: *GIndex* [22], *TreePi* [25] and *Tree+Δ* [26]. In these approaches data mining methods are firstly applied to extract the frequent subgraphs (*features*) and identify the graphs in the database which contain those subgraphs. Clearly, the effectiveness of these approaches depends on the quality of the selected features. In addition, the index construction time of these approach requires an additional high space cost and time overhead for enumerating all the graph fragments and performing the graph mining techniques. Moreover, all of these approaches deal with relatively small graph databases where they assume either implicitly or explicitly that the graph databases can completely or the major part of them fit into the main memory. None of them have presented a persistent storage mechanism of the large graph databases. In [27] Jiang et al. proposed another graph indexing scheme called *GString*. GString approach focus on decomposing chemical compounds into basic structures that have semantic meaning in the context of organic chemistry. In this approach the graph search problem is converted into a string matching problem and specific string indices is built to support the efficient string matching process. We believe that converting sub-graph search queries into sting matching problem could be an inefficient approach specially if the size of the graph database or the sub-graph query is large. Additionally, it is not trivial to extend GString approach to support processing of graph queries in other domain of applications.

6 Conclusions

Efficient sub-graph query processing plays a critical role in many applications related to different domains which involve complex structures such as: bioinformatics, chemistry and social networks. In this paper, we introduced *GraphRel* as a purely relational framework to store and query graph data. Our approach converts a graph into an intuitive relational schema and then uses powerful indexing techniques and advanced selectivity annotations of RDBMSs to achieve an efficient SQL execution plans for evaluating subgraph queries. In principle GraphREL has the following advantages:

1) It employs purely relational techniques for encoding graph data and processing the sub-graph search query. Hence, it can reside on any relational database system and exploits its well known matured query optimization techniques as well as its efficient and scalable query processing techniques.
2) It has *no* required time cost for offline or pre-processing steps.
3) It can handle static and dynamic (with frequent updates) graph databases very well. It is easy to maintain the graph database members and *no* special processing is required to insert new graphs, delete or update the structure of existing graphs.
4) The *selectivity* annotations for the SQL evaluation scripts provide the relational query optimizers with the ability to select the most efficient execution plans and apply an efficient pruning for the non-required graph database members.
5) As we have demonstrated in our experiments, using the well-known *scalability* feature of the relational database engine, GraphREL can achieve a very high scalability and ensure its good performance over very large graph databases and large sub-graph queries.

In the future, we will experiment our approach with other types of graphs and will explore the feasibility of extending our approach to deal with similarity queries and the general subgraph isomorphism problem as well.

Acknowledgments

The author would like to thank Jeffery XY Yu for his valuable comments on earlier drafts of this paper.

References

1. DBLP XML Records, http://dblp.uni-trier.de/xml/
2. Aboulnaga, A., Alameldeen, A., Naughton, J.: Estimating the Selectivity of XML Path Expressions for Internet Scale Applications. In: VLDB (2001)
3. Cai, D., Shao, Z., He, X., Yan, X., Han, J.: Community Mining from Multi-relational Networks. In: PPKDD (2005)
4. Clark, J., DeRose, S.: XPath 1.0: XML Path Language (XPath) (November 1999), http://www.w3.org/TR/xpath
5. Cooper, B., Sample, N., Franklin, M., Hjaltason, G., Shadmon, M.: A Fast Index for Semistructured Data. In: VLDB (2001)
6. Yoshikawa, M., et al.: XRel: a path-based approach to storage and retrieval of XML documents using relational databases. TOIT 1(1) (2001)
7. Cohen, S., et al.: Scientific formats for object-relational database systems: a study of suitability and performance. SIGMOD Record 35(2) (2006)
8. Botea, V., et al.: PIST: An Efficient and Practical Indexing Technique for Historical Spatio-Temporal Point Data. GeoInformatica 12(2) (2008)
9. Giugno, R., Shasha, D.: GraphGrep: A Fast and Universal Method for Querying Graphs. In: International Conference in Pattern recognition (2002)
10. Graefe, G.: Sorting And Indexing With Partitioned B-Trees. In: CIDR (2003)
11. Grust, T.: Accelerating XPath location steps. In: SIGMOD, pp. 109–120 (2002)
12. Grust, T., Mayr, M., Rittinger, J., Sakr, S., Teubner, J.: A SQL: 1999 code generator for the pathfinder xquery compiler. In: SIGMOD (2007)
13. Grust, T., Rittinger, J., Teubner, J.: Why Off-The-Shelf RDBMSs are Better at XPath Than You Might Expect. In: SIGMOD (2007)
14. Grust, T., Sakr, S., Teubner, J.: XQuery on SQL Hosts. In: VLDB (2004)
15. He, H., Singh, A.: Closure-Tree: An Index Structure for Graph Queries. In: ICDE (2006)
16. Klinger, S., Austin, J.: Chemical similarity searching using a neural graph matcher. In: European Symposium on Artificial Neural Networks (2005)
17. Kuramochi, M., Karypis, G.: Frequent Subgraph Discovery. In: ICDM (2001)
18. Lee, J., Oh, J., Hwang, S.e.: STRG-Index: Spatio-Temporal Region Graph Indexing for Large Video Databases. In: SIGMOD (2005)
19. Sakr, S.: Algebraic-Based XQuery Cardinality Estimation. IJWIS 4(1) (2008)
20. Teubner, J., Grust, T., Maneth, S., Sakr, S.: Dependable Cardinality Forecasts in an Algebraic XQuery Compiler. In: VLDB (2008)
21. Türker, C., Gertz, M.: Semantic integrity support in SQL: 1999 and commercial (object-)relational database management systems. VLDB J. 10(4) (2001)
22. Yan, X., Yu, P., Han, J.: Graph indexing: a frequent structure-based approach. In: SIGMOD (2004)
23. Yuen, L., Poon, C.: Relational Index Support for XPath Axes. In: International XML Database Symposium (2005)
24. Zhang, N., Özsu, T., Ilyas, I., Aboulnaga, A.: FIX: Feature-based Indexing Technique for XML Documents. In: VLDB (2006)
25. Zhang, S., Hu, M., Yang, J.: TreePi: A Novel Graph Indexing Method. In: ICDE (2007)
26. Zhao, P., Xu Yu, J., Yu, P.: Graph indexing: tree + delta = graph. In: VLDB (2007)
27. Zou, L., Chen, L., Xu Yu, J., Lu, Y.: GString: A novel spectral coding in a large graph database. In: EDBT (2008)

A Uniform Framework for Ad-Hoc Indexes to Answer Reachability Queries on Large Graphs

Linhong Zhu[1], Byron Choi[2], Bingsheng He[3], Jeffrey Xu Yu[4], and Wee Keong Ng[1]

[1] Nanyang Technological University, Singapore
{ZHUL0003,AWKNG}@ntu.edu.sg
[2] Hong Kong Baptist University, China
choi@hkbu.edu.hk
[3] Microsoft Research Asia
savenhe@microsoft.com
[4] Chinese University of Hong Kong, China
yu@se.cuhk.edu.hk

Abstract. Graph-structured databases and related problems such as reachability query processing have been increasingly relevant to many applications such as XML databases, biological databases, social network analysis and the Semantic Web. To efficiently evaluate reachability queries on large graph-structured databases, there has been a host of recent research on graph indexing. To date, reachability indexes are generally applied to the entire graph. This can often be suboptimal if the graph is large or/and its subgraphs are diverse in structure. In this paper, we propose a uniform framework to support existing reachability indexing for subgraphs of a given graph. This in turn supports fast reachability query processing in large graph-structured databases. The contributions of our uniform framework are as follows: (1) We formally define a graph framework that facilitates indexing subgraphs, as opposed to the entire graph. (2) We propose a heuristic algorithm to partition a given graph into subgraphs for indexing. (3) We demonstrate how reachability queries are evaluated in the graph framework. Our preliminary experimental results showed that the framework yields a smaller total index size and is more efficient in processing reachability queries on large graphs than a fixed index scheme on the entire graphs.

1 Introduction

Recent interests on XML, biological databases, social network analysis, the Semantic Web, Web ontology and many other emerging applications have sparked renewed interests on graph-structured databases (or simply *graphs*) and related problems (e.g., query processing and optimization). In this paper, we focus on querying large graphs. In particular, we are interested in a kind of fundamental queries from classical graph-structured databases – reachability query. Specifically, given two vertices u and v, a reachability query returns true if and only if there is a directed path from u and v (denoted $u \rightsquigarrow v$); otherwise, the query returns false.

Reachability queries have many emerging applications in graph-structured databases. For example, in XML, the ancestor and descendant axes of XPATH can be implemented with reachability queries on the graph representation of XML. Reachability queries are

X. Zhou et al. (Eds.): DASFAA 2009, LNCS 5463, pp. 138–152, 2009.

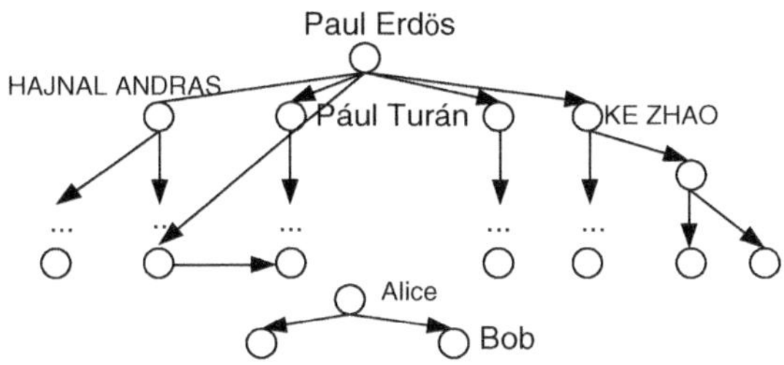

Fig. 1. An example of a real graph

also useful for building a query language [1] for the Semantic Web. As required by these applications, it is desirable to have efficient reachability query processing.

Example 1. Consider a directed graph that represents a social network of a set of researchers, where the vertices correspond to researchers and the edges correspond to the co-authorship relationship between two researchers, as shown in Figure 1. Social network analysis may often require reachability queries. For example, we may ask whether a researcher "Alice" has an *Erdös number*. A simple way to answer this query is to check whether "Alice" is reachable from "Paul Erdös".

To provide some background on reachability queries, we review existing naïve evaluation algorithms for reachability queries and the indexes for different kinds of graphs. There are two naïve alternatives for evaluating reachability queries on a graph: (1) A reachability query can be evaluated using a traversal of the graph. The runtime is $O(|G|)$, where $|G|$ denotes the size of the graph G. (2) A reachability query can also be evaluated by precomputing the transitive closure of the graph, whose size is quadratic to the graph size in the worst case. A reachability query can then be a simple selection on the transitive closure. It is clear that these two approaches are not scalable. Much indexing technique has been proposed for optimizing reachability queries on trees [2,3], directed acyclic graphs (DAGs) [4,5,6,7,8], and arbitrary graphs [9,10,11] (see Section 5). These indexes have demonstrated some performance improvement on the graphs with certain structural characteristics.

Unlike relational data, graph-structured data may vary greatly in its structure; e.g., trees, sparse/dense DAGs and sparse/dense cyclic graphs. It is evident that the structure of the graphs has an impact on the performance of reachability indexes on graphs. For instance, dual labeling [11] works best for sparse graphs but performs suboptimally on dense graphs. Hence, a single reachability index is sometimes not ideal to graphs that have different structures. Given these, we raise the following issues and propose some solutions for these issues in this paper:

1. A notion of data granularity is missing in graph-structured databases. Let us consider an example from relational databases. One may build a B+ tree on a *subset* of the attributes of a relation for range queries and a hash index on some other subsets of attributes for equi-joins. In comparison, to date, a graph index (such as dual labeling [11]) is either applied to the entire graph, or not applied at all. Is there a

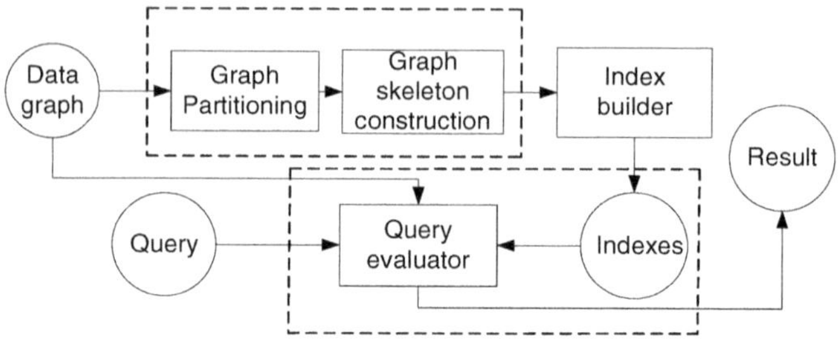

Fig. 2. Overview of our uniform framework for querying large graphs

general approach to seamlessly support multiple indexes? For example, one may apply dual labeling [11] to one subgraph and 2-hop [10] to another?

2. Structural differences between subgraphs have not been considered by state-of-the-art reachability indexes. Many real graphs such as Web graphs, telephone call graphs, and global social network graphs have different structures, both locally and globally. Real graphs, as a whole, are typically sparse with a constant average degree [12]. However, there may be local dense subgraphs. Hence, we need to address structural differences and determine suitable indexing techniques for subgraphs. Is it possible to detect different substructures from a graph and apply suitable indexes to these substructures?
3. Different reachability indexing techniques require different query evaluation algorithms. Is it possible to support multiple indexes and yet reuse existing query evaluation algorithms without modifying the indexes?

To address the above issues, we propose a uniform framework for indexing graphs. With the framework, we can flexibly use any existing index for reachability queries on subgraphs of a graph. An overview of our framework is shown in Figure 2. Our framework consists of two components: (1) Graph framework construction through graph partitioning and (2) reachability query evaluation on the graph framework with different indexes. As a proof of concept, our current prototype supports two state-of-the-art indexes for reachability queries, namely Interval [7] and HOPI [13].

In summary, we define a graph framework to represent a graph as a set of partitions and a graph skeleton. Each partition can use any existing reachability index. In conjunction with the framework, we define a cost function and propose a heuristic algorithm for graph partitioning. We illustrate how existing query evaluation techniques can be extended to our graph framework. In particular, a reachability query is casted into inter-partition and intra-partition reachability queries on indexes. We present our experimental evaluation on our framework with both synthetic and real graphs.

The remainder of the paper is organized as follows: In Section 2, we define notations used in this paper, our graph framework and the evaluation of reachability queries using the graph framework. The graph framework construction is presented in Section 3. In Section 4, we present an experimental study to evaluate the effectiveness and efficiency of our approach. Related work is discussed in Section 5. We conclude this work and present future work in Section 6.

2 Labeling Graph Framework for Reachability Query

In this section, we define our *graph framework* to represent reachability information of an arbitrary graph. In brief, the graph framework comprises information of strongly connected components, partitions of a graph and connections between partitions. We evaluate reachability queries efficiently by applying multiple indexing techniques to the graph framework. We defer the details of the construction of the graph framework to Section 3.

We describe the notations used in this work in Section 2.1. In Section 2.2, we present the definition of our graph framework and its properties. In Section 2.3, we show how to apply multiple indexes to the graph framework and how to process reachability queries in the graph framework.

2.1 Preliminaries

We denote a directed graph to be $G = (V, E)$, where V is a (finite) set of vertices, and E is a (finite) set of directed edges representing the connection between two vertices. We define an auxiliary function $\texttt{reach}(v_1, v_2)$ that returns true *iff* v_1 can reach v_2.

Definition 1. *The* condensed graph *of G is denoted as* $G^* = (V^*, E^*)$*, where a vertex* $v_i^* \in V^*$ *represents a strongly connected component* C_i *in G and each edge* $(v_i, v_j) \in E^*$ iff *there is at least one edge* $(u, v) \in E$ *such that* $u \in C_i$ *and* $v \in C_j$*.*

The condensed graph can be computed efficiently using Tarjan's algorithm with time complexity $O(|V|+|E|)$ [19].

We use G_i to denote a subgraph of G and V_i to denote the set of vertices in G_i. We define a (non-overlapping) partitioning of graph as follows:

Definition 2. *A partitioning of graph G P(G) is* $\{G_1, G_2, ..., G_k\}$*, where* $\forall\, i \in [1...k]$*,* $k \leq |V|$*,* $\cup_{i=1}^{k} V_i = V$*,* $V_i \cap V_j = \emptyset$*, where* $i \neq j$*.*

Example 2. Consider the graph shown in Figure 3(a). We partition the graph on the left into three partitions V_1=$\{0, 1, 2, 3, 4, 5\}$, V_2=$\{6, 7, 8, 9\}$ and V_3=$\{10, 11, 12\}$. G_1, G_2 and G_3 is a dense subgraph, a subtree and a sparse subgraph, respectively.

Based on this partitioning, we define a *partition-level graph* as follows.

Definition 3. *Given a partitioning P(G), the* partition-level graph $G_p(G)$ *is* (V_p, E_p)*, where each vertex* $v_i \in V_p$ *represents a partition* G_i *in P(G) and an edge* $(v_i, v_j) \in E_p$ if *there is an edge* $(u, v) \in E$ *such that* u *and* v *are vertices in Partitions* G_i *and* G_j*, respectively.*

Example 3. Consider the graph G and its partitioning $P(G)$ in Example 2. The partition-level graph $G_p(G)$ is shown in Figure 3(b). Vertices 1, 2 and 3 in G_p represent Partitions G_1, G_2 and G_3, respectively.

Next, let us consider the relationships between two partitions G_i and G_j that has not been captured by the partition-level graph. We define a *partition-level skeleton graph* $G_{ps}(G)$ to capture the connecting vertices of partitions of a graph G.

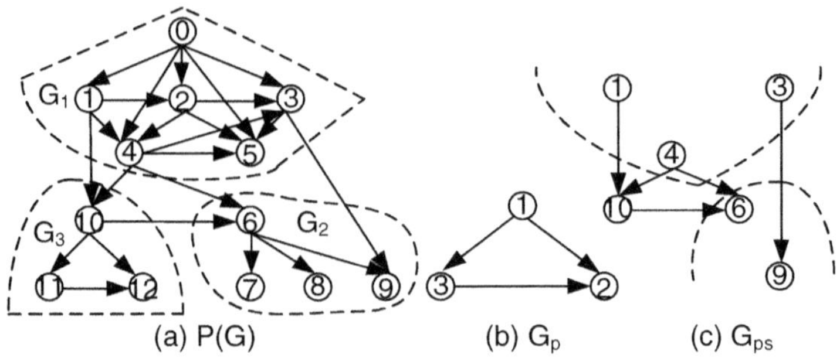

pid	$\mathcal{M}$ (pid)
1	$\{1, 3, 4\}$
2	$\{6, 9\}$
3	$\{10\}$

Fig. 3. A partitioning of a graph and its graph skeleton

Definition 4. *Given a partition-level graph $G_p(G)$, the* partition-level skeleton graph *$G_{ps}(G)$ is (V_{ps}, E_{ps}), where $V_{ps} \subseteq V$, $v \in V_{ps}$ is either a source or a target vertex of an inter-partition edge and E_{ps} is the set of inter-partition edges.*

With the above, a graph skeleton is defined as follows:

Definition 5. *Given a graph G and a partitioning of G, $P(G)$, the* graph skeleton *$S(G)$ is a 3-ary tuple (G_p, G_{ps}, $\mathcal{M}$), where G_p=(V_p, E_p) is the partition-level graph, G_{ps}=(V_{ps}, E_{ps}) is the partition-skeleton graph and $\mathcal{M} : V_p \rightarrow 2^{V_{ps}}$ is a mapping function that takes a partition v_i as input and gives a subset of V_{ps}: $\{v \mid v \in V_{ps}$ and $v \in G_i\}$ as output.*

To illustrate the mapping function $\mathcal{M}$, we give an example in Figure 3.

2.2 Graph Framework

Given the previous preliminary definitions, we propose our graph framework.

Definition 6. *A* graph framework *of a graph G, denoted as $H(G)$, consists of a strongly connected component index $C(V)$, a partitioning of the condensed graph $P(G^*)$ and a graph skeleton $S(G^*)$: $H(G)$= ($C(V)$,$P(G^*)$, $S(G^*)$).*

We remark that the proposed graph framework has the following properties:

- The graph framework supports multiple reachability indexing techniques. Different partitions may use different indexing techniques.
- The graph skeleton, which contains the inter-partition information, can be indexed in order to facilitate query processing. For example, we could apply the hash index to accelerate the search on the mapping $\mathcal{M}$ and any reachability indexes to the graph G_p and G_{ps} in $S(G^*)$.
- The graph framework consists of indexed subgraphs. Hence, the query evaluation on the graph framework can be transformed into the query evaluation on relevant indexed subgraphs.

2.3 Query Processing on Graph Framework with Reachability Indexing

In this subsection, we present an algorithm for evaluating a reachability query on our graph framework with multiple reachability indexes, as shown in Algorithm 1.

Algorithm 1. A query evaluation algorithm on the graph framework `evaluate-query`

Input: a graph framework $H(G)$: (C(V), $P(G^*)$, $S(G^*)$) with indexes, two input vertices x and y, where $x, y \in V$
Output: `true` if x can reach y, `false` otherwise
1: **if** C(x)==C(y)
2:     **return** `true`
3: denote $G_i(G_j)$ to be the partition of $x(y)$
4: **if** $G_i = G_j$
5:     **return** `true` if $x \rightsquigarrow y$ in G_i, `false` otherwise
6: **if** $G_i \not\rightsquigarrow G_j$ with respect to G_p in $S(G^*)$
7:     **return** `false`
8: **else** V_i=$\mathcal{M}(i)$, V_j=$\mathcal{M}(j)$
9:     for each vertex $v_i \in V_i$
10:         for each vertex $v_j \in V_j$
11:             if $v_i \rightsquigarrow v_j$ with respect to G_s in $S(G^*)$
12:                 **return** `true` if $x \rightsquigarrow v_i$ in G_i and $v_j \rightsquigarrow y$ in G_j
13: **return** `false`

Algorithm `evaluate-query` returns true if vertex x is able to reach vertex y. It returns false otherwise. The first step is to obtain the strongly connected component for vertices x and y. If they are in the same strongly connected component, the query returns true (Lines 1-2). Next, we compute the partitions where x and y reside, i.e., G_i and G_j (Line 3). If the two input vertices are in the same partition, we use the index of the partition to answer the reachability query (Lines 4-5). Next, if we find that G_i cannot reach G_j by using the index of G_p in $S(G^*)$, then u is not able to reach v and the query returns false (Lines 6-7). Otherwise, we apply the mapping $\mathcal{M}$ to obtain the set of vertices in G_s related to partitions G_i and G_j, i.e., V_i and V_j (Line 8). We test whether there is a vertex $v_i \in V_i$ that is able to reach V_j. If so, we return true if x is able to reach v_i in G_i and v_j reaches y in G_j (Lines 9-12). Otherwise, we return false (Line 13). The correctness of this algorithm can be easily derived from the definition of the graph framework.

Complexity. The time complexity of query processing on graph framework is index-dependant, i.e., it is determined by the reachability indexes applied to the graph framework. For example, assume that vertices x and y are not in the same partition and partition i, containing x, is indexed with interval labeling [7] with query time complexity $O(|V_i|)$ and partition j, containing y, is indexed with HOPI labeling [20] with query time complexity $O(|E_j|^{1/2})$. The partition-level graph G_p and partition-level graph G_s are indexed with dual labeling [11] with constant query time complexity. Hence, in this particular example, the overall complexity of `evaluate-query` is

$O(|E_{ij}|(|V_i|+|E_j|^{1/2}))$, where E_{ij} denotes the set of inter-partition edges connecting the partition i and partition j.

3 Graph Framework Construction

Given an input graph G, we construct a graph framework as follows: (1) compute condensed graph G^*; (2) compute a partitioning of G^*, $P(G^*)$, using a heuristic algorithm proposed in Section 3.1; (3) based on $P(G^*)$, compute the graph skeleton $S(G^*)$. The key in graph framework construction is to compute $P(G^*)$. Once $P(G)$ is computed, the graph skeleton $S(G)$ can simply be constructed as follows:

1. compute the partition-level graph G_p;
2. compute the set of inter-partition edges E_I;
3. compute the subgraph G_s which is induced by the edge set E_I; and
4. compute the mapping $\mathcal{M}$ between a partition and a set of vertices in G_s.

Hence, in this section, we focus on the details of determining $P(G^*)$ of a given graph G. First, we propose the objective function of our graph partitioning problem and the heuristic algorithmic strategy to solve the problem. Next, we present how to estimate the query cost of a graph and a graph framework, which is used in the heuristic graph partitioning.

3.1 Heuristics for Graph Partitioning

In this subsection, we present the objectivity of our graph partitioning problem. Next, we present the overall procedure for constructing the graph framework.

The graph partitioning problem considered in this work can be formulated as follows: Given a graph G=(V, E), determine k non-overlapping subsets of V_1,..., V_k such that:

1. $\cup_{i=1}^{k} V_i = V$, $V_i \cap V_j = \emptyset$ where $i \neq j$; and
2. the estimation of query time; i.e., the number of labels accessed during query processing, is minimal,

where k is determined during partitioning.

Overall algorithm. Let $E(G)$ and $E(H(G))$ denote the query costs on a graph G and a graph framework $H(G)$, respectively. Our heuristic graph partition algorithm works as follows: Given a k-partitioning of a graph $P(G)$ and its corresponding graph framework $H(G)$, we create a new $(k+1)^{\text{th}}$ partition for a set of vertices $P'(G)$ if the resulting new graph framework $H'(G)$ reduces the query cost; i.e., $E(H'(G)) \leq E(H(G))$ (shown in Algorithm 2). Hence, in Algorithm 2, the main strategy is to determine whether further partitioning reduces the cost.

More specifically, we start with a partitioning of G that all the vertices are in the same partition (Line 1). Consider the current k-partition $P(G)$. We find the partition G_{max} whose cost is maximum among all other partitions in $P(G)$ (Line 3). Next, we assign a set of vertices in G' to a new partition G_{k+1}: for every vertex v in G_{max}, we

Algorithm 2. The heuristic graph partition algorithm `partition`(G)

Input: a digraph G
Output: a partitioning of G, $P(G)$
1: let k=1, $C_{\mathtt{pre}}$=$E(G)$, $C_{\mathtt{decre}}$=0, $P(G)$=$\{G\}$
2: **do**
3: G_{max}=$\mathtt{arg}\max_{1 \le i \le k} \{E(G_i)\}$, $G_i \in P(G)$
4: $V_{k+1} = \emptyset$
5: **for** each vertex v of G_{max}
6: if $E(G_{max} \setminus v) < E(G_{max})$
7: $V_{k+1} = V_{k+1} \cup v$
8: $k+1$-partitioning $P'(G)$=$\{P(G) \setminus V_{k+1}, G_{k+1}\}$
/* where G_{k+1} is the subgraph induced by V_{k+1} */
9: call refinement procedure, $P(G)$=`Refinement`(G, $P'(G)$)
10: $C_{\mathtt{decre}}$=$E(P(G))$-$C_{\mathtt{pre}}$, $C_{\mathtt{pre}}$=$E(P(G))$, $k = k + 1$
11: **while** $C_{\mathtt{decre}}$ <0
12: **return** $P(G)$

place it into the new partition *if* its removal from G_{max} decreases its cost (Lines 5-7). This results in a $k+1$-partitioning $P'(G)$ (Line 8). In order to optimize the quality of the $k+1$ partitions, we invoke the partition refinement procedure `Refinement`(G, $P'(G)$) (Line 9) to obtain $P_r(G)$. We proceed to the next partitioning iteration if the cost has been reduced in the current iteration. Otherwise, Algorithm 2 terminates and returns $P_r(G)$ (Lines 10-12).

Partition refinement. Next, we present the details of the refinement procedure (Algorithm 3) used in Algorithm 2. In the refinement procedure, we improve the quality of a given k-partitioning. We apply a search technique to find a better k-partitioning with a lower cost. Initially, Algorithm 3 starts with a partitioning of graph $P(G)$ (Line 1). Then, for each vertex, we search for a good assignment that minimizes the cost: We find the best target partition `pid` for a vertex v such that the new k-partitioning produced by assigning v to Partition `pid` has the minimal cost among all other k-partitionings produced by other assignments (Lines 3-6). The partitioning algorithm terminates if the new partitioning found does not decrease the cost or the iteration number is up to a user input value m; otherwise, Algorithm 2 search for other possible assignments (Lines 7-9).

Complexity. Let the number of iterations needed in the refinement procedure be m. The complexity of the whole partition procedure is $O(mk^2(|V| + |E|))$, where k is the number of partitions. We remark that the values of m and k are often small in real applications.

3.2 Query Cost Estimation

In this subsection, we discuss how to model the costs, $E(G)$ and $E(H(G))$. One of the important properties of our framework is that it is able to support multiple reachability indexes, where any single specified reachability index is a special case of our

Algorithm 3. Partition Refinement `Refinement`$(G, P(G))$

Input: a digraph G, initial k-partitioning $P(G)$, iteration number m
Output: a new k-partitioning of G, $P_r(G)$
$C_{\texttt{pre}}$=$E(P(G))$, $C_{\texttt{decre}}$=0, i=0
do
  for each vertex $v \in V$ of G
    let $P_j(G)$ denotes a new partitioning resulted by removing v into partition j in $P(G)$
    $\texttt{pid}(v)$=$\texttt{arg} \min_{1 \le j \le k} \{E(P_j(G)) - E(P(G))\}$
    $P(G)$=$P_{\texttt{pid}}(G)$
  $C_{\texttt{decre}}$=E($P(G)$)-$C_{\texttt{pre}}$, $C_{\texttt{pre}}$=E($P(G)$), i++
while $C_{\texttt{decre}}$ <0 and $i < m$
return $P(G)$

framework. However, the accuracy of $E(G)$ or $E(H(G))$ is highly dependant on the cost model of reachability indexes involved. Hence, to compute the value of $E(G)$ and $E(H(G))$, the pre-condition is that involved reachability indexes have a reasonable cost model. As an illustration, we implement two state-of-the-art indexes, Interval [7] and HOPI [13] in our prototype of the graph framework. It has been known that the time complexity for a reachability query on the HOPI and Interval indexes are $O(|E|^{\frac{1}{2}})$ and $O(|V|)$, respectively. Therefore, the query time estimation for $E(G)$ can be modeled by Equation 1.

$$E(G) = \min(C_1|V|, C_2|E|^{\frac{1}{2}}), \tag{1}$$

where C_1 and C_2 are the unit cost in real measurements.

Based on $E(G)$, the estimated query time on $H(G)$, which consists of a k-partitioning $P(G)$ and a graph skeleton $S(G)$, is defined as follows:

$$E(H(G)) = \sum_{i=1}^{k} E(G_i) + E(G_p) + E(G_s), \tag{2}$$

where k is the number of partitions.

4 Experimental Evaluation

In this section, we perform an experimental study to evaluate the effectiveness and efficiency of our proposed techniques. All experiments were run on a machine with a 3.4GHZ CPU. The run-times reported are the CPU times. We implemented all the proposed techniques in C++. Regarding graph indexes, we used the HOPI implementation from [13] and we implemented the Interval scheme for DAGs [7]. We also used the implementation of a recent path-tree approach from Jin *et al.* [8].

4.1 Index Size and Query Performance Evaluation on Real Data

We used a collection of real graphs in this experiment. We report the statistics of the real graphs in Table 1. Among them, "days", "hep-th-new" and "eatRS" are obtained from a Web graph repository [21]; and the other real graphs are provided by Jin *et al.* [8].

Table 1. Statistics of real graphs

Tree-like graphs	$\|V\|$	$\|E\|$	$\|V\|^*$	$\|E\|^*$	Other graphs	$\|V\|$	$\|E\|$	$\|V\|^*$	$\|E\|^*$
kegg	14271	35170	3617	3908	days	13332	243447	13332	148038
vchocyc	10694	14207	9491	10143	hep-th-new	27770	352807	20086	130469
mtbrv	10697	13922	9602	10245	eatRS	23219	325624	15466	19916
agrocyc	13969	17694	12684	13408	hpycyc	5565	8474	4771	5859
anthra	13736	17307	12499	13104	nasa	5704	7942	5605	6537
human	40051	43879	38811	39576	xmark	6483	7654	6080	7025

Table 2. Index size comparison on real graphs.

Tree-like graphs	HOPI	Interval	Ptree	Our	Other graphs	HOPI	Interval	Ptree	Our
kegg	9488	10078	1703	2884	days	199788	227364	–	199826
vchocyc	33920	20196	830	1216	hep-th-new	268524	244748	–	204802
mtbrv	34312	20406	812	1204	eatRS	31032	67952	–	31034
agrocyc	43664	26728	962	1362	hpycyc	16576	11658	4224	**2118**
anthra	42888	26146	733	1160	nasa	49954	12852	5063	**1644**
human	84916	79058	965	1438	xmark	44112	14038	2356	**1880**

In the first set of experiments, we compared the index size of our graph framework with two popular approaches – Interval [7] and HOPI [20] and a recent path-tree approach [8] (PTree). The results are shown in Table 2. The reported index size is the number of integers in the indexes.

From the results presented in Table 2, we found that the index size of our approach is clearly smaller than that of the HOPI and Interval approaches. The reason is simple: for each subgraph in our framework, our approach chose a relatively better one between the two approaches. In addition, our approach is comparable to the Ptree approach. For graphs that are not "tree-like", our approach achieved a much smaller index size than the Ptree approach (the bold numbers in Table 2). Although our approach is sometimes worse than the Ptree approach for "tree-like" graphs, our approach is more general than the Ptree approach since the Ptree approach could be a special case of our approach (where the Ptree approach is applied to all partitions). However, since an accurate cost model for the Ptree approach has not been available, we did not apply the Ptree approach to any partition (subgraph) in our framework.

Next, we investigate the time for index construction and query processing. Here we only compared our methods with HOPI and the Interval approach. This is because the CPU time measurer and the query evaluator of these three approaches are all implemented by us while the Ptree approach is provided by its authors. All three approaches run on MS Windows while the Ptree approach runs on Linux. Hence, a comparison of indexing and query time between our approach and the Ptree approach may be affected by many implementation issues. When comparing the index construction time and the query time, we used two metrics, *i.e.*, S_L, and S_H to evaluate our approach, where $S_L = \frac{\min(\text{HOPI, Interval})}{Our}$, and $S_H = \frac{\max(\text{HOPI, Interval})}{Our}$. S_L measures the performance

Table 3. Construction time comparison on real graphs (ms)

Tree-like graphs	HOPI	Interval	Our	S_L	S_H	Other graphs	HOPI	Interval	Our	S_L	S_H
kegg	473.694	3.92742	4.19078	0.94	113	days	9484.38	93.113	9096.3502	0.01	1.04
vchocyc	1125	7.22463	7.3448	0.98	153	hep-th-new	56281.3	66.7182	71.7037	0.93	785
mtbrv	1140.63	5.47139	8.26619	0.66	138	eatRS	1718.75	33.1035	1728.2479	0.02	0.995
agrocyc	1500	7.18841	9.66451	0.74	155	hpycyc	609.375	4.52743	5.76455	0.79	106
anthra	1453.13	7.01311	9.72699	0.72	149	nasa	1203.13	3.46655	4.4054	0.79	273
human	4343.75	38.9534	59.8715	0.65	73	xmark	1000	3.80507	4.83549	0.79	207

Table 4. Total query time comparison on real graphs (ms)

Tree-like graphs	HOPI	Interval	Our	S_L	S_H	Other graphs	HOPI	Interval	Our	S_L	S_H
kegg	1315.07	1107.55	1097.85	1.008	1.2	days	7854.39	6645.64	7929.75	0.84	0.99
vchocyc	4136.98	2667.57	2674.55	0.997	1.5	hep-th-new	5576.43	7265.8	5641.19	0.99	1.3
mtbrv	4128.25	2672.27	2738.87	0.98	1.5	eatRS	3397.71	3514.07	3442.64	0.987	1.02
agrocyc	4286.65	2839.21	2797.93	1.015	1.5	hpycyc	3818.74	2599.91	2520.07	1.03	1.5
anthra	4254.57	2782.02	2767.95	1.005	1.5	nasa	5672.01	2739.5	2720.18	1.007	2.08
human	5003.2	3957.09	3901.01	1.014	1.28	xmark	5178.59	2677.69	2698.7	0.99	1.91

comparison between our method and the best of the two static methods. S_H measures the performance gain if an inefficient static method is chosen.

Table 3 illustrates that the Interval approach requires two traversals to construct the index and therefore always has the smallest indexing construction time. As known from previous work [9], the HOPI indexing time for large graphs can often be costly. Since a combination of HOPI and Interval is applied to our prototype implementation, the construction time of our approach is roughly between these two. In particular, the time depends on the percentage of the partitions using each of these indexing approaches. If most of the partitions are using HOPI, the construction time is closer to HOPI, such as the graph "days" and "eatRS", as shown in Table 3.

To study query performance, we issue one million random reachability queries on the indexes constructed for the real graphs. We used an in-memory IO simulation to estimate the IO cost. The IO simulation performs the following: Reachability labels (i.e., indexes) are stored in pages with the size 4KB. During query processing, we maintain a buffer with the size 4MB. When we check whether two vertices are reachable, we first obtain the ID of pages where the labels of two vertices are kept. Then we access the buffer to read the labels from required pages. If those pages are in buffer, we read the labels from pages directly. Otherwise, before reading labels from those pages, we insert each page into buffer or replace an old page in buffer using LRU replacement policy. Finally, we report the total query time and the number of labels accessed during query processing in Table 4 and 5, respectively. The average values of S_L in the total query time

Table 5. Number of labels accessed during query processing on real graphs

Tree-like graphs	HOPI	Interval	Our	S_L	S_H	Other graphs	HOPI	Interval	Our	S_L	S_H
kegg	663830	415144	407876	1.02	1.63	days	14986610	32996064	14957887	1.0	2.2
vchocyc	3251476	246004	242602	1.02	13.4	hep-th-new	8869499	47682220	8749328	1.01	5.45
mtbrv	3285414	248322	245036	1.01	13.4	eatRS	1333936	3181514	1333963	1.0	2.39
agrocyc	3292830	241414	237890	1.01	13.8	hpycyc	3017708	787830	770372	1.02	3.92
anthra	3298963	205832	206216	0.99	16	nasa	8993648	655730	655716	1.0	13.72
human	2160820	78618	80188	0.98	27	xmark	6838694	583146	583262	1.0	11.72

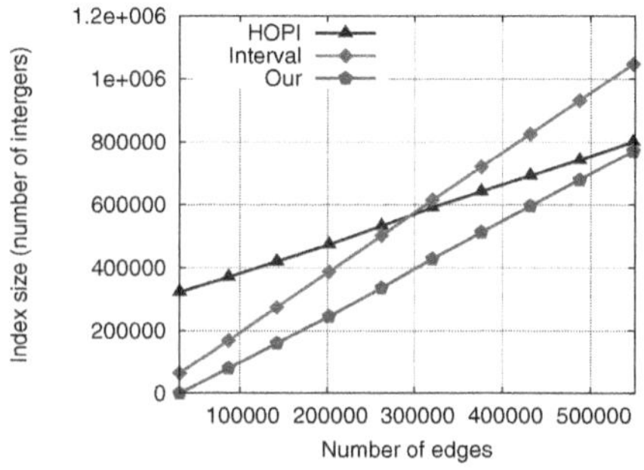

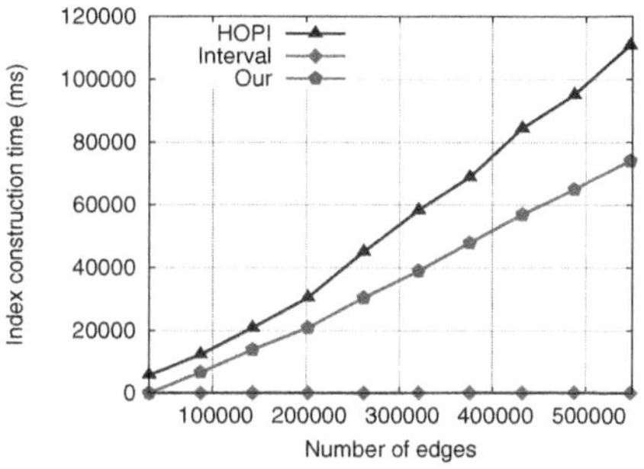

Fig. 4. Index size and construction time comparison on synthetic data

and the number of labels accessed are 0.99 and 1.006, respectively. These indicate that our approach is comparable to (or slightly better than) the best of Interval and HOPI approaches in the total query time and the number of labels accessed. Moreover, the average values of S_H in the total query time and the number of labels accessed are 1.44 and 10.38, respectively. This supports that our approach avoids the cost of selecting an inefficient static method. In all, our approach is both IO efficient and time efficient.

4.2 Index Size and Query Performance Evaluation on Synthetic Data

In Section 4.1, we compared our approach with Interval, HOPI and the Ptree approach on real graphs. In order to have a full control over graph structures, we implemented our own graph generator which controls the percentage of tree-like components and graph-like components. The generator works as follows: First, we generate a set of tree-like DAGs and a set of dense DAGs with the maximum fan-out of spanning tree F=6, the branch depth of spanning tree D=6. Then, we generate a large graph with n vertices by connecting p ($\times 100\%$) dense graphs to 1-p ($\times 100\%$) of tree-like graphs, where n and p are the two input parameters.

We generate a set of random graphs with n around 30k, and varying p from 0 to 0.9. The indexing size and index construction time comparison of Interval, HOPI, and our

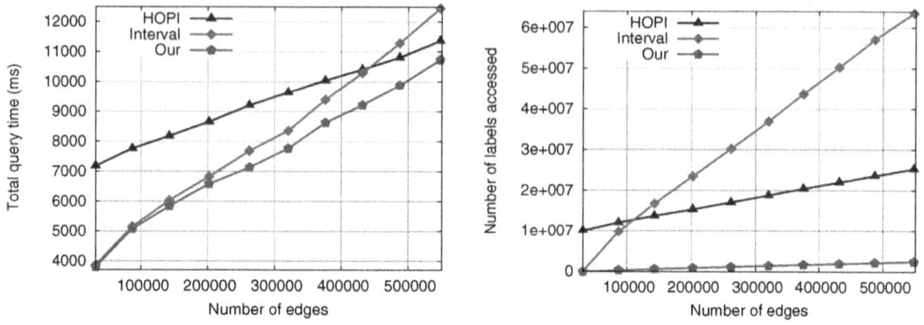

Fig. 5. Total query time and number of labels accessed comparison on synthetic data

approach is shown in Figure 4[1]. Similar to the previous experiment, the index size of our approach is significantly better than HOPI and Interval. Regarding the construction time, we find that the construction time of our approach is comparable to the Interval approach when the graph is sparse and slightly worse than the Interval approach when the graph is dense. However, the construction time of our approach is much smaller than that of the HOPI approach.

Figure 5 shows the total query time and the number of labels accessed of one million random queries. It is clear that our approach has a better query performance and is more IO efficient than the best of HOPI and Interval.

5 Related Work

There has been a large body of work on indexing for reachability queries on graphs. Due to space constraints, we list a few (non-exhaustive) examples in this section.

Dietz [3] assigns an interval to each vertex in a tree. A vertex can reach another vertex *iff* its interval is properly contained in the interval of the other vertex. There has been a host of work that demonstrates good query performance of the interval approach. Wu *et al.* [2] propose to use prime numbers to encode reachability information of a tree. A vertex is labeled with a product of prime numbers. A vertex is reachable from another vertex *iff* its label is divisible by the label of the other vertex. Wu and Zhang [5] extend this work [2] to support DAGs. Wang *et al.* [11] combine the interval approach for trees and a technique for indexing non-tree edges of a graph. The technique has a constant query time and small index size. Schenkel [10] and Cheng [13] extend *2-hop* labeling scheme, originally proposed by Cohen *et al.* [9], to efficiently index a large collection of XML documents. Trißl *et al.* [14] propose an efficient relational-based implementation that bases on the interval and *2-hop* labelings to index directed graphs.

All the aforementioned techniques index an entire graph. In contrast, our work focuses on a framework that supports applying different indexing techniques to different subgraphs. Hence, this work is orthogonal to any specific indexes.

[1] In this experiment, we did not compare the index size of our approach with Ptree [8], as our files storing the random graphs cannot be recognized by the Ptree implementation in Linux.

Perhaps [15,8] are the most relevant works. In [15], it proposes a hierarchical labeling scheme that identifies spanning trees for the interval labelings [3] and dense subgraphs in remainder graphs for 2-hop labeling [9]. While the problem being studied is similar to ours, their approach is tightly coupled with interval labelings and compression of the reachability matrix. Hence, it is not straightforward to incorporate arbitrary graph indexes into their technique. In addition, our method for identifying substructures for indexing is different. In [8], it applies a path-decomposition method to partition a DAG into paths. Next, a path-path graph is proposed to capture the path relationship in a DAG and is indexed with interval labelings. Each node u in a DAG is assigned with an X label denoting the DFS order, a Y label denoting the path order and the interval labels I of u's corresponding path in the path-path graph. The reachability query between two nodes u and v can be answered by comparing their X, Y, and I labels. Although we are working on building graph framework through graph partitioning, our partitioning method is to decompose the input graph into arbitrary subgraphs. That is, the structure of each partition is more general than paths.

Graph partitioning has been one of the classical problems in combinatorial optimization. The problem optimizes an input objective function. In general, this is an NP-complete problem. Various heuristics, *e.g.*, [16,17,18], have been proposed to find an optimal partition, with respect to the objective function. In this paper, our objective function is different from those solved by previous algorithms.

6 Conclusions and Future Work

In this paper, we proposed a uniform framework for efficiently processing reachability query on large graphs. Specifically, a graph is represented by a set of partitions and inter-partition connections. Subsequently, (possibly different) graph indexes can be applied to each partition. This facilitates a seamless application of the state-of-the-art of graph indexing on subgraphs represented in the graph framework. Our experimental study verified the effectiveness and efficiency of our framework. In our experiment with a large variety of synthetic graphs and real graphs, our framework consistently produced relatively small indexes when compared to the best index of a non-partition approach. In addition, our experiment showed that the framework improves the query processing performance over the non-partitioning methods.

We would like to point out that our proposed method has some limitations. We plan to extend our work in the future: First, our framework is proposed to enhance reachability query performance. Yet, reachability queries can be a part of other query formalisms. We are studying the connection between reachability queries and other query formalisms. Second, our query evaluation algorithm is proposed to evaluate one query at a time on a single machine. We plan to study distributed reachability query evaluation on a graph framework.

Acknowledgements. We are grateful to Jiefeng Cheng and Dr. Ruoming Jin for providing us the implementation for 2-hop labelings and Path tree labeling. We thank R. Bramandia for his insightful discussions.

References

1. W3C: OWL web ontology language overview, http://www.w3.org/TR/owl-features
2. Wu, X., Lee, M.L., Hsu, W.: A prime number labeling scheme for dynamic ordered xml trees. In: ICDE, pp. 66–78 (2004)
3. Dietz, P.F.: Maintaining order in a linked list. In: STOC, pp. 122–127 (1982)
4. Chen, L., Gupta, A., Kurul, M.E.: Efficient algorithms for pattern matching on directed acyclic graphs. In: ICDE, pp. 384–385 (2005)
5. Wu, G., Zhang, K., Liu, C., Li, J.: Adapting prime number labeling scheme for directed acyclic graphs. In: Li Lee, M., Tan, K.-L., Wuwongse, V. (eds.) DASFAA 2006. LNCS, vol. 3882, pp. 787–796. Springer, Heidelberg (2006)
6. Chen, L., Gupta, A., Kurul, M.E.: Stack-based algorithms for pattern matching on dags. In: VLDB, pp. 493–504 (2005)
7. Agrawal, R., Borgida, A., Jagadish, H.V.: Efficient management of transitive relationships in large data and knowledge bases. In: SIGMOD, pp. 253–262 (1989)
8. Jin, R., Xiang, Y., Ruan, N., Wang, H.: Efficiently answering reachability queries on very large directed graphs. In: SIGMOD, pp. 595–608 (2008)
9. Cohen, E., Halperin, E., Kaplan, H., Zwick, U.: Reachability and distance queries via 2-hop labels. SIAM J. Comput. 32(5), 1338–1355 (2003)
10. Schenkel, R., Theobald, A., Weikum, G.: Efficient creation and incremental maintenance of the hopi index for complex xml document collections. In: ICDE, pp. 360–371 (2005)
11. Wang, H., He, H., Yang, J., Yu, P.S., Yu, J.X.: Dual labeling: Answering graph reachability queries in constant time. In: ICDE, pp. 75–75 (2006)
12. Gibson, D., Kumar, R., Tomkins, A.: Discovering large dense subgraphs in massive graphs. In: VLDB, pp. 721–732 (2005)
13. Cheng, J., Yu, J.X., Lin, X., Wang, H., Yu, P.S.: Fast computation of reachability labeling for large graphs. In: Ioannidis, Y., Scholl, M.H., Schmidt, J.W., Matthes, F., Hatzopoulos, M., Böhm, K., Kemper, A., Grust, T., Böhm, C. (eds.) EDBT 2006. LNCS, vol. 3896, pp. 961–979. Springer, Heidelberg (2006)
14. Trißl, S., Leser, U.: Fast and practical indexing and querying of very large graphs. In: SIGMOD, pp. 845–856 (2007)
15. He, H., Wang, H., Yang, J., Yu, P.S.: Compact reachability labeling for graph-structured data. In: CIKM (2005)
16. Kernighan, B.W., Lin, S.: An efficient heuristic procedure for partitioning graphs. The Bell system technical journal 49(1), 291–307 (1970)
17. Karypis lab: Family of Multilevel Partitioning Algorithms, http://glaros.dtc.umn.edu/gkhome/metis/metis/overview
18. Pothen, A., Simon, H.D., Liou, K.P.: Partitioning sparse matrices with eigenvectors of graphs. SIAM J. Matrix Anal. Appl. 11(3), 430–452 (1990)
19. Tarjan, R.E.: Depth-first search and linear graph algorithms. SIAM J. Comput. 1(2), 146–160 (1972)
20. Schenkel, R., Theobald, A., Weikum, G.: Hopi: An efficient connection index for complex xml document collections. In: Bertino, E., Christodoulakis, S., Plexousakis, D., Christophides, V., Koubarakis, M., Böhm, K., Ferrari, E. (eds.) EDBT 2004. LNCS, vol. 2992, pp. 237–255. Springer, Heidelberg (2004)
21. Batagelj, V., Mrvar, A.: Pajek datasets, http://vlado.fmf.uni-lj.si/pub/networks/data/

Ricochet: A Family of Unconstrained Algorithms for Graph Clustering

Derry Tanti Wijaya and Stéphane Bressan

School of Computing, National University of Singapore,
Computing 1, Law Link, Singapore 117590, Republic of Singapore
derry.wijaya@gmail.com, steph@nus.edu.sg

Abstract. Partitional graph clustering algorithms like K-means and Star necessitate a priori decisions on the number of clusters and threshold for the weight of edges to be considered, respectively. These decisions are difficult to make and their impact on clustering performance is significant. We propose a family of algorithms for weighted graph clustering that neither requires a predefined number of clusters, unlike K-means, nor a threshold for the weight of edges, unlike Star. To do so, we use re-assignment of vertices as a halting criterion, as in K-means, and a metric for selecting clusters' seeds, as in Star. Pictorially, the algorithms' strategy resembles the rippling of stones thrown in a pond, thus the name 'Ricochet'. We evaluate the performance of our proposed algorithms using standard datasets and evaluate the impact of removing constraints by comparing the performance of our algorithms with constrained algorithms: K-means and Star and unconstrained algorithm: Markov clustering.

Keywords: Clustering, Weighted Graph clustering, Document clustering, K-Means Clustering, Star Clustering.

1 Introduction

Clustering algorithms partition a set of objects into subsets or clusters. Objects within a cluster should be similar while objects in different clusters should be dissimilar, in general. With a scalar similarity metric, the problem can be modeled as the partitioning of a weighted graph whose vertices represent the objects to be clustered and whose weighted edges represent the similarity values. For instance, in a document clustering problem (we use instances of this problem for performance evaluation) vertices are documents (vectors in a vector space model), pairs of vertices are connected (the graph is a clique) and edges are weighted with the value of the similarity of the corresponding documents (cosine similarity) [1, 2]. Partitional clustering graph algorithms, as the name indicates, partition the graph into subsets or regions. It is trying to identify and separate dense regions from sparse regions in order to maximize intra-cluster density and inter-cluster sparseness [3].

Partitional graph clustering algorithms like K-means and Star require crucial a priori decisions on key parameters. The K-means clustering algorithm [4] requires the number of clusters to be provided before clustering. The Star clustering algorithm [5]

X. Zhou et al. (Eds.): DASFAA 2009, LNCS 5463, pp. 153–167, 2009.

requires a threshold for the weight of edges to be fixed before clustering. The choice of the value of these parameters can greatly influence the effectiveness of the clustering algorithms.

In this paper, we propose a family of novel graph clustering algorithms that require neither the number of clusters nor a threshold to be determined before clustering. To do so, we combine ideas from K-means and Star. The proposed algorithms build sub graphs, assigning and dynamically reassigning vertices. We call the first vertex assigned to a sub graph the seed. The algorithms use degree of vertices and weights of adjacent edges for assignment and they use weights of adjacent edges for reassignment. During reassignment, some sub graphs disappear. The algorithms stop when there is no more reassignment. We call the intermediary and resulting sub graphs, clusters (for the sake of simplicity we may use cluster and sub graph interchangeably in the remainder of this paper). Pictorially, the algorithms' strategy resembles the rippling (iterative assignment) caused by stones (seeds) thrown in a pond, thus the name 'Ricochet'. Our contribution is the presentation of this family of novel clustering algorithms and their comparative performance analysis with state-of-the-art algorithms using real world and standard corpora for document clustering.

In the next section we discuss the main state-of-the-art weighted graph clustering algorithms, namely, K-means, Star and Markov Clustering. In section 3, we show how we devise unconstrained algorithms spinning the ricochet and rippling metaphor. In section 4, we empirically evaluate and compare the performance of our proposed algorithms. Finally, we synthesize our results and contribution in section 5.

2 Related Works

K-means [4], Star [5] and Markov Clustering (or MCL) [6] are typical partitional clustering algorithms. They all can solve weighted graph clustering problems. K-means and Star are constrained by the a priori setting of a parameter. Markov Clustering is not only unconstrained but also probably the state-of-the-art in graph clustering algorithms. Our proposal attempts to create an unconstrained algorithm by combining idea from K-means and Star. We review the three algorithms and their variants.

K-means clustering [4] divides the set of vertices into K clusters by choosing randomly K seeds or candidate centroids. The number of clusters, K, is provided a priori and does not change. K-means then assigns each vertex to the cluster whose centroid is the closest. K-means iteratively re-computes the position of the centroid based on the current members of each cluster. K-means converges because the average distance between vertices and their centroids monotonically decreases at each iteration. A variant of K-means efficient for document clustering is K-medoids [7]. We use K-medoids to compare with our proposed algorithms.

Unlike K-means, Star clustering [5], does not require the indication of an a priori number of clusters. It also allows the clusters produced to overlap. This is a generally desirable feature in information retrieval applications.

For document clustering, Star clustering analytically guarantees a lower bound on the topic similarity between the documents in each cluster and computes more accurate clusters than either the older single link [8] or average link [9] hierarchical clustering. The drawback of Star clustering is that the lower bound guarantee on the quality of each cluster depends on the choice of a threshold σ on the weight of edges in the graph.

To produce reliable document clusters of similarity σ, Star algorithm prunes the similarity graph of the document collection, removing edges whose cosine similarity in a vector space is less than σ. Star clustering then formalizes clustering by performing a minimum clique cover with maximal cliques on this σ-similarity graph. Since covering by cliques is an NP-complete problem [10, 11], Star clustering approximates a clique cover greedily by dense sub-graphs that are star shaped, consisting of a single Star center and its satellite vertices.

The selection of Star centers determines the Star cover of the graph and ultimately the quality of the clusters. [12] experimented with various metrics for the selection of Star centers to maximize the 'goodness' of the greedy vertex cover. The average metric (i.e. selecting Star centers in order of the average similarity between a potential Star center and the vertices connected to it) is a fast and good approximation to the expensive lower bound metric [5] that maximizes intra-cluster density in all variants of the Star algorithm. The average metric [12] is closely related to the notion of average similarity between vertices and their medoids in K-medoids [7].

Markov Clustering tries and simulates a (stochastic) flow (or random walks) in graphs [6]. From a stochastic view point, once inside a region, a random walker should have little chance to walk out [13]. The graph is first represented as stochastic (Markov) matrices where edges between vertices indicate the amount of flow between the vertices. MCL algorithm simulates flow using two alternating operations on the matrices: expansion and inflation. The flow is eventually separated into different regions, yielding a cluster interpretation of the initial graph. MCL does not require an a priori number of expected clusters nor a threshold for the similarity values. However, it requires a fine tuning inflation parameter that influences the coarseness and possibly the quality of the result clusters. We nevertheless consider it as an unconstrained algorithm, as, our experience suggests, optimal values for the parameter seem to be rather stable across applications.

Chameleon [14] is a hierarchical clustering algorithm that stems from the same motivation as the one that prompted our proposal: the need for dynamic decisions in place of a priori static parameters. However, in effect, Chameleon uses three parameters: the number of nearest neighbours, the minimum size of the sub graphs, and the relative weightage of inter-connectivity and closeness. Although the authors [14] observed that the parameters have a mild effect on 2D data, the users always need to preset the values of parameters and their effects are unknown for other types of data such as documents (high dimensional data). In a more recent paper [15], the author of Chameleon combines agglomerative hierarchical clustering and partitional clustering. Their experimental results suggest that the combined methods improve the clustering solution.

Our proposed family of algorithms, like Chameleon, is dynamic (we dynamically reassign vertices to clusters). However, our algorithms need no parameters. In spite of the absence of parameters, we hope to achieve effectiveness comparable to the best settings of K-means, Star and MCL. We also hope to produce more efficient algorithms than K-means, Star and MCL.

3 A Family of Unconstrained Algorithms

Ricochet algorithms, like other partitional graph clustering algorithms, alternate two phases: the choosing of vertices to be the seeds of clusters and the assignment of vertices to existing clusters. The motivation underlying our work lies in the observation that: (1) Star clustering algorithm provides a metric of selecting Star centers that are potentially good cluster seeds for maximizing intra-cluster density, while (2) K-means provides an excellent vertices assignment and reassignment, and a convergence criterion that increases intra-cluster density at each iteration. By using Star clustering metric for selecting Star centers, we can find potential cluster seeds without having to supply the number of clusters. By using K-means re-assignment of vertices, we can update and improve the quality of these clusters and reach a termination condition without having to determine any threshold.

Hence, similar to Star and Star-ave algorithms [12], Ricochet chooses seeds in descending order of the value of a metric combining degree with the weight of adjacent edges. Similar to K-means, Ricochet assigns and reassigns vertices; and the iterative assignment of vertices is stopped once these conditions are met: (1) no vertex is left unassigned and (2) no vertex is candidate for re-assignment.

The Ricochet family is twofold. In the first Ricochet sub-family, seeds are chosen one after the other ('stones are thrown one by one'). In the second Ricochet sub-family, seeds are chosen at the same time ('stones are thrown together'). We call the former algorithms Sequential Rippling, and the latter Concurrent Rippling. The algorithms in the Sequential Rippling, because of the way they select seeds and assign or re-assign vertices, are intrinsically hard clustering algorithms, i.e. they produce disjoint clusters. The algorithms in the Concurrent Rippling are soft clustering algorithms, i.e. they produce possibly overlapping clusters.

The algorithms in the Sequential Rippling can be perceived as somewhat straightforward extensions to the K-means clustering. We nevertheless present them in this paper for the purpose of completeness and comparison with the more interesting algorithms we propose in the Concurrent Rippling.

3.1 Sequential Rippling

3.1.1 Sequential Rippling (SR)

The first algorithm of the subfamily is call Sequential Rippling (or SR). In this algorithm, *vertices* are ordered in descending order of the average weight of their adjacent edges (later referred to as the weight of a vertex). The vertex with the highest weight is chosen to be the first seed and a cluster is formed by assigning all other vertices to the cluster of this first seed. Subsequently, new seeds are chosen one by one from the ordered list of vertices. When a new seed is added, vertices are re-assigned to a new cluster if they are closer to the new seed than they were to the seed of their current cluster (if no vertex is closer to the new seed, no new cluster is created). If clusters are reduced to singletons during re-assignment, they are assigned to the nearest non-singleton cluster. The algorithm stops when all vertices

```
Given a Graph G = (V, E). V contains vertices, |V| = N. Each vertex has a weight which is
the average similarity between the vertex and its adjacent vertices. E contains edges in G
(self-loops removed) with similarity as weights.

Algorithm: SR ( )
Sort V in order of vertices' weights
Take the heaviest vertex v from V
listCentroid.add (v)
Reassign all other vertices to v's cluster
While (V is not empty)
   Take the next heaviest vertex v from V
   Reassign vertices which are more similar to v than to other centroid
   If there are re-assignments
      listCentroid.add (v)
      Reassign singleton clusters to its nearest centroid
For all i ε listCentroid return i and its associated cluster
```

Fig. 1. Sequential Rippling Algorithm

```
Algorithm: BSR ( )
Sort V in order of vertices' weights
Take the heaviest vertex v from V
listCentroid.add (v)
Reassign all other vertices to v's cluster
Reassignment = true
While (Reassignment and V is not empty)
   Reassignment = false
   Take a vertex v ∉listCentroid from V whose ratio of its weight to
    the sum of its similarity  to existing centroids is the maximum
   Reassign vertices which are more similar to v than to other centroid
   If there are re-assignments
      Reassignment = true
      listCentroid.add (v)
      Reassign singleton clusters to its nearest centroid
For all i ε listCentroid return i and its associated cluster
```

Fig. 2. Balanced Sequential Rippling Algorithm

have been considered. The pseudocode of the Sequential Rippling algorithm is given in figure 1. The worst case complexity of Sequential Rippling algorithm is O (N^3) because in the worst case the algorithm has to iterate through at most N vertices, each time comparing the distance of N vertices to at most N centroids.

3.1.2 Balanced Sequential Rippling (BSR)

The second algorithm of the subfamily is called Balanced Sequential Rippling (BSR). The difference between BSR and SR is in its choice of subsequent seed. In order to balance the distribution of seeds in the graph, BSR chooses a next seed that is both a reasonable centroid for a new cluster (i.e. has large value of weight) as well as sufficiently far from the previous seeds. Subsequent seed is chosen to maximize the ratio of its weight to the sum of its similarity to the centroids of already existing clusters. This is a compromise between weight and similarity. We use here the simplest possible formula to achieve such compromise. It could clearly be refined and fine-tuned.

As in SR, when a new seed is added, vertices are re-assigned to a new cluster if they are closer to the new seed than they were to the seed of their current cluster. The algorithm terminates when there is no re-assignment of vertices. The pseudocode of Balanced Sequential Rippling algorithm is given in figure 2. The worst case complexity of Balanced Sequential Rippling algorithm is O (N^3) because in the worst case the algorithm has to iterate through at most N vertices, each time comparing the distance of N vertices to at most N centroids.

3.2 Concurrent Rippling

Unlike sequential rippling which chooses centroids one after another, concurrent rippling treats all vertices as centroids of their own singleton clusters initially. Then, each centroid concurrently 'ripples its influence' using its adjacent edges to other vertices. As the rippling progresses, some centroids can lose their centroid status (i.e. become non-centroid) as the cluster of smaller weight centroid is 'engulfed' by the cluster of bigger weight centroid.

3.2.1 Concurrent Rippling (CR)

The first algorithm of the sub-family is called Concurrent Rippling (CR). In this algorithm, for each vertex, the adjacent *edges* are ordered in descending order of weights. Iteratively, the next heaviest edge is considered. Two cases are possible: (1) if the edge connects a centroid to a non-centroid, the non-centroid is added to the cluster of the centroid (notice that at this point the non-centroid belongs to at least two clusters), (2) if the edge connects two centroids, the cluster of one centroid is assigned to the cluster of the other centroid (i.e. it is 'engulfed' by the other centroid), if and only if its weight is smaller than that of the other centroid. The two clusters are merged and the smaller weight centroid becomes a non-centroid. The algorithm terminates when the centroids no longer change. The pseudocode of Concurrent Rippling algorithm is given in figure 3.

Making sure that all centroids propagate their ripples at equal speed (lines 8 - 10 of Algorithm: CR () in figure 3), the algorithm requires the sorting of a list whose total size is the square of the number of vertices. Concurrent Rippling algorithm requires O (N^2logN) complexity to sort the N-1 neighbors of the N vertices. It requires another O (N^2logN) to sort the N^2 number of edges. In the worst case, the algorithm has to iterate through all the N^2 edges. Hence, in the worst case the complexity of the algorithm is O (N^2logN).

3.2.2 Ordered Concurrent Rippling (OCR)

The second algorithm of the sub-family is called Ordered Concurrent Rippling (OCR). In this algorithm, the constant speed of rippling is abandoned to be approximated by a simple ordering of adjacent edges according to their weights to the vertex (i.e. OCR abandons line 1, and lines 8 – 10 of Algorithm: CR () in figure 3).

The method allows not only to improve efficiency (although worst case complexity is the same) but also to process only the best 'ripple' (i.e. heaviest adjacent edge) for each vertex each time. The pseudocode of Ordered Concurrent Rippling algorithm is given in figure 4. The complexity of the algorithm is O ($N^2 \log N$) to sort the N-1 neighbors of the N vertices. The algorithm then iterates at most N^2 times. Hence the overall worst case complexity of the algorithm is O ($N^2 \log N$).

```
For each vertex v, v.neighbor is the list of v's adjacent vertices sorted by their similarity
to v from highest to lowest.
If v is a centroid (v.centroid == 1); v.cluster contains the list of vertices ≠ v assigned to v

Algorithm: CR ( )
1. Sort E in order of the edge weights
2. public CentroidChange = true
3. index = 0
4. While (CentroidChange && index < N-1 && E is not empty)
   5. CentroidChange = false
   6. For each vertex v, take its edge e_vw connecting v to its next
      closest neighbor w; i.e. w = v.neighbor [index]
   7. Store these edges in S
   8. Find the lowest edge weight in S, say low, and empty S
   9. Take all edges from E whose weight >= low
   10. Store these edges in S
   11. PropagateRipple (S)
   12. index ++
13. For all i ε V, if i is a centroid, return i and i.cluster

Sub Procedure: PropagateRipple (list S)
/* This sub procedure is to propagate ripples for all the centroids. If the ripple of one
centroid touches another, the heavier weight centroid will engulf the lighter centroid and
its cluster. If the ripple of a centroid touches a non-centroid, the non-centroid is assigned
to the centroid. A non-centroid can be assigned to more than one centroid, allowing
overlapping between clusters, a generally desirable feature */

While (S is not empty)
   Take the next heaviest edge, say e_vw, from S
   If v ∉x.cluster for all x ε V
         If w is a centroid, compare v's weight to w's weight
            If (w.weight > v.weight)
               add v and v.cluster into w.cluster
               Empty v.cluster
               If v is a centroid
                  v.centroid = 0
                  CentroidChange = true
            Else
               add w and w.cluster into v.cluster
               Empty w.cluster
               w.centroid = 0
               CentroidChange = true
         Else if w is not a centroid
            v.cluster.add (w)
            If v is not a centroid
               v.centroid = 1
               CentroidChange = true
```

Fig. 3. Concurrent Rippling Algorithm

For each vertex v, v.neighbor is the list of v's adjacent vertices sorted by their similarity to v from highest to lowest. If v is a centroid (i.e. v.centroid == 1); v.cluster contains the list of vertices $\neq v$ assigned to v

```
Algorithm: OCR ( )
public CentroidChange = true
index = 0
While (CentroidChange && index < N-1)
   CentroidChange = false
   For each vertex v, take its edge e_vw connecting v to its next closest
   neighbor w; i.e. w = v.neighbor [index]
   Store these edges in S
   PropagateRipple (S)
   index ++
For all i ε V, if i is a centroid, return i and i.cluster
```

Fig. 4. Ordered Concurrent Rippling Algorithm

3.3 Maximizing Intra-cluster Similarity

The key point of Ricochet is that at each step it tries to maximize the intra-cluster similarity: the average similarity between vertices and the centroid in the cluster.

Sequential Rippling (SR and BSR) try to maximize the average similarity between vertices and their centroids by (1) selecting centroids in order of their weights and (2) iteratively reassigning vertices to the nearest centroids. As in [12], selecting centroids in order of weights is a fast approximation to maximizing the expected intra-cluster similarity. As in K-means, iteratively reassigning vertices to nearest centroids decreases the distance (thus maximizing similarity) between vertices and their centroids.

Concurrent Rippling (CR and OCR) try to maximize the average similarity between vertices and their centroids by (1) processing adjacent edges for each vertex in order of their weights from highest to lowest and (2) choosing the bigger weight vertex as a centroid whenever two centroids are adjacent to one another. (1) ensures that at each step, the best possible merger for each vertex v is found (i.e. after merging a vertex to v with similarity s, we can be sure that we have already found and merged all vertices whose similarity is better than s to v); while (2) ensures that the centroid of a cluster is always the point with the biggest weight. Since we define a vertex weight as the average similarity between the vertex and its adjacent vertices; choosing a centroid with bigger weight is an approximation to maximizing the average similarity between the centroid and its vertices.

Further, although OCR is, like its family, a partitional graph clustering algorithm; its decision to only process the heaviest adjacent edge (hence the best possible merger) of each vertex at each step is compatible to the steps of agglomerative single-link hierarchical clustering. We believe therefore that OCR combines concepts from both partitional and agglomerative approaches. As in [15], such combination has been experimentally found to be successful than either of the methods alone.

4 Performance Analysis

In order to evaluate our proposed algorithms, we empirically compare, the performance our algorithms with the constrained algorithms: (1) K-medoids (that is given the optimum/correct number of clusters as obtained from the data set); (2) Star clustering algorithm, and (3) the improved version of Star (i.e. Star Ave) that uses average metric to pick star centers [12]. This is to investigate the impact of removing the constraints on the number of clusters (K-Medoids) and threshold (Star) on the result of clustering. We then compare, the performance of our algorithms with the state-of-the-art unconstrained algorithm, (4) Markov Clustering (MCL), varying MCL's fine-tuning inflation parameter.

We use data from Reuters-21578 [16], TIPSTER–AP [17] and a collection of web documents constructed using the Google News search engine [18] and referred to as Google. The Reuters-21578 collection contains 21,578 documents that appeared in Reuter's newswire in 1987. The documents are partitioned into 22 sub-collections. For each sub-collection, we cluster only documents that have at least one explicit topic (i.e. documents that have some topic categories within its <TOPICS> tags). The TIPSTER–AP collection contains AP newswire from the TIPSTER collection. For the purpose of our experiments, we have partitioned TIPSTER-AP into 2 separate sub-collections. Our original collection: Google contains news documents obtained from Google News in December 2006. This collection is partitioned into 2 separate sub-collections. The documents have been labeled manually. In total we have 26 sub-collections. The sub-collections, their number of documents and topics/clusters are reported in Table 1.

By default and unless otherwise specified, we set the value of threshold σ for Star clustering algorithm to be the average similarity of documents in the given collection.

In each experiment, we apply the clustering algorithms to a sub-collection. We study effectiveness (recall, r, precision, p, and F1 measure, $F1 = (2 * p * r) / (p + r)$), and efficiency in terms of running time.

In each experiment, for each topic, we return the cluster which best approximates the topic. Each topic is mapped to the cluster that produces the maximum F1-measure with respect to the topic:

$$\text{topic}\ (i) = \max_j \{F1\ (i, j)\}\ . \tag{1}$$

where F1 (i, j) is the F1 measure of the cluster number j with respect to the topic number i. The weighted average of F1 measure for a sub-collection is calculated as:

$$F1 = \Sigma\ (t_i/S) * F1\ (i, \text{topic}\ (i));\ \text{for}\ 0 \le i \le T\ . \tag{2}$$

$$S = \Sigma\ t_i;\ \text{for}\ 0 \le i \le T \tag{3}$$

where T is the number of topics in the sub-collection; t_i is the number of documents belonging to topic i in the sub-collection. For each sub-collection, we calculate the weighted-average of precision, recall and F1-measure produced by each algorithm.

Table 1. Description of Collections

Sub-collection	# of docs	# of topic	Sub-collection	# of docs	# of topic
reut2-000.sgm	981	48	Reut2-001.sgm	990	41
reut2-002.sgm	991	38	Reut2-003.sgm	995	46
reut2-004.sgm	990	42	Reut2-005.sgm	997	50
reut2-006.sgm	990	38	Reut2-007.sgm	988	44
reut2-008.sgm	991	42	Reut2-009.sgm	495	24
reut2-010.sgm	989	39	Reut2-011.sgm	987	42
reut2-012.sgm	987	50	Reut2-013.sgm	658	35
reut2-014.sgm	693	34	Reut2-015.sgm	992	45
reut2-016.sgm	488	34	Reut2-017.sgm	994	61
reut2-018.sgm	994	50	Reut2-019.sgm	398	24
reut2-020.sgm	988	28	Reut2-021.sgm	573	24
Tipster-AP1	1787	47	Tipster-AP2	1721	48
Google1	1019	15	Google2	1010	14

For each collection, we then present the average (using micro-averaging) of results produced by each algorithm.

Whenever we perform efficiency comparison with other graph-based clustering algorithms (Markov), we do not include the pre-processing time to construct the graph and compute all pair wise cosine-similarities. The pre-processing time of other graph-based clustering algorithms is the same as our proposed algorithm that is also graph-based. Furthermore, the complexity of pre-processing, O (n^2), may undermine the actual running time of the algorithms themselves. We however include this pre-processing time when comparing with non graph-based clustering algorithms (K-medoids) that do not require graph or all pair wise similarities to be computed. This is to illustrate the effect of pre-processing on the efficiency of graph-based clustering algorithms.

4.1 Performance Results

4.1.1 Comparison with Constrained Algorithms

We first compare the effectiveness and efficiency of our proposed algorithms with those of K-medoids, Star, and (an improved version of Star) Star-ave, in order to determine the consequences of combining ideas from both algorithms to obtain an unconstrained family. There is, of course, a significant benefit per se in removing the need for parameter setting. Yet the subsequent experiments show that this can be done, only in the some cases, at a minor cost in effectiveness.

In figure 5, we can see that the effect of combining ideas from both K-means and Star in our algorithms improves precision on Google data. Our algorithms also maintain recall therefore improving the F1-value. In particular, CR is better than K-medoids, Star and Star-ave in terms of F1-value. BSR and OCR are better than K-medoids and Star in terms of F1-value. In terms of efficiency (cf. figure 6), CR and OCR are faster than Star and Star-ave. K-medoids is much faster than all the graph-based clustering algorithms because it does not require computation of pair wise similarities between all the documents.

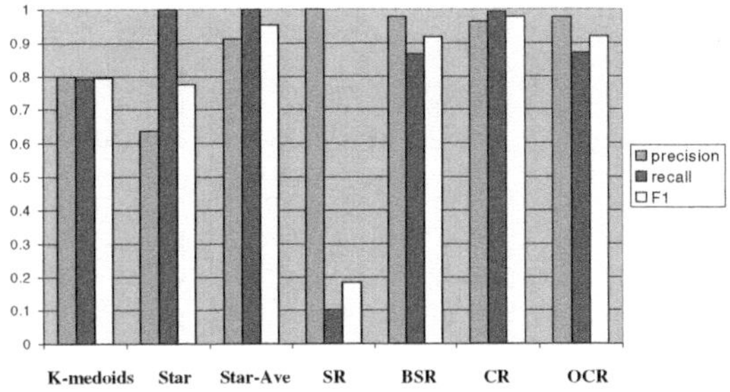

Fig. 5. Effectiveness on Google

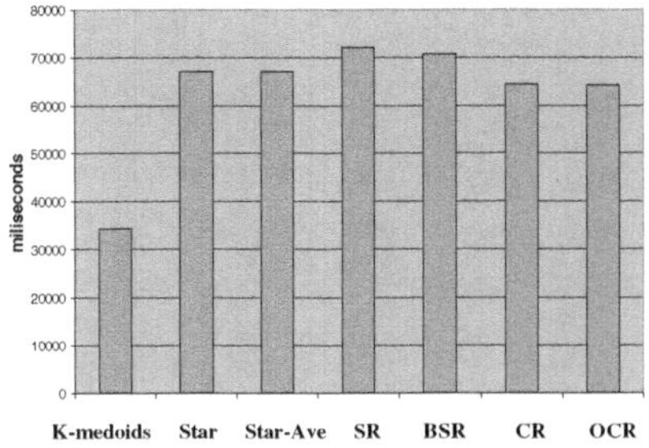

Fig. 6. Efficiency on Google

On Tipster-AP data (cf. figure 7), BSR and OCR yield a higher precision and F1-value than Star and Star-Ave. On Tipster-AP data, OCR performs the best among our proposed algorithms and its effectiveness is comparable to that of a K-medoids supplied with the correct number of clusters K. CR and OCR are also faster than Star and Star-ave (cf. figure 8).

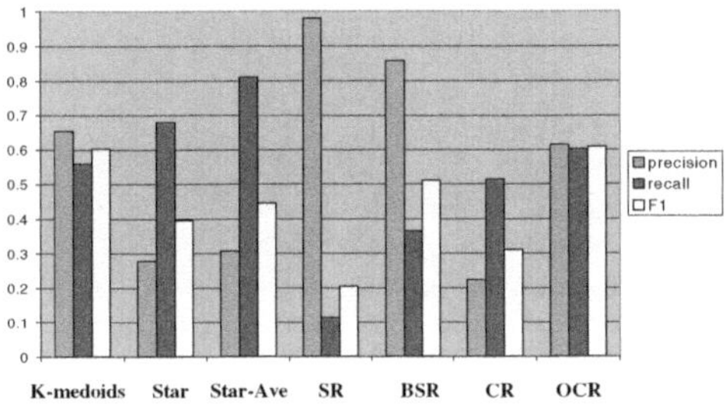

Fig. 7. Effectiveness on Tipster-AP

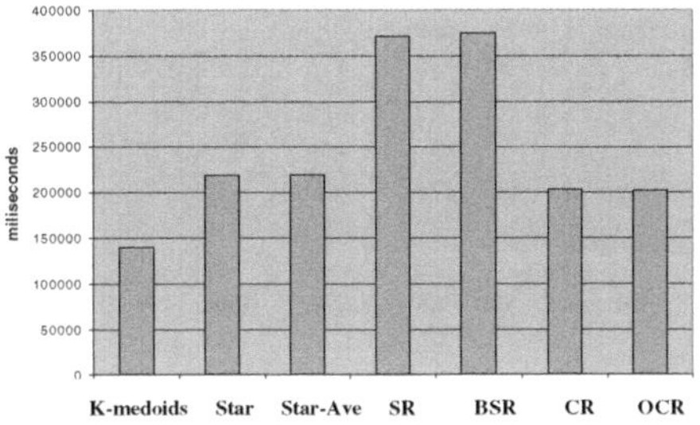

Fig. 8. Efficiency on Tipster-AP

On Reuters data (cf. figure 9), in terms of F1-value, OCR performs the best among our algorithms. OCR performance is better than K-medoids and is comparable to Star and Star-Ave. In terms of efficiency (cf. figure 10), OCR is faster than Star and Star-Ave but slower than K-medoids which does not require pair wise similarity computation between all the documents.

In summary, BSR and OCR are the most effective among our proposed algorithms. BSR achieves higher precision than K-medoids, Star and Star-Ave on all data sets. OCR achieves a balance between high precision and recall, and obtains higher or comparable F1-value than K-medoids, Star and Star-Ave on all data sets. Since our algorithms try to maximize intra-cluster similarity, it is precision that is mostly improved.

In particular, OCR is the most effective and efficient of our proposed algorithms. In terms of F1, OCR is 8.7% better than K-medoids, 23.5% better than Star, and 7.9% better than Star-Ave. In terms of efficiency, OCR is 56.5% slower than K-medoids (due to pre-processing time), 6.3% faster than Star and 6.5% faster than Star-Ave. When pre-processing time is not factored in (cf. Figure 11 and 12), the graph-based algorithms: Star, Star-Ave, CR, OCR are all faster than K-medoids (results on Reuters are consistent but not shown due to space constraints). This shows that it is mostly

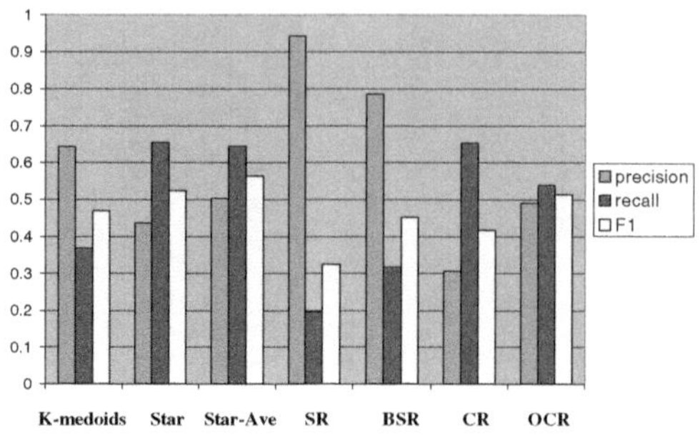

Fig. 9. Effectiveness on Reuters

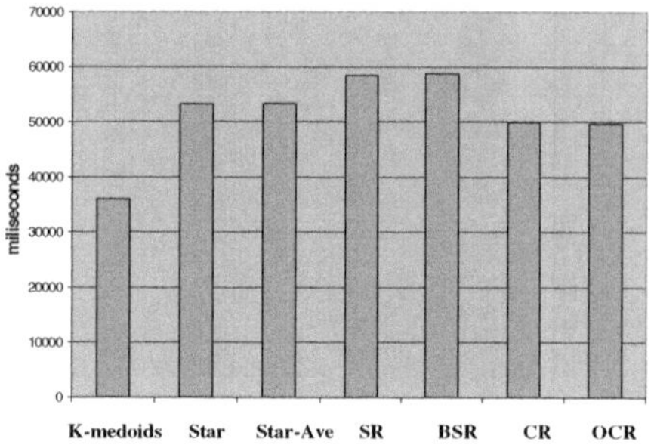

Fig. 10. Efficiency on Reuters

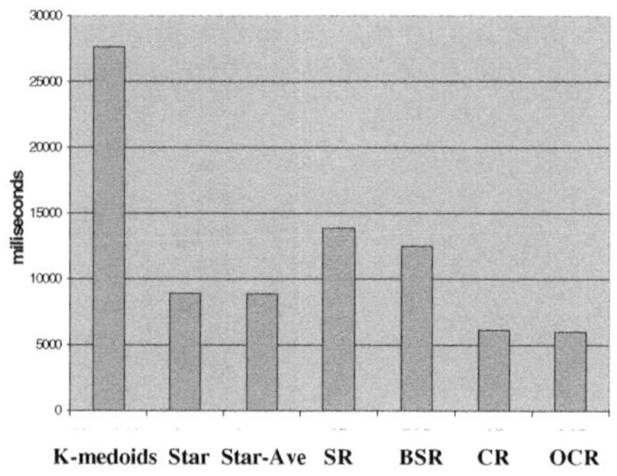

Fig. 11. Efficiency on Google

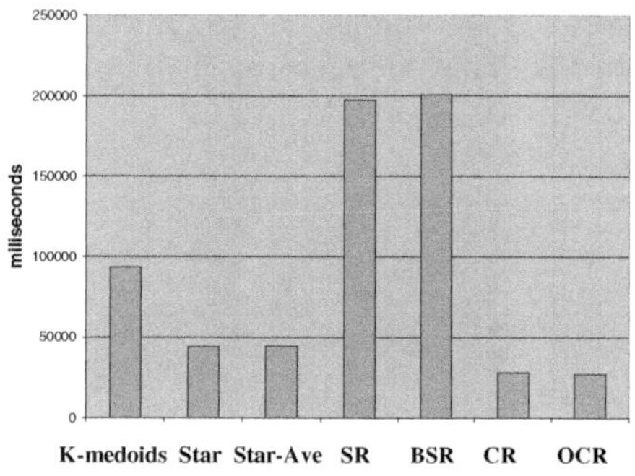

Fig. 12. Efficiency on Tipster-AP

pre-processing that has adverse effect on the efficiency of graph-based clustering algorithms. In the next section, we compare our best performing algorithms: BSR and OCR, with the unconstrained algorithm: Markov Clustering.

4.1.2 Comparison with Unconstrained Algorithms

We first illustrate the influence of MCL's inflation parameter on the algorithm's performance. We vary it between 0.1 and 30.0 (we have empirically verified that this range is representative of MCL performance on our data sets) and report results for representative values.

As shown in figure 13, at a value of 0.1, the resulting clusters have high recall and low precision. As the inflation parameter increases, the recall drops and precision improves, resulting in higher F1-value. At the other end of the spectrum, at a value of 30.0, the resulting clusters are back to having high recall and low precision again. In terms of efficiency (cf. figure 14), as the inflation parameter increases, the running time decreases, indicating that MCL is more efficient at higher inflation value. From figure 13 and 14, we have shown empirically that the choice of inflation value indeed affects the effectiveness and efficiency of MCL algorithm. Both MCL's effectiveness and efficiency vary significantly at different inflation values. The optimal value seems however to always be around 3.0.

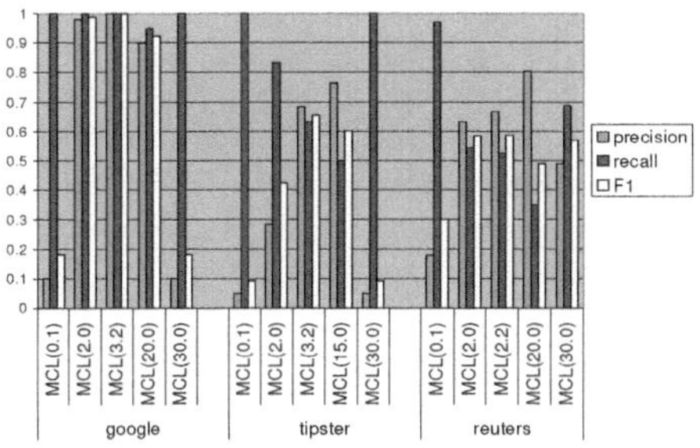

Fig. 13. Effectiveness of MCL

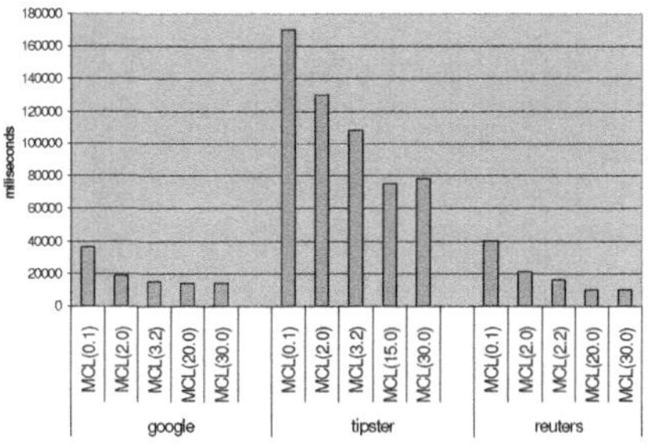

Fig. 14. Efficiency of MCL

We now compare the performance of our best performing algorithms, BSR and OCR, to the performance of MCL algorithm at its best inflation value as well as at its minimum and maximum inflation values, for each collection.

On Google data (cf. figure 15), the effectiveness of BSR and OCR is competitive (although not equal) to that of MCL at its best inflation value. Yet, BSR and OCR are much more effective than MCL at the minimum and maximum inflation values. In terms of efficiency, both BSR and OCR are significantly faster than MCL at all inflation values (cf. figure 16).

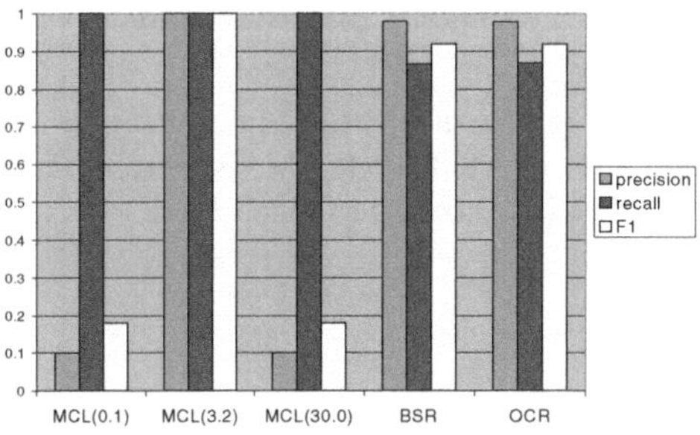

Fig. 15. Effectiveness on Google

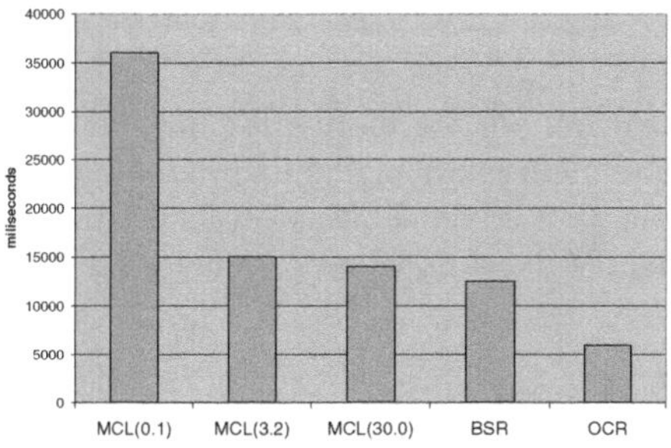

Fig. 16. Efficiency on Google

On Tipster-AP data (cf. figure 17), BSR and OCR are slightly less effective than MCL at its best inflation value. However, both BSR and OCR are much more effective than MCL at the minimum and maximum inflation values. In terms of efficiency (cf. figure 18), OCR is also much faster than MCL at all inflation values.

The same trend is noticeable on Reuters data (cf. figure 19). BSR and OCR are slightly less effective than MCL at its best inflation value. However, BSR and OCR are more effective than MCL at the minimum and maximum inflation values. In terms of efficiency (cf. figure 20), OCR is much faster than MCL at all inflation values.

In summary, although MCL can be slightly more effective than our proposed algorithms at its best settings of inflation parameter (9% more effective than OCR), one of our algorithms, OCR, is not only respectably effective but is also significantly more

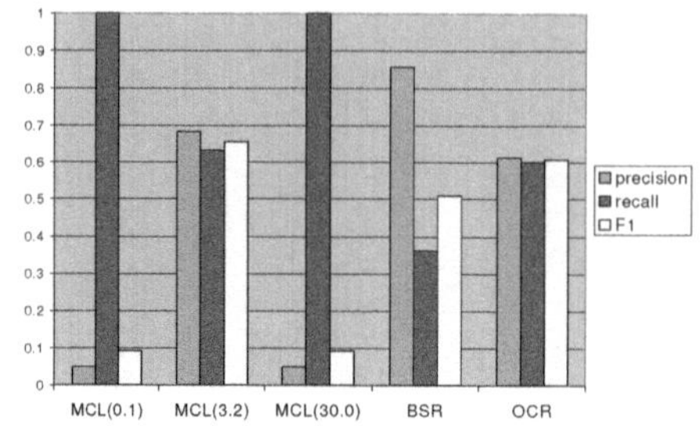

Fig. 17. Effectiveness on Tipster-AP

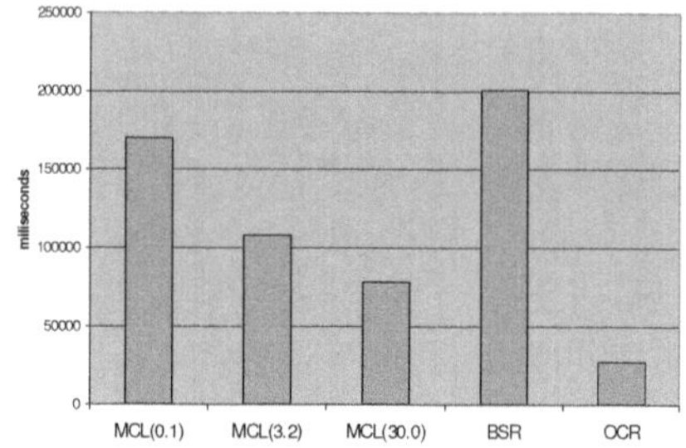

Fig. 18. Efficiency on Tipster-AP

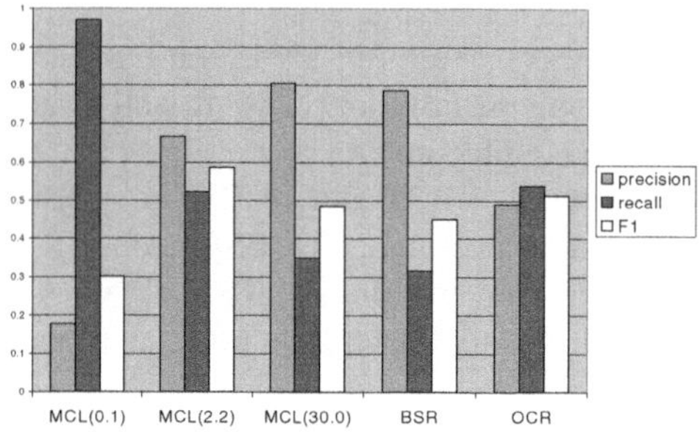

Fig. 19. Effectiveness on Reuters

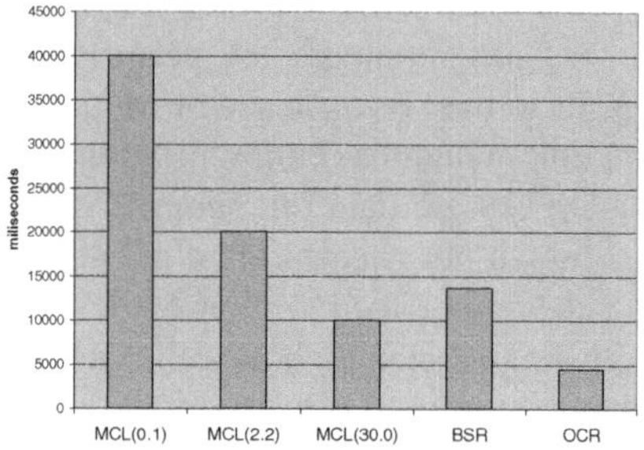

Fig. 20. Efficiency on Reuters

efficient (70.8% more efficient). Furthermore OCR does not require the setting of any fine tuning parameters like the inflation parameter of MCL that, when set incorrectly, can have adverse effect on its performance (OCR is in average, 334% more effective than MCL at its worst parameter setup).

5 Conclusion

We have proposed a family of algorithms for the clustering of weighted graphs. Unlike state-of-the-art K-means and Star clustering algorithms, our algorithms do not require the a priori setting of extrinsic parameters. Unlike state-of-the-art MCL clustering algorithm, our algorithms do not require the a priori setting of intrinsic fine tuning parameters. We call our algorithms 'unconstrained'.

Our algorithms have been devised by spinning the metaphor of ripples created by the throwing of stones in a pond. Clusters' centroids are stones; and rippling is the iterative spread of the centroids' influence and the assignment and reassignment of vertices to the centroids' clusters. For the sake of completeness, we have proposed both sequential (in which centroids are chosen one by one) and concurrent (in which every vertex is initially a centroid) versions of the algorithms and variants.

After a comprehensive comparative performance analysis with reference data sets in the domain of document clustering, we conclude that, while all our algorithms are competitive, one of them, Ordered Concurrent Rippling (OCR), yields a very respectable effectiveness while being efficient. Since all our algorithms try to maximize intra-cluster similarity at each step, it is mostly precision that is improved.

However, like other graph clustering algorithms, the pre-processing time to build the graph and compute all pair-wise similarities in the graph remains a bottleneck when compared to non graph-based clustering algorithms like K-means. We are exploring other ways to reduce this pre-processing time using indexing, stream processing or randomized methods.

We have therefore proposed a novel family of algorithms, called Ricochet algorithms, and, in particular, one new effective and efficient algorithm for weighted graph clustering, called Ordered Concurrent Rippling or OCR.

References

1. Salton, G.: Automatic Text Processing: The Transformation, Analysis, and Retrieval of Information by Computer. Addison-Wesley, Reading (1989)
2. Salton, G.: The Smart Document Retrieval Project. In: 14th Annual International ACM/SIGIR Conference on Research and Development in Information Retrieval, pp. 356–358 (1991)
3. Brandes, U., Gaertler, M., Wagner, D.: Experiments on graph clustering algorithms. In: Di Battista, G., Zwick, U. (eds.) ESA 2003. LNCS, vol. 2832, pp. 568–579. Springer, Heidelberg (2003)
4. MacQueen, J.B.: Some Methods for Classification and Analysis of Multivariate Observations. In: 5th Berkeley Symposium on Mathematical Statistics and Probability, Berkeley, vol. 1, pp. 281–297. University of California Press (1967)
5. Aslam, J., Pelekhov, K., Rus, D.: The Star Clustering Algorithm. Journal of Graph Algorithms and Applications 8(1), 95–129 (2004)
6. Van Dongen, S.M.: Graph Clustering by Flow Simulation. In: Tekst. Proefschrift Universiteit Utrecht (2000)
7. Kaufman, L., Rousseeuw, P.: Finding Groups in Data: An Introduction to Cluster Analysis. John Wiley and Sons, New York (1990)
8. Croft, W.B.: Clustering Large Files of Documents using the Single-link Method. Journal of the American Society for Information Science, 189–195 (1977)
9. Voorhees, E.: The Cluster Hypothesis Revisited. In: 8th SIGIR, pp. 95–104
10. Lund, C., Yannakakis, M.: On the Hardness of Approximating Minimization Problems. Journal of the ACM 41, 960–981 (1994)
11. Press, W., Flannery, B., Teukolsky, S., Vetterling, W.: Numerical Recipes in C: The Art of Scientific Computing. Cambridge University Press, Cambridge (1988)
12. Wijaya, D., Bressan, S.: Journey to the centre of the star: Various ways of finding star centers in star clustering. In: Wagner, R., Revell, N., Pernul, G. (eds.) DEXA 2007. LNCS, vol. 4653, pp. 660–670. Springer, Heidelberg (2007)
13. Nieland, H.: Fast Graph Clustering Algorithm by Flow Simulation. Research and Development ERCIM News 42 (2000)
14. Karypis, G., Han, E., Kumar, V.: CHAMELEON: A Hierarchical Clustering Algorithm Using Dynamic Modeling. IEEE Computer 32(8), 68–75 (1999)
15. Zhao, Y., Karypis, G.: Hierarchical Clustering Algorithms for Document Datasets. Data Mining and Knowledge Discovery 10(2), 141–168 (2005)
16. http://www.daviddlewis.com/resources/testcollections/reuters21578/ (visited on December 2006)
17. http://trec.nist.gov/data.html (visited on December 2006)
18. Google News, http://news.google.com.sg

Top-K Correlation Sub-graph Search in Graph Databases

Lei Zou[1], Lei Chen[2], and Yansheng Lu[1]

[1] Huazhong University of Science and Technology, Wuhan, China
{zoulei,lys}@mail.hust.edu.cn
[2] Hong Kong of Science and Technology, Hong Kong, China
leichen@cse.ust.hk

Abstract. Recently, due to its wide applications, (similar) subgraph search has attracted a lot of attentions from database and data mining community, such as [13,18,19,5]. In [8], Ke et al. first proposed correlation sub-graph search problem (CGSearch for short) to capture the underlying dependency between sub-graphs in a graph database, that is CGS algorithm. However, CGS algorithm requires the specification of a minimum correlation threshold θ to perform computation. In practice, it may not be trivial for users to provide an appropriate threshold θ, since different graph databases typically have different characteristics. Therefore, we propose an alternative mining task: ***top-K correlation sub-graph search***(TOP-CGSearh for short). The new problem itself does not require setting a correlation threshold, which leads the previous proposed CGS algorithm inefficient if we apply it directly to TOP-CGSearch problem. To conduct TOP-CGSearch efficiently, we develop a ***pattern-growth*** algorithm (that is PG-search algorithm) and utilize graph indexing methods to speed up the mining task. Extensive experiment results evaluate the efficiency of our methods.

1 Introduction

As a popular data structure, graphs have been used to model many complex data objects and their relationships, such as, chemical compounds [14], entities in images [12] and social networks [3]. Recently, the problems related to graph database have attracted much attentions from database and data mining community, such as frequent sub-graph mining [7,17], (similar) sub-graph search [13,18,19,5]. On the other hand, correlation mining [15] is always used to discovery the underlying dependency between objects, such as market-basket databases [2,10], multimedia databases [11] and so on. Ke et al. first propose correlation sub-graph search in graph databases [8]. Formally, correlation sub-graph search (CGSearch for short) is defined as follows [8]:

Given a query graph Q and a minimum correlation threshold θ, we need to report all sub-graphs S_i in graph database, where the Pearson's correlation coefficient between Q and S_i (that is $\phi(Q, S_i)$, see Definition 4) is no less than θ.

Actually, the larger $\phi(Q, S_i)$ is, the more often sub-graphs Q and S_i appear together in the same data graphs. Take molecule databases for example. If two sub-structures S_1 and S_2 often appear together in the same molecules, S_1 and S_2 may share some similar chemical properties. In Fig. 1, we have 5 data graphs in database D and a query

X. Zhou et al. (Eds.): DASFAA 2009, LNCS 5463, pp. 168–185, 2009.

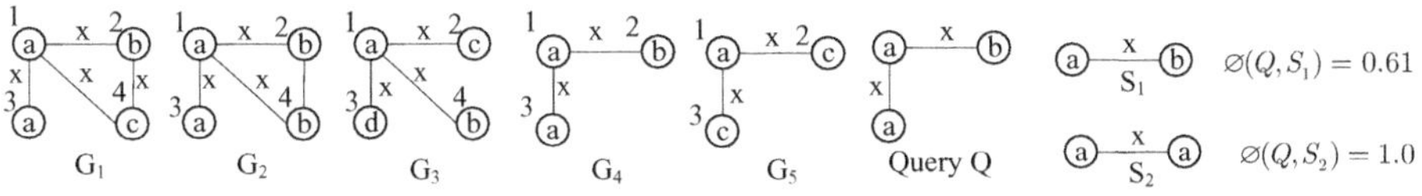

Fig. 1. A Graph Database

sub-graph Q and the threshold $\theta = 1.0$. In Fig. 1, for presentation convenience, we assume that all edges have the same label 'x'. The number beside the vertex is vertex ID. The letter inside the vertex is vertex label. Since only $\phi(Q, S_2) = 1.0 \geq \theta$, we report S_2 in Fig. 1. $\phi(Q, S_i) = 1.0$ means S_i (or Q) always appears in the same data graph g, if Q (or S_i) appears in data graph g. In CGSearch problem, each sub-graph S_i of data graphs in database D is a candidate, which is quite different from (similar) sub-graph search in previous work. In (similar) sub-graph search problem [13,18], it reports all data graphs that (closely) contain query graph Q. Therefore, only data graphs are candidates. Thus, the search space of CGSearch problem is much larger than that of (similar) sub-graph search, which leads CGSearch to be a more challenging task.

Furthermore, CGSearch is also different from frequent sub-graph mining. As we all know, all frequent sub-graph mining algorithms are based on "down-closure property"(that is Apriori property)[**?**] no matter they adopt bread-first search [9] or depth-first search [17]. However, Apriori property does not hold for $\phi(Q, S_i)$. It means that we cannot guarantee that $\phi(Q, S_i) \geq \phi(Q, S_j)$, if S_i is a sub-graph of S_j. Therefore, for CGSearch problem, we cannot employ the same pruning strategies used in frequent sub-graph mining algorithms.

In order to address CGSearch, Ke et al. propose CGS Algorithm [8], which adopts "filter-and-verification" framework to perform the search. First, according to a given minimum correlation coefficient threshold θ and a query graph Q, they derive the minimum support $lowerbound(sup(g))$. Any sub-graph g_i whose support is less than $lowerbound(sup(g))$ cannot be a correlation sub-graph w.r.t [1] query Q. Then, in the projected database (all data graphs containing query graph Q are collected to form the projected database, denoted as D_q), frequent sub-graph mining algorithms are used to find candidates g_i by setting minimum support $\frac{lowerbound(sup(g))}{sup(Q)}$ (that is *filter* process). After that, for each candidate g_i, the correlation coefficient is computed to find all true answers (that is *verification* process).

However, CGSearch in [8] requires users or applications to specify a minimum correlation threshold θ. In practice, it may not be trivial for users to provide an appropriate threshold θ, since different graph databases typically have different characteristics. Therefore, in this paper, we propose an alternative mining task:***top-K correlation sub-graph search***, *TOP-CGSearch*, which is defined:

Given a query graph Q, we report top-K sub-graphs S_i according to the Pearson's correlation coefficient between Q and S_i (that is $\phi(Q, S_i)$).

Obviously, the above methods in CGSearh [8] cannot be directly applied to the TOP-CGSearch problem, since we do not have the threshold θ. Simply setting a large

[1] With regard to.

threshold and then decreasing it step by step based on the number of answers obtained from the previous step is out of practice, since we need to perform frequent sub-graph mining algorithms to obtain candidates in each iterative step. Furthermore, we also need to perform verification of CGSearch [8] in each iterative step.

Can we derive top-K answers directly without performing frequent sub-graph mining algorithms for candidates and expensive verification several times? Our answer in this paper is YES. For TOP-CGSearch problem, we first derive a upper bound of $\phi(Q, S_i)$ (that is $upperbound(\phi(Q, S_i))$). Since $upperbound(\phi(Q, S_i))$ is a monotone increasing function w.r.t.$sup(S_i)$, we develop a novel **p**attern-**g**rowth algorithm, called *PG-search* algorithm to conduct effective filtering through the upper bound. During mining process, we always maintain β to be the K-largest $\phi(Q, S_i)$ by now. The algorithm reports top-K answers until all un-checked sub-graphs cannot be in top-K answers. Furthermore, in pattern-growth process, we only grow a pattern P if and only if we have checked all its parents P_i (see Lemma 2). Therefore, we significantly reduce the search space (the number of sub-graphs that need to be checked in *PG-search* algorithm). Finally, we utilize existing graph indexing techniques [13,18] to speed up the mining task. In summary, we made the following contributions:

1. We propose a new mining task TOP-CGSearch problem. For TOP-CGSearch, we develop an efficient "pattern-growth" algorithm, called PG-search algorithm. In PG-search, we introduce two novel concepts, **G**rowth **E**lement(GE for short) and Correct Growth to determine the correct growth and avoid duplicate generation.
2. To speed up the mining task, we utilize graph indexing techniques to improve the performance in PG-search algorithm.
3. We conduct extensive experiments to verify the efficiency of the proposed methods.

The rest of the paper is organized as follows: Section 2 discuss preliminary background. PG-search algorithm is proposed in Section 3. We evaluate our methods in experiments in Section 4. Section 5 discusses related work. Section 6 concludes this work.

2 Preliminary

2.1 Problem Definition

Definition 1. ***Graph Isomorphism.*** *Assume that we have two graphs* $G_1 < V_1, E_1, L_{1v}, L_{1e}, F_{1v}, F_{1e} >$ *and* $G_2 < V_2, E_2, L_{2v}, L_{2e}, F_{2v}, F_{2e} >$*.* G_1 *is graph isomorphism to* G_2*, if and only if there exists at least one bijective function* $f : V_1 \rightarrow V_2$ *such that: 1) for any edge* $uv \in E_1$*, there is an edge* $f(u)f(v) \in E_2$*; 2)* $F_{1v}(u) = F_{2v}(f(u))$ *and* $F_{1v}(v) = F_{2v}(f(v))$*; 3)* $F_{1e}(uv) = F_{2e}(f(u)f(v))$*.*

Definition 2. ***Sub-graph Isomorphism.*** *Assume that we have two graphs* G' *and* G*, if* G' *is graph isomorphism to at least one sub-graph of* G *under bijective function* f*,* G' *is sub-graph isomorphism to* G *under injective function* f*.*

If a graph S is sub-graph isomorphism to another graph G, we always say that graph G contains S.

Definition 3. ***Graph Support.*** *Given a graph database D with N data graphs, and a sub-graph S, there are M data graphs containing S. The support of S is denoted as $sup(S) = \frac{M}{N}$. Obviously, $0 \leq sup(S) \leq 1$.*

Definition 4. ***Pearsons Correlation Coefficient***
Given two sub-graphs S_1 and S_2, Pearson's Correlation Coefficient of S_1 and S_2, denoted as $\phi(S_1, S_2)$, is defined as:

$$\phi(S_1, S_2) = \frac{sup(S_1, S_2) - sup(S_1)sup(S_2)}{\sqrt{sup(S_1)sup(S_2)(1 - sup(S_1))(1 - sup(S_2))}} \tag{1}$$

Here, $-1 \leq \phi(S_1, S_2) \leq 1$

Given two graphs S_1 and S_2, if $\phi(S_1, S_2) > 0$, then S_1 and S_2 are positively correlated; otherwise, they are negatively correlated. In this paper, we focus on positively correlated sub-graph search.

Definition 5. *(Problem Definition)* ***Top-K Correlation Sub-graph Search*** *(TOP-CGSearch for short) Given a query graph Q and a graph database D, according to Definition 4, we need to find K sub-graphs S_i (i=1...K) in the graph database D, where $\phi(Q, S_i)$ are the first K largest.*

Note that, in this work, we only focus on positively correlated sub-graphs, if there exist less than K positively correlated sub-graphs, we report all positively correlated sub-graphs. Now we demonstrate top-K correlation sub-graph search with a running example.

EXAMPLE 1(Running Example). Given the same graph database D with 5 data graphs and a query graph Q in Fig. 1, we want to report top-2 correlated sub-graphs in D according to Definition 5. In this example, the top-2 answers are S_1 and S_2, where $\phi(Q, S_1) = 1.0$ and $\phi(Q, S_2) = 0.61$ respectively. For the other sub-graphs S_i, $\phi(Q, S_i) < 0.61$.

2.2 Canonical Labeling

In order to determine whether two graphs are isomorphism to each other, we can assign each graph a unique code. Such a code is referred to as **c**anonical label of a graph [4]. Furthermore, benefiting from canonical labels, we can define a total order for a set of graphs in a unique and deterministic way, regardless of the original vertex and edge ordering. The property will be used in Definition 8 and 12. It is known that the hardness of determining the canonical label of a graph is equivalent to determining isomorphism between graphs. So far, these two problems are not known to be either in P or in NP-complete [4]. There are many canonical labeling proposals in the literature [17]. In fact, any canonical labeling method can be used in our algorithm, which is orthogonal to our method. A simple way of defining the canonical label of a graph is to concatenate the upper-triangular entries of graph's adjacency matrix when this matrix has been symmetrically permuted so that the obtained string is the lexicographically smallest over the strings that can be obtained from all such permutations.

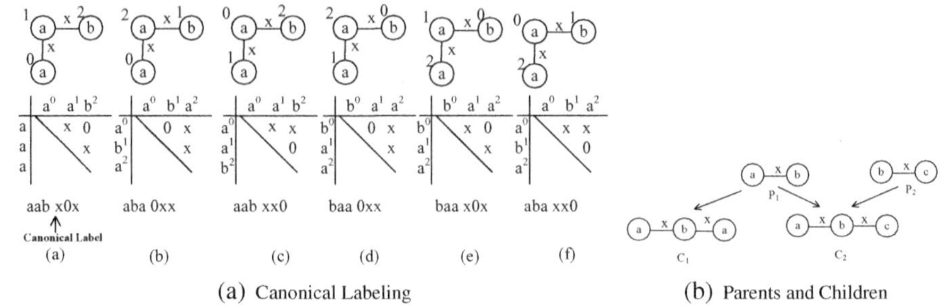

Fig. 2. Canonical Labels and Pattern Growth

Fig. 2(a) shows an example about obtaining the canonical label of a graph. In Fig. 2(a), we permutate all vertex orderings. For each vertex ordering, we can obtain a code by concatenating the upper-triangular entries of the graphs adjacency matrix. Among all codes, the code in Fig. 2(a)a is the lexicographically smallest ('0' is smaller than all letters). Thus, in Fig. 2(a), "aab x0x" is the canonical code of the graph. Fig. 2(a)a shows the canonical form of the graph.

Definition 6. ***Canonical Vertex Ordering and ID.*** *Given a graph S, some vertex ordering is called* **Canonical Vertex Ordering**, *if and only if this ordering leads to the canonical code of graph S. According to canonical vertex ordering, we define* **Canonical Vertex ID** *for each vertex in S.*

The vertex ordering in Fig. 2(a)a leads to canonical code, thus, the vertex ordering in Fig. 2(a)a is called *canonical vertex ordering*. Vertex ID in Fig. 2(a)a is called *canonical vertex ID*.

2.3 Pattern Growth

Definition 7. ***Parent and Child.*** *Given two* **connected** *graphes P and C, P and C have n and $n+1$ edges respectively. If P is a sub-graph of C, P is called a* **parent** *of C, and C is called a* **child** *of P.*

For a graph, it can have more than one child or parent. For example, in Fig. 2(b), C_2 has two parents that are P_1 and P_2. P_1 have two children that are C_1 and C_2.

During mining process, we always "grow" a sub-graph from its parent, which is called **pattern growth**. During pattern growth, we introduce an extra edge into a parent to obtain a child. The introduced extra edge is called **Growth Element** (GE for short). The formal definition about GE is given in Section 3.1.

3 PG-Search Algorithm

As mentioned in the Section 1, we can transfer a TOP-CGSearch problem into a threshold-based CGSearch. First, we set threshold θ to be a large value. According to CGS algorithm [8], if we can find the correct top-K correlation sub-graphs, then we report them and terminate the algorithm. Otherwise, we decrease θ and repeat the above process until we obtain the top-K answers. We refer this method as "Decreasing Threshold-Based CGS" in the rest of this paper. Obviously, when K is large, we have to set θ to be a small value to find the correct top-K answers, which is quite inefficient since we need to perform CGS algorithm [8] several times and there are more candidates needed to be investigated. In order to avoid this shortcoming, we propose a **p**attern-**g**rowth **search** algorithm (PG-search for short) in this section.

3.1 Top-K Correlation Sub-graph Query

The search space (the number of candidates) of TOP-CGSearch problem is large, since each sub-graph S_i in graph database is a candidate. Obviously, it is impossible to enumerate all sub-graphs in the database, since it will result in a combinational explosion problem. Therefore, an effective pruning strategy is needed to reduce the search space. In our algorithm, we conduct effective filtering through the upper bound. Specifically, if the largest upper bound for all un-checked sub-graphs cannot get in the top-K answers, the algorithm can stop. Generally speaking, our algorithms work as follows: We first generate all size-1 sub-graph (the size of a graph is defined in terms of number of edges, i.e. $|E|$) S_i, and insert them into heap H in non-increasing order of $upperbound(\phi(Q, S_i))$, which is the upper bound of $\phi(Q, S_i)$. The upper bound is a monotone increasing function w.r.t $sup(S_i)$. The head of the heap H is h. We set $\alpha = upperbound(\phi(Q, h))$. We compute $\phi(Q, h)$ and insert h into result set RS. β is set to be the Kth largest value in RS by now. Then, we delete the heap head h, and find all children C_i(see Definition 7) of h through pattern growth. We insert these C_i into heap H. For the heap head h, we set $\alpha = upperbound(\phi(Q, h))$. We also compute $\phi(Q, h)$, and then insert h into result set RS. The above processes are iterated until $\beta \geq \alpha$. At last, we report top-K answers in RS.

In order to implement the above algorithm, we have to solve the following technical challenges: 1)how to define $upperbound(\phi(Q, S_i))$; 2)how to find all h's children; 3)Since a child may be generated from different parents (see Fig. 2(b)), how to avoid generating duplicate patterns; and 4)how to reduce the search space as much as possible.

Lemma 1. *Given a query graph Q and a sub-graph S_i in the graph database, the upper bound of $\phi(Q, S_i)$ is denoted as:*

$$\phi(Q, S_i) \leq upperbound(\phi(Q, S_i)) = \sqrt{\frac{1 - sup(Q)}{sup(Q)}} * \sqrt{\frac{sup(S_i)}{1 - sup(S_i)}} \quad (2)$$

Proof. Obviously, $sup(Q, S_i) \leq sup(S_i)$, thus:

$$\begin{aligned}
\phi(Q, S_i) &= \frac{sup(Q,S_i) - sup(Q)sup(S_i)}{\sqrt{sup(Q)sup(S_i)(1-sup(Q))(1-sup(S_i))}} \\
&\leq \frac{sup(S_i) - sup(Q)sup(S_i)}{\sqrt{sup(Q)sup(S_i)(1-sup(Q))(1-sup(S_i))}} \\
&= \frac{sup(S_i)(1-sup(Q))}{\sqrt{sup(Q)sup(S_i)(1-sup(Q))(1-sup(S_i))}} \\
&= \sqrt{\frac{1-sup(Q)}{sup(Q)}} * \sqrt{\frac{sup(S_i)}{1-sup(S_i)}}
\end{aligned}$$

Theorem 1. *Given a query graph Q and a sub-graph S_i in a graph database D, the upper bound of $\phi(Q, S_i)$ (that is $upperbound(\phi(Q, S_i))$) is a monotone increasing function w.r.t $sup(S_i)$.*

Proof. Given two sub-graphs S_1 and S_2, where $sup(S_1) > sup(S_2)$, we can know that the following formulation holds.

$$\frac{upperbound(\phi(Q, S_1))}{upperbound(\phi(Q, S_2))} = \sqrt{\frac{sup(S_1)}{sup(S_2)}}\sqrt{\frac{1 - sup(S_2)}{1 - supp(S_1)}} > 1$$

Therefore, Theorem 1 holds.

Property 1. (Apriori Property) Given two sub-graph S_1 and S_2, if S_1 is a parent of S_2 (see Definition 7), then $Sup(S_2) < Sup(S_1)$.

Lemma 2. *Assume that β^* is the K-largest $\phi(Q, S_i)$ in the final results. Given a sub-graph S with $|S| > 1$, the necessary condition for $upperbound(\phi(Q, S)) > \beta^*$ is that, for all of its parents P_i, $upperbound(\phi(Q, P_i)) > \beta^*$.*

Proof. According to Property 1, we know that $Sup(S) \leq Sup(P_i)$. Since $upperbound$ $(\phi(Q, S))$ is a monotone increasing function w.r.t $sup(S)$, it is clear that $upperbound$ $(\phi(Q, P_i)) \geq upperbound(\phi(Q, S)) > \beta^*$. Therefore, Lemma 2 holds.

According to Lemma 2, we only need to check a sub-graph, if and only if we have checked all its parents P_i. It means that we can adopt "pattern-growth" framework in the mining process, that is to generate and check a size-(n+1) sub-graph S from its parent P_i after we have verified all S's parents. Furthermore, according to monotone increasing property of $upperbound(\phi(Q, S))$, we always "first" grow a sub-graph P with the highest support. We refer the method as "best-first". In order to combine "pattern-growth" and "best-first" in the mining process, we define Total Order in the following definition.

Definition 8. ***(Total Order)*** *Given two sub-graphs S_1 and S_2 in graph database D, where S_1 is not isomorphism to S_2, we can say that $S_1 < S_2$ if either of the following conditions holds:*

1) $sup(S_1) > sup(S_2)$;
2) $sup(S_1) == sup(S_2)$ AND $size(S_1) < size(S_2)$, where $size(S_1)$ denotes the number of edges in S_1.
3) $sup(S_1) == sup(S_2)$ AND $size(S_1) == size(S_2)$ AND $label(S_1) < label(S_2)$, where $label(S_1)$ and $label(S_2)$ are the canonical labeling of sub-graphs S_1 and S_2 respectively.

Furthermore, as mentioned in Section 2, we only focus on searching positively correlated graphs in this paper. Thus, according to Definition 4, if $\phi(Q, S_i) > 0$, then $sup(Q, S_i) > 0$, which indicates that Q and S_i must appear together in at least one graph in D. Therefore, we only need to conduct the correlation search over all data graphs that contain query graph Q, which is called *projected graph database* and denoted as D_Q. A projected database of the running example is given in Fig. 3(a). Similar with [8], we focus the mining task on the projected database D_Q. It means that we always generate candidate sub-graphs from the projected graph database through pattern growth. Since $|D_Q| < |D|$, it is efficient to perform mining task on D_Q.

At the beginning of PG-Search algorithm, we enumerate all size-1 sub-graph S_i in the projected graph database and insert them into the heap H in increasing order of *total order*. We also record their occurrences in each data graph. Fig. 3(b) shows all size-1 sub-graph S_i in the projected graph database in the running example. "$G_1 < 1, 2 >$" in Fig. 3(b) shows that there is one occurrence of S_1 in data graph G_1 and the corresponding vertex IDs are 1 and 2 in graph G_1 (see Fig. 1). In Fig. 3(b), the head of the heap H is sub-graph S_1. Therefore, we first grow S_1, since it has the highest support. For each occurrence of S_1 in the projected database D_Q, we find all **G**rowth **E**lements (GE for short) around the occurrence.

Definition 9. ***Growth Element.*** *Given a connected graph C having n edges and another connected graph P having $n-1$ edges, P is a sub-graph of C (namely, C is a child of P). For some occurrence O_i of P in C, we use $(C \setminus O_i)$ to represent the edge in C that is not in O_i. $(C \setminus O_i)$ is called* ***G****rowth* ***E****lement (GE for short) w.r.t O_i. Usually, GE is represented as a five-tuple $(< (vid1, vlable1), (elabel), (vid2, vlabel2) >)$, where $vid1$ and $vid2$ are* ***canonical vertex IDs*** *(defined in Definition 6) of P ($vid1 < vid2$). $vlable1$ and $vlable2$ are corresponding vertex labels, and $elabel$ is the edge label of GE. Notice that $vid2$ can also be "*", and "*" means the additional vertex for P.*

To facilitate understanding GE, we illustrate it with an example. Given a sub-graph S_1 in Fig. 4(a), there is one occurrence of S_1 in data graph G_1, which is denoted as the shaded area in Fig. 4(a). For the occurrence, there are three GEs that are also shown in Fig. 4(a). Growing S_1 (parent) through each GE, we will obtain another size-2 sub-graph (child) in Fig. 4(b). The numbers beside vertexes in each sub-graph S_i are *canonical vertex IDs* in S_i. The numbers beside vertexes in data graph G are just vertex IDs, which are assigned arbitrarily. Notice that, canonical vertex IDs in a parent may not be preserved in its child. For example, in Fig. 4(b), the canonical vertex IDs in S_1 are different from that in S_6. Given a size-(n+1) sub-graph S_i (child), it can be obtained by growing from different size-n sub-graphs (parents). For example, in Fig. 5(a)a, we can obtain sub-graph S_6 by growing S_1 or S_5. Obviously, duplicate patterns will decrease

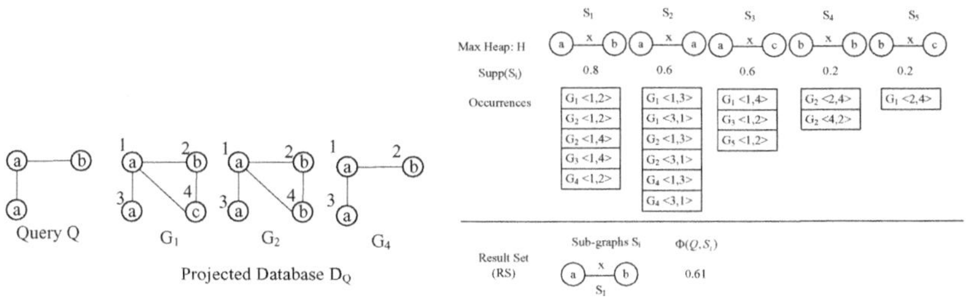

(a) Projected Database D_Q (b) The First Step in Mining Process

Fig. 3. Projected Database and The First Step

the performance of mining algorithm. Therefore, we propose the following definition about *correct growth* to avoid the possible duplicate generation.

Definition 10. ***Correct Growth****. Given a size-(n+1) sub-graph C(child), it can be obtained by growing some size-n sub-graphs P_i (parent), i=1...m. Among all growthes, the growth from P_j to C is called* **correct growth***, where P_j is the largest one among all parents according to Total Order in Definition 8. Otherwise, the growth is incorrect one.*

For example, in Fig. 5(a), S_9 has two parents, which are S_1 and S_5. Since S_5 is larger than S_1 (see Definition 8), the growth from S_5 to S_6 is a correct growth, and the growth from S_1 to S_6 is an incorrect growth. Obviously, according to Definition 8, we can determine whether the growth from P to C is correct or incorrect. We will propose another efficient method to determine the correct growth in Section 3.2.

Now, we illustrate the mining process with the running example. In the running example, we always maintain β to be the K largest $\phi(Q, S_i)$ by now ($K = 2$ in the running example). Initially, we set $\beta = -\infty$. First, we conduct sub-graph query to obtain projected graph database D_Q, which is shown in Fig. 3(a). Then, we find all size-1 sub-graphs (having one edge) in the projected database D_Q. In Fig. 3(b), we insert them into heap H in increasing order according to Total Order in Definition 8. We also record all their occurrences in the graph database. Since S_1 is the head of heap H, we compute $\alpha = \mathit{upperbound}(\phi(Q, S_1)) = 1.63$ and $\phi(Q, S_1) = 0.61$. We inset S_1 into result set RS. Now, $\beta = -\infty < \alpha$, the algorithm continues.

We find all growth elements (GEs for short) around S_1 in D_Q, which are shown in Fig. 5(b). Since G_3 is not in the projected database, we do not consider the occurrences in G_3 when we find GEs. For each GE, we check whether it leads to a correct growth. In Fig. 5(b), only the last one is a correct growth. We insert the last one (that is S_{10}) into the max-heap H, as shown in Fig. 6. Now, the sub-graph S_2 is the heap head. We re-compute $\alpha = \mathit{upperbound}(\phi(Q, S_2)) = 1.0$, $\phi(Q, S_2) = 1.0$, and insert S_2 into RS. We update β to be Top-2 answer, that is $\beta = 0.61$. Since $\beta < \alpha$, the algorithm still continues. The above processes are iterated until $\beta \geq \alpha$. At last, we report top-2 answers in RS.

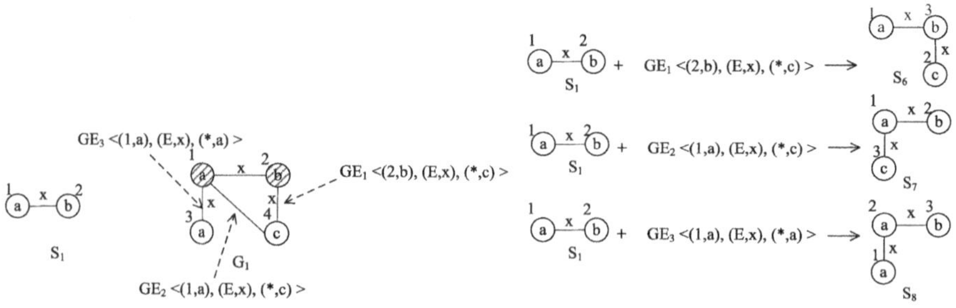

(a) How to Find GE (b) Grow Sub-graph by GE

Fig. 4. Finding GE and Grow Sub-graph by GE

3.2 PG-Search Algorithm

As discussed in Section 3.1, the heap H is ranked according to Total Order in Definition 8 (see Fig. 3(b) and 6). In Definition 8, we need $sup(S_i)$. Therefore, we maintain all occurrences of S_i in heap H. According to the occurrence list, it is easy to obtain $sup(S_i)$. However, it is expensive to maintain all occurrences in the whole database D, especially when $|D|$ is large. On the other hand, we need to find GE around each occurrence in the projected database D_Q (see Fig. 5(b)). Therefore, we need to maintain occurrences in the projected database. Furthermore, the projected database D_Q is always smaller than the whole database D. In order to improve the performance, in our PG-search algorithm, we only maintain the occurrence list in the projected database instead of the whole database.

However, Definition 8 needs $sup(S_i)$. Therefore, we propose to use $indexsup(S_i)$ to define Total Order instead of $sup(S_i)$, where $indexsup(S_i)$ is derived from sub-graph query technique.

Recently, a lot of graph indexing structures have been proposed in database literature [13,18] for sub-graph query. Sub-graph query is defined as: *given a sub-graph Q, we report all data graphs containing Q as a sub-graph from the graph database.* Because sub-graph isomorphism is a NP-complete problem [4], we always employ a *filter-and-verification* framework to speed up the search process. In filtering process, we identify the candidate answers by graph indexing structures. Then, in verification process, we check each candidate by sub-graph isomorphism. Generally speaking, the filtering process is much faster than verification process.

Definition 11. ***Index Support.*** *For a sub-graph S_i, index support is size of candidates in filtering process of sub-graph S_i query, which is denoted as $indexsup(S_i)$.*

Actually, $sup(S_i)$ is the size of answers for sub-graph S_i query, which is obtained in verification process. According to graph indexing techniques [13,18], we can identify the candidates without false negatives. Therefore, Lemma 3 holds.

Lemma 3. *For any sub-graph S_i in the graph database, $sup(S_i) \leq indexsup(S_i)$, where $indexsup(S_i)$ is the size of candidates for sub-graph S_i query.*

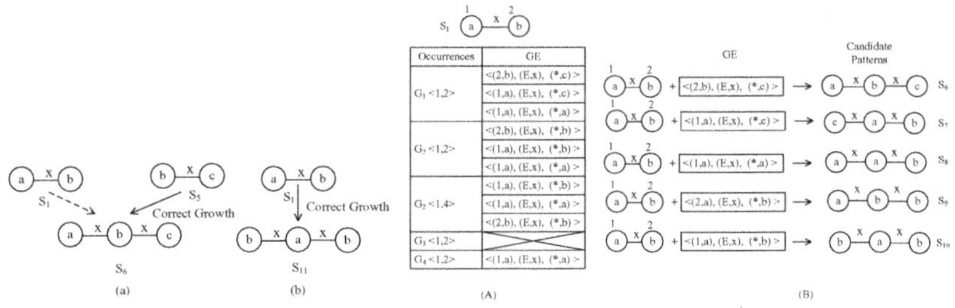

(a) Correct Growth (b) GEs for S_1

Fig. 5. Correct Growth and S_1's Growth Elements

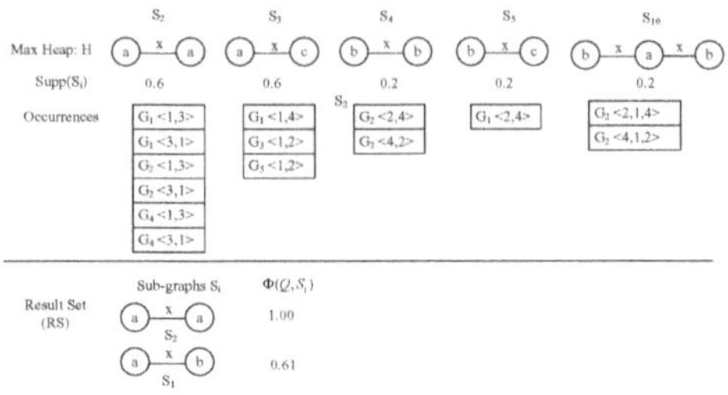

Fig. 6. The Second Step in Mining Process

Furthermore, $indexsup(S_i)$ also satisfy Apriori Property. For example, in gIndex [18], if Q_1 is a sub-graph of Q_2, the candidates for Q_2 query is a subset of candidates for Q_1. It means the following property holds.

Property 2. (Apriori Property) Given two sub-graph S_1 and S_2, if S_1 is a parent of S_2 (see Definition 7), then $indexsup(S_2) < indexsup(S_1)$.

Lemma 4. *Given a query graph Q and a sub-graph S_i in the graph database, the upper bound of $\phi(Q, S_i)$ is denoted as:*

$$\begin{aligned} \phi(Q, S_i) &\le upperbound2(\phi(Q, S_i)) \\ &= \sqrt{\frac{1-sup(Q)}{sup(Q)}} * \sqrt{\frac{indexsup(S_i)}{1-indexsup(S_i)}} \end{aligned} \tag{3}$$

Proof. According to Lemma 1 and 3:

$$\begin{aligned} \phi(Q, S_i) &\le \sqrt{\frac{1-sup(Q)}{sup(Q)}} * \sqrt{\frac{\sup p(S_i)}{1-\sup p(S_i)}} \\ &\le \sqrt{\frac{1-\sup p(Q)}{sup(Q)}} * \sqrt{\frac{index \sup p(S_i)}{1-index \sup p(S_i)}} \end{aligned}$$

Theorem 2. *Given a query graph Q and a sub-graph S_i in a graph database D, the upper bound of $\phi(Q, S_i)$ (that is $upperbound2(\phi(Q, S_i))$) is a monotone increasing function w.r.t $indexsup(S_i)$.*

Proof. It is omitted due to space limited.

In order to use graph index in the PG-Search algorithm, we need to re-define Total Order and Correct Growth according to *Index Support* as follows.

Definition 12. ***(Total Order)*** *Given two sub-graphs S_1 and S_2 in graph database D, where S_1 is not isomorphism to S_2, we can say that $S_1 < S_2$ if either of the following conditions holds:*

1) $indexsup(S_1) > indexsup(S_2)$;
2) $indexsup(S_1) == indexsup(S_2)$ AND $size(S_1) < size(S_2)$, where $size(S_1)$ denotes the number of edges in S_1.
3) $indexsup(S_1) == indexsup(S_2)$ AND $size(S_1) == size(S_2)$ AND $label(s_1) < label(s_2)$, where $label(S_1)$ and $label(S_2)$ are the canonical labeling of sub-graphs S_1 and S_2 respectively.

Lemma 5. *According to Definition 12, if sub-graphs* $S_1 < S_2$*, then* $upperbound2(\phi(Q, S_1)) \geq upperbound2(\phi(Q, S_2))$.

Definition 13. ***Correct Growth****. Given a size-(n+1) sub-graph C(child), it can be obtained by growing some size-n sub-graphs* P_i *(parent),* i*=1...*m*. Among all growthes, the growth from* P_j *to* C *is called the* **correct growth***, where* P_j *is the largest parent among all parents according to Total Order in Definition 12. Otherwise, the growth is incorrect one.*

Algorithm 1. PG-search algorithm

Require: Input: a graph database D and a query graph Q and K
Output: the top-K correlated sub-graph.

1: We conduct sub-graph Q search to obtain projected graph database D_Q.
2: Find all size-1 sub-graph S_i (having one edge), and insert them in max-heap H in order of Total Order in Definition 8. We also main their occurrences in projected graph database D_Q.
3: Set threshold $\beta = -\infty$.
4: For heap head h in max-heap H, we perform sub-graph query to obtain $sup(h)$. Then, we set $\alpha = upperbound2(\phi(h, Q))$ where $upperbound2(\phi(h, Q))$ is computed according to Equation 3.
5: For heap head h in max-heap H, we compute $\phi(h, Q)$ and insert it into answer set RS in non-increasing order of $\phi(h, Q)$.
6: Update β to be the Top-K answer in RS.
7: **while** ($\beta < \alpha$) **do**
8:     Pop the current head h into the table T.
9:     Find all GEs around the occurrences of h in the projected graph D_Q.
10:     **for** each GE g **do**
11:         Assume that we can get a sub-graph C through growing h by growth element g.
12:         **if** The growth from h to C is correct growth **then**
13:             Insert P into max-heap H according to Total Order in Definition 12. We also maintain the occurrence list for C in the projected graph database D_Q.
14:         **else**
15:             continue
16:     Set $\alpha = upperbound2(\phi(Q, h))$, where h is the new head in H.
17:     For the new heap head h in max-heap H, we perform sub-graph query to obtain $sup(h)$, and compute $\phi(h, Q)$. Then insert it into answer set RS in non-increasing order of $\phi(h, Q)$.
18:     Update β to be the Top-K answer in RS.
19: Report top-K results in RS

We show the pseudo codes of PG-search algorithm in Algorithm 1. In PG-search algorithm, first, given a query Q, we obtain projected graph database D_Q by performing sub-graph Q query (Line 1). Then, we find all size-1 sub-graphs S_i(having one edge) in D_Q, and insert them into max-heap H in increasing order according to Definition 12. We also need to maintain the occurrences of sub-graphs S_i in the projected database D_Q(Line 2). The threshold β is recorded as the top-K $\phi(Q, S_i)$ in result set RS by now. Initially, we set $\beta = -\infty$ (Line 3). We set $\alpha = upperbound2(Q, h)$, where h is the heap head in H (Line 4). For heap head h in max-heap H, we perform sub-graph search to obtain $sup(h)$, and compute $\phi(h, Q)$. We also insert h into result set RS (Line 5). Then, we update β to be the Top-K answer in RS (Line 6). If $\beta \geq \alpha$, the algorithm reports top-K results (Line 7). Otherwise, we pop the heap head h (Line 8). We find all GEs with regard with h in the projected graph database D_Q (Line 9). For each GE g, if growing h through g is not a correct growth, it is ignored (Line 15). If the growth is correct, we obtain a new sub-graph C. We insert it into heap H according to Total Order in Definition 12 (Line 13). Then, we set $\alpha = upperbound2(\phi(Q, h))$, where h is the new head in H (Line 16). For the new heap head h, we perform sub-graph search to obtain $sup(h)$ and compute $\phi(h, Q)$. We insert h into result set RS (Line 17). Update β to be the Top-K answer in RS (Line 18). We iterate Lines 7-21 until $\beta \geq \alpha$ (Line 7). As last, we report top-K answers (Line 19).

Notice that, in Line 4 and 16 of PG-search algorithm, we utilize graph indexing technique to support sub-graph query to obtain $sup(h)$. In Line 13, we also utilize indexes to obtain $indexsup(h)$, which is obtained in filtering process.

In order to determine the growth from h to C is correct or not (Line 12), we need to generate all parents P_i of h, and then perform sub-graph P_i query (only filtering process) to obtain $indexsup(P_i)$. At last, we determine whether h is the largest one among all P_i according to Definition 12. Obviously, the above method leads to numerous sub-graph query operations. We propose another efficient method for the determination as follows:

As we know, at each mining step, we always pop the heap head h_i from the heap H (Line 8). We can maintain all h_i into a table T, which is maintained in memory. For a growth from size-n sub-graph P to size-$(n+1)$ sub-graphs C, we generate all parents P_i of C. If all parents P_i except for P have existed in the table T, the parent P must be the largest one among all parents P_i (see Lemma 8 and proof). Therefore, the growth from P to C is correct one. Otherwise, the growth is incorrect. In the method, we only perform graph isomorphism test in table T. Furthermore, graph canonical labeling will facilitate the graph isomorphism. For example, we can record canonical labels of all sub-graphs in table T. The hash table built on these canonical labels will speed up the determination.

We first discuss Lemma 6 and Lemma 7, which are used to prove the correctness of Lemma 8 and Theorem 3.

Lemma 6. *Assume that h_1 and h_2 are heads of heap H (Line 8) in PG-Search algorithm in $i-$th and $(i+1)-$th iteration step (Lines 8-18). According to Total Order in Definition 12, we know that $h_1 < h_2$.*

Proof. 1)If h_2 exists in heap H in $i-$th iteration step: Since the heap H is in increasing order according to Definition 12 and h_1 is the heap head in $i-$th iteration step, it is straightforward to know $h_1 < h_2$.
2)If h_2 does not exist in heap H in $i-$th iteration step: According to PG-search algorithm, h_2 must be generated by growing the heap head h_1 in $i-$step. It means that h_1 is a parent of h_2. According to Property 2, $indexsup(h_1) \geq indexsup(h_2)$. Furthermore, $size(h_1) < size(h_2)$. Therefore, according to Definition 12, we know that $h_1 < h_2$.

Total Order in Definition 12 is "transitive ", thus, Lemma 7 holds.

Lemma 7. *Assume that h_1 and h_2 are heads of heap H in PG-Search Algorithm in $i-$th and $j-$th iteration step (Lines 8-18), where $i < j$. According to Definition 12, we know that $h_1 < h_2$.*

As discussed before, in PG-search algorithm, we maintain the heap head h of each step into the table T. The following lemma holds.

Lemma 8. *For a growth from size-n sub-graph P to size-$(n+1)$ sub-graphs C, we generate all parents P_i of C. If all parents P_i except for P have existed in the table T, the parent P must be the largest one among all parents P_i.*

Proof. (Sketch) Since all parents P_i except for P have existed in the table T, it means all P_i exist before P in heap H. According to Lemma 7, we know $P_i < P$. Therefore, P is the largest one among all parents P_i.

Theorem 3. ***Algorithm Correctness***. *PG-search algorithm can find the correct top-K correlated sub-graphs.*

Proof. (Sketch) Assume that $\beta \geq \alpha$ at nth iteration step of PG-search algorithm, where $\alpha = upperbound2(\phi(Q, h))$ and h is the heap head at nth step. According to Lemma 7, $h' < h$, where h' is the heap head at jth step, where $j > n$. According to Lemma 5, we know that $upperbound2(\phi(Q, h')) < upperbound2(\phi(Q, h)) = \alpha$. Since $\beta \geq \alpha$, $\alpha > upperbound2\ (\phi(Q, h'))$. It means that, in latter iteration steps, we cannot find top-K answers. Therefore, PG-search algorithm can find all correct top-K answers, if the algorithm terminates when $\beta \geq \alpha$ (Line 7). The correctness of PG-search algorithm is proved.

4 Experiments

In this section, we evaluate our methods in both real dataset and synthetic dataset. As far as we know, there is no existing work on top-K correlation sub-graph search problem. In [8], Ke et al. proposed threshold-based correlation sub-graph search. As discussed in Section 1, we can transfer top-K correlation sub-graph search into threshold-based one with decreasing threshold. In the following section, we refer the method as "Decreasing Threshold-Based CGS"(DT-CGS for short). We implement "DT-CGS" by ourselves and optimize it according to [8]. All experiments coded by ourself are implemented by standard C++ and conducted on a P4 1.7G machine of 1024M RAM running Windows XP.

1) Datasets. We use both real dataset (i.e. AIDS dataset) and synthetic daset for performance evaluation.
AIDS Dataset. This dataset is available publicly on the website of the Developmental Therapeutics Program. We generate 10,000 connected and labeled graphs from the molecule structures and omit Hydrogen atoms. The graphs have an average number of 24.80 vertices and 26.80 edges, and a maximum number of 214 vertices and 217 edges. A major portion of the vertices are C, O and N. The total number of distinct vertex labels is 62, and the total number of distinct edge labels is 3. We refer to this dataset as

AIDS dataset. We randomly generate four query sets, F_1, F_2, F_3 and F_4, each of which contains 100 queries. The support ranges for the queries in F_1 to F_4 are [0.02, 0.05], (0.05, 0.07], (0.07, 0.1] and (0.1, 1) respectively.

Synthetic Dataset. The synthetic dataset is generated by a synthetic graph generator provided by authors of [9]. The synthetic graph dataset is generated as follows: First, a set of *S* seed fragments are generated randomly, whose size is determined by a Poisson distribution with mean *I*. The size of each graph is a Poisson random variable with mean *T*. Seed fragments are then randomly selected and inserted into a graph one by one until the graph reaches its size. Parameter *V* and *E* denote the number of distinct vertex labels and edge labels respectively. The cardinality of the graph dataset is denoted by *D*. We generate the graph database using the following parameters with gIndex in [18]: D=10,000, S=200, I=10, T=50, V=4, E=1.

2) Experiment 1. Benefiting from graph indexing techniques, we can only maintain the occurrences in the projected database, as discussed in Section 3.2. In the experiment, we evaluate the indexing technique in PG-search. We use GCoding indexing technique [20] in the experiment. First, in Fig. 7(a) and 7(c), we fix the query sets F_1 and vary K from 10 to 100. Fig. 7(a) and Fig. 7(c) show the running time and memory consumption respectively. Second, in Fig. 7(b) and 7(d), we fix K to be 50 and use different query sets F_1 to F_4. In "PG-search without Index", we need to maintain the occurrence list in the whole database. Therefore, it needs more memory consumption, which is shown in Fig. 7(c) and 7(d). Observed from Fig. 7(a) and 7(b), it is clear that "PG-search" is faster than "PG-search without Index".

3) Experiment 2. In the experiment, we compare PG-search with decreasing threshold-based CGS (DT-CGS for short). We fix query sets F_1 and vary K from 2 to 100. Fig. 9(a) reports running time. We also report the final threshold β for top-K query in Fig. 9(b). Observed from Fig. 9(a), PG-search is faster than DT-CGS in the most cases. In DT-CGS, it is difficult to set an appropriate threshold θ. Therefore, in the experiments, we always first set $\theta = 0.9$, and then decrease it by "0.02" in each following step based on the number of answers obtained from the previous step. For example, in Fig. 9(a), when $K = 2, 5$, the running time in DT-CGS keeps the same, because the final threshold are $\beta = 1.0$ and $\beta = 0.95$ respectively (see Fig. 9(b)). It means that we can find the top-2 and top-5 answers, if we set $\theta = 0.9$ in the first step in DT-CGS. As the increasing of K, the performance of DT-CGS decreases greatly. Fig. 9(b) explains the cause. When K is large, the final threshold β is small. Thus, we have to perform CGS algorithm [8] many times according to decreasing threshold θ. Thus, it leads to more candidates in DT-CGS.

Furthermore, we observed that the increasing trend of running time in PG-search is much slower than that in DT-CGS algorithm. In PG-search algorithm, we need to compute $\phi(Q, h)$ in each iteration step and h is the head of heap H. Therefore, we can regard h as a candidate answer. Fig. 9(c) reports candidate size in PG-search and DT-CGS algorithm. Observed from Fig. 9(c), we also find the increasing trend of candidate size in PG-Search is much slower than that in DT-CGS. We have the similar results on other query sets in the experiments. Due to space limit, we do not report them here. We also compare PG-search with DT-CGS on synthetic datasets in Fig. 9. The experiment results also confirm the superiority of our method.

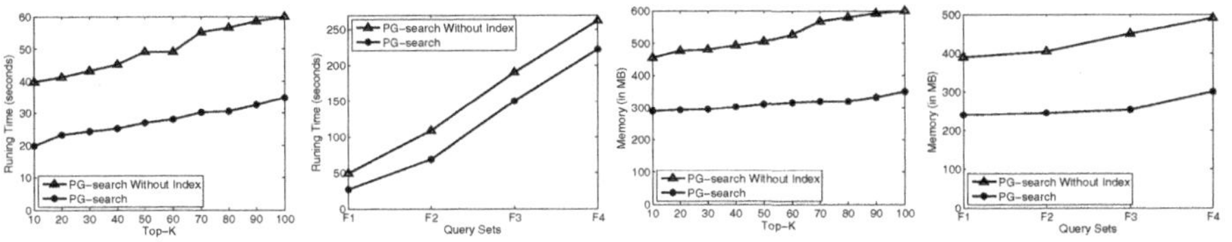

(a) Running Time in F_1 Query Sets (b) Running Time of Top-50 Query in different Query Sets (c) Memory Consumption in F_1 Query Sets (d) Memory Consumption of Top-50 Query in different Query Sets

Fig. 7. Evaluating Indexing Technique in PG-search Algorithm in AIDS datasets

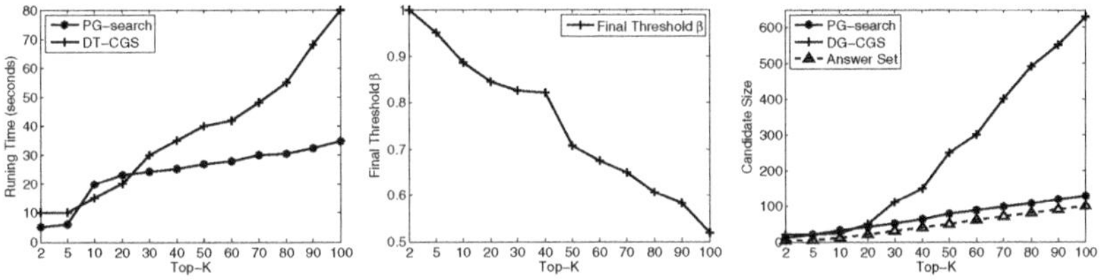

(a) Running Time in F_1 Query Sets (b) Final Threshold β (c) Candidate Size

Fig. 8. PG-search VS. Decreasing Threshold-based CGS on AIDS Dataset

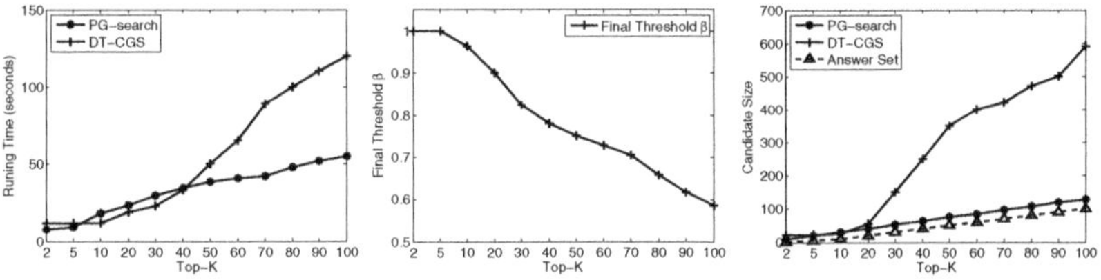

(a) Running Time in F_1 Query Sets (b) Final Threshold β (c) Candidate Size

Fig. 9. PG-search VS. Decreasing Threshold-based CGS on Synthetic Dataset

5 Related Work

Given a query graph Q, retrieving related graphs from a large graph database is a key problem in many graph-based applications. Most existing work is about sub-graph search [13,18]. Due to NP-complete hardness of sub-graph isomorphism, we have to employ *filtering-and-verification* framework. First, we use some pruning strategies to filter out false positives as many as possible, second, we perform sub-graph isomorphism for each candidate to fix the correct answers. Furthermore, similar sub-graph search is also important, which is defined as: *find all graphs that "closely contain" the query graph Q.*

However, the above similar subgraph search is structure-based. Ke et al. first propose correlation mining task in graph database, that is correlation graph search (CGSearch for short)[8]. CGSearch can be regarded as *statistical similarity* search in graph database, which is different from structural similarity search. Therefore, CGSearch provides an orthogonal search problem of structural similarity search in graph database. CGSearch in [8] requires the specification of a minimum correlation threshold θ to perform the computation. In practice, it may be not trivial for users to provide an appropriate threshold θ, since different graph databases typically have different characteristics. Therefore, we propose an alternative mining task: ***top-K** correlation sub-**g**raph **search**, TOP-CGS*.

Correlation mining is always used to discovery the underlying dependency between objects. The correlation mining task finds many applications, such as market-basket databases [2,10], multimedia databases [11]. In [16], Xiong et al. propose an all-strong-pairs correlation query in a market-basket database. In [6], Ilyas et al. use correlation discovery to find soft functional dependencies between columns in relational database.

6 Conclusions

In this paper, we propose a novel graph mining task, called as TOP-CGS. To address the problem, we propose a **p**attern-**g**rowth algorithm, called PG-search algorithm. In PG-search, we first grow the sub-graph S_i with the highest support and maintain the threshold β to be the K-largest correlation value $\phi(Q, S_i)$ by now. We increase β in need until we find the correct top-K answers. Furthermore, in order to improve the performance, we utilize graph indexing technique to speed up the mining task. Extensive experiments on real and synthetic datasets confirm the efficiency of our methods.

References

1. Agrawal, R., Srikant, R.: Fast algorithms for mining association rules in large databases. In: VLDB (1994)
2. Brin, S., Motwani, R., Silverstein, C.: Beyond market baskets: Generalizing association rules to correlations. In: SIGMOD (1997)
3. Cai, D., Shao, Z., He, X., Yan, X., Han, J.: Community mining from multi-relational networks. In: Jorge, A.M., Torgo, L., Brazdil, P.B., Camacho, R., Gama, J. (eds.) PKDD 2005. LNCS, vol. 3721, pp. 445–452. Springer, Heidelberg (2005)
4. Fortin, S.: The graph isomorphism problem. Department of Computing Science, University of Alberta (1996)
5. He, H., Singh, A.K.: Closure-tree: An index structure for graph queries. In: ICDE (2006)
6. Ilyas, I.F., Markl, V., Haas, P.J., Brown, P., Aboulnaga, A.: Cords: Automatic discovery of correlations and soft functional dependencies. In: SIGMOD (2004)
7. Inokuchi, A., Washio, T., Motoda, H.: An apriori-based algorithm for mining frequent substructures from graph data. In: Zighed, D.A., Komorowski, J., Żytkow, J.M. (eds.) PKDD 2000. LNCS, vol. 1910, pp. 13–23. Springer, Heidelberg (2000)
8. Ke, Y., Cheng, J., Ng, W.: Correlation search in graph databases. In: SIGKDD (2007)
9. Kuramochi, M., Karypis, G.: Frequent subgraph discovery. In: ICDM (2001)
10. Omiecinski, E.: Alternative interest measures for mining associations in databases. IEEE TKDE 15(1) (2003)
11. Pan, J.-Y., Yang, H.-J., Faloutsos, C., Duygulu, P.: Automatic multimedia cross-modal correlation discovery. In: KDD (2004)

12. Petrakis, E.G.M., Faloutsos, C.: Similarity searching in medical image databases. IEEE Transactions on Knowledge and Data Enginnering 9(3) (1997)
13. Shasha, D., Wang, J.T.-L., Giugno, R.: Algorithmics and applications of tree and graph searching. In: PODS (2002)
14. Willett, P.: Chemical similarity searching. J. Chem. Inf. Comput. Sci. 38(6) (1998)
15. Xiong, H., Brodie, M., Ma, S.: Top-cop: Mining top-k strongly correlated pairs in large databases. In: Perner, P. (ed.) ICDM 2006. LNCS, vol. 4065. Springer, Heidelberg (2006)
16. Xiong, H., Shekhar, S., Tan, P.-N., Kumar, V.: Exploiting a support-based upper bound of pearson's correlation coefficient for efficiently identifying strongly correlated pairs. In: KDD (2004)
17. Yan, X., Han, J.: gspan: Graph-based substructure pattern mining. In: ICDM (2002)
18. Yan, X., Yu, P.S., Han, J.: Graph indexing: A frequent structure-based approach. In: SIGMOD (2004)
19. Yan, X., Yu, P.S., Han, J.: Substructure similarity search in graph databases, pp. 766– 777 (2005)
20. Zou, L., Chen, L., Yu, J.X., Lu, Y.: A novel spectral coding in a large graph database. In: EDBT (2008)

Efficient RFID Data Imputation by Analyzing the Correlations of Monitored Objects

Yu Gu[1], Ge Yu[1], Yueguo Chen[2], and Beng Chin Ooi[2]

[1] Northeastern University, China
{guyu,yuge}@ise.neu.edu.cn
[2] National University of Singapore, Singapore
{chenyueg,ooibc}@comp.nus.edu.sg

Abstract. As a promising technology for tracing the product and human flows, Radio Frequency Identification (RFID) has received much attention within database community. However, the problem of missing readings restricts the application of RFID. Some RFID data cleaning algorithms have therefore been proposed to address this problem. Nevertheless, most of them fill up missing readings simply based on the historical readings of independent monitored objects. While, the correlations (spatio-temporal closeness) among the monitored objects are ignored. We observe that the spatio-temporal correlations of monitored objects are very useful for imputing the missing RFID readings. In this paper, we propose a data imputation model for RFID by efficiently maintaining and analyzing the correlations of the monitored objects. Optimized data structures and imputation strategies are developed. Extensive simulated experiments have demonstrated the effectiveness of the proposed algorithms.

1 Introduction

Radio Frequency Identification (RFID) [1,2] is experiencing fast developments in recent years. Today, an increasing number of products in our everyday life are attached with RFIDs, which provide fast accessibility of specifications as well as spatio-temporal information of the products. A RFID application is mainly composed of readers and tags. Readers are transponders capable of detecting tags within a distance from them. Tags are attached to the monitored objects. They can either actively transmit or passively respond to a unique identification code when they are close enough to a reader. By means of RFID technology, objects in the physical world can be easily identified, catalogued and tracked. With the tracking function, RFID can be widely applied in applications such as supply chain management [3], human activity monitoring and control [4], etc.

In RFID tracking applications, a huge amount of RFID readings generate a large number of rapid data streams containing spatio-temporal information of the monitored objects. However, there are many dirty data within the spatio-temporal RFID streams. Especially, missing readings are very common because

X. Zhou et al. (Eds.): DASFAA 2009, LNCS 5463, pp. 186–200, 2009.

of various factors such as RF collision, environmental interruption, and metal disturbance. Therefore, the RFID data streams usually contain a large percentage of defective spatial-temporal information. Data imputation, as a kind of data cleaning techniques, is quite important to improve the quality of RFID readings.

Existing studies [5,6] on physical RFID data imputation focus mainly on the analysis of historical readings of independent monitored objects, and ignore the correlations of trajectories of monitored objects. We observe that in many RFID applications, objects are being perceived as moving in swarms. Groups of the monitored objects such as humans, vehicles, herds and commodities are often moving together within many local tracks. For example, in a smart RFID museum scenario, many visitors are expected to move around the premises with their friends or family. Obviously, the positions of partners within a group are very useful for estimating the positions of missing RFID tags within the group. However, there are three important challenges in applying such group moving information to RFID data imputation:

- **Mutation.** The partnership of the monitored objects may change over time. For example, visitors may re-choose their partners because of the mutative interests about exhibition in the museum scenario.
- **Chaos.** People or objects in different groups may accidentally merge at the same location simultaneously. For example, a special show will attract unacquainted visitors at the same time.
- **Ambiguity.** When readers cannot obtain a reading from a tagged object in the monitoring space, we say an **empty reading** occurs. An empty reading can be generated from two cases: **missing reading** (readers fail to read a tag although the tag is covered by their sensing regions) or **vacant reading** (happens when the tagged object is not covered by any sensing region of a reader). It is difficult to identify whether an empty reading is a missing reading or a vacant one.

Figure 1 illustrates the above challenges. The circle and rectangle represent people from different organizations . *Chaos* can be found in region1 where o_1 and o_3 gather. When o_1 moves to region2, a *mutation* occurs because o_1's partner will change to o_5. *Ambiguity* exists between o_6 and o_7 when they both are not read by readers since we cannot distinguish whether it is because the tags have been moved out the sensing regions of readers (vacant readings) or they are disturbed (missing readings).

To our knowledge, this is the first paper that seeks to utilize the grouped trajectory information of monitored objects for RFID data imputation. Algorithms on handling the above challenges are proposed. The paper is organized as follows. Section 2 outlines the related work. Section 3 introduces the correlation model of RFID monitored objects. Section 4 proposes the techniques to efficiently maintain the correlations. Section 5 illustrates the optimized data imputation methods for improving accuracy. Section 6 provides the experimental analysis and Section 7 concludes the paper.

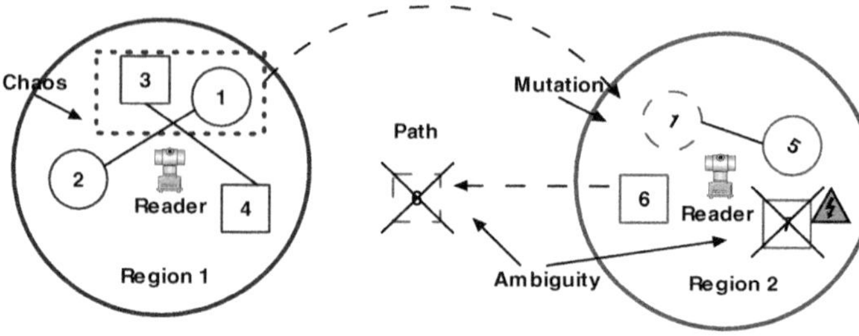

Fig. 1. Challenges for using group moving information

2 Related Work

The proliferation of RFID has generated new problems for database community. Many data cleaning techniques [5,6,8,9] have been proposed to improve the accuracy of RFID data collected from noisy environments. One important goal of RFID data cleaning is to fill up the missing readings in noisy communication conditions. For such data imputation based on physical readings, one commonly used technique is smoothing filter. All missing readings will be filled up if one reading is successfully read within one smoothing window. For example, a pipeline framework, which allows users to set the size of the smoothing window, is used to clean various dirty data in [5]. However, a fixed smoothing window size cannot well capture the dynamics of tag motion. A statistics-based approach is proposed in [6] to dynamically adjust the size of smoothing windows. Whereas, the method is only applicable to the situation of a single reader due to the limitation of the statistics model. For a general RFID application with multiple regions across the space, the statistics model will be ineffective. The cost optimization problem for historical readings based methods is discussed in [7], when additional context features and transition models are supposed to be predicted.

Different from the physical data imputation, some work has focused on cleaning data related to specific application semantics. For example, Khoussainova et al. [8] propose a cleaning strategy based on a probabilistic model. Rao et al. [9] propose to delay the data imputation to the time when queries are issued. The definition of dirty data is associated with specific applications. Moreover, it can only be applied to static data in a database, and not to the RFID streaming applications. Our spatio-temporal closeness model is also substantially different from the one in general sensor networks [10] as our model is used in the context of RFID, where moving objects are tracked.

3 RFID Monitored Objects Correlation Model

For simplicity, we assume that an RFID reader r periodically senses readings from a tagged object o if o is in the sensing region of r. Suppose R, O represent the sets of reader IDs and object IDs respectively. A RFID reading is modelled as a ternary tuple $p =< i \in R, j \in O, t >$, representing an object o_i is detected by a reader r_j at time stamp t. In our RFID model, we assume that RFID readers are independent, i.e., no sensing regions of two readers cover with each other. The sensing region of a reader r_i is called a logic region Υ_i. The space that is not covered by any logic region is called the dead region, denoted as $\tilde{\Upsilon}$. Those tagged objects in any logic region are supposed to be detected by a reader if no missing readings occur. While, those tagged objects will not be sensed by any readers if they are in $\tilde{\Upsilon}$.

All the detected objects within Υ_i at time stamp t is denoted as $\Upsilon_i(t) = \{o_j | \exists p =< i, j, t >\}$, While $\tilde{\Upsilon}(t)$ represents all the tagged objects in the dead region at time stamp t. We define all the objects with empty readings at time t as $\emptyset(t)$ and all the objects with missing readings at time t as $\phi(t)$. Therefore, we have $\emptyset(t) = \phi(t) \bigcup \tilde{\Upsilon}(t)$. If we use $\Delta(t)$ to represent all the tagged objects in the RFID working space at time t, we have $\Delta(t) = \emptyset(t) \bigcup (\bigcup_i \Upsilon_i(t))$. Note that $\Delta(t)$ is dynamic because tagged objects join and leave the RFID working space dynamically. Readers deployed at the entrances and exits can identify the changes of $\Delta(t)$. We define a function $\mathcal{R}^t(o_i)$ specifying the logic region that an object belongs to:

$$\mathcal{R}^t(o_i) = \begin{cases} k & \text{iff } o_i \in \Upsilon_k(t) \\ * & \text{iff } o_i \in \emptyset(t) \end{cases} \tag{1}$$

We can use a discrete stream $\mathcal{S}(o_i)$ specifying the logic regions of an object locating at a series of time stamps. For example, $\mathcal{S}(o_1) = 1111***222****33$. Note that a $*$ in a discrete stream is an empty reading. It can either be a missing reading or a vacant reading. During the estimation process, $*$ is replaced with 0 if it is estimated as a vacant reading. Otherwise, $k \in R$ replaces with $*$ when it is estimated as a missing reading and o_i is estimated at the logic region Υ_k. A discrete stream $\mathcal{S}(o_i)$ is transformed to a fully estimated stream, noted as $\tilde{\mathcal{S}}(o_i)$, if all $*$s in $\mathcal{S}(o_i)$ have been estimated. An example of fully estimated stream is $\tilde{\mathcal{S}}(o_1) = 1111002222000033$. We also define the actual stream $\bar{\mathcal{S}}(o_i)$ of an object as a sequence of the actual positions (k for Υ_k and 0 for $\tilde{\Upsilon}$) of o_i in different time stamps. For example, $\bar{\mathcal{S}}(o_1) = 1111000222000333$. If we ignore those consecutive repetitive readings and those empty readings within a discrete data stream $\mathcal{S}(o_i)$, we get a logic stream $\hat{\mathcal{S}}(o_i)$. For example, $\hat{\mathcal{S}}(o_1) = 123$.

We can measure the error number $e(\tilde{\mathcal{S}}(o_i))$ of an estimated stream by counting the number of incorrect estimated entries in $\tilde{\mathcal{S}}(o_i)$, comparing with the actual stream $\bar{\mathcal{S}}(o_i)$. For example, $e(\tilde{\mathcal{S}}(o_1)) = 2$. Given a number of observed streams $\mathcal{S}(o_i)$ $(i \in O)$, our problem is to achieve the following two level goals:

- Level1: Recover $\hat{\mathcal{S}}(o_i)$ from $\mathcal{S}(o_i)$, which is enough for some simple location sequence query in RFID applications.
- Level2: Estimate $\tilde{\mathcal{S}}(o_i)$ from $\mathcal{S}(o_i)$, so that $e(\tilde{\mathcal{S}}(o_i))$ can be as small as possible. $\tilde{\mathcal{S}}(o_i)$ is quite useful for the exact spatio-temporal queries of tagged objects.

Note that the pure temporal smoothing method or its variance cannot even be applicable for the first level goal. For example, for $\mathcal{S}(o_1) = 1111{*}{*}{*}{*}{*}{*}{*}{*}{*}{*}33$, we may get $\bar{\mathcal{S}}(o_1) = 13$ if those empty readings are simply ignored. For the level2 goal, because the staying periods of different objects in some regions may be quite different, the size of smoothing window cannot be efficiently defined. Furthermore, the dynamic smoothing window technique will not work due to the reader number limitation [6].

To utilize the group moving information for estimating the empty readings, we need define the correlation of moving tagged objects effectively. One intuition is that if two objects stay in the same region for a longer time, they will have higher probability to be the partners. We first assume that there are no missing readings, the correlation of two tagged objects o_i and o_j, denoted as $\lambda_1^t(o_i, o_j)$ can be defined as follows:

$$\lambda_1^t(o_i, o_j) = \begin{cases} \lambda_1^{t-1}(o_i, o_j) + 1 & \text{iff } \exists k, o_i, o_j \in \Upsilon_k(t) \\ \lambda_1^{t-1}(o_i, o_j) & \text{iff } o_i, o_j \in \tilde{\Upsilon}(t) \\ 0 & \mathit{otherwise.} \end{cases} \tag{2}$$

The above definition handle both the *chaos* and the *mutation* problems. The correlation of two objects increases if they stay in the same logic region longer. While, the correlation drops to zero if two objects are probably in different logic regions, which happens in *mutation* scenarios. However, the *ambiguity* is still a problem for such a definition because an empty reading in $S(o_i)$ can also be a missing reading. If we can distinguish between $\tilde{\Upsilon}(t)$ and $\phi(t)$, the correlation can be modified as:

$$\lambda_2^t(o_i, o_j) = \begin{cases} \lambda_2^{t-1}(o_i, o_j) + 1 & \text{iff } \exists k, o_i, o_j \in \Upsilon_k(t) \\ \lambda_2^{t-1}(o_i, o_j) & \text{iff } o_i, o_j \in \tilde{\Upsilon}(t) \\ & \quad \bigvee(o_i \in \phi(t) \bigwedge o_j \notin \tilde{\Upsilon}(t)) \bigvee(o_j \in \phi(t) \bigwedge o_i \notin \tilde{\Upsilon}(t)) \\ 0 & \mathit{otherwise.} \end{cases} \tag{3}$$

However, in real cases, what we can see is $\emptyset(t)$ instead of $\tilde{\Upsilon}(t)$. When $o_i \in \emptyset(t)$, it may be in the same or different places with o_j, keeping the correlation stable is the tradeoff choice according to the three challenges.

Thus ,we present our basic correlation definition as:

$$\lambda^t(o_i, o_j) = \begin{cases} \lambda^{t-1}(o_i, o_j) + 1 & \text{iff } \exists k, o_i, o_j \in \Upsilon_k(t) \\ \lambda^{t-1}(o_i, o_j) & \text{iff } o_i \in \emptyset(t) \bigvee o_j \in \emptyset(t) \\ 0 & \mathit{otherwise.} \end{cases} \tag{4}$$

Based on the correlations among objects, given a tagged object o_i, the reference partner of o_i can be defined as $\mathcal{P}^t(o_i)$, such that $\not\exists o_j, \lambda^t(o_i, o_j) > \lambda^t(o_i, \mathcal{P}^t(o_i))$. Note that there may be multiple objects maximizing the correlation to an object o_i. We randomly choose one of them as $\mathcal{P}^t(o_i)$.

In order to incrementally maintain $\lambda^t(o_i, o_j)$ from $\lambda^{t-1}(o_i, o_j)$, a complete graph is required, namely $\mathrm{G}(t) = (V, E)$, in which the vertex set is $V = \Delta(t)$ and the edge set is $E = \{(o_i, o_j, \kappa) | \forall o_i, o_j \in \Delta(t) \wedge i < j, \kappa = \lambda^t(o_i, o_j)\}$. The vertex is fully linked with weights and it cannot be divided into sub-graphs. A matrix $\mathcal{M}^t$, $\mathcal{M}^t[i][j] = \lambda^t(o_i, o_j)$ can be employed for dynamic maintenance of such a graph. The algorithm is easy to be implemented, but the time and space complexity are both $O(N^2)$, where $N = |\Delta(t)|$. The cost is quite high in RFID applications when the number of monitored objects is huge. Using the adjacency list to maintain the graph will not reduce the time complexity either. Therefore, we need to re-consider the $\lambda^t(o_i, o_j)$ definition and therefore compress the graph structure based on the the concept of reference partner.

Obviously, reference partner can be utilized in the correlation definition when an empty reading occurs. The optimized correlation definition is given in equation(5). Note that λ will be replaced with $\bar{\lambda}$ in computing $\mathcal{P}^t(o_i)$ in this case. Furthermore, the optimized correlation definition can lead to improved maintenance model which will be illustrated in the next section.

$$\bar{\lambda}^t(o_i, o_j) = \begin{cases} \bar{\lambda}^{t-1}(o_i, o_j) + 1 \text{ iff } (\exists k, o_i, o_j \in \Upsilon_k(t)) \\ \quad \bigvee(o_i \in \emptyset(t) \bigwedge o_j \notin \emptyset(t) \bigwedge \mathcal{R}^t(\mathcal{P}^{t-1}(o_i)) = \mathcal{R}^t(o_j)) \\ \quad \bigvee(o_j \in \emptyset(t) \bigwedge o_i \notin \emptyset(t) \bigwedge \mathcal{R}^t(\mathcal{P}^{t-1}(o_j)) = \mathcal{R}^t(o_i)) \\ \quad \bigvee(o_i \in \emptyset(t) \bigwedge o_j \in \emptyset(t) \bigwedge \mathcal{R}^t(\mathcal{P}^{t-1}(o_i)) = \mathcal{R}^t(\mathcal{P}^{t-1}(o_j))) \\ 0 \qquad \qquad \qquad \textit{otherwise.} \end{cases} \tag{5}$$

4 Optimized Maintenance of Correlation

4.1 Properties of Optimized Correlation

The optimized correlation has some desired properties which can be utilized for reducing the maintenance cost. They are: (1) $\bar{\lambda}^t(o_i, o_i) \geq \bar{\lambda}^t(o_i, o_j)$; (2) $\bar{\lambda}^t(o_i, o_j) = \bar{\lambda}^t(o_j, o_i)$; (3) $o_i \in \Upsilon_k(t) \wedge o_j \in \Upsilon_l(t) \wedge k \neq l \Rightarrow \bar{\lambda}^t(o_i, o_j) = 0$. Furthermore, an advanced property for $\bar{\lambda}^t$ can be inferred as follows:

Theorem 1. $\forall i, j, k$, Suppose $\bar{\lambda}^t(o_i, o_k) = \max\{\bar{\lambda}^t(o_i, o_k), \bar{\lambda}^t(o_i, o_j), \bar{\lambda}^t(o_j, o_k)\}$ $\Rightarrow \bar{\lambda}^t(o_i, o_j) = \bar{\lambda}^t(o_j, o_k)$

Proof. Suppose $\bar{\lambda}^t(o_i, o_k) = \kappa_1 > \bar{\lambda}^t(o_j, o_k) = \kappa_2$, according to equation (5), $\bar{\lambda}^{t-\kappa_2}(o_i, o_k) = \kappa_2 - \kappa_1$ and $\bar{\lambda}^{t-\kappa_2}(o_j, o_k) = 0$. At the time point $t - \kappa_2$, if $o_k, o_i, o_j \notin \emptyset(t)$, $\mathcal{A}^{t-\kappa_2}(o_i) = \mathcal{A}^{t-\kappa_2}(o_k) \neq \mathcal{A}^{t-\kappa_2}(o_j)$. Hence, $\bar{\lambda}^{t-\kappa_2}(o_i, o_j) = 0$. If $o_k \in \emptyset(t), o_i, o_j \notin \emptyset(t)$, $\mathcal{A}^{t-\kappa_2}(o_i) = \mathcal{A}^{t-\kappa_2}(\mathcal{P}^{t-\kappa_2-1}(o_k)) \neq \mathcal{A}^{t-\kappa_2}(o_j)$. By analogy of other situations, we can prove $\bar{\lambda}^{t-\kappa_2}(o_i, o_j) = 0$. Also, $\forall t - \kappa_2 < \tau \leq t$, we can infer $\mathcal{A}^\tau(o_i)$(or $\mathcal{A}^\tau(\mathcal{P}^{\tau-1}(o_i))$) $= \mathcal{A}^\tau(o_k)$(or $\mathcal{A}^\tau(\mathcal{P}^{\tau-1}(o_k))$) $= \mathcal{A}^\tau(o_j)$(or $\mathcal{A}^\tau(\mathcal{P}^{\tau-1}(o_j))$). So $\bar{\lambda}^\tau(o_i, o_j) = \bar{\lambda}^{\tau-1}(o_i, o_j) + 1$ and $\bar{\lambda}^t(o_i, o_j) = \kappa_2 = \lambda^t(o_j, o_k)$. Also, the situations for $\bar{\lambda}^t(o_i, o_k) = \bar{\lambda}^t(o_k, o_j)$ and $\bar{\lambda}^t(o_i, o_k) > \bar{\lambda}^t(o_k, o_j)$ can be inferred in the similar way.

Based on the properties of $\bar{\lambda}$, we can get a sub-graph $\mathrm{G}_k(t) = (V, E)$ for each $\Upsilon_k(t)$,where the vertex set is $V = \Upsilon_k(t)$ and the edge set is $E = \{(o_i, o_j, \kappa) | \forall o_i, o_j \in \Upsilon_i(t) \wedge i < j, \kappa = \bar{\lambda}^t(o_i, o_j)\}$. Also, $\emptyset(t)$ will correspond to a special sub-graph $G_*(t)$.We have $\forall o_i, o_j \in G_l(t) \Rightarrow \bar{\lambda}^t(o_i, o_j) \neq 0$. Likewise,

$o_i \in G_l(t), o_j \in G_k(t), l \neq k \Rightarrow \bar{\lambda}^t(o_i, o_j) = 0$. We name this kind of sub-graph as $\bar{\lambda} - Graph$ in this paper.

According to Theorem 1, we can compress a $\bar{\lambda} - Graph$ into a corresponding $\bar{\lambda} - list$ where $\bar{\lambda}$ values of consecutive objects in the list are recorded.

Theorem 2. *A $\bar{\lambda}$-Graph can be transformed into a $\bar{\lambda}$-List without any loss.*

Proof. We prove it by mathematical induction. For $|G_l(t)| = 1$ or $|G_l(t)| = 2$, it is a list. Suppose $|G_l(t)| = n$ can be represented by a list $|L_l(t)| = n$. For a new object o_i, $\exists o_j \in L_l(t)(|L_l(t)| = n)$, $\bar{\lambda}^t(o_i, o_j) = \max\{\forall o_k \in L_l(t), \bar{\lambda}^t(o_j, o_k)\}$. Furthermore, we can get a tree $|T_l(t)| = n+1$ by linking o_i with o_j. Suppose o_k is one neighbor of o_j in $|L_l(t)| = n$. According to Theorem 1, $\bar{\lambda}^t(o_i, o_k) \leq \bar{\lambda}^t(o_j, o_k)$. If $\bar{\lambda}^t(o_i, o_k) < \bar{\lambda}^t(o_j, o_k)$, by cutting the link of o_i and o_k and re-linking o_j and o_k, we can get a list $|L_l(t)| = n + 1$ (an example is shown in Figure 2 for $|L_k(t)| = 4$). If $\bar{\lambda}^t(o_i, o_k) = \bar{\lambda}^t(o_j, o_k)$, replace the edge between o_i and o_j with edge between o_i and o_k. $|L_k(t)| = n + 1$ can be obtained as well by repetitively executing the above rule until it reaches the head or tail of the list (an example is shown in Figure 2 for $|L_k(t)| = 5$). Hence, $|G_k(t)| = n + 1$ can be represented by $|L_k(t)| = n + 1$.

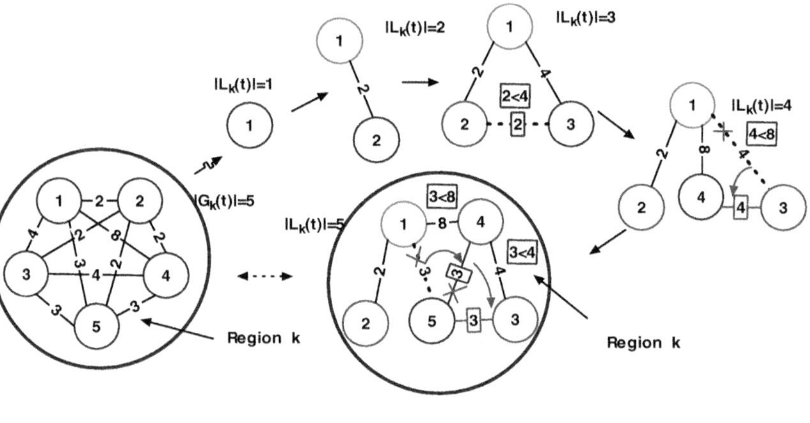

Fig. 2. An illustration of transforming $\bar{\lambda}$-Graph to $\bar{\lambda}$-List

When the detailed implementation is concerned, a linked list L_i^t can be designed to represent a $\bar{\lambda}$ list $L_i(t)$. The element e_{id} of the linked list is denoted as $< id, \bar{\lambda} >$, where $\bar{\lambda}$ represents the correlation of o_{id} and its previous node. In fact, we can build a $\bar{\lambda}$-List from the very beginning and transform L^{t-1} to L^t in the incremental manner based on the following principles.

- **Regrouping principle.** For any two objects o_i, o_j of different sub-lists, $\bar{\lambda}^t(o_i, o_j)$ =0. So when they are combined into a new $\bar{\lambda}$-list. We just need to link them directly and assign the $\bar{\lambda}$ value at the linking position as 1 according to our $\bar{\lambda}$ definition.

- **Splitting principle.** Suppose e_k, e_m and e_n are three consecutive elements in a $\bar{\lambda}$-List and $\bar{\lambda}^t(o_k, o_m) = \bar{\lambda}^t(o_m, o_n) = \kappa$. According to Theorem 1, $\bar{\lambda}^t(o_k, o_n) \geq \kappa$. However, if $\bar{\lambda}^t(o_k, o_n) > \kappa$, e_m can be inferred to be inserted into a $\bar{\lambda}$-List having contained e_k and e_n, at $t - \kappa + 1$. Whereas according to the regrouping principle, e_m will not lie between e_k and e_n. Therefore, we can testify $\bar{\lambda}^t(o_k, o_n) = \kappa$. Furthermore, combining it with Theorem 1, for any two objects o_i, o_j in $\bar{\lambda}$-List, $\bar{\lambda}^t(o_i, o_j)$ can be obtained by calculating the minimal $\bar{\lambda}$ between o_i and o_j. Thus, a list can be split into some sub-lists when some $\bar{\lambda}$ value drops to 0. Note that $\mathcal{P}^t(o_i)$ will be the o_i's neighbor in a $\bar{\lambda}$-list according to the analysis above.

4.2 Optimized Maintenance Strategies

In this section, we will propose three incremental $\bar{\lambda}$ maintenance strategies based on different splitting and regrouping principles. The proposed maintenance strategies show different performances in various conditions. Suppose $N = |\Delta(t)|$, N_r is the number of logic regions plus a dead region and $N_o = N/N_r$ represents the average number of objects in each region. Besides, there is a hidden metric P_c reflecting the average probability that a object will change its partners at the next time stamp.

Multiple Scans (MS) Strategy. In order to compute L^t from L^{t-1}, we can scan each L_i^{t-1} for multiple times. At each time, we abstract the objects in the same L_j^t. We illustrate the procedure in Figure 3. Suppose the input is $\Upsilon_1(t) = \{o_8, o_9, o_{10}, ...\}, \Upsilon_2(t) = \{o_{14}, o_{22}, ...\}$ and $\emptyset(t) = \{o_5, o_{20}, o_{26}, ...\}$. 1-1 represents the first step in the first loop. $min\lambda_i$ representing the minimum λ_i in L_i^{t-1} for each loop will be maintained in an incremental manner. For the objects in $\emptyset(t)$, reference partner is required to determine which list to insert. For example, $\mathcal{P}^{t-1}(o_5) = o_{20} \wedge \mathcal{R}^t(o_{20}) = *$, therefore, o_5 will be inserted into the special list L_*^t for the dead region. While $\mathcal{P}^{t-1}(o_{26}) = o_{22} \wedge \mathcal{R}^t(o_{22}) = 2$, so o_{26} will be inserted into the region list L_2^t.

The key steps of the MS algorithm is illustrated in Algorithm 1. The processing cost is related to P_c. If $P_c = 0$, the average complexity to maintain a L_i^{t-1} is just $O(N_o)$. If $P_c = 1$, the average complexity to maintain a L_i^{t-1} will be $O(N_o^2)$. Therefore, the total time complexity to transform L^{t-1} to L^t will be between $O(N_o N_r)$ and $O(N_o N)$.

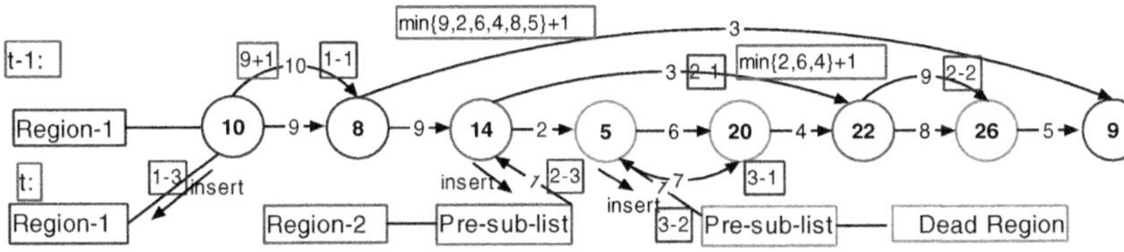

Fig. 3. An illustration of the MS procedure

Algorithm 1. The Multiple Scans Algorithm

Input: L^{t-1}
Output: L^t
for $L_i^{t-1} \in L^{t-1}$ **do**
  while $L_i^{t-1} \neq \emptyset$ **do**
    reset $min\lambda_i$
    $\alpha = L_i^{t-1}.head$
    if $\alpha \in \emptyset(t)$ **then**
      $\alpha = \mathcal{P}^{t-1}(\alpha)$
    for $\beta = o_i \in L_i^{t-1}$ **do**
      maintain $min\lambda_i$
      if $\beta \in \emptyset(t)$ **then**
        $\beta = \mathcal{P}^{t-1}(\beta)$
      if $\mathcal{R}^t(\alpha) = \mathcal{R}^t(\beta)$ **then**
        $\beta.\lambda = min\lambda_i + 1$
        $\hat{L}^{t-1}_{\mathcal{R}^t(\alpha)}.add(\beta)$
        $L_i^{t-1}.remove(\beta)$
    $L^t_{\mathcal{R}^t(\alpha)}.link(\hat{L}^{t-1}_{\mathcal{R}^t(\alpha)})$

Single Scan (SS) Strategy. Compared to MS, SS scans each sub-list L_i^{t-1} for only once. It inserts each o_i in the sub-list into the corresponding L_j^t. However, in order to maintain the $\bar{\lambda}$, for each L_j^t, the current $min\lambda_j$ must be maintained. All the L_j^t must be scanned to maintain the $min\lambda_j$ with each insertion. We illustrate the basic procedure in Figure 4 with the same input of Figure 3.

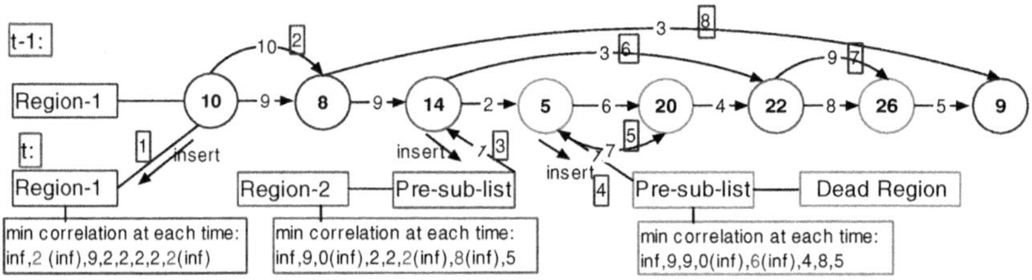

Fig. 4. An illustration of the SS procedure

The key steps of the SS algorithm is illustrated in Algorithm 2. The average complexity to maintain a list is $O(N_o N_r)$. The total complexity to transform from L^{t-1} to L^t will be $O(N_r N)$.

Interval Tree-Based Scan (ITS) Strategy. When computing the $\bar{\lambda}$ of any two objects, we can construct an interval tree to calculate the $\bar{\lambda}$ directly, instead of incrementally maintaining $min\lambda_i$. First, all $\bar{\lambda}$ values will be indexed as an ordered array. The leaf nodes of the tree are all the objects, and the internal

Algorithm 2. The Single Scan Algorithm

Input: L^{t-1}
Output: L^t

for $L_i^{t-1} \in L^{t-1}$ **do**
  for $\beta = o_i \in L_i^{t-1}$ **do**
    if $\beta \in \emptyset(t)$ **then**
      $\beta = \mathcal{P}^{t-1}(\beta)$
    for each L_i^t **do**
      if $min\lambda_i > \beta.\lambda$ **then**
        $min\lambda_i = \beta.\lambda$
    if β is the first in L_i^{t-1} to insert $L^t_{\mathcal{R}^t(\beta)}$ **then**
      $min\lambda_{\mathcal{R}^t(\beta)} = 0$
    $\beta.\lambda = min\lambda_{\mathcal{R}^t(\beta)} + 1$
    $min\lambda_{\mathcal{R}^t(\beta)} = \infty$
    $L^t_{\mathcal{R}^t(\beta)}.add(\beta)$
    $L_i^{t-1}.remove(\beta)$

nodes are the index of the children nodes with smaller $\bar{\lambda}$. The bounds of the index interval for each parent node is recorded. The complexity to construct such a tree for a sub-list will be $O(N_o log N_o)$. However, the extra space cost for the inter nodes is required. To calculate $\bar{\lambda}^t(o_i, o_j)$, from the root node, dichotomy is used until finding the corresponding left bound node and the right bound node. Then, the two correlation scores will be compared. The smaller one is the result. Note that in the list structure, $\bar{\lambda}^{t-1}(o_i, o_j)$ is stored in the node representing o_j. Therefore, the corresponding left bound index should be incremented by 1. For example, in Figure 5 which illustrates the procedure of ITS, to compute $\bar{\lambda}^{t-1}(o_8, o_9)$, namely $\bar{\lambda}^{t-1}(o_{index=2}, o_{index=8})$, the left index bound should be 3 and the right index bound should be 8. The scan of the tree needs $O(log N_o)$ time complexity. In this way, total time complexity to transform L^{t-1} to L^t will be $O(N log N_o)$.

MS, SS and ITS are suitable for different scenarios. MS is suitable for the situation when P_c is small and SS is suitable for the situation when N_r is small. ITS can be efficient when P_c and N_r are both large. Because no algorithm can be dominant over others in all cases, we can make the choice adaptively according to the applications.

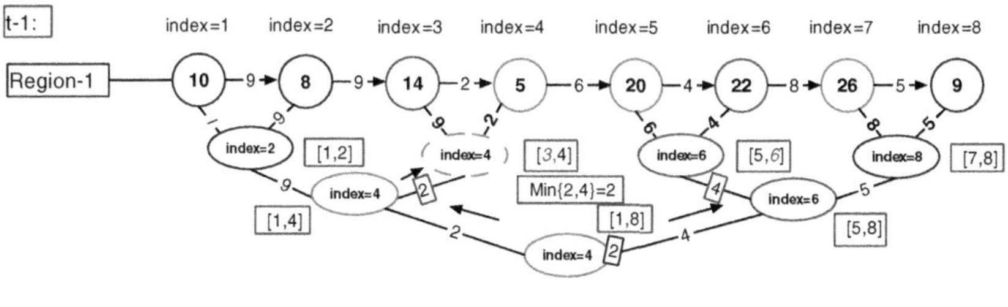

Fig. 5. An illustration of correlation computation based on interval tree

5 Remedy-Based RFID Data Imputation Strategy

Defining and maintaining the correlations are only the basis of the data imputation strategy. One way of data imputation is to construct the missing readings directly based on $\mathcal{P}^{t-1}$. We call it the direct imputation. However, the direct imputation may cause regular errors due to the change of partners. For example, for $\mathcal{S}(o_1)$=111***222 and $\mathcal{S}(o_2)$=11*****33, if $\mathcal{P}^3(o_2) = o_1$ and $\mathcal{P}^7(o_2) = o_1$, the direct imputation may result in $\tilde{\mathcal{S}}(o_2) = 111000233$. But obviously, the real case is much more likely $\tilde{\mathcal{S}}(o_2) = 110000033$, interpreted as o_1 leaves o_2 and leads to a different region. Therefore, we introduce a remedy-based RFID data imputation strategy. Instead of implementing imputation right away, the remedy-based RFID data imputation will check the states of two objects in the next logic destination region. Although the imputation is delayed, it is very useful to improve the accuracy of the results. Especially, the remedy-based RFID data imputation is quite useful for those objects having no substantial partners with them. We illustrate the strategy in Algorithm 3.

Algorithm 3. Remedy-based RFID Data Imputation

if $o_i \in \emptyset(t) \wedge \mathcal{P}^{t-1}(o_i) \notin \emptyset(t)$ **then**
  $o_i \rightarrow \mathcal{P}List.add(\mathcal{P}^{t-1}(o_i))$
if $o_i \notin \emptyset(t) \wedge \mathcal{P}^{t}(o_i) \notin \emptyset(t)$ **then**
  if $o_i \rightarrow \mathcal{P}List \neq \emptyset$ **then**
    $\mathcal{S}_\alpha = o_i \rightarrow \mathcal{P}List.searchbyID(\mathcal{P}^t(o_i))$
    if $\mathcal{S}_\alpha \neq \emptyset$ **then**
      for $\alpha \in \mathcal{S}_\alpha$ **do**
        $o_i.interpolate(\alpha.t, \alpha.r)$
  $o_i \rightarrow \mathcal{P}List.removeALL()$

The algorithm can be combined with the correlation maintenance strategy directly by introducing an adjunctive data structure $\mathcal{P}List$, which is used to reserve the possible reference partners. We denote the size of $\mathcal{P}List$ as $N_\mathcal{P}$ and the additional search cost $O(N_\mathcal{P})$ will be incurred when the imputation is trigged. However, it is trivial compared to the maintenance of correlations. The response delay of imputation is potentially determined by the time spent during the dead region for the corresponding objects.

As a pre-processing method, data imputation will serve the user's query. However, remedy-based imputation may lead to inaccurate results for the upper query because of the delay. RFID-oriented query can be based on database [11] or data stream [12]. For database query [11], the RFID readings will be inserted into a table, e.g., R-Table, and SQL query can be executed over the table. In this way, as long as the query is not incurred before the imputation judgement is finished, there will be no problems. Correspondingly, $o_i.interpolate(\alpha.t, \alpha.r)$ in the database context will be interpreted the following SQL:

UPDATE R-Table Set Region=$\alpha.r$ WHERE Object=$o_i.id$ AND time=$\alpha.t$.

But for the RFID data stream processing, automata model is usually employed to detect complex event pattern [12]. The data will not be stored into the database and query is executed in the continuous manner. For the common queries involving sequence, simultaneous, conjunction and disjunction, although the response will be delayed but the correct results can be guaranteed. However, for the query when negation operation is involved, there may be false positive result produced. For example, for the query $SEQ(A, NEG(B))_w$, which represents some object doesn't come to region B after leaving region A within w period, if the imputation cannot be finished within w, a false composite event will be notified. Fortunately, we can employ a similar version of algorithm 3 as the alternative method to solve the problem. The main idea is to first utilize direct imputation method. Then, delete the imputation data later once it is judged to be unreasonable. It can be obtained by simply modifying algorithm 3. Note that this alternative method is not suitable for the occasions when algorithm 3 is applicable. Therefore, we need to choose the proper remedy-based RFID data imputation strategy according to the query categories.

6 Experiments

6.1 Experiment Settings

In this section, we report the evaluation of our proposed model by simulated experiments, in terms of the efficiency and accuracy. All experiments were conducted on a PC of 2.6G Hz with 1G memory. The algorithms were implemented with C++. We designed three kinds of simulated data sets in a museum scenario.

- **DataSet1.** The locations of the monitored objects (visitors) are totally random at any time. In this case, objects hardly have stable partners, and P_c will be very high.
- **DataSet2.** Most monitored objects will have partners at each time stamp. The partners may change casually.
- **DataSet3.** All the monitored objects will have their partners all the time. The partners seldom change during the whole simulation period.

6.2 Maintenance Cost

We evaluate the processing time of three correlation maintenance strategies in this section. In order to better illustrate the flexibility in complex situation, the results of tests on dataset1 are given in Figure 6. In this case, P_c can be inferred by computing $\min(N_o, N_r)$, which helps us to observe its impacts on computational cost. The granularity of simulation time represented by T is second and the total cost during the simulation is used to illustrate the efficiency.

In Figure 6(a), we tested the efficiency of different correlation maintenance models. We can find that the list-based model based on the correlation $\bar{\lambda}$ will be much more efficient than the graph-based model based on the correlation λ. The matrix-based graph maintenance is very time-consuming. The advantage of ITS

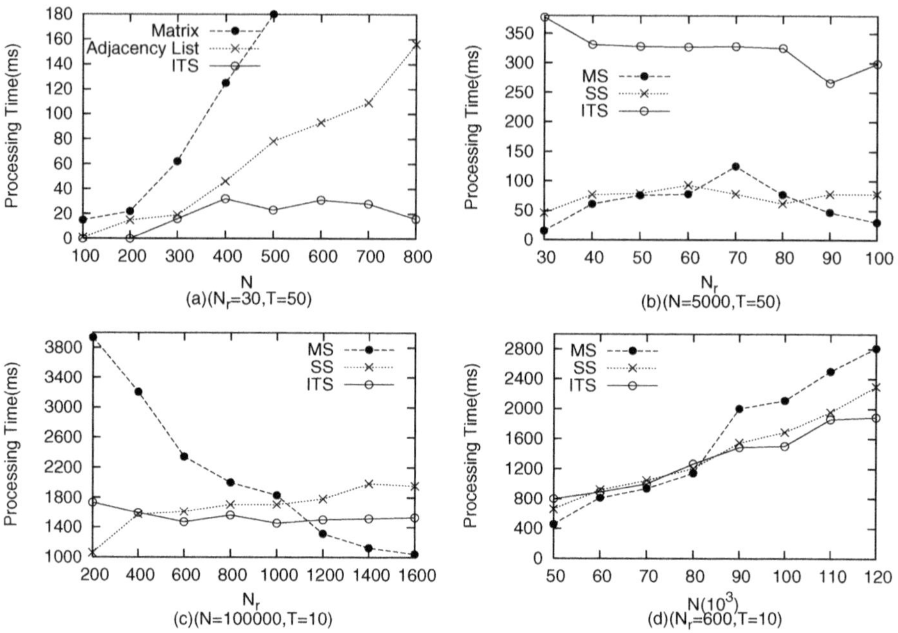

Fig. 6. Processing time comparison of different correlation algorithms

over adjaceny-list-based method will become quite prominent with the increase of N. In Figure 6(b), we compared the different optimization maintenance methods when $N = 5000$. ITS will consume some computational costs in constructing the tree and index, which is comparable to the maintenance of correlations in this scale of N. Therefore, compared to MS and SS, the cost of ITS will be higher. The contrast of MS and SS is not very obvious in this situation.

In Figure 6(c) and Figure 6(d), we studied the impacts of N and N_r on the computational cost of various correlation maintenance methods when a huge number of objects are involved. As we can see from these two figures, neither MS, SS nor ITS can dominate the others under different values of N and N_r, which is substantially accordant with our theory analysis. For dataset2 and dataset3, similar results can be found for the computational time of different correlation maintenance methods.

6.3 Accuracy

Because the direct imputation and remedy-based imputation methods will incur very low computational cost compared to the maintenance of correlations, we simply studied the accuracy of different correlation definitions and data imputation strategies. The compared methods include direct imputation and remedy-based imputation based on λ and $\bar{\lambda}$ (opt-λ in Figure 7). The error rates for our level1 goal and level2 goal of different imputation methods are illustrated in

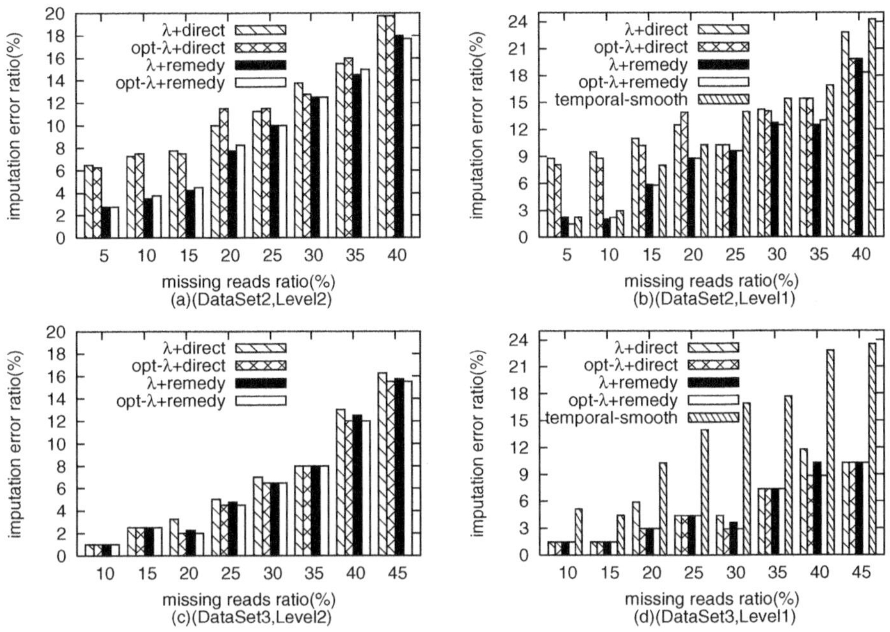

Fig. 7. Error ratio comparison of different data imputation strategies

Figure 7. For the level1 goal, we compared with the temporal smoothing method, which is the ideal method in this case if imputation is simply based on the temporal information. We take Dataset2 and Dataset3 as the testing datasets to illustrate the accuracy evaluation because partnerships in Dataset1 can hardly be preserved.

From Figure 7(a)-(d), we can see that the accuracy under the $\bar{\lambda}$ definition and the λ definition is quite close. While, the remedy-based method will improve the accuracy in Dataset2 where more mutations exist. Even for the level1 goal, compared to the temporal-smoothing method, the correlation-based imputation method will be more accurate, especially when the missing ratio is large. Compared to dateset2, the accuracy of our proposed imputation model is better in dataset3 where partners seldom change. Moveover, our methods can also gain relatively higher accuracy in the very noisy environment(e.g., with 40% missing ratio).

7 Conclusion

In this paper, we have proposed a novel RFID data imputation mechanism based on analyzing the correlations of monitored objects. Basic correlation model is first inferred to solve the three key challenges: mutation, chaos and ambiguity. The $\bar{\lambda}$ correlation is proposed to gain important correlation relationship between objects. Using this relationship, related optimization strategies about correlation maintenance are discussed. A remedy-based data imputation strategy is introduced to improve the accuracy. Finally, we have identified the effectiveness of the proposed models and methods through extensive experimental studies.

Acknowledgement

This research was partially supported by the National Natural Science Foundation of China under Grant No.60773220 and 60873009.

References

1. Want, R.: An introduction to RFID technology. IEEE Pervasive Computing 5(1), 25–33 (2006)
2. Want, R.: The Magic of RFID. ACM Queue 2(7), 40–48 (2004)
3. Asif, Z., Mandviwalla, M.: Integrating the supply chain with RFID: A Technical and Business Analysis. Communications of the Assiciation for Information Systems 15, 393–427 (2005)
4. Philipose, M., Fishkin, K.P., Perkowitz, M., Patterson, D.J., Fox, D., Kautz, H., Hahnel, D.: Inferring activities from interactions with objects. IEEE Pervasive Computing 3(4), 10–17 (2004)
5. Jeffery, S.R., Alonso, G., Franklin, M.J.: A pipelined framework for online cleaning of sensor data streams. In: Proceedings of ICDE, pp. 140–142 (2006)
6. Jeffery, S.R., Garofalakis, M., Franklin, M.J.: Adaptive cleaning for RFID data streams. In: Proceedings of VLDB, pp. 163–174 (2006)
7. Gonzalez, H., Han, J., Shen, X.: Cost-conscious cleaning of massive RFID data sets. In: Proceedings of ICDE, pp. 1628–1272 (2007)
8. Khoussainova, N., Balazinska, M., Suciu, D.: Towards correcting input data errors probabilistically using integrity constraints. In: Proceedings of MobiDE, pp. 43–50 (2006)
9. Rao, j., Doraiswamy, S., Thakkar, H., Colby, L.S.: A deferred cleansing method for RFID data analytics. In: Proceedings of VLDB, pp. 175–186 (2006)
10. Vuran, M.C., Akyildiz, I.F.: Spatial correlation-based collaborative medium access control in wireless sensor networks. IEEE/ACM Transactions on Networking (TON) 14(2), 316–329 (2006)
11. Wang, F.S., Liu, P.Y.: Temporal management of RFID data. In: Proceedings of VLDB, pp. 1128–1139 (2005)
12. Wang, F.S., Liu, S., Liu, P.Y.: Bridge physical and virtual worlds: complex event processing for RFID data streams. In: Ioannidis, Y., Scholl, M.H., Schmidt, J.W., Matthes, F., Hatzopoulos, M., Böhm, K., Kemper, A., Grust, T., Böhm, C. (eds.) EDBT 2006. LNCS, vol. 3896, pp. 588–607. Springer, Heidelberg (2006)

A Reprocessing Model Based on Continuous Queries for Writing Data to RFID Tag Memory*

Wooseok Ryu and Bonghee Hong

Department of Computer Engineering, Pusan National University,
San 30, Jangjeon-dong, Geumjeong-gu, Busan 609-735, Republic of Korea
{wsryu,bhhong}@pusan.ac.kr

Abstract. This paper investigates the problem of writing data to passive RFID tag memory and proposes a reprocessing model for assuring the atomicity and durability of writing transactions in volatile RF communications. This study is motivated by the need to support persistent writing transactions of RFID tag data despite asynchronous and intermittent RF connections. The problem arises when the tag disappears from the interrogation area of the reader before all data is written completely. The disconnection during the process of a write operation causes the tag data to be inconsistent. To solve the problem, we propose an asynchronous wake-up reprocessing model for resuming unsuccessful write operations. The basic idea is to use a continuous query scheme for asynchronously detecting re-observations of unsuccessfully written tags. To achieve this, we employ a continuous query index, which monitors the tag stream continuously and finds unsure tags efficiently. We also discuss implementation issues for the proposed model. This shows that the reprocessing model processes unsuccessful operations completely and avoids inconsistent data, despite volatile characteristics in the RF communications.

Keywords: RFID, Write Transaction, Tag Memory, Continuous Query.

1 Introduction

Radio Frequency IDentification (RFID) has been evolving into a distributed memory space [1]. Beyond simple identification, some RFID applications need to keep additional information about the product, such as price, expiry dates, and ownership in the passive RFID tag memory. As the tag memory enables self-description of tagged items, it is possible to access information about a tagged item without connecting to an information system.

Many industrial processes, such as supply-chain management, asset management, and manufacturing process management, require additional memory in the RFID tags [2]. In the automotive industry, for example, production information is maintained in passive tags attached to automotive components. The information in the tag refers to

* "This work was supported by the Grant of the Korean Ministry of Education, Science and Technology" (The Regional Core Research Program/Institute of Logistics Information Technology).

X. Zhou et al. (Eds.): DASFAA 2009, LNCS 5463, pp. 201–214, 2009.

manufacturing status and is accessed for quality assurance purposes, such as recalling defective components during the components' journey through the assembly process.

Accessing the data in a passive RFID tag's memory requires following two steps [3]. Firstly, the middleware requests a tag inventory operation to a specified reader. The reader broadcasts an RF signal and waits for the tags' reply signals. And then, the reader lists up the inventoried tags and return them to the middleware. Secondly, the middleware investigates inventoried tags. If a matching tag is found, it sends access operations to the reader. These access operations include read, write, lock, and kill.

An RFID transaction is a unit of logical work comprising tag access operations. To preserve the consistency of data in the tag's memory, the RFID transaction should execute the access operations correctly. RFID transaction processing is similar to mobile transaction processing [4] because they both involve disconnection of wireless communication. However, the RF communication has different characteristics. Firstly, the connection between the reader and the passive tag is volatile. The communication range between the reader and the tag is limited to less than 10 meters in the case of passive RFID tags [5]. Tags frequently move out of the range of a reader during the writing of data to the tag's memory. At that time, therefore, the data in the tag's memory might not be written correctly. Second characteristic is uncertainty. The RFID reader cannot guarantee 100% accuracy for tag access because of various interferences such as tag orientation, packing materials, and other obstacles [6]. A reader might not receive a reply message after sending a write message. If the tag disappears or the message is lost during processing of the write operation, the reader cannot determine whether the write operation was successfully processed or not.

The problem of processing RFID write operations is that these characteristics of RF communication will lead to inconsistent states of the tag data. A mobile client can handle these situations by immediately reconnecting to the host, because it maintains the operational status of the session. However, because a passive RFID tag has no internal power, it receives operational energy from the RF signal sent by readers. If it moves out of the field-of-view of the reader, the tag becomes unpowered and session information will be re-initialized. Data in the tag's memory will remain inconsistent. Therefore, a reprocessing mechanism is required that guarantees the correct execution of write operations. VTMS [7] is proposed to handle this problem. However, it is a backup storage service and a recovery mechanism for the operation is not provided.

We may consider that the middleware re-executes the write operation to the reader until the operation is processed completely. However, we do not know when the tag will be observed again and which reader will observe the tag. While a reader is performing an inventory operation for re-executing a write operation, another application may access the inconsistent data of the tag via another reader. Therefore, the middleware should re-execute the write operation to all readers. However, this approach would lead to middleware overload because the number of operations will be increased significantly. For example, assume that 1000 write operations are requested by the middleware, with 10% of them being unsuccessfully processed. This would mean that 5000 operations are required to handle 100 unsuccessful operations if 50 readers are connected to the middleware.

To solve this problem, we propose a reprocessing model for write operations, which handles suspended write operations caused by volatile and uncertain RF communications. Our approach is the asynchronous detection of re-observation of unsuccessful tags using a continuous query scheme when the tags are observed again. We employ a Continuous Query Index (CQI) to enhance the search for unsure tags. Inventoried tags from readers are stabbed into the CQI to check whether the tag requires re-execution or not. To do this, we specify an asynchronous wake-up reprocessing model based on the CQI scheme.

The main contribution of our work is that we develop a complete reprocessing model for overcoming the limitations of RF communications. Our model eliminates incompleteness in write operations by reprocessing unsuccessful operations. By using a continuous query scheme, our model enhances the performance of detecting unsuccessfully written tags. Our model gives consistent data access to passive RFID tags without requiring consideration of RF characteristics.

The remainder of this paper is organized as follows. We analyze the characteristics of RF communications and specify our work in Section 2. In Section 3, we present an asynchronous wake-up reprocessing model based on a CQI scheme for ensuring complete execution. In Section 4, we present middleware architecture for the proposed model with implementation results. A summary is presented in Section 5.

2 Problem Definition

We first describe the access mechanism of RFID tag memory. Then, we analyze the characteristics of RF communications and discuss problems in processing write operations.

2.1 Accessing the Tag Memory Data

A request-to-write operation from a user application specifies a target tag and target readers to detect the tag, in addition to the data. When the middleware receives the request, it requests a tag inventory operation to the target readers. The reader identifies a population of tags using a sequence of air-protocol commands [3]. Then, it reports the inventoried tags list to the middleware. The middleware compares the target tag with the inventoried tags and performs a write operation if the tag is found. The reader sends the write commands to the tag and receives a reply from the tag. This access to tag memory by the middleware is illustrated in Fig. 1.

Low-Level Reader Protocol specification (LLRP) [8] defines APIs for the reader. In the LLRP, the write operation is specified by the C1G2Write parameter. It specifies an OpSpecID, a memory bank, a word pointer, the data, and an access password. Because a single write command from the reader to the tag can write only one word in the tag's memory [3], a C1G2Write performs multiple write commands to the tag.

The result of a C1G2Write includes the number of words successfully written to the tag and the result of the operation, as shown in Table 1. If *NumWordsWritten* is not equal to the length of the data to be written, the result will be nonzero and will include information about the error state of execution of the write operation. Based on the LLRP, we may classify the result of an operation as one of four cases:

- *Successful*, where the operation has been processed successfully.
- *Temporarily Failed*, where the operation has not been processed successfully because of environmental reasons such as insufficient power. The operation may succeed after a retrial of the request.
- *Permanently Failed*, where the operation has failed because of memory overrun or access privileges. The operation will not succeed by simple retrial. It will be necessary to abort the operation.
- *No Response*, where the reader has failed to receive a reply to a write command from a tag. The middleware cannot determine whether the data has been written successfully or not, because the tag's state remains unknown.

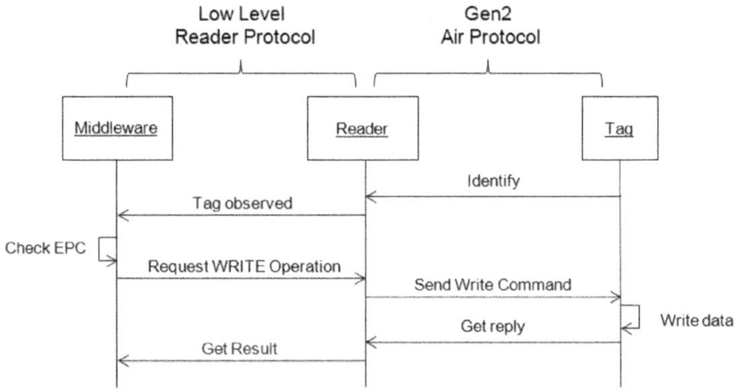

Fig. 1. Protocol for RFID operations [3][8]

Table 1. C1G2WriteOpSpecResult parameter [8]

Parameter	Description
OpSpecID	Identifier of OpSpec
NumWordsWritten	The number of words written as a result of this OpSpec
Result	0 Success 1 Tag memory overrun error 2 Tag memory locked error 3 Insufficient power to perform memory-write operation 4 Non-specific tag error 5 No response from tag 6 Non-specific reader error

2.2 A Problem with Processing Write Operations

The tag is called inconsistent when the write operation is not successfully executed, leaving the data indeterminate or partially written. Let us explain this state using the result of a write operation. Figure 2 shows three possible scenarios for a *No Response* result that causes tag data inconsistency. Figure 2(a) illustrates tag disappearance caused by volatile RF communications. If the tag moves outside the range of the reader after the reader has sent the write command, the reader cannot determine

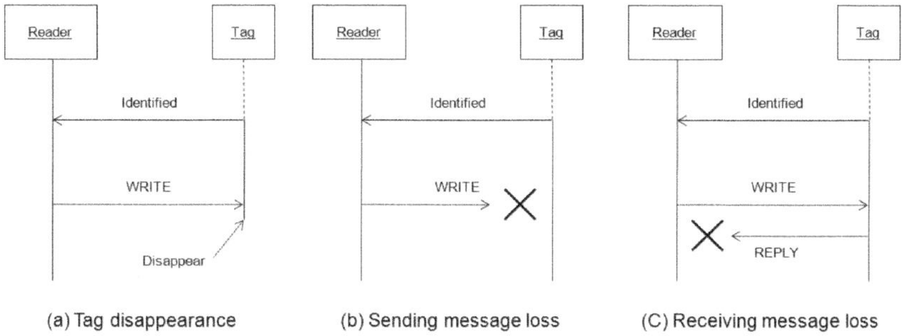

Fig. 2. Error cases for *No Response*

whether the command has been executed or not. Figure 2(b) shows that a message sent by the reader is lost and the operation is not processed. Figure 2(c) shows that the write operation is successfully processed in the tag and the tag data is successfully updated, but the reply message is lost.

The problem is that these characteristics will cause the tag state to be inconsistent. For all cases in Fig. 2, the reader does not receive a reply message and simply returns *No Response*. However, the tag memory may be already written or may be unchanged. Because a passive tag is not self-powered, it cannot maintain session and control information, and it will be re-initialized when the tag moves outside the interrogation area. As a result, its inconsistent data may be accessed by other applications.

To ensure the atomicity of the operation, tag inconsistency should be eliminated by processing the write operation again. It is possible to consider reader-level reprocessing to handle this problem. Currently, smart readers have been developed that perform retrial of operations as well as single execution of access operations [8]. However, the tag is floating in the reader space, and we cannot anticipate which reader will observe the tag next. If the tag is observed by another reader, it is inevitable that inconsistent data will be accessed.

3 Asynchronous Wake-up Reprocessing Model

In this section, we present an Asynchronous Wake-up (AW) reprocessing model for finalizing unsuccessful operations. We start with an introduction to the proposed model and discuss an asynchronous detection mechanism using a CQI scheme. Then, we present the execution mechanism of our model to complete the operations.

3.1 Introduction of AW Reprocessing Model

The objective of the AW reprocessing model is to enable write operation completion in volatile RF environments. Our model suspends execution of unsuccessful operations and separates them from other operations that are registered in the middleware for normal execution. When the tag's re-observation is detected, our model wakes up the execution of the unsuccessful operation. We start the discussion of our model with some definitions.

Definition 1. A write operation w is called *unsure* if a result of w is *No Response*. A w is called *partially completed* if the *NumWordsWritten* in the result is less than w.*length*. A w is called *incomplete* if it is either *unsure* or *partially completed*.

Definition 2. An *unsure tag* is a target tag of an *incomplete operation*.

If an operation is determined to be incomplete after execution, our model starts by inserting the operation into the unsure list. Each record in the unsure list is defined as a tuple of seven predicates (id, tid, offset, length, data, nww, pid), where the tid is the identifier of unsure tag, offset, length and data describe the write operation, nww is the number of words written successfully, and pid is the identifier of the parent operation which has resulted as incomplete.

The unsure list is used for the asynchronous detection of an unsure tag's re-observation. Inventoried tags from all readers should be checked whether they are unsure or not. Simple processing flow in the AW reprocessing model is illustrated in Fig. 3. When a tag is identified by Reader#B, its presence in the unsure list is checked. In this example, it matches the second record, where it is found that one word has already been written and one word remains to be written. Therefore, a write request for the word "34" is sent to Reader#B.

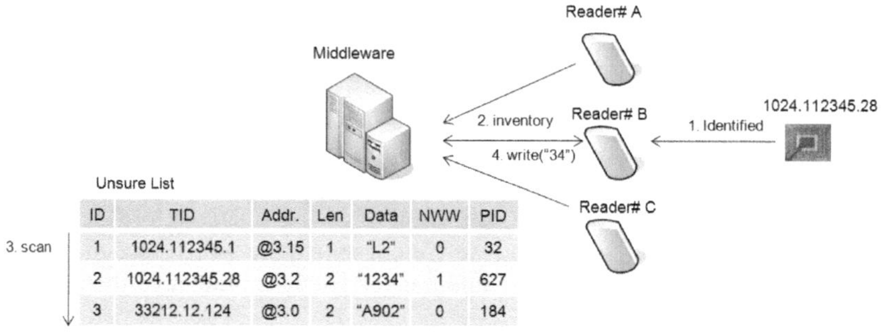

Fig. 3. Processing flow in the AW Reprocessing Model

3.2 Detection of the Tag's Re-Observation Using CQI

Detecting the tag's re-observation using the unsure list takes role of the continuous query processing. Inventoried tags from all readers are flowed to the middleware as stream, and each record in the unsure list should be compared continuously with the stream to determine whether the inventoried tag is an unsure tag or not. Therefore, detecting the tag's re-observation is the same as the continuous query processing where the records in the unsure list represents queries and each tag in the inventoried tags is a data to the continuous query processing.

Obviously, sequential scanning of the unsure list will be costly when there are many unsure tags. The purpose of the CQI [9] is to find a matching tag in the unsure list for an input tag stream more efficiently than via sequential scanning. The index stores records in the unsure list. The middleware can search for a matching tag by stabbing the index using inventoried tags. The role of the CQI is depicted in Fig. 4(a).

The CQI for the unsure list can be constructed as a one-dimensional index. Index key of the unsure list is the *tid* field. The *rid* dimension is not required because detecting the re-observation of an unsure tag does not consider which reader will be involved. However, there is a consideration in designing a CQI. Normal tags should be filtered out as soon as possible. The number of records in the unsure list is very small relative to the huge volume of the input tag stream. Only a few of the tags in the stream are expected to be unsure tags.

The extendable hashing method is suitable as a structure for the CQI. If bucket overflow occurs during insertion, this method doubles the directory size without creating and linking the new bucket. Even if a query result does not exist, the number of I/O operations is bounded by one directory access and n item accesses, where n is the bucket size. Therefore, this method can quickly filter out nonexistent tags in the index. A CQI based on the extendable hashing is illustrated in Fig. 4(b), where a reverse raw hex value of the tid is used as the hash key.

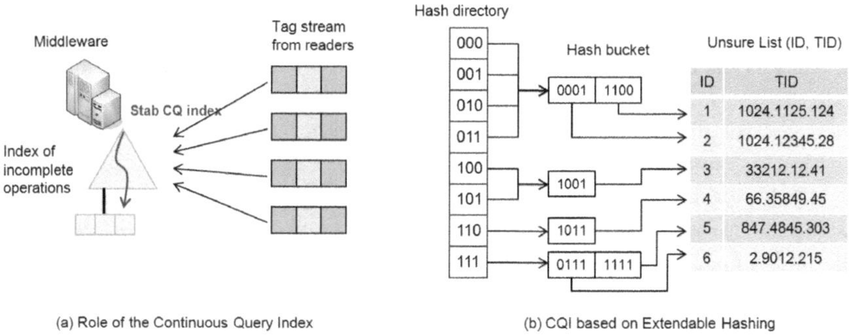

Fig. 4. CQI for the unsure list

3.3 Resuming Incomplete Operation After Re-observation

A flow chart of this reprocessing is shown in Fig 5. After a tag in the unsure list is found via the CQI in the Step 1, the next step is to wake up and resume the operation. For ease of explanation, we first discuss reprocessing of unsure operations for a simple case, which assumes that only single-word data is to be written and the reply result is either *Success* or *No Response*. Then, we will discuss more generalized cases.

In Step 2, the middleware verifies tag data with respect to the success or otherwise of the previous execution of the operation by sending a read command to the tag. When the reply is *Success* then checking of the returned data is required. If the data is verified as correct, then the operation is completed. If not, Step 3 is invoked to execute a write operation. If the middleware receives *No Response*, then Step 4 is invoked. It is also possible to force a rewrite of the data without verifying. However, the write command is less reliable than the read command because it requires more energy and a shorter communication range [10]. Therefore, read-before-write is more efficient than a simple rewrite.

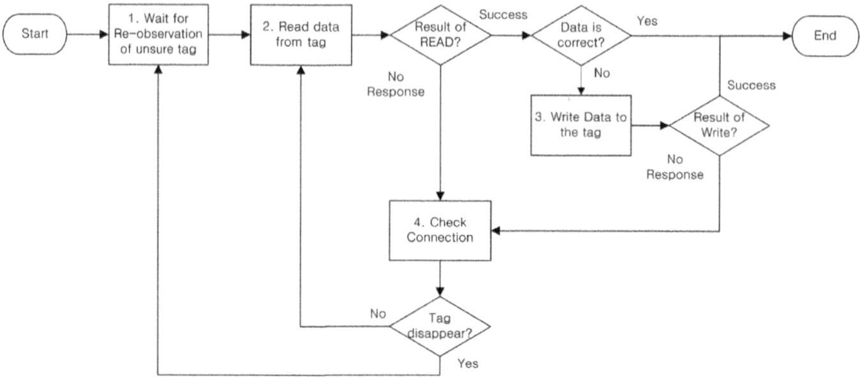

Fig. 5. Control flow for reprocessing unsure operations

In Step 3, the middleware sends a write command to reprocess the write operation. If the reply is *Success* then the reprocessing is complete. If the reader returns *No Response*, the operation is still incomplete, and so Step 4 is invoked. In Step 4, the middleware checks the connection to determine whether the tag is disappeared or the connection is temporarily unavailable. This check is achieved by sending an REQ_RN [3] command to the tag. The REQ_RN command requests a handle for accessing tag memory. If the tag is in the field-of-view of the reader, it returns a valid handle and Step 2 should be invoked for verification. If return of a handle fails, the middleware determines that the tag has disappeared and Step 1 should be invoked.

Let us consider an extension of our model from the unsure operation to the incomplete operation. A single write operation may require execution of multiple write commands, and the operation can be resulted to be *partially completed.* Resuming mechanism for the incomplete operation is basically the same as that of the unsure operation. Since we can find out how many words are written successfully, resuming the operation continues until the number of words written is equal to the length of data by investigating the replies to the write operation.

An algorithm for resuming incomplete operation is described in Algorithm 1. In the *Resume()* function, tag data is verified only when the previous write operation has resulted in *No Response*. The write operation is requested only for the remaining data and w.*nww* is increased by checking the reply. If the write operation results in *Temporarily Failed*, a forced re-write without verification is required because the tag memory is unchanged. The function *Check_Connection_and_Resume()* checks the connection between rid and w.tid. If the connection is still alive, this function calls *Resume()* again recursively. When the connection is lost, this function returns false and the operation will sleep again until the tag is re-observed.

Rollback() is called when the write operation results in *Permanently Failed.* Because the *Permanently Failed* error is the logical error, the operation cannot be processed further. It is required to roll-back the operation to keep consistency of the tag data. The role of *Rollback()* is to recover already written memory area to the original data. Rolling back the incomplete operation is similar to *Resume()* because it requires

Algorithm 1. Resume an incomplete operation

```
Algorithm Resume (w, rid)
w is a record of unsure list
rid is a reader identifier which re-observed w.tid
begin
    if w.isUnsure = true then
        reply = read(w.tid, w.offset+w.nww, 1)
        if reply.result = Success then
            w.isUnsure = false
            if reply.data is equal to w.data[w.nww] then
                w.nww = w.nww + 1;
                if w.nww = w.length then return Success
            end if
        else if reply.result is No Response then
            return Check_Connection_and_Resume (w, rid)
        end if
    end if

    reply = write(w.tid, w.offset+w.nww,
                  w.length-w.nww, w.data + w.nww)
    w.nww = w.nww + reply.nww;
    if reply.result = Success then
        return Success;
    else if reply.result = No Response then
        w.isUnsure = true
        return Check_Connection_and_Resume (w, rid)
    else if reply.result = Temporarily Failed then
        return Check_Connection_and_Resume (w, rid)
    else if reply.result = Permanently Failed then
        return Rollback (w, rid)
    end if
End
```

internal reprocessing to write the original data recursively. When the tag is re-observed, *Rollback()* is called again to roll back the operation completely. Therefore, the operation remains in the unsure list.

3.4 Reporting to the Application

When a write operation is processed in the middleware, a result of the operation is reported to the client as either *Success* or *Fail* with error code. When a write operation is determined to be incomplete, the operation cannot be reported as *Success* or *Fail.* A new return message is required to notify that the operation is incomplete.

A *Presumed Completed* message is sent to the application to indicate this status when the operation is resulted to be incomplete. The *Presumed Completed* means that the operation is not yet successfully processed but it will be internally processed completely. When the operation is processed successfully, the operation returns a *Success* message to the application. Interaction between the middleware and the client is illustrated in Fig. 6, which shows an example of the reprocessing sequences for one read command execution and one write command execution.

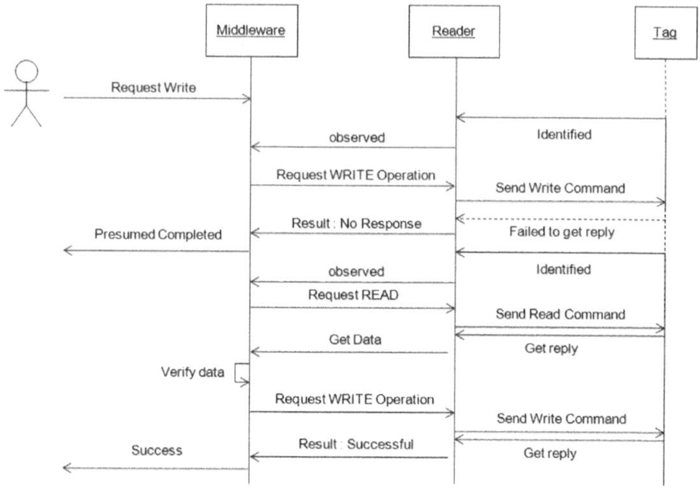

Fig. 6. Example sequence diagram for incomplete operation reprocessing

Table 2. List of message sent to the client

Message	Description
Presumed Completed	A Write operation is determined to be incomplete.
Success	Reprocessing of a operation is completed successfully
Presumed Failed	A logical error has occurred during reprocessing an operation and the operation is remained as inconsistent
Failed	A write operation is completely rolled back, or timeout has occurred during reprocessing the operation

Table 2 lists up notification messages that are sent to the client as a result of the reprocessing. When the incomplete operation results in *Permanently Failed*, the middleware sends a *Presumed Failed* message to the application to notify that an error has occurred and the operation will be rolled-back automatically. When the rolling-back of the operation is completed, a *Failed* message is sent to the client.

3.5 Completing the Operation

The main purpose of the proposed model is to guarantee complete writing to the tag memory despite volatile RF communications. In our model, incomplete operations will not be removed from the unsure list until the operation is processed completely. However, there is one exception to this rule, for which an incomplete operation cannot be reprocessed completely. This is when re-observation of a tag does not occur. Sometimes, an unsure tag is not observed again for reasons as malfunction of the tag, physical corruption, or movement to another site. We call this situation *starvation* of re-observation of an unsure tag. It may also cause middleware overload, because the increased size of the unsure list will affect middleware performance. Therefore, the middleware should handle tags that are not observed during a long period.

To achieve this, our model includes a timeout value for each write-request from the application, and the middleware will remove an unsure tag from the unsure list and send a *Failed* message to the application if the operation's timeout period expires. Detecting expiration for incomplete operations can be achieved by using a priority queue structure. The priority queue is constructed by using the expiration time for each unsure tag in the unsure list. The middleware then periodically checks the priority queue to enable detection of any expired unsure tag.

4 RFID Middleware Design and Implementation

Figure 7 shows a block diagram of the RFID middleware architecture for implementing the proposed reprocessing model. We add three new components to the middleware for the proposed model, which are shown in the highlighted area of Fig. 7. The detailed description of each component is as follows.

- Unsure Tag Filter: the component that receives the inventoried tag stream and checks whether each tag is unsure or not. To achieve this, it generates a stabbing query to the unsure tag index for each inventoried tag.
- Unsure Operation Manager: the component that manages incomplete operations. This maintains two modules: the unsure list, and the unsure tag index described in Section 3.2.
- Unsure Operation Processor: the component that performs the resume function for a re-observed tag. The resume function is described in Algorithm 1.

If an operation's execution is assessed as incomplete by the Operation Executor, the Operation Processor passes the incomplete operation request to the Unsure Operation Manager, which will insert the operation into the unsure list and the unsure tag index. The Unsure Tag Filter receives the input tag stream before any filtering by the Tag Filter. The Unsure Tag Filter checks if the input tag is unsure by stabbing into the unsure tag index. If the tag is not found in the index, it is passed to the Tag Filter for normal execution. If the tag is found in the index, it is reported to the Unsure Operation Manager, which will resume the unsure operation by a request to the Unsure Operation Processor. The Unsure Operation Processor performs the resume algorithm by requesting relevant operations of the Operation Executor.

We have implemented a prototype of the RFID middleware and simulated using virtual reader which supports various tag access operations. Our RFID middleware provides reprocessing of incomplete operations as well as processing of access operations and filtering of tag events. It is designed to be fully compatible with the ALE 1.1 standard specification [11], which is the standard interface for RFID middleware proposed by EPCglobal Inc. [12].

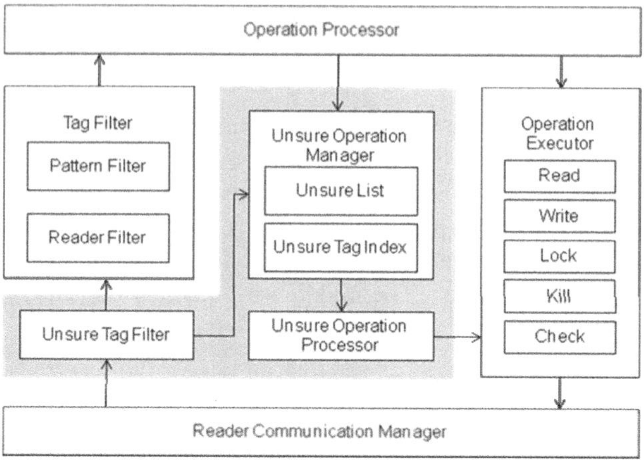

Fig. 7. Middleware architecture for the AW reprocessing model

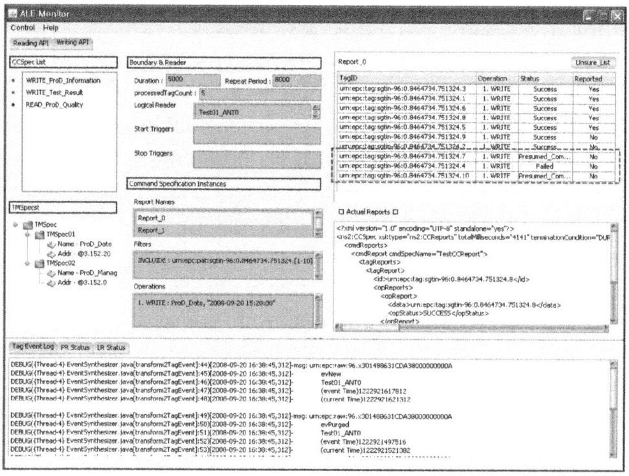

Fig. 8. A Screenshot of the ALE monitoring program

Figure 8 shows a snapshot of an ALE monitoring program which monitors running queries registered to the middleware. All user requests are listed in the left side of windows and detailed description of the selected request is shown in the middle. Matched tags are shown at right side of the window with operation results. If an operation is resulted to be incomplete, the operation is marked as *Presumed Completed.* In Fig. 8, two write operations are resulted to be incomplete.

Incomplete operations are immediately registered to the unsure list and are monitored separately. Figure 9 shows the state of incomplete operations with related information. When the tag is re-observed, the operation is resumed and the state information of the operation is updated. A tag highlighted by dotted box is processed successfully after a few minutes later, and the state of the tag is changed from *Presumed Completed* to *Success*.

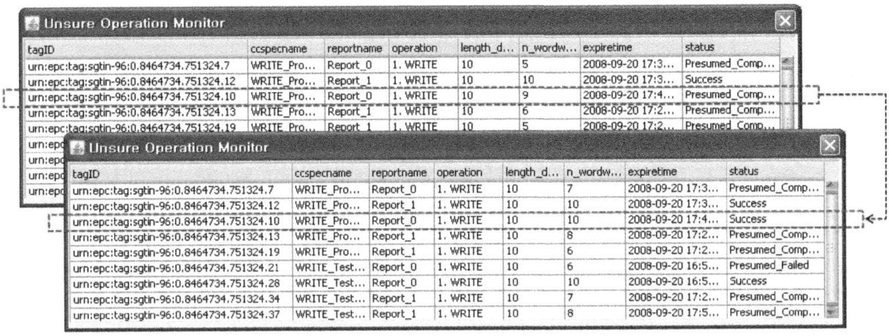

Fig. 9. Monitoring incomplete operations

5 Conclusion

The disconnection problem of RF communication during processing of a write operation from an RFID reader to a passive RFID tag will cause inconsistency of data in the tag memory. The volatile and uncertain characteristics of RF communications are caused by tag operations not being self-powered, limited reader range, and surrounding environments that interfere with communication. Unsuccessful write operations can cause data inconsistency, and so the atomicity and durability of writing transactions cannot be guaranteed.

To solve this problem, we have proposed an asynchronous wake-up reprocessing model for write operations. Our model suspends the execution of incomplete operations and separates them from other operations. Using a CQI scheme, re-observation of an unsure tag is detected asynchronously, following which the middleware wakes up the incomplete operation and resumes it internally. We define a step-by-step mechanism for reprocessing incomplete operations as well as a resumption algorithm. Our model processes incomplete operations completely without any inconsistent data access. We have verified this by implementing the proposed model within the RFID middleware.

The main contribution of our work is that we have developed a complete reprocessing model for overcoming the limitations of RF communications. In future work, we aim to extend our work to RFID transactions that support concurrent executions, and to enhance the CQI method that provides better performance.

References

1. Roy, W.: An Introduction to RFID Technology. IEEE Pervasive Computing 5(1), 25–33 (2006)
2. Banks, J., Hanny, D., Pachano, M., Thompson, L.: RFID APPLIED, pp. 328–329. Wiley, Chichester (2007)
3. EPCglobal Inc: EPC™ Radio-Frequency Identity Protocols Class-1 Generation-2 UHF RFID Protocol for Communications at 860-960 MHz, Version 1.1.0, `http://www.epcglobalinc.org/standards/uhfc1g2/uhfc1g2_1_1_0-standard-20071017.pdf`

4. Sanjay, K.M., Mukesh, K.M., Sourav, S.B., Bharat, K.B.: Mobile data and transaction management. Inf. Sci. 141(3-4), 279–309 (2002)
5. Ron, W.: RFID: A Technical Overview and Its Application to the Enterprise. IT Professional 7(3), 27–33 (2005)
6. Kenneth, P.F., Bing, J., Matthai, P., Sumit, R.: I sense a disturbance in the force: Unobtrusive detection of interactions with RFID-tagged objects. In: Davies, N., Mynatt, E.D., Siio, I. (eds.) UbiComp 2004. LNCS, vol. 3205, pp. 268–282. Springer, Heidelberg (2004)
7. Christian, F., Christof, R., Matthias, L.: RFID Application Development with the Accada Middleware Platform, IEEE Systems Journal, Special Issue RFID (January 2008)
8. EPCglobal Inc: Low Level Reader Protocol (LLRP), Version1.0.1, http://www.epcglobalinc.org/standards/llrp/llrp_1_0_1-standard-20070813.pdf
9. Lukasz, G., Tamer Özsu, M.: Issues in data stream management. SIGMOD Record 32(2), 5–14 (2003)
10. Karthaus, U., Fischer, M.: Fully Integrated Passive UHF RFID Transponder IC With 16.7-W Minimum RF Input Power. IEEE Journal of Solid-State Circuits 38(10), 1602–1608 (2003)
11. EPCglobal Inc.: The Application Level Events (ALE) Specification, Version 1.1, http://www.epcglobalinc.org/standards/ale/ale_1_1-standard-core-20080227.pdf
12. Epcglobal Inc., http://www.epcglobalinc.org

QoS-Oriented Multi-query Scheduling over Data Streams

Ji Wu[1], Kian-Lee Tan[1], and Yongluan Zhou[2]

[1] School of Computing, National University of Singapore
{wuji, tankl}@comp.nus.edu.sg
[2] Dept. of Mathematics & Computer Science, University of Southern Denmark
zhou@imada.sdu.dk

Abstract. Existing query scheduling strategies over data streams mainly focus on metrics in terms of system performance, such as processing time or memory overhead. However, for commercial stream applications, what actually matters most is the users' satisfaction about the Quality of Service (QoS) they perceive. Unfortunately, a system-oriented optimization strategy does not necessarily lead to a high degree of QoS. Motivated by this, we study QoS-oriented query scheduling in this paper. One important contribution of this work is that we correlate the operator scheduling problem with the classical job scheduling problem. This not only offers a new angle in viewing the issue but also allows techniques for the well studied job scheduling problems to be adapted in this new context. We show how these two problems can be related and propose a novel operator scheduling strategy inspired by job scheduling algorithms. The performance study demonstrates a promising result for our proposed strategy.

1 Introduction

Many typical applications of Data Stream Management System (DSMS) involve time-critical tasks such as disaster early-warning, network monitoring, on-line financial analysis. In these applications, output latency (as the main QoS measure) is extremely crucial. Managing system resources to maintain a high QoS is particularly important for applications that have Service Level Agreements (SLA) with the clients, where each client may have its own QoS requirement as to when query answers should be delivered.

In traditional DBMS, where data are pull-based, the output delay depends only on the query cost. However, in a DSMS, where data are pushed from scattered sources and their arrivals are out of the DSMS's control, the output latency becomes *tuple-dependent*. Input tuples may have experienced different degrees of delay (due to, for instance, the varying data transmission condition) before arriving at the system. Therefore, the query executor has to continuously adapt to the ever changing initial input delay to ensure results to be produced in a timely manner. Given the unpredictable workload and the limited resource, a DSMS may not always be able to meet all the QoS requirements. When that happens, efforts should be made to satisfy as many clients as possible to maximize the profits or minimize the loss.

Strategies proposed in this paper can be viewed as our initial effort towards QoS-oriented adaptive query processing for data streams. In this work, we take the output

X. Zhou et al. (Eds.): DASFAA 2009, LNCS 5463, pp. 215–229, 2009.

latency as the main QoS metric since it is the key parameter for typical online applications. Given that each input tuple is attached with a timestamp indicating when it is generated, a result tuple is said to meet the QoS if the time difference between the output time and the input timestamp is no more than the user-specified delay threshold. It is important to note that the output latency defined here embraces both query processing time and various delays incurred during query processing and data transmission. The latter are variables that fluctuate over time beyond query engine's control.

Several query scheduling strategies have been proposed to improve the query performance in a DSMS. However, the main objective of these algorithms is to minimize the average query processing time [5,10,12] or memory consumption [1]. Few of them deal with the issue of optimizing the total user satisfaction. Our proposed QoS-oriented metric complements the above work by considering a more realistic scenario. As can be seen later, a QoS-oriented perspective leads to an entirely different scheduling strategy from the existing work. In summary, contributions of this paper are:

1. Proposition of QoS-oriented query scheduling in a system-oriented context for continuous query processing;
2. An in-depth analysis of how the operator scheduling problem can be transformed into a job scheduling problem;
3. A novel planning-based scheduling strategy which is designed based on the above transformation and addresses the QoS-oriented multi-query optimization issue;
4. Extensive experimental studies that verify the effectiveness of the solution.

The rest of the paper is organized as follows: Section 2 introduces the problem to be solved and surveys the related work. Section 3 transforms the operator scheduling issue to a job scheduling problem to facilitate our problem analysis. Section 4 details the proposed scheduling algorithm. Section 5 demonstrates the effectiveness of the proposed strategy through our experimental study. Finally we conclude the paper in Section 6.

2 Preliminaries

2.1 Metric Definition

As mentioned before, the quality requirement is defined as the maximum tolerable delay of output tuples. To evaluate the QoS performance of the system, we define the following QoS penalty function.

Definition 1. *Given a query Q, let T^i_{out} denote the time when an output tuple i is produced and T^i_{in} be the maximal timestamp of all the input tuples that contribute to the tuple i. And L is the predefined QoS threshold for query Q. The penalty function for tuple i is:*

$$U_i = \begin{cases} 0 \; if \quad T^i_{out} - T^i_{in} \leq L \\ 1 \; otherwise \end{cases} \tag{1}$$

Accordingly, the query level QoS can be evaluated by taking the normalized aggregation of the tuple level penalties:

$$\frac{\sum_{i=1}^{n} U_i}{n}, \; n \text{ is the total number of output tuples} \tag{2}$$

Intuitively, a query's quality is inversely proportional to the value in Equation 2.

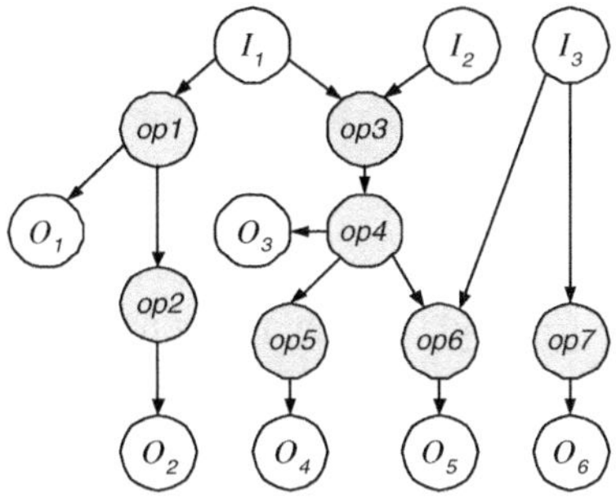

Fig. 1. A query graph example

In a multi-query environment, we seek to achieve high output quality across all participating queries. Each query q_i is assumed to be associated with a weight w_i to indicate its importance: A higher w_i implies a higher priority of q_i. Now, the penalty function U over all queries is the weighted sum of those of the individual queries:

$$U = \sum_{j=1}^{m} \frac{w_j \sum_{i=1}^{n_j} U_i}{n_j} \tag{3}$$

where, m is the number of participating queries.

For unbounded input streams, the objective function should be defined within an observation period. Then the parameter n in the equation refers to the total number of output tuples produced in the recent observation period (say last five hours). Note that the length of the period does not affect our algorithm. It should be meaningful to the application. For example, if its length is 5 hours, then our algorithm is optimizing the objective function defined on the last 5 hours.

2.2 System Models

Similar to existing work on stream processing, we model the entire Continuous Query (CQ) plan with a Directed Acyclic Graph (DAG). Vertices with only outgoing edges represent input streams and those with only incoming edges represent output streams. Other vertices are query operators. Edges connecting vertices are tuple queues that link the adjacent operators. Data flows are indicated by arrows. For example, Figure 1 shows a query graph with three input streams (I_1, I_2, I_3) and six output streams ($O_1, O_2, ..., O_6$). Each output stream corresponds to exactly one registered query in the system (O_1 is the output for query Q_1, and O_2 for query Q_2 ... so on and so forth). Also there are seven query operators ($op1, op2, ..., op7$) in this plan. Some operators are dedicated to a single query (such as $op7$ for query Q_6) while others are shared among several queries (such as $op4$ for query Q_3, Q_4 and Q_5).

In this problem setting, we assume complete and ordered query results are desired. For each input stream, tuple arrivals are ordered by their timestamps. Each query operator can only process tuples from its input queue in a First-Come First-Served (FCFS) manner so that the tuple order is preserved throughout the query execution. Since each input tuple has a timestamp indicating the tuple creation time and each query has a predefined QoS threshold, whenever a new input tuple arrives, the system can compute the deadline for producing the corresponding output in order to satisfy the QoS requirement. For example, given Q_1's QoS threshold L_1, if an input tuple $p \in I_1$ with timestamp T_{in}^p arrives at time t, then the deadline to produce the corresponding output tuple for query Q_1 is $T_{in}^p + L_1$. And the available time left for query processing, which is called *Remaining Available Time* (RAT), will be $T_{in}^p + L_1 - t$. If it actually requires C_p amount of CPU time, which is called *Remaining Processing Time* (RPT), for the system to process the tuple, then qualified output can be possibly produced only when $C_p \leq T_{in}^p + L_1 - t$ (i.e. RPT $\leq$ RAT).

However, it is not always easy to find out the deadlines of producing the qualified output especially when a query involves more than one input stream. For example, tuples from I_1 alone cannot determine the deadlines for the resultant output tuples of query Q_3 since their deadlines also depend on timestamps of inputs from I_2. Queries involving multiple streams will be discussed in detail in Section 4.3.

2.3 Problem Statement

The formal problem statement can be put as follows: *Given the query operator graph, continuously allocate a time slot for each operator to process each of its input tuples such that the objective function U, defined in Equation 3, is minimized.*

2.4 Related Work

Our work mainly concerns two areas: 1) operator scheduling in stream systems; 2) job scheduling algorithms that minimize the number of late jobs.

The issue of operator scheduling has been studied with different objectives. For example, The Chain algorithm [1] schedules the operators in a way such that runtime memory overhead can be minimized. In the rate-based scheduling algorithm proposed by Urhan *et al.* [12], the objective is to maximize the output rate at the early stage of query execution. There are also scheduling algorithms proposed for optimizing query response time (a.k.a. output latency) [5] or its variant metric (such as slowdown in [10]). However, the objectives in these work are more system-oriented in the sense that the optimal solution is the one that maximizes the system performance, but not users' satisfaction. In contrast, we adopt a QoS-oriented view, which brings in the user requirements as another dimension of the issue. Such slight difference, however, renders totally different problem settings, and hence completely different solutions. Probably the most relevant work to us is the one done by Carney *et al.* [3]. They provide interesting scheduling solutions to account for QoS-oriented requirements. However, their approach is only targeted at the scenario where operators are not shared among the queries. This is a strong assumption because stream applications often involve multi-query processing with operators shared among different queries. Scheduling over shared operators, as illustrated in this paper, can be far more complicated.

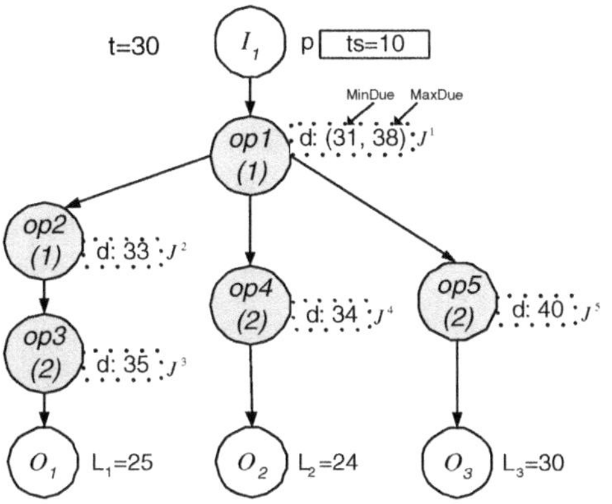

Fig. 2. Transform operator scheduling to job scheduling

The job scheduling problem (particularly, the problem of minimizing late jobs) has been studied over the years. Various algorithms were proposed to address this class of problem with different constraints. Karp [6] proved the weighted number of late jobs problem in general, denoted as $1||\sum w_j U_j$, is NP-hard. But it is solvable in pseudopolynomial time [8]. Polynomial algorithm is available if the processing time and the job weight can all be oppositely ordered [7]. In the more recent work [9], solutions were proposed to solve the same class of problem with the condition that job release time is not equal. However, to the best of our knowledge, no work has been done that can directly address the multi-query scheduling problem due to the complication of operator-sharing and precedence constraints that are unique in DSMS. Existing approaches are either too general or too restrictive to be applied directly in our context.

3 From Operator Scheduling to Job Scheduling

We show in this section how operator scheduling problem can be approximated by a job scheduling model. This provides a new angle to view the issue and allows us to borrow ideas from a well studied subject to develop low-cost operator scheduling algorithms.

In a typical single machine job scheduling problem, people look for a plan that allocates each job the appropriate time slot for execution so that the objective function is optimized. Each participating job J_i is associated with a processing cost c_i, a deadline d_i and a penalty value u_i. J_i is *on time* if its completion time $t_i \leq d_i$; otherwise, the job is *late* and the penalty u_i will be incurred.

Analogously, in continuous query processing, we can treat the work done by a query operator in response to the arrival of a new input tuple as a job. For example, for the query plan shown in Figure 2, the arrival of a tuple $p \in I_1$ (indicated by the box with solid lines) with timestamp $ts = 10$ triggers five jobs to be created in the system, each corresponds to one involved operator (Let J^x denote the job performed by operator op$x$ in the figure). The estimation of jobs' processing cost, deadline and penalty value are explained in the rest of this section.

Processing Cost. The job processing cost is the product of two parameters: unit processing cost and cardinality of the input size. Unit processing cost is the time taken for the operator to process one tuple from its input queue. Cardinality of the input size is determined by the multiplicity (or selectivity) of all upstream operators along the path from the input stream node to the current operator. For the example in Figure 2, let $\rho 1$ and $\rho 2$ denote the multiplicity of $op1$ and $op2$ respectively, then the input cardinality for $op3$ is simply $\rho 1 \rho 2$. If c_3 is the unit processing cost for $op3$, the job cost of J^3 would be $\rho 1 \rho 2 c_3$.

Deadline. Unlike traditional job scheduling problem, not all jobs are given explicit deadlines here. Firstly, it is important to distinguish two types of jobs in this context: *Leaf-Job* (L-Job) and *NonLeaf-Job* (NL-Job). L-Job refers to jobs performed by the last operator in a query tree. Examples of L-Jobs are J^3, J^4 and J^5 in Figure 2. Intuitively, for L-Job, its deadline coincides with the due date by which the query output should be produced. The value can be calculated by adding the input tuple timestamp with the respective query QoS threshold. For example, the deadline for J^3 would be 35 (input timestamp $ts = 10$ plus Q_1's QoS threshold $L_1 = 25$).

NL-Job refers to jobs performed by non-leaf operators. The output of an NL-Job becomes the input of some other NL-Job or L-Job in a query plan. J^1 and J^2 are examples of NL-Jobs in Figure 2. Computing deadline for NL-Job with fan-out equal to 1 is relatively easy. Basically, it can be derived backwards from its only immediate downstream job. For example, to compute J^2's deadline, we just need to know the deadline and processing cost of job J^3. For simplicity, assume the multiplicity of all operators in the example is 1 and the unit cost of each operator is indicated by the number in the corresponding bracket shown in the figure. Hence J^3's processing cost is $\rho 1 \rho 2 c_3 = 1 \times 1 \times 2 = 2$. And J^2's deadline is simply J^3's deadline minus J^3's processing cost, $35 - 2 = 33$. This essentially gives the latest date by which J^2 has to finish such that it is possible for the downstream job J^3 to complete on time.

Unfortunately, to define deadline for NL-Job with fan-out greater than 1 is much more involved. This is because different downstream jobs have different QoS requirements and due dates. Consider job J^1 in the figure whose fan-out is 3. The three immediate downstream jobs J^2, J^4 and J^5 require J^1 to complete latest by 32, 32 and 38 respectively in order for *themselves* to complete on time. There is no single definite deadline that can be determined for J^1 in this case. Here, we use two due dates, namely *MinDue* and *MaxDue*, to characterize the situation. The definition of *MinDue* and *MaxDue* are given as follows:

Definition 2. *The MinDue of job i is the latest time by which i has to complete such that there exists a feasible plan with all its downstream jobs to be scheduled on time.*

Definition 3. *The MaxDue of job i is the latest time by which i has to complete such that there exists a feasible plan with at least one of its downstream jobs to be scheduled on time.*

The significance of MinDue and MaxDue can be viewed as follows: If an NL-Job completes before its MinDue, then no downstream jobs will be overdue because this NL-Job cannot complete early enough. On the other hand, if the NL-Job completes after its

MaxDue, then no downstream jobs can possibly complete on time. Section 4 will show how MinDue and MaxDue are used in the proposed scheduling algorithm.

The MaxDue for an NL-Job is in fact just the maximum date among the deadlines derived from each of its downstream jobs. In the example, J^1's MaxDue would be $\max\{32, 32, 38\} = 38$. However, to compute MinDue is a bit complicated because of the schedule feasibility check. Due to space constraints, we omit the detail here. Interested reads can refer to [13] for the algorithm to compute MinDue. In the above example, the result MinDue for J^1 would be 31 (not 32). And the corresponding feasible schedule for its downstream jobs to complete on time is $J^2 \rightarrow J^4 \rightarrow J^5$. All the job deadlines in the example are shown inside boxes with dotted lines in Figure 2.

Job Penalty. To determine the penalty value associated with each job is another issue. In our objective function, the (weighted) penalty is only defined over each output tuple. That can be seen as the penalty applied to a late L-Job. However, for NL-Job, such penalty value is undefined since its completion does not affect the objective function in a direct way. Nevertheless, a late NL-Job, which causes its downstream L-Jobs to be late, does influence the overall QoS. We shall see how this subtlety is handled in our proposed scheduling algorithm in Section 4.

4 Scheduling Algorithm

The distinct features about multi-query scheduling as described in the previous section reveal that the problem is much harder to solve than a traditional job scheduling issue. This mainly stems from two reasons: 1) Jobs may be shared among different queries. 2) Job precedence constraints (given by the query plan tree) have to be observed. In what follows, we propose a novel dynamic planning-based heuristic, which effectively schedules the query operators to achieve a good overall QoS in polynomial time.

4.1 Job Set for Scheduling

Since the system deals with continuous queries running over potentially unbounded data streams, the scheduling strategy has to be an online algorithm. Here we propose a planning-based online strategy. Instead of selecting individual job, the scheduler selects *a set of jobs* from the received input tuples for each scheduling round. As we shall see, the planning-based approach enables the scheduler to take a holistic view when making the current scheduling decision, rendering a better output quality. However, the question that immediately follows is how large the job set should be. In other words, how far the scheduler should look ahead. If it looks too far, the plan may be outdated if later the subsequent arrivals of input tuples trigger jobs with earlier deadlines to be generated. This causes rescheduling that inflicts unnecessary overhead. On the other hand, if the lookahead is too short, poor decision may be reached due to lack of global deliberation. Hence, the ideal scenario is to include the minimum number of jobs enough to construct the global optimal (or quasi-optimal) plan for a near future.

To this end, we propose the following job selection criterion: All jobs to be included for scheduling must have their deadlines earlier than those of future jobs to be generated

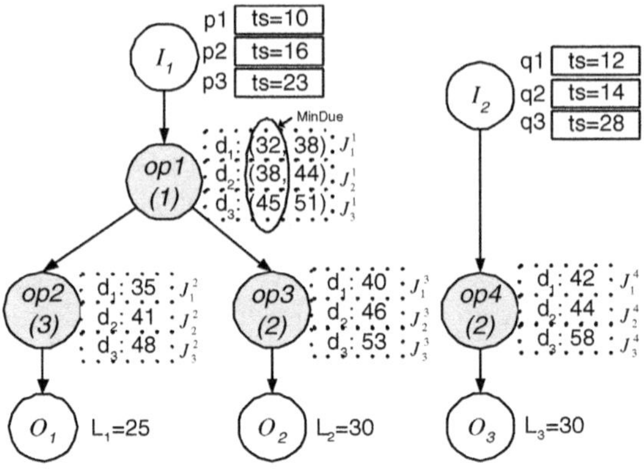

Fig. 3. Choosing the appropriate job set for scheduling

by the system. This rule can be enforced because input tuples are ordered according to their timestamps and processed in a FCFS manner by the operators. The monotonicity allows us to find a point in time which all future jobs' deadlines are bound to be beyond.

This point can be obtained by finding the earliest deadline among all the latest generated jobs associated with each operator. For example, Figure 3 shows a snapshot of a query graph with two input streams and four operators that produce three output streams. The graph shows there are three tuples buffered at each of the input queue ($p1, p2, p3$ at I_1 and $q1, q2, q3$ at I_2). These tuples virtually generate 12 jobs (boxes with dotted lines) for the system. Job performed by operator op$x$ with reference to input tuple py (or qy) is denoted as J_y^x in the figure. (Again, assume the query operators selectivity or multiplicity is 1 for simplicity) Their deadlines are indicated in the corresponding boxes. For NL-Job with fan-out greater than 1, we take its MinDue for our consideration. Therefore, among all the latest jobs for each operator (i.e. J_3^1, J_3^2, J_3^3, J_3^4), the one associated with $op1$ (i.e. J_3^1) has the earliest deadline $d = 45$, which is essentially the cut-off point: All jobs with deadlines less than or equal to 45 will be included for scheduling (i.e. those indicated as shaded boxes) while others are left for the next round of scheduling. The selected job set will be considered for scheduling in the algorithms depicted in the next section.

4.2 Scheduling Heuristic

In Section 3, we show that the uncertainties about NL-Job, in terms of both deadline and job penalty, greatly increase the problem complexity. In addition, precedence constraints among jobs with reference to the same input tuple further complicates the issue. For example, in Figure 3, any upstream job $J_y^1, y \in \{1, 2, 3\}$ has to be scheduled before the downstream job J_y^2 and J_y^3 to make the plan meaningful. In fact, the entire scheduling problem can be categorized as $1|prec|\sum w_j U_j$ [1] in standard notation [4]. It is an

[1] Denoting the problem of finding a non-preemptive schedule on a single machine such that the job precedence constraints are satisfied and the total weighted penalty function is minimized.

NP-hard problem and generally no good solutions or heuristics are known. To design a heuristic for this problem faces two challenges: 1) The produced plan must be feasible with all precedence constraints observed. 2) The benefits of executing NL-Jobs must be assessed in an efficient and intelligent way. We show how these two challenges are tackled in our proposed algorithm.

Evaluating Job Value. Denoting jobs with values (or utilities) is the essential step in a scheduling algorithm. In our problem setting, a duly completed job avoids penalty being applied to the objective function. Therefore, its value can be quantified as the amount of *Penalty Reductions* (PR) contributed to the system provided it is completed on time.

For L-Jobs, which are located at the bottom of the query tree, their PR values are calculated as follows: At the beginning of each scheduling cycle, the system calculates the current QoS of the L-Job's corresponding query according to equation 2. The amount of PR by completing the current L-Job, say job j, is therefore approximated as

$$w \times \left(\frac{\sum_{i=1}^{n} U_i + \prod \rho_j}{n + \prod \rho_j} - \frac{\sum_{i=1}^{n} U_i}{n + \prod \rho_j}\right) = \frac{w \prod \rho_j}{n + \prod \rho_j} \tag{4}$$

In the above equation, w is the weighting factor assigned to the given query, U_i is the penalty function defined in equation 1, and $\prod \rho_j$ is the production of multiplicities of all operators along the path from the first operator that takes the input stream to the leaf operator (i.e. the one performing the L-Job).

Unfortunately, computing PR for NL-Jobs is not straightforward. With the premise that the system will not be overloaded, we may reasonably assume that the number of late jobs in each scheduling cycle does not constitute a significant portion of the total number of jobs in that cycle. In other words, most NL-Jobs should be finished around their MinDues since a delayed NL-Job will seriously affect all the downstream jobs. In view of this, we adopt an optimistic approach in our heuristic: We assign an NL-Job's PR to be the sum of PRs of its immediate downstream jobs. Experiments show that this gives a very good estimation for the PR of an NL-Job.

Algorithm Sketch. The heuristic consists of two phases. Line 1 to 7 of Algorithm 1 sketch the first phase. As our intention is to maximize the potential NL-Job value (the optimistic view), the algorithm allocates all NL-Jobs at the earliest possible time without considering any L-Job in this phase. The sequence among NL-Jobs are determined according to their MinDue. By doing so, it implicitly enforces the precedence constraints because an upstream job always has an earlier MinDue than its downstream jobs. When an NL-Job is allocated a time slot, its scheduled completion time is recorded. This information will be used to check how much laxity is left between the current scheduled completion time and the due date.

In phase II of the heuristic (Line 8 to 30), L-Jobs are inserted into the schedule produced in phase I in a greedy manner. At the very outset, it is necessary to introduce the concept of Penalty Reduction Density (PRD) for NL-Job. The idea is similar to value density in job scheduling [11]. Essentially, it is the ratio between the value and the cost for a given job. Since for NL-Job, its value (in terms of PR) drops from maximum to

Algorithm 1. Job Scheduling Algorithm

Notations:
N: the set of all NL-Jobs
L: the set of all L-Jobs
P: Queue that records the final schedule
$x.cost$: the time cost for processing job x
$x.time$: current scheduled completion time for job x
$x.due$: deadline of job x if x is an L-Job; MinDue if x is an NL-Job
$x.maxdue$: MaxDue of NL-Job x
$x.pr$: the value (in terms of Penalty Reduction) of a job x
$x.prd$: Penalty Reduction Density of an NL-Job x

```
t:=0
while N is not empty do
    Find job i in N with the earliest MinDue
    N := N\{i}
    i.time := t + i.cost
    Append i at the end of P
    t := i.time
while L is not empty do
    Find job j in L with the earliest deadline
    L := L\{j}
    if CheckAncestor(j) returns FALSE then
        Append j at the end of P   /* j will miss the deadline */
    else
        Search from the beginning of P and find the first job k s.t. k.time ≥ j.due − j.cost
        Let Q denote the set that consists of job k and all jobs after k in P
        pen := 0   /* accumulated penalty */
        for all job m ∈ Q do
            if m.time < m.due then
                p := min{m.cost, max{0, m.time + j.cost − m.due}}
            else if m.time ≥ m.due AND m.time < m.maxdue then
                p := min{m.maxdue − m.time, j.cost}
            else
                p := 0
            pen := pen + m.prd × p
        if j.pr > pen then
            Insert j into P right before job k
            for all job n ∈ Q do
                n.time := n.time + j.cost
        else
            Append j at the end of P   /* j will miss the deadline */
```

zero as the completion time moves from below MinDue to above MaxDue, we can evaluate the PRD of an NL-Job using the following formula:

$$PRD = \frac{\text{PR}}{\text{MaxDue} - \text{MinDue}} \tag{5}$$

With PRD, we are able to estimate the potential penalty incurred with respect to the tardiness of an NL-Job completion.

Phase II of the algorithm goes as follows: All L-Jobs are considered for scheduling one by one ordered by their due dates. Firstly, function CheckAncestor() verifies the ancestor jobs are scheduled early enough for the current L-Job to complete on time (Line 11). This checking is straightforward and runs in $O(\log n)$ time. If the checking is passed, the L-Job will be tentatively allocated a latest possible time slot such that it can be completed by the due date. For all NL-Jobs whose completion times are deferred due to the insertion of the new L-Job, their potential losses, in terms of loss in PR, are assessed by multiplying the job tardiness with the job PRD (Line 17 to 24). And their aggregate value is compared against the new L-Job's PR. The new job can secure the tentative time slot only if its PR is greater than the total loss of PR of all NL-Jobs affected; otherwise, the L-Job is appended at the end of the schedule and treated as a late job. The total runtime of the heuristic is $O(n^2)$.

4.3 Multi-stream Queries

As mentioned at the beginning, for queries taking multiple input streams, the response time is defined as the difference between the result tuple delivery time and the maximal timestamp among all the contributing input tuples. However, quite often the system has no clue as to which particular contributing input tuple carries the largest timestamp value when they just arrive at the query engine. In the sequel, jobs triggered by the arrivals of these tuples will be assigned earlier deadlines than what is necessary (if those tuples indeed are not the one that carries the largest timestamp in the final output). Although such miscalculation only means some jobs will be unnecessarily scheduled in advance and no serious consequence will result most of time, it would be better to identify the delay or time difference among the input streams such that an offset can be applied to improve the deadline prediction. This is an issue related to input stream coordination. Strategies [2,14] have been proposed to identify the synchronization issue among the streams. However, this is an area where ideal solution has not been found. We will study this problem further in our future work.

4.4 Batch Processing

Tuple based operator scheduling offers fine-grained control over query execution. However, it leads to substantial overhead due to frequent context switches among operators. Batch based strategy effectively reduces such overhead by processing a series of tuples in one shot. The batch processing discussed in this section refers to grouping input tuples (from the same data source) such that they collectively trigger one job to be created for each involved operator. (as opposed to tuple based scheduling where every input tuple triggers one job for an involved operator) This not only helps cut down the number of context switches as mentioned, but also helps reduce the runtime of the scheduling algorithm since fewer jobs need to be scheduled in the system.

An important issue to consider here is the appropriate size for each batch. Given the dynamic input characteristics, we propose a dynamic criterion to determine the batch size as follows: Sequential tuples from the same input may form a single batch if 1)

The timestamp difference between the head tuple and the tail tuple in the batch does not exceed $a\mu$, where μ is the average laxity (defined as RAT-RPT) of jobs currently in the system and a is a runtime coefficient. 2) The timestamp difference of any two consecutive tuples is no more than $b\tau$, where τ is the average inter-arrival time of the corresponding input stream and b is a runtime coefficient.

Criteria 1 essentially constrains the length of the batch. The reference metric is the average laxity of the jobs currently in the system. The intuition here is job's laxity should be positively related to the length of delay that input tuples can tolerate. (Consider the delay experienced by the first tuple in the batch.) In our experiment, we set $a = 1$. It means the timestamp difference between the head tuple and tail tuple in the batch cannot be greater than the average jobs' laxity value in the system. Criteria 2 essentially determines the point that can mark the end of a batch. In our experiment, we set $b = 2$. It means if timestamp difference of two consecutive tuples is greater than two times of the average inter-arrive time, they will be separated into two batches.

Another issue here is to set the appropriate deadline of a job batch. The easiest way is to find a representative tuple from the batch and set the batch deadline according to that tuple's deadline. In our experiment, we choose the first tuple in the batch (which has the earliest timestamp value) as the representative tuple because it produces the most stringent deadline that guarantees no output tuples generated from the batch will be late if that deadline can be met.

5 Experimental Evaluation

5.1 Experimental Setup

We implemented our proposed algorithm as part of the QoS-aware DSMS prototype system. The system mainly consists of three components: query engine, statistical manager and query scheduler. The query engine is able to process queries involving selection, projection, join and aggregation. The statistical manager monitors information such as unit processing cost of each operator, input data rate as well as current QoS of each registered query and reports them to the scheduler, which then makes scheduling decisions based on these information.

The multi-query plan used for the experiment is generated randomly. The number of involved operators ranges from 24 to 48 and the number of involved queries are between 12 and 32. Each query has been given a QoS threshold, whose value is an integer between 500 to 10000 (ms). Also a weighting factor (integer between 1 and 10) is assigned for each query. We use three different data streams for our experiment. They are generated by a data generator which produces input streams following Poisson process with customizable mean inter-arrival time. Each produced tuple is given a timestamp indicating its generation time. Input tuples have to go through a "delay" operator before they can be processed. The "delay" operator essentially simulates the input transmission delay from data source to the query engine.

For comparison, we also implemented three other scheduling strategies: Earliest Deadline First (EDF), Least Laxity First (LLF)[2] and a random approach. All

[2] The strategy schedules the job with minimum slack time (i.e. min{RAT-RPT} among all jobs with RAT $\geq$ RPT) to execute.

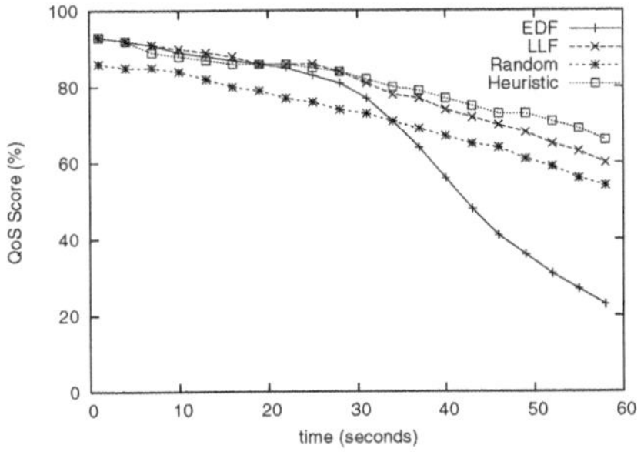

Fig. 4. QoS Score with increasing data rate (tuple-based)

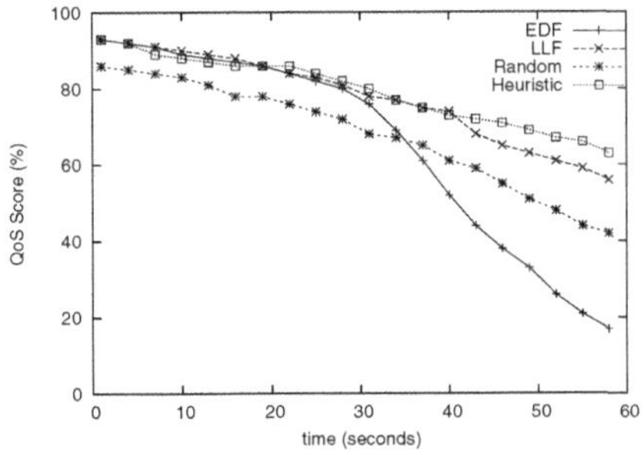

Fig. 5. QoS Score with increasing input transmission delay (tuple-based)

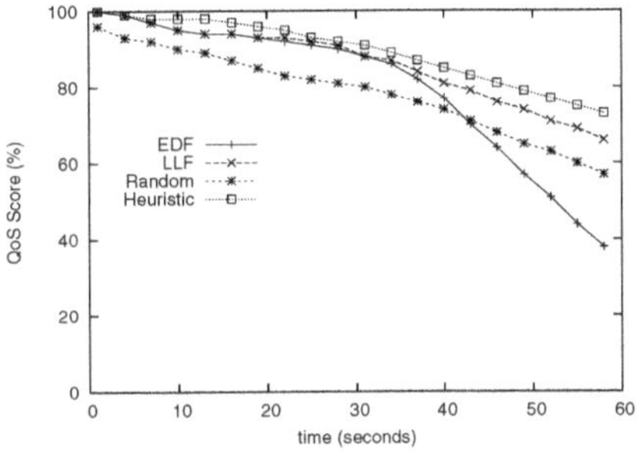

Fig. 6. QoS Score with increasing data rate (batch-based)

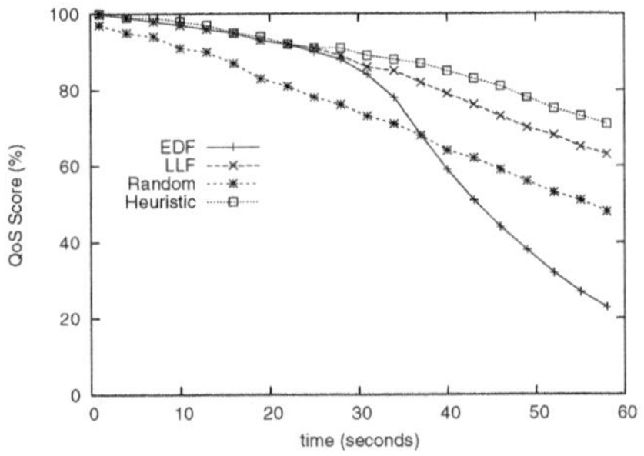

Fig. 7. QoS Score with increasing input transmission delay (batch-based)

experiments were conducted on an IBM x255 server running Linux with four Intel Xeon MP 3.00GHz/400MHz processors and 18G DDR main memory.

5.2 Performance Study

For ease of presentation, we convert the value of the objective function (defined in equation 3 in terms of weighted penalty aggregation) to a percentile QoS score as follows:

$$\text{QoS score} = \frac{U_{worst} - U}{U_{worst}} \qquad (6)$$

where U denotes the penalty value obtained from the experiment and U_{worst} denotes the worst possible penalty value. (i.e. the penalty value when all output tuples are late)

Strategy Comparison. We use the same set of queries and data to evaluate the performance of all four scheduling strategies. The experiment is designed as follows: We use two different ways to slowly increase the system workload over time and observe the resultant QoS score. In the first case, we achieve this by improving the data rate until the workload is equivalent to 80% of the system capacity. In the second approach,

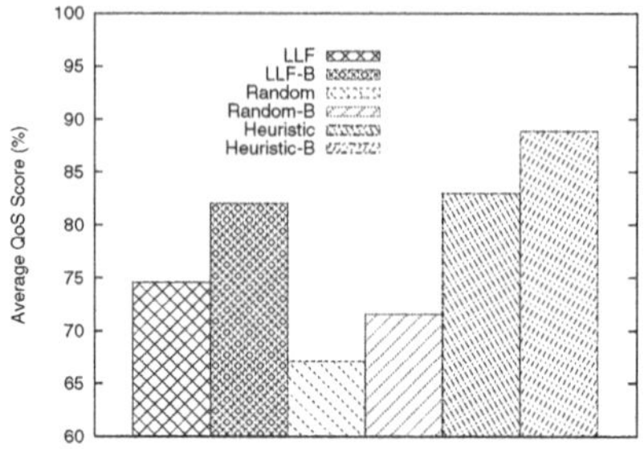

Fig. 8. Average QoS score

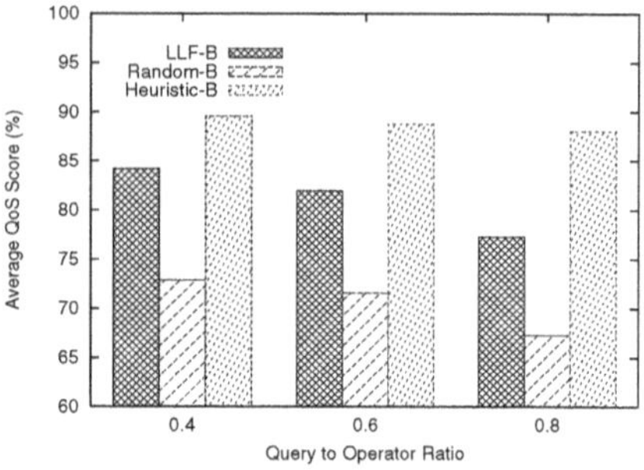

Fig. 9. Multi-query scalability

we keep the workload constant while slowly increasing the transmission delay of input tuples (achieved through the "delay" operator). This essentially reduces the RAT for input tuples leading to scheduling contentions. Results from Figure 4 and 5 clearly indicate that in both cases our heuristic approach performs better than other strategies. In particular, we can see that for EDF, the QoS score starts to drop significantly when the input load exceeds certain level. This can be explained by the Domino Effect: When system load becomes heavy, an EDF scheduler can perform arbitrarily badly because it consistently picks the most urgent task to execute, which may be already hopeless to meet the deadline. The LLF also performs worse than our heuristic due to its lack of vision to foresee the potential gain of scheduling NL-Jobs. The same set of experiments is also performed when all four strategies are running in batch-based scheduling mode (refer to Figure 6 and Figure 7). And similar conclusions can be reached.

Tuple-Based vs. Batch-Based Scheduling. We focus on performance comparison between tuple-based scheduling and batch-based scheduling in this section. Since the EDF strategy may perform arbitrarily badly, we do not include it for our consideration. Figure 8 plots the average QoS score achieved by the other three strategies for both tuple level scheduling and batch level scheduling. It turns out that batch-based scheduling outperforms tuple-based scheduling for all three strategies. This is mainly attributed to reduced number of context switches among query operators as well as decreased scheduling overhead. We also conducted experiments on input data with bursty nature. The performance contrast is even more obvious between tuple-based strategy and batch-based strategy. Due to space constraints, the details are not reported here.

Multi-query Scalability. A commercial stream system often runs a large number of similar queries, each for one subscribed user. That means within the query engine, operators are largely shared by different queries. This experiment examines the scalability of different scheduling strategies (excluding EDF for the same reason as above) in terms of query sharing. The degree of sharing is measured by the ratio between the number of queries to the number operators in the system. As depicted in Figure 9, with the increase of the query sharing, the superiority of our heuristic over other scheduling approaches become more and more evident. This is because our algorithm adopts a planning-based approach, which is able to look ahead to assess the potential value of each job for the entire system, while other strategies do not possess such clairvoyance when making the scheduling decision.

6 Conclusions

For a service-oriented data stream system, QoS-based query scheduling is an indispensable component. We propose a new multi-query scheduling strategy that aims to turn a DSMS into a true real-time system that can meet application-defined deadlines. The strategy is based on a novel transformation of our query scheduling problem to a job scheduling problem. Experimental study demonstrates a promising result of our proposed strategy. As part of the future plan, we will extend the current work to cater for applications with more generalized QoS specifications.

References

1. Babcock, B., Babu, S., Datar, M., Motwani, R.: Chain: Operator Scheduling for Memory Minimization in Data. In: SIGMOD, pp. 253–264 (2003)
2. Babu, S., Srivastava, U., Widom, J.: Exploiting k-constraints to reduce memory overhead in continuous queries over data streams. ACM Trans. Database Syst. 29(3), 545–580 (2004)
3. Carney, D., Çetintemel, U., Rasin, A., Zdonik, S.B., Cherniack, M., Stonebraker, M.: Operator Scheduling in a Data Stream Manager. In: VLDB, pp. 838–849 (2003)
4. Graham, R.L., Lawler, E.L., Lenstra, J.K., Rinnooy Kan, A.H.G.: Optimization and Approximation in Deterministic Sequencing and Scheduling: A Survey. Annals of Discrete Mathematics 5, 287–326 (1979)
5. Jiang, Q., Chakravarthy, S.: Scheduling strategies for processing continuous queries over streams. In: Williams, H., MacKinnon, L.M. (eds.) BNCOD 2004. LNCS, vol. 3112, pp. 16–30. Springer, Heidelberg (2004)
6. Karp, R.M.: Reducibility among Combinatorial Problems. In: Complexity of Computer Computations, pp. 85–103. Plenum, New York (1972)
7. Lawler, E.L.: Sequencing to minimize the weighted number of tardy jobs. RAIRO Operations Research 10, 27–33 (1976)
8. Lawler, E.L., Moore, J.: A functional equation and its application to resource allocation and sequencing problems. Management Science 16, 77–84 (1969)
9. Péridy, L., Pinson, E., Rivreau, D.: Using short-term memory to minimize the weighted number of late jobs on a single machine. European Journal of Operational Research 148(3), 591–603 (2003)
10. Sharaf, M.A., Chrysanthis, P.K., Labrinidis, A., Pruhs, K.: Efficient Scheduling of Heterogeneous Continuous Queries. In: VLDB, pp. 511–522 (2006)
11. Stankovic, J.A., Spuri, M., Rmamritham, K., Buttazzo, G.C.: Deadline Scheduling for Real-time Systems - EDF and Related Algorithms. Kluwer Academic Publishers, Norwell (1998)
12. Urhan, T., Franklin, M.J.: Dynamic Pipeline Scheduling for Improving Interactive Query Performance. In: VLDB, pp. 501–510 (2001)
13. Wu, J., Tan, K.L., Zhou, Y.: QoS-Oriented Multi-Query Scheduling over Data Streams. Technical Report (2008), `http://www.comp.nus.edu.sg/~wuji/TR/QoS.pdf`
14. Wu, J., Tan, K.L., Zhou, Y.: Window-Oblivious Join: A Data-Driven Memory Management Scheme for Stream Join. In: SSDBM Conference, 21 (2007)

An EM-Based Algorithm for Clustering Data Streams in Sliding Windows

Xuan Hong Dang[1], Vincent Lee[1], Wee Keong Ng[2], Arridhana Ciptadi[2], and Kok Leong Ong[3]

[1] Monash University, Australia
{xhdang;vincent.lee}@infotech.monash.edu
[2] Nanyang Technological University, Singapore
{awkng,arri0001}@ntu.edu.sg
[3] Deakin University, Australia
leong@deakin.edu.au

Abstract. Cluster analysis has played a key role in data understanding. When such an important data mining task is extended to the context of data streams, it becomes more challenging since the data arrive at a mining system in one-pass manner. The problem is even more difficult when the clustering task is considered in a sliding window model which requiring the elimination of outdated data must be dealt with properly. We propose SWEM algorithm that exploits the Expectation Maximization technique to address these challenges. SWEM is not only able to process the stream in an incremental manner, but also capable to adapt to changes happened in the underlying stream distribution.

1 Introduction

In recent years, we are seeing a new class of applications that changed the traditional view of databases as a static store of information. These applications are commonly characterized by the unbounded data streams they generate (or receive), and the need to analyze them in a continuous manner over limited computing resources [2,4]. This makes it imperative to design algorithms that compute the answer in a continuous fashion with only one scan of the data stream, whilst operating under the resource limitations. Among various data mining tasks, clustering is one of the most important tasks. Research in data stream clustering reported so far has mostly focused on two mining models, the landmark window [1,5] and the forgetful window [3,4]. While these two mining models are useful in some data stream applications, there is a strong demand to devise novel techniques that are able to cluster the data streams in a sliding window model which is the most general and also the most challenging mining model since it considers the elimination of outdated data.

We propose in this paper algorithm SWEM (clustering data streams in a time-based Sliding Window with Expectation Maximization technique) to address the above challenges. SWEM consists of two stages which are designed to strictly address the problem of constrained memory usage and one-pass processing over

X. Zhou et al. (Eds.): DASFAA 2009, LNCS 5463, pp. 230–235, 2009.

data streams. Furthermore, we develop in SWEM two important operations, namely splitting and merging micro components, in order to automatically adapt to changes happened frequently in stream distributions. Various experimental results confirm the feasibility of our proposed algorithm.

2 Problem Formulation

We focus on the time-based sliding window model. Let $TS_0, TS_1, \ldots, TS_i$ denote the time periods elapsed so far in the stream. Each time period contains multiple data points $x_i = \{x_i^1, x_i^2, ..., x_i^d\}$ (in d-dimensional space) arriving in that interval. Given an integer b, a time-based sliding window $\mathcal{SW}$ is defined as the set of records arriving in the last b time periods $\mathcal{SW} = \{TS_{i-b+1}, \ldots, TS_{i-1}, TS_i\}$. TS_i is called the latest time slot and TS_{i-b} is the expiring one. We also assume that the stream evolves with time and data points are generated as a result of a dynamic statistical process which consists of k mixture models. Each model corresponds to a cluster that follows a multivariate normal distribution. Consequently, any cluster $C_h, 1 \leq h \leq k$, is characterized by a parameter: $\phi_h = \{\alpha_h, \mu_h, \Sigma_h\}$ where α_h is the cluster weight, μ_h is its vector mean and Σ_h is its covariance matrix. Accordingly, our clustering problem is defined as the process of identifying parameters $\Phi_G = \{\phi_1, ..., \phi_k\}$ that optimally fit the current set of data points arriving in the last b time periods in the stream.

3 Algorithm Description

Initial Phase: We compute m distributions (also called micro components) modelling the data within each time slot of the sliding window. Let $\Phi_L = \{\phi_1, ..., \phi_m\}$ be the set of parameters of these local components where each MC_ℓ, $1 \leq \ell \leq m$, is assumed to follow a Gaussian distribution characterized by $\phi_\ell = \{\alpha_\ell, \mu_\ell, \Sigma_\ell\}$. For the initial phase where $SW = \{TS_0\}$, the initial values for these parameters will be randomly chosen.

In our framework, each data point belongs to all components yet with different probabilities. Given x, its probability (or weight) in a component ℓ^{th} is: $p(\phi_\ell|x) = \frac{\alpha_\ell \times p_\ell(x|\phi_\ell)}{p(x)} = \frac{\alpha_\ell \times p_\ell(x|\phi_\ell)}{\sum_{i=1}^{m} \alpha_i \times p_i(x|\phi_i)}$, in which $p_\ell(x|\phi_\ell) = (2\pi)^{-d/2}|\Sigma_\ell|^{-1/2} \exp\left[-\frac{1}{2}(x-\mu_\ell)^T \Sigma_\ell^{-1}(x-\mu_\ell)\right]$. We also assume data points are generated independently and thus, the probability of n records in TS_0 is computed by the product: $p(TS_0|\Phi_L) = \prod_{x_i \in TS_0} p(x_i|\Phi_L) = \prod_{i=1}^{n} \sum_{\ell=1}^{m} \alpha_\ell \times p_\ell(x_i|\phi_\ell)$, and in log likelihood form $Q(\Phi_L) = |TS_0|^{-1} \log \prod_{x \in TS_0} \sum_{h=1}^{m} \alpha_\ell \times p_\ell(x_i|\phi_\ell)$ which defines the average log likelihood measure.

In the first stage, SWEM employs the EM technique to maximize $Q(\Phi_L)$. Once the algorithm converges, the set of micro components are approximated by keeping a triple $T_\ell = \{N_\ell = |S_\ell|, \theta_\ell = \Sigma_{x_i \in S_\ell} x_i, \Gamma_\ell = \Sigma_{x_i \in S_\ell} x_i x_i^T\}$ for each MC_ℓ (where S_ℓ is the set of data points assigned to MC_ℓ to which they have the highest probability). The important property of T_ℓ is that it is sufficient to compute the mean and covariance of MC_ℓ. Concretely, $\mu_\ell = N_\ell^{-1}\theta_\ell$ and

$\Sigma_\ell = N_\ell^{-1}\Gamma_\ell - N_\ell^{-2}\theta_\ell \times \theta_\ell^T$. Furthermore, its additive property guarantees the mean and covariance matrix of a merged component can be easily computed from the triples of each member component. With these sufficient statistics, SWEM computes the k global clusters $\phi_h \in \Phi_G$ in the second stage:

$$\text{E-step:} \quad p(\phi_h|T_\ell) = \frac{\alpha_h^{(t)} \times p_h(\frac{1}{N_\ell}\theta_\ell|\mu_h^{(t)}, \Sigma_h^{(t)})}{\sum_{i=1}^{k} \alpha_i^{(t)} \times p_i(\frac{1}{N_\ell}\theta_\ell|\mu_i^{(t)}, \Sigma_i^{(t)})}$$

$$\text{M-step:} \quad \alpha_h^{(t+1)} = \frac{1}{n}\sum_{\ell=1}^{m} N_\ell \times p(\phi_h|T_\ell); \quad \mu_h^{(t+1)} = \frac{1}{n_h}\sum_{\ell=1}^{m} p(\phi_h|T_\ell) \times \theta_\ell;$$

$$\Sigma_h^{(t+1)} = \frac{1}{n_h}\left[\sum_{\ell=1}^{m} p(\phi_h|T_\ell)\Gamma_\ell - \frac{1}{n_h}\sum_{\ell=1}^{m} \left(p(\phi_h|T_\ell)\theta_\ell\right)\left(p(\phi_h|T_\ell)\theta_\ell\right)^T\right]$$

where $n_h = \sum_{\ell=1}^{m} N_\ell \times p(\phi_h|T_\ell)$.

Incremental Phase: In this phase SWEM utilizes the converged parameters in the previous time slot as the initial values for the mixture models' parameters. This helps minimize the number of iterations if the stream's characteristic does not vary much. However, in case the stream's distribution significantly changes, it is necessary to re-locate components. We develop splitting and merging operations in order to *discretely* re-distribute components in the entire data space.

An MC_ℓ is split if it is large enough and has the highest variance sum (i.e., its data are most spread). Assume dimension e having largest variance is chosen for splitting, parameters for two resulting components MC_{ℓ_1} and MC_{ℓ_2} are approximated by: $\mu_{\ell_1}^e = \int_{\mu_\ell^e - 3\sigma_\ell^e}^{\mu_\ell^e} x \times p_{\ell,e}(x|\phi_\ell)dx$; $\mu_{\ell_2}^e = \int_{\mu_\ell^e}^{\mu_\ell^e + 3\sigma_\ell^e} x \times p_{\ell,e}(x|\phi_\ell)dx$; $(\sigma_{\ell_1}^e)^2 = \int_{\mu_\ell^e - 3\sigma_\ell^e}^{\mu_\ell^e} x^2 \times p_{\ell,e}(x|\phi_\ell)dx - (\mu_{\ell_1}^e)^2$; $(\sigma_{\ell_2}^e)^2 = \int_{\mu_\ell^e}^{\mu_\ell^e + 3\sigma_\ell^e} x^2 \times p_{\ell,e}(x|\phi_\ell)dx - (\mu_{\ell_2}^e)^2$. For other dimensions, their means and variances are kept unchanged. On the other hand, two components are merged if they are small and close enough. The closeness is measured based on Mahalanobis distance: $Avg(D_{i,j}) = \frac{1}{2}[(\mu_i - \mu_j)^T \Sigma_j^{-1}(\mu_i - \mu_j) + (\mu_j - \mu_i)^T \Sigma_i^{-1}(\mu_j - \mu_i)]$. Parameters for the merging component is computed relied on the additive property: $\alpha_\ell = \alpha_{\ell_1} + \alpha_{\ell_2}$; $\mu_\ell = \frac{\alpha_{\ell_1}}{\alpha_{\ell_1}+\alpha_{\ell_2}} \times \mu_{\ell_1} + \frac{\alpha_{\ell_2}}{\alpha_{\ell_1}+\alpha_{\ell_2}} \times \mu_{\ell_2}$; $\Sigma_\ell = \frac{\Gamma_\ell}{n(\alpha_{\ell_1}+\alpha_{\ell_2})} - \frac{\theta_\ell \times \theta_\ell^T}{(n(\alpha_{\ell_1}+\alpha_{\ell_2}))^2}$. In that, $\theta_\ell = n \times (\alpha_{\ell_1} \times \mu_{\ell_1} + \alpha_{\ell_2} \times \mu_{\ell_2})$; $\Gamma_\ell = n[\alpha_{\ell_1}(\Sigma_{\ell_1} + \mu_{\ell_1}\mu_{\ell_1}^T) + \alpha_{\ell_2}(\Sigma_{\ell_2} + \mu_{\ell_2}\mu_{\ell_2}^T)]$.

Expiring Phase: This phase is applied when the window slides and the oldest time slot is expired from the mining model. $\Phi_G = \{\phi_1, \ldots, \phi_k\}$ is updated by subtracting the statistics summarized in $\Phi_L = \{\phi_1, \phi_2, ..., \phi_m\}$ of the expiring time slot. SWEM controls this process by using a fading factor λ $(0 < \lambda < 1)$ to gradually remove these statistics. At each iteration t, it reduces the weight of each expiring MC_ℓ by $N_\ell^{(t)} = \lambda^{(t)} N_\ell$. Equivalently, the reducing amount denoted by $r_\ell^{(t)}$ is $r_\ell^{(t)} = (1-\lambda)\lambda^{(t-1)} N_\ell$. The following theorem guarantees the number of iterations t can be any arbitrary integer while the total reducing weight on each expiring component approaches its original value.

Theorem 1. *Let t be an arbitrary number of iterations used by the SWEM algorithm. Then for each expiring micro component MC_ℓ: $\lim_{t\to\infty} \sum_t r_\ell^{(t)} = N_\ell$*

In specific, E-step updates: $p(\phi_h|T_\ell) = \frac{\alpha_h^{(t)} \times p_h(\frac{1}{N_\ell}\theta_\ell|\mu_h^{(t)}, \Sigma_h^{(t)})}{\sum_{i=1}^{k} \alpha_i^{(t)} \times p_i(\frac{1}{N_\ell}\theta_\ell|\mu_i^{(t)}, \Sigma_i^{(t)})}$

Table 1. Average log likelihood values returned by stdEM, SWEMw/oG and SWEM

	D2.K10.N100k			D4.K5.N100k			D10.K4.N100k		
TS	stdEM	w/oG	SWEM	stdEM	w/oG	SWEM	stdEM	w/oG	SWEM
2	-10.436	-10.512	-10.512	-19.252	-19.276	-19.276	-47.846	-47.869	-47.869
4	-10.427	-10.446	-10.446	-19.192	-19.215	-19.215	-47.933	-48.010	-48.010
6	-10.451	-10.604	-10.716	-19.164	-19.220	-19.326	-47.702	-47.712	-47.726
8	-10.444	-10.700	-10.735	-19.188	-19.226	-19.245	-47.859	-47.884	-47.886
10	-10.439	-10.523	-10.579	-19.202	-19.247	-19.258	-47.759	-47.820	-47.873

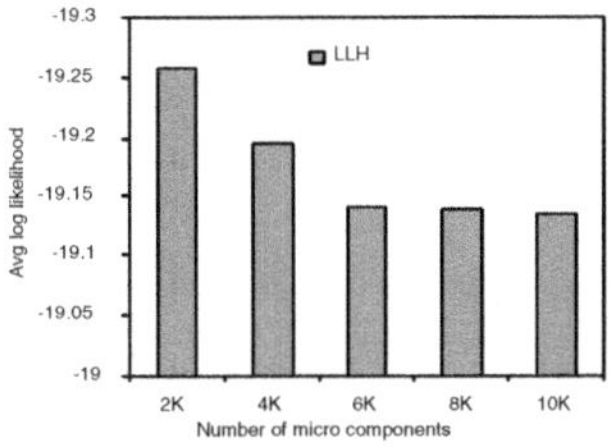

Fig. 1. Micro Components vs. Accuracy

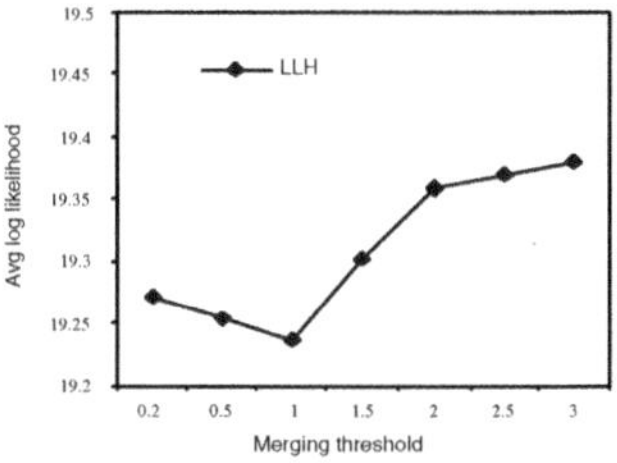

Fig. 2. Merging threshold $Avg(D_{i,j})$ vs. Accuracy

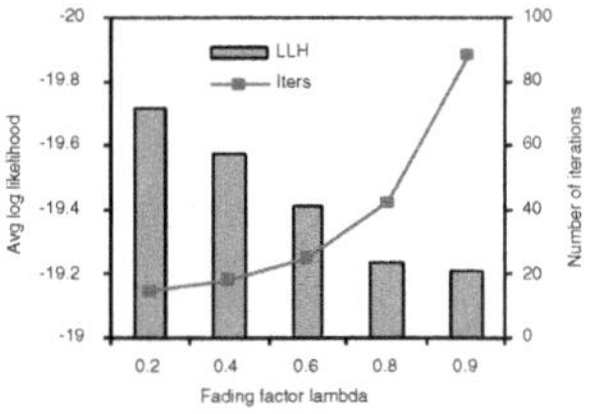

Fig. 3. Fading Factor λ vs. Accuracy

M-step: $n_G^{(t+1)} = n_G^{(t)} - \sum_{\ell=1}^{m} r_\ell^{(t+1)}; n_h^{(t+1)} = \alpha_h^{(t)} \times n_G^{(t)} - \sum_{\ell=1}^{m} p(\phi_h|T_\ell) \times r_\ell^{(t+1)};$
$\alpha_h^{(t+1)} = \frac{n_h^{(t+1)}}{n_G^{(t+1)}}$ where $\mu_h^{(t+1)} = \frac{\theta_h^{(t+1)}}{n_h^{(t+1)}};$ $\Sigma_h^{(t+1)} = \frac{1}{n_h^{(t+1)}}\left[\Gamma_h^{(t+1)} - \frac{1}{n_h^{(t+1)}}\theta_h^{(t+1)}\theta_h^{(t+1)T}\right],$
where $\theta_h^{(t+1)} = \theta_h^{(t)} - \sum_{\ell=1}^{m} p(\phi_h|T_\ell) \times r_\ell^{(t+1)} \times \frac{\theta_\ell}{N_\ell}$.

4 Experimental Results

Our algorithms are implemented using Visual C++ and the experiments are conducted on a 1.9GHz Pentium IV PC with 1GB memory space running Windows XP platform.

Clustering Quality Evaluation: Using the notations and method described in [1], three datasets $D2.K10.N100k$, $D4.K5.N100k$ and $D10.K4.N100k$ are generated. Unless otherwise indicated, the following parameters are used: $b = 5$; $m = 6K$ (where K is the number of global clusters), $Avg(D_{i,j}) = 1$, $\lambda = 0.8$, and each $TS_i = 10k$ data points. Table 1 shows the average log likelihood results returned by our experiments. It is observed that the clustering results of SWEM are very close to those of stdEM (a standard EM algorithm working without any stream constraints). Analogically, SWEM's clustering quality is almost identical to that of SWEMw/oG (a algorithm that derives global models by simply re-clustering all sets of micro components). This result verifies the efficiency of our gradually reducing weight technique.

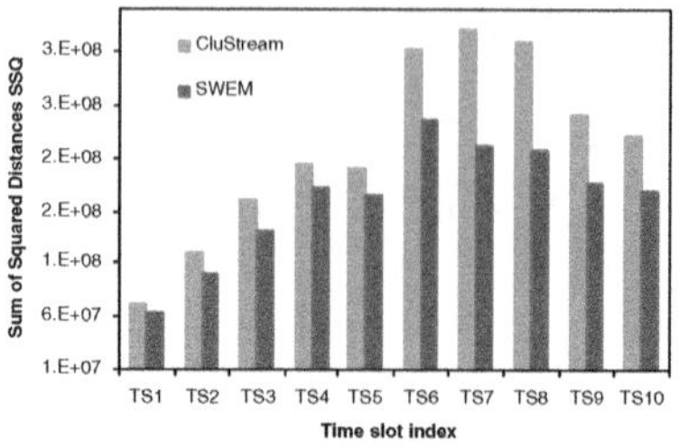

Fig. 4. Clustering quality comparison

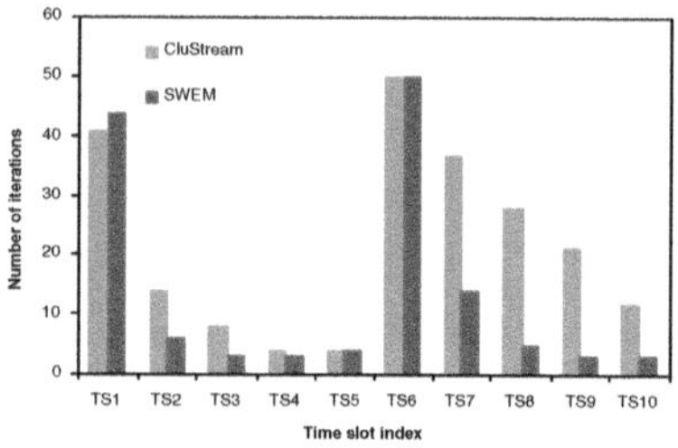

Fig. 5. Execution time comparison

Parameter Sensitivity: Using $D4.K5.N100k.AB$, a combination of two completely different stream distributions, we test the sensitivity of SWEM to various parameters. Figures 1, 2 and 3 report the accuracy of SWEM when the number of MC_ℓ, merging threshold $Avg(D_{i,j})$, and fading factor λ are respectively varied. We make the following observations. The average log likelihood becomes stable when $m = 6K$ (indicating that m need not be set too large, yet SWEM is still able to achieve high clustering quality); $Avg(D_{i,j})$ is set around 1 and λ is between 0.8 and 0.9.

Comparison with CluStream: It is observed from figures 4 and 5 that both the quality and execution time of SWEM are better than those of CluStream. This is understood by the difference in two algorithms' design. Whenever a significant change happens in the stream's distribution, SWEM re-distributes the set of micro components in the entire data space by using split and merge operations, CluStream, instead, simply creates a new cluster for a new point which cannot be absorbed by any clusters and merge other old ones. This causes the clusters' weights very imbalance and usually leads to a poor approximation of CluStream. Analogously, while SWEM minimizes the number of iterations by using the converged parameters of the previous time interval, CluStream always tries to update new data points into the set of micro clusters being maintained so far in the stream. This degrades the performance of Clustream and requires it more time to converge.

5 Conclusions

In this paper, we have addressed the problem of clustering data streams in one of the most challenging mining model, the time-based sliding window. We proposed SWEM algorithm that is able to compute clusters with a strictly single scan over the stream and work within confined memory space. Importantly, two techniques of splitting and merging components were developed to address the problem of time-varying data streams. SWEM has a solid mathematical background as it is designed based on the EM technique. Such a mathematically sound tool has been shown to be stable and effective in many domains despite the mixture models it employs being assumed to follow Gaussian distributions.

References

1. Aggarwal, C.C., Han, J., Wang, J., Yu, P.S.: A framework for clustering evolving data streams. In: Aberer, K., Koubarakis, M., Kalogeraki, V. (eds.) VLDB 2003. LNCS, vol. 2944, pp. 81–92. Springer, Heidelberg (2004)
2. Babcock, B., Babu, S., Datar, M., Motwani, R., Widom, J.: Models and issues in data stream systems. In: PODS, pp. 1–16 (2002)
3. Cao, F., Ester, M., Qian, W., Zhou, A.: Density-based clustering over an evolving data stream with noise. In: Jonker, W., Petković, M. (eds.) SDM 2006. LNCS, vol. 4165. Springer, Heidelberg (2006)
4. Chen, Y., Tu, L.: Density-based clustering for real-time stream data. In: SIGKDD Conference, pp. 133–142 (2007)
5. Aoying, Z., Feng, C., Ying, Y., Chaofeng, S., Xiaofeng, H.: Distributed data stream clustering: A fast em-based approach. In: ICDE Conference, pp. 736–745 (2007)

Two-Dimensional Retrieval in Synchronized Media Streams

Sujeet Pradhan[1,⋆] and Nimit Pattanasri[2,⋆]

[1] Department of Computer Science and Mathematics,
Kurashiki University of Science and the Arts, Kurashiki, Japan
sujeet@cs.kusa.ac.jp
[2] Department of Digital Content Research,
Academic Center for Computing and Media Studies,
Kyoto University, Kyoto, Japan
nimit@mm.media.kyoto-u.ac.jp

Abstract. We propose a two-dimensional information retrieval model for querying inside multimedia database where numerous media streams are semantically synchronized. The model exploits spatio-temporal arrangement of synchronized media streams during query evaluation phase. The output is a list of on-demand, dynamically integrated multimedia presentations satisfying a set of query keywords.

Keywords: comprehensive multimedia databases, multiple synchronized media streams, 2D information retrieval model.

1 Introduction

Multiple synchronized media streams have a wide range of applications including business, entertainment and education. Several commercial systems such as MPmeister[4] are already available for creating archives of high quality synchronized media stream contents. However, search capability inside synchronized streams is supported only at a coarse-grained level by these conventional systems[1]. The objective of this paper is to provide the basic essences of a formal query framework that is capable of extracting a list of *substreams* of synchronized media matching a set of user-specified keywords.

The fundamental problem of partial stream retrieval can be attributed to the fact that a single media fragment may not contain all the keywords specified by a user. Furthermore, even if a fragment does contain all the keywords, it may still lack the surrounding context, which would make the answer incomprehensible to the user. Therefore, traditional approaches of query processing with predetermined answer boundaries are not applicable. It should be noted that these boundaries may vary depending upon the keywords specified in the query. Therefore, it is impractical to index all such possible boundaries beforehand as the volume of index would grow exponentially. We propose a new query mechanism that determines the boundary at the query time, thus eliminating this

⋆ Both authors have equally contributed to this paper.

X. Zhou et al. (Eds.): DASFAA 2009, LNCS 5463, pp. 236–240, 2009.

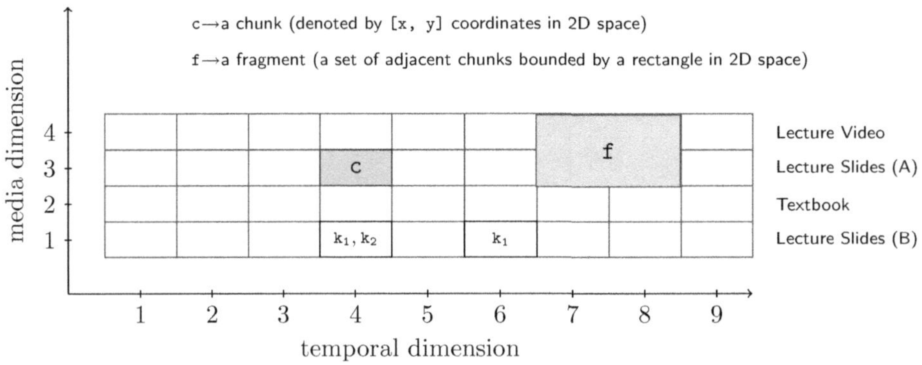

Fig. 1. Two-dimensional representation of a comprehensive multimedia database

problem of exponential number of indexing. The following section explains the formal description of our query model.

2 2D Retrieval in Synchronized Media Streams

2.1 Comprehensive Multimedia Database

A comprehensive multimedia database (CMD) may comprise several multiple synchronized media streams. However, for the sake of simplicity, here we shall consider only one such multiple synchronized media streams. Intuitively, each media stream is divided into a set of non-overlapped *chunks* — the smallest conceptual unit that is feasible to be indexed. A chunk thus may represent an image, a video shot, a PowerPoint/pdf slide, or a book page. Each chunk can be referenced by two dimensional coordinates of a rectangular coordinate system in which `[x, y]` refers to a chunk of a particular media stream. For example, `[4, 3]` represents Slide 4 of Lecture Slides (A) (i.e. Media Stream 3) in Fig. 1. The CMD thus stores a set of two dimensional coordinates for representing synchronized media streams as well as links to physical media chunks the coordinates reference to. The CMD also stores metadata associated with each chunk. For example, keywords k_1, k_2 are associated with the chunk `[4, 1]`, whereas keyword k_1 is associated with the chunk `[6, 1]`.

A *fragment* on the other hand, is a set of adjacent chunks bounded by a rectangle in the 2D coordinate space. For example, in Fig. 1, a fragment `f` is denoted by {`[7, 3]`, `[8, 3]`, `[7, 4]`, `[8, 4]`}. A chunk can therefore be considered as a singleton fragment. We use the term 'fragment' to denote a chunk unless a clear distinction is required.

2.2 Query Algebra

In order to realize an *answer model* that we described in the previous section, we require certain algebraic operations. These operations are basically classified as (1) *selection* and (2) *join* operations.

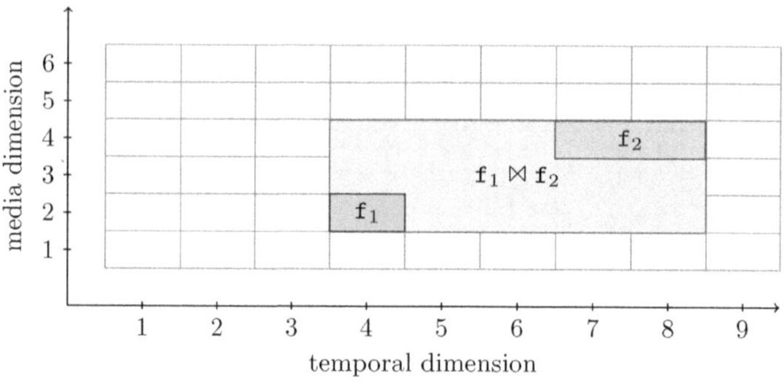

Fig. 2. Fragment Join operation between two fragments

Definition 1 (Selection). *Supposing* $\mathtt{F}$ *be a set of fragments in a CMD, and* $\mathtt{P}$ *be a predicate which maps a document fragment into true or* $false$, *a* selection *from* $\mathtt{F}$ *by the predicate* $\mathtt{P}$, *denoted by* $\sigma_{\mathtt{P}}$, *is defined as a subset* $\mathtt{F}'$ *of* $\mathtt{F}$ *such that* $\mathtt{F}'$ *includes all and only fragments satisfying* $\mathtt{P}$. *Formally,* $\sigma_{\mathtt{P}}(\mathtt{F}) = \{\mathtt{f} \mid \mathtt{f} \in \mathtt{F}, \mathtt{P}(\mathtt{f}) = true\}$.

Hereafter, the predicate P is also called a filter of the *selection* $\sigma_{\mathtt{P}}$.

The simplest filter is for the keyword selection of the type 'keyword $= k$' which selects only those fragments in CMD having the metadata 'k'.

Definition 2 (Fragment Join). *Let* $\mathtt{f}_1, \mathtt{f}_2, \mathtt{f}$ *be fragments in CMD. Then, fragment join between* $\mathtt{f}_1$ *and* $\mathtt{f}_2$ *denoted by* $\mathtt{f}_1 \bowtie \mathtt{f}_2$ *is the minimum bounding rectangle (MBR) in a 2D space that contains the two input fragments* $\mathtt{f}_1$ *and* $\mathtt{f}_2$.

Figure 2 shows the operation between two fragments $\{$[4, 2]$\}$ and $\{$[7, 4], [8, 4]$\}$. Note that the resulting fragment may include those chunks not contained in either of the input fragments.

Next, we extend this operation to a set of fragments. called *pairwise fragment join*, which is the set-variant of fragment join.

Definition 3 (Pairwise Fragment Join). *Let* $\mathtt{F}_1$ *and* $\mathtt{F}_2$ *be two sets of fragments in CMD,* pairwise fragment join *of* $\mathtt{F}_1$ *and* $\mathtt{F}_2$, *denoted by* $\mathtt{F}_1 \bowtie \mathtt{F}_2$, *is defined as a set of fragments yielded by taking* fragment join *of every combination of an element in* $\mathtt{F}_1$ *and an element in* $\mathtt{F}_2$ *in a pairwise manner. Formally,*

$$\mathtt{F}_1 \bowtie \mathtt{F}_2 = \{\mathtt{f}_1 \bowtie \mathtt{f}_2 \mid \mathtt{f}_1 \in \mathtt{F}_1,\ \mathtt{f}_2 \in \mathtt{F}_2\}.$$

Given two fragment sets $\mathtt{F}_1 = \{\mathtt{f}_{11}, \mathtt{f}_{12}\}$ and $\mathtt{F}_2 = \{\mathtt{f}_{21}, \mathtt{f}_{22}\}$, $\mathtt{F}_1 \bowtie \mathtt{F}_2$ produces a set of fragments $\{\mathtt{f}_{11} \bowtie \mathtt{f}_{21}, \mathtt{f}_{11} \bowtie \mathtt{f}_{22}, \mathtt{f}_{12} \bowtie \mathtt{f}_{21}, \mathtt{f}_{12} \bowtie \mathtt{f}_{22}\}$.

We now define *powerset fragment join* — another variant of the *fragment join* operation.

Definition 4 (Powerset Fragment Join). *Let* $\mathtt{F}_1$ *and* $\mathtt{F}_2$ *be two sets of fragments in a CMD,* powerset fragment join *between* $\mathtt{F}_1$ *and* $\mathtt{F}_2$, *denoted by* $\mathtt{F}_1 \bowtie^* \mathtt{F}_2$, *is defined as a set of fragments produced by applying* fragment join *operation to an arbitrary number (but not 0) of elements in* $\mathtt{F}_1$ *and* $\mathtt{F}_2$*. Formally,*

$$\mathtt{F}_1 \bowtie^* \mathtt{F}_2 = \{\bowtie (\mathtt{F}'_1 \cup \mathtt{F}'_2) \mid \mathtt{F}'_1 \subseteq \mathtt{F}_1, \mathtt{F}'_2 \subseteq \mathtt{F}_2, \mathtt{F}'_1 \neq \phi, \mathtt{F}'_2 \neq \phi\}$$
$$\textit{where } \bowtie \{\mathtt{f}_1, \mathtt{f}_2, \ldots, \mathtt{f}_n\} = \mathtt{f}_1 \bowtie \ldots \bowtie \mathtt{f}_n.$$

Obviously, *powerset fragment join* grows exponentially and becomes very expensive as the sizes of the input fragment sets get larger.

However, *powerset fragment join* can be logically transformed into the following equivalent algebraic expression.

$$\mathtt{F}_1 \bowtie^* \mathtt{F}_2 = (\mathtt{F}_1 \bowtie \mathtt{F}_1 \bowtie \mathtt{F}_1 \bowtie \mathtt{F}_1) \bowtie (\mathtt{F}_2 \bowtie \mathtt{F}_2 \bowtie \mathtt{F}_2 \bowtie \mathtt{F}_2)$$

where the expression $(\mathtt{F} \bowtie \mathtt{F} \bowtie \mathtt{F} \bowtie \mathtt{F})$ denotes the fixed point of a fragment set $\mathtt{F}$. Obviously, this new expression not only looks much simpler but also less costly to evaluate.

2.3 Query Evaluation

Definition 5 (Query). *A query is denoted by* $\mathtt{Q}_\mathsf{P}\{\mathtt{k}_1, \mathtt{k}_2, ..., \mathtt{k}_m\}$ *where* $\mathtt{k}_j$ *is called a query term for all* $j = 1, 2, \ldots, m$ *and* P *is a selection predicate.*

We write $\mathtt{k} \in \mathtt{keywords(c)}$ to denote that query term $\mathtt{k}$ appears in the metadata associated with a chunk $\mathtt{c}$.

Definition 6 (Query Answer). *Given a query* $\mathtt{Q}_\mathsf{P}\{\mathtt{k}_1, \mathtt{k}_2, ..., \mathtt{k}_m\}$*, answer* $\mathtt{A}$ *to this query is a set of media fragments defined to be*
$\{\ \mathtt{f} \mid (\forall \mathtt{k} \in \mathtt{Q}) \exists \mathtt{c} \in \mathtt{f} : \mathtt{c}$ *is a boundary chunk of* $\mathtt{f}\ \wedge$
$\mathtt{k} \in \mathtt{keywords(c)} \wedge \mathsf{P}(\mathtt{f}) = \textit{true}\ \}.$

Intuitively, a potential answer candidate to a query is a fragment consisting of several consecutive chunks in one or more media streams and each keyword in the query must appear in at least one chunk that constitutes the fragment. In addition, the fragment must satisfy the selection predicate(s) specified in the query.

A query represented by $\{\mathtt{k}_1, \mathtt{k}_2\}$ and a selection predicate P against a CMD can be evaluated by the following formula.

$$\begin{aligned}\mathtt{Q}_\mathsf{P}\{\mathtt{k}_1, \mathtt{k}_2\} &= \sigma_\mathsf{P}(\mathtt{F}_1 \bowtie^* \mathtt{F}_2) \\ &= \sigma_\mathsf{P}((\mathtt{F}_1 \bowtie \mathtt{F}_1 \bowtie \mathtt{F}_1 \bowtie \mathtt{F}_1) \bowtie (\mathtt{F}_2 \bowtie \mathtt{F}_2 \bowtie \mathtt{F}_2 \bowtie \mathtt{F}_2))\end{aligned}$$

where $\mathtt{F}_1 = \sigma_{\mathsf{keyword}=k_1}(\mathrm{F})$, $\mathtt{F}_2 = \sigma_{,\mathsf{keyword}=k_2}(\mathrm{F})$ and $\mathrm{F} = \{\mathrm{c} \mid \mathrm{c} \text{ is a chunk in CMD}\}$.

Fig. 3 shows an example query and an intuitive answer to this query. It is not difficult to see that the definitions given above are enough to precisely generate this answer (colored rectangle).

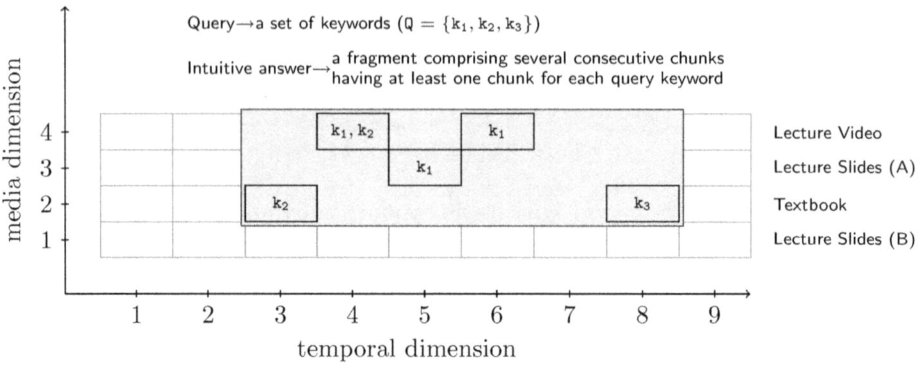

Fig. 3. A keyword query and a corresponding potential answer fragment

3 Conclusions

In this paper, we presented our preliminary work on partial retrieval of synchronized media streams. Our future work includes filtering and ranking of search results.

Acknowledgements

This project has been partially supported by Grant-in-Aid for Scientific Research (C) (20500107).

References

1. Hunter, J., Little, S.: Building and indexing a distributed multimedia presentation archive using SMIL. In: Constantopoulos, P., Sølvberg, I.T. (eds.) ECDL 2001. LNCS, vol. 2163, pp. 415–428. Springer, Heidelberg (2001)
2. Pradhan, S.: An algebraic query model for effective and efficient retrieval of XML fragments. In: VLDB, pp. 295–306 (2006)
3. Pradhan, S., Tajima, K., Tanaka, K.: A query model to synthesize answer intervals from indexed video units. IEEE Trans. on Knowledge and Data Engineering 13(5), 824–838 (2001)
4. Ricoh Corporation: MPmeister, `http://www.ricoh.co.jp/mpmeister/`

Eager Evaluation of Partial Tree-Pattern Queries on XML Streams

Dimitri Theodoratos and Xiaoying Wu

Department of Computer Science,
New Jersey Institute of Technology, USA
{dth,xw43}@njit.edu

Abstract. Current streaming applications have stringent requirements on query response time and memory consumption because of the large (possibly unbounded) size of data they handle. Further, known query evaluation algorithms on streaming XML documents focus almost exclusively on tree-pattern queries (TPQs). However recently, requirements for flexible querying of XML data have motivated the introduction of query languages that are more general and flexible than TPQs. These languages are not supported by known algorithms.

In this paper, we consider a language which generalizes and strictly contains TPQs. Queries in this language can be represented as dags enhanced with constraints. We explore this representation to design an original polynomial time streaming algorithm for these queries. Our algorithm avoids storing and processing matches of the query dag that do not contribute to new solutions (redundant matches). Its key feature is that it applies an eager evaluation strategy to quickly determine when node matches should be returned as solutions to the user and also to proactively detect redundant matches. We experimentally test its time and space performance. The results show the superiority of the eager algorithm compared to the only known algorithm for this class of queries which is a lazy algorithm.

1 Introduction

Current streaming applications require efficient algorithms for processing complex and ad hoc queries over large volumes of XML streams. Unfortunately, existing algorithms on XML streams focus almost exclusively on tree-pattern queries (TPQs). A distinguishing restrictive characteristic of TPQs is that they impose a total order for the nodes in every path of the query pattern.

We consider a query language for XML, called partial tree-pattern query (PTPQ) language. The PTPQ language generalizes and strictly contains TPQs. PTPQs are not restricted by a total order for the nodes in a path of the query pattern since they can constrain a number of (possibly unrelated) nodes to lie on the same path (*same-path* constraint). They are flexible enough to allow on the one side keyword-style queries with no structure, and on the other side fully specified TPQs.

In this paper, we address the problem of designing an efficient streaming algorithm for PTPQs. Our focus is on demanding streaming applications that require low response time and memory consumption.

X. Zhou et al. (Eds.): DASFAA 2009, LNCS 5463, pp. 241–246, 2009.

2 Query Language

A partial tree-pattern query (PTPQ) specifies a pattern which partially determines a tree. PTPQs comprise nodes and child and descendant relationships among them. Their nodes are grouped into disjoint sets called *partial paths*. PTPQs are embedded to XML trees. The nodes of a partial path are embedded to nodes on the same XML tree path. However, unlike paths in TPQs the child and descendant relationships in partial paths do not necessarily form a total order. This is the reason for qualifying these paths as partial. PTPQs also comprise node sharing expressions. A node sharing expression indicates that two nodes from different partial paths are to be embedded to the same XML tree node. That is, the image of these two nodes is the same – *shared* – node in the XML tree.

We represent PTPQs as node and edge labeled directed graphs. The graph of a PTPQ can be constructed by collapsing every two nodes that participate in node sharing expressions into a single node. The nodes in the graph are annotated by the (possibly many) PPs they belong to. The output node of a PTPQ is shown as a black circle. Figure 1 shows the query graph of PTPQ Q_1. For simplicity of presentation, the annotations of some nodes might be omitted and it is assumed that a node inherits all the annotating PPs of its descendant nodes. For example, in the graph of Figure 1, node C is assumed to be annotated by the PPs p_2 and p_3 inherited from its descendant nodes D and F.

The answer of a query on an XML tree is a set of results, where each result is the image of the output node in a match of the query on the XML tree. The formal definitions of a PTPQ and its embedding to an XML tree as well as the expressiveness results on PTPQs can be found in the full version of the paper [1].

3 Eager Evaluation Algorithm

In this section, we describe our streaming evaluation algorithm for PTPQs. The algorithm is called **Eager P**artial TPQ **S**treaming evaluation on **X**ML ($EagerPSX$). Let Q be the input query to be evaluated on a stream of events for an XML tree T. Algorithm $EagerPSX$ is event-driven: as events arrive, event handlers (which are the procedures *startEval* or *endEval*), are called on a sequence of query nodes whose label match the label of the tree node under consideration. Algorithm $EagerPSX$ is stack-based. With every query node X in Q, it associates a stack S_X.

3.1 Open Event Handler

Procedure $startEval$ is invoked every time an open event for a tree node x arrives. Let X be a query node whose label matches that of x. Procedure *startEval* checks if x qualifies for being pushed on stack S_X: node X is the root of Q or the stacks of all the parent nodes of X in Q are not empty and for each parent Y of X, the top entry of stack S_Y and x satisfy the structural relationship ('//' or '/') between X and Y in Q. If it does, x is called an *ancestor match* of X.

Avoiding redundant matches. Because the answer of a query comprises only the embeddings of the output node of the query, we might not need to identify all the matches

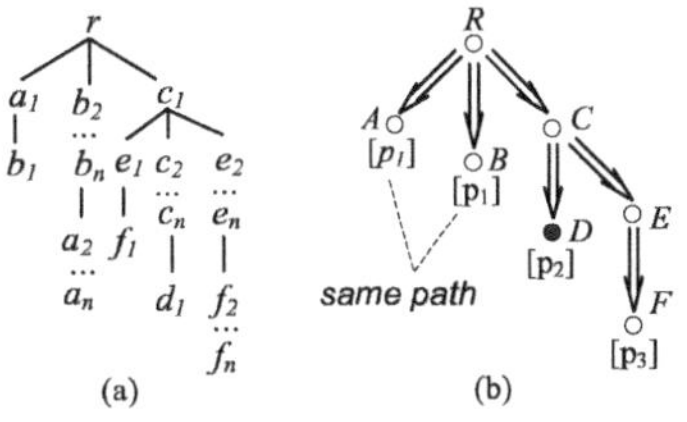

Fig. 1. (a) XML Tree, (b) Query Q_1

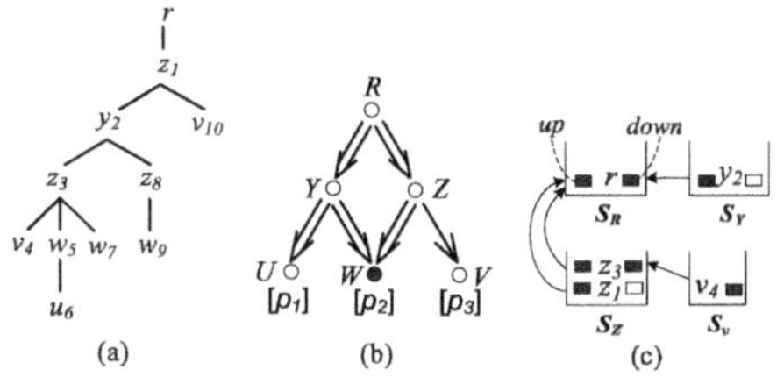

Fig. 2. (a) XML Tree, (b) Query Q_2, (c) Snapshots of stacks

of the query pattern when computing the answer of the query. Whenever a matching of a query node is found, other matches of the same node that do not contribute to a possible new matching for the output node are redundant and can be ignored.

A match of a query node X can be redundant in two cases. In the first case, X is a predicate node (a non-ancestor node of the output node of Q). Consider, for instance, evaluating the query Q_1 of Figure 1(b) on the XML tree T_1 of Figure 1(a). The nodes a_1, b_1, e_1 and f_1 which are matches for the predicate nodes A, B, E and F, respectively, contribute to the match d_1 of the output node D. The nodes $a_2, \ldots, a_n$, $b_2, \ldots, b_n$, $e_2, \ldots, e_n$, $f_2, \ldots, f_n$ which are also matches of the predicate nodes can be ignored since they all contribute to the same match d_1 of the output node. Note that these nodes correspond to $O(n^4)$ embeddings of the query with the same match for the output node. Avoiding their computation saves substantial time and space. By avoiding redundant predicate matches, our evaluation algorithm exploits the existential semantics of queries during evaluation.

In the second case, X is a backbone node (an ancestor node of the output node of Q). Consider again the example of Figure 1. Backbone node C has n matches $\{c_1, \ldots, c_n\}$. Note that before nodes $c_2, \ldots, c_n$ are read, both r and c_1 have already satisfied their predicates. Therefore, any match of the output node D that is a descendant of c_1 (e.g., d_1) can be identified as a solution and thus should be returned to the user right away. The nodes $\{c_2, \ldots, c_n\}$ need not be stored. Storing these nodes unnecessarily delays the output of query solutions, and wastes time and memory space. It is important to note that redundant backbone node matches contribute to a number of pattern matches which in the worst case can be exponential on the size of the query.

Our algorithm exploits the previous observations using the concept of *redundant* match of a query node. Redundant matches are not stored and processed by our algorithm. Note that most previous streaming algorithms either do not handle redundant matches [2,3] or avoid redundant matches for TPQs only [4].

Traversing the query dags. When node X is a sink node of query Q, Procedure *startEval* examines whether there are matches that become solutions by traversing two sub-dags G_P and G_B in that sequence. Dags G_P and G_B are a sub-dag of Q rooted at an ancestor node of X that consist of predicate nodes and backbone nodes, respectively. Procedure *startEval* traverses dag G_P in a bottom-up way and evaluates for each query node the predicate matches encoded in the stacks. After then, it traverses dag G_B in a top-down manner. It examines for each node its matches encoded in the stack, checking if there are candidate outputs (associated with the matches) become solutions and can be returned to the user. Redundant matches will be detected and pruned during each traversal. A traversal is terminated when either there is no more matches to examine, or some match that has been examined before is encountered.

Example. Consider evaluating query Q_2 of Figure 2(b) on the XML tree of Figure 2(a). When $\langle v_4 \rangle$ (the start event of v_4) is read, procedure *startEval* traverses the sub-dag (path in this case) Z/V starting with V. Subsequently, *startEval* goes up to Z. Then, it evaluates the predicates for the entry z_3 in stack S_Z. Procedure *startEval* ends its traversal at z_1. Figure 2(c) shows the snapshot of the query stacks at this time. Procedure *startEval* proceeds to traverse the sub-dag (path in this case) $Z//W$ starting with Z. It terminates its traversal on W since stack S_W is empty. Later, when $\langle u_6 \rangle$ is read, *startEval* traverses the sub-dags $Y//U$ and $Y//W$ in that sequence. As a result, node w_5 is found to be a solution and is returned to the user. Note that when $\langle w_7 \rangle$ is read, node w_7 is returned as a solution right away.

3.2 Close Event Handler

When a close event for a tree node x arrives, for each query node X whose label matches that of x, Procedure *endEval* is invoked to pop out the entry of x from S_X and checks if x is a *candidate match* of X. Node x is called a candidate match of X if x is the image of X under an embedding of the sub-dag rooted at X to the subtree rooted at x in T. If this is the case and X is a backbone node, each candidate output (a candidate match of the output node) stored in the entry of x is propagated to an ancestor of x in a stack, if X is not the root of Q, or is returned to the user, otherwise. If x is not a candidate match of X, the list of candidate output stored in the entry of x is either propagated to an ancestor of x in a stack, or is discarded, depending on whether there exists a path in stacks which consists of nodes that are candidate matches of the query nodes.

3.3 Analysis

We can show that Algorithm $EagerPSX$ correctly evaluates a query Q on a streaming XML document T. Under the reasonable assumption that the size of Q is bounded by constant, Algorithm $EagerPSX$ uses $O(|T|)$ space and $O(|T| \times H)$ time, where $|T|$ and H denote the size and the height of T, respectively. The detailed analysis can be found in [1].

4 Experimental Evaluation

We have implemented Algorithm $EagerPSX$ in order to experimentally study its execution time, response time, and memory usage. We compare $EagerPSX$ with PSX [5] and X_{aos} [2]. Algorithm PSX is the only known streaming algorithm for PTPQs and it is a lazy algorithm. Algorithm X_{aos} supports TPQs extended with reverse axes. All the three algorithms were implemented in Java. The experiments were conducted on an Intel Core 2 CPU 2.13 GHz processor with 2GB memory running JVM 1.6.0 in Windows XP Professional 2002. We used three datasets: a benchmark dataset, a real

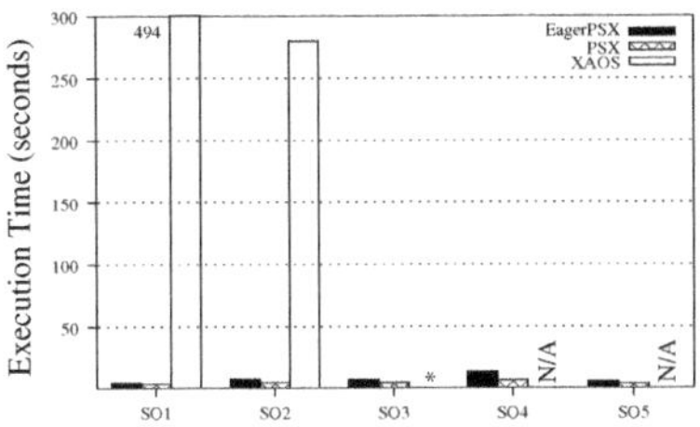

(a) Execution time

Query	$EagerPSX$	PSX	$XAOS$
SQ_1	0.02	3.83	493.8
SQ_2	0.01	4.6	279.5
SQ_3	0.012	4.64	*
SQ_4	0.09	6.43	N/A
SQ_5	0.015	4.19	N/A

(b) Time to 1st solution (seconds)

'*' denotes an execution that didn't finish within 7 hours; 'N/A' denotes incapacity of the algorithm to support the query.

Fig. 3. Query execution time and response time on synthetic dataset

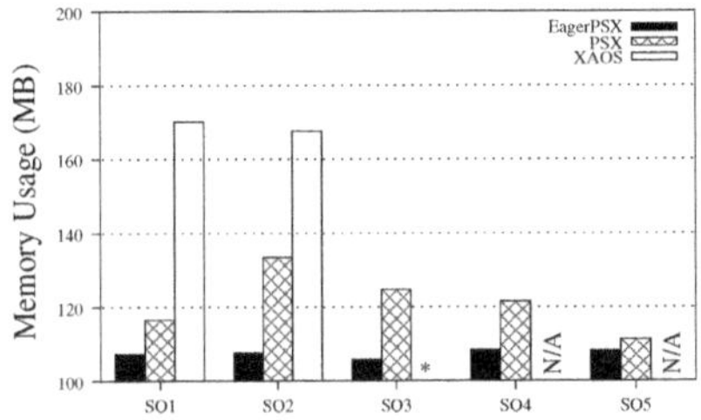

(a) Runtime memory usage

Query	$EagerPSX$	PSX	$XAOS$
SQ_1	592	63864	94696
SQ_2	0	198458	198458
SQ_3	2050	90068	*
SQ_4	4783	113696	N/A
SQ_5	519	7271	N/A

(b) Max number of stored candidate outputs

Fig. 4. Memory usage on synthetic dataset

dataset, and a synthetic dataset. The synthetic dataset is generated by IBM's XML Generator [1] with $NumberLevels = 8$ and $MaxRepeats = 4$. It has the size of 20.3MB and includes highly recursive structures. In the interest of space, following we only report the results on the synthetic dataset. Our experiments on the other two datasets confirm the results presented here. Additional experimental results can be found in [1].

On each one of the three datasets, we tested 5 PTPQs ranging from simple TPQs to complex dags. Both $EagerPSX$ and PSX supports all five queries but X_{aos} only supports the first three.

The execution time consists of the data and query parsing time and the query evaluation time. Figure 3(a) shows the results. As we can see, PSX has the best time performance, and in most cases it outperforms X_{aos} by at least one order of magnitude. $EagerPSX$ uses slightly more time than PSX, due to the overhead incurred by eagerly traversing the query dags.

The query response time is the time between issuing the query and receiving the first solution. Figure 3(b) shows the results. As we can see, $EagerPSX$ gives the best query response time for both simple and complext queries. Compared to PSX and X_{aos}, $EagerPSX$ shortens the response time by orders of magnitude. $EagerPSX$ starts to deliver query solutions almost immediately after a query is being posed.

[1] http://www.alphaworks.ibm.com/tech/xmlgenerator

Figure 4 shows the memory consumption of the three algorithms. As we can see, the memory usage of $EagerPSX$ is stable for both simple and complex queries. Figure 4(b) shows the maximal number of stored candidate outputs of the three algorithms. Among the three algorithms, $EagerPSX$ always stores the lowest number of candidate outputs. This is expected, since $EagerPSX$ uses an eager evaluation strategy that allows query solutions to be returned as soon as possible.

References

1. http://web.njit.edu/~xw43/paper/eagerTech.pdf
2. Barton, C., Charles, P., Goyal, D., Raghavachari, M., Fontoura, M., Josifovski, V.: Streaming xpath processing with forward and backward axes. In: ICDE (2003)
3. Chen, Y., Davidson, S.B., Zheng, Y.: An efficient XPath query processor for XML streams. In: ICDE (2006)
4. Gou, G., Chirkova, R.: Efficient algorithms for evaluating XPath over streams. In: SIGMOD (2007)
5. Wu, X., Theodoratos, D.: Evaluating partial tree-pattern queries on XML streams. In: CIKM (2008)

Energy-Efficient Evaluation of Multiple Skyline Queries over a Wireless Sensor Network

Junchang Xin[1], Guoren Wang[1], Lei Chen[2], and Vincent Oria[3]

[1] College of Information Science and Engineering, Northeastern University, China
{xinjunchang,wanggr}@ise.neu.edu.cn
[2] Department of Computer Science and Engineering, HKUST, Hong Kong, China
leichen@cse.ust.hk
[3] Department of Computer Science, New Jersey Institute of Technology, USA
oria@njit.edu

Abstract. Though skyline queries in wireless sensor networks have been intensively studied in recent years, existing solutions are not optimized for multiple skyline queries as they focus on single full space skyline queries. It is not efficient to individually evaluate skyline queries especially in a wireless sensor network environment where power consumption should be minimized. In this paper, we propose an energy-efficient multi-skyline evaluation (EMSE) algorithm to effectively evaluate multiple skyline queries in wireless sensor networks. EMSE first utilizes a global optimization mechanism to reduce the number of skyline queries and save on query propagation cost and parts of redundant result transmission cost as a consequence. Then, it utilizes a local optimization mechanism to share the skyline results among skyline queries and uses some filtering policies to further eliminate unnecessary data transmission and save the skyline result transmission cost as a consequence. The experimental results show that the proposed algorithm is energy-efficient when evaluating multiple skyline queries over wireless sensor networks.

1 Introduction

Wireless sensor networks (WSNs) are being widely deployed in many environment monitoring applications because of the decrease in hardware prices and the increase in reliability of wireless communications. For example when studying the behavior of various bird species, ornithologists utilize feeder sensors to determine the places where birds appear most frequently. The same feeder sensor can be equipped with a sound sensor, which can be used to detect the sound of birds. Therefore, the ornithologists can also monitor the sounds of birds. Other than environmental monitoring, WSNs have been successfully used in health care, home automation, traffic control, and battlefield surveillance, etc. WSNs indeed bring new opportunities as they allow to safely collect data from hazardous environment. But due to the limitation of current hardware, most sensors are battery powered and wireless communication is one of the major consumer of the sensor energy. Thus, the challenge for WSNs is to fulfill various monitoring tasks with a minimum communication cost which will also extend the life of WSNs.

X. Zhou et al. (Eds.): DASFAA 2009, LNCS 5463, pp. 247–262, 2009.

In response to the energy-saving challenge raised by WSNs, many proposals have come out to answer various types of energy-efficient query processing like average [7], top-k [10] and median [4]. In this paper, we investigate a new concept, called *subspace skylines*, over WSNs. As the new proposal is based on *skyline* [1], we will recall its definition:

Definition 1. *Given a set T of tuples in a $|D|$-dimensional space D, the skyline of D returns the tuples that are not dominated by the others. Here, tuple t dominates tuple t', if it holds that: 1) $t[i] \leq t'[i]$ (no worse than) for each dimension $i \in D$, and 2) $t[j] < t'[j]$ (better than) for at least one dimension $j \in D$.*

As mentioned above, energy is the critical resource of WSNs and wireless communication is a major consumer. A skyline approach can help find the critical nodes in the network and minimize the volume of traffic in the network. Based on the knowledge obtained from skyline, we can reduce the sampling rate of the skyline sensors or retrieve data from nearby sensors in order to save energy and extend the lifetime of the whole WSN. Here, a node n_i dominates a node n_j if $n_i.en \leq n_j.en$, $n_i.tr \geq n_j.tr$ and at least one of the following inequalities holds: $n_i.en < n_j.en$, $n_i.tr > n_j.tr$. The nodes not dominated by any others are the skyline nodes. Given a set of objects, the skyline query retrieves the complete set of skyline objects. Figure 1 shows an example of skyline in such an application, here a, c, d and h are skyline nodes, and the skyline is $\{a, c, d, h\}$.

In general, in environment monitoring applications, multiple sensing devices are installed on each sensor node and sensor readings consist in acquiring multiple attributes, each corresponding to a type of the physical measurement from a sensing device. Under this scenario, each sensor reading can be modeled as a multi-dimensional vector and each entry in the vectors represents one type of sensing value. In addition, users may be interested in retrieving data in various subspaces (*subspace skylines.* Formal definition is given in Section 3) since different users have different preferences. For example, we are currently involved in the process of building a real-time in-situ beach water monitoring system using WSNs. In this system, we use sensors to monitor the quality of water by sampling several parameters, such as E. coli bacteria, pH, salinity, and turbidity. The researchers from civil engineering department would like to ask various of subspace skyline queries over the system for different analysis purposes: "give me the places where E.coli is high and PH. is low" or "give me the places where E.Coli bacteria level is high, salinity is low and turbidity is low". Therefore, for

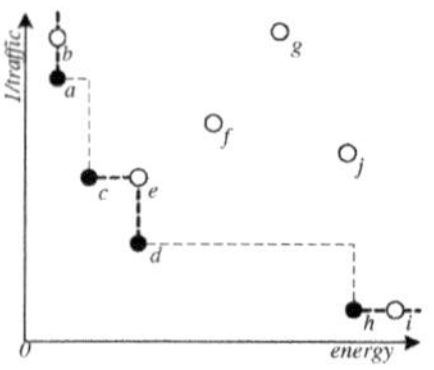

(a) Objects

Subspace	Skyline	XSkyline
energy,trafic	*a, c, d, h*	*a, b, c, d, e, h*
energy	*a, b*	*a, b*
trafic	*h, i*	*h, i*

(b) Skylines

Fig. 1. An example of skyline query

an environmental monitoring application like the above-mentioned example, it is quite common to have multiple skyline queries in difference subspaces which are posed into the WSN simultaneously to gather interesting information.

Skyline queries in WSNs have received considerable attention [14,15,3] in recent years, most existing work has focused on energy-efficient evaluation of only single skyline query in WSNs. We can naively apply the previous approaches to evaluate multiple subspace skylines one by one. Obviously, this will no longer be an energy-efficient approach because it will neglect the consanguineous correlations among the subspace skylines, which is exactly the base of our energy saving approach. Thus, in this paper, we propose a novel approach, called Energy-efficient Multi-Skyline Evaluation (EMSE), to evaluate multiple skyline queries in WSNs. EMSE employs two kinds of optimization mechanisms (global optimization and local optimization) to reduce both the skyline query propagation cost and skyline result transmission cost. The contributions of this paper are summarized as follows:

1. Through the comprehensive analysis of the relationship between a subspace skyline and its parent space skyline, we propose a global optimization mechanism on the base station.
2. Based the analysis of the skylines on different subspaces, we propose a local optimization mechanism on intermediate nodes.
3. The extensive simulations show that the proposed algorithm performs effectively on reducing the communication cost and saves energy on evaluating multiple skyline queries in WSNs.

2 Related Work

Chen et al. [3] address the problem of continuous skylines monitoring in sensor networks and presented a hierarchical threshold-based approach (MINMAX) to minimize the transmission traffic in sensor networks. Our previous work [14,15] investigated the problem of maintaining sliding windows skylines in WSNs, which seek the skylines over the latest data that are constrained by a sliding window, in which several types of filter are installed within each sensor to reduce the amount of data transmitted and save the energy consumption as a consequence. All existing work has focused on a single skyline query execution in WSNs, whereas we focus on multiple skyline queries.

Trigoni et al. [8] studied the multi-query optimization problem of region-based aggregation queries in WSNs and proposed an equivalence class based on a merging algorithm in [13] to merge partial aggregate values of multi-queries. Recently, Xiang et al. [12,11] studied the multi-query optimization problem in WSNs, and proposed a Two-Tier Multiple Query Optimization (TTMQO) scheme to enable similar queries to share both communication and computational resources in WSNs. However, they focus on how to optimize multiple aggregations in WSNs, while we are more concerned about how to optimize more complex queries, multiple skyline queries. Because dominance in the definition of skyline is a partial-order relationship and quite different with aggregations, all the traditional algorithms are not appropriate.

3 Preliminaries

The subspace skyline can be formally defined as following.

Definition 2. *Given a set T of tuples in a $|D|$-dimensional space D, a subset of dimensions $U \subseteq D$ ($U \neq \phi$) forms a $|U|$-dimensional subspace of D. The subspace skyline of U (denoted $Skyline(T, U)$) returns the tuples that are not dominated by any other tuple in the subspace U. Here, tuple t dominates tuple t' in the subspace U (denoted $t \succeq_U t'$) if: 1) $t[i] \leq t'[i]$ ($t[i]$ not worse than $t'[i]$) in each dimension $i \in U$, and 2) $t[i] < t'[i]$ ($t[i]$ better than $t'[i]$) in at least one dimension $i \in U$.*

Yuan et al. [16] has thoroughly studied the subspace skyline and pointed out the relationships between subspace skyline and its super space skyline.

Theorem 1. *[16] Given a set T of tuples in a $|D|$-dimensional space D, U and V are two subspaces of D ($U, V \subseteq D$), where $U \subseteq V$. Each tuple $t \in Skyline(T, U)$ is either a subspace skyline tuple in $Skyline(T, V)$ or shares the same values with another tuple $t' \in Skyline(T, V)$ in subspace U.*

Corollary 1. *[16] Given a set T of tuples in a $|D|$-dimensional space D, if for any two tuples t and t', $t[i] \neq t'[i]$ holds, then for two subspaces U, V of D ($U, V \subseteq D$), where $U \subseteq V \Rightarrow Skyline(T, U) \subseteq Skyline(T, V)$.*

Corollary 1 shows that the subspace skylines are contained in its super-space skyline when some special case (*distinct value condition*) holds. Once the super space skyline is calculated, all its subspace skylines can be answered facilely. If the containing relationship is hold in a more common condition, the number of subspace skylines executed in the network will be reduced remarkablely.

According to Theorem 1, besides super space skyline tuples, if all tuples having the same value with them in some dimensionalities are found, all the subspace skyline queries could be answered. Extended skyline [9], shown as Definition 3, just contains the above two kinds of tuples.

Definition 3. *Given a set T of tuples in a $|D|$-dimensional space D, a subset of dimensions $U \subseteq D$ ($U \neq \phi$) forms a $|U|$-dimensional subspace of D. The extended subspace skyline of U returns the tuples that are not strictly dominated by the others in subspace U, denoted as $XSkyline(T, U)$. Here, tuple t strictly dominates tuple t' in subspace U, if it holds that $t[i] < t'[i]$ (better than) for each dimension $i \in U$, denoted as $t \succ_U t'$.*

The extended skyline contains not only tuples of the traditional skyline, but also those tuples not strictly dominated by traditional skyline tuples. As shown in Figure 1, the extended skyline in $\langle en, tr \rangle$ is $\{a, b, c, d, e, h, i\}$. It not only contains all the skyline results of $\langle en \rangle$, $\langle tr \rangle$ and $\langle en, tr \rangle$, but also the extended skyline results of $\langle en \rangle$ and $\langle tr \rangle$.

Theorem 2. *[9] Given a set T of tuples in a $|D|$-dimensional space D, U and V are two subspaces of D ($U, V \subseteq D$), if $U \subseteq V$ holds, then $Skyline(T, U) \subseteq XSkyline(T, V)$.*

However, not all the extened skyline tuples are subspace skyline tuples, and these tuples have no contribution for any subspace skyline. Shown in Figure 1, e belongs to the extended skyline, $\{a, b, c, d, e, h, i\}$, of $\langle en, tr\rangle$, but it does not belong to any subspace skyline. So it is not appropriate to send e to the base station.

Lemma 1. *Given a set T of tuples in a $|D|$-dimensional space D, U and V are two subspaces of D ($U, V \subseteq D$), it satisfies that $U \subseteq V$. t and t' are two tuples belong to $XSkyline(T, V)$, they satisfy that $(t' \succeq_V t) \wedge (t' \succ_U t)$, if for any subspace $W \subseteq V$, $W \cap U \neq \phi$ holds, then $t \notin Skyline(T, W)$.*

Proof. $t' \succeq_V t \Rightarrow \forall i \in V, t'[i] \leq t[i]$, $W \subseteq V \Rightarrow \forall i \in W, t'[i] \leq t[i]$.
$t' \succ_U t \Rightarrow \forall i \in U, t'[i] < t[i]$.
$(W \cap U \neq \phi) \wedge (\forall i \in U, t'[i] < t[i]) \Rightarrow \exists j \in W, t'[j] < t[j]$.
Therefore, $t \notin Skyline(T, W)$. □

Lemma 1 shows that if tuple t is dominated by another tuple t' in V, t would not belong to skyline in any subspace having dimensionalities where the value of t and t' are not equal. Only in the subspace having equal value with t' in all its dimensionalities, t has the probability to be a skyline tuple.

Definition 4. *Given a set T of tuples in a $|D|$-dimensional space D, U is a subspace of D ($U \subseteq D$), t and t' is two tuples in T. If $\forall i \in U, t[i] \leq t'[i]$ and $\exists j \in U, t[i] < t'[i]$ hold, U is the dominated space of tuple t' corresponding to tuple t. If there is no such space $V \subseteq D$, satisfying $U \subset V$ and V is the dominated space of tuple t', U is the maximum dominated space of tuple t' corresponding to tuple t, denoted as $MDS(t, t')$.*

Definition 5. *Given a set T of tuples in a $|D|$-dimensional space D, U and V are two subspaces of D ($U, V \subseteq D$) and $U \subseteq V$. t and t' are two tuples in T. If $\forall i \in U, t[i] = t'[i]$ holds, U is called equivalent subspace between t and t' in V. If there is no such space $W \subseteq D$, satisfying $U \subset W \subset V$ and W is the equivalent subspace between t and t', U is called the maximum equivalent subspace between t and t' in V, denoted as $MES_V(t, t')$.*

In Figure 1, $MDS(c, e) = MDS(d, e) = \langle en, tr\rangle$, $MES_{en,tr}(c, e) = \langle tr\rangle$, $MES_{en,tr}(d, e) = \langle en\rangle$. According to $MDS(c, e) = \langle en, tr\rangle$ and $MES_{en,tr}(c, e) = \langle tr\rangle$, it can be concluded that e can possibly be a skyline tuple only on $\langle tr\rangle$, but according to $MDS(d, e) = \langle en, tr\rangle$ and $MES_{en,tr}(d, e) = \langle tr\rangle$, e can possibly be a skyline tuple only on $\langle en\rangle$. The intersection of the two subspace is null, so, e does not belong to the skyline of any subspace of $\langle en, tr\rangle$. On the basis of this conclusion, Theorem 3 is obtained.

Theorem 3. *Given a set T of tuples in a $|D|$-dimensional space D, U is subspace of D ($U \subseteq D$). And t is a tuple in T, there exists a set of tuple*

$\{t_1, t_2, ..., t_m\}$ satisfing $MDS(t_1,t) \supseteq U \wedge MES_U(t_1,t) = U_1, MDS(t_2,t) \supseteq U_1 \wedge MES_{U_1}(t_1,t) = U_2, \cdots, MDS(t_m,t) \supseteq U_{m-1} \wedge MES_{U_{m-1}}(t_1,t) = U_m$, if $W \cap (U - U_m) \neq \emptyset$ holds, then $t \notin Skyline(T, W)$ for any subspace $W \subseteq U$.

Proof. According to Definition 4, $MDS(t_1,t) \supseteq U \Rightarrow t_1 \succeq_U t$.

On the basis of Definition 4 and 5, $MDS(t_1,t) \supseteq U \wedge MES_U(t_1,t) = U_1 \Rightarrow t_1 \succ_{U-U_1} t$.

Thus, according to Lemma 1, in any subspace W, if $W \cap (U - U_1) \neq \emptyset$, $t \notin Skyline(T, W)$;

Similarly, $MDS(t_2,t) \supseteq U_1 \Rightarrow t_2 \succeq_{U_1} t$, $MDS(t_2,t) \supseteq U_1 \wedge MES_{U_1}(t_2,t) = U_2 \Rightarrow t_1 \succ_{U_1-U_2} t$. So, in any subspace W, if $W \cap (U - U_2) \neq \emptyset$, $t \notin Skyline(T, W)$;

Finally, in any subspace W, if $W \cap (U - U_m) \neq \emptyset$, $t \notin Skyline(T, W)$. □

4 Energy-Efficient Multi-skyline Evaluation

4.1 Global Multi-skyline Optimization

According Theorem 2, a subspace skyline result is the subset of extended skyline result in its super parent space. Once extended skyline result in a super space is known, all the skyline queries in its subspaces will be answered without retrieving any information from the sensor network. So the execution of several extended skyline queries in their super parent spaces will resolve all the original skyline queries, in the sensor network.

Can the execution of the full space extended skyline resolve all the subspace skyline queries, in the network? The answer is yes, it does. But it is not economical. Because the expected complexity of skyline query result is $O(\ln^{k-1} n/(k-1)!)$ [2] and the extended skyline query is only a simple extension of skyline query, the expected complexity of the result is also $O(\ln^{k-1} n/(k-1)!)$. It is concluded that replacing skyline queries with extended skyline queries would only slightly increase the number of results, while the increase of dimensions would sharply increase the number of the result. So we should merge skyline queries in subspaces into existing extended skyline queries in its super parent space, not create an extended skyline query in the space where no corresponding skyline query exists.

The original skyline queries, whose results are contained in the synthetic skyline query x, can be divided into two kinds: one has the same spaces with x, the other is in the proper subspace of x. If the original skyline query having the same spaces does not exist, executing x query is extravagant. This kind of skyline query determines whether x should be executed, so they are called *decisive query* of x; But the other kind of skyline query is answered by the results of the result of x. If the first kind of query exists, they have no influence to the execution of x, so we call them *covered query* of x. Their definition is introduced as follows.

Definition 6. *For a synthetic extended skyline query x, the original skyline query who is in the same query space with x is called decisive query of x.*

Definition 7. *For a synthetic extended skyline query x, the original skyline query whose query space is the proper subset of that of x is called covered query of x.*

	A	B	C	D
t_1	5	6	9	7
t_2	4	3	7	6
t_3	4	4	7	6
t_4	5	4	6	3
t_5	2	4	8	9

(a) Objects

Original Queries

s_1	A	x_1
s_2	AB	x_1
s_3	BD	x_2
s_4	C	x_2
s_5	BCD	x_2

Synthetic Queries

x_1	AB	s_2	s_1
x_2	BCD	s_5	s_3 s_4

(b) Queries

Original query results		GMSO results		SRT results	INF results
s_1	$\{t_5\}$				
s_2	$\{t_2, t_5\}$	x_1	$\{t_2, t_3, t_5\}$		
s_3	$\{t_2, t_4\}$			$01\{t_5\}, 10\{t_4\}, 11\{t_2, t_3\}$	$01\{t_5\}, 10\{t_4\}, 11\{t_2\}$
s_4	$\{t_4\}$	x_2	$\{t_2, t_3, t_4\}$		
s_5	$\{t_2, t_4\}$				

(c) Results

Fig. 2. An running example

Consider the set of 5 tuples shown in Figure 2(a). Original queries and corresponding synthetic queries are shown in Figure 2(b). From the query results in Figure 2(c), we can conclude that all the results of original queries are included in the synthetic query results produced by GMSO. The query number decreases from 5 to 2 and the cost of transmission decreases from 8 to 6 $tuple \cdot times$. Through transforming skyline queries in subspaces into extended skyline queries in several super parent spaces, not only the query propagation cost is reduced, but also parts of duplicate transmissions of tuples belonging to multiple subspace skyline results are avoided.

Basic Data Structure. All original skyline queries that need to be responded are stored in the set $S = \{s_1, s_2, \cdots, s_n\}$ and the corresponding synthetic extended skyline queries executed in the sensor network are stored in the set $X = \{x_1, x_2, \cdots, x_m\}$. For each original skyline query, we only need to record the subspace it inquires and the extended skyline query which responds it. They are stored in the form of: $s_i = \langle sid, querySpace, xid \rangle$. Here, sid is the unique identifier of the skyline query; $querySpace$ contains all the dimensions that skyline query s has; xid is the identifier of the synthesized extended skyline query responding the original skyline query.

For each synthetic query x, the information of the corresponding original skyline queries is also needed to be recorded, beside the subspace of x. Facilitating the insertion and termination processing of the skyline query, the relative skyline queries are stored into two sets, decisive query set and covered query set, separately. $x_i = \langle xid, querySpace, decisiveQueries, coveredQueries \rangle$. Here, xid is the unique identifier of the synthetic extended skyline query; $querySpace$ contains all the dimensions that the extended skyline query x has; $decisiveQueries$ and $coveredQueries$ fields contain respectively all the decisive queries and covered queries of the synthetic extended skyline query.

Data structures of s_i and x_i are shown in Figure 2(b). For example, the $querySpace$ of s_2 is AB and its xid is x_1. The $querySpace$ of x_2 is BCD, the $decisiveQueries$ is $\{s_5\}$ and $coveredQueries$ is $\{s_3, s_4\}$.

Skyline Query Insertion. The process of skyline query insertion is shown in Algorithm 1. When a new skyline query s arrives, scan the synthetic query list (Line 1-16). If the *querySpace* of s is the subset of the *querySpace* of a synthetic query x, merge s into x (Line2-9). At the same time, set the $s.xid$ with $x.xid$. In the process of merging, if the *querySpace* of s equals to that of x, add s into the decisive query list of x (Line 4) and if the *querySpace* of s is the proper set of that of x, add s into the covered query list of x (Line 6). Then the algorithm is terminated (Line8-9). If the attribute list of synthetic query x is the subset of that of s, put its decisive query list and covered query list into a temporary covered query list and terminate x (Line11-15). At last, according to s and the temporary covered query list, a new synthetic query is created (Line 17).

Algorithm 1. InsertSkyline

```
 1: for each element x in X do
 2:    if s.querySpace ⊆ x.querySpace then
 3:       if s.querySpace == x.querySpace then
 4:          x.decisiveQueries.add(s);
 5:       else
 6:          x.coveredQueries.add(s);
 7:       end if
 8:       s.xid = x.xid;
 9:       return;
10:    end if
11:    if s.querySpace ⊃ x.querySpace then
12:       tempSet.addAll(x.coveredQueries);
13:       tempSet.addAll(x.decisiveQueries);
14:       Terminate(x);
15:    end if
16: end for
17: X.newSyntheticQuery(s.querySpace, {s}, tempSet);
```

Consider the example shown in Figure 2. When a query $\langle s_6, ABC, null\rangle$ arrives, synthetic query x_1 would be terminated and a new synthetic query $\langle x_3, ABC, \{s_6\}, \{s_1, s_2\}\rangle$ is created, simultaneity. If another query $\langle s_7, CD, null\rangle$ arrives, the set of synthetic queries is not influenced, because s_7 is covered by existing synthetic query x_2.

Usually, the larger the dimensionality of the space is, the greater the probability that the subspace is contained is. In order to find out the super parent space extended skyline of a new skyline query as soon as possible, we sort synthetic queries in a descending order of the dimensionality, which improves the searching efficiency.

Because of the limitation of the number of sensors on each sensor node, the dimensionality would not be too high, bit operations can be used to improve the efficiency of the algorithm. Therefore, *querySpace* can be stored as a bit array of $32bit$, in which each bit indicates one dimension. If the corresponding dimension belongs to the query space, the bit would be set to be '1', otherwise set to be

'0'. The space with higher dimensionality can not be included by the space with lower dimensionality, so when judging their relationship, we first compare the dimensionalities of two spaces s_1 and s_2. Let's assume that $|s_1| > |s_2|$. And then "or" operation is used between the bit arrays of s_1 and s_2. If its result equals to the bit array of $s1$, it shows that s_1 includes s_2. It would greatly reduce the number of comparisons of judging the relationship between sets and sharply improve the efficiency of the algorithm.

In real applications, it is quite normal to insert a batch of subspace skyline queries and not economic to insert them respectively. As mentioned above, if a subspace query is inserted before its parents space query, it would lead to repeatedly creation and termination of the synthetic query. Thus queries can be sorted in a descending order of inclusion relation and those queries not having inclusion relation with any other query could be sorted in a descending order of dimensionality. Each query records all its subspace queries, and once finds out its parent space query or creates a new synthetic query, all its subspace queries could be directly added to the *coveredQueries*, which reduces the cost of scanning synthetic queries.

Skyline Query Termination. When a user skyline query s is terminated, the base station not only needs to delete s from S but also needs to judge whether the corresponding synthetic query x should be modified or not. After deleting s, if the decisive query set of x is null, it means that there is no other original skyline query whose subspace is equivalent to x. As a result, it is not economic to keep on executing x. Therefore, it should be terminated and the skyline queries covered by it should be inserted again. On the other hand, if decisive query set is not null, there still are some original skyline queries whose subspaces are equivalent to that of x, so the execution of x is necessary and no more other operations. The process of skyline query termination is shown in Algorithm 2. When the query s is terminated, according to the $s.xid$, find out the corresponding synthetic query x firstly (Line 1). If s is contained in the covered query list of x, delete it from $x.coveredQueries$ (Line 2-4), and if s is contained in the decisive query list of x, delete it from $x.decisiveQueries$ (Line 5-7). After the deletion, if the decisive query list of x is null, x it should be terminated and skyline queries in the covered query list should be inserted again as new queries. As a result, new synthetic queries are created(Line 8-13). At last, delete s from the original skyline list and the algorithm is terminated(Line 14).

In Figure 2, the decisive query set of the x_1 is $\{s_2\}$ and its covered query set is $\{s_1\}$, but the decisive query set of x_2 is $\{s_5\}$ and its covered query set is $\{s_3, s_4\}$. If s_2 is terminated, we must delete it from the set of decisive queries of corresponding synthetic query x_1. After that, we will find that the *decisiveQueries* become $\emptyset$, so x_1 should be terminated and the original query in its *coveredQueries* would be inserted as a new query. The termination of s_3 only need delete itself from the covered query set of synthetic query x_2. Since there is no modification for decisive query set of it, x_2 could be still executed.

Similarly, original queries in $x.coveredQueires$ are sorted in a descending order of inclusion. It avoids that the subspace query is inserted before the insertion

Algorithm 2. TerminateSkyline

```
x = X.getByID(s.xid);
if s ∈ x.coveredQueires then
    x.coveredQueires.delete(s);
end if
if s ∈ x.decisiveQueires then
    x.decisiveQueires.delete(s);
end if
if |x.decisiveQueires| == 0 then
    X.delete(x);
    for each element s' in x.coveredQueries do
        InsertSkyline(s');
    end for
end if
S.delete(s);
```

of its super parent space queries, so the number of the creation and termination of synthetic queries is reduced.

4.2 Local Multi-skyline Optimization

Sharing Result Transmission. Through the GMSO, original skyline queries are replaced by a group of synthetic extended skyline queries. However, there are always intersections among the query spaces of synthetic extended skyline queries, extended skyline results often have intersections. If each query result is transmitted respectively, the energy of sensor nodes would be wasted. So if transmission of the same tuple belonging to different extended skylines is shared, the transmission cost will be reduced sharply. In order to achieve this, we assign each tuple a signature that indicates which query the tuple belongs to. Each bit of the signature is related to a synthetic extended skyline query. After that, we classify the extended skyline tuples by these signatures, that is, tuples belonging to the same queries will be inserted into the same group and assigned the same signature. Then all the tuples would be transmitted only once and all the redundant results will be suppressed, with a little increase of signature transmission.

Shown in Figure 2(c), because t_5 and t_4 belong to x_1 and x_2 respectively, signatures of them are 01 and 10. While t_2 and t_3 both belong to x_1 and x_2, so their signature is 11. Comparing with GMSO, the cost of transmission is further reduced to 4 tuples and 6 bits signature.

In-Network Filtering. Not all of the extended skyline tuples are subspace skyline tuple. Theorem 3 illuminates that when extended skyline tuple t is dominated by a set of tuples, its maximum dominate space will shrink. Once its maximum dominated space becomes null, illustrating t would not be the skyline tuple of any space, so it is unnecessary to transform it in the network. Filtering these useless data, each extended skyline tuple is assigned an attribute, *possibleSpace*, to record the maximum space where the tuple would be skyline, while in the processing of calculating the extended skyline. In order to save the memory space, if the *possibleSpace* indicates the whole data space, it would be omitted. Without loss of generality, assume that t' dominates t in the extended skyline space. Then *possibleSpace* of t, where it possibly belongs to the skyline, is $MES_U(t, t')$. If another tuple t'' dominates t, the *possibleSpace* of t, where it possibly belongs to the skyline, shrinks to $MES_{possibleSpace}(t, t'')$. Once the *possibleSpace* of is null, delete t from the candidate results set R.

In the example shown in Figure 2, in the processing of the query x_1, when calculate the relationships between t_2 and $t3$, we find that $t_2 \succeq_{AB} t_3$ and $MES_{AB}(t_2, t_3) = A$, so the *possibleSpace* is initialized to be A; while calculate the relationships between t_3 and t_5, $t_5 \succeq_A t_3$ and $MES_A(t_3, t_5) = \emptyset$, then $possibleSpace = \emptyset$, meaning that t_3 does not belong to the skyline of any subspace of extended skyline query x_1. For the same reason, $t_2 \succeq_{BCD} t_3$ and $MES_{BCD}(t_2, t_3) = CD$, $t_4 \succeq_{CD} t_3$ and $MES_{CD}(t_3, t_5) = \emptyset$, indicating that t_3 does not belong to the skyline of any subspace of extended skyline query x_2. So it is unnecessary to transmit t_3 to its parent node. Comparing with the front method, the total cost is further reduced to 3 tuples and 6 bits signatures.

Influenced by the hierarchical routing structure of WSNs, it can not be avoid that transmitting the local skyline tuples, which does not belong to the global skyline, to the ancestor node. Such condition also needs to be considered. When there is a query insertion or termination, commands will be hierarchically transmitted to all the sensor nodes. In the transmission process, if MinMax filter [3] is sent together, some unnecessary local results transmission would be avoided and the cost will be reduced. Let the query subspace of synthetic query x_i is $|U_i|$. The calculating process of MinMax filter is shown in equation 1, 2 and 3.

$$max_j = MAX_{k=1}^{|U_i|} t_j[k] (1 \leq j \leq |T|) \tag{1}$$

$$minmax_i = MIN_{j=1}^{|T|} max_j \tag{2}$$

$$MinMax_i = \{minmax_i, minmax_i, \cdots, minmax_i\} \tag{3}$$

Now we set a MinMax filter for each query. When the number of synthetic queries is very large, they will still cost too much. Sometimes, only one byte will lead to a new package transmission. Moreover, because each MinMax filter only takes charge of the results of the responding skyline query, a tuple t may be dominated by the MinMax filter of a query x_i, but not dominated by the MinMax filter of another query x_j. The tuple t is still needed to transport to its parent node. In order to reduce these useless judgments, $MaxMinMax$ filter is proposed.

$$maxminmax_i = MAX_{i=1}^{|X|} minmax_i \tag{4}$$

$$MaxMinMax_i = \{maxminmax_i, \cdots, maxminmax_i\} \tag{5}$$

Next, the properties of $MaxMinMax$ is discussed to show its effectiveness.

Lemma 2. *For a set* $M = \{MinMax_1, MinMax_2, \cdots, MinMax_{|X|}\}$ *of MinMax filters and* $MaxMinMax$ *is the MaxMinMax filter of them. If a tuple* t *is dominated by* $MaxMinMax$*, it must dominated by any* $MinMax_i \in M$.

Proof. Immediate deducted from the definitions of MinMax filter and MaxMinMax filter. □

Theorem 4. *All tuples filtered by MaxMinMax filter would not belong to any synthetic query result.*

Proof. Due to the page limit, the proof has been omitted. □

Therefore, it is guaranteed that each tuple has been judged only once and the transmission cost is reduced.

5 Performance Evaluation

We have developed a simulator to evaluate the performance of our proposed EMSE approach. The nodes are set in to Crossbow Mica2 motes model [5], the generic MAC-layer protocol is used and only the energy consumption of wireless communication is taken in to account. For the sake of experimental unification, we randomly place n sensors in an area of $\sqrt{n} \times \sqrt{n}$ units, then each node holds one unit space averagely, and the communication radius is set to $2\sqrt{n}$ units. Meanwhile, we regulate that the maximal length of packets transmitted in the network is limited to 48 bytes. The followings are algorithms that have been evaluated.

-SMSE: The algorithm evaluating multiple skyline queries one by one.
-GMSO: The algorithm only using global optimization mechanism.
-EMSE: Our energy-efficient multiple skyline queries evaluation approach.

The simulated data are all generated by standard dataset generator [1], including independent (IND) and anti-correlated (ANT). Due to the page limitation, the results of IND is omitted, since it is similar with the performance of ANT data. In the experiments, we mainly consider two kinds of query distributions. One is uniform query (UQ) where users concern all the dimensions equally. The other is skew query (SQ) where users pay more attention to some dimensions but less attention to the others. We use uniform distribution data to simulate uniform query and use zipf distribution data to simulate skew query, in the experiments. Table 1 summarizes parameters under investigation, along with their ranges and default values. In each experiment, we vary only one single parameter each time, while setting the remainders to their default values. All simulations are run on a PC with 2.8GHz CPU and 512M of memory.

First, we investigate the performances of the algorithms, when the number of nodes changes. Figure 3 shows that the cost of all the algorithms is raised when the number of nodes increased. It is because the increment of sensor nodes directly results in the increment of data, which leads to the increment of the skyline results number. So the communication cost is also increased. It is obvious

Table 1. Simulation parameters

Parameter	Default	Range
Number of nodes	800	600, 700, 800, 900, 1000
Dimensionality	8	6, 7, 8, 9, 10
Number of queries	15	5, 10, 15, 20, 25
duplicate ratio	6%	2%, 4%, 6%, 8%, 10%

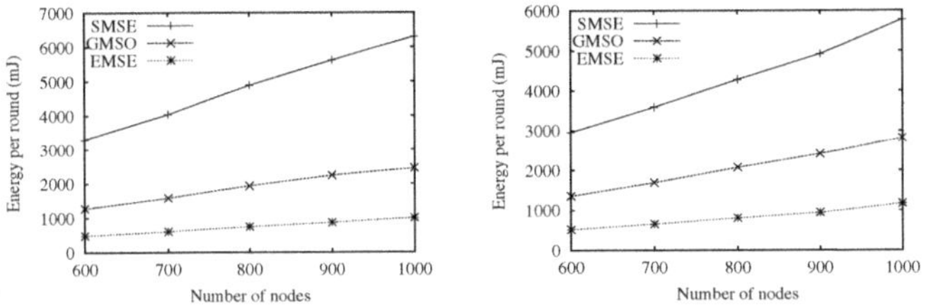

(a) UQ & ANT (b) SQ & ANT

Fig. 3. Communication cost vs. Number of sensor nodes

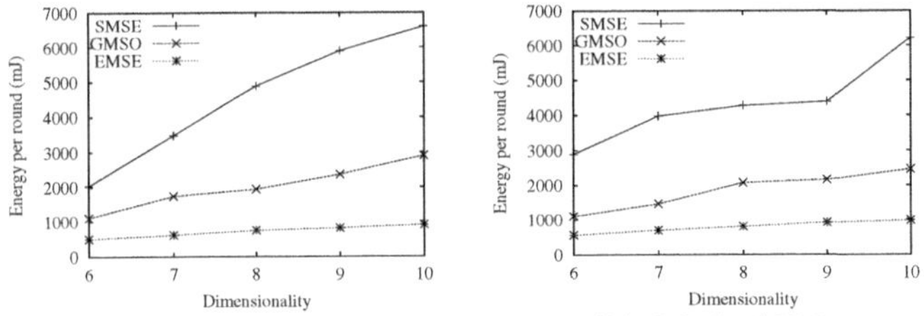

(a) UQ & ANT (b) SQ & ANT

Fig. 4. Communication cost vs. Data dimensionality

that the cost of SMSE is highest, GMSO's is lower, and the cost of EMSE is the lowest. It indicates the effectiveness of both our global optimization and local optimization mechanism. The cost of all the algorithms under independent distribution is lower than those under anti-correlated distribution, because skyline results in anti-correlated distribution are much more than those in independent distribution.

Second, we consider the influence of dimensionality to the algorithms. Figure 4 shows that the cost of all algorithms is increased when the dimensionality is raised. It is because that the increment of dimensionality leads to decrease of the probability of one tuple dominated by others, which makes skyline results enhanced and the cost of communication increase. The relation of performances of these three algorithms is similar with those shown in Figure 3, which further proves the utility of global optimization and local optimization mechanism.

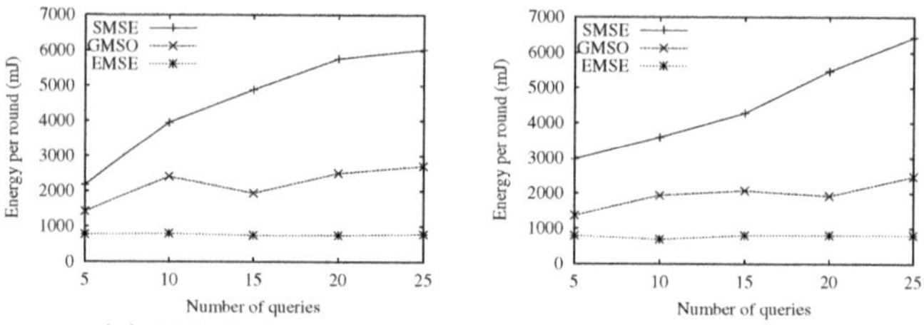

(a) UQ & ANT (b) SQ & ANT

Fig. 5. Communication cost vs. Number of skyline query

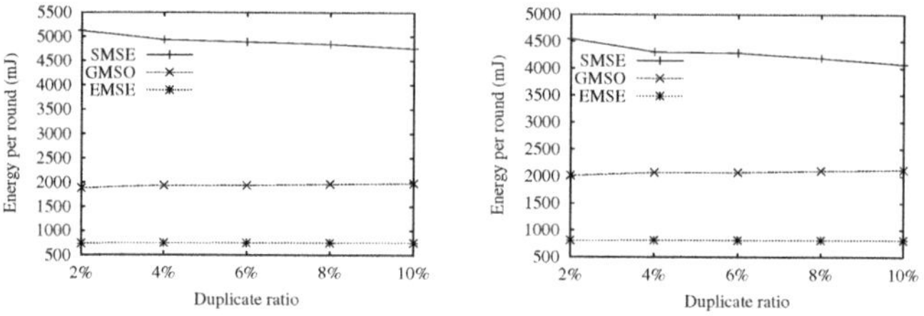

(a) UQ & ANT (b) SQ & ANT

Fig. 6. Communication cost vs. Data duplicate ratio

Then we consider the performances of the algorithms, when the query number is changed. While the query number increases, the communication cost of SMSE is also raised, shown in Figure 5. The reason is that the number of queries is enhanced, which leads to the increase of the number of query results, that also increases the number of data transmitted among sensor nodes. However, the cost of GMSO is lower than that of SMSE, but fluctuates, which indicates that GMSO is more effective, but is also affected by the diversity of queries. But the performance of EMSE is quite good, not influenced by queries. Through the two-step optimizations, the affection of queries is obviously decreased.

Finally, we discuss the influence of the data duplicate ratio on the algorithms. As shown in Figure 6, while the data duplicate ratio increases, the cost of SMSE is reduced a little. Because the increase of the ratio means the cardinality in the data domain is reduced, the probability of tuples being dominated is also reduced and so is the cost of the algorithm. However, the cost of EMSE and GMSO is gradually increases due to the increase of the ratio of duplicate data results and the increase of the number of extended skyline results. And the reason why the cost of EMSE is lower than GMSO is that EMSE avoids most unnecessary data transmissions. It transmits only data that belong to the extended skyline but not to any subspace skyline.

Through the experiments above, we can conclude that no matter what kind of distribution query obey, uniform query distribution or skew query distribution, and what kind of distribution data obey, independent distribution or anti-correlated distribution, the performance of GMSO is improved about 50% of that of SMSE. It indicates the utility of global optimization using query rewriting. However the performance of EMSE is 30-50% higher than that of GMSO, which indicates the utility of our proposed local optimization. After the two step optimizations, the number of data transmissions is reduced, which results in the decrease of energy consumption, so the life-span of WSNs is prolonged.

6 Conclusions

When multiple skyline queries are posed into WSNs, computing each skyline query individually is not energy-efficient. Because it ignores the essential relations among multiple skylines. In this paper, we proposed a novel and energy-efficient approach: the Energy-efficient Multiple Skyline Evaluation (EMSE). It aims at reducing the energy consumption on evaluating multiple skyline queries in wireless sensor networks. First, a global optimization mechanism which reduces the number of skyline queries is proposed to save the query propagation cost and the redundant results transmission cost. Then, a local optimization mechanism is employed to further reduce the communication cost through sharing the results transmission and filtering unnecessary data transmission. The experimental results prove that EMSE is an energy-efficient approach for calculating multiple skyline queries in wireless sensor networks.

Acknowledgement. This research is supported by the National Natural Science Foundation of China (Grant No. 60873011, 60773221, 60773219 and 60803026), National Basic Research Program of China (Grant No. 2006CB303103), and the 863 High Technique Program (Grant No. 2007AA01Z192).

References

1. Borzsonyi, S., Stocker, K., Kossmann, D.: The skyline operator. In: Proc. of ICDE, pp. 421–430 (2001)
2. Chaudhuri, S., Dalvi, N.N., Kaushik, R.: Robust cardinality and cost estimation for skyline operator. In: Proc. of ICDE, page 64 (2006)
3. Chen, H., Zhou, S., Guan, J.: Towards energy-efficient skyline monitoring in wireless sensor networks. In: Langendoen, K.G., Voigt, T. (eds.) EWSN 2007. LNCS, vol. 4373, pp. 101–116. Springer, Heidelberg (2007)
4. Greenwald, M.B., Khanna, S.: Power-conserving computation of order-statistics over sensor networks. In: Proc. of PODS, pp. 275–285 (2004)
5. C. Inc. MPR-Mote Processor Radio Board User's Manual
6. Kyriakos Mouratidis, D.P., Bakiras, S.: Continuous monitoring of top-k queries over sliding windows. In: Proc. of SIGMOD, pp. 635–646 (2006)
7. Madden, S., Franklin, M.J., Hellerstein, J.M., Hong, W.: TAG: A tiny aggregation service for ad-hoc sensor networks. In: Proc. of OSDI, pp. 131–146 (2002)
8. Trigoni, N., Yao, Y., Demers, A., Gehrke, J., Rajaraman, R.: Multi-query optimization for sensor networks. In: Prasanna, V.K., Iyengar, S.S., Spirakis, P.G., Welsh, M. (eds.) DCOSS 2005. LNCS, vol. 3560, pp. 307–321. Springer, Heidelberg (2005)

9. Vlachou, A., Doulkeridis, C., Kotidis, Y., Vazirguannis, M.: Skypeer: Efficient subspace skyline computation over distributed data. In: Proc. of ICDE, pp. 416–425 (2007)
10. Wu, M., Xu, J., Tang, X., Lee, W.-C.: Top-k monitoring in wireless sensor networks. TKDE 19(7), 962–976 (2007)
11. Xiang, S., Lim, H.B., Tan, K.-L., Zhou, Y.: Multiple query optimization for sensor network. In: Proc. of ICDE, pp. 1339–1341 (2007)
12. Xiang, S., Lim, H.B., Tan, K.-L., Zhou, Y.: Two-tier multiple query optimization for sensor network. In: Proc. of ICDCS, p. 39 (2007)
13. Xie, L., Chen, L.-j., Lu, S., Xie, L., Chen, D.-x.: Energy-efficient multi-query optimization over large-scale sensor networks. In: Cheng, X., Li, W., Znati, T. (eds.) WASA 2006. LNCS, vol. 4138, pp. 127–139. Springer, Heidelberg (2006)
14. Xin, J., Wang, G., Chen, L., Zhang, X., Wang, Z.: Continuously maintaining sliding window skylines in a sensor network. In: Kotagiri, R., Radha Krishna, P., Mohania, M., Nantajeewarawat, E. (eds.) DASFAA 2007. LNCS, vol. 4443, pp. 509–521. Springer, Heidelberg (2007)
15. Xin, J., Wang, G., Zhang, X.: Energy-efficient skyline queries over sensor network using mapped skyline filters. In: Proc. of APWeb/WAIM, pp. 144–156 (2007)
16. Yuan, Y., Lin, X., Liu, Q., Wang, W., Yu, J.X., Zhang, Q.: Efficient computation of the skyline cube. In: Proc. of VLDB, pp. 241–252 (2005)

Dominant and K Nearest Probabilistic Skylines*

Gabriel Pui Cheong Fung[1], Wei Lu[2,3], and Xiaoyong Du[2,3]

[1] School of ITEE, The University of Queensland, Australia
g.fung@uq.edu.au
[2] Key Labs of Data Engineering and Knowledge Engineering, Ministry of Education, China
[3] School of Information, Renmin University of China, China
{uqwlu,duyong}@ruc.edu.cn

Abstract. By definition, objects that are skyline points cannot be compared with each other. Yet, thanks to the probabilistic skyline model, skyline points with repeated observations can now be compared. In this model, each object will be assigned a value to denote for its probability of being a skyline point. When we are using this model, some questions will naturally be asked: (1) Which of the objects have skyline probabilities larger than a given object? (2) Which of the objects are the K nearest neighbors to a given object according to their skyline probabilities? (3) What is the ranking of these objects based on their skyline probabilities? Up to our knowledge, no existing work answers any of these questions. Yet, answering them is not trivial. For just a medium-size dataset, it may take more than an hour to obtain the skyline probabilities of all the objects in there. In this paper, we propose a tree called SPTree that answers all these queries efficiently. SPTree is based on the idea of space partition. We partition the dataspace into several subspaces so that we do not need to compute the skyline probabilities of all objects. Extensive experiments are conducted. The encouraging results show that our work is highly feasible.

1 Introduction

Given a set of objects, X, let U and V be two different objects in X. U is said to dominate V if U performs better than V in at least one dimension and not worse than V in all other dimensions. An object that cannot be dominated by any other object in X is called a skyline point. A collection of skyline points formulates a skyline. Traditionally, skyline points cannot be compared with each other because each skyline point performs superior than the other skyline points in some, but not all, dimensions, and different dimension is incomparable.

Yet, thanks to the idea of probabilistic skyline [1], skyline points can now be compared with each others if each skyline point contains repeated observations. For example, if we want to evaluate a basketball team, we will first collect the games (observations) that this team is involved, then identify the number of games that this team performed at skyline point, and finally obtain a probability to denote how likely this team can be a skyline point (i.e. skyline probability). Accordingly, we can compare the performances of all teams, as they should have different skyline probabilities.

* Wei Lu's and Xiaoyong Du's research is partially supported by National Natural Science Foundation of China under Grant 60573092 and Grant 60873017.

X. Zhou et al. (Eds.): DASFAA 2009, LNCS 5463, pp. 263–277, 2009.

When we are using the probabilistic skyline model, three questions will naturally be asked: (1) Which of the objects have skyline probabilities larger than a given object (Extract dominant probabilistic skyline points)? (2) Which of the objects are the K nearest neighbors to a given object according to their skyline probabilities (Identify K nearest probabilistic skyline points)? (3) What is the ranking of these objects based on their skyline probabilities (Rank the extracted probabilistic skyline points)? Up to our knowledge, no existing work can effectively answer any of the above questions. Yet, answering them is not trivial. It can take more than an hour to compute the skyline probabilities of all the objects in a 3 dimensional dataset with just 100 objects and each object contains 2,000 observations

In this paper, we propose a novel tree structure, called Space-Partitioning Tree (SP-Tree), that can answer all of the above questions efficiently. Our motivations and contributions are both solid. The idea of SPTree is came from space partitioning. Extensive experiments are conducted to evaluate the efficiency of our work. The results are highly encouraging. The rest of the paper is organized as follows – Section 2 defines our problem; Section 3 presents SPTree in details; Section 4 reports the experimental results; Section 5 discusses some important related works; Section 6 concludes this paper.

2 Problem Definition

Table 1 shows the symbols that would be used in this paper. Let $D = \{d_1, d_2, \ldots, d_N\}$ be a N-dimensional space, where d_i is a dimension. Given d_i, the smaller the value it is, the more preferable it will be. Let X be a set of objects in D.

By definition, given two objects, $U \in X$ and $V \in X$, and their observations, $u \in U$ and $v \in V$, u is said to dominate v, $u \prec v$, iff the following two conditions hold: (1) $\forall d_i \in D, u.d_i \leq v.d_i$; (2) $\exists d_i \in D, u.d_i < v.d_i$. The skyline probability of U, $P(U)$ is [1]:

$$P(U) = \frac{1}{|U|} \sum_{u \in U} P(u), \quad P(u) = \prod_{V} \left(1 - \frac{\|v \in V | v \prec u\|}{|V|}\right). \tag{1}$$

In this paper, we define the following new concepts:

Definition 1. p-dominant and p-dominant object *An object, V, is said to p-dominant another object, U, iff $P(V) > p$ and $P(U) = p$. Here, V is a p-dominant object.*

Table 1. Symbols and their meanings

Symbols	Meanings
D, d_i	D is an N dimensional space; d_i is a dimension of D, $d_i \in D$
X	A set of objects in D
U, u	U is a target object, $U \in X$; u is an observation, $u \in U$
V, v	V is any object other than U, $V \in X, V \neq U$; v is an observation of V, $v \in V$
$P(U)$ $(P(V))$	The skyline probability of U (V)
$P(u)$ $(P(v))$	The skyline probability of u (v)
$P_u(U)$ $(P_u(V))$	The upper bound skyline probability of U (V)
$P_l(U)$ $(P_l(V))$	The lower bound skyline probability of U (V)
S, V^s	S is a subspace after space partition; V^s is a subset of V drops in S, $V^s \subseteq V$
V^s_{min} (V^s_{max})	The min (max) point of the minimum bounding rectangle of V^s

Definition 2. ***K*-nearest *p*-skyline, S_K** *Given U with $P(U) = p$, K-nearest p-skyline is a set of K objects that is most nearest to U according to the skyline probabilities.*

Given an object, U, we try to solve the following three problems: (1) Extract objects that p-dominant U; (2) Rank p-dominant objects; and (3) Identify K-nearest p-skyline with respect to U. Note that computing $P(U)$ is trivial if we obtained all $P(u)$. Unfortunately, computing $P(u)$ is very time consuming because $\forall u \in U$ we have to identify how many $v \in V$ cannot dominate it. This is the major bottleneck.

3 Proposed Work

We propose a novel tree, Space-Partitioning Tree (SPTree), to answer the three questions stated in the previous section. In the followings, we first present the motivation of SPTree, then give an overview of it, and finally describe its implementation detail.

3.1 Motivation

Since computing the skyline probabilities of all the objects in a dataset is very expensive, we therefore try to think whether it is possible for not to compute their exact values, but to identify their range only, for solving our problems. Given U and $P(U)$, We observe that our problems have some interesting properties:

First, we only need to return the objects that satisfy the queries, but not the skyline probabilities. Second, let $P_l(V)$ and $P_u(V)$ be two probabilities, such that $P_l(V) \leq P(V) \leq P_u(V)$. If $P_l(V) \geq P(U)$, then V must be a p-dominant object. Similarly, if $P_u(V) < P(U)$, then V will never be a p-dominant object. Third, if we can obtain all $P_l(V)$ and all $P_u(V)$, without computing any $P(V)$, then we can immediately conclude some of their relationships with U. We only need to compute $P(V)$ for those V that their relationships with U are ambiguous. Forth, let $V' \neq V$ be another object in X. For ranking the p-dominant objects, if $\exists V', P_l(V) \geq P_u(V')$, then V must be ranked higher than V'. Furthermore, if $P_l(V) \geq P_u(V'), \forall V'$, then V must have the highest skyline probability. Similar for the case of $P_l(V) < P_u(V')$. Fifth, let $D_{min}(V,U)$ and $D_{max}(V,U)$ be the min difference and max difference between $P(V)$ and $P(U)$, respectively. For identifying K-nearest p-skyline, if there are K or more objects with max differences to $P(U)$ smaller than $D_{min}(V,U)$, then V can be pruned away. Similarly, $\forall V' \in X, V' \neq V$, if $D_{max}(V,U) \leq D_{min}(V',U)$, then V must belong to the K-nearest p-skyline of U.

To conclude, we use "range values" ($P_l(V)$ and $P_u(V)$) instead of "exact value" ($P(V)$) to solve our problem. Now our question is: how to compute $P_l(V)$ and $P_u(V)$?

3.2 Space Partitioning Approach

Given U, without any prior information, if we want to extract the objects which are p-dominant to U or are K-nearest to U, the first step is certainly to compute $P(U)$. Once we have obtained $P(U)$, we can utilize it to facilitate other processes. Thus, we build a tree (SPTree) according to U to index all the other objects.

Fig. 1 respectively show a sample dataset and the corresponding SPTree. In Fig. 1, there are three objects, A, B and U. To construct the SPTree, we first need to partition

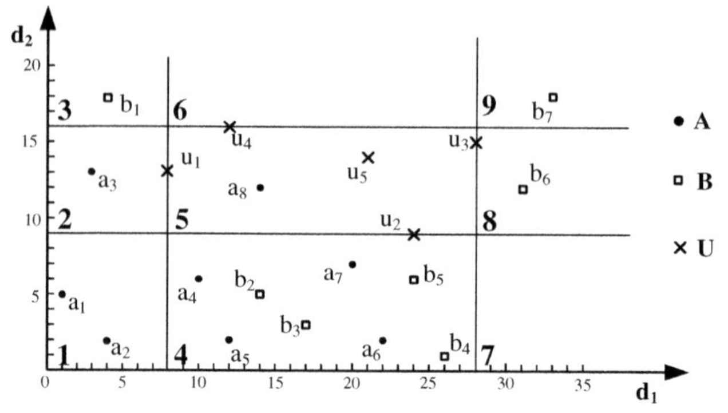

Fig. 1. An example dataset

the dataspace into subspaces based on the min and max values of U in each dimension. E.g. There are 9 subspaces in Fig. 1. Let S be a subspace. In Fig. 1, $S \in \{1,2,\ldots,9\}$. Only subspace 5 contains u. Observations of other objects may drop into more than one subspaces. Let $V^S \subset V$ be a set of observations drops in S. E.g. $A^1 = \{a_1, a_2\}$. We conduct the space partition because of the following reasons:

Recall that $P(u)$ is already computed. According to Theorem 2 in Appendix, if $\exists u \prec v, \forall v \in V^s$, then we can identify $P_u(V^s)$ by u, where

$$P_u(V^s) \leq p\left(\frac{|V^s| \times (|U| - |\{u \in U | u \prec v\}|)}{|U| \times (|V^s| - |\{v \in V^s | v \prec u\}|)}\right), \; p = \min_{u \in U}\{P(u) \mid u \prec v, \forall v \in V^s\}. \quad (2)$$

This process is very efficient because identifying the dominant relationship is simple [2] and we only need to consider U and a particular V. Similarly, we can obtain:

$$P_l(V^s) \geq p\left(\frac{|V^s| \times (|U| - |u \in U | u \prec v|)}{|U| \times (|V^s| - |\{v \in V^s | v \prec u\}|)}\right), \; p = \max_{u \in U}\{P(u) \mid v \prec u, \forall v \in V^s\}. \quad (3)$$

Hence, partitioning the dataspace can compute $P_u(V^S)$ and $P_l(V^S)$ efficiently.

There are two cases where we cannot use the above equations to compute $P_l(V^s)$ and $P_u(V^s)$: (1) We cannot find any pair of (u, v) that satisfies Eq. (2) or Eq. (3); and (2) We cannot further refine $P_l(V^s)$ and $P_u(V^s)$ by *recursively partitioning* the subspaces. We will explain case (2) shortly. If we cannot use Eq. (2) or Eq. (3) to compute $P_l(V^s)$ and $P_u(V^s)$, we have to compute them with the help of other objects. Yet, we do not need to consider all observations in the dataset but just need to consider the observations in the subspaces that may dominate U. E.g. in Fig. 1, if we compute $P_l(A^8)$ and $P_u(A^8)$, we only need to consider the observations of A and B in the subspaces 2, 5, 7 and 8 because objects in the subspace 1 and 4 will always dominate objects in subspace 8, and no objects in subspaces 3, 6 and 9 can dominate objects in subspace 8.

Let V^s_{min} and V^s_{max} be two points represent the min point and the max point of the minimum bounding box of V^s. Let $V'^s \neq V^s$ be another subspace. If either $V'^s_{max} \prec V^s_{min}$ or $V'^s_{min} \not\prec V^s_{max}$, then we do not need to compare the observations in V'^s with V^s. Hence, with space partition, the number of comparisons among objects is reduced dramatically,

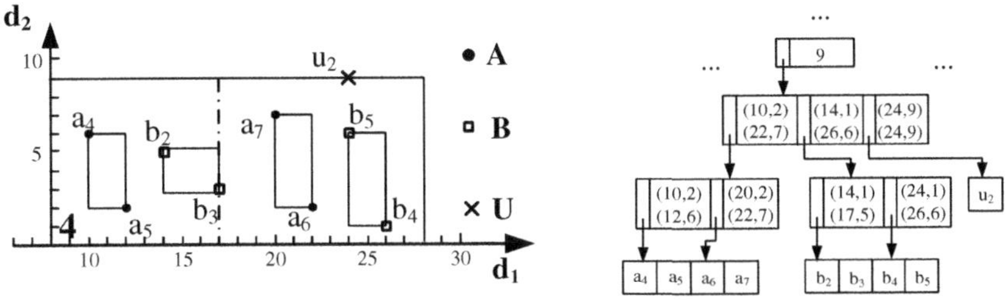

(a) Objects split in Subspace 4 (b) The corresponding nodes split in SPTree

Fig. 2. Nodes split in Subspace 4

whereas identifying V^s_{max} and V^s_{min} are trivial. Mathematically,

$$P_u(V^s) = \prod_{O \in X, O \neq V} \frac{\sum_{\forall O^{s'} \in O} |O^{s'}, O^{s'}_{max} \prec V^s_{min}|}{|O|}, \ P_l(V^s) = \prod_{O \in X, O \neq V} \frac{\sum_{\forall O^{s'} \in O} |O^{s'}_{min} \prec V^s_{max}|}{|O|}. \quad (4)$$

Once we obtained $P_l(V^s)$ and $P_u(V^s)$, we can update $P_l(V)$ and $P_u(V)$ incrementally:

$$P_l(V) = \sum_{V^s \in V} P_l(V^s) \times \frac{|V^s|}{|V|}, \quad P_u(V) = \sum_{V^s \in V} P_u(V^s) \times \frac{|V^s|}{|V|}. \quad (5)$$

Note that computing $P_l(V^s)$ and $P_u(V^s)$ is far more efficient than computing $P_l(V)$ and $P_u(V)$ because the search space in a subspace is far less than the whole space.

Let us refer to Fig 1 again. Assume that $P(U) = 0.4$. Suppose we obtained $0.6 \leq P(A) \leq 0.8$ and $0.45 \leq P(B) \leq 0.7$ after the above computation. Since $P_l(B) > P(U)$, B p-dominant U. For A, its relationship with U is undefined as its range overlap with $P(U)$. We therefore need to refine $P_l(A)$ and $P_u(A)$. The refinement is based on *recursively partitioning* each subspace further into two disjoint subspaces and compute $P_l(V^s)$ and $P_u(V^s)$ for each resulting subspace until we can distinguish the relationship between U and V. Two rules governed how we the partition the subspace. Rule 1: the total number of observations for all objects whose relationship with U are unknown have to be the same in both subspaces. By maintaining the subspaces with equal number of observations, we can usually converge to a solution much faster. Rule 2: partition a subspace depends solely on one particular dimension, d_i, where d_i varies for each iteration. Specifically, given two consecutive subspaces, S_1 and S_2 where $S_1 \cup S_2 = S$ and $S_1 \cap S_2 = \emptyset$, if $v.d_i < \delta, v \in V^s$, then v will be classified into S_1, otherwise it will be classified into S_2. Finally, let us use an example to illustrate how d_i is chosen. Assume $D = \{d_1, d_2\}$. The first and second iteration of the recursive partition will respectively depend on d_1 and d_2. The third iteration will depend back to d_1 again. For identifying δ, it is trivial once we constructed the SPTree as we will discuss it shortly. With these two rules, we can guarantee the size of the subspaces will not easily be biased by some dimensions, meanwhile the computational cost for each iteration will be less as we only need to partition the data according to one dimension at a time.

Let us refer to Fig. 1 again. Suppose the relationship of both A and B with U are ambiguous, then we need to partition the subspaces. Let us take Subspace 4 as an example. Fig. 2 shows the SPTree after nodes split in Subspace 4. According to Rule 1,

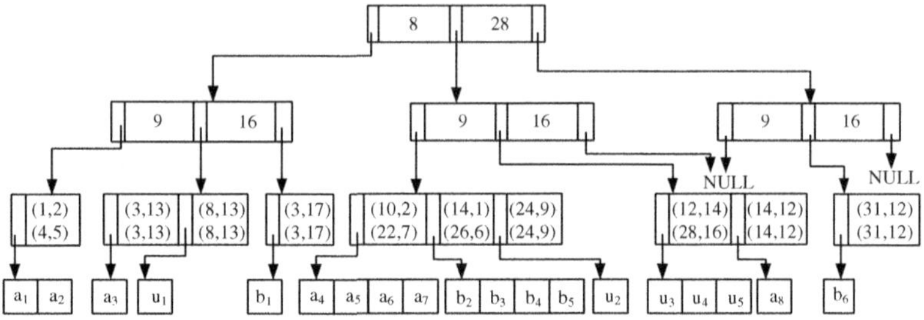

Fig. 3. An SPTree

we should partition Subspace 4 into two and each partition should have 4 observations. According to Rule 2, we partition the subspace based on d_1.

Let $\mathcal{V} \subseteq X$ be a set of objects p-dominant U. For the problem of extracting $\mathcal{V}$, it can be solved by applying the aforementioned framework. For ranking $\mathcal{V}$, we can identify the relationships among all $V \in \mathcal{V}$ by applying the same framework for each V. Finally, for identifying the K-nearest p-skyline, we can apply this framework to identify the relationship among all $V \in X$, and changing the ranking criterion from skyline probabilities to distance between $P(U)$ and $P(V)$. For implementing this framework, unfortunately, it is not simple. Deriving an appropriate data structure for efficient processing and identifying some special rules to further enhance the computational time are two non-trivial problems. In this paper, a Space Partitioning Tree (SPTree) is proposed.

3.3 SPTree Construction

Fig. 3 shows the SPTree for the dataset in Figure 1. Each node in the first N levels (N is the number of dimensions, i.e. $N = 2$ in this case) of a SPTree contains 2 keys and 3 pointers. Each pointer points to a node in the next level. For the nodes at the same level, they store the same pair of keys which denote the minimum and the maximum values of U at a particular dimension, i.e. at level i, the left key is $\min_{u \in U} u.d_i$ and the right key is $\max_{u \in U} u.d_i$. Each node at the $N+1$ level of SPTree corresponds to a subspace, S. The number of entries in a leave node depends on the number of objects that has observation in the corresponding S. Each entry corresponds to a V^s. It contains a pair of coordinations which denote V^s_{min} and V^s_{max} (i.e. the minimum bounding rectangle of V^s) and a pointer pointing to the first observation, v, in a leave node that stores all $v \in V^s$.

The SPTree may grow dynamically according to the case if a subspace requires recursively partition. Nodes would be added between the $N+1$ level of SPTree and the leave nodes. Fig. 2 (b) shows what would happen to the SPTree if we need to partition Subspace 4 in Fig. 1 (a). Fig. 2 (a) magnifies this subspace to include more details. In Subspace 4, there are 8 observations: a_4, a_5, a_6, a_7, b_2, b_3, b_4 and b_5. According to Rule 1, we should partition the subspace into two such that each of them contains 4 observations. According to Rule 2, we should partition the subspace based on only one of the dimensions. Suppose we based our decision on dimension d_1. Then, a_4, a_5, b_2 and b_3 will be grouped into one partition, and a_6, a_7, b_4 and b_5 will be grouped into another one. Hence, we will attach two nodes into the forth node (from left to right) at the 3rd level of the SPTree in Fig. 1. The result is shown in Fig. 2 (b). For the two attached nodes, each

Algorithm 1. `createSPTree(U, X)`

input : A target (U) and a set of objects ($X, U \notin X$)
output: SPTree, T

1 create first N level of T; // N is the number of dimensions
2 **foreach** $V \in X$ **do**
3   **foreach** $v \in V$ **do**
4         $S \leftarrow T.root$; // $T.root$ is the root node of T. S is a node
5   **for** $i \leftarrow 1$ **to** N **do**
6   **if** $v.d_i < S.key_1$ **then** $S \leftarrow S.pointer_1$;
7   **else if** $S.key_1 \leq v.d_i < S.key_2$ **then** $S \leftarrow S.pointer_2$;
8   **else** $S \leftarrow S.pointer_3$;
9   **end**
10   insert v into S group by (V, d_1);
11   incremental update V^s_{min} and V^s_{max} according to $v \in V$;
12   **end**
13 **end**
14 **return** T;

of them contains two entries. The meanings of these entries are the same as the entries of their immediate parent nodes. Note that the observations in the leave nodes are group by S, such that the pointers in their parents node can point to the correct entries.

Algorithm 1 outlines how SPTree is constructed. Line 1 creates the first N levels of the SPTree according to the minimum and the maximum values of U in each dimension. This process is trivial, so we will not elaborate it further. Line $2-13$ show how each observation, $v \in V$, is being inserted into the correct leave node in the SPTree, which is self-explained. Algorithm 2 outlines the major steps for identifying the p-dominant objects, $\mathcal{V}$. First, we compute $P(U)$ and $P(u)$ (line 1). All u are indexed by R-tree for efficient computation. All $P_u(V^s)$ and $P_l(V^s)$ are initialized to 1 and 0, respectively (line 2). After some parameters initialize (line 3), the main loop begins from line 4 to 23. In line $6-17$, for each $v \in V^s, V \in X$, we try to update the values of $P_u(V)$ and $P_l(V)$ with the help of $P_u(V^s)$ and $P_l(V^s)$ according to equations Eq. (2) to Eq. (5). Once we identify the relationship between V and U, V will be removed from X (line $11-16$). Finally, if the size of V^s is larger than 1 (i.e. the number of observations of an object that drops in S) and the relationship between V and U is not yet identified (i.e. $V \in X$), then we will partition S into two subspaces (line 20). Algorithm 3 outlines the partition process. It tries to get a value, δ (line 2), that guide us how the observations in S should be assigned to the two new nodes (line 3). δ is obtained by first sorting all the observations in S by a given dimension, d_i, (line 1) and then extract the observation, v, which is at the median position of S after sorting, and δ is simply the value of $v.d_i$ (line 2). Line $4-7$ try to assign the observations to the correct node, which are same explained. Finally, we link all of them together (line $8-10$).

Up to now, although we have only shown how we can extract the p-dominant objects, $\mathcal{V}$, from X, we can still use this similar framework to rank $\mathcal{V}$ or to identify the K-nearest p-skyline, which has already been discussed in the last paragraph of the last section. As

Algorithm 2. `overview(`U`,` X`)`

input : A target (U) and a set of objects ($X, U \notin X$)
output: p-dominant skyline points ($\mathcal{V}$)

compute $P(U)$ and $P(u), \forall u \in U$; // Base on Eq. (1)
$P_u(V^S) \leftarrow 1, \forall V^s$; $P_l(V^S) \leftarrow 0, \forall V^s$;
$\mathcal{V} \leftarrow \emptyset$; $T \leftarrow$ `createSPTree(`U, X`);` $i \leftarrow 1$;
repeat
    foreach $P \in T$ *(P is a parent node of a leave node)* **do**
        foreach $S \in T$ *(S is an entry)* **do**
            foreach $v \in V^S, V \in X$ **do**
                update $P_u(V)$ and $P_l(V)$ according to Eq. (2) and Eq. (3);
                if $P_u(V)$ *is not changed* **then** update $P_u(V)$ according to Eq. (4);
                if $P_l(V)$ *is not changed* **then** update $P_l(V)$ according to Eq. (4);
                update $P_l(V)$ and $P_u(V)$ according to Eq. (5);
                if V *is a p-dominant object* **then**
                    remove V from X;
                    add V into $\mathcal{V}$;
                else if V *can be pruned (see Section 3.4)* **then**
                    remove V from X;
                end
            end
        end
    end
    foreach $P \in T$ *(P is a parent node of a leave node)* **do**
        foreach $S \in L$ *(S is an entry)* **do**
            if $\exists V \in X$ *and* $|V^s| > 1$ **then** `partition(`S, d_i`);`
        end
    end
    $i \leftarrow (i = N)\ ?\ 1 : i+1$; // N is the number of dimensions
until $X = \emptyset$;
return $\mathcal{V}$;

the modification is rather simple, we do not attempt to further elaborate it due to the space limitation.

3.4 Pruning Strategies

Given a SPTree, T, after we compute $P_l(V^s)$ and $P_u(V^s)$, if $P_u(V^s) = 0$, then all the observations in V^s must be dominated by some other objects, we can then prune away V^s from T. If $P_l(V^s)$ equals to $P_u(V^s)$, then all $P(v)$ for $v \in V^s$ will be the same. So the offspring of V^s do not need to compute. For the problem of ranking p-dominant objects, if $P_u(V) < P(U)$, then V can be pruned away. However, the reverse is not true. If $P_u(V) > P(U)$, then even V can dominate U, we cannot prune it because we have to rank it. Finally, if $P_l(V) \geq P_u(V')$ for all $V' \in X, V' \neq V, V' \neq U$, then V would be the object with the highest skyline probability. Once we have identified this object, then

Algorithm 3. `partition`(S, d_i)

input : An entry, S, and a dimension, d_i

sort all $v \in V^s$ at the leaf-node by d_i;
$\delta \leftarrow v.d_i$ where v is the observation at the median position;
create a new node, L, with two entries, S_1 and S_2;
foreach $v \in V^s$ **do**
  if $v.d_i \leq \delta$ **then** incremental update $V_{min}^{S_1}$ and $V_{max}^{S_1}$ by v;
  else incremental update $V_{min}^{S_2}$ and $V_{max}^{S_2}$ by v;
end
$S.pointer \leftarrow L$;
$S_1.pointer \leftarrow$ first position of v in V^s at the leaf-node;
$S_2.pointer \leftarrow$ first v with $v.d_i = \delta$ in V^s at the leaf-node;

we can recursively use this rule to identify the object with the second highest skyline probability, and so on so forth.

3.5 Complexity

Let us call the nodes in the first N level of a SPTree as space nodes and all the other nodes as data nodes. For an N dimensional space, there will be at most 3^N space nodes. Each node contains one pointer and one value. For the space complexity, the best case is that all observations drop within one subspace. The worst case is that each subspace has at least one observation. Hence, the space complexities for space nodes are $O(1)$ and $O(3^N)$ for the best case and the worse case, respectively. Since V^s will be partitioned if we cannot identify the relationship between V and U, the best case is that no object is split and the worst case is that each object will be split until each data node has only one observation. Let n be the total number of observations in the dataset. The space complexities for data nodes are $O(n)$ and $O(4n-1)$ for the best case and the worse case, respectively. Combining all together, the space complexity for the average case would be $(3^N + n)$.

4 Experimental Study

All experiments are conducted with Intel 2.20GHz dual core CPU and 1GB memory using Java. We employed four datasets: (1) Uniformly Distributed (*UD*) – We generate the center of 200 objects randomly, and randomly generate 2,000 observations for each object. None of the objects are overlapped. (2) Slightly Overlapped (*SO*) – We generate this dataset in the way similar to that of *UD* except that the objects can be overlapped. On average, an object will be overlapped with 9 objects. (3) Highly Overlapped (*HO*) – We generate this dataset in the way similar to that of *SO* except that an object will be overlapped with 34 objects on average. (4) NBA – It is downloaded from `http://www.databasebasketball.com`. It contains 339,721 records of 1,313 players (from 1991 to 2005). Each player is treated as an object, and the records of the same player represent his observations. Each record has three attributes: points, assistants and rebounds.

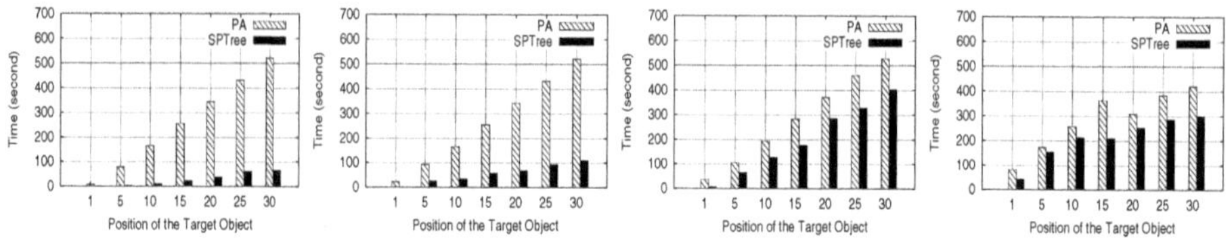

(a) UD (b) SO (c) HO (d) NBA

Fig. 4. Retrieval and ranking performances of the dominant probabilistic skylines

4.1 Dominant Skylines Identification

For each dataset, we conduct seven experiments. The first experiment takes the object with the highest skyline probability as U, and then extract the objects, V, with $P(V) > P(U)$. Since U has the highest skyline probability, no object will be extracted. The rest of the experiments are conducted in the way similar to the first experiment, except that we respectively take the object with the 5th, 10th, 15th, 20th, 25th and 30th highest skyline probability as U in each different experiment.

For comparison, we implement this approach as described in [1][1]. We call it as baseline approach (BA). This approach identifies all V with $P(V)$ larger than a *user-defined value* without computing the exact $P(V)$. Note that this approach cannot rank V. We slightly modify this algorithm for comparison as follows: (1) Given U, we first compute $P(U)$ to denote the user-defined value; (2) Use BA to extract all V with $P(V) > P(U)$; (3) We compute the exact $P(V), \forall V$ in order to rank them.

Figure 4 (a), 4 (b), 4 (c) and 4 (d) show the performances of extracting and ranking the dominant probabilistic skylines. In the figures, x-axes denote the ranking of U and y-axes denote the computational time (in sec). The black charts denote the performances of our SPTree and the shaded charts denote the baseline approach (BA).

At the first glance, SPTree excels BA for all datasets and all situations, especially for the datasets *UD* and *SO*. SPTree takes $0-65$ sec for *UD*, $1-110$ sec for *SO* and $6-400$ sec for *HO*. For a reference, computing the skyline probabilities of all objects requires more than 3,600 sec. The computational time of SPTree on the synthetic datasets is roughly: $(UD > SO > HO)$.

For the NBA dataset, SPTree seems does not have a clear pattern, i.e. even the given U is at a lower ranking, the computational time may not necessary be longer. For example, the time for extracting and ranking the dominant probabilistic skylines when U is at the 15th rank is faster than that at 10th. This is interesting because, intuitively, a higher rank object should take less computational time, as there are fewer objects dominating it, so that we only need to extract and rank fewer objects. But this is not the case. As such, we analysis the time spent on the most time consuming part of the algorithm – split and partition. As expected, more number of split and partition is needed if the computational time is higher. In order to understand why we have to perform more number of split and partition, we take a closer look at the distribution of the objects. We found out the number of split and partition we need depends on the overlapping between U

[1] In their paper, they have proposed two approaches, namely: (1) Top-Down; and (2) Bottom-Up. We have implemented the former one, because its excels the later one in most situations.

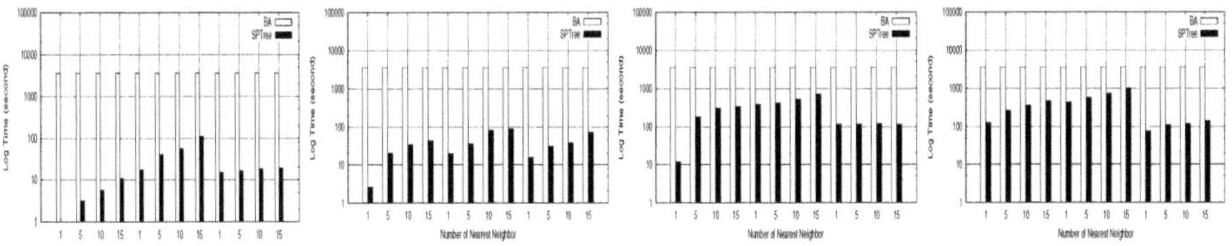

(a) UD (b) SO (c) HO (d) NBA

Fig. 5. Performances of the K nearest neighbor p-skylines

and the other objects. As such, we draw the following conclusion: (1) It is merely the distribution of the observations that affect the algorithm performance; and (2) There is no direct relationship between the skyline probabilities and the computational time.

4.2 K Nearest Skylines Identification

For each dataset, we first identify the objects with: (1) Highest skyline probability; (2) Lowest skyline probability; and (3) Median skyline probability, so as to serve as our target object, U. Then, we try to extract K objects that are nearest to each of these three objects. Here, $K = \{1,5,10,15\}$. As a result, for each dataset, we have to conduct $3 \times 4 = 12$ experiments. Since no existing work tries to address this problem, we have implemented the following approach: Given a set of objects, we compute all of their skyline probabilities, and then identify the K nearest neighbor of U accoriding to the skyline probabilities. We call it as Baseline Approach (BA).

Figure 5 (a), 5 (b), 5 (c) and 5 (d) show the retrieval performances of obtaining the K nearest probabilistic skylines. In the figures, x-axes denote different K, and y-axes is the computational time (*in log scale*) measured in seconds. The black charts denote our proposed work (SPTree) and the white charts denote the Baseline Approach (BA). The first four pairs, middle fours pairs, and last fours pairs of charts represent the time required for identifying the K nearest probabilistic skylines of U, where U is the object at the top, middle and bottom ranking according to the skyline probability.

From the figures, it is obvious that SPTree outperforms BA significantly (note that the y-axes are all in log scale). This is not surprise as we have several efficiency strategies to improve our identification process. The computational efficiency also follows the same conclusion as in the previous section: (UD $>$ SO $>$ HO). In general, the computational time increases linearly according to the number of nearest probabilistic skylines, K, which we have to extract. In addition, for SPTree, the middle four charts usually perform inferior than the first four charts and the last four charts. This finding suggests that the position of the target object, U, may affect the computational time. The reason is that overlapping among objects with skyline probabilities dropped in the middle range is usually highest. Moreover, the skyline probabilities of these objects are usually very similar. This will increase the computational cost significantly. For the NBA dataset, the best case is the last four charts, which is the objects with the lowest skyline probabilities. This is not surprise if we understand the data distribution in there. In NBA dataset, the lowest overlapping is the area where those players have

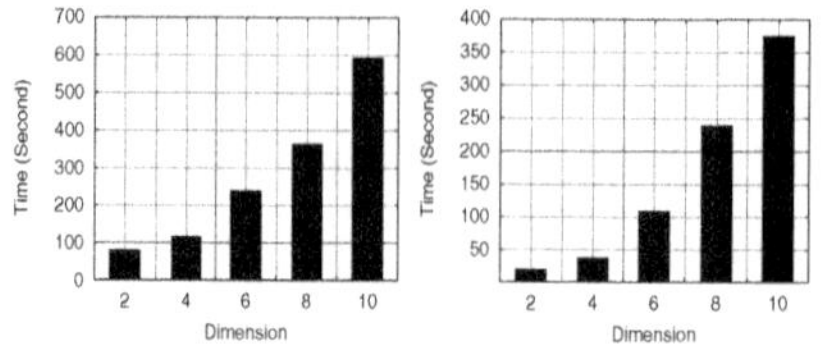

(a) (b)

Fig. 6. Effect of the dimensionality. (a) Ranking dominant p-skyline. (b) Identifying K nearest p-skyline.

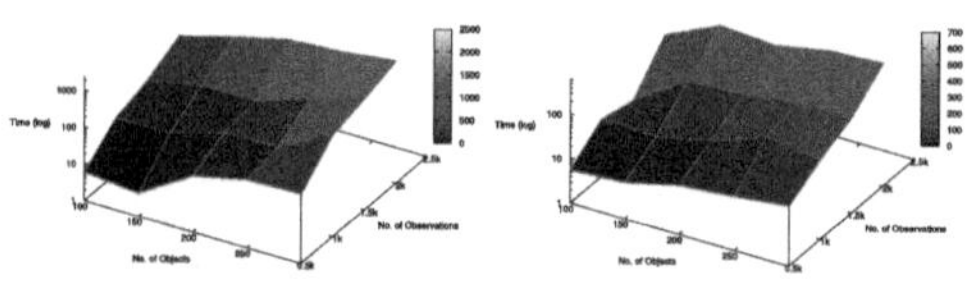

(a) (b)

Fig. 7. Relationship between no. of instances and no. of objects. (a) Ranking dominant p-skyline (b) Identifying K nearest p-skyline.

the lowest skyline probabilities, whereas both the top and middle range are heavily overlapped.

4.3 Further Discussion and Analysis

In this section, we analyze the following two issues: (1) The effect of dimensionality, and (2) The relationship between the number of objects and the number of instances. We conduct the following analysis: (1) Case 1 (Dimensionality) – We conduct two set of experiments, one for ranking the dominant probabilistic skylines, and the other one for identifying $K = 10$ nearest probabilistic skylines. The dataset that we use is similar to that of *SO* except that the dimensionality varies from $\{2,4,6,8,10\}$. For both experiments, U is the one with the 10th highest skyline probability. (2) Case 2 (Number of Objects and Instances) – We conduct two sets of experiments with the parameters and settings similar to Case 1, except that $N = \{100,150,200,250,300\}$ and $M = \{500,1000,1500,2,000,2,500\}$.

Figure 6 (a) and 6 (b) show the evaluation results of Case 1. The x-axes represent the dimensions of the dataspace, and the y-axes represent the time in seconds. When the dimensionality increases linearly, the computational time increases exponentially. This is expected because when the number of dimension increases linearly, the number of skyline point will be increased exponentially. As the number of skyline points increase dramatically, the number of comparison among points must also increase dramatically. Furthermore, if the dimensionality increases linearly, the number of subspaces will increase exponentially. This further contributions to the increase of computational cost.

Figure 7 (a) and 7 (b) show the evaluation results about the relationship between objects and observations (Case 2). In the figures, x-axes denote the total number of instances, y-axes represent the total number of observations, and z-axes are the computational time measured in second. From the figures, we have two important findings. The first finding is: the computational cost increases slowly if the number of objects increases while keeping the number of observations belong to an object is the same. Yet, the computational cost increases quickly if the number of observations increases while keeping the number of objects are the same. The reason is that the time required for splitting the nodes in the SPTree will be fewer if the number of observations is fewer. The second finding in this experiment is: the computational time required for identifying the K nearest p-skyline is much more stable and much less than that of ranking the dominant p-skyline. This is because if we rank the dominant p-skylines, we need to distinguish the partial relationship between every two skyline points, which is not the case for the K nearest p-skyline identification.

5 Related Work

Skyline query has been proposed for several years. Borzsonyi et.al [3] are the first to introduce skyline query into database community and extend the SQL statement to support the skyline query. In their work, four algorithms, BNL (Block-nested-loop) based nested-loop, DC (divide and conquer), B-tree based and R-tree based index, are proposed to compute the skyline. DC algorithm recursively partitions the data set into several parts and identifies skyline in each part. Then the separated parts will be recursively merged into one part. Sort-filter-skyline (SFS) algorithm, based on nested-loop as well, is proposed by Chomicki et al. [4]. SFS presorts the dataset in non-decreasing order with the key value of each point such that each point only needs to be compared with the points ranking before itself. Tan et al. [5] propose progressive skyline computation methods based on Bitmap and B-tree index. Kossmann et al. [6] propose a new online algorithm based on the nearest neighbor (NN) search to improve their methods. However, the NN algorithm needs to traverse the R-tree for several times and eliminate duplicates if dimension is larger than 2. Papadias et al. [7] propose a branch-and-bound algorithm (BBS) based on nearest neighbor search as well and indexed by R-tree. This algorithm is IO optimal and traverses the R-tree for only once.

Recently, researchers mainly focus on the following three aspects. Firstly, decrease the number of skyline points [8,9] when the dimensionality is high or the dataset is anti-correlated. Secondly, some variations of skyline query are proposed, including reverse skyline [10], multi-source skyline query [11], subspace skyline [12,13,14,15]. Thirdly, combine skyline query with other techniques, including continuously computing skyline in data streams [16,17,18], computing skyline in a distributed environment [19,20] and computing probabilistic skyline in uncertain data [1].

Most previous work focuses on computing skyline on certain data except the probabilistic skyline on uncertain data. Here, our work focuses on uncertain data as well, but we have different motivations with probabilistic skyline as explained before. We extend the probabilistic skyline to dominant and K nearest probabilistic skyline.

6 Conclusion and Future Work

In this paper, we proposed a novel algorithm, called SPTree, to answer the following questions: (1) Which of the objects have the skyline probabilities that dominate (larger than) the given object? (2) What is the skyline probability ranking of the objects that dominate the given object? (3) Which of the objects are the K nearest neighbors to the given object according to the skyline probabilities? Extensive experiments are conducted to evaluate it. The encouraging results indicated that SPTree is highly practical. However, SPTree is far from perfect. It may not be very efficiency if the dimensionality is high meanwhile the overlapping among object is huge. So, our future work will investigate this issue.

References

1. Pei, J., Jiang, B., Lin, X., Yuan, Y.: Probabilistic skylines on uncertain data. In: Proceedings of the 33rd International Conference on Very Large Data Bases (VLDB 2007) (2007)
2. Papadias, D., Tao, Y., Fu, G., Seeger, B.: Progressive skyline computation in database systems. ACM Trans. Database Syst (TODS) 30(1), 41–82 (2005)
3. Borzsonyi, S., Kossmann, D., Stocker, K.: The skyline operator. In: Proceedings of the 17th International Conference on Data Engineering (ICDE 2001) (2001)
4. Chomicki, J., Godfrey, P., Gryz, J., Liang, D.: Skyline with presorting. In: Proceedings of the 19th International Conference on Data Engineering (ICDE 2003) (2003)
5. Tan, K.L., Eng, P.K., Ooi, B.C.: Efficient progressive skyline computation. In: VLDB 2001, Proceedings of 27th International Conference on Very Large Data Bases (VLDB 2001) (2001)
6. Kossmann, D., Ramsak, F., Rost, S.: Shooting stars in the sky: An online algorithm for skyline queries. In: Proceedings of the 28th International Conference on Very Large Data Bases (VLDB 2002) (2002)
7. Papadias, D., Tao, Y., Fu, G., Seeger, B.: An optimal and progressive algorithm for skyline queries. In: Proceedings of the 2003 ACM SIGMOD International Conference on Management of Data (SIGMOD 2003) (2003)
8. Chan, C.Y., Jagadish, H.V., Tan, K.L., Tung, A.K.H., Zhang, Z.: Finding k-dominant skylines in high dimensional space. In: Proceedings of the ACM SIGMOD International Conference on Management of Data (SIGMOD 2006) (2006)
9. Lin, X., Yuan, Y., Zhang, Q., Zhang, Y.: Selecting stars: The k most representative skyline operator. In: Proceedings of the 23rd International Conference on Data Engineering (ICDE 2007) (2007)
10. Dellis, E., Seeger, B.: Efficient computation of reverse skyline queries. In: Proceedings of the 33rd International Conference on Very Large Data Bases (VLDB 2007) (2007)
11. Deng, K., Zhou, X., Shen, H.T.: Multi-source skyline query processing in road networks. In: Proceedings of the 23rd International Conference on Data Engineering (ICDE 2007) (2007)
12. Yuan, Y., Lin, X., Liu, Q., Wang, W., Yu, J.X., Zhang, Q.: Efficient computation of the skyline cube. In: Proceedings of the 31st International Conference on Very Large Data Bases (VLDB 2005) (2005)
13. Pei, J., Jin, W., Ester, M., Tao, Y.: Catching the best views of skyline: A semantic approach based on decisive subspaces. In: Proceedings of the 31st International Conference on Very Large Data Bases (VLDB 2005) (2005)
14. Tao, Y., Xiaokui Xiao, J.P.: Subsky: Efficient computation of skylines in subspaces. In: Proceedings of the 22nd International Conference on Data Engineering (ICDE 2006) (2006)
15. Tian Xia, D.Z.: Refreshing the sky: the compressed skycube with efficient support for frequent updates. In: Proceedings of the ACM SIGMOD International Conference on Management of Data (SIGMOD 2006) (2006)
16. Mouratidis, K., Bakiras, S., Papadias, D.: Continuous monitoring of top-k queries over sliding windows. In: Proceedings of the ACM SIGMOD International Conference on Management of Data (SIGMOD 2006) (2006)
17. Lin, X., Yuan, Y., Wang, W., Lu, H.: Stabbing the sky: Efficient skyline computation over sliding windows. In: Proceedings of the 21st International Conference on Data Engineering (ICDE 2005) (2005)
18. Tao, Y., Papadias, D.: Maintaining sliding window skylines on data streams. IEEE Trans. Knowl. Data Eng. (TKDE) 18(2), 377–391 (2006)
19. Cui, B., Lu, H., Xu, Q., Chen, L., Dai, Y., Zhou, Y.: Stabbing the sky: Efficient skyline computation over sliding windows. In: Proceedings of the 24th International Conference on Data Engineering (ICDE 2008) (2008)

20. Huang, Z., Jensen, C.S., Lu, H., Ooi, B.C.: Skyline queries against mobile lightweight devices in manets. In: Proceedings of the 22nd International Conference on Data Engineering (ICDE 2006) (2006)

Appendix

Theorem 1 ([1]). *Let* $V = \{v_1, ..., v_n\}$ *be the object where* $v_1, ..., v_n$ *are the observations of* V*. Then* $P(V_{min}) \geq P(V) \geq P(V_{max})$*.*

According to Theorem 1, we can use $P(V_{min})$ and $P(V_{max})$ to compute $P_u(V)$ and $P_l(V)$ respectively.

However, when computing the values of $P(V_{min})$ and $P(V_{max})$, we still need to traverse all the observations of the other objects. Practically, based on the observations of U and any other object $V \in X$, we deduce the following two important theorems, by which we can compute $P_u(V)$ and $P_l(V)$ only using the two objects V and U.

Theorem 2. *Given* $v \in V$*, if* $\exists u \in U$ *s.t.* $u \prec v$*, then*

$$P(v) \leq P(u) \times \left(1 - \frac{|\{u \in U | u \prec v\}|}{|U|}\right)\left(\frac{|V|}{|V| - |\{v \in V | v \prec u\}|}\right)$$

Proof. *$\forall O \in X (O \neq V,\ O \neq U)$, $\forall o \in O$ satisfying $o \prec u$, then by the definition of skyline, $o \prec v$. Thus, $|\{o \in O | o \prec v\}| \geq |\{o \in O | o \prec u\}|$. Let $R = X - \{U, V\}$, then:*

$$\begin{aligned} P(v) &= \prod_{O \in R} (1 - \frac{|\{o \in O | o \prec v\}|}{|O|}) \times (1 - \frac{|\{u \in U, u \prec v\}|}{|U|}) \\ &\leq \prod_{O \in R} (1 - \frac{|\{o \in O | o \prec u\}|}{|O|}) \times (1 - \frac{|\{u \in U, u \prec v\}|}{|U|}) \\ &= P(u) \times (\frac{|V|}{|V| - |v \in V | v \prec u|}) \times (1 - \frac{|\{u \in U | u \prec v\}|}{|U|}). \end{aligned}$$

The result follows. □

When identifying the value of $P(V_{min})$, if there exists an observation $u \in U$ and $u \prec V_{min}$, according to Theorem 2, only U and V will be searched.

Theorem 3. *Given* $v \in V$*, if* $\exists u \in U$ *s.t.* $v \prec u$*, then*

$$P(v) \geq P(u) \times \left(1 - \frac{|u \in U | u \prec v|}{|U|}\right)\left(\frac{|V|}{|V| - |v \in V | v \prec u|}\right)$$

Proof. *Trivial. Similar to the proof of Theorem 2* □

Similarly, when identifying the value of $P(V_{max})$, according to the Theorem 3, if there exist an observation $u \in U$ and $V_{max} \prec u$, only U and V will be considered.

Predictive Skyline Queries for Moving Objects

Nan Chen[1], Lidan Shou[1], Gang Chen[1], Yunjun Gao[2], and Jinxiang Dong[1]

[1] College of Computer Science, Zhejiang University, China
{cnasd715,should,cg,djx}@cs.zju.edu.cn
[2] School of Information Systems, Singapore Management University, Singapore
yjgao@smu.edu.sg

Abstract. The existing works on skyline queries mainly assume static datasets. However, querying skylines on a database of moving objects is of equal importance, if not more. We propose a framework for processing predictive skyline queries for moving objects which also contain other dynamic and static attributes. We present two schemes, namely RBBS and TPBBS, to answer predictive skyline queries. Experimental results show that TPBBS considerably outperforms RBBS.

Keywords: spatio-temporal database, moving object, skyline.

1 Introduction

Skyline query [1] is an important operation for applications involving multi-criteria decision making. Given a set of multi-dimensional data points, a skyline query retrieves a set of data points that are not dominated by any other points in all dimensions. A point dominates another if it is as good or better in all dimensions and better in at least one dimension. The existing works on skyline queries so far mainly assume static datasets. There are lots of algorithms for static skyline queries in the literature. For example, the BBS (branch-and-bound skyline) algorithm [4] is a widely cited one.

Meanwhile, management of moving objects has emerged as an active topic of spatial access methods. In moving-object-database (MOD) applications, indexing and querying for the current and near-future locations of moving objects is one of the key challenges. Various indexes have been proposed to index the locations of moving object, which continuously change by time in a 2D or 3D space. The Time-Parameterized R-tree (TPR-tree) [5] and the TPR*-tree [6] are typical examples of such indexes. There are however few reports on skyline algorithms for moving objects.

A moving object may contain spatial attributes such as the location, the velocity, and the distance to a specified point etc. Besides these spatial attributes, a moving object may also be associated with *non-spatial attributes* which can be either *time-parameterized* or *static*. We refer to the non-spatial time-parameterized attributes as NSTP. Regarding data combining the above three types of attributes, namely *spatial*, *NSTP*, and *static*, there are needs in querying the skyline of such data at some future time to support decision making. For example, in a digital battle, an aid worker may query a skyline in 5 minutes from now on for the injured soldiers, in order to choose the ones to prepare for the rescue. Both the aid worker and the injured soldiers may be on move. The attributes to be considered in the query may involve the distance from the

X. Zhou et al. (Eds.): DASFAA 2009, LNCS 5463, pp. 278–282, 2009.

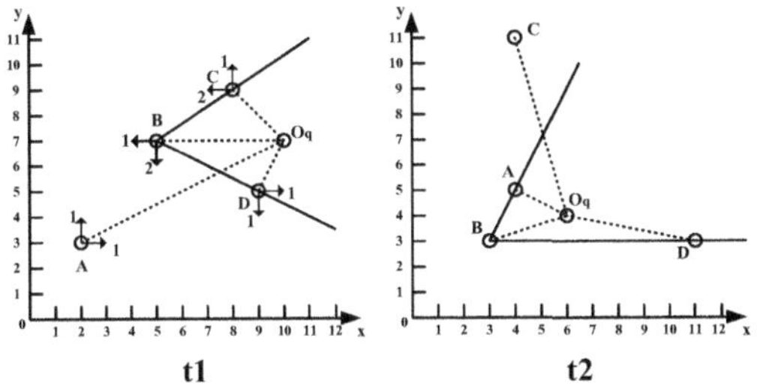

Fig. 1. Skyline for Moving Objects

injured soldiers to the aid worker (a spatial attribute), the severity of the injury which may change by time (a NSTP attribute), and the military rank (a static attribute).

Figure 1 shows the 2D layout of two predictive skyline queries on a dataset of four moving objects (A, B, C, and D) at two different time instances, namely t_1 and $t_2 = t_1+2$. Point O_q indicates the position of the query point, which is also a moving object. The solid lines in both subfigures indicate the skylines, while the dashed lines indicate the distances from the moving objects to the query points. We can see that the results of predictive skyline queries vary at different time instances and for different query points. The NSTP and static attributes are not visible in this figure.

In this work, we address the problem of answering predictive skyline queries for moving objects and their associated attributes. Our proposed solution is implemented in a prototype called PRISMO (PRedIctive Skyline queries for Moving Objects). In PRISMO, we design two schemes, namely RBBS and TPBBS, to answer predictive skyline queries. RBBS is a brute-force method, while TPBBS is based on an index structure called TPRNS-tree which indexes moving objects on all three types of attributes.

2 Problem Statement

Like some previous works [5] [6], PRISMO models moving objects as linear motion functions of time. Given a specific time t, which is no earlier than the current time, the spatial location of a moving object in a d-dimensional space is given by $O(t) = (\boldsymbol{x}, \boldsymbol{v}, t_{ref}) = \boldsymbol{x}(t) = (x_1(t), x_2(t), ..., x_d(t))$, where $\boldsymbol{x}$ is the position vector of the object at a reference time t_{ref}, $\boldsymbol{v}$ is a velocity vector $\boldsymbol{v} = (v_1, v_2, ..., v_d)$. Thus, we have $\boldsymbol{x}(t) = \boldsymbol{x}(t_{ref})+\boldsymbol{v}(t-t_{ref})$. Given a query point q which may also be a moving object, its location at t is $O_q(t) = (x_{q1}(t), x_{q2}(t), ..., x_{qd}(t))$. So the distance between the moving object and the query point is given by

$$distance(O(t), O_q(t)) = \sqrt{(x_1(t) - x_{q1}(t))^2 + (x_2(t) - x_{q2}(t))^2 + ... + (x_d(t) - x_{qd}(t))^2}.$$

Meanwhile, the NSTP attributes of a moving object can be represented as $y_1(t), y_2(t), \ldots, y_i(t)$, where i is the number of NSTP attributes. Each NSTP attribute y_k ($k = 1, \ldots, i$) is modeled similarly as a linear function of time $y_k(t) = y_k(t_{ref}) + c_k * (t - t_{ref})$, where c_k is a coefficient. A moving object may also have some static attributes $z_1, z_2, \ldots, z_j$, which do not vary by time. Therefore, given a moving object, its respective data point in vector form is $< x_1, x_2, ..., x_d, v_1, v_2, ..., v_d, y_1, y_2, ..., y_i, c_1, c_2, ..., c_i, z_1, z_2, ..., z_j >$.

Given a moving query object o_q and time instance t_q which does not precede the current time, the predictive skyline query, denoted by $Q(o_q, t_q)$, retrieves the set of skyline in the target space T at time t_q. The target space T contains the distance of the moving objects to the query point, the i-dimensional NSTP attributes, and the j-dimensional static attributes of the moving objects. Without loss of generality, we use the *min* condition to compute the skyline, in which smaller values are preferred.

3 Algorithms for Computing Predictive Skylines

3.1 RBBS

A natural approach to answering predictive skyline is to scan and compute the dataset, to rebuild an R-tree and then to process a standard BBS algorithm for each query. We call this brute-force solution as RBBS (BBS with Rescanning and Repacking).

When a point query $Q(o_q, t_q)$ is issued, we first obtain the position o'_q of the query point o_q at query time t_q, from the moving function of o_q. Second, we scan the whole dataset of moving objects. For each moving object, we compute the distance between its position at t_q and o'_q, and compute the values of the NSTP attributes at t_q. After scanning, we rebuild an R-tree whose dimensions include the distance, the NSTP attributes and the static attributes. Note that the R-tree is static and does not require insertion or deletion. In order to optimize the cost of rebuilding the R-tree, we use the packing method of STR (Sort-Tile-Recursive) [3]. After repacking an R-tree, RBBS processes a standard BBS algorithm to obtain the skyline set.

RBBS is straightforward and simple. However it needs to scan the dataset and repack the R-tree for each query. Although the I/O costs of repacking and BBS are both optimal, RBBS is a naive algorithm.

3.2 TPBBS

We advocate an index called the TPRNS-tree (time-parameterized R-tree with Non-Spatial dimensions) which integrates all three types of attributes of moving objects. Based on the TPRNS-tree, we propose a predictive skyline query scheme called TPBBS (time-parameterized BBS) which does not require dataset-scanning and index-repacking between consecutive queries.

The TPRNS-tree has similar index structure and insertion/deletion algorithms as the TPR-tree. However, a major difference between the two is that the TPR-tree indexes locations only, while the TPRNS-tree captures NSTP and static dimensions as well. The NSTP dimensions are treated similarly as the location dimensions in MBRs and VBRs, and the static dimensions are only recorded in MBRs. Therefore, the number of dimensions in MBRs and VBRs are different in TPRNS-tree: a MBR contains $D_m = d + i + j$ dimensions, while a VBR contains $D_v = d + i$. For a MBR, the bounds of the first $D_v = d + i$ dimensions move according to the corresponding VBR and are not necessarily compact, but those of the rest j static dimensions are always compact. The compactness of bounds has considerable impact on the update and query performance.

```
Algorithm TPBBS Q(o_q, t_q)
Input: o_q is the query point, t_q is the query time, R is the tree root.
Output: S is the list of the the predictive skyline.
S ← Ø, H ← Ø;
Compute the position o'_q of o_q at t_q;
For each child entry e_i of R
    Compute the position of point at t_q (or expand MBR to t_q);
    Form its te_point (or te_MBR) and compute its mv_point (or mv_MBR);
    Insert te_point (or te_MBR) into H according to its mv_point (or mv_MBR);
While H is not empty do;
    Remove the top entry E from H;
    If is a moving object
        Get te_point of E;
        If te_point is not dominated by any point in S
            Insert te_point into S;
        Else discard E;
    Else E is an index MBR
        Get te_MBR of E;
        If ce_MBR is dominated by any point in S at t_q
            discard E;
        Else: For each child entry e_i of E
            Compute the position of point at t_q (or expand MBR to t_q);
            Form its te_point (or te_MBR) and compute its mv_point (or mv_MBR);
            If te_point (or te_MBR)is not dominated by any point in S
                Insert te_point (or te_MBR) into H according to its mv_point (or mv_MBR);
            Else discard e_i;
End while;
Return S;
```

Fig. 2. Query Algorithm

Figure 2 shows the query algorithm of TPBBS. In TPBBS, we maintain a skyline set S, which becomes populated as the algorithm proceeds, and a sorted heap H containing entries of unexamined moving objects (leaf entries) and index MBRs (internal entries). Note that the entries in H and S are different from those stored in the TPRNS-tree. They need to be transformed and are therefore called *transformed entries*, or *te* in short. A transformed *leaf entry* is denoted by

$$te_{point}(o_q, t_q) =< ppd(o_q, t_q), NSTP(t_q), static >$$

which contains (i) the *point-to-point distance* (*ppd* in short) from its position to the position o'_q of the query point o_q at time t_q, (ii) the values of the NSTP attributes at time t_q and (iii) the values of the static attributes. A transformed *internal entry* is denoted by

$$te_{MBR}(o_q, t_q) =< prd(o_q, t_q), NSTP_{bottom-left}(t_q), static_{bottom-left}, pointer >$$

which contains (i) the *minimum point-to-region distance* (*prd* in short) between its expanded MBR in location space and o'_q at time t_q , (ii) the values of NSTP attributes of its expanded bottom-left corner at time t_q, (iii) the static attributes of its bottom-left corner, and (iv) a pointer to its corresponding internal node in the TPRNS-tree for getting the detail information of its children. It is also worthwhile to mention that the entries in H are sorted by a key called *minimum value* (*mv* in short). For a moving object (or an expanded MBR), its *mv* is computed as the sum of the coordinates of the first three members in its te_{point} (or te_{MBR}). The domination comparison between two points (or between a point and an expanded MBR), is to compare the coordinates of the first three members in their te_{point} (or te_{MBR}) one by one.

4 Experiments

We implement RBBS and TPBBS in C++, and conduct the experiments on a 2.6GHz Pentium 4 PC with 1GB memory, running Windows XP Professional. The page size and index node size are both set to 4 KB. Figure 3 shows the average page accesses and CPU time for each query while varying the cardinality from $10K$ to $1M$. The location attributes of the moving objects are generated similarly to that in [2] in the space domain of $1000 * 1000$, randomly choosing the moving velocity in each dimension from -3 to 3. For each NSTP dimension, the initial values range from 1 to 1000 randomly. The coefficients of their dynamic linear functions are chosen randomly from -5 to 5. We employ independent and anti-correlated distribution in the static dimensions, varying the values from 1 to 10000. There are 2 location dimensions, 1 NSTP dimension and 2 static dimensions in the datasets. The query point is generated as a moving object, and the query time ranges from 1 to 60 time units. As expected, TPBBS performs much better than RBBS. We also study the effect of the dimensionality, the data distribution and the moving velocity, which all lead to the same conclusion.

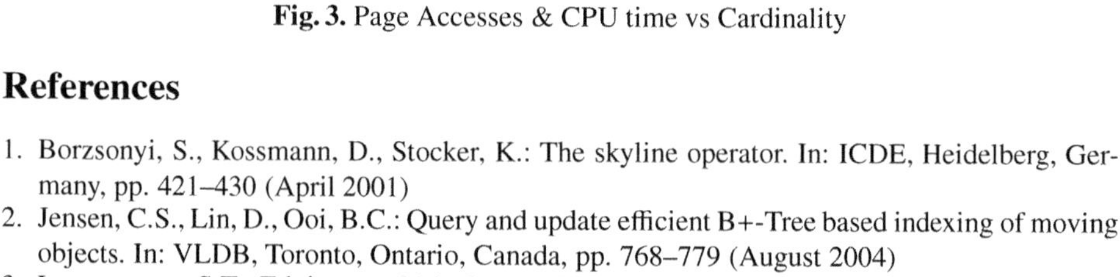

Fig. 3. Page Accesses & CPU time vs Cardinality

References

1. Borzsonyi, S., Kossmann, D., Stocker, K.: The skyline operator. In: ICDE, Heidelberg, Germany, pp. 421–430 (April 2001)
2. Jensen, C.S., Lin, D., Ooi, B.C.: Query and update efficient B+-Tree based indexing of moving objects. In: VLDB, Toronto, Ontario, Canada, pp. 768–779 (August 2004)
3. Leutenegger, S.T., Edgington, J.M., Lopez, M.A.: Str: A simple and efficient algorithm for r-tree packing. In: ICDE, Birmingham U.K, pp. 497–506 (April 1997)
4. Papadias, D., Tao, Y., Fu, G., Seeger, B.: Progressive skyline computation in database systems. ACM Trans. Database Syst(TODS) 30(11), 41–82 (2005)
5. Saltenis, S., Jensen, C.S., Leutenegger, S.T., Lopez, M.A.: Indexing the positions of continuously moving objects. In: SIGMOD Conference, Dallas, Texas, USA, pp. 331–342 (May 2000)
6. Tao, Y., Papadias, D., Sun, J.: The TPR*-Tree: An optimized spatio-temporal access method for predictive queries. In: VLDB, Berlin, Germany, pp. 790–801 (September 2003)

Efficient Algorithms for Skyline Top-K Keyword Queries on XML Streams*

Lingli Li , Hongzhi Wang, Jianzhong Li, and Hong Gao

Harbin Institute of Technology
lwsbrr@gmail.com, {wangzh,lijzh,honggao}@hit.edu.cn

Abstract. Keywords are suitable for query XML streams without schema information. In current forms of keywords search on XML streams, rank functions do not always represent users' intentions. This paper addresses this problem in another aspect. In this paper, the skyline top-K keyword queries, a novel kind of keyword queries on XML streams, are presented. For such queries, skyline is used to choose results on XML streams without considering the complicated factors influencing the relevance to queries. With skyline query processing techniques, algorithms are presented to process skyline top-K keyword queries on XML streams efficiently. Extensive experiments are performed to verify the effectiveness and efficiency of the algorithms presented.

Keywords: XML streams, keyword search, top-K, skyline.

1 Introduction

In some application, XML data is accessible only in streaming form, which is termed **XML Stream.** Querying XML streams with keywords without schema information is in great demand. The keyword search on XML streams is that given a set of keywords Q, the set of all the XML fragments in the XML streams with each of them containing all the keywords is retrieved. We call a node in XML fragment a *keyword match* if its value contains any keyword in Q.

Since in XML streams, infinite XML fragments will come and there are too many XML fragments satisfying the keyword search, to find the most relevant results and to limit the number of results are both required during the processing of keyword queries. How to identify the most relevant results for users is still an open problem. There are many techniques proposed to identify and rank the relevance of query results on XML data[1-4]. But sometimes, results are incomparable and it is hard to find a rank function of relevance. In some cases even though users submit the same query, different results may be required. In such cases, to find a definite rank function to rank the relevance of results is improper. Without rank functions, a problem comes how to identify the top-K results the most relevant to the requirements of users.

* Support by the Key Program of the National Natural Science Foundation of China under Grant No.60533110; the National Grand Fundamental Research 973 Program of China under Grant No.2006CB303000; the National Natural Science Foundation of China under Grant No.60773068 and No.60773063.

X. Zhou et al. (Eds.): DASFAA 2009, LNCS 5463, pp. 283–287, 2009.

Inspired by [5] , skyline query is a good tool to effectively solve this problem. For the applying skyline on keyword search, the distances between keyword matches are used as properties in the skyline. We propose a new kind of queries, loose skyline top-K keyword queries on XML streams. An efficient algorithm is presented to process such queries on XML streams.

The rest of this paper is organized as follows. In Section 2, we present the definition and processing algorithm of loose skyline top-K keyword queries on XML streams. In Section 3 we evaluate the performance of these algorithms by experiments. Section 4 concludes the whole paper.

2 Loose Skyline top-K Query on XML Streams

In this section, skyline is applied to improve the result selection for keyword queries on XML streams.

In an XML tree, the distance between nodes *n* and *e* is the number of edges on the path of *n-e*, denoted as $d(n, e)$. If *d* is unreachable to *e*, then $d(n, e) = \infty$.

Definition 2.1(**Keyword Match**). For a keyword set $W = \{K_1, \ldots, K_N\}$ and a leaf node *u* in a XML segment, if the value of *u* contains some keyword $K_i \in W$, node *u* is said to match keyword K_i and *u* is a keyword match of K_i, denoted by $K_i \subseteq u$.

Definition 2.2 (**Smallest Lowest Common Ancestor (SLCA)**). Given a keyword query *Q* on XML data, an XML node *i* in the XML data is called SLCA if it contains all the keywords in *Q* in its subtree, and none of its descendants does.

Definition 2.3 (**Query Result**). Given a keyword query *Q*, an XML segment *t* on the XML data is called a *query result* if it satisfies (i) the root of *t* is a SLCA of *Q* and (ii) each leaf of *t* contains at least one keyword not contained by other leaves of *t*.

For any two leaves *a* and *b* in a query result *T* and their minimum common ancestor *s*, a smaller sum $d(a, s)+d(s, b)$ implies larger relevance. By the definition of distance, the smaller the distance between two leaves is, the larger the relevance between them is. For every keyword K_i in a keyword query *W*, the set $A_i=\{a|a$ is a leave of a query result *T* of W and *a* contains $K_i\}$. The *relevance* of keyword K_i and keyword K_j in query result *T* is $\min_{a_i \in A_i, a_j \in A_j} d(a_i, a_j)$, denoted by $R_T(i, j)$.

For a keyword query $Q=\{K_1, \ldots, K_N\}$, $P_Q=\{(i, j)|K_i \in Q, K_j \in Q \text{ and } i<j\}$. All pairs in P_Q are sorted in the order of two dimensions. The order of a pair (i, j) in P_Q is k_{ij}. A result *T* can be considered as a vector $(R_1(T), \ldots, R_{|P_Q|}(T))$, where $R_{k_{ij}}(T) = R_T(i, j)$.

Definition 2.4 (**Keyword Dominate**). For a keyword query $Q=\{K_1, \ldots, K_N\}$ on XML streams, a result *T* dominates *T'*, denoted by $T < T'$, if $\forall$ k ($1 \le k \le |P_Q|$), $R_k(T) \le R_k(T')$.

From the definition of relevance, for two results of a keyword query *Q*, T_1 and T_2, T_1 dominating T_2 means that T_1 is more relevant than T_2. For all results of a keyword query *Q*, the most relevant results must be those dominated by none of other result. Therefore, we have the definition of *skyline point* of a keyword query *Q*.

Definition 2.5 (**Skyline Point**). Given a keyword query $Q = \{K_1, K_2, \ldots, K_N\}$ on XML streams, the query result *T* is called the skyline point of the received XML fragments, if *T* is not dominated by any other query result.

With the consideration that more results than skyline points for a keyword query may be required, inspired by [7], *loose skyline top-K keyword query* is defined.

Definition 2.6 (**Loose Skyline top-K Keyword Query**). A loose skyline top-K keyword query(LSK query in brief) is a query $Q= \{W, K\}$, where $W=\{K_1,K_2..., K_N\}$ is a keyword query, K is a constraint number. The results of Q on XML streams are a subset of the query results D of keyword query W, denoted as LS_D satisfying i) $\forall u \in LS_D$ is not dominated by any query result in D/ LS_D ii) $|LS_D|=K$.

To process the LSK queries on XML streams, we propose efficient algorithms. The basic idea of the algorithms is that for a LSK query $Q=\{W, K\}$, a set of query results of W containing the results of Q is maintained as *intermediate results* and such set is updated when new query results of W are generated.

The intermediate results should contain the results of a LSK query $Q=\{W, K\}$ and be as small as possible for the convenience of processing. At first, the *skyline layer* as the unit of *intermediate result* is defined.

Definition 2.7 (**Skyline Layer**). For a keyword query $Q =\{K_1,K_2,\ldots ,K_N\}$ and a set of query results M on XML streams, a subset of M is called *skyline layer* i(i≥1), denoted as L_i if it satisfies: i) $\forall a,b\in L_i$, a is not dominated by b, b is not dominated by a; ii) $\forall a\in L_i$, if $i\geq 2$, $\exists b\in L_{i-1}$, $b < a$; if $i=1$, $\forall b\in L_j$ (j≥1) a is not dominated by b; iii) $\forall a\in L_i$, if $L_{i+1}\neq\emptyset$, $\exists b\in L_{i+1}$, $a < b$

By this way, M can be divided into skyline layers from L_1 to $L_n(n\geq 1)$. By using this definition, we present the *garlic rule* to select the *intermediate results* denoted as L, of a LSK query $Q=\{W, K\}$ from the result set M of keyword query W.

Garlic Rule: Initially, $L=\emptyset$. For the query result set M of Q, L_1 is added to L at first. If $|L|$ is larger than K or all query results have been processed, the selection is finished. Otherwise the next skyline layer L_2 is chosen to be added to L. Such steps are processed until a layer L_t is added to L and following constraints are satisfied: i)$|L_1\cup L_2\cup\ldots\cup L_t|\geq K$; ii)if $t>1$, $|L_1\cup L_2\cup\ldots\cup L_{t-1}|< K$.

With the intermediate result set L selected by garlic rule, the results of Q can be easily generated by randomly selecting some nodes from the layer L_t. That is, if t=1 and $|L_1|>K$, K results are selected randomly from L_1; otherwise, $K-|L_1\cup L_2\cup\ldots\cup L_{t-1}|$ results are randomly selected from L_t as L'_t and the results of Q is $L_1\cup L_2\cup\ldots\cup L_{t-1}\cup L'_t$. Such selection rule is called **the lowest layer rule**. Therefore, for a LSK query Q on XML streams, the result of Q can be easily generated from the intermediate results. The update algorithm for intermediate result is presented in *Algorithm1*. In this algorithm, a new result m is compared with intermediate results in L from the lowest layer L_t to the higher layers until an element u in some layer L_i dominates m or none of elements in the top layer dominates m. If m is dominated by some element in L_i, m should be in lower layer than L_i. If L_i is the lowest layer and $|L|\geq K$, m is discarded. Otherwise, m is inserted to L_{i+1} and the elements in L_{i+1} dominated by m are inserted to the lower layer L_{i+2}. Such process is continued until element set E is to be inserted to the lowest layer L_t. Once E is inserted to L_t, the element set E' with each element dominated by some element in E should be pushed to lower layer than L_t. If $|L|+1-|E'|\geq$K, E' is discarded; otherwise, a new layer is created as the lowest layer with elements in E'. If none of elements in the top layer dominates m, it means that none of elements in L dominates m. m should be inserted to L_1 and the elements dominated by m are processed similarly as above steps. In order to accelerate the pushing elements to lower levels, for each elements e in $L_i\subseteq L$, pointers to the elements in L_{i+1} dominated by e is maintained.

Algorithm 1

```
Input: Q={W={K1,K2,...KN},K}, intermediate result set L, new query result m
Isdominated=false
for i=t to 1 do
    for each r ∈ Li do
    if r dominates m then
            Isdominated=true; Break;
    if Isdominated=true then
         l=i+1; Break;
if Isdominated=false then
    l=1
if l=t+1 then
    if |L|<K then
       add a new layer Lt+1={m} to L
    else
         add m to Ll
         E={u|u∈Ll∧m <u}
         while l≠t do
             Ll=Ll-E
             E={u|u∈Ll+1∧∃v∈E,v <u }
             Lt=Lt-E
       if |L|-|E|<K then
         add a new layer Lt+1=E to L
```

Algorithm 2

```
Input: Q={W,K}
Output: LS  // the results of query Q
BEGIN(n)
    n.state=00…0
    Stk.push(n)
endfunction
TEXT(n)
    for each Ki∈W do
        if Ki⊆n.v then
          n.state= n.state OR 2^i
          P[m][i].insert(n.id)
            //m is the parent of node n
endfunction
END(n)
//the descendants of node n contain all keywords
    if n.state =2^N-1 then
       Generate query result set M with root n
       for each result m in M do
          Update_intermediate(Q, L, m)
       Select LS from L with the lowest layer rule
    else
         Stk.top().state= Stk.top().state OR n.state
    Stk.pop()
endfunction
```

With intermediate result update algorithm just discussed, LSK query processing algorithm on XML streams is presented. To represent the structural relationship among nodes in XML streams, they are encoded by Dewey Code[6]. The state of each active node n is represented by an N-bits binary signature s_N. If any descendant of n contains keyword K_i, the ith position of s_N is set to 1; otherwise, it is set to 0.

In our algorithm, a stack is used to store the coding and states of all active nodes in the incoming sequence. An array P is used to store the keyword lists of all the non-leaf active nodes. For each non-leaf node i, the entry $P[i][j]$ stores all the descendants of node i containing keyword K_j. L stores current intermediate result set and LS stores the current results for the query.

The details of the algorithms are shown in *Algorithm2*. Node n is initialized by coding n with Dewey Code and pushing node n into the stack when opening tag of n is encountered (in Begin()). When the closing tag of an element n is received (in End()), the temporal results with n as root are generated by every keyword list K_i of node n stored in $P[n][i]$. The intermediate result L is updated with elements in the result $P[n][1]\times P[n][2]\times\ldots\times P[n][N]$ and current results are generated from L with the *lowest layer rule*. The details of the algorithms are shown in *Algorithm2*.

3 Experiments

We have implemented all the algorithms we proposed with reading XML files once from the disk to simulate XML streams. Two datasets, XMark and DBLP, are used. For each dataset, we selected a set of queries in order to test the efficiencies on different queries, we used both frequent keyword sets and rare keyword sets.

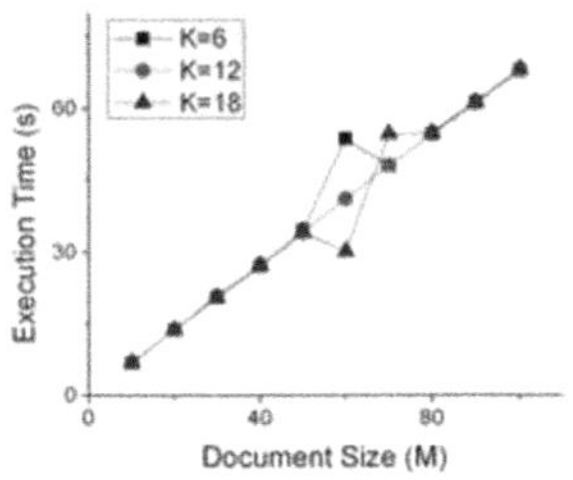

Fig. 1. Run time VS Doc. Size

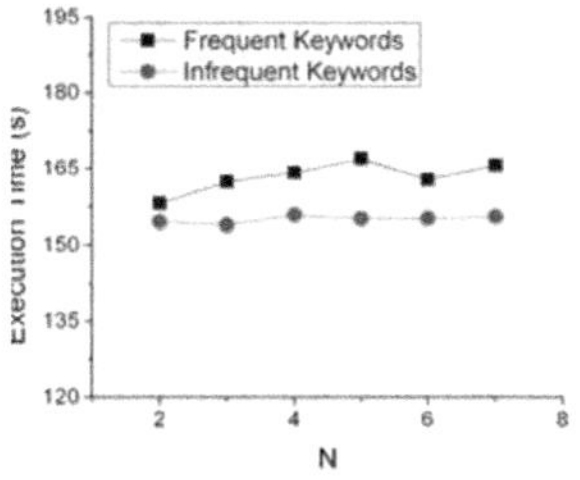

Fig. 2. Run time VS N

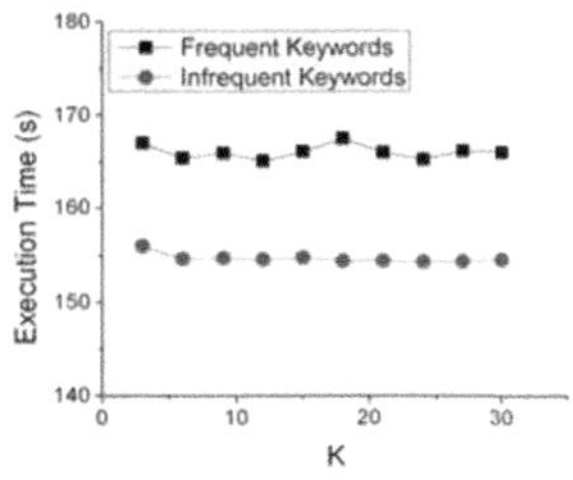

Fig. 3. Run time VS K

Scalability Experiments. In order to test the scalability, we generate XMark files with different parameters from 0.1 to 1.0 and the number of result *K* from 6 to 18. The experimental results are shown in Fig.1. From the results, the execution time is nearly linear to the number of elements.

Changing N. From Fig.2, it can be seen that both rare keyword sets and frequent keyword sets are insensitive to the value of *N*. That is because that in most cases the SLCA nodes has a small number of keyword matches.

Changing K. From Fig.3 we observe that, neither frequent nor infrequent query set are sensitive to the value of *K*. That is because in most cases, only a small number of query results need to be inserted to *L* instead of all of them.

4 Conclusion

With the broader application of XML data streams, top-k keyword search has become an important query on XML streams. Considering the different demands under the same keyword set query, in this paper, we propose the loose skyline top-K keyword query on XML streams. And we also propose effective algorithms for such kind of queries. Both analysis and experiments show that the techniques in this paper can obtain the search results efficiently and effectively with good scalability.

References

1. Hristidis, V., Papakonstantinou, Y., Balmin, A.: Keyword Proximity Search on XML Graphs. In: ICDE (2003)
2. Barg, M., Wong, R.K.: Structural proximity searching for large collections of semi-structured data. In: CIKM (2001)
3. Cohen, S., Mamou, J., Kanza, Y., Sagiv, Y.: XSEarch: A Semantic Search Engine for XML. In: VLDB (2003)
4. Guo, L., Shao, F., Botev, C., Shanmugasundaram, J.: XRANK: Ranked Keyword Search over XML Documents. In: SIGMOD (2003)
5. Borzsonyi, S., Kossmann, D., Stocker, K.: The skyline operator. In: Proc. 17th Int. Conf. on Data Engineering (2001)
6. Tatarinov, I., Viglas, S., Beyer, K.S., et al.: Storing and Querying Ordered XmlUsing a Relational Database System. In: SIGMOD (2002)
7. Pei, J., Jiang, B., Lin, X., Yuan, Y.: Probabilistic Skylines on Uncertain Data. In: VLDB (2007)

Searching for Rising Stars in Bibliography Networks

Xiao-Li Li, Chuan Sheng Foo, Kar Leong Tew, and See-Kiong Ng

Institute for Infocomm Research, Singapore 138632
{xlli,csfoo,kltew,skng}@i2r.a-star.edu.sg

Abstract. Identifying the rising stars is an important but difficult human resource exercise in all organizations. Rising stars are those who currently have relatively low profiles but may eventually emerge as prominent contributors to the organizations. In this paper, we propose a novel PubRank algorithm to identify rising stars in research communities by mining the social networks of researchers in terms of their co-authorship relationships. Experimental results show that PubRank algorithm can be used to effectively mine the bibliography networks to search for rising stars in the research communities.

Keywords: Rising Stars, Social Network Mining, Bibliography Networks.

1 Introduction

Many organizations are concerned with identifying "rising stars" — those who have relatively low profiles currently but who may subsequently emerge as prominent contributors to their organizations. However, there has been little work on this important task. In this paper, we investigate the possibility of discovering such rising stars from the social networks of researchers constructed using interactions such as research collaborations.

Most of the related social network mining research has focused on discovering groups or communities from social networks [1-2] and on the study of how these communities grow, overlap and change over time [3]. In this work, we consider the problem of detecting individual "stars" who rise above their peers over time in the evolving social networks that profile the underlying landscape of mutual influence. In universities and research institutions, it is possible to model the social network of researchers by the bibliography network constructed from their publications, where the nodes represent individual researchers, and the links denote co-author relationships.

From such a bibliography network, we aim to discover "rising stars". To do so, we consider the following factors: 1) *The mutual influence among researchers in the network*. For example, a junior researcher who is able to influence the work of his seniors and effectively collaborate with them, leveraging on their expertise, is far more likely to succeed in a research career. We model the degree of mutual influence using a novel link weighting strategy. 2) *The track record of a researcher.* We can measure this in terms of the average quality of the researcher's current publications. A researcher who publishes in top-tier journals and conferences is more likely to be an influential researcher as compared to another who publishes at less significant venues. This is accounted for by placing different weights on different nodes in the network model. 3) *The chronological changes in the networks*. Each researcher may work with different groups

X. Zhou et al. (Eds.): DASFAA 2009, LNCS 5463, pp. 288–292, 2009.

of people at different points in time. A researcher who can build up a strong collaborative network more rapidly than others is more likely to become a rising star.

In this work, we design a novel PubRank algorithm to mine rising stars from bibliography networks which incorporates the factors described above. Our algorithm derives information from the out-links of nodes, which is fundamentally different from many related node analysis algorithms that use information from the in-links.

Our technique is potentially useful for academics and research institutions in their recruitment and grooming of junior researchers in their organizations. It may also be useful to fresh PhDs and postdocs for selecting promising supervisors. Finally, it can be useful for tracking one's relative performance in the research community, and for deciding whom to collaborate (more) with.

We have also implemented a graphical interactive system RStar for public access to our results on the DBLP data (http://rstar.i2r.a-star.edu.sg/).

2 The Proposed Technique

Constructing the bibliography network. A bibliography network is a directed, weighted network where the nodes represent authors and the edges denote co-author relationships. When two authors v_i and v_j co-author a publication, there is mutual influence between them as the collaboration is typically beneficial to both parties. We model this mutual influence using the number of publications co-authored as a proxy for the strength of their collaboration relationship. We set the weight of the edge (v_i , v_j) to be the fraction of author $\boldsymbol{v}_j$'s publications that were co-authored with author $\boldsymbol{v}_i$, and the weight of the edge (v_j, v_i) to be the fraction of author $\boldsymbol{v}_i$'s publications that were co-authored with author $\boldsymbol{v}_j$. Researchers are then modeled to influence each other according to the strength of this relationship. Our weighting scheme captures the intuition that an expert researcher will tend to influence a junior researcher more than the junior influences the expert, as the expert will tend to have more publications, thus reducing the fraction of co-authored work with the junior researcher.

Accounting for the quality of publications: assigning node weights. The reputation and impact of a researcher is decided by the quality of his/her work. We incorporate this information by assigning node weights using the quality of a researcher's publications. While the citation count of a paper is commonly used as a measure of its quality, it is biased towards earlier publications because articles need time to accumulate citations. Rising stars, being junior researchers, are thus unlikely to have many highly cited papers. We therefore opted for an alternative measure based on the prestige of its publication venue. Numerous ranking schemas are available for this purpose. A commonly used system is as follows: rank 1 (premium), rank 2 (leading), rank 3 (reputable) and unranked [4].

Given a paper, we compute a measure of its quality based on the rank of the corresponding conference or journal where it was published. Then, given an author v_i who has a publication set P, we define his/her publication quality score $\lambda(v_i)$ as

$$\lambda(v_i) = \frac{1}{|P|} * \sum_{i=1}^{|P|} \frac{1}{\alpha^{r(pub_i)-1}}, \tag{1}$$

where pub_i is the i-th publication, $r(pub_i)$ is the rank of publication pub_i, and α ($0 < \alpha < 1$) is a damping factor so that lower ranked publications have lower scores. The larger $\lambda(v_i)$ is, the higher the average quality of papers published by researcher v_i.

Propagating influence in the bibliography network. The benefit of having a co-author is mutual. A young researcher will stand to gain by working with a more experienced and established collaborator, while the experienced researcher is far more productive by teaming up with like-minded researchers (both experts and promising novices) to do good work. This feedback nature has also been famously observed in the "social network" of co-referencing web-pages, and is exploited by Google's PageRank algorithm [5], where the PageRank of a page is defined in terms of the PageRanks of pages that link to it. We adapted the PageRank algorithm to compute a similar score for each node based on the propagation of influence in the bibliography network. We account for the mutual influence between authors and the quality of each author's publications to compute a similar PubRank score for each author (node):

$$PubRank(p_i) = \frac{1-d}{N} + d * \sum_{j=1}^{|V|} \frac{w(p_i, p_j) * \lambda(p_i) * PubRank(p_j)}{\sum_{k=1}^{|V|} w(p_k, p_j) * \lambda(p_k)} \tag{2}$$

In equation 2, N is the number of authors in the network, $w(p_i, p_j)$ is the weight of the edge (p_i, p_j) and $\lambda(p_i)$ is the publication quality score defined in equation 1.

A key difference between the PageRank and PubRank scores is that the PageRank score of a node is influenced by the scores of *nodes that link to it*, while the PubRank score of a node is dependent on the *nodes to which it links to*. In other words, unlike the PageRank algorithm and other link analysis algorithms which use *in-links* to derive information about a node, our algorithm uses a node's *out-links* to compute its score. This difference reflects the reality of the situation, as a researcher who has high quality publications and is able to contribute to the work of other influential researchers is likely to be a rising star.

Discovering rising stars from the evolving networks. The bibliography network grows larger each year as more papers are published. To account for this evolution of the network, we compute a series of PubRank scores for each author over several years. We hypothesize that if a researcher demonstrates an increase in his/her annual PubRank scores that are significantly larger than those of an average researcher, he/she will probably do very well in the coming years. We thus use a linear regression model to compute a gradient for each author, regressing his/her PubRank scores during a past time period against time (as measured in years). We then assess the significance of the gradient by computing its Z-score. Assuming that the gradients of the researchers have a Gaussian distribution, a critical region typically covers 10% of the area in the tail of the distribution curve. Thus, for a researcher v_i, if his/her Z-score is larger than 1.282, we regard v_i as statistically significant — v_i will be predicted as a rising star. In addition, we also require the researcher's PubRank score at the start of the time period to be lower than the average PubRank score of all researchers. This allows us to search for the "hidden" rising stars.

3 Experimental Evaluation

We performed two experiments using publication data from the Digital Bibliography and Library Project (DBLP). Our first experiment used all the DBLP data. This large data set with over one million publications tests the scalability of our algorithm. In our second experiment, we evaluate our algorithm on a subset of DBLP data from the Database domain. This is because one is often more interested in the performance of one's peers in the same technical domain than the entire field of computer science. The Database domain was chosen due to its long pedigree and relevance to our work in data mining. In our experiments the damping factor α was set to 2.

Results on the entire DBLP dataset. We used data from 1990-1995 to predict the rising stars, then look at their eventual PubRank scores a decade later in 2006 to verify if they have indeed realized their predicted potentials. We normalized the PubRank scores of all researchers using the Z-score measure as described in our method. Out of the 64,752 researchers with high PubRank scores (Z-score > 0), our method identified 4,459 rising stars. We compared the rising stars with researchers in general. On average, the rising stars continued to have significantly higher gradients in the period after 1995: the average gradient for the rising stars is 0.497 while the average gradient for all researchers is 0 (Z-score property). The predicted stars have indeed increased their PubRank scores significantly faster than researchers in general. In fact, although the rising stars all started out as relatively unknown researchers in 1990 (with PubRank scores lower than average), their final average Z-score in 2006 was 2.92, which means that they score well above that of the average researchers.

We performed a more in-depth analysis of the citation count of the top ten predicted rising stars, comparing them to the citation counts for 100 randomly selected non-rising star researchers. We found that the rising stars obtained significantly higher citation counts for their most cited papers, obtaining 440 citations on average as compared to 18.9 citations for randomly selected researchers.

We also ran our PubRank algorithm to mine the rising stars using the publication data from 1950 (1950–1955) to 2002 (2002–2007). In order to validate our predictions, we chose the *h*-index list (http://www.cs.ucla.edu/~palsberg/h-number.html) which is used to quantify the cumulative impact and relevance of an individual's scientific research output. The *h*-index, defined as the number of papers with citation count higher or equal to *h*, is a useful index to characterize the scientific output of a researcher [6]. Out of the 131 researchers with 40 or higher *h*-index score according to Google Scholar, 116 researchers (88.5%) are identified as rising stars by our algorithm across different years.

Results on the Database domain. A list of database conferences was obtained from schema [4] and we retrieved 19474 papers published at these venues from the DBLP data. Our PubRank algorithm is then used to identify the rising stars from 1990 to 1994 (rising stars in year n are predicted using historical data from n-5 to n-1). Note that a researcher can be predicted as a rising star in multiple years if their scores are always increasing significantly. To validate the results of our algorithm, we choose the top 20 rising stars for each year from 1990 to 1994. Out of the 100 rising stars, there are 63 unique individuals. Manual evaluation of the achievements of the 63 individuals showed that 43 (68.3%) have been appointed full professors at renowned universities, 7 (11.1%) of them are key appointment holders at established research laboratories and companies, and the remaining 13 are either Associate Professors or hold important positions in industry.

Table 1. Top 10 predicted rising stars from the database domain from years 1990-1994

Name	Position and Organization	Awards	Top Citation
Bharat K. Bhargava	Professor, Purdue University	IEEE Technical Achievement Award, IETE Fellow	143
H. V. Jagadish	Professor, University of Michigan, Ann Arbor	ACM Fellow	457
Hamid Pirahesh	Manager, IBM Almaden Research Center	IBM Fellow, IBM Master Inventor	1428
Ming-Syan Chen	Professor, Nat. Taiwan U	ACM Fellow, IEEE Fellow	1260
Philip S. Yu	Professor, UIC	ACM Fellow, IEEE Fellow	1260
Rajeev Rastogi	Director, Bell Labs Research Center, Bangalore	Bell Labs Fellow	1178
Rakesh Agrawal	Head, Microsoft Search Labs	ACM Fellow, IEEE Fellow, a Member of the National Academy of Engineering	6285
Richard R. Muntz	Professor, UCLA	ACM Fellow, IEEE Fellow	1191
Shi-Kuo Chang	Professor, U of Pittsburgh	IEEE fellow	171
Jiawei Han	Professor, UIUC	ACM fellow	6158

Table 1 shows the achievements of a selection of 10 outstanding individuals from the 63 we earlier identified. Their most highly cited publications all have over 100 citations (as found using Google Scholar) and 7 of them have been recognized as ACM and IEEE fellows (or both). The other individuals that we identified also have remarkable achievements such as being appointed editor-in-chief for prestigious journals or winning (10 year) best papers at major database conferences (*e.g.*, SIGMOD, PODS, VLDB, ICDE, KDD etc). Such achievements clearly show that they have indeed become the shining stars in the database domain, as we have predicted with our algorithm with publication data from more than a decade ago.

References

[1] Newman, M.E.J.: Detecting community structure in networks. European Physical Journal B 38, 321–330 (2004)
[2] Palla, G., Derenyi, I., Farkas, I., Vicsek, T.: Uncovering the overlapping community structure of complex networks in nature and society. Nature 435, 814–818 (2005)
[3] Backstrom, L., Huttenlocher, D., Kleinberg, J., Lan, X.: Group Formation in Large Social Networks: Membership, Growth, and Evolution. In: Proceedings of Twelfth ACM SIGKDD International Conference on Knowledge Discovery and Data Mining (2006)
[4] Long, P.M., Lee, T.K., Jaffar, J.: Benchmarking Research Performance in Department of Computer Science, School of Computing, National University of Singapore (1999), `http://www.comp.nus.edu.sg/~tankl/bench.html`
[5] Brin, S., Page, L.: The anatomy of a large-scale hypertextual Web search engine. In: Proceedings of international conference on World Wide Web, pp. 107–117 (1999)
[6] Hirsch, J.E.: An index to quantify an individual's scientific research output. In: PNAS (2005)

Hash Join Optimization Based on Shared Cache Chip Multi-processor

Deng Yadan, Jing Ning, Xiong Wei, Chen Luo, and Chen Hongsheng

College of Electronic Science and Engineering, National University of Defense Technology, Changsha 410073, China
dengyadan2008@yahoo.com.cn, {ningjing,xiongwei,luochen, chenhs}@nudt.edu.cn

Abstract. Chip Multi-Processor(CMP) allows multiple threads to execute simultaneously. Because threads share various resources of CMP, such as L2-Cache, CMP system is inherently different from multiprocessors system and, CMP is also different from simultaneous multithreading (SMT). It could support more than two threads to execute simultaneously, and some executing units are owned by each core. We present hash join optimization based on shared cache CMP. Firstly, we propose multithreaded hash join execution framework based on Radix-Join algorithm, then we analyze the factors which affect performance of multithreaded Radix-Join algorithm in CMP. Basing on this analysis, we optimize the performance of various threads and their shared-cache access performance in the framework, and then theoretic analysis of speedup in multithreaded cluster partition phase is presents which could give some advices to cluster partition thread optimization. All of our algorithms are implemented in EaseDB. In the experiments, we evaluate performance of the multithreaded hash join execution framework, and the results show that our algorithm could effectively resolve cache access conflict and load imbalance in multithreaded environment. Hash join performance is improved.

Keywords: Radix-Join,Shared L2-Cache,Cache Conflict, Chip Multi-Processor.

1 Introduction

As the number of transistors in processors continues to increase, processor frequency has reached the limit under current technology, and also the energy consumption. Therefore, the development trend of processors[9] is transforming from high-speed single-core to Chip Multi-Processor(CMP), and from instructions parallel to thread level parallel. At present, two cores and four cores processor are becoming the mainstream, and core number is increasing every year. Comparing to traditional multiprocessors system, off-chip communication speed is much lower than on-chip's and, parallel efficiency of CMP is obvious higher than multiprocessors. For DBMS, shared cache could reduce data redundancy and improve cache hit ratio. However, CMP generally share some resources between cores, which may lead to access conflict when using these shared resources, such as shared L2-Cache. Therefore, CMP has brought the database with both opportunities and challenges[7]. Many scholars[5][6][7][14]

X. Zhou et al. (Eds.): DASFAA 2009, LNCS 5463, pp. 293–307, 2009.

have researched database algorithm optimization based on CMP. Comparing to parallel optimization between queries, parallelism of internal query requires further study[14].

Current computer structures adopt multi-level cache architecture to reduce data access delay. More and more researches have proved that in multi-level cache architecture, with the decreasing price and increasing capacity of main memory, database servers with large main memory is economical and desirable, and disk I/O delay is gradually eased[1]. However, the speed gap between memory and on-chip cache is increasing, and the delay which processor waits for data from main memory to on-chip cache is becoming the main bottleneck for database[7][11][12][13][8][9]. Currently, database performance could be evidently improved either through cache access optimization to reduce cache delays or by thread-level parallelism to hide cache delay. For hash join, many scholars have presented some optimization algorithms, divided into two categories:

1. Cache access optimization. Such methods can be divided into prefetching[11] and partition[2][8].[11] utilizes software prefetching instructions to hide cache access delay and proposes two prefetching algorithm: group prefetching and pipeline technology; [2] presents Radix-Join algorithm which adopts multi-way partition in cluster partition process to reduce cache miss when cluster partition, and it can be used as basic hash join in many hash join optimization researches[10][12], also in our research; In [10], they present cache-oblivious hash join, setting optimal cache parameters in runtime, which could get better performance than adopt inherent parameters of processor cache in some cases.

2. Multithreaded parallelism optimization. [12] optimizes Radix-Join algorithm in MTA-2 CMP system, and the results show that speedup and throughput of hash join are improved with many threads execute concurrently; [13] optimize hash join in SMT (Simultaneous Multithreading) system which implements each operator in a two-threaded pattern. One thread processes even tuples on the (probe) input stream, and the other processes odd tuples; In [14], they divide hash join into some operations, then hash join could be executed according to pipeline composed by these operations, and every operation is executed by multiple threads.

However, these researches have not involved the performance optimization of shared cache which accessed by multithread in CMP system. Current CMP generally shares L2-Cache between cores which could effectively reduce data duplication and improve cache utilization, but the competition of shared resources may bring some negative effects on multithread execution, such as cache access conflict[4]. Cores in CMP could access shared cache at the same time. Core1 may replace cache lines which other cores will access in the future. When other cores access these cache lines, cache miss occurs, even resulting in cache thrashing which could bring serious performance decline.

In order to make full use of CMP, it is necessary to effectively solve performance bottlenecks caused by shared cache. Therefore, in this paper, we focus on improving shared cache access performance of multithreaded hash join algorithm, and also fully utilizing parallel computing resources of CMP. The contributions of this paper are as follows:

1. According to characteristic of shared L2-Cache CMP, we present a multithreaded hash join execution framework, which bases on Radix-Join algorithm, to execute hash join.

2. We analyze factors which effect multithreaded Radix-Join performance in shared cache CMP through two instances execution. Then on the basis of the analysis, by setting reasonable number of thread, thread start time and working sets of threads, we optimize performance of multithreaded hash join execution framework and analyze speedup of multithreaded cluster partition.

3. We implement our framework in EaseDB[8]. By analyzing the experiment results, we can prove our algorithm to be shared-cache-friendly and fully utilize computation resources of CMP. The hash join execution performance is also improved.

2 Multithreaded Hash Join Execution Framework

This paper presents how to optimize multithread and its cache performance of hash join in shared cache CMP. Therefore, it is more applicable for in-memory database or disk database which has large main memory and could cache enough pages to greatly reduce disk I/O delay. As following, we will present the details of multithreaded hash join execution framework which bases on Radix-Join[2] algorithm.

Radix-Join splits a table into H clusters using multiple passes. Cluster partition on lower $D=\sum_1^P D_p$ bits of the integer hash-value of a column is done in P sequential passes, in which each pass clusters tuples on D_p bits, starting with the leftmost bits. The number of clusters created by the cluster partition is $H=\prod_1^P H_p$, where each pass subdivides each cluster into $H_i=2^{D_p}$ new ones. The clusters of two tables are joined by hash join(Radix-Hash-Join) or nested loop join(Radix-NL-Join), while we only aim at Radix-Hash-Join.

This paper optimized Radix-Join algorithm based on shared L2-Cache CMP. Radix-Join algorithm is composed by two phases, cluster partition and cluster join. The multithreaded execution framework of Radix-Hash-Join is shown in Fig 1:

In Fig1, the temp table adopts *Parallel Buffer*[5] to manage memory. *Parallel Buffer* divides a big chunk memory into small chunks. Thread reads or writes memory in a unit of chunk, which could greatly reduce exclusive access cost of multithread. There are three kinds of thread in Fig 1:

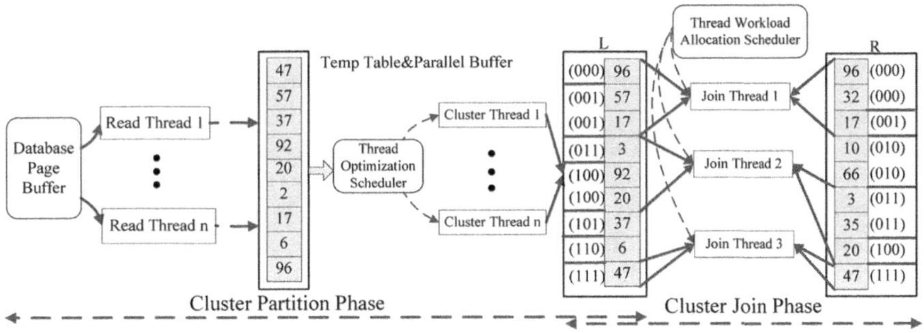

Fig.1. Radix-Hash-Join multithreaded framework

1. Page processing thread. Read Thread(RT) reads pages and executes project operator in *Query Execution Plan*(QEP) to convert tuples which satisfies query conditions to *Hash Cell*, and then, puts *Hash Cell* into temp table. *Hash Cell* is composed as <HashValue, KeyValue, TID>. In this expression, "HashValue" stands for hash value, "KeyValue" for value(s) of join column, and "TID" for tuple ID.

2. Cluster partition thread. After all pages in table have been processed, Cluster Thread(CT) partition temp table into small clusters. The start time of CT must be optimized(detailed analysis in Section 3.2.2), which is very important to performance of multithreaded cluster partition. Every CT will be assigned some clusters. For example, there are H_j clusters when m CTs are started, and then the H_j clusters are evenly assigned to the m threads. For every cluster, CT completes remaining cluster partition in P-j sequential passes. The *thread optimization scheduler* module corresponds to optimization of cluster partition thread in Section 3.2.2.

3. Cluster Join Thread. The *thread workload allocation scheduler* module corresponds to cluster join thread optimization in Section 3.2.3. Cluster partition generates two cluster tables(L and R). The key points of JT performance are workload balance of CMP and low cache conflict of JT.

3 Multithreaded Hash Join Execution Framework Optimization

In this section, we present optimization of multithreaded hash join execution framework according to characteristic of shared cache CMP. The two objects of optimization are: (1) low shared L2-Cache conflict; (2) load balance of CMP cores when cluster join.

3.1 Performance Analysis and Comparison of Multithread Radix-Join

Through performance comparison of multithreaded Radix-Join execution, we analyzed the factors which affect performance of multithreaded Radix-Join execution in CMP. The analysis could provide gist for the optimization in the following sections. The experiments were performed on Dual-Core CMP with 1M shared L2-Cache, and using EaseDB as experiment platform. The experiment data are L.tuple=2.72e+10^6 and R.tuple=2.73+10^6, *Cellsize*=12B(*CellSize* denotes Hash Cell size). Data distribution among clusters is uniform. Parameter P is in {7,6,5,4}, and corresponding *ClusterSize* is in {256k, 512k, 1024k, 2048k}.

Case 1. Execution time comparison of cluster joins which adopts different number of JT to execute cluster join.

Case 2. Execution time comparison of Radix-Hash-Join which adopts different number of CT and JT to execute respective cluster partition and cluster join.

From the multithreaded view, the CMP's efficiency is very low when single-threaded program runs on it. In Fig2(a), when *ClusterSize* equals 1024KB, the cluster join execution time of single thread is more expensive than the one of multithread, although cache conflict of single thread is greatly lower. However, only two threads could be executed simultaneously in Dual-Core CMP. For single database operation, such as cluster join, there is no proportional relationship between performance and the thread number. When thread number greatly exceed the number of processor core,

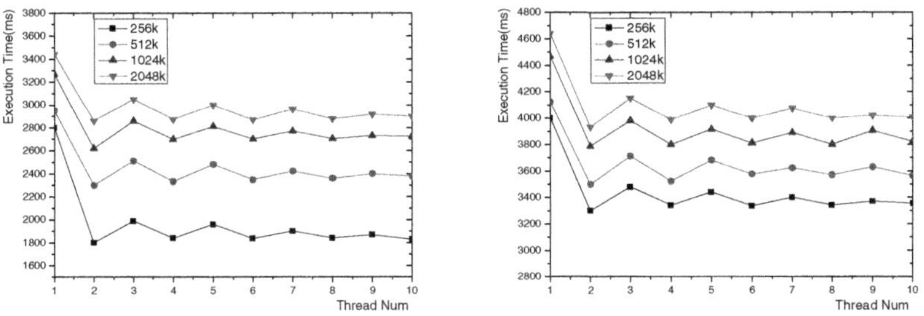

(a) Cluster join (b) Cluster partition and cluster join

Fig. 2. Radix-Join performance comparison

performance even have little decline caused by thread switch. Generally, thread time slice of modern operation system is 10-20ms, and the bandwidth between DDR2 main memory and L2-Cache is 25G/s. Considering delay cost of cache hit and cache miss, in a thread execution slice, data processed by one thread will greatly exceed L2-Cache capacity. Therefore, when thread number exceed processor core, there is not serious cache conflict which lead to biggish performance decline. In Fig2(a) and Fig2(b), When *ThreadNum* is bigger than *N* and odd(*N* denotes processor core number), performance has little decline caused by load imbalance of processor cores. While *ThreadNum* increases, *ClusterSize* decreases, and the degree of load imbalance is reduced, bringing in some performance improvement.

From the cache view to analyze, in Fig2(a), when *ClusterSize* equals 256KB, total data processed by two CTs are close to *C*(C denotes L2-Cache capacity), leading to very low cache conflict. When *ClusterSize* equals 512k, two JTs must access about 2MB data, and it is bigger than *C*. There will have a high probability which cache lines belong to hash table are replaced from L2-Cache to main memory, which is so-called cache conflict. Therefore, we could conclude that cache conflict seriously affects performance of multithread in shared cache CMP. However, compared when *ClusterSize* equals 1024KB, the performance does not decline severely, although there may be more serious cache conflict in theory when *ClusterSize* equals 1024KB or 2048KB. From this, we can conclude that there is no strict proportional relationship between data and performance decline when the cache capacity is much less than the amount of data to be processed. In Fig2(b), when *ClusterSize* equals 512 KB, the parameter *P* is six. Compared when *ClusterSize* equals 256KB, there is one less temp table access, but the advantage of less temp table access is counteracted by its serious cache conflict.

3.2 Multithreaded Hash Join Execution Framework Optimization

In shared L2-Cache CMP, basing on the conclusions in section 3.1, we must take two functions into account: (1) affection of shared L2-Cache to multithread execution; (2) data access characteristic of various threads. Then we decide thread execution parameters, such as thread number, thread start time and working sets allocation. In the multithreaded hash join execution framework, the objects of optimization are cluster partition and cluster join. The detail analyses are as follows.

3.2.1 Radix-Join Algorithm Parameter Optimization

In [2], they present three settings of parameter D about Radix-Hash-Join, respectively aiming at L1-Cache, L2-Cache and TLB. Because the main concern of this paper is shared L2-Cache, and Radix-Hash-Join performance using the two other settings would not be influenced in shared cache CMP, we only concern for the settings of L2-Cache in this paper. [2] presents that D equals $\log_2(L.\text{tuple}*CellSize/C)$, and this will lead to biggish cluster size. Since the situation of multithreaded cluster join is not under consideration, cache conflict would be serious when multithreaded cluster join execution. But from the analysis in Section 3.1, serious cache conflict will lead to severe performance decline. Therefore, we should add two restrictions: cluster join thread number in cluster join phase and processor core number of CMP, then setting $D=\log_2(2N*L.\text{tuple}* CellSize/C)$ to make sure the total data processed by N JTs less than C in cluster join phase.

3.2.2 Cluster Partition Thread Optimization

[2] presents that D_i is generally larger than one which must use swap table to realize *Hash Cell* exchange. When there are a large number of data, this setting could reduce cluster partition time and improve performance of Radix-Join. But CT must access additional swap table, leading to more possibility of cache conflict in a pass of cluster partition. From the performance analysis based on Fig 2(b), serious cache conflict may counteract the advantage of lesser cluster partition time. Therefore, we present a new cluster partition strategy. Firstly, the value of *DiffData* is computed through this equitation: $CPSize(D_i=1)-CPSize(D_i=D/P)$, in which $CPSize(D_i=1)$ and $CPSize(D_i=D/P)$ respectively present total data processed in cluster partition phase adopting $D_i=1$ and $D_i=D/P$. Then according to the critical value of *DiffData*(presented as *MaxData*), the strategy of cluster partition to be adopted is decided.

1. $DiffData<MaxData$, this denotes less cluster partition time, we could set $D_i=1$, namely, every cluster partition in P cluster partition time only process one bit of the hash value. Thereby, cluster partition could be done without swap table;
2. $DiffData \geq MaxData$, this denotes biggish cluster partition time. To reduce cluster partition cost, we adopt cluster partition strategy in [2].

How to set the value of *MaxData*? There is detailed analysis in Experiment 1. From the analysis in Section 3.1, we can deduce that if CTs are started prematurely, *ClusterSize* will be biggish, and resulting in high cache conflict in cluster partition. But if CTs are started late, the proportion of *(serial execution time)/(total execution time)* in whole cluster partition execution will be large, and weaken performance of multithreaded cluster partition. Therefore, it is important to get balance between the two cases.

1. When $DiffData<MaxData$, $H_j=\prod_{i=1}^{j} H_i$ denotes clusters number after jth cluster partition have finished. $TTSize=L.\text{tuple}*CellSize$ is the size of temp table. In the process of cluster partition, according to H_j $(1 \leq j \leq P)$, there are three cases:

Case 1. $H_j \leq N$ and $TTSize \leq C$. This implies that data are very little, when 1th cluster partition has finished, JTs could be started;

Case 2. $H_j \le N$ and *TTSize*>*C*. We judge whether $\frac{i*TTSize}{H_j} \le C$ when every cluster partition pass has finished. If exist *i* satisfying the condition, choose the maximum value of *i*(presented as *Max*(*i*)) and start *Max*(*i*) CTs. Otherwise, go to Case 3;

Case 3. H_j>*N*. After every cluster partition has finished, we also judge whether $\frac{i*TTSize}{H_j} \le C$ is satisfied ($2 \le i \le N$). If there are some values of *i* which satisfy the condition and *Max*(*i*)<*N*, then we have two feasible strategies to choice: (1)start *Max*(*i*) CTs, but this may reduce processor utilization;(2)delay CT start until $\frac{N*TTSize}{H_j} \le C$, then *N* CTs could be started, but this may reduce *Speedup* in Definition 1. Which one could get better performance? From the theoretic analysis in Section 3.3 and Experiment 6, we conclude that, under low cache conflict condition, the sooner CT starts, the less serial data process time and better performance we could get.

In Case 2 and Case 3, CT starting time(presented as *j*) equals to $\log_2 \frac{Max(i)}{C} * TTSize$. If *j* is close to *P*, then it is no sense to utilize multithread. For example, we can deduce that $\frac{N*TTSize}{H_{P-1}} > C$ is not satisfied by $TTSize = \frac{2^D * C}{2N}$ and $H_P = 2^D$. It denotes that, despite *N* CTs could be started when *P*-1th cluster partition finished with low cache conflict, performance will not be improved caused by *j* equals *P*-1. Therefore, before cluster partition phase begin, we should base on $j = \log_2 \frac{i}{C} * TTSize$ ($2 \le i \le N$) to compute the values of (*i, j*), then use these (*i, j*) as parameters to compute speedup according to forum *S*(*i, P, j*) in Section3.3. A certain value of (*i, j*) is chosen according to maximum speedup value. If the value of *j* is bigger than *MaxJ*, well then it is needless to satisfy $\frac{i*TTSize}{H_j} \le C$, we could start *N* CTs only when $H_j \ge N$. Otherwise, we could start CT according to strategies in Case 2 and Case 3. The value of *MaxJ* is analyzed in Experiment 2.

2. When $DiffData \ge MaxData$, *P* is small. We should start CTs as early as possible under the precondition of reducing cache conflict. Due to cluster partition needs access temp table, if *N* CTs are started after *P*-1th cluster partition finished, the average data processed by every CT is $\frac{2N*TTSize}{H_{P-1}} = C*2^{\frac{D}{P}}$, and total data accessed by *N* CTs is *TotalData*= $N*C*2^{\frac{D}{P}}$. $TotalData = \frac{4*TTSize}{H_{P-1}} = C*(2^{\frac{D}{P}+1}/N)$ when two CTs are started. $C*(2^{\frac{D}{P}+1}/N)$ is general bigger than *C* if there are a large number of data and small cores number of CMP. So, before cluster partition start, we should judge whether $C*(2^{\frac{D}{P}+1}/N)$>*C*. According to the result, there are two choices of cluster

partition: (1) $C*(2^{\frac{D}{P}+1}/N) > C$, the result implies that only two CTs are started when P-1th cluster partition finished, however, there also has serious cache conflict. Therefore, $Min(H_1,N)$ CTs could be started after the 1th cluster partition pass has finished; (2) $C*(2^{\frac{D}{P}+1}/N) \le C$, after every cluster partition has finished, we judge whether $\frac{2i*TTSize}{H_j} \le C$ is satisfied ($2 \le i \le N$). If there are values of i which satisfy the condition, $Max(i)$ CTs could be started. If there is not any i satisfying the condition when j equals $\lfloor P/2 \rfloor$, we should start N CTs immediately.

3.2.3 Cluster Join Thread Optimization

JT access cluster table(temp table) generated by cluster partition and output buffer. Cluster number in cluster table is bigger than core number of CMP in most cases, and according to the analysis in Section 3.1, when thread number is greater than N, performance would not have a large improvement. Therefore, the max number of JT equal to N. Owing to hash table is repeatedly accessed, cluster join is easy to occur serious cache conflict. Data are evenly distributed between clusters in normal circumstances after cluster partition. But when data are serious uneven, data distribution between clusters would be skew. Cluster join performance may suffer for two reasons.

1. The number of JT which should be started couldn't be accurately controlled. In some conditions, this may lead to severe cache conflict. For example, there are table L and R, and corresponding cluster sets are $ClusterSetL= (LCL_1,LCL_2,\ldots,LCL_A)$ and $ClusterSetR= (RCL_1,RCL_2,\ldots,RCL_B)$. N JTs are started. If we just allocate clusters between JTs evenly, this will bring serious cache conflict possibly, caused by data accessed by N JTs may be bigger than C, namely $\sum_{i=1}^{N} Sizeof(LCL_i + RCL_i) > C$. We couldn't accurately adjust JT number to avoid cache conflict.

2. Serious discrepancy of working sets size allocated to JTs would lead to load imbalance between cores in CMP and reduce performance of cluster join.

Therefore, to reduce cache conflict and load imbalance, we present a new multi-threaded cluster join execution strategy basing on cluster size classification. The cluster classification algorithm is as follows:

Algorithm 1. Cluster Classification Algorithm

Input: *ClusterSetL*[0..*A*-1], *ClusterSetR*[0..*B*-1]

Output: *WorkSet*[0.. *JTNum*]

Procedure: WSallocation(*ClusterSetL*, *ClusterSetR*)

```
i = 0; j = 0;k = 0;
While(i < A and j <B)
{
  While(ClusterSetL[i].radixvalue != ClusterSetR[j].radixvalue)
    j++;
  TotalClusteSize = Add(ClusterSetL[i].size, ClusterSetL[j].size);
```

```
for(k=1; k<= JTNum-1; k++)
  if(C /(k+1) <TotalClusteSize<= C /k)
  {
    add (i, j) to WorkSet[k];
    break;
  }
if(0< TotalClusteSize<= C /JTNum)
  {add (i, j) to WorkSet[JTNum];}
if(TotalClusteSize >C)
  {add (i, j) to WorkSet[0];}
i++;
}
```

Except for *WorkSet*[0], when deal with *WorkSet*[*i*], there are two cases according to whether $2i \le N$ is satisfied:

1. If $2i \le N$, JTs are divided into *i* groups, and each group includes $1+\left\lfloor \frac{N-i}{i} \right\rfloor$ JTs and joins a pair of clusters in *WorkSet*[*i*], such as (LCL_i, RCL_j). Hash cells in LCL_j (probe table) are evenly allocated to JTs in the group. A certain thread is specified to construct hash table, and other threads in this group must wait for the thread to finish hash table construction.

2. If $2i>N$, only *i* JTs need to be started. The pair of clusters in *WorkSet*[*i*] are evenly allocated to *i* JTs as working sets.

Firstly, *WorkSet*[1] is processed, then *WorkSet*[2], until *WorkSet*[*JTNum*]. Cluster pairs in *WorkSet*[0] are processed lastly. Although *WorkSet*[0] includes cluster pairs which are larger than *C*, but from performance analysis in Section 3.1, the performance of multithread is better than single thread in any cases. Therefore, *N* JTs are started. JT access output buffer once time after accessing relative large number of data. Therefore, output buffer access has little influence to JT, and output buffer is not included in above analysis.

In above analysis, thread number is bounded by core number in CMP which is suited for current CMP with small quantity of cores. When processor has a large number of cores, for example twenty or eighty cores, there is no need to start *N* threads for cluster partition and cluster join, and we could set the upper bound of thread number to eight. When CMPs with more cores appear, more appropriate upper bound could be set. Currently, shared L2-Cache is unified, namely, used as both data cache and instruction cache. Because hash join is typical data intensive computation, we could not take instruction data into account.

3.3 Multithreaded Cluster Partition Performance Analysis

Definition 1 *Speedup*. According to Amdahl law, *Speedup* is defined as $S(A, f)= \frac{A}{1+(A-1)f}$. *A* is physical thread number of computation, *f* denotes the proportion of (serial computation time)/(total computation time).

In shared L2-Cache CMP, due to cache conflict, high *Speedup* doesn't imply high performance. But it is still an important scale of performance, and we could analyze optimization strategy in cluster partition phase. The proposition and its proof are as follows.

Proposition 1. In cluster partition phase, the *speedup* $s=S(A_i, P, j)=\dfrac{PA_i}{P+j(A_i-1)}$, and the values of f in Definition 1 is proportional to j. A_i is thread number of cluster partition, and j denotes the jth cluster partition in P cluster partition passes.

Proof. Suppose that data is evenly distributed between clusters and execution time of a cluster partition pass is t. When the jth cluster partition has finished and *DiffData*<*MaxData*, A_i CTs are started, and serial execution time which from cluster partition start to now is $j*t$. H_j clusters are generated. The time which a thread deals with one cluster is

$$\frac{t}{H_j}(P-j) \qquad \text{Eq(1)}$$

then we could deduce the residual cluster partition time is

$$\frac{t}{H_j}(P-j)*\frac{H_j}{A_i} \Rightarrow \frac{t}{A_i}(P-j) \qquad \text{Eq(2)}$$

so the total execution time of cluster partition which adopts multithreaded execution is Eq(2)+ $j*t$

$$\frac{t}{A_i}(P-j)+j*t \qquad \text{Eq(3)}$$

and serial execution time is $P*t$. According to speedup's basic meaning, we could deduce the *Speedup* s is $P*t/$ Eq(3)

$$s=S(A_i,P,j)= Pt \Big/ \left(\frac{t}{A_i}(P-j)+jt\right) \Rightarrow PA_i \Big/ P+j(A_i-1) \qquad \text{Eq (4)}$$

by $S(A_i,P,j)=S(A_i,f)$, we could do following deduce

$$\frac{PA_i}{P+j(A_i-1)}=\frac{A_i}{1+(A_i-1)f} \Rightarrow f=j/P \qquad \text{Eq (5)}$$

through $f=j/P$, we can conclude f is proportional to j. when $DiffData \geq MaxData$, the proof is the same with *DiffData*<*MaxData*. □

The following is *Speedup* computation and optimization analysis of CT in Section 3.2.2. Processor is 4M shared L2-Cache Quad-Core CMP. Experiment data is *L*.tuple=670000, *CellSize*=12B, then $D=\log_2(2N*L.\text{tuple}*CellSize/C)=4$ and *DiffData*<*MaxData*. When j=2, the max values of i which satisfies $\dfrac{i*TTSize}{2^2} \leq C$ is two, then we could start two CTs. According to Eq(4), s=1.33. If we delay CT start to

j=3, and $\frac{4*TTSize}{2^3}<C$ is satisfied. Four CTs could be started, then s=1.23. Therefore, under low cache conflict, we can conclude that the sooner CT stars, the smaller value of f, fewer serial execution time, and better cluster partition performance we could get. The Experiment 6 confirms our conclusion. The formula of *Speedup* doesn't consider cache conflict influence. Therefore, it is only suitable for low cache conflict case.

4 Experimental Evaluation

4.1 Experiment Setup

We ran our tests on both shared L2-Cache CMP of Intel Dual-Core, as well as Quad-Core CMP. There are two kinds of L2-Cache, 1M and 4M. Database platform is EaseDB[8]. Hash join in our experiments are equal join of two tables. The setting of experiment data is the same with [10], and data in table are generated by random function.

4.2 Multithreaded Hash Join Performance Analysis

In this section, the multithreaded hash join execution framework is implemented based on EaseDB. Firstly, through performance analysis and comparison, we analyzed reasonable value of *MaxData*, then tested the performance of framework under different parameters and validated effect of optimization strategies in our paper. Performance comparison between our algorithm and latest hash join optimization algorithms is presented. Time count begins from total data pages have already been converted to temp table.

Experiment 1. In this experiment, we investigated reasonable value of *MaxData*. Processor is 1M shared L2-Cache Dual-Core CMP. The *L*.tuple values of different tables are ($3.5e+10^5$,$7.0e+10^5$,$1.4e+10^6$,$2.8e+10^6$,$5.57e+10^6$,$1.1185e+10^7$), *CellSize*= 12B, and the corresponding values of D are (4,5,6,7,8,9). To test execution time of cluster partition, we respectively adopted two cluster partition strategies to each table: where D_i equals 1 and the other D/P. When $DiffData<3TTSize$, from performance comparison of multithreaded cluster partition execution in Fig 3(a), due to serious L2-Cache conflict of each cluster partition pass in $D_i=D/P$ caused by swap table access, D_i=1 has better performance than $D_i=D/P$, despite $D_i=D/P$ has the advantage of lesser cluster partition time. When D=4, D_i=1 only need 69% execution time of $D_i=D/P$, and when D=5, D_i=1 need 57%. But with D increasing, the value of *DiffData* is increasing. When D respectively equal 4,5,6,7,8,9, the corresponding values of *DiffData* are *2TTSize*,*TTSize*,*2TTSize*,*3TTSize*,*4TTSize*,3TTSize. D_i=1's advantage of better cache access is counteracted by large cluster partition time. In Fig3(a), after D=7 and $DiffData \geq 3TTSize$, $D_i=D/P$ has better performance than D_i=1. Therefore, we can deduce $MaxData=3TTSize$.

Experiment 2. We studied the performance of cluster partition thread. The experiments were performed on 4M shared L2-Cache Quad-Core CMP and execute multithreaded cluster partition on table L to test performance of cluster partition thread in different start time. When *DiffData<MaxData*, L_i.tuple(i=1,2,3,4) are (6.67e+10^5, 1.0e+10^6, 1.33e+10^6, 1.67e+10^6), and corresponding values of D are (4,5,5,6), *CellSize*=12B. The (j, *Max*(i)) values which satisfy condition of $\frac{i*TTSize}{H_j} \le C$ are ((2,2),(3,2), (3,2),(4,3)). The *Start Time* in Fig 3(b) denotes CTs are started when |*Start Time*|th cluster partition finished. *Start Time*=0 implies CT is started when cluster partition phase beginning. Due to there is only one cluster, namely, the temp table, only one CT could be started. As shown in Fig 3(b), according to the (j, *Max*(i)) of L_1, two cluster threads could be started while 2th cluster partition pass finished. Cluster partition performance has been improved, and it is the best performance in all of *Start Time*. The other tables (L_2, L_3, L_4) also adopt (j, *Max*(i)) to start partition thread, but their serial execution time takes high proportion of total cluster partition time. Their performance when adopt parameter (j, *Max*(i)) are not the best in all of *Start Time* despite they also have low cache conflict. Therefore, we conclude that if $\log_2 \frac{i}{C} * TTSize \le \frac{1}{2} D$, cluster thread could be started according to Case 2 and Case 3 in Section 3.2.2. Otherwise, N cluster threads could be started while $H_j \ge N$. When $DiffData \ge MaxData$, we ran the test on 1M shared L2-Cache Dual-Core CMP. R.tuple=7.744e+10^6, *CellSize*=12B, D=9, P=3. As shown in Fig 3(c), due to $C*(2^{\frac{D}{P}+1}/N)>C$, CTs are started when the first cluster partition finished where we could get the best performance.

Experiment 3. The performance of JT was tested in this experiment. There are three execution models of cluster join: (a) Allocating working sets according to cluster classification; (b) Allocating working sets evenly; (c) Single thread. The experimental data (L.tuple,R.tuple) are A(1.0e+10^5,1.76e+10^5), B(2.5e+10^5, 3.666e+10^5),C(5.2667e+10^5,7.0e+10^5),D(1.167e+10^6,1.4e+10^6),E(1.867e+10^6,2.8e+10^6),F(4.0e+10^6,5.6e+10^6),G(7.433e+10^6,1.12e+10^7), *CellSize*=12B. Processor is 4M shared L2-Cache Quad-Core CMP. As shown Fig 4, with data increasing, the advantage of model (a) is more obvious. When *Table Size*=E, because data distribution is even, resulting in little performance difference between the three models. Model (c) couldn't fully utilize computation resources of CMP, and its performance is greatly lesser than the other two models.

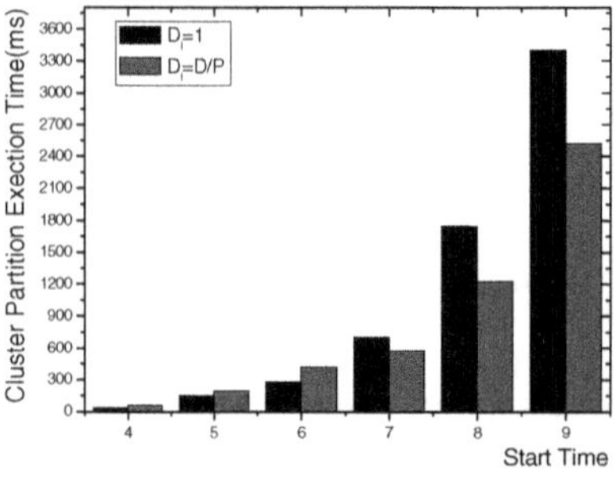

Fig. 3(a). Cluster partition performance

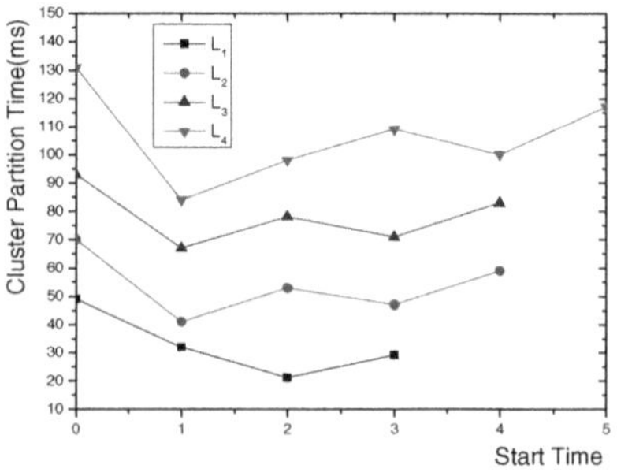

Fig. 3(b). Cluster partition performance

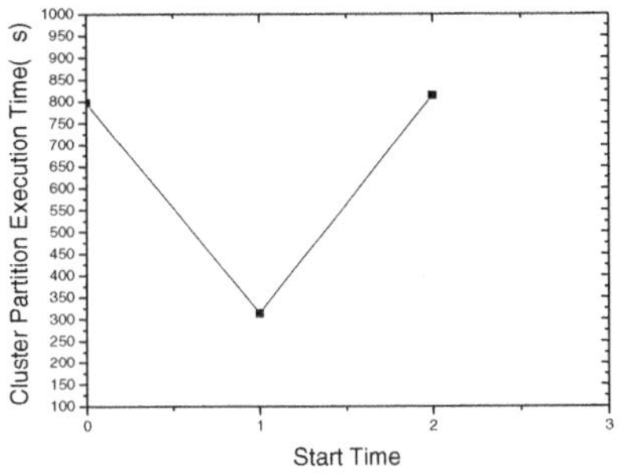

Fig. 3(c). Cluster partition performance

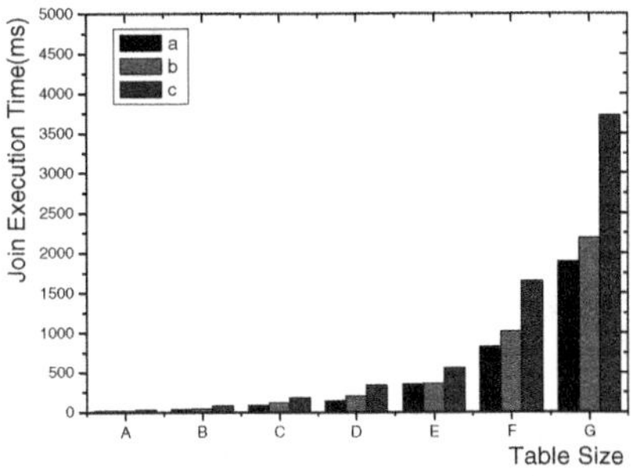

Fig. 4. Cluster join performance

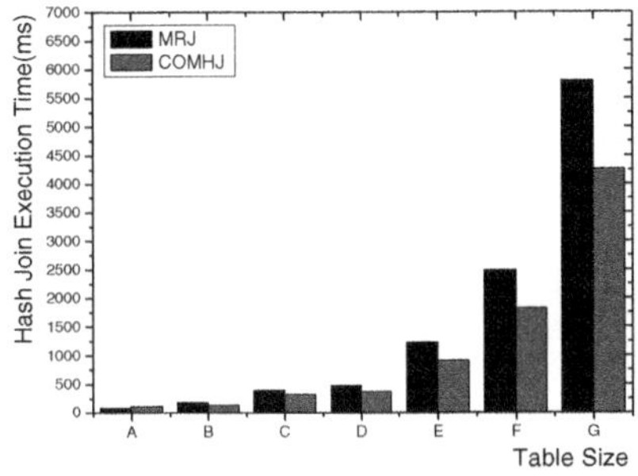

Fig. 5(a). Hash join performance

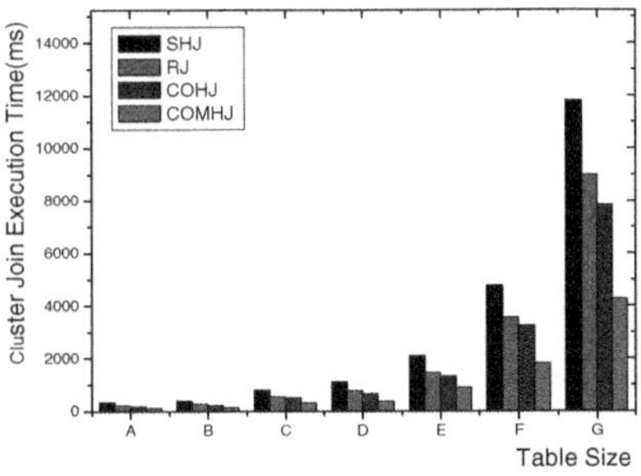

Fig. 5(b). Hash join performance

Experiment 4. To compare performance of various hash join algorithms, we tested some other hash join algorithms with our algorithm(COMHJ, cache optimized multi-threaded hash join), such as simple hash join(SHJ), Radix-Join(RJ)[2], multithreaded Radix-Join(MRJ), Cache-Oblivious hash join(COHJ)[8][10]. The experimental data and processor are the same with Experiment 3. In Fig 5(a), we compared performance of COMHJ with MRJ. If MRJ is just only optimized by multithread, and the influence of shared L2-Cache is not considered. The serious cache conflict would severely reduce performance. In Fig 5(b), we compared performance of COMHJ with other cache-optimized algorithm. In CMP, if hash join adopts single thread model, MRJ even has better performance than all of cache-optimized algorithm. Therefore, from the above analysis, we conclude that it is necessary to consider multithread and cache optimization in shared L2-Cache CMP.

Experiment 5. In this experiment, we investigated performance of our algorithm in Dual-Core and Quad-Core CMP. The L2-Cache capacity of Dual-Core CMP are 1M and 4M, and Quad-Core CMP is 4M. Firstly, we compared performance of COMHJ in the three kinds of CMP. Experiment data are the same with Experiment 3. Dual-Core with bigger L2-Cache not only starts cluster partition thread earlier, but also leading to bigger cluster size. Therefore, cluster partition time is decreased, and performance is improved as shown in Fig6. The Quad-Core CMP, comparing to 4M L2-Cache Dual-Core CMP, thread number which could be started to execute hash join is increased by one or two, and our optimization strategy is to start cluster partition threads as early as possible. Therefore, although the value of *N* is increased, but the cluster threads start time are not influenced basically. For example, when *L*.tuple= 1200000, if processor is Dual-Core CMP, the start time(*j*) of cluster partition thread is 2, and thread number is 2. When Quad-Core CMP, *j* is also 2, but three threads could be started. Its performance is better than Dual-Core CMP.

Experiment 6. Following, we would validate the result of theoretical analysis in Section 3.3. The processor is 4M shared L2-Cache Quad-Core CMP. L_1.tuple=670000, *CellSize1*=12B and L_2.tuple=408000, *CellSize2*=12B. As shown in Fig 7, if table is L_1, and *j*=2. According to the analysis in Section 3.2.2, two cluster partition threads could be started. If start time is delayed to *j*=3, and four threads could be started. Although *j*=3 could start more thread than *j*=2, but performance is reduced caused by more serial execution time. If table is L_2, and *j* also equals 2. Three threads could be started. If start time is delayed, the same with L_1, performance is decreased. Therefore, on the condition of few cache conflict, the sooner start time of cluster partition thread, the smaller serial execution time and better performance.

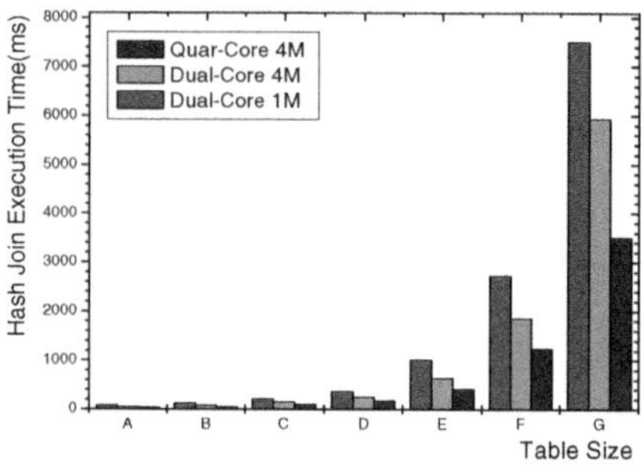

Fig. 6. Hash join performance

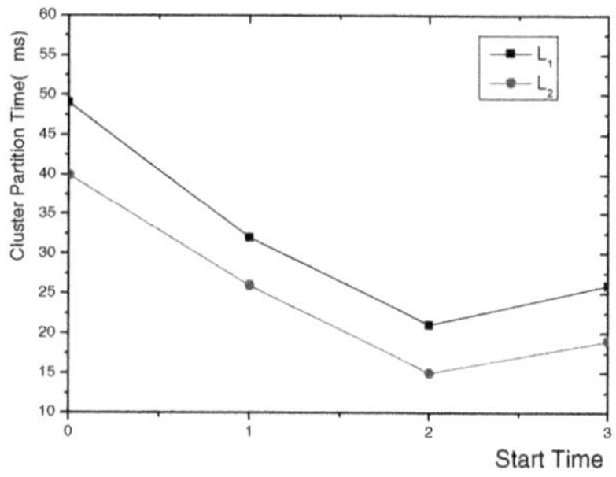

Fig. 7. Cluster partition performance

5 Conclusions and Future Work

Basing on Radix-Join algorithm, we presents multithreaded hash join execution framework to fully utilize CMP. According to the characters of share L2-Cache, we optimized various kind of threads in the framework, and the main contributions are: (1) presenting a new cluster partition algorithm based on multithread which could adopt appropriate cluster partition strategies according to data size at runtime, and cache access performance in cluster partition is improved; (2) basing on cluster classification, we present a new multithreaded cluster join execution strategy to achieve load balance between CMP cores and reduce cache conflict in cluster join execution. In the experiments, the multithreaded hash join execution framework is realized in EaseDB. The experiment results and analysis show that the framework in our paper has good performance of cache access, and it could fully utilize computing resources of CMP to improve performance. Its multithread and cache access optimization strategies could also be used in other database join optimization, such as nested loop join.

Acknowledgments. This research is supported by the National High-Tech Research and Development Plan of China(863) No.2007AA120400,No.2007AA12Z208,No 2006AA12Z205, NSFC China, No. 40801160.

References

[1] Hankins, R.A., Patel, J.M.: Effect of node size on the performance of cache-conscious B+-Tree. In: SIGMETRICS 2003 (2003)

[2] Boncz, P., Manegold, S., Kersten, M.L.: Database architecture optimized for the new bottleneck: Memory Access. In: Proceeding of the 25th VLDB conference, Edinburgh, Scotland (1999)

[3] Song, F., Moore, S., Dongarra, J.: L2 cache modeling for scientific application on chip Multi-Processors. In: Proceedings of the 2007 International Conference on Parallel Processing (2007)

[4] Chen, S., Gibbons, P.B., Kozuch, M.: Scheduling threads for constructive cache sharing on CMPs. In: Proceedings of the nineteenth annual ACM symposium on Parallel algorithms and architectures (2007)

[5] Cieslewicz, J., Ross, K.A., Giannakakis, I.: Parallel buffer for chip multiprocessors. In: DaMoN (2007)

[6] Cieslewicz, J., Ross, K.A.: Adaptive aggregation on chip multiprocessors. In: VLDB 2007 (2007)

[7] Hardavellas, N., Pandis, I., Johnson, R.: Database servers on chip multipr-ocessors limitations and opportunities. In: CIDR 2007 (2007)

[8] He, B., Luo, Q.: Cache-oblivious Database: Limitations and Opportunities. In: TODS 2008 (2008)

[9] Hennessy, J.L., Patterson, D.A.: Computer Architecture, 4th edn. (2007)

[10] He, B., Luo, Q.: Cache-Oblivious hash joins, HKU technical report (2006)

[11] Chen, S., Ailamaki, A., Gibbons, P.B.: Improving Hash Join Performance Through Prefetching. In: ICDE (2004)

[12] Cieslewicz, J., Berry, J., Hendrickson, B., Ross, K.A.: Realizing parallelism in database operations: Insights from a massively multithreaded architecture. In: DaMoN (2006)

[13] Zhou, J., Cieslewicz, J., Ross, K.A., Shah, M.: Improving database performance on simultaneous multithreaded processors. In: VLDB 2005 (2005)

[14] Garcia, P., Korth, h.F.: Pipelined Hash-Join on multithreaded architectures. In: DaMoN (2007)

[15] Garcia, P., Korth, H.F.: Database hash-join algorithms on multithreaded computer architectures. In: Proceedings of the 3rd conference on Computing frontiers, Ischia, Italy, pp. 241–252 (2006)

Traverse: Simplified Indexing on Large Map-Reduce-Merge Clusters

Hung-chih Yang and D. Stott Parker*

UCLA Computer Science Department
hcyang@cs.ucla.edu, stott@cs.ucla.edu

Abstract. The search engines that index the World Wide Web today use access methods based primarily on scanning, sorting, hashing, and partitioning (SSHP) techniques. The MapReduce framework is a distinguished example. Unlike DBMS, this search engine infrastructure provides few general tools for indexing user datasets. In particular, it does not include order-preserving tree indexes, even though they might have been built using such indexing components. Thus, data processing on these infrastructures is linearly scalable at best, while index-based techniques can be logarithmically scalable. DBMS have been using indexes to improve performance, especially on low-selectivity queries and joins. Therefore, it is natural to incorporate indexing into search-engine infrastructure.

Recently, we proposed an extension of MapReduce called *Map-Reduce-Merge* to efficiently join heterogeneous datasets and executes relational algebra operations. Its vision was to extend search engine infrastructure so as to permit generic relational operations, expanding the scope of analysis of search engine content.

In this paper we advocate incorporating yet another database primitive, indexing, into search engine data processing. We explore ways to build tree indexes using Hadoop MapReduce. We also incorporate a new primitive, *Traverse*, into the Map-Reduce-Merge framework. It can efficiently traverse index files, select data partitions, and limit the number of input partitions for a follow-up step of map, reduce, or merge.

1 Introduction

Search engines index the World Wide Web! As the WWW is still growing at an amazing speed and already contains a significant portion of all digitized information ever published, its (partial) indexes have grown extremely large. To build an index for the WWW, search engines usually employ clusters of thousands of nodes and operate a series of massive parallel data-processing tasks. An architecture of these search engine clusters is shown in Fig. 1. Intriguingly, DBMS are rarely deployed in major search engine subsystems [3].

To execute data-processing tasks on these clusters, search engines utilize generic parallel programming infrastructures such as MapReduce [5,13], a simple

* Work supported by NIH grants 1U54RR021813 and 1 RL1 LM009833-01.

X. Zhou et al. (Eds.): DASFAA 2009, LNCS 5463, pp. 308–322, 2009.

programming model that enables parallel data processing on large clusters built with cheap commodity hardware. This programming model is easy to use, very scalable, fault tolerant, and cost effective, so it has been proposed [13] to use it to build parallel data warehousing systems. On the other hand, search engine data processing can enhance its analytic power by applying database principles (such as the relational data model) in its operations [13]. Based on these two ideas, Map-Reduce-Merge [13] was proposed to efficiently join heterogeneous datasets on top of MapReduce.

Joining heterogeneous datasets has become important in search engine operations, as analyzing search engine algorithms and evaluating subsystems usually requires aggregating and comparing data from multiple sources. Fig. 1 depicts such an analytic cluster that takes data from every possible source through ETL (extract, transform, load) processes. Running these analytic tasks is difficult: not only is the data volume huge, but also the fundamental access method in search engines is based on hashing (not on hash indexing). Hashing is very effective in maintaining balanced loads on a large cluster, and it can efficiently aggregate data on *unique* keys. Since it disseminates data randomly over a cluster, processing and aggregating *related* data can be expensive.

For example, a search log dataset can contain URLs and some metrics. In which node a URL entry is stored in a cluster is usually determined by hashing the URL key. Therefore, aggregating metrics of the URLs that belong to a small host may require scanning the whole cluster! This scan, of course, can be easily

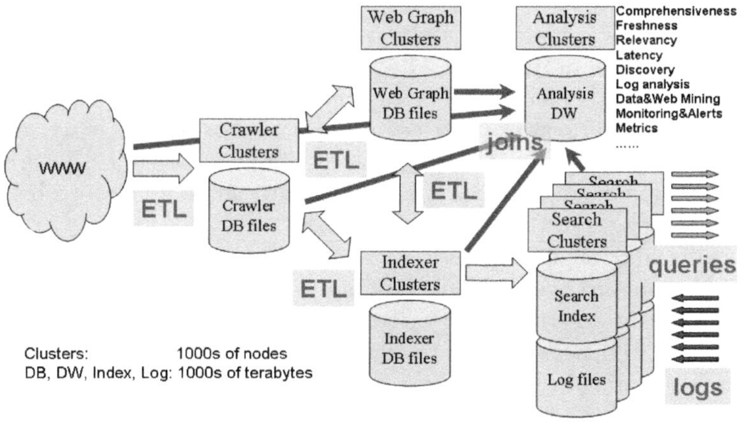

Fig. 1. A typical search engine architecture. Crawler clusters traverse the WWW and collect web pages into crawler databases. Webgraph clusters run link analysis, compute webpage ranks, and build webgraph databases. Indexer clusters build inverted indexes that are deployed into front-end search clusters to serve user queries. Search clusters also collect user query logs for advertisement, billing, and data mining purposes. WWW pages, crawler, webgraph, indexer databases, and query logs can be extracted, transformed, and loaded into analytic clusters for evaluating search engine algorithms, implementations, configurations, etc.

done in parallel with MapReduce. However, for a low selectivity task like this, the wisdom of database research prefers an index scan, rather than a full table scan. An index can be built using tree, hash, or array structures. In the aforementioned example, URLs from the same host can be clustered in a totally-ordered tree index, and then aggregated with a straightforward index scan.

Beyond providing a performance boost, index scans can reduce disk accesses, and thus extend hard-drive lifetime. Search engines usually utilize cheap, thus relatively unreliable, commodity components to build their clusters. At any point in time, for a large cluster of thousands of nodes, a certain percentage of these nodes will be down or have crashed, this being attributed mainly to disk failure. One way to reduce this failure rate is to lessen the load on hard drives. Using indexes can help reduce this load.

In this paper, we propose extending Map-Reduce-Merge to include some indexing capabilities. This idea proposes not only an alternative access method (besides hashing) for search engines, but also proposes the step of applying search engine technology in building generic, parallel data warehousing systems [13].

This paper starts by reviewing various access methods in search engines and databases in Sec. 1. In Sec. 2, we describe how to build indexes in map and reduce functions. We also describe how Map-Reduce-Merge is further enhanced by adding a *Traverse* primitive. We then use Hadoop to make the case of how indexes can improve performance for low-selectivity *ad hoc* queries in Sec. 3.

1.1 Access Methods in Search Engine Clusters

The core of a keyword-based search engine is an inverted index. This inverted index is built with data from several search engine subsystems deployed on large clusters of thousands of nodes. For disseminating data evenly on these nodes, hashing is a straightforward and economical approach. It is used extensively in cluster data processing, including search engine subsystems. In crawler and webgraph databases, a URL hash is used as key, while a keyword hash is used to look for webpages in these inverted indexes.

Though hashing is efficient to locate a specific webpage (as well as keyword, host, site, or domain) over thousands of nodes, it cannot help much in retrieving URLs that satisfy a certain querying condition. For example, a webserver can contain several webpages. To locate all the webpages belonging to a small server may require scanning every node in a webgraph cluster, because these webpages are placed all over the cluster by their URL hashes. This *webpage aggregation* problem can be easily solved by using a straightforward MapReduce task to scan the whole dataset. Though this scan is usually performed in parallel, true parallelism can be achieved only if there are enough processors to handle all data partitions; otherwise MapReduce tasks process data partitions in *waves*.

Hashing is sufficient to support search engine core operations – crawling, executing link analysis, and building inverted indexes. However, as we move to utilize search engine clusters as data warehouses for decision-support tasks, we hit limitations of hashing for linking related data entries. This problem applies to hashing-based parallel DBMS as well. One major goal of this research is to explore alternative access methods in hashing-dominated infrastructures.

1.2 Indexing in DBMS

Access methods (i.e., indexes) are essential in any data management or processing system, including database management systems (DBMS) and search engines. Where a DBMS builds indexes on structured datasets, a search engine builds inverted indexes over semi-structured webpages. A variety of indexes have been invented over the years, including tree indexes (B-Tree [2,4], R-Tree [8], GiST [9]), hashing, extendible hashing[6], bitmap [11], or join indexes, etc. Search engines can incorporate this trove of indexing structures and use them in analytic tasks. As we have indicated, indexing is an essential part of DBMS, and it can be a natural part of DBMS-extended search engine infrastructure.

For simplicity "partition", "part", "chunk", and "block" are used interchangeably in this paper. They were defined in different contexts; we use them to describe a block of disk space used to store a collection of records stored persistently in DFS (distributed file system).

1.3 Scanning, Sorting, Hashing vs. Indexing

Graefe pointed out in [7] that "index navigation plans scale truly gracefully, i.e., perform equally well on large and on very large databases, whereas scanning, sorting, and hashing scale at best linearly." This statement also applies to scalable parallel data processing systems, such as the ones used in search engines.

MapReduce systems, like MapReduce [5] and Hadoop [1], as well as extensions like Map-Reduce-Merge [13], provide a generic but fixed data-processing framework of scanning, hashing, partitioning, shuffling, sorting, and grouping. They have not taken advantage of the "graceful scalability" of indexing, because most search engine core operations need only batch scanning and grouping of webpages in order to build inverted indexes. However, decision-support operations on individual webpages, hosts, or query terms have become important in improving system performance and search relevancy. Sustained scanning of an entire cluster for these low-selectivity tasks is a waste of time and resources, not to mention that it can incur unnecessary hardware attrition.

These decision-support tasks can be executed in a so-called "research" analytic cluster. Webgraph databases can be viewed as such an analytic cluster implemented in search engines as it is used to do link analysis and evaluate the quality of webpages and algorithms. However, as this evaluation has become more *ad hoc*, we propose adopting yet another database experience in search engines: building an online analytic processing (OLAP) database, in which indexes are essential.

1.4 Search Engine Analytic Cluster

An analytic cluster is the OLAP term for a search engine. It is loaded with data extracted and transformed from various sources (see Fig. 1). Its mission is typically in research and decision support. Users can execute exploratory data analysis (EDA) tasks by running scripts in simple query languages like Pig [10] or Sawzall [12]. These EDA operations can be complicated or *ad hoc*, so it is inefficient to code them directly in MapReduce primitives. They usually require joining datasets, while MapReduce does not directly support relational joins [13]. Some analytic processing can benefit from storing data in structures like trees or nested tables [10], but MapReduce tasks iterate through datasets as lists. Thus, users need to emulate these sophisticated structures in lists.

Some examples of search-engine analytic challenges that can benefit from data structures beyond lists include:

1. **Finding all the URLs that fit a certain criterion**. As we have indicated, finding all URLs belonging to a small host can incur a full scan. Problems like this can be solved by building search trees similar to the B-Tree.
2. **Finding semi-duplicated URLs**. Marking duplicated webpages is an important task in search engines. Duplicated pages can take up a large chunk of space in subsystems. Also, returning URLs with highly similar content will not impress search engine users. Building a similarity tree can organize URLs based on similarity distances, and thus help in selecting pages with unique content.
3. **Joining two search-engine databases on a low-selectivity condition**. Indexes can help point- or range-queries on a single data source. They can also facilitate join operations over two data sources. For example, a Map-Reduce-Merge sort-merge join operator can scan indexes, which are sorted versions of the source datasets.

To build a search engine analytic system, engineers can elect to (a) install a DBMS-based data warehouse and analytic software, or (b) improve existing search engine infrastructures to satisfy these new requirements. Traditional warehousing systems are expensive and they may not be as scalable as search engine infrastructures due to overhead (e.g., transaction management) [3].

Therefore, in this paper we focus on the second approach — building and traversing tree structures on top of the MapReduce architecture. We use Hadoop for implementation.

1.5 Adding Indexing to MapReduce

It is well-known that choosing access strategies between indexing or scanning usually depends on selectivity, while there are many forms of index structures to choose from: tree, hash, or array, etc. DBMS usually provide a gamut of these access methods for a variety of requirements. On the other hand, search engines rely mostly on hashing as the primary document locating mechanism, although it can be deficient for analytic data processing tasks.

Since indexing is an important part of an analytic data processing system, in this paper we discuss how to extend search-engine infrastructures (e.g., MapReduce) to support it. There are many challenges for this transformation as many

MapReduce assumptions do not fit well with index building: large block sizes, read-only/append-oriented, iterative primitives, and batch processing, etc. Even so, from our benchmarks of index building on Hadoop MapReduce clusters, we found that indexing can still save a lot. In short, our focus is: (a) using map, reduce, or merge primitives to build indexes, and (b) introducing a new primitive called *Traverse* to facilitate a follow-up map, reduce, or merge step.

2 Traverse: Indexing in Map-Reduce-Merge

Index building is a process of organizing data into traversable structures, such as trees. Sorting can play an important role when building indexes in a bulk process. Since MapReduce is an effective parallel sorter [5], it is a natural tool to build database indexes, while it was originally designed for building inverted search-engine indexes.

2.1 Index as a Collection of Layered Distributed Files

The most popular tree index used in databases is the B-Tree and its variants. To build a B-Tree-like search structure in MapReduce, we can follow the steps shown in Fig. 2. Each layer of this search structure is a collection of partitions grouped into a distributed file. Each partition works like a B-Tree node that stores keys and pointers that point to partitions of a lower-level file.

A leaf-level index builder is simply a MapReduce sorter. It sorts a file over a user-defined key attribute and produces partition pointers as values. These pointers can simply be the names of the source file partitions. An internal-level index builder is also a MapReduce job that scans a sorted leaf-level index file, collects the boundary keys (the largest and smallest) of each leaf partition, and binds them into a key-range record. The same step of index building can be repeated until the root index file has only a small number of partitions. Usually one internal layer of indexing is sufficient, because MapReduce and Hadoop usually hold data in 64MB chunks or 128MB parts, and only around 1 million partitions are sufficient for a hundreds-of-terabyte DFS file. A small number of partitions can easily hold indexing information for 1 million partitions.

Like the B-Tree, the search tree we have described is one-dimensional. However the same principle can be applied in building tree indexes like the R-Tree for multi-dimensional datasets. In fact, besides keys and pointers, index files can even store aggregates associated with sub-trees [14].

2.2 Primitive Signatures

Once layers of index files are created as described in Sec. 2.1, users can write programs to traverse these layers of partitions. Since indexing is a common need, we propose to introduce a new primitive, *Traverse*, that works alongside Map, Reduce, and Merge. From [13], the signatures of map, reduce, and merge are reproduced here:

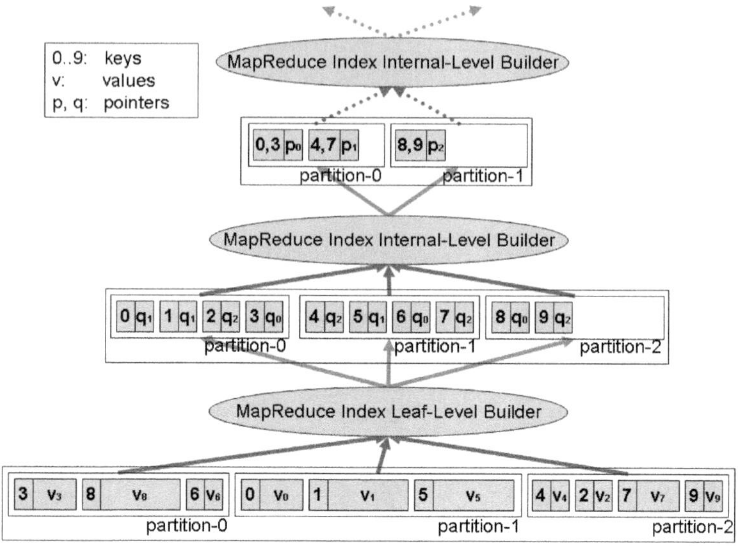

Fig. 2. Building Index Files Using MapReduce. At the bottom is an unsorted data file with keys from 0 to 9. From this data file, the first MapReduce index builder extracts keys and their partition information as values, sorts them, and produces a leaf-level index file shown in the middle of the figure. Over the sorted leaf-level index file, an index-building MapReduce job then scans over all partitions in parallel, finds the smallest and largest keys from each partition, and builds an internal-level index file, shown at the top of the figure. This process can be repeated until the root index file is small enough that one MapReduce job with a small number of nodes can scan it fully in parallel in one wave. Partitions of these internal- and leaf-level index files form a search forest of internal and leaf nodes.

$$\text{map: } (k_1, v_1)_\alpha \rightarrow [(k_2, v_2)]_\alpha$$
$$\text{reduce: } (k_2, [v_2])_\alpha \rightarrow (k_2, [v_3])_\alpha$$
$$\text{merge: } ((k_2, [v_3])_\alpha, (k_3, [v_4])_\beta) \rightarrow [(k_4, v_5)]_\gamma$$

These primitives scan DFS files (i.e., they do *table scans*), so they are mostly efficient for high selectivity jobs. For low selectivity tasks, using index files can reduce disk accesses and decrease the number of waves of operations for a large dataset stored in a small cluster. To incorporate *index scans* with these table-scan-oriented primitives, each can be paired up with a new primitive called *Traverse*. Below is the signature of the Traverse primitive:

$$\text{traverse: } (p, [ptr]) \rightarrow [ptr]$$

In this signature, p represents *predicates*, and ptr represents *pointers*. In one-dimensional search trees (e.g., the B-Tree) p is just the indexing key. In an R-Tree, p is a multi-dimensional key tuple (minimum bounding rectangle – MBR) that is used as a predicate to determine whether a key is inside the MBR or not. If it is, then this key might be stored in the sub-tree pointed by the pointer associated with this MBR predicate. In some indexes, a predicate can be associated with multiple pointers, so we have designed the signature with a collection of pointers.

The design of the Traverse signature follows the original reduce primitive defined in [5], which is reproduced here:

$$\text{map: } (k_1, v_1) \rightarrow [(k_2, v_2)]$$
$$\text{reduce: } (k_2, [v_2]) \rightarrow [v_3]$$

The value parameters (v_2 and v_3) are replaced by *ptr*. This design suggests that a traverse can be implemented as a *parallel* MapReduce job over an index file. Users can define traversal logic by applying a filter on keys and emitting pointers. A follow-up *Traverser* job takes the pointers emitted, scans, then filters the index partitions they point at. This traversal process ends when it finishes filtering leaf index partitions. File information of selected data partitions is then handed over to map, reduce, or merge tasks.

In implementing a traverse function, user-defined logic usually emits pointers passed in as arguments. Users can be creative by emitting pointers *generated* on the fly based on these arguments. However, these generated pointers can be invalid, and the system can either ignore them or throw an exception to warn users about a logic error.

An indexing predicate can be associated with more than one pointer for a sequence of traverses. An index file can also contain no key but a collection of pointers in each index entry. Some indexes can rely on external information, such as a key density estimate (or histogram), to determine how to traverse an index hierarchy. This kind of index can achieve maximal fan-out, because all space is used for storing index pointers.

2.3 Parallel Layered Traversal

In databases, the B-Tree is a very popular access method. A common access pattern is to traverse from the root down to a leaf node, and do a scan at the leaf level for a point or range query. For database structures like the B-Tree, complexity is measured by the number of disk blocks accessed. Thus each B-Tree node usually takes up one disk block of space. However, as alluded to in Sec. 2.2, in the parallel MapReduce environment, each index level can be visited by a collection of concurrent mappers, so it makes sense to allocate several blocks (file partitions) at the root level. Each layer of an index tree can be stored as a DFS file. A MapReduce job can scan an index DFS file at the root level, and collect key-pointer pairs that pass a user-defined predicate. These pointers point to a subset of nodes in the next layer of the index tree, and collectively these nodes can form another DFS file, which can be processed by a subsequent MapReduce job to advance the parallel traversal procedure until reaching data partitions.

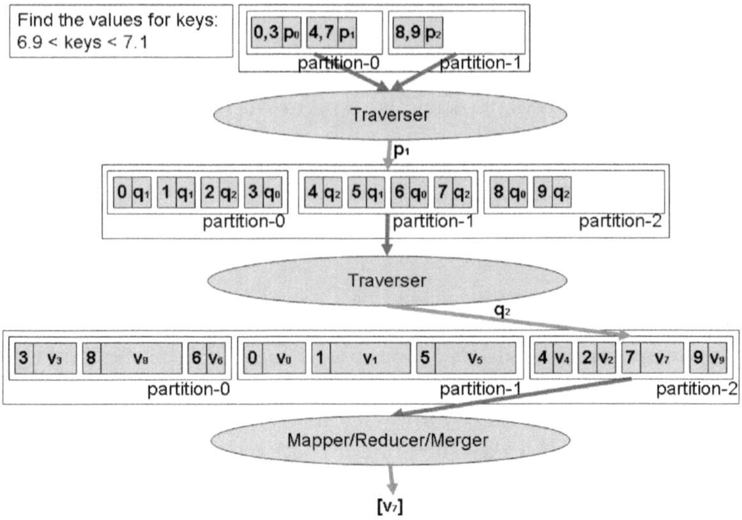

Fig. 3. Traversing Index Files Using MapReduce. This example is to find the values for keys that are greater than 6.9 and less than 7.1. From the top, a traverser function is applied to every partitions of an internal-level (root) index file. The output has one pointer, p_1, and it is passed to the next traversal iteration. The output has one pointer (can be with offset), q_2. It points to the data partition that hold the records that satisfy the querying condition.

3 Examples of Indexing Using Hadoop MapReduce

It is never trite to emphasize the advantage of using indexes in low selectivity queries. Hadoop and MapReduce are good at batch data processing, but their lack of infrastructural support for indexing makes them unsuitable for low-selectivity *ad hoc* queries. This motivates us to build index structures in Hadoop. In this section, we demonstrate examples of building and querying Hadoop index files. We also demonstrate that this indexing approach can be superior to the brute-force approach of scanning the whole dataset, in terms of footprint and even performance. Scan is *the approach of design* in the Hadoop and MapReduce infrastructures, thus traverse may not be an intuitive strategy. We implemented these examples based on the indexing mechanism described in Sec. 2.1 and Sec. 2.3, while we used the Hadoop system "as is" without any modifications. The first example is a point query to

$$\textit{look up a URL from a datatset} \qquad (1)$$

This example can actually be implemented using a hash-partitioning scheme without the help of an extra indexing structure. Hash-partitioning is convenient, because its mechanism is already present in the MapReduce architecture. However, a complete infrastructural support to implement this scheme still requires managing hash-partitioning meta-data in order to map keys to source data locations. This meta-data is equivalent to a hash index. An alternative is to build a search tree index over URLs as keys. Details of implementing this point query are described in Sec. 3.4. The second example is a range query to

$$\textit{locate all the URLs that belong to a small host} \quad (2)$$

This query can reuse the URL-based index file built for the first example, as long as the index's URL ordering function compares the URL's host component first. See Sec. 3.2 about ordering URLs. Notice that, unlike the first example, this range query cannot be resolved by simply hashing and partitioning, as a host's URLs are not clustered — their locations are determined by hashing over the whole URL as key and can be disseminated anywhere in a cluster. This query needs a tree index in order to avoid a comprehensive scan. Details of implementing this query are described in Sec. 3.5.

3.1 Experiment Setup

Our test dataset is a log file in which each record has a URL as key and several log/page quality metrics as attributes. We randomly choose a URL and use it in an equality predicate for Example 1. The host part of this URL is then used as key for Example 2. For each example, we implement and compare the performance of two pipelines of Hadoop MapReduce look-up processes: one takes a *comprehensive scan* over the source dataset, and the other executes an *index scan.* Steps of these two approaches are described below:

- **Comprehensive Scan:**
  This is a straightforward MapReduce approach. Mappers check keys against a querying predicate, and only emit them to a sole reducer if they pass the predicate. The reducer aggregates intermediate records based on the URL key. If nothing is passed from the mappers, then the querying URL does not exist in the source dataset.
- **Index Scan:**
  A more sophisticated scheme is to scan a pre-built index file, extract source partitions that satisfy the querying predicate, then scan only these partitions. Hadoop MapReduce does not have an infrastructure to directly support this approach, so users need to build one to get it done. It involves building an index and scanning a small number of index and data partitions:
  - Sort the source dataset using a user-defined key comparator. See Sec. 3.2 for details of ordering URLs. Notice that, before sorting, we may want to build a histogram on the source dataset in order to help the sorting MapReduce job to balance its load among tasks. This histogram building procedure is very important for load balancing, but requires extra MapReduce steps, and this goes beyond the scope of this paper.
  - Build a small index file by picking up records with the smallest and largest *keys* (the first and the last records) from each sorted data partition. The filename of the data partitions are emitted as *values.* The resulting index is a Hadoop DFS file with (<*start-key*, *end-key*>, *filename*) pairs as records.

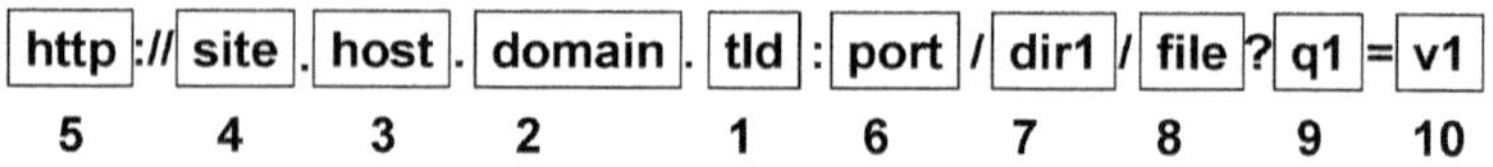

Fig. 4. Ordering URLs by Its Components. The component with the largest coverage, TLD, is compared first, then domain, host, site, and protocol, etc. The numbers in the figure indicate such a comparison sequence.

- Once an index file is built, it can be reused by many queries. Follow-up index building processes can build higher-level and subsequently smaller index files over an existing one. At the end, we get a series of index files whose records form a hierarchy of search keys and filenames as pointers. In our examples, one layer of index files is sufficient.
- A MapReduce querying task can then scan and traverse index files layer by layer to filter the candidate index/data partitions and eventually find in which sorted data partitions the querying key could be located. In this querying MapReduce job, only (*filenames*) are passed from mappers to reducers. Reducers group these candidate index/data partition filenames, scan them, and check out every key predicates. This process can be recursive if there are multiple layers of index files.

3.2 Ordering URLs

Inserting URLs into a search tree requires a comparison function. Search engines usually do not need a URL comparator, because they often rely on hashing to locate a URL from a data cluster. Unlike numbers that have a natural total order and strings that can be ordered lexicographically, there is no natural order on URLs. Therefore, we need to artificially define a URL comparator.

URLs can be considered as a compound key formed from their components — protocol, host, port, path, file, and arguments, etc. Therefore, the problem of comparing two URLs can be decomposed into comparing their components in a certain sequence. Fig. 4 shows one such sequence. It compares the component with the largest coverage, TLD, first, then proceeds to ones with smaller coverage. The benefit of this comparison sequence is that it can cluster URLs in the same domain, host, or site together. Thus, an index for locating a unique URL can also be used for locating URLs in the same domain, host, etc.

People may also be interested in querying URLs based on file formats or file extensions. To build an index for this purpose, URLs can be clustered based on their file extensions. This requires a different URL ordering strategy. Since most files are in html format, an index file can exclude html entries and focus on indexing entries with non-html file extensions. For html queries, an optimizer can apply a comprehensive scan on the source dataset. For other file extensions, it can apply an index scan on a html-excluded index file.

3.3 Building URL Index Files

The overall indexing scheme was described in Sec. 3.1. In our experiments, we use a data file with 128 data partitions, of which each is about 16MB. Altogether, there are 31 million records and the total size is about 2GB. We then build a histogram on the source dataset, use the histogram to sort it, and create an index file.

In this example, the leaf-level index file is small, so there is no need to create a higher internal-level index file. The index scan process is configured with one mapper and one reducer. Only 3 nodes (the least number of nodes that need to be allocated for a virtual Hadoop cluster) are allocated for this index scan process. The sole mapper scans this index file and emits the data partition filenames that pass the querying predicate. The reducer then groups and scans these candidate data partition files and emits qualified records.

3.4 Point Query

Example 1 is to look up a URL from a log dataset. A comprehensive scan is executed several times and configured with a variety of cluster sizes: 3, 4, 5, 6, 7, 8, 9, 10, 12, 14, 16, 18, 20, 25, 30, 40, 50, 60, and 70. We also ran one index scan job, but used only the least number of nodes (3) required to start up a Hadoop job.

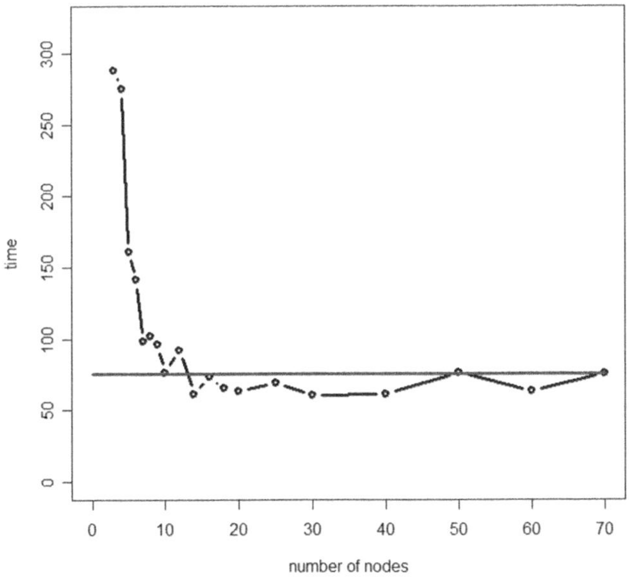

Fig. 5. Point Query to Look Up a URL from a Search Log. The blue line represents the times spent by a series of comprehensive scans with an increasing number of nodes. The red line represents the time spent by an index scan with the minimum number nodes required (3) to allocate a cluster.

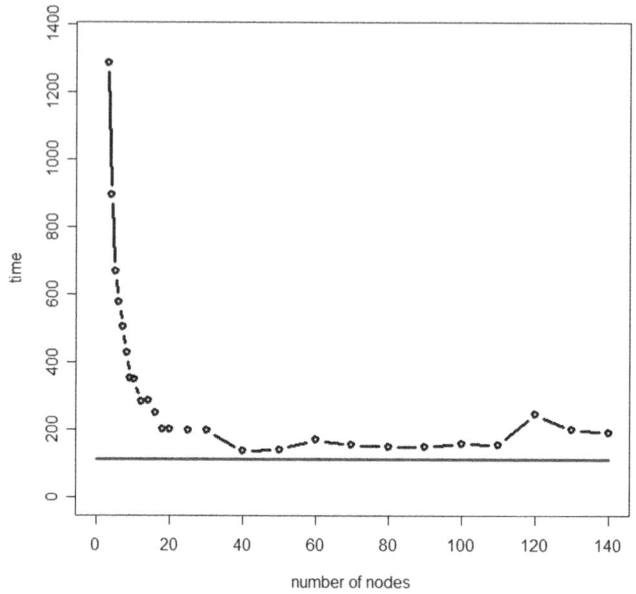

Fig. 6. Range Query to Look up a Small Host's URLs from a Search Log. The blue line represents a series of comprehensive scans with an increasing number of nodes. The red line represents an index scan with the minimum number of nodes required (3) to allocate a cluster.

Our benchmarks ignore the initiation time for a Hadoop job to locate input datasets and to initialize participating nodes. For a small scale test, this time is constant. Initiation involves communication between the Hadoop job tracker and name node, it can take longer for a comprehensive scan as it has more partitions. The overall result is shown in Fig. 5. It is not a surprise that the index scan gains an upper hand in terms of resource usage and time spent. The speedup of allocating more nodes for the comprehensive scans stops at around 14 nodes. The index scan is competitive even when compared with the best performance of the comprehensive scans. For 128 partitions and 70 nodes, the comprehensive scan can take up to two waves of scanning, while the index scan reads one index partition in the map phase, and one data partition in the reduce phase.

The index scan also has a smaller scanning footprint. The comprehensive process scans 128 data partitions and some number of intermediate files, while the index scan accesses one index partition and only one data partition. For both cases, the intermediate files passed from mappers to reducers are negligible in size.

Notice that the index-scan task executes a sequential scan on data partitions in the reducer. This can be further optimized, since the data partition is sorted already. In other words, this index scan could have been more efficient if we have had implemented an advanced search algorithm beyond linear scan. However, as MapReduce is limited to linear iteration, infrastructure changes might be needed.

3.5 Range Query

As aforementioned, we can also utilize the URL index file built for the point query example to locate all the URLs that belong to a small host. The result of an index scan and a series of comprehensive scans is shown in Fig. 6. The index scan beats every comprehensive scan, no matter how many nodes are allocated. It takes 40 nodes for the comprehensive scan to reach the highest performance. Beyond 40, allocating more nodes does not help in improving performance, and the overhead of coordinating and shuffling data can cause performance to deteriorate.

4 Conclusions

Search engine data-processing infrastructures rely mainly on scanning, sorting, hashing, and partitioning techniques to maintain load balance in large clusters. These techniques are sufficient to support search engine core operations, such as link analysis, crawling, or building inverted indexes. However, hashing cannot help range-based *ad hoc* queries, and scanning is linearly scalable at best. Sustained scanning of whole datasets can hasten disk failure as well. To avoid these issues, we can apply yet another database primitive to search engines: indexing. Indexing can improve performance especially on *ad hoc* low-selectivity queries.

In this paper, we first explore building search tree indexes in bulk MapReduce operations, as MapReduce is a very effective sorter, and sorting is intimately related to building search trees. We then propose improving the Map-Reduce-Merge framework by adding a new primitive called *Traverse*. It can process index file entries recursively, select data partitions based on query conditions, and feed only selected partitions to other primitives. Users can define logic to determine the optimal way of selecting partitions from data files, and effectively refine index access algorithms. This new primitive adds parallel indexing to the MapReduce repertoire in addition to scanning, sorting, hashing, and partitioning.

It is common knowledge that using indexes can boost performance for low selectivity queries. For load balancing, search engines usually employ hashing in their infrastructure. Indexing can help extend this infrastructure for decision-support data warehouses, as decision-support capability has become extremely important for optimizing search engine operations.

References

1. Apache. Hadoop (2006), `http://lucene.apache.org/hadoop/`
2. Bayer, R., McCreight, E.M.: Organization and Maintenance of Large Ordered Indices. Acta Inf. 1(3), 173–189 (1972)
3. Brewer, E.A.: Combining Systems and Databases: A Search Engine Retrospective. In: Hellerstein, J.M., Stonebraker, M. (eds.) Readings in Database Systems, 4th edn. MIT Press, Cambridge (2005)
4. Comer, D.: The Ubiquitous B-tree. Comp. Surveys 11(2), 121–137 (1979)
5. Dean, J., Ghemawat, S.: MapReduce: Simplified Data Processing on Large Clusters.. In: OSDI, pp. 137–150 (2004)

6. Fagin, R., et al.: Extendible Hashing - A Fast Access Method for Dynamic Files. TODS 4(3), 315–344 (1979)
7. Graefe, G.: B-tree Indexes, Interpolation Search, and Skew. In: Ailamaki, A., Boncz, P.A., Manegold, S. (eds.) DaMoN, p. 5. ACM, New York (2006)
8. Guttman, A.: R-Trees: A Dynamic Index Structure for Spatial Searching. In: Yormark, B. (ed.) SIGMOD 1984, pp. 47–57. ACM Press, New York (1984)
9. Hellerstein, J.M., et al.: Generalized Search Trees for Database Systems. In: VLDB 1995, pp. 562–573. Morgan Kaufmann, San Francisco (1995)
10. Olston, C., Reed, B., Srivastava, U., Kumar, R., Tomkins, A.: Pig Latin: a Not-So-Foreign Language for Data Processing. In: SIGMOD 2008, pp. 1099–1110. ACM, New York (2008)
11. O'Neil, P.E., Graefe, G.: Multi-Table Joins Through Bitmapped Join Indices. SIGMOD Record 24(3), 8–11 (1995)
12. Pike, R., Dorward, S., Griesemer, R., Quinlan, S.: Interpreting the Data: Parallel Analysis with Sawzall. Scientific Programming Journal 13(4), 227–298 (2005)
13. Yang, H.-C., Dasdan, A., Hsiao, R.-L., Parker Jr., D.S.: Map-Reduce-Merge: Simplified Relational Data Processing on Large Clusters. In: Chan, C.Y., Ooi, B.C., Zhou, A. (eds.) SIGMOD Conference, pp. 1029–1040. ACM, New York (2007)
14. Yang, H.-C., Parker Jr., D.S., Hsiao, R.-L.: The Holodex: Integrating Summarization with the Index. In: SSDBM, pp. 23–32. IEEE Computer Society, Los Alamitos (2006)

"Pay-as-You-Go" Processing for Tracing Queries in a P2P Record Exchange System

Fengrong Li[1], Takuya Iida[1], and Yoshiharu Ishikawa[2]

[1] Graduate School of Information Science, Nagoya University
[2] Information Technology Center, Nagoya University
Furo-cho, Chikusa-ku, Nagoya 464-8601, Japan
{lifr,iida}@db.itc.nagoya-u.ac.jp, ishikawa@itc.nagoya-u.ac.jp

Abstract. In recent years, data provenance or lineage tracing has become an acute issue in the database research. Our target is the data provenance issue in peer-to-peer (P2P) networks where duplicates and modifications of data occur independently in autonomous peers. To ensure reliability among the exchanged data in P2P networks, we have proposed a reliable record exchange framework with tracing facilities based on database technologies in [5,6]. The framework is based on the "pay-as-you-go" approach in which the system maintains the minimum amount of information for tracing with low maintenance cost and a user pays the cost when he or she issues a tracing query to the system. This paper focuses on its two alternative query processing strategies and compare their characteristics according to the performance.

1 Traceable P2P Record Exchange System

Data provenance is a facility that helps database users to interpret database contents and enhances the reliability of data [2,3]. We focus on the data provenance issue in information exchanges in *peer-to-peer (P2P)* networks. Although there exist many P2P systems and related proposals, they do not support the notion of data provenance. Based on this background, we proposed the concept of a *traceable P2P record exchange system* in [5,6], in which tuple-structured *records* are exchanged in a P2P network. The system employs "pay-as-you-go" approach [4] for tracing. In this paper, we show two alternative query processing strategies and compare their properties.

As an example, assume that information about novels are shared among peers in a P2P network and that each peer maintains a `Novel` record set that has two attributes `title` and `author`. Each peer maintains its own records and keeps its historical data related to itself. To make the tracing process easy, the system provides an abstraction layer which virtually integrates all the distributed relations and a datalog-like query language for writing tracing queries in an intuitive manner. For the details, please refer to [5,6]. In the following, we present some examples of tracing queries.

X. Zhou et al. (Eds.): DASFAA 2009, LNCS 5463, pp. 323–327, 2009.

Query 1: Suppose that peer A holds a record with title `t1` and author `a1` and that peer A wants to know which peer originally created this record:

```
BReach(P, I1) :- Data[Novel]('t1', 'a1', 'A', I2),
                 Exchange[Novel](P, 'A', I1, I2, _)
BReach(P1, I1) :- BReach(P2, I2), Exchange[Novel](P1, P2, I1, I2, _)
Origin(P) :- BReach(P, I), NOT Exchange[Novel](_, P, _, I)
Query(P) :- Origin(P)
```

Relation `Data[Novel]` is used to represent the user-level record set `Novel` in the underlying system level (it is called the *logical layer*). A `Novel` record is embedded as the first two attribute values of a `Data[Novel]` tuple. The third attribute of `Data[Novel]` contains the peer name which actually manages the record, and the fourth attribute represents the record id, which is unique among the P2P network. Relation `Exchange[Novel]` represents the record exchange history. For example, a `Exchange[Novel]` tuple `('B', 'A', '#B001', '#A001', '3/2/08')` means that peer A copied a record from peer B, where it had the id value `#B001`, and peer A assigned a new id `#A001` for the record when it was registered at peer A. The last attribute value `3/2/08` represents the timestamp of the exchange.

Query 2: This query detects whether peer C copied the record `(t1, a1)` owned by peer B or not:

```
Reach(P, I1) :- Data[Novel]('t1', 'a1', 'B', I2),
                Exchange[Novel]('B', P, I2, I1, _)
Reach(P, I1) :- Reach(P1, I2), Exchange[Novel](P1, P, I2, I1, _)
Query(I) :- Reach('C', I)
```

Note that Queries 1 and 2 perform backward and forward traversals of provenance information, respectively.

2 Query Processing

Although we have decided to take the "pay-as-you-go" approach and pay the cost when we perform tracing queries, the efficiency of query processing is still a quite important factor. Query 1 can be easily executed using the seminaive strategy [5]; we omit the details here. In the following, we use Query 2 to compare the performance of the *seminaive method* and the *magic set method*.

2.1 Query Evaluation Based on Seminaive Method

The *seminaive method* is based on simple iterative processing, but ensures no redundant evaluations are performed to process a recursive datalog query [1]. A tracing query in our P2P record exchange framework is executed by the cooperation of distributed peers using query forwarding.

Consider Query 2 is issued at peer B. At first, we need to translate it into a query in the *physical layer*. In the physical layer, two *virtual* relations `Data[Novel]` and Exchange[Novel] in the logical layer are stored as actual relations in a distributed manner. Each peer only stores its corresponding parts of the logical relations. For example, Data[Novel]'B' relation of the physical layer maintains a subset of Data[Novel] relation in the logical layer which corresponds to peer B. Relation Exchange[Novel] in the logical layer is represented as two physical relations To[Novel] and From[Novel]. For example, To[Novel]@'B' (From[Novel]@'B') represents the information of the records provided (copied) by peer B. In the above query, we used the notation like To[Novel]@P, in which P is a peer variable. The transformation result is as follows.

```
Reach(P, I1) :- Data[Novel]@'B'('t1', 'a1', I2),
                To[Novel]@'B'(I2, P, I1, _)
Reach(P, I1) :- Reach(P1, I2), To[Novel]@P1(I2, P, I1, _)
Query(I) :- Reach('C', I)
```

After the transformation, the query is executed using the extension of the seminaive method. For Query 2, the seminaive method generally visits all the peers which copied the record (t1, a1) offered by peer B. Since record provided by a certain peer often copied by multiple peers, it should visit a number of peers. If we assume that the target record provided by a peer is copied by n peers and m forwarding are performed along every path started from peer B, the process should visit n^m peers in total.

2.2 Query Evaluation Based on Magic Set Method

The *magic set* technique is a well-known strategy for the efficient execution of datalog programs [1]. By modifying a given program, it simulates "selection pushdown" for the top-down evaluation approach within the bottom-up evaluation approach.

First, we transform Query 2 into the following query according to the magic set rewriting rules.

```
Reach(P, I1) :- magic_Reach(P, I1), Data[Novel]@'B'('t1', 'a1', I2),
                From[Novel]@P(I1, 'B', I2, _)
Reach(P, I1) :- magic_Reach(P, I1), Reach(P1, I2),
                From[Novel]@P(I1, P1, I2, _)
magic_Reach(P1, I2) :- magic_Reach(P, I1), From[Novel]@P(I1, P1, I2, _)
magic_Reach('C', I):-
Query(I) :- Reach('C', I)
```

Once a program is modified by the magic set-based rewriting, we can execute the program using the seminaive method. The behavior of the modified program is, however, quite different from the normal seminaive method. In this case, the fourth rule above defines the actual start point; it first triggers the evaluation of the third rule. The additional magic predicate magic_Reach requires the following: for each record in peer C, we should traverse the path from peer C to the origin of the record.

We roughly estimate the cost of the query. Assume that peer C has l records. For each record in peer C, we need to traverse its path to the source. Since we follow the path towards the ancestor, the path does not contain branches. If we assume that the path length is a constant value on average, the total fowarding cost would be $O(l)$. This simple analysis appeals that the magic set-based strategy would be a promising method for Query 2.

3 Experimental Results

The purpose of the experiments is to observe the behaviors of two query processing strategies using a simple P2P record exchange model. The simulation model is summarized as follows. We first create $N = 100$ peers and $M = 500$ records; each record is randomly assigned to one of the peers. We assume that records are consists of two classes — "hot" records (20%) and normal records (80%). Hot records are more likely to be exchanged; when a peer wants to get a record from other peer, a hot record is selected with the chance of 80%. We perform random record exchanges until each peer exchanges $L = 50$ records on average.

Figure 1 shows the result. In this figure, we added the experimental results for $N = 500$ and $N = 1000$. Their parameters are same except for N.

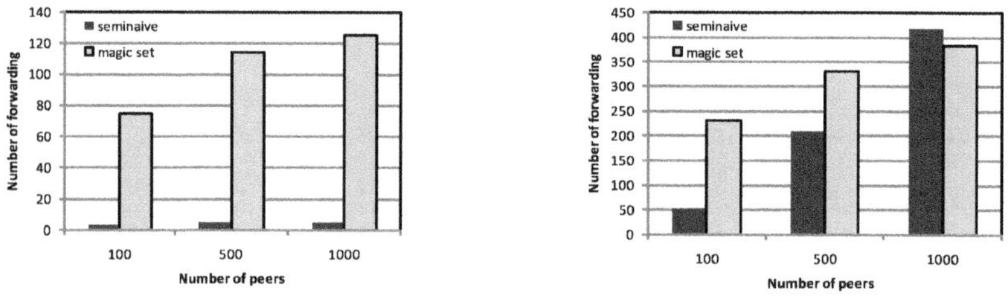

(a) Average case (b) Maximum case

Fig. 1. Query forwarding cost for Query 2

Figure 1(a) and Figure 1(b) shows the same experimental results, the magic set has the high cost. The main reason is that the average number of branches is not high in our simulation model. But the cost of the magic set method becomes better for $N = 1000$. In this case, a hot record is copied by a large number of peers so that the number of branches of forwarding paths become quite large.

Although the magic set has poor performance for the above experiment, it is quite effective for some situations. See the following Query 3, which is a modified version of Query 2.

Query 3: Is the record `(t1, a1)` in peer C a copy of `(t1, a1)` in peer B?

```
Reach(P, I1) :- Data[Novel]('t1', 'a1', 'B', I2),
                Exchange[Novel]('B', P, I2, I1, _)
Reach(P, I1) :- Reach(P1, I2), Exchange[Novel](P1, P, I2, I1, _)
Dup(I) :- Reach('C', I), Data[Novel]('t1', 'a1', 'C', I)
Query(I) :- Dup(I)
```

Figure 2 shows the results. The cost of the seminaive method is same as Query 2. On the other hand, the cost of magic set method becomes quite low, especially in the case of the maximal number of query forwarding. This is because, in contrast to Query 2, the number of forwarding path is only one due to the additional constraint of the third rule used for specifying the start record.

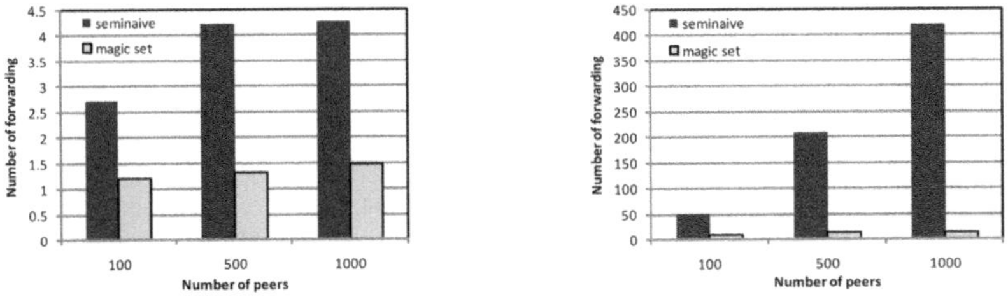

(a) Average case (b) Maximum case

Fig. 2. Query forwarding cost for Query 3

The experimental results indicate that we need to select an appropriate execution strategy depending on the situation.

4 Conclusions

In this paper, we compared two popular query processing methods, the seminaive method and the magic set method for our P2P record exchange framework by experiments: both methods have pros and cons; an appropriate execution strategy depends on the given query, the P2P network organization, the record exchange behaviors, etc. For the long version of this paper, please visit our homepage `http://www.db.itc.nagoya-u.ac.jp/`.

This research was partly supported by the Grant-in-Aid for Scientific Research (19300027, 19024037, 18200005) from MEXT and JSPS.

References

1. Abiteboul, S., Hull, R., Vianu, V.: Foundations of Databases. Addison-Wesley, Reading (1995)
2. Buneman, P., Cheney, J., Tan, W.-C., Vansummeren, S.: Curated databases. In: Proc. ACM PODS, pp. 1–12 (2008)
3. Buneman, P., Tan, W.-C.: Provenance in databases (tutorial). In: Proc. ACM SIGMOD, pp. 1171–1173 (2007)
4. Halevy, A., Franklin, M., Maier, D.: Principles of dataspace systems. In: Proc. ACM PODS, pp. 1–9 (2006)
5. Li, F., Iida, T., Ishikawa, Y.: Traceable P2P record exchange: A database-oriented approach. Frontiers of Computer Science in China 2(3), 257–267 (2008)
6. Li, F., Ishikawa, Y.: Traceable P2P record exchange based on database technologies. In: Zhang, Y., Yu, G., Bertino, E., Xu, G. (eds.) APWeb 2008. LNCS, vol. 4976, pp. 475–486. Springer, Heidelberg (2008)

Update Propagation Strategies Considering Degree of Data Update in Peer-to-Peer Networks

Toshiki Watanabe, Akimitsu Kanzaki, Takahiro Hara, and Shojiro Nishio

Dept. of Multimedia Eng., Grad. Sch. of Information Science and Technology, Osaka Univ.
{watanabe.toshiki,kanzaki,hara,nishio}@ist.osaka-u.ac.jp

Abstract. In a P2P network, it is common that data items are replicated on multiple peers for improving data availability. In such an environment, when a data item is updated, the update should be immediately propagated to other peers holding its replicas. In this paper, we propose two update propagation strategies considering the degree of change in data update in P2P networks.

1 Introduction

In a P2P network, it is common that data items are replicated on multiple peers for efficient data retrieval, improving data availability, and load balancing [1,2]. In many data sharing applications in a P2P network, a data update occurring on a particular peer should be immediately propagated to other peers holding its replicas in order to maintain consistency among replicas [3,4].

In [4], we proposed the UPT-FT (Update Propagation Tree with Fault Tolerance) strategy for delay reduction and node failure tolerance. This strategy creates an n-ary tree for each data item in a P2P network and propagates the update information according to the tree. The root of the tree is the owner of the original data item (*original node*), and the other nodes are peers holding its replicas. In this strategy, the updated data is certainly propagated to all replica holders every time when a data update occurs. Therefore, this strategy causes a heavy traffic for update propagation.

Here, some replica holders might need the updated data only when the degree of change in the data update is large. In such a case, in order to reduce the network load, it is effective to propagate the updated data only to replica holders that need.

In this paper, we propose two update propagation strategies considering the degree of change in data update that reduce the network load of update propagation.

2 Proposed Strategies

In this paper, we assume data that can express the degree of change in data update numerically. For example, numeric data (prices of some goods and stock prices) is suitable for this assumption. In our strategies, each replica holder decides the acceptable difference between the original data and its replica, which we call *update condition*. Replica holders need to receive the updated data when their update conditions are satisfied. In the following, all data are assumed as numeric data for simplicity of explanation.

X. Zhou et al. (Eds.): DASFAA 2009, LNCS 5463, pp. 328–333, 2009.

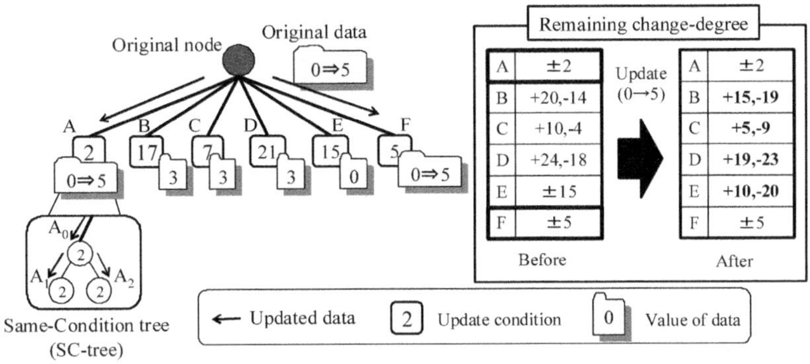

Fig. 1. Update propagation in UPDD-S

2.1 UPDD-S

To reduce needless update propagation, we propose UPDD-S (Update Propagation strategy considering Degree of Data update with Same-condition trees). This strategy reduces the network load by suppressing the update propagation to replica holders which do not need to receive the update. UPDD-S creates trees, each of which consists of replica holders with same update conditions, which we call *Same-Condition trees* (*SC-trees*). Fig. 1 shows the structure of a tree for update propagation in UPDD-S.

Update Propagation: In UPDD-S, when the original data is updated, the original node sends the updated data to root nodes of SC-trees whose update conditions are satisfied.

For update propagation, the original node manages the *remaining change-degrees* of root nodes of all SC-trees. Remaining change-degree indicates the current acceptable difference between original data and its replica. Therefore, when the degree of change in subsequent data update is larger than this value, the node needs to receive the updated data. In Fig. 1, before the update occurs the difference between the original data and replicas in tree A is 0, and peers in tree A set their update conditions as 2. Therefore, the original node records +2 and −2 as A's remaining change-degree.

When the original data is updated, the original node checks the remaining change-degree of each SC-tree and directly sends the updated data only to the root nodes whose update conditions are satisfied. At the same time, the original node updates the remaining change-degrees of all SC-trees because the differences between the original data and their replicas change. Finally, the nodes that receive the updated data from the original node propagate it according to the SC-trees.

In this strategy, a large number of SC-trees are created when replica holders set various kinds of update conditions. In such a case, the original node must manage a large number of information on the SC-trees and directly send the updated data to their root nodes. This leads to increase of the load of the original node.

2.2 UPDD-SO

To reduce the load of the original node, we propose another update propagation strategy, named UPDD-SO (UPDD with Same-condition trees and Ordered-condition trees).

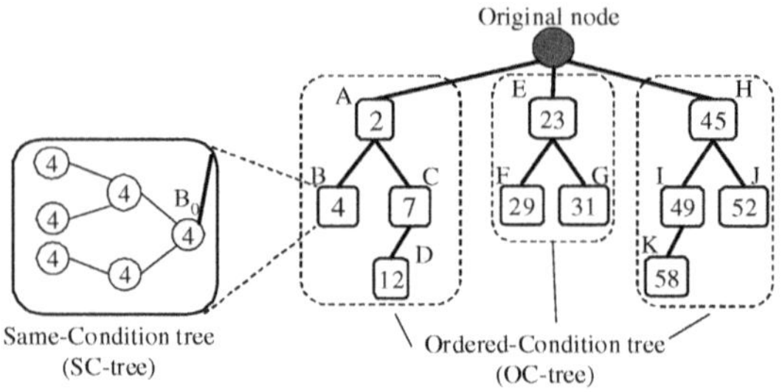

Fig. 2. Structure of a logical tree in UPDD-SO ($G = 3$)

Fig. 2 shows the structure of a logical tree for update propagation in UPDD-SO. UPDD-SO creates two types of trees. One is SC-tree that is same as UPDD-S. In addition, SC-trees are grouped into G groups so that peers in each group have similar update conditions. Then, the root nodes of all same condition trees in each group construct a logical tree, named *Ordered-Condition tree* (*OC-tree*), where the update condition of each node in an OC-tree is smaller than that of its children. This is based on the fact that the smaller the update condition is, the more frequent the peer needs to receive the updated data. Allocating replica holders with smaller update conditions at higher positions in the tree helps reducing unnecessary update propagations.

Update Propagation: The procedure of update propagation is as follows:

1. When the original data is updated, the original node sends the updated data to its children that need it. Then, the original node updates the remaining change-degrees of its descendants in the same way as in UPDD-S. In Fig. 3(a), when the original data changes from 0 to 5, the original node sends the updated data to peer A_0. At the same time, the original node updates the remaining change-degree of tree F.
2. Each node that receives the updated data updates its own replica, and propagates it according to the SC-tree that the node joins. Then, the node sends the updated data to its children that need it. After that, the node updates the remaining change-degrees of its descendants in the same way as the original node. If there is no descendant that needs to receive the updated data, the procedure goes to step 3. In Fig. 3(a), peer A_0 sends the data to peer B_0 and updates the remaining change-degree of tree C.
3. The node that stops the propagation of the updated data checks the remaining change-degree of its descendants. If the remaining change-degree includes a smaller absolute value than its own update condition, the node informs the information on the descendant to its parent. We call this message *control message*. By using the control messages, each node can recognize the minimum remaining change-degree of its descendants. In Fig. 3(b), peer B_0 stops the update propagation and informs its parent of D's remaining change-degree (i.e., +2 and -12) since D's remaining change-degree shows a smaller absolute value (i.e., 2) than B's update condition (i.e., 4). Peer A_0 records the remaining change-degree of tree D.

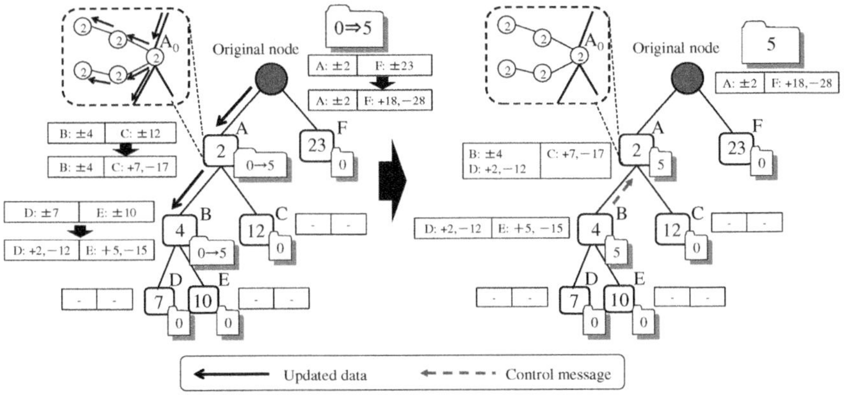

(a)Update propagation (b)Control message propagation

Fig. 3. Update propagation in UPDD-SO

3 Simulation Experiments

3.1 Simulation Model

In the simulation experiments, there are 1,000 peers and 100 kinds of data items in the entire system. At every peer, the query-issuing frequency at each time slot is 0.01.

Each replica holder randomly selects its update condition from D update conditions ($\frac{100}{D} * 1, \frac{100}{D} * 2, ..., \frac{100}{D} * D$). Initially, the value of the original data is 100 and it is updated as shown in Fig. 4. In our strategies, SC-trees and OC-trees are binary trees. The size of updated data is 10. The sizes of a message for maintenance of tree structures (*maintenance message*) and a control message described in Section 4.2 are 1.

3.2 Network Load

First, we examine the network load in UPT-FT, UPDD-S and UPDD-SO. The network load is defined as the total amount of transmitted data including updated data, maintenance messages and control messages. Here, we set D as 20.

From Fig. 5, we can see that our strategies drastically reduce the network load compared with UPT-FT. Here, of our proposed strategies, UPDD-SO increases the network load compared with UPDD-S. This is because UPDD-SO needs to exchange control messages for update propagation. Moreover, in UPDD-SO, it can happen that the updated data has to be propagated to nodes in a SC-tree along the OC-tree even when the update conditions of any intermediate nodes along the path are not satisfied, which results in increase of the load for the updated data.

3.3 Update Propagation Load

Next, we examine the following criteria in UPDD-S and UPDD-SO.

- Update propagation load of the original node
  The total amount of data propagated by the original node.

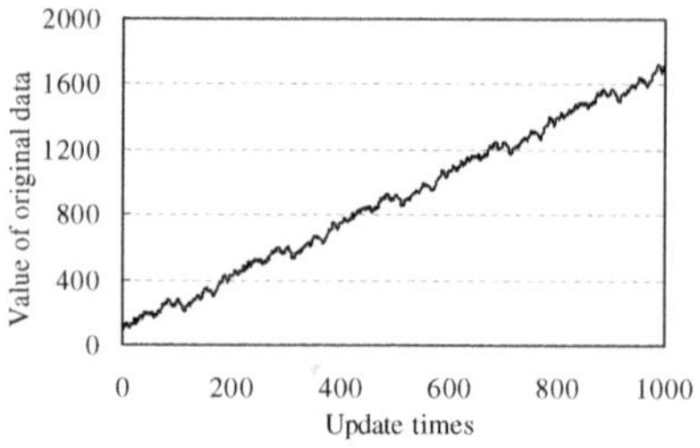

Fig. 4. Transition of original data

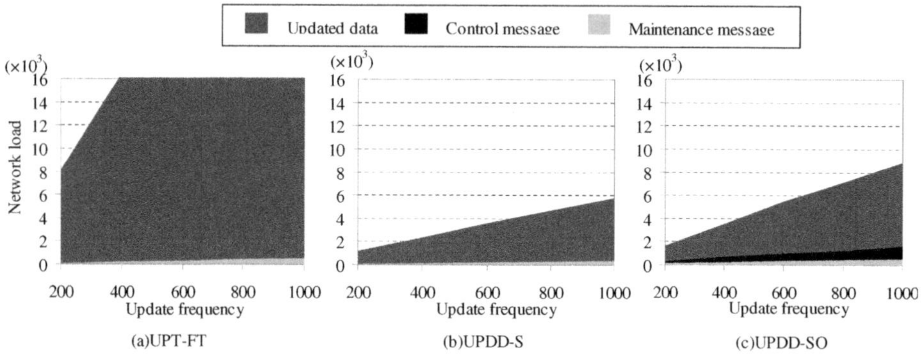

Fig. 5. Network load ($D = 20$)

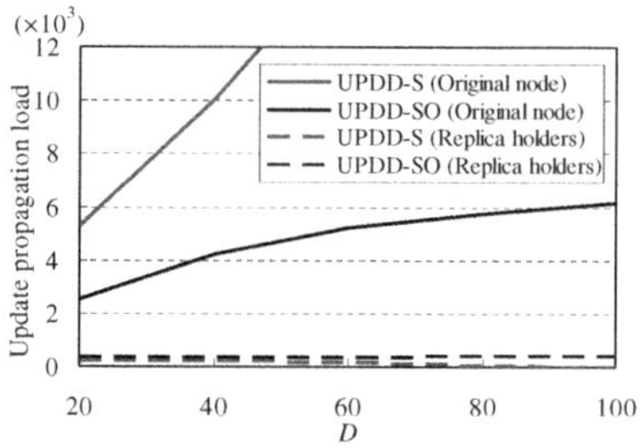

Fig. 6. Update load

- Update propagation load of replica holders
  The average of the amount of data propagated by each replica holder.

In the experiments, we set the update frequency as 200.

From Fig. 6, we can see that the load of the original node in UPDD-S becomes very large. On the other hand, UPDD-SO suppresses the load of the original node while keeping the load of replica holders low by introducing OC-trees.

4 Conclusions

In this paper, we proposed new update propagation strategies considering the degree of change in data update. These strategies reduce the network load for update propagation.

As part of our future work, we plan to consider a more efficient approach for maintaining the tree structures and further reduces the network load.

Acknowledgement

This research was partially supported by Special Coordination Funds for Promoting Science and Technology:"Yuragi Project", Grant-in-Aid for Scientific Research (18049050) of the Ministry of Education, Culture, Sports, Science and Technology, Japan, and the Ministry of Internal Affairs and Communications of Japan under the Research and Development Program of "Ubiquitous Service Platform".

References

1. Cohen, E., Shenker, S.: Replication strategies in unstructured peer-to-peer networks. In: Proc. SIGCOMM 2002, pp. 177–190 (2002)
2. Lv, Q., Cao, P., Cohen, E., Li, K., Shenker, S.: Search and replication in unstructured peer-to-peer networks. In: Proc. ICS 2002, pp. 84–95 (2002)
3. Wang, Z., Das, S.K., Kumar, M., Shen, H.: Update propagation through replica chain in decentralized and unstructured P2P systems. In: Proc. P2P 2004, pp. 64–71 (2004)
4. Watanabe, T., Kanzaki, A., Hara, T., Nishio, S.: An Update Propagation Strategy for Delay Reduction and Node Failure Tolerance in Peer-to-Peer Networks. In: Proc. FINA 2007, pp. 103–108 (2007)

XML Data Integration Using Fragment Join

Jian Gong, David W. Cheung, Nikos Mamoulis, and Ben Kao

Department of Computer Science, The University of Hong Kong
Pokfulam, Hong Kong, China
{jgong,dcheung,nikos,kao}@cs.hku.hk

Abstract. We study the problem of answering XML queries over multiple data sources under a schema-independent scenario where XML schemas and schema mappings are unavailable. We develop the fragment join operator — a general operator that merges two XML fragments based on their overlapping components. We formally define the operator and propose an efficient algorithm for implementing it. We define schema-independent query processing over multiple data sources and propose a novel framework to solve this problem. We provide theoretical analysis and experimental results that show that our approaches are both effective and efficient.

1 Introduction

Data integration allows global queries to be answered by data that is distributed among multiple heterogeneous data sources [1]. Through a unified query interface, global distributed queries are processed as if they were done on a single integrated data source. To achieve data integration, a schema mapping is often used, which consists of a set of mapping rules that define the semantic relationship between the global schema and the local schemas (at the data sources). In these systems, such as Clio [2], processing a global query typically involves two steps: query rewriting, and data merging.

While much work has been done on query rewriting, very little has been done on data merging. In most existing approaches, data merging is mostly an *ad hoc* computation — a special data merging routine is custom-coded for each mapping rule. This approach leads to inflexible system design. In this paper we propose a schema independent framework that allows data merging be processed without referring to any specific schema mapping rules.

Let us illustrate our idea by an example. Figure 1(a) shows two XML documents taken from UA Cinema website and IMDB website, respectively. Both UA and IMDB contain the *title* and the *director* of each *movie*. In addition, UA contains *venue* and *price*, while IMDB contains the movie's *reviews*. Consider a user who wants to find out the *title*, *director*, *price*, and *review* for each *movie*. This is expressed by the twig pattern query shown in Figure 1(b). Note that neither UA nor IMDB can answer the query alone because UA lacks *reviews* and IMDB lacks *pricing* information. The (global) query thus has to be broken into two query fragments, one for each site. The returned results from the two sites should then be merged based on their common components. Figure 1(c) shows an example of the query result. Our goal is to answer such twig pattern queries in a schema-independent fashion where mapping rules are not needed.

X. Zhou et al. (Eds.): DASFAA 2009, LNCS 5463, pp. 334–338, 2009.

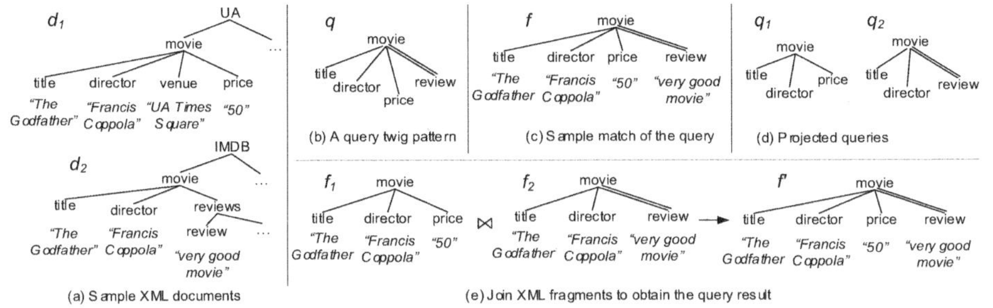

Fig. 1. Query on sample XML documents and the results

In our approach, we join data fragments based on their overlapping content in order to answer queries. For example, we first *project* the global query on the two XML documents and obtain two local queries (Figure 1(d)). Then, we retrieve XML fragments $m(t, d, p)$ from UA and $m(t, d, r)$ from IMDB. Afterward, we *join* these fragments based on their overlapping parts, which are title (t) and director (d) (Figure 1(e)).

2 Preliminaries

An XML document D is a rooted, node-labeled tree $D = \langle N, E, r \rangle$, wherein N is a node set, $E \subseteq N \times N$ is an edge set, and $r \in N$ is the root node. Each node in an XML document has a label and may contain some text. The *vocabulary* of an XML document d, denoted by $v(d)$, is the set of distinct node labels of d.

Definition 1. (XML FRAGMENT) *An XML fragment f is an edge-labeled XML document, where each edge is labeled by either "/" (parent-child edge) or "//" (ancestor-descendant edge). An XML fragment f is a fragment of an XML document d, denoted as $f \sqsubseteq d$, if there exists an injective mapping $\lambda : f.N \rightarrow d.N$, such that: (i) $\forall n \in f.N$, $n = \lambda(n)$, and (ii) $\forall e(n_1, n_2) \in f.E$ labeled as "/" (resp., "//"), $\lambda(n_1)$ is the parent (resp., ancestor) of $\lambda(n_2)$.*

Definition 2. (TWIG PATTERN AND MATCH) *A twig pattern is an XML fragment, where the text content of the nodes is disregarded. A fragment f is a* **match** *to a twig pattern q, denoted as $f \vdash q$, if there exists a mapping $\gamma : q.N \rightarrow f.N$, such that the node labels and edges of q are preserved in f. A fragment f_1 is contained in another fragment f_2, denoted as $f_1 \preceq f_2$, if all the nodes and edges of f_1 are contained in f_2.*

Definition 3. (PROJECTION) *Given a fragment f and a vocabulary $v(d)$ of a document d, the projection of f on $v(d)$, denoted as $\rho_{v(d)}(f)$, is obtained by removing from f all the nodes whose labels are not in $v(d)$ and the corresponding connecting edges.*

3 The Fragment Join Operator

Definition 4. (FRAGMENT JOIN) *Given a set of of fragments f_1, ..., f_n ($n \geq 2$), a fragment f is a join of f_1, ..., f_n, denoted as $(f_1, ..., f_n) \overset{\bowtie}{\rightarrow} f$, if $\exists f'_1 \preceq f, ..., f'_n \preceq f$,*

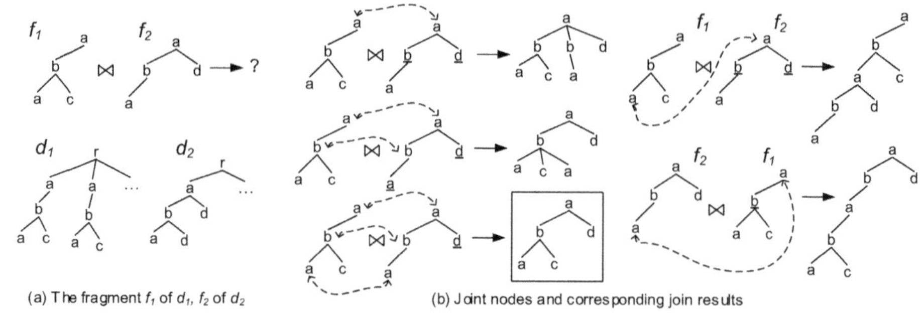

Fig. 2. XML fragment join on different joint sub-trees

such that: 1) $f'_i = f_i$, $1 \leq i \leq n$, *2)* $\forall n \in f.N$, $n \in f'_1.N \cup ... \cup f'_n.N$, *and 3)* $\forall e \in f.E$, $e \in f'_1.E \cup ... \cup f'_n.E$.

In addition, the join set of $f_1, ..., f_n$ *is a set of fragments* $F = \{f | (f_1, ..., f_n) \overset{\bowtie}{\rightarrow} f\}$, *denoted as* $(f_1, ..., f_n) \overset{\bowtie}{\Rightarrow} F$.

Definition 5. (JOINT SUB-TREE) *Given two fragments* f_1 *and* f_2*, a subtree* js *is a joint sub-tree of* f_1 *and* f_2 *if (1)* $js \preceq f_1$, $js \preceq f_2$, *(2) the root of* js *= the root of* f_2.

Figure 2(b) shows the five results of the fragment join between f_1 and f_2 shown in Figure 2(a). Each of these results is based on a joint sub-tree, whose nodes are pointed by double-arrowed dashed lines in the two fragments.

We propose Algorithm 1 for evaluating the fragment join of two fragments f_1 and f_2. For example, consider the first join result shown in Figure 2(b). The joint-subtree for this join result consists of a lone node a. The *boundary nodes* are the children of the root node a in f_2, which are labeled b and d (underlined). The subtrees of these boundary nodes are attached to the matching node a in f_1 forming the join result.

4 Schema-Independent, Query-Based Data Integration

Our research problem is formally stated as following: *given XML documents* d_1 *and* d_2*, and a twig pattern query* q*, compute* $F = \{f | f \vdash q; (f_1, f_2) \overset{\bowtie}{\rightarrow} f; f_1 \sqsubseteq d_1; f_2 \sqsubseteq d_2\}$. Our approach to solve this problem consists of the following phases.

Projection. The twig query q is rewritten into local queries $q_1 = \rho_{v(d_1)}(q)$ and $q_2 = \rho_{v(d_2)}(q)$ using the *project* operator (Section 2). We then apply the fragment join operator on q_1 and q_2 to find a joint sub-tree js for which the join result is q.

Matching. Two sets of fragments F_1 and F_2 are returned, which contains all matches to the local query q_1 in d_1 and all matches to the local query q_2 in d_2, respectively [1].

Join. For each pair of fragments $(f_1, f_2) \in F_1 \times F_2$, we compute the fragment join of f_1 and f_2 using the joint-subtree obtained in the *projection* phase. The join results are returned as the query's answer.

[1] We thank the authors of [3] for providing us with the implementation of TwigList, used as a module for evaluating twig queries in our work.

Algorithm 1. The join evaluation algorithm

Input: XML fragments f_1 and f_2
Output: a set of XML fragments F, with the join sub-trees used for each $f \in F$

1: $JS \leftarrow enumerateJointSubtrees(f_1, f_2)$
2: **for all** $js \in JS$ **do**
3: $\quad f \leftarrow join(f_1, f_2, js)$
4: $\quad$ output (f, js)
5: **end for**
6: repeat 1-6 with f_1 and f_2 exchanged, if necessary

function $join(f_1, f_2, js)$
1: $f \leftarrow copy(f_1)$
2: **for all** $x \in js.N$ **do**
3: $\quad$ let x_1, x_2 be the corresponding nodes of x in f_1 and f_2, respectively
4: $\quad$ **for all** x_2's child c **do**
5: $\quad\quad$ **if** $c \notin js.N$ **then**
6: $\quad\quad\quad sf \leftarrow constructFragment(f_2, c)$
7: $\quad\quad\quad addChild(f_1, x_1, sf)$
8: $\quad\quad$ **end if**
9: $\quad$ **end for**
10: **end for**
11: return f

Figure 3 illustrates our approach (the found joint sub-tree contains the underlined nodes). We note that projecting a global query onto local sources so that one single local query is applied to each source may not be sufficient to retrieve the complete set of query results. For example, consider again query q in Figure 3. We observe that joining q_{11}, a sub-twig pattern of q_1 containing nodes b and c and the edge between them, with q_2 also gives us q (using the joint sub-tree b). Therefore, in order to ensure that all valid query results are found, we should consider all pairs of sub-twig patterns of q_1 and q_2 that can form q.

Definition 6. (RECOVERABILITY) *Given a twig pattern q, a pair of twig patterns (q_i, q_j) is recoverable for q, denoted as* $(q_i, q_j) \overset{r}{\rightsquigarrow} q$, *if* $(q_i, q_j) \overset{\bowtie}{\underset{r}{\rightarrow}} q$ *using some joint sub-tree js; else, (q_i, q_j) is non-recoverable for q, denoted as* $(q_i, q_j) \overset{r}{\not\rightsquigarrow} q$.

We add two more schema-level phases to the Projection-Matching-Join framework, in order to ensure completeness of the query results.

Decomposition. After the projection phase in which local queries q_1 and q_2 are derived, the decomposition phase returns: $Q_1 = \{q_i | q_i \preceq q_1\}$, and $Q_2 = \{q_j | q_j \preceq q_2\}$.

Recoverability checking. After the decomposition phase, this phase returns: $\{(q_i, q_j) \,|(q_i, q_j) \in Q_1 \times Q_2 \wedge (q_i, q_j) \overset{r}{\rightsquigarrow} q\}$.

5 Experimental Evaluation and Conclusion

We use DBLP and CiteSeer datasets in our experiments. The raw CiteSeer data are in plain text BibTeX format. We converted them into an XML file having similar schema to that of DBLP data. The size of Citeseer dataset is 15MB. We randomly sample the

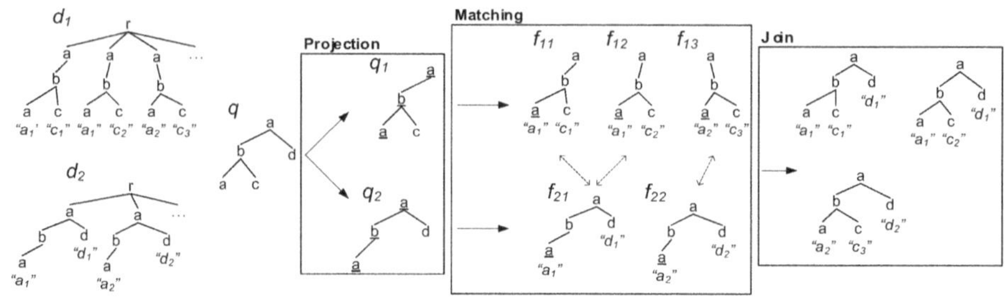

(a) Two XML documents

(b) The twig pattern query and *Projection-Matching-Join* query answering process

Fig. 3. Query answering from multiple data sources: projection, matching, and join

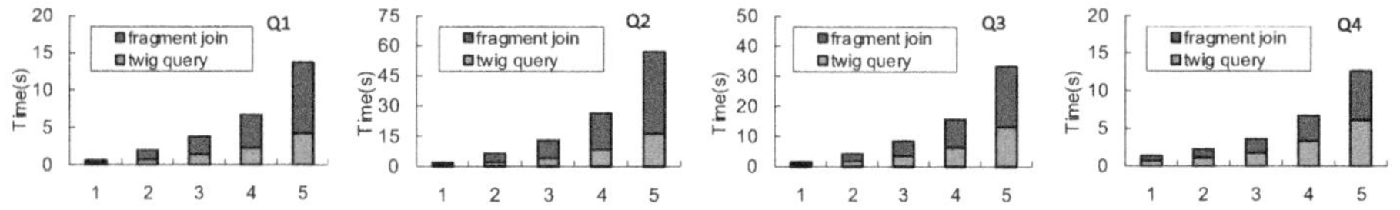

Fig. 4. Overall performance of PDRMJ for all queries and datasets

original DBLP (130MB) dataset to extract the publication records and attributes, and obtain five DBLP datasets, whose sizes are: 1MB, 10MB, 20MB, 40MB, and 80MB, respectively. Thus, we have five pairs of datasets used for the queries, each consisting of the Citeseer dataset plus one of the sampled DBLP datasets.

We manually created four test twig pattern queries, named Q1-Q4, each of which queries on a set of atrributes of papers, such as *title*. All these queries can only be answered using both DBLP and Citeseer datasets (but not one of the two datasets alone) by fragment join in our framework.

The overall performance of our complete, optimized approach (PDRMJ) is tested in Figure 4 for all queries Q1-Q7 on all datasets. The overall response time is broken down to two parts: (i) the time spent by all sub-twig pattern queries issued against the different sources, and (ii) the time spent by the fragment joins. We observe that the performance for all queries scales roughly linearly to the size of the DBLP dataset (recall that the size of the CiteSeer dataset is fixed). In addition, nearly half of the cost is due to the twig pattern queries against the sources.

In conclusion, we developed a fragment join operator for query-based data integration from multiple sources. We studied the problem of schema-independent data integration based on this operator. We conducted experiments to show the effectiveness of our approaches.

References

1. Lenzerini, M.: Data integration: a theoretical perspective. In: PODS (2002)
2. Yu, C., Popa, L.: Constraint-based XML query rewriting for data integration. In: SIGMOD (2004)
3. Qin, L., Yu, J.X., Ding, B.: TwigList: make twig pattern matching fast. In: Kotagiri, R., Radha Krishna, P., Mohania, M., Nantajeewarawat, E. (eds.) DASFAA 2007. LNCS, vol. 4443, pp. 850–862. Springer, Heidelberg (2007)

An Adaptive Method for the Efficient Similarity Calculation

Yuanzhe Cai[1,2], Hongyan Liu[3], Jun He[1,2], Xiaoyong Du[1,2], and Xu Jia[1,2]

[1] Key Labs of Data Engineering and Knowledge Engineering, MOE, P.R. China
[2] School of Information, Renmin University of China, P.R.China
{yzcai,hejun,duyong,jiaxu}@ruc.edu.cn
[3] Department of Management Science and Engineering, Tsinghua University, P.R. China
liuhy@sem.tsinghua.edu.cn

Abstract. *SimRank* is a well-known algorithm for similarity calculation based on object-to-object relationship. However, it suffers from high computation cost. In this paper, we find that the convergence behavior of different object pairs is different when we use *SimRank* to compute the similarity of objects. Many similarity scores converge fast, while others need more time before convergence. Based on this observation, we propose an adaptive method called *Adaptive-SimRank* to speed up similarity calculation. Using this method, we don't need to recalculate those converged pairs' similarity. The experiments conducted on web datasets and synthetic dataset show that our new method can reduce the running time by nearly 35%.

Keywords: Similarity Calculation, Linkage Mining.

1 Introduction

As a process of grouping a set of objects into classes of similar objects, clustering has been widely studied in many domains such as information retrieval, bioinformatics, market basket analysis, collaborative filtering and bioinformatics. In the information retrieval field, clustering document collections is based on an intuitive notion: documents are grouped by topics, and documents in the same topic tend to heavily use the same set of terms. In the bioinformatics field, Scientists want to cluster genes, proteins, and their behaviors in order to discover their functions. For market basket analysis, merchants want to find customer groups who have similar purchasing behaviors. In the collaborative filtering field, similar users and items are clustered based on user or item similarities. In the bioinformaticsfield, scientists cluster the scientific papers in the same topic by analyzing their reference relationship.

In these applications, the object-to-object relationship can be the most useful information for clustering. To use this kind of information, Jeh and Widom [1] proposed a new similarity measure called *SimRank* (for simplicity, we also call the algorithm proposed in [1] to compute this measure *SimRank*). Using this measure to compute similarity can obtain higher accuracy than using other methods [2] [3] [4]. But in the meantime, its computation cost is high. For a web of small size (about 4000 web pages), it took more than 17 hours to finish the computation in our experiment. Thus, to speed up the computation of *SimRank* is important.

X. Zhou et al. (Eds.): DASFAA 2009, LNCS 5463, pp. 339–353, 2009.

Improving the performance of the *SimRank* algorithm is a difficult task. First, many other fast methods are not available for this task because the real world graph is sparse and large. Further, there are few papers to discuss how to improve the *SimRank* algorithm. Finally, the convergence of *SimRank* algorithm is fast for real world graph. In our experiments, the average number of iterations for a relationship graph (its formal definition is in section 3) with about 4000 nodes is only about 50. It is difficult to improve this already fast convergence rate.

After study of the structure of many real world relationship graphs, we found that the convergence rate of *SimRank* for each pair of objects during each step of iteration is not uniform. Some pairs' *SimRank* scores become convergent quickly, but others slowly.

In this paper, based on the above observation, we design a simple and effective algorithm, called *Adaptive-SimRank*, to improve the performance of *SimRank* algorithm. In this algorithm, those *SimRank* scores which have already converged during the computation iterations will not be recomputed in the following iterations. Our experiment results show that this method can speed up the performance of *SimRank* by nearly 35%.

The main contributions of this paper are as follows:

- Based on the characteristics of real world object-to-object relationship graphs, we develop a new algorithm, *Adaptive-SimRank*, which improves the performance of *SimRank* by nearly 35%. This method can also be used for other algorithms [5] [6] to speed up the similarity calculation.
- We also prove the convergence of *Adaptive-SimRank* algorithm by both theoretical and empirical studies.
- We evaluate our method on real datasets to confirm its applicability in practice.

This paper is organized as follows. We introduce the related work in section 2 and define the graph models in section 3. Preliminaries are presented in section 4, and the *Adaptive-SimRank* algorithm is described in section 5. Our performance study is reported in section 6, and finally this study is concluded in section 7.

2 Related Work

We categorize existing work related to our study into two classes: clustering based on link analysis and Random walk on graph.

Clustering based on link analysis: The earliest research work for similarity calculation based on link analysis focuses on the citation patterns of scientific papers. The most common measures are *co-citation* [2] and *co-coupling* [3]. *Co-citation* means if two documents are often cited together by other documents, they may have the same topic. The meaning of *co-coupling* for scientific papers is that if two papers cite many papers in common, they may focus on the same topic. In [4], Amsler proposed to fuse bibliographic *co-citation* and *co-coupling* measures to determine the similarity between documents. However, all these methods compute the similarity only by considering their immediate neighbors. *SimRank* [1] is proposed to consider the entire relationship graph to determine the similarity between two nodes. But this method has a high time complexity, which limits its use to large datasets. Thus, there are some methods proposed to improve the performance of *SimRank*. In the paper *SimRank first proposed* [1], authors also proposed a algorithm called *Pruning-SimRank* [1], which computes similarities by a small scope of relationship graph around two nodes. However, because this method ignores much information, the accuracy of this method is not very good. Another method to enhance the performance of *SimRank* algorithm is *Fingerprint-SimRank* [5], which pre-computes several steps of random walk path for each object and uses these steps to calculate the similarity between objects. Although *Fingerprint-SimRank* improves the performance, this algorithm has a high space complexity. In addition, Xiaoxin Yin et al [6] proposed a hierarchical structure, *SimTree,* to represent the similarity between objects and develop an effective similarity calculation algorithm—*LinkClus*. However, in that paper, there is no sound theoretical proof of the convergence for that algorithm. Thus, it is difficult to decide in which iteration *LinkClus* will gain the best accuracy. Different from all these methods, our method focuses on the convergence of *SimRank*. All these above-mentioned techniques can combine with our method to improve the performance of *SimRank*.

Random walks on graphs: Theoretical basis of our work is hit times for two surfers walking randomly on the graph. We mainly refer to some research works about expected *f-meeting distance theory* [1, 7]. Some other research papers also help us to understand our research, which is the theory of random walk with restart. *PageRank* [8-10] is a famous algorithm using this theory.

3 Problem Definition

The object-to-object relationship can be represented by a graph. A relationship graph $G(V, E)$ is a directed graph, where each vertex in V represents an object in a particular domain and an edges in E describes the relationship between objects. Furthermore, we use $I(v)$ to represent all the neighbors of a node v . $|I(v)|$ is the number of neighbor nodes of v. For example, Fig.1 is a relationship graph that describes the web page relationship crawled from Cornell computer science department. In this graph, a directed edge <p, q> means a reference (hyperlink) from page p to page q. Our research work focuses on how to cluster the similar pages on the web graph. In this website relationship graph, these web pages have been manually classified into seven domains, course, department, faculty, project, staff, student, and others. Based on this classification, we will evaluate our method to cluster these pages automatically into seven fields.

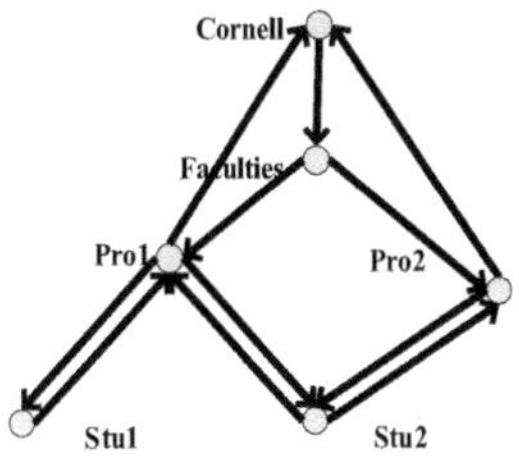

(a)

Name	URL
Cornell	http:\\www.cs.cornell.edu
Faculties	http:\\www.cs.cornell.edu\People\faculty\index.htm
Pro1	http:\\www.cs.cornell.edu\~professor1
Pro2	http:\\www.cs.cornell.edu\~professor2
Stu1	http:\\www.cs.cornell.edu\~stu1
Stu2	http:\\www.cs.cornell.edu\~stu2

(b)

Fig. 1. An example of a web page relationship graph

4 Review of *SimRank*

In this section, we will introduce *SimRank* and define the convergence of *SimRank*.

4.1 Introduction of *SimRank*

SimRank [1], a classical linkage-based similarity calculation algorithm, measures similarity of two objects based on the principle, "two objects are similar if they link to the similar objects" [1]. The recursive similarity computation equation is as follows:

$$\begin{cases} 1 & (V_a = V_b) \\ s(V_a, V_b) = \dfrac{c}{|I(V_a)||I(V_b)|} \displaystyle\sum_{i=1}^{|I(V_a)|} \sum_{j=1}^{|I(V_b)|} s(I_i(V_a), I_j(V_b)) & (V_a \neq V_b) \end{cases} \quad (1)$$

Where c is a decay value, which is a constant between 0 and 1. $I(V_a)$ is the set of neighbor nodes of a. $|I(V_a)|$ is the number of neighbors of node V_a.

At another point of view, $s(V_a, V_b)$ can be considered as how soon and how many times two surfers, a and b, will meet, when they walk starting from V_a and V_b respectively and travel randomly on the graph G. Thus, based on the *expected f-meeting distance* theory [1], Jeh and Widom gave another definition of *SimRank* score $s(V_a, V_b)$ as follows.

$$s(V_a, V_b) = \sum_{t:(V_a, V_b) \sim\sim> (V_x, V_x)} P[t] c^{l(t)} \quad (2)$$

Where c is also the decay value. And t=<$w_1, w_2, \ldots, w_k$> is a travel path from V_a to V_b and $l(t) = k/2$, which is the number of steps starting from V_a and V_x(the hit position). $P[t]$ of travel path t is $\prod_{i=1}^{k-1} \frac{1}{|I(w_i)|}$, which represents the probability of a surfer traveling along this path.

In order to explain formula (2), we use the graph shown in Fig.1 to show the calculation of the similarity between Pro1's web page and Pro2's web page. Let two surfers starting from node Pro1 and node Pro2 respectively walk only one step. Then, they may meet at node Cornell or node Stu2. If meeting at node Cornell, the travel path is <Pro1, Cornell, Pro2>. The probability of a surfer starting form Pro1 to Cornell is 1/3, because there are three web pages connected to node Pro1. Similarly, the probability of a surfer starting from Pro2 to Cornell is 1/2. Thus, in this path t, $P[t]$ = 1/3×1/2=1/6. We set c = 0.8. The number of steps is 1. Thus, for this path, the *SimRank* score is $1/6\times(0.8)^1 = 0.133$. Because there are only two paths between Pro1 and Pro2, < Pro1, Cornell, Pro2> and < Pro1, Stu2, Pro2>, s(Pro1, Pro2) is equal to 0.267 at the first step of iteration (i.e. walking one step). If walking two steps, there is only one path <Pro1, Stu1, Pro1, Stu2, Pro2> (i.e., Pro1→Stu1→Pro1, Pro2→Stu2→Pro1), the *SimRank* score on this path is $(1/3\times1\times1/2\times1/2) \times 0.8^2 =$ 0.0533. *SimRank* will search all paths and sum them. Finally, s(Pro1,Pro2) = 0.267 + 0.0533+ …= 0.404.

4.2 Similarity Calculation

Similarity calculation is an iteration process. Starting with $s_0(V_a, V_b)$, we can calculate $s_{k+1}(V_a, V_b)$ from $s_k(V_a, V_b)$ by equation (4).

$$s_0(V_a, V_b) = \begin{cases} 0 & (\text{if } V_a \neq V_b) \\ 1 & (\text{if } V_a = V_b) \end{cases} \tag{3}$$

$$s_{k+1}(V_a, V_b) = \frac{c}{|I(V_a)||I(V_b)|} \sum_{i=1}^{|I(V_a)|} \sum_{j=1}^{|I(V_b)|} s_k(I_i(V_a), I_j(V_b)) \tag{4}$$

The major steps of *SimRank* algorithm are as follows:

Algorithm 1. *SimRank*

Input: Decay Factor *c*, Transfer Probability Matrix *T* (the probability of moving from state *i* to state *j* in one step), Tolerance Factor ε (under normal case, $\varepsilon = 0.001$)

Output: Similarity Matrix s_k

$k \leftarrow 1$;

$s_0 \leftarrow$ identity matrix;

while($Max(|s_k(V_a, V_b) - s_{k-1}(V_a, V_b)| / |s_{k-1}(V_a, V_b)|) > \varepsilon$)

$k \leftarrow k+1$;

$s_{k-1} \leftarrow s_k$;

for each element $s_k(V_a, V_b)$

$$s_k(V_a, V_b) = c \cdot \sum_{i=1}^{|I(V_a)|} \sum_{j=1}^{|I(V_b)|} T_{V_a V_i} \cdot T_{V_b V_j} \cdot s_{k-1}(V_i, V_j);$$

end for

end While

return s_k

4.3 Convergence of SimRank

In paper [1], the author proposed that the *SimRank* had a rapid convergence, usually in five iterations. However, in our experiments, after five iterations, some *SimRank* scores are still increasing much. We define the convergence factor d as follows:

$$d = max\ (|s_{k+1}(V_a, V_b) - s_k(V_a, V_b)| / |s_k(V_a, V_b)|) \tag{5}$$

We say when $d < 10^{-3}$, *SimRank* scores become convergent. We calculate the convergence factor d for four web datasets as shown in table 1. As can be seen from table 1, these *SimRank* scores converge into fixed values after about 40 steps of iteration.

Table 1. Statistics about Web Dataset convergence

iteration / Web DataSet	1	2	3	4	5	6	Final Iteration
Wisconsin	20	114.9	207.9	59.7	33.11	4.44	***40 iter.***
Washington	48.9	181.6	593.6	633.8	250.9	29.7	***45 iter.***
Texas	47.4	313.6	352.1	49.2	4.1	1.8	***39 iter.***
Cornell	48.8	260.5	260.5	140.4	5.46	2.10	***41 iter.***

5 Adaptive-SimRank Algorithm

In this section, we first show our observation of *SimRank* calculation on real world datasets. Then, we describe our algorithm, *Adaptive-SimRank*. Finally, we analyze the reason of *Adaptive-SimRank*. The proof of its convergence is in Appendix B.

5.1 Adaptive-SimRank Intuition

According to the above introduction about *SimRank*, *SimRank* computation process is the process to search the even-step paths between two nodes. If all of the even-step paths between two nodes are not long, the convergence for two nodes is quickly. For example, in Fig.2 (a), because there are only one path between node V_a and V_b and the path length is 2, we can obtain the final *SimRank* score of node V_a and V_b for only one step of iteration. However, there are many long even-step paths between node V_c and V_d such as <V_c, V_e, V_d>, <V_c, V_f, V_d>, <V_c, V_e, V_c, V_f, V_d>, <V_c, V_f, V_d, V_f, V_d> and so on, thus we need to iterate for about nine steps before we obtain the final similarity score. Based on this analysis, we discover that because of different graph structures, some pairs obtain the convergence values faster than other pairs.

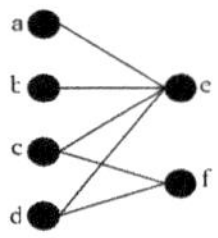

(a)

Calculation Nodes	Iteration Times	Similarity Score
s(V_a, V_b)	1 times	0.8
s(V_c, V_d)	9 times	0.6316

(b)

Fig. 2. Example of Iteration Calculation

In our observation on real world graphs, large graphs in real applications are usually sparse graphs with fairly dense blocks. In table 2, we list the statistic data of the four web sets. We use $\alpha = s / (n*(n-1)/2)$ to describe the density of a graph, where s is the number of edges and n is the number of vertices in the graph. We can see that α of each graph is very low, less than 0.008, which indicates that all of these real world graphs are sparse graphs. Fig. 3 illustrates that the links in Cornell data set are not homogeneous distribution. This is because in a group of websites, the links mainly distribute with the web host. In other words, links in each host are much more and links between hosts are less. Usually, each research lab will have their web host to introduce themselves and each host has their special research topic. Therefore, we can discover that in the same research area, the density of links are much high, but the links between research areas are much less. Because of the property of sparsity and non-uniform distribution of links for the real world graphs, some node pairs in the graph will converge faster than others.

Table 2. Statistics of Four Datasets

Data Set	**Vertexes#(*n*)**	**Edges#(*s*)**	α
Wisconsin	1263	5305	0.0067
Cornell	867	2691	0.0072
Texas	827	2667	0.0078
Washington	1205	3299	0.0045

Note: $\alpha = s / (n*(n-1)/2)$, describing the density of graph.

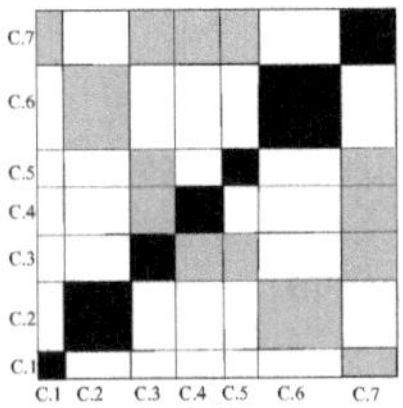

Fig. 3. Cornell Dataset analysis.(The web pages in cornell dataset have been divided into seven classes by research areas, such as Hardware, Artificial intelligence, Database & Information Retrieve, Operation System & Distribute System, Software Engineering & Programming Language, Theory of Computing and Others, which are marked from C.1 to C.7. The function, $R(A,B) = L(A,B) / L(A)$, discribe the density of links. Where, $L(A,B)$ is the number of links from block *A* to block *B* and $L(A)$ is the total number of links from *A* to any other blocks. If $R(A,B) \geq 0.4$, the block is drawed as black and $0.4 > R(A,B) > 0.1$, the block is gray and in other situations, the block is white.)

Fig.4 shows some statistics for the Cornell Dataset. Fig.4(a) shows the newly added proportion of pairs whose *SimRank* score converge to a relative tolerance of .001 in each iteration. Fig. 4(b) describes the cumulative version of the same data, which shows the percentage of pairs that have converged through a particular iteration steps. We see that in 30 iteration steps, the *SimRank* score for over three-forth of pairs have already converged. In our observation, it is also possible that some pairs' *Simrank* are not converged. For example, a pair had the same values for several iterations, and later on changed significantly. In Fig.4(a), the navigate values of the convergence property describe this situation. That is because there are some long paths between nodes which increases these pairs' *SimRank* score. For this problem, we will discuss it later. Furthermore, in Fig.4(a) and 4(b), we can discover that the distribution of the proportion of newly added converged node pairs is approximate normal distribution. In Appendix A, we show that other three datasets also have this character. Thus, based on this skewed distribution of convergence, we propose an approximate algorithm to calculate *SimRank*.

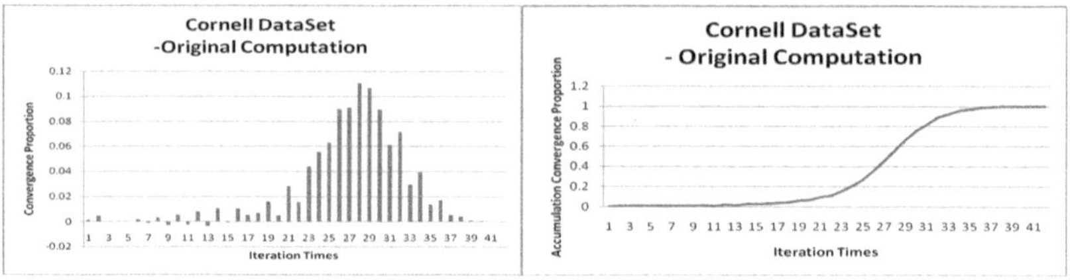

(a) (b)

Fig. 4. Experiment results on Cornell dataset. Results on other three datasets are given in Appendix A. (a) X-axis represents the iteration steps and the height of bar means the newly added proportion of convergence pairs at this step of iteration. The similarity score is computed by *SimRank*. (b) Cumulative convergence proportion at each step of iteration. The x-axis means the iteration step k and the y-axis give the proportion of convergent pairs that have a convergence steps less than or equal to k.

5.2 Adaptive-SimRank

Above-mentioned observation of non-uniform convergence indicates that the time of *SimRank* computation can be reduced by avoiding overlap computation. In particular, if a *SimRank* score has already converged, we don't need to recalculate it in the following iterations. Based on this, we propose a new algorithm, *Adaptive-SimRank*, whose major steps are shown in algorithm 2:

Input: Decay Factor c, Transfer Probability Matrix T, Tolerance Factor ε
Output: Similarity Matrix s

  $k \leftarrow 1$;
  $s_0 \leftarrow$ identity matrix;
  flagmatrix $\leftarrow 0$; // Record whether node pairs has converged :
      //0 represents not convergent. 1 convergent.
  while($Max(|s_k(V_a, V_b) - s_{k-1}(V_a, V_b)| / |s_{k-1}(V_a, V_b)|) > \varepsilon$)
    $k \leftarrow k+1$;
    $s_{k-1} \leftarrow s_k$;
    for each element $s_k(V_a, V_b)$
      if(flagmatrix(V_a, V_b) == 0)   // not convergent

$$s_k(V_a, V_b) = c \cdot \sum_{i=1}^{|I(V_a)|} \sum_{j=1}^{|I(V_b)|} T_{V_a V_i} \cdot T_{V_b V_j} \cdot s_{k-1}(V_i, V_j) \;;$$

    end for
    for each element $s_k(V_a, V_b)$
      If($|s_k(V_a, V_b) - s_{k-r}(V_a, V_b)| / |s_{k-r}(V_a, V_b)| < \varepsilon$)
        // r: an iteration number to control accuracy
        flagmatrix(V_a, V_b) = 1;
    end for
  end while
  return s_k

The time and space complexity of *Adaptive-SimRank* is the same as that of *SimRank*. r is used to decide whether or not this *SimRank* score has already converged. Commonly, r is equal to 1. We will discuss it in the next section.

5.3 Analysis of *Adaptive-SimRank*

As we have mentioned above, the process of node pair similarity calculation is the process to search the even-step paths between two nodes. In the kth step of iteration, algorithm *SimRank* searches for $2k$-step paths for each pair. For example, we calculate the *SimRank* score for the graph shown in Fig.5(a) and list their *SimRank* value in the right table. We can find that in the first iteration, 2-step paths, $<V_a, V_c, V_b>$ and $<V_a, V_d, V_b>$, have been searched and similarity score in this iteration is $(0.5\times0.5 + 0.5\times0.5) \times 0.8^1 = 0.4$. In the second iteration, 4-step paths, $<V_a, V_c, V_a, V_d, V_b>$, $<V_b, V_d, V_b, V_c, V_a>$, $<V_a, V_d, V_a, V_c, V_b>$, $<V_a, V_d, V_b, V_c, V_b>$, are discovered and added to the original similarity score. We can lock some node pairs' *SimRank* scores, which means that we will stop searching the path between these pairs. For example, if we fix the $s(V_c, V_d)$ value, 0.56, in the second step of iteration, we only find six paths, $<V_c, V_a, V_d>$, $<V_c, V_b, V_d>$, $<V_a, V_c, V_a, V_d, V_b>$, $<V_b, V_d, V_b, V_c, V_a>$, $<V_a, V_d, V_a, V_c, V_b>$ and $<V_a, V_d, V_b, V_c, V_b>$, between node V_c and V_d. Meanwhile, in the process of similarity calculation, node pairs reinforce each other to find paths, if we ignore some paths between node V_c and V_d, this will affect the final similarity score of pair (V_a, V_b). In Fig.5(b), after three steps of iteration, $s(V_a,V_b)$ will keep 0.62 forever. Based on the above analysis, if we lock those pairs which have not converged, that will affect accuracy. In Fig.4, the negative value shows that some pairs' *SimRank* scores in the kth iteration don't have a great change, but in $(k+1)^{th}$ iteration these pairs change again. However, in our experiments on four real datasets, which are shown in table 3, these pairs are less than 2 percent of total node pairs. Thus, it will not greatly affect accuracy. One way to increase accuracy is that we can increase the value of r, but that will affect the improvement of performance. In table 3, for these four datasets, we can check that when we use this judging formula, $|s_k(V_a,V_b) - s_{k-2}(V_a,V_b)| \,/\, |s_{k-2}(V_a,V_b)| < \varepsilon$, to predict the convergence of pairs, those converged pairs don't have a great change during following steps of iteration. In common situation, setting r to 1 is good for most datasets.

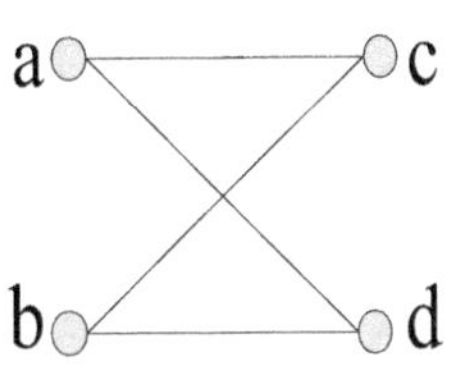

(a)

Iter. / Val.	1	2	3	4	5	6	...
s(a,b)	0.4	0.56	0.62	0.62	0.62	0.62	0.62
s(c,d)	0.4	0.56	0.56	0.56	0.56	0.56	0.56

(b)

Fig. 5. *SimRank* Calculation

Table 3. Accuracy Threshold VS Convergent Pairs

r / **Dataset**	**1**	**2**	**3**
Wisconsin	0.35%	0%	0%
Cornell	0.72%	0%	0%
Washington	0.11%	0%	0%
Texas	1.93%	0%	0%

6 Empirical Study

In this section, we conduct experiments to evaluate the performance of *Adaptive-SimRank*. We focus on the following questions:

***Q*1:** How accurate is *Adaptive-SimRank*?

***Q*2:** What is the computational cost of different algorithms?

***Q*3:** What is the worst situation about *Adaptive-SimRank*?

After describing the experimental setting in section 6.1, we will answer the question Q1 in section 6.2. In section 6.3, we will analyze the cost of *Adaptive-SimRank* on the website dataset. Finally, we will discuss on which dataset the performance of *Adaptive-SimRank* has the worst performance.

6.1 Experiment Setting

In this set of experiments, we use two different datasets, website datasets and synthetic datasets.

Website Dataset: We use the CMU four-university datasets [11]. These datasets contain web pages crawled from computer departments of four universities, Cornell, Texas, Washington, and Wisconsin. Web pages in these datasets have been manually divided into seven classes, student, faculty, staff, department, course, project and others. These classes will be used as the standard to evaluate the accuracy of our algorithm.

Synthetic Dataset: We use V_xE_y to represent the relationship graph with x nodes and y edges between these nodes. All of the edges have been randomly distributed between nodes. Clearly, this synthetic dataset can't be used to test the performance. However, because of the uniform distribution of links, this synthetic dataset can be used to represent the worst situation of the performance improvement by *Adaptive-SimRank*.

When we test the accuracy of these methods, we take PAM [12], a k-medoids clustering approach, to cluster objects based on similarity score calculated by these methods. We randomly select the initial centroids for 50 times when do clustering. We compare the clustering result with the real class and choose the most accurate results among the 50 results.

All our experiments are conducted on a PC with a 1.86G Intel Core 2 Processor, 2GB memory, under windows XP Professional environment. All of our programs are written in java.

6.2 Accuracy of *Adaptive-SimRank* Algorithm

To test accuracy, we compare *Adaptive-SimRank* with *SimRank* on the website datasets. We set r to 1. The result is shown in Fig.6.

From Fig.6 we can see that the accuracy of *Adaptive-SimRank* is slightly lower than *SimRank* algorithm, especially for the Texas dataset. The reason why the Texas dataset has a lowest accuracy is that in table 3, 1.93 percent of node pairs are misjudged to be convergent, which is the highest proportion among all the datasets. However, in sum, the loss of accuracy of *Adaptive-SimRank* is not much.

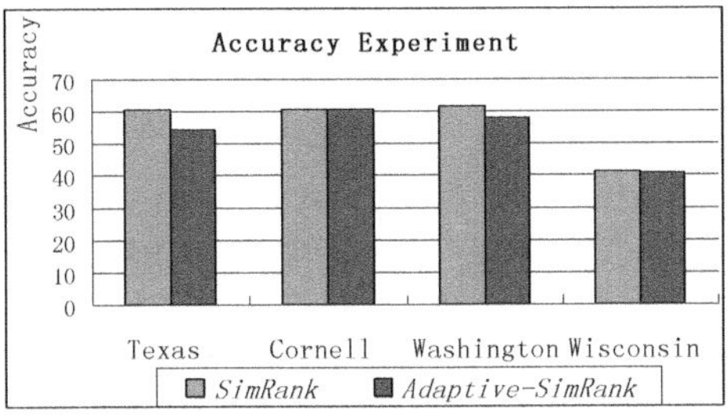

Fig. 6. Accuracy on web site datasets

6.3 Computation Cost of *Adaptive-SimRank* Algorithm

The computation cost for all these algorithms depends on two aspects, the time for each step of iteration and the number of iterations.

We test *Adaptive-SimRank* on the four website datasets. Fig.7 shows time cost at each step of iteration. We can find that from the first to the 17th step of iteration, the time cost of each iteration of *Ataptive-SimRank* is almost the same as that of *SimRank*, but after that the time cost of our algorithm drops quickly. Fig.4 and figures in appendix A show that between the 16th step of iteration and 35th step of iteration, a great number of pairs have converged. Thus, the time cost of our algorithm drops very quickly. Table 4 discribes the total time of these two methods, we can see that our method speeds up the orginal *SimRank* by nearly 35%.

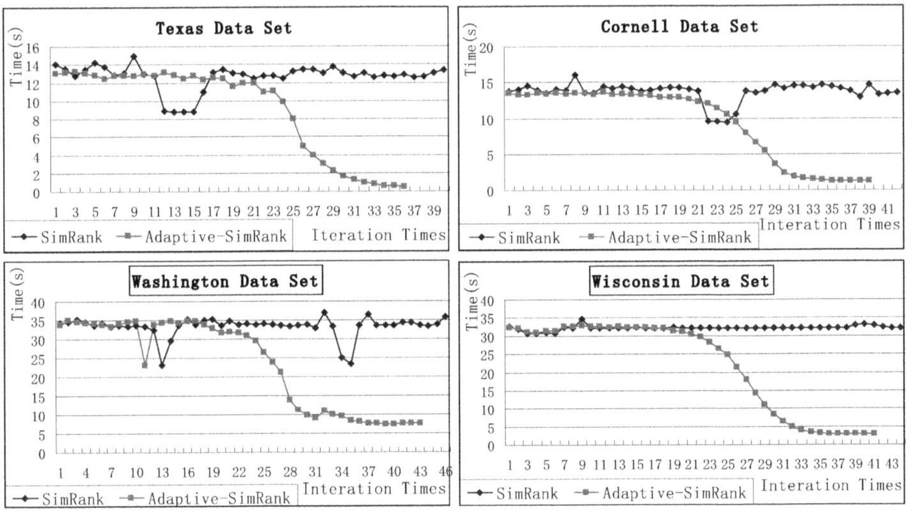

Fig. 7. Number of iterations VS Time(seconds)

Table 4. Total Time VS Algorithm

Alg. \ Data Set	Texas	Cornell	Washington	Wisconsin
SimRank	507.139s	576.199s	1535.818s	1417.959s
Adaptive-SimRank	**328.782s**	**362.531s**	**1013.637s**	**897.330s**

We also test the performance of our algorithm on graphs with different number of nodes. In table 5, we can see that with the increase of the number of nodes, the time cost of *SimRank* and *Adaptive-SimRank* increase quickly, while our method always improve the performance of *SimRank* by nearly 35%.

Table 5. Performance VS the Number of Nodes

Alg. \ **Node#**	**Wi. 1263**	**Wi.&Wa. 2468**	**Wi.&Wa.&T. 3295**	**Wi.&Wa.&T.&C. 4162**
SimRank	1417.959s	12336.839s	31385.259s	70025.662s
AdatptiveSimRank	**897.33s**	**7765.776s**	**18827.152s**	**38220.047s**

6.4 The Worst Situation of *Adaptive-SimRank*

In the end, we want to discuss the worst situation of *Adaptive-SimRank*. In our experiment, we find that the performance of *Adaptive-SimRank* does not keep very well for a graph whose links have a uniform distribution. Fig.8 shows the performance improvement proportion on the real datasets and uniform distribution graphs ($V_{827}E_{2667}$, $V_{867}E_{2691}$, $V_{1205}E_{3299}$, $V_{1263}E_{5305}$), which have the same number of nodes and links with the real datasets.

In Fig.8, the performance improvement proportion for the synthetic dataset is about half less than the real world dataset. Fig.9 shows the reason of this phenomenon. Because of uniform distribution of links, a great number of pairs converge in almost the same step, from the 26th step to the 28th step of iterations, not like the real datasets, whose link distribution is close to normal distribution. Thus, *Adaptive-SimRank* doesn't take a great advantage in this kind of graphs.

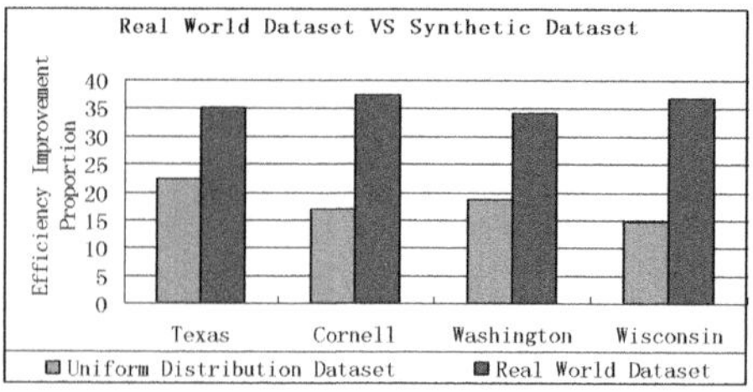

Fig. 8. Performance Improvement Proportion

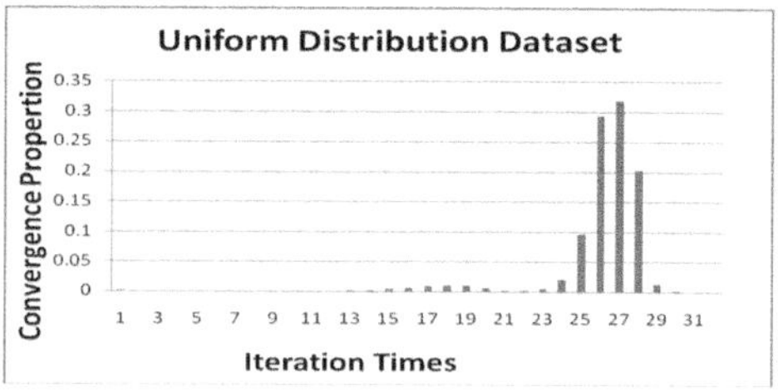

Fig. 9. Convergence Proportion on $V_{827}E_{2667}$

7 Conclusion

In this paper, we propose a new algorithm, *Adaptive-SimRank*, to calculate similarity measure *SimRank* based on object-to-object relationship efficiently. This work is based on our two observations. One is that most node pairs in the relationship graph converge to their final *SimRank* score quickly, while others need more time to converge. The other is that the distribution of the newly added converged node pairs in each step of iteration for the real graph is close to normal distribution. Algorithm *Adaptive-SimRank* exploits these observations to speed up the calculation of *SimRank* by avoiding unnecessary computation. The convergence of our algorithm is give in Appendix B. Empirical study demonstrate the efficiency and accuracy of our algorithm.

Acknowledge

This work was supported in part by the National Natural Science Foundation of China under Grant No. 70871068, 70621061, 70890083, 60873017, 60573092 and 60496325.

References

[1] Jeh, G., Widom, J.: SimRank: A measure of structural-context similarity. In: SIGKDD (2002)
[2] Small, H.: Co-citation in the scientific literature: A new measure of the relationship between two documents. Journal of the American Society for Information Science (1973)
[3] Kessler, M.M.: Bibliographic coupling between scientific papers. American Documentation (1963)
[4] Amsler, R.: Applications of citation-based automatic classification. Linguistic Research Center (1972)
[5] Fogaras, D., Racz, B.: Scaling link-base similarity search. In: WWW (2005)
[6] Yin, X.X., Han, J.W., Yu, P.S.: LinkClus: Efficient Clustering via Heterogeneous Semantic Links. In: VLDB (2006)
[7] Jeh, G., Widom, J.: Scaling personalized web search, Technical report (2001)
[8] Page, L., Brin, S., Motwani, R., References, T.: The PageRank citation ranking: Bringing order to the Web, Technical report (1998)
[9] Langville, A.N., Meyer, C.D.: Deeper inside PageRank. Internet Math. J. (2003)
[10] Kamvar, S., Haveliwala, T., Golub, G.: Adaptive Methods for the Computation of PageRank, Technical report (2003)
[11] CMU four university data set, `http://www.cs.cmu.edu/afs/cs/project/theo-20/www/data/`
[12] Han, J.W., Kamber, M.: Data Mining Concepts and Techniques. Morgan Kaufmann Publishers, San Francisco (2001)

Appendix A

Fig. 10 illustrates the convergence situation for each step of iteration on the other three datasets, Washington dataset, Texas dataset and Wisconsin dataset. Based on our observation, the distribution of the newly added converged node pair for the real world graph is close to normal distribution.

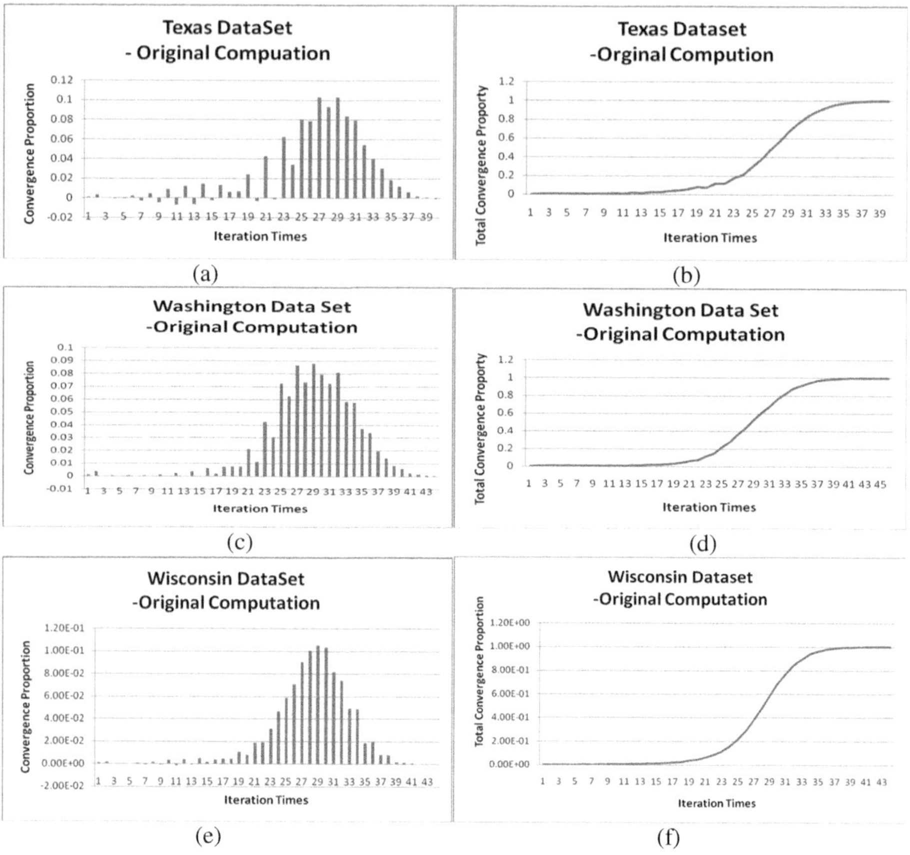

Fig. 10. Convergence on Texas dataset, Washington dataset, and Wisconsin dataset. The meanings of axes are the same as that of Fig.3.

Appendix B

Proof of convergence of *Adaptive-SimRank*

In this appendix, we prove the convergence of *Adaptive-SimRank*.

Lemma A: Let $s(V_a,V_b)$ be the similarity value calculated by *SimRank*, $s(V_a,V_b) = \sum_{t:(V_a,V_b)\sim\sim>(V_x,V_x)} P[t]c^{l(t)} = c\sum_{i=1}^{n_1} P[t] + c^2\sum_{i=1}^{n_2} P[t] + \ldots + c^k\sum_{i=1}^{n_k} P[t] + \ldots,\ s(V_a,V_b)$.

Let $s_a(V_a,V_b)$ be the similarity value calculated by *Adaptive-SimRank*, $s_a(V_a,V_b)$

$$= \sum_{t:(V_a,V_b)\sim\sim>(V_x,V_x)} P[t]c^{l(t)} = c\sum_{i=1}^{m_1} P[t] + c^2\sum_{i=1}^{m_2} P[t] + c^k\sum_{i=1}^{m_k} P[t] +\ldots,\ s_a(V_a,V_b).$$

Let $u_k = c^k \sum_{i=1}^{n_k} P[t]$, and $v_k = c^k \sum_{i=1}^{m_k} P[t]$, Then $0 \le v_k \le u_k$.

Proof: There are three situations for the kth step of iteration for *Adaptive-SimRank*.

Case 1: In the kth step of iteration, *Adaptive SimRank* locks the similarity value between V_a and V_b before the kth step of iteration. Thus, $0=v_k \le u_k$.

Case 2: In the kth step of iteration, *Adaptive-SimRank* searches some of the paths between V_a and V_b, but loses some paths, because that paths is ignored by the convergence of other pairs. Thus, $0 \le v_k < u_k$.

Case 3: In the kth step of iteration, *Adaptive-SimRank* searches all of the paths between V_a and V_b. Thus, $0 \le v_k = u_k$.
In sum, $0 \le v_k \le u_k$. . ■

Lemma B: $s_{ak}(V_a,V_b) \le s_k(V_a,V_b) \le Mc\dfrac{1-c^k}{1-c}$, $M \ge 0$.

Proof: According to Lemma A, it's easy to get $s_{ak}(V_a,V_b) \le s_k(V_a,V_b)$.

$$s_k(V_a,V_b) = c\sum_{i=1}^{n_1} P[t] + c^2\sum_{i=1}^{n_2} P[t] +\ldots+ c^k\sum_{i=1}^{n_k} P[t] +\ldots$$

Let M equal $max(\sum_{i=1}^{n_1} P[t], \sum_{i=1}^{n_2} P[t], \ldots, c^k\sum_{i=1}^{n_k} P[t], \ldots,)$, then

$$s_k(V_a,V_b) \le cM+c^2M + \ldots+c^kM+\ldots= Mc\frac{1-c^k}{1-c}$$

Thus, $s_{ak}(V_a,V_b) \le s_k(V_a,V_b) \le Mc\dfrac{1-c^k}{1-c}$. ■

Theorem: $s_a(V_a,V_b)$ will converge to a fixed value.

Proof: $s_a(V_a,V_b) = \lim_{k->+\infty} s_{ak}(V_a,V_b)$, $s(V_a,V_b) = \lim_{k->+\infty} s_k(V_a,V_b)$

$$\lim_{k->+\infty} s_{ak}(V_a,V_b) \le \lim_{k->+\infty} s_k(V_a,V_b) \le \lim_{k->+\infty} Mc\frac{1-c^k}{1-c} = \frac{Mc}{1-c}$$

Therefore, $s_a(V_a,V_b)$ has a upper bound.
In Lemma A, $0 \le v_k$, thus $s_a(V_a,V_b)$ is the positive term series.
Thus, *Adpative-SimRank* will converge. ■

Periodic Pattern Analysis in Time Series Databases

Johannes Assfalg, Thomas Bernecker, Hans-Peter Kriegel, Peer Kröger, and Matthias Renz

Institute for Informatics, Ludwig-Maximilians-Universität München, Germany
{assfalg,bernecker,kriegel,kroegerp,renz}@dbs.ifi.lmu.de

Abstract. Similarity search in time series data is used in diverse domains. The most prominent work has focused on similarity search considering either complete time series or certain subsequences of time series. Often, time series like temperature measurements consist of periodic patterns, i.e. patterns that repeatedly occur in defined periods over time. For example, the behavior of the temperature within one day is commonly correlated to that of the next day. Analysis of changes within the patterns and over consecutive patterns could be very valuable for many application domains, in particular finance, medicine, meteorology and ecology. In this paper, we present a framework that provides similarity search in time series databases regarding specific periodic patterns. In particular, an efficient threshold-based similarity search method is applied that is invariant against small distortions in time. Experiments on real-world data show that our novel similarity measure is more meaningful than established measures for many applications.

1 Introduction

In a large range of application domains, e.g. environmental analysis, evolution of stock charts, research on medical behavior of organisms, or analysis and detection of motion activities we are faced with time series data that feature cyclic activities composed of regularly repeating sequences of activity events. In particular for the recognition and analysis of activities of living organisms, cyclic activities play a key role. For example, human motions like walking, running, swimming and even working are composed of cyclic activities that correspond to significant motion events.

In this paper, we focus on similarity search on time series with a special focus on cyclic activities, in particular on the evolution of periodic patterns that repeatedly occur in specified periods over time. Examples of such time series are depicted in Figure 1(a). The upper time series shows the motion activity of a human, in particular the vertical acceleration force that repetitively occurs during a human motion like walking or running. Consecutive motion patterns show similar but distinct characteristics. We can observe changes in the shape of consecutive periodic patterns that are of significant importance if, for example, we want to analyze the motion behavior of any person. Many other applications that take advantage of the ability to examine the evolution of periodic patterns can be found in the medical or in the biological domain. For example, chronobiologists are highly interested in exploring the relationship between the activity of a cell or a complete organism and the amount as well as the duration of daylight affecting the

X. Zhou et al. (Eds.): DASFAA 2009, LNCS 5463, pp. 354–368, 2009.

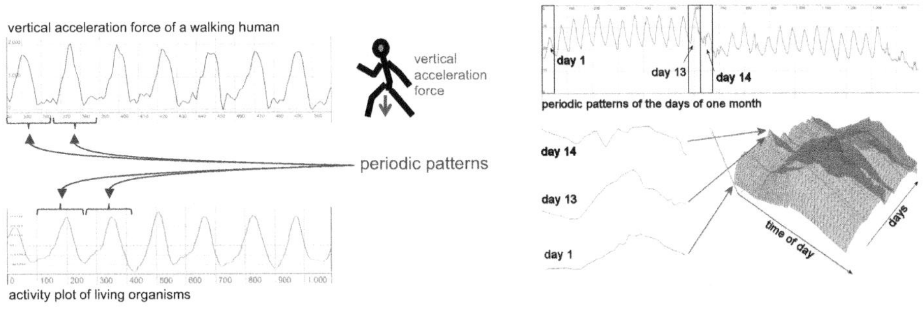

(a) Evolution of periodic patterns in medical and biological applications.

(b) Original and dual-domain representation of a time series.

Fig. 1. Origin, application and representation of periodic patterns

cell or organism. Obviously, an important task is the identification of similar periodic patterns in daylight cycles and biological responses like the concentration of hormones (cf. Figure 1(a)). Another important domain where we find lots of time series containing periodic patterns is the environmental research. Examples are time series that describe the change of temperature values measured several times a day for each day within a month. In this case, the periodic pattern is the temperature course of a day. In order to be able to track the evolution of such periodic patterns, we propose to string consecutive patterns together to a sequence of patterns as shown in the example depicted in Figure 1(b). A time series is then split into a sequence of subsequences, which we call dual-domain time series. It represents the temporal behavior along a "second" time axis, e.g. each hour of a day. The original time domain is thus made coarser, e.g. it now represents each day of the entire time period. This way, we are able to define structures modelling the characteristic of the evolution of periodic patterns. In our example application, the new time series model allows us to examine the evolution of global climatic changes by considering the summer or winter months of the last 20 years. Contrary to [2], where the focus lies on the determination of periodicity features or the detection of motion directly from the periodic patterns, we take our attention to methods that help us to analyze the evolution of periodic patterns.

Given the new time series model, we are now interested in the examination of things that happen at a certain time. Thereby, we have our focus on the relationship between the times of both time domains at which certain events occur. Here, we take special emphasis on events that refer to an exceeding of a given activity threshold. Similarity search methods based on events that refer to exceeding of a given activity threshold have been introduced in [3,4]. Given a certain threshold value τ, this approach reduces single-domain time series to a sequence of intervals corresponding to time periods where the amplitude value of a time series exceeds τ. Our approach represents periodic patterns as polygons, that analogously correspond to threshold-exceeding amplitude values. This approach is useful for a lot of application domains, where the exact value of a time series is less important than the fact whether a certain amplitude (activity) threshold is exceeded or not. Furthermore this approach is more robust to noise and errors in measurement. We are subsequently able to identify similar threshold-exceeding patterns by comparing polygons. In order to efficiently perform similarity queries, we extract relevant feature information from those polygons.

The main contributions of this paper are the following: We introduce a new similarity measure for time series that takes two time domains into account. For the similarity measures we propose feature-based representations of dual-domain time series and show how they can be organized in an efficient way. The rest of this paper is organized as follows: First we introduce a matrix representation of dual-domain time series. Afterwards we introduce the so-called intersection set, that consists of the polygons generated by a threshold plane intersecting the dual-domain time series at a given threshold value τ. Furthermore, we present an approach to efficiently process index-supported similarity search based on periodic patterns. For that purpose we employ different features that are extracted from the intersection sets. Finally we evaluate the efficiency as well as the effectiveness of our approach in a broad experimental section.

2 Related Work

There are a lot of existing approaches performing similarity search on time series. Searching patterns can be supported by the Dynamic Time Warping approach (DTW) that is introduced for data mining in [7] and that presents a possibility to match the most corresponding values of different time series. Since the length of time series is very often quite large, the DTW approach suffers from its quadratic complexity with respect to the length of the time series. Thus a number of dimensionality reduction techniques exist. For example the Discrete Wavelet Transform (DWT) [1], the Discrete Fourier Transform (DFT) [16], the Piecewise Aggregate Approximation (PAA) [15,23], the Singular Value Decomposition (SVD) [20], the Adaptive Piecewise Constant Approximation (APCA)[14], Chebyshev Polynomials [9], or the Piecewise Linear Approximation (PLA) [17] could be used. In [12], the authors propose the GEMINI framework, that allows to incorporate any dimensionality reduction method into efficient indexing, as long as the distance function on the reduced feature space fulfills the lower bounding property. However, those solutions are hardly applicable for searching similar patterns because in most cases, important temporal information is lost. In contrast to those solutions, in [21] the authors propose a bit sequence representation of time series. For each amplitude value, a corresponding bit is set if this value exceeds a certain threshold value. Similarity is finally computed based on those bits in an efficient way, since this approach lower bounds the Euclidean Distance or DTW. However, it is not possible to specify a certain threshold value at query time. This problem is addressed with inverse queries in [19].

Many approaches for similarity search on time series are based on features extracted from time series, i.e. in [18,10,13]. A similarity model for time series that considers the characteristics of the time series was proposed in [22], where a set of global features including periodicity, self-similarity, skewness and kurtosis among others is used to compute the similarity between time series. The features proposed in [5] are calculated over the whole amplitude spectrum. Thus, time-domain properties can be captured over the whole available amplitude range.

In this paper, we consider properties for the whole amplitude range of dual-domain time series which is novel to the best of our knowledge.

3 Time Series Representation

3.1 Dual-Domain Time Series

Intuitively, a dual-domain time series is a sequence of sequences, i.e. we have an amplitude spectrum and – in contrast to traditional single-domain time series – two time axes. More formally, a dual-domain time series is defined by

$$X_{dual} = \langle\langle x_{1,1}, \ldots, x_{1,N-1}, x_{1,N}\rangle, \ldots, \langle x_{M,1}, \ldots, x_{M,N-1}, x_{M,N}\rangle\rangle$$

where $x_{i,j}$ denotes the value of the time series at time slot i in the first (discrete) time domain $T = \{t_1, \ldots, t_N\}$ and at time slot j in the second (discrete) time domain $S = \{s_1, \ldots, s_M\}$. In the following, we call the $x_{i,j}$ *measurement configurations*. We assume $\forall i \in 1, \ldots, N-1 : t_i < t_{i+1}$ and $\forall j \in 1, \ldots, M-1 : s_j < s_{j+1}$.

Both axes T and S may also be any other ordered domain such as a spatial axis or a color spectrum, so that the concepts presented in this paper can also be applied to such types of data. The concepts can further be extended to the case of a multi-domain representation of time series. For the sake of presentation, we focus on dual-domain time series with two time domains, i.e. T and S are domains of discrete time slots.

3.2 Intersection Sets

As proposed in [3,4], time series considering a single time domain can be represented as a sequence of intervals according to a certain threshold value τ. For the recognition of relevant periodic patterns that are hidden in the matrix representation of dual-domain time series, we extend this approach to a novel abstract meaning. Hence, we consider an abstraction of the time series. In case of multiple domains, we speak of an *n-domain* time series, where the dual-domain case corresponds to $n = 2$. Adding the amplitude axis to the n-domain time series yields an $(n+1)$-dimensional surface.

The dual-domain time series can be structured using an elevation grid which is created by the grid squares of the measurement configurations $x_{i,j}$ where $1 \leq i \leq N$ and $1 \leq j \leq M$ (cf. Section 3.1). Each grid square of a time series X_{dual} (in the following

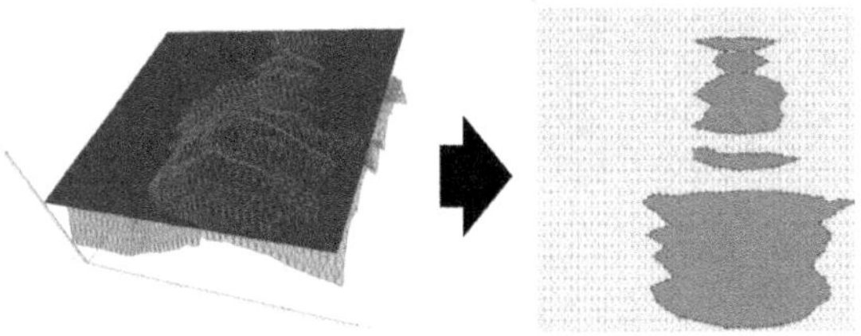

Fig. 2. Dual-domain time series with a threshold plane and the intersection polygon set

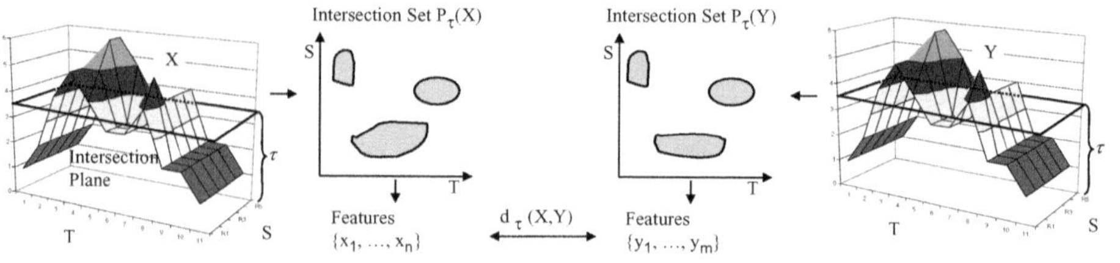

Fig. 3. Similarity measure for a given threshold τ

denoted as X for simplicity) can be denoted by $(x_{i,j}, x_{i+1,j}, x_{i+1,j+1}, x_{i,j+1})$ where $1 \leq i \leq N-1$ and $1 \leq j \leq M-1$. In our case, the threshold line for τ corresponds to an n-dimensional threshold hyperplane which intersects the time series. The result of this intersection is a set of n-dimensional polygons $P_\tau(X) = \{p_1, \ldots, p_K\}$ which we call the *intersection set*. The intersection set is created by intersecting the plane with the amplitudes of each of the grid squares. An example of an intersection set is depicted in Figure 2. The polygons of an intersection set with respect to a certain value of τ contain those amplitude values of the time series that are above the threshold plane τ, and thus, they deliver all the information about the periods of time during which the n-domain values of the time series exceed τ. With this abstraction, we are able to compare two time series with respect to coherences in time.

4 Similarity Query Processing

4.1 Similarity Measure and Feature Transformation

In order to analyze dual-domain time series based on periodic patterns with respect to a certain threshold τ, we have to define a distance value for such time series. As described above, the patterns of interest emerge as polygons forming intersection sets that are created. So the distance value $d_\tau(X, Y)$ of two dual-domain time series X and Y should reflect the dissimilarity of their corresponding intersection sets. In order to save computational cost and to allow for the usage of index structures like the R^**-tree* [6], we derive local or global features for the polygons and compare these features instead of the exact polygons (cf. Figure 3). In the following sections, we describe several local and global features suitable for capturing the characteristics of the intersection sets.

4.2 Similarity Measure Based on Local Features

Local features describe a polygon p belonging to an intersection set $P_\tau(X)$ for a given dual-domain time series X and a given threshold value τ. Let a polygon p consist of $|p|$ vertices $v_1, \ldots, v_{|p|}$. Let vertex v_i be defined by the tuple $(x_i, y_i) \in T \times S$. Then the *Polygon Centroid* feature (PC) describes the position of the vertices by calculating their central point. Formally, the PC feature of a polygon p is defined as

$$PC(p) = \frac{1}{|p|} \sum_{i=1}^{|p|} v_i = \left(\frac{1}{|p|} \sum_{i=1}^{|p|} x_i, \frac{1}{|p|} \sum_{i=1}^{|p|} y_i \right).$$

The *Polygon MBR* feature (PM) is a conservative approximation of a polygon. It describes a polygon by means of its minimal bounding rectangle (MBR). Formally, the 4-dimensional PM feature of a polygon p is defined as

$$PM(p) = \left(\min_{i=1..|p|}(x_i), \min_{i=1..|p|}(y_i), \max_{i=1..|p|}(x_i), \max_{i=1..|p|}(y_i) \right).$$

The number of polygons varies for different intersection sets and so does the number of local features. In order to calculate the distance value $d_\tau(X, Y)$ based on local features we employ the Sum of Minimal Distance (SMD) measure [11]. The SMD matches each polygon (i.e. the corresponding local feature) of $P_\tau(X)$ to its best matching partner of $P_\tau(Y)$ and vice versa:

$$d_\tau(X,Y) = \tfrac{1}{2}\left(\frac{1}{|P_\tau(X)|}\sum_{x \in P_\tau(X)}\left(\min_{y \in P_\tau(Y)} d(x,y)\right) + \frac{1}{|P_\tau(Y)|}\sum_{y \in P_\tau(Y)}\left(\min_{x \in P_\tau(X)} d(x,y)\right)\right)$$

where x and y are the features describing the elements of the intersection set and where $d(x, y)$ is a distance function defined on these features. In our case, this distance function is the Euclidean distance.

4.3 Similarity Measure Based on Global Features

Contrary to local features, global features try to capture the characteristics of an intersection set by a single feature value or feature vector. In this section, we present three examples for global features.

Let $P_\tau(X)$ be an intersection set as described above. Let furthermore K be the number of polygons $P_\tau(X)$ consists of. Then the *Intersection Set MBR* feature (ISM) is the global version of the local PM feature approximating the complete set of polygons by a minimal bounding rectangle. So, ISM is defined analogously as

$$ISM(P_\tau(X)) = \left(\min_{\substack{i=1..|p| \\ k=1..K}}(x_{k,i}), \min_{\substack{i=1..|p| \\ k=1..K}}(y_{k,i}), \max_{\substack{i=1..|p| \\ k=1..K}}(x_{k,i}), \max_{\substack{i=1..|p| \\ k=1..K}}(y_{k,i}) \right).$$

The *Intersection Set Centroid* feature (ISC) is the global variant of the local PC feature considering all polygon vertices of the intersection set. Let S be the overall number of all vertices of all polygons of $P_\tau(X)$. Then $ISC(P_\tau(X))$ analogously calculates the central point of all vertices of the intersection set:

$$ISC(P_\tau(X)) = \frac{1}{S}\sum_{i=1}^{S} v_i = \left(\frac{1}{S}\sum_{i=1}^{S} x_i, \frac{1}{S}\sum_{i=1}^{S} y_i \right).$$

A more sophisticated high-level feature is the *Fill Quota* feature (FQ). For each row and each column of the data matrix, the percentage of polygon coverage is computed. Hence, the horizontal and vertical values generate two single-domain time series that describe the position as well as the size of the polygons. The computation of the polygon coverage is processed based on the grid squares (cf. Section 3.1). Each grid square is tested for its contribution to the area of a polygon. For a dual-domain time series X that consists of N rows and M columns, we denote the coverage area of a grid square at the position (i, j) by $A_{i,j}, i = 1..N, j = 1..M$. For the i-th row and the j-th column, the values are computed as follows:

$$FQ(x_i) = \frac{1}{M} \sum_{j=1}^{M} A_{i,j} \text{ and } FQ(y_j) = \frac{1}{N} \sum_{i=1}^{N} A_{i,j}.$$

Afterwards we apply a standard technique for dimensionality reduction to the projected time series so that we store only n feature values $c_1, \ldots, c_n$ (for example Fourier coefficients) for each of the two projected time series $FQ(x)$ and $FQ(y)$. Note that $n \ll N$ and $n \ll M$. This leads to the following definition of the FQ feature:

$$FQ(P_\tau(X)) = (c_1(FQ(x)), \ldots, c_n(FQ(x)), c_1(FQ(y)), \ldots, c_n(FQ(y))).$$

The distance value for two intersection sets based on global features can be calculated without the SMD measure, as for each intersection set we derive the same amount of global features. So in this case, $d_\tau(X, Y)$ is calculated as the Euclidean distance between the associated global features.

5 Efficient Query Processing

In the previous section, we introduced similarity measures which are adequate to compare evolutions of periodic patterns in time series. In this section, we show how similarity queries based on the proposed similarity measures can be performed in an efficient way. In particular, we consider the ε-range query and the k-nearest-neighbor query which are the most prominent similarity query methods and are used as basic preprocessing steps for data mining tasks [8].

The proposed methods are based on the features extracted from the original time series as described in Section 4.1. Since, the feature extraction procedure, in particular the computation of the intersection sets, is very time consuming, it is not feasible to do at query time. For this reason, we propose to do the feature extraction in a preprocessing step and organize the precomputed features in an efficient way. For example, the features can be extracted during the insertion of the object into the database. Thereby we have to solve the problem that the features extracted from the objects are associated with certain amplitude-threshold values. For this reason, we either have to define a fixed threshold which is used for all similarity computations or we have to precompute the features for all possible threshold values. The former solution is too restrictive as it does not allow the option for an adequate threshold readjustment at query time. On the other hand there exists an unlimited number of thresholds which would have to be taken into account leading to immense storage overhead.

5.1 Feature Segments

In the following, we propose a method for the efficient management of preextracted features that allows for the specification of the threshold at query time. In fact, we have

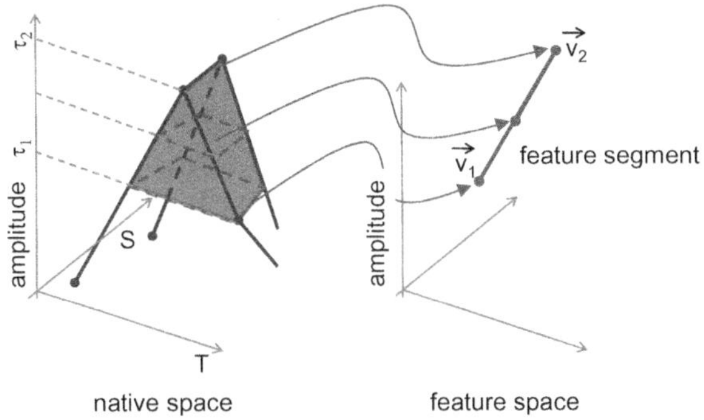

Fig. 4. Extracting feature segments from a time series

to take only a finite number of thresholds into account to derive features for all possible thresholds. In particular, for a dual-domain time series of size n (n different amplitude values), we need to extract only features for at most n different thresholds. The features for the remaining threshold values can be easily computed by means of linear interpolation. Let us assume that we are given all amplitude values of a dual-domain time series in ascending order of their amplitude values. The topology of the intersection sets associated with thresholds equal to two adjacent amplitude values does not change. Furthermore, also the change of the shape of all intersection sets associated with the corresponding thresholds is steady and homogeneous. As a consequence, we only need to extract the features at thresholds specified by all amplitude values occurring in a given time series. The features between two adjacent values of these amplitude values can be generated by linear interpolation, so we store each d-dimensional feature as a $(d+1)$-dimensional *feature segment*. An example is illustrated in Figure 4.

On the left hand side, there is a section of a dual-domain time series from which we extract polygons (dotted lines) for the intersection planes at the two thresholds τ_1 and τ_2. The corresponding features points (vectors) $\boldsymbol{v_1}$ and $\boldsymbol{v_2}$ are sketched on the right hand side. The features of all polygons that result from intersection planes at thresholds between τ_1 and τ_2 are represented by the line between $\boldsymbol{v_1}$ and $\boldsymbol{v_2}$ called *feature segment*. Consequently, we only need to extract and manage a set of feature segments corresponding to a finite set of thresholds which are bound by the size of the corresponding dual-domain time series. In order to calculate the features of all objects at query time, we have to intersect the feature segments with the intersection plane corresponding to the given query threshold τ. Obviously, at query time we only need to take those feature segments into account that intersect the query threshold τ. The query cost can be reduced if we use an adequate organization of the feature segments that allows us to access only the feature segments which are relevant for a certain query.

5.2 Feature Segment Organization

After extracting the feature segments from all time series objects, they are partitioned and stored in disc blocks of equal size. As mentioned above, for efficiency reasons it is necessary to organize the feature segments of all objects in such a way that given a query

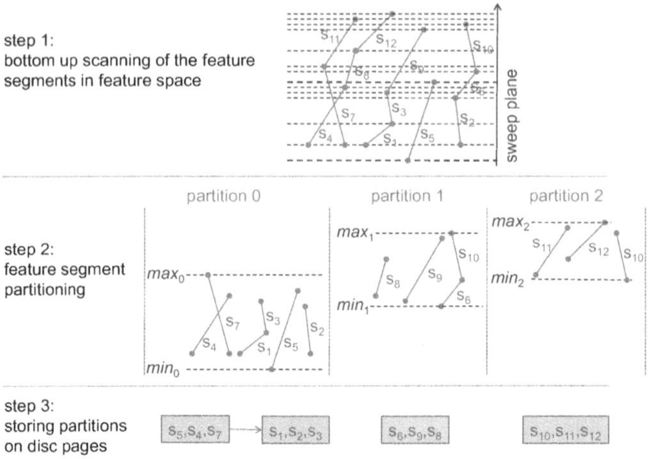

Fig. 5. Partitioning and storing of feature segments

threshold τ only those segments need to be accessed that intersect τ. For that purpose, we sort the feature segments of all objects according to the lower bound of the segments in ascending order and partition them into groups which are stored into equal-sized disc blocks. This is done using a sweep plane as illustrated in the example depicted in Figure 5 (step 1). During the sweep plane scan over the feature space, the feature segments which have been reached by the sweep plane are collected into a group. After a fixed number k of segments has been collected (the number k is based on the capacity of a disc block. In our example the disc block capacity k is set to 3), the minimal amplitude vale min_i and the maximal amplitude value max_i over all segments collected so far are computed (step 2). Then the algorithm proceeds with the next segments. If a new segment lies within min_i and max_i, it is added to group i using a new disc block. This new block is concatenated to the existing blocks of group i. In case group i is not suitable for the storage of a segment, a new group is created. The bounds of this new group are determined as soon as the first block of the new group is filled.

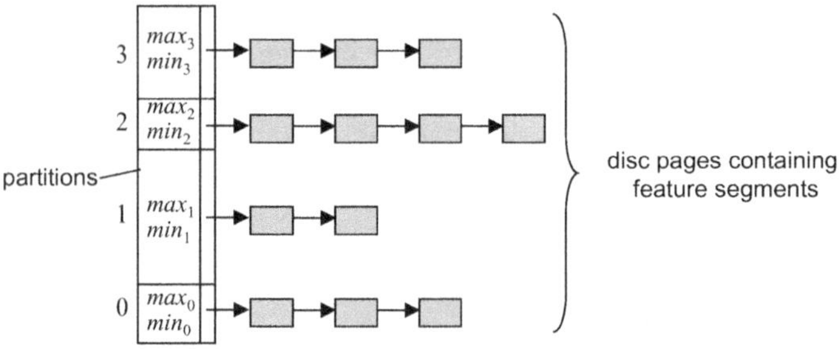

Fig. 6. Efficient organization of disc pages containing the feature segments

The resulting disc blocks are organized group-wise within the following indexing structure. As mentioned above, each feature segment group i consists of concatenated disc blocks containing the feature segments and the minimum and maximum value min_i and max_i. The two values min_i and max_i are used to index the feature segment groups. The index consists of an array of triples (min_i, max_i, ref) which correspond to the feature segment group i. The triple entity ref is a pointer to the disc blocks containing the feature segments of the corresponding group. The array entries are sorted in ascending order of the corresponding min_i value. The index structure is illustrated in Figure 6.

5.3 Query Processing

At query time, the introduced index allows us to efficiently search for the relevant feature segment group using binary search over the triple entries min_i. Here, we assume that the array fits into main memory. Otherwise, a secondary storage structure is required to index the feature segment groups w.r.t. the min_i value. Following the ref pointer all relevant feature segments can be sequentially accessed by scanning the corresponding disc pages. If a new object is inserted into the database, its feature segments are generated and sorted in ascending order w.r.t. their minimum amplitude value. The algorithm then tries to insert the new feature segments into existing groups. A new group is not generated until the insertion of new segment entries would require an enlargement of the max_i value of one of the existing group i. In case of a deletion of an object, we need a complete scan over the feature segment groups and remove the corresponding entries. Afterwards we try to merge disc blocks of a group that are not completely filled anymore. If a group consists of only one disc block which is less than half full, then the group is deleted and the remaining entries of the disc block are assigned to one of the neighboring groups.

6 Experimental Evaluation

6.1 Datasets

We evaluated the effectiveness of similarity search on dual-domain time series utilizing two real-world datasets. The *TEMP* dataset contains environmental time series data created by stationary measurements of several years[1]. It consists of 60 temperature time series from the year 2000 to 2004 that have been preclassified corresponding to the seasons summer and winter. Each object consists of up to 31 days and each day is represented by 48 measurements that have been normalized because of matching reasons on different ranges of the temperature measures that occur with different months. The *NSP* dataset is a chronobiologic dataset describing the cell activity of Neurospora[2] within sequences of day cycles. This dataset is used to investigate endogenous rhythms. We

[1] http://www.lfu.bayern.de/

[2] Neurospora is the name of a fungal genus containing several distinct species. For further information see *The Neurospora Home Page*: http://www.fgsc.net/Neurospora/neurospora.html.

converted single-domain time series that describe cell activities by splitting the measurements according to the artificial day cycle. The *NSP* dataset consists of 120 objects and has been classified into five classes according to the day cycle length (16, 18, 20, 22, and 26 hours). The efficiency evaluation was performed on an artificial dataset that contained 10-1000 objects. We created two subsets having different resolutions: Each time series of the dataset ART_{20} consisted of 20×20 measurements. Analogously, the dataset ART_{50} contained objects having a resolution of 50×50 measurements.

6.2 Effectiveness of the Time Series Representation

Considering the single-domain representation of time series, we performed similarity queries on the given datasets utilizing the techniques that are applicable for computing similarity on single-domain time series, such as the Euclidean distance (in the following denoted as *EUCL*), the DTW [7] and the threshold-based approach [3,4], in the following referred to as *THR*. Later in this section, we outline the obtained results of our newly introduced approach of measuring the distances for comparison. For an explanation of the tables and the curves that appear in this section, we give a short overview of the distance measures that have been considered for the experimental evaluation in Table 1.

Table 1. Distance measures considered for the evaluation

Abbreviation	Description
EUCL	Euclidean distance
DTW	Dynamic Time Warping
THR	Threshold Distance (single-domain approach)
ISC	Intersection Set Centroid feature
ISM	Intersection Set MBR feature
FQ	Fill Quota feature
PC	Polygon Centroid feature
PM	Polygon MBR feature

Table 2 lists the average precision values for the different similarity measures utilizing the single-domain representation of the datasets. In comparison to the Euclidean distance and the DTW approach, the threshold-based similarity measure hardly leads to higher average precision values. This can be observed for both datasets *TEMP* and *NSP*. As we outline in the following, with our newly introduced dual-domain representation approach in combination with a suitable threshold and the presented features, we are able to improve these results. We also compared the effectiveness of the local and global features that have been introduced in Section 4 using our threshold-based approach for different values of τ.

The results vary significantly with the threshold τ and also with the datasets. Depending on τ and on the used feature we outperform the Euclidean distance calculated on the time series. For the *TEMP* dataset and the local features (PC and PM) a threshold value of $\tau = 0.5$ yields the best results. The global features perform better for a threshold value of $\tau = 0.75$ (cf. Table 2(a)). The evaluation of the feature-based

Table 2. Average precision for different measures on the single-domain and the dual-domain time series representation for a given threshold τ

Measure	EUCL	DTW	THR	ISC	ISM	FQ	PC	PM
$\tau = 0.25$	0.58	0.62	0.55	0.54	0.55	0.55	**0.56**	0.55
$\tau = 0.5$	0.58	0.62	0.58	**0.65**	0.54	0.59	0.61	0.61
$\tau = 0.75$	0.58	0.62	0.56	0.60	**0.67**	0.58	0.61	0.60

(a) Average precision achieved on the *TEMP* dataset.

Measure	EUCL	DTW	THR	ISC	ISM	FQ	PC	PM
$\tau = 0.25$	0.56	0.52	0.56	0.73	**0.99**	0.74	0.41	0.53
$\tau = 0.5$	0.56	0.52	0.60	0.80	**0.93**	0.63	0.55	0.59
$\tau = 0.75$	0.56	0.52	0.47	0.51	**0.67**	0.46	0.52	0.52

(b) Average precision achieved on the *NSP* dataset.

approach using the *NSP* dataset leads to different results. Especially the utilization of the ISM feature leads to a high degree of effectiveness. In general, similarity search based on the dual-domain time series representation leads to better results with higher average precision in comparison to the single-domain approach. Figure 7 depicts two precision-recall plots that support this statement. In this figure we included the results for the single-domain representation (DTW and Euclidean distance) as well as for the dual-domain representation in combination with different features.

6.3 Efficiency of Threshold-Based Similarity Search on Dual-Domain Time Series

In order to evaluate the efficiency of our newly introduced approach we performed ϵ-range queries with a varying database size on the datasets ART_{20} and ART_{50} and measured the query time. The query objects were selected randomly and averaged the results. We examined the benefit of precalculating the features by storing them as segments as described in Section 5. The corresponding results are marked with "pre-calc". For comparison, we calculated the intersection sets and features at query time and labelled the results with "onl. calc". Furthermore, we performed similarity search using the traditional measures having a single-domain time series representation (i.e. Euclidean distance and DTW). Figure 8(a) depicts that calculating the required information at runtime is significantly more expensive than retrieving the information from our precalculated segments. However, the DTW can be outperformed anyway. Utilizing pre-calculation yields a better runtime than if the Euclidean distance is applied. Here, the threshold-based approach benefits from its reduction of dimensionality. Obviously, the runtime for the dataset ART_{50} is significantly higher than for the dataset ART_{20}, which is due to the complexity of the data and thus of the intersection sets. Figure 8(b) depicts a difference in the runtime comparing local and global features, representatively

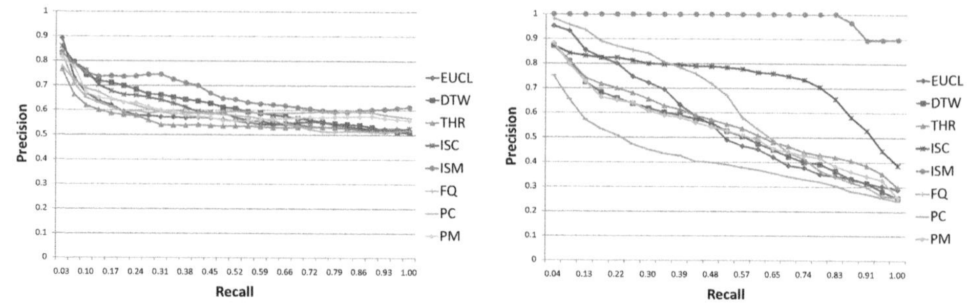

(a) Evaluation of the *TEMP* dataset having $\tau = 0.75$.

(b) Evaluation of the *NSP* dataset having $\tau = 0.25$.

Fig. 7. Precision-recall plots for different features and different representations

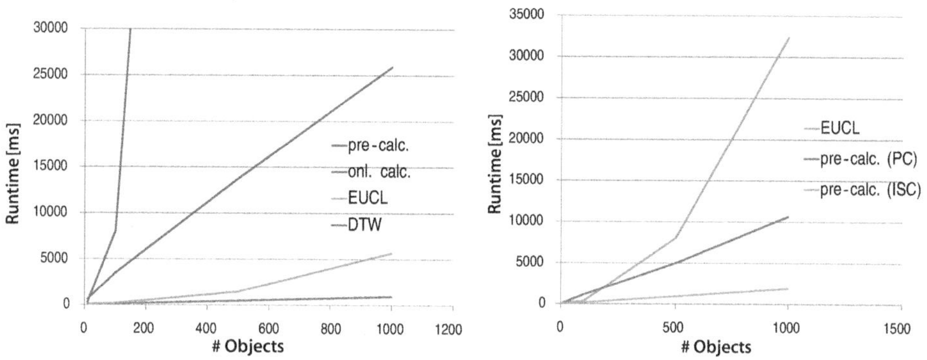

(a) Results for dataset ART_{20}.

(b) Results for dataset ART_{50}.

Fig. 8. Results of the efficiency evaluation having a threshold value of $\tau = 0.25$

performed using the PC and the ISC feature. This is due, on the one hand, to the SMD that has to be applied with the local features but not with the global features, and on the other hand to the number of local features that is significantly higher than that of the global features, since each polygon has to be described separately.

7 Conclusions

In this paper, we proposed a new approach to perform similarity search on time series having periodic patterns in an effective and efficient way. Regarding single-domain time series having periodic patterns, the threshold-based approach can hardly improve the results of similarity computations in comparison to traditional techniques like the Euclidean distance or the DTW. Transforming the time series into the dual-domain space and thus considering the periodicity, we can better observe how the patterns change in time. Furthermore, the ability to focus on relevant amplitude thresholds by utilizing the extraction of polygons from the time series and the usage of suitable, even simple features enables us to get better results for periodic pattern analysis. The quality of the results when utilizing an arbitrary feature however depends on the datasets. As a consequence, the effectiveness of global and local features varies with the type of data. The results with respect to the performance show a clear tendency. Regarding the traditional techniques and further a straightforward approach of computing the polygons and extracting the features from the time series at query time, we can reduce the runtime of similarity queries significantly by performing the computation and extraction in a preprocessing step. Furthermore, similarity computations using global features can be processed more efficiently than with local features.

References

1. Agrawal, R., Faloutsos, C., Swami, A.N.: Efficient similarity search in sequence databases. In: Lomet, D.B. (ed.) FODO 1993. LNCS, vol. 730, pp. 69–84. Springer, Heidelberg (1993)
2. Albu, A.B., Bergevin, R., Quirion, S.: Generic temporal segmentation of cyclic human motion. Pattern Recognition 41(1), 6–21 (2008)
3. Aßfalg, J., Kriegel, H.-P., Kröger, P., Kunath, P., Pryakhin, A., Renz, M.: Similarity search on time series based on threshold queries. In: Ioannidis, Y., Scholl, M.H., Schmidt, J.W., Matthes, F., Hatzopoulos, M., Böhm, K., Kemper, A., Grust, T., Böhm, C. (eds.) EDBT 2006. LNCS, vol. 3896, pp. 276–294. Springer, Heidelberg (2006)
4. Aßfalg, J., Kriegel, H.-P., Kröger, P., Kunath, P., Pryakhin, A., Renz, M.: Threshold similarity queries in large time series databases. In: Proceedings of the 22nd International Conference on Data Engineering, ICDE 2006, Atlanta, GA, USA, April 3-8, 2006, p. 149 (2006)
5. Aßfalg, J., Kriegel, H.-P., Kröger, P., Kunath, P., Pryakhin, A., Renz, M.: Similarity search in multimedia time series data using amplitude-level features. In: Satoh, S., Nack, F., Etoh, M. (eds.) MMM 2008. LNCS, vol. 4903, pp. 123–133. Springer, Heidelberg (2008)
6. Beckmann, N., Kriegel, H.-P., Schneider, R., Seeger, B.: The R*-tree: An efficient and robust access method for points and rectangles. In: Proceedings of the 1990 ACM SIGMOD International Conference on Management of Data, Atlantic City, NJ, May 23-25, 1990, pp. 322–331 (1990)
7. Berndt, D.J., Clifford, J.: Using dynamic time warping to find patterns in time series. In: KDD Workshop, pp. 359–370 (1994)
8. Böhm, F.K.C.: The k-nearest neighbor join: Turbo charging the kdd process. Knowledge and Information Systems (KAIS) 6(6) (2004)
9. Cai, Y., Ng, R.T.: Indexing spatio-temporal trajectories with chebyshev polynomials. In: SIGMOD Conference, pp. 599–610 (2004)
10. Deng, K., Moore, A., Nechyba, M.: Learning to recognize time series: Combining arma models with memory-based learning. In: IEEE Int. Symp. on Computational Intelligence in Robotics and Automation, vol. 1, pp. 246–250 (1997)
11. Eiter, T., Mannila, H.: Distance measures for point sets and their computation. Acta Informatica 34(2), 109–133 (1997)
12. Faloutsos, C., Ranganathan, M., Manolopoulos, Y.: Fast subsequence matching in time-series databases. In: Proceedings of the 1994 ACM SIGMOD International Conference on Management of Data, Minneapolis, Minnesota, May 24-27, 1994, pp. 419–429 (1994)
13. Ge, X., Smyth, P.: Deformable markov model templates for time-series pattern matching. In: KDD, pp. 81–90 (2000)

14. Keogh, E.J., Chakrabarti, K., Mehrotra, S., Pazzani, M.J.: Locally adaptive dimensionality reduction for indexing large time series databases. In: SIGMOD Conference, pp. 151–162 (2001)
15. Keogh, E.J., Chakrabarti, K., Pazzani, M.J., Mehrotra, S.: Dimensionality reduction for fast similarity search in large time series databases (2000)
16. Korn, F., Jagadish, H.V., Faloutsos, C.: Efficiently supporting ad hoc queries in large datasets of time sequences. In: SIGMOD 1997, Proceedings ACM SIGMOD International Conference on Management of Data, Tucson, Arizona, USA, May 13-15, 1997, pp. 289–300 (1997)
17. Morinaka, Y., Yoshikawa, M., Amagasa, T., Uemura, S.: The L-index: An indexing structure for efficient subsequence matching in time sequence databases. In: Cheung, D., Williams, G.J., Li, Q. (eds.) PAKDD 2001. LNCS, vol. 2035. Springer, Heidelberg (2001)
18. Nanopoulos, A., Alcock, R., Manolopoulos, Y.: Feature-based classification of time-series data. Information processing and technology, pp. 49–61 (2001) ISBN 1-59033-116-8
19. Nanopoulos, A., Manolopoulos, Y.: Indexing time-series databases for inverse queries. In: Quirchmayr, G., Bench-Capon, T.J.M., Schweighofer, E. (eds.) DEXA 1998. LNCS, vol. 1460, pp. 551–560. Springer, Heidelberg (1998)
20. Chan, K.-p., Fu, A.W.-C.: Efficient time series matching by wavelets. In: Proceedings of the 15th International Conference on Data Engineering, Sydney, Austrialia, March 23-26, 1999, pp. 126–133 (1999)
21. Ratanamahatana, C.A., Keogh, E.J., Bagnall, A.J., Lonardi, S.: A novel bit level time series representation with implication of similarity search and clustering. In: Ho, T.-B., Cheung, D., Liu, H. (eds.) PAKDD 2005. LNCS, vol. 3518, pp. 771–777. Springer, Heidelberg (2005)
22. Wang, X., Smith, K.A., Hyndman, R.J.: Characteristic-based clustering for time series data. Data Min. Knowl. Discov. 13(3), 335–364 (2006)
23. Yi, B.-K., Faloutsos, C.: Fast time sequence indexing for arbitrary lp norms. In: VLDB 2000, Proceedings of 26th International Conference on Very Large Data Bases, Cairo, Egypt, September 10-14, 2000, pp. 385–394 (2000)

Multi-level Frequent Pattern Mining

Todd Eavis and Xi Zheng

Concordia University, Montreal, Canada
eavis@cs.concordia.ca,
zhengxi96@yahoo.com

Abstract. Frequent pattern mining (FPM) has become one of the most popular data mining approaches for the analysis of purchasing patterns. Methods such as Apriori and FP-growth have been shown to work efficiently in this setting. However, these techniques are typically restricted to a single concept level. Since typical business databases support hierarchies that represent the relationships amongst many different concept levels, it is important that we extend our focus to discover frequent patterns in multi-level environments. Unfortunately, little attention has been paid to this research area. In this paper, we present two novel algorithms that efficiently discover multi-level frequent patterns. Adopting either a top-down or bottom-up approach, our algorithms exploit existing fp-tree structures, rather than excessively scanning the raw data set multiple times, as might be done with a naive implementation. In addition, we also introduce an algorithm to mine cross-level frequent patterns. Experimental results have shown that our new algorithms maintain their performance advantage across a broad spectrum of test environments.

1 Introduction

Frequent pattern mining (FPM) is a relatively recent but important data mining pursuit. Simply put, by identifying commonly occurring purchase or product combinations, FPM assists decision makers in improving their strategies for dealing with complex data environments. For the most part, previous research in the area has focused on *single level* frequent pattern mining (SLFPM). In practice, however, most practical business/retail databases consist of products or elements with multiple concept or classification levels. For example, a retailer might consider a carton of milk as (a) Sam's club skim milk (b) skim milk (c) milk (d) dairy product, depending upon the context of the analysis. To be useful, therefore, frequent pattern mining should be viable within multi-level domains.

One approach to multi-level mining (MLFPM) would be to directly exploit the standard algorithms in this area — Apriori [1] and FP-growth [6] — by iteratively applying them in a level by level manner to each concept level. In fact, for relatively small data sets, this approach is probably adequate. However, I/O costs in particular tend to increase significantly with larger data sets. With FP-growth, for example, the raw data set must be fully scanned each time a new fp-tree is constructed. This, of course, would occur once for each hierarchy level.

X. Zhou et al. (Eds.): DASFAA 2009, LNCS 5463, pp. 369–383, 2009.

In this paper, we propose two new algorithms that deal with the I/O problem described above. Specifically, we dramatically minimize I/O costs by using fp-trees themselves as input to subsequent iterations of the mining process. Since our methods extend FP-growth, we refer to them as *FPM-B* and *FPM-T* respectively. Essentially, FPM-B adopts a *bottom-up* approach, scanning previous fp-trees with a leaf-to-root traversal, then re-ordering and building a new tree for the next level. FPM-T, in contrast works *top-down*, scanning existing trees from root-to-leaf, building an intermediate tree at the same time and, after trimming and resizing, producing a new tree for the next level of the hierarchy. Both methods have been shown to work effectively under specific conditions. As an extension, we also apply our techniques to *cross-level* mining, so as to identify frequent patterns existing *between* arbitrary classification levels.

The remainder of the paper is organized as follows. Section 3 reviews related work in multi-level frequent pattern mining. The new bottom up (FPM-B), top down (FPM-T) and cross-level (FPM-cross) methods are then presented in Sections 4, 5, and 6 respectively. Experimental results are discussed in Section 7, with final conclusions offered in Section 8.

2 Related Work

In the domain of quantitative association rules, Srikant et al. first introduced the idea of *partial completeness*, specifically dealing with the information lost because of partitioning [12]. Miller et al. [8] explored distance-based quality and rule interest measures in order to find quantitative rules, while Aumann et al. adopted statistical theory for this same problem [2]. Moreover, Zhang et al. also investigated statistical theory to search for quantitative patterns [14]. Specifically, they introduced a re-sampling technique as the basis of their approach.

Pan et al. proposed CARPENTER to deal with high-dimensional bio-informatics data sets [9]. Essentially they transformed data sets into lists of transaction IDs that could eventually be used to build fp-trees. Zhu et al. presented an algorithm called Pattern-Fusion to work with colossal patterns [15]. The basic idea here is to fuse the small core patterns of the colossal patterns in one step. With this approach, they reduce the cost of generating mid-sized patterns, which is the general technique adopted by Apriori and FP-growth.

Generalized Sequential Patterns (GSP), an early algorithm in mining sequential patterns, presented by Srikant et al., also uses an Apriori-like approach [13]. PrefixSpan, proposed by Pei et al., employs a divide and conquer technique, treating each sequential pattern as a prefix and getting all patterns from the divided partitions [11].

Grahne et al. studied the problem of mining monotonic constraint based frequent itemsets [3]. Specifically, they identify all minimal answers that are valid and all minimal valid answers. Pei et al. also proposed algorithms to discover convertible constraint based frequent itemsets [10], in this case by exploiting the FP-growth algorithm. Finally, Liu et al. presented a noise-tolerant threshold for

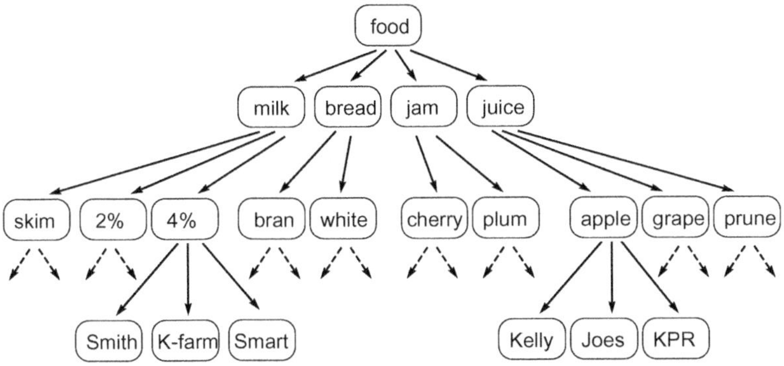

Fig. 1. A simple concept hierarchy

mining approximate frequent itemsets [7]. They also use rules to prune itemsets if their sub-itemsets are not frequent.

3 Multi-level Mining

3.1 Conceptual Overview

MLFPM builds upon the notion of a *concept hierarchy*, whereby each item in the data set can be associated with multiple meanings. For example, given the product hierarchy of Figure 1, "Smith" brand 4% milk bread may be viewed from three different perspectives. At the bottom level (Level 1), it literally means "Smith" brand 4% milk, a specific *product*. At Level 2, it refers to the *kind* of milk, while an Level 3, it simply represents milk. In the current context, we use the terms *higher level* and *lower level* to represent higher and lower levels of abstraction.

Given the hierarchy in Figure 1, we can investigate shopping cart queries that are executed against transaction databases. Individual entries in this data set would be of the form "4% K-farm milk, 2% Smart milk, Kelly apple juice." For simplicity, we encode the database in a concise numeric form that identifies not only the given product, but its classification level as well. For example, given the hierarchy of Figure 1, we might represent Level 3 by assigning "1" to milk, "2" to bread, "3" to jam and "4" to juice. In the sub-level for milk, Level 2, we assign "1" to skim milk, "2" to 2% milk and "3" to 4% milk. Furthermore, in the sub-level for 4% milk, we assign "1" to Smith, "2" to K-farm and "3" to Smart. As a result, "132" represents K-farm 4% milk, while "411" implies Kelly apple juice. Table 1 illustrates how a simple transaction database for the running example might be encoded. Finally, we note that when necessary, we can extend the encoding by adding the wildcard character "*" for lower levels. For example, "13*" would imply all brands of 4% milk.

Table 1. Encoded transaction table

TransactionID	Items
1	111 132 212 221 311
2	111 211 212 222 412 321
3	111 132 212 221 311
4	112 131 222 312 423 322
5	111 112 131 211 222 312
6	132 211 221 311 321

3.2 A Simple MLFPM Benchmark

It is certainly possible to adapt the original Apriori method proposed by Han et al. [5]. Briefly, the idea is to iteratively process the k levels of the hierarchy from bottom to top by generating *candidates* for k-itemsets from large $(k-1)$-itemsets. The candidates are subsequently tested against the data set to determine whether or not they meet the user-defined threshold. However, the cost of all of the candidate generation and checking quickly become prohibitive.

As such, a more appealing option is to exploit the non-candidate generation algorithm, FP-growth. Recall that FP-growth employs a data structure called an *fp-tree* to record counts of the most commonly occurring product combinations. Simply put, the idea for the benchmark is to first scan the data set once to generate frequent items for all levels, filtering out items whose counts are below the *threshold* (i.e, a user-defined frequency ratio). After this, we iteratively scan the database $d-1$ times to generate an fp-tree for each concept level. Finally, we mine the fp-tree at each level and discard the tree when mining is finished. Ultimately, the fp-growth approach is considerably less I/O-intensive that the Apriori model and will therefore serve as our baseline algorithm for comparative purposes.

4 FPM-B: A Bottom-Up Approach

The primary motivation for the new algorithms is that the benchmark is forced to repeatedly scan the full data set. Since an fp-tree stores complete information for a data set, we observe that it may be faster to scan the associated fp-trees instead of the underlying data set itself. At the same time, when scanning an fp-tree, we can directly create a new fp-tree for the subsequent level. Moreover, as the multi-level algorithms move from lower concept levels to higher concept levels, the size of the fp-trees becomes significantly smaller, allowing us to exploit this inherent form of tree compression.

In short, FPM-B starts from the leaf nodes of an existing fp-tree and traverses each branch upwards until it reaches its root. It then employs standard FP-growth tree-construction methods in order to create an fp-tree at the subsequent (i.e., higher) concept level. The process is described in Algorithm 1. In the remainder of this section, we review the primary steps or phases of the algorithm.

Algorithm 1. FPM-B

Input: an encoded transaction data set
Output: multi-level frequent pattern rules

```
scan the database once and build a header table for every level
filter out clearly infrequent items
build an fp-tree for the bottom level
mine the bottom level tree
for level k = 2 to level d do
    identify the leaf nodes for the next level
    generate the associated fp-tree
    mine this new fp-tree and identify frequent patterns
end for
```

Step 1: Initial data set scan. Items are initially inserted into a compact, hash-based storage structure called a *header table.* Every level-based fp-tree has its own header table to store frequent items. Unlike the benchmark, FPM-B only identifies frequent items at the *highest* level. For the rest of the levels, FPM-B identifies *all* items at the given concept level. For example, given Table 1, and thresholds 2, 4, and 5 at levels 1, 2, and 3 respectively, the first scan would identify as frequent "1**", "2**", and "3**" at level 3. "4**" is infrequent since its count is 2 and this is obviously below the Level 3 threshold. In addition, of course, we get all items, frequent and infrequent, at the lower levels.

Step 2: Filtering. With FPM-B, we scan lower level fp-trees in order to create higher level fp-trees. We have to careful, of course, not to eliminate an infrequent node at level $k-1$ that may in fact be frequent at level k. Therefore, items at lower levels should only be filtered relative to the frequency status at the highest level. In Figure 1, for example, since "4**" is infrequent at Level 3 (the highest level), all its *descendants* (412, 423, 41*, 42*) will be filtered out. But item "321" will not be filtered, even though item "32*" has a count of just 3, since "321" has a corresponding ancestor "3**" that is frequent at Level 3.

Step 3: Second data set scan. We are now ready to proceed with tree-building, following the logic of the standard fp-growth algorithm. In short, we scan each transaction, re-ordering items according to their *descending* frequency in the header table, and then inserting them into the new tree. After building the bottom level fp-tree (BLFPT), we perform the mining process as per the threshold. Note that although we may have slightly more items in the BLFPT than might be the case with the benchmark, the mining process is only performed on those items that are frequent at their level. As a concrete example, the BLFPT for Table 1 is shown in Figure 2(a). Note that the number listed after the colon in each label is an the occurrence count.

Step 4: Identify leaf nodes. Since the BLFPT is in fact a concise representation of the original data set, we may use it as a data set *proxy* to directly generate the next level fp-tree. We begin by scanning the tree and identifying

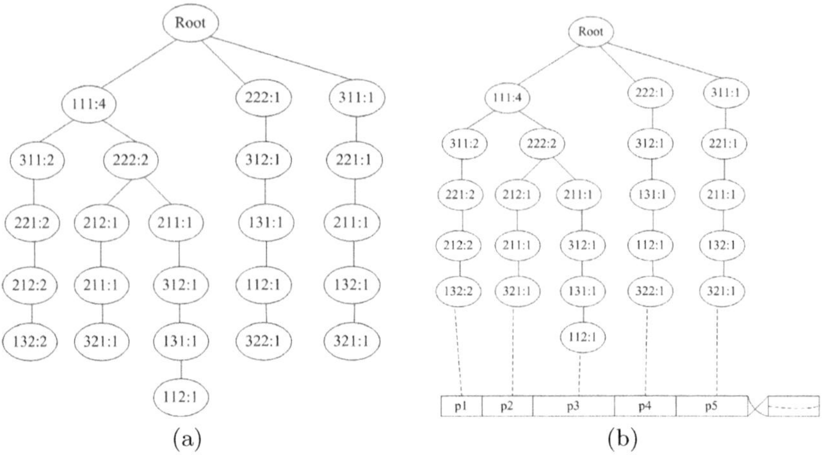

Fig. 2. (a) Bottom level fp-tree (BLFPT) (b) Identifying leaves

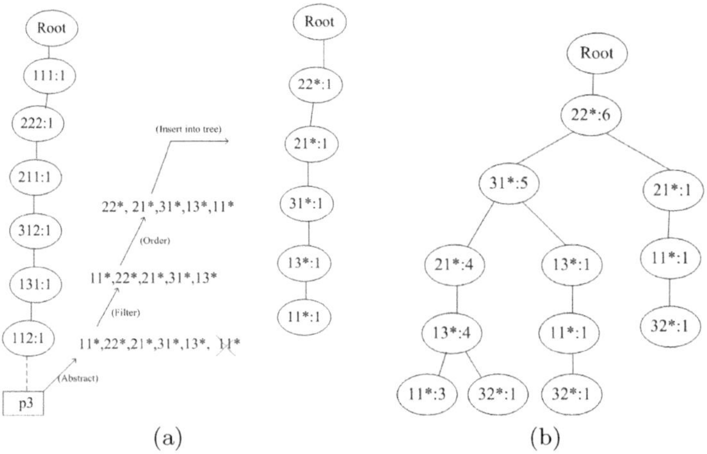

Fig. 3. (a) Filter duplicate items (b) Bottom-up construction of new fp-tree

its leaf nodes. A pointer to each leaf is then inserting into the *Leaf Node Array*. An illustration for the running example is shown in Figure 2(b).

Step 5: Generation of a new fp-tree. We now perform a bottom up scan of each leaf node until we reach the root. At the same time, we collect the labels of all nodes scanned in the process. We use "*" to replace digits of the labels in order to form new labels that, in turn, are used to create the nodes at the next level. Because multiple nodes at the current level may generate duplicate wildcarded labels at the next level, we adopt the idea of *boolean rules* described in [5] in order to do *eliminations*. That is, a repeated item in one transaction line will be treated as though it occurred only once. So, for example, for the third branch of the tree in Figure 2(a), we start from the leaf node "112" and subsequently identify "131", "312", "211", "222" and "111". The corresponding items at the next level are therefore "11*", "13*", "31*", "21*", "22*" and "11*". Since we now have two "11*"s in this branch, we keep just one of them, as shown in Figure 3(a). We then use the (descending) order defined in the header table for Level 2 to sort the names, getting "22*", "21*", "31*", "13*" and "11*". Using this order, we insert all the new nodes into the next fp-tree. The final result is illustrated in Figure 3(b).

Step 6: Mining. At the conclusion of the tree-construction process, we have a new compressed fp-tree for the next level. At this point, we may initiate the mining process in order to identify the appropriate frequent itemsets. This level-by-level process is very similar to the mining phase for single level frequent pattern identification (SLFPM).

5 FPM-T: A Top-Down Approach

The second approach to multi-level mining reverses the direction of the traversal through the fp-trees of the classification hierarchy. It is described in Algorithm 2. Though many of its steps are conceptually similar to those of the previous algorithm, the alternate traversal pattern creates unique processing requirements. We describe the main features of FPM-T below.

Algorithm 2. FPM-T

Input: an encoded transaction data set
Output: multi-level frequent pattern rules

```
scan the database build top and bottom header tables
filter out clearly infrequent items
build an fp-tree for the bottom level
mine the bottom level tree
for level k = 2 to level d do
    create a new tree from the tree of level k - 1
    generate the associated header table
    re-order nodes and combine sibling branches
    mine the current fp-tree and identify frequent patterns
end for
```

Step 1: Initial data set scan. As is the case with the benchmark and FPM-B, FPM-T needs to scan the database to generate frequent item listings. However, at this stage FPM-T only creates header tables for the highest and lowest levels. In fact, as we will see below, FPM-T actually traverses existing fp-trees in order to generate the remaining header tables.

Step 2: Filtering. This progress is virtually identical to the filtering process performed in FPM-B. In short, we only perform filtering on lower levels based on the values at the highest level.

Step 3: Second data set scan. Again, this phase is analogous to the one in FPM-B. Specifically, we perform a second scan of the data set in order to generate an fp-tree for the bottom level (BLFPT).

Step 4: Building derivative fp-trees. We begin scanning from the root of the current tree. As we do so, we create associated items for the next level. Figure 4(a) illustrates how the first branch might be created. If two items in the same branch have the same item label, we again exploit boolean rules to eliminate one. Duplicate elimination is depicted in Figure 4(b). In order to further compress the tree, we also combine child nodes of *sibling branches* if both have the same item label. An example can be found in Figure 4(c). Here, because the third branch of our running example contains a "211", it is merged with the "21*" node of the new tree, along with its relevant children.

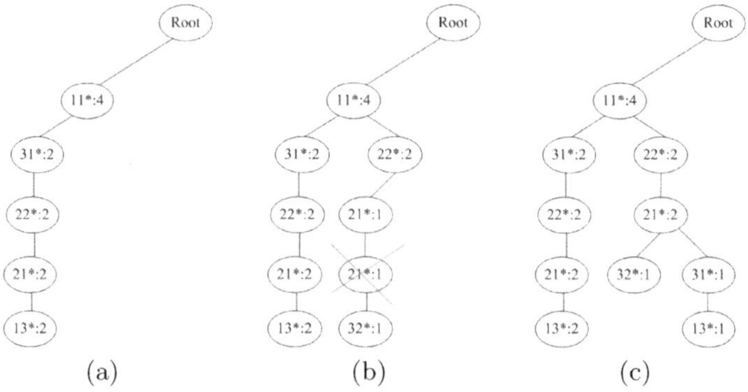

Fig. 4. (a) New branch created (b) Eliminate nodes in a top-down manner. (c) Combine nodes between siblings.

Recall that in the initial scan, we generate just the top and bottom header tables. We now build the intermediate table(s) by scanning the current fp-tree, which is typically very fast. As we scan the tree from the root downwards, every new item is inserted into the header table. If the item already exists, we simply update its count. We also use a linked-list to maintain a connection between items that have the same item label but are located in different branches. An example can be seen in Figure 5.

Step 5: Re-order the tree. In FP-growth, the most frequent item in the header table is deemed to have the "lowest" order, and consequently appears closer to the root in the associated fp-tree. Since we create nodes for the new level from the current tree, every new node follows an order relative to the structure of

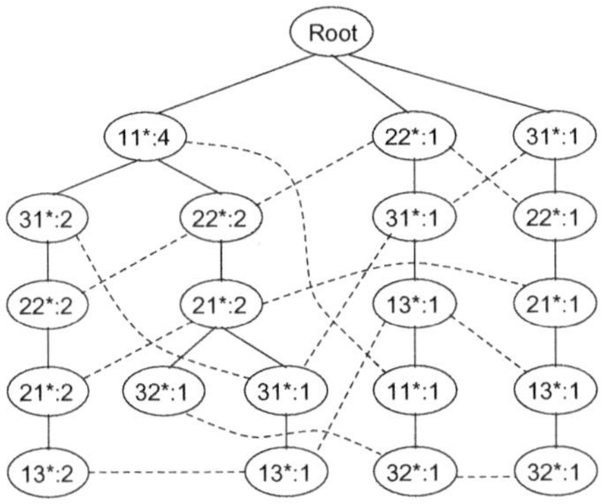

Fig. 5. Level 2 header table

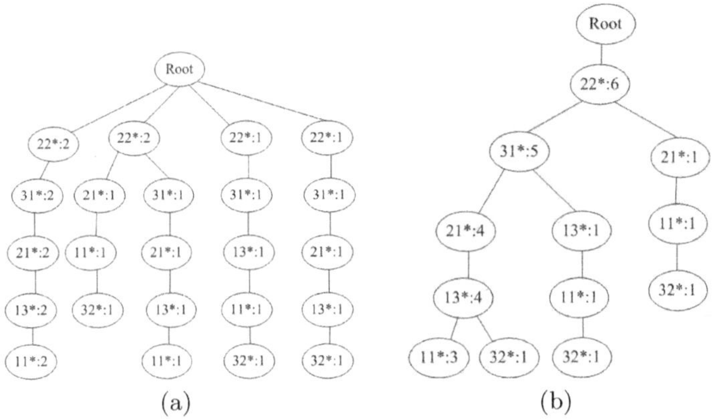

Fig. 6. (a) After re-ordering (b) After combining

the previous tree. However, in the new level, the order of items in the header table may be different from that in the previous level; thus, we must ensure that the new nodes also satisfy the appropriate descending order pattern. Figure 6(a) illustrates how this re-ordering would proceed. Note, however, that as we re-order the nodes, the tree may actually expand as a side effect. In order to minimize this growth, items sharing a common parent and having the same label are combined into a single node. A graphical example can be found in Figure 6(b).

Step 6: Mining. At this stage, we are left with a cleaned, compressed fp-tree for the next level. We can therefore proceed directly to the mining process, which again is very similar to that of single level frequent pattern mining (or FPM-B).

5.1 Complexity Considerations

We note that because the FPM algorithms build upon essentially linear traversals of the fp-trees (and the mining itself is unchanged), performance characteristics are similar to those of the underlying fp-growth algorithm (the re-ordering in FPM-T adds a real but modest overhead). For detailed performance analysis, see [4]. In terms of memory, we note that while multiple fp-trees are utilized, only two need to be used at any one time. In fact, since branches are never reprocessed, it would be possible to dynamically reclaim the memory of the parent tree during the generation of the child tree. As such, a judicious implementation would require no significant increase in memory.

6 FPM-Cross: Cross Level Mining

From time to time, users or decision makers may also want to understand associations *across* levels. For example, knowing that customers buy milk and *cherry* jam together may be more useful in making strategic decisions if the company in question is able to consistently obtain a discount from cherry jam wholesalers. In this case, buying milk and jam that is at the same level in the concept hierarchy becomes less important than buying milk and *cherry* jam that is at a different level, or *cross-level*, in the concept hierarchy. Therefore, we now turn our attention to the concept of cross-level frequent pattern mining (CLFPM).

6.1 Filtering

Due to the combinatorial explosion of item relationships in the cross-level context, it is not feasible to insert items at all levels into a single fp-tree. Clearly, some form of filtering is required. We first note that, in the addition to the d level thresholds of MLFPM, CLFPM mining is also associated with a single global cross-level threshold. Only items from all levels that are above this threshold can be considered as frequent. Consequently, we can use the first threshold as a filter to do level-by-level mining (as in MLFPM), and use both thresholds to do cross-level mining.

Recall that when building an fp-tree from the previous level, we can not use the first threshold to filter items in the tree except at the highest level, because we need to generate the items of the next level from the current level. But in cross-level, the situation changes, since we only generate a single "mixed" cross-level tree. As a result, we can use the first threshold to filter out items whose counts are either below the threshold for their level *or* the threshold for cross-level mining.

6.2 The Supporting MLFPM Method

In terms of supporting algorithms for FPM-cross, we could conceivably build upon either FPM-B or FPM-T. However, the new cross-level tree is deeper than before, which means that longer branches will be grown. Because FPM-T requires us to re-order and subsequently combine branch nodes, it is not ideally suited to the CLFPM context. Instead we utilize FPM-B. The integrated process is shown in Algorithm 3.

Algorithm 3. FPM-Cross

```
Input: base level fp-tree
Output: cross-level frequent pattern rules
filter nodes as per both mining thresholds
identify fp-tree leaf nodes
for each branch of the base level fp-tree do
  for every node of the branch do
    generate item names for all levels
    check each item i against the cross level header table
    if i exists in the header table then
      insert i into a temporary array T
    end if
    re-order the nodes of T
    initialize cross level nodes
    insert these nodes into the cross level fp-tree
  end for
end for
do mining and generate cross-level frequent pattern rules
```

When applying the bottom-up method, we start from the lowest user-requested level. For example, let us say the user asks for the cross-level relationship between Level 2 and Level 5. In this case, we refer to Level 2 to as the *base level*. We begin with the base fp-tree, scanning from leaf to root, identifying the label information for individual branches. We then generate corresponding nodes for *all* levels. When items do not meet the threshold at their concept level, or the threshold of the unique cross-level, their labels are deleted. We then take the filtered items, as we do in FPM-B, and insert them into the new cross-level fp-tree. This resulting fp-tree fulfills the requirements of cross-level frequent pattern mining.

An example of the base level fp-tree is shown in Figure 7(a). Thresholds for each level are 2, 2, and 2, while the threshold for cross-level analysis is 4. Consider the branch containing the leaf node “752:1”. Here we get: “752”, “236”, “456”, “235”, “45*”, “23*”, “75*”, “4**”, “7**”, and “2**”. We find “752” is infrequent (count = 1), and further that “236” and “75*” are also infrequent in the cross-level context. Thus, only “456”, “235”, “45*”, “23*”, “4**”, “7**”, and “2**” are inserted, leaving the final cross-level fp-tree (CLFP) depicted in Figure 7(b). Note that, while larger than a single FPM-B tree, CLFP is not significantly bigger due to pruning and filtering steps.

In terms of a viable benchmark in the cross-level domain, any proposed solution must be capable of adding items from the next level to each transaction line and then building some form of cross-level tree. For example: “235”, “456”, “786”, “457 → “235”, “456”, “786”, “457”, “23*”, “45*”, “78*”, “45*”, “2**”, “4**”, “7**”. The advantage of our method over a “naive” implementation is again the minimization of data set scanning. Specifically, we can do cross-level construction after building an initial fp-tree, and then using the bottom-up technique to generate the new cross-level tree. This allows us to directly exploit the methods developed for multi-level frequent pattern mining.

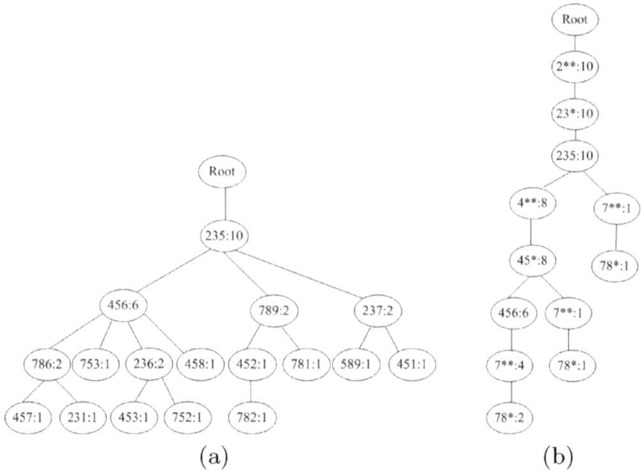

Fig. 7. (a) Base-level tree (b) Cross-level fp-tree

7 Experimental Results

We have performed extensive experimental evaluation of the new multi-level methods under a variety of test conditions. Due to space limitations, we will describe a sampling of the most illustrative tests in this section. In terms of the evaluation environment itself, all tests were conducted on a Dell workstation with a 1.8 GHz processor, 2 GB of memory, and a 250 GB SATA hard drive. The algorithms were implemented with Java 5.0 and run on a Linux OS (Fedora Core 5). With respect to data set generation, we employ a number of ideas taken from [1,5]. In particular, our data generation application allows us to dynamically tune the following parameters: T (Number of transactions), I (Average length per transaction line), L (Number of levels in the hierarchy), and F (Fan-outs from level to level). Thresholds are randomly set between .01% and 4% of the full transaction count. We also utilize a zipfian skew parameter to arbitrarily cluster data points.

In Figure 8(a), we see a comparison between FPM-T, FPM-B and the benchmark algorithm, using the parameters T = 250000, I = 3, L = 6, and F = 5 (data set D1). Essentially, this is a performance test for data sets with short, simple transactions. The graph clearly demonstrates an advantage for the FPM methods. As expected, the relative improvement increases with added dimensions since the cost of processing increasingly smaller fp-trees drops almost to zero. FPM-T, in particular, executes in just a little over half the time of the benchmark at six dimensions. In addition to the benefit of using fp-trees as input representations, FPM-T excels in short transaction environments due to the fact that it also uses fp-trees to build the header tables, thereby further reducing data set scan costs. This point is underscored in Figure 8(b), which demonstrates that the total cost of header table construction for FPM-T is about half that of the benchmark.

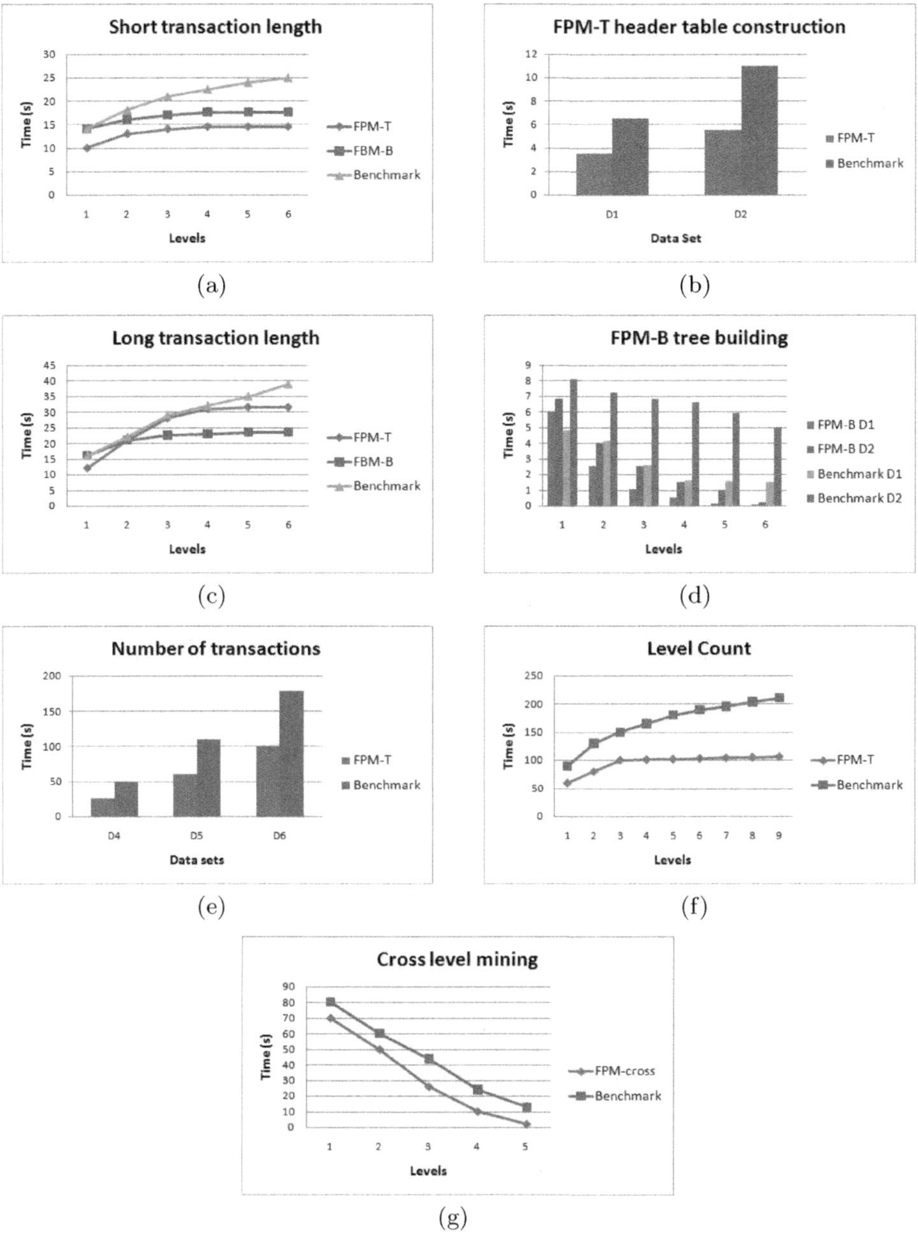

Fig. 8. (a) Baseline test with short transaction length (b) header table construction costs (c) longer transaction length (d) tree building costs (e) effect of data set size (f) number of concept levels (g) cross-level mining

For environments associated with longer average transaction lengths (data set D2, I = 6), the same general trend is evident; that is, performance of the benchmark deteriorates in the presence of increased classification levels. The results are depicted in Figure 8(c). Here, however, FPM-T is not as impressive due to the fact that its node re-ordering costs grow with longer transaction length. FPM-B does not require re-ordering and is able to more fully exploit the reduced cost of constructing fp-trees from existing trees. Figure 8(d) isolates this benefit in a direct comparison with the benchmark. Depending upon the combination of level count and dataset, we can see that tree building for FPM-B represents as little as one third the cost of the benchmark.

Figure 8(e) illustrates the effect of increasing the number of transactions in the input set. In this case, we set I = 3, L = 6, and F = 5, and generate data sets of 500000 (D4), 1 million (D5), and 1.5 million (D6) transactions. Since the average transaction length is short in this particular case, we have selected FPM-T for the comparison. As shown in the graph, data set size has little effect upon relative performance, with the FPM algorithm consistently requiring just 50% of the total processing time.

The effect of increasing the level count is also an important consideration for multi-level mining since the primary purpose of the new algorithms is to minimize excessive I/O associated with data set scans. As level count increases, the benefit of the new methods should be quite pronounced. Figure 8(f) shows just that, with the incremental cost of the FPM-T method effectively dropping to zero after four or five levels (almost no I/O or mining costs) on a data set with one million records (data set D5). Total cost is again about half the cost of the benchmark, which must do full I/O on every iteration.

Finally, we turn to cross-level mining. In particular, Figure 8(g) assesses the cost of cross level mining on the six-level data set D4 as we shift the base from level one to level five. We see that even the times for the benchmark drop consistently as we move the base since, of course, we have fewer items to insert into the cross level tree. The times for FPM-cross, however, range from 12% to 93% less than the benchmark at comparable points.

8 Conclusions

Frequent pattern discovery is one of the fundamental mining tasks found in contemporary decision support environments. While efficient algorithms exist for the somewhat artificial case of non-hierarchical product data sets, we are aware of no research that extends these methods to the much more practical multi-level classification context. In this paper, we present a pair of MLFPM algorithms that reduce run time to less than half that of more naive alternatives, largely by minimizing redundant disk accesses. We also extend the core work by incorporating one of the FPM methods into a cross-level mining framework. Experimental evaluation demonstrates the viability of the new methods across a variety of test scenarios.

References

1. Agrawal, R., Srikant, R.: Fast algorithm for mining association rules in large databases. In: VLDB, pp. 487–499 (1994)
2. Aumann, Y., Lindell, Y.: A statistical theory for quantitative association rules. In: ACM SIGKDD, pp. 261–270 (1999)
3. Grahne, G., Lakshmanan, L.V.S., Wang, X.: Efficient mining of constrained correlated sets. In: ICDE, pp. 512–521 (2000)
4. Grahne, G., Zhu, J.: Fast algorithms for frequent itemset mining using fp-trees. IEEE Transactions on Knowledge and Data Engineering 17(10) (2005)
5. Han, J., Fu, Y.: Discovery of multiple-level association rules from large databases. In: VLDB, pp. 420–431 (1995)
6. Han, J., Pei, J., Yin, Y.: Mining frequent patterns without candidate generation. In: ACM SIGMOD, pp. 1–12 (2000)
7. Liu, J., Paulsen, S., Sun, X., Wang, W., Nobel, A., Prins, J.: Mining approximate frequent itemsets in the presence of noise. In: SIAM international conference on data mining (2006)
8. Miller, R.J., Yang, Y.: Association rules over interval data. In: ACM SIGMOD, pp. 452–461 (1997)
9. Pan, F., Cong, G., Tung, A.K.H., Yang, J., Zaki, M.: Carpenter: Finding closed patterns in long biological datasets. In: ACM SIGKDD, pp. 637–646 (2003)
10. Pei, J., Han, J., Lakshmanan, L.V.S.: Mining frequent itemsets with convertible constraints. In: ICDE, pp. 433–442 (2001)
11. Pei, J., Han, J., Mortazavi-Asl, B., Wang, J., Pinto, H., Chen, Q., Dayal, U., Hsu, M.-C.: Mining sequential patterns by pattern-growth: the prefixspan approach. In: Transactions on Knowledge Data Engineering, pp. 1424–1440 (2004)
12. Srikant, R., Agrawal, R.: Mining quantitative association rules in large relational tables. In: ACM SIGMOD, pp. 1–12 (1996)
13. Srikant, R., Agrawal, R.: Mining sequential patterns: generalizations and performance improvements. In: Apers, P.M.G., Bouzeghoub, M., Gardarin, G. (eds.) EDBT 1996. LNCS, vol. 1057, pp. 3–17. Springer, Heidelberg (1996)
14. Zhang, H., Padmanabhan, B., Tuzhilin, A.: On the discovery of significant statistical quantitative rules. In: SIGKDD, pp. 374–383 (2004)
15. Zhu, F., Yan, X., Han, J., Yu, P.S., Cheng, H.: Mining colossal frequent patterns by core pattern fusion. In: ICDE, pp. 706–715 (2007)

Mining Entropy l-Diversity Patterns

Chaofeng Sha, Jian Gong, and Aoying Zhou

School of Computer Science
Fudan University, Shanghai 200433, China
{cfsha,jiangong,ayzhou}@fudan.edu.cn

Abstract. The discovery of diversity patterns from binary data is an important data mining task. This paper proposes entropy l-diversity patterns based on information theory, and develops techniques for discovering such diversity patterns. We study the properties of the entropy l-diversity patterns, and propose some pruning strategies to speed our mining algorithm. Experiments show that our mining algorithm is fast in practice. For real datesets the running time are improved by serval orders of magnitude over brute force method.

1 Introduction

Many real-life datasets can be modeled as binary data, in which an item may *appear* or not appear in a given record. The well-known transaction presentation of supermarket basket data in association rule mining is a good example of binary data, in which each transaction is a record. Other examples include documents (documents as records, and keywords as items), and different kinds of web data (e.g. users as records, and user's click history as items) or high-dimensional data. Therefore, binary data analysis has attracted much attention in data mining research community.

In this paper, we consider another kind of pattern: diversity pattern. Given a binary dataset, where each row is a user, and each column is a sample, the authors of [4] proposed the maximum diversity sample selection problem, based on subset coverage. In this paper, we used entropy to measure the diversity of itemsets, and proposed the problem of finding entropy l-diversity pattern from binary datasets. Some work in data mining and machine learning has adopted entropy as the diversity measure of the objects. Here we use the idea of [3] to measure the diversity of the sensitive attribute, and discuss some properties of the entropy l-diversity patterns. In our algorithm design, we use those properties to prune the pattern search space, which improve the efficiency of our mining algorithm.

The rest of this paper is organized as follows. Section 2 discusses related works. Section 3 presents problem description and some bounds of joint entropy that will be used to pruning the candidate patterns. Section 4 describes our mining algorithm and Section 5 presents our experimental evaluation on real datasets. We conclude our work in Section 6.

X. Zhou et al. (Eds.): DASFAA 2009, LNCS 5463, pp. 384–388, 2009.

2 Related Work

Joint entropy is proposed in [2] as a quality measure for itemsets, and serval efficient algorithms to mine those maximally informative k-itemsets. The main goal of their work is to select distinct items, as well as to minimize redundancy within the resulted itemsets. Their work is orthogonal to ours, here we aim to find those minimum patterns with entropy no less than threshold l, and we use more bounds of joint entropy to pruning the pattern search space.

Our algorithm is mainly based on the algorithmic framework proposed in [4]. They defined subset coverage diversity, discussed the properties of this diversity measure, and proposed an algorithm to find minimum sample set (here is item) with coverage diversity larger than a threshold ρ. While our work used the entropy l-diversity appeared in [3], which is a privacy preservation model there. This model requires that the values of sensitive attribute in each equivalence class is not too skew, i.e., the entropy of the sensitive attribute is not less than a threshold l.

3 Mining Minimum Entropy l-Diversity Patterns

In this section we introduce the entropy l-diversity patterns. For the content of information theory, the reader is referred to [1].

Definition 1. *Given an entropy threshold l and a binary dataset D, an itemset $X(X \subseteq I)$ is an entropy l-diversity pattern, if and only if $H(X) \geq l$.*

Mining Minimum Entropy l-Diversity Patterns: Given a binary dataset D and a threshold l, find all minimum patterns $X \subseteq I$ such that $H(X) \geq l$.

When mining entropy l-diversity patterns, the elementary sub-procedure is computing the entropy of any itemset. It's time consuming to scan whole dataset for computing entropy once. Therefore our aim is to reduce the number of computation of entropy of large itemsets. This is achieved by using serval lower bounds and upper bounds of those entropy, which just use the entropy of 1-itemsets and 2-itemsets, and conditional entropy between item pairs. We will develop some such bounds, which are listed as follows.

Theorem 1. *For any $n \geq 3$ random variables $X_1, X_2, \cdots, X_n$,*

$$H(X_1, X_2, \cdots, X_n) \leq \frac{H(X_1, \cdots, X_{n-1}) + H(X_1, \cdots, X_{n-2}, X_n) + H(X_{n-1}, X_n)}{2} \quad (1)$$

By the chain rule for entropy, we get: $H(X_1, X_2, \cdots, X_n) \leq H(X_1, \cdots, X_{n-1}) + \min_{1 \leq i \leq n-1} H(X_n|X_i)$, $H(X_1, \cdots, X_{n-2}, X_n) \leq H(X_1, \cdots, X_{n-1}) + \min_{1 \leq i \leq n-1} H(X_n|X_i)$, *and* $H(X_1, \cdots, X_{n-2}, X_n) \geq H(X_1, \cdots, X_{n-1}) - \min\left(\min_{1 \leq i \leq n-2} H(X_{n-1}|X_i), H(X_{n-1}|X_n)\right)$.

4 Our Mining Algorithm

In this section, we develop the Exhaustive Pattern Enumeration algorithm to find entropy l-diversity patterns. The algorithm also consists of two phases. In the first phase, we find a pattern C with $H(C) \geq l$ as the initialization itemset, which used the greedy idea. Starting from empty itemset, in each step, we add one item which maximizes the entropy gain of itemset. When the entropy of current itemset is lager than l, we stop this searching phase. This phase is similar to the algorithm 5 in [2], where they return the found itemset as the result. That algorithm is an approximate algorithm.

Now in the second phase, we should search the itemsets with size less than the size of itemset returned by the first phase, and then return the patterns with minimum number of items. In the worst case, the running time of the enumeration procedure in this phase is exponential. While we use those pruning strategies described following the algorithm, the running time of our algorithm is practicable. When searching the pattern space, in depth-first order, we visit the nodes dynamically and calculate entropy of the itemsets.

***K-l* Entropy Diversity Pattern Algorithm**
Input: Itemset I, Binary data set D, Threshold l, maximum size K, $K = |C|$.
Output: Patterns P having $H(P) \geq l$ with minimum size ($\leq K$).

1. Initialize.
   Candidate minimum itemset list, $cList = \phi$.
2. Current itemset, $cItemset = \phi$.
3. Remaining itemset, $rItemset = I$.
4. Enumerate($cItemset$,$rItemset$).
5. P = the minimum itemsets in $cList$.

Subroutine: Enumerate(*cItemset*,*rItemset*)

6. if $|cItemset| \geq K$ return;
7. for each item $X_i \in rItemset$
8. if $H(cItemset \cup \{X_i\}) \geq l$
9. Insert set $cItemset \cup \{X_i\}$ into $cList$
10. else
11. $cItemset = cItemset \cup \{X_i\}$.
12. $rItemset = rItemset - \{X_i\}$.
13. Enumerate($cItemset$,$rItemset$).
14. $rItemset = rItemset - \{X_i\}$.

Pruning Strategy 1. Our first pruning strategy is updating the entropy upper bound of candidate itemsets. We can use the size $K = |C|$ of the initialization itemset returned by ForwardSearch algorithm. The first two steps in sub-procedure Enumerate() uses this bound. It is not necessary to compute the entropy of itemsets with size larger than K, we just prune those itemsets directly. In addition, the 9th step in Enumerate() also check whether the itemset size is larger than k. Besides those two examinations, we can dynamically use the smallest size of candidate itemset in $cList$ as an upper bound. While this size is less than K, we update K to this value.

Pruning Strategy 2. In the enumeration procedure, we can estimate the incremental entropy gain when adding one item to the current candidate itemset. We could calculate $H(X_i)$ for any item X_i, joint entropy $H(X_i, X_j)$, and conditional entropy $H(X_i, X_j)$ between any item pair, before K-l-EDP algorithm. Now before the 8th step in Enumerate(), we can estimate an upper bound of joint entropy, which is $H(cItemset \cup \{X_i\}) \leq H(cItemset) + \min_{X_j \in cItemset} H(X_j|X_i)$. If the right hand of this inequality is less than l, we know that the left hand is also less than $l$then we don't need to compute $H(cItemset \cup \{X_i\})$.

On the other hand, if we have computed $H(cItemset \cup \{X_i\})$, then for $X_j \in rItemset$, we could get an upper bound and a lower bound of $H(cItemset \cup \{X_j\})$ respectively as: $H(cItemset \cup \{X_i\}) + \min_{X_k \in cItemset} H(X_j|X_k)$ and $H(cItemset \cup \{X_i\}) - \min\left(\min_{X_k \in cItemset} H(X_i|X_k), H(X_i|X_j)\right)$. If this upper bound is less than l, then we know that$cItemset \cup \{X_i\}$ is not an entropy l-diversity patter; and if this lower bound larger than l, then we know that$cItemset \cup \{X_i\}$ should be an entropy l-diversity pattern.

Pruning Strategy 3. We have another upper bound of entropy of the candidate itemset, before the 7th and 10th steps. Similarly, if the upper bounds is less than l, we need not to compute the joint entropy, due to the fact that this itemset should not be an entropy l-diversity pattern.

In addition, we assume that the visiting node is an itemset with p items, denoted as $cItemset = \{X_{i_1}, \cdots, X_{i_p}\}$, and the smallest size of candidate itemsets is K. Then we know that we could add at most $K - p$ items into the itemset.

5 Experimental Evaluation

In this section, we evaluate the performance of our solution to the entropy l-diversity pattern problem on real datasets. All experiments were run on a Windows XP machine with Intel Pentium4 2.4GHz CPU and 2GB RAM. Our algorithm is implemented using Java.

5.1 Scalability

First, we evaluate the performance of mining algorithm on the Mushroom dataset, which consists of 126 items and 8124 records. The left panel of Figure 1 shows the execution time of mining algorithm on the Mushroom dataset. As can be seen, the running time is small when the entropy threshold $l \leq 3.5$. While the execution time increases significantly with the increase of the threshold $l \geq 4$.

Next, we evaluate the performance of our mining algorithm on Chess dataset, which consists of 73 items and 3196 records. As can be seen from the right panel of Figure 1, the running time performance is similar to the one on Mushroom dataset. When the threshold $l \geq 4.5$, the execution time increases significantly with the increase of the threshold. Because the number of visited nodes increase significantly, and we find that the size of returned patterns is no less than 6.

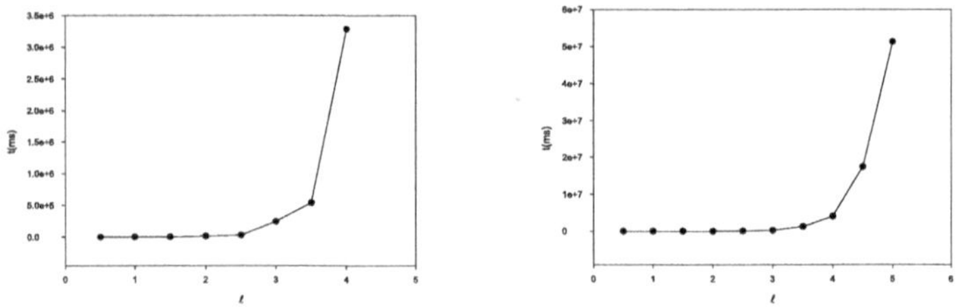

(1) Mushroom Dataset (2) Chess Dataset

Fig. 1. The execution time of mining algorithm on Mushroom and Chess dataset

5.2 Quality of Entropy Diversity Patterns

In this experiment, we examined the quality of entropy diversity patterns found by our mining algorithm. Here we present two interesting entropy diversity patterns found from Voting dataset, when we set the threshold $l = 6$: {handicapped-infants, water-project-cost-sharing, mx-missile, superfund-right-to-sue, synfuels-corporation-cutback, immigration, export-administration-act-south-africa}, and {handicapped-infants, water-project-cost-sharing, synfuels-corporation-cutback, immigration, mx-missile, crime, export-administration-act-south-africa}. As can be seen, all patterns are 7-itemsets. We find that the mined patterns almost have the first two bills, which maybe due to those bills are more controversial.

6 Conclusion

In this paper, we introduce the entropy l-diversity pattern mining problem. After presenting some properties of the entropy l-diversity patterns, we propose the mining algorithm based on depth-first search, which is powered by three pruning strategies. We evaluate the performance of our mining algorithm on serval datasets and demonstrate the quality of the mined patterns.

References

1. Cover, T., Thomas, J.: Elements of Information Theory. Wiley Interscience, Hoboken (1991)
2. Knobbe, A., Ho, E.: Maximally informative k-itemsets and their efficient discovery. In: KDD, pp. 237–244 (2006)
3. Machanavajjhala, A., et al.: l-Diversity: Privacy Beyond k-Anonymity. In: ICDE 2006 (2006)
4. Pan, F., Roberts, A., McMillan, L., de Villena, F., Threadgill, D., Wang, W.: Sample selection for maximal diversity. In: Perner, P. (ed.) ICDM 2007. LNCS, vol. 4597. Springer, Heidelberg (2007)

Query Optimization for Complex Path Queries on XML Data*

Hongzhi Wang, Jianzhong Li, Xianmin Liu, and Jizhou Luo

Harbin Institute of Technology
Harbin, China
{wangzh, lijzh, liuxianmin, luojizhou}@hit.edu.cn

Abstract. With the proliferation of XML data and applications on the Internet, efficiently XML query processing techniques are in great demand. Many path queries on XML data have complex structure with both structural and value constraints. Existing query processing techniques can only process some part of such queries efficiently. In this paper, a query optimization strategy for complex path queries is presented. Such strategy combines index on values, structural index and join on labeling scheme and generates effective query plan for complex queries. Experimental results show that the optimization strategy is efficient and effective; our method is suitable for various path queries with value constraints and our method scales up for large data size with good query performance.

Keywords: XML; Path Query; Query Optimization.

1 Introduction

XML has become the de facto standard for information representation and exchange over the Internet. An XML document can be naturally modeled as a tree, where elements are modeled as nodes in the tree and direct nesting relationships between elements are modeled as edges between nodes [28]. All standard XML query languages, e.g., XPath and XQuery, can retrieve a subset of the XML data nodes satisfying certain path constraints. For example, XPath query //book[author="aaa"]//figure[id=1] will retrieve all figure nodes with id equaling to 1 that appear under books with a name of author "aaa".

With the proliferation of XML data and applications on the Internet, efficient XML query processing techniques are in great demand. In many applications, path query can be complex with both structural and value constraints. The processing of such kind of queries (CXQ for brief) brings following challenges:

* Support by the Key Program of the National Natural Science Foundation of China under Grant No.60533110; the National Grand Fundamental Research 973 Program of China under Grant No.2006CB303000; the National Natural Science Foundation of China under Grant No.60773068 and No.60773063.

X. Zhou et al. (Eds.): DASFAA 2009, LNCS 5463, pp. 389–404, 2009.

- The selectivity of a value and structural constraint is possible to be large or small. When the selectivity is high, only a few nodes are useful to the query. If the query is processed with the method that is to process structural part at first and then the value constraint, many useless nodes are accessed. The efficiency is decreased. Currently, most techniques such as [3,4,13,30] and [2] for XML query evaluation focus on structural part. With processing many useless nodes, high efficiency will not be gained.
- A CXQ may contain multiple value constraints with various selectivities. Additionally, different parts of the structural part of a CXQ also have various selectivities. Therefore, the order of processing different parts of the query affects the processing efficiency of a CXQ. Current query processing techniques for XML data do not consider such problem.
- For a complex query, one kind of techniques can only process some part of it efficiently. Current work does not consider the problem of how to combine efficient techniques to process CXQ efficiently.

To address these problems, in this paper, with efficient storage structure, our method combines the advantages of tree search, join-based and index-based methods to process CXQ efficiently. To process CXQ with various structural and value constraint selectivities efficiently, we design cost-model-based query optimization strategies.

The major contributions of this paper include:

- A form of query plan for CXQ is presented.
- With effective support of storage structure and the query processing operators on it, a cost model is designed as the base of query optimization.
- In order to choose the most efficient query plan for a CXQ, a cost-based query optimization strategy is presented. Such techniques can be generalized for query optimization for other storage structure with similar cost model.

The rest of the paper is organized as follows: Section 2 introduces some background knowledge. Section 3 presents the form of query plan of CXQ. The cost model for CXQ is presented in Section 4. Cost-model-based query optimization strategies are presented in Section 5. We present our experimental results and analysis in Section 6. Related work is described in Section 7 and Section 8 concludes the paper.

2 Preliminaries

XML documents are usually modeled as labeled trees: elements and attributes are mapped to nodes in the trees and direct nesting relationships are mapped to edges in the trees. In this paper, we only focus on the element nodes; it is easy to generalize our methods to the other types of nodes defined in [28].

All structural indices for XML data take a path query as input and report exactly all those matching nodes as output, via searching within the indices. Equivalently, those indices are said to cover those queries. Existing XML indices differ in terms of the classes of queries they can cover. DataGuide [7] and 1-index [21] can cover all simple

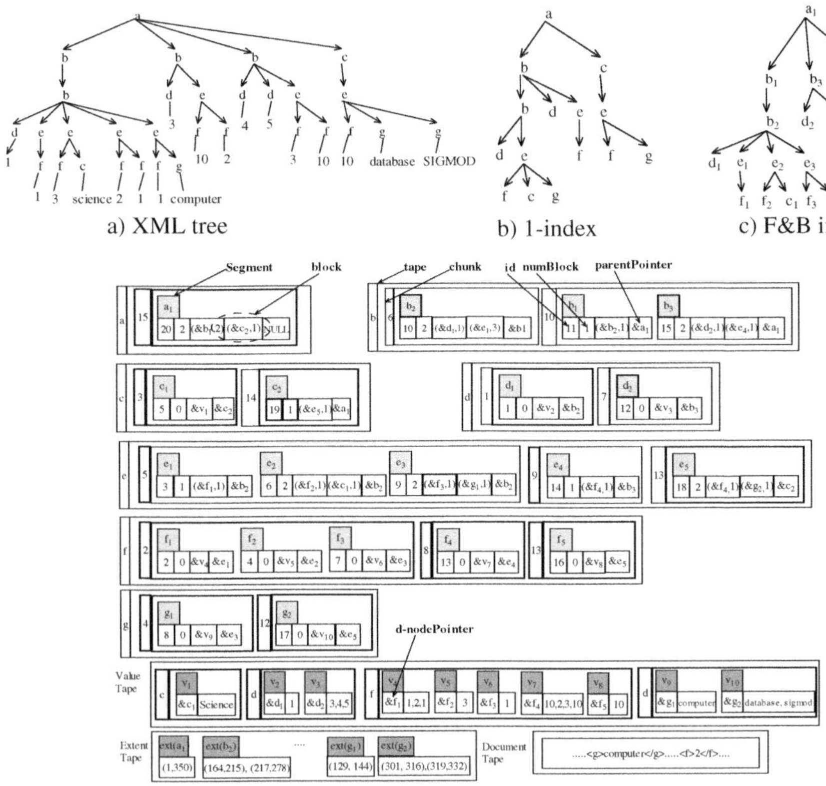

a) XML tree b) 1-index c) F&B index

d) Storage Structure

Fig. 1. Example of the Storage Structure

path queries, that is, path queries without branch. [13] showed that F&B Index is the minimum index that covers all branching path queries. Note that if XML data is modeled as a tree, its 1-index and F&B Index will also be a tree. Each index node n can be associated with its extent, which is a set of data nodes in the data tree that belongs to the index node *n*.

We show an example XML data tree, its 1-index and F&B Index in Figure 1(a), Figure 1(b) and Figure 1(c), respectively. In the 1-index, all the 2nd level b elements in the data tree are classified as the same tag b index node; this is because all those nodes cannot be distinguished by their incoming path, which is a/b. However, those b nodes are classified into three groups in the F&B Index; this is because branching path expressions, e.g., a/b[d] and a/b[b], can distinguish them.

XPath [6] is a path description language presented by W3C. An XPath expression can be modeled as a tree, in which node is tag or value constraint and edge has two kinds of tags, "/" and "//". We define the node with value constraint as VC-node. For example, the query *a/b[//e[f][g='computer']]//c* can be modeled as Figure 2(a), where single line represents "/" and double line represents "//". The part of

XPath matching the result to be returned is called trunk. The part in "[ ]" is called a branching constraint. In the example, a/b//c is the trunk; [//e[f][g='computer'] is a branching constraint. g is a VC-node.

We describe some notations used in the rest of the paper. We distinguish nodes in the original data tree and nodes in the 1-index (or F&B Index); the former is termed Dnodes and the latter is termed 1index nodes (or Inodes), respectively. Since the F&B Index is a refinement of the 1-index, each Inode n in the F&B Index corresponds to a unique 1-index node and thus is assigned the 1-index node number (denoted as n.1indexID). We term the child axis (i.e., /) as PC axis and term descendant-or-self axis (i.e., //) as AD axis.

Disk-based F&B [30] index is to store the F&B index on disk with reasonable cluster strategies. We choose the disk-based F&B index as the storage structure in this paper because it supports various search strategies and query operators on XML data. The cluster has three levels:

1. All Inodes with the same tag are clustered together on a logical unit called tape.
2. The child Inodes of an Inode is clustered as follows: the child Inodes from the same parent are grouped by their tag names. Each of such cluster is called a chunk.
3. Nodes with the same tag are clustered according to their 1indexID. This is because it can be shown that the answers to simple path queries are exactly those Inodes with the same 1indexID. Each of such clusters is called a block.

To support query processing efficiently, there are also some additional storage structure.

1. In order to support bottom-up search, with each node, a pointer to its parent is added. To support the query with value constraints, pointers from values to corresponding nodes are added.
2. Values of nodes with the same tag are stored together on the value tape. To support the value constraint efficiently, indices are built on the values. The start and end positions of all nodes are stored together on the extent tape.
3. To obtain a whole fragment of the XML document, the document tape is used to store the whole document. With the start and end positions in the extent tape, the contents of elements can be obtained from the document tape.

For example, the disk-based F&B index of the XML document in Figure 1(a) is shown in Figure 1(d). On disk-based F&B index, query operations of depth first search (DFS for brief), breadth first search (BFS) and bottom-up search (BUS) can be used to process path queries. Based on clustering by 1-index node, a path query without twig constraint can be processed by scanning 1-index instead of traverse the whole subtree. This operator is RangeFetch. With additional tapes containing value information, extents and the whole document, processing path queries with disk-based F&B index can return values or XML segments instead of ids only. With extents, structural join [3] can be applied to accelerate "//" operation (such operator is called SegSJ). The details of these algorithms are shown in [30]. The operator for seeking from value to its corresponding node is call value search (VS for brief).

3 Query Plans for CXQs

In this section, we present the model of the query plan for a CXQ based on the storage structure presented in Section 2.

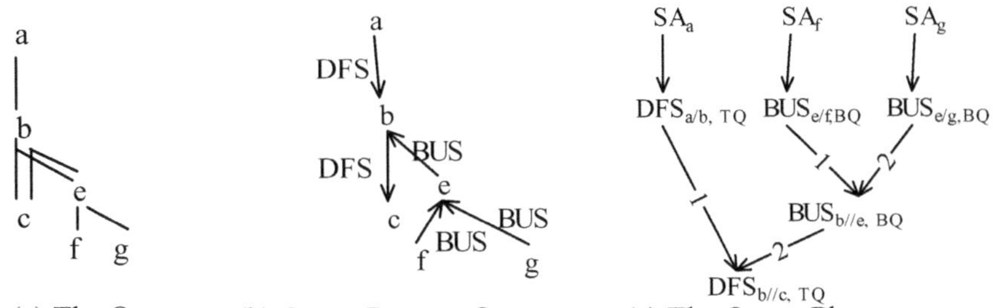

(a) The Query (b) Query Process Operators (c) The Query Plan

Fig. 2. An Example for Query Plan

As the fundamental components of query plans, we define *query operators* DFS, BFS, BUS, VS, SegSJ and RangeFetch, corresponding to the algorithm Depth-first Search, Breadth-first Search, Bottom-up Search, Value Search, Segment Join and Range Fetch, respectively. Here, VS operator is to perform the step of finding required nodes just satisfying single value constraint without considering other structural information.

Each query operation is a binary operator to process a sub-query with form a*b where *a* and *b* are tags and * is "/" or "//".Each operation is a triple (*op*, *axis*, *dir*), where *op* is the name of the operator; axis is whether the operation is used to evaluate the sub-query with "/" or "//"; *dir* is the operation which is used to evaluate the sub-query in trunk (TQ for brief) or the sub-query in branching constraint (BQ for brief). For an operation to answer sub-query *a*b*, if *dir=TQ*, the results of this operation are the elements matching *b*; if *dir=BQ*, the result of this operation are the elements matching *a*.

We also define three *access operators* that access tape directly. The operator for accessing some tape sequentially is defined as sequential access (SA for brief). When the next operator is to access nodes by value constraint with index, as discussed in Section 2, all nodes will not be accessed. Such accessing is defined as index access (IA for brief). The third case is that at the end of a series of RangeFetch operations, as discussed in [30], only a successive fragment of one tape is accessed. Accessing operator in such case is defined as RangeFetch access (RA for brief).

A CXQ can be represented as a query tree. Each edge of such a query tree can be processed with an operator. For example, the query *a/b[//e[f]][g]]//c* with structure in Figure 2(a) can be executed by the strategy shown in Figure 2(b) with query plan in Figure 2(c). where each edge in Figure 2(a) corresponds to an operator. Edge $a \to b$ corresponds to $DFS_{a/b,TQ}$.

A query plan P can be modeled as a DAG $G_p=(V_{Gp}, E_{Gp})$ with each vertex as an operator and each edge as the relationship between two operators. Each vertex $v\in V_{Gp}$ is an operator. And $e\in E_{Gp}$: $v_1\rightarrow v_2$ means that some input of v_2 is the result of v_1.

For each vertex v in G_p with multiple incoming edges, there is an order attached to each edge representing the order of evaluation of the operators. In Figure 2(c), the order is the number in the edges.

4 The Cost Model

In this section, the cost model for query optimization is presented. We use the logical I/O during query processing as cost of query processing. It is reasonable because I/O cost dominates total cost.

Table 1. Notations of Cost Model

notation	semantics
S_t	the set of candidate node with tag t
$\|S\|$	the number of nodes in set S
$blk(S)$	the number of blocks of nodes in set S
$Child_n(b)$	the set of b children of node n
$tree\_size(n)$	the number of blocks containing the subtree with n as root
$constraint(t,C)$	the constraint as a node with tag t satisfying constraint C
$number_{rt,st}(n, C)$	the number of blocks should be traversed to find a node satisfying constraint C with the relationship described with rt of n. rt is the PC or AD, representing parent-child and ancestor-descendant relationship, respectively. st is the traversal type with value DFS or BFS
$parent_n(t)$	the set of parents of node with tag t. If $t=*$, it means all the parent of n.
$path(n,p)$	the set of ancestors of node n in the path from n to its nearest ancestor with tag p
$ancestor_n(t)$	the set of ancestors of n with tag t. If t=*, it means all the parent of n.
$sel(S_t, r*t)$	the subset of S_t with t nodes selected by index with an r node as parent if *="/" or ancestor if *="//"
$l_{ind}(t)$	the height of B+index of the *id*s of nodes with tag t
$inter_S$	the number of disk containing the intervals corresponding to the index nodes in S
$cost_d(S,t)$	the cost of seeking the left most($d=L$) or right most($d=R$) descendant with tag t of all nodes in S
$cost_{value}(C,t)$	the cost of seeking all values satisfying constraint t in the set of values of t nodes
$V_{t[C]}$	the set of nodes with tag t containing value satisfying constraint C

Table 2. The Cost Model of Operators

		single slash	double slash
BFS	TQ	$\sum_{n_a \in S_a} blk(Children_{n_a}(b))$	$\sum_{n_a \in S_a} tree\_size(n_a)$
	BQ	$\sum_{n_a \in S_a} num_{PC,DFS}(n_a, constraint(b,C))$	$\sum_{n_a \in S_a} num_{AD,DFS}(n_a, constraint(b,C))$
DFS	TQ	$\sum_{n_a \in S_a} blk(Children_{n_a}(b))$	$\sum_{n_a \in S_a} tree\_size(n_a)$
	BQ	$\sum_{n_a \in S_a} num_{PC,BFS}(n_a, constraint(b,C))$	$\sum_{n_a \in S_a} num_{AD,BFS}(n_a, constraint(b,C))$
BUS	TQ	$blk(\bigcup_{n_b \in S_b} parent_{n_b}(*))$	$blk(\bigcup_{n_b \in S_b} path(n_b, a))$
	BQ	$blk(\bigcup_{n_b \in S_b} parent_{n_b}(*))$	$blk(\bigcup_{n_b \in S_b} ancestor_{n_b}(*))$
	index	$blk(\bigcup_{n_b \in sel(S_b, a/b)} parent_{n_b}(*)) + l_{ind(a)} \times \left\| \bigcup_{n_b \in S_b} parent_{n_b}(a) \right\|$	$blk(\bigcup_{n_b \in sel(S_b, a//b)} path(n_b, a)) + l_{ind(a)} \times \left\| \bigcup_{n_b \in sel(S_b, a//b)} n_b \right\|$
SegSJ	TQ	$inter_{S_a} + inter_{S_b} + blk(S_a) + blk(S_b)$	$inter_{S_a} + inter_{S_b} + blk(S_a) + blk(S_b)$
	BQ	$inter_{S_a} + inter_{S_b} + blk(S_a) + blk(S_b)$	$inter_{S_a} + inter_{S_b} + blk(S_a) + blk(S_b)$
RangeFetch		2	$cost_L(a, b) + cost_R(a,b)$
VS		$cost_{value}(C, a) + blk(V_{a[C]})$	

The framework of the estimation of the cost is to estimate the cost in each step and sum up all the costs in the steps. The parameters used in the description of the cost model are shown in Table 1. The cost of a query plan is the sum of the cost of all its operators. Their values can be estimated with techniques in [21-24].

The cost models of the operators are summarized in Table 2. The unit of the result of the cost model is the number of I/Os. For discussion, the operator VS is to process step *a[C]* where *C* is a constraint and other operators are to process step *a*b* where "*" means "/" or "//". The explanations of the cost model are as follows.

- Here, S_a and S_b are the sets of *a* nodes and *b* nodes obtained from previous steps, respectively.
- The cost models of BFS and DFS in case of TQ are the same. When the axis is '/', BFS and DFS are to access all the *b* children of *a* nodes. When the axis is '//', BFS and DFS are to traverse all sub-trees with nodes as roots.
- The cost models of BFS and DFS in case of BQ are different because these two methods may meet the nodes satisfying constraints in different position.
- BUS operator with axis '/' can obtain parents without accessing another nodes. Without index, BUS operator with axis '//' needs to access the ancestors of each node in S_a. For axis '//', with BUS operator for TQ and BQ, for each node in S_b,

the numbers of ancestors to be accessed are different. Therefore, we use different functions to describe them.

- During processing the estimation of BUS, if some nodes share same parent, their parent is accessed only once.
- The cost of SegSJ is the sum of number of blocks of F&B index node and the intervals of candidates.
- The cost of RangeFetch is the cost of locating of the start and end position in the tape of *b*.
- The cost of VS is the sum of the cost of its two steps, obtaining the values satisfying the constraint and seeking the index nodes corresponding to the values.

The cost models of operators SA, IA and RA are presented as follows, where selected($tape_a$) is the selected part during accessing.

$$\text{cost}_{\text{SA}}(a) = \text{size}(tape_a)$$
$$\text{cost}_{\text{IA}}(a) = \text{size}(\text{selected}(tape_a))$$
$$\text{cost}_{\text{RA}}(a) = \text{size}(\text{selected}(tape_a))$$

5 The Generation of Query Plan

For a CXQ, all possible query plans can be generated with following strategy. The generation of query plan is divided into steps. The number of steps equals to the number of edges in the query graph. In each step, one edge is contracted. Since there may be multiple edges in query graph, which edge to be contracted has multiple choices. Each step of contraction can be represented as an operator. In the last step, the graph is contracted into a point. The steps of contraction are represented as a digraph $G=(V, E)$, where $v \in V$ is a status of the contraction and each $e \in E$: $v_i \to v_j$ is a transmission from status v_i to v_j. For $e \in E$: $v_i \to v_j$, the edge in query graph contracted to transform v_1 to v_2 is called a *transform edge* of *e*, denoted by T_e. *G* is called the *transform graph* of corresponding query. In *G*, a node without incoming edge is called a *source* and a node without outgoing edge is called a *sink*. An example of the generation of the candidate query plans of the query a/b[e]/c is shown in Figure 3.

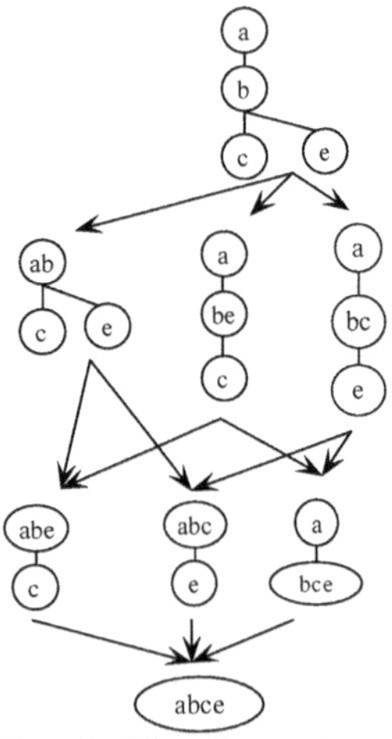

Fig. 3. The Generation of the Query Plan

Each edge $e \in E$ corresponds to an operator, denoted by *e.opt*. In *G*, the operators in each path from source to sink represent a feasible query plan. For the operator of each edge *e*, there may be several choices:

- In the case that T_e is in the trunk of the query and none of the nodes from root to the start node of Te have branching constraints, if Te represents"//", the choices of *e.opt* include DFS, BFS, SegSJ, $\text{BUS}_{\text{noindex}}$, $\text{BUS}_{\text{index}}$ or RangeFetch; otherwise Te may be DFS, BFS, SegSJ, BUSnoindex, or RangeFetch.

- If T_e is in branching and represents "//", *e.opt* may be DFS, BFS, SegSJ, BUSindex or BUSnoindex.
- If T_e points from node to value, *e.opt* is VS.
- In other cases, the choices of *e.opt* include DFS, BFS, SegSJ and BUSnoindex.

The operators for the edges in the transform graph are selected individually. It is because evaluation of each subquery with different operator will not affect the intermediate result. As the first stage of optimization, for each edge *e*, the operator with smallest cost is selected and the weight of *e*, denoted by *e.w*, is such cost. The search for optimal query plan is to find the path from source to sink with smallest sum of weights of edges.

n is the number of nodes in the query tree. Each choice in the search space corresponds to an arrangement of all the edges in the query tree. Therefore, the size of search space is $O((n-1)!)$. Since the status is generated in level, this problem can be solved with dynamic programming. The algorithm is shown in Algorithm 1. With such algorithm, the number of generated statuses is $O(n\times 2^n)$.

Algorithm 1 SearchOptSolution(G)
Input: G**:** The query graph;
Output: The query plan as the query operator list
$S_0 = \emptyset$
$n_{S_0}.cost = 0$
$n_{S_0}.path = NULL$
$n_{S_0}.piori = NULL$
for i = 1 to $|G.E|$ **do**
  T= $GenCombination(i,|G.V|)$
  for each $T \in$ T **do**
    $n_T.cost = \min_{j\in T}(n_{T/i}.cost + e_j.wight)$
    $r = arc \min_{j\in T}(n_{T/i}.cost + e_j.wight)$
    $n_T.path = n_{t-r}.path \cup e_r$
return $n_{|G.V|}.path$

In the algorithm, the subscript is a set. The subscripts of status in the *i*th phase are represented by the combination of *i* elements in $|G.V|$. The function GenCombination(i,n) is to generate all the combinations with *i* from *n*. For each transmission from one status to another, one element is added to its subscript. In the algorithm, The cost model is applied on the estimation of the cost of each status.

In most of cases, in the algorithm, n is small (less than 10) and the query optimization is fast. For the cases that n gets large, in order to accelerate the query optimization, we design some strategies to reduce the search space.

Initialization Cost Filter. The idea is that for the query, we design a query plan *p* as a seed by heuristic method. During the searching, when the cost of current steps is larger than *p*, further searching in this branching is stopped. *p* is called the *seed plan*. The seed plan is chosen by following heuristic strategy without any statistic information:

1. If a node is in trunk with all the nodes in its incoming path has no branching constraint, RangFetch is added to its incoming edge.
2. For each other single slash edge, a BFS operator is added.
3. For each other double slash edge, a SegSJ operator is added.

4. For a leaf f with value constraint, a VS operator is added and all the edges in the path from f to the node in trunk as the root of the sub-tree with f is converted into BUS.
5. The order of operators associating with a node with multiple incoming or outgoing edges is determined by following strategy: the priority is as (1)branching with BUS (2)branching for constraint (3) branching with more nodes.

Rule 1-3 of the heuristic strategy choose the fastest operator when the data is distributed uniformly. Rule 4 insures that the value query can be supported and Rule 5 is to process the operator that will filter more nodes.

For a structural query, the seed plan is better than the query plan with operator SegSJ in each step. The reasons are as following:

In a Dnode, the pointers to its children are stored. Processing single slash step with BFS accesses required children directly while SegSJ accesses extra extent tape. Therefore, the step of single slash, BFS is better than SegSJ. For the step of double slash, SegSJ is chosen in seed plan.

Optimal Query Plan Branch Cutting. This strategy is using the optimal cost of the searching in former steps as the filter to cut branches. When a query plan with smaller cost is obtained, the filter cost is changed to this value.

6 Experimental Results

6.1 Experimental Setup

All of our experiments were performed on a PC with Pentium 3GMHz, 2G memory and 100G IDE hard disk. The OS is Windows 2000 Server. We implemented the system using Microsoft Visual C++ 6.0. We also implement XSketch [22][23][24] by ourselves as well as query estimation on XSketch and query optimization strategy based on the estimation result of XSketch. We allocate 500K main memory for XSketch, including the graph Sketch for structural part and histograms for value part. We used LRU policy for buffer replacement. For comparison, we obtained the source code of NoK system [1] and TJFast [19] from the original authors. NoK is one of a native XML storage and query processing systems. TJFast is to using Dewey to improve TwigStack. TJFast is the best twig join algorithm to process all twig queries in the absence of indexes. We modify the code of TJFast to support value constraints.

In order to compare fairly with other systems, all the experiments are run in the warm buffer.

The dataset we tested is the standard 100M XMark benchmark dataset [27]. It has a fairly complicated schema, with several deeply recursive tags. We used a relatively small buffer size (less than 1%) to better simulate the case when the XML data size grows up to Gigabytes. Some statistics of the data, its 1-index and its F&B Index are shown in Table 3. We use the following metrics: elapsed time (time) and the number of physical I/Os (PIO).

Table 3. Size of the ED F&B Index

#Document Nodes	#Tags	#1-index Nodes	#F&B Index Nodes
2048193	77	550	528076

In order to better test and understand the characters of our system under different workload, we designed a set of queries with different characters. For a CXQ, there are two major dimensions: axis type, value constraint selectivity. Axis type includes *without AD axis* (i.e., PC axis only) or *with AD axis*. Value constraint includes *high selectivity* and *low selectively*. For example, query with AD axis and low selectivity is abbreviated as *AD-low*. We select two representative queries for each category and number those queries from Q?1 to Q?8, where if ? is 'X', the query is for XMark; if ? is 'D', the query is for DBLP. The semantics of the queries are shown in Table 4. The statistic information of queries for XMark is shown in Table 4.

Table 4. The Workload of Experiments

Queries	Type	Queries	Type
Q?1, Q?2	PC-high	Q?3, Q?4	PC-low
Q?4, Q?5	AD-high	Q?7, Q?8	AD-low

Table 5. The Query Set

QID	Type	Result	Query
QX1	PC-low	327	/site/open_auctions/open auction[bidder/increase = 42]/annotation
QX2	PC-low	313	/site/people/person[address[zipcode>=8 and zipcode<= 9]]/profile/interest
QX3	PC-high	3966	/site/regions/Europe/item[quantity = 1]/description/text
QX4	PC-high	10148	/site/open_auctions/open_auction[bidder/[increase>5.0 and increase< 100]/initial
QX5	AD-low	158	//site//closed_auction[//price> 30 and //price < 4][//annotation]//type
QX6	AD-low	299	/site//open_auction[//annotation//happiness = 2][//privacy]//reserve
QX7	AD-high	38634	/site/open_auctions/open_auction[//happiness][//quantity=1][//descriptio n/text]//date
QX8	AD-high	6770	/site//person[//age>20 and //age< 100][//emailaddress][//creditcard] //interest

6.2 Quality of Plans

In this subsection, we check the quality of query plans selected by our strategy. We randomly (but not exhaustively) generated ten query plans for each query of XMark. We executed these plans on 100M XMark document and picked up the best, the worst of these plans and computed average time of these query plans. Note that the worst and best plans in the Table 4 are not necessarily the worst and best plan for a query but only the worst and best plans among a randomly selected plan set. The execution time of the worst, best plan and average time of the query plan set on 100M XMark document are shown in MAX, MIN and AVG column in Table 5, respectively. The plans as seed of the search are shown in SED column in Table 5.

Table 6. Comparison of Run Time

Query	OPT(ms)	SED(ms)	MAX(ms)	MIN(ms)	AVG(ms)
QX1	5	38	151	41	88
QX2	8	9	12	8	9
QX3	126	161	538	174	334
QX4	73	73	95	73	88
QX5	13	17	770	13	273
QX6	30	198	995	632	851
QX7	173	305	794	645	745
QX8	28	145	20171	292	2852

The plan execution time on 100M XMark document is shown in Table 4, where time in OPT column is the execution time for the plan as the result of our optimization algorithm.

From the execution results of query plan, the query plan generated by our algorithm can always avoid really bad plans. The query plans of our optimizer are always better than the average efficiency of randomly generated query plan.

6.3 Optimization Time

In this subsection, we examine the time taken by optimizer. The time taken by optimizer is shown in Table 5.We compare the optimization time with and without branch-cutting, in the line of OPT and NON-OPT, respectively. We also compare the optimization time with execution time of query plan, which is in the line of TIME.

Table 7. Optimization Time

	QX1	QX2	QX3	QX4	QX5	QX6	QX7	QX8
NON-OPT(ms)	0	0	0	0	0	0	16	15
OPT(ms)	0	0	0	0	0	0	0	0
Execution(ms)	5	8	126	73	13	30	173	28

From the result, we can observe that when query is large, our heuristic strategy can improve query optimization time significantly. And comparing with execution time, query optimization is small.

6.4 Scalability

We tested our system on XMark document with different sizes, from 10M to 100M. The execution results of our system on Xmark documents with different sizes are shown in Figure 4. From the results, our system has good scalability. The physical I/Os and execution time of our system are about linear to document size. It shows our system has good scalability.

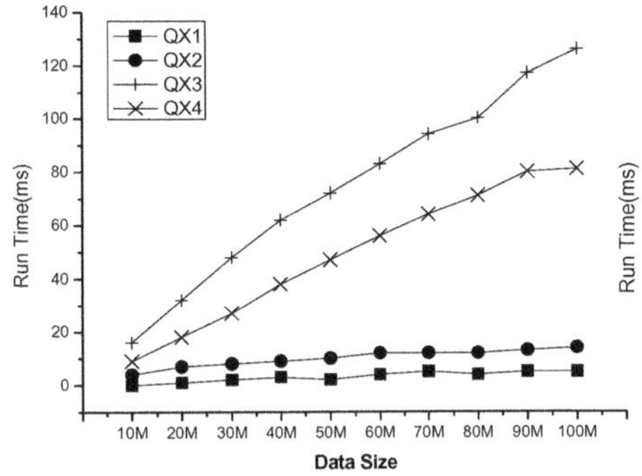

a) time VS. size: QX1-QX4

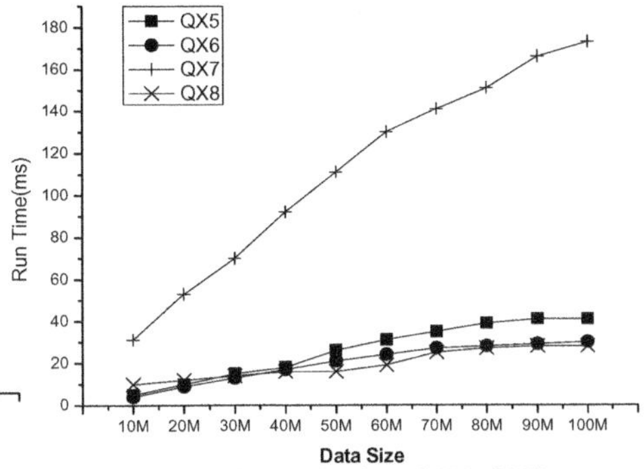

b) time VS. size: QX5-QX8

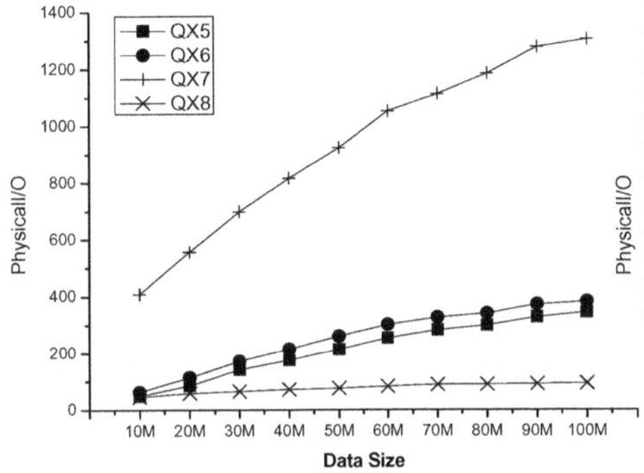

c) PIO VS. size: QX1-QX4

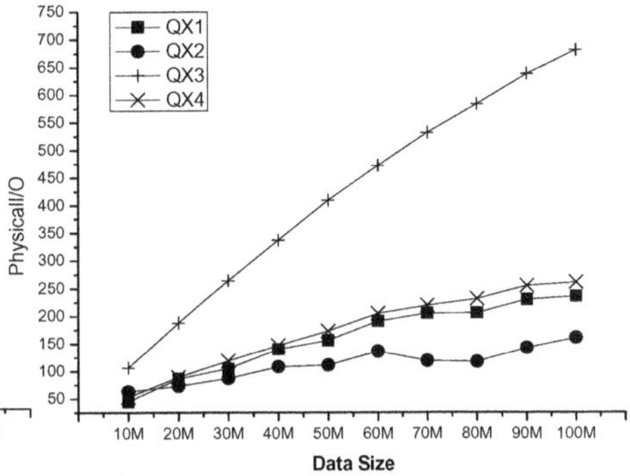

d) PIO VS. size: QX5-QX8

Fig. 4. Scalability Experimental Results

6.5 Comparing with Other Systems

In this subsection, we compare the performance of our system with another two systems: NoK [1] and TJFast [19]. We set buffer size 256K Bytes. Query optimization time and final sorting time is included in our algorithms such that the results are in the document order. We measured the evaluation time of all the applicable algorithms and compared them in Table 8.

Table 8. Comparison with Other Systems

	QX1	QX2	QX3	QX4	QX5	QX6	QX7	QX8
OPT(ms)	5	8	126	73	13	30	173	28
NoK(ms)	1572	n/a	750	n/a	n/a	n/a	n/a	n/a
TJFast(ms)	1094	17891	437	1094	297	391	860	1282
	QD1	QD2	QD3	QD4	QD5	QD6	QD7	QD8
OPT(ms)	5	17	54	29	4	13	97	87
NoK(ms)	172	n/a	17032	n/a	n/a	n/a	n/a	n/a
TJFast(ms)	3453	765	797	265	782	454	829	297

NoK. Since current version NoK can process query with only single slash and equality constraint, we only compare Nok with *Q?1* and *Q?3*, For queries on XMark and DBLP, our system outperforms Nok 6 to 300 times.

TJFast. From the comparison, our system outperforms TJFast 2 to 700 times. It is because even though in TJFast, the value constraint can filter some leaves, nodes corresponding to other leaves can not be filtered. In our system, since optimizer can choose the plan to begin at the part with smallest selectivity and search from such part only requires to access a small ratio of the nodes of corresponding to other nodes in the query.

7 Related Work

There are two major methods for XML structural query processing. One is index-based method; another is join-based method.

There have been many previous work on structure of XML data. Most of the structural indexes are based on the idea of considering XML document as a graph and partitioning the nodes into equivalent classes based on certain equivalence relationship. The resultant index can support (or cover) certain class of queries. A rigid discussion of the partitioning method and its supported query classes is presented in [26]. DataGuide [7] are based on *equivalent* relationship and 1-index [21] are based on the *backward bi-similarity* relationship. Both of them can answer all simple path queries. F&B Index [13] uses both backward and forward bisimulation and has been proved as the minimum index that supports all branching queries. A(k)-index [15], D(k)-index [25], M(k)-index and M*(k)-index [10] are approximations of 1-index. All these indices can not be applied to support CXQ directly.

XML queries can also be evaluated on-the-fly using the join-based approaches. Structural join and twig join are such operators and their efficient evaluation algorithms have been extensively studied [5, 8, 11, 18, 29, 31] [4, 12]. Their basic tool is the coding schemes that enable efficient checking of structural relationship of any two nodes. TwigStack [4] is the optimal twig join algorithm to answer twig queries with only double slash without using additional index. TJFast [19] improves TwigStack to support both twig queries with single slash and double slash efficiently. Hence it is chosen for comparison in our experiment. However, these methods can not support CXQ efficiently.

NoK [1] is a native XML data management system with CXQ supporting. It features a succinct XML storage scheme that essentially stores a string representation of the XML data tree obtained from pre-order traversal. We also choose Nok for comparison.

Most recently, the idea of integrating both tree traversal-based and join-based query processing has been proposed [9, 14]. The major difference of their systems with our proposal is that they either do not use a structural index [9] or they are only using the index as a filter to accelerate structural joins [14]. On the contrary, our focuses is more on using the structural index to its full power and integrate both join-based and index-based query processing methods to accelerate CXQ processing.

[16] and [17] present a method by indexing values and attaching bloom filter representing incoming path to the value. The query processing is to select value with and matching path approximately with the bloom filter. This method is designed for single path queries

with value constraint at leave and not suitable for queries with complex structure and multiple value constraints. [20] presents a disk-based system of XML query processing based on concurrency control. This system supports path query with value constraints by adding value filter operation on the candidates of twig join or structural join. As a comparison with their system, our method combines the benefits of structural join operation and structural index and gains efficiency, especially for the data with many duplicate structures such as DBLP.

8 Conclusions

Complex path queries (CXQ for brief) are an important kind of queries for XML data. Existing techniques can not process CXQs efficiently. With the supporting of storage structure, the query optimization strategies for CXQs are presented in this paper. In order to choose the optimal query plan for CXQs, a cost-based query optimization strategy is proposed. Extensive experiments show that our method outperforms existing methods significantly and has good scalability. Our query optimization strategy is effective and efficient. Our future work includes investigating XQuery query evaluation and optimization methods, as well as holistic methods for efficiently CXQ processing in our system.

Acknowledgement

The authors would like to thank Ning Zhang and Jiaheng Lu for their help in carrying out the experiments.

References

1. Zhang, N., Kacholia, V., Ozsu, M.T.Ä.: A succinct physical storage scheme for efficient evaluation of path queries in XML. In: ICDE 2004 (2004)
2. Zhang, N., ÄOzsu, M.T., Ilyas, I.F., Aboulnaga, A.: Fix: Feature-based indexing technique for xml documents. In: VLDB 2006 (2006)
3. Al-Khalifa, S., Jagadish, H.V., Patel, J.M., Wu, Y., Koudas, N., Srivastava, D.: Structural joins: A primitive for efficient XML query pattern matching. In: ICDE 2002 (2002)
4. Bruno, N., Koudas, N., Srivastava, D.: Holistic twig joins: Optimal XML pattern matching. In: SIGMOD 2002 (2002)
5. Chien, S.-Y., Vagena, Z., Zhang, D., Tsotras, V.J., Zaniolo, C.: Efficient structural joins on indexed XML documents. In: Bressan, S., Chaudhri, A.B., Li Lee, M., Yu, J.X., Lacroix, Z. (eds.) CAiSE 2002 and VLDB 2002. LNCS, vol. 2590. Springer, Heidelberg (2003)
6. Clark, J., DeRose, S.: XML path language (XPath). In: W3C Recommendation (November 16, 1999), `http://www.w3.org/TR/xpath`
7. Goldman, R., Widom, J.: Dataguides: Enabling query formulation and optimization in semistructured databases. In: VLDB 1997 (1997)
8. Grust, T.: Accelerating XPath location steps. In: SIGMOD 2002 (2002)
9. Halverson, A., Burger, J., Galanis, L., Kini, A., et al.: Mixed mode XML query processing. In: VLDB 2003 (2003)
10. He, H., Yang, J.: Multiresolution indexing of xml for frequent queries. In: ICDE 2004 (2004)

11. Jiang, H., Lu, H., Wang, W., Ooi, B.C.: XR-Tree: Indexing XML data for efficient structural join. In: ICDE 2003 (2003)
12. Jiang, W., Wang, H., Yu, J.X.: Holistic twig joins on indexed XML documents. In: VLDB 2003 (2003)
13. Kaushik, R., Bohannon, P., Naughton, J.F., Korth, H.F.: Covering indexes for branching path queries. In: SIGMOD 2002 (2002)
14. Kaushik, R., Krishnamurthy, R., Naughton, J.F., Ramakrishnan, R.: On the integration of structure indexes and inverted lists. In: SIGMOD 2004 (2004)
15. Kaushik, R., Shenoy, P., Bohannon, P., Gudes, E.: Exploiting local similarity for indexing paths in graph-structured data. In: ICDE 2002 (2002)
16. Li, H.-G., Aghili, S.A., Agrawal, D.P., El Abbadi, A.: FLUX: Content and structure matching of xPath queries with range predicates. In: Amer-Yahia, S., Bellahsène, Z., Hunt, E., Unland, R., Yu, J.X. (eds.) XSym 2006. LNCS, vol. 4156, pp. 61–76. Springer, Heidelberg (2006)
17. Li, H.-G., Aghili, S.A., Agrawal, D., Abbadi, A.E.: Flux: fuzzy content and structure matching of xml range queries. In: WWW 2006 (2006)
18. Li, Q., Moon, B.: Indexing and querying XML data for regular path expressions. In: VLDB 2001 (2001)
19. Lu, J., Ling, T.W., Chan, C.Y., Chen, T.: From region encoding to extended dewey: On efficient processing of XML twig pattern matching. In: VLDB 2005 (2005)
20. Mathis, C., HÄarder, T., Haustein, M.P.: Locking-aware structural join operators for XML query processing. In: SIGMOD 2006 (2006)
21. Milo, T., Suciu, D.: Index structures for path expressions. In: ICDE 1999 (1999)
22. Polyzotis, N., Garofalakis, M.N.: Statistical synopses for graph-structured XML databases. In: SIGMOD 2002 (2002)
23. Polyzotis, N., Garofalakis, M.N.: Structure and value synopses for XML data graphs. In: Bressan, S., Chaudhri, A.B., Li Lee, M., Yu, J.X., Lacroix, Z. (eds.) CAiSE 2002 and VLDB 2002. LNCS, vol. 2590. Springer, Heidelberg (2003)
24. Polyzotis, N., Garofalakis, M.N., Ioannidis, Y.E.: Selectivity estimation for XML twigs. In: ICDE 2004 (2004)
25. Qun, C., Lim, A., Ong, K.W.: D(k)-index: An adaptive structural summary for graph-structured data. In: SIGMOD 2003 (2003)
26. Ramanan, P.: Covering indexes for XML queries: Bisimulation - Simulation= Negation. In: VLDB 2003 (2003)
27. Schmidt, F.W., Kersten, M.L., Carey, M.J., Manolescu, I., Busse, R.: XMark: A benchmark for XML data management. In: Bressan, S., Chaudhri, A.B., Li Lee, M., Yu, J.X., Lacroix, Z. (eds.) CAiSE 2002 and VLDB 2002. LNCS, vol. 2590, Springer, Heidelberg (2003)
28. W3C. XML Query 1.0 and XPath 2.0 data model (2003), `http://www.w3.org/TR/xpath-datamodel`
29. Wang, W., Jiang, H., Lu, H., Yu, J.X.: PBiTree coding and efficient processing of containment joins. In: ICDE 2003 (2003)
30. Wang, W., Wang, H., Lu, H., Jiang, H., Lin, X., Li, J.: Efficient processing of xml path queries using the disk-based F&B index. In: VLDB 2005 (2005)
31. Zhang, C., Naughton, J.F., DeWitt, D.J., et al.: On supporting containment queries in relational database management systems. In: SIGMOD 2001 (2001)

Containment between Unions of XPath Queries

Rui Zhou[1], Chengfei Liu[1], Junhu Wang[2], and Jianxin Li[1]

[1] Swinburne University of Technology, Melbourne, Australia
{rzhou,cliu,jianxinli}@swin.edu.au
[2] Griffith University, Gold Coast, Australia
J.Wang@griffith.edu.au

Abstract. In this paper, we address the containment problem for unions of XPath queries with and without schema. We find the problem can be always reduced into containment problem between one single query and a union of queries. When schema is not available, the problem can be further reduced into checking containment between pairwise queries (each from one union), but this only holds for some XPath subsets, such as $XP^{\{/,//,[]\}}$, but not for $XP^{\{/,//,[],*\}}$. We then show the problem is still solvable in $XP^{\{/,//,[],*\}}$, though no efficient algorithm exists. When schema is at hand, we propose a strategy to rewrite a query into a union of simplified queries based on schema information, and then apply methods developed when schema is not taken into account. The problem is then reduced into checking containment between unions of queries in $XP^{\{/,[]\}}$ without schema.

1 Introduction

Testing query containment is a key technique in many database applications. In query optimization [1], it is a subtask to check if two formulations of a query are equivalent, and hence to choose the formulation with less evaluation cost. In data integration scenario, especially rewriting queries using views [2], it provides a means to find equivalent or contained rewritings, and also to detect redundant rewritten queries to save computation time. Query containment can also be used to maintain integrity constraints [3] and determine when queries are independent of updates to the database [4].

In relational context, containment has been studied for conjunctive queries[1] [1] and unions of conjunctive queries [6]. In [6], it shows, for two unions of conjunctive queries P and Q, P is contained in Q if and only if any query p in P is contained in some query q in Q. Therefore, containment for unions of conjunctive queries can be reduced into containment for (a number of) pairs of queries. While for XML queries, especially XPath [7] queries, the containment problem is mainly studied between two single queries [8, 9, 10, 11], but not for unions of XPath queries. Some questions still need to be answered: Can we draw a similar conclusion for unions of XPath queries as that for unions of conjunctive queries? If not, how can we determine containment between unions of XPath queries?

[1] A typical query that is extensively studied in relational database, see [5].

X. Zhou et al. (Eds.): DASFAA 2009, LNCS 5463, pp. 405–420, 2009.

Does it make any difference whether an XML schema is available? We will look into these questions in this paper. Before heading into the main part, we would like to stress again the importance of determining containment between unions of queries. One typical application is rewriting queries using views, where to detect a redundant rewriting means to detect if a rewritten query is contained in a union of other rewritten queries. This is a special case of union containment, for one of the unions contains only one query.

In this paper, we show that the pairwise comparison property for unions of conjunctive queries holds only for some XPath subsets, such as $XP^{\{/,//,[]\}}$ featuring child, descendant, and branch axes. For a larger subset $XP^{\{/,//,[],*\}}$, with wildcards added, we provide an example to show two queries can be combined together to contain a third query, though neither of them could solely contain the third one. Therefore we should devise some new strategy to detect containment relationship for unions of queries in this subset. To make the work comprehensive, we also discuss containment under schema information. A query p not contained in another query q in general, may be contained in q under schema constraints, because schema imposes some constraints, confining wildcards and descendant axes in the query being interpreted in some particular ways. In order to tackle this problem, we propagate schema constraints into queries to eliminate wildcards and descendant edges, and thus simplify the queries into queries in subset $XP^{\{/,[]\}}$. Then after chasing the simplified patterns in $XP^{\{/,[]\}}$, established methods for unions of queries without schema can be applied.

In this paper, Our contributions are highlighted as follows:

– We are the first to investigate the containment problem for unions of queries in XML context, particularly on XPath queries. We show the problem can be always reduced into containment problem between one single query and a union of queries.
– When schema is not available, the problem can be further reduced into checking containment between two single queries (each from one union). However, this result only holds for some simple XPath subsets like $XP^{\{/,//,[]\}}$, not for $XP^{\{/,//,[],*\}}$. But fortunately in $XP^{\{/,//,[],*\}}$, the problem is still solvable.
– When schema is available, we suggest a strategy to rewrite a query into a union of simplified queries based on schema information, and then apply the methods developed when schema is not considered. The problem is then reduced into checking containment between unions of queries in $XP^{\{/,[]\}}$ without schema.

The rest of this paper is organized as follows. In Section 2, we will give some notations and background knowledge. Then we propose two important theorems and tackle the containment problem without schema in Section 3. Schema information will be taken into account in Section 4 to eliminate wildcard nodes and expand descendant edges. In Section 5, we extend the discussion from XPath queries into general tree pattern queries. Related work is given in Section 6. Finally, we draw a conclusion and propose some future work in Section 7.

2 Preliminaries

In this section, we introduce some notations and background knowledge. XML documents and XPath queries are modelled as trees and tree patterns, and evaluating an XPath query on an XML document is modelled as matching a tree pattern to a tree. We also formulate the definition for containment between unions of XPath queries at the end of this section.

2.1 XML Trees

In the literature, an XML document is modeled as an unordered tree[2] with nodes labeled from an infinite alphabet Σ (Σ is finite, if a schema is available), the label of each node corresponds to an XML element, an attribute name or a data value, the root node of the tree represents the root element in the document. We slightly modify the model by adding a new root with a unique label $r \in \Sigma$ to the tree, serving as the *document node*. In this way, the root node in the previous model becomes a (single) child of this document node, and every XML document starts with a root node labeled r, see Definition 1. We will see the aim of this modification in the next subsection. We denote all possible trees over Σ as T_Σ.

Definition 1. *An XML document is a tree* $t = (V_t, E_t, r_t, \Sigma)$, *where*

- V_t *is the node set, and* $\forall v \in V_t$, v *has a label in the alphabet* Σ, *denoted as* $label(v)$;
- E_t *is the edge set;*
- $r_t \in V_t$ *is the root node of* t, *and* $label(r_t) = r$;

2.2 XPath Query

XPath is the core subclass of XML query languages. We consider a subset of XPath featuring child axes (/), descendant axes (//), branches ([ ]), and wildcards (*). It can be represented by the following grammar:

$$p \rightarrow .|l| * |p/p|p//p|p[p]$$

where "." denotes the current context node, "l" is a label from alphabet Σ and "*" represents a wildcard label. We denote this subset as $XP^{\{/,//,[],*\}}$. The result of evaluating an XPath expression $p \in XP^{\{/,//,[],*\}}$ on a tree $t \in T_\Sigma$, denoted as $p(t)$, is a set of nodes in t. The formalized semantics are given in [12] (omitted here), where the context node is fixed on the document node if the context node is not explicitly specified. Like in [8], besides allowing the usage of "." immediately inside a predicate [ ], we further allow "." to appear at the beginning of an expression to capture the XPath queries starting with / or //. For example, queries /a/b and //a//b, which do not conform to the above

[2] Order is ignored in most previous research works, and so it is in this work.

grammar, i.e. ignored in previous works (such as [13,9,8]), can be now rewritten into ./a/b and .//a//b, and therefore be captured. Since "." always inspects the document node (which we intentionally added in the former section) by default, the rewritten expressions correctly preserve the semantics. Moreover, XPath expressions starting with a label can be safely rewritten into some expressions starting with "." as well, eg, a/b equals to ./a/b. In this way, an XPath query corresponds to a unique tree pattern query, see the following definition.

Definition 2. *An XPath query q can be expressed as a tree pattern $(V_q, E_q, r_q, d_q, \Sigma_q)$, where*

- *V_q is the node set, and $\forall v \in V_q$, v has a label in a finite alphabet $\Sigma_q \cup \{*\}$, denoted as $label(v)$;*
- *E_q is the edge set, and $\forall e \in E_q, type(e) \in \{/, //\}$. We use the term "pc-edge"("ad-edge") to represent the type of an edge, "/"("//").*
- *r_q is the root node of the query, corresponding to the leading "." tag in q (if the current context node is not specified, then $label(r_q) = r$);*
- *d_q is the answer (also called distinguished or return) node of the query, identified with a circle;*

The result of evaluating an XPath query, equals to finding embeddings from a tree pattern query q to a tree t, which can be represented as $q(t) = \{f(d_q) | f$ is some embedding from p to $t\}$. Embedding is defined as follows:

Definition 3. *An embedding from a tree pattern $q = (V_q, E_q, r_q, d_q, \Sigma_q)$ to a tree $t = (V_t, E_t, r_t, \Sigma)$ is a mapping $f : V_q \to V_t$, satisfying:*

- *Root preserving: $f(r_q) = r_t$;*
- *Label preserving: $\forall v \in V_q$, $label(v) = *$ or $label(v) = label(f(v))$;*
- *Structure preserving: $\forall e = (v_1, v_2) \in E_q$, if e is a pc-edge, $f(v_1)$ is the parent node of $f(v_2)$; e is an ad-edge, $f(v_1)$ is an ancestor node of $f(v_2)$ including the case $f(v_1)$ being the parent of $f(v_2)$.*

In this work, all XPath queries are observed as tree patterns for ease of discussion, and we will provide some discussions on general tree patterns (queries with more than one distinguished nodes) in Section 5.

2.3 Containment Formulation

For any two tree pattern query p and q, p is said to be contained in q, denoted as $p \subseteq q$, iff $\forall t \in T_\Sigma, p(t) \subseteq q(t)$. We now extend the definition to unions of queries. We use a lowercase letter and an uppercase letter to reflect a single query and a union of queries respectively. Let $P = \{p_1, p_2, \cdots, p_m\}$ be a union of queries, the result of this set of queries on a tree t, denoted as $P(t)$, is defined as $p_1(t) \cup p_2(t) \cup \cdots \cup p_m(t)$. For two unions of queries $P = \{p_1, p_2, \cdots, p_m\}$ and $Q = \{q_1, q_2, \cdots q_n\}$, P is said to be contained in Q, iff $\forall t \in T_\Sigma, P(t) \subseteq Q(t)$ (i.e. $p_1(t) \cup p_2(t) \cup \cdots \cup p_m(t) \subseteq q_1(t) \cup q_2(t) \cup \cdots \cup q_n(t)$).

3 Containment without Schema

In this section, we will investigate containment problem between unions of XPath queries without schema information. We start with reducing the problem into a simplified form and introducing boolean tree pattern. After that, we endeavor to solve the problem for boolean tree patterns belonging to subset $P^{\{/,//,[]\}}$. Finally, a larger subset with wildcards, $P^{\{/,//,[],*\}}$, will be discussed.

3.1 Containment Reduction

To simplify the problem, we propose a theorem to reduce the containment checking between two unions of queries into containment checking between one single query and a union of queries. The theorem also holds when a schema is at hand.

Theorem 1. *For two unions of XPath queries $P = \{p_1, p_2, \cdots, p_m\}$ and Q, we have: $P \subseteq Q$, iff $\forall p_i \in P$, $p_i \subseteq Q$.*

Proof. If: Given $\forall p_i \in P$, $p_i \subseteq Q$, we have, from definition, for any tree $t \in T_\Sigma$, $p_i(t) \subseteq Q(t)$. Therefore, for any tree $t \in T_\Sigma$, $p_1(t) \cup p_2(t) \cup \cdots \cup p_m(t) \subseteq Q(t)$, that is $P \subseteq Q$;
Only if: Given $P \subseteq Q$, we have $\forall t \in T_\Sigma, P(t) \subseteq Q(t)$, i.e. $p_1(t) \cup p_2(t) \cup \cdots \cup p_m(t) \subseteq Q(t)$, thus $\forall p_i$, $p_i(t) \subseteq p_1(t) \cup p_2(t) \cup \cdots \cup p_m(t) \subseteq Q(t)$. □

The idea conveyed by the above theorem is simple, but it lays a foundation to check union containment, because we can always safely simplify the left part of the comparison into a single query. This leads to some further explorations and observations in Section 3.3 and 3.4.

3.2 Boolean Tree Pattern

Boolean pattern (short for boolean tree pattern) is a tree pattern query with no distinguished node. The result of evaluating a boolean pattern $\bar{p}$ on a tree t, $\bar{p}(t)$, is a boolean value, either true or false. $\bar{p}(t)$ is true, means there exists an embedding from $\bar{p}$ to t, otherwise $\bar{p}(t)$ is false. For two boolean patterns, $\bar{p}$ and $\bar{q}$, we say $\bar{p}$ is contained in $\bar{q}$, denoted as $\bar{p} \subseteq \bar{q}$, iff $\forall t \in T_\Sigma, \bar{p}(t) \Rightarrow \bar{q}(t)$.

Each XPath tree pattern corresponds to a unique boolean pattern, which can be obtained by adding a child node with a distinct label x to the distinguished node, and make the distinguished node not outstanding any more (shown in Fig. 1). Let the corresponding boolean patterns of XPath patterns p and q be $\bar{p}$ and $\bar{q}$ respectively. According to [14], $p \subseteq q$ iff $\bar{p} \subseteq \bar{q}$. Consequently, for ease of discussion, XPath tree pattern queries are considered as boolean patterns in the rest of the paper. Notations p and q will refer to boolean patterns from now on, and we no longer use $\bar{p}$ and $\bar{q}$. And the result of a union of boolean tree patterns $Q = \{q_1, q_2, \cdots, q_n\}$ on a tree t can be then expressed in the form of $Q(t) = q_1(t) \vee q_2(t) \vee \cdots \vee q_n(t)$, for $Q(t)$ is a boolean value. We also use $P^{\{/,//,[]\}}$ to denote the corresponding boolean pattern subset for XPath tree patterns in $XP^{\{/,//,[]\}}$.

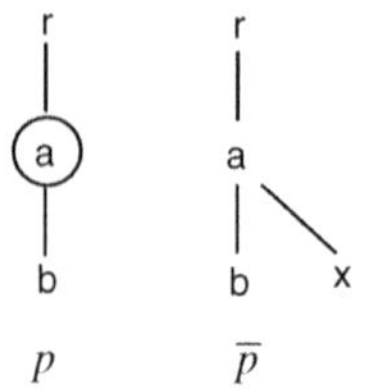

Fig. 1. A XPath Tree Pattern p and Its Corresponding Boolean Pattern $\bar{p}$

3.3 Patterns in $P^{\{/,//,[]\}}$

As is shown in Theorem 1, checking containment between unions of queries can always be reduced into checking containment between one single query and a union of queries. Then one question arises: Can the problem be further reduced into checking containment for a number of query pairs? That is, if a query is contained in a union of queries, does it mean that the single query is contained in some particular query from the union? The answer is yes, but the result is restricted to a certain subset of queries. For simplicity, we illustrate this result (expressed as Theorem 2) within a query subset $P^{\{/,//,[]\}}$, which has branches and descendant axes, but no wildcards. We will point out a larger query subset where the property still holds after proving the theorem.

Theorem 2. *For a boolean pattern p and a union of boolean patterns $Q = \{q_1, q_2, \cdots, q_n\}$ in $P^{\{/,//,[]\}}$, we have: $p \subseteq Q$, iff $\exists q_i \in Q$, such that $p \subseteq q_i$.*

Proof. The sufficient condition is obvious. Now we will prove the necessary condition by proving its contrapositive statement, i.e. to show that if p is not contained in any $q_i \in Q$, then p cannot be contained in Q. Before starting the proof, we first introduce a technique called homomorphism providing a sufficient and necessary condition to decide containment between two single patterns in $P^{\{/,//,[]\}}$.

A homomorphism from one pattern $p = (V_p, E_p, r_p, \Sigma_p)$ to another $q = (V_q, E_q, r_q, \Sigma_q)$, is a function $h : V_p \rightarrow V_q$ satisfying the definition of embedding (given in Definition 3). The only difference is that homomorphism is a mapping from one query pattern to another, while embedding is a mapping from a pattern to a data tree. According to Theorem 3 in [8], for two boolean patterns p and q in $P^{\{/,//,[]\}}$, $p \subseteq q$ iff there exists a homomorphism from V_q to V_p. In other words, if $p \not\subseteq q$, there must exist a node v_i in V_q, such that we cannot find any homomorphism h that has a corresponding node $h(v_i)$ in V_p, satisfying label preserving and structure preserving conditions w.r.t. nodes v_i and $h(v_i)$. We call such node v_i a *private* node of q against p. We also name, on some path in q (from root to leaf), the first private node as a *transitional* node.

To prove the contrapositive statement of the necessary condition in Theorem 2, given $\forall q_i \in Q, p \not\subseteq q_i$, we could construct a tree t, such that $p(t)$ holds while $q_i(t)$ is false. And hence $p(t)$ does not imply $Q(t) = q_1(t) \vee q_2(t) \vee \cdots \vee q_n(t)$, namely p is not contained in Q. The tree t can be constructed as follows: replace each ad-edge in p with two pc-edges and an additional distinct label z.

For instance, a//b can be transformed into a/z/b. Here label z does not appear in any Σ_{q_i}(i.e. $z \in \Sigma - \bigcup_{i=1}^{n} \Sigma_{q_i}$), where Σ_{q_i} is the alphabet of q_i. Since Σ is infinite (when there is no schema available) and Σ_{q_i} is finite (because the number of labels in a query is limited), this transformation is always possible. After the transformation, it is straightforward that the result tree t conforms to pattern p, and thus $p(t)$ is true. However, for any q_i, we can show $q_i(t)$ is false. The reason is: since $p \nsubseteq q_i$, there must be some transitional node v_i in q_i, such that for the transitional node, we cannot find a corresponding node $f(v_i)$ in t defined by any embedding f from q_i to t. Otherwise, if such embedding f existed, we could obtain a twin homomorphism h from q_i to p based on f. Here the twin homomorphism h would have the same mapping function as embedding f, because, in f, no nodes in q_i can be mapped onto z-nodes (nodes with distinct label z) in t. Therefore, a corresponding node $h(v_i)$ in p would exist for the homomorphism. This result contradicts with the assumption that v_i is a transitional node. Recall that, a transitional node in q_i could not map onto any node in p by any homomorphism, as a result, $\forall q_i \in Q$, $q_i(t)$ is false, i.e. $Q(t) = q_1(t) \vee q_2(t) \vee \cdots \vee q_n(t)$ is false. In addition, $p(t)$ is true, hence $p(t) \nRightarrow Q(t)$. The contrapositive statement of the necessary condition is proved. □

The complexity of testing containment between one pattern p and a union of patterns $Q = \{q_1, q_2, \cdots, q_n\}$ is $O(|p| \cdot \sum_{i=1}^{n} |q_i|)$, bounded by $O(n|p||q|_{max})$, where $|q|_{max}$ is $\max\{|q_i|\}$. This is an immediate result from that finding a homomorphism from a pattern q_i to p is of complexity $O(|p||q_i|)$, where $|p|$, $|q_i|$ are the size (number of nodes) of p and q_i respectively. However, Theorem 2 only holds in $P^{\{/,//,[]\}}$, or a large subset $\hat{P}^{\{/,//,[],*\}}$ mentioned in [15]. $\hat{P}^{\{/,//,[],*\}}$ refers to an XPath query subset further including wildcards, but with two additional restrictions: (i) no wildcard node is incident with ad-edges(//) and (ii) there is no wildcard leaf node. The reason that Theorem 2 holds for a limited query subset lies in the following aspects: (1) If there are wildcard nodes in the patterns, homomorphism only serves as a sufficient (but not necessary) condition to determine containment between two patterns. We will give an example and devise a strategy in the next subsection to deal with patterns in $P^{\{/,//,[],*\}}$. (2) Moreover, if there is a schema available, the theorem does not necessarily hold in $P^{\{/,//,[]\}}$ as well, because alphabet Σ becomes finite, and there may not always exist a distinct label z to transform an ad-edge a//b into a/z/b, and thus the current proof is not sufficient. We will consider queries conforming to a schema in Section 4.

3.4 Patterns in $P^{\{/,//,[],*\}}$

We first give a simple example to show that Theorem 2 is not true for subset $P^{\{/,//,[],*\}}$. See in Fig 2, $p \nsubseteq q_1$ and $p \nsubseteq q_2$, but it is obvious that p is equivalent to $Q = q_1 \vee q_2$, because if b is a descendant of a, then either b is a direct child of a or b is a descendant of a's child. And $p = Q$ implies $p \subseteq Q$. The example shows that several patterns may be combined together to contain a target pattern, though none of them could solely contain the target. This observation makes

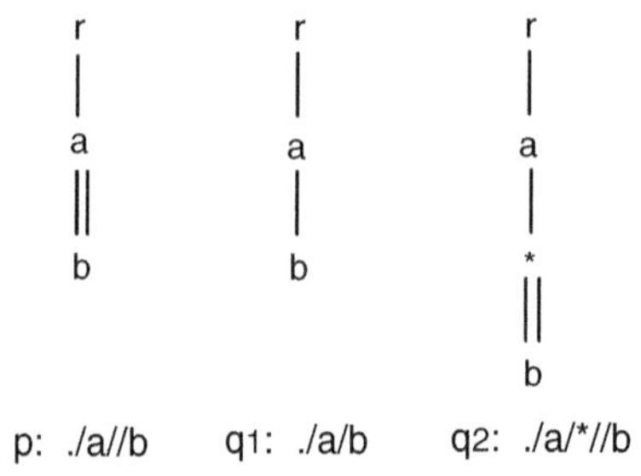

Fig. 2. Containment with Wildcard Node

the problem complicate in $P^{\{/,//,[],*\}}$, for it is difficult to know which patterns should be combined together to contain the target.

Fortunately, taking advantage of Lemma 3 in [8], the containment problem between one single pattern p and a union of patterns $Q == \{q_1, q_2, \cdots, q_n\}$ can be reduced to checking containment between two single patterns as well. The two single patterns can be constructed easily, as is shown in Fig. 3, where c in both p' and q' is a label in Σ, $T_0 \in T_\Sigma$ is a tree such that for any q_i, $q_i(T_0)$ is true. This can be achieved by fusing the roots of q_i (this is possible because they share the same root label r), and replacing all wildcards with an arbitrary label, and all ad-edges with pc-edges. It has been proved that $p \subseteq Q (Q = q_1 \vee q_2 \vee \cdots \vee q_n)$, iff $p' \subseteq q'$. Since $p', q' \in P^{\{/,//,[],*\}}$ and deciding $p' \subseteq q'$ is coNP-complete, the containment problem between unions of patterns in $P^{\{/,//,[],*\}}$ is coNP-complete. Despite we cannot break the intrinsic complexity result for subset $P^{\{/,//,[],*\}}$, we manage to convert the problem into one that we have a solving strategy.

One may realize that since $P^{\{/,//,[]\}}$ is a subset of $P^{\{/,//,[],*\}}$, we can, without lose of generality, use the construction method above, to check whether $p' \subseteq q'$ using homomorphism technique in order to determine containment relationship between p and Q for subset $P^{\{/,//,[]\}}$. This introduces another strategy to solve the problem for subset $P^{\{/,//,[]\}}$. The observation is true, but the drawback of the above method is that it is less efficient than the method implied by Theorem 2. We now illustrate it by analyzing the algorithm complexity. For pattern p', it

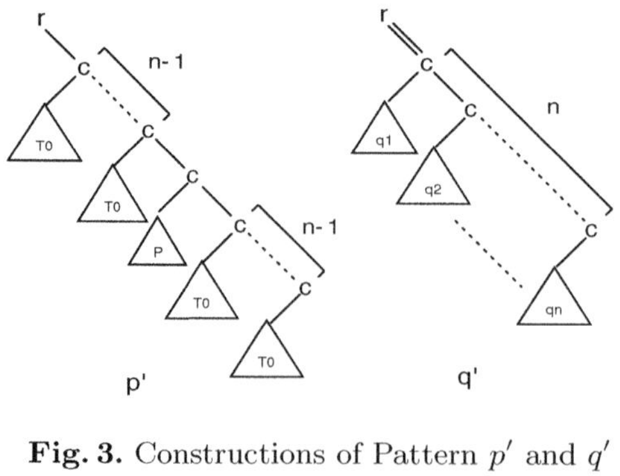

Fig. 3. Constructions of Pattern p' and q'

contains $2(n-1)$ number of T_0 trees, each of which has the size $\sum_{i=1}^{n}|q_i|$ (see how to construct T_0 in the previous paragraph), and thus $|p'|$ is $|p|+2(n-1)\sum_{i=1}^{n}|q_i|$. The size of $|q'|$ is easy to get as $\sum_{i=1}^{n}|q_i|$. The algorithm complexity $O(|p'||q'|)$ is $O((|p|+2(n-1)\sum_{i=1}^{n}|q_i|)\cdot\sum_{i=1}^{n}|q_i|)$, bounded by $O(n|p||q|_{max}+2(n-1)n^2|q|^2_{max})$, larger than $O(n|p||q|_{max})$ given in the last section. To conclude, if there is no wildcard node in the query pattern, it is better to compare p with queries in Q one by one.

4 Containment with Schema

Given a schema G, if for any tree t conforming to G, we have $p(t) \Rightarrow q(t)$, then we say p is contained in q under G, denoted as $p \subseteq_G q$. Schema provides a means to define or constrain XML data. A pattern p not contained in pattern q in general, may be contained in q for trees conforming to a certain schema. We show a simple example in Fig. 4, p_1 has a wildcard node, and p_2 has an ad-edge, they are not contained in $Q = q_1 \vee q_2$ in general. But if schema G is available, all the queries should conform to G. It is not hard to see $p_1 \subseteq_G p_2 \subseteq_G Q$, in fact $p_1 =_G p_2 =_G Q$.

Schema information is usually modelled as regular expressions or a few number of constraints. The works [10, 9] show that containment between patterns in $P^{\{/,//,[],*,DTD\}}$ is EXPTIME-complete, and some more theoretical results w.r.t. various pattern subsets can be found in [11]. Since the containment problem is already difficult for two single patterns, it is unlikely to have an efficient method to determine containment between unions of patterns in $P^{\{/,//,[],*,DTD\}}$. The aim of our work is not to break the proved EXPTIME-complete upperbound for two queries, nor to provide any exact complexity results for unions of queries, but to reexamine the problem from another angle and to suggest a strategy to check containment between two single patterns or unions of patterns with schema information. The idea is to propagate DTD constraints into queries so as to eliminate wildcards and descendant edges. Consequently, the problem could be converted into containment between unions of simplified queries in $P^{\{/,[]\}}$. Then, after chasing patterns in $P^{\{/,[]\}}$ with DTD constraints, we can apply Theorem 1 and 2 to evaluate the containment relationship.

In our paper, we model the schema as a directed graph G. (we don't consider disjunctions in the schema.) G is a DAG means the schema is not recursive,

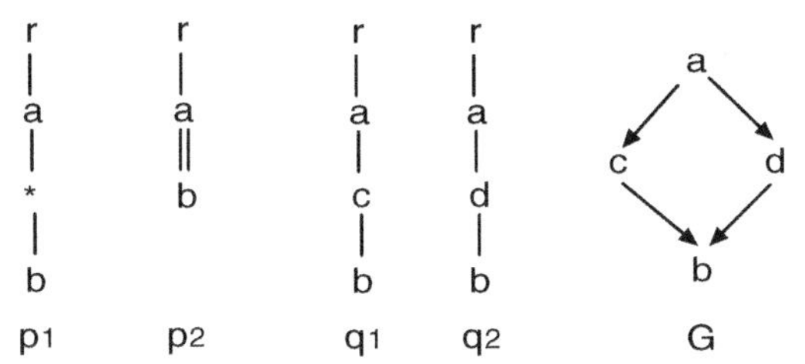

Fig. 4. Containment with Schema Information

otherwise G will have circles. We will consider G as a DAG first in Section 4.1, 4.2 and 4.3 and will discuss recursive schema in Section 4.4.

4.1 Eliminating Wildcards

With a schema available, a wildcard node can be replaced by specific labels in Σ, as long as the result pattern conforms to the given schema. A naive method is to pick an arbitrary label in Σ for each wildcard node, and then to verify if the clearly specified query complies with schema G. This method requires to verify

Algorithm 1. Algorithm for Eliminating Wildcard Nodes

for all node v that is not a $*$-node in pattern p **do**
  $L(v) \leftarrow \{label(v)\}$;
end for
for all leaf $*$-node v in p **do**
  $L(v) \leftarrow \Sigma$;
end for
Mark all leaf nodes and all non $*$-nodes;
repeat
  for each $*$-node x in p whose children $x_{c_1}, x_{c_2}, \cdots, x_{c_k}$ are all marked **do**
    for $i = 1$ to k **do**
      $S_i \leftarrow \phi$;
      for each $\beta \in L(x_{c_i})$ **do**
        if $((x, x_{c_i})$ is a pc-edge and there is some $\alpha \in L(x)$ such that (α, β) is an edge in $G)$ or $((x, x_{c_i})$ is an ad-edge and there is some $\alpha \in L(x)$ such that there is a path from α to β in $G)$ **then**
          $S_i \leftarrow S_i \cup \{\alpha\}$;
        end if
      end for
    end for
    $L(x) \leftarrow \bigcap_{i=1}^{k} S_i$;
    Mark x;
  end for
until all $*$-nodes in p are marked
Unmark root p_r and all non $*$-nodes;
repeat
  for each $*$-node x in p whose parent x_p is unmarked **do**
    for each $\beta \in L(x)$ **do**
      if $((x_p, x)$ is a pc-edge and there is some $\alpha \in L(x_p)$ such that (α, β) is an edge in $G)$ or $((x_p, x)$ is an ad-edge and there is some $\alpha \in L(x_p)$ such that there is a path from α to β in $G)$ **then**
        Add (β, α) into $P(x)$;
      else
        Remove β from $L(x)$;
      end if
    end for
    Unmark x;
  end for
until all $*$-nodes are unmarked

$|\Sigma|^k$ (where k is the number of wildcard nodes in a pattern) number of queries, and is obviously not efficient.

We propose an improved algorithm shown in Algorithm 1. The basic idea is to use existing structural information in the query to avoid wild guesses. It is inspired by [16] in which a similar idea was used to test the satisfiability of a tree pattern under schema G. Here, in our scenario, we need to record detailed label relationships (parent-child or ancestor-descendant) for adjacent node pairs, because these relationships could be further utilized to transform one query with wildcards into a union of queries without wildcards.

Algorithm 1 scans the wildcard nodes in p twice: bottom-up (line 8-21) and then top-down (line 23-34). The bottom-up phase calculates a set of possible labels $L(x)$ for each wildcard node x, using information about possible labels of its children. The top-down phase further refines the set $L(x)$ using information about the parent label of x. The parent-child and ancestor-descendant relationships are recorded as label pairs in $P(x)$. Note that in the algorithm, we omit, for brevity, the step to check if $L(x) = \phi$ after line 18 and 29 ($L(x) = \phi$ means pattern p does not conform to G), since it is not difficult to implement. The algorithm runs in $O(|p||\Sigma|^2)$: lines 1-7 runs in $O(|p|)$; the bottom-up phase visits each node in p at most twice, and within the loop, lines 12-16 can be done in $O(|\Sigma|^2)$. And thus lines 8-21 run in $O(|p||\Sigma|^2)$. The top-down phase also runs in $O(|p||\Sigma|^2)$. If there is an index on schema G, pc(α,β) or ad(α,β) can be checked efficiently.

4.2 Eliminating Ad-Edges

Now we have obtained unions of queries without wildcards. However, Theorem 2 is still not sufficient to decide containment between two sets of queries under schema (recollect the example in Fig. 4). We need to replace all the ad-edges with concrete paths compromising only pc-edges, because ad-edges must be interpreted in specific ways constrained by the schema.

A naive method to expand an ad-edge (v_1,v_2), similar to eliminating wildcard nodes, is to find all the paths between two labels $label(v_1)$ and $label(v_2)$ in schema G, and replace the ad-edges with one of these concrete paths. Obviously, there may be many ways to replace an ad-edge, and thus a pattern consisting ad-edges will be transformed into a union of a large number of patterns in $P^{\{/,[]\}}$. Then with the follow-up treatment in Section 4.3, one can determine the containment relationship for unions of queries under schema.

To avoid generating a possibly exponential number of patterns, a better solution is to wisely replace an ad-edge (v_1,v_2) with a subgraph between $label(v_1)$ and $label(v_2)$ in G, denoted as $G_s(label(v_1), label(v_2))$. To define $G_s(label(v_1), label(v_2))$ formally, $G_s(label(v_1), label(v_2)) = \{$ node $v|v \in G \wedge v$ is reachable from $label(v_1) \wedge label(v_2)$ is reachable from $v\}$. As long as the given pattern conforms to G, i.e. $label(v_2)$ is reachable from $label(v_1)$ in G, subgraph $G_s(label(v_1), label(v_2))$ will always exist, no matter G has circles or not. In addition, to find $G_s(label(v_1), label(v_2))$ is not expensive. It includes a top-down traverse in G from $label(v_1)$ to $label(v_2)$, and a bottom-up traverse from $label(v_2)$ to $label(v_1)$. The idea is similar to Algorithm 1, and hence we omit the details.

4.3 Chasing Patterns in $P\{/,[]\}$

Now we have wildcards and ad-edges eliminated, and all the patterns transformed into $P\{/,[]\}$. To reduce the problem into one without schema, we have the last step to chase the patterns in $P\{/,[]\}$ as much as possible with sibling constraints and functional constraints [9]. When G is not recursive, the process is not difficult, since the result patterns (after chasing) should be finite. The problem then converts into checking containment for unions of queries in $P\{/,[]\}$ without schema. Thereafter, we can apply Theorem 1 and 2 to solve it. Note that, after expanding ad-edges, a pattern may become a DAG rather than a rigorous tree pattern in $P\{/,[]\}$. but the chase process is the same except that we may not apply sibling constraints at some node whose child nodes are following or-semantics, because these child nodes are expanded from an ad-edge expressed as alternative paths (or subgraphs), making sibling constraints not satisfied on such node. On the other hand, when homomorphism is used to detect containment between such or-semantic patterns, a final step needs to be added. For example, when we conduct a mapping from pattern p to q, we can draw the conclusion $q \subseteq p$ with two further conditions holding: (1) for every subgraph chased from $G_s(label(v_1), label(v_2))$ in p, one of its subgraph connecting v_1 and v_2 must be mapped on to q; (2) for every two nodes v_1' and v_2' with an ad-edge in q, if v_2' is mapped, then every v_2''s ancestor (on all alternative subgraphs) should be mapped by some node in p.

4.4 Recursive Schema

One challenge arises: if the schema is recursive, a pattern can be chased continuously without stop, and a /-path (path only consisting of pc-edges) may contain a circle repeated for any times. In such cases, we allow the loop to appear once in the chased pattern to keep track of the nodes in a circle, and we also tag the loop start node and loop end node. Now we are able to rewrite a query in $P^{\{/,//,[],*,DTD\}}$ into a union of finite number of queries in $P^{\{/,[]\}}$. Theorem 2 will then be sufficient to decide the containment. A condition needs to be added when to find a homomorphism from pattern q to pattern p (p, q are in $P^{\{/,[]\}}$ with loop start node and loop end node tagged): if v_1 and v_2 are the loop start node and loop end node in q, there must exist a circle in G with labels $label(h(v_1))$, $label(h(v_2))$ as start and end respectively. Here, $h(v_1)$ and $h(v_2)$ may not be loop start and end nodes in pattern p. Fig. 5 shows an example. p_0 and q_0 are two queries involving ad-edges. G is a schema containing a circle. After expanding ad-edges and chasing with schema G for p_0 and q_0, we get patterns p and q. In pattern q, $a//c$ is expanded and chased into $a/b/c$ with node v_1 labelled a, node v_2 labelled c as the loop start and loop end. Similarly, $b//a$ is expanded and chased into $b/c/a$ in p. Considering the result patterns p and q, there is a homomorphism from q to p, and moreover, nodes $h(v_1)$ and $h(v_2)$, the corresponding

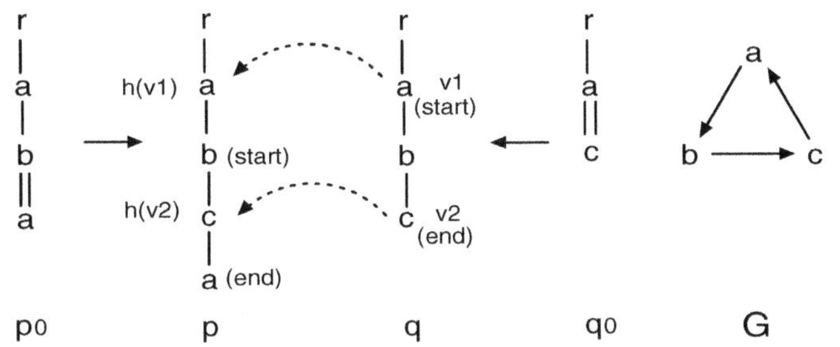

Fig. 5. Expanding Edges under Recursive Schema

nodes in p of loop start v_1 and loop end v_2 in q, have labels $a = label(h(v_1))$ and $c = label(h(v_2))$ that are start and end nodes of a circle in G. Therefore, pattern p is contained in pattern q, and thus the original pattern p_0 is contained in q_0. Note that the condition does not require $h(v_1)$ and $h(v_2)$ be the loop start and loop end nodes in p.

In the above discussion, we assume that there are no intersected circles in G, i.e. the recursive loops have no overlaps. This assumption obviously simplifies the problem, and it is still interesting and challenging to investigate the containment problem of unions of patterns under complex recursive schema.

5 Discussions on General Tree Pattern Queries

Different from XPath tree patterns, general tree pattern queries may contain more than one distinguished nodes. This may add difficulty to containment checking in some circumstance. In fact, it is due to multiple distinguished nodes that containment between general tree patterns is not the same as containment between boolean patterns. We illustrate the observation by firstly reviewing Proposition 1 in [8] and then show when the proposition is not correct and why.

To restate it: let two general tree patterns be p and q, we can obtain two boolean patterns p' and q' by adding distinct labels $l_1, \cdots, l_k$ to the k distinguished nodes in p and q respectively, then we have $p \subseteq q$ iff. $p' \subseteq q'$. However, one could discover that the proposition only holds for output-order-sensitive queries. In other words, the distinguished nodes should have a fixed order so that we can label them in a unique way. See Fig. 6 for an example. Pattern q has two distinguished nodes b and c. Suppose there is no predefined order for the distinguished nodes, to transform q into a boolean pattern, we have two labelling schemes shown as q_1 and q_2 respectively. Obviously, q_1 and q_2 are not identical. Therefore, in such situation, a general tree pattern query should be transformed into a unions of boolean patterns, rather than a single pattern. Hence if we have k distinguished nodes, we will have $k!$ ways to label them, resulting in $k!$ boolean patterns to represent a general tree pattern. This will significantly complicate containment detection if the number of distinguished nodes is large. Luckily,

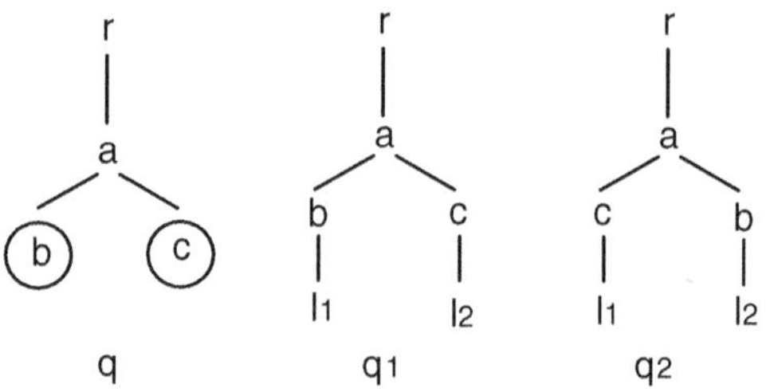

Fig. 6. Different Labeling Scheme for a General Tree Pattern Query

for XPath queries, there is only one distinguished node, so the boolean pattern evolved from its corresponding XPath pattern is unique. And the conclusions we got in the above sections are still correct. In some real applications, general tree pattern queries indeed organize the distinguished nodes in a fixed order, such as tree pattern queries induced from XQuery queries. But we should keep in mind that containment between general tree patterns and boolean patterns are not always the same.

6 Related Work

Query containment was first put forward together with query equivalence in order to optimize query evaluation in relational context [1], where containment problem is studied for two queries containing select, project and join operators. Later, containment of unions of queries is discussed in [6]. It provided a sufficient and necessary condition showing that containment between unions of queries can be reduced into containment between a number of pairwise queries. In [17], the authors showed if relations are modelled as multisets of tuples, the previous sufficient and necessary condition holds only for one type of label system, while for another type of label system, the containment problem is undecidable.

Unfortunately, the established theory for relational queries cannot be applied in XML context. The containment problem between unions of XPath queries is still open, though fruitful results have been produced for containment between two single queries. In some pioneer works, the problem was shown in PTIME for XPath subsets $XP^{\{/,[],*\}}$, $XP^{\{/,//,[]\}}$ and $XP^{\{/,//,*\}}$, and furthermore proved to be coNP-complete for $XP^{\{/,//,[],*\}}$ in [8]. When a schema is available, the problem turned out to be more difficult, because data trees are constrained according to a particular pattern, and thus XPath queries with wildcard nodes and ad-edges cannot be interpreted arbitrarily. Wood [9] and Neven and Schwentick [10] independently showed the containment between two XPath queries in $XP^{\{/,//,[],*,DTD\}}$ is decidable, in fact is EXPTIME-complete. Neven and Schwentick [10] also discussed disjunction and variables in XPath. More theoretical results with respect to various XPath query subsets are summarized in [11]. A richer XPath fragment, XPath2.0, is recently examined in [18].

7 Conclusions and Future Work

In this paper, we have addressed the containment problem between unions of XPath queries. We showed that the problem can be always reduced into containment between one query and a union of queries. We also proved that, for XPath subset $XP^{\{/,//,[]\}}$, the problem can be reduced into checking containment between two single queries, each from one union. For a larger subset $XP^{\{/,//,[],*\}}$, we utilize an existing technique to develop an effective strategy to solve the problem. When a schema is available, we could use the schema to eliminate wildcard nodes and expand ad-edges in the query so that our developed theorem could be applied thereafter to decide containment relationship between unions of queries under schema information.

One direction for future work is to consider more complicated recursive relationships in schema, eg. two circles may have intersections. This is always a difficult problem, may result in chasing patterns in $P^{\{/,[]\}}$ rather challenging.

Acknowledgments. This work was supported by the Australian Research Council Discovery Project under the grant number DP0878405.

References

1. Aho, A.V., Sagiv, Y., Ullman, J.D.: Equivalences among relational expressions. SIAM J. Comput. 8(2), 218–246 (1979)
2. Halevy, A.Y.: Answering queries using views: A survey. VLDB J. 10(4), 270–294 (2001)
3. Gupta, A., Sagiv, Y., Ullman, J.D., Widom, J.: Constraint checking with partial information. In: PODS, pp. 45–55 (1994)
4. Levy, A.Y., Sagiv, Y.: Queries independent of updates. In: VLDB, pp. 171–181 (1993)
5. Chandra, A.K., Merlin, P.M.: Optimal implementation of conjunctive queries in relational data bases. In: STOC, pp. 77–90 (1977)
6. Sagiv, Y., Yannakakis, M.: Equivalences among relational expressions with the union and difference operators. J. ACM 27(4), 633–655 (1980)
7. Clark, J., DeRose, S.: XML path language (XPath) 1.0. In: W3C Recommendation (November 1999), `http://www.w3.org/TR/xpath`
8. Miklau, G., Suciu, D.: Containment and equivalence for a fragment of XPath. J. ACM 51(1), 2–45 (2004)
9. Wood, P.T.: Containment for xPath fragments under DTD constraints. In: Calvanese, D., Lenzerini, M., Motwani, R. (eds.) ICDT 2003. LNCS, vol. 2572, pp. 297–311. Springer, Heidelberg (2002)
10. Neven, F., Schwentick, T.: XPath containment in the presence of disjunction, dTDs, and variables. In: Calvanese, D., Lenzerini, M., Motwani, R. (eds.) ICDT 2003. LNCS, vol. 2572, pp. 312–326. Springer, Heidelberg (2002)
11. Schwentick, T.: Xpath query containment. SIGMOD Rec. 33(1), 101–109 (2004)
12. Wadler, P.: A formal semantics of patterns in xslt and xpath. Markup Lang. 2(2), 183–202 (2000)

13. Amer-Yahia, S., Cho, S., Lakshmanan, L.V.S., Srivastava, D.: Tree pattern query minimization. The VLDB Journal 11(4), 315–331 (2002)
14. Wang, J., Yu, J.X., Liu, C.: On tree pattern query rewriting using views. In: Benatallah, B., Casati, F., Georgakopoulos, D., Bartolini, C., Sadiq, W., Godart, C. (eds.) WISE 2007. LNCS, vol. 4831, pp. 1–12. Springer, Heidelberg (2007)
15. Wang, J., Yu, J.X., Liu, C.: Contained rewritings of xPath queries using views revisited. In: Bailey, J., Maier, D., Schewe, K.-D., Thalheim, B., Wang, X.S. (eds.) WISE 2008. LNCS, vol. 5175, pp. 410–425. Springer, Heidelberg (2008)
16. Lakshmanan, L.V.S., Ramesh, G., Wang, H., Zhao, Z.: On testing satisfiability of tree pattern queries. In: VLDB, pp. 120–131 (2004)
17. Ioannidis, Y.E., Ramakrishnan, R.: Containment of conjunctive queries: beyond relations as sets. ACM Trans. Database Syst. 20(3), 288–324 (1995)
18. ten Cate, B., Lutz, C.: The complexity of query containment in expressive fragments of xpath 2.0. In: PODS, pp. 73–82 (2007)

FlexBench: A Flexible XML Query Benchmark*

Maroš Vranec and Irena Mlýnková

Department of Software Engineering, Charles University in Prague, Czech Republic
maros.vranec@gmail.com, mlynkova@ksi.mff.cuni.cz

Abstract. In this paper we propose a new approach to XML benchmarking – a flexible XML query benchmark called *FlexBench*. The flexibility is given by two aspects. Firstly, FlexBench involves a large set of testing data characteristics so that a user can precisely describe the application. And, secondly, FlexBench is able to adapt the set of testing query templates to the particular set of synthesized testing data. Hence, contrary to the existing works, the testing is not limited by the fixed set of queries and basic data characteristics (usually only size) to a single (and often simple) application. We depict the advantages of the proposed system using a set of preliminary experiments.

1 Introduction

Since XML [11] has become a de-facto standard for data representation and manipulation, there exists a huge amount of so-called *XML Management Systems* (XMLMSs) that enable one to store and query XML data. Hence, being users, we need to know which of the existing XMLMSs is the most sufficient for our particular application. On the other hand, being vendors, we need to test correctness and performance of our system and to compare its main advantages with competing SW. And, being analysts, we are especially interested in comparison of various aspects of existing systems from different points of view. Consequently the area of benchmarking XMLMSs has opened as well.

In general, a *benchmark* or a *test suite* is a set of testing scenarios or test cases, i.e. data and related operations which enable one to compare versatility, efficiency or behavior of *system(s) under test* (SUT). In our case the set of data involves XML documents, possibly with their XML schema(s), whereas the set of operations can involve any kind of XML-related data operations. Nevertheless, the key operations of an XMLMS are usually XML queries. Currently, there exist several XML query benchmarks which provide a set of testing XML data collections and respective XML operations that are publicly available and well-described. However, although each of the existing benchmarks brings certain interesting ideas, there are still open issues to be solved.

In this paper we focus especially on the key persisting disadvantage of all the existing approaches – the fact that the sets of both data and queries are

* This work was supported in part by the Czech Science Foundation (GAČR), grant number 201/09/P364.

X. Zhou et al. (Eds.): DASFAA 2009, LNCS 5463, pp. 421–435, 2009.

fixed or the only parameter that can be specified is the size or the number of XML documents. We propose a new approach to XML benchmarking – a flexible XML query benchmark called *FlexBench*. The flexibility is given by two aspects. Firstly, FlexBench involves a large set of testing data characteristics so that a user can precisely describe the tested application. But, to ensure user-friendliness and simplicity (i.e. one of the key requirements for a benchmark) we also provide a set of predefined settings which correspond to classical types of XML documents identified in an analysis of real-world XML data. And, secondly, FlexBench is able to adapt the set of testing queries to the particular set of synthesized testing data. Hence, contrary to the existing works, the benchmarking is not limited by the fixed set of queries and basic data characteristics to a single simple application. We depict the advantages of the proposed system using a set of preliminary experiments.

The paper is structured as follows: Section 2 overviews existing XML benchmarks, their classification and (dis)advantages. Section 3 describes FlexBench in detail. Section 4 involves the results of preliminary tests made using FlexBench. And, finally, Section 5 provides conclusions and outlines possible future work and open issues.

2 Related Work

In general, there exists a large set of XML query benchmarks. The seven best known representatives are XMark [13], XOO7 [17], XMach-1 [10], MBench [16], XBench [24], XPathMark [18] and TPoX [21].

From the point of view of purpose we can differentiate so-called *application-level* and *micro* benchmarks. While an application-level benchmark is created to compare and contrast various applications, a micro-benchmark should be used to evaluate performance of a single system in various situations. In the former case the queries are highly different trying to cover all the key situations, whereas in the latter case they can contain subsets of highly similar queries which differentiate, e.g., in selectivity. Most of the seven benchmarks are application-level; the only representative of micro-benchmarks is MBench.

Another set of benchmark characteristics involves the number of users it is intended for, the number of applications it simulates and the number of documents within its data set. Most of the benchmarks are single-user, single-application and involve only a single document. The only exception, XBench, involves (a fixed set of) four classes of XML applications with different requirements. On the other hand, XMach-1 and TPoX are multi-user benchmarks and enable one to test other XML management aspects, such as, e.g., indexing, schema validation, concurrency control, transaction processing, network characteristics, communication costs, etc.

Another important aspect of XML benchmarks are characteristics of the data sets. All the representatives involve a data generator, but in most cases the only parameter that can be specified is the size of the data. Most of the benchmarks involve own simple data generator, some of them (i.e. XBench and TPoX) exploit a more complex data generator, but pre-set most of its parameters. Expectably, all the benchmarks involve also one or more schemas of the data representing the simulated applications.

The last important set of characteristics describes the operation set of the benchmarks. All the benchmarks involve a set of queries, some of them (i.e. XMach-1, MBench and TPoX) also a set of update operations. The two multi-user benchmarks also support additional, less XML-like operations with the data. The most popular operations are XQuery [8] queries, whereas the benchmarks try to cover various aspects of the language, such as, e.g., ordering, casting, wildcard expressions, aggregations, references, constructors, joins, user-defined functions, etc. However, in all the cases the sets of queries are fixed.

3 FlexBench Benchmark

As we have mentioned in the Introduction, the aim of FlexBench is to deal with the problem of fixed parameters of the existing benchmarks. From one point of view it is an advantage since one of the general requirements for benchmarking is simplicity [9]. It is even proven by the analysis of existing XQuery benchmarks [5] that the most popular benchmark is XMark – a single-application and single-user benchmark with a fixed set of queries and a single data characteristic, i.e. size in bytes. On the other hand, such a simple fixed benchmark enables one to test only one specific situation, however according to statistical analysis of real-world XML data [20] there are multiple types of XML data, i.e. multiple applications. And, naturally, new ones occur every day.

Consequently, we state the following two aims of FlexBench:

1. We want to support as much characteristics of the tested application as possible.
2. The benchmark system should still be simple and easy-to-use.

To fulfill the first condition the FlexBench involves three parts – a data generator, a schema generator and a query generator. In general the synthesis of a benchmark can start with any of the three generators. An overview of possible strategies can be seen in Table 2.

All of the existing benchmarks exploit a restricted combination of the second and third approach, where the schema and the queries are fixed and the data are synthesized according to the schema. In FlexBench we use the first approach – the XML documents are synthesized first, then the schemas and, finally, the queries on top of them. In our opinion it is the most user-friendly approach and, in addition, it enables us to exploit XML data characteristics[1] from a statistical analysis of real-world XML data [20]. Not only do their characteristics describe the data from various points of view and very precisely, but the strategy enables

[1] Due to space limitations we will not repeat the definitions of the characteristics. Most of them (such as depth or fan-out) are quite common, whereas the definitions of the less known ones (such as relational or DNA patterns) can be found in [20].

Table 1. Three possible strategies when synthesizing a benchmark

Strategy	**Description**
data → schema & queries	XML documents are synthesized/provided first. Schema and queries are then synthesized on the top of them. It is also possible to synthesize the queries on top of the schema (instead of the XML documents). The schema can be synthesized by an external tool as well, since many suitable ones already exist.
schema → data → queries	The schema is synthesized/provided first. XML documents valid against it are synthesized next (possibly by an external tool). Queries are then synthesized on top of the XML documents or the schema.
queries → schema → data	Queries are synthesized (or provided by a user) first. Then the respective XML documents and their schemas are created on top of them.

us to exploit the particular results for pre-setting FlexBench (see Section 3.4). Since most of these characteristics cover properties of XML documents, not their schemas, mainly because not all XML documents have a schema, when deciding whether to synthesize XML documents on top of their schemas or vice versa, this fact convinced us to synthesize the documents first.

On the other hand, supporting a wide set of data characteristics may lead to some conflicts. There are many dependencies between them and a user may specify contradicting properties. For instance, the size (in bytes) of a document and the number of its elements are correlated and particular values may be in conflict. A related important aspect is that the generator never respects the specified values accurately. For example, the output files usually have a little percentage of bytes higher (or lower) than a user originally specified. However, these deviations have a minimum impact on the quality of synthesized data and it is a common feature of most of the existing data generators.

3.1 Data Generator

FlexBench data generator supports basic data types of its parameters, i.e. data characteristics (e.g. strings, integers, floating point numbers, etc.) as well as advanced ones – discrete statistical distributions. The supported data characteristics are listed in Table 2 which involves their classification, data types and the most important aspect – conflicting parameters.

The first set of so-called *basic* parameters involves the output directory and the amount of documents to be synthesized. Naturally they have no conflict with the other parameters and they can be specified as a constant value.

The set of *structural* parameters describes the structure of the document tree. In particular its depth, fan-out (i.e. a kind of width), number of attributes and total size in bytes. All these characteristics can be specified as a statistical distribution. The important aspects are their conflicts with other parameters.

Table 2. Parameters of the benchmark generator and their relations

Type	Parameter name	Conflicts with	Type
Basic	Output directory	None	Constant
	Number of documents	None	Constant
Structural	Size (in bytes)	Number of elements, fan-out, depth, percentage of text, attribute values	Statistical distribution
	Fan-out	Depth, size	Statistical distribution
	Depth	Fan-out, size	Statistical distribution
	Number of attributes	Size, percentage of text	Statistical distribution
Textual	Percentage of text	Size	Constant
	Percentage of mixed-content elements	Percentage of text	Constant
	Depth of mixed-content	Depth	Statistical distribution
	Percentage of simple mixed-content elements	Percentage of text	Constant
Patterns	Percentage of pure recursions	Other recursions	Constant
	Percentage of trivial recursion	Other recursions	Constant
	Percentage of linear recursion	Other recursions	Constant
	Percentage of general recursion	Other recursions	Constant
	Percentage of DNA patterns	None	Constant
	Percentage of relational patterns	None	Constant
	Percentage of shallow relational patterns	None	Constant
Schema	Percentage of DTDs	None	Constant
	Percentage of XSDs	None	Constant

Naturally, the worst situation is for the size in bytes which is correlated with almost all other characteristics. On the other hand, it is the most common parameter in all the existing benchmarks and it is also the most natural parameter to be specified by a user. Almost the same information would be specified by the number of elements, however, for the previously specified reasons, we have decided to support the total size instead. As for the fan-out and depth which specify the "shape" of the tree, they are correlated mutually as well as with the size. Also note that instead of depth and fan-out, we could use the distribution of levels of the document tree. However, similarly to the previous case, our choice seems to be more natural and user-friendly. Finally, the number of attributes deals with the special type of nodes of the document tree which are correlated only with size and textual parameters.

The third set of data characteristics – the *textual* ones – do not influence the shape of the document tree but the particular textual values. These may occur in three situations: as attribute values, within textual contents of elements or within mixed contents of elements. The most important and natural parameter is undoubtedly the total percentage of text in the XML document which is correlated with size. On the other hand, the remaining parameters specify the number of mixed-content and simple-mixed content elements, as well as the depth of mixed-content elements, i.e. their complexity. Naturally, they are correlated with respective characteristics, i.e. percentage of text or depth of the document.

The fourth set of data characteristics involves various types of data *patterns* as were defined in [20]. Since recursion is related to element names, there are almost no conflicts with other data characteristics for all its four types; they may conflict only mutually. Similarly, since the remaining patterns, i.e. relational patterns, shallow relational patterns and DNA patterns, specify pre-defined subtrees of the synthesized tree, they are not in direct conflict with any of the other parameters as well.

The last set of characteristics involves the percentage of XML documents for which a schema (a DTD [11] or an XSD [7,22]) should be inferred. Typically this will be 0% or 100%, however for special applications, such as those dealing with schema inference or schema evolution, it may be useful to infer a schema only for a subset of the synthesized XML documents. Naturally, these parameters are in no conflict with others.

3.2 Schema Generator

According to the statistical analysis [20], XML schemas of XML documents are used quite often. As long as some XMLMSs use schemas as an information how to create the internal representation of XML documents, FlexBench involves a schema generator as well. However, for the sake of simplicity it involves a third-party implementation, since there exists a plenty of suitable solutions [19]. Hence, since this is not the key aspect of FlexBench, we will omit further details.

3.3 Query Generator

Having synthesized a set of XML documents and having inferred their XML schema if required, the next important component of FlexBench is a query generator. Since, in general, it can have on input any kind of XML data, it must be able to synthesize queries over them, so that we can get reasonable results. Hence, the question is what kind of queries it should synthesize and how.

FlexBench query generator is based on the idea of exploitation of a set of XQuery templates with empty parts that are filled with XML document names and respective element/attribute names according to the given XML data.

For instance, consider the following query template:

```
for $a in doc("input.xml")//elem
order by $a
return <result>{$a}</result>
```

Its instance is created as follows: The `input.xml` is replaced with name of a synthesized XML document and the element name `elem` with any of its elements.

The apparent problem is that we need to tight the data with the queries, i.e. to enable a user to specify which of the elements/attributes should be used in the templates and, hence, queried. Under a closer observation we can see that most of them are directly represented by data generator parameters. For example, if the user tends to synthesize many mixed-content elements, the query generator can "guess" that (s)he wants to synthesize queries involving these elements as well. In general, the user may specify which of the elements should be queried. However, since this approach is useful only in case the user is acquainted with the data in detail, the FlexBench data generator outputs elements with "interesting" features (such as the most common element, the mostly used element in recursion, the mostly used mixed-content element, the mostly used trivial element, etc.) of which a user may choose.

The second question is how to select the query templates. As we have mentioned, the choice of the particular queries depends on the type of the benchmark. Since FlexBench is an application-level benchmark, our aim is to support as many distinct types of queries as possible. In [5] the queries used in the existing benchmarks are divided into several categories as depicted in Table 3.

Table 3. Categories and numbers of queries in existing benchmarks

Category	MBench	XMark	XOO7	XMach	XBench
Core XPath	12	3	1	0	1
XPath 1.0	4	3	8	3	12
Navigational XPath 2.0	22	5	6	1	22
XPath 2.0	5	8	6	2	23
Sorting	1	1	1	1	9
Recursive functions	2	0	0	1	0
Intermediate results	0	0	0	0	0

In the following sections we will go through all the categories, define them briefly and discuss the respective characteristics. As we have mentioned, the FlexBench query synthesis is based on the idea of query templates and creating their instances suitable for the given data. Hence, the characteristics are related mainly to the amount of synthesized queries. For the space limitations we do not list all the currently supported templates – the whole set can be found in [23].

Core XPath Queries. These queries test XMLMS performance when finding an XML element specified by the navigational part of the XPath [14]. The usage of position information, all functions and general comparisons are excluded.

FlexBench distinguishes between two special cases – the queried element at the first level and a nested element. From the point of view of the query generator it is useful to let the user specify how deep the queried element should occur.

Hence, FlexBench enables one to specify the number of the queries and statistical distribution of levels of queried elements.

Text Queries. Preserving the order of a text is one of the most important aspects of XMLMSs. The text queries test the performance of an XMLXS when searching for a text. They also belong to the category of Core XPath. FlexBench enables one to specify the number of these queries.

XPath 1.0 Queries. When storing a document-centric XML document, an XMLMS must store the elements in their original order (otherwise the meaning of the text would be lost). That is why there exist ways to query the ordered access. In general, there are two possibilities:

- *Absolute order* – queries that retrieve elements according to their absolute position in the XML tree
- *Relative order* – queries that return elements according to their neighboring elements in the XML tree

FlexBench enables one to specify the number of absolute and relative order access queries. Besides these obvious parameters, more customizable absolute ordered queries are useful, e.g. a user can specify what index should be used.

Navigational XPath 2.0 Queries. From navigational XPath 2.0 [6] the usage of position information and all aggregation and arithmetic functions are excluded. However, we include queries with `some` and `every` clauses to construct quantified expressions as well. A user is able to specify the amount of such queries; `some` and `every` versions of a query are synthesized randomly in rate 50:50.

XPath 2.0 Queries. An instance of this category can be queries which apply an aggregation function on the data. Examples of such functions are a sum of some set of numbers, an average value, etc. The aggregation-function queries suppose that there exist similar elements which can be used as a source for the aggregation. FlexBench enables one to specify the number of particular aggregate-function queries.

Sorting Queries. These queries use sorting of a set of the given attributes to get an ordered list of elements/attributes. FlexBench enables users to specify their amount.

Queries with Recursive Functions. XQuery enables a user to define own functions, possibly recursive. Since such a function is highly related to the data, as a template we have chosen a simple recursive function so generic that it can be used on any XML document. It computes the depth of an XML document.

```
declare function local:depth ( $root as node()? ) as xs:integer?
  {
  if ($root/*)
  then max($root/*/local:depth(.)) + 1
  else 1
  };

local:depth(doc("input.xml"))
```

Similarly to the previous cases, FlexBench enables users to specify the number of synthesized recursive queries.

Queries with Intermediate Results. Intermediate result queries contain `let` clause of XQuery. A user can specify the amount of such queries.

Currently, FlexBench supports a set of simple query templates for each of the previously described query class (see [23]). Naturally, there are much more complex constructs and queries that can be expressed in XQuery, such as various joins, complex user-defined functions, etc. As for the future work, we intent to involve them in FlexBench as well and to deal with their more precise binding with the synthesized data. Such complex queries will require looser XQuery templates and a kind of data analyzer capable of creating their correct instances. The current implementation of FlexBench is a preliminary version that enables one to demonstrate the basic advantages of the chosen approach. Hence, the templates were selected so that no complex data analysis was necessary.

3.4 Pre-defined Settings of Parameters

To fulfill the second aim of FlexBench stated at the beginning of this chapter, we provide a set of its predefined settings. Since we have created the set of supported data parameters on the basis of characteristics analyzed in [20], we can exploit the data categories and their real-world characteristics stated in the analysis as well. In particular, the XML data are divided into these six classes:

- *data-centric* documents, i.e. documents designed for database processing (e.g. database exports, lists of employees, lists of IMDb movies, etc.),
- *document-centric* documents, i.e. documents which were designed for human reading (e.g. Shakespeare's plays, XHTML documents, novels in XML, DocBook documents, etc.),
- documents for data *exchange*, e.g. medical information on patients and illnesses, etc.,
- *reports*, i.e. overviews or summaries of data (usually of database type),
- *research* documents, i.e. documents which contain special (scientific or technical) structures (e.g. protein sequences, DNA/RNA structures, etc.) and
- *semantic web* documents, i.e. RDF documents.

A user can use these pre-defined categories through command-line parameters. For instance, specifying:

```
java -jar flexbench.jar -data-exchange
```

is equivalent to specifying all the parameters like:

```
java -jar flexbench.jar NumberOfGeneratedFiles=9
                        PercentageOfTextGen=uniform(31,40)
                        PercentageOfFilesWithDTD=100 ...
```

4 Preliminary Experiments

To depict advantages of the proposal we have chosen basic typical use cases of FlexBench. We describe them and their results in this section.

4.1 Comparing XMLMSs

We start with a typical aim of XML benchmarking – to compare performance of several XMLMSs. We use the six pre-defined sets of parameters specified in Section 3.4 and three simple XMLMSs Qizx [3], Qexo [1] and Saxon [4].

First of all, the total execution times of all benchmark queries are depicted in Table 4.

Table 4. Total execution time on six different types of applications

	Qizx (s)	Qexo (s)	Saxon (s)
Data-centric	3.268	4.942	11.423
Document-centric	10.564	19.919 (but failed on text queries)	61.767
Exchange	8.116	15.438	18.290
Reports	55.324	failed	failed
Research	3.945	5.050	7.139
Semantic web	7.430	9.874	27.902

As we can see, Qizx and Qexo XMLMSs performed almost equally in data-centric, data exchange, research and semantic web scenarios, whereas Qizx sustained superior performance. Remaining two categories output more interesting results. Both Qexo and Saxon failed in case of reports – the cause of failure was low heap space available. (We will describe the scalability of FlexBench in Section 4.2 to determine maximum XML document size which Qexo can work with.) Saxon has the worst efficiency in general.

Interesting ones are also the document-centric results. It seems that Qizx is much faster than Qexo when querying document-centric documents. Moreover,

Table 5. Comparing XMLMSs in query categories of document-centric benchmark

Query category	**Number of of queries**	**Average time for query (ms)** Saxon	Qizx	Qexo	eXist
Core XPath (exact match) queries	137	328	25	67	405
XPath 1.0 (absolute and relative order) queries	86	157	26	52	250
Navigational XPath 2.0 (quantification) queries	9	109	64	failed	289
XPath 2.0 (aggregate function) queries	70	55	38	75	301
Sorting queries	9	873	437	failed	274
Recursive function queries	14	failed	failed	failed	1549
Intermediate result queries	12	170	162	234	413

Qexo does not support text queries because of function usage which it does not support. Note that such experiments could not be performed using any of the existing benchmarks, because they do not support so wide characteristics of the data to specify different applications.

To analyze the document-centric application more precisely we synthesized 55 absolute ordered queries, 70 aggregate function queries, 137 exact match queries, 12 intermediate result queries, 9 quantification queries, 14 recursive function queries, 31 relative order queries, 9 sorting queries and 72 text queries for the document-centric category. As we can see in Table 5, Qexo needs twice the time of Qizx and even fails in some of the cases. None of the categories shows extra deviation from this pattern, so we can conclude that Qexo has overall problems when working with the document-centric documents. On the other hand, Saxon has problems with basic queries (especially in exact match queries), but outperforms Qexo in more complex ones. And, finally, eXist [2] was added to depict how complex recursive function queries are.

4.2 Scalability of Benchmark Generator

FlexBench capabilities can be also used to detect the limits of an XMLMS. As we have seen in Table 4, Qexo has some problems with large files and its default allocated heap space. We will keep its default setting of heap space considering Qexo as an unconfigurable black box and we will try to determine the approximate size of the XML file that Qexo can process successfully. Firstly, we synthesize XML documents with various sizes ranging from 1MB to 13MB. Then we perform the respective tests. Finally, as we have seen from a log file, Qexo has problems with XML files bigger than 7MB.

Naturally, such testing can be done using any data generator. The advantage of FlexBench is that we can find the limits for distinct applications, i.e. data collections and operations.

4.3 Modifying Parameters

In the following tests we will show the advantage of the various parameters of FlexBench. We can use any of the parameters (or their combination) and selected type(s) of queries to study their impact on the selected XMLMSs. For illustration we choose influence of recursion and text queries on Saxon and Qizx.

Figure 1 shows how the percentage of recursion correlates with corresponding total execution times for text queries. As we can see, in both the selected cases the systems behave quite naturally – the more the percentage of recursion is, the higher the execution times are. However, Saxon overcomes Qizx in all three cases.

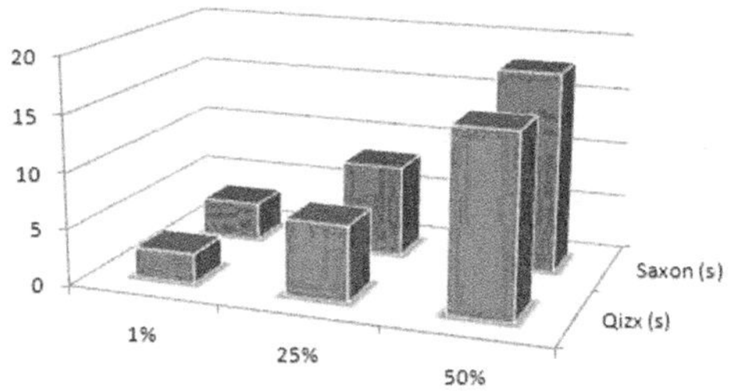

Fig. 1. Effect of percentage of recursion on text queries

4.4 Consistency of FlexBench Output

Last but not least, we will discuss the consistency of the results. One might argue why results over data and queries that are completely synthetic should be trustworthy. How big is the chance that the results will not be entirely different with the same parameters passed to FlexBench again? For the sake of result consistency, multiple tries were made when benchmarking. Each time the whole benchmark was re-synthesized and applied to Qizx. The results can be seen in Table 6.

Table 6. Stability of results of a randomly synthesized benchmark

Total execution time	**Qizx**
Document-centric benchmark #1 (s)	26.592
Document-centric benchmark #2 (s)	29.871
Document-centric benchmark #3 (s)	25.948
Document-centric benchmark #4 (s)	27.358
Document-centric benchmark #5 (s)	27.297
Semantic web benchmark #1 (s)	30.117
Semantic web benchmark #2 (s)	27.550
Semantic web benchmark #3 (s)	28.579
Semantic web benchmark #4 (s)	33.579
Semantic web benchmark #5 (s)	32.718

As we can see, the average total execution time for synthesized document-centric benchmark is 27.413s and the standard deviation is 1.333s. In case of synthesized semantic web benchmark, the average total time is 30.509s and standard deviation is 5.190s. Considering total randomness of synthesized data and queries (even the amount of queries is more-or-less randomized) the results are convincing.

5 Conclusion and Future Work

There are several major achievements of this paper. Firstly, the XML data generator supports numerous interesting parameters of XML documents, in fact so many that no third-party solution was suitable to be exploited and utilized. It is also a notable fact that the parameters were taken from available statistics about real-world XML data sets and, hence, the realistic results could be used for reasonable pre-setting. Secondly, the query generator can synthesize queries so that they can be applied on the given data. Every other XML benchmark has its query workload fixed and consequently, the respective data generators are highly restricted. Thus, in general, the main advantage of our approach is that we are much more flexible when synthesizing data, since we are not bound to any pre-defined queries.

Experimental results showed us how easy is to determine interesting insights about tested XML databases. Thanks to our different kinds of data and various queries we were able to show different behaviors of the benchmarked XMLMSs. This would not be possible with a fixed set of data and XQuery queries. Moreover, we have created pre-defined sets of benchmark parameters corresponding to the real use-cases of XML data.

Authors of [9] state that every benchmark should have the following four basic properties:

- *relevance*: FlexBench synthesizes XML queries that cover most constructs of XQuery.
- *portability*: FlexBench is implemented in Java and outputs portable XML format.
- *scalability*: FlexBench is scalable through lots of data/query parameters.
- *simplicity*: No parameter of FlexBench is mandatory and we provide a set of pre-defined data sets.

Another way to evaluate the quality of an XQuery workload was stated in [12]:

1. Is there a restriction on XML document structure?
2. Is there a database size and load volume scalability?
3. Is there a query type variability?
4. Is there an ad hoc and open interface for schema input and operation input?

FlexBench's answers are:

1. No, there are no restrictions. XML documents and their schemas are synthesized and a user can specify numerous parameters of the result.

2. Yes, there is. Moreover, every other parameter of synthesized XML documents is scalable as well (not only the size parameter).
3. Yes, there is. FlexBench supports more categories of queries than the rest of benchmarks.
4. If we consider FlexBench parameters as a form of a schema, then yes, FlexBench is also easily extensible by new parameters and new templates for synthesized queries.

Naturally, there is also a plenty of possible future improvements of FlexBench. Firstly, we intent to perform more elaborate experiments with various types of XMLMSs from XML-enabled to native XML ones and, especially, with commercial solutions. Secondly, we plan to extend the proposed idea as much as possible. In particular we will focus on the set of XML query templates that can be much wider and enable one to test various aspects of an XML application. As we have mentioned, this task opens a wide research area for creating reasonable instances of the templates. At the same time, we need to tighten the relation between the synthesized data and the queries so that the user can specify more precisely which items should be queried and how. Side but still important tasks involve implementation of a user-friendly interface for specifying the characteristics as well as a repository for their predefined settings. And, last but not least, we want to focus on benchmarking of stream processing [15], where the various characteristics of synthetic data can be widely exploited as well.

References

1. Qexo – The GNU Kawa implementation of XQuery. Kawa (2007), `http://www.gnu.org/software/qexo/`
2. eXist-db: Open Source Native XML Database. exist-db.org (2008), `http://exist.sourceforge.net/`
3. Qizx/db. Pixware (2008), `http://www.xmlmind.com/qizx/`
4. Saxon: The XSLT and XQuery Processor. SourceForge.net (2008), `http://saxon.sourceforge.net/`
5. Afanasiev, L., Marx, M.: An Analysis of the Current XQuery Benchmarks. In: ExpDB 2006: Proc. of the 1st Int. Workshop on Performance and Evaluation of Data Management Systems, Chicago, Illinois, USA, pp. 9–20. ACM, New York (2006)
6. Berglund, A., Boag, S., Chamberlin, D., Fernndez, M.F., Kay, M., Robie, J., Simeon, J.: XML Path Language (XPath) 2.0. W3C (January 2007)
7. Biron, P.V., Malhotra, A.: XML Schema Part 2: Datatypes, 2nd edn. W3C (October 2004)
8. Boag, S., Chamberlin, D., Fernndez, M.F., Florescu, D., Robie, J., Simeon, J.: XQuery 1.0: An XML Query Language. W3C (January 2007)
9. Bohme, T., Rahm, E.: Benchmarking XML Database Systems - First Experiences. In: HPTS 2001: Proc. of 9th Int. Workshop on High Performance Transaction Systems, Pacific Grove, California (2001)
10. Bohme, T., Rahm, E.: XMach-1: A Benchmark for XML Data Management. Database Group Leipzig (2002), `http://dbs.uni-leipzig.de/en/projekte/XML/XmlBenchmarking.html`

11. Bray, T., Paoli, J., Sperberg-McQueen, C.M., Maler, E., Yergeau, F.: Extensible Markup Language (XML) 1.0, 4th edn. W3C (September 2006)
12. Bressan, S., Li Lee, M., Li, Y.G., Lacroix, Z., Nambiar, U.: The XOO7 benchmark. In: Bressan, S., Chaudhri, A.B., Li Lee, M., Yu, J.X., Lacroix, Z. (eds.) CAiSE 2002 and VLDB 2002. LNCS, vol. 2590, pp. 146–147. Springer, Heidelberg (2003)
13. Busse, R., Carey, M., Florescu, D., Kersten, M., Manolescu, I., Schmidt, A., Waas, F.: XMark – An XML Benchmark Project. Centrum voor Wiskunde en Informatica (CWI), Amsterdam (2003), `http://www.xml-benchmark.org/`
14. Clark, J., DeRose, S.: XML Path Language (XPath) Version 1.0. W3C (November 1999)
15. Dvorakova, J., Zavoral, F.: Using Input Buffers for Streaming XSLT Processing. In: GlobeNet/DB 2009: Proc. of the 1st Int. Conf. on Advances in Databases, Guadeloupe, French Caribbean. IEEE, Los Alamitos (2009)
16. Runapongsa, K., et al.: The Michigan Benchmark. Department of Electrical Engineering and Computer Science, The University of Michigan (2006), `http://www.eecs.umich.edu/db/mbench/`
17. Bressan, S., et al.: The XOO7 Benchmark (2002), `http://www.comp.nus.edu.sg/~ebh/XOO7.html`
18. Franceschet, M.: XPathMark. University of Udine, Italy (2005), `http://users.dimi.uniud.it/~massimo.franceschet/xpathmark/`
19. Mlynkova, I.: An Analysis of Approaches to XML Schema Inference. In: SITIS 2008: Proc. of the 4th Int. Conf. on Signal-Image Technology and Internet-Based Systems, Bali, Indonesia. IEEE, Los Alamitos (2008)
20. Mlynkova, I., Toman, K., Pokorny, J.: Statistical Analysis of Real XML Data Collections. In: COMAD 2006: Proc. of the 13th Int. Conf. on Management of Data, New Delhi, India, pp. 20–31. Tata McGraw-Hill Publishing Ltd., New York (2006)
21. Nicola, M., Kogan, I., Raghu, R., Gonzalez, A., Liu, M., Schiefer, B., Xie, G.: Transaction Processing over XML (TPoX), `http://tpox.sourceforge.net/`
22. Thompson, H.S., Beech, D., Maloney, M., Mendelsohn, N.: XML Schema Part 1: Structures, 2nd edn. W3C (October 2004)
23. Vranec, M., Mlynkova, I.: FlexBench: A Flexible XML Query Benchmark (September 2008), `http://urtax.ms.mff.cuni.cz/~vranm3bm/dp/flexbench/`
24. Yao, B.B., Ozsu, M.T.: XBench – A Family of Benchmarks for XML DBMSs. University of Waterloo, School of Computer Science, Database Research Group (2003), `http://se.uwaterloo.ca/~ddbms/projects/xbench/`

Consistent Answers from Integrated XML Data*

Zijing Tan[1], Chengfei Liu[2], Wei Wang[1], and Baile Shi[1]

[1] School of Computer Science,
Fudan University, Shanghai, China
{zjtan,weiwang1,bshi}@fudan.edu.cn
[2] Faculty of Information and Communication Technologies,
Swinburne University of Technology, Australia
cliu@groupwise.swin.edu.au

Abstract. We consider the problem of using query transformation to compute consistent answers when queries are posed to virtual XML data integration systems, which are specified following the local-as-view approach. This is achieved in two steps. First the given query is transformed to a new query with global constraints considered, then the new query is rewritten to queries on the underlying data sources by reversing rules in view definitions. The XPath query on the global system can be transformed in XQuery. We implement prototypes of our method, and evaluate our framework and algorithms in the experiment.

1 Introduction

When data sources are integrated together, inconsistencies wrt global integrity constraints are likely to occur. The consistency of data derived from the data integration and query answering is an important issue of the consistent query answering(CQA) problem[5] and data quality. In this paper, we consider the problem of using query transformation to get consistent answers from virtually integrated XML data. The virtually integrated XML data is left as is, and queries are transformed when they arrive, as such the transformed query on the original, possibly inconsistent XML data will yield exactly the consistent answers.

In our discussion, the global system and each data sources use XML as their schemas. The mappings between the global schema and the data sources are defined in *local as view(LAV)* approach, and global integrity constraints are predefined. Given a query on the global system, we first transform it into a new query with global constraints considered for consistent answers. As the global XML data instance is not materialized, we then transform the new query to queries on the underlying data sources for the actual query answers. This is achieved by reversing rules in the view definitions. For a given XPath query on the global system, we illustrate that both transformations mentioned above

* This work is supported by the National Natural Science Foundation of China under grant No. 60603043, Australian Research Council Discovery Project under grant No. DP0878405.

X. Zhou et al. (Eds.): DASFAA 2009, LNCS 5463, pp. 436–440, 2009.

can be implemented by XQuery. We implement prototypes of our method, and evaluate our framework and algorithms in the experiments.

Related Work. For XML, absolute and relative keys[3], functional dependencies[1,9] are the most commonly discussed constraints. We give a more general constraint model in this paper, which can express functional dependencies and keys. It also naturally extends the recently discussed conditional functional dependencies[4] to XML. [8] investigates the existence of repairs with respect to a set of integrity constraints and a DTD. It differs from our goal to compute consistent answers from integrated XML data. In [6,7], view definition and query rewriting are also discussed. The difference is that, we need to answer queries on global system using the source instances in LAV approach. [10] studies the problem of answering queries through a target schema, and target constraints may be incorporated in the query rewriting. To the best of our knowledge, neither of the former work considers global integrity constraint violations with XML as global schema.

2 Constraint Model

A query(path) Q is evaluated at a *context* node v in an XML tree T, and its result is the set of nodes of T reachable via Q from v, denoted by $\lfloor v\{Q\}\rceil$. In particular, when there is only one node in $\lfloor v\{Q\}\rceil$, we use $v\{Q\}$to denote this node. If v is the *root* node, we write $\lfloor Q\rceil$ for $\lfloor v\{Q\}\rceil$.

We give a general form of XML constraint first. The constraint σ is of the form $(Q, Q', (Q_1, \ldots, Q_n))(\bar{X}_1, \ldots, \bar{X}_m)[x_{l1}{=}x_{k1}, \quad \ldots, \quad x_{lh}{=}x_{kh}{\Rightarrow}x_{l(h+1)}{=}x_{k(h+1)}]$. Here $Q, Q', Q_1, \ldots, Q_n$ are all simple paths, and Q is a root path, or $Q = \epsilon$. $\bar{X}_j (j \in [1, m])$ is a list of n variables, below we use x_{ji} to denote the ith$(i \in [1, n])$ variable in $\bar{X}_j$. $x_{l1},\ldots,x_{l(h+1)},x_{k1},\ldots,x_{k(h+1)}$ is either a variable or a constant; If it is a variable, it must be in $\bar{X}_j$.

We say a constraint σ is satisfied by an XML tree T, denoted as $T \models \sigma$ iff: For any symbol mapping h from σ to T, $\forall\ v{\in}\lfloor Q\rceil$, if (1)$\exists v_j{\in}\lfloor v\{Q'\}\rceil$ for $j \in [1, m]$, and (2)for $j, j' \in [1, m]$, if $j \neq j'$, $v_j \neq v'_j$, and (3)there is only one leaf node in $\lfloor v_j\{Q_i\}\rceil$ for $i \in [1, n]$, and (4)$val(v_j\{Q_i\}) = h(x_{ji})$, and (5)if $h(x_{l1}){=}h(x_{k1})$, $\ldots$, $h(x_{lh}){=}h(x_{kh})$ all hold, $h(x_{l(h+1)}) = h(x_{k(h+1)})$.

Please note that by $val(v_j\{Q_i\}) = h(x_{ji})$, all the variables in $x_{l1},\ldots,x_{l(h+1)}$, $x_{k1},\ldots,x_{k(h+1)}$ are bound by node values from T. We call Q context path, Q' target path, and $Q_1, \ldots, Q_n$ value paths. If the context path is ϵ, σ holds inside the whole document, called *absolute* constraint. Otherwise it holds inside every subtree rooted at a node in $\lfloor Q\rceil$, called *relative* constraint. Given a constraint σ in the form $(Q, Q', (Q_1, \ldots, Q_n))\ldots$, in a given XML T, we call $v{\in}\lfloor Q\rceil$ context nodes, $v_1{\in}\lfloor v\{Q'\}\rceil$ target nodes, and $v_1\{Q_i\}$ value nodes for constraint σ.

The given constraint model can express some commonly discussed constraints, for example, keys and functional dependencies. It also naturally extends the recently discussed conditional functional dependencies(CFD)[4] to XML. The CFD aims at capturing the consistency of data by incorporating bindings of semantically related values, which is more suitable for data cleaning and related works.

3 Constraint-Based Query Rewriting for Consistent Query Answers

Given a query Q on global system, we can rewrite Q to a new query Q' on global system with constraints considered. To answer Q' on the original, possibly inconsistent global system will generate the consistent query answers.

We use $\neg$ to denote *negation* required in the rewriting method. Given a constraint σ of the form $(Q, Q', (Q_1, \ldots, Q_n))(\bar{X}_1, \ldots, \bar{X}_m)[x_{l1}{=}x_{k1}, \ \ldots, \ x_{lh}{=}x_{kh} \Rightarrow x_{l(h+1)}{=}x_{k(h+1)}]$, it is converted to the form $\neg((Q, Q', (Q_1, \ldots, Q_n))$ $(\bar{X}_1, \ldots, \bar{X}_m)$ $[x_{l1}{=}x_{k1}, \ldots, x_{lh}{=}x_{kh}, \neg(x_{l(h+1)}{=}x_{k(h+1)})])$. Intuitively speaking, we first find the violating target nodes by using *negation* on the conditions, then calculate the consistent answers by excluding violating ones using another *negation*.

The approach of constraint based query rewriting is summarized: 1)The nodes $\{v_1, ..., v_k\}$ qualified by Q are selected; 2)Constraints in Σ should be verified if their target nodes have v_i as ancestor nodes. These constraints can be determined by considering Q, the context and target path of constraints; 3)Constraints are verified one by one, ordered by the concatenation of their context and target paths. Please note that v_i is the domain for constraint validations. However, since constraints may have context beyond v_i, the original document may be accessed again in constraint validations; 4)The resulting XML document is generated top-down from v_i, excluding the violating value nodes of constraints in Σ.

Algorithm 1. $Constraint - Based - Rewrite(Q, T, \Sigma)$

Input: query Q, virtual global instance T, global constraint set Σ.

for each $\sigma{=}(P, P', (\ldots))(\ldots)[\ldots]{\in}\Sigma$
    if Q is a prefix of P/P' **then**
        Insert σ into σ_list properly: /* σ_list is initially empty. */
            for $\sigma'{=}(S, S', (\ldots))(\ldots)[\ldots]{\in}\sigma\_list$, if S/S' is a prefix of P/P', ensure that σ' is before σ in σ_list.
        /* σ_list will be a list of constraints, which should be validated for query Q. */
Evaluate Q on T, and let the nodes qualified by Q on T be $\{v_1, ..., v_k\}$.
for each v_i $(i{\in}[1, k])$
    for each σ_j from σ_list
        Validate σ_j; during the validation, at least one target node for σ_j should be selected from T_{v_i}.
        /* We use T_{v_i} to denote the subtree rooted at v_i in T */
        if σ_j does not hold **then**
            Mark the violating value nodes of σ_j in T_{v_i}.
    For node u in T_{v_i}, if all the descendant nodes of u are marked, mark u as well.
    Output T_{v_i} top-down, excluding the marked nodes.

---

4 Query Rewriting in LAV

With the new query Q' produced based on target constraints in section 3, we need to further rewrite it to queries on the underlying data sources, because the global instance is not materialized. To define the mappings between the global schema and the data sources, we propose a definition of XML views, which supports edge-path mappings and data-value bindings. The given XML views are also flexible enough to incorporate bindings of semantically related values.

We rewrite the query on global system to queries on underlying data sources by reversing rules in the view definitions. We just illustrate some key points in this step: 1)The value bindings need to be reversed; 2)More than one paths in the global schema may match the given query in the reversed view definition; 3)When a node is extracted from the source instance, its child nodes must be extracted iteratively by the succeeding rules; 4) Some *necessary* default elements may be added to make the result conforming to the target schema.

5 Experimental Results

We implement a rewriting module that translates XPath queries on the global system to XQuery queries on the underlying data sources, based on the constraint rewriting and reversed view definitions. Then we run the generated XQuery queries on the source XML documents using an XQuery system built on SAXON.

We perform our experiments with artificial data sets generated by the IBM XML Generator. In the experiments, we compare the time of ordinary query answering with the time of consistent query answering.

We run various types of XPath queries, and below we give the evaluation time for a query with filters on data values. We consider *noise* of data, *number* of constraints, *scalability* of data set, and *selectivity* of query in the experiments. The *noise* factor has no effect on the running time of ordinary query. And for consistent query answering, it also takes almost the same time to process consistent query answering on a consistent XML document, or on a badly inconsistent XML document. With a fixed data set, the consistent query answering time increases with more constraints considered. But for two constraints, the time to get violating nodes wrt them is the sum of the time to get their own violating nodes. So the time growth is not explosive with increasing constraints.

In figure 1, we vary the size of source data sets. Each data source grows with the same pace, and here the size of total data sources is given. Please note that we use different measures for the two time. The time of ordinary answering and consistent answering all grows with larger data sets, and consistent answering needs more time, which implies that most of time is spent on getting violating nodes wrt target constraints. With larger data sets, the time for constraint validating is in quadratic growth, which grows faster than the time for ordinary query answering. In figure 2, with fixed data sources, we run queries with decreasing selectivity by using different conditions in filters. Because we have no index built, the time of ordinary query answering is not sensitive to the selectivity of queries. However, the set of qualified nodes is the domain for constraint

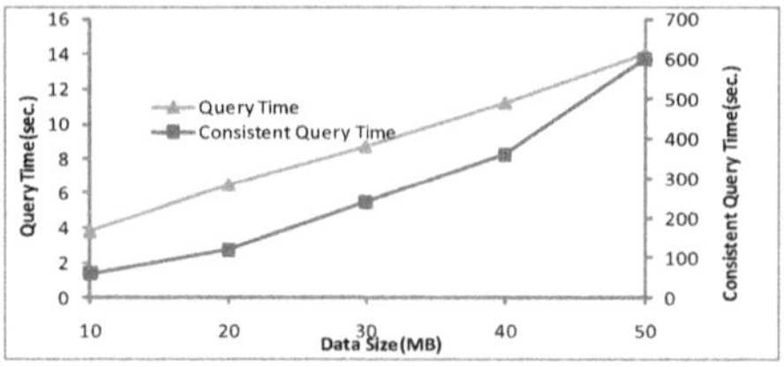

Fig. 1. Time with increasing data size

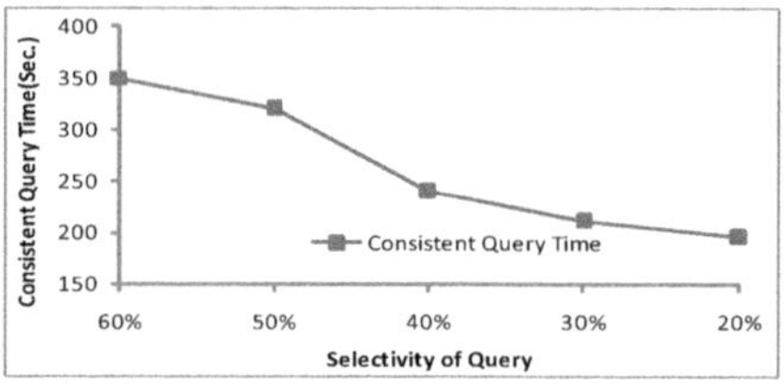

Fig. 2. Time with decreasing query selectivity

validations. There is no need to compare target nodes if neither of them is qualified by the query. So the time for consistent query answering decreases with lower selectivity dramatically.

6 Conclusions

We use query transformation to compute consistent answers when queries are posed to virtual XML data integration systems, based on global constraints rewriting and rules reversing in LAV setting. To run the transformed query on the original, possibly inconsistent global system will yield exactly the consistent answers. We also implement prototypes of our method, and evaluate our framework and algorithms in the experiment.

References

1. Arenas, M., Libkin, L.: A Normal Form for XML Documents. TODS 29(1), 195–232 (2004)
2. Arenas, M., Bertossi, L.E., Chomicki, J.: Consistent Query Answers in Inconsistent Databases. In: PODS 1999, pp. 68–79 (1999)
3. Buneman, P., Davidson, S., Fan, W., Hara, C., Tan, W.: Reasoning about Keys for XML. In: DBPL 2002, pp. 133–148 (2002)
4. Bohannon, P., Fan, W.F., Geerts, F., Jia, X., Kementsietsidis, A.: Conditional Functional Dependencies for Data Cleaning. In: ICDE 2007, pp. 746–755 (2007)
5. Chomicki, J.: Consistent query answering: Five easy pieces. In: Schwentick, T., Suciu, D. (eds.) ICDT 2007. LNCS, vol. 4353, pp. 1–17. Springer, Heidelberg (2006)
6. Fan, W.F., Bohannon, P.: Information Preserving XML Schema Embedding. TODS 33(1) (2008)
7. Fan, W.F., Geerts, F., Jia, X., Kementsietsidis, A.: Rewriting Regular XPath Queries on XML Views. In: ICDE 2007, pp. 666–675 (2007)
8. Flesca, S., Furfaro, F., Greco, S., Zumpano, E.: Querying and repairing inconsistent XML data. In: Ngu, A.H.H., Kitsuregawa, M., Neuhold, E.J., Chung, J.-Y., Sheng, Q.Z. (eds.) WISE 2005. LNCS, vol. 3806, pp. 175–188. Springer, Heidelberg (2005)
9. Vincent, M.W., Liu, J., Liu, C.: Strong Functional Dependencies and their Application to Normal Forms in XML. TODS 29(3), 445–462 (2004)
10. Yu, C., Popa, L.: Constraint-based XML query rewriting for data integration. In: SIGMOD 2004, pp. 371–382 (2004)

Optimal Privacy-Aware Path in Hippocratic Databases*

Min Li[1], Xiaoxun Sun[1], Hua Wang[1], and Yanchun Zhang[2]

[1] Department of Mathematics & Computing
University of Southern Queensland, Australia
{limin,sunx,wang}@usq.edu.au
[2] School of Computer Science & Mathematics
Victoria University, Australia
yzhang@csm.vu.edu.au

Abstract. Privacy becomes a major concern for both customers and enterprises in today's corporate marketing strategies, many research efforts have been put into developing new privacy-aware technologies. Among them, Hippocratic databases are one of the important mechanisms to guarantee the respect of privacy principles in data management, which adopt purpose as a central concept associated with each piece of data stored in the databases. The proposed mechanism provides basic principles for future database systems protecting privacy of data as a founding tenet. However, Hippocratic databases do not allow to distinguish which particular method is used for fulfilling a purpose. Especially, the issues like purpose hierarchies, task delegations and minimal privacy cost are missing from the proposed mechanism.

In this paper, we extend these mechanisms in order to support inter-organizational business processes in Hippocratic databases. A comprehensive approach for negotiation of personal information between customers and enterprises based on user preferences is developed when enterprises offer their clients a number of ways to fulfill a service. We organize purposes into purpose directed graphs through AND/OR decomposition, which supports task delegations and distributed authorizations. Specially, customers have controls of deciding how to get a service fulfilled on the basis of their personal feeling of trust for any service customization. Quantitative analysis is performed to characterize privacy penalties dealing with privacy cost and customer's trust. Finally, efficient algorithms are given to guarantee the minimal privacy cost and maximal customer's trust involved in a business process.

1 Introduction

With the widespread use of information technology in all walks of life, personal information is being collected, stored and used in various information systems. Achieving privacy preservation has become a major concern. Issues related to

* This research is funded by an ARC Discovery Project DP0663414.

X. Zhou et al. (Eds.): DASFAA 2009, LNCS 5463, pp. 441–455, 2009.

privacy have been widely investigated and several privacy protecting techniques have been developed. To our best knowledge, the most well known effort is the W3C's Platform for Privacy Preference (P3P) [10]. P3P allows websites to express their privacy policy in a machine readable format so that using a software agent, consumers can easily compare the published privacy policies against their privacy preferences. While P3P provides a mechanism for ensuring that users can be informed about privacy policies before they release personal information, some other approaches are proposed [4,8,16,17,22], where the notion of *purpose* plays an important role in order to capture the intended usage of information.

As enterprises collect and maintain increasing amounts of personal data, individuals are exposed to greater risks of privacy breaches and identity theft, many enterprises and organizations are deeply concerned about privacy issues as well. Many companies, such as IBM and the Royal Bank Financial Group, use privacy as a brand differentiator [3]. By demonstrating good privacy practices, lots of business try to build solid trust to customers, thereby attracting more customers [20,5,6,12]. Together with the notion of *purpose*, current privacy legislations also define the privacy principles that an information system has to meet in order to guarantee customer's privacy [11,1,2,21]. Mechanism for negotiation is presented by Tumer et al. [21]. Enterprises specify which information is mandatory for achieving a service and which is optional, while customers specify the type of access for each part of their personal information.

On the basis of the solution for the exchange between enterprises and customers, Hippocratic databases enforced fine-grained disclosure policies to an architecture at the data level [1]. In the proposed architecture, enterprises declared the purpose for which the data are collected, who can receive them, the length of time the data can be retained, and the authorized users who can access them. Hippocratic databases also created a privacy authorization table shared by all customers, but it does not allow to distinguish which particular method is used for fulfilling a service. Moreover, enterprises are able to provide their services in different ways, and each different method may require different data. For example, notification can be done by email or by mobile phone or by fax. Depending on the different kinds of methods, customers should provide different personal information. Asking for all personal information for different service methods as compulsory would clearly violate the principle of minimal disclosure.

On the server side, a single enterprise usually could not complete all procedures of a service by itself, rather a set of collaborating organizations participating in the service. Enterprises might need to decompose a generic purpose into more specific sub-purposes since they are not completely able to fulfill it by themselves, and so they may delegate the fulfillment of sub-purposes to third parties. It is up to customers to decide on a strategy of how to get a service fulfilled on the basis of their personal feeling of trust for different service components. A question that many customers have when interacting with a web server, with an application, or with an information source is "Can I trust this entity?". Different customizations may require different data for which considerations may vary; there might be different trust levels on different partners (sub-contractors). The choice of service customization has significant impact on the privacy of individual customers.

In this paper, we present an approach to automatically derive the optimal way of authorizations needed to achieve a service from enterprise privacy policies. In particular, we organize purposes into purpose directed graphs through AND/OR decompositions, which support the delegation of tasks and authorizations when a host of partners participating in the business service provides different ways to achieve the same service. Further, we allow customers to express their trust preferences associated with each partner of the business process. Thus, a weight combining privacy cost and customer trust is given on each arc of the graph in the form of privacy penalties, and the process for fulfilling a purpose can be customized at runtime and guarantees minimal privacy cost and maximal customer trust because it was selected with criterion of the optimal privacy penalty. Finally, an efficient algorithm is proposed to find optimal privacy-aware path in Hippocratic databases. Our work is grounded on modeling and analysis of purposes for Hippocratic databases and proposes enhancements to Hippocratic database systems in order to deal with inter-organizational business processes.

The remainder of the paper is structured as follows. Section 2 introduces the motivation of this paper based on a running example. Section 3 presents some background information on Hippocratic database systems. We introduce purpose directed graph with delegation in Section 4 and discuss how to characterize the privacy penalty and efficiently find the optimal solution in Section 5. We provide a brief survey of related work in Section 6. We conclude the paper in Section 7.

2 Motivation

We consider the following example throughout the paper.

Example: *Ebay is an online seller in Australia and provides an online catalogue to its customers who can search for the items they wish to buy. Once customers have decided to buy goods, Ebay needs to obtain certain personal information from customers to perform purchase transactions. This information includes name, shipping address, and credit card number. Ebay views purchase, its ultimate purpose, as a three-step process: credit assessment, delivery, and notification. Credit assessment relies on Credit Card Company (CCC). Delivery can be done either by a delivery company or the post office, while notification can be done by email or by mobile phone.*

Obviously, Ebay provides many ways to achieve the purchase service and each different method could require different data. An important principle is that enterprises should disclose to customers which data is collected and for what purpose. Also, enterprises should maintain minimal personal information necessary to fulfill the purpose for which the information was collected.

From the customers' point of view, they do not want to disclose more data than needed to get the desired service; rather, they want the process that best protects their privacy based on their preferences. Depending on the method of notification, Ebay needs either an email address or a mobile phone number. For example, Jimmy, a professor plagued by spam, may treasure his email address and give away his business mobile phone number. Bob, a doctor whose mobile phone is always ringing, may have the opposite preference. Therefore, it is up to customers to decide how to get a service fulfilled on the basis of their personal feeling of any service customization.

Furthermore, if we consider the delivery service, Ebay could not fulfill the service by itself, but rather relies on a delivery company or post office. That means Ebay may outsource a large part of data processing to third parties participating in a single business process. However, the more the data is used, the more likely it might be disclosed, since the personal information is transmitted from one to another. This requires that enterprises maintain minimal personal information necessary to fulfill the purpose. Moreover, the partners chosen by Ebay might also be trusted differently by its potential customers. The burden of choice is on the human who must decide what to do on the basis of his/her personal feeling of trust of the enterprises. For instance, Albert may prefer to delivery by a delivery company, since it is fast; whereas, Bob may chose delivery by post office because it is safe. Different partners (sub-contractors) chosen for the same purpose may be with different trust levels. The choice of service customization has significant impact on the privacy of individual customers.

If we consider these factors, both the privacy cost and customer's trust should be considered as important factors in privacy security system when enterprises publish comprehensive privacy policies involving hierarchies of purposes, possibly spanning multiple partners. Formally, it can be stated as follows:

Minimal privacy cost: Is there a way to fulfill the purpose with minimal privacy cost?

Maximal customer's trust: Is there a way to fulfill the purpose with maximal trust between enterprises and customers?

Classical privacy-aware database systems such as Hippocratic databases do not consider these issues, we are interested in solutions that support customers and companies alike, so that companies can publish comprehensive privacy policies involving multiple service methods, possibly delegation of tasks and authorizations. Moreover, the solutions will allow customers to personalize services based on their own privacy sensitivities and their trust of partners who might contribute to the requested service.

3 Overview of Hippocratic Databases

Hippocratic databases use *purpose* as a central concept [1]. A purpose describes the reason(s) for data collection and data access, which is stored in the database as a "special" attribute occurring in every table of the database. This attribute specifies the purpose (reason/goal) for which a piece of information can be used.

For example, Table 1 shows the schema of two tables, customer and order, that store the personal information including purposes. In particular, table customer

Table 1. Database schema

table	**attribute**
customer	purpose, customer-id, name, address, email, fax-number, credit-card-info
order	purpose, customer-id, transaction, book-info, status

Table 2. Privacy metadata schema

table	**attribute**
privacy-polices	purpose, table, attribute, {external-receipts}, {retention-period}
privacy-authorizations	purpose, table, attribute, {authorized-users}

stores personal information about customers, and table order stores information about the transactions between enterprises and their customers. Then, for each purpose and data item stored in the database, we have:

External-recipients: the actors to whom the data item is disclosed;

Retention-period: the period during which the data item should be maintained;

Authorized-users: the users entitled to access the data item.

Purpose, external recipients, authorized users, and retention period are stored in the database with respect to the metadata schema defined in Table 2. Specifically, the above information is split into separate tables: external-recipients and retention period are in the *privacy-policies table*, while authorized-users in the *privacy-authorizations table*. The purpose is stored in both of them. The privacy-policies table contains the privacy policies of the enterprise, while privacy-authorizations table contains the access control policies that implement the privacy policy and represents the actual disclosure of information. In particular, privacy-authorizations tables are derived from privacy-policies tables by instantiating each external recipient with the corresponding users. Therefore, Hippocratic database systems define one privacy-authorizations table for each privacy-policies table, and these tables represent what information is actually disclosed.

Hippocratic database system is an elegant and simple solution but does not allow for dynamic situations that could arise with web services and business process softwares. In such settings, enterprises may provide services in many different ways and may delegate the execution of parts of the service to third parties. This is indeed the case of a virtual organization based on business process for web service where different partners explicitly integrate their efforts into one process [15].

4 Purpose Directed Graph with Delegation

Agrawal et al.[1] proposed a structure to split a purpose into multiple purposes and then stored them in the database. Karjoth et al.[16] used a directory-like notation to represent purpose hierarchies, which loses the logic relation between a purpose and its sub-purposes. In particular, this notation does not distinguish if a sub-purpose is derived by AND or OR decomposition [19]. Assuming a purpose p is AND-decomposed into sub-purposes $p_1, \ldots, p_n$, then all of the sub-purposes must be satisfied in order to satisfy p. For example, Ebay AND-decomposes purchase into delivery, credit assessment, and notification, then all of the three sub-purposes have to be fulfilled for fulfilling purchase purpose. However, if a purpose p is OR-decomposed into sub-purposes $p_1, \ldots, p_n$, then one of the sub-purposes must be satisfied in order to satisfy p. For instance, Ebay further OR-decomposes delivery into direct delivery relying on delivery companies and delivery by post office (shown in Fig.1). In this way, only one of them could be necessary to fulfill the delivery purpose. In essence, AND-decomposition is used to define the process for achieving a purpose, while OR-decomposition defines alternatives for achieving a purpose.

Our approach is based on traditional goal analysis [18], and consists of decomposing purposes into sub-purposes through an AND/OR refinement. The idea is to represent purpose hierarchies with directed graphs.

Definition 1. *A purpose directed graph PDG is a pair (P, A), where P is a set of purposes and A is the set of arcs, each arc represents a hierarchical relation between the purposes.*

A purpose directed graph (PDG) can be used to represent goal models in goal-oriented requirements engineering approaches [7]. For our purposes, they represent the entire set of alternative ways for delivering a service required by customers. Such representations can also be used to model the delegations of tasks and authorizations in the security modeling methodology proposed by Giorgini et al.[14].

An enterprise could provide different methods to achieve a service or rely on different partners to achieve the same part of the service. In particular, Ebay relies on a delivery company, Worldwide Express (WWEx), for shipping books. Ebay needs to delegate customer's information, such as name and shipping address, to WWEx. In turn, WWEx depends on local delivery companies for door-to-door delivery. To this end, WWEx delegates customer information to the local delivery companies $\text{LDC}_1, \cdots, \text{LDC}_n$ for door-to-door delivery. Consequently, different processes can be used to fulfill the required service. To capture this insight, we introduce the notion of path.

Definition 2. *Let $PDG = (P, A)$ be a purpose directed graph. A path from v_0 to v_m is defined as a sequence $W = (v_0, a_1, v_1, ..., a_m, v_m)$, where a_i is an arc from v_{i-1} to v_i for $i = 1, ..., m$.*

A purpose directed graph PDG is rooted if it contains a vertex v, such that all the vertices of PDG are reachable from v through a directed path. The vertex v is called a root of PDG.

For example, consider the purpose directed graph depicted in Fig.1. Each vertex is composed by two parts: a purpose identifier and an enterprise needed

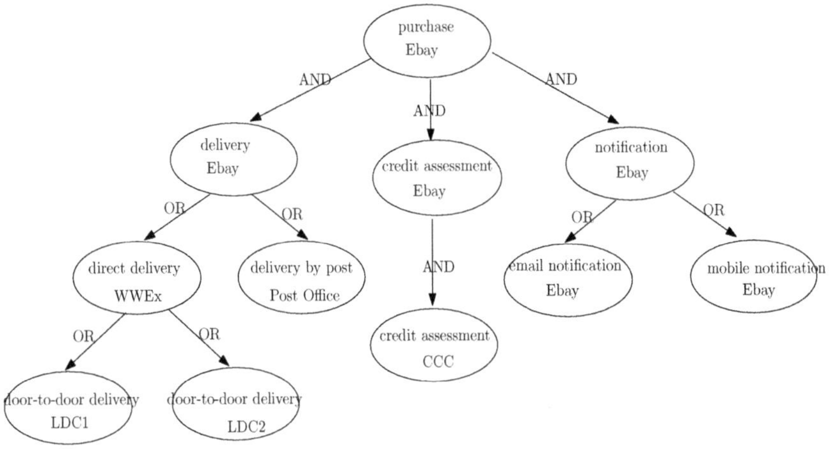

Fig. 1. Purpose directed graph

to fulfill the purpose, and each of the purposes represents the policies of a single enterprise. The vertex 'purchase' is the root of the graph and purchase is the root-level purpose. Essentially, if a path $W = (v_0, a_1, v_1, ..., a_m, v_m)$ satisfies that v_0 is the root purpose and there exists no downward paths from v_m, we say the path is an essential path. An essential path represents a possible process through which an enterprise can fulfill the root purpose.

The enterprise-wide privacy policies are derived by looking at the Hippocratic database of each partner involved in the business process and merging them into a single purpose. Therefore, purposes can be recognized as the outcome of a process of refinements of goals in security requirements modeling methodologies[13]. The task delegation is indeed the case of a virtual organization based on business process for web service where different partners explicitly integrate their efforts into one process.

5 Finding Optimal Privacy-Aware Path

Our goal is to decide which is the essential optimal privacy-aware path to fulfill the root purpose with respect to the customer's preference. This can be performed through the following quantitative analysis.

5.1 Objective Characterization

Since our reference business model is that of virtual organizations, we assume that there will often be more than one way to deliver a service. Yet, they may differ in an important aspect, notably they may require different private data items, which incurs different privacy cost. Further, depending on each customer's individual preferences, the same decomposition path might have significantly different trust values for different customers. In order to support quantitative analysis, we need to introduce the notion of privacy penalty.

Definition 3. *The privacy penalty of an arc a is defined as a pair* $w_a = (\alpha, \beta)$, *where* α *is the privacy cost and* β *is the customer's trust value on the arc a.*

Choice of α, β**:** The privacy penalty pair (α, β) on each arc can be pre-defined by asking the enterprises and customers to specify the level of privacy cost and trust they feel about the sub-suppliers. Since the personal information is transmitted from one to another, this may increase the danger of the leakage of personal information. Therefore, we use α to depict the privacy cost. Generally, we assume that there are different trust values based on the customer's personal feeling of the trust on different service custormizations. For example, Bob prefers mobile notification more than email notification because of the personal experience, so there is a high trust value on mobile notification.

Intuitively, the privacy penalty of a path should consist of two parts: one is the sum of the privacy cost on each arc and the other is the minimum trust among these arcs.

Definition 4. *Let* $\mathcal{P} = (v_0, a_1, v_1, \cdots, a_m, v_m)$ *be a path in the PDG. Then, the privacy penalty of the path* $\omega_{\mathcal{P}} = \omega_{a_1} + \ldots + \omega_{a_m} = (\sum_{i=1}^{m}(\alpha_i), min_{i=1}^{m}(\beta_i))$, *where* $\omega_{a_i} = (\alpha_i, \beta_i)$, $i = 1, ..., m$.

Essentially, a path represents a possible process through which an enterprise can fulfill a root purpose. For our purpose, we use the *sum* of the private cost of each arcs because we argue that the more a piece of data is used, the more likely it might be misused. The smaller the sum is, the less the privacy cost is. Therefore, *sum* measures are the ones that capture best one's intuitions on the cost of privacy. We also use the minimization function on trust values to get the smallest trust value on this path. The larger the value is, the more the trust is on this path. Our goal is to decide which is the process with the optimal privacy penalty (i.e., the minimal privacy cost and maximal trust value) to fulfill the root purpose with respect to the user's preferences. In order to describe the user's preference, we next introduce a flexible objective function.

Flexible objective function: If the privacy penalty on the arc a is defined as $w_a = (\alpha, \beta)$, we introduce the following objective function to balance the privacy cost and customer trust with a preference coefficient γ $(0 \leq \gamma \leq 1)$.

$$alt(a) = \gamma \times \alpha + (1 - \gamma) \times \beta \qquad (1)$$

The choice of parameter γ depends on the customer's preference. If the customer cares whether data are disclosed at all, then γ may be set with a value in the interval $0.5 \leq \gamma \leq 1$. On the other hand, if the customer stresses more on trust, then γ can be set with a value between 0 and 0.5.

In addition to the objective function, we propose to decompose purposes into sub-purposes through an AND/OR decomposition. In essence, AND-decomposition is used to define the process for achieving a purpose, while OR-decomposition defines alternatives for achieving a purpose. Normally, the node purpose can be either AND-decomposed or OR-decomposed. A decomposition arc is either an OR-arc or an AND-arc.

Definition 5. *Let $PDG = (P, A)$ be a purpose directed graph, for each vertex $v \in P$, we denote $OUT(v) = OUT_{or}(v) \cup OUT_{and}(v)$ as the set of all successors of v, where $OUT_{or}(v)$ refers to all successors connecting v with OR-arcs, and $OUT_{and}(v)$ stores all successors connecting v with AND-arcs. Especially, if $OUT(v) = \emptyset$, we say the vertex v is a leaf of PDG.*

For example, in Fig. 2 the root purpose r is AND-decomposed into three sub-purposes: delivery, credit assessment and notification, then $OUT(r) = OUT_{and}(r) = \{$delivery, credit assessment and notification$\}$. Further, considering the node v with purpose 'mobile notification', since $OUT(v) = \emptyset$, then the node 'mobile notification' is a leaf of the purpose directed graph.

5.2 The Algorithm

In this section, we present efficient algorithms to track the optimal path that the enterprises need to fulfill a purpose. Next, we analyze two situations in finding the optimal privacy-aware path.

Case 1: if the root purpose is OR-decomposed, the algorithm consists of following steps:

1. To contract each vertex v with all its successors in $OUT_{and}(v)$ to a compound vertex v_c; suppose $OUT_{and}(v) = \{v_1, \cdots, v_k\}$, we define $cost[v_c] = \sum_{i=1}^{k} \alpha(v, v_i)$, $trust[v_c] = \min_{i=1}^{k} \beta(v, v_i)$;
2. To transfer the purpose directed graph PDG into $\overline{PDG}$ with no AND-arcs and find the optimal path $\overline{p}$ using function $optimal\_path(\overline{PDG})$;
3. If the optimal path of $\overline{PDG}$ contains a compound vertex (or vertices), then expand the compound vertex (or vertices) on $\overline{p}$ to become the optimal solution of PDG.

In Algorithm 1, $\alpha(u, v)$ represents the privacy cost between the two nodes u and v, and $\beta(u, v)$ refers to the trust value on the arc (u, v). For each leaf vertex, Sum function is used to track the distance between the leaf and the root, while $predecessor[]$ records all predecessor vertices of the leaf, and $previous[]$ records the vertices on the optimal path from the leaf to the root. alt on line 10 is the objective function with the preference coefficient γ. If $\gamma \geq 0.5$, it means customers prefer more on privacy protection, then the minimal objective value is needed depending on the minimization function; while if $\gamma < 0.5$, it means customers prefer more on trust, then the maximal objective value is needed depending on the maximization procedure.

Algorithm 1. *optimal_path*(PDG, OR_r)

Input: a purpose directed graph PDG with OR-decomposed root r.
Output: The optimal path D

1. Contract each vertex v with all its successors in $OUT_{and}(v)$ to the compound vertex v_c
2. Transfer PDG into $\overline{PDG}$
3. $\quad \overline{p} = optimal\_path(\overline{PDG})$
4. If $\overline{p}$ contains compound vertex(vertices),
5. $\quad$ expand the compound vertex(vertices) on $\overline{p}$ to p,
6. $\quad\quad D = p$
7. else
8. $\quad\quad D = \overline{p}$

function: *optimal_path*$(\overline{PDG})$:

Input: $\overline{PDG}$ with root purpose r and leaves $v_1, \ldots, v_k$, pre-defined privacy cost and trust function $\alpha(*,*)$, $\beta(*,*)$, and preference coefficient $0 \leq \gamma \leq 1$

for each vertex v (not a compound vertex) in $\overline{PDG}$:
$\quad cost[v] := 0$
$\quad trust[v] := \infty$
$\quad$ for each leaf v_i $(i = 1, \cdots, k)$
$\quad Sum(v_i) := 0$, $previous[v_i] := \{v_i\}$
$\quad$ while $r \notin predecessor[v_i] = \{u_{i_1}, \cdots, u_{i_s}\}$
$\quad$ {
$\quad\quad$ for each u_{i_j} $(1 \leq j \leq s)$
$\quad\quad\quad cost(u_{i_j}, v_i) := cost[v_i] + cost[u_{i_j}] + \alpha(u_{i_j}, v_i)$
$\quad\quad\quad trust(u_{i_j}, v_i) := min\{trust[v_i], trust[u_{i_j}], \beta(u_{i_j}, v_i)\}$
$\quad\quad\quad alt(u_{i_j}, v_i) := \gamma \times cost(u_{i_j}, v_i) + (1 - \gamma) \times trust(u_{i_j}, v_i)$
$\quad\quad\quad$ if $\gamma \geq 0.5$ /* prefer cost */
$\quad\quad\quad\quad$ let $alt(u_{i_m}, v_i) = \min_{j=1}^{s} alt(u_{i_j}, v_i)$
$\quad\quad\quad\quad previous[v_i] := previous[v_i] \cup \{u_{i_m}\}$
$\quad\quad\quad\quad Sum(v_i) := Sum(v_i) + alt(u_{i_m}, v_i)$
$\quad\quad\quad\quad v_i := u_{i_m}$
$\quad\quad\quad$ if $\gamma < 0.5$ /* prefer trust */
$\quad\quad\quad\quad$ let $alt(u_{i_m}, v_i) = \max_{j=1}^{s} alt(u_{i_j}, v_i)$
$\quad\quad\quad\quad previous[v_i] := previous[v_i] \cup \{u_{i_m}\}$
$\quad\quad\quad\quad Sum(v_i) := Sum(v_i) + alt(u_{i_m}, v_i)$
$\quad\quad\quad\quad v_i := u_{i_m}$
$\quad\quad$ }
$\quad$ /*end while and all paths from the leaf to the root are found*/
$\quad$ if $\gamma \geq 0.5$
$\quad\quad$ assume $Sum(v_t) = min_{i=1}^{k} Sum(v_i), (1 \leq t \leq k)$
$\quad\quad$ output $previous[v_t]$
$\quad$ if $\gamma < 0.5$
$\quad\quad$ assume $Sum(v_t) = max_{i=1}^{k} Sum(v_i), (1 \leq t \leq k)$
$\quad\quad$ output $previous[v_t]$
end function

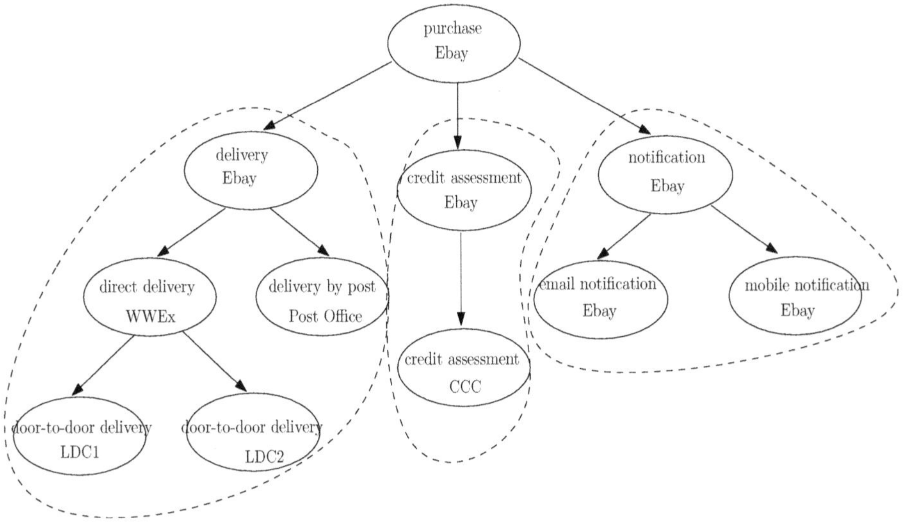

Fig. 2. *sub_PDG* in Purpose directed graph

Case 2: if the root purpose is AND-decomposed, in order to design efficient algorithms to determine the process by which a service can be delivered with optimal privacy penalties, we need the definition of sub-purpose directed graph.

Definition 6. *Let $PDG = (P, A)$ be a purpose directed graph, if the root purpose r is AND-decomposed into several sub-purposes, then each sub-purpose with all its descendants form a sub-purpose directed graph of PDG, and we denote it by sub_PDG. Essentially, if the root of the sub_PDG is further AND-decomposed into several sub-purposes, then each sub-purpose with all its descendants form a sub-purpose directed graph of sub_PDG, which is also a sub-sub-purpose directed graph of PDG, and we denote it by $sub\_(sub\_PDG)$.*

For example, in Fig. 2 Ebay AND-decomposes purpose purchase into three sub-purposes: delivery, credit assessment and notification. According to the definition of sub-purpose directed graph, the purpose delivery with all its decedents consist of a sub-purpose directed graph. The same situation applies to the other two sub-purposes, so there are three sub-purpose directed graphs as in Fig. 2 (circled in broken line). Since in each sub-purpose directed graph, the root is further OR-decomposed, there is no sub-sub-purpose directed graph in Fig. 2.

For the sake of simplicity, we assume that the root of each $sub\_(sub\_PDG)$ is OR-decomposed. In this case, the algorithm consists of following steps:

1. To decompose the purpose directed graph PDG into several sub-purpose directed graphs.
2. For each sub-purpose directed graph sub_PDG with root purpose r,
   (a) if the root purpose r is OR-decomposed, run algorithm $optimal\_path(sub\_PDG, OR\_r)$ to find the optimal path in sub_PDG;

(b) if the root purpose r is AND-decomposed, further decompose the *sub_PDG* into several sub-sub-purpose directed graphs, then run algorithm *optimal_path* ($sub\_(sub\_PDG), OR\_r'$) to find the optimal path in each $sub\_(sub\_PDG)$ with root r'. Combine all the optimal paths of each $sub\_(sub\_PDG)$ into the optimal solution of *sub_PDG*.

3. To combine all the optimal paths of each *sub_PDG* into the optimal solution of *PDG*.

Algorithm 2. $optimal\_path(PDG, AND\_r)$

Input: A purpose directed graph PDG with AND-decomposed root
Output: The optimal path D

decompose PDG into several sub_PDG
for each sub_PDG with root r
  if the root r is OR-decomposed in sub_PDG
    run algorithm $optimal\_path(sub\_PDG, OR\_r)$
    output $p_{or} = optimal\_path(sub\_PDG)$
  if the root r is AND-decomposed in sub_PDG
    further decompose the sub_PDG into several $sub_(sub_PDG)$s
    for each $sub\_(sub\_PDG)$ with root r'
      run algorithm $optimal\_path(sub\_(sub\_PDG), OR\_r')$
      output $p_{and} = optimal\_path(sub\_PDG)$
$D = (\cup p_{or}) \cup (\cup p_{and})$

In Algorithm 2, p_{or} refers to the optimal path of *sub_PDG* with an OR-decomposed root, while p_{and} refers to the optimal path of *sub_PDG* with an AND-decomposed root.

Until here, ether the root purpose is AND-decomposed or OR-decomposed, we can find the optimal path under any specific value of γ through our algorithms. If $\gamma \geq 0.5$, it means the customer stresses more on privacy cost. The optimal solution to satisfy the customer's preference is the optimal path with the minimal objective value when varying the value of γ. Our method is to search all the possible optimal path based on the value of γ from 0.5 to 1 by the interval of 0.01. Then, the optimal path with the minimal objective value will be chosen as the optimal solution. For the situation $\gamma < 0.5$, we search all the possible optimal path based on the value of γ from 0 to 0.5 by the interval of 0.01. Then, the optimal path with the maximal objective value will be chosen as the optimal solution, since the customer prefers more on trust.

6 Related Work

Our work is related to several topics in the area of privacy and security for data management, namely privacy policy specification, privacy-preserving data management systems and multilevel secure database systems. We now briefly survey the most relevant approaches in these areas and point out the differences of our work with respect to these approaches.

The W3Cs Platform for Privacy Preference (P3P) [23] allows web sites to encode their privacy practice, such as what information is collected, who can access the data for what purposes, and how long the data will be stored by the sites, in a machine-readable format. P3P enabled browsers can read this privacy policy automatically and compare it to the consumers set of privacy preferences which are specified in a privacy preference language such as a P3P preference exchange language (APPEL) [10], also designed by the W3C. Even though P3P provides a standard means for enterprises to make privacy promises to their users, P3P does not provide any mechanism to ensure that these promises are consistent with the internal data processing. By contrast, the work in our paper provides an effective strategy to maximize privacy protection. Further, we allow customers to express their trust preferences associated with each partner of the business process in order to achieve maximal customer trust.

Byun et al. presented a comprehensive approach for privacy preserving access control based on the notion of purpose [9,8]. In the model, purpose information associated with a given data element specifies the intended use of the data element, and the model allows multiple purposes to be associated with each data element. The granularity of data labeling is discussed in detail in [9], and a systematic approach to implement the notion of access purposes, using roles and role-attributes is presented in [8]. Similar to our approach, they introduce purpose hierarchies in order to reason on access control. Their hierarchies are based on the principles of generalization and specification and are not expressive enough to support complex strategies defined by enterprises. However, we organize purposes into purpose directed graph through AND/OR decomposition, which supports the delegation of tasks and authorizations when a host of partners participating in the business process provides different ways to achieve the same service. We also present an efficient method to automatically derive the optimal way of authorizations needed to achieve a service from enterprise privacy policies.

The concept of Hippocratic databases, incorporating privacy protection within relational database systems, was introduced by Agrawal et al.[1]. The proposed architecture uses privacy metadata, which consist of privacy policies and privacy authorizations stored in two tables. LeFevre et al.[2] enhance Hippocratic databases with mechanisms for enforcing queries to respect privacy policies stated by an enterprise and customer preferences. In essence, they propose to enforce the minimal disclosure principle by providing mechanisms to data owners that control as who can access their personal data and for which purpose. Although the work on the Hippocratic databases[1,2] is closely related to ours, our approach has some notable differences. First, we introduce more sophisticated concepts of purpose, i.e., purposes are organized in purpose directed graph through AND/OR decomposition. The second difference is that Hippocratic databases does not allow to distinguish which particular method is used; whereas, we discuss the situations that could arise with web services and business process software. Third, we provide an efficient method to automatically derive the optimal way of authorizations needed to achieve a service from enterprise privacy policies.

7 Conclusions

In this paper, we analyze the purposes behind the design of Hippocratic database systems, and organize them in hierarchal manner through AND/OR decomposition. We apply the purpose directed graph to characterize the ways the enterprised need to achieve a service which may rely on many different partners. Specially, the selection of the partners and the identification of a particular plan to fulfill a purpose is driven by the customer's preference. We use a goal-oriented approach to analyze privacy policies of the enterprises involved in a business process, in which one can determine the minimum disclosure of data for fulfilling the root purpose with respect to customer's maximum trust. On the basis of the purpose directed graph derived through a goal refinement process, we provide efficient algorithms to determine the optimal privacy-aware path for achieving a service. This allows to automatically derive access control policies for an inter-organizational business process from the collection of privacy policies associated with different participating enterprises.

References

1. Agrawal, R., Kiernan, J., Srikant, R., Xu, Y.: Hippocratic Databases. In: Proceedings of VLDB 2002, pp. 143–154. Morgan Kaufmann, San Francisco (2002)
2. Agrawal, R., Evfimievski, A., Srikant, R.: Information sharing across private databases. In: Proceedings of SIGMOD 2003, pp. 86–97. ACM Press, New York (2003)
3. Ashley, P., Powers, C.S., Schunter, M.: Privacy promises, access control, and privacy management. In: Third International Symposium on Electronic Commerce (2002)
4. Backes, M., Pfitzmann, B., Schunter, M.: A toolkit for managing enterprise privacy policies. In: Snekkenes, E., Gollmann, D. (eds.) ESORICS 2003. LNCS, vol. 2808, pp. 162–180. Springer, Heidelberg (2003)
5. Bertino, E., Ferrari, E., Squicciarini, A.C.: Trust-X: A Peer-to-Peer Framework for Trust Establishment. IEEE Trans. Knowl. Data Eng. 16(7), 827–842 (2004)
6. Blaze, M., Feigenbaum, J., Lacy, J.: Decentralized trust management. In: Proc. IEEE Symp. Security Privacy, pp. 164–173 (1996)
7. Bresciani, P., Giorgini, P., Giunchiglia, F., Mylopoulos, J., Perini, A.: TROPOS: An agent-oriented software development methodology. JAAMAS 8(3), 203–236 (2004)
8. Byun, J.W., Bertino, E., Li, N.: Purpose based access control of complex data for privacy protection. In: Proceedings of SACMAT 2005, pp. 102–110. ACM Press, New York (2005)
9. Byun, J.W., Bertino, E., Li, N.: Purpose based access control for privacy protection in relational database systems. Technical Report 2004-52, Purdue University
10. Cranor, L., Langheinrich, M., Marchiori, M., Reagle, J.: The platform for privacy preferences 1.0 (P3P1.0) specification. W3C recommendation (2002), `http://www.w3.org/TR/P3P/`

11. Ferrari, E., Thuraisingham, B.: Security and privacy for web databases and services. In: Bertino, E., Christodoulakis, S., Plexousakis, D., Christophides, V., Koubarakis, M., Böhm, K., Ferrari, E. (eds.) EDBT 2004. LNCS, vol. 2992, pp. 17–28. Springer, Heidelberg (2004)
12. Finin, T., Joshi, A.: Agents, trust, and information access on the semantic web. ACM SIGMODRec 31(4), 30–35 (2002)
13. Giorgini, P., Massacci, F., Mylopoulos, J., Zannone, N.: Requirements engineering meets trust management. In: Jensen, C., Poslad, S., Dimitrakos, T. (eds.) iTrust 2004. LNCS, vol. 2995, pp. 176–190. Springer, Heidelberg (2004)
14. Giorgini, P., Massacci, F., Mylopoulos, J., Zannone, N.: Modeling security requirements through ownership, permission and delegation. In: Proceedings of RE 2005, pp. 167–176. IEEE Press, Lausanne (2005)
15. Handy, C.: Trust and the virtual organization. Harv. Bus. Rev. 73, 40–50 (1995)
16. Karjoth, G., Schunter, M., Waidner, M.: Platform for enterprise privacy practices: Privacy-enabled management of customer data. In: Dingledine, R., Syverson, P.F. (eds.) PET 2002. LNCS, vol. 2482, pp. 69–84. Springer, Heidelberg (2003)
17. Massacci, F., Zannone, N.: Privacy is linking permission to purpose. In: Proceedings of the 12th International Workshop on Sec. protocols (2004)
18. Nilsson, N.J.: Problem solving methods in AI. McGraw-Hill, New York (1971)
19. Nilsson, N.J.: Principles of Artificial Intelligence. Morgan Kaufmann, San Francisco (1994)
20. Seamons, K.E., Winslett, M., Yu, T., Yu, L., Jarvis, R.: Protecting privacy during on-line trust negotiation. In: Dingledine, R., Syverson, P.F. (eds.) PET 2002. LNCS, vol. 2482, pp. 129–143. Springer, Heidelberg (2003)
21. Tumer, A., Dogac, A., Toroslu, H.: A Semantic based Privacy framework forweb services. In: Proceedings of ESSW 2003 (2003)
22. Yasuda, M., Tachikawa, T., Takizawa, M.: Information flow in a purpose-oriented access control model. In: Proceedings of ICPADS 1997, pp. 244–249. IEEE Press, Lausanne (1997)
23. World Wide Web Consortium (W3C). A P3P Preference Exchange Language 1.0 (APPEL 1.0), `www.w3.org/TR/P3P-preferences`

Privacy-Preserving Clustering with High Accuracy and Low Time Complexity

Yingjie Cui, W.K. Wong, and David W. Cheung

Department of Computer Science
The University of Hong Kong
Pokfulam, Hong Kong
{yjcui,wkwong2,dcheung}@cs.hku.hk

Abstract. This paper proposes an efficient solution with high accuracy to the problem of privacy-preserving clustering. This problem has been studied mainly using two approaches: data perturbation and secure multiparty computation. In our research, we focus on the data perturbation approach, and propose an algorithm of linear time complexity based on 1-d clustering to perturb the data. Performance study on real datasets from the UCI machine learning repository shows that our approach reaches better accuracy and hence lowers the distortion of clustering result than previous approaches.

1 Introduction

Nowadays, as information accumulates rapidly, data mining - the technology of discovering knowledge out of information, becomes more and more prevalent. Clustering analysis is a commonly used data mining tool. It is a process of grouping a set of physical or abstract objects into classes of similar objects [5]. Given a similarity measure, it tries to maximize the intra-cluster similarity as well as minimize the inter-cluster similarity of data. K-means clustering is one of the most famous clustering problems. The objective of k-means clustering is to cluster a set of n objects into k partitions such that the average (Euclidean) distance from objects to its assigned cluster center is minimized where k is given by the user. K-means clustering has been applied in many applications in real life, such as customer behavior analysis, biology research and pattern recognition, etc.. Many of these applications are done over very large datasets, for example, millions of transaction records. Besides, k-means clustering is shown to be an NP-hard problem [4]. So, the heuristic algorithm proposed by Lloyd [13] is usually used in real applications. The algorithm adopts the hill climbing technique and returns a local optimum solution to the user.

The data used in the clustering process may contain sensitive information of an individual. For example, the financial transaction records held by a bank. Privacy issues are concerned on how the bank uses the information collected. The data owner (the bank) should not let others (the general public) observe the data (the clients' transaction records). On the other hand, there are many situations that there are other parties involved in clustering analysis. First, consider the

X. Zhou et al. (Eds.): DASFAA 2009, LNCS 5463, pp. 456–470, 2009.

situation that a law enforcement agency needs to investigate people's financial transactions which are divided between different financial institutions, e.g. banks and credit card companies. These institutions cooperate together to compute a global k-means clustering. The database in one financial institution should not be revealed to others. Consider another situation: as the development in cloud computing introduces the idea of software as a service (SaaS), a data owner can outsource the clustering task to a service provider in order to gain the benefits such as cost relief and payment on consumed resources only. The data owner sends his data to the service provider and executes an application at service provider. Since the service provider is a third party, it is not trusted. Privacy of individuals should be protected against the service provider. So, there is a need in studying privacy-preserving clustering problem in which the clustering result is found without accessing the original data.

Some algorithms have been proposed to address the privacy-preserving clustering problem. We describe two main approaches as follows:

1. Secure multiparty computation (SMC) approach: this approach addresses the privacy-preserving clustering problem in the multiparty case [18,8]. A number of data owners, each of them owns a database, cooperate together to compute global clustering result. The result is computed through certain rounds of complex communications among the data owners.
2. Perturbation approach: the data owner generates a perturbed database from the original database by adding noise to it [15,6,11]. The perturbed database can be accessed by any other parties. Hence, one can collect the perturbed databases he wants to perform clustering on and performs data mining on his own.

Although SMC approach provides accurate result and is provably secure, it cannot be applied to the general case of privacy- preserving clustering, for example the outsourcing scenario. In addition, there are considerable communication

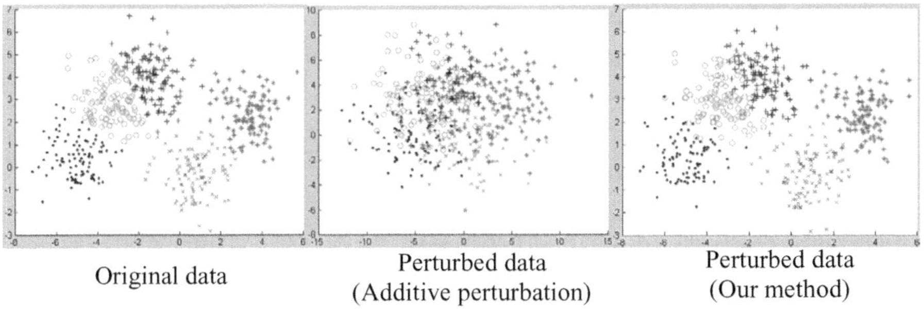

Fig. 1. An example illustrating the distortion of clustering result. The leftmost graph represents the clustering result on the original dataset. The clusters are represented in different colors. The graph in the center represents the disturbed points by using additive noise perturbation. The rightmost graph represents the disturbed points using our method.

overheads in the computation. So, we adopt the perturbation approach that can be applied in general case of privacy-preserving clustering and we can use existing algorithms for k-means clustering to compute the clusters. However, current perturbation approaches on privacy-preserving clustering are either insecure against a practical attacker [10] or destructive to the clustering result. Most of current studies focus on protecting the privacy of the data and do not consider the distortion to the mining result in the generation of noises. Figure 1 shows the distortion introduced to the clustering result by an existing perturbation method [6] and that by our method. If nosies are not added carefully, the noises can easily disturb the clustering result. In addition, a secure additive perturbation takes $O(nd^2)$ time, where n is the number of objects in the database, d is the number of dimensions of each object. It is a considerable high cost especially when the number of dimensions is high.

In this paper, we study the problem of privacy-preserving clustering and give a *secure* and *efficient* perturbation method in which the clustering result is *accurate*.

Contributions of this paper: we study the problem of privacy-preserving clustering using the data perturbation approach. Our contributions include: (1) we propose a new attack model to the distance-preserving perturbation and hence show that it is not secure even the third party obtains the distances between points from a black box oracle; (2) we propose a novel perturbation method using 1-d clustering to perturb the data in database which is secure, efficient and preserves accuracy of clustering result; (3) we give a theoretical study on the cost and security of the proposed technique; (4) we evaluate the proposed scheme with experiments on real datasets.

The rest of this paper is organized as follows. Section 2 mentions some related work. Section 3 defines the problem of privacy-preserving clustering and states the requirements on the solution. Then, we propose a new attack to distance-preserving perturbations in section 4 and hence show that distances between points cannot be revealed to attackers although distances alone do not reveal the location of points. Section 5 introduces our solution to this problem, and we perform experiments in Section 6 to compare the accuracy and efficiency of our solution with previous approaches. Section 7 concludes the paper and gives directions for future work.

2 Related Work

[15] first addressed the problem of privacy-preserving clustering. They proposed geometric transformations, which include shift, scaling and rotation, to disguise the original data. The geometric transformations used in [15] except scaling can be summarized as distance-preserving transformations. By distance-preserving transformations, we mean that $|x - y| = |T(x) - T(y)|$ for all x and y in the dataset, where T denotes the transformation.

Since all the pairwise distances are preserved, this kind of transformation always preserves the accuracy of the clustering result. The privacy of distance-preserving transformation has been studied by [10]. In algebra [2] this kind of transformation is called rigid motion, and can be represented by $T(x) = Mx + v$, where M is an orthogonal matrix and v is a vector. The *known input-output* attack assumes that some of the original data records are leaked and thus the attacker knows them. Besides, the attacker also has the released data, and knows the correspondence of the leaked original data and the transformed data. So the attacker's task is to solve M and v using some linearly independent *known input-output* points to set up linear regression equations. If $v = 0$, knowing d linearly independent points is enough for the attacker to solve M thus recover the whole original dataset. If v is not equal to 0, one more point is needed. This infers the safety bound of distance-preserving transformation against regression attack is not high.

[15] also proposed an additive random perturbation to improve the privacy by adding normally or uniformly distributed noise to the sensitive numerical attributes, which can be represented as $y_i = x_i + r_i$. Using this additive perturbation approach can raise the difficulty for the attacker to recover the original data. However, it can largely distort the clustering result. Besides, it is susceptible to data reconstruction methods such as PCA and Bayesian estimation [6], which can filter out much of the noise if the data is highly correlated. When the correlation of the data is high, the information of data concentrates in several dimensions that have higher variances than others. If the additive noise is evenly distributed in all dimensions, it can be largely filtered out if the attacker applies PCA to reduce dimensions that has small variances, while the information of the data is still approximately preserved. So [6] proposed a method to add noise for better resistance against this eigen-analysis attack, using the same covariance for the noise as that of the original data. However, since [6] mainly focused on the privacy problem of data publishing, its method of perturbation did not consider the need of accuracy in clustering. In other words, the clustering result can be largely distorted. And we have conducted experiments to show this. Besides, the calculation of a covariance matrix of the original data requires the time complexity of $O(nd^2)$, where n is the cardinality of the dataset and d is the number of dimensions. This is a considerable cost for high dimensional data.

Another kind of perturbation is the multiplicative perturbation, which is based on the Johnson-Lindenstrauss lemma [9]: if data points in a high dimensional space are projected onto a space of lower dimension, the distances between the data points can be approximately preserved. Also, after the projection, high dimensional data can not be recovered from low dimensional data because of the loss of information. Thus projections can be applied to distance-based operations such as privacy-preserving clustering. The best (in terms of accuracy) and most commonly used dimension reduction technique is PCA. It can be used to select the dimensions that have greater variances than others, and then the data can be projected onto these dimensions. Its costs which include calculating and decomposing the covariance matrix and matrix multiplication require a time complexity of $O(nd^2) + O(d^3)$, which is not suitable for high dimensional data. So a computationally simpler but less accurate dimension reduction method, the random projection, was proposed [3]. This projection can be represented as $Y = XM$, where X is the $n * d$ original dataset, M is a $d * e$ randomized matrix, e is the dimension after reduction. The elements of M are i.i.d. samples of a certain distribution, and Y is the $n * e$ projected dataset. This projection method reduced the time complexity to be $O(nde)$. [11] used this random projection to perturb the original data for distributed privacy-preserving data mining.

[18] proposed a solution to the privacy-preserving clustering problem using a different approach. Data are partitioned according to different attributes and distributed to several parties. These parties together find the clustering result, but do not share their original data with each other. The privacy of this approach is preserved using Yao's framework of secure multiparty computation [19]. Since all the computations are done through encryption and decryption, no distortion of data happens, so the accuracy is preserved. There has been other work [8] which partitions the data in a different way and uses the same secure framework with [18]. This approach assumes several non-colluding parties to do the clustering task, which may be difficult to find in realistic situation. Besides, the bit communication cost among different parties in each iteration of k-Means algorithm is $O(nrk)$, where n is the number of data points, r is the number of parties and k is the number of clusters, which is not low.

3 Problem Definition

A data owner owns a database DB, which consists of n objects. Each object is represented by a d-dimensional point. The distance between two points $x = (x_1, x_2, \ldots x_d)$ and $y = (y_1, y_2, \ldots, y_d)$ is measured by Euclidean distance[1], i.e., $|x - y| = \sqrt{\sum_{i=1}^{d}(x_i - y_i)^2}$. The problem of k-means clustering is, given k as input, to partition the n into k partitions: $P_1, P_2, \ldots, P_k$, such that the intra-cluster variance is minimized. The intra-cluster variance is measured by $\sum_{i=1}^{k}\sum_{x \in P_i}(x_i - \mu_i)^2$, where μ_i is the cluster center of P_i and $\mu_i = \frac{\sum_{x \in P_i} x_i}{|P_i|}$.

In privacy-preserving clustering, a third party data miner requires computing k-means clustering on DB. Due to privacy of information, the data owner releases a perturbed database DB' to the data miner. DB' is computed by perturbing the original objects in DB. We model the perturbation as $DB' = \{y \mid \forall x \in DB, y = T(x)\}$, where T represents the perturbation process[2]. Our objective in this paper is to develop a transformation process T such that

1. T is efficient. The cost at the data owner side is low and should be lower than the cost of performing clustering on his own. Otherwise, our technique is not suitable in privacy preservation of outsourcing scenarios.
2. T is secure. An attacker who obtains DB' and some background information on DB cannot recover DB.

---

[1] There are other distance measures used in the literature, e.g., Manhattan distance. We focus to Euclidean distance in this paper because it is more popular in practice.

[2] The perturbation function T can be irreversible and non-deterministic.

3. T preserves the accuracy of the clustering result. The clustering result on DB' should be similar to that on DB.

4 Distance Based Attack

An ideal perturbation method is that we can preserve clustering results for any datasets. The major part of clustering algorithms is to compute the distances between points and compare the distances to partition the points. If the data miner can compute the distances accurately, he can compute the correct mining result. Otherwise, if the distances are disturbed by random noises, the correctness of the mining result can not be ensured. So, it is an intuitive idea to explore the feasibility of a perturbation scheme which the data miner can accurately compute the distance between any two points[3]. However, in this section, we will show a negative result that such perturbation scheme cannot be secure by proposing an attack to it.

Theorem 1. *A perturbation scheme T is* ***insecure*** *against a known input-output attack given T allows an attacker to observe the distances between points in the original space.*

Proof. Suppose an attacker has retrieved the perturbed database DB' and a black box oracle G. G lets the attacker compute the distance between two points x and y by $|x - y| = G(x', y')$, where x' (y' resp.) is the perturbed point of x (y resp.). In a *known input-output* attack, the attacker obtains a number of m original points z_i and the corresponding perturbed points z'_i in the database. In order to recover a victim point v' in DB' to v in DB, the attacker first computes the distances between v and every known point z_i. So, we can set up m quadratic equations[4]: $|v - z_i| = G(v', z'_i)$. Each of the equation forms a d-hypersphere in the original space and v lies on the intersection of the hyperspheres. In general, if $m \geq d + 1$, the intersection is in fact a point. So, the attacker can conclude the original value of v. □

We use an example to illustrate the proposed attack in 2-D space as shown in figure 2. The attacker knows a set of three points A, B and C in the DB. He also knows the corresponding perturbed points A', B', and C' in DB' respectively.

[3] This scheme covers the distance-preserving transformation, but is not restricted to it. For example, this perturbation scheme can also include encryption, with the help of a trusted secure device[1]. The data owner can encrypt the data and then send it to the miner along with the secure device, and the data miner then uses the secure device to decrypt the data and calculate the pairwise distances in the process of clustering.

[4] Some previous work has also proposed scaling in perturbation which the distances are not exactly preserved. In such case, we may replace $G(v', z'_i)$ by $sG(v', z'_i)$ in the equation where s is the unknown scaling factor. In general, the attack can still recover the points given the RHS of the equation is a polynomial of $G(v', z'_i)$, though the attacker may need more known points for solving the increased number of unknowns.

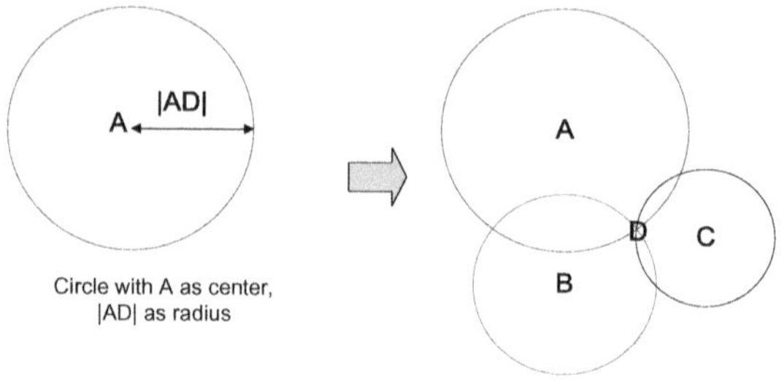

Fig. 2. An example illustrating the proposed attack. The attacker observes the original value of the point D by knowing A, B, C and the distances $|AD|$, $|BD|$, $|CD|$.

Suppose now he is trying to infer the original point D of a perturbed point D' in DB'. He calculates the distance of D to every known point using G. So, he gets $|AD|$, $|BD|$, $|CD|$. The attacker then draws three circles with A, B, C as the centers and $|AD|$, $|BD|$, $|CD|$ as the radii in the 2-D space. D must be on the perimeter of each circle. Now there is only one intersection of the circles. So, the attacker can conclude that the intersection in the example is D. Notice that even the attacker knows two points (say, A, B) only in the example, he can infer that D is one of the two intersections of the two circles. So, there is $\frac{1}{2}$ chance that the attacker obtains D by a random guess. This infers the safety bound of such perturbation scheme that preserves the pairwise distances is low, and leads us to develop a heuristic perturbation scheme in which distances between points are distorted but the clustering result is almost preserved.

5 Perturbation Based on 1-d Clustering

5.1 The Measure of Distortion

Previous perturbation methods use additive noise and multiplicative noise to improve the privacy. However, the additive noise can largely distort the clustering result, since it does not take into account the pairwise distances of the original data. The distortion of the clustering result can be measured by the difference between the clustering results of the original data and the perturbed data, and there are several metrics that have been proposed, such as Variation of Information (VI), Mirkin Metric and Van Dongen Metric. [14] studied these metrics, and proved that VI is a sensible metric for comparing clusters. As a metric of comparing the difference between two clustering results, VI is defined as the formula[5] below:

$$VI = -\sum_{i=1}^{k} p_i \log p_i - \sum_{j=1}^{k} q_j \log q_j - 2\sum_{i=1}^{k}\sum_{j=1}^{k} t_{ij} \log \frac{t_{ij}}{p_i q_j}$$

Here p_i ($i \in [1, k]$) is the proportion of the cluster i in one cluster result, i.e. p_i is the number of the elements in cluster i divided by the total number of data points. And q_j ($j \in [1, k]$) is the proportion of the cluster j in the other

[5] The base of the logarithm in this formula is 2.

cluster result. t_{ij} is the proportion of the same data points shared by p_i and q_j. This metric has a range [0, 2log$k$], and higher value means greater difference. Besides, since the k-means problem is NP-hard [4], much of the previous work has adopted the iterative approach proposed by Lloyd [13]. The result of this approach heavily depends on the initial choice of means, so the comparison of clustering results between original data and perturbed data should be restricted, for example, they should be with the same initial means set.

5.2 Our Solution

We observe that for multidimensional data, the clustering result of the whole dataset will not change too much if we perturb the data in such a way that the clustering of each single dimension is preserved and the range of each dimension is also preserved. Having this observation, we propose a randomized perturbation algorithm that is linear in time complexity and based on 1-d clustering.[6]

The details of the proposed algorithm is shown in Algorithm 1. We perform this algorithm on each dimension of the dataset, firstly calculate the 1-d clustering result and then perturb the data based on that clustering result.

The distance between two neighboring clusters C_1 and C_2 are calculated using the formula in [16]: $C_1.count * C_2.count * (C_1.mean - C_2.mean)^2/(C_1.count + C_2.count)$.

5.3 The Analysis of Time Complexity

Step 1 requires 1 scan of the data. Step 2 and step 3 require 1 scan of the data. After step 3, we get m clusters, and m is less than or equal to θ, the input number of initial intervals which is proportional to the cardinality of elements.

As for step 4 and step 5, we first look at the distance calculation and merging part. Since the number of clusters diminishes by a half after each round, the total step of these operations is the sum of a geometric progression with the proportion 1/2. So the total step of distance calculation is less than $2m - 1$, and the total step of merging is the half of that of distance calculation. Then we look at the finding of median of the distances. This can be done using a divide-and-conquer approach, and has been implemented in the STL of C++ as the function nth_element(), and its time complexity has been shown to be linear [7].

Step 6 is designed to control the number of final clusters, and its time complexity is $O(\mu\log\mu)$, where μ is the specified number of final clusters. The time complexity of step 7 is equal to one scan of the elements.

In conclusion, the time complexity of the above algorithm is linear with respect to the cardinality of data, i.e. $O(n)$. When applied to the whole dataset, the time complexity of this algorithm is $O(nd)$, where n is the count of the data points

[6] We use a heuristic to compute the clusters on one dimension but not the classic k-means clustering because we just need a approximate partitioning result and it is too expensive to execute classic k-means algorithms.

Algorithm 1. Perturbation Based on 1-d Clustering

Input: the data array of each dimension, the number of initial intervals θ, the number of final clusters μ.
Output: the perturbed data array.

PHASE 1, 1-d Clustering: partition the data on one dimension

1. Scan the data to find the maximum and minimum element of the data array.

2. Divide the range of data into θ intervals, with the length of each interval to be $(max - min)/\theta$, where max and min are the maximum and minimum elements of the array, and θ is an input number given by the user. Generally, θ is proportional to the data cardinality.

3. Project each data element to the intervals. Since elements are generally not evenly distributed, some intervals will have elements in them and some will not. Elements that belong to the same interval are treated as the initial clusters. The lower bound and upper bound of each interval that contains elements are used as the lower bound and upper bound of that cluster.

4. Calculate the distances between neighboring clusters.

5. Find the median of these distances, and merge the neighboring clusters that have distances smaller than the median. After merging, use the lower bound of the lower cluster as the new lower bound, and the upper bound of the upper cluster as the new upper bound.

6. Repeat step 4 and step 5 until the number of clusters is smaller than 2μ. Then execute step 4, sort the distances, merge the neighboring clusters that have the smallest distance and continue merging until the number of clusters is μ.

PHASE 2, Random Perturbation: Generate a random point in the partition as the perturbed point

7. For each cluster, calculate the distance between the lower bound and the mean and the distance between the upper bound and the mean. Use the smaller one of these two distances as the radius and the mean as the center to calculate a range. For each element in that cluster, generate a uniformly distributed random number in that range, and replace the original element with the random number.

---

and d is the dimension. So its scalability to large dataset with high dimension is good, compared with other approaches that require the time complexity $O(nd^2)$, $O(nd^2) + O(d^3)$ or $O(nde)$.

5.4 The Analysis of This Perturbation against Existing Attack Models

[6] proposed two attack models of the additive random perturbation. The first one is the eigen-analysis attack based on PCA, which tries to filter out the

random noise in case that the original data is highly correlated. And the second one is the Bayesian Attack which tries to maximize the probability $P(X|Y)$, where Y is the disguised data and X is the reconstructed data.

The key point of the first attack model is that the attacker can calculate the covariance matrix of the original data from the disguised data, which can be represented as the formula below:

$$Cov(X_i + R_i, X_j + R_j) = \begin{cases} Cov(X_i, X_j) + \delta^2, \text{ for } i = j \\ Cov(X_i, X_j), \text{ for } i \neq j \end{cases}$$

The assumption of applying this formula is that the attacker knows the distribution of the additive noise, i.e. he knows the variance or noise δ^2. However, in our approach, the original data is not additively perturbed, and the variance of the noise can not be separately learnt by the attacker.

The Bayesian attack is based on the assumption that both the original data and the randomized noise are subjected to multivariate normal distributions and the attacker knows this. This assumption is quite strong [12]. In our approach where the noise does not have such kind of distribution, this assumption can not be satisfied.

As for the *known input-output* attack and the distance based attack, since our approach changes the pairwise distances among data points, it can not be regarded as the orthogonal transformation, so the attacker can not use linear regression to recover the orthogonal transformation matrix, or use the pairwise distances to infer the original data.

In conclusion, our perturbation approach has good resistance against the above previous attack methods and our new attack model.

6 Experiments

In this section, we evaluate our proposed perturbation scheme in three aspects: privacy preserved, accuracy of clustering result, and execution cost. We compare the performance of our scheme against two existing approaches that are robust to all existing attacks: the additive random noise using the same covariance matrix as the original data [6], and the multiplicative perturbation, which includes the projection based on PCA dimension reduction and the random projection [11]. We denote the approaches as "Additive perturbation" and "Multiplicative perturbation" respectively.

There are different techniques in the multiplicative perturbation approach. The dimension reduction based on PCA is the most accurate dimension reduction technique, since the principle components captures the maximum possible variance [12]. In other words, PCA-based projection reaches the highest accuracy in the multiplicative perturbation approach. So we choose the PCA-based projection as the representative multiplicative perturbation to compare the distortion of clustering result with other approaches. Besides, as we have described in Section 2, the random projection outperforms PCA-based projection in time complexity. So we choose random projection to be the representative multiplicative perturbation when comparing the time complexity of different approaches.

6.1 Implementation Details

For the additive perturbation approach, we first calculate the covariance matrix of the original data, and then use the *mvnrnd* function of MATLAB to generate the multivariate random noise which has the same covariance matrix as the original data. There are no input parameters to this approach.

For the PCA-based projection for multiplicative perturbation, more noise will be introduced if more dimensions are reduced, but it also causes a higher distortion to the clustering result. We need to give the number of dimensions to preserve as the input to the algorithm. We pick the input such that the proportion of preserved variance is higher than 0.99.

For the random projection for multiplicative perturbation, the elements of the randomized projection matrix is chosen from a standard normal distribution $N(0,1)$. We use MATLAB to generate projection matrices of different numbers of dimensions, and then project the original dataset using these matrices. The algorithm requires the number of dimensions to preserve as the input. We will try different numbers of dimensions in the experiment for comparisons.

For the perturbation based on 1-d clustering, the number of initial intervals θ and the number of final clusters μ are the input to the algorithm. The number of final clusters is a trade off between the accuracy and privacy: the more final clusters in each dimension, the lower distortion can be reached, and the less noise can be introduced into the original data. Here we choose μ for different datasets in a way to make the accuracy of clustering result comparable with other approaches. The number of initial intervals θ is also specified by the user, and here we set the numbers to be half of the data cardinality for all the datasets.

6.2 Experiment Settings

We have implemented the algorithms using C++ and MATLAB. The version of C++ compiler is gcc 3.4.2. All the experiments are performed on a PC with Intel Core 2 Duo CPU E6750 2.66G and 2G RAM. The operating system is Windows XP Professional Version 2002 SP2.

6.3 Datasets

We use 4 real datasets from the UCI machine learning repository [17]: Wine, Breast Cancer Wisconsin (Original), Statlog (Shuttle) and Musk (Version 2). The details of the datasets are shown in the table.

We will evaluate the privacy preserved, and accuracy of clustering result of all approaches on the Wine, Shuttle and BCW datasets. The Musk dataset has a relatively high number of dimensions, so it is used to compare the execution time in perturbing the data.

Table 1. Datasets used in the experiment

Dataset Name	Number of Records	Dimensions	Classes
Wine	178	13	3
Shuttle	58000	9	7
BCW	683	10	2
Musk	6598	166	2

6.4 Measurements of Performance

For each of the dataset, we perturb it using the three approaches: additive perturbation and multiplicative perturbation and our approach using 1-d clustering perturbation. So, we obtain the perturbed datasets for each of the approach. Then, we perform k-means clustering algorithm by Lloyd [13] over the original dataset and different perturbed datasets. Since the initial cluster centers may affect the output of k-means clustering algorithm, we use the same initial centers for all cases. The measurement of the approaches are defined as follows:

Privacy Preserved. Privacy is measured as the amount of difference between the original dataset and the perturbed dataset [6]. We use the mean square error (MSE) as the measure of difference between original and perturbed dataset, which is the same as [6]. MSE can be calculated as MSE $= \sum_{x \in DB} |T(x) - x|^2$, where $T(x)$ represents the perturbed point of x. If an approach has a smaller MSE, the perturbed data is very similar to the original data. So, an attacker can approximately acquire the sensitive information. More privacy is preserved in the scheme that has a higher MSE.

Accuracy of Clustering Result. We use Variation of Information (VI) as described in section 5.1. A smaller VI represents less difference between the clustering results on original dataset and that on the perturbed dataset. An ideal situation is VI$=0$ which means the clustering results are the same.

Execution Cost. We measure the total execution time of each of perturbation approach in generating the perturbed dataset from the original dataset.

6.5 Experiment Results on Privacy Preserved and Accuracy of Clustering Result

The experiment results on the 3 datasets: Wine, BCW, Shuttle are shown in table 2, table 3, table 4 respectively.

In the experiments, our proposed algorithm outperforms both of the existing approach except that MSE of additive perturbation is larger (preserves more privacy) than our approach in BCW and Shuttle datasets. However, VI of addictive perturbation approach on these two datasets are extremely high for these

Table 2. Experiment Results on Dataset: Wine

	Parameters	VI	MSE
Multiplicative perturbation	1 dimension	0	14.5115
Addictive perturbation		0.0839841	16.5334
1-*d* Clustering Perturbation	$\mu = 8*3$, $\theta = 0.5*178$	0	53.0283

Table 3. Experiment Results on Dataset: BCW

	Parameters	VI	MSE
Multiplicative perturbation	1 dimension	0	7.06364
Addictive perturbation		1.708	4.00013e+010
1-*d* Clustering Perturbation	$\mu = 3*2$, $\theta = 0.5*683$	0	2.06604e+009

two datasets. Note that VI represents the accuracy of clustering result and is bounded by $2logk$. In BCW dataset, VI is bounded by $2log2 = 2$. So, additive perturbation gives nearly the worst mining result of clustering (1.708). In Shuttle dataset, VI is bounded by $2log7 = 5.61$. So, addictive perturbation gives a very bad mining result too (3.7063). Similarly, VI of addictive perturbation is poor in the Wine dataset. So, although addictive perturbation can preserve more privacy, the noises introduced are too large that it heavily damages the clustering result. We remark addictive perturbation is not able to preserve the clustering result and hence is not suitable in the problem of privacy-preserving clustering.

On the other hand, multiplicative perturbation has shown a comparable (a little bit higher in average) VI compared to our approach. It is because the principle components with large variants are preserved in the multiplicative perturbation approach. However, the obtained privacy in this approach, which is measured by MSE, is much smaller than the other approaches. If we try to raise the privacy by reducing more dimensions, the accuracy will be affected.

6.6 Experiment Results on Execution Time

The experiment results of perturbation time complexity on the Musk dataset is shown in table 5. It shows that our proposed approach is the fastest among the three approaches. It is because we have the lowest time complexity which is linear to the number of objects in the dataset and the number of dimensions. If the number of dimension is increased, the difference in execution time will be

Table 4. Experiment Results on Dataset: Shuttle

	Parameters	VI	MSE
Multiplicative perturbation	4 dimension	1.36994	53.8315
Addictive perturbation		3.7063	6317.08
1-*d* Clustering Perturbation	$\mu = 4*7$, $\theta = 0.5*58000$	0.990317	113.62

Table 5. Experiment Results on Dataset: Musk

	Parameters	Time(s)
Multiplicative perturbation	120 dimensions	2.828
	80 dimensions	2.344
	40 dimensions	1.875
	20 dimensions	1.626
	10 dimensions	1.497
Additive perturbation		5.031
1-d Clustering Perturbation	$\mu = 3*2$, $\theta = 0.5*6598$	1.341

further widened. Multiplicative perturbation has a comparable execution time when the number of dimensions preserved is reduced to a smaller value like 10. It is because the time complexity of random projection technique is $O(nde)$, where e is the number of dimensions preserved. So, when e is 10, the performance will be comparable to our proposed algorithm. However, as more dimensions are reduced, more information is lost in the original dataset, which will result in poorer clustering result. Note that the random projection itself is a less accurate projection method than the projection based on PCA, and in the previous section we have done experiments to show the accuracy of the PCA-based projection.

7 Conclusions and Future Work

In this paper we proposed a solution with high accuracy and low time complexity to the problem of privacy-preserving clustering. Besides, we proposed a new distance-based attack model to the distance-preserving perturbation, which strengthened our motivation to find solutions using perturbations that do not preserve pairwise distances. Previous approaches such as perturbation using additive random noises can largely distort the clustering result. In order to improve the accuracy, our approach takes into account the distribution of the original data by doing 1-d clustering on each dimension and then perturbs it using random noise. Another drawback of previous approaches is the high time complexity, especially when dealing with high dimensional data. Our approach is linear with respect to both the cardinality and dimensionality, i.e. $O(nd)$, thus its scalability to large and high dimensional dataset is good. The performance study on real datasets shows our approach reaches good accuracy and causes low time overhead compared with previous approaches.

As future work, we plan to study the applicability of our approach to general problems of data publishing, such as range queries and aggregate queries.

References

1. Agrawal, R., Asonov, D., Kantarcioglu, M., Li, Y.: Sovereign Joins. In: IEEE ICDE (2006)
2. Artin, A.: Algebra. Prentice Hall, New Jersey (1991)

3. Bingham, E., Mannila, H.: Random Projection in Dimensionality Reduction: Applications to Image and Text Data. In: ACM SIGKDD (2001)
4. Drineas, P., Frieze, A., Kannan, R., Vempala, S., Vinay, V.: Clustering Large Graphs via the Singular Value Decomposition. Machine Learning 56, 9–33 (2004)
5. Han, J., Kamber, M.: Data Mining: Concepts and Techniques. Morgan Kauffmann, San Francisco (2001)
6. Huang, Z., Du, W., Chen, B.: Deriving Private Information from Randomized Data. In: ACM SIGMOD (2005)
7. ISO/IEC 14882: 2003(E), p. 562 (2003)
8. Jagannathan, G., Wright, R.: Privacy-Preserving Distributed k-Means Clustering over Arbitrarily Partitioned Data. In: ACM KDD (2005)
9. Johnson, W., Lindenstrauss, J.: Extensions of Lipshitz Mapping Into Hilbert Space. In: Proc.of the Conference in Modern Analysis and Probability. Contemporary Mathematics, vol. 26, pp. 189–206 (1984)
10. Liu, K., Giannella, C.M., Kargupta, H.: An attacker's view of distance preserving maps for privacy preserving data mining. In: Fürnkranz, J., Scheffer, T., Spiliopoulou, M. (eds.) PKDD 2006. LNCS, vol. 4213, pp. 297–308. Springer, Heidelberg (2006)
11. Liu, K., Kargupta, H.: Random Projection-Based Multiplicative Data Perturbation for Privacy Preserving Distributed Data Mining. IEEE TKDE 18(1) (2006)
12. Liu, K., Giannella, C., Kargupta, H.: A survey of Attack Techniques on Privacy-Preserving Data Perturbation Methods. In: Privacy-Preserving Data Mining: Models and Algorithms (2007)
13. Lloyd, S.: Least Squares Quantization in Pcm. IEEE Transactions on Information Theory IT-28(2), 129–136 (1982)
14. Meila, M.: Comparing Clusterings - An Axiomatic View. In: ICML (2005)
15. Oliveira, S., Zaiane, O.: Privacy Preserving Clustering By Data Transformation. In: Proceedings of the 18th Brazilian Symposium on Databases, pp. 304–318 (2003)
16. Ward Jr., J.: Hierarchical Grouping to Optimize an Objective Function. Journal of the American Statistical Association 58(301) (1963)
17. UCI Machine Learning Repository, `http://archive.ics.uci.edu/ml/`
18. Vaidya, J., Clifton, C.: Privacy Preserving K-Means Clustering over Vertically Partitioned Data. In: ACM SIGKDD (2003)
19. Yao, A.: How to Generate and Exchange Secrets. In: Proceedings of the 27th IEEE FOCS, pp. 162–167 (1986)

Efficient and Anonymous Online Data Collection

Mafruz Zaman Ashrafi and See Kiong Ng

Institute for Infocomm Research, A*STAR, Singapore

Abstract. The wide adoption of the Internet has made it a convenient and low-cost platform for large-scale data collection. However, privacy has been the one issue that concerns Internet users much more than reduced costs and ease of use. When sensitive information are involved, respondents in online data collection are especially reluctant to provide truthful response, and the conventional practice to employ a trusted third party to collect the data is unacceptable in these situations. Researchers have proposed various anonymity-preserving data collection techniques in recent years, but the current methods are generally unable to resist malicious attacks adequately, and they are not sufficiently scalable for the potentially large numbers of respondents involved in online data collections. In this paper, we present an efficient anonymity-preserving data collection protocol that is suitable for mutually distrusting respondents to submit their responses to an untrusted data collector. Our protocol employs the onion route approach to unlink the responses from the respondents to preserve anonymity. Our experimental results show that the method is highly efficient and robust for online data collection scenarios that involve large numbers of respondents.

1 Introduction

Decision makers today are increasingly relying on large databases to guide their decision making. Many corporations now regularly collect large amounts of data in the context of their market research, as do government agencies for various national interests. This has resulted in many commercial entities that are engaged in the collection, packaging and merchandising of data as their principal business.

The phenomenal adoption of the Internet has made it possible for substantial amounts of data to be collected with incredible ease via online data collection [8,14]. However, in the light of the recent alarming rise in privacy breaches [14], the trust between the online data collector and the individual is diminishing. Individuals are understandably reluctant to share their data online, or when they are compelled to do so, simply provide incorrect data [2-4], leading to inaccurate data mining. In order to successfully exploit the internet as an effective platform for large-scale data collection, it is therefore necessary for online data collections to be anonymity-preserving, resistant to malicious attacks, and at the same time scalable to handle large numbers of respondents.

In this paper, we consider a data collection scenario in which a data miner (assume untrusted) wishes to collect potentially private and sensitive data from a large number of mutually distrusting individuals for use in a data mining experiment. For

X. Zhou et al. (Eds.): DASFAA 2009, LNCS 5463, pp. 471–485, 2009.

example, the human resource department of a large organization may wish to poll its employees for sensitive information (e.g. how satisfied they are with respect to their bosses' management skills) in a corporate climate survey, or a government agency may wish to poll the people's opinions on a newly introduced policy. While the individuals may be willing to provide their data truthfully because of the potentially beneficial outcomes of the data collection exercise, it is important to ascertain that their responses will be used in an aggregate study and that they cannot be linked back to them. In fact, they are understandably distrusting of the data collector and even their fellow respondents.

The conventional approach is to employ a trusted third party to collect the data. However, from the respondent's perspective, the notion of someone (no matter how trusted) who holds the links of the data back to the respondents [1] is too risky to be acceptable. To overcome this, cryptographic and random shuffling based techniques such as [2, 3] have previously been proposed to unlink the data from the respondents. Unfortunately, the current techniques are limited in their scalability for handling large-scale online data collection. For example, in the method proposed in [2], multiple rounds of communication and sequential processing are required: the data miner collects encrypted responses from each respondent and then sends the set encrypted responses back to each respondent for decryption and re-shuffling, one respondent by one respondent. Assuming a typical 1024-bit public key and that each original response is only 1000 bits long, it will take about 7.5 milliseconds to strip off an encryption layer [3]. This means that a data collection exercise of 100,000 respondents using the method could take up to 863 days, even without considering the time needed for additional modular exponentiations for larger responses, communication time, etc.

1.1 Our Solution

Both efficiency and anonymity are two key issues that should be taken into consideration when designing protocols for online data collection. The protocol should work well in a setup in which the respondents are mutually untrusting and the data collector is also untrusted. The protocol should be robust against malicious attacks, and allow the data collector to obtain all the responses exactly as sent by the honest respondents (ensures data accuracy) without linking them back to the respondents.

We propose a protocol based on the Onion Route technique and incorporate additional techniques to address the specific needs of anonymous online data collection mentioned above. Our major contributions can be outlined as follows:

1. *Collusion Resistance.* Our proposed protocol preserves a respondent's anonymity regardless of whether the data collector is malicious or not. Even if a malicious data collector collaborates with a set of malicious respondents, the actual respondent will still appear no more likely to be the originator of the response than the remaining honest respondents.
2. *Tamper Resistance.* Our proposed protocol maintains the integrity and secrecy of a respondent's response in a strong adversarial model. We incorporate an

anonymous digital signature to make it impossible (unless her key is compromised) for an adversary to change an honest respondent's response.
3. *Efficiency.* Our proposed protocol requires much less computation and communication than other anonymity preserving data collection protocols. The entire data collection process takes only a fraction of the time and require less communication cost in compare to the other state-of-the-art methods.

2 Related Works

As mentioned, the privacy of the data provider is an important issue and various anonymity-preserving data collection models have been proposed [2-5, 8-12, 15]. We discuss some of the current techniques below.

Randomized response: The respondents are required to perturb some of their sensitive response before submitting to the data miner. For example, for a query expecting a binary answer (e.g. what is a respondent's gender), the respondent could flip a bias coin and alters her response if the coin comes up with the head. The parameters of the perturbation process (e.g. the bias) are shared with the data miner, so the data miner could modify his algorithms accordingly so as to discover results comparable to that with the original (i.e. non-perturbed) responses. Privacy is protected indirectly as the data collector can never be certain about the truthfulness of a response [8,14]. The more perturbed the data are, the more privacy is guaranteed. However, there is an inadvertent tradeoff in the data mining accuracy because of the noise that had been introduced into the data. Our technique proposed in this paper is designed to provide the data miner with exact, unperturbed responses. This way, there is no need to modify any data mining algorithms, and there is no sacrifice of the accuracy of the results for preserving anonymity.

Cryptographic techniques: A typical cryptographic approach can be found in [2]. Each respondent encrypts her response using a set of public keys before sending it to the data collector. The data collector forwards the set of responses back to each respondent sequentially (i.e. one by one), who will strip off a layer of encryption from each of the responses, shuffle the responses and transmit them back to the collector who will then send them to the next responder for similar processing. At the end of this sequential randomization process, the data miner sends the randomized responses (those that data miner received from the final responder) to all responders for verification[1]. Such methods are not amenable to large-scale data collection because of the shuffling of the encrypted response set by each respondent sequentially. Moreover, despite the encryption and shuffling, the methods are not collusion resistant. A dishonest data miner could easily reveal the identity of an honest respondent if it collaborates with other dishonest respondents during the course of the protocol. In this work, we make sure that our proposed technique is efficient and at the same time collusion resistant.

[1] The protocol presented in [3] also employs similar techniques as [2] but it requires a higher communication overhead to achieve the same level of anonymity.

Mix networks: Recently, mix networks such as onion route [5, 11, 26], Crowds [13], k-anonymous message transmission [10], Hordes [11], and DC-nets [6] have been devised for anonymous communications over public networks by hiding the identities of the message senders. We believe that such anonymous communication networking techniques can also be deployed for our data collection scenarios, but there are additional concerns that need to be addressed. For example, a malicious respondent may exploit the anonymity of the communication channel to send multiple spurious responses or even replace the honest respondents' responses sent to the data miner. In this work, we employ an onion routing approach to unlink responses from the respondents and show how to incorporate various mechanisms to address the inherent weaknesses of mix networks for online data collection.

3 Preliminaries

Our online data collection scenario involves an untrusted data miner (i.e. data collector) who wishes to collect private or sensitive data from N mutually distrusting individuals. Each of the respondents R_i submits her private response m_i to the data miner whose purpose is to apply data mining algorithms on the collective data submitted to find useful patterns (we will henceforth refer to the data collector as the data miner). Our aim is to devise a setting in which to allow the data miner to collect the respondents' private data $(m_1, m_2, \cdots, m_N)$ accurately and without the risk of disclosing the link between a response m_i and the respondent R_i even in adversarial situations.

The Onion Route was an infrastructure designed to provide anonymous private connections over a public network for applications such as anonymous web browsing. An onion routing network consists of a set of onion routers (ORs) that provides a way for two parties to communicate anonymously without disclosing the address of the connection initiator. The data sent through the onion routers are encrypted using a layered data structure known as the *onion*. The onion layers are encrypted several times using a number of public keys, and only the intended recipients are able to decrypt the outermost layer with their private key to find the next recipient of that message. The recipient of an onion only knows the previous and next recipients (i.e. one hop). As such, any recipient along the path is unable to determine the originator of the onion. Each of the intermediate recipients decrypts only a layer of an onion, the content remains unknown until it reaches the final destination.

As the onion route is designed to enable anonymous communication between two parties over the public network, it is conceivable that we can use it for anonymous data collection. However, a simple protocol that directly uses the onion route for anonymous data collection is inadequate to meet the potential threats and challenges in online data collection scenarios. As the onion routing technique hides the identities of the respondents of all responses, if a malicious respondent sends multiple spurious responses, the data miner will not be able to distinguish hers from her honest peers' . The data miner is also vulnerable to the *man-in-the-middle* attack in which an adversary replaces an honest respondent's response as it is impossible for the data miner to refer the received response back to the rightful respondent to verify the inaccurate response. An honest respondent can also be handicapped in such a setting as it is impossible for her to know whether dishonest respondents are colluding with each other or are posing any passive attack which may subsequently expose her identity to the data miner.

Furthermore, the communication pattern in the data collection scenario is *many-to-one*, but the onion route is primarily designed for many-to-many anonymous communication scenarios. When an honest respondent sends an onion to a network path each of the intermediate hops of that path is aware of this honest respondent's intention. It is therefore prone to an eavesdropping attack. For example, an adversary may eavesdrop and discover the identity of an honest respondent if it manages to take control of a portion of the network and monitors the communication between the participants [10]. K-anonymous methods have been proposed to overcome such attacks, but the communication cost of k-anonymous message transmission is cubic to the number of respondents. This renders it infeasible for handling large numbers of participants [3], as in the case with our online data collection scenario.

4 Proposed Method

We propose a data collection method that can make use of the onion route to hide the respondents' identities effectively and efficiently. We introduce two techniques to address the shortcomings of the simple Onion route approach discussed in the previous section. Firstly, we incorporate a digital signature scheme to reduce the man-in-the-middle attack and to disallow spurious responses from malicious respondents by ensuring that no respondent is able to send more than one response. The data miner can be assured that all responses are coming from legitimate respondents, while at the same time, an honest respondent can also decide whether she will continue or abort the protocol before sending a response to the data miner, based on whether she has found any discrepancies herself. Secondly, we introduce a novel method for each respondent to evaluate the honesty of her peers before selecting a network path to send a response to the data miner. The method involves pre-computing a reputation matrix designed to identify adversarial peers (e.g. those who refuse to relay the onion) so that a respondent can exclude them from her subsequent choice of network path to send a response to the data miner.

4.1 Assumptions

Before we go into the details of our proposed protocol, let us outline our assumptions on the respondents and the data miner:

1. Prior to exchanging a response with the data collector, each of the participants (the respondents and the data miner) is aware of each other's primary

public keys. This is a standard assumption of any anonymity preserving data collection protocols [2-3].
2. The respondents are able to choose network paths with at least two intermediate ORs. A path with less than two ORs will not be sufficient to protect a respondent's privacy as each participant in an onion routing network path knows the IP addresses of its previous and next recipients.

4.2 The Techniques

As mentioned, we incorporate two techniques into our proposed protocol to enable the onion route for anonymous data collection. The next subsection describes our incorporation of a digital signature scheme to ensure anonymous authenticity in the responses, while the subsequent subsection will describe our novel use of circular onion paths to compute reputation matrices, and how the respondents can use these matrices to construct reliable linear onion paths in the network for sending their responses to the data miner.

Anonymous Authenticity. In the direct deployment of the basic onion route technique for anonymous data collection, the data miner is unable to recognize whether a malicious participant had modified an honest respondent's reply or not, since there is no way for it to refer the messages back to the originators for verification. We incorporate a digital signature scheme into our proposed protocol for authentication to avoid such *man-in-the-middle* attacks. Each respondent generates a secondary asymmetric key pair $P_{Secondary}\{x_i, y_i\}$ and sends the secondary public key x_i anonymously to the data miner before sending any response. The data miner can consider each of these public keys as a password and use it to verify a message's authenticity. Note that since the data miner receives all the secondary public keys anonymously, he does not know the ownerships of the secondary public keys. Each of the respondents generates a digital signature of its response by using secondary private key and submits both the response and the digital signature to the data miner. The data miner will able verify the digital signature with exactly one of the secondary public keys that she received from the respondents. For example, a respondent R_i shall always digitally sign her message m_i using her secondary private key y_i, and the recipient the data miner m_i will then verify the digital signature v_i. We denote $v = (m)_y$ as a digital signature of message m using a private key y and $V_x\{m, v\}$ is a verification function of a digital signature using a public key x associated with a private key y.

As we shall describe later (in Section 4.3), the data miner anonymously receives the secondary public key from each respondent. It is unable to know which particular public key $\{x_i, e_i\}$ belongs to a respondent R_i. Thus, the message authentication process does not compromise the anonymity of an honest respondent R_i as the data miner is unable to associate m_i with R_i.

Reputation Profiling: A dishonest respondent in a network path may attempt to sabotage the communication protocol by dropping or tampering with some of the data packets of an honest respondent before it reaches the data miner. Alternatively, a malicious peer who controls a portion of her network path may attempt to discover the originator of the data packets. However, if an honest respondent is able to know about its peer respondents' integrity, she can avoid choosing those dishonest respondents in her network paths. We propose a novel path selection process to assist an honest respondent for excluding such malicious peers from the network path chosen. The process allows an honest respondent to determine the reputation of the peers in her network path prior to relaying her data packets through the peer respondents as an anonymising network path to the data miner.

The path selection process works as follows: first, a respondent R_i builds a set of *circular* onion paths that consists of varying numbers of intermediate hops (i.e. intermediate hops who will relay the response along the path). To conceal the real intention of respondent R_i, she may also include the data miner as an intermediate hop. When R_i forwards its data packets along the circular paths, the data packets will eventually return to R_i intact if all the intermediate hops in the path are honest. If R_i finds that her data packets have been tampered with or even dropped, she can infer that some of the intermediate hops along that circular path are dishonest. Such information are stored in a *reputation matrix* with which the respondent would use to choose which of her peers to avoid. The respondent then transmits her actual response via a *linear* path consisting of multiple intermediate hops having high reputation ratings in the reputation matrix. We describe the details of circular and linear path construction below.

Circular Path: To develop its own perspective of some (or all) of its peers' reputation, each of the N respondents constructs a reputation profile (matrix) of her peers using circular paths. Algorithm 1 (Figure 1) shows how each respondent constructs circular paths for reputation profiling. In Steps 1 and 2, a respondent R_0 randomly pools N' number of peer respondents to construct n circular paths. To construct a circular path, R_0 randomly chooses k unique respondents from the set and then adds herself as the final hop (Steps 4 to 12). Note that the size of each circular path is randomly chosen (k in Step 4). This ensures that even if all the N' peer respondents were colluding, none of them will be able to figure out the path construction patterns. If the value of k were fixed for R_0, the colluding parties may figure out the value of k which may then be used to distinguish whether a current path is circular or linear later.

After constructing a set of circular paths in T, R_0 then generates a set of dummy responses $Q = \{q_i\,|1\ldots n\}$ of varying sizes to disguise R_0's real response which it will be submitting via the linear path (we will describe the construction of the linear path later). R_0 then generates the respective onions $Q_0 = \{q_{i_0}\,|1\ldots n\}$ for each dummy response q_i, and sends each onion q_{i_0} along the path t_i. The successes and failures of transmitting these onions are recorded in two $N' \times N'$ matrices M_S and M_F. The matrix M_S keeps track of the number of times R_0 successfully sends a response q_i through a circular path t_i. The matrix M_F keeps track of the corresponding number of times that a response q_j has failed to return to R_0 intact via a path t_j. Finally, R_0 then computes the reputation of a given peer $R_j \in \Psi$ using the conditional probability shown in the Equation 1 below.

Algorithm 1: Building Circular Path

1. $\Psi = \{R'_i \mid i = 1 \ldots N'\}$
2. T : Total Paths $T = \{t_1 \cdots t_n\}$
3. **for** $i = 1 \ldots \ldots n$
4. $\quad k = rand[2, N']$
5. $\quad t_i = \{\ \}$
6. $\quad$ **for** $j = 1 \ldots \ldots k$
7. $\quad\quad$ **do**
8. $\quad\quad\quad R_j = \Psi[rand[1, N']]$
9. $\quad\quad$ **while** $(R_j \notin t_i.prev)$
10. $\quad\quad t_i.\text{add}\{R_j\}$
11. $\quad$ **end**
12. $\quad t_i.\text{add}\{R_0\}$
13. **end**

Algorithm 2: Building Linear Path

1. $\Omega_{temp} = \Omega$
2. $M_{temp} = M_S$
3. $t_i = \operatorname{argmax}_j \Omega_{temp}(R_j)$
4. $\Omega_{temp}(R_{t_i}) = 0$
5. **for** $i = 2 \ldots \ldots k'$
6. $\quad C_{best} = \left\{ c \mid \Omega_{temp} R_c = \max_j \Omega_{temp}(R_j) \right\}$
7. $\quad t_i = \operatorname{argmax}_{c \in C_{best}} M_{temp}(t_{i-1}, c)$
8. $\quad \Omega_{temp}(R_{t_i}) = 0$
9. $\quad M_{temp}(t_{i-1}, t_i) = 0$
10. **end**

Fig. 1. Algorithms

$$\Omega(R_j) = \frac{\sum_{i=1}^{n} M_S[i,j]}{\left(\sum_{i=1}^{n} M_S[i,j] + \sum_{i=1}^{n} M_F[i,j]\right)} \tag{1}$$

Linear Path: To transmit its actual response, a respondent uses a linear path between the respondent and the data miner. Let us now discuss how each of the respondents constructs a linear path based on the reputation (Ω values) of N' of her peer respondents. Again, let us denote the original respondent (i.e. the initiator) as R_0 and let k' be the length of the linear path chosen randomly by R_0; $2 \leq k' < N'$. To construct the linear path for sending her response to the data miner, R_0 chooses the best peer respondents $\{R_1, R_2, ..., R_{k'}\}$ based on their reputation (Ω values) as follows. First, R_0 starts by selecting the best peer respondent R_j that has the maximum value in the Ω vector (Step 3). It then finds the subsequent $k' - 1$ peer respondents as follows. Let C_{best} be the set of candidates that having the next best Ω value (Step 6). As there may more than one such candidate, we pick the best peer $R_{j'}$ in the C_{best} based on the peer-to-peer relationship found in the M_S matrix (subsequently stored as M_{temp}), as shown in Step 7. We iterate this process until R_0 finds k' peer respondents for her linear path. R_0 can then create an onion using this linear path to transmit her response to the data miner.

4.3 The Protocol

Our proposed anonymous data collection protocol consists of four main phases. As mentioned, we assume that each respondent knows the data miner's public key and IP address prior to the execution of the protocol.

Phase 1: Registration

1. Each respondent R_i generates a primary key pair $\{P_i, S_i\}$. It sends P_i, its IP address A_i, and its digital signature v_i directly to the data miner for registration,where $v_i = \{P_i, A_i\}_{S_i}$.
2. The data miner then forwards the information to all other respondents.

Phase 2: Secondary Key Submission

3. Each respondent R_i generates a secondary public key $\{x_i, y_i\}$ and computes the hash of x_i as $h_i = h(x_i)$.
4. Respondent R_i forms a set of n_i circular paths $T = \{t_1, t_2, \cdots, t_{n_i}\}$ and $\forall t_j \in T$ it randomly chooses k_j intermediate hops as described in Algorithm 1.
5. For $j = 1 \cdots n_i$ respondent R_i encrypts a dummy data q_j and forms an onion using the primary public keys of the k_j intermediate respondents of a given network path $t_j \in T$:

   $m_i = E_{R_i}(A_i \parallel q_i)$
   $m_{i+1} = E_{R_1 \in t_j}(A_1 \parallel m_i)$
   ....
   $m_{k_j} = E_{R_{k_j} \in t_j}\left(A_{k_j} \parallel m_{k_j - 1}\right)$
6. R_i then forwards the onion m_{k_j} to the intermediate hop $R_{k_j} \in t_j$.
7. Upon receiving the message m_{k_j} the intermediate hop R_{k_j} uses her primary private key S_{k_j} to strip off the k_j^{th} layer encryption and obtains $m_{k_j-1} = D_{R_{k_j}}\left(m_{k_j}\right)$ which exposes the IP address A_{k_j-1} of the next intermediate hop. The intermediate hop R_{k_j} then forwards m_{k_j-1} to respondent R_{k_j-1} who is at IP address A_{k_j-1}. This process continues until the onion reaches the originating respondent R_i.
8. R_i evaluates $\Omega(R_j)$ s.t. $\forall R_j \in t$ where t is a circular path in T, using Equation 1. It then constructs a linear path t consisting of k intermediate hops based on the reputation profiles, constructs response packet α_i by concatenating a header ω_i (i.e. the random padding bits γ_i and its size, first 16 bits of ω_i represent the size of γ_i) and her secondary public key x_i, generates the onion and finally sends that onion to the data miner via the path t'.
9. The data miner uses its private key S_{DM} and decrypts: $\Psi_i = D_{S_{\rm DM}}[m_i]$. It then uses header information ω_i, unpacks the response packet α_i and finds the R_i's secondary public key x_i.

Phase 3: Verification

10. If all respondents have behaved honestly, then at the beginning of this phase, the data miner should be holding N secondary public keys from the N respondents. Otherwise the protocol aborts.
11. The data miner hashes the secondary public key set: $H_S[i] = h(x_i)$, $1 \leq i \leq N$. The data miner then forwards H_S to all respondents.
12. Each respondent R_i verifies whether its h_i as computed previously in Step 3 is included in H_S. If so, R_i sends an acknowledgment to the data miner. Otherwise, the protocol aborts.

Phase 4: Data Submission

13. Each respondent R_i constructs her response packet α_i by concatenating a header ω_i, her actual response r_i with the hashcode of its secondary public key h_i and a digital signature $v_i = \{r_i, h_i\}_{y_i}$. R_i also generates the hashcode $h(v_i)$ and stores it for future verification (Step 18).

14. R_i then forms a linear path and builds the onion in the same way as described in the Secondary Key Submission Phase (Phase 2, Steps 4 to 8) and sends the onion of her response packet α_i to the data miner.
15. The data miner decrypts the message using its private key S_{DM}, uses header information ω_i to unpacks the response packet α_i and then finds the corresponding secondary public key x_i using hashcode h_i.
16. The data miner verifies the digital signature: $V_{x_i}\{\alpha_i, y_i\}$=accept.
17. The data miner also verifies whether R_i has already submitted a valid response or not. If α_i is a valid first response, it updates x_i to the database of secondary keys received for verifying subsequent submissions.
18. Finally, the data miner broadcasts the hashed digital signature $h(v_i)$ to all respondents to let the originator R_i know that her response has been successfully received by the data miner.

4.4 Correctness and Integrity

Let us now show that how the data miner restricts a dishonest respondent R_i to submit more than one valid response to it to maintain correctness and integrity of the protocol.

Proposition: *At the end of the above protocol, the data miner collects N responses correctly.*

Proof: We show that the protocol allows exactly N valid secondary keys and exactly N valid responses are transmitted and received. First, the Verification phase (Phase 3) ensures that exactly N secondary keys are submitted and received. It achieves this by having the data miner first making sure that exactly N keys were received, and then forwarding H_S, the set of hashcodes of the secondary keys that it has received, to each respondent in $\{R_i \,|i = 1, 2, \cdots, N\}$ to verify her respective secondary key independently. If a respondent R_i finds that her key is absent, the protocol is aborted. This verifies that the N secondary keys that have been received by the data miner are valid. Next, the Data Submission phase (Phase 4) ensures that exactly N valid responses were submitted and received. First, the data miner ensures the correctness of the responses by accepting only a response packet α_i if and only if α_i has a valid digital signature v_i. This means that to submit a valid response to the data miner, a respondent has to use one of the N secondary keys that was previously transmitted. Let us suppose a dishonest respondent R_j tries to submit more than one response during this phase. Since the data miner also records each response against the secondary key x_i, any respondent submitting multiple responses using the same key would be detected. Thus, at the end of a successful protocol execution, the data miner will only have collected N valid responses correctly.

The integrity of an honest respondent's response may be compromised by malicious respondents. Let us consider a scenario in which some malicious respondents acting as intermediate hops attempt to modify an honest respondent's response. Recall that to send a response to the data miner, an honest respondent R_i initiates a linear path and then forwards her response along the path.

Suppose there exists a malicious respondent along $R_i's$ chosen linear path who wholly or partially replaces a data packet α_i before it forwards it to the data miner. In our protocol, such an attack will be detected because every data packet α_i has a digital signature v_i. As each response packet α_i is signed using a secret secondary private key y_i only known by R_i, a malicious respondent R_j will be unable to generate a valid digital signature for the tampered response. The only harm a malicious respondent R_j can do is to arbitrarily drop α_i. However, doing so does nothing to reveal the identity of R_i.

It is also possible that a malicious respondent may try to modify or even replace the secondary public key of an honest respondent during the Secondary Key Submission phase (Phase 2), since a respondent does not use any digital signatures when submitting her secondary public key to the data miner. However, in the subsequent Verification phase (Phase 3), each respondent is allowed to verify whether her secondary public key reaches to the data miner or not (i.e. treating the key as a password for authentication). As such, a malicious respondent forging an honest respondent's response using a modified or replaced secondary public key for the honest respondent would be detected.

4.5 Complexity

Let us now analyze the complexity of our proposed protocol in terms of computation and communication costs.

Computational complexity. In Phase 2, each respondent builds circular and linear paths that involve n number of peer respondents. Each respondent then constructs the onion and encrypts it n number of times. Similarly, upon receiving the onion, each of the respondents decrypts it once. While each respondent may be included as an intermediate hop in many network paths, on average, each of the respondents decrypts only n (i.e. total size of the network path $N * n$ divide by the number of respondent N) number of onions. Thus, the average computational complexity for each of the respondent is $O(n)$ where $n < N$.

In Phases 1 and 2 of our protocol, the data miner receives primary and secondary public keys from all respondents and it performs N decryptions. In Phase 3 every respondent verifies only the hashcode of her secondary primary key which does not require any decryption. Phase 4 of our proposed protocol works in exactly the same manner as Phase 2, with some extra computation that the data miner required for verifying the digital signature of each respondent's response. As each of the respondents generates one digital signature, the data miner verifies N digital signatures in total. Based on the above analysis, we can conclude the computational complexity of the data miner is $O(N)$.

Communication cost. In Phase 1, each respondent sends its public key and receives all the other participants' public keys and IP addresses. This requires 2 rounds of communication. Suppose each public key has k bits, each IP address has 32 bits and each digital signature has s bits, the total communication cost of this phase is $C_{P_1} = \left(kN^2 + 32N^2 + sN\right)$. In Phase 2, each respondent selects n respondents as intermediate hops and builds a set of network paths T to send its secondary public key to the data miner. It sends an onion to each of the elements of path T and each onion containing the secondary public key with size k', header information which is ω bits long, together with the n intermediate hop's IP addresses each of them is 32 bits. Thus, the total communication cost of this phase is equal to $C_{P_2} = \sum_{i=0}^{N} (k' + \omega + 32n) \times n$.

During the verification phase (Phase 3), the data miner sends the hashcode of each respondent's secondary public key to all respondents. The total number of bits transmitted during this phase is equal to $C_{P_3} = hN^2$, where h is length of the hashcode. Similarly, Phase 4 needs 1 round for data submission and 1 round for verification of successful data submission. If we assume header information is ω bits, each respondent's response as Y bits, the secondary hashcode as h bits, the corresponding digital signature as s' bits, then the total communication cost of data submission is equal to $\sum_{i=0}^{N} (\omega + Y + s' + h + 32n) \times n$ and the cost of successful verification (i.e. hashcode of digital signature) is equal to $h(s_i) \times N^2$, hence communication cost of this phase equal to $C_{P_4} = \sum_{i=0}^{N} (\omega + Y + s' + h + 32n \times n) + hN^2$. The total communication cost of our protocol is thus the aggregate communication cost of the four phases:

$$C_{total} = C_{P_1} + C_{P_2} + C_{P_3} + C_{P_4} \tag{2}$$

5 Performance Analysis

We perform two simulation experiments to evaluate the relative performance of our proposed protocol. First, we evaluate the robustness of our protocol by measuring how an honest respondent's degree of anonymity is affected against a set of malicious participants. Then, we compare the efficiency of our proposed protocol with the current state-of-the-art anonymity preserving data collection protocol.

5.1 Anonymity

We use information entropy [16] to measure the anonymity of an honest respondent (that is, how much information about an honest respondent and her response a dishonest data miner or a dishonest respondent may be able to obtain). Let N be the number respondents in the anonymity set A. To illustrate an honest respondent's anonymity in the worse case scenario, let us assume that for some reason the data miner was able to exclude k of the respondents from the initial anonymity set A such that it has a new anonymity set X where $X \subseteq A$ and $K = (N - k)$ size of the set X. The entropy of the system after the data miner has obtained this subset X is:

$$H(X) = \log_2(K) \tag{3}$$

We can now measure the degree of anonymity of an honest respondent using the above formula. The degree of anonymity in our data collection protocol should reflect how much information an adversary may gain with an attack for it to link a respondent with her response. We can measure this by calculating

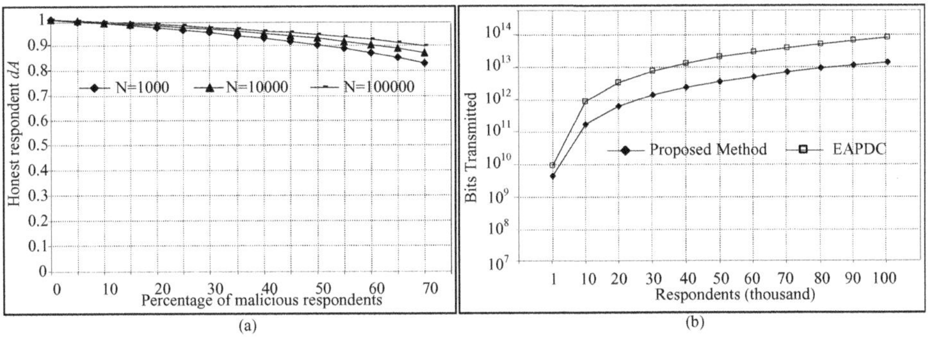

Fig. 2. (a)Anonymity Loss in Malicious Models (b)Comparative Communication Costs

the difference of entropy $|H_M - H(X)|$ before and after an adversary attack. We use the following formula to measure the degree of anonymity of an honest respondent on a [0-1] scale:

$$dA = 1 - \frac{H_M - H(X)}{H_M} = \frac{H(X)}{H_M} \tag{4}$$

We performed a simulation experiment to investigate how many corrupted respondents a data miner would need in order to guess the linkage between an honest respondent and its response with a significant probability. Figure 2(a) shows the effect on the degree of anonymity with increasing percentage of corruption (i.e. malicious collaborators) in varying numbers of total respondents. As shown in the figure, to reduce the degree of anonymity significantly, say to less than 0.9, the malicious data miner would need to collaborate with more than 45% of the total respondents when the respondent size N=1000. When the total size of respondents increases, the malicious data miner would need a significantly higher number of collaborators. For example, when there are 100,000 respondents, it would need at least 65% (that is, more than 65,000) malicious respondents to decrease the anonymity by only 15%, from 1.0 to near 0.85.

5.2 Efficiency

We compare our proposed protocol with the current best-known anonymity preserving data collection method, Efficient Anonymity-Preserving Data Collection (EAPDC) [2]. Note that although the protocol outlined in [3] is also similar to EAPDC, we do not include it in this comparison study because of the excessive communication overhead that it requires to achieve the similar level of anonymity as EAPDC. To compare the efficiency of the methods, we calculate the total number of bits each of those methods transmitted by varying the total number of respondents.

In our simulation experiment, we assume for both the protocols that the primary and secondary public keys that each respondent uses are 1024 bits, the data submitted by each respondent are 2000 bits long, and that all the respondents employ

128 bits long MD5 hashing. Each respondent in our proposed method adds random padding bits as header to her circular and linear path data packet. These bits are randomly chosen. In this study we assume the respondent header size is up to 1000 bits long, which implies that all of the responses the respondents send through circular and linear network paths are on average approximately 3000 bits long. For our proposed method we also vary the respondent's network path size (the total number of intermediate hops that each respondent uses to send her data to the data miner). We increase the total number of intermediate hops from 500 to 1500 as the respondents size increase from 1000 to 100000.

Figure 2(b) shows the comparison between the two methods. The results showed that our proposed onion-route based method transmitted a significantly lower number of bits of data(between 2 to 5.25 times less data) as compared to the EAPDC protocol.

Let us examine how we achieve the significant communication cost savings. For EAPDC, since each of the respondents receives all responses that they submitted to the data miner, who strips off a layer of encryption, shuffles each of the responses randomly and then sends these random responses back to the data miner, the respondents' responses are sent back and forth between the data miner and all respondents at least N times. This was not required in our proposed protocol. Furthermore, in EAPDC, once all respondents have performed their respective shuffling, the data miner broadcasts them to each of the respondents to verify those random responses. The data miner then receives a digital signature from each respondent if the verification succeeded and broadcasts those signatures to all respondents to collect the secondary secret keys from each of respondents. Our method requires neither digital signatures nor secret keys from each of the respondents. Instead, the data miner simply sends the hashcode of the responses to the respondents for verification. This significantly reduces the communication overheads.

Finally, it is worth mentioning that the data transmission ratio between our method and EAPDC increases as the number of respondents increases. For example, the ratio is 2.16 with 1000 participants, and it increases up to 5.85 when there are 100000 participants. This ascertains that our protocol is indeed much better suited for data mining scenarios involving large numbers of respondents.

6 Conclusion

Today's decision makers are increasingly turning to the internet as a convenient and low-cost option for large-scale data collection. Unfortunately, the inherent insecurity of the internet poses great privacy risks and does not encourage respondents to respond truthfully, especially when personal and sensitive information are involved. The current data collection methods are either inadequately resistant to malicious attacks, or not sufficiently scalable for the potentially large numbers of respondents involved in online data collections. Some of the methods even introduced inaccuracies into the data as a way to protect the respondents' anonymity.

As the output of data mining algorithms is highly dependent on both the quantity and quality of the input data collected, it is important to ensure that the respondents are able submit their responses truthfully and correctly. In this paper, we have described an onion-route based method for anonymity preserving online data collection. Our proposed protocol is able to maintain an honest respondent's anonymity and protect the integrity of her response robustly even in various malicious models involving dishonest data miner as well as colluding respondents. Our protocol is also highly efficient in terms of computation and communication costs. Our empirical performance studies showed that our method was able to reduce the message transmission requirements six-fold as compared to the current state-of-the-art protocol, making it highly amenable to meet the inevitable challenge of scale associated with online data collection.

References

1. Agrawal, R., Evfimievski, A., Srikant, R.: Information sharing across private databases. In: Proc. of the SIGMOD 2003 (2003)
2. Brickell, J., Shamatikov, V.: Efficient Anonymity-Preserving Data Collection. In: Proc. of the 12th ACM SIGKDD, pp. 76–85 (August 2006)
3. Yang, Z., Zhong, S., Wright, R.N.: Anonymity-preserving data collection. In: Proc. of the ACM SIGKDD, pp. 21–24 (August 2005)
4. Golle, P., McSherry, F., Mironov, I.: Data Collection With Self-Enforcing Privacy. In: Proc. of the ACM CCS, pp. 69–78 (2006)
5. Chaum, D.: Untraceable electronic mail, return addresses, and digital pseudonyms. Communications of the ACM 24, 84–90 (1981)
6. Chaum, D.: The dining cryptographers problem: unconditional sender and recipient untraceability. Journal of Cryptology 1, 65–75 (1988)
7. Warner, S.L.: Randomized response: A survey technique for eliminating evasive answer bias. The American Statistical Association 60, 63–69 (1965)
8. Evfimievski, A., Srikant, R., Agrawal, R., Gehrke, J.: Privacy preserving mining of association rules. In: Proc. of the ACM SIGKDD (July 2002)
9. Ambainis, A., Jakobsson, M., Lipmaa, H.: Cryptographic randomized response techniques. In: Bao, F., Deng, R., Zhou, J. (eds.) PKC 2004. LNCS, vol. 2947, pp. 425–438. Springer, Heidelberg (2004)
10. Ahn, L.V., Bortz, A., Hopper, N.J.: k-anonymous message transmission. In: Proceedings of the 10th ACM CCS (2003)
11. Levine, B.N., Shields, C.: Hordes: a multicast based protocol for anonymity. Journal of Computer Security 10, 213–240 (2002)
12. Reiter, M.K., Rubin, A.D.: Crowds: Anonymity for Web transactions. ACM Transactions on Information and System Security 1, 66–92 (1998)
13. Syverson, P.F., Goldschlag, D.M., Reed, M.G.: Anonymous Connections and Onion Routing. In: Proc. of the IEEE Symp. on S&P, p. 44 (1997)
14. Evfimievski, J.G., Srikant, R.: Limiting privacy breaches in privacy preserving data mining. In: Proc. of the 22nd ACM SIGMOD, pp. 211–222 (June 2003)
15. Dingledine, R., Mathewson, N., Syverson Tor, P.: Second Generation Data Mining Onion Route. In: Proc. of the 13th USENIX Security Symp. (2004)
16. Cover, T.M., Thomas, J.A.: Elements of Information Theory. John Wiley & Sons, Inc., Chichester (1991)

Decomposition: Privacy Preservation for Multiple Sensitive Attributes

Yang Ye[1], Yu Liu[2], Chi Wang[2], Dapeng Lv[2], and Jianhua Feng[2]

[1] Institute for Theoretical Computer Science, Tsinghua University
Beijing, 100084, P.R.China
yey05@mails.tsinghua.edu.cn
[2] Department of Computer Science, Tsinghua University
Beijing, 100084, P.R.China
{liuyu-05,wangchi05,lvdp05}@mails.tsinghua.edu.cn
fengjh@tsinghua.edu.cn

Abstract. Aiming at ensuring privacy preservation in personal data publishing, the topic of anonymization has been intensively studied in recent years. However, existing anonymization techniques all assume each tuple in the microdata table contains one single sensitive attribute (the *SSA* case), while none paid attention to the case of multiple sensitive attributes in a tuple (the *MSA* case). In this paper, we conduct the pioneering study on the MSA case, and propose a new framework, decomposition, to tackle privacy preservation in the MSA case.

1 Introduction

Anonymization[1] is the most popularly adopted approach for privacy-preserving data publishing. Anonymization techniques typically perform *generalization*[1,2] on QI attributes, as depicted in Table 3. Principles such as k-anonymity[1] put constraints on each QI-group. The widely adopted principle l-diversity[3] requires each group contains at least l "*well-represented*" sensitive values, and reduces the risk of *sensitive attribute disclosure* to no higher than *1/l*.

Current researches on anonymization all assume there is one single sensitive attributes (the SSA case) in the microdata table. This assumption is arbitrary. In the running example, two attributes, *Occupation* and $Salary$ are sensitive attributes. Consider an adversary who obtains the QI values $\{M, 10076, 1985/03/01\}$ of Carl. Given the published Table 3, s/he can locate Carl in the first QI-group. However, since the first two tuples of Group 1 have "$nurse$" as the occupation value and according to common sense, nurse is generally a female occupation, thereby the adversary can locate Carl in the last two tuples. S/he will be able to reveal with high confidence that Carl's monthly salary is 8000-10000 dollars (In tables of this paper, integer i in "salary" column means the monthly salary is between the range of $1000i - 1000(i + 1)$ dollars).

This paper provides the first study towards privacy preservation in the MSA case. We propose a new publishing methodology, decomposition, to achieve privacy preservation in the MSA case. Instead of performing generalization on QI

X. Zhou et al. (Eds.): DASFAA 2009, LNCS 5463, pp. 486–490, 2009.

Table 1. The Microdata Table

Tuple#	Gender	ZipCode	Birthday	Occupation	Salary
1(Alice)	F	10078	1988/04/17	nurse	1
2(Betty)	F	10077	1984/03/21	nurse	4
3(Carl)	M	10076	1985/03/01	police	8
4(Diana)	F	10075	1983/02/14	cook	9
5(Ella)	F	10085	1962/10/03	actor	2
6(Finch)	M	10085	1988/11/04	actor	7
7(Gavin)	M	20086	1958/06/06	clerk	8
8(Helen)	F	20087	1960/07/11	clerk	2

Table 2. Part of a Vote Register List

Name	Gender	ZipCode	Birthday
Alice	F	10078	1988/04/17
Betty	F	10077	1984/03/21
Carl	M	10076	1985/03/01
Diana	F	10075	1983/02/14
Ella	F	10085	1962/10/03
Finch	M	10085	1988/11/04
Gavin	M	20086	1958/06/06
Helen	F	20087	1960/07/11

Table 3. The Generalized Table

#	Gender	ZipCode	Birth.	Occ.	Sal.
1	*	1007*	1983-88	nurse	1
2	*	1007*	1983-88	nurse	4
3	*	1007*	1983-88	police	8
4	*	1007*	1983-88	cook	9
5	*	*008*	1958-88	actor	2
6	*	*008*	1958-88	actor	7
7	*	*008*	1958-88	clerk	8
8	*	*008*	1958-88	clerk	2

attributes and forming QI-groups, our technique decomposes the table into so-called *SA-groups*. To retain valuable information lost in the transformed sensitive attributes, the original *sensitive table* is also published without privacy leakage.

2 General Idea of Decomposition

We term our methodology "*decomposition*". Firstly, it publishes the decomposed sensitive table. Secondly, instead generalized on QI attributes, tuples are grouped properly. Their QI values remain unchanged while tuples within a group share the union of their sensitive values, as shown in Table 4 and Table 5.

Definition 1. (SA-group) *A SA-group G contains tuples with their original, non-transformed QI values and for each S^i, each tuple in G is associated with the set of $G.S^i$ values.*

We first assume there is one single sensitive attribute S and aim at achieving l-diversity. We shall research, given a diversity parameter l, how to decompose the table into SA-groups so that: (*i*) each group had better contains exactly l distinct sensitive values. (*ii*) the number of such SA-groups should be maximized. Following **Largest-l group forming Procedure** is adopted: first place tuples with identical sensitive values into a same "*bucket*". Let B_i denote the ith largest bucket and $\mathscr{B} = \{B_1, B_2, \ldots, B_m\}$ denote the set of buckets. We have:

Table 4. The Decomposed Table for Single Sensitive Attribute

#	Gender	Zip.	Birth.	Occ.
1	F	10078	1988/04/17	police
	F	10085	1962/10/03	nurse
	M	20086	1958/06/06	actor
	M	10076	1985/03/01	clerk
2	F	10077	1984/03/21	nurse
	M	10085	1988/11/04	actor
	F	10075	1983/02/14	cook
	F	20087	1960/07/11	clerk

Table 5. The Decomposed Table for Two Sensitive Attributes

#	Gender	Zip	Birth.	Occ.	Sal.
1	F	10078	1988/04/17	police	1 2 8
	F	10085	1962/10/03	nurse	
	M	20086	1958/06/06	actor	
	M	10076	1985/03/01	clerk	
2	F	10077	1984/03/21	nurse	2
	M	10085	1988/11/04	actor	4
	F	10075	1983/02/14	cook	7
	F	20087	1960/07/11	clerk	9

$n_i = |B_i|$, $n_1 \geq n_2 \geq \cdots \geq n_m$. In each iteration, one tuple is removed from each of the l largest buckets to form a new SA-group. Similar to [5], we can prove:

Theorem 1. *The Largest-l group forming procedure creates as many groups as possible.*

We shall also investigate: (i) in which case there will be not tuples left after the procedure; and (ii) what is the property of residual tuples, if any.

Theorem 2. *When the Largest-l group forming procedure terminates, there will be no residual tuples if and only if the buckets formed after the bucketizing step satisfy the following properties (we term it l-Property):*
(i) $\frac{n_i}{n} \leq \frac{1}{l}$, $i = 1, 2, \ldots, m$ *(Use the same notation:* n_i, m, n *as in Theorem 1);*
(ii) $n = kl$ *for some integer* k*.*

When the buckets formed through bucketization satisfy the first condition while do not satisfy the second condition of l-Property, we have following conclusion:

Corollary 1. *If the buckets satisfy:* $\frac{n_i}{n} \leq \frac{1}{l}$, *then when the Largest-l group forming terminates, each non-empty bucket contains just one tuple.*

Corollary 2. *The largest permissable assignment to the diversity parameter* l *is* $l_{per} = \lfloor \frac{n}{n_1} \rfloor$

The extension of Decomposition to the MSA case is intuitive. First, the sensitive table T^S is published. Next, one sensitive attribute (denoted S^{pri}), is chosen as the "*primary sensitive attribute*" and largest-l procedure is exerted on S^{pri} to form SA-groups.

Definition 2. *(*Primary Sensitive Attribute*) In the MSA case, the* primary sensitive attribute *is the sensitive attribute chosen by the publisher, according to which SA-groups are formed.*

Third, for each SA-group and each non-primary sensitive attribute, the original values are united up, as depicted in Table 5. Reduplicated values are counted once because multiple counts just increase the privacy disclosure risk. We should not assign a uniform l for all S^i. Instead, each S^i should have its own l_i.

Table 6. The Final Publishing of Decomposition

The Sensitive Table

Occupation	Salary
nurse	1
nurse	4
police	8
cook	9
actor	2
actor	7
clerk	8
clerk	2

The Decomposed Table after Adding Noise

Group#	Gender	ZipCode	Birthday	Occupation	Salary
1	F	10078	1988/04/17	police	1
	F	10085	1962/10/03	nurse	2
	M	20086	1958/06/06	actor	4
	M	10076	1985/03/01	clerk	8
2	F	10077	1984/03/21	nurse	2
	M	10085	1988/11/04	actor	4
	F	10075	1983/02/14	cook	7
	F	20087	1960/07/11	clerk	9

Definition 3. *($(l_1, l_2, \ldots, l_d)$-diversity) A decomposed table is said to satisfy $(l_1, l_2, \ldots, l_d)$-diversity, if for each of its SA-group G and each $i \in \{1, 2, \ldots, d\}$, $G.S^i$ contains at least l_i distinct sensitive values.*

As for some non-primary sensitive attribute S^i, there may be groups with less than l_i distinct S_i values, like in Group 1 of Table 5, $l_{per}(Salary) = \frac{8}{2} = 4$. For Group 1 to satisfy the privacy goal, some "*noise*" is added. In sum, the final publishing of *decomposition* is shown in Table 6.

3 Experiments

In the experiments, we utilized the "Adult" database from the UCI Machine Learning Repository (http://www.ics.uci.edu/mlearn/mlrepository.html) and the *KL-divergence* metric to measure data quality.

First we treat *Work-class* as the sensitive attribute and develop 4 tables from Adult: q-QI-Adult ($5 \leq q \leq 8$). q-QI-Adult takes the first d of other attributes as QI. We compare decomposition against the widely-adopted multi-dimensional generalization algorithm Mondrian[4] when achieving l-diversity. Figure 1 through Figure 4 depicts the KL-divergence of the anonymized datasets created by two algorithms. We also compare the execution time of both techniques. For lack of space, only the result on 8-QI-Adult is in Figure 5. Decomposition greatly outperforms generalization in both data quality and efficiency.

For the MSA case, we develop 4 tables: d-SA-Adult ($1 \leq d \leq 4$). d-SA-Adult uses the first 5 attributes as QI attributes and the subsequent d attributes as sensitive attributes. *Work-Class* is treated as primary sensitive attribute. Figure 6 depicts the KL-divergence of decompositiond d-SA-Adult tables where l_{pri} is set from 3 to $l_{per}(\textit{work-class}) = 7$. For each non-primary sensitive attribute S^i, l_i is set to $l_{per}(S^i)$. In Figure 6, the experimental result is quite close to the theoretical estimation of $\log(\prod_i l_i)$. Figure 7 depicts the execution time of decomposition on d-SA-Adult tables. Again, each non-primary sensitive attribute is set to its largest permissible diversity parameter while l_{pri} varies from 3 to 7.

We conduct a separate experiment to measure the number of noises in the MSA case. This experiment is on 2-SA-Adult, which takes *Work-Class* as the

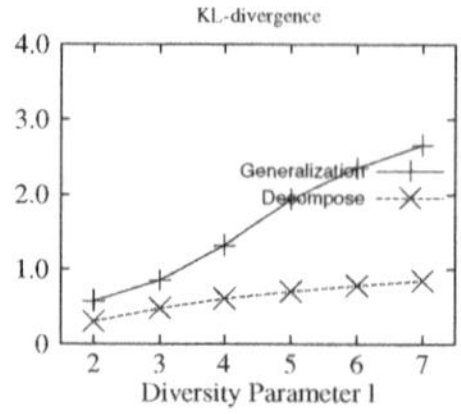

Fig. 1. 5-QI (SSA)

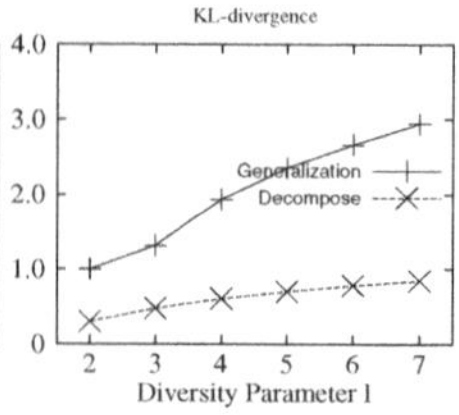

Fig. 2. 6-QI (SSA)

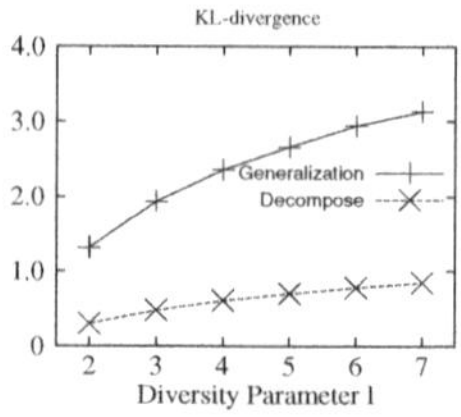

Fig. 3. 7-QI (SSA)

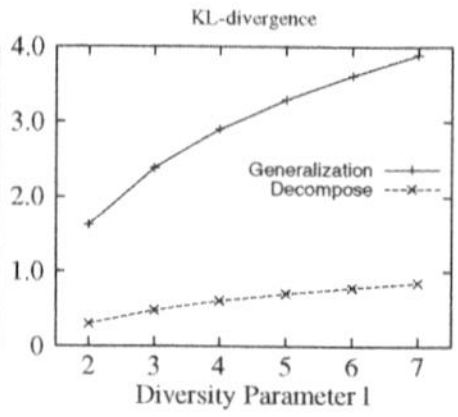

Fig. 4. 8-QI (SSA)

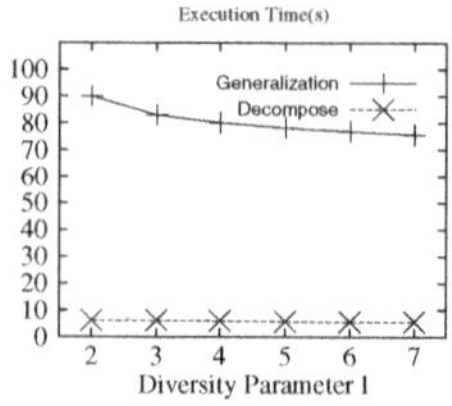

Fig. 5. Time (SSA)

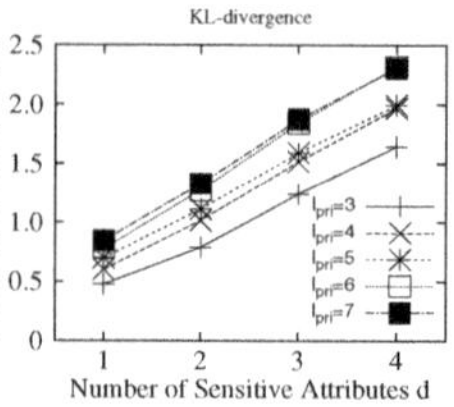

Fig. 6. MSA

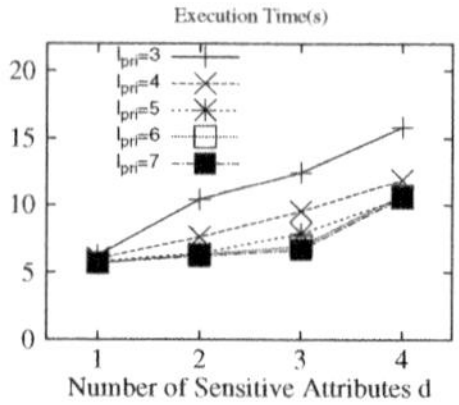

Fig. 7. Time (MSA)

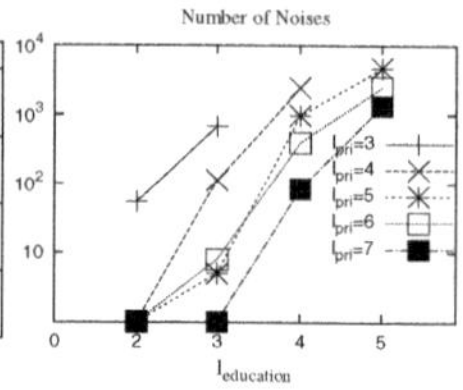

Fig. 8. Noise (MSA)

primary sensitive attribute and *Education* as the non-primary sensitive attribute. Figure 8 depicts the number of noises as the function of l_{pri} and $l_{Education}$.

4 Conclusions

This paper conducts the pioneering research towards privacy preservation in the MSA case, and lays down a foundation for future works, including combining categorical and numerical sensitive attributes, working on dynamic dataset, etc.

Acknowledgments. This work is partly supported by the National Natural Science Foundation of China Grant 60873065, 60553001, the National Basic Research Program of China Grant 2007CB807900, 2007CB807901, and the National High Technology Development 863 Program of China Grant 2007AA01Z152.

References

1. Sweeney, L.: Achieving k-anonymity privacy protection using generalization and suppression. International Journal on Uncertainty, Fuzziness and Knowledge-based Systems 10(5), 571–588 (2002)
2. LeFevre, K., DeWitt, D.J., Ramakrishnan, R.: Incognito: efficient full-domain k-anonymity. In: SIGMOD, pp. 49–60 (2005)
3. Machanavajjhala, A., Gehrke, J., Kifer, D., Venkitasubramaniam, M.: l-diversity: privacy beyond k-anonymity. In: ICDE, pp. 24–26 (2006)
4. LeFevre, K., DeWitt, D.J., Ramakrishnan, R.: Mondrian: multidimensional k-anonymity. In: ICDE, p. 25 (2006)
5. Ye, Y., Deng, Q., Wang, C., Lv, D., Liu, Y., Feng, J.-H.: BSGI: An Effective Algorithm towards Stronger l-Diversity. In: Bhowmick, S.S., Küng, J., Wagner, R. (eds.) DEXA 2008. LNCS, vol. 5181, pp. 19–32. Springer, Heidelberg (2008)

Privacy-Preserving Queries for a DAS Model Using Encrypted Bloom Filter

Chiemi Watanabe and Yuko Arai

Department of Information Science, Ochanomizu University
2-1-1, Otsuka, Bunkyo-ku, 112-8610 Tokyo, Japan
chiemi@is.ocha.ac.jp, yukoarai@db.is.ocha.ac.jp

Abstract. Recently, Database-As-a-Service (DAS) has attracted considerable attention. Users require protecting sensitive data from the DAS administrators. Most of previous studies, which proposed the solutions using cryptographic techniques, assume that a large amount of data will be inserted into the database and new data will be uploaded infrequently. We propose a secure query execution model for such an environment. Our approach is to represent all schemes of each tuple in a plaintext table as one Bloom filter index, and to replace queries with keyword searches of the Bloom filter index. Same values in each tuple are transformed into different values by two-phase encryption. DAS administrators cannot determine the schemes of the original table even if they look view the database and queries. Therefore, our approach is robust against the estimation of schemes in original tables.

Keywords: Privacy Preserving, Cloud Database, Database as a Service, Database Security, Encryption.

1 Introduction

Recently, Database-As-a-Service (DAS) has attracted considerable attention. DAS provides data management service in the cloud computing environment. Users may be concerned that DAS administrators are third parties to access their data. Traditional database systems function under the assumption that an administrator is trustworthy. Therefore, it is natural for clients to demand that their sensitive data be protected against the administrators.

To resolve this issue, cryptography is commonly employed. Many researchers have proposed several models for performing queries on encrypted data [1] [4] [5] [6]. In DAS model, user's queries are generally translated into two types of queries: a *first query* to be performed on the server and a *second query* to be performed on the client. A part of query processing, filtering, is performed directly on the encrypted data in the server using a *first query*; therefore, the computational cost to the DAS client is small. True results can be obtained by performing a *second query* on the client using the filtering results returned from the DAS server.

Most existing studies assumes a "write once read many (WORM)" environment, such as a data warehouse, in which a data owner uploads large amount of preformatted static data, and these studies extracts statistical information from the database for

X. Zhou et al. (Eds.): DASFAA 2009, LNCS 5463, pp. 491–495, 2009.

data encryption and query translation. Another major drawback of the existing approaches is that they limit their target data types. Also, the authors think that the situation, in which different encryption schemes are used for different attributes, may give hints to adversaries to attack the database.

In this paper, we propose a secure query execution model on the assumption of an environment where multiple users frequent upload and update data. Fig.1(a) shows an example of translating a table using our proposed approach. A plaintext table includes *id, date,* and *content* as attributes. In the translated table, only two attributes, namely, *etuple* and *bfindex,* are stored on the server. *etuple* is the encrypted value of a plaintext tuple. *bfindex* is a secure Bloom filter. *bfindex* is created by merging secure Bloom filters for all attributes of the original table, such as attributes {id, date, content}. It is impossible for adversaries to determine what attributes and values are stored in the *schedules* table from the *qx4k3rj* (encrypted value of the table name "schedules") table. Fig.1(b) shows how a query is translated. A range query condition for the *date* attribute and a text retrieval condition for the *content* attribute in a plaintext query are transformed to arguments of a function *smatch*() in a *first query* (translated query shown in Fig.1). The function realizes keyword searches for each *bfindex.* Adversaries will not be able to determine what attributes and values the *schedules* table has from the *translated query* (*first query*).

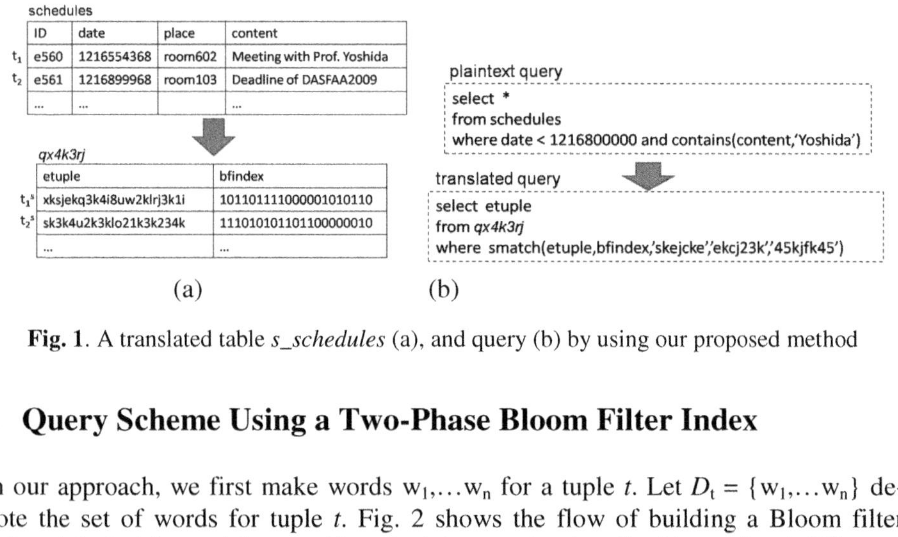

Fig. 1. A translated table *s_schedules* (a), and query (b) by using our proposed method

2 Query Scheme Using a Two-Phase Bloom Filter Index

In our approach, we first make words $w_1,\ldots w_n$ for a tuple *t*. Let $D_t = \{w_1,\ldots w_n\}$ denote the set of words for tuple *t.* Fig. 2 shows the flow of building a Bloom filter index (*bfindex*) for a plaintext table *schedules,* as also described in section 1. When an attribute value has a numeric type (e.g. id, date), the numeric data are translated into a set of words. It is described in detail in section 3. When an attribute value has a string type (e.g., id, content), the string arrangement is separated by word, and the attribute name is attached to the word. Furthermore, the attribute values are added to D_t after attaching the attribute name and the word "em" (this is used for exact matching query). The set of words D_t in Fig.2 is for tuple *t.* The words "date:lt:B0" and "date:lt:B4" are translated words for the attribute *date*'s value.

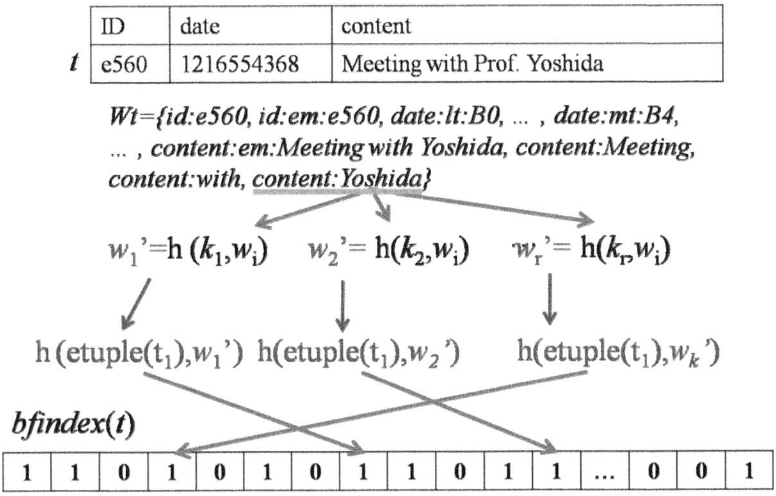

Fig. 2. Process flow for translating a tuple *t*

We apply the HMAC (Keyed-Hashing for Message Authentication) function in two phases. We obtain *r* values $w'_1,\ldots,w'_r$ using keys $K=\{k_1,..,k_r\}$ for a word *w* in the set of words D_t for a tuple *t*. Users share these *r* keys for the *first query* on the server. In the second phase, we again compute HMAC using *etuple* as a key for the result values of the first phase hashing $w'_1,\ldots,w'_r$. By applying the secure approach, even if some tuples have the same values in the original table, adversaries cannot obtain any useful features from *bfindex* because the same values are translated into different values using different *etuple* values for each of them.

Next we describe about the translation of queries. Suppose the following plaintext query which contains two conditions.

```
SELECT *
FROM schedules
 WHERE contains(content, 'Yoshida')  and  place = 'room602'
```

First, we create a set of words Q={"content:Yoshida", "place:em:room602"} according to the query conditions. The way of creating the set of words is the same as that for the set of words D_t for tuple *t*. After creating a set of query words *Q*, we apply the HMAC function for each words using r' pieces of keys $K'=\{k'_1,\ldots,k'_{r'}\}$, which are randomly selected from $K=\{k_1, \ldots, k_r\}$. The number of *K*' and the members are different for each word. We call these hashed values ***query elements***. Using these hashed values, we translate the plaintext query. The above query is translated as follows:

```
SELECT etuple
FROM qx4k3rj
WHERE smatch(etuple, bfindex, aljk34,1kjakm,akj3w4j,akj23kj,14kjshk),
```

where query elements are { aljk34,1kjakm,akj3w4j,akj23kj,14kjshk }.

By applying the randomly selected word set *K'* for each query word, the accuracy of filtering (*first query*) may decrease. However, the security level increases. This is because even if the adversary analyzes the query log, a sentence of *first query*, and the results, it will be difficult for an adversary to guess the number of query conditions that are described in the original query.

In the translated query, there is a function *smatch*() that includes some hash values as arguments. The translated query is sent to the DAS server.

In the DAS server, we compute the hash values again for each of the arguments in a the *smatch*() function using each *etuple* as a hash key, and we check the bits at the positions of the hashed values.

3 Range Query for Numerical Values Using Bucketization

For a numeric attribute value, we select three sets of bucket numbers: (1) a bucket number corresponding to a value *x*, (2) a set of bucket numbers whose maximum is less than the value *x*, and (3) a set of bucket numbers whose minimum is more than the value *x*. We denote these sets as $B_=(x)$, $B_<(x)$, and $B_>(x)$, respectively. Next, we attach "*atr*:em" to the word in $B_=(x)$, " *atr*:lt" to the words in $B_<(x)$ and "*atr*:mt" to the words in $B_>(x)$. In Fig.3, 304(= x) is translated into three sets of words; {"em:B5","lt:B1","lt:B2","lt:B3","lt:B4","mt:B6","mt:B7","mt:B8","mt:B9","mt:Ba"}.

Range queries also replace keyword matching using *bfindex*. If x is less than 193, $B_<(x)$ contains B4. Likewise, if x is greater than 650, $B_>(x)$ contains B7.

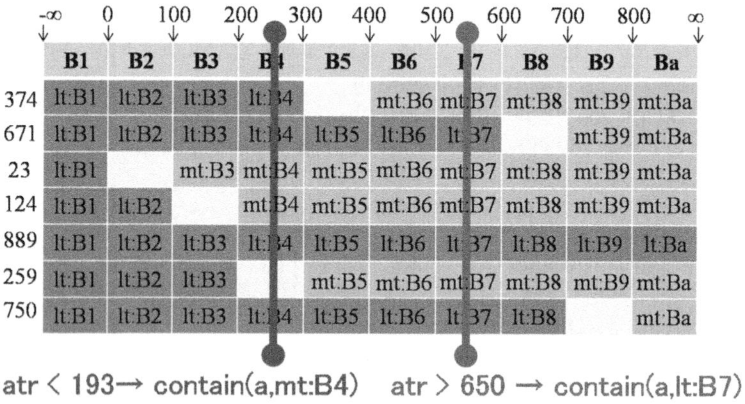

Fig. 3. Translation Numerical Data and the Range Query using Bucketization

4 Conclusion

In this paper, we presented a secure query execution model on the assumption of an environment where multiple users frequently upload and update data. The features of plaintext data are not revealed with frequent updates because static information about the uploaded data is not used when transforming the plaintext values.

In our approach, an original table is transformed into a table whose attributes are only *etuple* and the Bloom filter index. A plaintext query is replaced in keyword searches in the Bloom filter. Our approach is robust against the estimation of schemes in a plaintext table because all schemes in a plaintext table are replaced into the Bloom filter index, which only includes 1 and 0. Moreover, our approach is strong against the estimation of true values because the same values between each tuple in a plaintext table are encrypted to different values using two-phase encryption. Furthermore, it is difficult to obtain useful information even if an administrator sees t he query logs, a sentence of the *first query* and the results, because there are many query patterns used in our approach.

A major problem of our approach is query performance. Our approach should apply hash functions by the number of query words for each tuple. While the approach is effective for protecting against administrator's reasoning attacks, it is main reason why the speed of the search is not so high. In future work, we should evaluate the performance and we should propose the speed-up technique of our approach.

References

1. Hacigumus, H., Iyer, B., Li, C., Mehrotra, S.: Executing SQL over Encrypted Data in the Database-Service-Provider Model. In: Proceeding of the ACM SIGMOD International Conference on Management of Data, pp. 216–227 (June 2002)
2. Yang, Z., Zhong, S., Wright, R.N.: Privacy-preserving queries on encrypted data. In: Gollmann, D., Meier, J., Sabelfeld, A. (eds.) ESORICS 2006. LNCS, vol. 4189, pp. 479–495. Springer, Heidelberg (2006)
3. Boneh, D., Di Crescenzo, G., Ostrovsky, R., Persiano, G.: Public key encryption with keyword search. In: Cachin, C., Camenisch, J.L. (eds.) EUROCRYPT 2004. LNCS, vol. 3027, pp. 506–522. Springer, Heidelberg (2004)
4. Hore, B., Mehrotra, S., Tsudik, G.: A privacy-Preserving Index for Range Queries. In: Proceedings of the 30th VLDB Conference, pp. 720–731 (2004)
5. Agrawal, R., Kiernan, J., Srikant, R., Xu, Y.: Order Preserving Encryption for Numeric Data. In: Proceedings of the ACM SIGMOD, pp. 563–574 (2004)
6. Ge, T., Zdonik, S.B.: Answering Aggregation Queries in a Secure System Model. In: Proceedings of the 33rd VLDB Conference, pp. 519–530 (2007)

Hash-Search: An Efficient SLCA-Based Keyword Search Algorithm on XML Documents

Weiyan Wang[1], Xiaoling Wang[1,2], and Aoying Zhou[2]

[1] School of Computer Science and Technology, Fudan University, China
acmwwy@gmail.com
[2] Shanghai Key Laboratory of Trustworthy Computing, Software Engineering institute of East China Normal University
xlwang@sei.ecnu.edu.cn, ayzhou@sei.ecnu.edu.cn

Abstract. XML is a de-facto standard for exchanging and presenting information and keyword search over XML documents has become an interesting topic. However semi-structured XML data give rise to many challenges of conventional information retrieval technologies. In order to return highly-related data nodes and improve the quality of keyword search result, SLCA(*Smallest Lowest Common Ancestor*)-based keyword search on XML data is recently attracting more and more attention in the database community. In this paper, we design efficient index and propose hash-based method to answer SLCA-based keyword search queries. Our approach outperforms *Incremental Multiway-SLCA approach*, which is the most efficient algorithms in the literature. We demonstrate the effectiveness of our algorithms analytically and experimentally.

1 Introduction

Keyword search on XML data has become one of the most convenient and popular approach to search information from XML document collections. Consider a query Q including k keywords $w_1, \cdots, w_k$, conventional LCA semantics suggest simply returning all nodes in the XML tree whose subtrees contain all k keywords. However, such semantics will lead to lots of lowly-related data nodes that are remotely connected to some nodes with keyword labels. In order to return highly-related data nodes and improve the quality of keyword search result, a notion of *Smallest Lowest Common Ancestor (SLCA)* semantics is proposed in [2].

According to the SLCA semantics, only data nodes in the XML tree satisfying following two conditions are returned to users :(1) all the keywords appear in the subtrees rooted at the nodes, and (2) the nodes have no descendent nodes whose subtrees also contain all the keywords. The answer to the query "John, XRank" includes nodes 1.4.5, 1.4.10 and 1.20. Nodes 1 and 1.4 are discarded because they violate the second condition of the SLCA semantics. Similar to *SLCA* semantics, another semantics named *Meaningful LCA (MLCA)* is proposed in [4] to enhance the effectiveness of keyword search in XML data.

X. Zhou et al. (Eds.): DASFAA 2009, LNCS 5463, pp. 496–510, 2009.

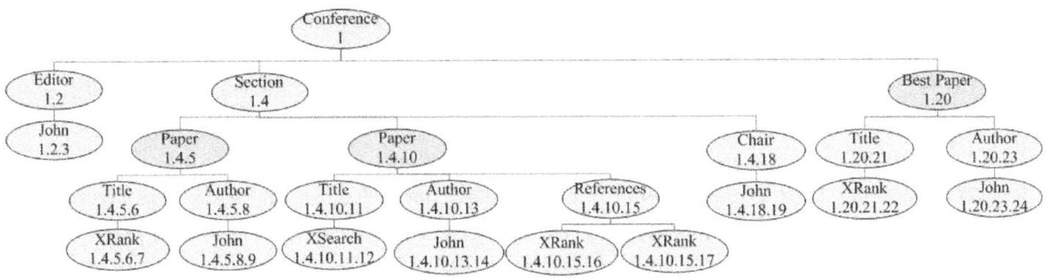

Fig. 1. Example Conference.xml(each node is associated with combined preorder id)

The most efficient algorithms for keyword search using SLCA semantics is the *Incremental Multiway-SLCA (IMS)* algorithm [1], which is proven to outperform the *Scan Eager (SE)* and the *Indexed Lookup Eager (ILE)* algorithms [2]. However, the IMS algorithm has two disadvantages. First, the IMS algorithm stores each data node with a Dewey label so that LCA node can be gained by comparing nodes with Dewey labels, which costs $O(d)$ time, where d is the depth of the XML tree. However, when the XML tree is deep, LCA computation would impair the efficiency of the IMS algorithm. Second, to determine the closest node of S_i to v in the XML tree, the IMS algorithm has to conduct a binary search on S_i, which costs $O(dlog(|S_i|))$ including the process of comparing with Dewey labels. For that reason, the size of S_i will limit the performance of the IMS algorithm.

The main contributions of our paper are as follows:

- We introduce a new notion of *domination* to indicate the relationship between a data node and a word and develop a hash-based index structure to effectively store all useful *domination* information. By doing so, we are able to eliminate LCA computation through simply examining *domination* information in the index, which is unbounded by the depth of the XML tree.
- We propose a novel approach for keyword search in XML databases according to SLCA semantics named *Hash-Search (HS)*. Different from the IMS algorithm, we initially select the smallest keyword list as the working node list (denoted by S_1). Then for each node v in S_1,we find the potential SLCA by checking the *domination* information between $v's$ ancestors and all keywords. Such process can be done efficiently by binary search all the nodes on the path from the XML tree root to v.
- We experimentally evaluate the Hash-Search algorithm by comparing with the Incremental Multiway-SLCA algorithm[1]. The experiments demonstrates that our approach outperforms former work.

The rest of this paper is organized as follows. Preliminaries are given in Section 2. In Section 3, we introduce our approach, including the definition of *domination*, Hash-Index and Hash-Search algorithm. The system architecture are presented in Section 4. The results of experiments are shown in Section 5. Section 6 presents related research works and it's followed by the conclusion in Section 7.

2 Preliminaries

We model the XML tree $\mathcal{T}$ with conventional labeled ordered tree model. Each element in the XML tree is represented by a node v whose label is $L(v)$. The root of the XML tree is denoted by *root*. Given a word w, we use $w \in L(v)$ to denote that the label of v directly contains w. Besides, each node has two unique ids to represent it. One is the preorder id $pre(v)$ which satisfies the condition of preorder numbering that if and only if a node u precedes a node v in the preorder left-to-right depth-first traversal of the tree the inequality $pre(u) < pre(v)$ holds. The other kind of id is the combination of all v's ancestors' preorder ids, called *combined preorder id* (denoted by $pres(v)$). The formal definition is as followed: (1)$pres(v) = pre(v)$ if $v = root$; (2)$pres(v) = pres(u).pre(u)$ if u is the father of v. Note that $pre(v)$ is the last element of $pres(v)$. Therefore, we only need to store $pres(v)$ for v. *Combined preorder id* enables us to efficiently determine the preorder ids of v's ancestors. Moreover, the preorder id of the LCA of two nodes u and v can be easily computed by comparing $pres(u)$ and $pres(v)$.

Consider a keyword search query K with k keywords $w_1, \ldots, w_k$ and a XML tree $\mathcal{T}$, for each keyword w_i, let S_i denote the list of data nodes in $\mathcal{T}$ whose labels directly contain keyword w_i. We define S_i to be the *node list* of w_i. The data nodes in S_i is sorted by preorder id. Without loss of generation, we assume w_1 to be the keyword with lowest frequency in $\mathcal{T}$ among k keywords. In other words, $|S_1| = min\{|S_1|, \ldots, |S_k|\}$. S_1 is considered the *working node list* of our approach.

A *match* of K is the set of data nodes $S = \{v_1, \ldots, v_k\}$ where $|S| = |K|$ and $v_i \in S_i, i \in [1, k]$. An *answer subtree* for K is a subtree of $\mathcal{T}$ that contain all the keywords; i.e., there is a *match* S for K that each node in S belongs to the *answer subtree*. The *smallest answer subtree* is an *answer subtree* that no subtree of it is an *answer subtree*. The SLCA problem is to return the set of roots of all *smallest answer subtree* for the query K. We use function $slca(w_1, \ldots, w_k)$ to denote the results.

Given two nodes u and v, $u <_\alpha v$ denotes that u is a proper ancestor of v and $u \leq_\alpha v$ denotes that u is an ancestor of v ($u = v$ or $u <_\alpha v$). Given a node $v \in \mathcal{T}$, we can find all its ancestor node through examining its *combined preorder id* with $O(d)$ time, where d is the maximum depth of $\mathcal{T}$.

Given k data nodes $v_1, \ldots, v_k$, we use the function $lca(v_1, \ldots, v_k)$ to compute the *lowest common ancestor* of them and return null if one of them doesn't exist in the XML trees. Specifically, given two nodes u and v, $lca(u, v)$ is the node that whose combined preorder id is the longest common prefix of the combined preorder ids of u and v. Given k sets of data nodes $S_1, \ldots, S_k$, a data node v belongs to $lca(S_1, \ldots, S_k)$ if there are k nodes $v_1, \ldots, v_k$ that $v = lca(v_1, \ldots, v_k)$, where $v_i \in S_i, i \in [1, k]$. A node v belongs to $slca(S_1, \ldots, S_k)$ if $v \in lca(S_1, \ldots, S_k)$ and no descendent of v is in $lca(S_1, \ldots, S_k)$. It is obvious that $slca(w_1, \ldots, w_k) = slca(S_1, \ldots, S_k)$. For brevity, in the rest sections, we use $lca(\{v\}, w_1, \ldots, w_k)$ to denote $lca(\{v\}, S_1, \ldots, S_k)$, and use $slca(\{v\}, w_1, \ldots, w_k)$ to represent $slca(\{v\}, S_1, \ldots, S_k)$. In addition, given a data node u, a slca node v is said to be *limited* to u if $u \leq_\alpha v$ and can be called $u - limited$ slca node.

The following example demonstrates the above definitions.

Example 2.1 Consider the XML tree shown in Fig. 1 and the keyword search query is "XRank, John". For each node in the XML tree, the *combined preorder id* is presented under its label. $S_1 = \{1.4.5.6.7, 1.4.10.15.16, 1.4.10.15.17, 1.20.21.22\}$,$S_2 = \{1.2.3, 1.4.5.8.9, 1.4.10.13.14, 1.4.18.19, 1.20.23.24\}$.Because $|S_1| < |S_2|$,S_1 is the *working node list* .Moreover, the ancestors of 1.4.18.19 are 1,1.4, 1.4.18 and 1.4.18.19 which can be obtained from its *combined preorder id.* $lca(\text{"}XRank\text{"}, \text{"}John\text{"}) = \{1,1.4, 1.4.5, 1.4.10, 1.20\}$, $slca(\text{"}XRank\text{"}, \text{"}John\text{"}) = \{1.4.5, 1.4.10, 1.20\}$. In addition, 1.4.5 is *1.4-limited* or *limited* to 1.4.

3 Our Approach

As discussed above, the efficiency of the IMS algorithm is restricted by two basic processes : LCA computation and the determination of the closest node of a set to a node. The key motivation of our approach is to avoid these time-consuming processes by using a designed elegant index named *Hash-Index* based on the concept of domination.

In this section, we propose our algorithm named *Hash-Search (HS)* to process SLCA-based keyword search queries. In Section 3.1, we introduce the notion of *domination*, which is essential for the HS algorithm. We also give some important properties. The description and construction Hash-Index are provided in Section 3.2. Finally, Section 3.3 presents the HS algorithm.

3.1 Domination

Given a data node v in the XML tree and a word w, w is said to be *dominated* by v if there is a data node p satisfying both conditions: $v \leq_\alpha p$ and $w \in L(p)$, which is denoted by $v \succ w$. Notice that for each word $w \in L(v)$, we have $v \succ w$. Furthermore, for a data node v in the XML tree, a word w is considered to belong to the *domination set* of v if $v \succ w$. We refer to H_v as the *domination set* of v.

Example 3.1 Consider the word "John", the node v where $pres(v) = 1.4.5$ satisfies $v \succ John$ and $v \succ Paper$. $H_v = \{$Paper, Title, XRank, Author, John$\}$.

We give several important properties that help us to improve the efficiency of SLCA computation by using *domination* concept. Because of space limitation, we omit the proof of these properities.

Lemma 1. *Let u and v be two nodes, if $u \leq_\alpha v$, then $pre(u) \leq pre(v)$.*

Lemma 2. *Let u and p be two nodes such that $p \leq_\alpha u$. If w is a word such that $u \succ w$, then $p \succ w$.*

Lemma 3. *Let u and p be two nodes such that $p \leq_\alpha u$. $H_u \subseteq H_p$.*

Lemma 4. *For k keywords $w_1, \ldots, w_k$, $lca(S_1, \ldots, S_k) = \{u | w_i \in H_u, i \in [1, k]\}$.*

Lemma 4 provides a useful property to determine whether a node is lca node without knowing any specific nodes whose labels contain keywords. We call the process of determination *lca node verification.*

Lemma 5. *Consider k keywords $w_1, \ldots, w_k$ and a node u such that $u \in lca(S_1, \ldots, S_k)$, then for each v where $v \leq_\alpha u$, $v \in lca(S_1, \ldots, S_k)$.*

Lemma 6. *Given a set of keywords $K = \{w_1, \ldots, w_k\}$, for any word $w_m \in K$, $slca(w_1, \ldots, w_k) =$*

$$RemoveAncestor(\bigcup_{v \in S_1} slca(\{v\}, w_2, \ldots, w_k))$$

Lemma 7. *Given any two nodes u, v and a set S, if $pre(u) < pre(v)$ and $pre(slca(\{u\}, S)) \geq pre(slca(\{v\}, S))$, then $slca(\{v\}, S) \leq_\alpha slca(\{u\}, S)$.*

Lemma 8. *Given any two nodes u, v and a set S, if $pre(u) < pre(v)$ and $pre(slca(\{u\}, S)) < pre(slca(\{v\}, S))$ and $slca(\{u\}, S)$ is not an ancestor of $slca(\{v\}, S)$, then $slca(\{u\}, S)$ is a slca node.*

3.2 Hash-Index

In this section, we discuss the detail of the Hash Index including the structure description and the construction of the index.

Structure of Hash-Index. Given a XML tree $\mathcal{T}$, Hash-Index consists of the following components:

- a hash table called *the Frequency Table (FT)* stores the frequencies of keywords in the $\mathcal{T}$. The FT is originally stored in the disk and read into memory during the preprocessing step. When a keyword search query comes, we use the FT to determine the keyword in the query with the smallest frequency.
- a disk-based B+ tree called *the Node Library (NL)* stores all the nodes in $\mathcal{T}$. The NL keys are the keywords of data nodes in $\mathcal{T}$. For each key in the NL, the data associated with it is a list of *combined preorder ids* of the data nodes whose labels directly contain that key. These ids are stored in ascending order.
- a disk-based hash table called *the Domination Set Pool (DSP)* stores domination sets of all nodes in $\mathcal{T}$. Consider a node v whose domination set is H_v. For each word $w \in H_v$, we generate a pair $(pre(v), w)$ as one entry stored in the DSP. The hash key is then determined from the pair. Using this method, to determine whether a word w is *dominated* by a data node v, we can simply check whether the pair $(pre(v), w)$ is contained in the hash table. The reason why we choose hash table to store domination information is that hash table can answer point query more efficiently than other indices.

Example 3.2 For the XML file in Fig. 1, we show the corresponding Hash-Index on $\mathcal{T}$ in Fig. 2 and Fig. 3. For example, in the Tree $\mathcal{T}$, The frequency of "John" is 5. Through hashing calculation, "John" and its frequency are stored

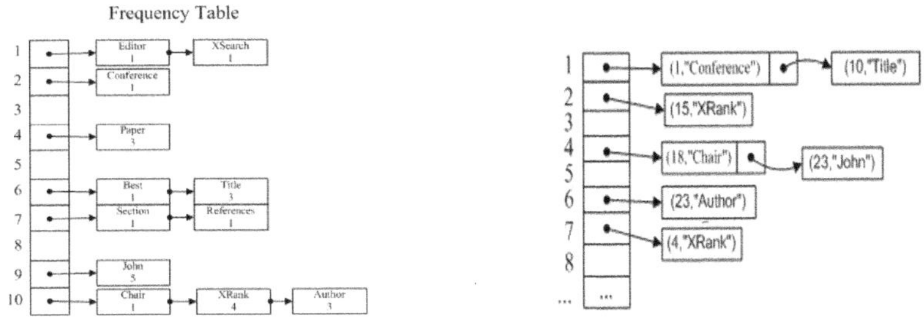

(a)Frequency Table (b)Domination Set Pool

Fig. 2. Frequency Table and Domination Set Pool of Example 3.2

in $FT[9]$. The NL given in the Fig. 3 records "John" and *combinedpreorderids* of nodes whose labels contain "John": 1.2.3 ,1.4.5.8.9,1.4.10.13.14, 1.4.18.19 and 1.20.23.24. Moreover, the DSP contains the pair (23,"John") but doesn't include (15,"John") , which means that the node with id 23 dominates the keyword "John", while the node with id 15 not.

Index Construction

An input XML document is parsed by a *SAX* parser to produce a stream of events. Such events include *Start-Document, End-Document, Start-Element, End-Element, and Text.* Notice that we regard other types of nodes in the XML document, such as attributes, as special element nodes in this paper. The Hash-Index is constructed during parsing the XML document. We initialize the environment of Hash-Index when a *Start-Document* event is met and finish parsing as well as Hash-Index construction when *End-Document* event occurs. *Start-Element* event recognizes the beginning of an element.

The algorithm for *Start-Element* events is given in Algorithm 1. In the algorithm, preorder counter *PC* and current combined preorder *CCP* is maintained to compute the preorder id and combined preorder id respectively for each node in the XML tree. Initially, *PC* is set to 0 and *CCP* is set to empty. When a start element E is met, we create a node v to represent the corresponding node in the XML tree. Line 1-2 updates *PC* and *pre(v)*. Before line 3, CCP is equal to the combined preorder id of v's parent (CCP is empty if v is the root of the XML tree). Lines 5 to 11 store the *domination* information between the element E and the ancestors of v to the DSP. Let u be any ancestor of v, according to the definition of *domination*, the pair $(pre(u), E)$ should be included in the DSP. At the same time, we can directly gain ids of ancestors of v by enumerating each element of *pres(v)*. According to Lemma 1, Line 5 ensures that any node is processed before its ancestors, which is the foundation of lines 6 to 7. The correctness of lines 6 to 7 stems from the fact that if the pair $(pre(u), E)$ has already been in the DSP, then for each ancestor of u (denoted by v), $(pre(v), E)$ must be in the DSP too, by Lemma 2. Therefore, it's not necessary to store the *domination information* between E and all ancestors of u. Line 12 inserts *pres(v)* to the data list associated with key E in the NL. And line 13 updates the FT.

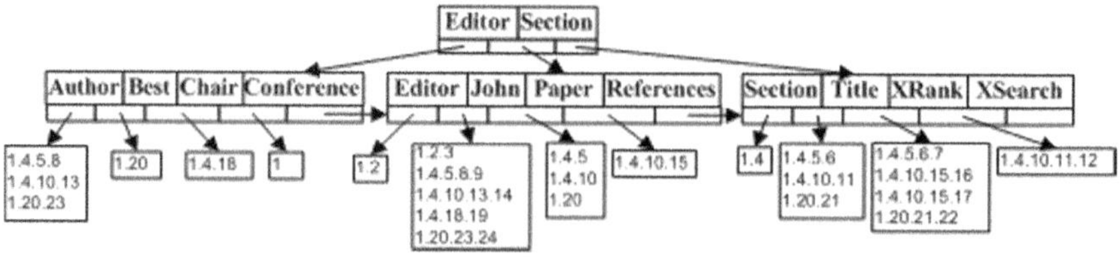

Fig. 3. The Node Library of Example 3.2

Algorithm 1. StartElement()

Input: Element Name E
Output:
Data: Preorder Counter PC, Current Combined Preorder CCP, Current Node v

1 $PC = PC + 1$;
2 $pre(v) = PC$;
3 Add $pre(v)$ to the tail of CCP;
4 $pres(v) = CCP$;
5 **for** *each preorder id* $pid \in pres(v)$ *in decreasing order* **do**
6   **if** *pair* $(pid, E) \in DSP$ **then**
7   break;
8   **else**
9   Add pair (pid, E) to DSP;
10   **end**
11 **end**
12 Add $pres(v)$ to NL with key E;
13 The frequency of E in FT is increased by 1.

Moreover, *End-Element* event identifies the end of an element, when the end of E is met, all nodes belong to the subtree of v has been processed.

For *Text* event, *text* is treated almost the same as the start element in lines 1 to 4 except that CCP is not updated by $pre(v)$ because *text* has no descendent to compute its combined preorder id through CCP. Because of the space limitation, we omit the detailed algorithm here.

Efficiency of Index Construction Hash-Index can be efficiently generated by scanning the XML document once. Let d be the height of the XML tree and n is the number of words in the tree. For Algorithm 1, the update of the DSP in lines 5 to 11 involve $O(d)$ time because the insertion complexity of hash table is $O(1)$ and we need to insert at most d entries into the DSP for each element. The cost of line 12 is $O(log(n))$ for the property of B+ Tree. In sum, the complexity of Algorithm 1 is $O(d + log(n))$. Moreover, the cost of processing *End-Element* is $O(1)$. Algorithm dealing with *Text()* is same as the algorithm to process start element in Algorithm 1,which costs $O(d + log(n))$. So the overall complexity of index construction is $O(nd + nlog(n))$ because each word in the XML data tree must be processed with $O(d+log(n))$ time whether it's a start element or a word in the text string.

3.3 Hash Search

In this section, firstly, we introduce Algorithm *LimitedSLCA()* which uses the Domination Set Pool (DSP) to obtain potential SLCA. Based on Algorithm *LimitedSLCA()*, we present our Hash-Index-based algorithm called *Hash Search (HS)*, whose details are shown in Algorithm 3.

Obtaining Potential SLCA

Given a data node v, an node *anc* which is an ancestor of v, and k keywords $w_1, \ldots, w_k$, Algorithm 2 computes $slca(\{v\}, w_1, \ldots, w_k)$ *limited* to *anc* based on the Hash-Index. This algorithm is frequently used by Hash-Search algorithm to determine potential slcas.

Initially line 1 organizes preorder ids of all ancestors of v into an array $IDArray$ in ascending order. These ids can be easily gained from $pres(v)$. Based on Lemma 1, $pre(anc) \leq pre(slca) \leq pre(v)$ always holds. So Lines 2 to 3 determine the range of the potential slca in $IDArray$ by setting $left$ and $right$ to the position of $pre(anc)$ and $pre(v)$ in the $IDArray$ respectively. Lines 4 to 13 test whether $IDArray[left + 1]$ belongs to $lca(\{v\}, w_1, \ldots, w_k)$.

Lines 14 to 28 present the process of binary search to find slca. In each loop, we examine the node *MidNode* whose preorder id is $IDArray[mid]$. If it is a lca node, all its ancestors must be lca nodes by lemma 5. While the slca node should satisfy $MidNode \leq_\alpha slca$, therefore, we only need to consider descendants of *MidNode*. In the similar way, if *MidNode* is not a lca node, based on lemma 3, all its descendants are not lca nodes, thus we can safely discard them. Lines 17 to 22 present the verification process using the DSP. Only when all keywords are in the *domination set* of *MidNode* can we consider *MidNode* an element of $lca(\{v\}, w_1, \ldots, w_k)$.

Efficiency of Algorithm 2 The determination of slca nodes limited to one node can be done efficiently. Let d denote the height of the XML data tree, and assume we need to find $slca(\{v\}, w_1, \ldots, w_k)$. The cost of Line 1 is $O(d)$ since we have to check each ancestor of v. After that, we determine the slca node through searching $IDArray$ with binary search technique. In each loop, the verification step in both lines 5 to 10 and lines 17 to 22 costs $O(k)$ because the search complexity of the *DSP* is $O(1)$ and we need to search *domination* information for each keyword. Therefore, the overall complexity of Algorithm 2 is $O(d + klog(d))$.

Example 3.3 In Example 3.2, let K = "John","References". We denote node v where $pres(v) = 1.4.10.15.16$ and *anc* where $pres(anc) = 1$. We demonstrate how we calculate the $slca(\{v\}, w_1, \ldots, w_k)$ *limited* to *anc*. We start to store $pres(v)$ into $IDArray$ so that $IDArray$ contains five elements : 1, 4, 10, 15 and 16. After that, we have $left = position = 1$, $right = 5$. Before binary search, we test whether the son of *anc* can dominate all keywords. Notice that (4,"John") and (4,"References") are in the DSP. So we initiate the binary search and finally return the node 1.4.10 because it dominates keywords.

Algorithm 2. LimitedSLCA()

```
   Input: Combined Preorder ID of a node v pres(v), Set of Keywords
          K = {w_1, ..., w_k}, an Ancestor of v anc
   Output: slca({v}, w_1, ..., w_k) if it's limited to anc and NONE otherwise
1  Split pres(v) into preorder ids and store them in an array IDArray in ascending
   order.
2  left = position = the position of pre(anc) in IDArray;
3  right = the position of last elements in IDArray;
4  LCA = true;
5  for each w_i ∈ K do
6      if pair(IDArray[left + 1], w_i) ∉ DSP then
7          LCA = false;
8          break;
9      end
10 end
11 if LCA = false then
12     return pres(anc);
13 end
14 while left < right do
15     mid = (left + right)/2;
16     LCA = true;
17     for each w_i ∈ K do
18         if pair(IDArray[mid], w_i) ∉ DSP then
19             LCA = false;
20             break;
21         end
22     end
23     if LCA = true then
24         right = mid -1;
25     else
26         left = mid +1;
27     end
28 end
29 if left < position then
30     return NONE;
31 else
32     return pres(IDArray[left]);
33 end
```

Querying Evaluation At the beginning, the algorithm determines the working set S_m by comparing frequencies of keywords in the XML data tree. After that, the algorithm enumerates each node in the working set and computes the potential slca node for it using Algorithm 2. The correctness of this algorithm is guaranteed by Lemma 6.

Initially, the FT is used in Lines 1 to 3 to determine the frequencies of keywords in the keyword query. Line 4 finds the keyword with lowest frequency and set it as w_1. In addition, we make use of NL to get the node list of

Algorithm 3. HashSearch()

Input: Keyword Query $K = \{w_1, \ldots, w_k\}$
Output: $Answer = slca(w_1, \ldots, w_k)$
Data: Potential slca Node v

1 **for** *each keyword* $w_i \in K$ **do**
2   let f_i = the frequency of w_i in FT;
3 **end**
4 without loss of generality, let w_1 be the word with smallest frequency $f_1 = \min\{f_1, \ldots, f_k\}$;
5 gain S_1 through the NL with key w_1;
6 $Answer = \{\}$;
7 $v = 1$;
8 **for** *each node* u *in* S_m **do**
9    $x = LimitedSLCA(u, K - \{w_1\}, root)$;
10   **if** $pre(v) < pre(x)$ **then**
11   **if** $v \not\leq_\alpha x$ **then**
12          $Answer = Answer \cup \{v\}$;
13   **end**
14       $v = x$;
15   **end**
16 **end**
17 $Answer = Answer \cup \{v\}$;
18 return $Answer$;

w_1 and work on this set in the following steps. Lines 6 to 17 compute the slca nodes according to Lemma 6. Recall that nodes in S_1 are sorted by preorder id. Consider two nodes u and v in S_1 where $pre(u) < pre(v)$, according to Lemma 7, if $pre(slca(\{u\}, S_2, \ldots, S_k)) \geq pre(slca(\{v\}, S_2, \ldots, S_k))$, then $slca(\{v\}, S_2, \ldots, S_k)$ should be discarded since it violates the second condition of the definition of slca. Moreover, under the condition that $pre(slca(\{u\}, S_2, \ldots, S_k)) < pre(slca(\{v\}, S_2, \ldots, S_k))$, if $slca(\{u\}, S_2, \ldots, S_k) \not\leq_\alpha slca(\{v\}, S_2, \ldots, S_k)$, then $slca(\{u\}, S_2, \ldots, S_k)$ is a slca node (by Lemma 8), we can insert it into the answer set . Otherwise, $slca(\{u\}, S_2, \ldots, S_k)$ should be removed and $slca(\{v\}, S_2, \ldots, S_k)$ will become the undecided slca node. Lines 10 to 15 apply Lemma 7 and Lemma 8 to determine the slca nodes.

Let n be the number of words in the XML data tree. Let d be the height of the XML data tree. Let k be the number of keywords in the keyword query. And let S_1 as the working set of the HS algorithm. Lines 1 to 4 require $O(k)$ time to get the frequencies and compare them. It takes $O(log(n))$ time to gain S_1 from the NL in line 5. After that, for each node in S_1, we compute the potential slca for it using Algorithm 2, whose complexity is $O(d + klog(d))$ in line 9. Notice that we can easily determine the ancestor-descendant relationship between two nodes by comparing their combined preorder ids. So the cost of line 11 is $O(d)$. In sum, the overall complexity of the HS algorithm is $O(k + log(n) + |S_1|(d + klog(d)))$, which is smaller than $O(kd|S_1|log(|S|))$, the complexity of the IMS algorithm, where S is the data node list with the highest frequency[4].

Example 3.4 Consider computing SLCAs for keywords {"XRank","John" } on the tree T shown in the Fig. 1.At first, through the FT, we know that $f_1 = 4$, $f_2 = 5$. Because $f_1 < f_2$, we let w_1 = "XRank". Then by querying the NL, we have the node list for w_1 {1.4.5.6.7,1.4.10.15.16,1.4.10.15.17,1.20.21.22}. With the $LimitedSLCA$ function, we determine the potential slcas for each node in the node list for w_1. At the first iteration, x = 1.4.5. Because $pre(v) = 1 < pre(x) = 5$, v = 1.4.5. At the second iteration, x = 1.4.10, which satisfies $pre(v) = 5 < pre(x) = 10$ and $v \nleq_{\alpha} x$,1.4.5 is inserted into $Answer$. While at the third iteration, since x is equal to v, v remains the same. At the last iteration, we have x = 1.20. Then $pre(v) = 10 < pre(x) = 20$ and $v \nleq_{\alpha} x$. So 1.4.10 is considered as an answer and v is replaced by 1.20. Line 17 finally adds 1.20 to $Answer$. Therefore, the answer to the keyword query is {1.4.5,1.4.5,1.20}.

4 System Architecture

In this section, we present the architecture of the Hash-Search implementation, which is shown in Fig. 4. The IndexBuilder parses a XML document T with SAX Parser and construct the Hash-Index, which includes Node Library, Frequency Table, and Domination Set Pool.

The Hash-Index plays an important role in Search Engine. When a keyword query is submitted, the Search Engine will use Frequency Table, which is loaded into memory before accepting any queries, to determine the keyword with smallest frequency. The smallest keyword list is later gained by indexing Node Library with that keyword. After that, through examining domination information between node and keyword in Domination Set Pool, Hash-Search can efficiently compute SLCA nodes and return them to users.

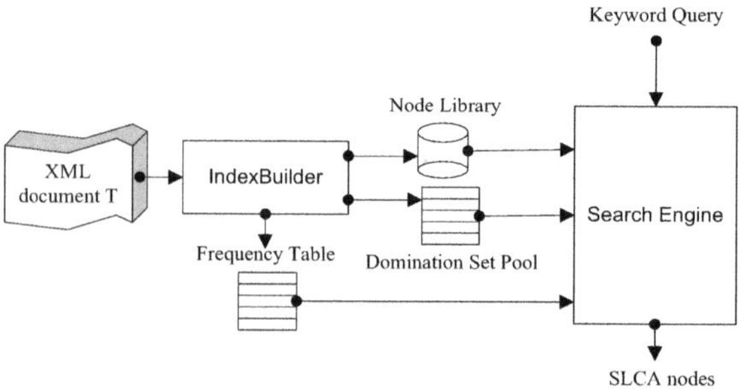

Fig. 4. HashSearch Architecture

5 Experimental Evaluation

In this section, we evaluate the performance of our approach. The experiments were conducted on a 2GHz dual-core laptop with 1GB of RAM. Our approach are tested on two real datasets, DBLP [14] and INEX [15]. DBLP documents are simple structures and the depth is about 10. INEX documents are attribute-abundant and structure-complex.

5.1 Experimental Setup

We have implemented in C++ the program called HS to evaluate the performance of our proposed algorithms and two versions of algorithms in [1] denoted by IMS and $IIMS$. IMS and $IIMS$ utilize B-tree in indexed form and non-indexed form respectively. All implementation are based on Berkeley DB [7]. The Node Libraries(NL) in HS use B+ tree to organize data nodes in the XML tree. Similar to the non-indexed B+ tree, the B+ tree in the NL utilizes keywords of data nodes as keys. But data in the NL of HS is the list of combined preorder ids of data nodes that directly contain the keyword. Moreover, the hash index in Berkeley DB is used in the Domination Set Pools(DSP) of HS. The Berkeley DB database was configured using a page size of 8KB and a cache size of 1GB.

We evaluated IMS, $IIMS$, HS with different classes of queries. We denoted the class of queries $kN - L - H$ in the same way as [1], where N stands for the number of keywords in the query, L and $H(L \leq H)$ are the lowest frequency and the highest frequency of N keywords respectively. We limit the query in the form that only one of the N keywords has the frequency L, the other $N-1$ keywords have the frequency H. We generated 10 random queries for each class of queries and executed each query 6 times. The average time of last five executions is reported.

The objectives of our experiments are twofold. The first is to study the performance of our HS algorithm when searching different sizes of documents. The second is to evaluate the advantage of processing queries with different keyword numbers and frequency, and our approach is compared with IMS and $IIMS$ [1]. The results are presented in the next subsection.

5.2 Performance of Query Evaluation

This experiment is designed to study the performance of the query evaluation with respect to different data sets and a variety of queries. The performance of the query evaluation on DBLP and INEX documents are tested against a range of significant parameters of document size, keyword number N, the lowest frequency L and the highest frequency H.

In the first experiment, we use DBLP data and the number of node is from 1M to 10M. We test 10 randomly generated queries, and the corresponding parameters for each query is (k2-100-1000).The results are given in Fig. 5. The processing time is roughly linearly scalable to the size of the XML nodes in the collection. In addition, the processing time is about 100 milliseconds, even with 10M XML nodes, about 200MB. Compared with the time needed by IMS and $IIMS$, our method saves about more than half of processing time.

The second experiment is to study the impact of query parameters on both DBLP and INEX documents. We compare the processing time of our approach with the processing time of IMS and $IIMS$ algorithms. Fig. 6 and Fig. 7 illustrate that our approach is found to be more efficient for both simple and complex document structures.

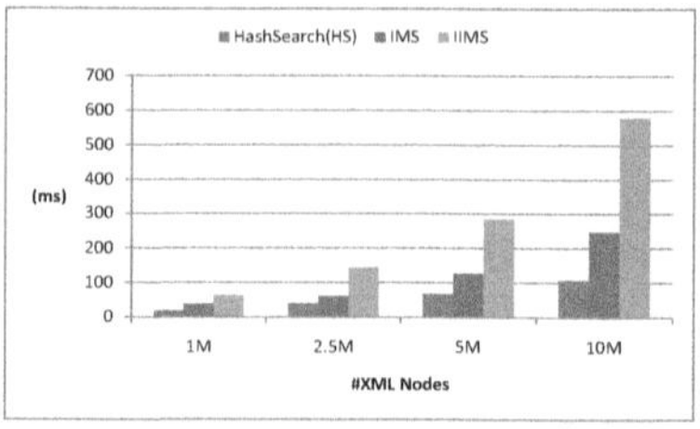

Fig. 5. Efficiency

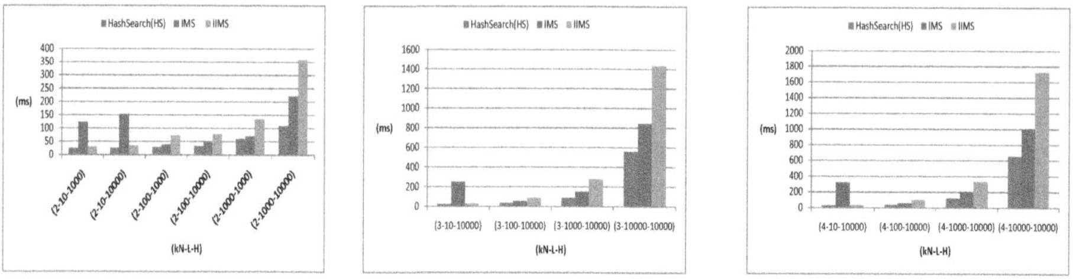

(a)two keywords (b)three keywords (c)four keywords

Fig. 6. Performance over DBLP data

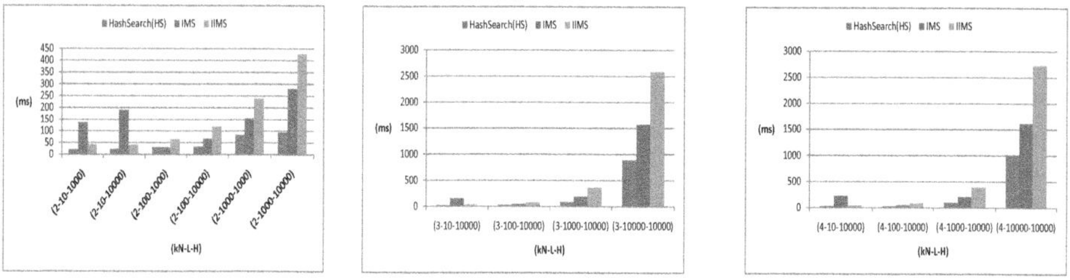

(a)two keywords (b)three keywords (c)four keywords

Fig. 7. Performance over INEX data

In Fig. 6 and 7, we use queries with 2, 3 and 4 keywords to evaluate both DBLP and INEX documents. The frequency parameters L and H range from 10 to 10000. The efficiency of our approach is affected less by the keyword number than $Multiway - SLCA$ methods.

When the keyword number is fixed and keyword frequency increases, Fig. 6 and 7 show the processing time of all algorithms. When low frequency is very small, $IIMS$ and HS perform better than IMS. HS is relatively faster than $IIMS$. When low frequency is large, IMS and HS take less time than $IIMS$. Moreover, the processing time of IMS is nearly twice as that of HS. The reason is that the performance of HS depends on the smallest frequency of keywords and the depth of the XML tree, while the performance of IMS and $IIMS$ algorithm is determined by not only these two factors but also the highest frequency of keywords. When the highest frequency of keywords become large, efficiency of $Multiway - SLCA$ approaches will be limited.

6 Related Work

The computation of the LCA of two nodes on trees has been solved by some early work [5,6]. Unfortunately, these algorithms are only intended to handle limited data that can be totally stored in the main-memory. To integrate the keyword search into XML query language, [8] introduces the *meet* operator whose definition is a special case of LCA problem for searching XML data. Such operator can be realized efficiently by running joins on relations. XRank [9] adopts a stack-based algorithm to answer Web-like keyword queries and ranks the results with extended Page-Rank hyperlink metric for XML. The ranking techniques are orthogonal to the retrieval and can be easily incorporated into our work.

Our work is closely related to three research work [1,2,4]. [4] enables users to retrieve XML data without any background knowledge about the document schema through a novel technique called Schema Free XQuery. In addition, a stack-based search algorithm is provided by [4] to answer queries using the *Meaningful LCA (MLCA)* semantics, which is similar to the SLCA semantics. XKSearch [2] proposes two efficient algorithms : the *Scan Eager (SE)* algorithm and the *Index Lookup Eager (ILE)* algorithm. MultiKSearch [1] proposes a more efficient algorithm named IMS than the ILE and SE algorithms by using several optimization strategies to reduce the number of LCA computation. Moreover, MultiKSearch extended its algorithm to support more general search queries involving the AND and OR semantics. In particular, the IMS algorithm is compared against in this paper.

XSeek [11] is an XML keyword search engine which can generate return nodes according to the analysis of both XML data structures and keyword match patterns. It's the first system that manages to automatically infer meaningful return nodes for keyword search results provided by XKSearch [2]. [12] introduces two intuitive and non-trivial properties for data and query : *monotonicity* and *consistency* to reason about keyword search strategies. Moreover, it proposes a novel semantics named MaxMatch which satisfies both porperties for identifying relevant matches and provides an efficient algorithm to implement it. MaxMatch is integrated into XSeek system[11].

7 Conclusions and Future Work

In this paper, we tackle the problem of how to process LCA queries efficiently over XML documents. We presents a novel approach called *Hash-Search approach* to answer SLCA-based keyword search queries on XML documents. Our approach outperforms the best-known *Incremental Multiway-SLCA approach.* Our experiments using different kinds of XML datasets demonstrate the efficiency of our approach. In the future work, we will focus on optimization technology for reducing the number of elements stored in the indexes without affecting the effectiveness of the algorithm.

Acknowledgement. This work is supported by NSFC grants (No. 60673137 and No. 60773075), National Hi-Tech 863 program under grant 2008AA01Z1470967 and Shanghai Leading Academic Discipline Project(Project NumberB412).

References

1. Sun, C., Chan, C.Y., Goenka, A.K.: Multiway SLCA-based Keyword Search in XML Data. In: WWW (2007)
2. Xu, Y., Papakonstantinou, Y.: Efficient Keyword Search for Smallest LCAs in XML Databases. In: SIGMOD Conference (2005)
3. Xu, Y., Papakonstantinou, Y.: Efficient LCA based Keyword Search in XML Data. In: CIKM (2007)
4. Li, Y., Yu, C., Jagadish, H.V.: Schema-Free XQuery. In: VLDB (2004)
5. Bender, M., Colton, M.F.: The LCA problem revisited. In: Latin American Theoretical Informatics (2000)
6. Harel, D., Tarjan, R.E.: Fast algorithm for finding nearest common ancestors. SIAM Journal on Computing (1984)
7. Sleepycat Software, The Berkeley Database (Berkeley DB), `http://www.sleepycat.com`
8. Schmidt, A., Kersten, M., Windhouwer, M.: Querying XML Document Made Easy:Nearest Concept Queries. In: ICDE (2001)
9. Guo, L., Shao, F., Botev, C., Shanmugasundaram, J.: XRANK: Ranked Keyword Search over XML Documents. In: SIGMOD (2003)
10. Botev, C., Amer-Yahia, S., Shanmugasundaram, J.: Expressiveness and performance of full-text search languages. In: Ioannidis, Y., Scholl, M.H., Schmidt, J.W., Matthes, F., Hatzopoulos, M., Böhm, K., Kemper, A., Grust, T., Böhm, C. (eds.) EDBT 2006. LNCS, vol. 3896, pp. 349–367. Springer, Heidelberg (2006)
11. Liu, Z., Chen, Y.: Identifying Meaningful Return Information for XML Keyword Search. In: SIGMOD (2007)
12. Liu, Z., Chen, Y.: Reasoning and Identifying Relevant Matches for XML Keyword Search. In: VLDB (2008)
13. Li, G., Ooi, B.C., Feng, J., Wang, J., Zhou, L.: EASE: An Effective 3-in-1 Keyword Search Method for Unstructured, Semi-structured and Structured Data. In: SIGMOD (2008)
14. DBLP XML records, `http://dblp.uni-trier.de/xml/`
15. INEX XML data sets

MCN: A New Semantics Towards Effective XML Keyword Search

Junfeng Zhou[1,2], Zhifeng Bao[3], Tok Wang Ling[3], and Xiaofeng Meng[1]

[1] School of Information, Renmin University of China
[2] Department of Computer Science and Technology, Yanshan University
{zhoujf2006,xfmeng}@ruc.edu.cn
[3] School of Computing, National University of Singapore
{baozhife,lingtw}@comp.nus.edu.sg

Abstract. We investigate the expressiveness of existing XML keyword search semantics and propose *Meaningful Connected Network (MCN)* as a new semantics to capture meaningful relationships of the given keywords from XML documents. Our evaluation method adopts a two-step strategy to compute all MCNs. In the first step, we identify a set of query patterns from a new schema summary; in the second step, all query patterns are processed based on two efficient indices, partial path index and entity path index. The experimental results show that our method is both effective and efficient.

1 Introduction

As an effective search method to retrieve useful information, keyword search has gotten a great success in IR field. However, the inherently hierarchical structure of XML data makes the task of retrieving the desired information from XML data more challenging than that from flat documents. An XML document can be modeled as either a rooted tree or a directed graph (if IDRefs are considered). For example, Fig. 1 shows an XML document D, where solid arrows denote the containment edge, dashed arrows denote the reference edge, and the number beside each node denotes the node id.

The critical issue of XML keyword search is how to find meaningful query results. In both tree model and graph model, the main idea of existing approaches is to find a set of *Connected Networks* (CNs) where each CN is an acyclic subgraph T of D, T contains all the given keywords while any proper subgraph of T does not. In particular, in tree data model, Lowest Common Ancestor (LCA) semantics [1] is first proposed, followed by SLCA (smallest LCA) [1] and MLCA [5] which apply additional constraints on LCA. In graph data model, methods proposed in [6,7,8,9,10] focused on finding matched CNs where IDRefs are considered.

In practice, however, most existing approaches [1,2,3,4,5,6,7,8,9,10] only take into account the structure information among the nodes in XML data, but neglect the *node categories*; thus they suffer from the *limited expressiveness*, which makes them fail to provide an effective mechanism to describe how each part in the returned data fragments are connected in a meaningful way, as shown in Example 1.

X. Zhou et al. (Eds.): DASFAA 2009, LNCS 5463, pp. 511–526, 2009.

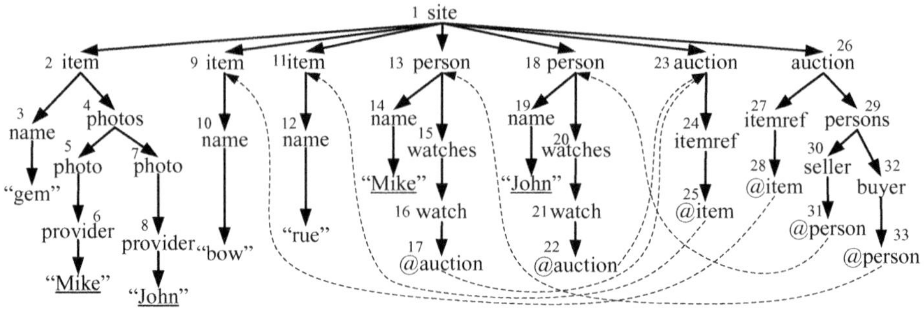

Fig. 1. An example XML document D

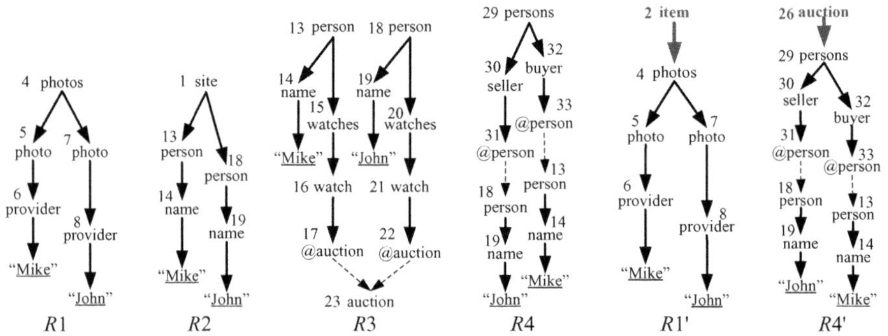

Fig. 2. $R1$ to $R4$ are four possible answers of existing methods for query Q={Mike, John} against D in Fig. 1. $R3$, $R1'$ and $R4'$ are three meaningful answers of our method.

Example 1. Consider a keyword query Q={Mike, John} issued on D in Fig. 1. If IDRefs in D are not considered, then $R1$ and $R2$ in Fig. 2 are two (not all) matched results according to the LCA semantics [1] where the LCA node are *photos* and *site*, respectively. If IDRefs in XML data are considered, we may find more results, e.g., $R1$ to $R4$ are four matched results of existing methods. However, we cannot identify any meaningful relationship between 'Mike' and 'John' from $R1$, $R2$ and $R4$, because the nodes (*photos, site, persons*) connecting 'Mike' and 'John' do not convey any useful information to explain the relationship between 'Mike' and 'John'. We observe when talking about relationships between data elements, users just care about relationships of those representative nodes (*photo, person, item* and *auction* in Fig. 1), which we call entities in ER model, and most of the time their query is based on the relationships of entities, rather than those meaningless ones (e.g., *photos*, *site* and *persons*), which are just used to organize data. With this observation, $R3$, $R1'$ and $R4'$ in Fig. 2 should be meaningful results. $R3$ means 'Mike' and 'John' watch the same *auction*; $R1'$ means both 'Mike' and 'John' provided a *photo* to the same *item*; $R4'$ means 'Mike' bought an *item* sold by 'John' in an *auction*. However, according to the semantics of existing works [1,2,3,4,5,6,7,8,9,10], $R1'$ and $R4'$ will be removed as answers because of $R1$ and $R4$, respectively.

Motivated by the above problem, we propose a new semantics called *Meaningful Connected Network (MCN)* to capture the *meaningful relationships* of the given keywords from *XML document graph* by considering reference relationship. A MCN is an acyclic subgraph T of the given XML document D and contains all the keywords at least once, which is defined based on the relationships among entities (or entity instances), rather than simply on data elements without considering their categories. For the above query, we first check which entity instances 'Mike' ('John') belongs to, then find the meaningful relationships of the two entity instances of 'Mike' and 'John'. Finally, $R3$, $R1'$ and $R4'$ (not $R1$, $R2$ and $R4$) as MCNs are considered as meaningful results and returned.

However, finding even the first CN is reducible to the classical group Steiner tree problem, which is known to be NP-complete [12]. As a MCN may contain nodes that are redundant to a CN (e.g., item in $R1'$ is redundant to $R1$), finding all MCNs from the given XML document is more difficult than finding all CNs. Note all MCNs can be classified into different groups according to their structure, we call the structure of a MCN together with all query keywords a *query pattern* (QP) (e.g., $R3$, $R1'$ and $R4'$ are three QPs after removing all node id), thus finding all occurrences of MCNs equals to solving the following two problems.

- ($P1$) Efficiently identify all QPs from the underlying schema.
- ($P2$) Efficiently evaluate all QPs against the given XML data.

For problem $P1$, we propose to use *entity graph* as a schema summary to capture the meaningful relationships of entity nodes from the original schema. Entity graph is generated from the original schema by removing the noisy information while preserving the connection relationships of entity nodes. Then we propose an algorithm based on entity graph to efficiently compute all QPs.

For problem $P2$, we design two efficient indices to improve the query performance. The first is *partial path index.* For each keyword k, partial path index stores a set of paths, each of which starts with an entity instance and ends at one of its attribute node that directly contains k. The second is *entity path index.* Entity path index stores all the matching instances of entity pairs, where each pair is joined by one edge of the entity graph.

In summary, our contributions are listed as follows:

- We propose an effective semantics, i.e., Meaningful Connected Network (MCN), to enhance the expressiveness of keyword search query.
- We propose to use an entity graph to reduce the cost of computing QPs.
- We propose an algorithm based on partial path index and entity path index to improve the performance of query evaluation. We proved the costly structural join operations[1] can be avoided from the evaluation of *all* QPs.
- Extensive experiments are conducted on datasets of various characteristics, and the results show our method is both effective and efficient.

2 Background and Related Work

Schema: We assume the schema is always available, as we can use the methods proposed in [13,14] to infer the schema (if unavailable). We use a node labeled

[1] Join operations based on ancestor-descendant or parent-child relationship.

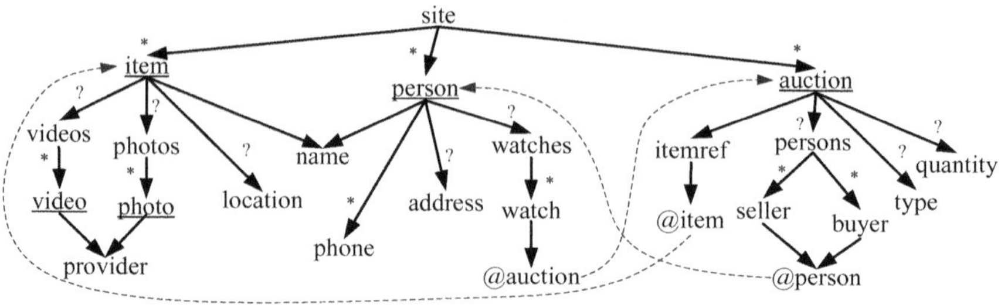

Fig. 3. An example schema S of XMark

directed graph to model a schema. Formally, $S = (V_S, E_S)$, where V_S denotes a set of schema elements each with a distinct tag name, E_S denotes a set of directed edges between schema elements. As shown in Fig. 3, there are two kinds of edges in S. The first is the *containment edge*, which is drawn as a solid arrow from an element to its child element. The second is the *reference edge*, which is drawn as a dashed arrow from the attribute of referrer element to referee element.

Node Categories: In the following discussion, whenever we mention *entity* and *attribute*, they refer to the notions defined in ER-model, rather than that defined in XML specification[2]. Generally speaking, two kinds of methods can be adopted to specify the category of each schema element, which are (1) automatic methods using heuristic inference rules [4,6] and (2) manual method done by users, DBA or domain expert. The inference rules used in [4,6] are as follows:

1. A node represents an entity if it corresponds to a $*$-node in the DTD.
2. A node denotes an attribute if it does not correspond to a $*$-node, and only has one child, which is a value.
3. A node is a connection node if it represents neither an entity nor an attribute. A connection node can have a child that is an entity, an attribute or another connection node.

The automatic method can avoid the cost of manual intervention, but it may not be quite correct. E.g., for the schema in Fig. 3, *person* is a $*$-node, thus by rule 1, *person* represents entity, instead of the attribute of *site*. *name* is considered as an attribute of *person* and *item* according to rule 2. *site* is not a $*$-node and has no value child, so it is a connection node by rule 3. However, according to the above rules, *phone* will be considered as entity, which is unreasonable, since *phone* is usually considered as a multi-value attribute of *person*. By using manual method from scratch, we can get accurate category of each schema node. However, it may impose great burden to users, DBA or domain experts.

Therefore, to achieve as accurate as possible node categories while paying minimum manual intervention, we first employ the above three inference rules to get an approximately accurate categorization, followed by a minor manual

[2] http://www.w3.org/TR/REC-xml/

Table 1. Summary of notations

Notation	Description	Notation	Description
MEW	meaningful entity walk	QP	query pattern
(M)CN	(meaningful) connected network	TP	tree (or twig) pattern
PPI	partial path index	EPI	entity path index

adjustment from users, DBA or domain expert. Using this method, we can consider *phone*, *watch*, *buyer* and *seller* as multi-value attributes of their entities, respectively. We take the underlined nodes as entities in Fig. 3.

To facilitate our discussion, whenever we use *entity*, it refers to the entity-type which corresponds to an entity node of a schema; the term *entity instance* is used to denote the instance of *entity* in XML document. In this paper, we mainly focus on how to provide effective and efficient mechanism to extract meaningful results based on the results of existing classification methods. The notations used in our discussion are shown in Table 1.

Discussion and Related Work: Among existing XML keyword search methods [1,2,3,4,5,6,7,8,9,10], The basic semantics [1,2,3,4,5] is based on the *tree* model, thus cannot capture the meaningful relationship conveyed by IDRefs.

For *graph* model (IDRefs considered) based methods [6,7,8,9,10], the first group [9,10] directly compute all *CN*s from the given XML document. However, finding even the minimal connected network is reducible to the classical NP-complete group Steiner tree problem [12]. Thus these methods [9,10] apply special constraints to CN and find only a subset of all CNs.

The second group [6,7,8] adopt a two-step strategy: (1) compute the set of QPs that are isomorphic to the set of CNs, (2) evaluate all QPs to get the matching results. However, when the schema graph becomes complex and when evaluating large amount of QPs, both the two steps are no longer a trivial task.

For step 1: the methods proposed in [6,8] focus on finding all QPs of schema elements, text values are attached to different schema elements, thus they cannot process queries involving text values attached to two schema elements of same name, e.g., {person:Mike, person:John}. [7] proposed a method to compute from the schema graph all QPs of keyword queries that allow both text values and schema elements. The main idea is to find all QPs from a new *expanded* graph G' that is generated from G and each QP is a subgraph of G'.

For a given keyword query $Q = \{k_1, k_2, ..., k_m\}$, [7] maintains for each keyword k an inverted list which stores the data elements directly containing k and works through the following steps. (1) For each keyword $k_i \in Q$, produce the element set S_{k_i} which consists of all data elements containing k_i. (2) Based on the m sets $S_{k_1}, S_{k_2}, ..., S_{k_m}$, produce data element set S_K for all subsets K of Q, where $S_K = \{d | d \in \cup_{k \in K} S_k \wedge \forall k \in K, d\ contains\ k \wedge \forall k \in Q - K, d\ does\ not\ contain\ k\}$ [11]. (3) For all elements of S_K, find the set of corresponding schema elements $S_{label(S_K)} = \{l | \exists d \in S_K, l = label(d)\}$. For each $l \in S_{label(S_K)}$, add a node named l to the original schema graph by attaching K to l to produce a new schema graph G'. (4) Find from G' all the QPs, where each QP q is a subgraph of G',

q contains all keywords at least once, while any proper subgraph of q does not. Thus the performance of the first step is affected by three factors: (1) I/O cost of accessing all data instances to construct G' in the first two steps; (2) the maximum size of QP, and (3) the size of the new expanded graph G'.

For step 2: as a keyword query may correspond to multiple QPs, and each QP consists of several tree (twig) patterns (*TPs*) connected together by reference edges, which imposes great challenges for subsequent query processing. Existing methods made improvements from the following four orthogonal aspects: (1) designing efficient algorithms [15,16], (2) reducing the size of the given QP [17], (3) designing efficient indices to reduce the size of input streams [15,16], and (4) reusing the intermediate results to improve the whole performance. However, all these methods suffer from costly structural join operation, which greatly affects the whole performance.

3 Semantics of Keyword Search Queries

From D in Fig. 1, we know person "Mike" bought the item sold by person "John", which can be expressed as answer $R4'$ in Fig. 2. $R4'$ manifests the meaningful relationship between entity instances may contain edges of different directions (*mixed directions* problem) and cannot be got by simply traversing the document. Even if we omit the direction of each edge, it is infeasible to XML document because of its large size. An alternative way is finding such relationship from schema graph, which has much smaller size. However, some relationships produced in such case may be meaningless (*meaningfulness* problem). For example, R' : "item→name←person" is a possible relationship produced by traversing undirected schema graph S_u of S in Fig. 3, such relationship is meaningless since according to S, person and item must have different child element of same element type name in XML documents, which contradicts R'. Thus, we need an effective way to capture both "*mixed directions*" and "*meaningfulness*" so as to avoid losing meaningful relationships (e.g., $R4'$) by traversing directed graph and producing meaningless relationships (e.g., R') by traversing undirected graph.

Definition 1 (Walk). *A v_0-v_k walk W : $v_0, e_1, v_1, e_2, v_2, \ldots, v_{k-1}, e_k, v_k$ of the undirected schema graph S_u is a sequence of vertices of S_u beginning with v_0 and ending at v_k, such that each two consecutive vertices v_{i-1} and v_i are joined by an edge e_i of S_u.*

The number of edges of W is called the length of W, which is denoted as $L(W)$. For any two nodes u and v of S_u, if there exists at most one edge joining u and v in S_u, W can be written as W : $v_0, v_1, \ldots, v_k$. Any $v_i - v_j$ walk W' : $v_i, e_{i+1}, v_{i+1}, \ldots, v_{j-1}, e_j, v_j (0 \le i \le j \le k)$ extracted from W is called a sub-walk of W, which is denoted as $W' \subseteq W$.

Intuitively, a walk denotes a possible connection relationship of two schema nodes where direction is not considered. Note Definition 1 doesn't require the listed vertices and edges to be distinct, there may be more than one walk between two nodes. We define walk so as to avoid the problem of "*mixed directions*", and

Definition 2 is used to avoid the problem of "*meaningfulness*" and capture the meaningful connection relationship of two entities.

Definition 2 (Meaningful Entity Walk (MEW)). *Let S be a schema graph, a $v_0 - v_k$ walk W of the undirected schema graph S_u is a meaningful entity walk of S if both v_0 and v_k are entity nodes and*

- *$L(W) \leq 1$, or,*
- *W doesn't contain a sub-walk W' that has the form $u \rightarrow v \leftarrow w$ in S, where "$\rightarrow$" denotes a solid arrow from $u(w)$ to v in S. Moreover, if W' has the form $u \leftarrow v \rightarrow w$ in S, v must be an entity node.*

Example 2. According to Definition 2, W_1:"person" is a MEW since person is an entity and $L(W_1) = 0 \leq 1$. W_2:"item→name←person" is not a MEW according to Definition 2, the reason is stated in the first paragraph of this section, where W_2 is denoted as R'. W_3:"person→watches→watch→@auction- -→auction" is a MEW according to Definition 2, which means a person is watching an auction. W_4:"person←- -@person←buyer←persons←auction→persons→seller →@person - -→person" is a MEW which means a person bought an item sold by another person. W_5:"item←site→person" is not a MEW according to Definition 2 as W_5 has the form "$u \leftarrow v \rightarrow w$" and site is not an entity, which means the relationship of item and person cannot be interpreted by a non-entity node, i.e., site.

Definition 3 (Meaningful Connected Network (MCN)). *Let $Q = \{k_1, k_2, ..., k_m\}$ be the given keyword query. A meaningful connected network of Q on the XML document D is a subgraph T of D, which holds all the following properties:*

1. *T contains $k_i (1 \leq i \leq m)$ at least once,*
2. *for any node u of T, if u is not an entity instance and joined by just one edge with other nodes of T, u contains at least one keyword,*
3. *for any two nodes u and v of T, if u and v are entity instances, there exists at least one $u - v$ MEW instance W on T,*
4. *no proper subgraph of T can hold for the above three properties.*

Example 3. Consider the six subgraphs in Fig. 2, where there are four entities, i.e., *person*, *photo*, *item* and *auction*. For $R1$, we cannot explain intuitively the relationships of the two *photo* nodes as they are connected together through a connection node, i.e., *photos*. Similarly, we cannot explain intuitively the relationships of the two *person* nodes in $R2$ and $R4$ as they are connected together through two connection nodes, i.e., *site* and *persons*. Since the walk "photo←photos→photo" in $R1$ is not a MEW (*photos* is not an entity node), according to Definition 3, $R1$ is not a MCN. $R2$ and $R4$ are not MCNs for the same reason. Thus they will be considered as meaningless answers. The intuitive meanings of $R3$, $R1'$ and $R4'$ are explained in Example 1, where each pair entity nodes in $R3$, $R1'$ and $R4'$ are connected through MEWs. According to Definition 3, $R3$, $R1'$ and $R4'$ are MCNs and considered as meaningful answers.

As shown in [11], the size of the joining sequence of two data elements in a given XML document is data bound (data bound means the size of a result may be as large as the number of nodes in an XML document), so is the MCN. Thus users are usually required to specify the maximum size C for all MCNs, which equals to the number of edges. However, such method may return results of very weak semantics or lose meaningful results. For example, each one of $R3$, $R1'$ and $R4'$ contains 3 entity instances, thus the semantic strongness of the three MCNs should be equal to each other, if all edges have the same weight. By specifying $C = 10$, $R3$ and $R4'$ will not be returned as matching results. On the other hand, a MCN may contain no connection node but an overwhelming number of entity instances, if its size equals to C, it may convey very weak semantics. Therefore in our method, the constraints imposed on MCN is not the maximum number of edges, but that of entity instances. This C is a user-specified variable with a default value of 3. As a result, a formal definition of keyword search is as below.

Definition 4 (Keyword Search Problem). *For a given keyword query Q, find all matched MCNs from the given XML document D, where each MCN contains at most C entity instances.*

4 Computation of Query Patterns

Definition 5 (Entity Path). *Path $p : v_1, v_2, ..., v_k (2 \leq k)$ of schema graph S is called an entity path if only v_1 and v_k are entity nodes, and for any $v_i(2 \leq i \leq k-1), v_i \neq v_j(1 \leq j \leq k \wedge i \neq j)$.*

Definition 6 (Partial Path). *A partial path p is a path of the given QP Q, which starts with an entity node n that is the only entity node of p.*

Intuitively, an entity path describes the direct relationship of two entities, a partial path denotes containment relationship of an entity and one of its attributes. As the definition of MCN is based on relationship of entity nodes, an important operation is for a given keyword k, finding the set of entity nodes that have entity instances containing k as their attribute or attribute values, which is denoted as $selfE(k)$. We maintain an *auxiliary index* H that stores the set of partial paths (*not their instances*) for each keyword k, where each partial path has database instances appearing in the given XML document D, which can be got after parsing D. For example, H stores "photo/provider, person/name" for 'Mike'. According to H, selfE('Mike')={photo,person}.

Definition 7 (Entity Graph). *Let S be a schema graph, $\mathcal{P}$ the set of entity paths of S. The* entity graph *of S is represented as $G = (V, E)$, which consists of only entity nodes of S, and for each entity path $p \in \mathcal{P}$ that connect u and v in S, there is an edge in G that joins u and v.*

As shown in Fig. 4 (A), in an entity graph, two entity nodes may be joined by two or more edges, e.g., person and auction are joined by e_4, e_5, e_6. Each edge

Algorithm 1. getQP(Q, G, C) /*$Q = \{k_1, k_2, ..., k_m\}$*/

$\mathcal{Q} \leftarrow \emptyset$ /*queue of QPs*/
if ($|Q| = 1$) **then** {$S_{QP} \leftarrow$ selfE(k_1); **return** S_{QP}}
foreach (combination $\mathcal{E} = (E'_1, E'_2, ..., E'_m)$, $E'_i \in selfE(k_i)$) **do**
  get the partition $P = \{S_{E_1}, S_{E_2}, ..., S_{E_q}\}$ of $\mathcal{E}$ according to entity names
  get the keyword set $\mathcal{K}_{E_i} = \{k | E \in S_{E_i} \wedge E \in selfE(k) \wedge k \in Q\}$ of S_{E_i}
  get the entity set $S_{\mathcal{K}_{E_i}} = \{E_i^K | K \subseteq \mathcal{K}_{E_i}\}$ of $E_i (1 \leq i \leq q)$
  put nodes of $S_{\mathcal{K}_{E_i}}$ into S_{QP} if they are QPs, otherwise put them into $\mathcal{Q}$
  while ($\neg$ empty($\mathcal{Q}$)) **do**
    $J \leftarrow$ RemoveHead($\mathcal{Q}$)
    foreach (edge $e = (E, u)$ in G_u that is incident with a node u of J) **do**
      if ($E \in \{E_1, E_2, ..., E_q\}$) **then**
        foreach (node $E' \in S_{\mathcal{K}_E}$) **do**
          $J' \leftarrow$ Add E' and (E', u) to J;
          if (isQP(J')$\wedge$nEntity(J') $\leq C$) **then** $S_{QP} \leftarrow S_{QP} \cup \{J'\}$
          else if (nEntity(J') $< C$) **then** Add J' to $\mathcal{Q}$
      else
        $J' \leftarrow$ Add E and (E, u) to J;
        if (nEntity(J') $< C$) **then** Add J' to $\mathcal{Q}$
return S_{QP}

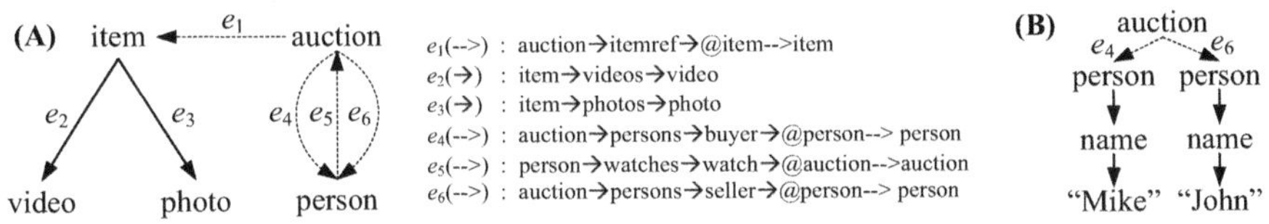

Fig. 4. The entity graph G (A) of S in Fig. 3, (B) is the QP of $R4'$ in Fig. 2.

of an entity graph may be a containment edge (solid arrow) or reference edge (dashed arrow). Each containment edge of G denotes an entity path that consists of just containment edges in S, and each reference edge denotes an entity path that consists of at least one reference edge in S. According to Definition 7, we have the following lemma.

Lemma 1 *There exists a one-to-one mapping between the edges of an entity graph and the entity paths of the original schema graph.*

Moreover, we observe that a QP consists of two kinds of relationships, (1) the relationship between entity nodes, i.e., *entity path*; (2) the relationship between entity nodes and attribute or attribute values, i.e., *partial path*. For example, after removing the node id, $R4'$ as a QP is shown in Fig. 4 (B), which contains two entity paths, i.e., e_4 and e_6, and a partial path, i.e., "person/name". According to Lemma 1, for a QP, the relationships of entity nodes can be got from entity graph, and the partial paths can be got from the auxiliary index H stated in the paragraph after Definition 6. For simplicity, we use the relationships of entity nodes of a QP to denote the QP in the following.

Algorithm 1 shows how to compute all QPs, which has three parameters, a keyword query Q, an entity graph G and the maximum number C of entity nodes. If Q contains only one keyword k, the set of QPs S_{QP} equals to selfE(k) (line 2); otherwise, for each combination $\mathcal{E}=(E'_1, E'_2, ..., E'_m)$ where $E'_i \in selfE(k_i)$, it computes the set of QPs of $\mathcal{E}$ (line 3-18). In particular, it first gets a partition $P=\{S_{E_1}, S_{E_2}, ..., S_{E_q}\}$ of $\mathcal{E}$ according to entity names, where $S_{E_i} = \{E|E \in \mathcal{E} \wedge label(E) = E_i\}$ (line 4). Then it gets the set of keywords $\mathcal{K}_{E_i}$ of S_{E_i} and gets all combinations of keywords contained by an entity node, i.e., multiple keywords may be contained by an entity node(line 5-6). Then adds all nodes of an entity set $S_{\mathcal{K}_{E_i}}$ to $\mathcal{Q}$ if they do not contain all keywords; otherwise, put them into S_{QP} (line 7). In line 8-18, while $\mathcal{Q}$ is not empty, a graph J is removed from $\mathcal{Q}$ in line 9 for further computing. isQP(J')=TRUE means that J' is a QP and nEntity(J') denotes the number of entity nodes in J'. Finally, S_{QP} is returned (line 19). Note a MCN is an instance of a QP, the checking of QP is similar to that of MCN except that QP is defined on schema graph.

Example 4. Assume C=3, G is the entity graph in Fig.4(A). As shown in Fig.5, according to D in Fig. 1, for $Q=\{M, J\}$,we have selfE('M')=selfE('J')=$\{P, PH\}$. There are three combinations of entities for Q, i.e., (PH, PH), (P, PH) and (P, P). For (PH, PH), according to line 4 of Algorithm 1, $P=\{S_{PH}\}$, where $S_{PH}= \{PH_1, PH_2\}$ (PH_1 and PH_2 denote they are two entity nodes of same name). In line 5, we know that a PH node may contain two keywords, i.e., $\mathcal{K}_{PH}=\{M, J\}$. According to line 6, there are three possible cases where a photo node contains these keywords, that is, $S_{\mathcal{K}_{PH}}=\{PH^{[M]}, PH^{[J]}, PH^{[M,J]}\}$. In line 7, $PH^{[M]}$ and $PH^{[J]}$ are put into $\mathcal{Q}$ and $PH^{[M,J]}$ is put into S_{QP} since it is already a QP. In line 9, $PH^{[M]}$ is first removed from $\mathcal{Q}$, since there is only one node, i.e., item, adjacent to photo in G and nEntity($PH^{[M]} \leftarrow^{e_4} I$)=2, it is put into $\mathcal{Q}$ in line 18. There are six newly generated graph for $PH^{[M]} \leftarrow^{e_4} I$, and only $PH^{[M]} \leftarrow^{e_4} I \rightarrow^{e_4} PH^{[J]}$ is a QP. We omit the computation for (P, PH) and (P, P) and just show them in Fig. 5.

Theorem 1. (Completeness) *For a given ATP query, Algorithm 1 produces all QPs that satisfy each QP has at most C entity nodes.*

The correctness of Theorem 1 is obvious according to Algorithm 1 and Example 4 and we omit the proof for limited space. Compared with the method of [7],

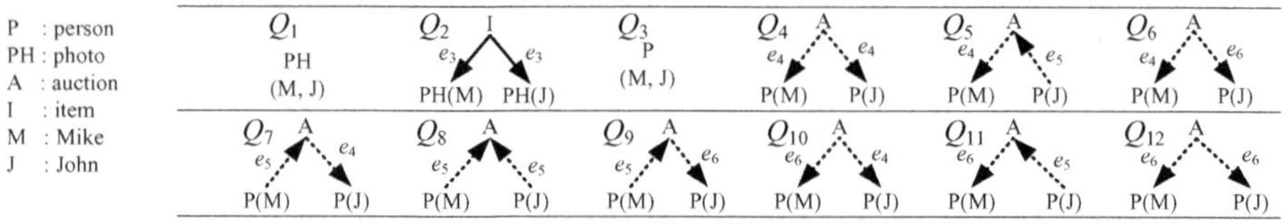

Fig. 5. The set of QPs of $Q=\{Mike, John\}$, partial paths are omitted, for Q_1 and Q_2, the partial path is 'photo/provider', for Q_3 to Q_{12}, partial path is 'person/name'

our method made improvements from three aspects: (1) by postponing checking the real dataset until evaluating QPs, Algorithm 1 avoids the costly I/O operation, (2) Algorithm 1 is based on entity graph, which is much smaller than the expanded schema graph, and (3) the value of C in Algorithm 1 is much smaller than that of [7], where C equals to the number of edges in a QP.

5 Query Processing

To accelerate the evaluation of QPs, we firstly introduce *partial path index* PPI. For each keyword k, we store in PPI a list of tuples of the form $< P_{ID}, Path >$, where P_{ID} is the ID of a partial path, and $Path$ is a database instance of P_{ID} in XML documents. All $Path$s are clustered together according to P_{ID} and sorted in document order. The PPI of D in Fig. 1 is shown in Table 2, where only partial content is presented. The second index is *entity path index EPI*, for each edge e of the given entity graph, we maintain in EPI a set of database instances of e. The EPI of D in Fig. 1 is shown in Table 3.

Theorem 2. *Let Q be a given keyword query. Using PPI and EPI, the structural join operations can be avoided from the evaluation of Q.*

Proof. **[Sketch]** According to Algorithm 1, any keyword query Q has a set of QPs S_{QP}. Each $Q' \in S_{QP}$ consists of two kinds of relationships, (1) the relationship between entity nodes and (2) the relationship between entity nodes and attribute or attribute values. The former can be computed by probing EPI and the latter can be computed by probing PPI, the final results can be got from the results of all selection operations on PPI and EPI.

As shown in Algorithm 2, we find the set of QPs S_{QP} in line 1. In line 2-3, all QPs that contain at least one selection operation that produces empty set is removed from S_{QP}. In line 4-12, we check for each node E_K of a QP Q', whether E_K has database instances that contain all keywords in K_E (K_E is the set of keywords attached to E_K). If E_K does not have such database instances, $R_{E_K} = \emptyset$. In line 13-17, for each QP $Q' \in S_{QP}$, if there is an entity node E_K and $R_{E_K} = \emptyset$, Q' is removed from S_{QP} in line 14; otherwise, Q' is evaluated in line 16. In line 17, the results $\mathcal{R}_{Q'}$ of Q' are put into $\mathcal{R}$. In line 18, $\mathcal{R}$ is returned.

Table 2. The partial path index of the XML document D in Fig. 1

Keyword	Tuple sets	Keyword	Tuple sets
Mike	$< photo/provider, 5/6 >$ $< person/name, 13/14 >$	John	$< photo/provider, 7/8 >$ $< person/name, 18/19 >$

Table 3. The entity path index of the XML document D in Fig. 1

Edge	Entity Paths	Edge	Entity Paths
e_1	{23/24/25/11, 26/27/28/9}	e_4	{26/29/32/33/13}
e_2	$\emptyset$	e_5	{13/15/16/17/23, 18/20/21/22/23}
e_3	{2/4/5, 2/4/7}	e_6	{26/29/30/31/18}

Table 4. The selection operations of the 12 QPs of Q in Example 5

Selection Op.	Result	Queries	Selection Op.	Result	Queries
$photo/provider{\sim}`Mike'$	$\neq \emptyset$	Q_1, Q_2	e_3	$\neq \emptyset$	Q_2
$photo/provider{\sim}`John'$	$\neq \emptyset$	Q_1, Q_2	e_4	$\neq \emptyset$	Q_4 to Q_7, Q_{10}
$person/name{\sim}`Mike'$	$\neq \emptyset$	Q_3 to Q_{12}	e_5	$\neq \emptyset$	Q_5, Q_7 to Q_9, Q_{11}
$person/name{\sim}`John'$	$\neq \emptyset$	Q_3 to Q_{12}	e_6	$\neq \emptyset$	Q_6, Q_9 to Q_{12}

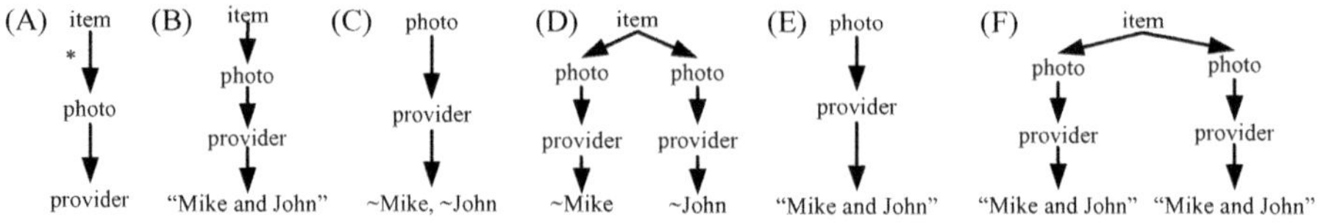

Fig. 6. Illustrating of redundant QP

Example 5. According to Example 4, $Q=\{Mike, John\}$ corresponds to 12 QPs shown in Fig. 5, i.e., $S_{QP}=\{Q_i|1\leq i\leq 12\}$. According to Table 3 and Table 2, we can get Table 4, which shows all selection operations of S_{QP}. Obviously, there are 8 selection operations involved in Q_1 to Q_{12}. According to line 4-12 and the PPI in Table 2, we have $\mathcal{E}=\{PH(M,J), PH(M), PH(J), P(M,J), P(M), P(J)\}$ for Q_1 to Q_{12} in Fig. 5. As $R_{PH(M,J)}=R_{P(M,J)}=\emptyset$, we can delete Q_1 and Q_3 from S_{QP} in line 14. In line 16, we evaluate the remainder QPs, among which three QPs have non-empty result sets, i.e., Q_2, Q_6, Q_8. $\mathcal{R}_{Q_2}=\sigma_{photo/provider\sim`Mike'}\bowtie\sigma_{e_3}\bowtie\sigma_{e_3}\bowtie\sigma_{photo/provider\sim`John'}=\{R1'\}$. Similarly, $\mathcal{R}_{Q_6}=\{R4'\}$, $\mathcal{R}_{Q_8}=\{R3\}$, where $R3, R1', R4'$ are the three MCNs in Fig. 2. Other QPs in S_{QP} have empty result sets. Therefore the final result set $\mathcal{R}_Q=\mathcal{R}_{Q_2}\cup\mathcal{R}_{Q_6}\cup\mathcal{R}_{Q_8}=\{R3, R1', R4'\}$.

Note Algorithm 1 may produce redundant QPs. For instance, consider the schema graph in Fig. 6 (A), where photo and item are entities, (B) is the XML document conforming to (A). Assume $C = 3$, i.e., each QP contains at most three entity nodes, Fig. 6 (C) and (D) are two QPs according to Algorithm 1. Obviously, Fig. 6 (E) and (F) are two matches of Fig. 6 (C) and (D), respectively. In fact, the QP in Fig. 6 (D) is redundant since in Fig. 6(E), both the two photo nodes and the two provider nodes represent same data element in Fig. 6 (B). Therefore before evaluating each QP Q' in $\mathcal{Q}$, we need to check in line 2-14 for each E_K, whether there exists entity instances of label E such that each entity instance d contains all keywords of K and no other $E'_{K'}$ exists such that $k \in (K' - K)$ and d contains k. Thus we have Theorem 3.

Theorem 3 (Non Redundancy). *If a QP evaluated in line 16 of Algorithm 2 produces a MCN d, then no other QPs can produce d as their instance.*

6 Experimental Evaluation

6.1 Experimental Setup

We used a PC with Pentium4 2.8 GHz CPU, 2G memory, 160 GB IDE hard disk, and Windows XP professional as the operating system. We implemented *IM*

Algorithm 2. indexMerge(Q, G, C)

$S_{QP} \leftarrow$ getQP(Q, G, C)
foreach (edge $e \in Q', where\ Q' \in S_{QP}$) **do**
  if $(R_{\sigma_e} = \emptyset)$ **then** $S_{QP} \leftarrow S_{QP} - \{Q'\}$
foreach ($node\ E_K \in Q' \wedge Q' \in S_{QP} \wedge E_K \notin \mathcal{E}$) **do**
  if (E_K *is attached with keywords set* K_E) **then**
    $R_{E_K} \leftarrow$merge the results of selection operations of each keyword of K_E
  foreach $(E'_{K'} \in \mathcal{E} \wedge label(E) = label(E'))$ **do**
    $R = R_{E_K} \cap R_{E'_{K'}}$
    if $(K_E \subset K'_{E'})$ **then** $R_{E_K} \leftarrow R_{E_K} - R$
    else if $(K_E \supset K'_{E'})$ **then** $R_{E'_{K'}} \leftarrow R_{E'_{K'}} - R$
    else $\{R_{E_K} \leftarrow R_{E_K} - R;\ R_{E'_{K'}} \leftarrow R_{E'_{K'}} - R\}$
  $\mathcal{E} \leftarrow \mathcal{E} \cup \{E_K\}$
foreach $(Q' \in S_{QP})$ **do**
  if $(\exists E_K \in \mathcal{E} \wedge E_K \in Q' \wedge R_{E_K} = \emptyset)$ **then** $S_{QP} \leftarrow S_{QP} - \{Q'\}$
  else
    $\mathcal{R}_{Q'} \leftarrow$ merge the results of all selection operations of Q'
    $\mathcal{R} \leftarrow \mathcal{R} \cup \mathcal{R}_{Q'}$
return $\mathcal{R}$

Table 5. Statistics of datasets, L denotes Length

Dataset	Size(M)	Nodes(M)	Max L	Avg L	Index/Doc.
XMark	115	1.7	12	5.5	5.3
SIGMOD	0.5	0.01	6	5.1	4.8

(short for indexMerge) algorithm using Microsoft Visual C++ 6.0. The compared methods include SLCA [1], XSEarch [3]. Further, we select two query engines, X-Hive[3] and MonetDB[4], to compare the performance of evaluating QPs.

6.2 Datasets, Indices and Queries

We use XMark[5] and SIDMOD[6](short for SIGMOD Record) datasets for our experiments. The main characteristics of the two datasets can be found from Table 5. The last column of Table 5 is *the ratio of index size to document size*, where index consists of (1) *PPI*, (2) *EPI*, and (3) assistant index used to get self entity nodes, of which *PPI* has much larger size than the other two.

We select 40 keyword queries (32 from XMark and 8 from SIGMOD), which are omitted for limited space and classified into 4 groups containing 2, 3, 4 and 5 keywords, respectively. Table 6 shows the statistics of our keyword queries.

[3] http://www.x-hive.com
[4] http://monetdb.cwi.nl/projects/monetdb/XQuery/index.html
[5] http://monetdb.cwi.nl/xml
[6] http://www.sigmod.org/record/xml/

Table 6. Statistics of keyword queries. The 2nd column is the average number of keywords of a keyword query, the 3rd column is the average number of QPs of a keyword query, the 4th column is the average number of entities in a QP, the 5th column is the average number of distinct selection operations for the set of QPs of a keyword query.

Keyword queries	# of Keywords	Avg. # of QPs	Avg. # of entities	Avg. # of Sel. Op.
1st group	2	5.7	1.77	11.8
2nd group	3	10.3	2.13	14.1
3rd group	4	16.5	2.42	19.4
4th group	5	19	2.61	21.2

In our experiment, the node category is assigned using the method discussed in Section 2, we assume each MCN has at most 3 entity nodes, i.e., C=3 for Algorithm 2. Note C=3 means there may have 17 edges in a MCN, which is large enough to find most meaningful relationships.

6.3 Evaluation Metrics

We consider the following performance metrics to compare the performance of different methods: (1) Running time, (2) Precision and (3) Recall.

We define the Precision and Recall using the following steps: (1) users submit their keyword queries, (2) by asking users' search intension, we formulate the XQuery expressions corresponding to their keyword queries, then let them select the XQuery expressions that meet their search intension. For a given keyword query Q, the result of the selected XQuery expressions is denoted as R. (3) evaluate all keyword queries using different methods, the result of a specific method on Q is denoted as R_Q. Then the Precision and Recall of this method are defined as: Precision=$(R_Q \cap R)/R_Q$, Recall=$(R_Q \cap R)/R$.

6.4 Performance Comparison and Analysis

Figure 7 (a) to (d) compare the Precision and Recall of different methods for the four group of keyword queries in Table 6, from which we know for XMark dataset, the Recall of our method is 100%, this is because all results that meet users' search intension are returned by our method. However, the Precision is a little worse than SLCA and XSEarch, this is because our method may return results involving IDREF. If the users' search intension involves IDREF, obviously, our method will be more effective; otherwise, SLCA has the highest Precision. The average figures in Figure 7 (a) and (b) shows that for XMark dataset, though the Precision of our method is not better than SLCA and XSEarch, it has the highest Recall. For SIGMOD dataset, as shown in Figure 7 (c) and (d), both Precision and Recall of our method are better than SLCA and XSEarch, since in such a case IDREF is not considered, thus the number of QP is very small, usually equals to 1, and our method always return results that meet the users' search intension.

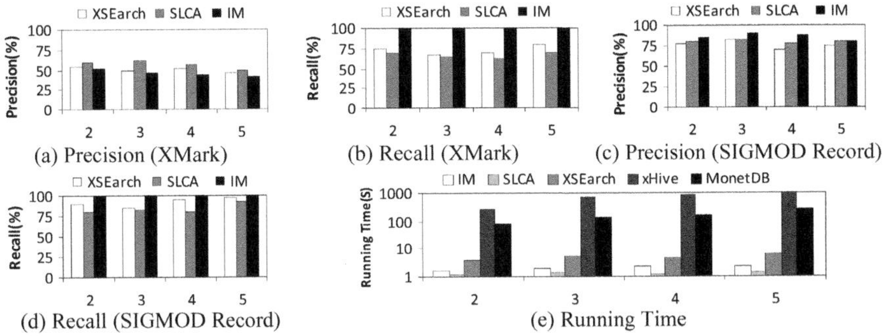

Fig. 7. The comparison of the average Precision (a and c), Recall (b and d) and running time (e). 2,3,4 and 5 denote the number of keywords in a query and the corresponding four query groups in Table 6, respectively.

Figure 7 (e) shows the running time of different methods, where xHive and MonetDB process all QPs as our method. From this figure we know existing query engines cannot work well for these queries as they need to process large number of complex QPs (we try to merge as many as possible query patterns to one XQuery expression so as to make full use of their optimization methods to achieve better performance). Further, except SLCA (which is based on tree model and thus has lower Recall), our method achieves best query performance.

As the demo and optimization techniques of [7] are not publicly available, we do not make comparison with it. Even though, the improvement of our method is obvious and predictable. In our experiment, (1) C=3, (2) our method is based on an entity graph that is much smaller than the original schema graph, (3) our method does not need to scan real data. However, C=3 in our method means C=12 in [7] for XMark dataset, such a value is unnegligible because [7] is based on an expanded schema graph and the cost of computing QPs grows exponentially, let alone scanning the real data to construct an expanded schema graph for each query.

7 Conclusion

In this paper, *Meaningful Connected Network (MCN)* was proposed to enhance the expressiveness of XML keyword search. Entity graph and two indices were then introduced to improve the performance of query evaluation. We proved our method is not only effective (*Completeness* and *No Redundancy*), but also efficient (the costly structural join operations can be equivalently transformed into just a few selection and value join operations). The experimental results verify the effectiveness and efficiency of our method in terms of various evaluation metrics. We will focus on designing effective automatic node classification method and ranking mechanism considering node categories in the future work to provide higher reliability to the effectiveness of our keyword search method.

Acknowledgements. This research was partially supported by the grants from the Natural Science Foundation of China under grant number 60833005, 60573091; China 863 High-Tech Program (No:2007AA01Z155); China National Basic Research and Development Program's Semantic Grid Project (No. 2003CB317000).

References

1. Yu, X., Yannis, P.: Efficient Keyword Search for Smallest LCAs in XML Databases. In: SIGMOD Conference, pp. 537–538 (2005)
2. Lin, G., Feng, S., Chavdar, B., Jayavel, S.: XRANK: Ranked Keyword Search over XML Documents. In: SIGMOD Conference, pp. 16–27 (2003)
3. Sara, C., Jonathan, M., Yaron, K., Yehoshua, S.: XSEarch: A Semantic Search Engine for XML. In: VLDB Conference, pp. 45–56 (2003)
4. Ziyang, L., Yi, C.: Identifying meaningful return information for XML keyword search. In: SIGMOD Conference, pp. 329–340 (2007)
5. Yunyao, L., Cong, Y., Jagadish, H.V.: Schema-Free XQuery. In: VLDB Conference, pp. 72–83 (2004)
6. Cong, Y., Jagadish, H.V.: Querying Complex Structured Databases. In: VLDB Conference, pp. 1010–1021 (2007)
7. Vagelis, H., Yannis, P., Andrey, B.: Keyword Proximity Search on XML Graphs. In: ICDE Conference, pp. 367–378 (2003)
8. Sara, C., Yaron, K., Benny, K., Yehoshua, S.: Interconnection semantics for keyword search in XML. In: CIKM Conference, pp. 389–396 (2005)
9. Konstantin, G., Benny, K., Yehoshua, S.: Keyword proximity search in complex data graphs. In: SIGMOD Conference, pp. 927–940 (2008)
10. Hao, H., Haixun, W., Jun, Y., Philip, S.Y.: BLINKS: ranked keyword searches on graphs. In: SIGMOD Conference, pp. 305–316 (2007)
11. Vagelis, H., Yannis, P.: DISCOVER: Keyword Search in Relational Databases. In: Bressan, S., Chaudhri, A.B., Li Lee, M., Yu, J.X., Lacroix, Z. (eds.) CAiSE 2002 and VLDB 2002. LNCS, vol. 2590, pp. 670–681. Springer, Heidelberg (2003)
12. Reich, G., Widmayer, P.: Beyond Steiner's problem: a VLSI oriented generalization. In: WG Workship (1990)
13. Geert, J., Frank, B., Stijn, N., Inferring, V.: XML Schema Definitions from XML Data. In: VLDB Conference, pp. 998–1009 (2007)
14. Geert, J.B., Frank, N., Thomas, S., Karl, T.: Inference of Concise DTDs from XML Data. In: VLDB Conference, pp. 115–126 (2006)
15. Nicolas, B., Nick, K., Divesh, S.: Holistic twig joins: optimal XML pattern matching. In: SIGMOD Conference, pp. 310–321 (2002)
16. Haifeng, J., Wei, W., Hongjun, L., Jeffrey, X.Y.: Holistic Twig Joins on Indexed XML Documents. In: VLDB Conference, pp. 273–284 (2003)
17. Sihem, A., SungRan, Y., Laks, C., Minimization, V.S.L.: of Tree Pattern Queries. In: SIGMOD Conference, pp. 497–508 (2001)

On the Discovery of Conserved XML Query Patterns for Evolution-Conscious Caching

Sourav S. Bhowmick

School of Computer Engineering, Nanyang Technological University, Singapore
assourav@ntu.edu.sg

Abstract. Existing XML query pattern-based caching strategies focus on extracting the set of frequently issued *query pattern trees* (QPT) based on the support of the QPTs in the history. These approaches ignore the *evolutionary* features of the QPTs. In this paper, we propose a novel type of query pattern called *conserved query paths* (CQP) for efficient caching by integrating the *support* and *evolutionary* features together. CQPs are paths in QPTs that never change or do not change *significantly* most of the time (if not always) in terms of their support values during a specific time period. We proposed a set of algorithms to extract *frequent* CQP*s* (FCQPs) and *infrequent* CQP*s* (ICQPs) and rank these query paths using evolution-conscious *ranking functions*. Then, these ranked query paths are used in *evolution-conscious caching* strategy for efficient XML query processing. Finally, we report our experimental results to show that our strategy is superior to previous QPT-based caching approaches.

1 Introduction

In a XML data repository, a collection of XML queries may be issued by different users over a period of time. These queries can be represented as a collection of *query pattern trees* (QPTs) [12]. Given such a query collection, a *frequent* XML *query pattern* refers to a rooted QPT that is a subtree of at least *minsup* fraction of XML queries. Recently, several algorithms [11,12,13] have been proposed to mine these frequent patterns from the historical query log and cache the corresponding query results to reduce the response time for future queries that are the same or similar. These techniques are primarily designed for static collection of XML queries and cannot handle evolution of query workload efficiently. Consequently, several incremental algorithms [4,6] have been proposed to address the issue of efficiently maintaining the frequent query patterns.

Our initial investigation revealed that existing frequent query pattern-based caching strategies are solely based on the concept of frequency without taking into account the temporal features of the evolving query workload. Every occurrence of a query subtree contributes equally to the caching strategy regardless of *when* the query was issued. Consequently, this may not always be an effective approach in many real-life applications. For instance, consider the two queries, QPT_2 and QPT_4, in Figure 1. Assume that QPT_2 had been issued many times in the past but rarely in recent times whereas QPT_4 is only formulated frequently in recent times. Interestingly, QPT_2 may still remain as a frequent query over

X. Zhou et al. (Eds.): DASFAA 2009, LNCS 5463, pp. 527–542, 2009.

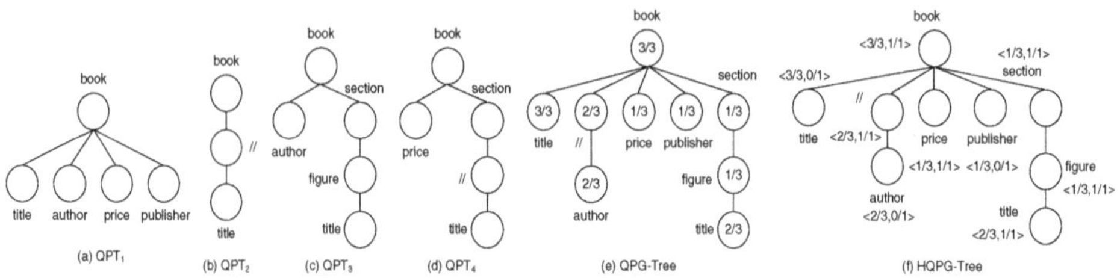

Fig. 1. QPTs, QPG-tree and HQPG-tree

the entire query collection due to its popularity in the past. On the other hand, in spite of its recent popularity, QPT_4 may be considered as *infrequent* with respect to the *entire* query collection in the history due to its lack of popularity in the past. Note that, it is indeed possible that more queries similar to QPT_4 are expected to be issued in the near future compared to queries similar to QPT_2.

***In this paper, we propose a more effective and novel caching strategy that incorporates the evolutionary patterns of* XML *queries*.** In our approach, each QPT consists of a set of *rooted query paths* (RQPs). Informally, a RQP in a QPT is a path starting from the root. For example, */book/section/figure* is a RQP of the XML query shown in Figure 1(c). In our approach, we first discover two groups of RQPs, the *frequent conserved query paths (*FCQP*)* and the *infrequent conserved query paths (*ICQP*)*, from the historical XML queries. Intuitively, *conserved query paths* (CQP) are RQPs whose *support* values never change or do not change significantly most of the times (if not always) during a time period. Here *support* represents the fraction of QPTs in the query collection that *includes* a specific RQP. *Hereafter, whenever we say changes to the* RQP*s/*QPT*s, we refer to the changes to the support of the* RQP*s.*

The second step of our approach is to build a more efficient evolution-conscious caching strategy using the discovered CQPs (FCQPs and ICQPs). Our strategy is based on the following principles. For RQPs that are FCQP, the corresponding query results should have higher priority to be cached since the support values of the RQPs is not expected to change significantly in the near future and the RQPs will be issued frequently in the future as well. Similarly, for RQPs that are ICQP, the corresponding query results should have lower caching priority.

We adopt a *path-level* caching strategy for XML queries instead of *twig-level* (subtree-level) caching. However, it does not hinder us in evaluating twig queries as such queries can be decomposed into query paths. In fact, decomposing twig queries into constituent paths has been used by several *holistic twig join* algorithms. Our focus in this paper is to explore how evolutionary characteristics of XML queries can enable us to design more efficient caching strategies. Our path-level, evolution-conscious caching approach can easily be extended to twig-level caching and we leave this as our future work. Importantly, we shall show later that our caching strategy can outperform a state-of-the-art twig-level, evolution-unaware caching approach [13].

Compared to existing caching strategies for XML data [5,2,12,13], our work differs as follows. Firstly, we use frequent and infrequent conserved RQPs instead of frequent QPTs for caching strategies. Secondly, not only the frequency of the RQPs is considered, but also the evolution patterns of their support values are incorporated to make the caching strategy evolution-conscious. In summary, the main contributions of this paper are as follows. (a) We propose a set of metrics to measure the evolutionary features of QPTs (Section 2). (b) Based on the *evolution metrics*, two algorithms (D-CQP-MINER and R-CQP-MINER) are presented in Section 3 to discover novel patterns, namely *frequent* and *infrequent conserved query paths.* (c) A novel path-level evolution-conscious caching strategy is proposed in Section 4 that is based on the discovered CQPs. (d) Extensive experiments are conducted in Section 5 to show the efficiency and scalability of the CQP-MINER algorithms as well as effectiveness of our caching strategy.

2 Modeling Historical XML Queries

We begin by defining some terminology that we shall use later for representing historical XML queries. A *calendar schema* is a relational schema R with a constraint C, where $R = (f_n : D_n, f_{n-1} : D_{n-1}, \cdots, f_1 : D_1)$, f_i is the name for a calendar unit such as year, month, and day, D_i is a finite subset of positive integers for f_i, C is a Boolean-valued constraint on $D_n \times D_{n-1} \times \cdots \times D_1$ that specifies which combinations of the values in $D_n \times D_{n-1} \cdots D_1$ are valid. For example, suppose we have a calendar schema *(year: {2000, 2001, 2002}, month: {1, 2, 3, ⋯ ,12}, day: {1, 2, 3, ⋯ , 31})* with the constraint that evaluate $\langle y, m, d\rangle$ to be "true" only if the combination gives a valid date. Then, it is evident that $\langle 2000,\ 2,\ 15\rangle$ is valid while $\langle 2000,\ 2,\ 30\rangle$ is invalid. Hereafter, we use $*$ to represent any integer value that is valid based on the constraint.

Given a calendar schema R with the constraints C, a calendar pattern, denoted as $\mathbb{P}$, is a valid tuple on R of the form $\langle d_n, d_{n-1}, \cdots, d_1\rangle$ where $d_i \in D_i \cup \{*\}$. For example, given a calendar schema $\langle$*year, month, day*$\rangle$, the calendar pattern $\langle *,\ 1,\ 1\rangle$ refers to the time intervals "the first day of the first month of every year". Next we introduce the notion of *temporal containment.* Given a calendar pattern $\langle d_n, d_{n-1}, \cdots, d_1\rangle$ denoted as $\mathbb{P}_i$ with the corresponding calendar schema R, a timestamp t_j is represented as $\langle d'_n, d'_{n-1}, \cdots, d'_1\rangle$ according to R. The timestamp t_j is *contained* in $\mathbb{P}_i$, denoted as $t_j \simeq \mathbb{P}_i$, if and only if $\forall$ $1 \leq l \leq n$, $d'_l \in d_l$.

2.1 Representation of an XML Query

We adopt the *query pattern trees* (QPT*s*) [12,13] representation method in this paper. A *query pattern tree* is a rooted tree $QPT = \langle V, E\rangle$, where V is a set of vertex and E is the edge set. The root of the tree is denoted by $root(QPT)$. Each edge $e = (v_1, v_2)$ indicates node v_1 is the parent of node v_2. Each vertex v's label, denoted as $v.label$, is a tag value such that $v.label$ is in {"//", "*"} $\cup$ $tagSet$, where $tagSet$ is the set of all element and attribute names in the schema. Furthermore, if $v \in V$ and $v.label \in$ {"//", "∗"} then there must be a $v' \in V$ such that $v' \in tagSet$ and is a child of v if $v.label =$ "//".

A QPT is a tree structure that represents the hierarchy structure of the predicates, result elements, and attributes in the XML query. Based on the definition of QPT, in existing approaches the *rooted subtree* of a QPT is defined to capture the common subtrees in a collection of XML queries [11,13]. However, in this paper, we are interested in *rooted query paths*, which can provide a finer granularity for caching than *rooted subtrees*. Rooted query paths are special cases of rooted subtrees. Given a QPT $QPT = \langle V, E \rangle$, $RQP = \langle V', E' \rangle$ is a *rooted query path* of QPT, denoted as $RQP \subseteq QPT$, such that (1) $Root(QPT) = Root(RQP)$ and (2) $V' \subseteq V$, $E' \subseteq E$, and RQP is a path in QPT. For example, */book/section/figure* is a RQP in Figure 1(c).

2.2 Representation of XML Query History

Each QPT is represented as a pair (QPT_i, t_i), where t_i is the timestamp recording the time when QPT_i was issued. As a result, the collection of queries (QPTs) can be represented as a sequence $\langle\ (QPT_1, t_1), (QPT_2, t_2), \cdots, (QPT_n, t_n)\ \rangle$, where $t_1 \leq t_2 \leq \ldots \leq t_n$. Then, a *Query Pattern Group* (QPG) is a bag of QPTs $[(QPT_i, t_i), (QPT_{i+k}, t_{i+k}), \cdots, (QPT_j, t_j)]$ such that $1 \leq (i,j) \leq n$ and $\forall\ m\ (i \leq m \leq j)$, $t_m \simeq \mathbb{P}_x$ where $\mathbb{P}_x$ is the user-defined calendar pattern. Observe that the QPTs in a specific QPG are issued within the same calendar pattern according to the calendar schema. Users can define their own time granularity according to the workload and application-specific requirements.

The sequence of QPTs can now be partitioned into a sequence of query pattern groups denoted as $\langle QPG_1, QPG_2, \cdots, QPG_k \rangle$. The occurrences of all QPTs in a QPG are considered to be *equally* important. In our approach, we compactly represent each QPG as a *query pattern group tree (*QPG*-tree)*.

Definition 1. Query Pattern Group Tree (QPG-tree):*Let $QPG = [QPT_i, QPT_{i+1}, \cdots, QPT_j]$ be a query pattern group. A query pattern group tree is a 3-tuple tree, denoted as $T_G = \langle\ V, E, \aleph\ \rangle$, where V is the vertex set, E is the edge set, and $\aleph$ is a function that maps each vertex to the support value of the corresponding rooted query path (*RQP*), such that $\forall\ RQP \subseteq QPT_k$, $i \leq k \leq j$, there exists a rooted query path, $RQP' \subseteq T_G$, that is extended included to RQP.*

Consider the three QPTs in Figures 1(a), (b), and (c). The corresponding QPG-tree is shown in Figure 1(e). The QPG-tree includes all RQPs and records the *support* values (the values inside the nodes of the RQPs in the figure). Given a query pattern group QPG_i, the *support* of a RQP in QPG_i is defined as Φ_i(RQP) $= K\ /\ L$, where K denotes the number of times the RQP is *extended included* in the QPTs in QPG_i and L denotes the number of QPTs in QPG_i. When the RQP is obvious from the context, the support is denoted as Φ_i. Note that the traditional notion of *subtree inclusion* [9] is too restrictive for QPTs where handling of wildcards and relative paths are necessary. Hence, the concept of *extended subtree inclusion*, a sound approach to testing containment of query pattern trees, was proposed by Yang et al. [11] to count the occurrence of a tree pattern in the database. Here, we adopt this concept in the context of RQPs. Given two rooted query paths, RQP_1 and RQP_2, $RQP_1 \prec RQP_2$ denotes that RQP_1 is *extended included* in RQP_2. Our definition of extended inclusion is similar to that of Yang et al. [11]. The only difference is that we assume the subtrees are RQPs. The formal definition is given in [15].

Since there can be a sequence of QPGs in the history, the mean support value of a RQP is represented as *Group Support Mean* (GSM). That is, let $\langle QPG_1, QPG_2, \cdots, QPG_n \rangle$ be a sequence of QPGs in the history. The GSM of a rooted query path, $RQP \subseteq QPG_i$ $(0 \leq i \leq n)$, denoted as $\overline{\Phi}(RQP)$, is defined as $\frac{1}{n}\sum_{i=1}^{n} \Phi_i$.

To facilitate discovery of specific patterns from the evolution history of the RQPs in the QPG-trees, we propose to merge the sequence of QPG-trees into a "global" tree called *historical* QPG*-tree* (HQPG*-tree*). It is similar to the idea of QPG-tree except for the function $\aleph$. In QPG-tree, the $\aleph$ function is used to map each vertex to a single support value of the rooted path at that vertex. In the definition of HQPG-tree, $\aleph$ is replaced by Ψ function which is used to map each vertex to a *sequence* of support values. For example, Figure 1(f) shows an example HQPG-tree by partitioning the QPTs in Figures 1(a), (b), (c), and (d) into two QPGs. The first three QPTs are in one group, while the last is in another group. The sequence of values associated with each vertex in Figure 1(f) corresponds to the support values. The formal definition is given in [15].

2.3 Evolution Metrics

Given a sequence of historical support values of a RQP, we can undertake two approaches to measure its evolutionary characteristics. First, in the *regression-based* approach, the evolution metric computes the "degree" of evolution (or conservation) from the sequence directly. Second, in the *delta-based* approach, we first compute the changes to consecutive support values in the sequence and then quantify the evolution characteristics of the RQP using a set of *delta-based* evolution metrics.

Regression-based Evolution Metric: Intuitively, the evolutionary pattern of a RQP can be modeled using regression models [10]. We propose a metric called *query conservation rate* to monitor the changes to supports of query paths using the linear regression model:$\Phi_t(RQP) = \Phi_0(RQP) + \lambda t,\ where\ 1 \leq t \leq n$. Here the idea is to find a "best-fit" straight line through a set of n data points $\{(\Phi_1(RQP),\ 1),\ (\Phi_2(RQP),\ 2),\ \cdots,\ (\Phi_n(RQP),\ n)\}$, where $\Phi_0(RQP)$ and λ are constants called *support intercept* and *support slope*, respectively. The most common method for fitting a regression line is the method of least-squares [10]. By applying the statistical treatment known as linear regression to the data points, the two constants, $\Phi_0(RQP)$ and λ, can be determined. The correlation coefficient, denoted as r, can then be used to evaluate how the regression fits the data points actually.

Definition 2. Query Conservation Rate: *Let* $\langle \Phi_1, \Phi_2, \cdots, \Phi_n \rangle$ *be the sequence of historical support values of the rooted query path* RQP. *The query conservation rate of* RQP *is defined as* $\mathbb{R}(RQP) = r^2 - |\lambda|$ *where* $\lambda = \frac{\sum_{i=1}^{n} i\Phi_i - \sum_{i=1}^{n}\Phi_i \sum_{i=1}^{n} i}{n\sum_{i=1}^{n} i^2 - (\sum_{i=1}^{n} i)^2}$ *and* $r = \frac{n\sum_{i=1}^{n}(\Phi_i * i) - (\sum_{i=1}^{n}\Phi_i)(\sum_{i=1}^{n} i)}{\sqrt{[n\sum_{i=1}^{n}(\Phi_i)^2 - (\sum_{i=1}^{n}\Phi_i)^2][n\sum_{i=1}^{n} i^2 - (\sum_{i=1}^{n} i)^2]}}$.

Note that the larger the absolute value of the support slope, the more significantly the support changes over time. At the same time, the larger the value of r^2, the more accurate is the regression model. Hence, the larger the query conservation rate $\mathbb{R}(RQP)$, the support values of the RQP change less significantly or are more *conserved.* Also it can be inferred that $0 \leq \mathbb{R}(RQP) \leq 1$.

Delta-based Evolution Metrics: We now define a set of evolution metrics that are defined based on the changes to the support values of a RQP in consecutive QPG pairs. We begin by defining the notion of *support delta.* Let QPG_i and QPG_{i+1} be any two consecutive QPGs. For any rooted query path, RQP, the *support delta* of RQP from ith QPG to $(i+1)th$ QPG, denoted as $\delta_i(RQP)$, is defined as $\delta_i(RQP) = |\Phi_{i+1}(RQP) - \Phi_i(RQP)|$.

The *support delta* measures the changes to support of a RQP between any two consecutive QPGs. Obviously, a low δ_i is important for a RQP to be conserved. Hence, we define the *support conservation factor* metric to measure the percentage of QPGs where the support of a specific RQP changes *significantly* from the preceding QPG.

Definition 3. Support Conservation Factor: *Let* $\langle\ QPG_1, \cdots, QPG_n\ \rangle$ *be a sequence of* QPGs *. For any rooted query path,* RQP*, the support conservation factor in this sequence, denoted as* $\mathbb{S}(\alpha,\ RQP)$*, where* α *is the user-defined threshold for support delta, is defined as* $\mathbb{S}(\alpha, RQP) = \frac{\sum_{i=1}^{n-1} d_i}{n-1}$ *where (a) if* $\delta_i(RQP) \geq \alpha$ *then* $d_i = 1$*; (b) if* $\delta_i(RQP) < \alpha$ *then* $d_i = 0$.

Observe that the smaller the value of $\mathbb{S}(\alpha, RQP)$ is, the less significant is the change to the support values of the RQP. Consequently, at first glance, it may seem that a low $\mathbb{S}(\alpha, RQP)$ implies that the RQP is conserved. However, this may not be always true as small changes to the support values in the history may have significant effect on the evolutionary behavior of a RQP over time. We define the *aggregated support delta* metric to address this.

Definition 4. Aggregated Support Delta: *Let* $\langle QPG_1, QPG_2, \cdots, QPG_n\rangle$ *be a sequence of* QPGs *in the history. The aggregated support delta of* RQP*, denoted as* $\Delta(RQP)$*, is defined as:* $\Delta(RQP) = \sqrt{\frac{1}{n-1}\sum_{i=1}^{n-1}(\Phi_i - \Phi_{i+1})^2}$.

3 CQP-Miner Algorithms

We begin by formally presenting two definitions for CQPs by using the regression-based metric and delta-based metrics, respectively.

Definition 5. Conserved Query Path (CQP): *A* RQP *is a conserved query path in a sequence of* QPGs *iff any one of the following conditions is true: (a)* $\mathbb{R}(RQP) \leq \zeta$ *where* ζ *is the threshold for query conservation rate; (b)* $\mathbb{S}(\alpha, RQP) \leq \beta$ *and* $\Delta(RQP) \leq \gamma$ *where* α*,* β*, and* γ *are the thresholds for support delta, support conservation factor, and aggregated support delta, respectively.*

```
Algorithm 1: QPG-tree Construction
Input:  A bag of QPTs:
        [QPT_1, QPT_2, ..., QPT_n]
Output: The QPG tree: T_G

Description
Initialize T_G as the first QPT QPT_1
for all 2 ≤ I ≤ n do
    for all RQP ⊆ QPT_i do
        for all RQP' ⊆ T_G do
            if RQP' ≺ RQP then
                update the support of RQP'
            if RQP ⊈ T_G then
                Insert RQP to T_G
        end for
    end for
end for
Return(T_G)
```

(a) QPG-tree Contruction Algorithm

```
Algorithm 2: D-CQP-Extract
Input:  An HQPQ tree: T_H
        The user-defined thresholds
        α, β, γ, ξ, ξ'.
Output: Sets of FCQPs and ICPQs: F and I

Description
for all RQP ⊆ T_H (top-down)
    if ξ' < Φ̄ < ξ then
      prune all the children of this RQP
    if Φ̄ ≥ ξ, S(α, RQP) ≤ β
      and Δ(RQP) ≤ γ then F = F ∪ RQP
    if Φ̄ ≤ ξ', S(α, RQP) ≤ β
      and Δ(RQP) ≤ γ then I = I ∪ RQP
end for
Return (F, I )
```

(b) D-CQP-Extract Algorithm

```
Algorithm 3: R-CQP-Extract
Input:  An HQPQ-tree: T_H
        The user-defined thresholds
        ζ, ξ, ξ'.
Output: Sets of FCQPs and ICPQs: F and I

Description
for all RQP ⊆ T_H (top-down)
   if ξ' < Φ̄ < ξ then
      prune all the children of this RQP
   if Φ̄ ≥ ξ and R(RQP) ≤ ζ then
      F = F ∪ RQP
   if Φ̄ ≤ ξ' and R(RQP) ≤ ζ then
      I = I ∪ RQP
end for
Return (F, I )
```

(c) R-CQP-Exxtract Algorithm

Fig. 2. Algorithms for CQP mining

There are two variants of CQPs, *frequent conserved query paths* (FCQPs) and *infrequent conserved query paths*(ICQPs), which are important for our caching strategy. Both of them have the following characteristics: (a) the support values of the RQPs are either large enough or small enough; and (b) their support values do not evolve significantly in the history.

Definition 6. FCQP and ICQP: *Let* RQP *be a conserved query path. Let* ξ *and* ξ' *be the minimum and maximum group support mean (*GSM*) thresholds, respectively. Also,* $\xi > \xi'$*. Then, (a)* RQP *is a Frequent Conserved Query Path (*FCQP*) iff* $\overline{\Phi}(RQP) \geq \xi$*; (b)* RQP *is an Infrequent Conserved Query Path (*ICQP*) iff* $\overline{\Phi}(RQP) \leq \xi'$*.*

3.1 Mining Algorithms

Given a collection of historical XML queries, the objective of conserved query paths mining problem is to extract the frequent and infrequent CQPs. Using the delta-based and regression-based evolution metrics, we present two algorithms to extract the sets of FCQPs and ICQPs. We refer to these algorithms as D-CQP-MINER and R-CQP-MINER, respectively. Each algorithm consists of the following two major phases.

HQPG-tree Construction Phase: Given a collection of XML queries, an HQPG-tree is constructed in the following way. Firstly, the queries are transformed into QPTs. Then, the QPTs are partitioned into groups based on the timestamps and user-defined calendar pattern, where each QPG is represented as a QPG-tree. Next, the sequence of QPG-trees are merged together into an HQPG-tree. We elaborate on the construction of the QPG-tree and merging QPG-trees. The algorithm of constructing the QPG-tree is shown in Figure 2(a).

The algorithm of merging the sequence of QPG-trees into the HQPG-tree is similar to the above algorithm. The only difference is that rather than increasing the support values of the corresponding RQPs, a vector that represents the historical support values is created for each RQP. If the RQP does not exist in the

HQPG-tree, then the vector of supports for this RQP should be a vector starting with *i-1* number of *0*s, where i is the ID of the current query pattern group.

CQP Extraction Phase: Given the HQPG-tree, the FCQPs and ICQPs are extracted based on the user-defined thresholds for the corresponding evolution metric(s). Corresponding to the two definitions of CQPs, two algorithms are presented. The first algorithm is based on the delta-based evolution metrics and the second one is based on the regression-based evolution metric. We refer to these two algorithms as D-CQP-*Extract* and R-CQP-*Extract*, respectively. In both algorithms, the top-down traversal strategy is used to enumerate all candidates of both frequent and infrequent CQPs. We use the top-down traversal strategy based on the downward closure property of the GSM values for RQPs.

Lemma 1. *Let RQP_1 and RQP_2 be two rooted query paths in an* HQPG*-tree. If RQP_1 is included in RQP_2, then $\overline{\Phi}(RQP_1) \geq \overline{\Phi}(RQP_2)$.*

Due to space constraints, the proof is given in [15]. Based on the above lemma, we can prune the HQPG-tree during the top-down traversal. That is, for RQPs whose $\overline{\Phi}$ are smaller than ξ, no extensions of the RQPs can be FCQPs. Similarly, for RQPs whose $\overline{\Phi}$ are smaller than ξ', all of their extensions also satisfy this condition to be ICQPs. The D-CQP-*Extract* algorithm is shown in Figure 2(b). We first compare the values of $\overline{\Phi}$ with the thresholds of GSM. In this case, some candidates can be pruned. After that, the value of $\mathbb{S}(\alpha, RQP)$ is calculated and compared with β. Note that as $\mathbb{S}(RQP)$ is expensive to compute, it is only calculated for the candidates that satisfy all other constraints. The R-CQP-*Extract* algorithm (Figure 2(c)) is similar to the D-CQP-*Extract*, the only difference being the usage of different metrics.

4 Evolution-Conscious Caching

We now present how to utilize the discovered CQPs to build the evolution-conscious cache strategy. There are two major phases, the CQP *ranking phase* and the *evolution-conscious caching (ECC) strategy phase.*

4.1 The CQP Ranking Phase

In this phase, we rank the CQPs discovered by the CQP-MINER algorithm using a *ranking function.* The intuitive idea is to assign high rank scores to query paths that are expected to be issued frequently. Note that there are other factors such as the query evaluation cost and the query result size that are important for designing effective caching strategy [11,12].

Definition 7. Ranking Functions: *Let the cost of evaluating a* RQP *(denoted as $Cost_{eval}(RQP)$) is the time to execute this query against the* XML *data source without any caching strategy, while the size of the result (denoted as $|result(RQP)|$) is the actual size of the view that stores the result. Then the ranking function, $\mathcal{R}$, is defined as: (a) If* D-CQP-MINER *is used to extract* ICQPs *and*

Algorithm 4: Cache-Conscious Query Evaluation
Input: A new XML query: q_x ,
Ranked FCQPs and ICQPs in descending order: F_p and I_p

Description:
$M = \{RQP_i | RQP_i \prec q_x\} \cap F_p$
  if $M \neq \emptyset$
    choose a sequence of ordered $RQP_i \in M$ based on their ranking
    decompose $q_x = RQP_i \circ \cdots \circ RQP_j \circ q'_x$
  end if
  evaluate the query by combining the results
  for all $RQP_i \cdots RQP_j \in M$
    update $\mathcal{R}(RQP_i)$
    if $\mathcal{R}(RQP_i) < Min\{\mathcal{R}(RQP)\}$
      evict the cached result of RQP_i from caching
    end if
  end for

(a) Cache-Conscious Query Evaluation Algorithm

Algorithm 5: Evolution-Conscious Cache Maintenance Policy
Input: Q, ΔQ, K be the set of queries that have been cached

Description:
  Compute $q = \frac{|\Delta Q|}{|Q|}$
  if $q \geq \epsilon$
    Regenerate I_p and F_p.
    if $M' = K \cap I_p \neq \emptyset$
      evict $RQP \in M'$
    end if
    while there is space left in the cache
      cache the RQP with maximum rank but not in the cache
    end while
  end if

(b) Evolution-Conscious Cache Maintenance Policy Algorithm

Fig. 3. Algorithms for evolution-conscious caching

FCQPs, *then* $\mathcal{R}(RQP) = \frac{Cost_{eval}(RQP) \times \overline{\Phi}(RQP)}{\mathbb{S}(\alpha, RQP) \times \Delta(RQP) \times |result(RQP)|}$; *(b) If* R-CQP-MINER *is used to extract* ICQPs *and* FCQPs, *then* $\mathcal{R}(RQP) = \frac{Cost_{eval}(RQP) \times \overline{\Phi}(RQP)}{\mathbb{R}(RQP) \times |result(RQP)|}$.

Observe that we have two variants of the ranking function as our ranking strategy depends on the two sets of evolution metrics used in the regression-based (R-CQP-MINER) and delta-based (D-CQP-MINER) CQPs discovery approaches. Particularly, these evolution metrics are used to estimate the expected number of occurrences of the query paths. The remaining factors are used in the similar way as they are used in other cache strategies [3,8,12].

4.2 The ECC Strategy Phase

The goal of this phase is to construct an evolution-conscious caching strategy that utilizes the ranked FCQPs and ICQPs in such a way that the query processing cost for future incoming queries is minimized. As the cache space is limited, the basic strategy is to cache the results for the FCQPs with the *largest* rank scores by replacing the cached results of the RQPs with *smaller* rank scores.

We first introduce the notion of *composing query*. Suppose at time t_1, the cache contains a set of views $V = \{V_1, V_2, \cdots, V_n\}$ and the corresponding queries are $Q = \{Q_1, Q_2, \cdots, Q_n\}$. When a new query Q_{n+1} comes, it inspects each view V_i in V and determines whether it is possible to answer Q_{n+1} from V_i. View V_i answers query Q_{n+1} if there exists another query C which, when executed on the result of Q_i, gives the result of Q_{n+1}. It is denoted by $C \circ Q_i = Q_{n+1}$, where C is called the *composing query* (CQ). When a view answers the new query, we have a *hit*, otherwise we have a *miss*.

Cache-conscious query evaluation: Figure 3(a) describes the query evaluation strategy. When a new query q_x appears, it may match to more than one of the RQPs in the set of FCQPs (which are denoted as M). Hence, q_x can be considered to be the join of many $RQPs$ and the composing query q'_x. Formally, $q_x = RQP_1 \circ RQP_2 \cdots, RQP_j \circ q'_x$, where $RQP_1, RQP_2, \cdots, RQP_j$ are the cached RQPs with the highest rank scores and are contained in q_x, q'_x is the composing query that does not contain any of the RQPs in the cache. The answers are obtained by evaluating the composing query and joining the corresponding results (Lines 2-7). Next, for all $RQPs$ that are contained in M, the corresponding ranks are updated with respect to the changes of $\overline{\Phi}$ (Lines 8-9). If the rank for any of these $RQPs$ falls below the minimum value of these RQPs in the cache, then the corresponding query results will be evicted (Lines 10-12). Note that we do not update the values of evolution metrics of ICQPs and FCQPs during the caching process. Rather, it is done off-line as discussed below.

Evolution-conscious cache maintenance policy: One can observe that under heavy query workload, mining FCQPs and ICQPs frequently during evaluation of every new query can be impractical. Hence, rather than computing new sets of FCQPs and ICQPs whenever a new query appears, we recompute these CQPs *only when the number of new queries that have been issued, in comparison with the set of historical queries, is larger than some factor* q. Note that this mining process can be performed off-line.

Formally, let t_p be the most recent time when we computed the sets of FCQPs and ICQPs in the history. Let $|Q|$ denote the number of XML queries in the collection at t_p. Assume that we recompute the sets of FCQPs and ICQPs at time t_n where $t_n > t_p$. Let $|\Delta Q|$ be the set of new queries that are added during t_p and t_n. Then, $q = \frac{|\Delta Q|}{|Q|}$.

The algorithm for query evaluation is shown in Figure 3(b). First, it computes the q value. If q is greater than or equal to some threshold ϵ then the FCQPs and ICQPs are updated off-line. In Section 5.2, we shall empirically show that $\epsilon = 0.5$ produces good results. If the RQPs that have been cached are in the list of regenerated ICQPs, then the corresponding results in the cache have to be evicted (Lines 5-7). Consequently, there may be some space in the cache available that can be utilized. If the space is enough, then cache those RQPs in F_p having maximum rank but have not been cached yet (Lines 8-10).

5 Performance Evaluation

The mining algorithms and the caching strategy are implemented in Java. All the experiments were conducted on a Pentium IV PC with a 1.7Ghz CPU and 512MB RAM, running MS Windows 2000. We use two set of synthetic datasets generated based on the DBLP.DTD (http://dblp.uni-trier.de/xml/dblp.dtd) and SSPLAY.DTD (http://www.kelschindexing.com/shakesDTD.html). Firstly, a DTD graph is converted into a DTD tree by introducing some "//" and "*" nodes. Then, all possible rooted query paths are enumerated. Similar to [6,12,13], the collection of QPTs is generated based on the set of RQPs using the Zipfian distribution and these QPTs are randomly distributed in the temporal dimension. Example of two sets of QPTs in the DBLP and SSPLAY datasets is given in [15]. Each basic dataset consists of up to 3,000,000 QPTs, which are divided into 1000 QPGs. The characteristics of the datasets are shown in Figure 4(a).

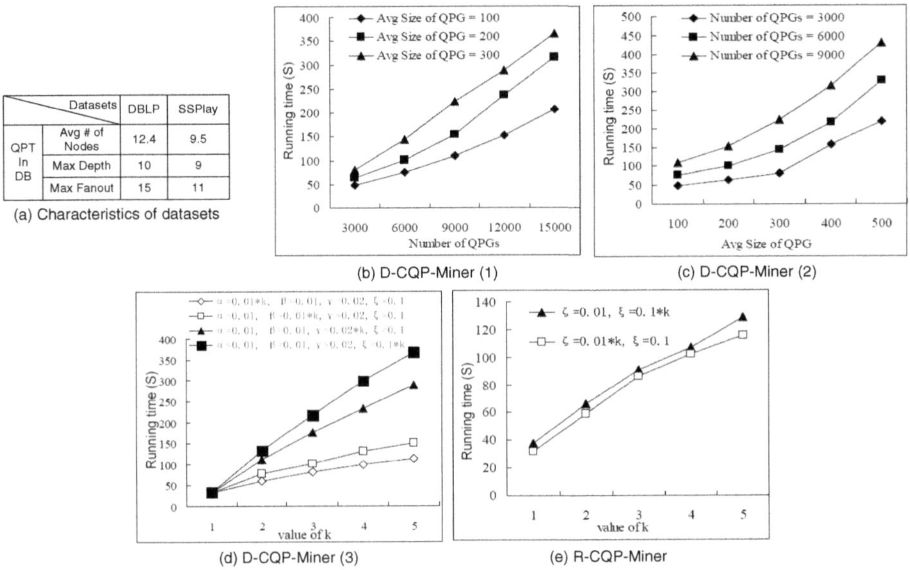

	Datasets	DBLP	SSPlay
QPT In DB	Avg # of Nodes	12.4	9.5
	Max Depth	10	9
	Max Fanout	15	11

(a) Characteristics of datasets

Fig. 4. Datasets and performance of CQP-Miner

5.1 CQP-Miner

Algorithm Efficiency: We evaluate the efficiency by varying the average size of QPGs and the number of QPGs (the size of the time window). Figures 4(b) and (c) show the running time of the D-CQP-Miner when the size of the dataset increases. In the first case, the number of QPGs is increased while the average size of each QPG is fixed. In the second case, the average size of each QPG increases while the number of QPGs is fixed. The DBLP dataset is used and the parameters are fixed as follows: $\alpha = 0.02$, $\beta = 0.05$, $\gamma = 0.02$, and $\xi = 0.25$. Also, we set $\xi' = \xi/10$. It can be observed that when the size of the dataset increases, the running time increases as well. The reason is intuitive as the size of the HQPG-tree becomes larger, it requires more time for the tree construction and handling large number of candidate CQPs. The running time of the R-CQP-Miner shows a similar trend. Due to space constraints, the reader may refer to [15] for details.

Effects of Thresholds: As there are four thresholds: α, β, γ, and ζ for the D-CQP-Miner, experiments are conducted by varying one of the them and fixing the others. For instance, in Figure 4(d), "$\alpha = 0.01 * k$, β=0.01, γ=0.02, ξ=0.1" means that we fix the values of β, γ and ξ, and vary α from 0.01 to 0.05 by varying k from 1 to 5. In this experiment, the DBLP dataset with 300,000 queries is used. The results in Figure 4(d) show that the running time of D-CQP-Miner increases with the threshold values. Observed that the changes to ξ and α have more significant effect on the running time than the changes to β and γ. This is because ξ affect the total number of FCQPs and ICQPs and the values of α affect both support deltas and support conservation factors.

α	β	γ	ζ	overlap	o@10	o@30	o@60
0.01	0.01	0.01	0.01	0.66	1	0.83	0.77
0.01	0.01	0.02	0.01	0.71	1	0.87	0.83
0.01	0.02	0.01	0.01	0.85	1	1	0.88
0.02	0.01	0.01	0.01	0.79	1	0.93	0.79
0.02	**0.02**	**0.02**	**0.01**	**0.93**	**1**	**1**	**1**
0.02	**0.02**	**0.01**	**0.01**	**0.97**	**1**	**1**	**1**
0.02	**0.01**	**0.02**	**0.01**	**0.94**	**1**	**0.98**	**1**
0.01	0.02	0.02	0.01	0.91	1	1	0.91
0.03	0.02	0.01	0.01	0.88	1	1	0.89

(a) Mining Results

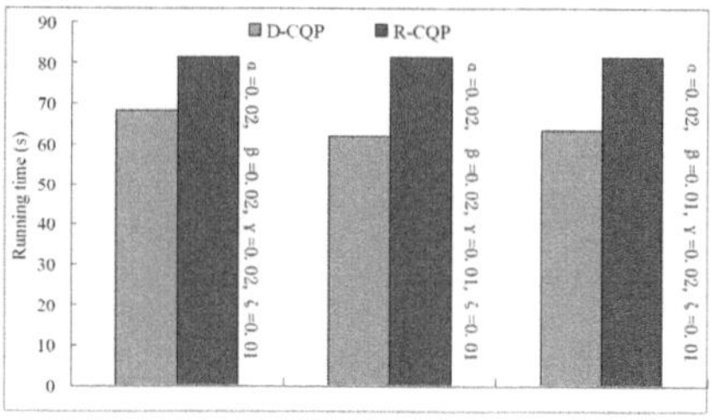

(b) Running Times

Fig. 5. Comparison of mining algorithms

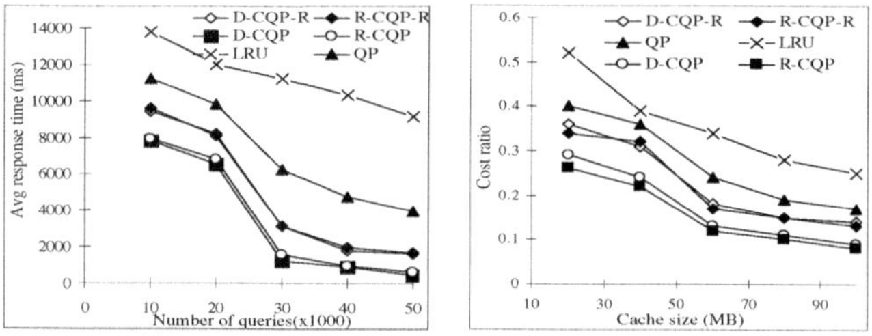

(a) Avg response time (b) Cost ratio

Fig. 6. Performance of caching strategies

Similarly, the thresholds, ζ and ξ, are varied to evaluate their effects on the running time of the R-CQP-MINER. The results are shown in Figure 4(e). The SSPLAY dataset with 900,000 queries is used. It can be observed that the running time increases with the thresholds. The reason is that when the values for any of the two parameters increase, the number of CQPs increases.

Comparison of Mining Results: As the two algorithms use different evolution metrics, to compare the mining results, we define the notion of *overlap* metric. Let F_D and I_D be the sets of FCQPs and ICQPs, respectively, in the D-CQP-MINER mining results. Let F_R and I_R be the sets of FCQPs and ICQPs in the R-CQP-MINER mining results. The *overlap* between the two sets of mining results is defined as: $Overlap = \frac{1}{2} \times (\frac{|F_D \cap F_R|}{|F_D \cup F_R|} + \frac{|I_D \cap I_R|}{|I_D \cup I_R|})$.

Basically, the *overlap* value is defined as the number of shared CQPs divided by the total number of unique CQPs in both mining results. Based on this definition, it is evident that the larger the *overlap* value, the more similar the mining results are. In this definition all the CQPs in the mining results are taken into consideration. However, in caching, only the top-k frequent/infrequent CQPs in the results are important. Hence, we define the notion of *overlap*@k metric. Let $C_D(k)$ and $C_R(k)$ be the sets of top-k conserved query paths in the D-CQP-MINER and R-CQP-MINER results, respectively, where $C_D(k) \subseteq F_D \cup I_D$ and $C_R(k) \subseteq F_R \cup I_R$. The *overlap*@k (denoted as $o@k$) is defined as: $o@k = \frac{|C_D(k) \cap C_R(k)|}{|C_D(k) \cup C_R(k)|}$.

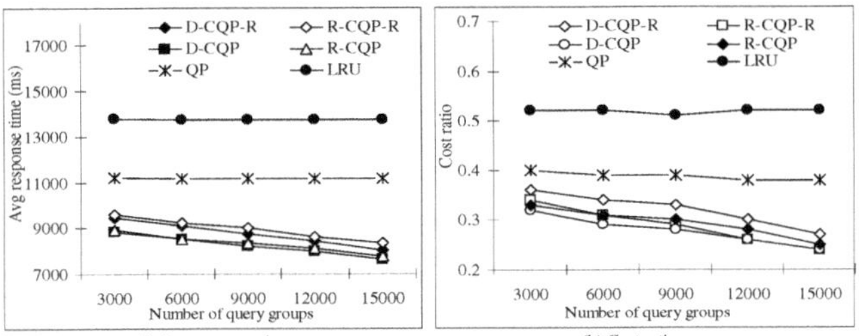

(a) Avg response time (b) Cost ratio

Fig. 7. Effect of QPG size

The experimental results with the SSPLAY dataset is shown in Figure 5(a). We vary the thresholds of the evolution metrics and compute the *overall* and *overall@k*. Interestingly, the *overlap* value can be very close to *1* when the threshold values are appropriately set. This indicates that both algorithms share a large number of CQPs even though they are based on different evolution metrics. Moreover, it can be observed that the top-*10* CQPs are exactly the same. Even for the top-*60* CQPs, the two categories of evolution metrics can produce identical sets of CQPs under appropriate threshold values. This is indeed encouraging as it indicates that both the regression-based and delta-based evolution metrics can effectively identify the top-k CQPs that are important for our caching strategy.

Comparison of Running Times: We now compare the running times of the two algorithms when they produce identical top-k CQPs under appropriate thresholds. We choose the three sets of threshold values shown in Figure 5(a) that can produce identical top-60 CQPs (shaded region in the table). Figure 5(b) shows the comparison of the running time. The DBLP dataset is used and ξ is set to *0.1*. It can be observed that D-CQP-MINER is faster than R-CQP-MINER when they produce the same top-60 CQPs.

5.2 Evolution-Conscious Caching

We have implemented the caching strategy by modifying the replacement policies of LRU with the knowledge of FCQPs and ICQPs as stated in the previous section. From the original collections of QPTs, some QPTs are chosen as the basic query paths and are extended to form the future queries. To select the basic query paths, queries that are issued more recently have a higher possibility of being chosen. That is, given a sequence of n QPGs, $\frac{n-i}{2^{i+1-n}}$ QPTs are selected from the ith group. Then, the set of selected queries are extended according to the corresponding DTD. The future queries are generated by extending the previous query paths with the randomly selected query paths. Note that for each of the following experiments, *10* sets of queries are generated for evaluation and the figures show the average performance. The QPTs used for generating examples of the *10* sets of queries are given in [15].

We use the same storage scheme as in [12]. That is, we use the index scheme of [7] to populate the SQL Server 2000 database and create the corresponding indexes. The system accepts tree-patterns as its queries, and utilizes structural join method [1] to produce the result. No optimization techniques are used.

Basically, six caching strategies are implemented: the D-CQP-MINER and R-CQP-MINER-based strategies (denoted as DCQP and RCQP, respectively), D-CQP-MINER and R-CQP-MINER-based strategies without a ranking function (denoted as DCQP-R and RCQP-R, respectively), the original LRU-based caching strategy (denoted as LRU), and the state-of-the-art frequent query pattern-based caching strategy (2PX-MINER [13] based caching strategy denoted as QP). Note that the FCQPs and ICQPs used in the following experiments are discovered using the D-CQP-MINER and R-CQP-MINER, by setting $\alpha = 0.02$, β=0.02, γ=0.01, ζ=0.01, and $\xi = 0.2$.

Average Response Time: The *average response time* is the average time taken to answer a query. It is defined as the ratio of total response time for answering a set of queries to the total number of queries in this set. Note that the query response time includes the time for ranking the CQPs (The CQP ranking phase). Figure 6(a) shows the average response time of the six approaches while varying the number of queries from 10,000 to 50,000 with the cache size fixed at 40MB. We make the following observations. First, as the number of queries increases, the average response time decreases. This is because when the number of queries increases, more historical behaviors can be incorporated and the frequent query patterns and conserved query paths can be more accurate. Second, DCQP, DCQP-R, RCQP, and RCQP-R perform better than QP and LRU. Particularly, when the number of queries increases, the gaps between our approaches and the existing approaches increases as well. For instance, our caching strategies can be up to 5 times faster than the QP approach and 10 times faster than the LRU approach when the number of queries is up to 50,000. Third, the rank-based evolution-conscious caching strategies outperform the rank-unconscious caching strategies highlighting the benefits of using the ranking functions.

Cost ratio: The *cost ratio* represents the query response time using different types of caching strategies against the response time without any caching strategy for all query examples. Figure 6(b) shows the performance of the six caching strategies in terms of the cost ratio measure. The number of queries is fixed at 2000, while the cache size varies from 20MB to 100MB (for the SSPLAY dataset). It can be observed that DCQP, DCQP-R, RCQP, and RCQP-R perform better than QP and LRU. Particularly, observe that the ratio difference between state-of-the-art QP approach and LRU is between 0.09 $\sim$ 0.12. If we consider this as the benchmark then observed further difference of 0.1 $\sim$ 0.13 between our approach and QP is significant. In other words, the idea of including evolutionary feature of queries for caching is an effective strategy.

Number of QPGs: Figures 7(a) and (b) show how the average response time and cost ratio change when the number of QPGs increases. The SSPLAY dataset is used and the average size of each QPG is 300. We vary the number of QPGs from 3,000 to 15,000. Observe that the evolution-conscious caching strategies perform better when there are more QPGs. This is because when the number of QPGs is large, our CQPs are more accurate.

q	LRU	R-CQP-R	D-CQP-R	QP
0.1	0.52	0.29	0.36	0.39
0.2	0.52	0.32	0.41	0.45
0.3	0.52	0.37	0.48	0.53
0.4	0.52	0.43	0.56	0.61
0.5	0.52	0.51	0.63	0.74
0.6	0.52	0.68	0.71	0.86

(a) Effects of q on cost ratio (1)

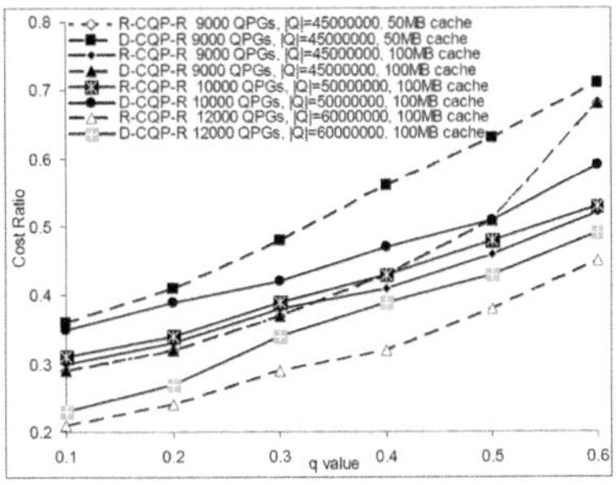

(b) Effects of q on cost ratio (2)

Fig. 8. Effects of q on cost ratio

Maintenance cost of ICQPs and FCQPs: As mention in Section 4.2, the sets of FCQPs and ICQPs need to be updated after certain number of queries are issued. In this experiment, we empirically determine the threshold value ϵ such that as long as $q < \epsilon$ we do not need to update the ICQPs and FCQPs. We first vary q to study its effect on the quality of our caching strategy. Note that from the running cost point of view, the larger the value of q, the lesser is the overhead. Figure 8(a) shows the performance of our proposed approaches compared to the LRU and QP approaches (in terms of cost ratio). We set $|Q| = 45000000$ (9000 QPGs) and the cache size is fixed to 50MB. It can be observed that the cost ratio increases with the increase in q for all approaches except the LRU-based approach. For the QP approach, rather than repeatedly updating the frequent query patterns whenever new queries are issued, the same strategy of periodically updating the mining results is used. It can be observed that the performance of our proposed RCQP-R and DCQP-R are better than the QP approach for any q value. Furthermore, RCQP-R and DCQP-R are better than the LRU approach in most cases when $q < 0.5$.

In Figure 8(b) we vary $|Q|$ and the cache size to study the effect of q on the cost ratio. It can be observed that our approaches produce good performance in most cases when $q < 0.5$ ($\epsilon = 0.5$). That is, our approach can improve the query performance without updating the FCQPs and ICQPs as long as $|\Delta Q| < \frac{|Q|}{2}$.

6 Conclusions and Future Work

In this paper, we proposed a novel type of XML query pattern named conserved query paths (CQPs) for efficient caching. To the best of our knowledge, this is the first approach that integrates evolutionary features of XML queries along with frequency of occurrences for building an efficient caching strategy. Conserved query paths are rooted query paths (RQPs) in QPTs that never change or do not change significantly most of the time in terms of their support values during a specific time period. Based on two evolution metrics, we presented two algorithms (D-CQP-MINER and R-CQP-MINER) that extract frequent and infrequent CQPs from the historical collection of QPTs. These CQPs are ranked according to our proposed ranking function and used to build the evolution-conscious caching strategy. Experimental results showed that the proposed algorithms can be effectively used to build more efficient caching strategies compared to state-of-the-art caching strategies. In future, we wish to explore how calendar pattern selection can be automated. Also, we would like to extend our framework to provide a more sophisticated probabilistic ranking function. Finally, we plan to investigate strategies to automate the maintenance of ICQPs and FCQPs.

Acknowledgement. The author wishes to acknowledge and thank Dr Qiankun Zhao for implementing the ideas discussed in this paper.

References

1. Al-Khalifa, S., Jagadish, H.V., et al.: Structural Joins: A Primitive for Efficient XML Query Pattern Matching. In: ICDE (2002)
2. Chen, L., Rundensteiner, E.A., Wang, S.: Xcache: A Semantic Caching System for XML Queries. In: SIGMOD, p. 618 (2002)
3. Chen, L., Wang, S., Rundensteiner, E.: Replacement Strategies for XQuery Caching Systems. Data Knowl. Eng. 49(2), 145–175 (2004)
4. Chen, Y., Yang, L., et al.: Incremental Mining of Frequent XML Query Patterns. In: Perner, P. (ed.) ICDM 2004. LNCS, vol. 3275. Springer, Heidelberg (2004)
5. Hristidis, V., Petropoulos, M.: Semantic Caching of XML Databases. In: WebDB (2002)
6. Li, G., Feng, J., et al.: Incremental Mining of Frequent Query Patterns from XML Queries for Caching. In: ICDM (2006)
7. Li, Q., Moon, B.: Indexing and Querying XML Data for Regular Path Expressions. In: VLDB (2001)
8. Mandhani, B., Suciu, D.: Query Caching and View Selection for XML Databases. In: VLDB (2005)
9. Ramesh, R., Ramakrishnan, L.V.: Nonlinear Pattern Matching in Trees. JACM 39(2), 295–316 (1992)
10. Weisberg, S.: Applied Linear Regression, 2nd edn. Wiley, Chichester (1985)
11. Yang, L., Lee, M., et al.: Mining Frequent Query Patterns from XML Queries. In: DASFAA (2003)
12. Yang, L., Lee, M., et al.: Efficient Mining of XML Query Patterns for Caching. In: VLDB (2003)
13. Yang, L., Lee, M., et al.: 2pxminer: An Efficient Two Pass Mining of Frequent XML Query patterns. In: SIGKDD (2004)
14. Zaki, M.J.: Efficiently Mining Frequent Trees in a Forest. In: SIGKDD, pp. 71–80 (2002)
15. Bhowmick, S.S.: cqp-Miner: Mining Conserved XML Query Patterns For Evolution-Conscious Caching. Technical Report, CAIS-12-2007 (2007), `http://www.cais.ntu.edu.sg/~assourav/TechReports/CQPMiner-TR.pdf`

ValidMatch: Retrieving More Reasonable SLCA-Based Result for XML Keyword Search

Lingbo Kong, Rémi Gilleron, and Aurélien Lemay

Mostrare, INRIA Futurs,
Villeneuve d'Ascq, Lille, 59650 France
mLinking@gmail.com,
{remi.gilleron,aurelien.lemay}@univ-lille3.fr

Abstract. This paper proposes a new effective filtering mechanism for pruning the uninteresting nodes implied in the SLCA-based (Smallest LCA – Lowest Common Ancestor) fragments for XML keyword search. Its fundamental concept is the valid contributor. Given two nodes v and u, and u is v's parent, the child v is a valid contributor to its parent u, if (1) v's label is unique among all u's children; or (2) for the siblings with same label as v, v's content is not covered by any of them. The new filtering mechanism can be described as following: every node in each retrieved fragment should be valid contributor to its parent.

1 Introduction

With the widespread use of XML, adapting keyword search to XML data has become attractive, generalized as XML keyword search (XKS). The latest work [1] proposes not only the axiomatic properties that an XKS technique should satisfy, but also a concrete algorithm returning the fragments rooted at the SLCA nodes. To ensure the meaningfulness of the fragments, [1] proposes a contributor-based filtering mechanism, which requires that every node n in a fragment does not have any sibling n_2 satisfying $dMatch(n) \subset dMatch(n_2)$, where the function $dMatch(n)$ represents all the keywords contained in the subtree rooted at the node n.

However, the above filtering mechanism has the false positive problem (discarding interesting nodes), and the redundancy problem (keeping uninteresting nodes). Fig.1 demonstrates these two problems. Fig. 1(a) *corresponds to a SLCA-based fragment for the keyword query* Q_1 = "Li Zhou top-k XML query", *which is interested to check if there is some paper on* "top-k XML query" *written by authors named* "Li" *and* "Zhou". *According to the contributor concept, the node* "0.2.1.1 (title)" *will be discarded for* $dMatch(0.2.1.1)$ = {top-k, XML} $\subset$ $dMatch(0.2.1.2)$ = {top-k, XML, query}. *However, it should not be, as it is the title of the paper. As for* Fig. 1(b), *it is a SLCA-based fragment for the keyword query* Q_2 = "DASFAA 2008 author name", *whose intuition is to collect all the distinct authors who have published papers in DASFAA 2008. We can see that the information in it is duplicate, but the contributor-based filtering is unable to prune it, for the relationship of the* $dMatchs$ *of the two nodes* "0.2.0.0" *and* "0.2.1.0" *is* $dMatch(0.2.0.0) = dMatch(0.2.1.0)$ = {author, name}.

To overcome those above two problems, this paper first proposes a new filtering mechanism in Section 2, whose kernel is the concept of valid contributor. Then, it

X. Zhou et al. (Eds.): DASFAA 2009, LNCS 5463, pp. 543–548, 2009.

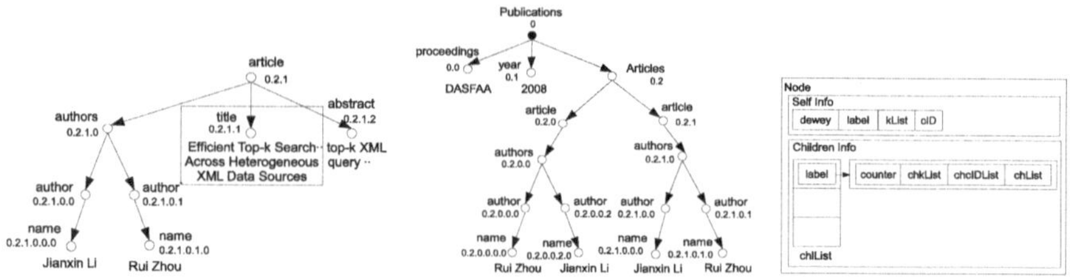

(a) (b) (c)

Fig. 1. (a): False positive example for contributor-based filtering with Q_1; (b): Redundancy example for contributor-based filtering with Q_2; (c): Node description used in Section 3. The integer sequence beside each node in (a) and (b) is the Dewey code of that node.

introduces a concrete algorithm – ValidMatch in Section 3. Section 4 illustrates the experimental result, which confirms the efficiency and effectiveness of our new filtering mechanism.

2 Valid Contributor

In this section, we first propose the tree content set and tree keyword set in Definition 1. Based on it, we formalize the concept of valid contributor in Definition 2. Then it is easy to construct the filtering mechanism as following: *all the nodes in a SLCA-based fragment should be valid contributors for their parents.*

Definition 1 (Tree Content Set and Tree Keyword Set of a node). *Given an XML tree T, the content C_v of a node v in T is the word set implied in v's label, text and attributes. Given a keyword query $Q = \{w_1, \ldots, w_k\}$, the node v is a keyword node if $C_v \cap Q \neq \emptyset$ where $\emptyset$ is the empty set. The tree content set TC_v of a node v is the content union of all the keyword nodes rooted at the node v, that is $TC_v = \bigcup_{v' \in T_v} C_{v'}$, where T_v represents the subtree rooted at the node v, and $v' \in T_v$ stands for a keyword node in T_v. The tree keyword set of the node v is defined simply as $TC_v \cap Q$, denoted as TK_v.*

Definition 2 (Valid Contributor). *Given an XML tree T, and the keyword query $Q = \{w_1, \ldots, w_k\}$, S is a SLCA-based fragment in T. u, v are two nodes in S, and u is the parent of v. The child v is a valid contributor of u if either of the following two conditions holds:*

1. *v is the unique child of u with label $\lambda(v)$;*
2. *v has several siblings $v_1, \ldots, v_m$ $(m \geq 1)$ with same label as $\lambda(v)$, but the following conditions hold:*
   (a) $\nexists v_i$, $TK_v \subset TK_{v_i}$;
   (b) $\forall v_i \wedge TK_v = TK_{v_i}$, $TC_v \neq TC_{v_i}$.

Now let's see how our valid contributor-based mechanism overcomes the two problems in the contributor-based mechanism. *For* Fig. 1(a), *even though* $TK_{0.2.1.1} \subset TK_{0.2.1.2}$, *we still should keep the node* "0.2.1.1" *in the final result, because their labels are different from each other* (according to the rule 1 in Definition 2). *As for* Fig. 1(b), *we should delete one of the two nodes* "0.2.0" *and* "0.2.1", *because their tree content sets are same* (according to the rule 2.(b) in Definition 2).

3 ValidMatch

Algorithm 1 describes the implementation of the above ideas. Since its first three stages are similar with those in MaxMatch in [1], we pay our attention here mainly to the fourth stage of ValidMatch. To support the basic computations, such as $TK_v = TK_{v_i}$, $TK_v \subset TK_{v_i}$ and $TC_v = TC_{v_i}$, we design the node data structure as shown in Fig. 1(c).

Algorithm 1. VALIDMATCH ALGORITHM

VALIDMATCH (T, Q)

Input: The XML data T and the keyword query $Q = \{w_1, w_2, \ldots, w_k\}$
Output: All the fragments containing only reasonable nodes
$D_i \leftarrow$ getKeywordNodes(T, Q); /* Collect the keyword nodes of the keyword w_i $(1 \leq i \leq k)$
SLCAs $\leftarrow$ getSLCA($D_1, \ldots, D_k$); /* Return all the interesting SLCA nodes, same with that in [1]
Matches $\leftarrow$ getMatch(SLCAs, $D_1, \ldots, D_k$); /* Collect all the related keyword nodes for every corresponding SLCA
For (each match S in Matches) /* For each SLCA-based fragment, prune the uninteresting
  pruneMatch(S)

PRUNEMATCH (S)

/* CONSTRUCTING STEP */
$i \leftarrow |S.knodes| - 1$;
$start \leftarrow S.a$;
While $(i \geq 0)$ **do**
  $n_i \leftarrow S.knodes$'s i^{th} keyword node
  $n_i.ancestors$ stores its ancestors from n_i to $start$ /* sorted according to their lengths;
  For (each ancestor x of n_i on the path from n_i to $start$)
    Construct a node X following the node data structure described in Fig. 1(c), and fill X based on the information in x;
    If (the tree $S.r$ contains a node with same Dewey code as X)
      Update the corresponding node using X;
    Else
      Add X into $S.r$;
  /* The following two lines are added to guarantee that the node information of n_i could be transferred to all its ancestors.
  If $(start \neq S.a)$
    Update the nodes from $start$ to $S.a$ using the information of the node corresponding to $start$;
  $i \leftarrow i - 1$;
  If $(i < 0)$ break;
  Else $start \leftarrow$ LCA(n_i, n_{i+1});
/* PRUNING STEP */
Breadth-first traverse the tree, and **For** (each node n)
  If ($n.chlList$ is not *null*)
    If $(n.chlList[j].counter != 1)$
      $usedKNums \leftarrow \{\}$, $usedCIDs \leftarrow \{\}$;
      For (each child ch in $n.chlList[j].chList$)
        If $(knum(ch.kList) \in usedKNums)$
          Keep ch when $ch.cID \notin usedCIDs$;
        Else If (no element, which is larger than $knum(ch.kList)$ in $n.chlList[j].chkList$, covers $knum(ch.kList)$)
          Keep ch in the result, **Add** $knum(ch.kList)$ into $usedKNums$ and add $ch.cID$ into $usedCIDs$;
    Else Keep that child node in $n.chlList[j]$ whose $counter$ is 1.

A node comprises two parts: (1) the information of the node itself in "Self Info" frame; and (2) the information of its children in "Children Info" frame. The former stores the basic information of the node itself, including its Dewey code (*dewey*), label (*label*), keyword list (kList, which corresponds to the tree keyword set of the node) and the content ID (*cID*, which represents the tree content set, and only records the word pair (min, max) of the tree content set of the node. The min and max are the smallest and the largest word in the tree content set according to the lexical order.). The latter stores the information of a node's children, maintained in a list *chlList* according to their distinct labels. For each distinct label, there is an item, which stores *counter* (the number of the children with that distinct label), *chkList* (records the sorted distinct keyword list numbers), *chcIDList* (stores the cIDs of the children), and *chList* (records the references to those children).

Since the $kList$ of a node directly corresponds to its tree keyword set, it is easy to compute $TK_v = TK_{v_i}$ and $TK_v \subset TK_{v_i}$ based on the $kLists$ of v and v_i. Here is an illustration for the computation of $TK_v \subset TK_{v_i}$. In Fig. 1(a), $TK_{0.2.1.1} = \{$top-k, XML$\}$ for Q_1, and its $kList$ is $\begin{array}{|c|c|c|c|c|}\hline 0&0&1&1&0\\\hline\end{array}$; while, for the node "0.2.1.2", its $TK_{0.2.1.2} = \{$top-k, XML, query$\}$, and its $kList$ is $\begin{array}{|c|c|c|c|c|}\hline 0&0&1&1&1\\\hline\end{array}$. Clearly, the latter $kList$ covers the former $kList$[1]. As for the computation of $TC_v = TC_{v_i}$, we use an approximate method based on $(\min, \max)$. If $v_{\min} = v_{i\min} \wedge v_{\max} = v_{i\max}$, we conceive that their tree content sets are same.

Now let us briefly introduce the pruneMatch stage, especially about its pruning step, for it is easy to understand the constructing step following the node description[2]. In pruning step, all the nodes of a SLCA-based fragment are checked, and the final result only keeps the valid contributors. Line 26 corresponds to the rule 1 in Definition 2, for the counter of the n's j^{th} label equals 1 ($n.chlList[j].counter = 1$) means that n has only one child with that label. Lines 22 and 23 correspond to the rule 2.(b), while lines 24 and 25 correspond to the rule 2.(a) with the help of $usedKNums$ (used key numbers) and $usedCIDs$ (used cIDs).

4 Experiments

We implement our ValidMatch and the MaxMatch in Java, and compare them using four XML datasets: "dblp20040213" (197.6MB) and three XMark datasets – standard (111.1MB), data1 (334.9MB) and data2 (669.6MB). They are parsed with Xerces 2.9.0, and the shredded tuples are stored in PostgreSQL 8.2.4 with three simple tables – "label (label, number)", "element (node's label, Dewey, level, kList, cID)" and "value (node's label, Dewey, attribute, keyword)". During the parsing, the stop-words [2] are filtered with Lucence, and some words are randomly selected to compose keyword queries.

There are two parameters to evaluate the two algorithms: efficiency and effectiveness. The efficiency is illustrated by recording the elapsed time of running a keyword query on a dataset. As for the effectiveness, we take MaxMatch as the baseline, and record two ratios – common fragment ratio (CFR) and average pruning ratio (APR). The average pruning ratio is defined as $APR = \frac{\sum_{a \in A} \frac{|x_a - v_a|}{|x_a|}}{|V - V \cap X|}$. The sets A, V and X of a

[1] We can also construct the keyword number (*knum*) for each kList, for instance knum($\begin{array}{|c|c|c|c|c|}\hline 0&0&1&1&0\\\hline\end{array}$) $= 0 * 2^4 + 0 * 2^3 + 1 * 2^2 + 1 * 2^1 + 0 * 2^0 = 6$. With the help of the AND computation between two *knums*, we can also determine the equality and subset relationships between two nodes (refer to [1] for more detail).

[2] From the knowledge of Graph theory, the LCA operation in this step is to locate the lowest common ancestor of the given two nodes.

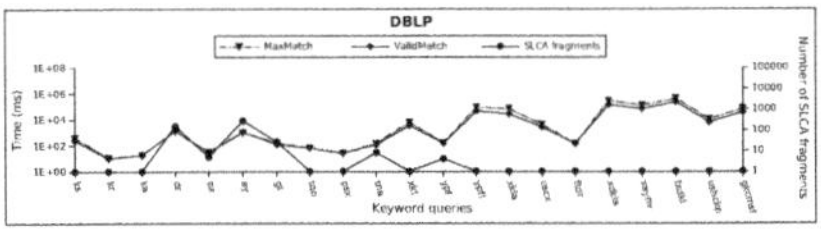

(a) dblp20040213

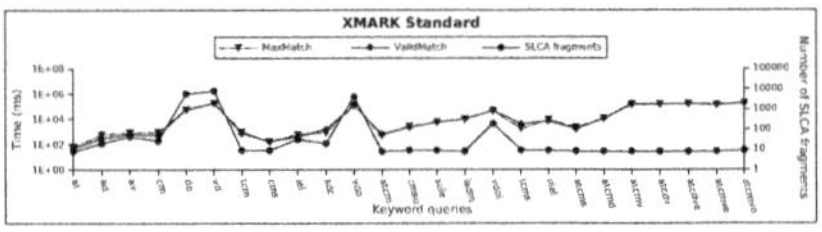

(b) xmark standard

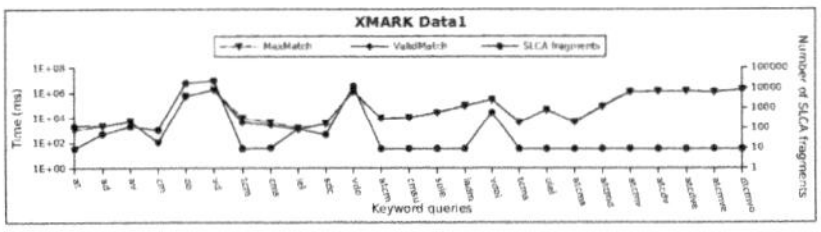

(c) xmark data1

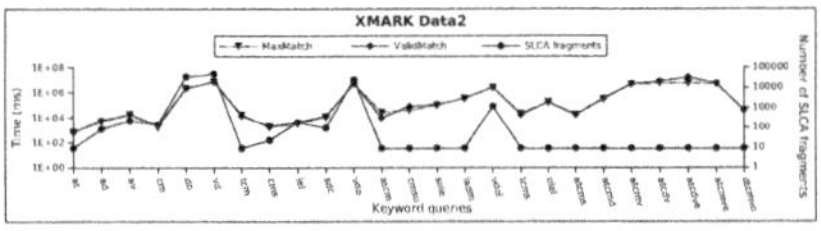

(d) xmark data2

Fig. 2. Performance of ValidMatch and MaxMatch on the datasets with different keyword queries

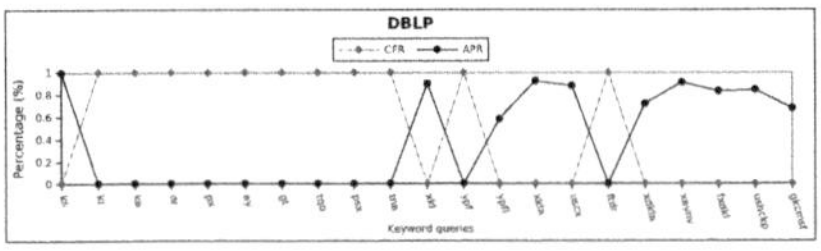

(a) dblp20040213

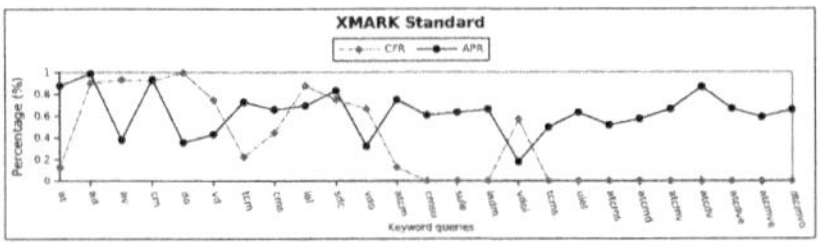

(b) xmark standard

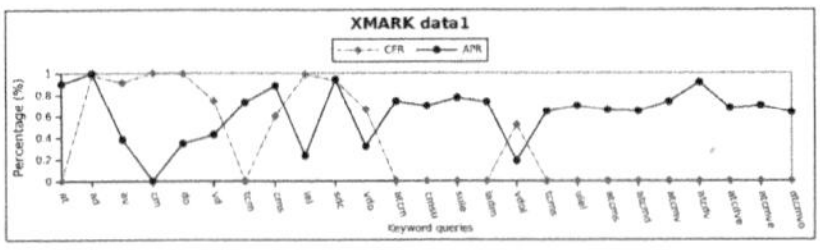

(c) xmark data1

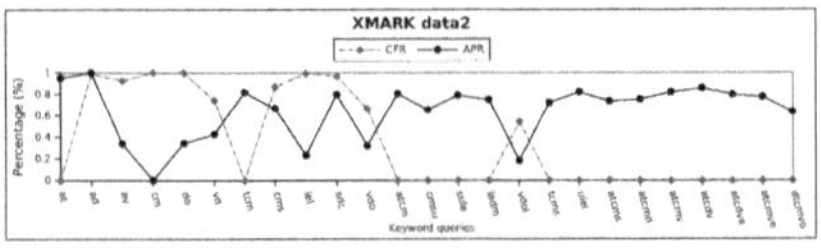

(d) xmark data2

Fig. 3. Ratio of the retrieved results by ValidMatch and MaxMatch

query Q stand respectively for the SLCA node set, the two fragment sets computed by our ValidMatch and MaxMatch. The v_a and x_a represent the corresponding meaningful fragments in V and X with the same SLCA node a, and we conceive that v_a and x_a are same if their node sets are same. The $v_a - x_a$ stands for the nodes further discarded by our ValidMatch, and $V \cap X$ stands for the same fragments in V and X. And the CFR is defined as $CFR = \frac{|V \cap X|}{|A|}$. It is easy to infer that the smaller the CFR for a query is, the more fragments our ValidMatch needs to prune in the result retrieved by MaxMatch for that query. As for the APR, the larger the APR for a query is, the more uninteresting nodes our ValidMatch discards from the fragments retrieved by MaxMatch.

• **Performance:** Figure 2 shows the performance of ValidMatch and MaxMatch. From those figures we can see that ValidMatch has competent performance as MaxMatch. Besides, we also record the number of the fragments for each query in those figures, shown as "SLCA fragments" lines for better understanding the CFR and APR lines in Fig. 3. If there are 10 SLCA fragments for Q, and the CFR is 20%, we can directly know that there are $10 * 20\% = 2$ same fragments returned by both ValidMatch and MaxMatch. We also can get $|V - V \cap X|$ for the Q used in APR is $10 - 10 * 20\% = 8$.

• **Effectiveness:** Figure 3 illustrates the effectiveness of our valid-contributor-based ValidMatch algorithm. For DBLP data, the distribution of the CFR values seems obeying some practical law, either 1 or 0. And it heavily depends on whether the SLCA node is the root (CFR=0) or not (CFR=1). This is rational, because, for the real XML data, the SLCA-based fragments not rooted at the root node are self-complete. As for the APR, all its values where CFR=0 are larger than 40%. As for Fig. 3(b)~3(d), most of their APR values are larger than 0, and their CFR values are seldom 1. This is caused, we infer, by the synthetic property of XMark series, for the keywords are distributed randomly in the data.

5 Conclusion

This paper proposes valid contributor-based filtering mechanism to prune the uninteresting nodes in the SLCA-based fragments for XML keyword search. Compared with the contributor-based one in [1], ours is better, which overcomes both the false positive problem and the redundancy problem.

References

1. Liu, Z., Chen, Y.: Reasoning and identifying relevant matches for xml keyword search. In: 34th International Conference on Very Large Data Bases (VLDB) (2008)
2. www.syger.com/jsc/docs/stopwords/english.htm

Efficient Data Structure for XML Keyword Search

Ryan H. Choi[1,2] and Raymond K. Wong[1,2]

[1] The University of New South Wales, Sydney, NSW, Australia
[2] National ICT Australia, Sydney, NSW, Australia
{ryanc,wong}@cse.unsw.edu.au

Abstract. In this paper, we present a compression technique for inverted lists that are specifically designed for XML keyword search algorithms. In addition, we support all navigational operators required by such algorithms and these operators run faster than on traditional inverted lists. Our technique requires small memory footprint and consumption, thus it is also ideal for many resource constraint applications. Experiments show that our technique is efficient and scales well.

1 Introduction

Keyword search on XML has lately received much attention from the community for a simple and schema free way of querying XML data. Many previous works [1,2,3,4] focus on what nodes should be returned as answer nodes (e.g., SLCA), and how they can be returned efficiently. In addition to the problems discussed by previous works, we have identified a few more problems. First, all previous works heavily rely on inverted lists of an XML document, but the size of these inverted lists are several times larger than the raw XML document. One of the reasons is the large physical size of Dewey labels. Since all keyword search algorithms rely on fast ancestor/descendant and preceding/following node comparators, the use of Dewey cannot be avoided. Second, many algorithms call a number of navigational operators on inverted lists very frequently, and these operators become the bottleneck, which decrease the overall performance of these algorithms.

In this paper, we present a compression technique for inverted lists that are specifically designed for XML keyword search algorithms. Dewey labels are difficult to compress, as the length of a label for a node is variable. There are several numeric compression algorithms (see Zhang et al. [5]) that can compress a sequence of numbers, but it is not clear how they can be used, as Dewey labels can be seen as sequences of variable sized numbers. To solve this problem, we adopted ISX Storage Scheme [6]. Using this scheme, we use an integer value to represent the location of a node, and we use this location as a label for a node in an XML document. We then organize these labels in sequences of fixed sized numbers, and compress them by using smaller number of bits. To provide fast node navigational operators, we generate compact summaries of compressed numbers, and build additional search indexes on top of these summaries.

X. Zhou et al. (Eds.): DASFAA 2009, LNCS 5463, pp. 549–554, 2009.

The main contributions of this paper are summarized as follows.

- To the best of our knowledge, this is the first work that addresses the problem of compressing inverted lists specifically designed for XML keyword search.
- We present an efficient technique for compressing inverted lists while providing fast navigational operators required by XML keyword search algorithms.
- We evaluate our technique on various sizes of dataset, and run existing keyword search algorithms to evaluate performance improvements.

Organization: Section 2 presents related work, operators in inverted lists and ISX Storage Scheme. Section 3 describes our compression technique. Section 4 presents experiment results. Finally, Section 5 concludes the paper.

2 Background

XML keyword search algorithms: XRank [7] uses a stack to find a set of answer nodes in an XML tree such that, either the answer node or its descendant nodes contain at least one occurrence of all keywords. XKSearch [1] and Multiway SLCA [2] present efficient algorithms based on Smallest Lowest Common Ancestor (SLCA) semantics. Some works [3,4] try to improve query quality. While the above works provide solutions for returning answer nodes, none of them addresses issues about the size and access time of inverted lists.

XML compressions and differential encoding: XPRESS [8] presents how to process queries on compressed data. ISX Storage Scheme [6] compactly represents an XML document structure and utilizes indexes to support fast node navigational operators. However, it is not clear how they can be used to compress inverted lists. Some works (e.g., Zhang et al. [5]) present efficient ways of storing a sequence of numbers by storing the difference between a number and its preceding number. However, they cannot be directly used to compress inverted lists, as a sequential scan is required to access the contents of inverted lists. Moreover, it is not clear how to store variable length values such as Dewey.

Model: We model XML data T as a rooted, labeled and ordered tree. Every node v in T has a tag name, and every leaf node has a text node. A text node is represented by a child of v. An internal node contains one or more text nodes.

Supporting Keyword Search Algorithms: In order to support existing XML keyword search algorithms, we add supports for two additional functions: $lm(L,\ \lambda(v))$, and $rm(L,\ \lambda(v)))$. Given an inverted list $L = [\lambda(v_1), ..., \lambda(v_j), \lambda(v_{j+1}), ..., \lambda(v_n)]$, and $\lambda(v_j) \leq \lambda(v) \leq \lambda(v_{j+1})$, $lm(L,\ \lambda(v))$ returns $\lambda(v_j)$. Similarly, $rm(L,\ \lambda(v))$ returns $\lambda(v_{j+1})$. Both functions return null if such $\lambda(v)$s do not exist. Many previous works [1,2,3,4] use these two functions extensively.

ISX Storage Scheme: We use the topology layer from ISX Storage Scheme [6] to represent the data structure of an XML tree T. It uses a balanced parenthesis encoding scheme [9], in which a single bit 0 or 1 is used to represent an opening and closing tag, respectively. We define a label $\lambda(v)$ for a node v in an XML tree T to be the number of open parenthesises between the root node and v in T. We extend ISX Storage Scheme such that, given a label $\lambda(v)$, the ith bit representing $\lambda(v)$ can be found efficiently. This additional index contains the total number of open parenthesises between the root and an ith block in Tier 1.

3 Approach

3.1 Compressing Inverted Lists

Our system uses a hash table to store keywords and their hash values. Labels in an inverted list L for a keyword k are treated as a sorted sequence of numbers. This sequence of numbers is compressed by storing the sequence of differences between a number and its preceding number. Next, we make groups of numbers in L for k called *segments*, and use b bits to encode the numbers in each segment.

Compressed inverted lists are stored across a number of *Base Data* blocks. Each Base Data block contains a header and a group of segments. A header consists of four parts. The first part contains an integer, which counts the number of segments stored in a block. We use $\lceil \lg \frac{|B|}{8} \rceil$ bits, where $|B|$ is the size of a block in bits. The second, third and fourth parts are used to represent $[(S_s, |S|, b)]$, where S_s and $|S|$ is the start position and the size of a segment in a block, respectively. Each element in the tuple uses $\lceil \lg \frac{|B|}{8} \rceil$, $\lceil \lg s \rceil$ and $\lceil \lg z \rceil$ bits, where s is the maximum number of labels that a segment can contain; and z is the largest label in a segment.

An inverted list L for a keyword k is compressed as follows. First, labels in L are preprocessed such that the difference between a label and its preceding label is calculated for $\forall \lambda(v) \in L$. Second, these labels are split into segments of size s, and the biggest label z from each segment is selected. We then use $b = \lceil \lg z \rceil$ bits to encode the labels that are in the same segment as z. The above process is repeated for all segments. While performing the process, we record the absolute values of the first and last labels in each segment, as they are used for indexes.

3.2 Main and Aux Indexes

In order to avoid linear scans of inverted lists, Main and Aux Indexes are built.

Main Index (MI) summarizes segments in Base Data. It is mandatory to access Base Data, and is partitioned into a number of disk blocks. Each MI block contains a header and a group of MI entries. A header contains one integer, which counts the number of entries in a block. A MI entry is created for each segment in Base Data, and is in form of (k, d, BId, SId), where k is the keyword that a segment represents; d is the absolute value of the last label in the previous segment, or the first label if the segment is the first one; BId is a block ID of Base Data where this segment can be found; and SId is the position where this segment is located within the block BId. We use 32 bits for each k, d and BId, and $\lceil \lg \frac{|B|}{8} \rceil$ bits for SId. All entries in MI are sorted in ascending order. To do binary search on MI efficiently, we build Aux Indexes. With Aux Indexes, we only need 1 disk read to find a MI block, as Aux Indexes are stored in memory.

Aux Index (AI) is an optional index, and one or more AIs are built as required. Similar to MI, AI is partitioned into a number of disk blocks. Each AI block contains a header and a group of AI entries. Similarly, a header contains a counter. An AI entry is created for each MI block, and is in form of (k, d, BId), where k and d are the first two values appeared in the first MI entry in a MI block; and BId is the block ID of AI, where (k, d) pair is extracted. We use 32 bits for each value in the tuple. Lastly, all entries are sorted in ascending order.

3.3 Searching

With MI and AIs, we can search for a label $\lambda(v)$ for a keyword k in an inverted index $I = \{L_1, ..., L_n\}$ efficiently. We perform $contains(I,\ k,\ \lambda(v))$ search as follows. First, the topmost Aux Index (AI_n) is read, and the header is decoded to get the number of AI entries stored in the block. Second, we do binary search within a block to get ith entry in the block. Once we find two entries such that, $(k, d, BId)_i \leq (k, \lambda(v)) < (k, d, BId)_{i+1}$, we read $(k, d, BId)_i$. Third, we read the block with *BId* of the next Aux Index (AI_{n-1}), and do binary search similarly. We iteratively keep searching until we reach the first Aux Index (AI_1). We now read a block of MI, and do binary search within the MI block similarly. Fourth, when we find the matching MI entry, we read *BId* and *SId* from the entry, and read the block with *BId* of Base Data. We then decode the block to retrieve $(S_s, |S|, b)$. After that, we jump to the segment and do linear scan. While doing linear scan, labels in the segment are reconstructed by iteratively adding consecutive labels. The first absolute value in the segment can be calculated by adding d, which is found in the MI entry. $contains(I,\ k,\ \lambda(v))$ can be easily modified to implement $get(I, k)$. $lm(L, \lambda(v))$ and $rm(L, \lambda(v))$ are implemented similarly, but we return two labels, $\lambda(v_i)$ and $\lambda(v_{i+1})$, whose values are $\lambda(v_i) \leq \lambda(v) \leq \lambda(v_{i+1})$.

4 Experiments

The experiment settings were as follows. OS=Mac OS X, CPU=Core 2 Duo 2.4 GHz, Ram=2 GB, Harddisk=5400 RPM, and implementation=Java 1.5.

Datasets: We used XMark to generate XML documents of size 112, 223, 447, 670, 896 and 1100 MB, respectively. Then, two sets of inverted lists were generated: one with Dewey and one with our labeling scheme.

Comparison: We implemented ILE [1] and Stack [7] algorithms. Inverted lists were implemented in Oracle Berkeley DB as described by Xu et al. [1].

Queries: The following queries, which were adopted from Liu & Chen [3], were used. *Q1: closed_auction,price*; *Q2: closed_auction,price,date,itemref, quantity,type,seller,buyer*; *Q3: open_auction,person257*; and *Q8: seller,04/02/1999.*

4.1 Results

Figure 1(a) and 1(b) show the runtime performance when various numbers of keywords were used. In this experiment, we used one keyword at a time from

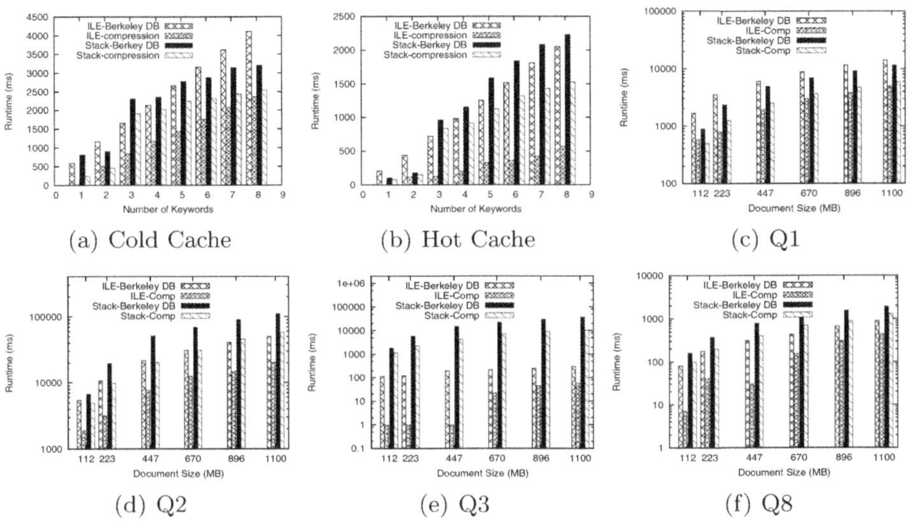

Fig. 1. Query Performance

query Q2, and used 112 MB document. In addition, cold and hot buffers were also tested. The figures show that, as the number of keywords increases, the runtime for both algorithms increases. This is because the number of join operations and labels to process are increased for ILE and Stack algorithm, respectively. The rate of increase of both algorithms when performed on our compressed inverted lists is lower than when the same algorithms were performed on Berkeley DB. This is because we can retrieve the right set of labels for a keyword faster than Berkeley DB, and comparing two labels takes constant time unlike Dewey labels.

Figure 1(c)–1(f) show the query performance when the queries Q1–Q3 and Q8 were executed on various sizes of dataset. Due to limited space, we omit figures for Q4–Q6 and Q7. These are similar to Figure 1(e). The runtime of both algorithms increase as the size of dataset increases. This is because the number of answer nodes are increased, and thus the number of join operations and labels to process are also increased. Both algorithms ran faster on our inverted lists, and the reasons are similar to the experiments in Figure 1(a) and 1(b).

5 Conclusion

We have presented a compression technique for inverted lists that are specifically designed for XML keyword search algorithms. Our data structure supports all navigational operators that many keyword search algorithms require. Our experiments show that existing keyword search algorithms run faster on our compressed inverted lists than traditional Dewey based inverted lists.

References

1. Xu, Y., Papakonstantinou, Y.: Efficient keyword search for smallest lcas in xml databases. In: SIGMOD (2005)
2. Sun, C., Chan, C.Y., Goenka, A.K.: Multiway slca-based keyword search in xml data. In: WWW (2007)
3. Liu, Z., Chen, Y.: Identifying meaningful return information for xml keyword search. In: SIGMOD (2007)
4. Xu, Y., Papakonstantinou, Y.: Efficient lca based keyword search in xml data. In: EDBT (2008)
5. Zhang, J., Long, X., Suel, T.: Performance of compressed inverted list caching in search engines. In: WWW (2008)
6. Wong, R.K., Lam, F., Shui, W.M.: Querying and maintaining a compact xml storage. In: WWW (2007)
7. Guo, L., Shao, F., Botev, C., Shanmugasundaram, J.: Xrank: Ranked keyword search over xml documents. In: SIGMOD (2003)
8. Min, J.K., Park, M.J., Chung, C.W.: Xpress: A queriable compression for xml data. In: SIGMOD (2003)
9. Zhang, N., Kacholia, V., Özsu, M.T.: A succinct physical storage scheme for efficient evaluation of path queries in xml. In: ICDE (2004)

WS-BPEL Business Process Abstraction and Concretisation

Xiaohui Zhao[1], Chengfei Liu[1], Wasim Sadiq[2], Marek Kowalkiewicz[2], and Sira Yongchareon[1]

[1] Centre for Complex Software Systems and Services
Swinburne University of Technology
Melbourne, Victoria, Australia
{xzhao,cliu}@groupwise.swin.edu.au,
575318X@student.swin.edu.au
[2] SAP Research
Brisbane, Australia
{wasim.sadiq,marek.kowalkiewicz}@sap.com

Abstract. Business process management is tightly coupled with service-oriented architecture, as business processes orchestrate services for business collaboration at logical level. Given the complexity of business processes and the variety of users, it is a sought-after feature to show a business process with different views, so as to cater for the diverse interests, authority levels, etc., of users. This paper presents a framework named *FlexView* to support process abstraction and concretisation. A novel model is proposed to characterise the structural components of a business process and describe the relations between these components. Two algorithms are developed to formally illustrate the realisation of process abstraction and concretisation in compliance with the defined consistency rules. A prototype is also implemented with WS-BPEL to prove the applicability of the approach.

1 Introduction

In service-oriented architecture (SOA), business processes are widely applied to organise service composition and service orchestration [1-4]. As one of the leading SOA advocators, the Web service community formally adopted business process technology in 2001 by establishing the Business Process Execution Language for Web Services (WS-BPEL) [5]. WS-BPEL supports the specification of both composition schemas and coordination protocols to fulfil complicated B2B interactions.

Reluctantly, most of current business process modelling languages, including WS-BPEL, stick to a fixed description of business processes. Although WS-BPEL can be used to define both abstract processes and executable processes, WS-BPEL is in lack of mechanisms to automatically represent a business process with different views on demand. The concept of "process view" has emerged recently to support for flexible views on business process representation, and thereby separate the process representation from executable business process models. This feature has been longed for in the practical business process application environment, for the purpose of authority control, privacy protection, process analysis, etc. [6, 7] For instance, users may prefer to see part of the process details at a time, due to the complexity of the business process.

X. Zhou et al. (Eds.): DASFAA 2009, LNCS 5463, pp. 555–569, 2009.

Users with different interests or different authority levels, may be interested to or be allowed to see different views of the same business process. For another instance, in a graphical displaying tool for business processes, the flexibility on showing a reduced version of business process at a time is highly expected, due to the limit of screen size. Similar functions can be found in other application areas. A good example is *google maps*, which allows users to zoom in or zoom out a map, while the displayed details on map automatically adapt to the scale level, for instance, streets and roads are shown on a large scale map, yet a small scale map only shows suburbs and towns.

To realise such "smart zooming" functions towards business process representation, this paper proposes a framework named *FlexView* to support flexible process abstraction and concretisation. With FlexView, users are allowed to define and switch among the different views for a business process. A comprehensive model defines the structural constructs of a business process and the relations between them. Two algorithms formally illustrate how to enforce the process abstraction and concretisation operations in compliance with structural consistency.

The remainder of this paper is organised as follows: Section 2 discusses the requirements for supporting flexible views with a motivating example; Section 3 introduces a process component model with a set of rules on structural consistency, and the algorithms for realising abstraction and concretisation; Section 4 addresses the incorporation of FlexView into WS-BPEL, and also introduces the implementation of a prototype; Section 5 reviews the related work and discusses the advantages of our framework; concluding remarks are given in Section 6 with an indication for the future work.

2 Motivating Example

Figure 1 (a) shows all details of the business process for a simplified sales management service, where the process starts from receiving purchase orders, and then handles the production, cost analysis and shipping planning concurrently, and finally terminates by sending the invoice. Each task may interact with proper Web service(s) to fulfil the assigned mission. The dashed arrows represent the synchronisation dependencies between tasks, for example, the arrow between "production" and "dispatch products" denotes that task "dispatch products" can only start after the completion of "production".

This business process mainly involves four departments, viz., sales department, workshop, accounting department, and distribution centre. For a user from the distribution centre, the user may only care about the shipping details, and thus the user may choose to "zoom out" the details for production and cost handling. The user can obtain the view for this business process as shown in Figure 1 (b). In this view, the details for production and cost handling are abstracted into two new tasks, i.e., "handle production at workshop" and "handle cost calculation at accounting department". These two tasks hide the details yet preserve the existence of the production and cost handling procedures. In this transformation, the related links are hidden automatically, as well as the synchronisation link from "schedule production" to "cost analysis". The synchronisation link from "production" to "dispatch products" is converted to connect "handle production at workshop" to "dispatch products", as these two tasks inherit the synchronisation dependency of the former one.

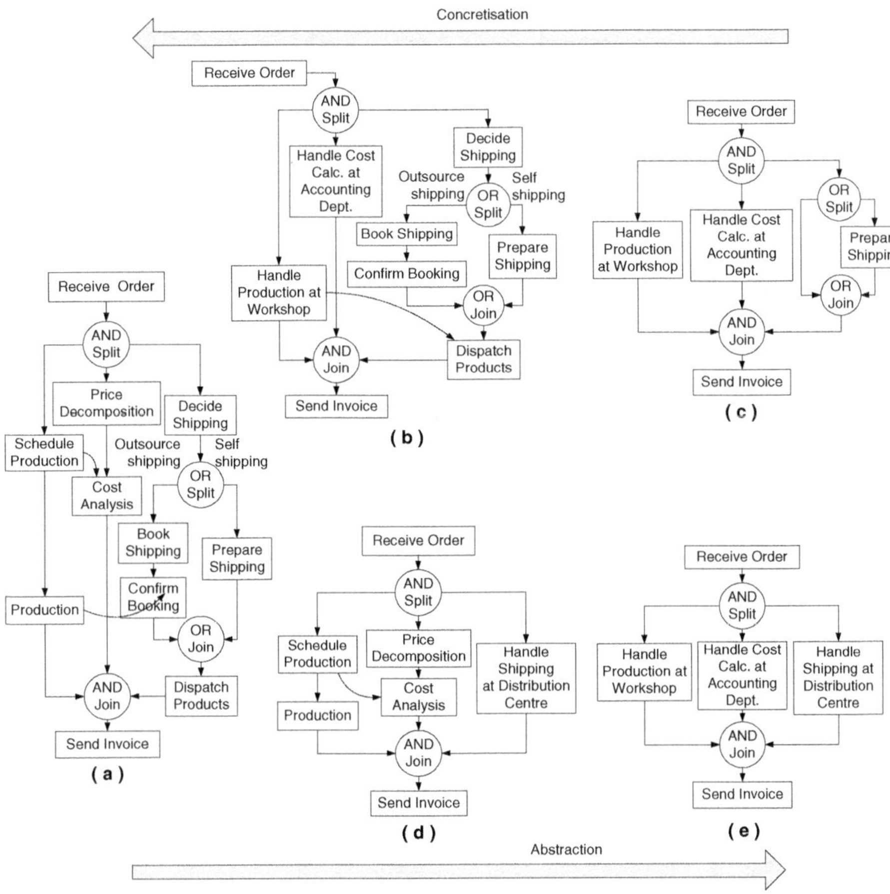

Fig. 1. Motivating example business process

Due to the screen size of the user's computer, the user may want to check the details of a single shipping option at a time. In this case, the process should change to the view shown in Figure 1 (c). In this view, the shipping procedure is represented as an Or-split/join structure, which contains a branch with task "prepare shipping", and an empty branch standing for the existence of an alternative shipping option. The user may later on select to "zoom in" this empty branch to see the details for the alternative option.

The users from other departments may not be authorised to see the shipping details. Such users may only see the view shown in Figure 1 (d), where all the shipping details are hidden in a new task "handle shipping at distribution centre". The synchronisation link from "production" to "dispatch products" is also hidden, as the underlying synchronisation dependency is not effective for this view. Figure 1 (e) displays a further abstracted view of the business process, which only outlines the core part of the business process with three parallel tasks. The authorised users can choose to concretise the interested part to see more details. In either process concretisation (from the right to the left in Figure 1) or abstraction (from the left to the right), the structure of the new views keeps consistent with the previous one.

To support process abstraction and concretisation functions, new mechanisms are on demand to allow wrapping a sub process into a specific task or link, and releasing the sub process back from a task or link. In details, we list the following technical requirements:

- Maintain the relations between the hidden sub processes and the corresponding tasks/links.
- Preserve the structural information of a business process, such as split/join structures and synchronisation links, during process abstraction/concretisation.
- Support cascading abstraction/concretisation operations.
- Keep the structural consistency of process views during transformations.

Towards these requirements, our FlexView framework employs a process component model to describe the structure of process views and structural components, and maintain the relations between structural components. The algorithms are designed to enable the procedure of abstraction and concretisation. A set of defined rules regulate the structural consistency during the procedure. The framework is implemented with WS-BPEL as a proof-of-concept.

3 Framework of FlexView

3.1 Process Component Model

To well describe the structure of a process view and maintain the relations between structural components, we define a process component model. This model provides the foundation for process abstraction and concretisation functions, and particularly takes into account the characteristics of WS-BPEL.

Definition 1. (Gateway) Gateways are used to represent the structure of a control flow. Here we define five types of gateways, namely *Or-Split*, *Or-Join*, *And-Split*, *And-Join*, and *Loop*. Figure 2 shows the samples of these gateways, respectively. *Or-Split/Join* and *Loop* gateways may attach conditions to restrict the control flow.

Definition 2. (Synchronisation Link) In an *And-Split/Join* structure, synchronisation links are used to represent the synchronisation dependency between the tasks belonging to different branches. For example, in Figure 2 (b), the synchronisation link between t_i and t_j, represented as a dashed arrow, denotes that t_j can only start after the completion of t_i. In WS-BPEL, such a synchronisation dependency is supported with a <link> element.

Functions *ind*(m) and *outd*(m) define the number of edges which take m as the terminating node and the starting node, respectively. Note *ind* and *outd* only count the number of edges but not synchronisation links.

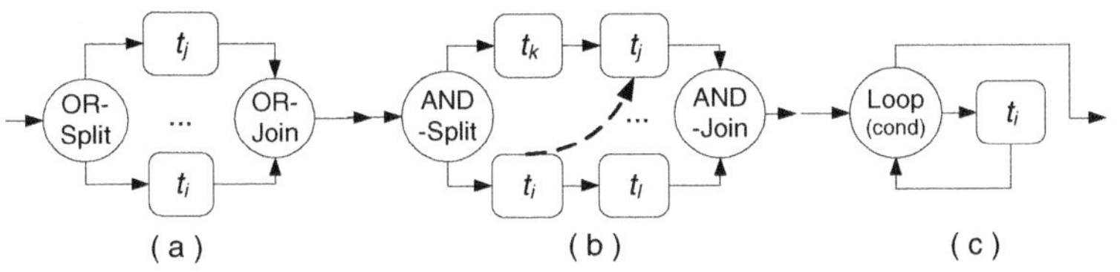

Fig. 2. Gateway samples

Function *type*: $G \rightarrow Type$ is used to specify gateway types, where $Type=\{Loop, And\text{-}Join, And\text{-}Split, Or\text{-}Join, Or\text{-}Split\}$. According to the natural characteristics of these gateways, we can define the following rules in terms of the incoming and outgoing degrees:

$$\text{if } type(g) = \text{“}Loop\text{”} \begin{cases} ind(g)=1,\ outd(g)=2 & \text{if } g \text{ is at the starting position;} \\ N/A & g \text{ is not allowed at the ending position;} \\ ind(g)=2,\ outd(g)=2 & \text{Otherwise.} \end{cases}$$

$$\text{if } type(g) = \text{“}And\text{-}Split\text{” or “}Or\text{-}Split\text{”} \begin{cases} ind(g)=0,\ outd(g)>1 & \text{if } g \text{ is at the starting position;} \\ N/A & g \text{ is not allowed at the ending position;} \\ ind(g)=1,\ outd(g)>1 & \text{Otherwise.} \end{cases}$$

$$\text{if } type(g) = \text{“}And\text{-}Join\text{” or “}Or\text{-}Join\text{”} \begin{cases} N/A & g \text{ is not allowed at the starting position;} \\ ind(g)>1,\ outd(g)=0 & \text{if } g \text{ is at the ending position} \\ ind(g)>1,\ outd(g)=1 & \text{Otherwise.} \end{cases}$$

Definition 3. (Sub Process) A sub process is a structural component of a business process, and it also maintains the necessary information for sub process composition. The structure of a sub process s can be modelled as an extended directed graph in the form of tuple $(N, G, E, L, m_s, m_t, L_O)$, where

- $N=\{n_1, n_2, \ldots, n_x\}$, $n_i \in N$ $(1 \le i \le x)$ represents a task of s.
- $G=\{g_1, g_2, \ldots, g_y\}$, $g_i \in G$ $(1 \le i \le y)$ represents a gateway of s.
- E is a set of directed edges. An edge $e=(m_1, m_2) \in E$ corresponds to the control dependency between m_1 and m_2, where $m_1 \in N \cup G$, $m_2 \in N \cup G$.
- L is a set of synchronisation links. A synchronisation link $l = (m_1, m_2) \in L$ corresponds to the synchronisation dependency between m_1 and m_2, where $m_1 \in N \cup G$, $m_2 \in N \cup G$.
- m_s is the starting node of s, which satisfies that $m_s \in N \cup G$ and $ind(m_s)=0$.
- m_t is the terminating node of s, which satisfies that $m_t \in N \cup G$ and $outd(m_t)=0$.
- $\forall n \in N \setminus \{ m_s, m_t \}$, $ind(n)=outd(n)=1$. This property is guaranteed by the usage of gateways.
- L_O is a set of hidden synchronisation links, i.e., the synchronisation links that have a node not included in N or G. $\forall l=(m_1, m_2) \in L_O$, $(m_1 \in N \cup G) \wedge (m_2 \notin N \cup G)$ or $(m_1 \notin N \cup G) \wedge (m_2 \in N \cup G)$. The synchronisation links in L_O are not displayable as they connect to foreign nodes, but such synchronisation link information is preserved for sub process composition.

Definition 4. (Sub Process Hierarchy) A sub process hierarchy $\Gamma(p)$ for business process p maintains all the related sub processes and mapping information for process representation. $\Gamma(p)$ can be represented as tuple (S, δ, γ), where

- S is a finite set of distinct sub processes.
  For a sub process $s \in S$,
  $$\forall l=(n_1, n_2) \in s.L \cup s.L_0 \cup s.G\ \exists s_1 \in S\ (n_1 \in s_1.N \cup s_1.G) \wedge (n_2 \in s_1.N \cup s_1.G),$$
  $$\text{or } \exists s_1 \in S,\ s_2 \in S,\ (n_1 \in s_1.N \cup s_1.G) \wedge (n_2 \in s_2.N \cup s_2.G).$$
- $s_0 \in S$ is the root sub process, which shows the most abstracted view of p.
- δ: $E' \rightarrow S'$ ($E' \subseteq \bigcup_{s \in S} s.E$ and $S' \subseteq S \backslash \{s_0\}$) is a bijection describing the relations between edges and sub processes. Correspondingly, we have the inverse function δ^{-1}: $S' \rightarrow E'$.
- γ: $N' \rightarrow S'$ ($N' \subseteq \bigcup_{s \in S} s.N$ and $S' \subseteq S \backslash \{s_0\}$) is a bijection describing the relations between nodes and sub process. Correspondingly, we have the inverse function γ^{-1}: $S' \rightarrow N'$.
- A sub process can only occur in one of the two inverse functions. This denotes that $\forall s \in S \backslash \{s_0\}$, if $\delta^{-1}(s) \neq$null then $\gamma^{-1}(s)$=null; if $\gamma^{-1}(s) \neq$null then $\delta^{-1}(s)$=null.

The sub processes of a sub process hierarchy can be defined in a nested way. Figure 3 shows a sub process hierarchy example, where $subp_1$, …, $subp_5$ are five sub processes in this hierarchy, and $subp_1$ is the root sub process.

In this example, tasks t_i and t_j of sub process $subp_1$ can be mapped to sub processes $subp_2$ and $subp_5$ by functions $\gamma(t_i)$ and $\gamma(t_j)$, respectively. Further, task t_k and edge e_m of $subp_2$ can be mapped to sub processes $subp_3$ and $subp_4$ by functions $\gamma(t_k)$ and $\delta(e_m)$, respectively. A *concretisation* operation denotes the extension by replacing a task or edge with the mapped sub process. Therefore, root sub process $subp_1$ can be concretised into combination $subp_1$+$subp_2$, $subp_1$+$subp_5$, or $subp_1$+$subp_2$+$subp_5$, where tasks t_i and t_j are replaced by the corresponding sub processes. Correspondingly, the *abstraction* operation can be realised by wrapping a sub process back into a task or edge with functions γ^{-1} and δ^{-1}. Each result combination denotes a partial view of the business process.

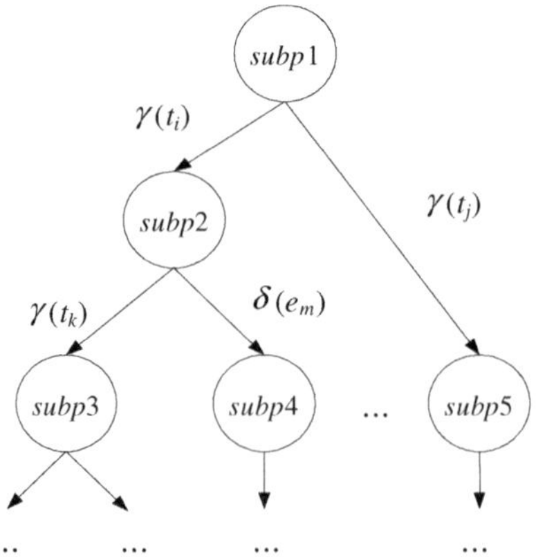

Fig. 3. Sub process hierarchy example

Such a sub process hierarchy is fully customisable for users, and thereby enables the adaptation to user-defined partitions and categorisations according to different levels of BPEL abstraction and concretisation.

Definition 5. (Process View) A process view represents the viewable part for a business process at a time. In the sub process hierarchy, each process view corresponds to a sub tree including the root sub process, where the mapped tasks/edges are concretised with corresponding sub processes. A fully concretised view, i.e., the view containing all the sub processes in this hierarchy, is equivalent to the base business process, and the view containing only the root sub process is the most abstracted view.

3.2 Consistency and Validity Rules

As explained in the motivating example, some rules are defined to guarantee the structural consistency of process views during view abstraction and concretisation. This section is to discuss the rules on preserving execution orders, branch subjection, synchronisation dependencies, and so on.

- *Preliminary*

- A *dummy branch* denotes a branch in a split/join structure such that the branch contains nothing but only one edge.
- A common split gateway predecessor (CSP), x, of a set of tasks, T, denotes a split gateway such that x is the predecessor of each task in T.
- $before(t_1, t_2)$ denotes that task t_1 will be executed earlier than task t_2. This means that there exists a path from starting t_1 to t_2 in the corresponding directed graph, while the path does not contain any go-back edge of a loop structure. Apparently, *before* is a transitive binary relation.
- $CSP(t_1, t_2)$ returns the set of common split gateway predecessors of t_1 and t_2, or returns null if the two tasks have no common split gateway predecessors.
- $branch(g, t_1, t_2)$ is a boolean function, which returns true if t_1 and t_2 lie in the same branch led from split gateway g, otherwise returns false.

- *Structural Consistency and Validity Rules*

In regard to an abstraction/concretisation operation, the original process view v_1 and the result view v_2 are required to comply with the following rules:

Rule 1. (Order preservation) As for the tasks belonging to v_1 and v_2, the execution sequences of these tasks should be consistent, i.e.,

If $t_1, t_2 \in v_1.N \cap v_2.N$ such that $before(t_1, t_2)$ exists in v_1, then $before(t_1, t_2)$ also exists in v_2.

Rule 2. (Branch preservation) As for the tasks belonging to v_1 and v_2, the branch subjection relationship of these tasks should be consistent, i.e.,

If $t_1, t_2 \in v_1.N \cap v_2.N$ and $g \in CSP(t_1, t_2)$ in v_1, $g \in CSP(t_1, t_2)$ in v_2 such that $X(g, t_1, t_2)$ in v_1, then $X(g, t_1, t_2)$ in v_2, where $X \in \{branch, \neg branch\}$.

Rule 3. (Synchronisation dependency preservation) If an abstraction operation involves any tasks with synchronisation links, the synchronisation links should be rearranged to preserve the synchronisation dependency. Assume that sub process s comprising tasks t_1 and t_2 is to be abstracted into a compound task t_c as shown in Figure 4,

– for task $t_x \in s.N$ and t_x has an outgoing synchronisation link l,

If $\forall t \in s.N$, $before(t, t_x)$ then the source task of l should be changed to t_c, otherwise l should be hidden.

– for task $t_x \in s.N$ and t_x has an incoming synchronisation link l,

If $\forall t \in s.N$, $\neg before(t, t_x)$ then the target task of l should be changed to t_c, otherwise l should be hidden.

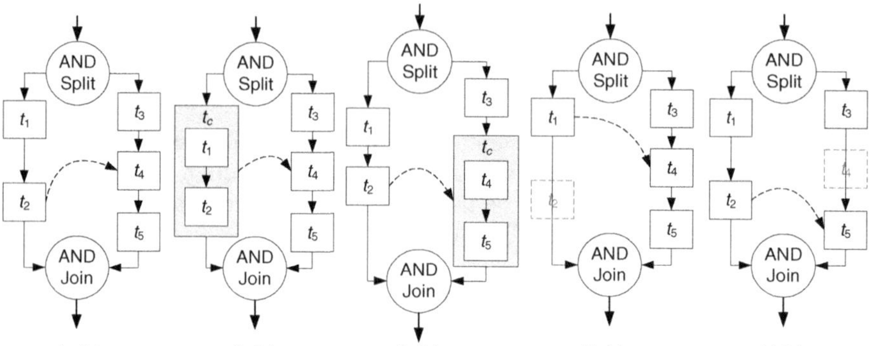

Fig. 4. Synchronisation link handling

In Figure 4, the transformation from (a1) to (a2), where t_1 and t_2 are hidden in task t_c, and the transformation from (a1) to (a3), where t_4 and t_5 are hidden in task t_c, illustrate the two mentioned scenarios, respectively.

In the case that a task involving a synchronisation link is abstracted into an edge, the re-arrangement of synchronisation links is subject to Rule 1. For example, Figure 4 (b1) and (b2) illustrate the re-arrangements in cases that t_2 and t_4 are hidden in edges.

Rule 4. (No empty Split/Join or Loop structures) If a loop structure contains no tasks, or if a split/join structure contains only dummy branches, then the loop or split/join structure should be hidden.

Rule 5. (No dummy or single branch in And-Split/Join structures) If an And-split/join structure contains both dummy and non-dummy branches, then the dummy branch(es) should be hidden. If the And-split/join structure contains only one non-dummy branch, then the And-split/join structure will be degraded into a sequential structure.

Rule 6. (Dummy branch in Or-Split/Join structures) If an Or-split/join structure contains a dummy branch, then this dummy branch should remain to indicate the existence of an alternative execution path. If an Or-split/join structure contains multiple dummy branches, these branches should merge into one dummy branch.

3.3 Process Abstraction and Concretisation

To realise the view abstraction and concretisation under the restriction of structural consistency, two algorithms are developed to formalise the procedures of process view transformation.

Given a sub process hierarchy $\Gamma(p)=(S, \delta, \gamma)$, the following functions are to be used in the algorithms: *addEdge*(s, e) inserts edge e into set E of sub process s. *addTask*(s, t) inserts task t into set N of sub process s. *addLink*(s, l) inserts synchronisation link l to set L of sub process s. *removeLink*(s, l) deletes synchronisation link l from set L of sub process s. *removeTask*(s, t) deletes task t from set N of sub process s. *removeEdge*(s, e) deletes edge e from set E of sub process s. *combineSubProc*(s_1, s_2) combines the constitute sets, i.e., N, G, E, L and L_0, of sub process s_2 into sub process s_1. *removeSubProc*(s_1, s_2) removes the constitute sets of sub process s_2 from sub process s_1. *toSequence*(s, g_1, g_2) flats a single branch split/join structure scoped by gateways g_1 and g_2 in sub process s into a sequence structure, i.e., removes the two gateways and re-connects the gateways' adjacent nodes to the single branch.

Algorithm 1.

taskZoomIn(s, t) transforms sub process s into a more concrete sub process s', by concretising task t.

```
1   s'=s; subp=γ(t);
2   if t=s'.m_s then s.m_s=subp.m_s;
3   if t=s'.m_t then s.m_t=subp.m_t;
4   do while (∃e=(m_x, t)∈s'.E)
5       removeEdge(s', e); addEdge(s', (m_x, subp.m_s));
6   loop
7   do while (∃e=(t, m_y)∈s'.E)
8       removeEdge(s', e); addEdge(s', (subp.m_t, m_y));
9   loop
10  removeTask(s', t); combineSubProc(s', subp);
11  for each sync link l=(m_1, m_2)∈s'.L_0
12     if (m_1∈s'.N)∧(m_2∈s'.N) then
13        addLink(s', l); s'.L_0=s'.L_0\{l};
14     end if
15  for each sync link l=(m_1, m_2)∈s'.L\subp.L
16    if (m_1=t) or (m_2=t) then
17       removeLink(s', l);
18  for each link l=(m_1, m_2)∈s'.L_0\subp.L_0
19     if (m_1=t) or (m_2=t) then s'.L_0=s'.L_0\{l};
20  return s';
```

Lines 2-3 handle the connection in case that the task to concretise is the starting or ending node. Lines 4-9 connect edges according to Rule1 and Rule 2. Line 10 replaces task t with sub process *subp*. Lines 11-14 reveal the hidden synchronisation links if their source and target nodes are both visible during the concretisation. Lines 15-17 delete the synchronisation links that are involved with task t, because the newly revealed synchronisation links from *subp* will replaces these links. Lines 18-19 sort the hidden synchronisation links that are involved with task t.

The procedure of zooming in an edge is similar to Algorithm 1, and we here do not detail it due to space limit.

Algorithm 2.
zoomOut(*s*, *x*) transforms sub process *s* into a more abstract sub process *s′* by abstracting the part containing task or edge *x*.

```
1   s'=s; ∃subp∈S such that x belongs to subp.
2   if δ⁻¹(subp)≠null then
3     addEdge(s', δ⁻¹(subp));
4   else if γ⁻¹(subp)≠null then addTask(s', γ⁻¹(subp));
5   end if
6   for each sync link l=(m1, m2)∈s'.L
7     if (m1∈subp.N∪subp.G) and (∀t∈subp.N, before(t, m1)) then
8       if γ⁻¹(subp)≠null then
9         addLink(s', (γ⁻¹(subp), m2)); removeLink(s', l);
10      else if δ⁻¹(subp)≠null then
11        e= δ⁻¹(subp)=(m3, m4); addLink(s', (m3, m2)); removeLink(s', l);
12      end if
13    else if (m2∈subp.N∪subp.G) and (∀t∈subp.N, ¬before(t, m2)) then
14      if γ⁻¹(subp)≠null then
15        addLink(s', (m1, γ⁻¹(subp))); removeLink(s', l);
16      else if δ⁻¹(subp)≠null then
17        e= δ⁻¹(subp)=(m3, m4); addLink(s', (m1, m4)); removeLink(s', l);
18      end if
19    end if
20  end for
21  s'=removeSubProc(s', subp);
22  do
23    for each loop structure with loop gateway g in s'
24      if ∃e=(g, g)∈s'.E then removeLoop(s', g); // remove empty loop structure
25    for each split/join structure scoped by split gateway g1 and join gateway g2, in s'
26      flag=0;
27      if (outd(g1)=ind(g2)=1) and (∃e=( g1, g2)∈s'.E) then
28        removeEdge(s', e); toSequence(s', g1, g2); flag=1;
29      end if
30    if (s'.type(g1)=And-split) AND (∃e=(g1, g2)∈s'.E) then removeEdge(s', e);
31    if outd(g1)=ind(g2)=1 then
32      toSeqence(s', g1, g2); flag=1;
33      end if
34    end for
35  loop until (flag=0)
36  return s';
```

Lines 2-5 replace the sub process to abstract with the mapped task or edge. Lines 6-20 handle the synchronisation links according to Rule 3. If *subp* has an outgoing link and the link leaves from the last node of *subp*, lines 7-12 rearrange the link to preserve the synchronisation dependency. Similarly, lines 13-19 do the arrangement, if *subp* has an incoming link and the link joins to the first node of *subp*.

Lines 22-35 iteratively check the structural consistency according to Rules 4-6, until no conflicts exist. According to Rule 4, lines 23-24 and lines 27-29 delete empty loop structures and empty split/join structures, respectively. According to Rules 5 and 6, line 30 deletes dummy branches in an And-split/join structure, and lines 31-33 flat any split/join structures with single branches into sequential structures. Note, due to the set definition, the dummy branches in an Or-split/join structure are already combined together.

The result sub process from these algorithms can be easily converted to a process view for representation, by discarding set L_0.

4 Incorporation into WS-BPEL

To enable abstraction and concretisation for WS-BPEL processes, we first need to incorporate the proposed model into WS-BPEL. As listed in Table 1, the main structural constructs of our model correspond to proper WS-BPEL elements. In WS-BPEL, every edge is implicitly represented, i.e., the execution sequence is determined by the occurrence sequence of elements nested in <sequence>, <pick>, <flow>, <while>, <switch> elements.

Table 1. WS-BPEL elements and our structural constructs

Structural construct	*WS-BPEL element*	*Description*
Task sequence	<sequence>	Allow for sequential execution of tasks.
A pair of *Or-Split/Join* gateways with conditions	<pick>	Perform the non-deterministic execution of one of several paths depending on an external event.
A *Loop* gateway with conditions	<while>	Perform a specific iterative task repeatedly until the given condition becomes false.
A pair of *Or-Split/Join* gateways with conditions	<switch>	Perform a conditional behaviour with a set of branches.
A pair of *And-Split/Join* gateways	<flow>	Perform parallel execution of a set of branches.
A synchronisation link	<link>	Support the synchronisation between tasks or gateways on the branches inside a <flow> element.
A sub process	<scope>	Originally used for defining the compensation scope for fault handling in WS-BPEL, yet here we use it to store the structural content and contextual information (such as variables and declarations), for sub processes.
A dummy branch of a split/join structure	<empty>	Originally used to denote a dummy task, yet here we use it to stand for a dummy branch of a split/join structure.

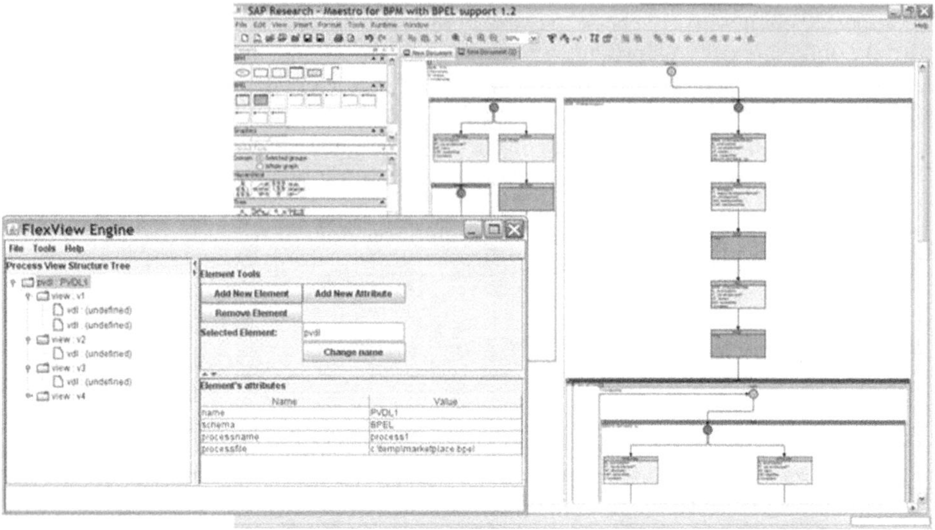

Fig. 5. View generation system architecture

We have developed a prototype for the proof-of-concept purpose. This prototype is based on SAP Research Maestro for BPEL, with extension on process views. This prototype is purely programmed in Java, and utilises some packages from Tensegrity Software [8] for user interface design. Extensible Stylesheet Language Transformation (XSLT) [9] is selected as the technical tool to enforce process abstraction and concretisation. The FlexView engine is responsible for handling the generation of process views according to the user's requests and the pre-defined sub process hierarchy, while the Maestro is used as the displaying tool to represent process views graphically. Users send requests for "zooming in" or "zooming out" the representation of a business process through the FlexView engine, and see the result views in the Maestro. The user interfaces of FlexView and Meastro are given in Figure 5.

In current version, users have to define the sub process hierarchy in advance. However, we are developing necessary mining techniques to identify typical process patterns for defining sub processes. With such support, our FlexView system can automatically or semi-automatically create the sub process hierarchy.

5 Related Work and Discussion

Works on workflow/process views are related to ours. In regard to structural consistency during the process transformations, Liu and Shen [10] proposed an order-preserving approach for deriving a structurally consistent process view from a base process. In their approach, the generation of "virtual activities" (compound tasks) needs to follow their proposed membership rule, atomicity rule, and order preservation rule. Recently, Eshuis and Grefen [11] formalised the operations of task aggregation and process customisation, and they also proposed a series of construction rules for validating the structural consistency. Martens [12] discussed the verification on the structural consistency between a locally defined executable WS-BPEL process and a globally specified abstract process based on Petri net semantics. Compared with these work, first of all, our approach focused more on realising the process transformation at technical level rather than theoretical level. Secondly, in the mentioned works, the customisation process actually lost some tasks. Yet, our approach preserved the hidden tasks and necessary mapping relations, and thus supported both abstraction and concretisation operations. Finally, synchronisation links were considered in our approach.

To support process privacy and interoperability, many works targeted at applying workflow/process views in the inter-organisational collaboration environment. van der Aalst and Weske [13] proposed a "top-down" workflow modelling scheme in their public-to-private approach. Organisations first agree on a public workflow, and later each organisation refines the part it is involved in, and thereafter generates its private workflow. This work reflected a primitive idea of workflow view. In [14], Schulz and Orlowska focused on the cross-organisational interactions, and proposed to deploy coalition workflows to compose private workflows and workflow views together to enable interoperability. Issam, Dustdar et al. [15] extracted an abstract workflow view to describe the choreography of a collaboration scenario and compose individual workflows into a collaborative business process. By deploying workflow views in the workflow interconnection and cooperation stages, their approach allows partial visibility of workflows and resources. Our previous works [16, 17] also established a relative workflow model for collaborative business process modelling. A relative workflow for an organisation comprises the local workflow processes of the organisation and the filtered workflow process views from its partner organisations. In this way, this approach can provide a relative collaboration context for each participating organisation. Some follow-up work targeted at the instance correspondence [18] and the process evolvement [19] in collaborative business processes, as well as role-based process view derivation and composition [7]. In supplement to these works, our approach provided a practical implementation solution by incorporating the view concept into a popular standard business process modelling language. The abstraction and concretisation functions were naturally applicable to support privacy protection or perception control in the collaboration environment.

Proviado project [20] adopted process views for personalised visualisation of large business processes, and they allowed some trade-off between the structural consistency and the adequate visualisation. Our work firmly complied with the proposed structural consistency and validity rules, and supported bi-directional process view operations.

Our work is motivated by practical requirements from areas of process visualisation, process analysis, user friendly process representation, and so on. The work brings the process view concept to the technical level, and incorporates it into a standard process modelling language. In summary, this work contributes to the following aspects:

(1) Abstraction and concretisation functions towards process representations. With these two operations, users are allowed to choose and switch among different views of the same business process. In this way, our approach caters for the diversity of users' interests, authority levels, and so on. Although this paper chooses WS-BPEL as the candidate model to apply abstraction and concretisation functions, the essential

idea of the proposed approach is applicable to most process models, like Business Process Modelling Notations (BPMN) [21], Petri net based workflow models, etc.

(2) Information preservation and structural consistency during transformation. The proposed model and developed algorithms guarantee that our process abstraction and concretisation are lossless in information and consistent in structure. Consequently, the two operations can be performed back and forth rather than one way only.

(3) Deployment in Web service domain using WS-BPEL language. The whole framework is completely incorporated into WS-BPEL via a prototype, which applies XSLT techniques and external repositories to realise WS-BPEL process abstraction and concretisation.

6 Conclusions and Future Work

This paper proposed a framework to support abstraction/concretisation functions towards flexible process view representation. A model was defined to describe the process components and their relations, while a set of algorithms were developed to enforce the abstraction/concretisation operations in compliance with the defined structural consistency rules. The whole framework was incorporated into WS-BPEL language, and a prototype was also developed for the proof-of-concept purpose.

Our future work is to refine the process component model, and further investigate the techniques to automat the generation of the sub process hierarchy. Besides, we plan to investigate similar use cases using BPMN as graphical representation.

Acknowledgement

The research work reported in this paper is supported by Australian Research Council and SAP Research under Linkage Grant LP0669660.

References

1. Smith, H., Fingar, P.: Business Process Management - The Third Wave. Mehan-Kiffer Press (2003)
2. Khoshafian, S.: Service Oriented Enterprise. Auerbach Publisher (2006)
3. Papazoglou, M.: Web Services: Principles and Technology. Prentice-Hall, Englewood Cliffs (2007)
4. Alonso, G., Casati, F., Kuno, H., Machiraju, V.: Web Services - Concepts, Architectures and Applications. Springer, Heidelberg (2004)
5. Andrews, T., Curbera, F., Dholakia, H., Goland, Y., Klein, J., Leymann, F., Liu, K., Roller, D., Smith, D., Thatte, S., Trickovic, I., Weerawarana, S.: Business Process Execution Language for Web Services (BPEL4WS) 1.1 (2003)
6. Zhao, X., Liu, C., Li, Q.: Challenges and Opportunities in Collaborative Business Process Management. Information System Frontiers (2008)
7. Zhao, X., Liu, C., Sadiq, W., Kowalkiewicz, M.: Process View Derivation and Composition in a Dynamic Collaboration Environment. In: The 16th International Conference on Cooperative Information Systems, Monterrey, Mexico, pp. 82–99 (2008)

8. Tensegrity Software, http://www.tensegrity-software.com
9. XSLT, http://www.w3.org/TR/xslt
10. Liu, D.-R., Shen, M.: Workflow Modeling for Virtual Processes: an Order-Preserving Process-View Approach. Information Systems 28, 505–532 (2003)
11. Eshuis, R., Grefen, P.: Constructing Customized Process Views. Data & Knowledge Engineering 64, 419–438 (2008)
12. Martens, A.: Consistency between Executable and Abstract Processes. In: The 7th IEEE International Conference on e-Technology, e-Commerce, and e-Services, Hong Kong, China, pp. 60–67 (2005)
13. van der Aalst, W.M.P., Weske, M.: The P2P approach to interorganizational workflows. In: Dittrich, K.R., Geppert, A., Norrie, M.C. (eds.) CAiSE 2001. LNCS, vol. 2068, pp. 140–156. Springer, Heidelberg (2001)
14. Schulz, K.A., Orlowska, M.E.: Facilitating Cross-organisational Workflows with a Workflow View Approach. Data & Knowledge Engineering 51, 109–147 (2004)
15. Issam, C., Schahram, D., Samir, T.: The View-Based Approach to Dynamic Inter-Organizational Workflow Cooperation. Data & Knowledge Engineering 56, 139–173 (2006)
16. Zhao, X., Liu, C., Yang, Y.: An organisational perspective on collaborative business processes. In: van der Aalst, W.M.P., Benatallah, B., Casati, F., Curbera, F. (eds.) BPM 2005. LNCS, vol. 3649, pp. 17–31. Springer, Heidelberg (2005)
17. Zhao, X., Liu, C.: Tracking over collaborative business processes. In: Dustdar, S., Fiadeiro, J.L., Sheth, A.P. (eds.) BPM 2006. LNCS, vol. 4102, pp. 33–48. Springer, Heidelberg (2006)
18. Zhao, X., Liu, C., Yang, Y., Sadiq, W.: Handling instance correspondence in inter-organisational workflows. In: Krogstie, J., Opdahl, A.L., Sindre, G. (eds.) CAiSE 2007 and WES 2007. LNCS, vol. 4495, pp. 51–65. Springer, Heidelberg (2007)
19. Zhao, X., Liu, C.: Version management in the business process change context. In: Alonso, G., Dadam, P., Rosemann, M. (eds.) BPM 2007. LNCS, vol. 4714, pp. 198–213. Springer, Heidelberg (2007)
20. Bobrik, R., Reichert, M., Bauer, T.: View-based process visualization. In: Alonso, G., Dadam, P., Rosemann, M. (eds.) BPM 2007. LNCS, vol. 4714, pp. 88–95. Springer, Heidelberg (2007)
21. OMG: Business Process Modeling Notation (BPMN 1.1) (2008)

Quality Evaluation of Search Results by Typicality and Speciality of Terms Extracted from Wikipedia

Makoto Nakatani, Adam Jatowt, Hiroaki Ohshima, and Katsumi Tanaka

Department of Social Informatics, Graduate School of Informatics, Kyoto University
Yoshida-honmachi, Sakyo, Kyoto, 606-8501 Japan
{nakatani,adam,ohshima,tanaka}@dl.kuis.kyoto-u.ac.jp

Abstract. In Web search, it is often difficult for users to judge which page they should choose among search results and which page provides high quality and credible content. For example, some results may describe query topics from narrow or inclined viewpoints or they may contain only shallow information. While there are many factors influencing quality perception of search results, we propose two important aspects that determine their usefulness, "topic coverage" and "topic detailedness". "Topic coverage" means the extent to which a page covers typical topics related to query terms. On the other hand, "topic detailedness" measures how many special topics are discussed in a Web page. We propose a method to discover typical topic terms and special topics terms for a search query by using the information gained from the structural features of Wikipedia, the free encyclopedia. Moreover, we propose an application to calculate topic coverage and topic detailedness of Web search results by using terms extracted from Wikipedia.

Keywords: Search results quality, Wikipedia mining, Term extraction, Term typicality, Term speciality.

1 Introduction

Web search engines have become frequently used for acquiring information over the Internet. Web search results given by search engines are usually composed of a list of Web pages with some information such as titles, snippets and urls. However, it is often difficult for users to judge which page they should choose among search results. In many cases, users require Web pages including credible and comprehensive information about the search query. According to the online survey that we have recently conducted on 1000 respondents in Japan [1], users search the Web mostly because they require basic (46%) or detailed (36.8%) information about their search queries. Yet, conventional search engines usually do not provide users with any detailed information about the extent to which search results cover typical query topics. Some Web pages in search results may be regarded as being of low quality because they contain information related to query topics that are described from a narrow or an inclined viewpoint. In this

X. Zhou et al. (Eds.): DASFAA 2009, LNCS 5463, pp. 570–584, 2009.

sense, a page is deemed to be of high quality if it covers as many typical topics about a query term as possible. On the other hand, if a Web page conveys only shallow information in spite of covering many typical topics, the page will be also regarded as low quality one. In this paper, we propose the notion of the "topic coverage" and "topic detailedness" of Web pages for evaluating their quality. Topic coverage of a Web page means how many typical topics about the search query are covered by the Web page. On the other hand, topic detailedness of a Web page intuitively means how many special topics are included in the page. We believe that it will become easier for users to judge which page they should choose by showing them the above two measurements.

We would like to emphasize here that the complete quality evaluation of web pages is actually a complex, multi-dimensional issue. In order to find high quality pages one would generally have to analyze many aspects such as information accuracy and freshness, content organization, completeness, readability and so on. In this research we focus only on two aspects of quality evaluation of Web pages, topic coverage and topic detailedness. Both are actually query-dependent quality measures and can thus fit well into a search scenario in which users seek high quality pages for their queries. The proposed measures are also user-dependent to some extent. For example, users who are experts within certain topics would probably search for highly-specialized, detailed pages while non-experts users should generally prefer documents covering broad and typical topics related to their queries.

For calculating the topic coverage and topic detailedness of Web search results it is first necessary to extract typical and special terms for a search query used for generating these results. In this paper, we define typical and special terms for a search query as follows:

- Typical terms of a search query are terms that frequently appear in the domain of the search query.
- Special terms of a search query are terms that appear mostly in the domain of the search query.

Typical terms are used for measuring topic coverage of Web pages, and special terms are used for measuring topic detailedness.

We propose a method to discover typical topic terms and special topics terms for a search query by using the information gained from the structural features of Wikipedia. Wikipedia, the free online encyclopedia that anyone can edit, provides a huge number of interlinked entries. It started in 2001 becoming a prominent example of successful collaboration of thousands users on the Web. According to the statistics which Wikipedia has released as of June 2008 [1], the English Wikipedia contains about 2.4 million articles, and there are about 7 million registered user accounts. According to the Nature Journal, Wikipedia is about as accurate in covering scientific topics as the Encyclopedia Britannica [2]. In this work, we focus on category and link structure of Wikipedia for the purpose of measuring typicality and speciality of terms. In Wikipedia, each

[1] http://en.wikipedia.org/wiki/Special:Statistics

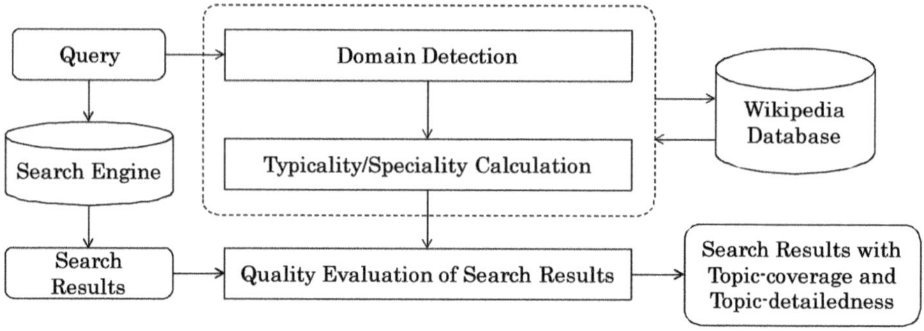

Fig. 1. Overview of our proposed system

article is assigned to one or more categories, and it links to and is linked by other related articles. In our approach, the category structure is used for detecting the domain of query term, and we calculate typicality and speciality of topic terms by analyzing the link structure in Wikipedia.

We have also implemented a system that presents Web search results with the scores of topic coverage and topic detailedness. The overview of the proposed system is illustrated in Fig. 1. Given a query, (i) the domain of a search query is detected, and (ii) typicality and speciality scores of terms are calculated by using the category and link structure of Wikipedia. At the last, (iii) topic coverage and topic detailedness of each Web page acquired from a search engine are measured by using typical and special terms extracted from Wikipedia and pages are annotated with these both measures.

The remainder of the paper is organized as follows: Section 2 discusses related work. In Section 3, we propose the method of measuring typicality and speciality of terms by using the structural features of Wikipedia. In Section 4, we present the approach of measuring topic coverage and topic detailedness of Web search results by using typicality and speciality of terms. Section 5 provides conclusion and discusses our future work.

2 Related Work

2.1 Quality Evaluation of Web Pages

The quality of Web pages has been evaluated so far from various viewpoints. Link analysis has been probably the most frequently exploited approach for the quality evaluation in information retrieval. PageRank [3] and HITS [4] are well-known algorithms in which the number of in-links of a Web page are used as a rough measure for the popularity and, indirectly, the quality of the page. Following the success of PageRank, Haveliwala [5] proposed a topic-sensitive PageRank measure, which separately determines a set of popularity scores for predetermined topics. Cho et al. [6] discovered that page ranking by link analysis causes the "rich-get-richer" phenomenon, and they proposed the method of measuring quality from Web snapshots by analyzing the changes in PageRank values over time. While link analysis considers the perspective of Web page authors, the information extracted from social annotations generated by users has recently attracted much attention for evaluating Web contents. The possibility of evaluating the quality of Web pages by using the information extracted from social bookmarking sites such as Del.icio.us[2] is described in [7][8].

Some researchers also proposed machine learning approaches for evaluating the quality of Web pages [9][10][11]. In these approaches, HTML structure, the number of links and language features such as number of unique words and so on are used as parameters for machine learning. Mandl et al. [11] implemented AQUAINT, a quality-based search engine, using a machine learning method. Our method is different from these works in that it uses Wikipedia, as a knowledge base constructed by the collaborative effort of multiple users. We also propose two query-dependent factors for page quality measurement, topic coverage and topic detailedness by which our method analyzes Web pages.

2.2 Term Extraction

Large text corpora have been successfully used for knowledge extraction. For example, Hearst [12] proposed a method for the automatic acquisition of the hyponymy lexical relations from unrestricted text. Several researchers have begun to seek effective ways for mining huge data collections since the detailed analysis of large content repositories is often impossible or prohibitively costly. Bollegala [13] proposed a semantic similarity measure that uses page counts and text snippets returned by a Web search engine for computing the similarity between terms or entities. In a similar fashion, Cilibrasi and Vitanyi [14] introduced a semantic distance measure called Google Normalized Distance between query terms based on the returned Web count values.

Wikipedia has recently attracted much attention as a large-scale, semi-structured corpus for data mining; and "Wikipedia mining" has become a new research area [15][16][17]. Strube [15] proposed a method of measuring semantic relatedness by using category data of Wikipedia articles. In addition, several applications based on knowledge extracted from Wikipedia have been demonstrated [18][19]. For example, *Koru* [18] is a new search engine that uses knowledge from Wikipedia for automatic query expansion. The Wikify! system proposed by Mihalcea [19] attaches links to entry articles of terms selected in Web pages by using the keyword extraction and word sense disambiguation based on Wikipedia data. These systems help users with search and learning, while our system aims at evaluating quality of Web pages by using the information gained from the Wikipedia.

Our work is also related to detecting domain-specific knowledge. Some methods for extracting domain-specific terms from online documents have been proposed in various domains, such as, the field of medical informatics [20][21]. Eibe et al. [22] described a method for finding domain-specific phrases by using

[2] http://del.icio.us

machine learning techniques. The above solutions, however, usually require a large number of manually labeled training data. Bing et al. [23] introduced a non-supervised method for detecting topic-specific concepts and definitions from Web pages. Their techniques first identify sub-topics or salient concepts of the topic, and then find and organize informative pages to be presented for users. Our approach differs from these methods in that, first, it extracts domain-specific terms from Wikipedia for the purpose of quality evaluation and, second, it is based on unsupervised and domain-independent algorithm.

3 Typicality and Speciality of Terms

There are many studies about term extraction as mentioned in the above section. Yet, typicality or speciality of extracted terms have not been referred in those works. They are prerequisite for assessing coverage and detailedness of Web pages returned for a query.

Our proposed method is composed of the following steps. To extract typicality and speciality of topic terms we first detect the domain of a search query by using the link and category structure of Wikipedia. The next step is to extract terms from Wikipedia articles included in the detected domain. After term extraction, we calculate typicality and speciality of terms by analyzing the distribution of links to the article of each term.

3.1 Detecting a Domain of Search Query

First, we describe the method for detecting a domain of a search query when there exists a Wikipedia article about the original query term. Suppose that q is a search query and a_q is a Wikipedia article about q. The first step is to acquire the set of categories that a_q belongs to. We express it as $C_{direct}(q)$. Each category that $C_{direct}(q)$ contains is intuitively a hypernym of the query term. For example, an article about "iPod" belongs to ten Wikipedia categories such as "Portable Media Player", "2001 introductions" and "Semi-protected" etc. "Portable Media Player" is regarded as an appropriate category for $C_{direct}($"iPod"$)$. Although "2001 introductions" can be regarded as a hypernym of "iPod", we remove it from direct categories since other articles contained in this category are hardly related to "iPod". "Semi-protected" is also removed from direct categories because it is a label that expresses the temporal state of articles. In the proposed method, we do not deal with categories that are classified by time axis such as "20xx_yyyy" and with the categories which are only labels about article state such as "Protected" and "All articles with unsourced statements" etc.

We consider that not only direct categories but also indirect categories are required for expressing in what kind of context a search query is used. In the example of "iPod", categories such as "iPod software" and "iPod accessories" are not directly combined with the article about "iPod". However, these categories are strongly related to "iPod". In order to add such indirect categories to the domain of a search query, we gather the category data of articles that are linking

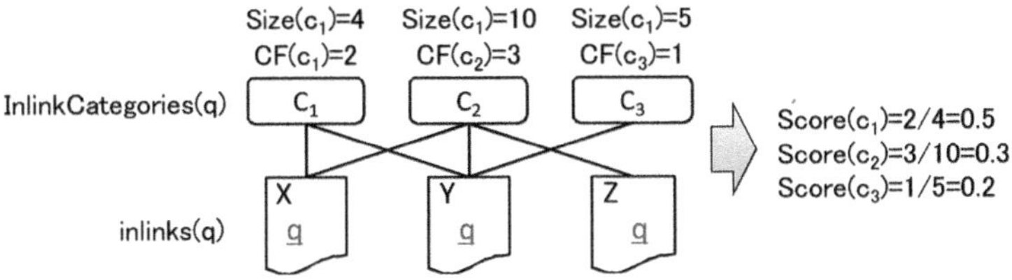

Fig. 2. Explanation of indirect category scoring where inlinks(q) is the articles linking to the article of q

to an article about the query. A basic idea about how to find indirect categories is described in Fig. 2. First, articles linking to an article about a search query, expressed as $inlinks(q)$, and categories that those articles belong to (denoted as $InlinkCategories(q)$) are acquired:

$$InlinkCategories(q) = \cup_{a_i \in inlinks(q)} Categories(a_i) \tag{1}$$

Next, the categories containing many articles from $inlinks(q)$ could be regarded as indirect categories. However we need to consider the size of each category. The score of each category in $InlinkCategories(q)$ is measured by the following equation:

$$Score(c) = \frac{CF(c)}{Size(c)} \tag{2}$$

where $CF(c)$ is the number of articles which are contained both in a category c and in $inlinks(q)$, and $Size(c)$ means the number of articles contained in the category c. If $Score(c)$ is larger than a threshold α and the category contains more than β articles, the category c can be regarded as belonging to $C_{indirect}(q)$, indirect categories of query. In this paper, α is set as 0.5 and β is set as 5. Next, $Domain(q)$, a domain of search query, is determined by calculating the union of direct categories and indirect categories:

$$Domain(q) = C_{direct}(q) \cup C_{indirect}(q) \tag{3}$$

In case when a Wikipedia article about the original query does not exist, the query is divided into as long word sequences ($q = \{q_1, q_2, ..., q_n\}$) as possible for which Wikipedia articles exist. For example, if "iPod nano battery" is given as a query, it can be divided into "iPod nano" and "battery". Although the articles of "iPod" and "nano" exist, "iPod nano" is a longer word sequence than each of them and there is also a Wikipedia article about "iPod nano". Note that our proposed method cannot deal with a query for which none of the terms exist on Wikipedia articles. However, such situation rarely happens according to our study. For each of divided terms, direct categories can be extracted and category scores can be calculated in the same way as the domain detection of a single term that was described above. Direct categories for each divided terms are just adopted to as total direct categories. In order to acquire total indirect

Table 1. Examples of detecting a domain of a search query

Query: q	$C_{direct}(q)$	$C_{indirect}(q)$
iPod	iPod Portable media players	iPod games Macintosh all-in-ones Directors of Apple Inc. Macintosh computers by product line iMac series iPod accessories iPod software X86 Macintosh computers
parkinson's disease	Aging associated diseases Geriatrics Parkinson's disease	Pervasive developmental disorders Motor neuron disease Cognitive disorders Tardive dyskinesia Dopamine agonists Antiparkinsonian agents Heterocyclic compounds (4 or more rings)
iPod headphone	Headgear Headphones iPod Portable media players	iPod accessories iMac series

categories, category scores for each divided term are linearly combined:

$$TotalScore(c) = \sum_{q_i \in q} weight(q_i) \cdot Score_{q_i}(c) \tag{4}$$

where $weight(q_i)$ is calculated as follows:

$$weight(q_i) = \frac{w_{q_i}}{\sum_{q_i \in q} w_{q_i}} \tag{5}$$

$$w_{q_i} = \frac{\log(1 + |outlinks(q_i)|)}{\log(1 + |inlinks(q_i)|)} \tag{6}$$

Wikipedia articles for which the number of out-links is low and the number of in-links is high tend to be abstract words with broader concepts. The above weighting scheme assigns a relatively small weight value for such words. For example, if a given query is "iPhone Japan", categories of "Japan" are not important and *weight*("Japan") has a low value. To the contrary, given "iPod Zune" as a query, both terms are important and almost equivalent weights are given. The rest is the same as that in the case of a single word.

Some examples of detecting a domain of a search query are described in Table 1.

3.2 Calculating Typicality and Speciality of Terms

We describe here the way for calculating typicality and speciality scores of terms by using the link structure of Wikipedia. Intuitively, typical terms should frequently occur in the domain of a search query, while special terms should occur mostly in the domain and rarely outside of it. We explain our idea in Fig. 3. Given q as a search query, the domain of q is detected by the method described in Section 3.1. We regard terms linked by many articles included in the domain

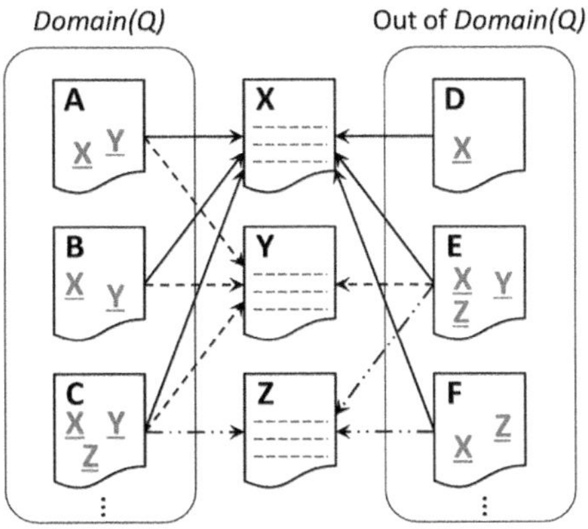

Fig. 3. Analysis of the distribution of link frequency for measuring typicality and speciality of terms

of query as typical terms, and terms linked by mostly articles included in the domain as special terms. Intuitively, a typical term is frequently used in the domain of a query, and a special term is hardly used out of the domain of query. In Fig. 3, X and Y are typical terms and Z is not a typical term. On the other hand, only Y is a special term, while X and Z are not special terms because the articles of X and Z are linked by many articles not included in the domain of the query.

Details of our proposed method are described as follows. For each category included in $Domain(q)$, we acquire all articles in it and express these articles as D_q which means domain pages of a search query q. Next, we define link frequency (LF). $LF(t, D)$ is the number of articles which link to the article of t and are included in the article set D. Typicality and speciality of a given term t when q is a search query are calculated by using $LF(t, D_q)$. Typicality score is calculated by dividing $LF(t, D_q)$ by the number of articles included in D_q, and speciality score is $LF(t, D_q)$ divided by $LF(t, D_W)$ where D_W means all articles of Wikipedia. Each equation is shown as follows:

$$Typicality(t, q) = \frac{LF(t, D_q)}{|D_q|} \tag{7}$$

$$Speciality(t, q) = \frac{LF(t, D_q)}{LF(t, D_W)} \tag{8}$$

3.3 Experiments

Experimental Setting. We prepared 20 queries in total for evaluating our proposed method of measuring typicality and speciality of terms. 10 of these queries are common terms chosen from the most popular Wikipedia articles [3].

[3] http://en.wikipedia.org/w/index.php?title=Wikipedia:Popular_pages

Table 2. Examples of typicality and speciality of terms extracted from Wikipedia

Query:q	Term:t	$Typicality(q, t)$	$Speciality(q, t)$
iPod	Apple Inc.	0.4046	0.0367
	IPod shuffle	0.2948	0.3953
	IPod Camera Connector	0.1445	0.4717
Global warming	Carbon dioxide	0.2702	0.0848
	Greenhouse gas	0.2561	0.2740
	Bali roadmap	0.1293	0.8279
Support vector machine	Machine learning	0.3755	0.3032
	Algorithm	0.1119	0.0203
	Kernel trick	0.0361	0.4545
Query expansion	Information retrieval	0.3947	0.2616
	Recall	0.0526	0.6667
	Relevance feedback	0.0439	0.833

The other 10 queries are technical terms about data mining and information retrieval, and so on. In this experiment, we do not deal with queries for which a Wikipedia article does not exist such as "iPod Zune" although in Section 3.1 we described how to process such queries. Wikipedia can be downloaded [4], and in our experiment we used the English Wikipedia database dumped in July 2008. For each query, we calculated typicality and speciality of terms by using structural features of Wikipedia. We took top 10 typical terms together with another 10 terms selected at random for each query and we showed these terms to 5 evaluators who are graduate students in informatics. Evaluators rated each term on a scale of 1 (low) to 7 (high) to reflect its typicality and speciality levels. The purpose of this evaluation is investigating the following two points:

- Accuracy of top 10 typical terms.
- Validity of speciality of each term.

Results. First, we describe some examples of typicality and speciality of terms extracted from Wikipedia in Table 2. In the example of "global warming", "carbon dioxide" and "greenhouse gas", which are generally considered as the cause of global warming, had high typicality. "Carbon dioxide" which is a relatively general term had low speciality, but "greenhouse gas" which conceptually contains "carbon dioxide" showed high speciality. This is because "greenhouse gas" is a special term only used in the domain of global warming. "Bali roadmap", a roadmap adopted after a climate change conference held in Bali, showed much higher speciality.

Next, we describe the accuracy of top 10 typical terms for our queries. We calculated an average precision for evaluating the accuracy of typicality. Here we regarded terms of which typicality score provided by evaluators were more than 5.0 as the correct terms and those with the score lower than 5.0 as incorrect ones. The result is shown in Table 3. If a technical term was given as a search query, the average precision was about 70%. On the other hand, the average precision for common term queries was only about 50%.

[4] http://download.wikimedia.org

Table 3. Accuracy of the top 10 typical terms (left table) and validity of speciality of each term (right table)

Query Type	Avg. Precision (top 10)
Common Terms	0.49
Technical Terms	0.71
Total	0.60

Query Type	Spearman's coefficient
Common Terms	0.40
Technical Terms	0.45
Total	0.42

Last, we describe the validity of the speciality calculation of the terms. We calculated Spearman's rank correlation coefficient between speciality scores measured by our proposed method and average scores given by user evaluation. As shown in Table 3, it turned out as a result that they have a positive correlation.

4 Quality Evaluation of Search Results

In this section, we propose two measurements, "topic coverage" and "topic detailedness", for facilitating the evaluation of the quality of Web pages by using typicality and speciality scores of included terms.

4.1 Topic Coverage

Topic coverage of a Web page means how many typical topics about a search query are covered by the Web page. We consider that a Web page containing more typical terms has higher coverage. Topic coverage of a Web page p about a query q is calculated by the following equation:

$$TopicCoverage(p,q) = \sum_{t \in terms(q)} C(t,p) \cdot Typicality(t,q) \tag{9}$$

where $terms(q)$ are extracted terms and $C(t,p)$ is an indicator taking values 0 or 1 depending whether a Web page p contains a given term t. A Web page with high topic coverage can be regarded as of high quality while a Web page with low coverage may be written from a narrow viewpoint or an inclined viewpoint. Note that we use here $C(t,p)$ instead of the term frequency of t in a Web page for measuring topic coverage because term frequency is not directly related to how many topics the page includes.

4.2 Topic Detailedness

Topic detailedness of a Web page means how many special topics about a search query are included in the Web page. We consider that a Web page containing more special terms is more detailed. Topic detailedness of a Web page p about a query q is calculated by the following equation:

$$TopicDetailedness(p,q) = \sum_{t \in terms(q)} TF(t,p) \cdot Speciality(t,q) \tag{10}$$

where $terms(q)$ are extracted terms and $TF(t,p)$ means term frequency of t in the Web page p. In this case, term frequency of t should be taken into consideration because Web pages with special terms that are repeatedly used are more likely to be about detailed topics.

Even if a Web page shows a high coverage, it may contain only shallow information. In general, we consider a Web page with low coverage and low detailedness as a low quality Web page.

4.3 Application

We implemented a system that presents Web search results with the scores of topic coverage and topic detailedness. The objective of this system is to make it easier for users to choose the right pages when exploring search results. Fig. 4 shows a screen shot of our system. The interface for inputting queries is the same as the one in a standard search engine. The system gives users search results which contain the scores of topic coverage and topic detailedness of each Web page in addition to titles, snippets and urls. We used Yahoo! Web search API service [5] for acquiring the search results.

Currently, our system downloads each Web page for calculating topic coverage and topic detailedness. Response speed of the system could be improved by using only title and summary for calculating the two measurements. However, we need to be aware of that it is difficult to evaluate the quality of a Web page by using only titles and snippets contained in Web search results as the length of available text is limited.

4.4 Experiments

For showing the effectiveness of our proposal, we have to clarify the following two points:

- Accuracy of our proposed method for measuring topic coverage and topic detailedness.
- Relation between the overall quality of Web pages, and topic coverage or topic detailedness.

We prepared 10 search queries for clarifying the above points. For each query, we imposed the following tasks on evaluators:

1. Read the top 10 Web pages acquired by a Web search engine.
2. Rate on a scale of 1 to 10 the topic coverage and topic detailedness of each page.
3. Rate on a scale of 1 to 10 the overall quality of each page.

In this experiment, we regarded the average of each score given by 5 evaluators as the answers of coverage, detailedness and overall quality for each page.

[5] http://developer.yahoo.com/search/web/V1/webSearch.html

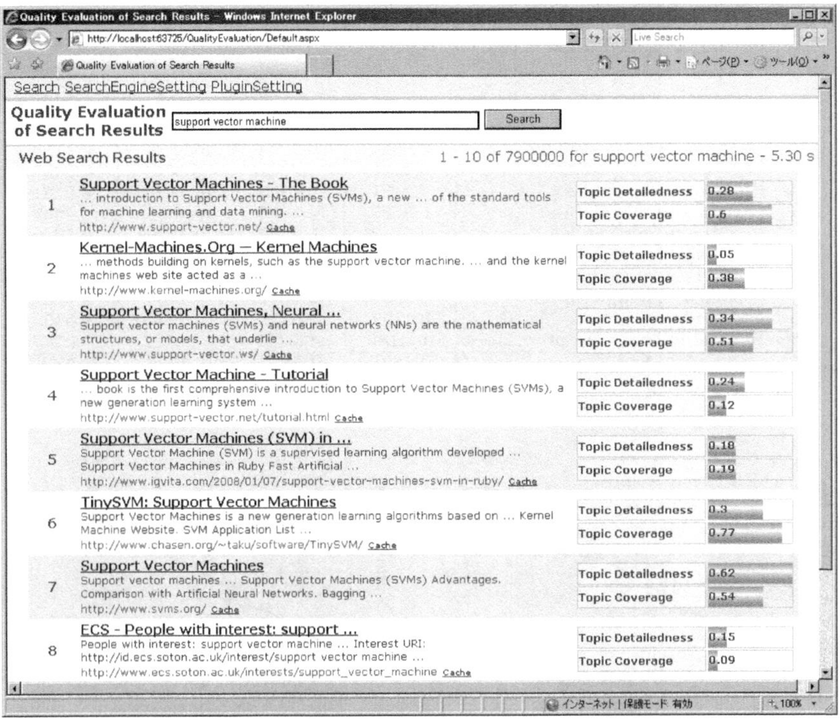

Fig. 4. A screenshot of the proposed system which presents Web search result with topic coverage and topic detailedness

We first discuss the effectiveness of our proposed method for measuring topic coverage and topic detailedness of Web pages. We used the following three rankings:

Original Ranking (OR) : An original ranking by Web search engine.
System Ranking (SR) : A ranking sorted by topic coverage or topic detailedness which we proposed.
User Ranking (UR) : A ranking sorted by topic coverage or topic detailedness scored by evaluators.

For each query, we calculated Spearman's rank correlation coefficient between the original ranking and user ranking, and between the system ranking and user ranking. The result is shown in Table 4. In general, the system ranking has more strongly positive correlation with the user ranking in comparison with the original ranking. This indicates that the proposed method for measuring topic coverage and topic detailedness is appropriate and these two measurements should help users with judging which pages they choose among search results.

Next, we investigate the correlation between topic coverage / topic detailedness and overall quality of Web pages that were assigned by the evaluators. As shown in Fig. 5, we found out that topic coverage and topic detailedness of Web

Table 4. Spearman's rank correlation coefficient between original ranking (OR) or system ranking (SR) and user ranking (UR)

Query	topic coverage		topic detailedness	
	OR v.s UR	SR v.s. UR	OR v.s. UR	SR v.s. UR
support vector machine	0.6364	0.2970	0.5758	0.1758
Eric Clapton	0.2121	0.8061	-0.0061	0.6485
subprime lending	0.2121	0.4303	0.1152	0.6121
parkinson's disease	0.3697	0.3576	0.3333	0.4909
Hurricane Katrina	0.2242	0.4303	0.3576	0.5030
global warming carbon dioxide	0.2970	0.6848	0.3576	0.8182
iPod Zune comparison	0.1030	0.5879	0.1515	0.4424
ancient Olympics event	0.1394	0.5273	0.1030	0.6727
obesity causes	0.1152	0.6121	0.2242	0.5152
PageRank search engine optimization	-0.4182	-0.0424	-0.4182	0.0182
Avg.	0.1891	**0.4691**	0.1794	**0.4897**

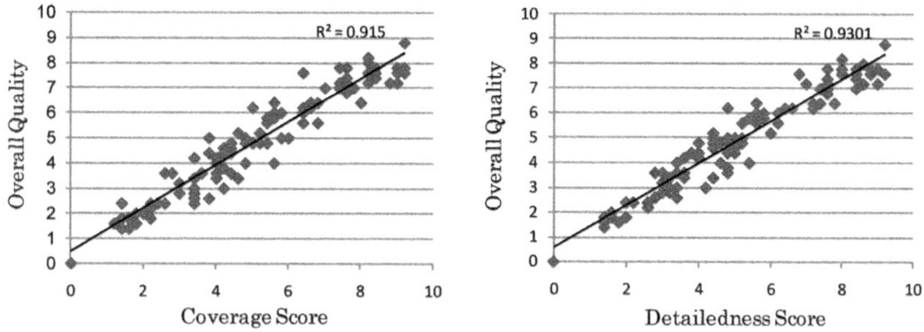

Fig. 5. Relationships between overall quality and topic coverage, topic detailedness

pages are strongly correlated to their overall quality. This means that topic coverage or topic detailedness are important factors for evaluating the quality of Web pages.

5 Conclusion and Future Work

In this paper, we introduced the notions of topic coverage and topic detailedness of Web pages which are important factors in evaluating their quality. Topic coverage and topic detailedness are calculated by using typical terms and special terms. We have proposed a method of measuring typicality and speciality of terms by utilizing structural features of Wikipedia. We also implemented a system that presents Web search results with the scores of topic coverage and topic detailedness of pages. The experimental results suggested that our proposed methods are effective in classifying topic terms and that automatic evaluation of Web page quality by topic coverage and topic detailedness has a positive correlation with a manual evaluation.

Our proposed methods using Wikipedia data still have some disadvantageous that need to be approached. One problem is the word sense disambiguation like in the example of "Java". We think that this problem can be solved by applying disambiguation methods proposed, for example in [18][19]. Another problem is certain, inherent limitation of Wikipedia. Although Wikipedia contains a huge amount of content, it does not necessarily cover all the possible topics and the quality and scope of its articles may actually differ for different topics. However, currently Wikipedia is the largest, manually edited knowledge base available online. It is also frequently and promptly updated according to real world changes.

We focused only on two aspects of quality evaluations of Web pages, topic coverage and topic detailedness. Both are actually query-dependent and user-dependent quality measures. In our future work, we plan to combine the proposed methods with other query-independent quality measures such as a readability for more precisely evaluating quality of Web pages. We also intend to propose a system for re-ranking search results by user's knowledge level about a search query or its domain.

Acknowledgments. This work was supported in part by the following projects and institutions: Grants-in-Aid for Scientific Research (Nos. 18049041 and 18049073) from MEXT of Japan, a MEXT project entitled "Software Technologies for Search and Integration across Heterogeneous- Media Archives," a Kyoto University GCOE Program entitled "Informatics Education and Research for Knowledge- Circulating Society," and the National Institute of Information and Communications Technology.

References

1. Nakamura, S., Konishi, S., Jatowt, A., Ohshima, H., Kondo, H., Tezuka, T., Oyama, S., Tanaka, K.: Trustworthiness analysis of web search results. In: Kovács, L., Fuhr, N., Meghini, C. (eds.) ECDL 2007. LNCS, vol. 4675, pp. 38–49. Springer, Heidelberg (2007)
2. Giles, J.: Internet encyclopedia go head to head. Nature 438 (2005)
3. Brin, S., Page, L.: The anatomy of a large-scale hypertextual web search engine. Computer Networks and ISDN Systems 30, 107–117
4. Kleinberg, J.M.: Authoritative sources in a hyperlinked environment. J. ACM 46(5), 604–632 (1999)
5. Haveliwala, T.H.: Topic-sensitive pagerank. In: WWW 2002: Proceedings of the 11th international conference on World Wide Web, pp. 517–526. ACM, New York (2002)
6. Cho, J., Roy, S., Adams, R.E.: Page quality: in search of an unbiased web ranking. In: SIGMOD 2005: Proceedings of the 2005 ACM SIGMOD international conference on Management of data, pp. 551–562. ACM, New York (2005)
7. Yanbe, Y., Jatowt, A., Nakamura, S., Tanaka, K.: Can social bookmarking enhance search in the web? In: JCDL 2007: Proceedings of the 7th ACM/IEEE-CS joint conference on Digital libraries, pp. 107–116. ACM, New York (2007)
8. Bao, S., Xue, G., Wu, X., Yu, Y., Fei, B., Su, Z.: Optimizing web search using social annotations. In: WWW 2007: Proceedings of the 16th international conference on World Wide Web, pp. 501–510. ACM, New York (2007)

9. Amento, B., Terveen, L., Hill, W.: Does "authority" mean quality? predicting expert quality ratings of web documents. In: SIGIR 2000: Proceedings of the 23rd annual international ACM SIGIR conference on Research and development in information retrieval, pp. 296–303. ACM, New York (2000)
10. Ivory, M.Y., Hearst, M.A.: Statistical profiles of highly-rated web sites. In: CHI 2002: Proceedings of the SIGCHI conference on Human factors in computing systems, pp. 367–374. ACM, New York (2002)
11. Mandl, T.: Implementation and evaluation of a quality-based search engine. In: HYPERTEXT 2006: Proceedings of the seventeenth conference on Hypertext and hypermedia, pp. 73–84. ACM, New York (2006)
12. Hearst, M.A.: Automatic acquisition of hyponyms from large text corpora. In: Proceedings of the 14th conference on Computational linguistics, Morristown, NJ, USA, Association for Computational Linguistics (1992)
13. Bollegala, D., Matsuo, Y., Ishizuka, M.: Measuring semantic similarity between words using web search engines. In: Proceedings of the 16th international conference on WWW. ACM, New York (2007)
14. Cilibrasi, R.L., Vitanyi, P.M.B.: The google similarity distance. IEEE TKDE 19(3) (2007)
15. Strube, M., Ponzetto, S.P.: Wikirelate! computing semantic relatedness using wikipedia. In: Proceedings of National Conference for Artificial Intelligence (2006)
16. Milne, D., Medelyan, O., Witten, I.H.: Mining domain-specific thesauri from wikipedia: A case study. In: International Conference on Web Intelligence (2006)
17. Erdmann, M., Nakayama, K., Hara, T., Nishio, S.: An approach for extracting bilingual terminology from wikipedia. In: Haritsa, J.R., Kotagiri, R., Pudi, V. (eds.) DASFAA 2008. LNCS, vol. 4947, pp. 380–392. Springer, Heidelberg (2008)
18. Milne, D.N., Witten, I.H., Nichols, D.M.: A knowledge-based search engine powered by wikipedia. In: Proceedings of the sixteenth ACM conference on CIKM. ACM, New York (2007)
19. Mihalcea, R., Csomai, A.: Wikify!: linking documents to encyclopedic knowledge. In: Proceedings of the sixteenth ACM conference on CIKM. ACM, New York (2007)
20. Bennett, N.A., Qin He, K.P., Schatz, B.R.: Extracting noun phrases for all of medline. In: Proceedings of the American Medical Informatics Association (1999)
21. Klavans, J.L., Muresan, S.: Definder: Rule-based methods for the extraction of medical terminology and their associated definitions from on-line text. In: Proceeding of the American Medical Informatics Association (2000)
22. Frank, E., Paynter, G.W., Witten, I.H., Gutwin, C., Nevill-Manning, C.G.: Domain-specific keyphrase extraction. In: Proceedings of the Sixteenth IJCAI. Morgan Kaufmann Publishers Inc., San Francisco (1999)
23. Liu, B., Chin, C.W., Ng, H.T.: Mining topic-specific concepts and definitions on the web. In: Proceedings of the 12th international conference on WWW. ACM, New York (2003)

A Ranking Method for Web Search Using Social Bookmarks

Tsubasa Takahashi[1] and Hiroyuki Kitagawa[1,2]

[1] Graduate School of Systems and Information Engineering
tsubasa@kde.cs.tsukuba.ac.jp
[2] Center for Computational Sciences
University of Tsukuba
1-1-1 Tennodai, Tsukuba, Ibaraki 305–8573, Japan
kitagawa@cs.tsukuba.ac.jp

Abstract. Recently, Social Bookmark, which allows us to register and share our own bookmarks on the web, is attracting attention. Social Bookmark makes it possible to retrieve structured data such as (URL, Username, Timestamp, Set of tags). More importantly, the retrieved data represents user interests. There are two aspects of bookmark usage: data for reuse and data for hot issues. This paper, focusing on timestamps of bookmarks, proposes a way to measure the freshness of a web page. It further proposes a page evaluation method that improves S-BITS, our previously proposed method to evaluate informativeness of web pages using social bookmarks, using the freshness evaluation. Finally, it demonstrates the effectiveness of the proposed method through experiments.

1 Introduction

Social Bookmarks are attracting attention recently. They belong to new information sharing services and allow individuals to bookmark and annotate web pages. A user can manage bookmarks by annotating web pages with sets of tags. A bookmarked page has valuable attributes for its user. A bookmarking action from a user to a page is a vote from the user for the page. People differ in expertise in a typical region. In a typical region, opinions from people who have high expertise will be trustworthy. In our previous work [6], we proposed S-BITS, which uses Social Bookmarks to evaluate and rank web pages based on user expertise. It showed higher precision than existing web search engines and an existing method that looks only at bookmark counts.

Social Bookmarks have timestamps that indicate when a user bookmarks a page. Some web pages are fresh only for a short time, such as news scripts; others are fresh for a long time, such as Wikipedia scripts and manuals. The principle difference is the spread of timestamps. Short-lived pages are bookmarked a lot but loose their popularity in a short time. Long-life pages are bookmarked over a protracted period of time. This paper, taking into account the lifetime of a web page's content based on the spread of timestamps, proposes a method named FS-BITS to evaluate freshness of the web page. FS-BITS is an extension of our previous method S-BITS.

X. Zhou et al. (Eds.): DASFAA 2009, LNCS 5463, pp. 585–589, 2009.

2 Social Bookmark

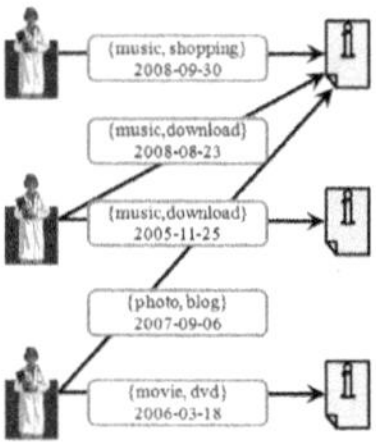

Fig. 1. SBM

Social Bookmarks (shortly SBMs) are new information sharing services that allow individuals to bookmark and annotate web pages. A user can manage bookmarks by deciding a set of tags. *delicious*[1], at the top of SBM services, now has more than a million users and continues to grow. SBM services have been popular, and many services like *Hatena Bookmark*[2] (Japanese Top Service Provider) have appeared. The bookmark data, which is structured as a tuple (url, username, timestamp, set of tags), is available in SBM services (Fig. 1). The data shows user interests for certain pages. A user, using personal knowledge and intuition, can annotate a page with tags that show keywords and/or categories. Timestamps indicate when users get interested in the information.

3 Related Work

With the spread of SBM services, research into Folksonomy, including SBMs, has increased. Bao et al. [4] demonstrated that keyword associations based on social annotations can improve web searches. Yanbe et al. [3] proposed SBRank, which indicates how many users bookmarked a page, and estimated effectiveness of SBRank as an indicator of web search. They further suggested integrating SBRank and PageRank [2] into a ranking method.

Research into temporal data is also abundant. Hotho et al. [5] presented an approach for discovering topic-specific trends within folksonomies.

Our previous approach [6] features introduction of concepts of authorities and hubs as in HITS [1] into the page evaluation framework using SBM. We deal with the freshness of all pages in SBM by extending our previous work.

4 Freshness of Web Page

In SBM services, whether or not a page has attracted attention and has been freshened is determined by whether or not the page has been recently bookmarked. Since pages have diverse contents, however, the lifetime of freshness for each page is also diverse. For that reason, we cannot simply say that freshness is lost even if the page has not been bookmarked for a week or a month.

Many pages go unnoticed after the initial hot peak. Typical examples are pages that include highly-temporary topics, such as news scripts. In contrast, there are also many pages that people have evaluated over a long period . Examples are manuals and reference pages. Freshness of the former is lost after its hot peak and

[1] http://delicious.com/

[2] http://b.hatena.ne.jp/

the latter is longer. We define the duration of freshness as $lifetime$. Considering their $lifetimes$, the former's is short and the latter's is long.

The main difference between the two is the spread of bookmark timestamps. We assume that lifetime δ_{life} is associated with *spread*, which is the spread of bookmark timestamps. Two indicators express *spread*: One is standard deviation (SD); the other is interquartile range (IQR). SD has features that reflect the shape of the population and its changes. But it is sensitive to small changes. IQR has robust features for small changes. It is weak, however, to changed values when the population is large. This paper, dealing with SD and IQR, defines lifetime δ_{life} as follows (Formula 1):

$$\delta_{life} = C_1\ SD + C_2\ IQR + C_0 \tag{1}$$

where C_0, C_1 and C_2 are constants.

We consider the page keeps current freshness when one of the following two conditions is met.

1. Enough time has passed since the first bookmark to the page, but it is still getting new bookmarks and can be considered to be within its lifetime.
2. The page just starts being bookmarked recently.

For the former pages, we can observe the spread of bookmark timestamps, and estimate lifetime δ_{life}. If the time passed since the last bookmark time is less than δ_{life}, we regard the page is fresh. While for the latter pages, we cannot estimate lifetime δ_{life} because it is highly possible to change its population and distribution dramatically when the new bookmarks are appeared. Therefore, we regard that a page is fresh if less than δ_0 times has passed since the occurrence of the last bookmark. To formulate the above concepts, freshness of page p is evaluated as follows (Formulas 2 to 4):

$$dt(p) = t_{today} - t_{last}(p) \tag{2}$$

$$freshness(p) = \begin{cases} true & if\ dt(p) < max(\delta_{life}(p), \delta_0) & (3) \\ false & otherwise & (4) \end{cases}$$

where $t_{last}(p)$ is the latest bookmark timestamp of p. Note that $\delta_{life}(p)$ differs page to page, while δ_0 is a common parameter for all pages. In the experiment in section 7, we set δ_0=7[days], and $C_1 = C2 = 0.5$, $C_0 = 0$.

5 Ranking Web Pages

We now propose FS-BITS, which is a ranking method for web search considering user expertise and page authoritativeness and freshness.

In FS-BITS, we consider page freshness. A fresh page is informative, and it is highly likely that non-fresh one is redundant. We therefore prune pages without freshness. If freshness of page p_i (denoted by $freshness(p_i)$) is considered non-fresh, it is pruned. When page p_i is pruned, the bookmarks and users connected with page p_i are also pruned.

In this way, based on the page freshness, a pruned graph is constructed. After pruning, we can create rankings that target only fresh pages. FS-BITS operation consists of 4 steps as follows:

1. Based on the query, select the result page set P_0, relevant user set and tag set.
2. Merge another page set P_1 to P_0. P_1 is extracted based on co-user and co-tag relationship.
3. Truncate the pages which are out of date (use the freshness measurement).
4. Rank the pages (just like the HITS algorithm).

Our previous ranking method S-BITS [6] consists of steps 1, 2 and 4. Since S-BITS does not consider the freshness of web pages, it regards pages that received good evaluations in the past as still being authoritative pages. Solving this matter by inserting step 3 which evaluates page freshness, we improve the precision of page evaluation.

6 Experiments

This section presents and discusses experimental results. We measure effectiveness of the proposed method. We compared and analyzed propriety for 3 ranks: the original Yahoo! rank, the S-BITS rank and the FS-BITS rank. We collected SBM data in Hatena Bookmark from July to August 2008. The collected data we used in this experiments included around 300 thousand pages and 5 million bookmarks. The initial page set P_0 for S-BITS and FS-BITS was obtained from the Yahoo! Web Search API [3]. We took the top 200 pages from that API. To evaluate each method manually, we recruited 6 examinees. Then we evaluated 30 cases including 10 queries. We asked examinees to distinguish whether or not the page is fresh and whether or not it is informative. We then asked them to score satisfaction for each ranking using a five-grade evaluation. Because there is a limit to what can be done manually, the evaluation is based on only the top 20 pages.

In precision of informativeness (Fig. 2), FS-BITS, the newly proposed method, shows higher precision than the others. FS-BITS removes pages estimated as non-fresh, related bookmarks and users. With this removal, FS-BITS can compose a more sophisticated graph than S-BITS. In precision of freshness (Fig. 3) as well, FS-BITS shows higher precision than the others. Because S-BITS has no way to include freshness concept, it cannot out-perform the others. By demonstrating higher precision in FS-BITS, we can say that the distinction of page freshness is effective. Regarding satisfaction for each ranking (Fig. 4), examinees expressed good satisfaction by awarding high scores. Both S-BITS and FS-BITS, our proposed methods, received higher scores than Yahoo!.

As the above results show, by taking page freshness into account and confining search targets to fresh pages, it is possible to improve precision and satisfaction of retrieval.

[3] http://developer.yahoo.co.jp/search/web/V1/webSearch.html

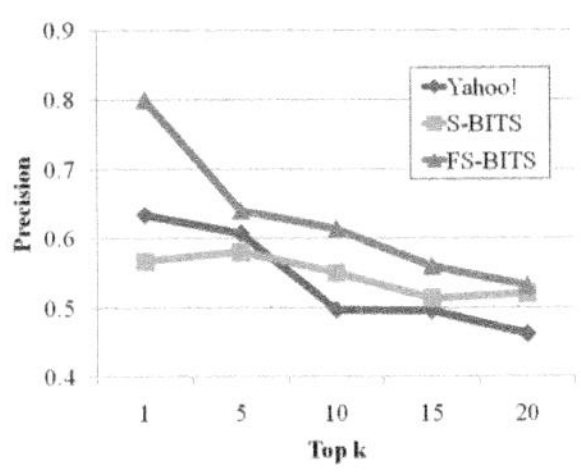

Fig. 2. Average precision for informativeness

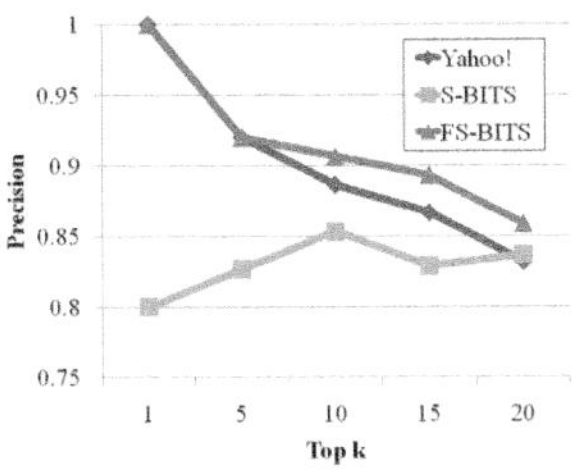

Fig. 3. Average precision for freshness

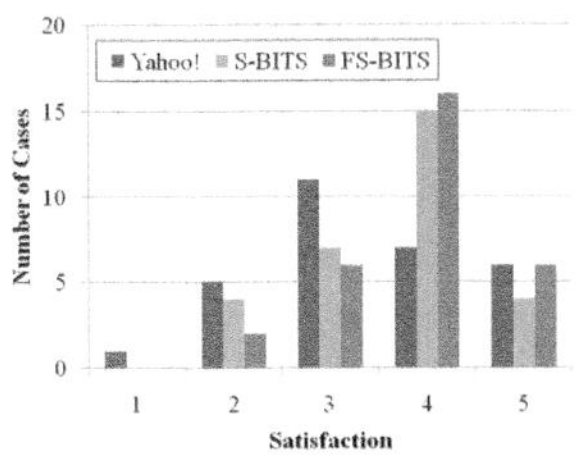

Fig. 4. Satisfaction

7 Conclusion

Focusing on timestamps, which show bookmarked times in SBM, we proposed a way to evaluate page freshness. Additionally, combining S-BITS and the freshness evaluation, we proposed FS-BITS, which includes the time-line. Experiments confirmed the effectiveness of FS-BITS to evaluate page authoritativeness based on user expertise considering page freshness. Future work will focus on refining the model to define freshness. Moreover, by constructing more sophisticated models and discovering rules or knowledge by looking at various elements of SBM services, we will propose interactive and effective web search models.

Acknowledgement. This research has been supported in part by the Grant-in-Aid for Scientific Research from MEXT (#1924006).

References

1. Kleinberg, J.: Authoritative Sources in a Hyperlinked Environment. In: Proc. of the 9th ACM SIAM Symposium on Discrete Algorithms (SODA 1998), pp. 668–677 (1998)
2. Page, L., Brin, S., Motwani, R., Winograd, T.: The pagerank citation ranking: Bringing order to the Web. Technical report, Stanford Digital Library Technologies Project (1998)
3. Yanbe, Y., Jatowt, A., Nakamura, S., Tanaka, K.: Towards improving web search by utilizing social bookmarks. In: Baresi, L., Fraternali, P., Houben, G.-J. (eds.) ICWE 2007. LNCS, vol. 4607, pp. 343–357. Springer, Heidelberg (2007)
4. Bao, S., Wu, X., Fei, B., Xue, G., Su, Z., Yu, Y.: Optimizing Web Search Using Social Annotations. In: Proc. of the 16th World Wide Web Conference, pp. 501–510 (2007)
5. Hotho, A., Jäschke, R., Schmitz, C., Stumme, G.: Trend detection in folksonomies. In: Avrithis, Y., Kompatsiaris, Y., Staab, S., O'Connor, N.E. (eds.) SAMT 2006. LNCS, vol. 4306, pp. 56–70. Springer, Heidelberg (2006)
6. Takahashi, T., Kitagawa, H.: S-BITS: Social-Bookmarking Induced Topic Search. In: Proc. of the 9th International Conference on Web-Age Information Management (WAIM 2008), pp. 25–30 (2008)

Fractional PageRank Crawler: Prioritizing URLs Efficiently for Crawling Important Pages Early*

Md. Hijbul Alam, JongWoo Ha, and SangKeun Lee

College of Information and Communications
Korea University, Seoul, Republic of Korea
{hijbul,okcomputer,yalphy}@korea.ac.kr

Abstract. Crawling important pages early is a well studied problem. However, the availability of different types of framework for publishing web content greatly increases the number of web pages. Therefore, the crawler should be fast enough to prioritize and download the important pages. As the importance of a page is not known before or during its download, the crawler needs a great deal of time to approximate the importance to prioritize the download of the web pages. In this research, we propose Fractional PageRank crawlers that prioritize the downloaded pages for the purpose of discovering important URLs early during the crawl. Our experiments demonstrate that they improve the running time dramatically while crawling the important pages early.

1 Introduction

The content of search engines is fed by crawlers which download web pages recursively by following the links. However, the number of web pages is growing at an astonishing rate and most of the time users only view the top ranked pages in the search results. Therefore, a crawler needs to prioritize the download of the web pages. In spite of the existence of state of the art crawling techniques, crawling important pages early poses a great challenge, because crawlers have to determine the importance of web pages before downloading them. The existing strategies require a very large running time to prioritize the URLs.

In this research, we propose three algorithms to reduce the running time of the process of prioritizing URLs for the purpose of crawling important pages early. We model a special random surfer that visits web pages only once. The order in which the web pages are visited depends on their Fractional PageRank. The Fractional PageRank of a page is the summation of the probabilities of all of the paths from the seed pages to this page during a certain phase. We assume that pages with high Fractional PageRank values are likely to have high PageRank values and we give high priority to their download. We compare the results of our algorithms with that of the Windowed RankMass crawlers [2] that use the PageRank (PR) lower bound of a page to prioritize the discovered URLs.

* This work was supported by grant No. RO1-2006-000-10510-0 from the Basic Research Program of the Korea Science & Engineering Foundation.

X. Zhou et al. (Eds.): DASFAA 2009, LNCS 5463, pp. 590–594, 2009.

2 Proposed Methodology

In perspective of our proposed crawlers, we group web pages into two broad types; Downloaded pages and Undownloaded pages. Among the undownloaded pages whose URLs are known to or discovered by the crawlers are called Discovered pages. A downloaded page is called Explored page if the URLs from that page are discovered or downloaded by the crawlers. Otherwise the downloaded pages is called Unexplored pages. In the random surfer model of the PageRank, every edge in a webgraph has some probability that a random surfer will traverse and the PageRank of a page is the summation of all of the paths probabilities a page receives through its incoming edges [1][2]. Fractional PageRank (FPR) of a page is the summation of the probabilities of all of the paths from the seed pages to this page in a certain phase during crawling. Therefore, the formulation of Fractional PageRank is almost as same as the PageRank except it is computed considering only certain phase of crawling. Because of the space limitation we discuss the formulation of Fractional PageRank for **Fractional PageRank Discovered Crawler** only in the following.

$$fpr_v = (1-d) * t_v + \sum_{u \to v} \frac{fpr_u * d}{o_u}; v \in V, u \in U \tag{1}$$

$$fpr_u = 0; u \notin U, u \in E \tag{2}$$

When a random surfer is on a page u, with the probability d which is set to 0.85, the random surfer will click on one of the o_u links on the page with equal probability of $\frac{1}{o_u}$. Here, t_v is trusted values for the seed pages and V, U, E denote the Discovered, Unexplored and Explored set respectively and all the sets are disjoint. We differentiate the FPR from the PageRank by considering the path probabilities during a certain phase. In the case of FPR Discovered crawler, the certain phase means the FPR is computed only for the undownloaded but discovered pages ($v \in V$) through the incoming paths from the Unexplored Pages ($u \in U$). When all of the outlinks from the Unexplored page are extracted the FPR of that page will set to 0 permanently and u becomes one of the elements of Explored set E.

Algorithm 1. Fractional PageRank Discovered Crawler

Foreach p_i in the set of trusted seed pages **do**
    $fpr_i = (1-d) * t_i;$
While (Queue is not empty) **do**
    Pick p_i with largest fpr_i;
    Download p_i;
    Foreach p_j linked to by p_i and p_j is not downloaded **do**
        $fpr_j = fpr_j + (d * fpr_i)/o_i;$
    $fpr_i = 0;$

Algorithm 2. Fractional PageRank Explored Crawler

Foreach p_i in the set of trusted seed pages **do**
  $fpr_i = (1 - d) * t_i$;
  Download p_i;
While (Queue is not empty) **do**
  Pick p_i with largest fpr_i;
  Foreach p_j linked to by p_i and p_j is not in Explored **do**
    Download p_j if not downloaded yet;
    $fpr_j = fpr_j + (d * fpr_i)/o_i$;
  $fpr_i = 0$;
  mark p_i as Explored;

Algorithm 3. Two Layer Fractional PageRank Crawler

Foreach p_i in the set of trusted seed pages **do**
  $fpr_i = (1 - d) * t_i$;
  Download p_i;
Assign all the seed pages in the parentQ;
While (parentQ is not empty) **do**
  Pick p_i with largest fpr_i from parentQ;
  Foreach p_j linked to by p_i and p_j is neither in Explored nor in parentQ **do**
    Download p_j if not downloaded yet;
    $fpr_j = fpr_j + (d * fpr_i)/o_i;//$ insert or update fpr of j in childQ
  $fpr_i = 0$;
  mark p_i as Explored;
  if parentQ is empty assign all the urls of the childQ to parentQ

We propose two other novel algorithms which directly prioritize the Unexplored pages. First one is the **Fractional PageRank Explored Crawler**; we accumulate the FPR of Unexplored page until the page is explored. Therefore, here prioritization is done for ordering the exploration of the Unexplored pages. All the outlinks of the highest FPR Unexplored page will be downloaded together. As this algorithm deals with the downloaded Unexplored pages, prioritizing scheme runs on the small amount of pages and hence it is faster.

The last algorithm we propose is **Two Layer Fractional PageRank Crawler**. To decrease the overhead of prioritizing the URLs further, the queue used in the FPR Explored crawler is divided into two levels. The top level is called the parent queue and the bottom level is called the child queue. The FPR for this crawler is defined as the summation of the path probabilities it receives while a page stays in the child queue. The downloaded pages are first placed into the child queue and prioritized here according to FPR. When the parent queue becomes empty, all of the pages from the child queue will be placed into the parent queue. In the parent queue, the page with the highest FPR will be explored first and the URLs pointed to by that page will be downloaded together.

In the proposed three algorithms we define and use the Fractional PageRank instead of the PR lower bound as computing PR lower bounds is very expensive [2]. In the computation of PR lower bounds, if any Explored node receives a new inlink, the PR lower bounds of all its descendent nodes will be computed repeatedly which costs a great deal of time and memory. Moreover, in the experiment we found that the FPR crawlers are also able to crawl important pages early.

3 Experimental Results

We simulate our algorithms on a sub graph of the Web consisting of 80 million pages and about 2.4 billion links from the .uk top domain, namely the uk-2006-06 graph, which was crawled by University of Millan [3]. For our simulation, we used only a single Intel Core 2 Quad 2.4 Ghz CPU with 4 GB RAM in a Solaris machine. We perform all of the processing in the main memory. In this experiment, the top 160 pages according to PageRank among the collection are selected as the seed pages and the trusted values are distributed equally among them.

The proposed crawlers outperform the 100% Windowed RankMass crawler, not only in terms of the running time, but also crawling important pages earlier. The 100% Windowed RankMass crawler runs faster than all of the other RankMass crawlers [2]. Besides that we also compare our algorithms with 20% Windowed RankMass crawler (Figure 1 & 2). To visualize the effectiveness of the

Table 1. Algorithm Performance by Running Time

Name of Algorithm	Hours	No of pages
Two Layer Fractional PageRank crawler	40 mins	76688586
Fractional PageRank Explored crawler	54 mins	76688586
Fractional PageRank Discovered crawler	71 mins	76688586
100% Windowed RankMass crawler	4:11 hours	69000000
20% Windowed RankMass crawler	20 hours	70000000

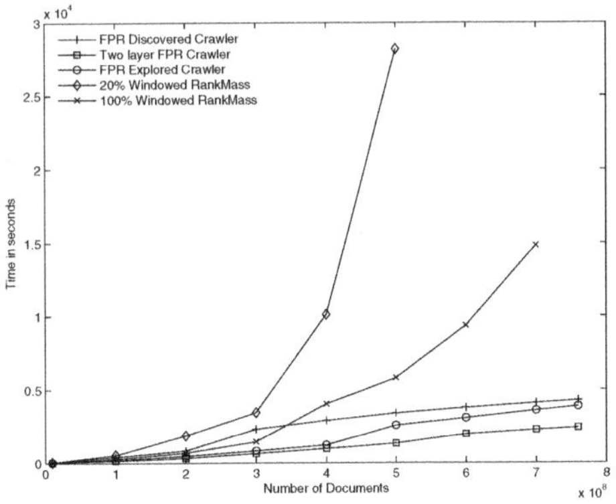

Fig. 1. Running Time of Algorithms

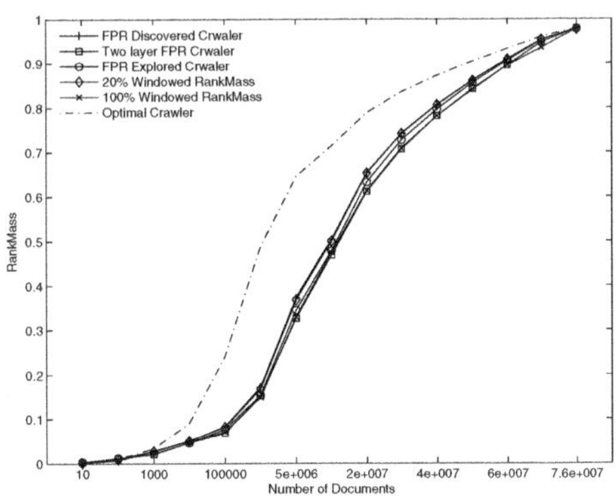

Fig. 2. Effectiveness of Algorithms

proposed crawlers, we compute the PageRank values for all of the pages in the uk-2006-06 graph and plotted the sum of the PageRank values of the downloaded pages at different points of the crawling process. The strategy which makes the cumulative PageRank higher downloading less documents is an effective one. The upper bound corresponds to the value of the cumulative PageRank collected by the optimal crawler that is assumed to know all of the URLs and the PageRank values of the pages and to greedily download the highest PageRank page first from the frontier.

It can be seen from Figure 2 that all of the three proposed FPR crawlers provide better ordering than the 100% Windowed RankMass crawler. The Two Layer FPR crawler behaves similarly to the 100% Windowed RankMass crawler. However, the FPR Explored crawler clearly provides better ordering than the 100% Windowed RankMass crawler. The experimental results (Figure 2 & Table 1) show that the FPR Discovered crawler produces the same ordering with the 20% Windowed RankMass crawler within 70 minutes whereas the 20% Windowed RankMass crawler takes about 20 hours.

The FPR Explored crawler performs better than the FPR Discovered crawler in terms of the running time. However, the ordering quality is degraded because of its mixed mode of sequential and random access according to the FPR. This means that although it picks the web pages with the highest FPR to be explored, it downloads all of the URLs pointed to by the explored pages together.

4 Conclusions

We propose three crawling algorithms that are scalable and able to prioritize the web scale frontier effectively and efficiently. In the FPR Explored and Two Layer FPR algorithms we prioritize the downloaded unexplored pages and both of these algorithms demonstrate the effectiveness of choosing important pages early. In the future, we will study about the combination of Fractional PageRank and other available information in the downloaded unexplored pages to design an effective algorithm for focused crawler.

References

1. Brin, S., Page, L.: The anatomy of a large-scale hypertextual web search engine. Comput. Netw. ISDN Syst. 30, 107–117 (1998)
2. Cho, J., Schonfeld, U.: Rankmass crawler: a crawler with high personalized pagerank coverage guarantee. In: Proceedings of the 33rd International Conference on Very Large Data Bases, pp. 375–386 (2007)
3. Web Collection Uk-2006-06. Crawled by the Laboratory for Web Algorithmics, University of Millan, `http://law.dsi.unimi.it/`

Selectivity Estimation for Exclusive Query Translation in Deep Web Data Integration

Fangjiao Jiang[1,2], Weiyi Meng[3], and Xiaofeng Meng[1]

[1] School of Information, Renmin University of China
{jiangfj,xfmeng2006}@gmail.com
[2] College of Physics and Electronic Engineering, Xuzhou Normal University
[3] Computer Science Dept, SUNY at Binghamton
meng@cs.binghamton.edu

Abstract. In Deep Web data integration, some Web database interfaces express exclusive predicates of the form $Q_e = P_i(P_i \in P_1, P_2, \ldots, P_m)$, which permits only one predicate to be selected at a time. Accurately and efficiently estimating the selectivity of each Q_e is of critical importance to optimal query translation. In this paper, we mainly focus on the selectivity estimation on infinite-value attribute which is more difficult than that on key attribute and categorical attribute. Firstly, we compute the attribute correlation and retrieve approximate random attribute-level samples through submitting queries on the least correlative attribute to the actual Web database. Then we estimate Zipf equation based on the word rank of the sample and the actual selectivity of several words from the actual Web database. Finally, the selectivity of any word on the infinite-value attribute can be derived by the Zipf equation. An experimental evaluation of the proposed selectivity estimation method is provided and experimental results are highly accurate.

1 Introduction

The Deep Web continues to grow rapidly [1], which makes exploiting useful information a remarkable challenge. Metaquerier, which provides a uniform integrated interface to the users and can query multiple databases simultaneously, is becoming the main trend for Deep Web data integration.

Query translation plays an important role in a metaquerier. However, due to the large-scale, heterogeneity and autonomy of the Web databases, automatic query translation is challenging. One of the important aspects is that Web database interfaces may express different predicate logics. The integrated query interface and many Web database interfaces express conjunctive predicates of the form $Q_c = P_1 \wedge P_2 \wedge \ldots \wedge P_m$, where P_i is a simple predicate on single attribute. While some Web database interfaces express exclusive predicates of the form $Q_e = P_i(P_i \in P_1, P_2, \ldots, P_m)$, which means any given query can only include one of these predicates. Exclusive attributes are often represented on a Web database interface as a selection list of attribute names or a group of radio buttons each of which is an attribute. A very interesting problem is, among all the $Q_e s$ on an interface, which one has the lowest selectivity? It is of critical

X. Zhou et al. (Eds.): DASFAA 2009, LNCS 5463, pp. 595–600, 2009.

importance to optimal query translation. In this paper, we mainly focus on the selectivity estimation of infinite-value attribute for exclusive query translation.

Before we carry out our study, we have two important observations: 1) there exist different correlations between different attribute pairs, and 2) the word frequency of the values on an infinite-value attribute usually has a Zipf-like distribution. Based on these observations, we propose a correlation-based sampling approach to obtain the approximate random attribute-level sample and a Zipf-based approach that can estimate the selectivity of any word by Zipf equation.

The rest of paper is organized as follows. Section 2 gives the overview of query selectivity estimation. Section 3 proposes the correlation-based sampling approach. Section 4 proposes a Zipf-based selectivity estimation approach. Section 5 reports the results of experiments. Section 6 introduces the related work. Section 7 concludes the paper.

2 An Overview of Query Selectivity Estimation

The overall flow chart of our approach is given in Fig.1.

Attribute correlation calculation for a domain. For any given domain (e.g., Books), we first calculate attribute correlation for each pair of attributes (***Attribute Correlation calculation)*** and identify the least correlative attribute $Attr_i$ for each specific attribute $Attr_u$. Because attribute correlation of each attribute pair in a domain is usually independent of the Web databases, the attribute correlation can be used for all the Web databases in the same domain.

Selectivity estimation for a Web database. Given an infinite-value attribute $Attr_u$ and a specific Web database, we use a series of query probes on $Attr_i$ in the Web database interface to obtain an approximate random attribute-level sample on $Attr_u$ (***Correlation-based sampling)***. The word rank on $Attr_u$ can be calculated from the sample, which is viewed as the actual word rank on $Attr_u$ of the Web database due to the randomness of the sample. Then several words on $Attr_u$ are used to probe the actual Web database and the frequencies

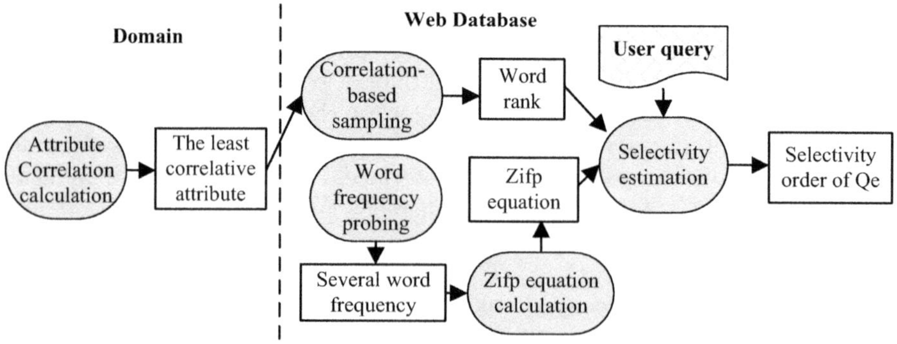

Fig. 1. The processing flow of our approach

of these words are returned (***Word frequency probing***). Zipf equation can be estimated using the word ranks and the actual frequencies of several words (***Zipf equation calculation***). Finally, for any word on $Attr_u$, we can estimate its frequency by the Zipf equation and its rank (***Selectivity estimation***).

3 Correlation-Based Sampling for Word Rank

In this paper, we use *Attribute Word Distribution* of different attributes to define the concept of attribute correlation.

Definition 1 Attribute Word Distribution (AWD). *Given all the words $w_1, w_2, \ldots, w_m$ of the values of attribute A in a database D, the Attribute Word Distribution for A is a vector $\overrightarrow{v}(v_1, v_2, \ldots, v_m)$, each component of which v_i is the frequency of the word w_i. Under the assumption that no word appears more than once in an attribute value, the frequency of the word w_i is the number of tuples returned by the query $\sigma_{A=w_i}D$.*

Definition 2 Attribute Correlation. *Attribute Correlation is the dependence between any attribute pair($Attr_u$, $Attr_v$) and is measured by the difference of the Attribute Word Distributions of the returned results on an attribute ($Attr_u$).*

A measure of the distribution difference is Kullback-Leibler(KL) divergence. If we submit different queries Q_1, Q_2, ..., Q_s on $Attr_v$, we will gain the corresponding result sets S_1, S_2, ..., S_s on $Attr_u$. Suppose that S is the union of S_1, S_2, ..., S_s and S consists of a set of words w_1, w_2, ..., w_k. Then the KL-divergence of $Attr_u$ from S to S_j is:

$$D_{KL}(S||S_j) = \sum_{l=1}^{k} prob(Attr_u = w_l|S) log \frac{prob(Attr_u = w_l|S)}{prob(Attr_u = w_l|S_j)}$$

where prob($Attr_u$ =w_l| S) refers to the probability that $Attr_u$=w_l in S and prob($Attr_u$=w_l| S_j) refers to the probability that $Attr_u$=w_l in S_j.

Attribute correlation is the average of the KL divergence of $Attr_u$ from S to S_j:

$$Correlation(Attr_u, Attr_v) = \frac{1}{s} \sum_{j=1}^{s} D_{KL}(S||S_j)$$

After discovering the least correlative attribute $Attr_i$, we submit some query probes on $Attr_i$ to the Web databases and collect the returned results on attribute $Attr_u$ as the attribute-level sample of $Attr_u$, which is the approximate random sample. Then we order the words of the sample by their frequencies and the word rank can be viewed as the actual one due to the randomness of the sample.

4 Zipf-Based Selectivity Estimation

It is well known that English words of a general corpus satisfy the Zipf distribution. However, it is not clear if the words of text attributes in different domains also follow this distribution. Our experiments indicate that they do.

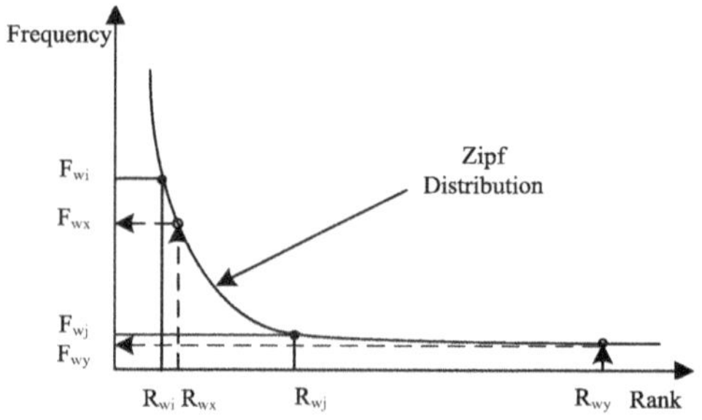

Fig. 2. Zipf-based Selectivity Estimation

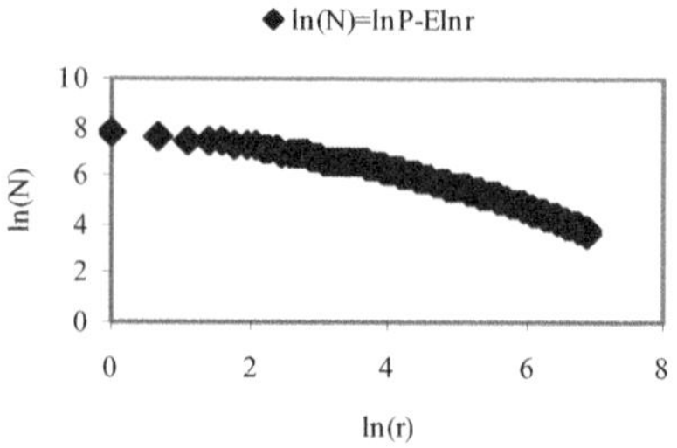

Fig. 3. Distribution transformation

Zipf distribution can be represented by $N = P(r+p)^{-E}$[4], where N represents the frequency of the word, r represents the rank of the word and P, p and E are the positive parameters. As Fig.2 shows, we submit word i, word j to the Web database and obtain their frequencies F_{wi}(i.e., N_i) and F_{wj}(i.e., N_j), respectively. And we know the ranks of these two words (i.e., r_i and r_j) from the sample obtained in section 3. Then, we can estimate the parameters P, p and E as follows.

- Equation Transformation: After the logarithm transformation, the Zipf equation is changed to $ln(N) = lnP - Eln(r+p)$. Because the parameter p $(0 < p < 1)$ is usually much smaller than word rank r(i.e., some applications even assume $p = 0$), the parameter E is approximately viewed as the slope of the line $ln(N) = lnP - Eln(r)$ as shown in Fig.3.
- Parameter E: E can be calculate by the equation $E \approx \frac{ln(N_i)-ln(N_j)}{ln(r_j)-ln(r_i)}$.
- Parameter p: When E is estimated, parameter p can be derived from the equation $\frac{N_i}{N_j}=\frac{P*(r_i+p)^{-E}}{P*(r_j+p)^{-E}}$. So we have $p \approx \frac{r_j - r_i * e^m}{e^m - 1}$ $(m = \frac{1}{E} * \ln \frac{N_i}{N_j})$.
- Parameter P: Finally, the parameter P is derived. $P \approx N_j * (r_j + p)^E$.

Consequently, we can use the Zipf equation and the word ranks to compute the selectivity of any word on the attribute.

It is worth noticing that the parameters P, p and E are not unique. We study the relationships among the precision, word ranks and rank distances. The results show that the precision will go down when the rank increases and to keep the precision stable, the distance of two word ranks should be increased with the increase of the word ranks.

5 Experiments

We evaluate our approach with the precision measure which is defined as follows.

$$Precision = \frac{1}{N} \sum_{n} \left| \frac{N_r - E_s}{N_r} \right|$$

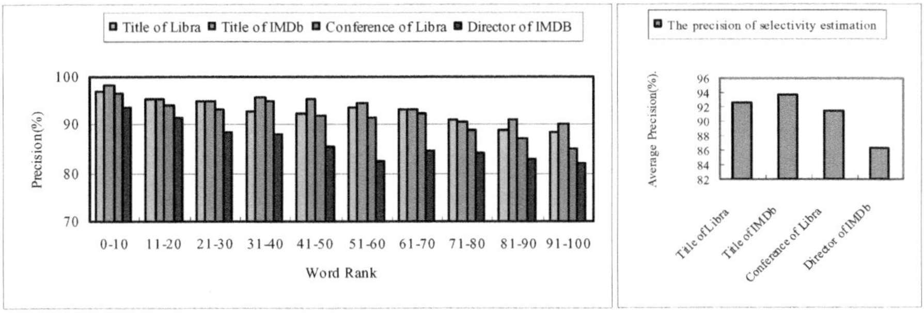

Fig. 4. The precision of selectivity estimation

where N_r is the number of results when submitting the word on the attribute to the actual Web database, E_s is the selectivity of the word on the same attribute estimated by our approach, and n is the number of the words that we test in the experiments.

We select the top 100 words on *Title*, *Conference* attribute of Libra, *Title*, *Director* attribute of IMDb and submit them to actual Web databases. Meanwhile, we estimate the selectivity of these words using our approach. Overall, as we can see from Fig.4, the precision of our approach is generally good.

However, there is still some deviation on estimation values. The reasons are that any two attributes are somehow correlative with each other and the words on some infinite-value attributes do not satisfy Zipf distribution perfectly.

Given that our approach can cope with selectivity estimation of all the infinite-value attributes and it is domain independent, it is generally feasible to be applied in query translation for exclusive query interface.

6 Related Work

The problem of selectivity estimation through uniform random sampling has received considerable attention [2,5]. [2] cannot be applied as we do not have full access to the Web databases. [5] proposes a random walk approach to sampling the hidden databases, which is a database-level sampling and relatively complex compared with our attribute-level sampling. [3] focuses on the selectivity estimation of the text type attribute with several constraints (e.g., *any*, *all* or *exactly*, etc.) in Web database interfaces.

7 Conclusions

In this paper, we study the query translation problem of the exclusive query interface and present a novel Zipf-based selectivity estimation approach for infinite-value attribute. Experimental results on several large-scale Web databases indicate that our approach can achieve high precision on selectivity estimation of infinite-value attribute for exclusive query translation.

Acknowledgments. This work is supported in part by China 863 High-Tech Program(No.2007AA01Z155); NSF of China (No.60833005, 60573091); China National Basic Research and Development Program's Semantic Grid Project (No.2003CB317000). US NSF grants IIS-0414981 and CNS-0454298.

References

1. The Deep Web: Surfacing Hidden Value, `http://www.completeplanet.com/Tutorials/`
2. Oliken. F.: Random Sampling from databases. PhD Thesis, University of California, Berkeley (1993)
3. Zhang. Z., He. B., Chang. K. C. C.: On-the-fly Constraint Mapping across Web Query Interfaces. In: IIWEB (2004)
4. Mandelbrot, B.B.: Fractal Geometry of Nature. W. H. Freeman and Co., New York (1988)
5. Dasgupta. A., Das. G., Mannila. H.: A random walk approach to sampling hidden databases. In: SIGMOD, pp. 629–640 (2007)

Detecting Aggregate Incongruities in XML

Wynne Hsu, Qiangfeng Peter Lau, and Mong Li Lee

Department of Computer Science, National University of Singapore
{whsu,plau,leeml}@comp.nus.edu.sg

Abstract. The problem of identifying deviating patterns in XML repositories has important applications in data cleaning, fraud detection, and stock market analysis. Current methods determine data discrepancies by assessing whether the data conforms to the expected distribution of its immediate neighborhood. This approach may miss interesting deviations involving aggregated information. For example, the average number of transactions of a particular bank account may be exceptionally high as compared to other accounts with similar profiles. Such incongruity could only be revealed through aggregating appropriate data and analyzing the aggregated results in the associated neighborhood. This neighborhood is implicitly encapsulated in the XML structure. In addition, the hierarchical nature of the XML structure reflects the different levels of abstractions in the real world. This work presents a framework that detects incongruities in aggregate information. It utilizes the inherent characteristics of the XML structure to systematically aggregate leaf-level data and propagate the aggregated information up the hierarchy. The aggregated information is analyzed using a novel method by first clustering similar data, then, assuming a statistical distribution and identifying aggregate incongruity within the clusters. Experiments results indicate that the proposed approach is effective in detecting interesting discrepancies in a real world bank data set.

1 Introduction

Research efforts in data cleaning have been steadily gaining momentum. Despite this, many data quality problems have yet to be tackled. One such problem involves discrepancies that can only be uncovered based on aggregated information. Consider the example XML document of bank accounts with their transactions given in Figure 1. An initial data exploratory analysis within the left account reveals nothing abnormal. Individually, the customer transaction details appear normal. None of the transactions involves extraordinary amounts (⟨Amt⟩) that will raise concerns. However, when we aggregate the number of transactions per customer (⟨Trans_Count⟩) and average transaction amount (⟨Avg_Trans_Amt⟩) at the account level, an interesting discrepancy is immediately apparent. By considering the values of elements ⟨Trans_Count⟩, ⟨Avg_Trans_Amt⟩, and ⟨Country⟩ of an account, we can identify accounts with unusually low (or high) transaction averages or counts, compared to other accounts from the same country. In fact, a transaction average of 200 Japanese Yen, the equivalence of less than USD$3, would have been very uncommon in Japan. Highlighting such discrepancies is important for formulating useful risk management policy. We introduce the notion of *aggregate incongruity* to capture this class of data discrepancies.

X. Zhou et al. (Eds.): DASFAA 2009, LNCS 5463, pp. 601–615, 2009.

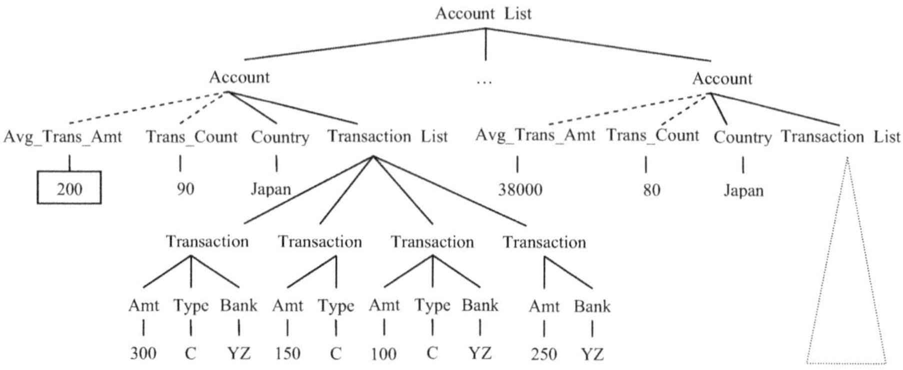

Fig. 1. Example of aggregate incongruity in average transaction amount

Aggregated information plays an important role in many real world application domains. For instance, the World Health Organization (WHO) constantly monitors the reports of infectious cases all over the world. Finding large deviation in the total number of infectious cases in one region as compared to other regions will alert WHO to the possibility of new outbreaks in that region. In the industry, many businesses use aggregated information to formulate their strategies to increase competitiveness. A senior manager would be interested to know which are the branch locations that exhibit large deviations in their key performance indicators, as these signal the need for some change in the business strategy for these branches. With globalization, the complexity of data analysis has increased significantly. Finding discrepancies from the summarized data represents a promising approach to effectively reduce this complexity.

Previous works in finding discrepancies in data have been explored for relational data [1] and more recently, for XML [2]. These approaches use the notion of support counting and a scoring metric based on the supports to determine the discrepancy of data. These methods also rely on discretization of continuous data before counting supports of the discretized intervals. However, aggregated data is often continuous instead of categorical and, the method used to discretize continuous attributes unduly influences the detection process by distorting the distance between continuous data values. This constitutes the motivation for developing an approach that naturally handles continuous data in the presence of categorical data without discretization.

In this work, we propose a framework that pro-actively detects discrepancies based on aggregated data. The framework utilizes the XML hierarchical structure to systematically perform aggregation on a set of identified attributes. Deviation analysis of the aggregated information is then carried out. Three issues need to be resolved.

First, we need to select an appropriate *space* within which the deviation analysis of the aggregated information is to be carried out. For example, when analyzing the average transaction amounts, the number of bank employees and opening hours of the bank will be irrelevant, however the customer's age and income is likely to be useful. Determining the appropriate space for deviation analysis is an open question. We take advantage of the inherent nesting structure of XML to derive the appropriate spaces.

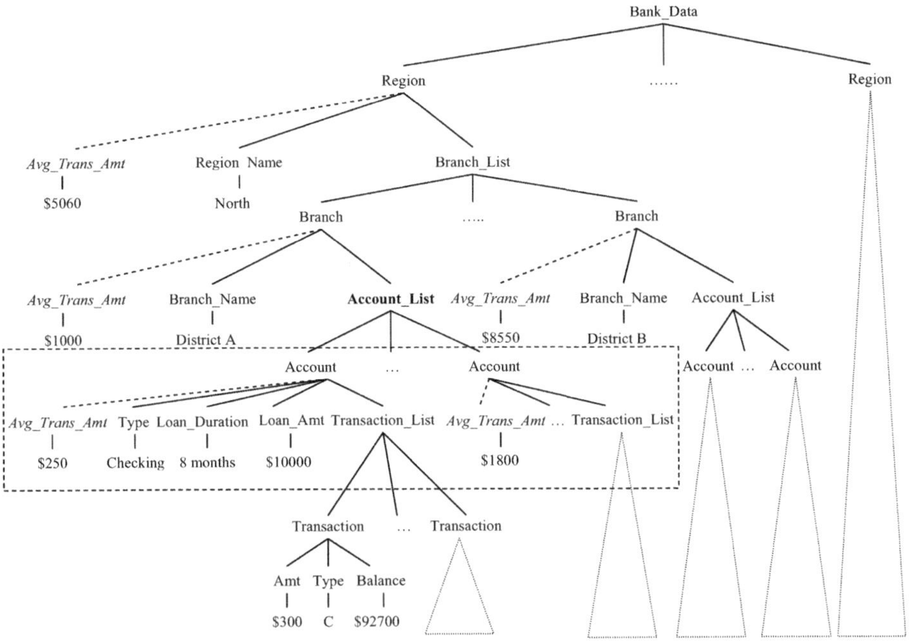

Fig. 2. Example Bank XML data with aggregates inserted at different levels and some details omitted

This makes sense as the encoding of relationships between XML elements are in the form of a hierarchy, i.e. elements in the same sub-tree tend to be related to one another.

Second, is finding the *context* in which to analyze aggregate incongruities within a given space. Figure 1 compared average transaction amount in the context of a group of accounts in Japan. If accounts from other countries were analyzed in the same context, the unusually low ⟨Avg_Trans_Amt⟩ may not be detected. To find these groupings, we use clustering to partition the space such that aggregate incongruity can be discovered.

Third, is determining which set of values should be aggregated. In Figure 2, a list of transactions is contained in an account and a branch has a list of accounts. If we consider the average transaction amount at the account level, we may discover aggregate incongruities in the average transaction amount between different accounts. At the branch level, we may discover aggregate incongruities of average transaction amount between different branches. Aggregation at different levels of the XML hierarchy reveals different incongruities. Hence, we need a mechanism that automatically propagates the aggregated information from the leaf nodes of an XML tree up its hierarchy.

To the best of our knowledge, this is the first work that examines how we can systematically employ aggregation in XML to identify interesting aggregate incongruities. We present a statistical approach to identify data discrepancies within a context. This allows us to handle continuous data naturally. Experiments on synthetic and a real-world bank data set show that our framework is able to reveal data discrepancies at different levels of aggregation and performs well on continuous data.

2 Concepts and Definitions

An XML document is modeled as a tree, $T = (V, E)$, where V is the set of vertices (nodes) and E is the set of edges. The leaf nodes[1] of an XML document contains values as text. We denote this value of a leaf node, v, as $val(v)$. Let $label(v)$ denote the unique element type label for a node v. The XML hierarchical structure organizes data according to their relevance – the closer the nearest shared ancestor of two elements is, the more relevant they are, where closeness is measured by the difference in ancestor-descendant element depth.

Definition 1. (Nearest Common Ancestor) *Let $T = (V, E)$ be a XML tree, and $U = \{v_1, v_2, \cdots, v_n\}$, $U \subset V$. Further let $desc(u, v)$ be the predicate that is true if v is the descendant of u and $dist(u, v)$ is the number of edges traversed by the shortest path from u to v. Suppose $A = \{u \in V | desc(u, v_1) \wedge desc(u, v_2) \cdots \wedge desc(u, v_n)\}$ is a set of nodes that are the common ancestors of all $v_i \in U$. We define the nearest common ancestor of U, denoted as $NCA_U = v_c \in V$ if and only if $v_c \in A$, and $\forall v_i, v_j \in U, \forall w \in A - \{v_c\}, dist(v_i, v_c) + dist(v_c, v_j) \leq dist(v_i, w) + dist(w, v_j)$.*

In Figure 2, the NCA of ⟨Transaction⟩ nodes within the sub-tree of an ⟨Account⟩ node is the corresponding ⟨Transaction_List⟩ node that is a child of ⟨Account⟩ node. For ⟨Transaction⟩ nodes across all accounts in the sub-tree of a ⟨Branch⟩ node, the NCA is the corresponding ⟨Account_List⟩ node that is the child of ⟨Branch⟩ node.

Definition 2. (Aggregate Node) *Given an XML tree $T = (V, E)$, an aggregate function $f \in F$ (the set of aggregate functions), and set of leaf nodes $U \subset V$ such that $\forall u, v \in U, label(u) = label(v)$, we can create an aggregate node $v_{f,U}$ in T such that $v_{f,U}$ is a leaf node and a sibling node of NCA_U. The value of $v_{f,U}$ is the result of applying aggregate function f to the set of values of the nodes in U.*

Figure 2 shows the aggregate nodes of average transaction amount inserted at various levels of the XML hierarchy (denoted by dashed lines). At the lowest level, the aggregate nodes ⟨Avg_Trans_Amt/$250⟩ and ⟨Avg_Trans_Amt/$1800⟩ summarize the average transaction amount for different bank accounts. At the branch level, ⟨Avg_Trans_Amt/ $1000⟩ and ⟨Avg_Trans_Amt/$8550⟩ summarize the average transaction amounts for each branch. Finally, ⟨Avg_Trans_ Amt/$5060⟩ summarizes the average transaction amounts for different regions. Note that these aggregate nodes are siblings of the nearest common ancestor of the nodes that have been aggregated, i.e. ⟨Transaction_List⟩, ⟨Account_List⟩, and ⟨Branch_List⟩ for account, branch and region levels respectively.

Definition 3. (Object) *Given an XML tree $T = (V, E)$, an object, denoted as $Obj(v)$, rooted at node $v \in V$ is a non-empty set of leaf nodes that are children of v, i.e. $Obj(v) = \{u \in V | desc(v, u) \wedge dist(u, v) = 1 \wedge val(u) \neq \emptyset\}$.*

The ⟨Account⟩ object in Figure 2 consists of nodes: ⟨Avg_Trans_Amt⟩, ⟨Type⟩, ⟨Loan_Amt⟩, and ⟨Loan_Duration⟩. We analyze if a particular aggregate value in an

[1] A node refers to an element and the value of a node is the corresponding text node in XML

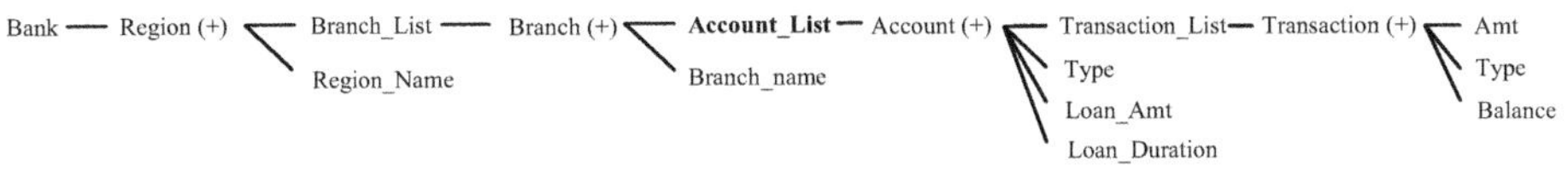

(+) denotes *many*, $[0, n]$, multiplicity

Fig. 3. Example Bank XML schema without inserted aggregates

object is an aggregate incongruity w.r.t. a group of objects. This group of objects is identified via a *pivot node.*

Definition 4. (Pivot Node) *Given an XML tree,* $T = (V, E)$ *and object* $Obj(v)$*, suppose* u *is the first occurrence of a node, on the path from node* v *to the root of* T*, whose corresponding element in the XML schema has a multiplicity of many (i.e. [0,n]), then the parent node of* u *is the pivot of* $Obj(v)$*, denoted by* $pivot_v$*.*

Figure 3 shows the schema of the bank account data set (Figure 2). The objects represented by ⟨Account⟩ nodes have their respective ⟨Account_List⟩ nodes (in bold) as their pivot node. In this particular case, the pivot node in the XML tree (Figure 2) of each ⟨Account⟩ object node is their parent ⟨Account_List⟩ node.

Definition 5. (Pivoted Space) *Let* v *be a node in* V*. Given an object node* $Obj(v)$ *and its pivot node* $pivot_v$*, we define a pivoted space* $PS(v)$ *as the set of nodes such that:* $\forall v_i \in PS(v), Obj(v_i) \neq \emptyset \wedge label(v_i) = label(v)$*, and* $NCA_{PS(v)} = pivot_v$*.*

In Figure 2, the nodes enclosed in a box denotes a pivoted space whose pivot node is the ⟨Account_List⟩ node indicated in bold.

3 Aggregate Incongruity Detection (AID) Framework

There are four key processes in our Aggregate Incongruity Detection (AID) framework to identify potential aggregate incongruities:

1. Aggregate Nodes Insertion – Given an XML schema and a list of aggregate functions, we automatically augment the XML tree with the appropriate aggregate nodes at various levels of the XML hierarchy.
2. Pivoted Space Identification – Given the augmented XML document, we identify all pivot nodes and determine all pivoted spaces.
3. Neighborhood Identification – Within each pivoted space, we identify groupings of data that form the context for detecting incongruity.
4. Aggregate Incongruity Identification – Given the context, a statistical method is used to identify aggregate incongruity with respect to the context.

3.1 Aggregate Nodes Insertion

We first augment the XML tree with aggregate nodes at various levels of the hierarchy. Any aggregate function may be applied. Here, we focus on AVG (mean of a set of continuous values), COUNT (cardinality of a set of objects) and SUM.

Given an XML schema and a set of aggregate functions to be applied, for each leaf node, we determine the applicability of the aggregate function with respect to the domain of the leaf node. For example, SUM and AVG functions require the input set of values to be continuous while the COUNT function does not. For each applicable function on each leaf node, an aggregate node is inserted. Since aggregate nodes at different levels of an XML tree can reveal different discrepancies, our algorithm must automatically propagate these aggregates up the XML hierarchy.

Algorithm 1(a) gives the details for augmenting an XML schema tree (Figure 3) with aggregate nodes (elements). It traverses the tree, T_s, in a breadth first manner and looks for aggregate functions whose function domain matches the current node's value domain (Line 5). For example, for a node, v, if $Obj(v) \neq \emptyset$, COUNT will be applicable to the node. Similarly, if $val(v)$ is numeric, then SUM and AVG functions will be applicable. Next, the path to root from the current node is obtained (Line 6) and for each node, u in that path such that $pivot_u$ exists and has a parent (Line 8), an aggregate node, a, is created using the CREATEAGG method. This method returns an aggregate node that will contain the aggregate value of applying aggregate function f to all values of nodes corresponding to $label(v)$ in the sub-tree rooted at $pivot_u$. Then, if a has not been previously inserted into the tree (Line 10), it will be inserted as a sibling of $pivot_u$. For example, using the schema in Figure 3, assuming the current node is ⟨Amt⟩ and the aggregate function is AVG. The path to root is ⟨Amt⟩, ⟨Transaction⟩, ⟨Transaction_List⟩, ⟨Account⟩, ⟨Account_List⟩, ... , ⟨Bank_Data⟩. Starting from the first node in the path, ⟨Amt⟩ is not an object so $hasPivot(u)$ is false. The next node, ⟨Transaction⟩, is an object, has pivot node, ⟨Transaction_List⟩, and does not have a sibling node of average transactions amounts. Hence ⟨Avg_Trans_Amt⟩ is added as a sibling of ⟨Transaction_List⟩. Continuing along the path to root, the next object is ⟨Account⟩ with pivot node, ⟨Account_List⟩, therefore ⟨Avg_Trans_Amt⟩ is added as a sibling node to it. This continues until the root node. Finally, the ⟨Bank_Data⟩ node has no parent so AVG aggregates for transaction amount are not propagated further. After augmenting the XML schema with the specification of aggregate nodes using Algorithm 1(a), the aggregate values can be computed in a single parse of the XML document.

3.2 Pivoted Space Identification

The inserted aggregate nodes have to be analyzed within their pivoted spaces. Algorithm 1(b) details how objects and their respective pivoted spaces are identified. Given an XML document tree, T and its schema tree, T_s, a depth first traversal of the XML tree is performed. For example in Figure 2, assume the ⟨Transactions⟩ nodes appear from left to right sequence in the XML tree, i.e. $t_1, t_2, ...t_n$, where t_1 is the leftmost node and t_n is the rightmost node. The ⟨Transaction⟩ nodes below the t_i node will be traversed before the ⟨Transaction⟩ nodes below t_{i+1}.

Let the label of an XML tree node, $label(v)$, be the corresponding node, v_s, in the schema tree. The hash table H_p, keyed by labels, maintains the current pivot node of a particular pivoted space being constructed and H_s maintains the current space that contains the identified objects. Lines 8 to 12 handles the case where the current space is completed and a new pivoted space is to be identified – the old space is added to A and the hash tables updated. Lines 13 to 14 adds the current object into the correct space.

Algorithm 1. Algorithms for AID

(a) Aggregate Propagation

Input: F – set of functions, $T_s = (V, E)$ – schema tree

Output: $T_s = (V, E)$ – schema tree with aggregate specification

```
1: Q ← root(T_s)
       ▷ queue for Breadth First Traversal
2: while Q ≠ ∅ do
3:    v ← DEQUEUE(Q)
4:    ENQUEUE(Q, children(v))
          ▷ enqueue all children of v
5:    for each f ∈ F s.t. ValueDom(v) ∈ Dom(f) do
6:       P ← PATHTOROOT(v)
7:       for each u ∈ P do
8:          if hasPivot(u) ∧ hasParent(pivot_u) then
9:             a ← CREATEAGG(label(v), f, u)
10:            if ¬hasSibling(pivot_u, a) then
11:               ADDCHILD(parent(pivot_u), a)
12:            end if
13:         end if
14:      end for
15:   end for
16: end while
17: return T_s
```

(b) Pivoted Space Identification

Input: $T = (V, E)$ – XML tree, $T_s = (V_s, E_s)$ – schema tree

Output: A – a set of pivoted spaces

```
1: H_p ← V_s × {∅}   ▷ hash table to map object label to
   pivot node init. to null value for each element in XML
2: H_s ← V_s × {∅}   ▷ hash table to map object label to
   pivot space init. to null set for each element in XML
3: S ← {root(T)}   ▷ stack for depth first traversal
4: while S ≠ ∅ do
5:    v ← POP(S)
6:    MULTIPUSH(S, children(v))
7:    if isObject(v) then
8:       if pivot(v) ≠ GET(H_p, label(v)) then
9:          A ← A ∪ {GET(H_s, label(v))}
               ▷ store prev. space
10:         PUT(H_p, label(v), pivot(v))
11:         PUT(H_s, label(v), ∅)
12:      end if
13:      P ← GET(H_s, label(v)) ∪ {v}
14:      PUT(H_s, label(v), P)
15:   end if
16: end while
17:    ▷ store remaining spaces
18: for each α ∈ range(H_s) s.t. α ≠ ∅ do
19:    A ← A ∪ {α}
20: end for
21: return A
```

Lines 17 to 20 outputs the remaining spaces in H_s. Note that instead of storing A in main memory, at Line 9, the completed pivot space can be output to disk if there are space limitations.

3.3 Neighborhood Identification

After identifying the pivot spaces, we derive the local contexts in which to analyze potential aggregate incongruities. This is achieved by enumerating each subspace of lower dimensions from the pivoted space and clustering the subspace.

DBScan [3] is able to find arbitrary shaped clusters in data using the notions of core point, density reachable and density connected. Given an integer for $MinPts$ and a radius ϵ, a point is said to be a core point if there are $MinPts$ within distance ϵ from it. Any point within ϵ distance of a core point is directly-density reachable from the core point. Then, density reachable is a transitive asymmetric relation of directly-density reachable. For example, suppose point a and b are core points and c is not a core point. Further suppose that c is directly density reachable from b and b is directly density reachable from a. Then, c is density reachable from a. Lastly, density connected is the symmetric relation between two points, a and b such that there exists a point c where a and b are each density reachable from c. DBScan clustering groups all objects that are density connected as being in the same cluster. We apply DBScan on each subspace, $S(v)$, of the pivoted space, $PS(v)$, with the following modifications: (1) each point not belonging to any cluster is considered a cluster of size one, (2) the values in each dimension are transformed using min/max normalization to the range $[0, 1]$, and (3) the distance between k-dimensional data points is computed using a modification of the Euclidean distance metric (L2-norm) that also handles categorical data,

$$d(\mathbf{x}, \mathbf{y}) = \sqrt{\sum_{i=1}^{k} \Delta(x_i, y_i)^2} \quad , \quad \Delta(x_i, y_i) = \begin{cases} 0 & \text{if attribute } i \text{ is categorical} \\ & \text{and } x_i \neq y_i \\ 1 & \text{if attribute } i \text{ is categorical} \\ & \text{and } x_i = y_i \\ x_i - y_i & \text{otherwise} \end{cases}$$

DBScan partitions $S(v)$ into a set of clusters. Each cluster is called a *neighborhood.*

Incongruities in high dimensional space are inherently difficult for the user to comprehend. Therefore, for practical purposes, it is typically sufficient to limit the enumeration of subspaces up to five dimensions. In addition, although we employ DBScan for its simplicity and ability to find arbitrary shaped clusters, other clustering techniques can also be used to find neighborhoods.

3.4 Aggregate Incongruity Identification

After identifying the neighborhoods, the values of each dimension are analyzed using *Chauvenet's criterion. Chauvenet's criterion* identifies a value from the sample as a statistical outlier if the value's deviation from the sample mean given the sample standard deviation has a low probability of occurring. We use the significance level (denoted by α) to determine the critical region for such statistical outliers that are termed aggregate incongruity in the AID framework.

Definition 6. (Aggregate Incongruity) *Given a set of values, X, and a significance level, α, a value $x \in X$ is an aggregate incongruity with respect to X if,*

$$P(z \geq \frac{|x - \mu_X|}{\sigma_X}) \leq \frac{\alpha}{2}$$

where μ_X and σ_X is the mean and standard deviation of the values in X respectively, and the random variable z is assumed to be: (1) a normal distribution if $|X| \geq 30$, (2) a t-distribution with $|X| - 1$ degrees of freedom otherwise. The degree of incongruity of $x \in X$ is denoted by p-value $= 2 \cdot P(z \geq \frac{|x-\mu_X|}{\sigma_X})$. A smaller p-value indicates a higher degree of incongruity.

In practice, we use an inverse cumulative distribution function $D^{-1}(a)$ of the distribution z (that returns b from the equation $P(z \geq b) = a$) to determine the range of values in a neighborhood dimension beyond which are potential aggregate incongruities. The range of values is given by,

$$\left[D^{-1}(1 - \frac{\alpha}{2}) \cdot \sigma_X + \mu_X, D^{-1}(\frac{\alpha}{2}) \cdot \sigma_X + \mu_X\right]$$

where X is the set of values from a dimension of a neighborhood N.

4 Experiments

A series of experiments were conducted to investigate the performance of AID in handling different scenarios created from synthetic data generation. We also compare AID to XODDS [2] which utilizes discretization to handle continuous values. Finally, AID is run on a real world Bank Account data set to uncover potential aggregate incongruity. Experiments are performed on an Intel Core 2 dual 2 GHz computer. Programs are written in Java and ran with 1GB heap size.

4.1 Synthetic Data Set

We create a synthetic XML document with a pivoted space of N objects, each with 3 elements containing continuous values in two stages: clean data generation, and noisy data addition.

In the clean data generation stage, we choose a seed point (x, y, z) at random where $x, y, z \in [0, 1]$. From this seed point (x, y, z), we generate a neighborhood that is normally distributed with mean (x, y, z) and standard deviation $(\sigma_x, \sigma_y, \sigma_z)$. For easy analysis of results, we set $\sigma_x = \sigma_y = \sigma_z = \sigma$ where σ is a user supplied standard deviation. This neighborhood is obtained by taking repeated random samples that are within 1.96σ for each x, y, z dimension, i.e. 95% confidence level. Sampling stops when the neighborhood has reached a user specified size. After each desired neighborhood has been generated the "clean" synthetic data set is output.

In the noise addition stage, we insert varying numbers of incongruous data points into the neighborhoods until the user specified percentage of noise is reached. Noise points are distributed proportionally among the neighborhoods. To generate noise points for a neighborhood, the same normal distributions used to generate the clean neighborhood with means, x, y, z, and user specified standard deviation σ are used to sample noise points that are beyond 1.96σ of the means.

Performance is measured based on the F-measure, that is the harmonic mean between Precision and Recall, i.e. $F = \frac{2PR}{P+R}$, where F, P, R are the F-measure, precision and recall respectively[2]. Positive instances in our experiments are equivalent to noise points. We perform four experiments on the synthetic data set.

Effect of Varying Noise on Single Neighborhood. First, we investigate how noise levels affect performance for different settings of ϵ. A clean data set of 1000 points within a single neighborhood using $\sigma = 0.05$ is generated and various levels of noise is added to generate the individual noisy data sets. Then, AID is run using different values for ϵ on each data set with $\alpha = 0.03$. Figure 4 shows the plots of F-measure versus noise level for different settings of ϵ. The results show that if a suitable value of ϵ is selected, the performance is marginally affected by noise level. Also, using too small a value of ϵ severely impedes the performance of AID under low levels of noise (2% – 6%) but less so at higher levels (8% – 10%).

Effect Varying Neighborhood Overlap. Second, we investigate the effect of neighborhood overlap on performance. Five data sets are generated with various values of

[2] $P = \frac{\#\ true\ positive}{\#\ true\ positive\ +\ \#\ false\ positive}$ and $R = \frac{\#\ true\ positive}{\#\ true\ positive\ +\ \#\ false\ negative}$.

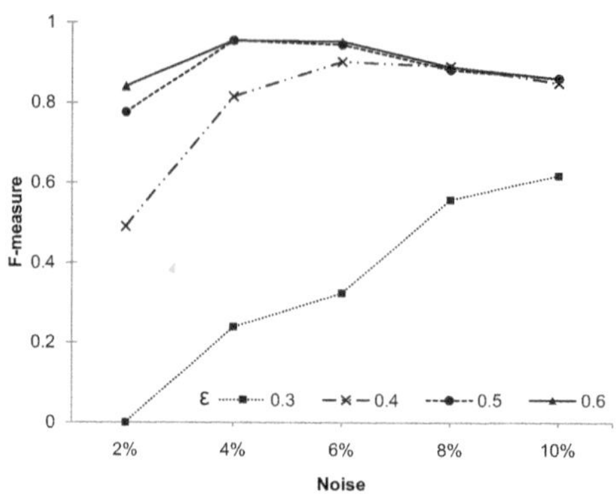

Fig. 4. Plot of F-measure vs. noise on single neighborhood

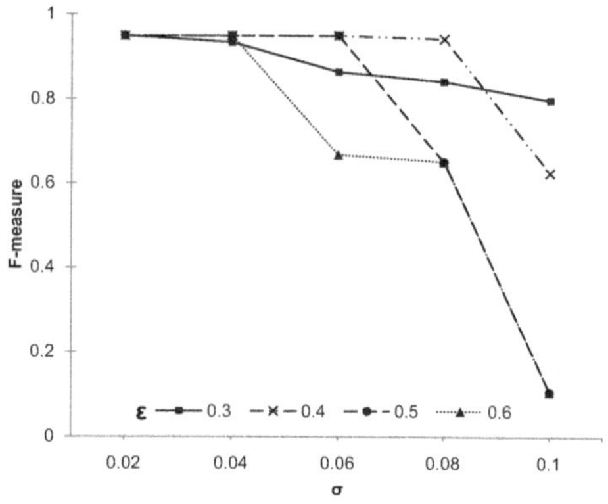

Fig. 5. Plot of F-measure vs. σ on multiple neighborhoods

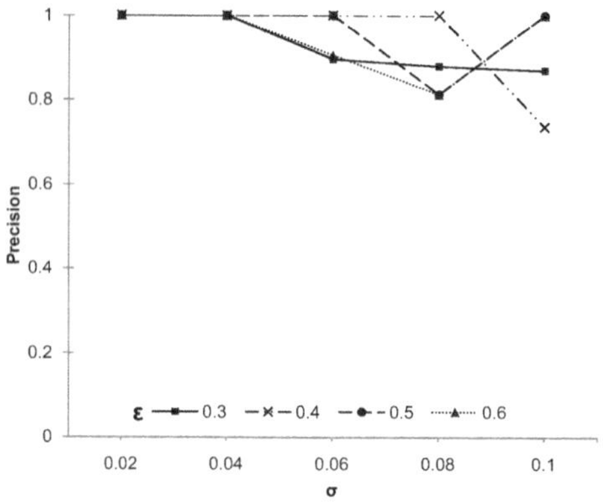

Fig. 6. Plot of Precision vs. σ on multiple neighborhoods

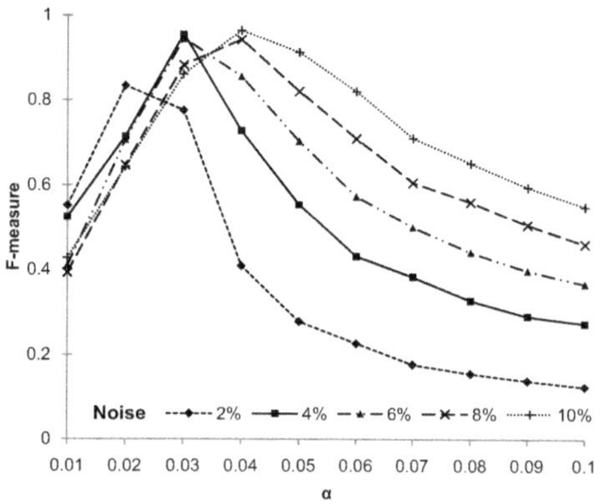

Fig. 7. Plot of F-measure vs. α

σ, each consists of three neighborhoods. Each neighborhood has 400 points. Then, 5% noise is added to each data set. The value of σ in this process controls the degree of neighborhood overlap. Figure 5 depicts the performance across varying values for σ for each of the five data sets with different settings for ϵ. Higher levels of overlap ($\sigma > 0.04$) degrade performance quickly regardless of ϵ. Conversely, if the inherent neighborhoods were well defined, i.e. $\sigma \leq 0.04$, performance is less affected by ϵ. Further analysis indicated that recall degrades sharply at high levels of overlap but precision remains marginally affected (Figure 6). Consequently, this experiment shows that even with high degree of overlap, the detected aggregate incongruities will be true positives.

Effect of Varying α. Third, we investigate the effect of varying α on performance, a clean data set of 1000 points containing a single neighborhood was generated. Then noise was added at various levels to give five data sets. For each data set, AID was run with various values for α and $\epsilon = 0.5$. The results are shown in Figure 7. Performance on each data set is plotted versus α. In general, low values of α (< 0.03) degrade

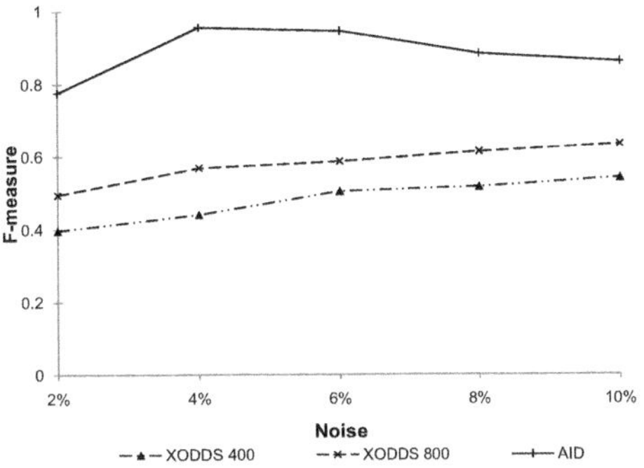

Fig. 8. F-measure comparison of XODDS with AID on single neighborhood

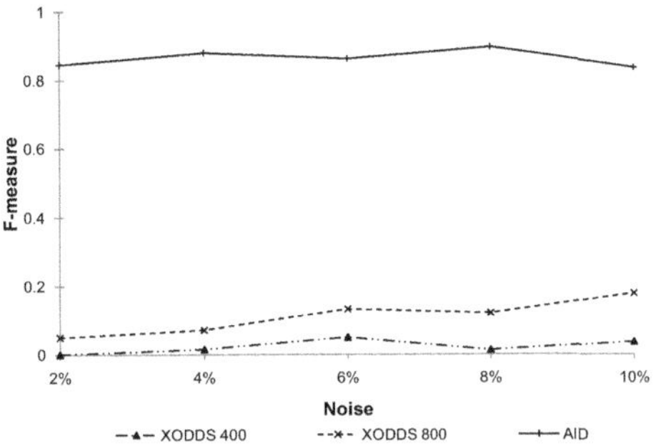

Fig. 9. F-measure comparison of XODDS with AID on multiple neighborhoods

performance regardless of the noise level. At higher values of α, performance degrades with higher noise levels being less affected when compared to lower noise levels.

Comparison with XODDS. One of the motivations for developing the AID framework is to naturally handle continuous values common among aggregated data, thus reducing performance dependency on discretization algorithms. We compare AID with the XODDS [2] framework that uses a support counting approach to identify deviations in data. Equi-width binning is used to discretize continuous attributes before detection. Projections are made over each pivoted space, in the process the supports for co-occuring data is accumulated. A scoring metric is employed to score each candidate with respect to data that co-occurs with it. One such metric used is the xQ-measure that is the proportion of the support for a target value divided by the support of data co-occuring with it. A lower score indicates higher degree of deviation. After scoring, the candidates are sorted in increasing order of score. Lastly, an adaptive threshold is determined based on the score and used to identify deviations.

We use XODDS in two variants differentiated by the number of bins used for discretization, namely, 400 and 800. The XODDS parameters were set as $MinSup = 10$ and $SoftCut = 20\%$. Both variants used the xQ-measure as the scoring metric. AID is used with the following parameter settings: $\epsilon = 0.5$, and $\alpha = 0.03$.

First, we compare the performance using a single neighborhood. 1000 clean points were generated within a single neighborhood using $\sigma = 0.05$. Then varying levels of noise is added to give five data sets. Figure 8 shows the plots of the variants of XODDS and AID for F-measure versus noise level. For each technique, the noise level affects performance marginally. The results show that the performance of XODDS is affected by the number of bins used during discretization. Higher number of bins offer greater distinction of discrepancies for XODDS. However, AID systematically performs better than both XODDS variants across various noise levels in terms of F-measure.

Next, we compare the performance on multiple neighborhoods. Five neighborhoods of 200 clean points each were generated using $\sigma = 0.02$. Then, varying levels of noise is added to give five data sets. Note that the smaller value of σ compared to the first part gives more well-defined neighborhoods. Figures 9, illustrates the performance plots of the variants of XODDS and AID for F-measure. The results show that the performance of XODDS has severely decreased. This is likely to be due to equi-width binning distorting the inherent neighborhoods in each of the data sets. Conversely, performance of AID is relatively unaffected and systematically out-performs XODDS.

This experiment shows that discretization of continuous attributes can significantly degrade the performance of detecting discrepancies in data. Hence justifying the need to develop a framework that handles continuous attributes naturally.

4.2 Real World Data Set

We apply the AID framework to a real world data set of bank accounts from the Czech Republic and report a few meaningful aggregate incongruities that were discovered at different hierarchical levels. This data set was used in the Discovery Challenge, held as a part of the 3rd European Conference on Principles and Practice of Knowledge Discovery in Databases (PKDD 1999). We extract from the bank account data set 4,500 accounts, 207,989 transactional records, and 670 loan records.

We first convert the data set to a flat XML schema that contains a single list of accounts. Under each ⟨Account⟩ element is a list of transactions, branch name, and region name of account. Each ⟨Transaction⟩ contains ⟨Amt⟩ and ⟨Balance⟩ elements that hold the values for transaction amount and post-transaction balance respectively. The aggregates AVG and COUNT were propagated on transaction ⟨Amt⟩ and ⟨Balance⟩ values to the ⟨Account⟩ level. We investigate if aggregate incongruities can be identified with respect to other categorical values of branch and region names. As the structure of the XML data is flat, we expect the single pivoted space of accounts to be dense. Hence AID was used with $\epsilon = 0.1$ and $\alpha = 0.05$. We ignore neighborhoods that have less than four tuples since these neighborhoods are too small for meaningful incongruities.

Many of the aggregate incongruities detected have large neighborhoods as the XML data consists of a single pivoted space of 4,500 accounts. We illustrate the smaller examples here. Table 1 lists three aggregate incongruities and their neighborhoods that provide a context for them. In Table 1(a), an account with low average transaction balance is highlighted with respect to other accounts in the South Moravia region with similar average transaction amount. Table 1(b) shows an account with high average transaction amount with respect to the Central Bohemia region and average transaction balance. The results shows that the AID framework is capable of identifying aggregate incongruities with respect to other categorical values.

Next, we investigate if aggregate incongruities can be detected at higher hierarchical levels. The bank data set is converted to a schema with two additional hierarchical levels of branch and region, similar to Figure 3. Accounts are now placed under their respective ⟨Region⟩ and ⟨Branch⟩ elements in which they belong to. For the aggregate nodes insertion process we use the aggregate functions AVG and COUNT. The process inserts aggregates of the continuous values such as transaction ⟨Amt⟩ and ⟨Balance⟩ at the different hierarchical levels of: Account, Branch, and Region.

One aspect of this organization of data in XML is that pivoted spaces at higher levels in the hierarchy are more sparse than in the lower level. For example, there may be thousands of ⟨Transaction⟩ objects within a particular bank account, but there are only eight ⟨Region⟩ objects within the bank. In light of this, instead of using a fixed value for ϵ during the neighborhood identification process, we use a different value of

Table 1. Various Aggregate Incongruity In Accounts

Region name	Avg transaction amount	Avg transaction balance
South Moravia	2270.75	16878.74
South Moravia	2204.24	16661.81
South Moravia	2079.93	16383.27
South Moravia	2045.21	16885.06
South Moravia	1983.33	16929.55
South Moravia	2140.96	***15838.56**
South Moravia	2105.89	16475.28
South Moravia	2053.53	16717.17
South Moravia	2281.22	16758.93

(a) *low average transaction balance with p-value $= 0.044 < \alpha = 0.05$.

Region name	Avg transaction amount	Avg transaction balance
Central Bohemia	2231.37	17341.64
Central Bohemia	2154.57	17376.11
Central Bohemia	2076.12	17767.96
Central Bohemia	2096.63	18068.14
Central Bohemia	2332.83	17607.68
Central Bohemia	***2471.29**	17569.89
Central Bohemia	2243.81	16721.47
Central Bohemia	2186.86	16277.11
Central Bohemia	2130.68	17559.80
Central Bohemia	2085.77	17297.51

(b) *high average transaction amount with p-value $= 0.048 < \alpha = 0.05$.

Table 2. Aggregate Incongruity Among Regions of the Bank

Avg loan amount	Avg loan payment	Avg transaction balance
153957.29	4338.95	38883.76
155392.27	4418.90	38444.00
156235.60	4539.18	37847.71
152549.21	4046.09	38264.74
136480.42	3785.39	38161.15
***122731.48**	3678.90	38535.49
165996.71	3916.35	38294.26
154541.13	4550.52	39179.12

*low average loan amount in North Bohemia Region with p-value $= 0.07 < \alpha = 0.1$.

Table 3. Aggregate Incongruity Among Branches in South Moravia region

Avg loan amount	Avg loan duration (Months)	Avg loan payment
170146.29	36.0	4098.14
168725.00	40.0	4083.63
168204.00	44.4	3725.80
168113.14	44.6	3731.00
157396.00	38.0	4401.50
177221.65	38.8	4503.12
***198584.00**	40.0	4616.17

*high average loan amount in Hodonin branch with p-value $= 0.072 < \alpha = 0.1$.

ϵ for different pivoted spaces based on their depth in the XML tree with the function, $E(PS(v), \epsilon_0) = \frac{\epsilon_0}{\ln(depth(v)+2)}$, where ϵ_0 is an initial value for ϵ. The function E reduces the value of ϵ for the neighborhood identification process for pivoted spaces that are deeper in the XML tree. Using this modification, the last three processes of AID are run with $\epsilon_0 = 0.8$ and $\alpha = 0.1$ on the Bank data set.

Table 2 illustrates a neighborhood among ⟨Region⟩ objects that contains an aggregate incongruity. Based on the neighborhood as a context, the average loan amount in the North Bohemia Region is significantly less than the other regions of the bank. This incongruity may be interesting to management as it highlights that with regards to the average loan payment and transaction balance of the accounts in North Bohemia the average amount of loans an account has is less than usual. Consequently, action may be taken to increase the consumption of loan related products in North Bohemia.

Table 3 depicts the neighborhood within the South Moravia region where the average loan amount for Hodonin branch is much higher compared to the average loan duration and average loan payment. Note that although the average loan payment for Hodonin branch may seem higher than the rest, it is not detected as an aggregate incongruity as it is not statistically significant. This aggregate incongruity may indicate that the Hodonin branch is issuing loans that are larger than usual.

Lastly, Table 4 shows two aggregate incongruities among bank accounts in different neighborhoods at the Cheb Branch of West Bohemia Region. The neighborhoods can be

Table 4. Aggregate Incongruity Among Accounts in Two Neighborhoods of Cheb Branch, West Bohemia Region

Avg transaction amount	Avg transaction balance	Count transaction
2100.00	17496.67	30
2285.64	17661.31	59
1974.61	18150.17	132
2090.50	20040.20	159
1321.14	19163.24	153
1300.45	15124.87	195
2181.58	15483.12	148
2079.64	16908.20	67
1844.09	* **11578.52**	85
2437.25	17237.17	117
1195.48	15417.96	97
1354.27	16232.52	103

(a) * low average transaction balance in low transaction count neighborhood with p-value $= 0.033 < \alpha = 0.1$.

Avg transaction amount	Avg transaction balance	Count transaction
951.10	19589.26	278
2554.83	22543.31	299
2018.47	28643.10	341
2899.85	31863.26	316
2484.33	* **38291.42**	376
2415.75	22954.26	294
1061.15	21477.48	364
2696.50	17325.22	260
1339.82	17912.48	291
1028.48	16740.92	304
2174.68	24304.43	254
2988.13	29862.17	204

(b) * high average transaction balance in high transaction count neighborhood with p-value $= 0.05 < \alpha = 0.1$.

distinguished by the count of transactions. In Table 4(a), one account has a significantly low average balance when compared to other accounts of similar average transaction amount and number of transactions. In Table 4(b) another account has a significantly high average balance instead. These may indicate unusual banking activity.

5 Related Work

Finding deviations is often viewed as outlier detection. There are two types of outliers, the class outliers and the attribute outliers. Class outliers are multivariate data points (tuples) which do not fit into any cluster formed by the remaining data. The majority of existing works has focused on class outliers and associated outlier-ness with the complete dimensions or attributes of the input data set [4,5,6,7]. Methods for identifying class outliers can be broadly classified into distribution-based, clustering-based, distance-based and density-based approaches.

Methods for finding class outliers cannot be applied to discover the class of data discrepancies obtained via aggregation. Class outlier detection typically only finds rare attribute values. However, using aggregation to find discrepancies involves finding values that seldom occur together, which is a special case of attribute outliers. Attribute outliers are univariate points that exhibits deviating correlation behavior with respect to other attributes [1,8,2] thus giving a context to the deviation. Attribute outlier detection has received little attention to date. The work in [1] proposes a method to identify attribute outliers in relational databases. This involves exhaustive enumeration of sets of attribute values to find discrepancies.

With the rapid proliferation of XML repositories, research on XML data quality is gaining momentum. The current focus is on XML duplicate detection by matching values of the corresponding fields of two objects and the structural similarity between the XML objects [8,9,10]. Recently, the work in [2] extended [1] to find attribute outliers in XML. However, it does not consider the role of aggregation in detecting additional data discrepancies and the detection method relies on discretization of continuous data that affects detection performance.

6 Conclusion

In conclusion, we presented a complete framework for the detection of aggregate incongruities in XML documents. The AID framework is able to automatically propagate aggregates throughout the XML document with minimal user intervention. Furthermore, the XML document is split into meaningful pivoted spaces based on its structure for incongruity identification. To provide meaningful contexts for the user to assess incongruities, we identify neighborhoods using subspace clustering. Then, within these neighborhoods we apply a well-known statistical method for identifying deviating data.

Our experiments on synthetic data show how AID performs with respect to noise, overlapping neighborhoods, and different parameter settings. Comparison with the recent XODDS [2] framework that uses discretization to handle continuous data shows that the approach used in AID to handle continuous data is better. On a real-world data set of Bank accounts, we report aggregate incongruities that are meaningful to the user.

Although the AID framework is used to analyze aggregates, it can also be used to analyze any continuous data in XML with discrete data as part of its context. Future work includes: investigating adaptive ways to tune the detection parameters based on data, evaluating the use of other clustering techniques, evaluating other statistical approaches to detect deviating data, and developing suitable result visualization to aid user action.

References

1. Koh, J.L.Y., Li Lee, M., Hsu, W., Lam, K.-T.: Correlation-based detection of attribute outliers. In: Kotagiri, R., Radha Krishna, P., Mohania, M., Nantajeewarawat, E. (eds.) DASFAA 2007. LNCS, vol. 4443, pp. 164–175. Springer, Heidelberg (2007)
2. Koh, J.L.Y., Lee, M., Hsu, W., Ang, W.T.: Correlation-based attribute outlier detection in XML. In: Proceedings of the 24th International Conference on Data Engineering, Cancun, Mexico, pp. 1522–1524 (2008)
3. Ester, M., Kriegel, H.P., Sander, J., Xu, X.: A density-based algorithm for discovering clusters in large spatial databases with noise. In: Proc. of 2nd International Conference on Knowledge Discovery and Data Mining (KDD 1996), pp. 226–231 (1996)
4. Aggarwal, C., Yu, S.: An effective and efficient algorithm for high-dimensional outlier detection. The VLDB Journal 14(2), 211–221 (2005)
5. Knorr, E.M., Ng, R.T.: Finding intensional knowledge of distance-based outliers. In: VLDB 1999: Proceedings of the 25th International Conference on Very Large Data Bases, pp. 211–222. Morgan Kaufmann Publishers Inc., San Francisco (1999)
6. Teng, C.M.: Polishing blemishes: Issues in data correction. IEEE Intelligent Systems 19(2), 34–39 (2004)
7. Zhu, X., Wu, X.: Class noise vs. attribute noise: a quantitative study of their impacts. Artif. Intell. Rev. 22(3), 177–210 (2004)
8. Low, W.L., Tok, W.H., Lee, M.L., Ling, T.W.: Data Cleaning and XML: The DBLP Experience. In: ICDE, p. 269. IEEE Computer Society, Los Alamitos (2002)
9. Puhlmann, S., Weis, M., Naumann, F.: XML duplicate detection using sorted neighborhoods. In: Ioannidis, Y., Scholl, M.H., Schmidt, J.W., Matthes, F., Hatzopoulos, M., Böhm, K., Kemper, A., Grust, T., Böhm, C. (eds.) EDBT 2006. LNCS, vol. 3896, pp. 773–791. Springer, Heidelberg (2006)
10. Weis, M., Naumann, F.: Dogmatix tracks down duplicates in XML. In: SIGMOD 2005: Proceedings of the 2005 ACM SIGMOD international conference on Management of data, pp. 431–442. ACM Press, New York (2005)

Materialized View Selection in XML Databases

Nan Tang[1], Jeffrey Xu Yu[2], Hao Tang[3], M. Tamer Özsu[4], and Peter Boncz[1]

[1] CWI, Amsterdam, The Netherlands
{N.Tang,P.Boncz}@cwi.nl
[2] The Chinese University of Hong Kong, Hong Kong
yu@se.cuhk.edu.hk
[3] Renmin University, Peking, China
haotang@ruc.edu.cn
[4] University of Waterloo, Ontario, Canada
tozsu@cs.uwaterloo.ca

Abstract. Materialized views, a RDBMS silver bullet, demonstrate its efficacy in many applications, especially as a data warehousing/decison support system tool. The pivot of playing materialized views efficiently is view selection. Though studied for over thirty years in RDBMS, the selection is hard to make in the context of XML databases, where both the semi-structured data and the expressiveness of XML query languages add challenges to the view selection problem. We start our discussion on producing minimal XML views (in terms of size) as candidates for a given workload (a query set). To facilitate intuitionistic view selection, we present a view graph (called VCUBE) to structurally maintain all generated views. By basing our selection on VCUBE for materialization, we propose two view selection strategies, targeting at space-optimized and space-time tradeoff, respectively. We built our implementation on top of Berkeley DB XML, demonstrating that significant performance improvement could be obtained using our proposed approaches.

1 Introduction

Materialized views, a RDBMS silver bullet, increase by orders of magnitude the speed of queries by allowing pre-computed summaries. In the context of XML databases, both the semi-structured data and the expressiveness of XML query languages complicate the selection of views. Answering XML queries using materialized views [4,24,15,16,7,22] and XML view maintenance [21,20] have been addressed recently, however, the problem of materialized view selection, which plays an important role in profiting from the above notable achievements of answering XML queries using views, is not well discussed. In this work, we describe a framework of selecting XML views to materialize. Broadly speaking, the problem of view selection in XML databases is the following: given XML databases $\mathcal{X}$, storage space B and a set of queries $\mathcal{Q}$, find a set of views $\mathcal{V}$ over $\mathcal{X}$ to materialize, whose combined size is at most B.

The problem of view selection is well studied in on-line analytical processing (OLAP) in data warehouses [9,10,25,5,13,6], where multiple aggregated views

X. Zhou et al. (Eds.): DASFAA 2009, LNCS 5463, pp. 616–630, 2009.

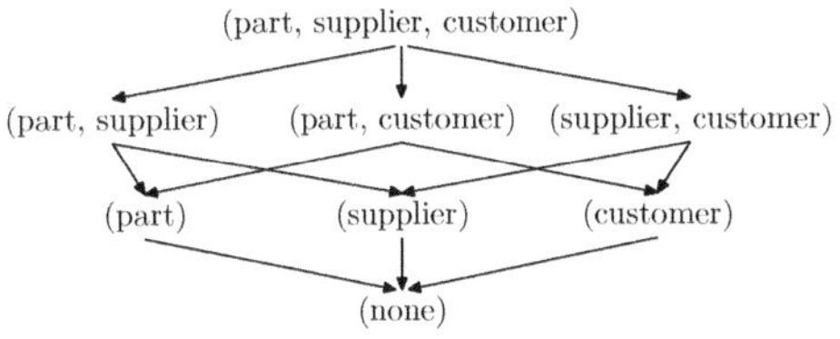

Fig. 1. Sample Data Cube

Table 1. Sample XML Queries

Q_1	//conference[@booktitle = "SIGMOD" or "VLDB"] /author[@name = "Venky Harinarayan" or "Jeffrey D. Ullman"]
Q_2	//conference[@keyword = "view" or "data cube"] /author[@name = "Jeffrey D. Ullman"]
Q_3	//conference[[@keyword = "view" or "data cube"] and[@booktitle = "SIGMOD" or "VLDB"]]

are organized as a data cube [11,18] for pre-computed summaries. This problem is generalized to network environments e.g. distributed databases [14] and P2P networks [8], as both computation cost and net communication could be saved. XML is a new standard for data exchange over the Internet, with standardized query languages (e.g. XPATH and XQUERY). There are similarities, and differences (in syntactic level) for the *view selection* problem between XML and relational databases: (*i*) there is inherent impedance mismatch between the relational (sets of tuples) and XML (ordered tree) data model; (*ii*) SQL is for relational data which is flat, regular, homogeneous and unordered. XPATH and XQUERY are for XML data which is nested, irregular, heterogeneous and ordered. Both items add to the complexity of the view selection problem.

We consider fundamental properties of view selection for optimizing a given query workload, with the objective in accordance with the formal perspective of view selection in RDBMS [6]. There are two conventional solutions in RDBMS: *common subexpressions exploitation* and *query aggregation*. The main problem of adopting common subexpressions is that the *deep copied* XML fragments lose the structural information required for performing the subsequent structural joins. While in RDBMS, the relational algebra is straightforwardly operated on intermediate tuples.

Next we discuss whether *query aggregation* can be applied. We review this technique in RDBMS first. Consider a TPC-D database with three tables: *part, supplier* and *customer*. The 8 possible groupings of tables are depicted in Figure 1. Computing each of the 8 groups of tables can be derived from the cells that are reachable to it, e.g., (*part*) can be derived from (*part, supplier*), (*part, customer*) and (*part, supplier, customer*). There are two difficulties of applying this technique in XML, considering the three XML queries in Table 1. (1) Enumerating all views that may contribute to answering a query is expensive. Assume that, given a query, we can find all views to answer it by relaxing queries [3]. For example, relax query axis e.g. `//conference/author` to `//conference//author` in Q_1 and Q_2; generalize label to wildcard $*$ e.g. `//conference` to //$*$ in Q_{1-3}, etc. The number of queries in this problem space is prohibitively large, which makes it infeasible. (2) Given candidate views, what relationships among them are to be maintained and in which way to organize them such that efficient algorithms might be devised, like data cube for RDBMS. Note that query/view answerability is closely related to query/view containment [17,16]. On the surface, it is natural to maintain the containment relationship between views. However, checking containment of XPATH queries is coNP-complete [17], with only a restricted syntax set (/, //, $*$ and branches $[\cdots]$). This indicates that it is impractical to find all containment relationships among candidate views.

Contributions & Roadmap. We start with an introduction of XML query and problem statement in Section 2. In Section 3, we prove the existence and uniqueness of a minimal view to answer two given queries, followed by an algorithm to generate such a view. We then describe, given multiple queries, how to generate a minimal view set while ensuring optimality. This solves the first difficulty. Note that computing the containment relationship between all candidate views ($\mathcal{V}$) requires $\mathsf{P}_2^{|\mathcal{V}|}$ (order 2 permutation of $|\mathcal{V}|$) comparisons. In Section 4, we propose to maintain the minimal view set using a directed acyclic graph, called VCUBE, to solve the second difficulty. VCUBE maintains the answerability relationship between views as edges, along with the process of generating views. This avoids computing the relationship between views pairwise, which is expensive. In Section 5, we describe two algorithms on top of VCUBE to select a set of views to materialize. Moreover, extensive experiments are conducted in Section 6 to show the efficiency of our proposed algorithms in boosting query performance. We conclude this paper in Section 7 with outlook on future works.

2 XML View Selection

2.1 Preliminaries

XPATH ***query.*** The set of XPATH queries studied in this paper is given in Table 2. It may involve the child-axis (/), descendant-axis (//), wildcards ($*$) and predicates. Predicates can be any of these: equalities with string, comparisons with numeric constants, or an arbitrary XPATH expression with optional operators "and" and "or". Sample XPATH queries may reference to Table 1.

Query containment. The *containment* of XPATH fragments involving /, //, $*$ and [] is studied in [17]. The extension of XPATH fragments containing additional operators "and" and "or" in predicates is studied in [4]. We borrow the definition of *containment* of queries from [17]. For an XML query P, $P(\mathcal{X})$ denotes the *boolean* result of P over database $\mathcal{X}$. We say $P(\mathcal{X})$ true if there exists at least one result; it is false otherwise. For two queries P and Q, P is *contained* in Q, denoted as $P \sqsubseteq Q$, iff $P(\mathcal{X})$ implies $Q(\mathcal{X})$, in every XML database $\mathcal{X}$. Containment is a partial order, i.e., $V \sqsubseteq P, P \sqsubseteq Q \Rightarrow V \sqsubseteq Q$. Equivalence is a two-way containment. Two queries P, Q are equivalent, denoted as $P \equiv Q$, iff $P \sqsubseteq Q$ and $Q \sqsubseteq P$.

Table 2. Query Definition

Path ::=	Step$^+$
Step ::=	Axis NameTest Predicates?
Axis ::=	"/" | "//"
NameTest ::=	elementName | " $*$ "
Predicates ::=	"[" Expr "]"
Expr ::=	Step | OrExpr
OrExpr ::=	AndExpr | OrExpr "or" AndExpr
AndExpr ::=	CompExpr | AndExpr "and" CompExpr | Step
CompExpr ::=	"@" attributeName Comp Value
Comp ::=	" = " | " > " | " < " | " >= " | " <= " | "!= "
Value ::=	Number | String

Example. (1) $//a/b$ is contained in $//a//b$, since the //-axis is more general than /-axis; (2) $//a/b$ is contained in $//a/*$, since the wildcard $*$ matches any label;

(3) $//a[@n < 10]$ is contained in $//a[@n < 100]$; (4) $//a[@s = \text{“}x\text{”}]$ is contained in $//a[@s = \text{“}x\text{”}\ or\ @s = \text{“}y\text{”}]$.

Answering queries using views. Existing works focus on rewriting a given query using materialized view with/without accessing the base data. We attempt to use the materialized view only to answer a query, without accessing the base data. $V \models Q$ denotes that a view V can be used to answer a query Q. $V \models \mathcal{Q}$ denotes that a view V can be used to answer each query in $\mathcal{Q} = \{Q_1, Q_2, \cdots, Q_n\}$.

2.2 Problem Statement

XML ***view selection.*** The problem of XML view selection is formally defined as follows: let $\mathcal{Q} = \{Q_1, Q_2, \cdots, Q_n\}$ (each Q_i is associated with a non-negative weight w_i) be a workload, $\mathcal{X}$ be an XML database, B be the available storage space, and COST() be a cost estimation function for query processing. The problem is to find a set of views $\mathcal{V}$ whose total size is at most B that minimizes:

$$\text{COST}(\mathcal{X}, \mathcal{V}, \mathcal{Q}) = \sum_{Q_i \in \mathcal{Q}} \text{COST}(\mathcal{X}, \mathcal{V}, Q_i) \times w_i \tag{1}$$

here, $\text{COST}(\mathcal{X}, \mathcal{V}, Q_i)$ denotes the cost of evaluating query Q_i using some view in $\mathcal{V}$, which is materialized over the XML database $\mathcal{X}$. This object function is in accordance with the formal perspective of view selection in RDBMS [6].

The optimal solution. For a query Q_1, the *complete* set of candidates $\mathcal{C}_1$ is all views V that may answer Q_1, i.e., $\mathcal{C}_1 = \{V | V \models Q_1\}$. The complete set of candidates $\mathcal{C}$ for a workload $\mathcal{Q} = \{Q_1, Q_2, \cdots, Q_n\}$ is $\mathcal{C} = \bigcup_{i=1}^{n} \mathcal{C}_i$. The naïve way is to enumerate all candidate views $\mathcal{C}$. Then, by exhaustively searching $\mathcal{C}$ to identify the optimal solution (i.e. a view set $\mathcal{O}$) that minimizes Equation 1.

There is no existing work to enumerate all complete candidate views. Assume that for a query Q, we can define a set of rules, such as an edge relaxation (e.g., $/a$ to $//a$), a subtree promotion (e.g., $//a/b[Q_1]//Q_2$ to $//a[.//Q_2]/b[Q_1]$), a leaf node deletion (e.g., $//a/b/c$ to $//a/b$) [2], and label generation (e.g., $//a$ to $//*$), etc. With above rules, we may (possibly) enumerate the complete candidate views, while we still face the two difficulties addressed in Introduction. (1) The number of candidate views is prohibitively large. The basic relaxation rules [2] already generate an exponential amount of candidates. Taking label generation into account, the problem space is exponentially exploded. (2) It is infeasible to identify the relationships among all candidate views, which requires $\mathsf{P}_2^{|\mathcal{C}|}$ comparisons, and per containment check is NP-hard. We describe a new way to generate a small view set $\mathcal{V}$, which is a subset of the complete set (i.e., $\mathcal{V} \subseteq \mathcal{C}$) but ensures to be a superset of the optimal solution (i.e., $\mathcal{O} \subseteq \mathcal{V}$).

3 Candidate View Generation

We first discuss the problem of answering a query using materialized views only, i.e., without accessing the base data. We then describe how to generate a minimal set of views as candidates for materialization.

3.1 Query/View Answerability Criteria

Recall that in Table 2, each query Q is represented as a sequence of location steps $Q = s_{Q_1} s_{Q_2} \cdots s_{Q_n}$. Here, each step s_{Q_i} has the form $a_{Q_i} l_{Q_i} [p_{Q_i}]$, where $a_{Q_i} \in \{/, //\}$ is an `Axis`, l_{Q_i} is an element name or wildcard $*$ for `NameTest`, and p_{Q_i} is a predicate that can be any XPATH fragment or empty. Figure 2 shows the three steps of query $Q : //a[@n < 1998$ or $@s =$ "str"$]/*//b[c/*//d]$. For materialization, the XML fragments that satisfy the query and rooted at the label of the last step (i.e. b) will be materialized as *deep copies* of XML fragments.

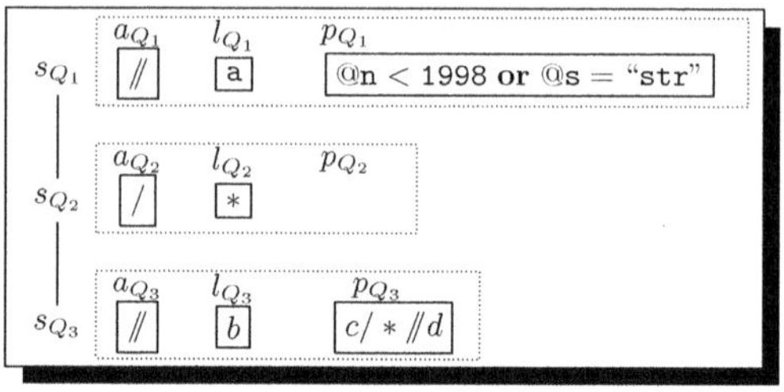

Fig. 2. Query Steps

Criteria for answerability. Given a view V and a query Q, where $V : s_{V_1} s_{V_2} \cdots s_{V_m}$ and $Q : s_{Q_1} s_{Q_2} \cdots s_{Q_n}$, V answers Q, denoted as $V \models Q$, iff the following conditions are satisfied: (1) $m \leq n$; (2) $s_{Q_i} \equiv s_{V_i}$ (equivalent) for $1 \leq i \leq m-1$; and (3) $s_{Q_m} s_{Q_{m+1}} \cdots s_{Q_n} \sqsubseteq s_{V_m}$.

Item 1 states that the number of view steps must be no more than the number of query steps. Item 2 declares that the first $m-1$ query steps must be exactly the same as corresponding view steps. Therefore, we need not access the base data to refine the materialized view fragments. Two steps are equivalent, denoted as $s_{Q_i} \equiv s_{V_i}$, iff $a_{Q_i} = a_{V_i}$ (the same axis), $l_{Q_i} = l_{V_i}$ (the same label) and $p_{Q_i} \equiv p_{V_i}$ (equivalent predicates). Though being NP-hard for testing the equivalence/containment of two predicates in theory, the predicates are typically not complicated such that it could be handled in real applications. Item 3 claims that the XPATH fragment with the form $s_{Q_m} s_{Q_{m+1}} \cdots s_{Q_n}$ is contained in the XPATH fragment s_{V_m}, which guarantees that the query result can be extracted from the materialized view result. This criteria are similar to the ones used in [16].

3.2 Constructing A Minimal Query

Given two queries Q_1 and Q_2, we say Q is a minimal query that answers Q_1 and Q_2, iff $Q \models Q_1, Q \models Q_2$, and there does not exist another query Q' where $Q' \models Q_1, Q' \models Q_2$ and $Q' \models Q$. We have the following theorem.

Theorem 1. *Given two queries P, Q, the minimal query that answers both P and Q exists and is unique.*

Proof. There exists a query $/*$ (abbreviation of /child::$*$) that answers both P and Q. The query $/*$ actually materializes the very first *root* element of an XML document, which carries the entire information of an XML document (the virtual *document root* is exclusive).

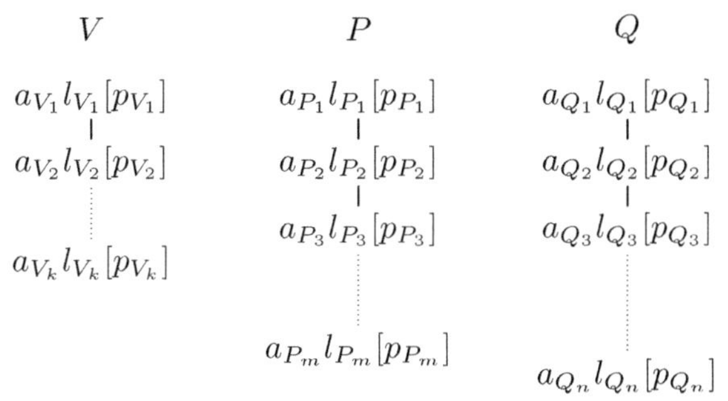

Fig. 3. Representation of Queries

Next we prove by construction that there exists a minimal query V, where $V \vDash P$, $V \vDash Q$. For any query V', if $V' \vDash P$ and $V' \vDash Q$, then $V' \vDash V$. We depict the view V and two queries P, Q in Figure 3.

Based on the criteria for query/view answerability, for any V that answers P and Q, we have the followings:

$$
\begin{aligned}
V \vDash P \Rightarrow\ & a_{V_i} = a_{P_i} && 1 \le i \le k-1; \\
& l_{V_i} = l_{P_i} && 1 \le i \le k-1; \\
& p_{V_i} \equiv p_{P_i} && 1 \le i \le k-1; \\
& a_{P_k} l_{P_k} [p_{P_k}] \cdots a_{P_m} l_{P_m} [p_{P_m}] \sqsubseteq a_{V_k} l_{V_k} [p_{V_k}] \\
V \vDash Q \Rightarrow\ & a_{V_i} = a_{Q_i} && 1 \le i \le k-1; \\
& l_{V_i} = l_{Q_i} && 1 \le i \le k-1; \\
& p_{V_i} \equiv p_{Q_i} && 1 \le i \le k-1; \\
& a_{Q_k} l_{Q_k} [p_{Q_k}] \cdots a_{Q_n} l_{Q_n} [p_{P_n}] \sqsubseteq a_{V_k} l_{V_k} [p_{V_k}]
\end{aligned}
$$

With regards to the minimal query, the number of steps should be as long as possible (e.g. $a[./b/c] \vDash a/b[./c] \vDash a/b/c$), and the predicates should be as restrictive as possible. Therefore, for the minimal query V that answers both P and Q, k should be the first position where $s_{P_k} \not\equiv s_{Q_k}$, i.e., some of the following conditions are not satisfied: $a_{P_k} = a_{Q_k}$, $l_{P_k} = l_{Q_k}$ or $p_{P_k} \equiv p_{Q_k}$.

Next step is to find a predicate that minimally contains both predicates: $a_{P_k} l_{P_k} [p_{P_k}] \cdots a_{P_m} l_{P_m} [p_{P_m}]$ and $a_{Q_k} l_{Q_k} [p_{Q_k}] \cdots a_{Q_n} l_{Q_n} [p_{Q_n}]$. Logically, the minimal predicate containing two predicates that are defined on the same schema is the union of them, i.e., using "or". Above construction is unique, and any other query V' that answers both P and Q must answer V based on the query/view answerability, which proves that V is minimal. ■

We denote by $Q_1 \circ Q_2$ the minimal query that may answer both Q_1 and Q_2. Note that we say the minimal query is unique in terms of equivalence, e.g., a/b and $a[./b]/b$ are different in syntax but always produce the same result. Therefore, $Q_1 \circ Q_2$ is a singleton (i.e. only one view instead of a set of views). Furthermore, we have the following proposition.

Proposition 1. *If a query P answers a query Q, then the minimal query that answers P and Q is P, i.e., $P \circ Q = P$ iff $P \vDash Q$.*

Computing the minimal query. We describe an algorithm to compute the operation $Q_1 \circ Q_2$. Recall that a query Q can be represented as a sequence of steps as: $Q = s_{Q_1}\ s_{Q_2} \cdots s_{Q_n}$, and each step is represented in the form: Axis NameTest Predicates? $(a_{Q_i}\ l_{Q_i}\ p_{Q_i})$. Two steps are *equivalent*, if their Axis, Name and Predicates? are equivalent, correspondingly. Algorithm 1 shows how to compute the minimal query. The correctness of this algorithm can be directly verified from the proof of Theorem 1 (by construction).

Algorithm 1 MinQuery(P, Q)

1: $P = s_{P_1}\ s_{P_2} \cdots s_{P_m}$;
2: $Q = s_{Q_1}\ s_{Q_2} \cdots s_{Q_n}$;
3: **if** $s_{P_1} \not\equiv s_{Q_1}$ **then**
4: $\quad V = /*$, or P if $Q \sqsubseteq P$, or Q if $P \sqsubseteq Q$;
5: $\quad$ **return** V;
6: **else**
7: $\quad V = s_{P_1}$;
8: **end if**
9: $k = \min(m, n)$;
10: **for** i **from** 2 **to** k
11: $\quad$ **if** $s_{P_i} \equiv s_{Q_i}$
12: $\quad\quad V+ = s_{P_i}$;
13: $\quad$ **else**
14: $\quad\quad$ **break**
15: $\quad$ **end if**
16: **end for**
17: $V+ = [s_{P_{i+1}} \cdots s_{P_m}$ or $s_{Q_{i+1}} \cdots s_{Q_n}]$;
18: **return** V

Next we illustrate why the joined view $V = /*$ if $s_{P_1} \not\equiv s_{Q_1}$ and P, Q are not contained by each other (lines 4-5). Here, we may safely omit some predicates. For example, given $P : a[./c/d/e]$ and $Q : b[./c/d/e]$, the minimal query $V = P \circ Q = /*$ while not $V' = / * [./c/d/e]$ which seems more restrictive. There are only two cases for predicates: `true` or `false`. (1) $[./c/d/e]$ is `true`, to materialize V' is equivalent to materialize V, the *root* element. (2) $[./c/d/e]$ is `false`, the result of V' is empty. We do not materialize a query with empty result. Both bases are normalized to $/*$.

We consider the following cases for minimizing the predicates. (1) Comparison predicates. The minimization of comparisons with numeric constant is straightforward. For example, $n < x$ and $n < y$ can be minimized to $n < x$ iff $x \leq y$; $n > x$ or $n > y$ can be minimized to $n > x$ iff $x \geq y$, etc. (2) Path minimization. We have P or $Q = Q$ if $P \sqsubseteq Q$; P and $Q = P$ if $P \sqsubseteq Q$. Algorithms to find the minimized query are applicable to our approach, which are omitted here due to space constraints. [12] and [1] investigate to minimize comparison predicates and minimize tree pattern queries, respectively.

3.3 Optimality of Candidate Views

In this section, we first describe the cost model, as a criterion for measuring views to answer a set of queries. We then discuss how to generate a view set as candidates, which is guaranteed to be a superset of the optimal solution and safely avoids enumerating all potential views.

We use $\text{SIZE}(\mathcal{X}, V)$ to denote the result size of applying view V over an XML database $\mathcal{X}$. When the XML database $\mathcal{X}$ is clear from the context, we use $\text{SIZE}(V)$ as a simplification. Without loss of generality, we assume that the materialized views do not have index support. Evaluating a query over a materialized view requires one scan of the materialized XML fragments. We use $\text{CARD}(V)$ to denote the number of labels in an XML view V. We have the following general cost estimation model of evaluating query Q based on view V materialized over an XML database $\mathcal{X}$:

$$\text{COST}(\mathcal{X}, V, Q) = \alpha \cdot \text{SIZE}(\mathcal{X}, V) + \beta \cdot \text{SIZE}(\mathcal{X}, V) \cdot \text{CARD}(Q) \quad (2)$$

When the view is materialized on a disk, the overhead is dominated by disk I/O, in which case $\alpha \gg \beta$. Otherwise, if the views are materialized in a semantic cache, the overhead is dominated by the computational cost, which is determined by the materialized view fragments and the query size, thus $\alpha \ll \beta$.

Naturally, we have $\text{SIZE}(Q') \geq \text{SIZE}(Q)$ if $Q' \models Q$. For two views V and V', we say V is *better* than V' if V can be used to obtain a smaller value of Equation 1. Given a single query Q and two views V and V', where $V \models Q$ and $V' \models Q$, V is better than V' in answering Q iff $\text{SIZE}(V) < \text{SIZE}(V')$. Furthermore, given a query set $\mathcal{Q}$ and two views V and V', where $V \models \mathcal{Q}$ and $V' \models \mathcal{Q}$, V is better than V' in answering $\mathcal{Q}$ iff $\text{SIZE}(V) < \text{SIZE}(V')$. Recall that $V \models \mathcal{Q}$ means that V may answer each query Q in $\mathcal{Q}$.

There are other factors that might influence the cost model, which are not considered here. These factors may include the storage model of the materialized XML fragments, and different indices that may be exploited, which may lead to a more complicated model. However, we believe that our cost model, being simple but general, enables us to investigate representative algorithms.

Finding a candidate set. Next we discuss how to generate a candidate view set that is a superset of the optimal solution. We show some properties of the minimal query first. The minimal query, computed over the operator "$\circ$", satisfies the following identities:

$$\begin{array}{ll} \text{(L1) } P \circ Q = Q \circ P & \text{(commutative law)} \\ \text{(L2) } Q \circ Q = Q & \text{(idempotent law)} \\ \text{(L3) } (P \circ Q) \circ V = P \circ (Q \circ V) & \text{(associative law)} \end{array}$$

The first two equivalences can be verified directly through Algorithm 1. We simply illustrate L3 next. Based on Algorithm 1, assume that P, Q, V have k, m, n steps, respectively, and the ith step is the first step that s_{P_i}, s_{Q_i} and s_{V_i} are not equivalent. Both $(P \circ Q) \circ V$ and $P \circ (Q \circ V)$ are equivalent to the form: $s_{P_1} \cdots s_{P_{i-1}} [s_{P_i} \cdots s_{P_k} \texttt{ or } s_{Q_i} \cdots s_{Q_m} \texttt{ or } s_{V_i} \cdots s_{V_n}]$ such that identity L3 holds. Thus, the computation of the minimal view of a query set is order independent.

Given a workload $\mathcal{Q} = \{Q_1, Q_2, \cdots, Q_n\}$, we first aggregate them pairwise to get $\binom{n}{2}$ views, which are obtained by computing $Q_1 \circ Q_2$, $Q_1 \circ Q_3$, etc. We then can aggregate $Q_1 \circ Q_2$ and $Q_1 \circ Q_3$ to get $Q_1 \circ Q_2 \circ Q_3$, and so on. We get $\mathcal{V} = \{Q_1, \cdots, Q_n, Q_1 \circ Q_2, Q_1 \circ Q_3, \cdots, Q_{n-1} \circ Q_n, \cdots, Q_1 \circ Q_2 \circ \cdots \circ Q_n\}$. There are $O(2^n)$ candidate views generated, which is deduced by $\binom{n}{1} + \binom{n}{2} + \cdots + \binom{n}{n} = 2^n - 1$. Although the worst case is still exponential, the size of candidate views is much smaller than all potential views. Furthermore, we will introduce a simple bound to significantly reduce the number of generated views.

Next we illustrate why we can safely ignore any view $V' \in \mathcal{C}$ ($\mathcal{C}$ is all views that can be generated) if $V' \notin \mathcal{V}$, while still ensuring optimality. This question comes from the fact that, for two queries Q_1, Q_2 and $V = Q_1 \circ Q_2$, there may exist another view V', where $V' \models V$ which is not considered for materialization. Here, V' can be potentially used to answer some query Q_3 but is not generated.

We prove this by showing that there exists another candidate view $V'' \in \mathcal{V}$ that is better than V'. Assume that $\mathcal{Q}'$ is a subset of $\mathcal{Q}$ that V' can answer each query in $\mathcal{Q}'$, i.e., $\mathcal{Q}' = \{Q | Q \in \mathcal{V} \wedge V' \models Q\}$. V' is not in $\mathcal{V}$, therefore, V' is not the minimal view to answer $\mathcal{Q}'$. Suppose that the minimal view to answer $\mathcal{Q}'$ is V''. Naturally, we have $V' \models V''$ and V'' is better than V' for the given workload $\mathcal{Q}$ to minimize Equation 1.

4 Candidate View Organization

View organization. We organize the candidate views $\mathcal{V}$ as a directed acyclic graph (named VCUBE), which is denoted as $\mathcal{G}(N, E)$. Each node $u \in N(\mathcal{G})$ represents a view u_V in $\mathcal{V}$. Each node u has a level, denoted as LEVEL(u), which is the number of queries in the original workload $\mathcal{Q}$ that are used to generate the view u_V. There is an edge $(u, v) \in E(\mathcal{G})$, iff LEVEL$(u)$ = LEVEL$(v) + 1$ and v_V can be used to generate u_V, i.e., the query set used to generate v_V is a subset of the query set used to generate u_V.

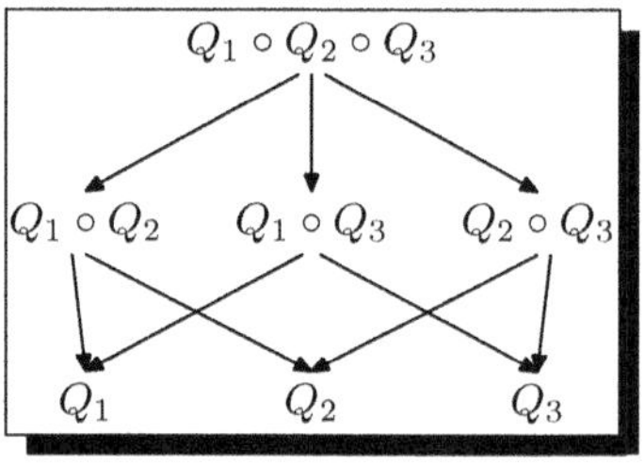

Fig. 4. Sample VCUBE

Example. Given a workload $\mathcal{Q}_e = \{Q_1, Q_2, Q_3\}$, the level of $Q_1 \circ Q_2$ is 2, i.e., LEVEL$(Q_1 \circ Q_2) = 2$. There is an edge from $Q_1 \circ Q_2 \circ Q_3$ to $Q_1 \circ Q_2$, since LEVEL$(Q_1 \circ Q_2 \circ Q_3) = 3$, LEVEL$(Q_1 \circ Q_2) = 2$ and $\{Q_1, Q_2\}$ is a subset of $\{Q_1, Q_2, Q_3\}$. The VCUBE for workload $\mathcal{Q}_e$ is shown in Figure 4.

It deserves noting that we only maintain edges between adjacent levels of views e.g. no edge from $Q_1 \circ Q_2 \circ Q_3$ to Q_1, although the former answers the latter. The nice property is that, the answerability is traced not only by direct edges, but also by reachability relationships. Based on VCUBE, our problem is reduced to finding a minimum weight set of nodes that covers all leaves and minimizes Equation 1. We say VCUBE guarantees optimality since the optimal view set is subsumed in all candidates and answerability is verifiable. Next we illustrate why we could safely avoid checking the answerability between two unrelated views e.g. $Q_1 \circ Q_2$ and Q_3.

In Figure 4, there are two edges from $Q_1 \circ Q_2$, to Q_1 and Q_2, respectively. Assume that $Q_1 \circ Q_2$ can also answer Q_3 but there is no edge from $Q_1 \circ Q_2$ to Q_3, whether we would be underestimating $Q_1 \circ Q_2$, since we consider it to answer Q_1 and Q_2 only. According to Proposition 1, $Q_1 \circ Q_2 \circ Q_3 = Q_1 \circ Q_2$ if $Q_1 \circ Q_2 \models Q_3$, and there should have existed a path from $Q_1 \circ Q_2 \circ Q_3$ to Q_3. Therefore, the optimal solution can always be identified. More specifically, assume that there are two nodes $u, v \in V(\mathcal{G})$ where $u_V \models v_V$ but there is no path from u to v. The nearest common ancestor (node w) of u and v satisfies $w_V = u_V \circ v_V = u_V$ since $u_V \models v_V$. The path from u to v could thus be omitted.

The benefits of edge construction in VCUBE are twofold. (1) The edges can be directly constructed when computing the minimal views, without checking the answerability between views, which is expensive. (2) The number of edges maintained is much smaller than the edges computed depending on answerability. This can reduce both the graph size (the number of edges) and the overhead of searching optimal solution over it.

View graph optimization. We describe to optimize the view graph by safely omitting some views to be generated. We introduce a special view as an upper bound. The special view is $\Delta : /*$, whose result is just the *root* element of $\mathcal{X}$ and would never be materialized. For any query Q, we have $Q \circ \Delta = \Delta$. Therefore, if some node in the bottom-up construction is Δ, we do not generate any other nodes that may reach it, which can greatly reduce the candidates to be generated. Recall that in Equation 1, we have a size constraint B. Therefore, if some view V has the estimated materialized size larger than B, we do not materialized V and can safely set V to Δ.

5 Materialized View Selection Algorithms

Finding the optimal solution for Equation 1 is NP-hard, which can be reduced to the set cover problem[1]. In this section, we describe heuristic methods to identify approximate solutions.

Estimating view size. View selection algorithms require knowledge of the size of each view. There are many ways of estimating the sizes of views [19,23]. The approach we adopt is sampling, i.e., running views on a representative portion of the base data and multiplied by a scaling factor.

We have to consider space-time tradeoffs in view selection. In the space-optimized perspective, we would like to select the views with the smallest size to answer as many queries as possible. In the time-optimized perspective, we will materialize the query workload only, which can answer any given query directly but with a large size. We will first discuss an algorithm targeting space-optimized, followed by an algorithm for space-time consideration.

5.1 Space-Optimized Algorithm

The goal of space-optimized is to material the smallest XML fragments to answer a given query workload. We adopt a bottom-up, level-by-level dynamic programming strategy over the VCUBE.

Given a candidate view set $\mathcal{V}$ generated from a workload $\mathcal{Q}$, its VCUBE $\mathcal{G}(N, E)$ and a node $u \in N(\mathcal{G})$, we use $\textsc{views}(u)$ to represent the set of queries that generate the view u_V e.g. $\textsc{views}(Q_1 \circ Q_2) = \{Q_1, Q_2\}$. Conversely, for a subset of queries $\mathcal{Q}'$, we denote by $\textsc{genv}(\mathcal{Q}')$ the views generated by aggregating each query in $\mathcal{Q}'$ e.g. $\textsc{genv}(Q_1, Q_2) = Q_1 \circ Q_2$. Our recursive definition of the minimum cost of computing $\textsc{size}(u_V)$ for each node $u \in N(\mathcal{G})$ is as follows:

$$\textsc{size}(u_V) = \textsc{min}\{\textsc{size}(\textsc{genv}(\textsc{views}(u_V) - \textsc{views}(v_V))) + \textsc{size}(v_V), \textsc{size}(u_V)\} \quad (3)$$

here, node $v \in N(\mathcal{G})$ is any graph node that is reachable from u. For instance, when computing the size of $Q_1 \circ Q_2 \circ Q_3$, we compare its size with the sum of sizes of each combination in $(Q_1,\ Q_2 \circ Q_3)$, $(Q_2,\ Q_1 \circ Q_3)$ and $(Q_3,\ Q_1 \circ Q_2)$, and record the smallest one.

[1] http://en.wikipedia.org/wiki/Set_cover_problem

Algorithm 2 SPACEOPTIMAL($\mathcal{G}$)

1: **for** i **from** 1 **to** $|N(\mathcal{G})|$ **do**
2: estimate SIZE(n_i)
3: **end for**
4: **for** i **from** $|\mathcal{Q}|+1$ **to** $|N(\mathcal{G})|$ **do**
5: **for** j **from** 1 **to** $|N(\mathcal{G})|$ **do**
6: **if** GENV(VIEWS(n_i) $\cup$ VIEWS(n_j)) doesn't exist **or** (VIEWS(n_k) $=$ VIEWS(n_i) $\cup$ VIEWS(n_j) **and** SIZE(n_k) $>$ SIZE(n_i) $+$ SIZE(n_j))
7: SIZE(n_k) $\leftarrow$ $size(n_i) + size(n_j)$
8: **end if**
9: **end for**
10: **end for**

The intuition of Equation 3 is to compute the smallest size for each graph node from all the views that it might answer. Algorithm 2 shows a dynamic programming based approach with time complexity $O(n^2)$ where n is the number of graph nodes. Here, we use n_i to represent the i-th node in the bottom-up, level-by-level traversal mode. Initially, each view has an estimated size (lines 1-3). In each iteration (leaf views could be skipped in line 4), we compute the size of one view V by counting all the views that are answerable by V, but marking the smallest size only. Note that a view might not exist in graph originally, if its size exceeds a given upper bound, while this view will be used (not materialized) in selecting views in our dynamic program. After computing the last view (the graph root), we could find the smallest XML fragments to materialize to answer the given query workload $\mathcal{Q}$. We illustrate this algorithm by an example next.

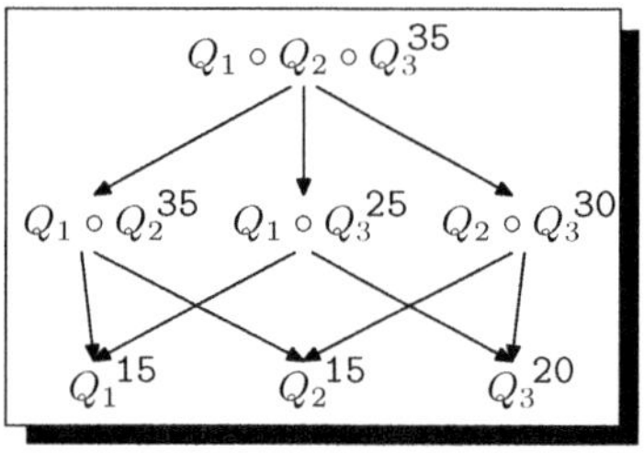

Fig. 5. VCUBE with Sizes

Example. Consider the VCUBE in Figure 5, each view is associated with an estimated size. Assume that the size constraint is $B = 40$. We start with an initial view set $\mathcal{V} = \{Q_1, Q_2, Q_3\}$. Consider $Q_1 \circ Q_2$, which is omitted since SIZE($Q_1 \circ Q_2$) $= 35 >$ SIZE(Q_1)$+$SIZE(Q_2) $= 30$. However, $Q_1 \circ Q_3$ and $Q_2 \circ Q_3$ are considered to be materialized since SIZE($Q_1 \circ Q_3$) $<$ SIZE(Q_1)$+$SIZE(Q_3) and SIZE($Q_2 \circ Q_3$) $<$ SIZE(Q_2)$+$SIZE(Q_3). To answer all queries $\mathcal{Q}$, materializing $Q_1 \circ Q_3$ and Q_2 requires the size 40, while for $Q_2 \circ Q_3$ and Q_1, it requires $45 > 40$. Therefore, we replace Q_1, Q_3 in $\mathcal{V}$ by $Q_1 \circ Q_3$ and $\mathcal{V} = \{Q_1 \circ Q_3, Q_2\}$.

Next we consider $Q_1 \circ Q_2 \circ Q_3$, whose estimated size is less than the summation of $Q_1 \circ Q_3$ and Q_2, i.e., SIZE($Q_1 \circ Q_2 \circ Q_3$) $= 35 <$ SIZE($Q_1 \circ Q_3$)$+$SIZE(Q_2) $= 40$. Furthermore, SIZE($Q_1 \circ Q_2 \circ Q_3$) $= 35 < B = 40$, thus we have $\mathcal{V} = \{Q_1 \circ Q_2 \circ Q_3\}$.

5.2 Space-Time Algorithm: A Greedy Solution

In above example, the space-optimized approach only considers to use the smallest sized views to answer the given workload. However, materializing $Q_1 \circ Q_3$ and Q_2, whose total size is 40, requires a larger size but may have a lower query execution time.

Next we consider a space-time optimized approach. We define a utility function on the view graph, for each candidate view V and a workload $\mathcal{Q}$, where each query $Q_i \in \mathcal{Q}$ is associated with a weight w_i.

$$\text{UTIL}(V) = \frac{\Sigma_{V \models Q_i} w_i}{\text{SIZE}(V)} \tag{4}$$

here, we compute the utility of a view V by considering how it can be used to improve the query response time. The utility value is in inverse proportion to the size of view, which means that the smaller the materialized size of a view, the larger utility value it has. The utility value is in direct proportion to the summated weights of queries the view may answer, which means that the more queries the view can answer, the larger the utility value will be. The weight of a view could be query independent as query frequency or query dependant like *full-resp-time/view-resp-time*.

Algorithm description. We simply describe the algorithm of heuristically selecting a set of views. (1) Compute the utility value of each view in the view graph, and select the one with the largest utility value. (2) Remove the selected view and all queries it may answer. Recompute the utility values of remaining views. (3) Repeat this procedure until all queries can be answered or the total size exceeds the size constraint B. This greedy strategy is similar to the one used in [11] for computing data cube.

6 Performance Study

We report on empirical results in two orientations. Firstly, we measure the two algorithms for selecting materialized views. Secondly, we study the performance gain using materialized views, against the evaluation over the base data (BD for short). For simplicity, we represent our space-optimized algorithm as SO, and the greedy strategy as GR.

The experiments were conducted on a PC 1.6GHz machine with 512MB of memory, running Windows XP. The algorithms were coded in C++. We used Berkeley DB XML[2] for managing XML data with basic index support and storing views. We used XMark as our test benchmark, with 16 queries (not listed for space consideration). Q_{1-4} are in the form /site/people/person[@id="person#"]//? where person# represents a person id (e.g. person10) and ? a NameTest (e.g., name); queries Q_{5-8} are in the form /site/regions//item[@id="item#"]//?; queries Q_{9-12} are in the form /site/closed_auctions/closed_auction//?; and queries Q_{13-16} are in the form //open_auctions//open_auction[Expr1][Expr2]//? where Expr1 is a path predicate.

6.1 Selecting Views in Practice

We mainly measure the effect of size parameters in two selection algorithms. The first group of experiments is to fix the size limitation for each single view large enough (e.g. the document size), while varying the total size upper bound (B). Figure 6 shows the number the views selected when varying the upper bound B for different documents. Here in x-axis, 0.25 means 0.25MB and 1 for 1MB, etc. Both sub-figures show that, along with the enlarging size constraint, the number

[2] http://www.oracle.com/database/berkeley-db/xml/index.html

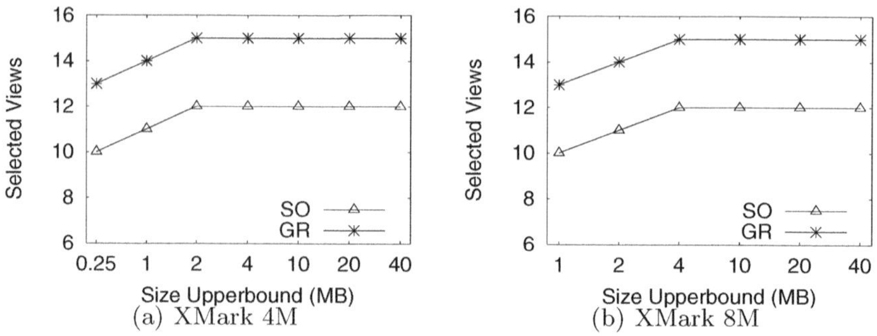

Fig. 6. Varying Size Upperbound B

of views for both algorithms increase. For example, for 4MB document, the SO selects 10 to 12 views and GR picks 13 to 15 views. When the constraint B exceeds some threshold (e.g. 2MB for 4MB document), the size constraint cannot affect the result of view selection. This group of experiments also tells that we could use a relative small size constraint, e.g., half of document size for XMark, to get the optimal solution. The benefit to select a small size bound is that, answering queries over base data could be accelerated by underlying indices. This could be faster than a materialized view without index support.

We also examine how the size limit of each view (parameter b) affects the size of VCUBE in terms of number of graph nodes. Recall that in a VCUBE, if the size of some view exceeds b, we don't generate this graph node. Theoretically, the VCUBE size of a workload of 16 queries is $2^{16} = 65536$. Take an Xmark 4MB document and fixed B =1M. If we set b to be 4MB, there are 24 nodes in VCUBE. When we change b to 1MB, there are 16 nodes generated. The cause of this result comes from two facts: (1) the views that might answer queries in different groups will have a large size and thus are not materialized; (2) the number of views materialized for the same group of queries decreases when we restrict b.

Due to the small size of VCUBE, the costs for both selection strategies are surprisedly small. Note that the size of each view is pre-estimated using a small sample. In our case, we use an XMark 1MB document and estimate with a scaling factor (e.g. 4 for an XMark 4MB document). Therefore, the selection is only affected by the number of graph nodes but not the document size. It takes 10ms for SO strategy and 50ms for GR strategy, and this basically keeps the same for different sized XMark documents.

In the worst case, the estimated view size via aggregating two queries exceeds b. This case is reflected in above tests where no views generated for two groups of queries. However, the query might be materialized if its size is below b.

6.2 Answering Queries Using Materialized Views

We compare the response time of SO, GR and BD (without using views), on top of Berkeley DB XML. We test XMark documents for different factors, from 1, 2 to 32. The weights of queries are all set to 1 as a normalized frequency in

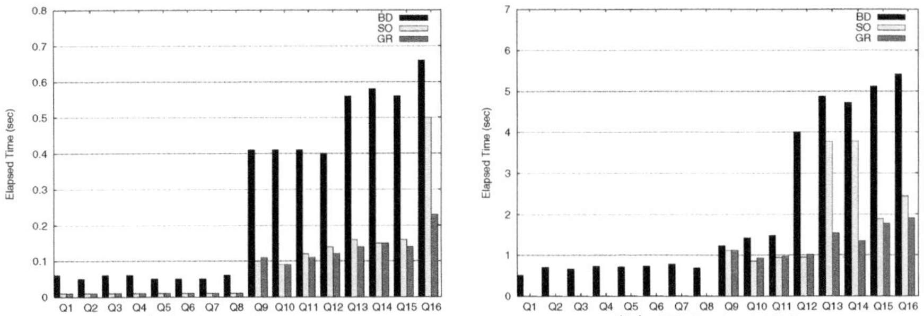

(a) XMark 4M (b) XMark 32M

Fig. 7. Answering Queries using Views

equivalent weights. All tests give similar results, so we only report two documents in Figure 7, where x-axis carries all test queries and y-axis its response time in millisecond. This group of experiments verifies two important goals as expected: (1) Answering queries using materialized views is much faster than evaluating over the base data, even if the base data is well indexed. The reason is simple, as the materialized views have smaller size and many structural joins of queries have been pre-evaluated by views. (2) GR outperforms SO in terms of response time. This comes from the different aims of algorithm designs. SO aims at an optimized space consumption, while GR balances the space overhead and cost estimation model for query processing, which is simple yet general and effective.

7 Conclusion and Future Work

We have described a framework to select XML views for materialization. For a given workload, we present a new way to generate a small number of candidates while ensuring optimality. Based on a well organized graph structure (VCUBE) for maintaining a minimal set of candidate views, we present two heuristic algorithms for view selection. We experimentally demonstrate that query response time could be significant reduced using materialized views. Moreover, the VCUBE gives full possibilities to develop other algorithms with different aims and for further optimization.

In the future: (*i*) We plan to further develop algorithms over VCUBE, for both better selected views and faster computation; (*ii*) In this paper, we store materialized views using a disk-based database. We would like to cache views using main-memory databases, to examine the advantage of selecting materialized views in different environments. (*iii*) We want to implement view selection as a transparent optimization strategy, which is self-tuning and works off-line by analyzing query logs. (*iv*) Incremental (or lazy) materialized view maintenance is an interesting topic, compared with re-materialization each time for changed base data and query logs. (*v*) We also plan to investigate, as a relaxed version, the problem of view selection that allows to access the base data.

References

1. Amer-Yahia, S., Cho, S., Lakshmanan, L.V. S., Srivastava, D.: Minimization of tree pattern queries. In: SIGMOD (2001)
2. Amer-Yahia, S., Koudas, N., Marian, A., Srivastava, D., Toman, D.: Structure and content scoring for XML. In: VLDB (2005)
3. Amer-Yahia, S., Lakshmanan, L.V. S., Pandit, S.: FleXPath: Flexible structure and full-text querying for XML. In: SIGMOD (2004)
4. Balmin, A., Özcan, F., Beyer, K.S., Cochrane, R., Pirahesh, H.: A framework for using materialized XPath views in XML query processing. In: VLDB (2004)
5. Baralis, E., Paraboschi, S., Teniente, E.: Materialized views selection in a multidimensional database. In: VLDB (1997)
6. Chirkova, R., Halevy, A.Y., Suciu, D.: A formal perspective on the view selection problem. In: VLDB (2001)
7. Fan, W., Geerts, F., Jia, X., Kementsietsidis, A.: Rewriting regular xpath queries on xml views. In: ICDE (2007)
8. Gribble, S.D., Halevy, A.Y., Ives, Z.G., Rodrig, M., Suciu, D.: What can database do for peer-to-peer? In: WebDB (2001)
9. Gupta, H.: Selection of views to materialize in a data warehouse. In: ICDT (1997)
10. Gupta, H., Mumick, I.S.: Selection of views to materialize under a maintenance cost constraint. In: ICDT (1999)
11. Harinarayan, V., Rajaraman, A., Ullman, J.D.: Implementing data cubes efficiently. In: SIGMOD Conference (1996)
12. Hopcroft, J.E., Ullman, J.D.: Set merging algorithms. SIAM J. Comput. 2(4) (1973)
13. Karloff, H.J., Mihail, M.: On the complexity of the view-selection problem. In: PODS (1999)
14. Kossmann, D.: The state of the art in distributed query processing. ACM Comput. Surv. 32(4) (2000)
15. Lakshmanan, L.V. S., Wang, H., Zhao, Z.J.: Answering tree pattern queries using views. In: VLDB (2006)
16. Mandhani, B., Suciu, D.: Query caching and view selection for XML databases. In: VLDB (2005)
17. Miklau, G., Suciu, D.: Containment and equivalence for an XPath fragment. In: PODS (2002)
18. Mumick, I.S., Quass, D., Mumick, B.S.: Maintenance of data cubes and summary tables in a warehouse. In: SIGMOD Conference (1997)
19. Polyzotis, N., Garofalakis, M.N., Ioannidis, Y.E.: Selectivity estimation for xml twigs. In: ICDE (2004)
20. Sawires, A., Tatemura, J., Po, O., Agrawal, D., Abbadi, A.E., Candan, K.S.: Maintaining XPath views in loosely coupled systems. In: VLDB (2006)
21. Sawires, A., Tatemura, J., Po, O., Agrawal, D., Candan, K.S.: Incremental maintenance of path expression views. In: SIGMOD Conference (2005)
22. Tang, N., Yu, J.X., Özsu, M.T., Choi, B., Wong, K.-F.: Multiple materialized view selection for XPath query rewriting. In: ICDE (2008)
23. Wang, W., Jiang, H., Lu, H., Yu, J.X.: Bloom histogram: Path selectivity estimation for xml data with updates. In: VLDB (2004)
24. Xu, W., Özsoyoglu, Z.M.: Rewriting XPath queries using materialized views. In: VLDB (2005)
25. Yang, J., Karlapalem, K., Li, Q.: Algorithms for materialized view design in data warehousing environment. In: VLDB (1997)

Implementing and Optimizing Fine-Granular Lock Management for XML Document Trees

Sebastian Bächle[1], Theo Härder[1], and Michael P. Haustein[2]

[1] Department of Computer Science,
University of Kaiserslautern, Germany
{baechle,haerder}@informatik.uni-kl.de
[2] SAP AG
Hopp-Allee 16, 69190 Walldorf, Germany
haustein@h-ms.de

Abstract. Fine-grained lock protocols with lock modes and lock granules adjusted to the various XML processing models, allow for highly concurrent transaction processing on XML trees, but require locking facilities that efficiently support large and deep hierarchies with varying fan-out characteristics. We discuss these and also further requirements like prefix-based node labels, and present a lock management design that fulfills all these requirements and allows us to perfectly exploit the advantages of our tailor-made lock protocols for XML trees. Our design also supports the flexible use of heuristics for dynamic lock escalation to enhance workload adaptivity. Benchmark runs convincingly illustrate flexibility and performance benefits of this approach and reveal that careful lock protocol optimization pays off.

1 Motivation

Native XML database systems (XDBMS) promise tailored processing of XML documents, but most of the systems published in the DB literature are designed for efficient document retrieval [13,19]. This "retrieval-only" focus was probably caused by the first proposals of XQuery respectively XPath where the update part was left out [21]. With the advent of the update extension [22] and its support in commercial systems, however, the requirement for concurrent and transaction-safe document modifications became widely accepted.

Currently, all vendors of XML(-enabled) database management systems support updates only at document granularity. Although this approach reduces some of the complexity of updating XML, it is only feasible if the database consists of collections of small documents. Efficient and effective transaction-protected collaboration on XML documents, however, becomes a pressing issue for medium- to large-sized documents, especially, in the face of a growing number of use cases for XML as the central data representation format in business environments. Therefore, our goal is to provide a suitable solution for highly concurrent transaction processing on XML documents that fulfills all requirements of an XDBMS in terms of applicability and flexibility.

X. Zhou et al. (Eds.): DASFAA 2009, LNCS 5463, pp. 631–645, 2009.

In recent years, a great variety of XML storage models, indexing approaches and sophisticated query processing algorithms have been proposed. Unfortunately, they all make different assumptions about the availability of specific data access operators, or even do not touch the problem of an embedding in a real system environment at all. This diversity in conjunction with the various language models and programming interfaces for XML is the most crucial problem in XML concurrency control. All those different ways of accessing the same pieces of data makes it impossible to guarantee that only serializable schedules of operations occur. Hence, we strive for a general solution that is even applicable for a whole spectrum of XML language models (e.g., XPath, XQuery, SAX, or DOM) in a multi-lingual XDBMS environment.

Because of the superiority of locking in other areas, we also focus on lock protocols for XML. We have already developed a family consisting of four DOM-based lock protocols called the taDOM group by adjusting the idea of multi-granularity locking [8] to the specific needs of XML trees. Here, we discuss mechanisms how such protocols can be efficiently implemented in a native XDBMS or any other system that requires concurrent access to XML documents.

In Sect. 2, we emphasize the need and advantage of lock protocols tailored to the specific characteristics of XML processing, sketching the properties of our taDOM protocols, and discuss the role of prefix-based node labeling for efficient lock management. Sect. 3 gives an overview of related work. In Sect. 4, we demonstrate how to embed XML lock protocols in a system environment and describe some implementation details of a lock manager that simplifies the use of a hierarchical lock protocol for large and deep document trees. In Sect. 5, we introduce an approach to dynamically balance benefit and lock overhead, outline the problem of conversion deadlocks, and demonstrate the feasibility of our approach with experimental results in Sect. 6. Finally, Sect. 7 concludes the paper and gives an outlook on future work.

2 Preliminaries

XML lock protocols aim to enable XDBMS concurrent read and write accesses of different transactions to the same documents, and thus increasing the overall performance of the system. Hence, regarding the tree structure of XML, the protocols have to synchronize read and write operations inside of a tree structure at different levels and in different granularities.

Hierarchical lock protocols [9]—also denoted as multi-granularity locking—were designed for hierarchies of objects like tables and tuples and are used "everywhere" in the relational world. The allow for fine-grained access by setting R (read) or X (exclusive) locks on objects at the lower levels in the hierarchy and coarse grained access by setting the locks at higher levels in the hierarchy, implicitly locking the whole subtree of objects at smaller granules. To avoid lock conflicts when objects at different levels are locked, so-called intention locks with modes IR (intention read) or IX (intention exclusive) have to be acquired along the path from the root to the object to be isolated and vice versa when the locks are released [9].

Although the MGL protocol can also be applied to XML document trees, it is in most cases too strict, because both R and X mode on a node, would always lock the whole subtree below, too. While this is the desired semantics for part-of object hierarchies as in relational databases, these restrictions do not apply to XML where transactions must not necessarily be guaranteed to have no writers in the subtree of their current node. Hence, MGL does not provide the degrees of concurrency, that could be achieved on XML documents.

In the following, we will give a brief introduction into our TaDOM lock protocols, which refine the ideas of the MGL approach and provide tailored lock modes for high concurrency in XML trees.

2.1 TaDOM Protocol Family

To develop true DOM-based XML lock protocols, we introduced a far richer set of locking concepts, beyond simple intention locks and, in our terms, subtree locks. We differentiate read and write operations thereby renaming the well-known (IR, R) and (IX, X) lock modes with (IR, SR) and (IX, SX) modes, respectively. We introduced new lock modes for single nodes called NR (node read) and NX (node exclusive), and levels of siblings called LR (level read). As in the MGL scheme, the U mode (SU in our protocol) plays a special role, because it permits lock conversion. The novelty of the NR and LR modes is that they allow, in contrast to MGL, to read-lock only a node or all nodes at a level (under the same parent), but not the corresponding subtrees.

The LR mode required further a new intention mode CX (child exclusive). It indicates the existence of an SX or NX lock on some direct child nodes and prohibits inconsistent locking states by preventing LR and SR locks. It does not prohibit other CX locks on a context node c, because separate child nodes of c may be exclusively locked by other transactions (compatibility is then decided on the child nodes themselves). Alltogether these new lock modes enable serializable transaction schedules with read operations on inner tree nodes, while concurrent updates may occur in their subtrees.

For phantom protection, edge locks are used as a secondary type of locks. They have only three different modes (read, update, and exclusive) and are requested for the so-called virtual navigation edges of elements (previous/next sibling, first/last child) and text nodes (previous/next sibling). Transactions have to request shared edge locks when they navigate "over" such an edge to another node, and exclusive edge locks when they want to insert new child/sibling nodes in the tree. This mechanism signals readers node deletions of uncommitted transactions and hinders writers from inserting nodes at positions where they would appear as phantoms for others.[1]

Continuous improvement of this basic concepts lead to a whole family of lock protocols, the taDOM family, and finally ended in a protocol called taDOM3+,

[1] Although edge locks are an integral part of taDOM, they are not in the focus of this work and will not be further regarded due to space restrictions.

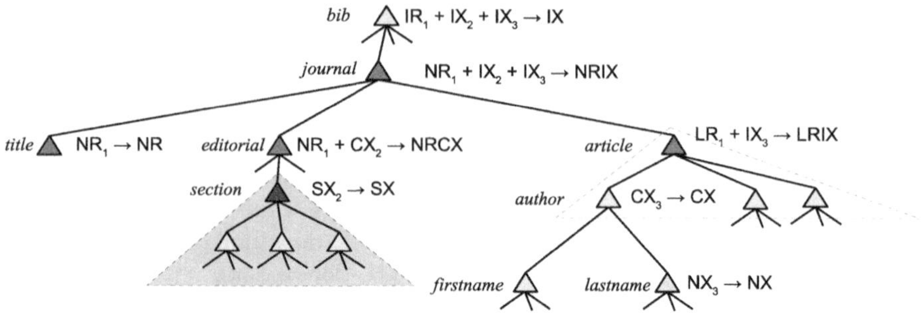

Fig. 1. Example of the taDOM3+ protocol

which consists of 20 different lock modes and allows highest degrees of parallelism on XML document trees. Correctness and, especially, serializability of the taDOM protocol family was shown in [11,20].

To illustrate the general principles of these protocols, let us assume that transaction *T1* navigates from the context node *journal* in Fig. 1 to the first child *title* and proceeds to the *editoral* sibling. This requires *T1* to request NR locks for all visited nodes and IR locks for all nodes on the ancestor path from root to leaf. Then, *T1* navigates to the first *article* to read all child nodes and locks the *article* node and all children at once with the perfectly fitting mode LR. Then, transaction *T2* deletes a *section* from the *editorial*, and acquires an SX lock for the corresponding node, CX for the *editorial* parent and IX locks for all further ancestors. Simultaneously, transaction *T3* is able to update the *firstname* node, because the LR lock of *T1* is compatible with the required IX intention modes.

2.2 Node Labeling

XML operations often address nodes somewhere in subtrees of a document and these often require direct jumps "out of the blue" to a particular inner tree node. Efficient processing of all kinds of language models [6,21] implies such label-guided jumps, because scan-based search should be avoided for direct node access and navigational node-oriented evaluation (e.g., getElementById() or getNextSibling()) as well as for set-oriented evaluation of declarative requests (e.g., via indexes).

Because each operation on a context node requires the appropriate isolation of its path to the root, not only the node itself has to be locked in a sufficient mode, but also the corresponding intention locks on all ancestor nodes have to be acquired. Therefore, the lock manager often has to procure the labels for nodes and their contexts (e.g., ancestor paths) requested. No matter what labeling scheme is used, document access cannot always be avoided (e.g., getNextSibling()). If label detection or identification, however, mostly need access to the document (usually stored on disk), a dramatic cost factor may burden concurrency control. Therefore, the node labeling scheme used may critically influence lock management overhead [10].

In the literature, range-based and prefix-based node labeling [4] are considered the prime competitive methods for implementation in XDBMSs. A comparison and evaluation of those schemes in [10] recommends prefix-based node labeling based on the Dewey Decimal Classification [5]. As a property of Dewey order encoding, each label represents the path from the document's root to the related node and the local order w.r.t. the parent node; in addition, sparse numbering facilitates node insertions and deletions. Refining this idea, a number of similar labeling schemes were proposed differing in some aspects such as overflow technique for dynamically inserted nodes, attribute node labeling, or encoding mechanism. Examples of these schemes are ORDPATHs [14], DeweyIDs [10], or DLNs [2]. Because all of them are adequate and equivalent for our processing tasks, we prefer to use the substitutional name stable path labeling identifiers (SPLIDs) for them.

Here, we can only summarize the benefits of the SPLID concept; for details, see [10,14]. Existing SPLIDs are immutable, that is, they allow the assignment of new IDs without the need to reorganize the IDs of nodes present – an important property for stable lock management. As opposed to competing schemes, SPLIDs greatly support lock placement in trees, e.g., for intention locking, because they carry the node labels of all ancestors. Hence, access to the document is not needed to determine the path from a context node to the document root. Furthermore, comparison of two SPLIDs allows ordering of the related nodes in document order and computation of all XPath axes without accessing the document, i.e., this concept provides holistic query evaluation support which is important for lock management, too.

3 Related Work

To the best of our knowledge, we are not aware of contributions in the open literature dealing with XML locking in the detail and completeness presented here. So far, most publications just sketch ideas of specific problem aspects and are less compelling and of limited expressiveness, because they are not implemented and, hence, cannot provide empirical performance results. As our taDOM protocols, four lock protocols developed in the Natix context [12] focus on DOM operations and acquire appropriate locks for document nodes to be visited. In contrast to our approach, however, they lack support for direct jumps to inner document nodes as well as effective escalation mechanisms for large documents. Furthermore, only a few high-level simulation results are reported which indicate that they are not competitive to the taDOM throughput performance.

DGLOCK [7], proposed by Grabs et al., is a coarse-grained lock protocol for a subset of XPath that locks the nodes of a structural summary of the document instead of the document nodes themselves. These "semantic locks" allow to cover large parts of a document with relatively few locks, but require annotated content predicates to achieve satisfying concurrency. Hence, even if only simple predicates are used, a compatibility check of a lock request may require physical access to all document nodes affected by a lock, respectively by its predicate. Furthermore, the protocol does not support the important descendant axis.

XDGL [15] works in a similar way, but provides higher concurrency due to a richer set of lock modes, and introduces logical locks to support also the descendant axis. The general problem of locks with annotated predicates, however, remains unsolved. SXDGL [16] is snapshot-based enhancement of XDGL that uses additional lock modes to capture also the semantics of XQuery/XUpdate. It employs a multi-version mechanism for read-only transactions to deliver a snapshot-consistent view of the document without requesting any locks.

OptiX and SnaX [17] are two akin approaches, which make also extensive use of a multi-version architecture. OptiX is the only optimistic concurrency control approach adapted to the characteristics of XML so far. SnaX is a variant of OptiX that relaxes serializability and only guarantees snapshot consistency for readers.

4 Infrastructure

As shown in Sect. 2 the taDOM protocols are applied only on the level of logical operations on prefix-labeled trees. In a real system environment, however, different data access methods might be used for performance reasons or because the underlying physical representation of a document can only support a subset of these operations.

The rationale of our approach is to map all types of data access requests—at least logically—to one or a series of operations of a DOM-like access model. Hence, our approach can be used in many different system environments, as long as they natively support or at least additionally provide a prefix-based node addressing. Note again, that this is a crucial requirement for XML lock protocols, but anyway also worthwhile for many processing algorithms and indexing methods. For the mapped operation primitives, we can apply the lock protocol independently of the actual access model or physical data representation. So, we can profit from both, the performance of fast, native data access operations as well as the concurrency benefits gained from the use of the fine-grained taDOM protocols.

Database management systems typically follow the principle of a layered architecture to encapsulate the logic for disk and buffer management, physical and logical data representation, query evaluation and so on. Except from cross-cutting system services like the transaction management, monitoring or error handling, it is relatively easy to exchange such a layer as long as it provides the same functionality to the next higher system layer. In our approach, we use this property to introduce a new layer in this hierarchy, which realizes the mapping of data access operations to logical taDOM operations and back to physical data access operations. Fig. 2 shows the two core parts of our design—a stack of three thin layers, placed between the physical representation and the higher engine services, and the lock manager implementation itself.

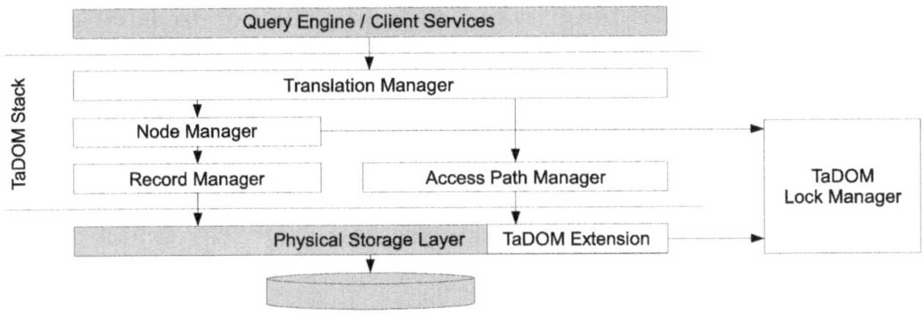

Fig. 2. Embedding of the lock protocols in a system infrastructure

4.1 Data Access Layer

The taDOM stack provides the query engine or the client transparently with a transaction consistent, "single-user view" on the data. It is located on top of the physical access layer, which provides access primitives for the underlying storage structures. Hence, the stack replaces the classic, monolithic logical access layer, responsible for the translation of high-level data access requests to the corresponding data access primitives.

At the top is the so-called *translation manager*. It maps all different kinds of access requests to the one or a series of calls of the node manager, which provides a set of operations which can be easily covered by the lock modes of the taDOM protocols. These are the navigation and modification operations as they are known from DOM, efficient bulk operations for subtrees like scans, insertions, and deletions, and, of course, direct jumps to nodes via their SPLID. The latter is certainly the biggest strength of the whole protocol as we will see later. For our explanations, we consider first the most simple case, where the translation manager only has to provide node-wise access and manipulation operations, because they directly correspond to DOM operations.

The node manager is responsible for the acquisition of the appropriate node and edge locks according to the taDOM protocols. When the required locks for an operation are granted, it simply passes the request directly to the *record manager*, which provides the same set of operations as the node manager.

The record manager translates the operations finally into the actual physical data access operations of the physical storage layer. Depending on the chosen storage model, this can be, for example, a simple access to a main memory representation of the document tree, or a B*-tree access. Independent of the chosen storage model, however, the physical storage layer only has to preserve *structural consistency*, e.g., if multiple threads representing different transactions access and modify the shared data structures concurrently. The *transactional consistency* will always be preserved, because the taDOM protocol applied at the node manager level already ensures that the operations will create a serializable schedule.

Depending on the needs of the higher system layers and the capabilities of the underlying storage structures, it may be desirable or necessary to use efficient indexes, e.g., to fetch attribute nodes by their value, or to fetch all element nodes of a certain name. Of course, these are complex operations, which can not be easily mapped to the navigational base model. From the point of view of the taDOM model, however, this can be also considered as "random" jumps to specific document nodes, which, in turn, are supported very well through the SPLID-labeling. Hence, with a small extension of the physical storage layer, our approach can also support efficient index accesses. The diverse index structures only have to lock every node with a node lock, before they return it as a result. In contrast to the previous case, however, index structures have to take additional care of phantoms themselves.[2]

4.2 Lock Manager

Although taDOM is based on the idea of widely used concepts of multi-granularity locking, hierarchical locking on XML trees is fairly different from the common way. While multi-granularity locking in relational databases usually requires only four or less granules of locks, e.g., for single tuples and whole tables, XML trees have a varying and dynamic depth and also varying and much smaller fanouts than a table with millions of tuples.

So far, hardly anything was reported in the literature about the implementation of XML lock managers. Without having a reference solution, we had to develop such a component from scratch where the generic guidelines given in [9] were used.

The central part of the lock manager is the lock table. It coordinates all required data structures and is also responsible for granting lock requests. For each locked object, a lock *header* is created, which contains name and current mode of the locked object together with a pointer to the lock *queue* where all lock requests for the object are attached to. Each lock request carries the ID of the respective transaction, the requested/granted mode. All lock requests of a transaction are doubly chained to speed-up lock release at transaction end.

Further necessary data structures are transaction entries to store housekeeping information for each transaction, as well as two hash tables h_{ta} and h_{lock} for fast lookups of lock headers and transaction entries. The hash tables, lock headers and the transaction entries use a fast latching mechanism to minimize synchronization points between concurrent threads.

A lock request is generally processed as follows. When a transaction T requests lock mode m for object o, h_{ta} is used to find the transaction entry te of T. If it does not exist, a new one is created. Then h_{lock} is used to find the lock header h of o. If o is not locked, a new lock header for o is registered in h_{lock}. If T already holds a lock for o, it tries to replace the currently granted mode with a mode that covers both the old granted and the requested mode. If it is T's first lock request for o it creates a new request r and appends it to the request queue of

[2] Mechanisms for the prevention of phantoms are specific to the index structures, but, because this is also a problem in relational indexes, similar solutions are usually applicable.

h. If the requested mode is compatible with all requests of other transactions, the lock is granted. Otherwise, *T* must wait until either all incompatible locks of other transactions are released, the request timed out or the transaction is aborted by the deadlock detector due to a circular wait-for-relationship.

Because we need to synchronize objects of varying types occurring at diverse system layers (e.g., pinning pages by the buffer manager and locking XML-related objects such as nodes, edges, and indexes), which exhibit incomparable lock compatibilities, very short to very long lock durations, we encapsulated everything in so-called lock services, which provide a convenient interface to the various system components [1].

To simplify and speed up lock management for our hierarchical lock protocols, we made our lock implementation "tree-aware". First, the node lock service automatically infers from a request, e.g., mode NR for SPLID 1.25.3.7 directly the ancestor path and the required intention locks on this path. These locks are acquired from the lock table in a single request. The lock table itself gets the transaction entry for the requesting transaction and starts granting the requests along the ancestor path from root to the leaf generally in the similar manner as sketched above.

When an intention lock on an ancestor node is granted that has been locked by the requestor before, it checks if the granted mode on the ancestor is strong enough to cover also the actual leaf request, e.g., when the node with SPLID 1.25 was already locked with an SR lock. If the ancestor lock is strong enough to satisfy the actual leaf request, we can directly stop the acquisition of further locks along the path.

5 Protocol Optimization

As it turned out by empirical experiments, lock depth is the most important and performance-critical parameter of an XML lock protocol. Lock depth *n* specifies that individual locks isolating a transaction are only acquired for nodes down to level *n*. Operations accessing nodes at deeper levels are isolated by subtree locks at level *n*. Note, choosing lock depth 0 corresponds to the case where only document locks are available. In the average, the higher the lock depth parameter is chosen, the finer are the lock granules, but the higher is the lock administration overhead, because the number of locks to be managed increases. On the other hand, conflicting operations often occur at levels closer to the document root (at lower levels) such that fine-grained locks (and their increased management) at levels deeper in the tree do not always pay off. Obviously, taDOM can easily be adjusted to the lock-depth parameter. A general reduction of the lock depth, however, would jeopardize the benefits of our tailored lock protocols.

5.1 Dynamic Lock Depth Adjustment

Obviously, the optimal choice of lock depth depends on document properties, workload characteristics and other runtime parameters like the number of concurrent users etc., and cannot be decided statically. The most effective solution to reduce lock management overhead at runtime is lock escalation: The fine-grained resolution of a lock protocol is—preferably in a step-wise manner—reduced by acquiring coarser lock granules. Applied to our case, we have to dynamically reduce lock depth and lock subtrees closer to the document root using single subtree locks instead of separately locking each descendant node visited. We aim at running transactions initially at a high lock depth to benefit from the fine-grained resolution of our lock protocols in hot-spot regions, but reserve the option to dynamically reduce the lock depth in low-traffic regions encountered to save system resources.

We use a counter for each request object, which is incremented everytime a node is locked intentionally as the direct parent for a lock request. If this counter reaches a certain threshold, indicating that—depending on the level, which we know from the tree-aware lock table—relatively much children of this node are already locked[3], it seems very likely that the transaction will access further child nodes, and it would be beneficial to escalate the intention lock request either with an LR lock if a simple NR lock was requested for the child, or even with a shared or exclusive lock for the whole subtree depending on the requested mode for the child. To avoid blocking situations, we exploit the context knowledge from lock table about the current lock mode and the requests of all transactions for that node again, and check whether concurrent transactions already hold incompatible locks, before we finally decide about the subtree-local lock escalation.

The escalation thresholds are computed from the simple formula $threshold = k * 2^{-level}$, which takes into account that typically the fanout as well as the conflict potential decreases on deep levels. The parameter k can be adjusted according to current runtime properties.

5.2 Use of Update Locks

Update locks are special lock modes used to avoid so-called conversion deadlocks. These deadlocks arise if two transactions read the same object and then both attempt to upgrade the lock for modification. In relational systems, update locks are mainly used for update cursors. They allow for a direct upgrade to exclusive lock mode when the transaction decides to modify the current record, or for a downgrade to a shared lock when the cursor is moved to the next record without any changes. Transactions in XDBMS do not follow such easy access patterns. Instead, they often perform arbitrary navigation steps in the document tree, e.g., to check the content child elements, before modifying a previously visited node.

Tests with our XDBMS prototype XTC revealed that many deadlocks result from the conversion of edge locks and not from the conversion of node locks. In a typical situation of such deadlocks, for example, two transactions try to append a new fragment under a node when they have already acquired a shared lock for its last-child edge while checking the value of the ID attribute of the current

[3] The actual number of locked child nodes may be less if the same child node is locked several times.

last child. As the insertion of a new last child requires conversion to an exclusive lock for the current last-child edge, both transactions form a deadlock. Hence, we have to carefully enrich our access plans with hints when to use update locks for accessing nodes, subtrees, or edges.

6 Experimental Results

We evaluated the described optimizations in our XDBMS prototype XTC with a mix of eight transaction types, which access and modify a generated XMark document at varying levels and in different granules. Three transaction types are read-only and access the document in various ways. The first one simply reconstructs the subtree of an item, the second iterates over the person siblings and item siblings to find the seller of an item, and the third one fetches the mails in the mailbox of an item. The update transactions examine smaller parts of the document through navigation, before they change the structure and the content of the document, e.g., by placing bids on items, by changing user data, or by inserting new users, items, or mails.

In all cases, we used an additional element index to randomly select a jump-in point for the transaction. To place a bid, for example, we first perform some navigation in a randomly selected open_auction subtree to check the current highest bid and to determine which content nodes have to be modified, before the new bid is actually inserted.

As we wanted to examine only the behaviour of the locking facilities in situations with a high blocking potential and not the influence of other system parameters, we chose an initial document size of only 8 MB and used a buffer size large enough for the document and the additional element index. In our experiments, we only used the taDOM3+ protocol, because it outperforms all other protocols of the taDOM family, and focused on lock-depth optimization.

In the first experiment, we evaluated the influence of the escalation heuristics and the parameter k on the effectiveness and efficiency of the lock protocol. The benchmark load was produced by 50 client instances, which continuously started transactions of a randomly selected type. We weighted the transaction types to achieve a balanced workload that evenly accesses and modifies the document at lower and deeper levels. For each escalation heuristics, we varied the initial lock depth from 0 to 8 and measured throughput, response time, abort rate, and the number of locks required for each configuration in several benchmark runs of one minute. For the escalation heuristics, we chose the parameter k equal to 2048 (called *moderate*) and 1536 (*eager*) respectively 512 (*aggressive*).

To document the general benefit of an XML lock protocol, we run the benchmark also in a "single-user mode", which allowed scheduled transactions only exclusive access to the documents.

The results in Fig. 3(a) reveal that the reduced lock overhead of our dynamic escalation mechanism has a positive influence on transaction throughput. The highest transaction rates were already achieved at lock depth 2, which seems fitted best to our test workload. For all higher lock depths, throughput remains

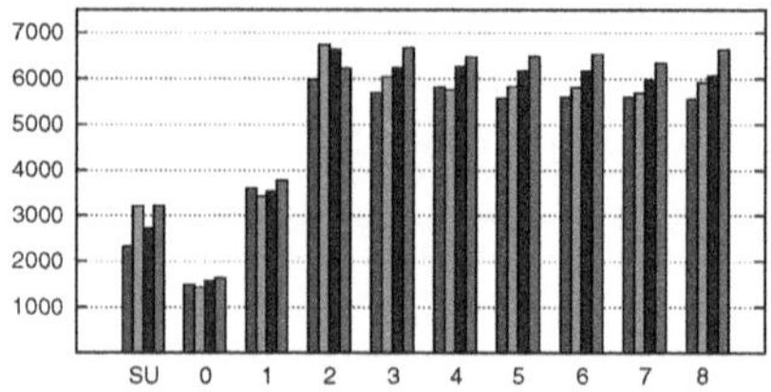

(a) Committed transactions per minute

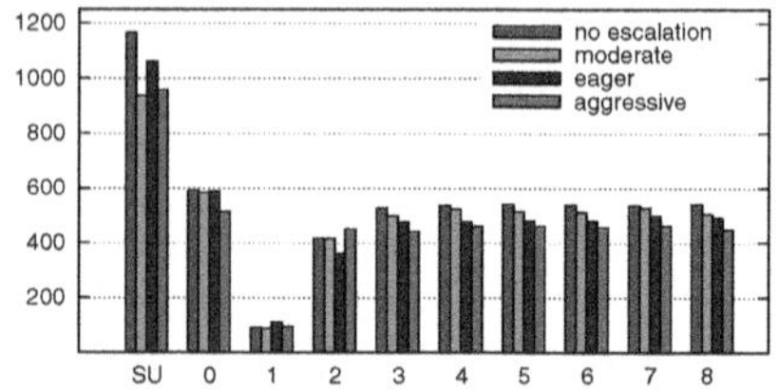

(b) Avg. response time of committed transactions in ms

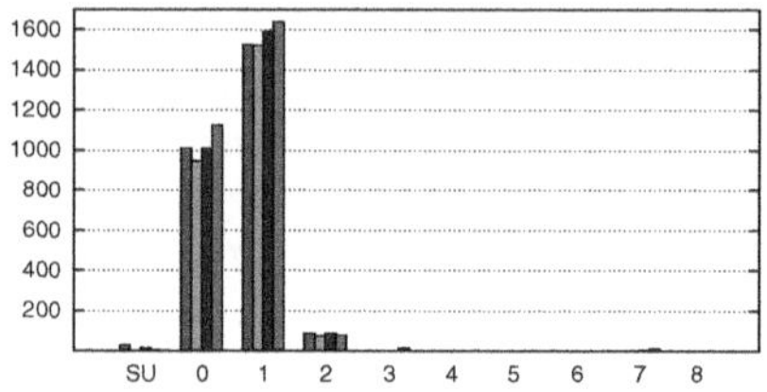

(c) Aborted transactions per minute

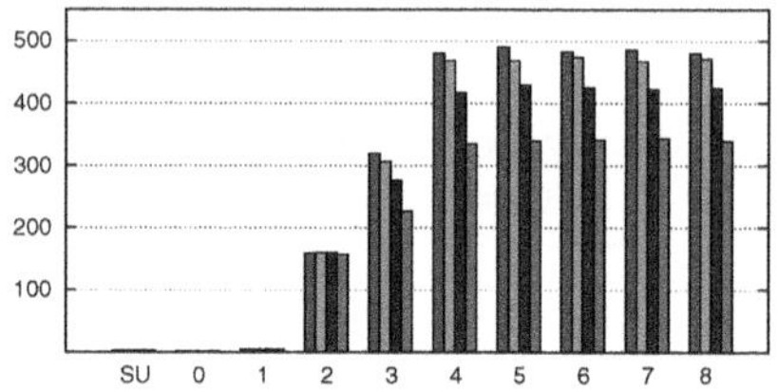

(d) Avg. number of locks per transaction

Fig. 3. Effect of lock escalation on transaction throughput and response times

relativley constant for all heuristics and also remarkably higher than in single user mode. The comparison of the escalation heuristics proves that less lock overhead directly leads to higher throughput.

The average response times of successful transactions (Fig. 3(b)) mirror the results shown in Fig. 3(a) for the lock depth 2 and higher. Shortest response times correspond to highest transaction throughput.

Fig. 3(c) shows that the lock depths 0 and 1 suffered from extremely high abort rates, caused by a high amount of conversion deadlocks, which immediately arise when a write transaction first uses shared subtree locks at root level or level 1, and then tries to convert this lock into an exclusive subtree lock the perform an update operation. Fig. 3(c) also clearly indicates that the escalation heuristics do not lead to higher deadlock rates when the maximum lock depth is chosen appropriately.

Fig. 3(d) demonstrates how effective simple escalation heuristics could reduce overhead of lock management. The number of required locks grows with higher lock depths and then saturates at a certain level. As expected, the aggressive heuristics achieved the best results in this category and saved in the average 30%[4] compared to the plain protocol variant without any escalation. Also the eager heuristics could save a mentionable amount of locks, whereas the benefit of the moderate heuristics was only marginal.

For our second experiment series, we took the balanced workload of the previous experiment (denoted *default*) and changed the weights of the transaction

[4] The actual savings potential is in fact even considerably higher, because acquired locks can be removed as soon as they become obsolete after an escalation.

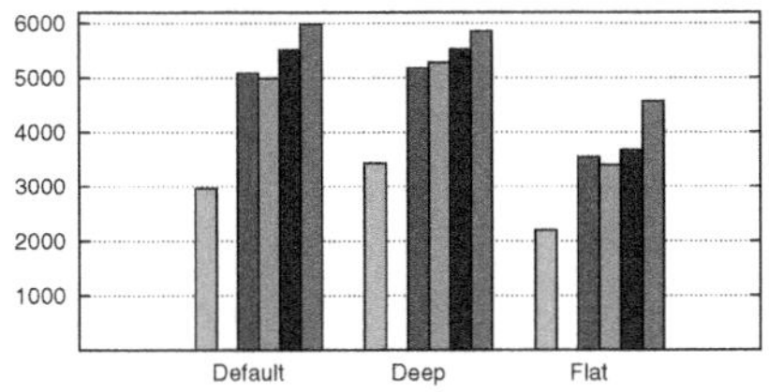

(a) Committed transactions per minute

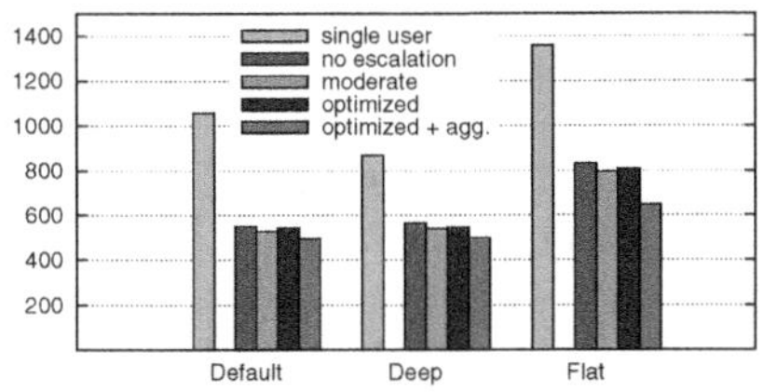

(b) Avg. response time in ms

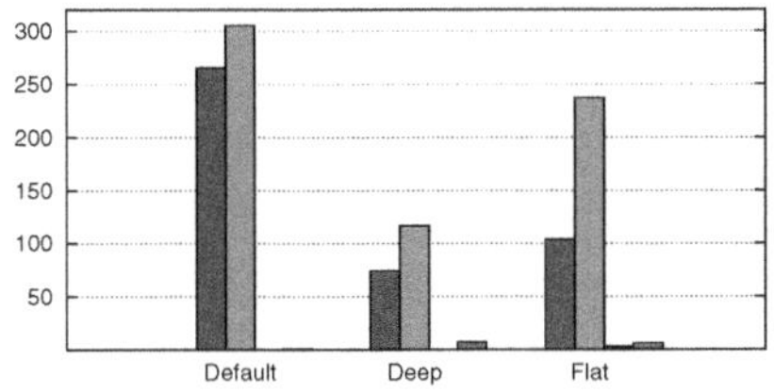

(c) Aborted transactions per minute

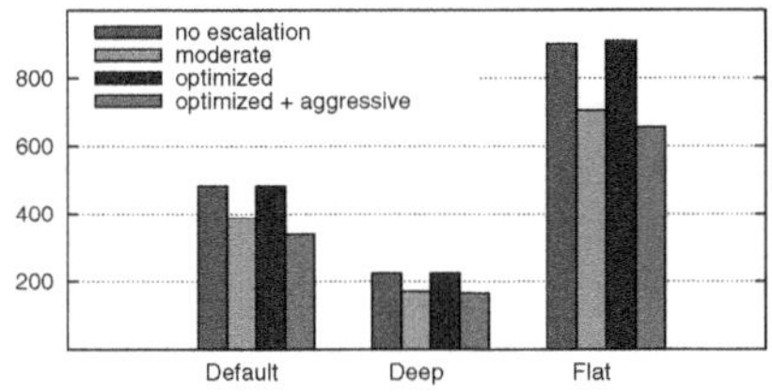

(d) Avg. number of locks per transaction

Fig. 4. Results of workload variations and adjusted escalation strategies

types to create a workload mix that mainly consists of transaction types that access and modify the document at deeper levels (*deep*) and another one that mainly operates on higher document levels (*flat*). We ran the benchmark with maximum lock depth 8, which allows fine-grained locks even deep in the document, and again in single-user mode. For these configurations, we also evaluated variants with moderate escalation heuristics from the previous experiment, a variant where we modified the transaction types to make careful use of update locks (*optimized*) to avoid conversion-induced deadlocks, and a fourth variant (*optimized + aggressive*) that combined the use of update locks with the aggressive heuristics, which produced the best results in the previous experiment.

The results in Fig. 4(a) proof again that all variants of taDOM3+ achieve higher transaction throughput than the solution with exclusive document access. The optimized version with aggressive escalation nearly achieves a gain of 20% in throughput as compared to the plain taDOM version. Of course, we observe similar results for the response times in Fig. 4(b), too.

The abort rates in Fig. 4(c) show the benefit of carfully set update lock modes during processing. The deadlock rate decreases to nearly zero, which in turn explains the throughput gain in Fig. 4(a).

Finally, Fig. 4(d) illustrates that our optimizations complement each other. On the one hand, correct application of update lock modes is helpful if lock escalations are used, because this increases danger of deadlocks otherwise. On the other hand, lock escalations help to reduce the overhead of lock management.

Altogether, the experimental results demonstrate well that a fine-grained locking approach pays off and provides higher throughput and shorter response times than exclusive document locks. The experiments also confirmed that taDOM3+ in combination with our adaptations is able to provide constantly high transaction throughput for higher lock depths, and that it can effectively and efficiently be adjusted to varying workloads to achieve high concurrency without a waste of system resources.

A closer analysis of the shown results revealed that our protocols would allow even much higher concurrency in the XML tree for lock depths higher than 2. Here, however, the data structures of the physical storage layer—which is a B*-tree in our prototype—became the bottleneck.

7 Conclusions and Outlook

In this paper, we explained the realization of fine-grained concurrency control for XML. We started with an introduction into the basics of our tailor-made lock protocols, which are perfectly eligible for a fine-grained transaction isolation on XML document trees, and emphasized the advantages of prefix-based node labeling schemes for lock management. Thereafter, we turned on general implementation aspects, where we showed how the XML protocols can be integrated in a layered architecture and how even different storage models and indexes can be incorporated into our concept. We also explained how we adapted the a widely used lock manager architecture for our needs and presented ways to optimize the runtime behaviours of the lock protocols.

In our future work, we will focus on the integration of advanced XML indexes, which make use of structural document summaries, in our isolation concept, and the interplay between XML concurrency control on the one hand and efficient query evaluation algorithms for declarative queries based on XQuery on the other hand.

References

1. Bächle, S., Härder, T.: Tailor-made Lock Protocols and their DBMS Integration. In: Proc. EDBT 2008 Workshop on Software Engineering for Tailor-made Data Management, pp. 18–23 (2008)
2. Bähme, T., Rahm, E.: Supporting Efficient Streaming and Insertion of XML Data in RDBMS. In: Proc. 3rd Int. Workshop Data Integration over the Web, Riga, Latvia, pp. 70–81 (2004)
3. Chen, Q., Lim, A., Ong, K.W., Tang, J.: Indexing XML documents for XPath query processing in external memory. Data & Knowledge Engineering 59(3), 681–699 (2006)
4. Christophides, W., Plexousakis, D., Scholl, M., Tourtounis, S.: On Labeling Schemes for the Semantic Web. In: Proc. 12th Int. WWW Conf., Budapest, pp. 544–555 (2003)
5. Dewey, M.: Dewey Decimal Classification System, `http://frank.mtsu.edu/~vvesper/dewey2.htm`
6. Document Object Model (DOM) Level 2 / Level 3 Core Specific., W3C Recommendation
7. Grabs, T., Bähm, K., Schek, H.-J.: XMLTM: Efficient transaction management for XML documents. In: Proc. CIKM Conf., pp. 142–152 (2002)

8. Gray, J.: Notes on Database Operating Systems. In: Flynn, M.J., Jones, A.K., Opderbeck, H., Randell, B., Wiehle, H.R., Gray, J.N., Lagally, K., Popek, G.J., Saltzer, J.H. (eds.) Operating Systems. LNCS, vol. 60, pp. 393–481. Springer, Heidelberg (1978)
9. Gray, J., Reuter, A.: Transaction Processing: Concepts and Techniques. Morgan Kaufmann, San Francisco (1993)
10. Härder, T., Haustein, M.P., Mathis, C., Wagner, M.: Node Labeling Schemes for Dynamic XML Documents Reconsidered. Data & Knowledge Engineering 60(1), 126–149 (2007)
11. Haustein, M.P., Härder, T.: Optimizing lock protocols for native XML processing. Data & Knowledge Engineering 65(1), 147–173 (2008)
12. Helmer, S., Kanne, C.-C., Moerkotte, G.: Evaluating Lock-Based Protocols for Cooperation on XML Documents. SIGMOD Record 33(1), 58–63 (2004)
13. Jagadish, H.V., Al-Khalifa, S., Chapman, A.: TIMBER: A native XML database. The VLDB Journal 11(4), 274–291 (2002)
14. O'Neil, P., O'Neil, E., Pal, S., Cseri, I., Schaller, G., Westbury, N.: ORDPATHs: Insert-Friendly XML Node Labels. In: Proc. SIGMOD Conf., pp. 903–908 (2004)
15. Pleshachkov, P., Chardin, P., Kuznetsov, S.O.: XDGL: XPath-Based Concurrency Control Protocol for XML Data. In: Jackson, M., Nelson, D., Stirk, S. (eds.) BNCOD 2005. LNCS, vol. 3567, pp. 145–154. Springer, Heidelberg (2005)
16. Pleshachkov, P., Chardin, P., Kusnetzov, S.: SXDGL: Snapshot Based Concurrency Control Protocol for XML Data. In: Barbosa, D., Bonifati, A., Bellahsène, Z., Hunt, E., Unland, R. (eds.) XSym 2007. LNCS, vol. 4704, pp. 122–136. Springer, Heidelberg (2007)
17. Sardar, Z., Kemme, B.: Don't be a Pessimist: Use Snapshot based Concurrency Control for XML. In: Proc. 22nd Int. Conf. on Data Engineering, p. 130 (2006)
18. Schmidt, A., Waas, F., Kersten, M.L., Carey, M.J., Manolescu, I., Busse, R., et al.: XMark: A Benchmark for XML Data Management. In: Proc. VLDB Conf., pp. 974–985 (2002)
19. Schöning, H.: Tamino – A DBMS designed for XML. In: Proc. 17th Int. Conf. on Data Engineering, pp. 149–154 (2001)
20. Siirtola, A., Valenta, M.: Verifying Parameterized taDOM+ Lock Managers. In: Geffert, V., Karhumäki, J., Bertoni, A., Preneel, B., Návrat, P., Bieliková, M. (eds.) SOFSEM 2008. LNCS, vol. 4910, pp. 460–472. Springer, Heidelberg (2008)
21. XQuery 1.0: An XML Query Language, `http://www.w3.org/XML/XQuery`
22. XQuery Update Facility, `http://www.w3.org/TR/xqupdate`
23. Yu, J.X., Luo, D., Meng, X., Lu, H.: Dynamically Updating XML Data: Numbering Scheme Revisited. World Wide Web 8(1), 5–26 (2005)

On Multidimensional Wavelet Synopses for Maximum Error Bounds

Qing Zhang, Chaoyi Pang, and David Hansen

Australia E-Health Research Center
firstname.lastname@csiro.au

Abstract. Having been extensively used to summarize massive data sets, wavelet synopses can be classified into two types: space-bounded and error-bounded synopses. Although various research efforts have been made for the space-bounded synopses construction, the constructions of error-bounded synopses are yet to be thoroughly studied. The state-of-the-art approaches on error-bounded synopses mainly focus on building one-dimensional wavelet synopses, while efficient algorithms on constructing multidimensional error-bounded wavelet synopses still need to be investigated. In this paper, we propose a first linear approximate algorithm to construct multidimensional error-bounded L_∞-synopses. Our algorithm constructs a synopsis that has $O(\log n)$ approximation ratio to the size of the optimal solution. Experiments on two-dimensional array data have been conducted to support the theoretical aspects of our algorithm. Our method can build two-dimensional wavelet synopses in less than 1 second for a large data set up to 1024×1024 data array under given error bounds. The advantages of our algorithm is further demonstrated through other comparisons in terms of synopses construction time and synopses sizes.

1 Introduction

Wavelet techniques have been successfully used in image and signal processing for decades. Haar wavelet is the simplest yet powerful wavelet tool that has been extensively investigated during recent years. Studies have demonstrated the applicability of Haar wavelet analysis on image compressing [1], approximate query processing [2] and data streams [3]. Briefly, the idea is to apply Haar wavelet transformation on input data to obtain a compact data synopsis, containing only a set of selected wavelet coefficients. There has been an explosion of publications on how to choose coefficients (a.k.a coefficient thresholding) in recent years, which forms an active research topic in the literature.

One common thresholding criterion is to minimize approximation errors under a given number of coefficients. The synopses thus created are named *space-bounded* synopses. Various error metrics, collectively known as L_p-norm error, have been applied on synopses construction. Among them, L_2-norm error metrics, i.e. root-mean-squared, are well-understood and widely adopted. However, it is pointed out by many researchers that the choice of using L_2-norm error is

X. Zhou et al. (Eds.): DASFAA 2009, LNCS 5463, pp. 646–661, 2009.

natural but yet to be sufficient [4]. A major shortcoming of L_2 synopses is that users have no way to control approximation error of individual items, which can result in wide variance as well as severe bias in the quality of the approximation [5]. To alleviate this, researches have made efforts on building non-L_2 synopses, such as L_∞-synopses (i.e. maximum error synopses) that aim at minimizing maximum approximation error. Literature has shown those synopses outperform L_2 in many cases [2,6]. Garofalakis et al proposed the first algorithm to build space-bounded L_∞-synopses, based on randomized rounding techniques [7]. However its probabilistic nature makes it still possible to generate poor synopses in certain cases. Thus Garofalakis et al proposed a deterministic algorithm, that explicitly minimizes the maximum error and can always guarantee better synopses [8]. The main drawback of this approach is that it imposes large resources requirements on systems, which renders itself impractical in many real situations, such as large or streaming data processing. Karras et al thus proposed a one-pass greedy based algorithm to build L_∞-synopses for large data sets [9]. Note that so far all the mentioned techniques are restricted to store a subset of wavelet coefficients and are thus classified further as restricted synopses. Guha et al argued that from the intrinsic purpose of synopses, this restriction is neither necessary nor guaranteeing best quality synopses. They thus proposed a greedy algorithm and a fully polynomial-time approximation scheme (FPTAS) to build unrestricted L_∞-synopses [4,10,11]. This FPTAS, like many other algorithms in L_∞-synopses construction, features high time and space complexity. Aiming at better unrestricted L_∞-synopses construction, Karras et al seek the 'help' from error-bounded synopses and proposed an indirect approach. Specifically, instead of computing space-bounded synopses, they suggest a detour through a series of error-bounded synopses constructions [12]. They demonstrate theoretically and practically that this detour actually improves the overall performance.

Another thresholding criterion is to minimize synopses size under a given approximation error. These synopses are thus named *error-bounded* synopses. However compared with the thriving research fruits on spaced-bounded L_∞-synopses construction, error-bounded L_∞-synopses, only received few attentions in the literature. In fact, these synopses are also commonly used in practice as space-bounded synopses. Many data processing applications, such as time series data analyses [13], statistical data for hypothesis testing [14], location management with spatiao-temporal data reduction [15], require compressing data with guaranteed approximation error. Thus error-bounded synopses are favored in these scenarios. As dualities of space-bounded L_∞-synopses, error-bounded synopses are only discussed in few recent papers. Muthukrishnan proposed the first algorithm to construct maximum error-bounded synopses in [16]. It is a natural extension of the optimal algorithm proposed in [8] and not surprisingly suffers similar performance penalties. It has subquadratic time complexity and quadratic space complexity, which is obviously impractical to large data processing. A greedy improvement of that method was proposed in [17]. Another investigation, as mentioned before, was proposed by Karras et al in [12]. Interestingly, it was a by-product of their space-bounded synopses construction algorithm. For a data set with n data, it takes $O((\frac{\varepsilon}{\delta})^2 n)$ time and $O(\frac{\varepsilon}{\delta}\log n + n)$ space to build a one-dimensional wavelet synopses with maximum error bounded by ε. The performance of this method relies however heavily on the resolution value, δ, which is a double-sided blade for the algorithm's efficiency and effectiveness. Large δ renders a large size synopses, but can be built within 1 second. Small δ leads to a succinct or even near-optimal synopses (in terms of synopses size), with the building time ranging from hours to days. The authors, in the experiment section of their paper [12], also mentioned that they need to run several trials before they can find a suitable δ value.

Despite this process, the existing two approaches for error-bounded L_∞-synopses construction are still inefficient for summarizing large data sets. One potential problem is that they both rooted from examining possible coefficient values in every node of the Haar wavelet error tree. Thus they need to at least maintain such a tree structure in the system's memory, which requires $O(n)$ space complexity. Also the repeatedly checking on every node makes it not easy to be tuned as a pure linear algorithm, which is an utmost prerequisite for large and streaming data processing. Another issue is that both two algorithms are focusing on building one-dimensional synopses. How to efficiently construct multidimensional error-bounded L_∞-synopses is still not well-understood. Simply extend existing solutions to build multidimensional synopses will inevitably exacerbate the performance. These motivates the work presented in [18] and this paper. In [18], we focused on building synopses for one-dimensional data and proposed a couple of approximate algorithms. While this paper extends and completes that work by proposing algorithms for multidimensional synopses. Table 1 summarizes the above reviews by enumerating various algorithms' complexity as well as their specifications. It also lists the result of this paper for comparison purpose. n is the data size. B and ε are the space and error bound, respectively. R is the cardinality of possible coefficient values. δ is the value of resolution. k is the dimensionality of data.

Table 1. Summary of various wavelet synopses

	Ref.	**Time**	**Space**	**Spec.**
Space Bound	[5]	$O(n^2 B \log B)$	$O(n^2 B)$	Optimal Restrict
	[9]	$O(n \log^3 n)$	$O(n \log n)$	Greedy Restrict
	[12]	$O(R^2 n(\log \varepsilon^* + \log n))$	$O(R \log n + n)$	Indirect Unrestrict
	[10,11]	$O(R^2 n \log^2 B)$	$O(R \min\{B^2 \log \frac{n}{B}, n \log B)$	FPTAS Unrestrict
	[4]	$O(n)$	$O(B + \log n)$	Approx. Unrestrict
Error Bound	[16]	$O(n^2/\log n)$	$O(n^2)$	Optimal Restrict
	[12]	$O((\frac{\varepsilon}{\delta})^2 n)$	$O(\frac{\varepsilon}{\delta}\log n + n)$	Approx. Unrestrict
	[18]+ **This Work**	$O(n)$	$\frac{2^k}{k}\log n$	Approx. Unrestrict

Our contributions: (1) We present an approximate algorithm on multidimensional Haar wavelet error-bounded L_∞-synopses construction. To the extend of our knowledge, this is the first algorithm on multidimensional error-bounded L_∞-synopses that has ever been presented in literature. (2) Our algorithm can build synopses satisfying both absolute and relative error requirements. It is an approximate algorithm that guarantees the quality of constructed synopses with a $\frac{2^k}{k}\log n$ approximation ratio to the size of the optimal solution (k is the number of dimensions). (3) Our algorithm has linear time complexity and sublinear space complexity, which make it naturally applicable for large data processing. (4) We also experimentally show that our algorithm can be indirectly used to construct space-bounded L_∞-synopses. The performance is even better than the greedy space-bounded L_∞-synopses construction algorithm proposed in [4].

The rest of this paper are organized as follows. Section 2 presents preliminary knowledge of Haar wavelet transformation and synopses construction. Readers familiar with wavelet techniques can easily skip this part. Section 3 explains details of our algorithm, theoretical analyses of the algorithm's performance and possible extensions. Section 4 reports experiment results and section 5 concludes this paper.

2 Preliminary

In this section we first briefly introduce one-dimensional Haar wavelet transformation. Then we discuss two approaches on multidimensional Haar wavelet transformation, namely standard and non-standard transformation. Specifically we focus on one method of non-standard transformation, proposed by Chakrabarti et al [19]. Finally we present formal definitions of wavelet synopses.

2.1 One-Dimensional Haar Wavelet Transformation:

The Haar wavelet transformation is a hierarchical decomposition of the original data. In each step, it filters the input data into two parts: the approximation part as the result of low-pass filters; and the detail part, a.k.a. coefficients, as the result of high-pass filters. Due to the special properties of Haar wavelet mother function, we can simplify one-dimensional transformation through a series of summing and differentiating. Specifically, for every two adjacent data, we sum them and then divide $\sqrt{2}$ to get the approximations. We compute the difference between them and also divide $\sqrt{2}$ to get the coefficients. Note that by using $\sqrt{2}$, instead of the commonly used 2, we can directly get the normalized wavelet coefficients. This procedure is named one-step transformation. Recursively, we apply this one-step transformation on the computed approximations till we only have one data in the approximation part. This one approximation data plus all the computed coefficients form the final result of wavelet transformation, i.e. wavelet coefficients. Suppose we are given an original data set: $[1, 2, 1, 2]$. After applying the above techniques, we compute the one-dimensional wavelet transformation as $\left[3, 0, \frac{-1}{\sqrt{2}}, \frac{-1}{\sqrt{2}}\right]$.

2.2 Multidimensional Haar Wavelet Transformation:

There exist two methods to perform multidimensional wavelet transformation: the standard and non-standard decomposition [1]. The *standard decomposition*

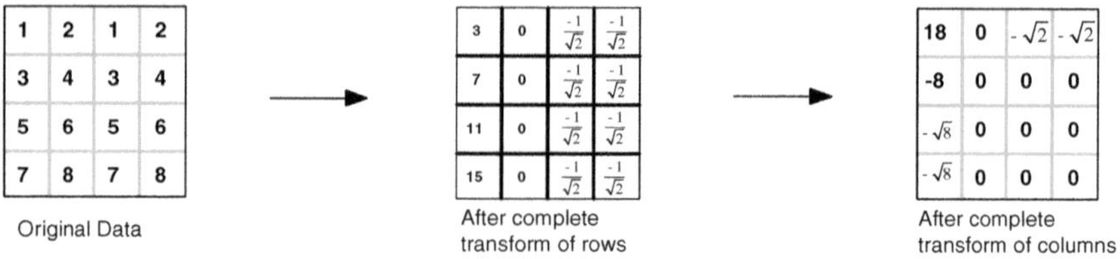

(a) Standard transformation

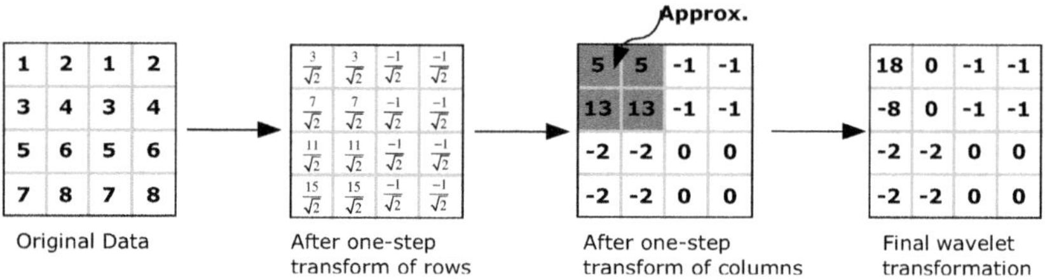

(b) Non-standard transformation

Fig. 1. Two-dimensional wavelet transformation

is to apply one-dimensional wavelet transformations on every dimension of a k-dimensional data array repeatedly. Figure 1 (a) presents an example of such decomposition on a two-dimensional data array. As described, one-dimensional wavelet transformations are applied firstly on rows, followed by another set of one-dimensional transformations on columns. The *non-standard decomposition* is to apply a one-step transformation alternately on each dimension. After those one-step transformations, we get the approximation and the detail parts. Then we apply the one-step transformations on the approximation part recursively, until we have only one data in the approximation part. Figure 1 (b) presents an example of such decomposition. After one-step transformation on rows, we start one-step transformation on columns. Then we get the upper-left four data as the approximations while others as the details. Recursively applying one-step operations on those four data, we finally reach the one data approximation, 18. This data with other computed details form the non-standard wavelet coefficients. Note that both methods have been extensively used in a variety of applications and none of them show consistently superior [19].

In this paper, we construct error-bounded synopses based on non-standard wavelet transformation. Our choice is mostly motivated by the fact that in non-standard transformation, after every one-step transformation on all dimensions we can directly get part of the final wavelet coefficients. This enables a similar tree structure of wavelet transformation, as that used in one-dimensional transformation. Figure 2 (a) illustrates a multidimensional transformation tree with 3 levels transformation. Figure 2 (b) presents a corresponding 3 levels Haar transformation on a MRI image of human brain. All dark images represent details and the only bright one at the upper left corner represents the overall approximation.

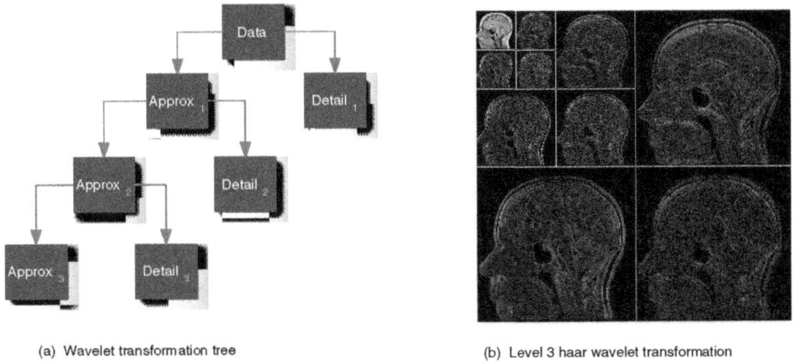

Fig. 2. Hierarchical non-standard wavelet transformation

To clearly present our synopses construction algorithms, we will not use the above method of non-standard wavelet transformation. Instead, we apply the *box operation* method proposed by Chakarbarti et al: the procedure of this method can be thought conceptually as a $2 \times 2 \times \cdots \times 2 (= 2^k)$ hyper-box being shifted across the k-dimensional data array, performing pairwise computation, distributing the results to appropriate locations [19]. We use a two-dimensional transformation to illustrate the box operation. Consider a 2×2 square shifted across the data, the four data values, e, f, g, h, in a square, are transformed to an approximation t and three details p, q, r, as presented in Figure 3 (a). Note that the equations used in pairwise computation are slightly different to that in [19]. This is again because we want to generate normalized coefficients directly. Figure 3 (b) presents an example of the non-standard transformation through box operations on the data array used in Figure 1. Readers can refer to [19] for more details of the box operation.

2.3 Wavelet Synopses

To construct wavelet synopses, we choose to store certain *non-zero* values as wavelet coefficients. These will be used to approximate the original data. Given a data set $D = \{d_i : 1 \leq i \leq n\}$, we define the wavelet transformation as $\mathcal{W}$ and the reconstruction (i.e. inverse wavelet transformation) as $\mathcal{W}^{-1}$. The wavelet synopsis is defined as S. Then by definition, we have $\mathcal{W}(D)$ represents the set of wavelet coefficients. And $\mathcal{W}^{-1}(S)$ represents the approximation generated by S. The difference between D and $\mathcal{W}^{-1}(S)$ is measured by L_p-norm error metrics. When $p = \infty$, it implies the maximum error metrics. Existing wavelet synopses can be categorized into two groups as follows:

Definition 1. *Space-bounded synopses: Given a data set D, a positive integer B and L_p-norm error metrics, find a synopsis S, such that $|S| = B$ and $\left\|D - \mathcal{W}^{-1}(S)\right\|_p$ is minimized.*

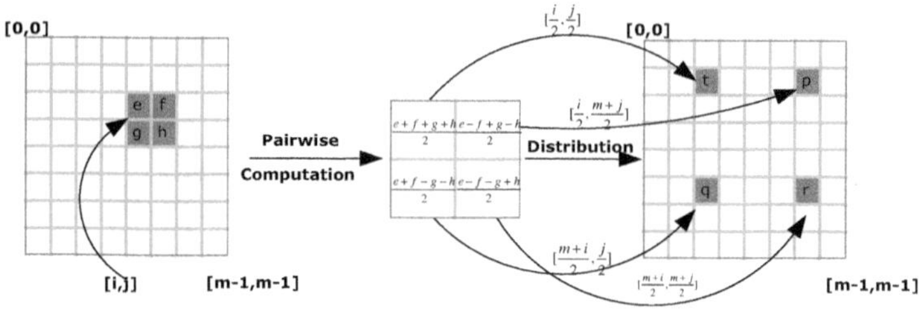

(a) Box operation: computation and distribution

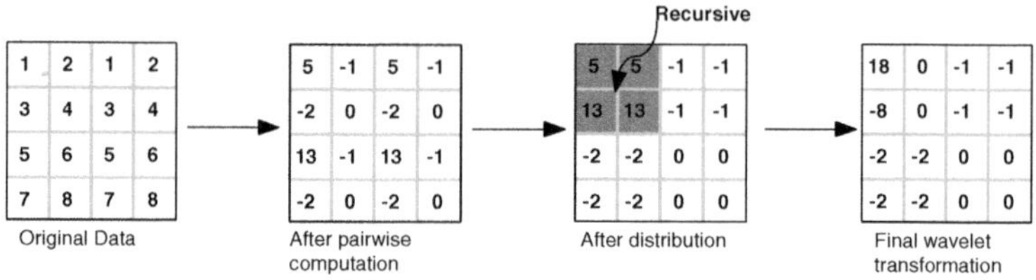

(b) Non-standard transformation with box operations

Fig. 3. Box operation

Definition 2. *Error-bounded synopses: Given a data set D and a non-negative real number* ε*, find a synopsis S, such that* $\left\|D - \mathcal{W}^{-1}(S)\right\|_p \leq \varepsilon$ *and* $|S|$ *is minimized.*

Furthermore, S can be broadly defined as unrestricted synopses where S is not suggested to be a subset of $\mathcal{W}(D)$. In the following section, we will discuss how to build a multidimensional error-bounded L_∞-synopsis, with the synopsis's size is guaranteed to be a $O(\log n)$ approximation ratio to the size of the optimal solution.

3 Building Multidimensional Error-Bounded L_∞-Synopses

In this section, we first present the sketch of our error-bounded L_∞-synopses construction algorithm. Then we explain details of synopses computation. This is followed by theoretical analyses of our algorithm. We also discuss possible extensions of the algorithm at the end of this section. According to Definition 2, for any approximation values generated by error-bounded L_∞-synopsis S, we have the following inequations:

$$d_i - \varepsilon \leq \mathcal{W}^{-1}(S)_i \leq d_i + \varepsilon \tag{1}$$

Thus for any synopses S, it is easy to see that as long as Equation 1 holds, S is a L_∞-synopsis bounded by ε and vice versa. This simple yet mighty observation motivates our following synopses construction algorithm.

3.1 Algorithm Sketch

Intuitively, our algorithm starts from the top level of the wavelet transformation tree in Figure 2 (a) (i.e. the data node). For each data d_i of the original data set, we relax it to a range $[d_i - \varepsilon, d_i + \varepsilon]$ according to Equation 1. This range is a collection of all valid approximation values of d_i. Our algorithm will then use the ranges, instead of the original data set to build S. Specifically, for every transformation that decomposes a parent node into two child nodes, the approximation and detail, we greedily find smallest number of non-zero values that are necessary to make all ranges of the parent node valid. As long as we fix all the values, we can compute the valid ranges of the approximation child node, which will be used recursively for next level decomposition. This process will be continued until we reach the bottom of the transformation tree. Algorithm 1 presents our synopses construction algorithm.

Input: D, k-dimensional data array$[1 \ldots n, \ldots, 1 \ldots n]$; ε, error bound
Output: S, k-dimensional error-bounded L_∞-synopsis
1 Relax each d_i of D to a range: $[d_i - \varepsilon, d_i + \varepsilon]$;
2 **while** $n \neq 1$ **do**
3 Shift a 2^k hyper-box in the k-dimensional array $[1 \ldots n, \ldots, 1 \ldots n]$;
4 In each box, compute the values of the detail part according to ranges in the hyper-box;
5 Determine the range of approximation part according to computed values of detail;
6 Output values as part of S;
7 $n = \frac{n}{2}$;
8 **end**
9 Compute the last value from the range stored in cell $(1, \ldots, 1)$;

Algorithm 1. Building error-bounded L_∞-synopses

3.2 Compute Approximation Ranges and Details in a Hyper-Box

Now we present how to compute details and new approximation ranges from ranges in a hyper-box i.e. step 4 and 5 of algorithm 1. For simplicity reasons, we will always use two-dimensional synopses to explain our techniques hereafter. Computation for higher dimension wavelet synopses can be easily derived based on the ideas presented here. Consider a 2×2 box shifting in a two-dimensional data array. We use e, f, g, h to represent the four data values and t, p, q, r to represent the computed approximation and details, as depicted in Figure 3 (a). Those four data values are bound by ranges as:

$$\begin{cases} e_1 \leq e \leq e_2 \\ f_1 \leq f \leq f_2 \\ g_1 \leq g \leq g_2 \\ h_1 \leq h \leq h_2 \end{cases} \quad (2)$$

The box operation , as described in Figure 3 (a), can be presented as:

$$\begin{bmatrix} .5 & .5 & .5 & .5 \\ .5 & -.5 & .5 & -.5 \\ .5 & .5 & -.5 & -.5 \\ .5 & -.5 & -.5 & .5 \end{bmatrix} \begin{bmatrix} e \\ f \\ g \\ h \end{bmatrix} = \begin{bmatrix} t \\ p \\ q \\ r \end{bmatrix} \rightarrow A \begin{bmatrix} e \\ f \\ g \\ h \end{bmatrix} = \begin{bmatrix} t \\ p \\ q \\ r \end{bmatrix} \tag{3}$$

Note that A is the Haar wavelet transformation matrix and $A = A^{-1}$. Combining Equations 2 and 3, we have Equation 4. Based on this, we can decide values of t, p, q, r through two following cases greedily.

$$A \begin{bmatrix} e_1 \\ f_1 \\ g_1 \\ h_1 \end{bmatrix} \leq \begin{bmatrix} t \\ p \\ q \\ r \end{bmatrix} \leq A \begin{bmatrix} e_2 \\ f_2 \\ g_2 \\ h_2 \end{bmatrix} \tag{4}$$

Case 1: If there exists a common range, $[c_1, c_2]$, that bounds e, f, g, h simultaneously in Equation 2, we can simply set the three details, p, q, r as 0. Then t can be computed as:

$$\begin{bmatrix} e_1 \\ f_1 \\ g_1 \\ h_1 \end{bmatrix} \leq A \begin{bmatrix} t \\ 0 \\ 0 \\ 0 \end{bmatrix} \leq \begin{bmatrix} e_2 \\ f_2 \\ g_2 \\ h_2 \end{bmatrix} \Rightarrow 2c_1 \leq t \leq 2c_2$$

Case 2: Otherwise, we set the values of three details as the middle values of their valid ranges. For instance, according to Equation 4, we compute p's range as $\left[\frac{e_1-f_1+g_1-h_1}{2}, \frac{e_2-f_2+g_2-h_2}{2}\right]$. Thus we set $p = \frac{e_1+e_2}{4} - \frac{f_1+f_2}{4} + \frac{g_1+g_2}{4} - \frac{h_1+h_2}{4}$. Similar computations exist for q and r. Then from Equation 4, we can compute the range of t from the following inequations.

$$\begin{bmatrix} e_1 \\ f_1 \\ g_1 \\ h_1 \end{bmatrix} - A \begin{bmatrix} 0 \\ p \\ q \\ r \end{bmatrix} \leq A \begin{bmatrix} t \\ 0 \\ 0 \\ 0 \end{bmatrix} \leq \begin{bmatrix} e_2 \\ f_2 \\ g_2 \\ h_2 \end{bmatrix} - A \begin{bmatrix} 0 \\ p \\ q \\ r \end{bmatrix}$$

The last value, i.e. in step 9 of Algorithm 1, can be easily set as 0, if valid, or the middle value of its valid range.

3.3 Algorithm Analyses

In this part, we analysis the time and space complexity of our synopses construction algorithm. We also prove that our algorithm can generate quality guaranteed synopses. That is the size of constructed synopsis is bounded in $O(\log n)$ to the size of optimal solution.

Complexity Analysis: As described in Algorithm 1, we output details from the computations on data of every 2^k hyper-box directly, to build k-dimensional wavelet synopses. The wavelet transformation matrix A is fixed for a given dimension number k, and can thus be pre-computed. Then the time used to compute details and approximation's range for each hyper-box is $O(2^k)$. Suppose we have n data in a k-dimensional array, the synopses construction time T is:

$$T = 2^k\left(\frac{n}{2^k} + \frac{n}{2^{2k}} + \cdots + 1\right) = O(n)$$

Also during our synopses construction, we can directly output details and only store the approximation's range for each computation in a 2^k hyper-box. To get any range of an approximation node in the transformation tree, we need a 2^k hyper-box of its parent node. This requirement will be propagated to the top of the transformation tree. Thus the running space, used on computing synopses is:

$$(2^k - 1)\log_{2^k} n = O\left(\frac{2^k}{k}\log n\right)$$

Theorem 1. *Algorithm 1 can build a k-dimensional error-bounded L_∞-synopses on n data with time complexity $O(n)$, space complexity $O(\frac{2^k}{k}\log n)$.*

Quality Analysis: The quality of synopses, i.e. the sizes of the synopses, is closely related to our greedy mechanisms. Before analyzing the sizes of synopses generated by Algorithm 1, we need to first understand the error tree structure used to compute approximations from synopses. Figure 4 illustrates an error tree for generating approximations from a two-dimensional synopsis. Leaf nodes contains the computed values from the synopsis. Non-leaf nodes contains the details. Specifically the root contains the last computed valued in step 9 of Algorithm 1. Every approximate value is computed through traversing from the root to its parent node. The example in Figure 4 computes approximations of the data used in Figure 3 (b), with the maximum error bounded by 1.5.

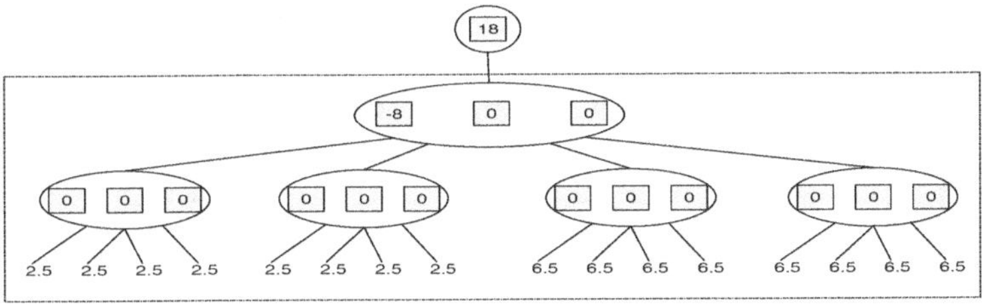

Fig. 4. Error tree for two-dimensional nonstandard reconstruction

Our greedy method can be conceptually thought of constructing an error tree upwardly, starting from valid ranges at the leaf nodes. Each computation inside the hyper-box generates details stored in a non-leaf node. It also generates approximation ranges of the node. Here we define a special subtree that only has non-zero details at it's root node as *pivot twig.* As in Figure 4, the subtree located inside the big rectangle is a pivot twig. The following lemma presents an important property of a pivot twig.

Lemma 1. *Given a pivot twig t of an error tree constructed by Algorithm 1, Let α represent the number of non-zero details inside t, that would be set by the optimal solution. Then $\alpha \geq 1$.*

Proof. Let t_r represent the root node of t. According to the definition of t, there must exist no common range from all ranges of the child nodes of t_r, before we set non-zero details in t_r. Also any non-zero detail outside t will not help to make a common range for t_r's children. Thus even the optimal solution must also has at least one non-zero detail in t. Otherwise the approximation values in the leaf nodes of t will against the maximum error bound. This proves $\alpha \geq 1$. □

Intuitively, Lemma 1 tells us that whenever there exists a pivot twig in our approximate solution, there must be some non-zero coefficients, constructed by the optimal solution, in that twig. Thus we can bound the sizes of approximate synopses through the following theorem:

Theorem 2. *Let the size of an optimal k-dimensional error-bounded L_∞-synopsis be $S(o)$. Let the size of the greedy synopsis constructed by Algorithm 1 be $S(g)$. Then $\frac{S(g)}{S(o)} \leq \frac{2^k}{k} \log n$.*

Proof. Suppose the number of pivot twigs in the error tree, constructed by Algorithm 1, is n_{pt}. Since our algorithm only generates non-zero coefficients on the root or root's ancestors of pivot twigs. And each root of a pivot twig has at most $log_{2^k}n$ ancestors. Thus we have $S(g) \leq n_{pt} * log_{2^k}n * (2^k - 1)$. According to Lemma 1, we have $n_{pt} \leq S(o)$. This proves $\frac{S(g)}{S(o)} \leq \frac{2^k}{k} \log n$. □

3.4 Discussions

Our greedy method can be applied on construction of other types of synopses, with few, if not none, modifications. Here we briefly discuss two interesting synopses: maximum relative-error-bounded synopses and space-bounded synopses.

Building maximum relative-error-bounded synopses: The relative error is arguably having the best practical performance. We can apply similar methods used for processing one-dimensional data [20]. Given a relative error bound ε_{rel} and a sanity bound α, we compute the valid approximation range of a data d_i as: $[d_i - \varepsilon_{rel} \cdot \max\{|d_i|, \alpha\}, d_i + \varepsilon_{rel} \cdot \max\{|d_i|, \alpha\}]$. Algorithm 1 can then be applied to build a multidimensional maximum relative-error-bounded synopses without any modification!

Building space-bounded synopses: Using indirect method, we also can compute space-bounded synopses by our greedy algorithm. The idea is simple and straightforward. Given a data set D, to compute a multidimensional space-bounded L_∞-synopses with space B, we run several trials with various guessed error bound ε until we find a synopses whose size equals or less than B and minimizes ε. The lower and upper bound of ε can be simply set as 0 and $\max_i\{|d_i|\}$ ($d_i \in D$) at the very beginning of trials.

4 Experiment

In this section, we test the performance of our greedy algorithm on multidimensional error-bounded synopses construction. For simplicity reason, all our tests are based on two-dimensional wavelet synopses construction. Three gray scaled images downloaded from internet are tested as data sets. Specifically, a 1024×1024 array represents a MRI image of human brain, a 512×512 array represents a fractal image, a 256×256 array represents a hieroglyph alphabets image. We implement our algorithms using GNU C++ 4.0. All experiments reported in this section are performed on an Intel Core 2 Duo 2.8 GHz CPU with 2GB main memory, running Mac OS X.

4.1 Constructing Synopses

We test our greedy method on absolute/relative error-bounded synopses construction in this part. The experimental data sets used include the MRI image and the fractal image. These are two-dimensional arrays with 1M and 262K data respectively. Each data represents a pixel intensity and thus ranges from 0 (total absence, black) to 1 (total presence, white).

We build absolute error-bounded, two-dimensional synopses for the MRI image. Figure 5 (a) presents the original data. $|D|$ represents the data size. Two synopses, based on two absolute error bounds 0.05 and 0.2, are constructed for the MRI image. Figure 5 (b) and (c) present the approximations generated by those two synopses. ε represents the error bound and $|S|$ represents the synopses size.

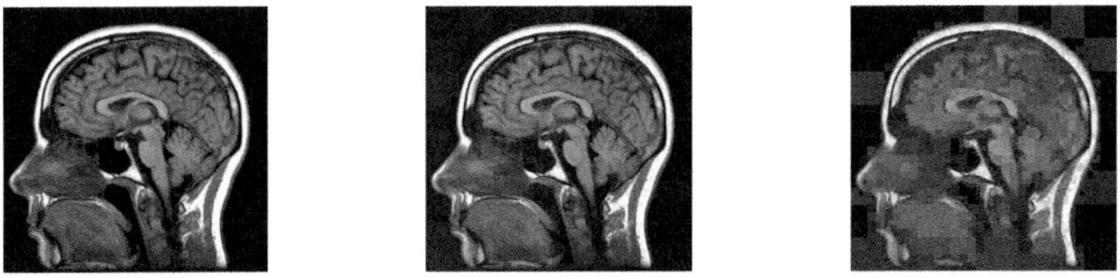

(a) $|D| = 1048576\ (2^{20})$ (b) $\varepsilon = 0.05, |S| = 96574$ (c) $\varepsilon = 0.2, |S| = 10293$

Fig. 5. Absolute error-bounded synopses for the MRI image

(a) $|D| = 262144\ (2^{18})$ (b) $\varepsilon_{rel} = 20\%, |S| = 85183$ (c) $\varepsilon_{rel} = 60\%, |S| = 26565$

Fig. 6. Relative error-bounded synopses for the fractal image

We build relative error-bounded, two-dimensional synopses for the fractal image. Figure 6 (a) presents the original data. Two relative error bounds, 20% and 60%, are set to build synopses. Figure 6 (b) and (c) present them separately.

Both Figure 5 and 6 demonstrate that higher error bound leads to higher compression ratio, but with sacrifices on approximation quality.

4.2 Time Efficiency

In this part we evaluate the time used on building error-bounded synopses. Besides the MRI and fractal images we build absolute and relative error-bounded synopses for another data set, the hieroglyph alphabet image, to form a complete test on various data sizes. Figure 7 presents the experiment result by setting absolute error 0.05 and relative error 20% in the synopses construction. As claimed in Theorem 1, the synopses construction time only relates to the original data size. Our experiment result supports that claim. Hieroglyph alphabet image is the smallest data set that can be summarized by error-bounded synopses in less than 0.1 second. While the MRI image, the largest data set, takes around 1 second to be summarized.

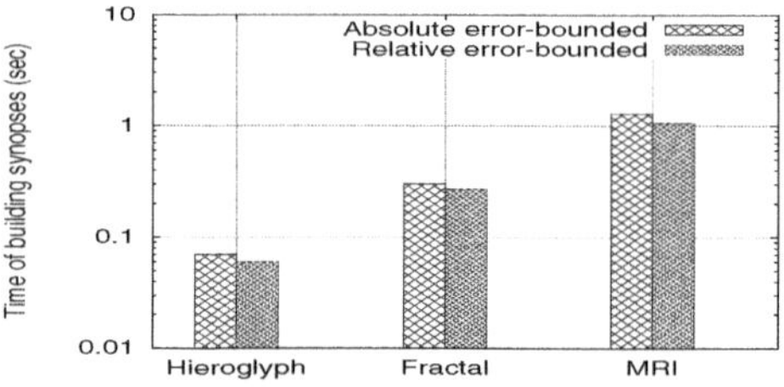

Fig. 7. Building time of error-bounded synopses

4.3 Synopses Quality Comparison

Our last experiment focuses on measuring the quality of synopses. For space-bounded synopses, the quality is measured by the maximum approximation

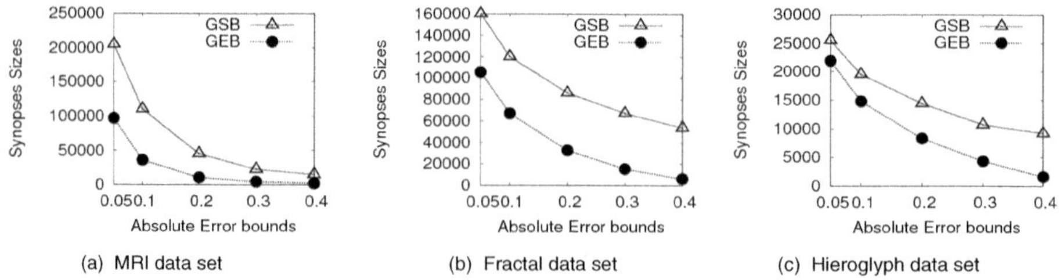

Fig. 8. Size comparison of error-bounded synopses generated by GSB and GEB

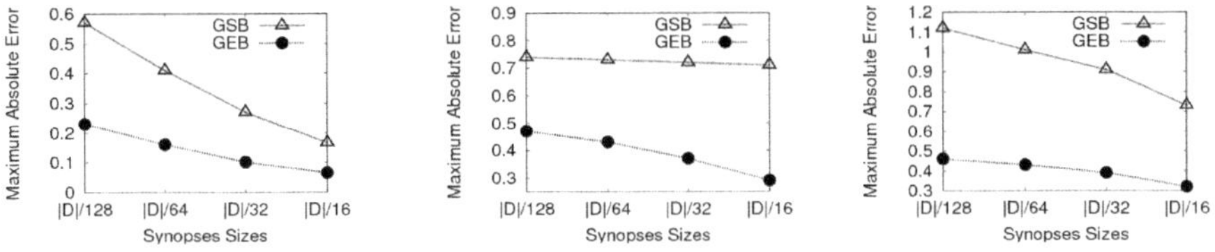

(a) MRI data set (b) Fractal data set (c) Hieroglyph data set

Fig. 9. Error comparison of spaced-bounded synopses generated by GSB and GEB

(a) original image (b) approx. by GEB (c) approx. by GSB

Fig. 10. Approx. comparison of space-bounded synopses generated by GSB and GEB

error. Similar to that, we use the size of synopses to measure the quality of error-bounded synopses. However since our algorithm is the only known one that can build multidimensional error-bounded L_∞-synopses, we measure the synopses quality through indirect comparisons with the quality of synopses constructed by an algorithm originally aiming for multidimensional space-bounded L_∞-synopses. Specifically, we use both algorithms to generate space-bounded and error-bounded synopses and compare their qualities. Section 3.4 already presents basic ideas of apply our algorithm on building space-bounded synopses indirectly. Similar ideas can be easily derived for algorithms, originally aiming for space-bounded synopses, to build error-bounded synopses.

We choose the greedy algorithm, proposed by Guha et al [4], to be the indirectly compared algorithm. The reason of our choice is threefold. First it is a state-of-the-art research result of building multidimensional space-bounded synopses. Second it is an approximate algorithm with $O(\log n)$ approximation ratio. Third it has linear time complexity which makes it highly applicable for large data sets. Let GEB denotes our greedy algorithm for error-bounded synopses. Let GSB denotes the greedy algorithm for space-bounded synopses [4]. Figure 8 and 9 report comparison results on building various error-bounded and space-bounded synopses. Note that in Figure 9, $|D|$ represents the size of original data.

When compared on building error-bounded synopses it is not surprise that GEB achieves better quality of synopses, compared to those indirectly constructed by GSB. Interestingly, on building space-bounded synopses, GEB also achieves synopses with smaller maximum approximation error, compared to those directly constructed by GSB. We present visual comparisons of the approximations produced by same sizes space-bounded synopses, generated by GSB and GEB, in Figure 10.

5 Conclusion

In this paper, we examined building multidimensional wavelet synopses. We proposed an efficient and effective greedy algorithm that can construct absolute as well as relative error-bounded L_∞-synopses. Our method is a one-pass linear algorithm with $O(\log n)$ space complexity of fixed dimension number, this makes it naturally suitable for processing large or streaming data sets. It also has a provable $O(\log n)$ approximation ratio to guarantee synopses quality. In addition, as demonstrated by experiments, our greedy method can be used to indirectly build space-bounded synopses with good quality. As our next step research, we will investigate building multidimensional error-bounded synopses for generic L_p error metrics.

References

1. Stollnitz, E.J., Derose, T.D., Salesin, D.H.: Wavelets for computer graphics: theory and applications, 1st edn. Morgan Kaufmann, San Francisco (1996)
2. Matias, Y., Vitter, J.S., Wang, M.: Wavelet-based histograms for selectivity estimation. In: ACM SIGMOD, pp. 448–459 (1998)
3. Gilbert, A.C., Kotidis, Y., Muthukrishnan, S., Strauss, M.: Surfing wavelets on streams: One-pass summaries for approximate aggregate queries. The International Journal on Very Large Data Bases, 79–88 (2001)
4. Guha, S., Harb, B.: Approximation algorithms for wavelet transform coding of data streams. In: IEEE TOIT, vol. 54(2) (February 2008)
5. Garofalakis, M., Kumar, A.: Wavelet synopses for general error metrics. ACM Trans. Database Syst. 30(4), 888–928 (2005)
6. Chapelle, O., Haffner, P., Vapnik, V.N.: Support vector machines for histogram-based image classification. IEEE TONN 10(5), 1055–1064 (1999)
7. Garofalakis, M., Gibbons, P.B.: Wavelet synopses with error guarantees. In: ACM SIGMOD, pp. 476–487 (2002)
8. Garofalakis, M., Kumar, A.: Deterministic wavelet thresholding for maximum-error metrics. In: ACM PODS, pp. 166–176 (2004)
9. Karras, P., Mamoulis, N.: One-pass wavelet synopses for maximum-error metrics. In: VLDB, pp. 421–432 (2005)
10. Guha, S., Harb, B.: Approximation algorithms for wavelet transform coding of data streams. In: SODA, pp. 698–707 (2006)
11. Guha, S., Harb, B.: Wavelet synopsis for data streams: minimizing non-euclidean error. In: ACM SIGKDD, pp. 88–97 (2005)
12. Karras, P., Sacharidis, D., Mamoulis, N.: Exploiting duality in summarization with deterministic guarantees. In: ACM SIGKDD, pp. 380–389 (2007)
13. Chen, H., Li, J., Mohapatra, P.: Race: time series compression with rate adaptively and error bound for sensor networks. In: IEEE International conference on Mobile Ad-hoc and Sensor systems (2004)

14. Shimokawa, H., Te Han, S., Amari, S.: Error bound of hypothesis testing with data compression. In: Proceedings of IEEE International Symposium on Information Theory, p. 114 (1994)
15. Cao, H., Wolfson, O., Trajcevski, G.: Spatio-temporal data reduction with deterministic error bounds. In: DIALM-POMC, pp. 33–42 (2003)
16. Muthukrishnan, S.: Subquadratic algorithms for workload-aware haar wavelet synopses. In: Ramanujam, R., Sen, S. (eds.) FSTTCS 2005. LNCS, vol. 3821, pp. 285–296. Springer, Heidelberg (2005)
17. Pang, C., Zhang, Q., Hansen, D., Maeder, A.: Building data synopses within a known maximum error bound. In: APWeb/ WAIM 2007 (2007)
18. Pang, C., Zhang, Q., Hansen, D., Maeder, A.: Unrestricted wavelet synopses under maximum error bound. In: EDBT/ ICDT (2009)
19. Chakrabarti, K., Garofalakis, M., Rastogi, R., Shim, K.: Approximate query processing using wavelets. VLDB Journal 10(2-3), 199–223 (2001)
20. Pang, C., Zhang, Q., Hansen, D., Maeder, A.: Constructing unrestricted wavelet synopses under maximum error bound. Technical report, ICT CSIRO (08/209)

A Revisit of Query Expansion with Different Semantic Levels*

Ce Zhang[1], Bin Cui[1], Gao Cong[2], and Yu-Jing Wang[1]

[1] Key Laboratory of High Confidence Software Technologies (Peking University), Ministry of Education, CHINA & School of EECS, Peking University
{zhangce,bin.cui,wangyujing}@pku.edu.cn
[2] Aalborg University, Denmark
gaocong@cs.aau.dk

Abstract. Query expansion has received extensive attention in information retrieval community. Although semantic based query expansion appears to be promising in improving retrieval performance, previous research has shown that it cannot consistently improve retrieval performance. It is a tricky problem to automatically determine whether to do query expansion for a given query. In this paper, we introduce Compact Concept Ontology (CCO) and provide users the option of exploring different semantic levels by using different CCOs. Experimental results show our approach is superior to previous work in many cases. Additionally, we integrate the proposed methods into a text-based video search system (*iVSearcher*), to improve the user's experience and retrieval performance significantly. To the best of our knowledge, this is the first system that integrates semantic information into video search and explores different semantic levels.

1 Introduction

Query extension has been extensively studied in information retrieval community to address the word mismatch or semantic gap between queries and documents. Word mismatch is a fundamental and serious problem in information retrieval since it is common that users may use different words to retrieve document from the words used by authors to describe the concepts in documents [7]. There are a host of work on Query Expansion (QE), e.g., [15,8]. Existing query expansion methods usually need to extract terms' similarity based on distinct sources, e.g., documents and relevance feedback made by users. What is more related to our work is ontology-based query expansion to alleviate semantic gap. For example, user query is $bunny$ but the desired documents may be described by $rabbit$ or $lapin$; keyword-based search will fail to find the desired documents. On the other hand, a query expansion method can expand the user's query $bunny$ to other terms related to $rabbit$ or $lapin$, so that the search can capture semantic features.

Semantic based query expansion is expected to help discover the semantic similarity between query and document, and hence improve the performance of text retrieval.

* This research was supported by the National Natural Science Foundation of China under Grant No.60603045.

X. Zhou et al. (Eds.): DASFAA 2009, LNCS 5463, pp. 662–676, 2009.

However, experiments in previous work have shown that query expansion is often not a robust technique. Previous work showed that semantic-based query expansion can ameliorate some of the problems of mismatched words on small collections while degrade performance when the query is relatively complete and long. For example, Voorhees [15] used WordNet to expand query terms and provided a comprehensive discussion and found that such expansions make little performance improvement for common queries, however, significantly improves the performance on short ones [7,15]. Following the work by Voorhees, more works focused on similar topics but unfortunately failed to improve retrieval performance in general.

On the other hand, the existing expansion strategies for query expansion methods are quite rigid and hardly to be tuned by users. For example, individual query performance differs more for more aggressive expansion strategies (i.e., expanding using longer chains of links in WordNet or weighting certain terms more heavily), while the overall performance of a query set may not be optimal by aggressively expanding queries [7]. Some methods use human judgement, e.g., relevant feedback [12], to obtain useful information. However, such approaches are not convenient due to the large-amount feedback [8].

In spite of the negative news on semantic query expansion, we believe it is still a valuable complementarity for text retrieval after carefully examining the existing query expansion methods with different query expanding strategies. Semantic query expansion can achieve better performance than methods without considering semantic if users are involved to determine whether to do semantic expansion.

In this paper, we propose a novel query expansion method for effective text retrieval. First, we represent each text with a weighted tree which encodes the semantic of the whole text based on WordNet, and then calculate the similarity between two texts globally using their weighted trees. Second, to explore different semantic levels for text search, we propose an algorithm for merging WordNet to generate Compact Concept Ontologies (CCOs) with different scales and replace WordNet with them when calculating semantic similarity. Different semantic expanding levels on CCO can be used for effective query/text expansion.

To evaluate the performance of our approach, we compare cosine similarity and query expansion method on both standard dataset NPL corpus [17] and video text dataset which was collected from YouTube. Our experimental results on NPL corpus show that different approaches yield different performance on individual query; experimental results on video text show our method outperforms existing methods. Experimental results also show that semantic expansion work better on text-based online video search area, where the text description is usually short. We also implement a demo system of text-based video search engine based on our CCO mechanism, and the system can improve users' experience of video search.

The rest of this paper is organized as follow: In section 2, we introduce some related work; In section 3, we present the algorithm for calculating semantic similarity and Compact Concept Ontology (CCO), which are the two core components of our approach; we evaluate these algorithms in Section 4, and demonstrate our system in Section 5; and finally we conclude our work in Section 6.

2 Related Work

In this section, we present a brief introduction to previous work in the area of query expansion, especially those related to our methods. The existing query expansion approaches can be divided into two classes, i.e., 1) global analysis and 2) local analysis [9]. Global analysis seriously relies on the statistical information of corpus and is relatively robust, while local analysis uses only some initially retrieved documents for further query expansion.

There are some well-known global analysis algorithms, e.g., term clustering [10] and Latent Semantic Indexing (LSI) [11]. A trivial global method mentioned in [21] uses the co-occurrence relation between concepts and obtains relatively good result. Moreover, Crouch et al. [22] clusters the document and carries out query expansion. Although these methods can obtain good performance sometimes, they still suffers from the problem caused by corpus-wide statistical analysis that consumes a considerable amount of computing resources [8].

On the other hand, local expansion methods use only some initially retrieved documents for further query expansion. For example, the relevance feedback method [12] modifies a query based on users' relevance judgments of the retrieved documents [8] and experimental results indicate that these kinds of methods yield satisfactory performance when users provide adequate feedbacks. Considering the obstacle of collecting such feedback information, some methods tend to use approximate information, e.g., pseudo-relevance feedback [13], which assumes that the top-ranked documents to be relevant, however the performance is significantly affected by the quality of the initial retrieval. It is found in [9] that local expansion methods outperform global ones in some circumstances and there are also methods that combine them together, e.g., [14].

Last but not at least, some researchers have built semantic search engines based on WordNet for text documents. For example, Voorhees investigated the problem of query expansion using semantic knowledge and presented four expansion strategies [6] :

1. by synonyms only;
2. by synonyms plus all descendants in an *isa* hierarchy;
3. by synonyms plus parents and all descendants in an *isa* hierarchy;
4. by synonyms plus any synset directly related to the given synset.

Though Voorhees found that those methods does not make significant improvement on TREC corpus, she indicated that those approaches work well on some queries, especially the short ones. This phenomenon is quite trivial because it is not difficult for us to image that, for a long and complete query, WordNet-based expansion brings more noise than useful information. While for short queries, WordNet-based expansion can capture more semantic information of query and enhance the query performance.

3 Calculating Semantic Similarity

In this section, we discuss the algorithm for calculating semantic similarity and semantic level determined by Compact Concept Ontology (CCO).

Although WordNet-based query expansion discussed in Sec. 2 can utilize the semantic information of the query, it may introduce undesirable expansion, especially when expanding the query word with its descendants in an *isa* hierarchy because the number of descendants is generally too large to obtain a desirable searching result [7]. For example, we found that on the video text dataset used in our experiment if we expand the query *animal dance* by their descendants using WordNet, the top 100 videos in the result list are only related to *animal* but not *dance* since the expanded query contains many words of animal names, but few words related to *dance*.

To solve this problem, one possible solution is to expand both the query and text descriptions of videos to their parents in the WordNet to prevent the large number of expanding. But how to determine the expanding degree – which we call *semantic level* – is a nontrivial task. As the hypernym (is-a-kind-of) relationships in WordNet of different words do not contain equal semantic information, determining a consistent semantic level for each word is infeasible. For example, if we want to expand the word *bunny* to *animal*, we must expand it to its 8th superconcept, while expanding the word *dance* with 8th level will exceed the root node of WordNet! Therefore, we need a new strategy to determine the expanding degree of each word. To this end, we will build a new ontology in which the hypernym relationships can almost equally convey semantic information and thus we can fix on an universal expansion layer and change semantic level by change different CCO. We also propose a new strategy for query expansion as discussed above. Moreover, the new ontology also allows users to explore different semantic levels easily by just selecting different CCO to use.

Note that, though it is like that we change the structure of documents, this problem can be solved easily by some appropriate indexing methods. For example, when we create the invert index of documents, we can index not only the words themselves, but also their hypernyms in different ontology. Based on this reason, we still classify our methods into the area of *query* expansion, despite of that it use the information of detailed documents while searching.

3.1 Algorithm for Calculating Semantic Similarity

We first discuss the approach to compute semantic similarity between text features, which is also a strategy of query expansion. There is a host of work on computing semantic similarity between two words using WordNet. However, there is a gap from the semantic similarity between words to the semantic similarity between texts. Although it seems that a straightforward extension from word-level semantic similarity to text-level semantic similarity, e.g. [1], will work well, it is disproved in [5]. The algorithm in [1] treats each word pair independently, and thus cannot capture the semantic of the text as a whole. For example, a word may have multiple senses and its context in text will help to identify the real sense in a text. To this end, we next present a novel method of computing semantic similarity between two texts based on WordNet.

The main idea of our approach is to first represent each text with a weighted tree which can encode the overall semantic of the whole text based on WordNet, and then calculate the semantic similarity between two texts using their weighted trees.

Given a text P, we build its weighted tree T_P as follows. For every word w_i in P, we find the corresponding node in WordNet for each sense of w_i to get a set

$SC_i = \{C_1, ..., C_r\}$ of nodes, where r is the number of senses of w_i. Each node C_j in SC_i and the corresponding path from C_j to the root of WordNet tree will be added to T_P (if the path is not in T_P); for each node from C_j to the root, we increase its weight by one. In this way, we will get the complete weighted tree T_P for P after we process all words in P. The weighted tree T_P represents the semantic characteristics of the whole text P. The similarity of two texts can be calculated by comparing the similarity of their weighted trees. The similarity between two weighted trees T_1 and T_2 can be defined as the straight-forward dot product of the two trees or as follows:

$$Sim(T1, T2) = m - \sum_{node_i} \left| \frac{W_1(node_i)}{W_1(root)} - \frac{W_2(node_i)}{W_2(root)} \right|,$$

where $W_1(.)$ and $W_2(.)$ are weights for nodes in tree T_1 and T_2 respectively, and $node_i$ represents every node appearing in either T1 or T2. If a certain node only appears in one tree, its weight in the other tree is set as zero.

The above similarity definition is a natural extension of the similarity in [4] which computes the similarity between two words using WordNet. Compared with the method in [1], we compute similarity as a whole. Considering the time complexity, suppose the depth of ontology tree and the number of tree nodes are constant, when generating the weighted tree, we need to find all corresponding paths for each word, thus the time complexity is $O(c * s * m)$. When comparing two weighted tree, we just need $O(c * s * (m + n))$. Thus, calculating the semantic similarity between two given texts cost $O(c * s * (m + n))$. Consider [1], It cost $O(c * s * m * n * O(WordSim))$. Where n and m are the number of words in two texts, c is the average height of WordNet, s is the average number of senses of each word in WordNet, $O(WordSim)$ is the time complexity of semantic similarity methods between words. Thus, our method is more efficient.

The algorithm discussed above calculates the semantic similarity between texts by comparing their weighted trees. While trying to consider semantic features, it ignores the fact that sometimes we need not expand each word to the root of WordNet. So here we define the concept of $expanding - layer$ (k), which will be helpful when we discuss the semantic level later. When founding the corresponding path of each concept, we do not use the whole path from root, but a shorter path from each concept to the kth superconcept and later we link all these short paths to the original root. Fig. 1 shows an example of the corresponding path of node $bus\#3$ with expanding layer k=2.

3.2 Semantic Level and Compact Concept Ontology (CCO)

The proposed algorithm for computing semantic similarity uses WordNet as the concept ontology. However, we find that WordNet does not fit very well with our algorithm while searching. The nature of our algorithm is to expand each word to its superconcepts, but in WordNet, the proper number of layers to expand is generally different for different words. For example, when we want to expand the word $bunny$ to $animal$ (e.g. the query $animal\ dance$), we find that the $expanding\ layer$ for $bunny$ is k=8, which is too large for the word $dance$. Thus, when using WordNet as ontology, it is difficult for us to create a system which has a suitable semantic level. We conjecture that we can

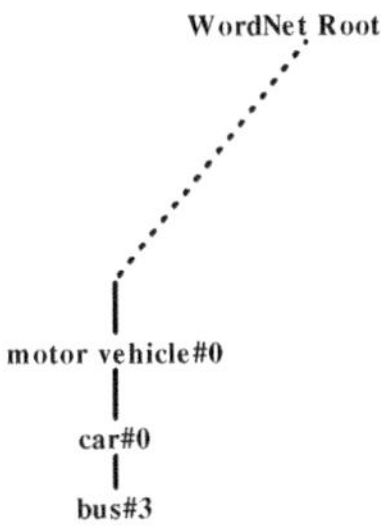

Fig. 1. Example of the path of bus#3 with expanding level k=2

solve this problem by merging different nodes of WordNet together – when we merge those semantically similar nodes together, we can get an ontology focusing on this given semantic level. Thus, given a searching query, we can fixate the *expanding layer* and change the different ontology to use in order to achieve different semantic levels. Therefore, we design a new ontology from WordNet, named Compact Concept Ontology (CCO),which has the following properties:

1. CCO inherits the hierarchical relationship of WordNet structure.
2. CCO combines different senses of the same word into a single concept if they are very similar.
3. CCO highlights semantic factors and considers combination when two concepts have similar semantic meanings.
4. Compared with WordNet, CCO is much more compact and yields better time efficiency.

Where 1) ensures the methods based on WordNet can be migrated to CCO without modification; 2), 3) and 4) result in larger synsets of each word.

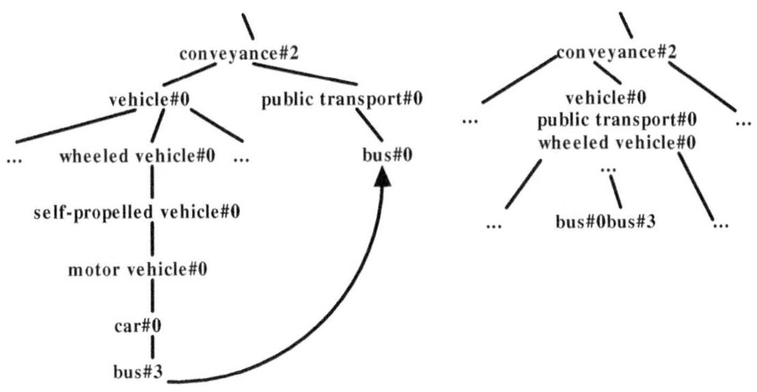

Fig. 2. Process of Union

We next present the WordNet Union Algorithm to build CCO from WordNet by combining semantically similar nodes. More specifically, we find semantically similar node pairs (i, j) and the nearest common ancestor node a in WordNet. Then, we combine all the nodes on the paths between a and i,j to form a new node k, which is child node of a, and also combine i and j to form a new node, which is child node of node k. Figure 2(a) and (b) shows an example. In Figure 2(a), nodes bus#0 and bus#3 are semantically similar nodes and the nodes between them and their common ancestor will be combined into one node in Figure 2(b) and nodes bus#0 and bus#3 are also combined into one node.

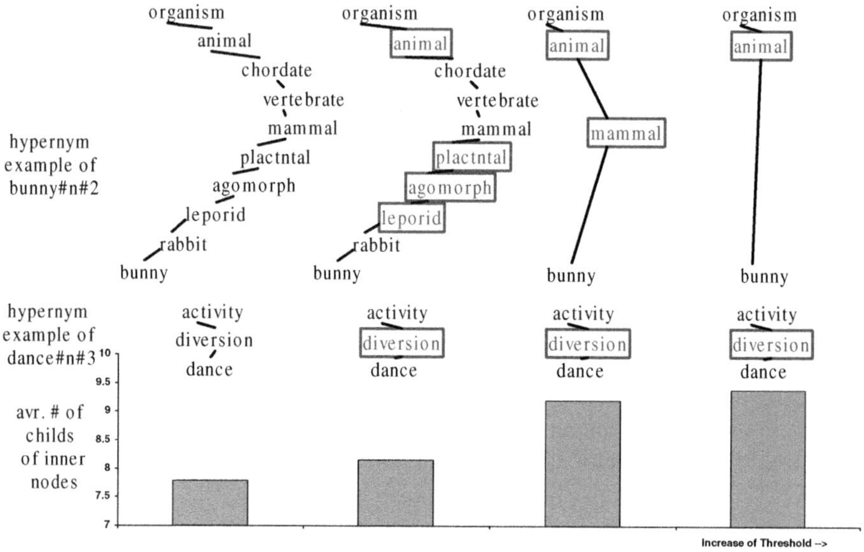

Fig. 3. Comparison between Different CCOs

The open problem is to find the semantically similar node pairs (i, j). They need to satisfy the following two conditions: 1) they represents two senses of the same word, and 2) they are highly semantically similar, i.e. $Sim(i, j) > threshold$, where $Sim(.,.)$ is computed using the WUP method in [3]. By determining the different thresholds while merging, we can get CCOs with different scales, which tends to convey different semantic levels. We show some CCOs in Fig. 3, where the boxed nodes mean new nodes generated by our merging methods and the smallest CCO only has approximately 2000 nodes which is much smaller than WordNet. Fig. 3 illustrates the difference between CCOs, from which we can easily see that, by choosing different CCO as ontology we can expand the word *bunny* to different levels of superconcepts, while the word *dance* has little change. This fact somehow proves that, CCOs can provide optional semantic levels for search task.

4 Performance Study

4.1 Overview

We conduct extensive experiments to evaluate the performance of the proposed methods in both computing semantic similarity and document/video retrieval by comparing with other competitive techniques. All experiments were conducted on a PC running Windows XP with 1.5 GHz CPU and 2G memory.

To evaluate the performance of the proposed retrieval approach, we compare our method with the classic WordNet-based query expanding [15] and cosine similarity (without considering semantics) both in text corpus and our Web-video corpus. Here we fixate the *expanding layer* to two.

Data: We conduct experiments on two data sets, including a standard text retrieval set NPL corpus [17] and a Web-video set.

NPL corpus is a widely used standard set for text retrieval, in which there are 11429 documents and each of them is annotated by the relevance to different queries. We use all of these documents after processed by a WordNet stemmer. NPL also provides 93 queries for evaluating and we use all of them in our experiment.

Our second data set is a real Web-video data set. Text description for video and textual query from user are usually short and incomplete, therefore we expect that semantic query expansion will be appropriate. There is no benchmark dataset available for Web-video retrieval. Thus, we create a video set manually. We randomly collected 20,000 videos Web pages from YouTube website. The surrounding texts including description, tags and title were extracted via parsing the Web pages. All the videos are also attached with their original YouTube categories, including: 1) Entertainment; 2) Nonprofits; 3) Sports; 4)How To; 5) News; 6) Pets; 7) Travel; 8) Science. A WordNet stemmer is used to do stemming. We note that there is standard video collections, e.g., TRECVID [19], but the TRECVID data set does not fit Web video well, as argued in [18].

Metrics: The performance of retrieval on NPL corpus is measured by the wildly used standard 11-point Recall/Precision table [15], where *precision* is the number of correct results divided by the number of all returned results and *recall* is the number of correct results divided by the number of results that should have been returned. By varying the recall values, we can obtain different precision values, and then a curve (or table) described the relation between recall and precision can characterize the performance of methods clearly. Detailed information can be found in Hersh's book [16]. Moreover, we also report the performance of *Average Performance*, that is, as introduced in [20], the average of 11 precisions in the aforementioned 11-point Recall/Precision table.

It is very difficult to manually annotate the relevance of the whole Web-video set to a given user query. Hence, we cannot use 11-point Recall/Precision as we did on NPL corpus. Instead, we use Precision@10, one of the popular metrics used in information retrieval area, which is used to measure the fraction of the top 10 videos retrieved that are relevant to the user's query.

4.2 Performance Evaluation

Text Retrieval. We evaluate the performance of our methods on text retrieval. We use NPL, a standard text corpus for text retrieval, to carry out our experiment and employ standard pre-process, including elimination of stop words and stemming. We first report the performance of individual methods in Fig. 4, including 1) non-semantic cosine similarity, 2) WordNet-based query expansion [6] and 3) our methods using different CCOs (CCO0(WordNet), CCO1, CCO2 and CCO3).

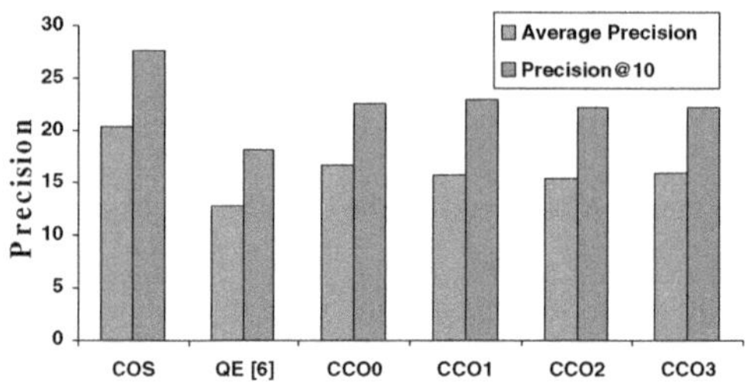

Fig. 4. Performance of Text Retrieval on NPL corpus

It is not surprising to see that each expansion method obtains worse performance than the non-semantic methods, because as argued in Voorhees's work [7], such expansion methods do not fit when query is relatively complete and long. This is understandable because query expansion for complete and long query often brings more noise than useful information. Although query expansion does not improve performance in general, we notice that semantic expansion provides useful complementary information for some queries.

If we allow users to manually choose semantic levels (each semantic level corresponds to a CCO) to be expanded, the query expansion using CCOs can improve the performance significantly as illustrated in Fig. 5. The line with label COS+QE shows the result of the method of choosing the better one between cosine similarity and query expansion [6]. The line with label COS+WN shows the optimal result of our method that can be achieved if users can manually select the best semantic levels using Wordnet. The line with label COS+CCOs shows the optimal result that our approach can achieve if users can manually select the best semantic levels using CCOs. The experimental results show that, combining WordNet-based query expansion methods [6] with cosine similarity brings little improvement, the average precision for 11-point recalls only increases from 20.29% to 20.51%. Meanwhile, our methods using WordNet as optional ontology can improve precision from 20.29% to 22.88%, and this performance can further improve to 23.38% if we use 4 CCOs as ontology. The improvement is quite significant. In practice, users may care more about precision. If we set recall at 10%, our approach using Wordnet can improve the precision of cosine from 42.63% to 48.75% and our approach using CCOs can further improve the performance to 50.33%, while the approach of selecting the best of cosine and the query expansion [6] can only slightly improve performance from 42.63% to 43.36%.

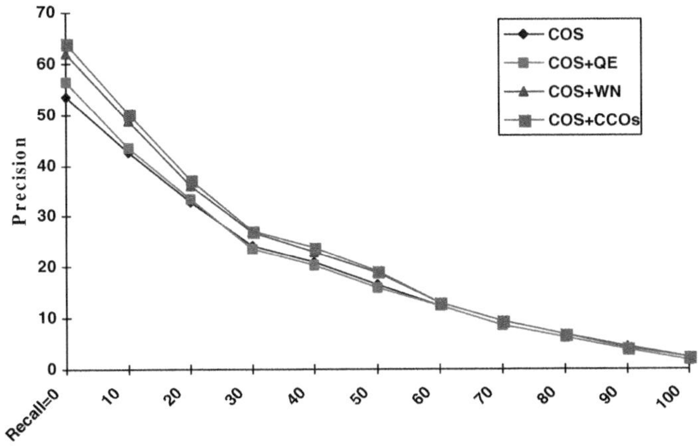

Fig. 5. Optimal Performance on NPL Corpus

As shown in Fig. 5, our query expansion methods are complemental to the non-semantic cosine similarity. Readers may wonder whether the WordNet-query expansion methods [6] can also provide the useful semantic levels. We tried to adjust the semantic levels of WordNet by increasing the expanding layers. The experimental results show that its performance decrease sharply when the number of expanding layers increase as shown in Fig. 6. The reason could be that there are already too much useless information added to the query.

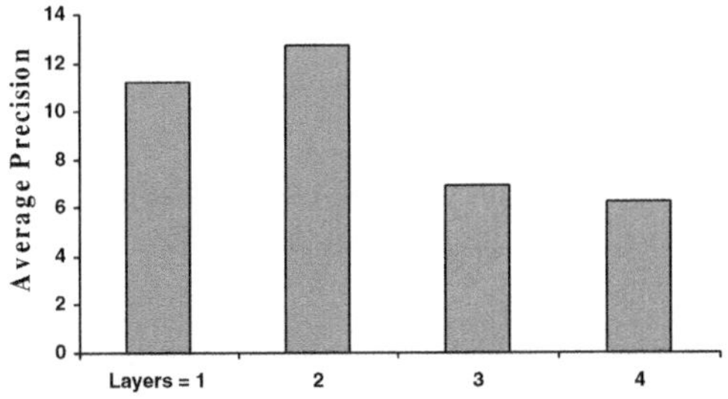

Fig. 6. The Decrease Trend of the approach [6] on NPL corpus

On the other hand, our CCO-based methods can meet the need of different semantic levels of different kinds of queries. We next show using examples in Fig. 7 that for different queries the best performance can be obtained at different semantic levels based on different CCOs. In this paper, we set 4 semantic levels which is easy for users to adjust the semantic levels.

As shown in the 5 sample queries in Fig. 7, the best performance can be obtained by selecting different ontologies, i.e. CCOs, which indicates the usefulness of semantic

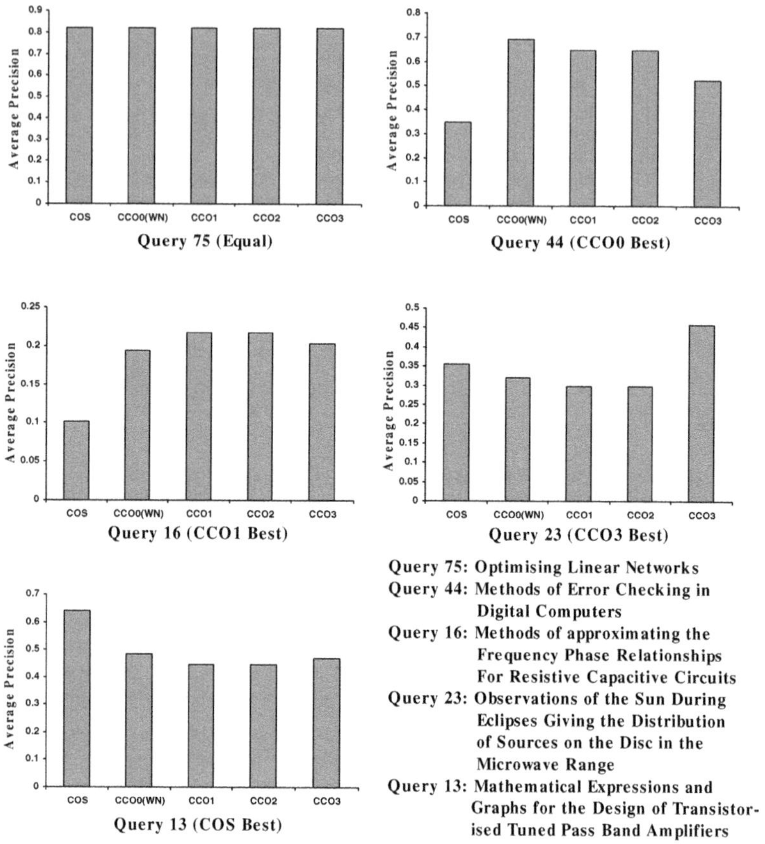

Fig. 7. Sample Query Results of CCOs-based Methods

levels of our methods. For different queries, user can adjust the ontology and thus different semantic level to enhance the performance of the retrieval system.

Text-Based Video Search. We next evaluate our methods on a real-world video data set. We randomly selected 10 noun phrases appeared in our dataset as queries, and checked whether the videos in top 10 returned results are relevant to the query. We first report the performance of each single method, as showed in Fig. 8.

WordNet-based query expansion method [6] outperforms cosine similarity, which is inconsistent with the results on NPL corpus. As discussed in Sec. 1, the text description of video is generally short and semantic expansion would be more appropriate. Additionally, the queries that we used are quite short. The results clearly show that query expansion based on CCOs provides superior performance on video text corpus. Our CCO-based methods are robust when query is short, because we obtain the same expansion pattern for the same word, in both documents and queries.

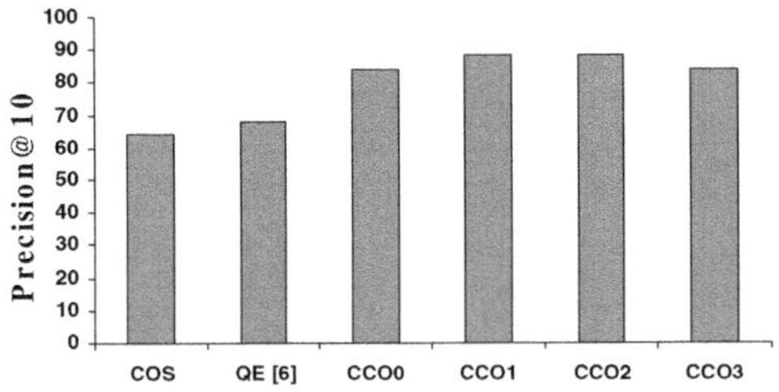

Fig. 8. Performance on Web-video Corpus

We next report the optimal results in Table 1, i.e., for each query we choose the best semantic level based on human judgement. The column COS represents the system using cosine similarity without semantic expansion, $COS + QE$ represents the system using query expansion method in [6] and we report the better results between COS and the method in [6], and $COS + CCOs$ is our system based on $CCOs$. We can see that $COS + CCOs$ significantly improves the accuracy of video search.

Table 1. Result of Experiment

	COS	COS+QE[6]	COS+CCOs
Precision@10	64%	72%	88%

5 A System Demonstration Based on Our Approach

We use our query expansion approach to create a text-based video searching system, i.e., *iVSearcher*, because a high-performance query expansion technology is even more needed in video area because of the normally short and incomplete descriptions and queries. With the explosive growth of the Internet and hardware, techniques for multimedia information retrieval are receiving increasing attention in multimedia research community. In recent years, there is massive volume of the online video information, empowered by the permeation of Internet into daily life. However, the technology for video retrieval has not kept pace with the fast growth of video data. Indeed, most of the existing video search engines, such as Google Video, apply the keyword search for video retrieval. Given the massive online video resources, it is becoming increasingly difficult to find an interested video, because the text description may not precisely capture the semantic information of a video.

The system has 20,000 text descriptions for online videos from YouTube. In this system, an end-user only needs to select the wanted $semantic\ level$ for his/her query and our system can determine the query expanding automatically with these given semantical level using our Compact Concept Ontologies (CCOs). All of these process need neither statistical information of corpus nor users' feedback of relevant, which is distinct from most existed methods.

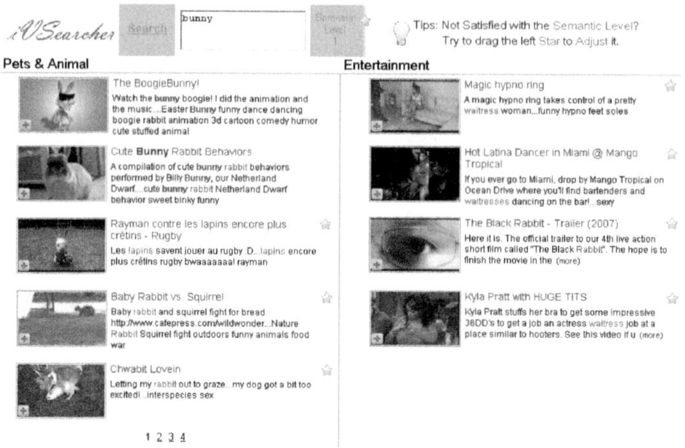

Fig. 9. Example of search result of query "bunny"

Fig. 9 gives the main interface of *iVSearcher* system. In contrast with other video search engines, *iVSearcher* integrates semantic information and allows the users to adjust semantic levels of his/her query by dragging the star mark, which will present different sets of search results. By setting semantic level to the lowest, our system simply works like a key-word based search engine, while a higher semantic level can return videos with similar semantic meanings. For example in Fig. 9, although the user query only contains *bunny*, all the videos described by *rabbit* or *lapin* are also returned, even the videos containing the word *waitress* – an unfamiliar sense of *bunny*. All the results are showed with category information and the videos marked with star cannot be returned by the classic keyword search method, which clearly shows the advantage of our approach. Note that, the user can select whether to show results in category or not, while we suggest to show the category information because of various semantic senses a query word may convey.

Fig. 10 shows another example to demonstrate the effectiveness of *iVSearcher* by comparing with the traditional keyword based approach and the query expansion approach in [6] using different queries. Our semantic-based search is very effective for the query *animal dance*; all returned videos by the classic bag-of-words model are not related to *animal dance*; the query expansion method [6] only returns those related to *animal*.

We can also see that the users can easily adjust the semantic levels to find promising results and the semantic levels are corresponding to the CCOs we generated, e.g., the four levels given in Fig. 3. We generate the weighted trees for each video's description using different CCOs. Given a query Q, we first generate the weighted tree of the query using the CCO determined by the semantic level given by user, then calculate the similarity with each video's weighted tree respectively, and finally return the most similar results. Fig. 9 and 10 give examples of our system. In Fig. 9 we use the query *bunny* with semantic level 1, which just uses WordNet (the largest CCO) as ontology. In Fig. 10, we use the query *animal dance* with semantic level 4, which uses the smallest

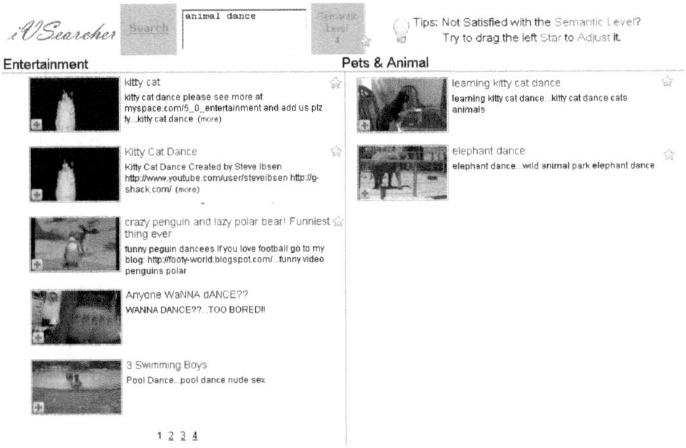

Fig. 10. Example of search result of query "animal dance"

CCO, which directly expands all kinds of animals to the node *animal* and holds the *dance* almost unchanged.

6 Conclusion

In this paper, we presented our novel query expansion mechanism for effective text retrieval. Specially, we proposed a strategy of WordNet-based query expanding and Compact Concept Ontologies (CCOs) to render different semantic levels accessible. Our experimental results show that the scheme retains a high degree of accuracy as compared to text retrieval method without query expansion. We demonstrated that demo system *iVSearcher* based on a new semantic similarity algorithm and CCOs is more effective than benchmark approaches for text/video retrieval. Moreover, with the proposed CCO mechanism, users can easily adjust the semantic levels, which provides better user experience.

In the future, we plan to explore the algorithm for automatically determining semantic levels based on CCO. A possible method is making semantic levels of an query expansion rely on the information provided by an expansion, which could be captured by models in information theory. Moreover, we are trying to collect more text descriptions of online videos to reflect the textual feature of current Web-video systems.

References

1. Boni, M.D., Manandhar, S.: Implementing clarification dialogues in open domain question answering. In: Natural Language Engineering (2005)
2. Fellbaum, C.D.: WordNet: An Electronic Lexical Database. MIT Press, Cambridge (1998)
3. Wu, Z., Palmer, M.: Verb semantic and lexical selection. In: ACL 1994 (1994)
4. Leacock, C., Chodorow, M.: Combining local context andwordnet similarity for word sense identification. MIT Press, Cambridge (1998)

5. Zhang, C., Wang, Y.J., Cui, B., Cong, G.: Semantic similarity based on compact concept ontology. In: WWW 2008 (2008)
6. Voorhees, E.M.: Using WordNet for Text Retrieval. In: Fellbaum, C. (ed.) WordNet An Electronic and Lexical Database. MIT Press, Cambridge (2008)
7. Mandala, R., Takenobu, T., Hozumi, T.: The Use of WordNet in Information Retrieval. In: COLING/ACL Workshop on Usage of WordNet in Natural Language Processing Systems (1998)
8. Aly, A.A.: Using a Query Expansion Technique To Improve Document Retrieval. In: Information Technologies and Knowledge, vol. 2 (2008)
9. Xu, J., Croft, W.B.: Query Expansion Using Local and Global Document Analysis. In: SIGIR 1996 (1996)
10. Jones, S.K.: Automatic keyword classification for information retrieval. Butterworth's (1971)
11. Deerwester, S., Dumai, S.T., Furnas, G.W., Landauer, T.K., Harshman, R.: Indexing by latent semantic analysis. Journal of the American Society for Information Science 41(4) (1994)
12. Salton, G., Buckley, C.: Improving retrieval performance by relevance feedback. Journal of the American Society for Information Science 41(4) (1990)
13. Buckley, C., Salton, G., Allan, J., Singhal, A.: Automatic query expansion using SMART. In: Proceedings of the 3rd Text REtrieval Conference (1994)
14. Xu, J., Croft, W.B.: Improving the effectiveness of information retrieval with local context analysis. In: ACM Transactions on Information System (2000)
15. Voorhees, E.M.: Query Expansion using Lexical-Semantic Relations. In: SIGIR 1994 (1994)
16. Hersh, W.R.: Information Retrieval: A Health and Biomedical Perspective. Springer, Heidelberg (2003)
17. Sparck-Jones, K., Webster, C.A.: Research in relevance weighting. British Library Research and Development Report 5553, University of Cambridge (1979)
18. Chow, J.H., Dai, W., Zhang, R.F., Sarukkai, R., Zhang, Z.F.: Joint categorization of queries and clips for Web-based video search. In: MIR 2006 (2006)
19. Smeaton, A.F., Over, P., Kraaij, W.: Evaluation campaigns and TRECVid. In: MIR 2006 (2006)
20. Croft, W.B., Harding, S.: An Evaluation of Information Retrieval Accuracy with Simulated OCR Output. In: In Proceedings of the Third Annual Symposium on Document Analysis and Information Retrieval (1994)
21. Qiu, Y., Frei, H.P.: Concept based query expansion. In: SIGIR 1993 (1993)
22. Crouch, C.J., Yang, B.: Experiments in automatic statistical thesaurus construction. In: SIGIR 1992 (1992)

A Unified Indexing Structure for Efficient Cross-Media Retrieval*

Yi Zhuang[1,2], Qing Li[3], and Lei Chen[4]

[1] College of Computer & Information Engineering, Zhejiang Gongshang University, P.R.China
[2] Zhejiang Provincial Key Laboratory of Information Network Technology, P.R.China
[3] Dept of Computer Science, City University of Hong Kong, HKSAR, P.R.China
[4] Dept of Computer Science & Engineering, HKUST, HKSAR, P.R.China
zhuang@zjgsu.edu.cn, itqli@cityu.hk.edu, leichen@cse.ust.hk

Abstract. An important trend in web information processing is the support of content-based multimedia retrieval (CBMR). However, the most prevailing paradigm of CBMR, such as content-based image retrieval, content-based audio retrieval, etc, is rather conservative. It can only retrieve media objects of single modality. With the rapid development of Internet, there is a great deal of media objects of different modalities in the multimedia documents such as webpages, which exhibit latent semantic correlation. Cross-media retrieval, as a new multi- media retrieval method, is to retrieve all the related media objects with multi- modalities via submitting a query media object. To the best of our knowledge, this is the first study on how to speed up the cross-media retrieval via indexes. In this paper, based on a Cross-Reference-Graph(*CRG*)-based similarity retrieval method, we propose a novel unified high-dimensional indexing scheme called *CIndex*, which is specifically designed to effectively speedup the retrieval performance of the large cross-media databases. In addition, we have conducted comprehensive experiments to testify the effectiveness and efficiency of our proposed method.

1 Introduction

With the rapid development of Internet and multimedia technology, content-based multi- media retrieval and indexing issues have been extensively studied for many years. The existing approaches focus on the content-based retrieval for single-modality-based media objects, such as content-based image retrieval [2, 3], content-based audio retrieval [7] and content-based video retrieval [5], etc. However, as shown in Figure 1, for the media objects of different modalities (e.g., image, audio and video, etc) appeared in webpages, there exists to some extent latent semantic

* This work is partially supported by the Program of National Natural Science Foundation of China under Grant No. 60873022; The Program of Zhejiang Provincial Natural Science Foundation of China under Grant No. Y1080148; The Key Program of Science and Technology of Zhejiang Province under Grant No. 2008C13082; The Open Project of Zhejiang Provincial Key Laboratory of Information Network Technology; The Key Project of Special Foundation for Young Scientists in Zhejiang Gongshang University under Grant No. Q09-7.

X. Zhou et al. (Eds.): DASFAA 2009, LNCS 5463, pp. 677–692, 2009.

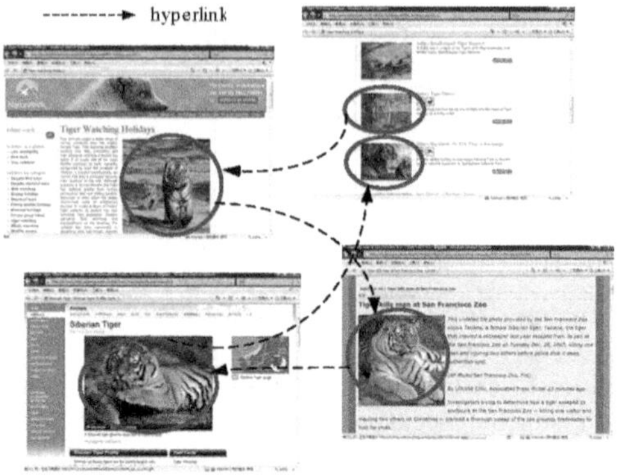

Fig. 1. The latent semantic correlation in the webpages

correlation among them. Cross-media retrieval [11, 12] is a new multimedia retrieval method which has received increasing attention due to its practical usage. It is perceived as a novel retrieval method which returns the media objects of multi-modalities in response to a query media object of single modality. For example, when user submits a "tiger" image, the retrieval system may return some "tiger"-related audio clips and video clips via using the so called *Cross Reference Graph*, as to be introduced in Section 3.

Compared with the traditional single-modality retrieval [2, 3, 4, 5], the cross-media retrieval, also known as multi-modality search in other work such as [10], tries to breakthrough the restriction of media modality in multimedia retrieval. The fundamental challenges of the cross- media retrieval lie in two folds:

- For an image and an audio clip, it is hard to measure the correlation between them with the same semantic information due to their heterogeneity of modalities. So how to model the correlation of media objects of different modalities(e.g., image, audio and video, etc) via mining the semantic correlation of media objects among the multimedia documents such as webpages? In addition, to effectively facilitate the cross-media retrieval with different modalities, how to generate and refine a cross reference graph(*CRG*) which is the representation of the correlation of different media objects?

- To efficiently facilitate the large-scale cross-media retrieval, a naïve solution is to create an index for each media and aggregate the results from each media indexes. However, this is definitely not a practical approach due to the space and I/O cost, so how to design a novel integrated indexing scheme to index such a large *CRG* is an important issue.

In this paper, we address both the effective and efficient issues of cross-media retrieval together and mainly focus on the efficient retrieval techniques. We first briefly review a novel cross-media retrieval method based on *CRG* from our previous work [12], then we propose a unified indexing structure called CIndex, which is specifically designed for indexing the cross- media retrieval over large multi-modality media databases. With the aid of the CIndex, a cross- media retrieval of query example in high-dimensional spaces is transformed into a range query in the single dimensional space.

The primary contributions of this paper are as follows:

1. Based on the novel cross reference graph(*CRG*) [12] as a foundation of cross-media retrieval, we propose a novel unified multi-feature-based

high-dimensional indexing structure called CIndex to facilitate the efficient cross-media retrieval.
2. We present a theoretical analysis and comparison on the search and storage cost of the proposed method.
3. We perform an extensive experimental study using some 50,000 images, 2000 audio clips and 5000 video clips to evaluate the efficiency of our approach.

The rest of this paper is arranged as follows. We briefly review the related work in Section 2. In Section 3, the preliminaries are given. In Section 4, for the purpose of self-containment, we briefly summarize our previous work on the cross-media retrieval [12]. In Section 5, a unified index structure called *CIndex* is proposed to dramatically speed up the retrieval performance. In Section 6, we report the results of extensive experiments which are designed to evaluate the effectiveness and efficiency of the proposed approach. Finally, we conclude the paper in Section 7.

2 Background

Our work is motivated by, and based on, previous research on multi-media retrieval, high- dimensional indexing and multi-feature indexing, as reviewed in this section.

Previous work addressing the multimedia retrieval can be classified into two groups: single modality and multi-modality fusion.

The retrieval approach in single modality deals with a single type of media, and most content-based retrieval approaches (e.g., [2, 3, 4]) fall into this group. Among them, the QBIC system [2], MARS project [3], VisualSEEK system [4] focus on image retrieval, VideoQ system [5] is for video retrieval, and WebSEEK [6] system is a Web-oriented search engine that can retrieve both images and video clips. These approaches differ from each other in either the low-level features extracted from the data, or the distance functions used for similarity calculation. Despite the differences, all of them are similar in the following fundamental aspects: (1) they all rely on low-level features; (2) they all use the query- by-example paradigm; (3) they are single-modality-based retrieval systems.

Recently, with the rapid development of Internet and multimedia technology, a great deal of multimedia information has been produced. Especially for the large-scale webpage repository downloaded from Internet(cf. Figure 1), there exists some latent semantic correlation among different media objects extracted from the webpages. Multi-modality retrieval has received much attention increasingly. The semantic-based correlation analysis of different types of media objects of different modalities is becoming a hot research topic. *Octopus* [10] is an early prototype system which can support multi-modality retrieval. However, no index has been considered in that system, so the retrieval performance of the *Octopus* is not satisfactory when the number of media objects becomes large. More recently, of the work on Cross-media retrieval in [11, 12], as an extension to traditional multimedia retrieval, has been conducted which tries to fuse media objects of different modalities.

The cross-media indexing issue belongs to high-dimensional indexing and multi-feature indexing method. There is a long stream of research on solving the high-dimensional indexing problems [13]. Existing techniques can be divided into three main categories. The first category is based on data & space partitioning, hierarchical tree index structures such as the R-tree [14] and its variants [13], etc. Their performance deteriorates rapidly as the dimensionality increases due to the "*curse of dimensionality* "[13]. The second category is to represent original feature vectors using smaller, approximate representations, such as VA-file [15]. Although it can accelerate the sequential scan by using data compression, it incurs higher computational cost to decode the bit-string. The last category is using a distance-based indexing method such as iDistance [16], etc. iDistance is a distance-based scheme, in which high-dimensional points are mapped to a single-dimension distance values by computing their distances from the centroid respectively, which are then indexed by a B^+-tree. The drawback of iDistance is that it can not significantly prune the search region and especially when the dimensionality becomes larger. In [8], Jagadish proposed a multi-feature indexing method. Meanwhile, Shen, et al [9] proposed a multi-scale indexing (MSI) structure for web image retrieval, which nicely combines two modality information of image, such as its semantic and visual features. However the usability of the MSI structure is very limited and it can only support two kinds of media objects (e.g., text and image), thus not truly supporting search of multi-modality media objects.

3 Preliminaries

As we know, to measure two media objects with the same modality, primitive/low-level similarity as the metric. However, for two media objects with different modalities, the correlation between such two media objects needs to be introduced.

Definition 1. *A multimedia document(MD) can be modeled as a five-tuple:*

MD := <DocID, URISet, KeywList, ElemSet, LinkSet>

where DocID is denoted as an identifier of a document; URISet is the identifiers of the uniform resource location of different media objects; KeyList includes the semantic information of this document; ElemSet includes the features (low-level visual or audio features and high-level semantic features) of the media objects, denoting the set of the media objects in the corresponding document; LinkSet is the set of (hyper)links extracted from documents(e.g., webpages).

For media objects of the single modality, Table 1 shows the similarity metrics of each type. The symbols to be used in the rest of the paper are shown in Table 2.

Table 1. Primitive features and similarity metric adopted

Media	***Features***	***Similarity Metrics***
Image	256 HSV 32-d Tamura texture	Euclidean Distance
Audio	4 spectral characteristics: (1). *Spectral Centroid*; (2). *Rolloff*; (3). *Spectral Flux*; (4). *RMS*	To extract the centroid of a audio clip using the fuzzy clustering algorithm [7], and use the cosine function as a similarity metric
Video	shot boundary detection, using first frame of each shot as key-frame	key-frame similarity as shot similarity, average pair-wise shot similarity as video similarity

Table 2. The symbols used

Symbol	Meaning
Ω	The media objects database, $\Omega=\{X_1,X_2,\ldots,X_n\}$
X_i	The i-th media object, where the media object X can image(I), audio(A) and video(V), etc.
O_j	The centroid of the j-th cluster sphere
CR_j	The class radius for the j-th cluster sphere
$d(X_i,X_j)$	The similarity distance between two media objects (X_i and X_j), which is defined in Table 1, where X can be I, A and V, etc.
$\lvert\bullet\rvert$	The number of $\bullet$
n	The total number of media objects and $n=\lvert\Omega\rvert$
MD_j	The j-th multimedia document

4 CRG-Based Cross-Media Retrieval: *A Review*

To support cross-media retrieval, in our previous work [12], we have studied the retrieval method for cross-media through mining the co-existence information of the heterogeneous media objects and users' relevance feedbacks. In this section, we provide a summary review of the relevant materials which are essential for our subsequent discussions.

As mentioned before, different media objects correspond to heterogeneous low level features (i.e., the features extracted are different and the dimensionalities of features are different as well), so it is hard to directly measure the similarity between two media objects of different modalities. For example, an image is represented with some visual perceptual features such as color histogram, texture and shape, etc. An audio clip can be represented with some intrinsic acoustic features such as timbre, rhythm, pitch and daubechies wavelet coefficient histograms (DWCHs), etc. Therefore, it is not easy to establish the correlation between these different media objects by only using their low-level features. Instead of the low-level features, we can use the coexistence information such as the corresponding hyperlink information of the media objects in webpages to establish their relationships. This is because to some extent, some latent semantic correlations between different media objects may exist in a webpage or different webpages. Hence, in order to support the effective cross-media retrieval, it is critically important to evaluate the correlation among media objects of different modalities. In this section, we introduce a *Cross Reference Graph*(*CRG*) model in which the media object of different modalities are uniformly represented as vertices and the correlations among them are the weights of the weighted edges in it.

Definition 2 (Cross Reference Graph). *A Cross Reference Graph(CRG) is denoted as CRG=(V, E), where V is a set of media objects, E is a set of edges. Note that, for two media objects of the same modality, E refers to the similarity of them. However, for two media objects of different modalities, E refers to the correlation between them.*

Figure 2 shows an example of the Cross Reference Graph model. Note that, the solid line is denoted as the similarity metric and the dash line is represented as the correlation between two different media objects. Assume that there are three kinds of media objects in an MD, namely image, audio and video, and there is at least one media object in each MD. However, the proposed method can be easily extended to more kinds of media objects. Let **W** be an *n*×*n* affinity matrix with $Wij=d(Xi,Xj)$, representing the *CRG*, we construct the *CRG* via the flowing five steps.

Step 1. The affinity matrix **W** is initialized by:

$$Wij=\infty \ (0<i,\ j<n) \tag{1}$$

Step 2. The **W** is constructed by Eq.(2) defined in Table 2. The low-level feature distance of image, audio and video must be normalized respectively.

$$\begin{cases} \forall X_i, X_j \in I, & W_{ij} = \left\| X_i^f - X_j^f \right\| \\ \forall X_i, X_j \in A, & W_{ij} = \left\| X_i^f - X_j^f \right\| \\ \forall X_i, X_j \in V, & W_{ij} = \left\| X_i^f - X_j^f \right\| \end{cases} \tag{2}$$

Step 3. Learn from the media object co-existence information. The graph is modified as Eq.(3) in which ε is a small constant and $X_i^h = X_j^h$ means that Xi and Xj are in the same *MD*, where $Xi, Xj \in \Omega$.

$$W_{ij} = \varepsilon \quad if \quad (X_i, X_j \in \Omega \wedge X_i^h = X_j^h) \tag{3}$$

Step 4. In this step, we first model the local geometrical structure as Eq.(4) where σ is a small constant reflecting the view of locality:

$$W_{ij} = \begin{cases} W_{ij} & if\,(W_{ij} < \sigma) \\ \infty & otherwise \end{cases} \tag{4}$$

We define the length of a path as the sum of the weights along the path. To model the global geometrical structure of the *CRG*, we reconstruct the graph by finding the shortest paths for all pairs of vertices in it, and then replace the weights Wij by the length of shortest paths between each pairs of vertices.

Step 5. In the last step, we use a correlation matrix **C** to measure the correlation among the media objects which is shown in Eq.(5), where $wij \in \mathbf{W}$ and θ is a smoothing factor:

$$C_{ij} = \begin{cases} \exp\left({-w_{ij}^2}\big/{2\theta^2} \right) & if \quad i \neq j \\ 0 & if \quad i = j \end{cases} \tag{5}$$

The proposed *CRG* construction approach actually nonlinearly fuses the information carried by media objects of different modalities according to their low-level feature and co-existence. In other words, the *CRG* is a multimedia manifold representation which can reflect the multimedia correlation more precisely [19]. As media objects of different modalities are represented uniformly in the *CRG*, hence the cross-media retrieval can be facilitated more easily.

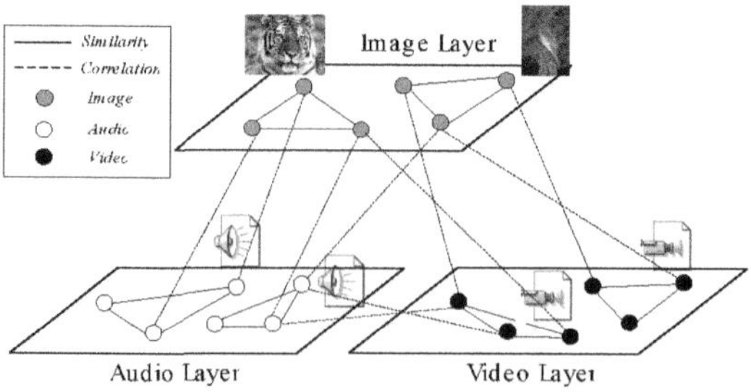

Fig. 2. An example of the Cross Reference Graph model

Although the *CRG* can model the correlations among media objects of different modalities to some extent, there still exist two problems. First, due to the semantic gap, although the proposed manifold-learning-based fusion method can reduce the ambiguity at a semantic level, the geodesic distance defined in the *CRG* is just an approximation of the correlations of media objects since they are estimated based on low-level feature distances. Second, it is difficult for the *CRG* to recover the multimedia manifold accurately if the number of media objects is not large enough. Hence, the user feedbacks need to be adopted to refine the *CRG* and make it consistent with the users' perception, which is locally isometric to the multimedia semantics. To enhance the retrieval accuracy, a practical approach, a log-based learning of feedback is proposed to refine the *CRG*, which is described in [12] in detail.

5 The CIndex Structure

When the cross-media database size becomes large, its corresponding *CRG* will become large. Retrieving related media objects in such a large graph is a CPU and I/O intensive operation. So it is inefficient to retrieve over large cross-media databases by only using linear scan. In this section, we propose a novel unified index structure, which is called CIndex, to facilitate the efficient cross-media retrieval.

5.1 Pruning Heuristics

As a preliminary step, three media objects such as image, audio and video are first clustered by adopting a hierarchical clustering algorithm such as BIRCH [14]. Figure 3 illustrates the three subspaces for the three different type media objects (i.e., image, audio and video).

To support the cross-media retrieval, a cross reference graph is next obtained through learning the co-existence of media objects in the webpages, as discussed in Section 4. For example, for an image, its corresponding cross reference graph can be modeled as an adjacency list. As shown in Figure 4, in particular, for the image whose ID is 21, the ID numbers of the semantically correlated audio clips are 3, 9, 18 and 26, and its corresponding video clips are 7 and 39, respectively. Notice that the values below the IDs are the correlation values of two media objects with different modality.

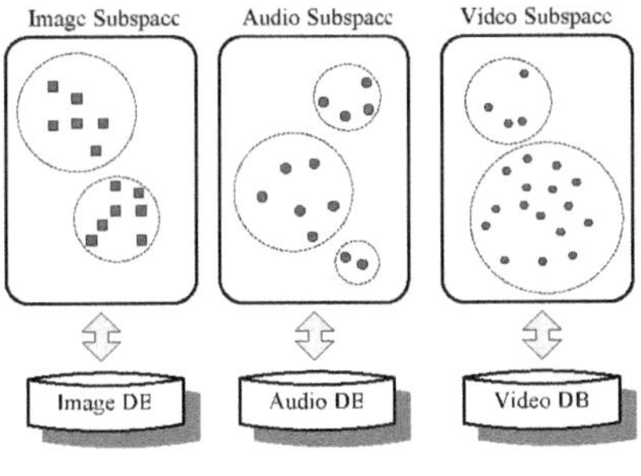

Fig. 3. The three subspace clustering

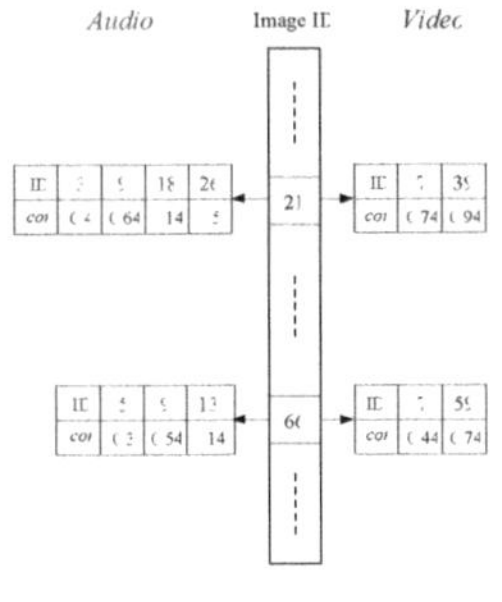

Fig. 4. The adjacency-list-based *CRG* representation

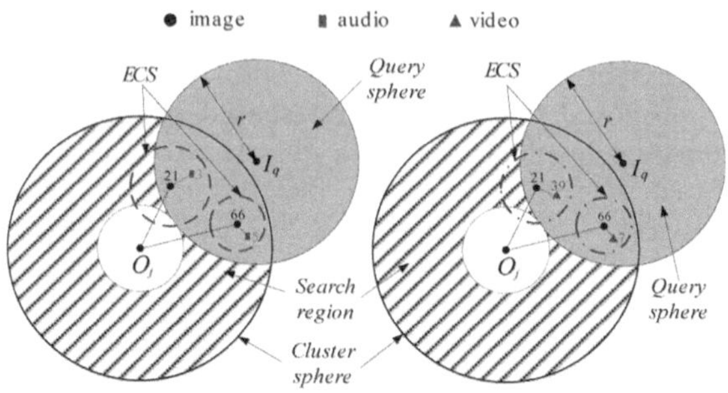

(a). The ECS for Audio (b). The ECS for Video

Fig. 5. The two corresponding ECSs in the high-dimensional feature space of images

Definition 3 (Embedded Correlation Subspaces). *For a media object X_i, the media objects whose modalities are different from X_i are in its embedded correlation subspaces (ECS) if their semantics are the same as X_i's, where X can be an image(I), audio(A) or video(V), etc.*

Based on Definition 3, without loss of generality, we choose image as an example. In Figure 5, each image corresponds to two *Embedded Correlation Subspaces*[1] which are organized as two adjacency lists(ref. Figure 4). That is to say, the media objects in the same ECS is semantically similar/identical. For example, in the image's feature space of Figure 5(a), the (*red*) dash circle refers to the ECS for an audio whose semantics is identical to that of the image whose ID number is 21. Similarly, in Figure 5(b), the (*blue*) dash circle represents the ECS for a video.

Like the iDistance [16], the basic idea of the pruning method of the CIndex is to transform the cross-media similarity search in high-dimensional spaces into the range searches in one- dimensional space.

5.2 The Data Structure

In order to efficiently facilitate the pruning operation, we propose the CIndex-support method in which some index keys are derived below.

Specifically, for a query image I_i, to efficiently retrieve the audio clips relevant to I_i, the index key of I_i can be represented as follows:

$$key(I_i)=\beta*\langle d(I_i,O_j),\theta\rangle+c(I_i,A_k)/_{MAX} \quad (6)$$

where $d(I_i,O_j)$ denotes the similarity distance between I_i and O_j, $c(I_i,A_k)$ refers to the correlation between I_i and audio object A_k , as defined by Eq.(5). $\langle \bullet,\theta\rangle$ denotes the rounded • value with θ decimal places and θ={1,2,3,…}. The symbol β is a large constant (e.g., 10^3) which makes $\langle d(I_i,O_j),\theta\rangle$ an integer. The value of $c(I_i,A_k)$ should be normalized into [0,1] through division by a constant *MAX*, thus the value range of $d(I_i,O_j)$ will not overlap with that of $c(I_i,A_k)$.

[1] Note that, for a media object X_i, its corresponding embedded correlation subspace is assumed as a virtual one in which the semantics of the correlated media objects of different modalities is the same to that of X_i .

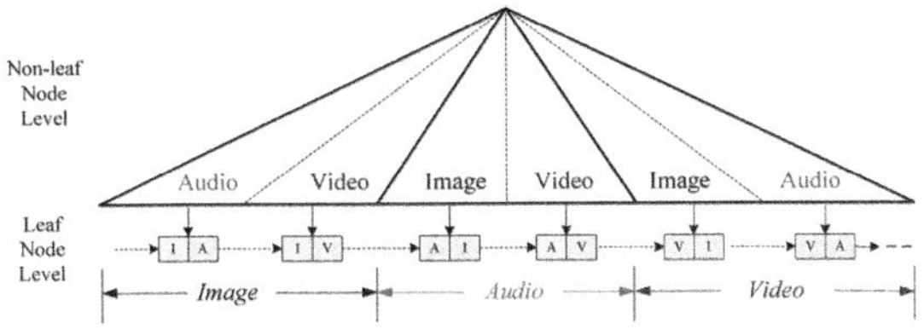

Fig. 6. The CIndex index structures

Since the images are grouped into T clusters, to get a uniform index key of image in different clusters, the index key in Eq. (6) can be rewritten by Eq. (7):

$$key(I_i)=\alpha*CID+\beta*\langle d(I_i,O_j),\theta\rangle+c(I_i,A_k)/MAX \tag{7}$$

where the CID is the ID number of cluster that I_i falls in, and α is a stretch constant which is set 5×10^3 empirically.

Similarly, to facilitate retrieving videos via submitting an image I_i, the index key of I_i can be represented by Eq. (8):

$$key(I_i)=\alpha*CID+\beta*\langle d(I_i,O_j),\theta\rangle+c(I_i,V_w)/MAX \tag{8}$$

where $c(I_i,V_w)$ refers to the correlation between I_i and video object V_w.

To further obtain a uniform index key expression, we combine Eq.(7) and Eq.(8) together by adding two constants (i.e., $S\_A=0$ and $S\_B=5\times10^5$). Therefore, a uniform cross-media index key for the image I_i can be rewritten below:

$$key(I_i)=\begin{cases} S\_A+\alpha*CID+\beta*\langle d(I_i,O_j),\theta\rangle+c(I_i,A_k)/MAX \\ \quad \text{if the correlated audio of } I_i \text{ is } A_k \\ S\_B+\alpha*CID+\beta*\langle d(I_i,O_j),\theta\rangle+c(I_i,V_w)/MAX \\ \quad \text{if the correlated video of } I_i \text{ is } V_w \end{cases} \tag{9}$$

Analogously, for the index key of an audio(A_k) and video (V_w), their corresponding uniform cross-media index keys can be derived as follows:

$$key(A_k)=\begin{cases} S\_C+\alpha*CID+\beta*\langle d(A_k,O_j),\theta\rangle+c(A_k,I_i)/MAX \\ \quad \text{if the correlated image of } A_k \text{ is } I_i \\ S\_D+\alpha*CID+\beta*\langle d(A_k,O_j),\theta\rangle+c(A_k,V_w)/MAX \\ \quad \text{if the correlated video of } A_k \text{ is } V_w \end{cases} \tag{10}$$

$$key(V_w)=\begin{cases} S\_E+\alpha*CID+\beta*\langle d(V_w,O_j),\theta\rangle+c(V_w,I_i)/MAX \\ \quad \text{if the correlated image of } V_w \text{ is } I_i \\ S\_F+\alpha*CID+\beta*\langle d(V_w,O_j),\theta\rangle+c(V_w,A_k)/MAX \\ \quad \text{if the correlated audio of } V_w \text{ is } A_k \end{cases} \tag{11}$$

where $S\_C=10^6$, $S\_D=1.5\times10^6$, $S\_E=2\times10^6$ and $S\_F=2.5\times10^6$.

Eqs. (9), (10) and (11) represent the cross-media index keys of image, audio and video respectively, which correspond to three independent indexes. In order to incorporate them into an integral index, we derive a new uniform index key expression as shown in Eq. (12).

$$key(X_i) = \begin{cases} SCALE\_I + key(I_i), & if\ X_i = I_i \\ SCALE\_A + key(A_i), & if\ X_i = A_i \\ SCALE\_V + key(V_i), & if\ X_i = V_i \end{cases} \tag{12}$$

where Xi denotes a media object(i.e., Xi can be an image, a audio clip or a video clip). The three constants $SCALE\_I$=0, $SCALE\_A$= 5×10^6 and $SCALE\_V$=10^7 should be set large enough to stretch the value ranges of the index keys so that they do not overlap with each other.

Algorithm 1. CIndexBuild(Ω, *CRG*)

Input: Ω: media object repository, *CRG:* the Cross reference graph;
Output: *bt*: the CIndex
1. initialize;
2. while ($X_i \in \Omega$) /* X_i *can be an image, audio or video* */
3. locate the X_i in the *CRG*;
4. get the media objects semantically related to X_i;
5. *bt*←***InsertBtree***(**key**(X_i)); /* key(X_i) is shown in Eq.(12) */
6. end while
7. return *bt*

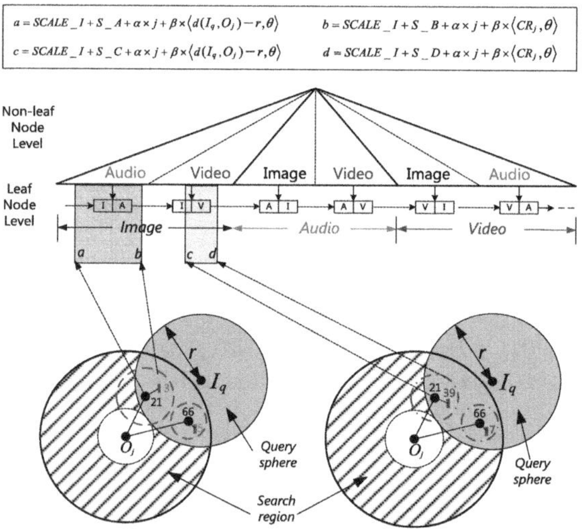

Fig. 7. The CIndex-based cross-media retrieval

5.3 Query Algorithm

To utilize the CIndex to support cross-media retrieval with different modality media object, the query process is composed of two stages (cf. Figure 7): the media object filtering and the result refinement. When user submits a query object Xq[2] as an input,

[2] In Figure 7, the query object is the image Iq. The aim of its cross- media retrieval is to return the related media objects of different modalities (e.g., *audio* and *video*) with respect to Iq.

we first get the correlated media objects of *Xq* via a CIndex-based searching of the cross reference graph(*CRG*). Then, a refinement process is conducted. Note that, the query object *Xq* can either be an image(*Iq*), audio clip(*Aq*) *or* a video clip(*Vq*). The details are given by **Algorithm 2**. It is worth mentioning in function ***Search***(*Xq,r,j*), given a query (i.e., *Xq* and *r*), for its *j*-th affected (i.e., *intersected* or *contained*) cluster, its corresponding search range [*left, right*] can be derived as follows:

$$left = SCALE\_X + S\_Y + \alpha \times j + \beta \times \langle d(X_q, O_j) - r, \theta \rangle \text{, and}$$

$$right = SCALE\_X + S\_Y + \alpha \times j + \beta \times \langle CR_j, \theta \rangle$$

where *X* in *Xq* is the same as the *X* in *SCALE_X*, *SCALE_X* and *S_Y* are two constants mentioned in Section 4.1 in which *X* can be *I*, *A* and *V*, *Y* can be *A*, *B*, *C*, *D*, *E* and *F*.

Algorithm 2. CrossSearch(X_q, r)

```
Input: a query media object Xq, a query radius r
Output: the query result S
S←Φ;                /* initialization */
for each other media types except itself(Xq) do
  for j:=1 to T do     /* T is the number of clusters */
    if Θ(Oj,CRj) dose not intersects Θ(Xq,r) then
      next loop;
    else
      S1←Search(Xq,r, j);
      S←S∪S1;
      if Θ(Oj,CRj) contains Θ(Xq,r) then end loop;
    end if
  end for
end for
return S;           /* return the candidate media objects */

Search(Xq,r, j)
left ← SCALE_X + S_Y + α×j + β×⟨d(Xq,Oj) − r,θ⟩;
right ← SCALE_X + S_Y + α×j + β×⟨CRj,θ⟩;
S←BRSearch[left, right, j];
for each media object Xi∈S
  if d(Xq,Xi)>r then S←S-Xi;  /* Xi is deleted from the candidate set S */
end for
return S;
```

5.4 Aggregating the Query Results

In **Algorithm 2** above, the query result *S* includes the different modality media objects which are semantically similar/identical to the query example. It is thus imperative to sort the media objects in terms of the similarity or correlation values between the media objects in *S* and the query one. In this regard, we propose a fusion-based aggregating algorithm (**Algorithm 3**) for the answer set, with the following formula for calculating the ranking score:

$$RankScore = \beta * \langle d(X_q, X_i), \theta \rangle + c(X_i, Y_j) / MAX \qquad (13)$$

where *Xi* and *Yj* are two semantically related media objects of different modality.

Algorithm 3. SearchRank(S, X_q, r)

Input: The query result S, a media object X_q, query radius r
Output: the ranked query result S'

1. $S' \leftarrow \Phi$; /* *initializtion* */
2. for X_i, $X_j \in S$ do
3. compute the similarities between X_i and X_q, and sorted;
4. compute the correlation of X_i and Y_j and sorted;
5. a comprehensive ranking score of X_i can be obtained based on Eq. (13);
6. end for
7. return S'; /* *return the ranked query results* */

5.5 Extensibility of CIndex

In anticipation of the continuous advancement of multimedia technology, for which new media types may emerge, our proposed CIndex shows a good characteristic of extensibility. When a new media type comes, the CIndex can be easily extended to support such new media type without re-organizing the original index data, due to that the CIndex is a distance-based indexing scheme. For the purpose of exposition, Figure 6 which is the adopted CIndex to support cross-media retrieval over image, audio and video data, is transformed into the new one as shown in Figure 8 when a new type of media data such as Flash movies [15] need to be accommodated. Note that the shadow parts of the CIndex in Figure 8 represent the newly inserted ones. Consequently, the new CIndex can support more powerful cross-media retrieval over not only image, audio and video data, but also Flash movies.

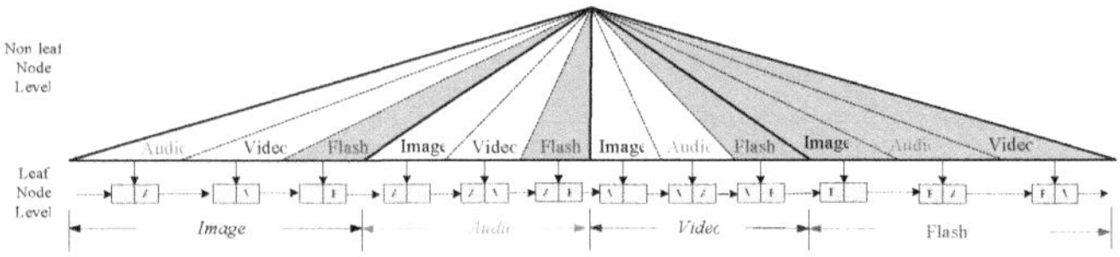

Fig. 8. The extensibility of the CIndex Structure

6 Experimental Results

To test the effectiveness and efficiency of the proposed cross-media retrieval and indexing method, we have implemented a prototype system, called *CMR*, to facilitate the cross-media retrieval over image, audio and video databases. The experimental data includes 50000 images, 2000 audio clips and 5000 video clips, which are randomly downloaded from the Internet and collected from Microsoft Encarta [20] – a multimedia encyclopedia. Note that, compared with our previous work in [12], the data size of the experiments is much larger than that of [12]. All the experiments are executed on a Pentium IV CPU at 2.0GHz with 256 Mbytes memory and index page size is fixed to 4096 Bytes.

6.1 Effectiveness of the Retrieval Method

In this experiment, we testify the effectiveness and efficiency of our retrieval method. As shown in Figure 9, when user submits an audio clip of "bird", several candidate

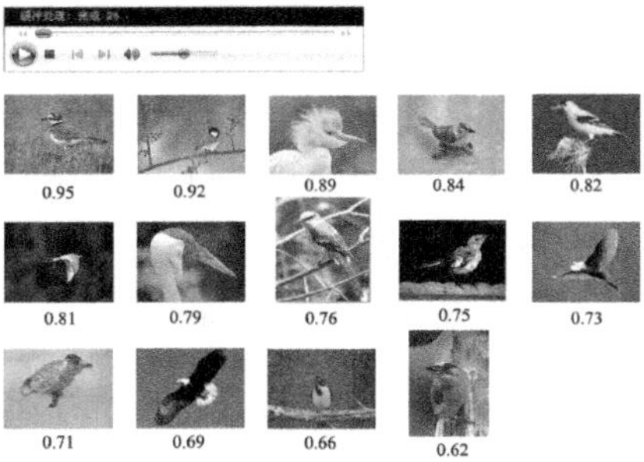

Fig. 9. One retrieval example

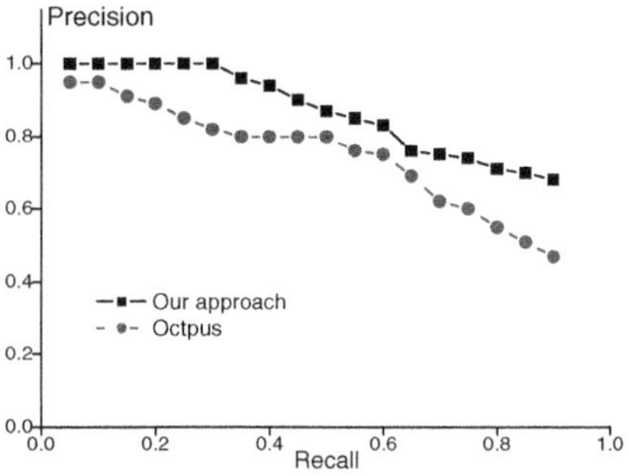

Fig. 10. Recall *vs.* precision

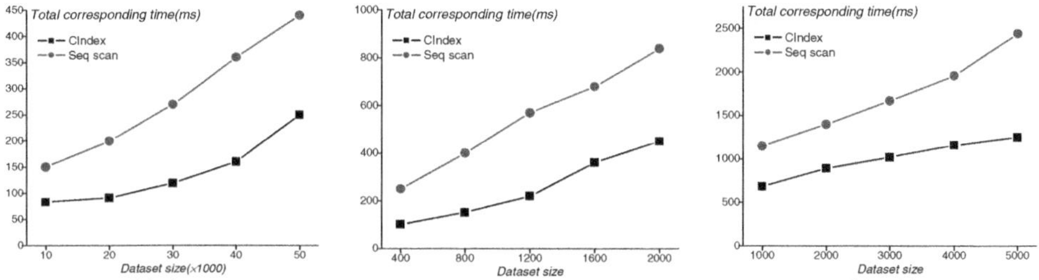

(a). image (b). audio (c). video

Fig. 11. Effect of data size

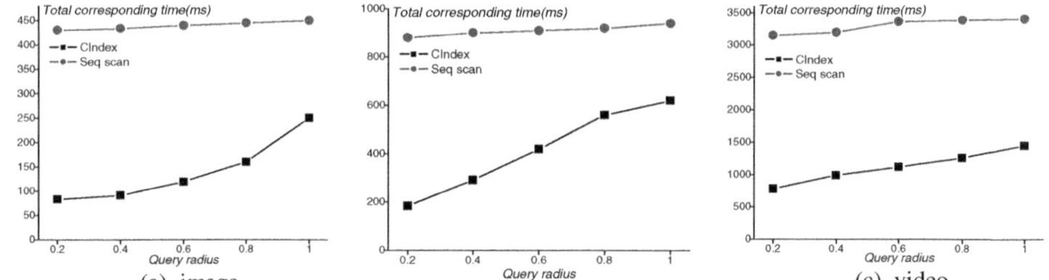

(a). image (b). audio (c). video

Fig. 12. Effect of query radius

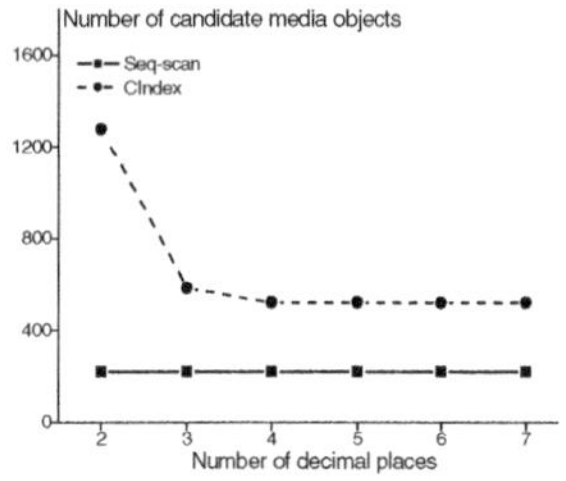

Fig. 13. Effect of θ on the efficiency of range search

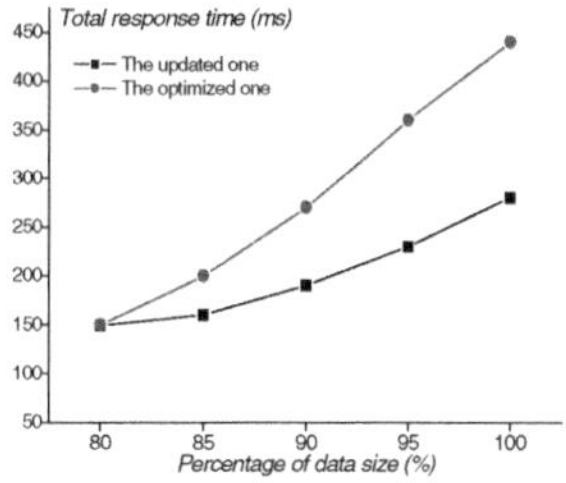

Fig. 14. Effect of Dynamic Insertion

media objects are retrieved by the system with the aid of the *CRG*. The number below the candidate images are similarity metric values.

Denoting the set of ground-truth as *rel*, and the set of results returned by a *k*-NN search as *ret*, the recall and precision achieved by this retrieval method are defined as:

$$recall = \frac{|rel \cap ret|}{|ret|}, \quad precision = \frac{|rel \cap ret|}{|rel|} \tag{14}$$

Figure 10 illustrates a *recall-precision* curve for the performance comparisons between the *CRG* method and that of Octopus [10]. In particular, it compares the average retrieval result (the average precision rate under the average recall rate) of 20 media objects queries randomly chosen from the database. The figure shows that the retrieval effectiveness of our proposed method is better than that of the *Octopus* by a large margin. This is because during the training process, our algorithm not only learns from the low-level features of the media objects but also learns from the co-existence. Moreover, different from the *Octopus* [10], such information is represented and reinforced by a manifold way which can make a positive impact on the improvement of the retrieval accuracy.

6.2 Efficiency of the CIndex

In the following, we test the performance of our proposed CIndex under different sizes of databases and different selectivity. We also study the effect of tunable parameters T and θ. Finally, we test the performance of CIndex under the situations of dynamic update.

6.2.1 Effect of Data Size

In this experiment, we measure the performance behavior with varying number of media objects. The comparison of the CIndex and sequential scan is conducted in terms of range search with the number of media objects varying from 2000 to 10000. Figure 11 shows the performance of query processing (for all the three media types) in terms of CPU cost. It is evident that CIndex outperforms the sequential scan methods significantly. The CPU cost of CIndex increases slowly as the data size grows. We also notice that the gap between CIndex and the sequential scan is large since the sequential scan is a CPU-intensive operation.

6.2.2 Effect of Query Radius

In this experiment, we proceed to evaluate the effect of query radius on the performance of a similarity search using the CIndex. Figure 12 shows that when *query* radius ranges from 0.2 to 1, the CIndex is always superior to the sequential scan in terms of page access and the CPU cost. This is because the sequential scan is a CPU-expensive retrieval operation.

6.2.3 Effect of θ on Search Efficiency

In this experiment, we study the effect of θ – the number of decimal place – on the efficiency of range search with an identical search radius. Figure 13 illustrates that the number of candidate media objects retrieved by the CIndex is decreasing gradually as θ increases, since the increase of θ will result in the precision enhancement of the index key. It is interesting to note that the search efficiency of the CIndex can not improve any more when θ exceeds a threshold, e.g., θ=3. This is because with the increase of θ, the distance ($\langle d(X_i,O_j),\theta\rangle$) difference, alternatively the precision of index key, is getting smaller when comparing with $d(X_i,O_j)$. The efficiency of the

CIndex does not improve anymore once θ reaches an optimal value. Therefore, θ is also a turning factor to the search optimization. Based on our experiment results, θ=3 is an optimal value.

6.2.4 Effect of Dynamic Insertion

In the last experiment, we investigate the effect of dynamic insertion on our indexing method. We initialize the CIndex by randomly choosing the first batch of images which consists of about 2000 images, and then insert other batches of images, with each batch containing about 2000 images also. After each insertion of a batch of images, we conduct a range search to see how the insertion affects the performance. Figure 14 shows the changing trends of total response time between sequential scan and our method. We notice that the performance of our method only degrades slightly due to the dynamic insertions on the CIndex, which suggests that the proposed method can be used for online processing.

7 Conclusions

In this paper, we have proposed a novel unified cross- media index scheme called *CIndex*, to boost up the retrieval performance over the large-scale cross-media retrieval. To the best of our knowledge, this is the first study on the indexing issue of cross-media search. Based on the cross reference graph(*CRG*) model [12], three steps are designed in building the CIndex: first, all media objects of the same modality are grouped into some clusters. Secondly, the centroid distance of each media object(X_i) is computed, and the correlations between X_i and its semantically related media objects whose modalities are different from X_i are obtained by mining the co-existence(*links*) in webpages. Finally, these two are combined to get the index key based on a B+-tree to effectively prune the search region and dramatically boost up the search performance in sequel. We have shown by experimental studies that our CIndex is effective in "understanding" the latent multimedia semantics, and much more efficient than sequential scan.

Acknowledgements

This work was done while Yi Zhuang was a Ph.D student in Zhejiang University. The authors would like to thank Prof. Yueting Zhuang and Dr. Fei Wu for their advising and comments.

References

[1] McGurk, H., MacDonald, J.: Hearing Lips and Seeing Voices. Nature 264, 746–748 (1976)
[2] Flickner, M., Sawthney, H., Niblack, W., et al.: Query by image and video content: The QBIC system. IEEE Computers 28(9), 23–31 (1995)
[3] Rui, Y., Huang, T.-S., Chang, S.-F.: Image Retrieval: Current Techniques, Promising Directions and Open Issues. JVCIR 10, 39–62 (1999)

[4] Smith, J.R., Chang, S.-F.: VisualSEEK: a fully automated content-based image query system. In: ACM Multimedia (1996)
[5] Chang, S.F., Chen, W., Meng, H.J., Sundaram, H., Zhong, D.: VideoQ: An automated content based video search system using visual cues. In: ACM Multimedia (1997)
[6] Smith, J., Chang, S.-F.: WebSEEK, a content-based image and video search and catalog tool for the Web. IEEE Multimedia (1997)
[7] Zhao, X.Y., Zhuang, Y.-T., Wu, F.: Audio Clip Retrieval with Fast Relevance Feedback based on Constrained Fuzzy Clustering and Stored Index Table. In: Chen, Y.-C., Chang, L.-W., Hsu, C.-T. (eds.) PCM 2002. LNCS, vol. 2532. Springer, Heidelberg (2002)
[8] Jagadish, H.V., Ooi, B.-C., Shen, H.T., Tan, K.-L.: Towards efficient multi-feature query processing. IEEE TKDE 18(3), 350–362 (2006)
[9] Shen, H.T., Zhou, X.F., Cui, B.: Indexing and Integrating Multiple Features for WWW images. World Wide Web Journal 9(3), 343–364 (2006)
[10] Yang, J., Li, Q., Zhuang, Y.-T.: Octopus: Aggressive Search of Multi-Modality Data Using Multifaceted Knowledge Base. In: WWW (2002)
[11] Wu, F., Zhang, H., Zhuang, Y.-T.: Learning Semantic Correlations for Cross-Media Retrieval. In: ICIP (2006)
[12] Zhuang, Y.-T., Yang, Y., Wu, F.: Mining Semantic Correlation of Heterogeneous Multimedia Data for Cross-media Retrieval. IEEE TMM 10(2), 221–229 (2008)
[13] Böhm, C., Berchtold, S., Keim, D.: Searching in High-dimensional Spaces: Index Structures for Improving the Performance of Multimedia Databases. ACM Computing Surveys 33(3) (2001)
[14] Guttman, R.: R-tree: A dynamic index structure for spatial searching. In: SIGMOD (1984)
[15] Weber, R., Schek, H., Blott, S.: A quantitative analysis and performance study for similarity-search methods in high-dimensional spaces. In: VLDB (1998)
[16] Jagadish, H.V., Ooi, B.C., Tan, K.L., Yu, C., Zhang, R.: iDistance: An Adaptive B+-tree Based Indexing Method for Nearest Neighbor Search. ACM TODS 30(2), 364–397 (2005)
[17] Zhang, T., Ramakrishnan, R., Livny, M.: BIRCH: An efficient data clustering method for very large databases. In: SIGMOD (1996)
[18] Yang, J., Li, Q., Liu, L.W.-Y., Zhuang, Y.-T.: Searching for Flash Movies on the Web: a Content and Context Based Framework. World Wide Web Journal 8(4), 495–517 (2004)
[19] Tenenbaum, J.B., de Silva, V., Langford, J.C.: A Global Geometric Framework for Nonlinear Dimensionality Reduction. Science 290(5500), 2319–2323 (2000)
[20] Microsoft Encarta (2006), http://encarta.msn.com/

Dimension-Specific Search for Multimedia Retrieval

Zi Huang[1], Heng Tao Shen[1], Dawei Song[2], Xue Li[1], and Stefan Rueger[3]

[1] School of ITEE, The University of Queensland, Australia
[2] School of Computing, The Robert Gordon University, UK
[3] Knowledge Media institute, The Open University, UK

Abstract. Observing that current Global Similarity Measures (GSM) which average the effect of few significant differences on all dimensions may cause possible performance limitation, we propose the first Dimension-specific Similarity Measure (DSM) to take local dimension-specific constraints into consideration. The rationale for DSM is that significant differences on some individual dimensions may lead to different semantics. An efficient search algorithm is proposed to achieve fast Dimension-specific KNN (DKNN) retrieval. Experiment results show that our methods outperform traditional methods by large gaps.

1 Introduction

Given an image feature database, the similarity measure determines the retrieval effectiveness. All existing similarity measures compute the *global similarity* which is aggregated from all dimensions of the feature space without exploring the impact of local distance along each individual dimension. We refer them as Global Similarity Measures (GSM) [7]. Although the distances along some dimensions are significant, such differences become non-discriminative in global similarity computation which averages the differences to all dimensions. Intuitively, properly utilizing the distance along each individual dimension might help to improve retrieval by filtering more irrelevant objects from top-K results.

On the other hand, high dimensionality of the feature space drives existing indexing structures [2,3,5,6] deficient due to the 'curse of dimensionality'. One important reason is that they maintain all dimensions as a whole for global similarity computations and all dimensions need to be accessed during query processing. [4] approaches high-dimensional indexing as a physical database design problem and proposes to vertically decompose the data by maintaining a separate table for each dimension (vertical decomposition for short). Some data points can be pruned away from full distance computations by accessing fewer dimensions/tables based on global upper and lower bounds estimated from all dimensions.

In this paper, we propose the first Dimension-specific. Different from conventional GSM, DSM takes dimension-specific constraints into consideration, where two feature vectors must match within a certain tolerance threshold along

X. Zhou et al. (Eds.): DASFAA 2009, LNCS 5463, pp. 693–698, 2009.

each individual dimension. In other words, DSM computes the global similarity as in conventional GSM, subject to the maximum allowable variation in each dimension. The rationale for DSM is that significant differences on individual dimensions may often lead to different human perception (i.e., semantics).

Corresponding to DSM, we introduce a new type of query called Dimension-specific KNN (DKNN) query to find KNN with dimension-specific constraints. To achieve fast retrieval for DKNN query, we propose an search algorithm to coordinate efficient dimension-specific pruning and global KNN pruning concurrently based on derived pruning rules. It accesses the data in a dimension-by-dimension manner and the intermediate candidate set obtained on the current dimension is propagated to the next dimension for further lookup and continuous reduction. It avoids most of high-dimensional distance computations and is robust to increasing dimensionality.

An extensive empirical performance study is conducted on the widely used Getty image datasets. The experimental results showed that DSM generated better MAP over GSM based on classical Euclidean distance by very large gaps. Furthermore, our DKNN search algorithm achieves great pruning power and outperforms traditional KNN method extended for DKNN query by an order of magnitude.

2 Dimension-Specific KNN Search

2.1 Dimension-Specific Similarity Measure (DSM)

Definition 1 (Dimension-specific Similarity Measure). *Given a query object represented by its feature vector* $x_q = (x_q^1, x_q^2, ..., x_q^D)$ *and a database object represented by its feature vector* $x=(x^1, x^2, ..., x^D)$*, their dimension-specific dissimilarity* $DS(x_q, x)$ *is defined as:*

$$DS(x_q, x) = \begin{cases} d(x_q, x), & if\ \forall i \in \{1..D\}, x_q^i \cong_{\varepsilon^i} x^i \\ +\infty, & otherwise \end{cases}$$

where $d(x_q, x)$ *is the global dissimilarity measure applied if the dimension-specific constraints (i.e.,* $\forall i \in [1..D], x_q^i \cong_{\varepsilon^i} x^i$*) are satisfied, and D is the dimensionality of the feature space.* [1]

DSM is able to avoid individual significant differences being potentially neglected in global measures. Given a query, the problem we investigate here is to find the *top-K most similar results* from the image database. By applying DSM, the similarity between two images is measured by the Euclidean distance as its *global similarity*, subject to the *dimension-specific conditions.* We define this type of query as Dimension-specific KNN (DKNN) query.

In DSM, ε is an important parameter which determines the qualification of a data point to compute its global distance to a query. To assign the value of ε, one

---

[1] $x_q^i \cong_{\varepsilon^i} x^i$ if $|x_q^i - x^i| \leq \varepsilon^i$.

simple way is to analyze historical query results for a proper fixed ε value to all dimensions. This may work for the feature space whose data along all dimension are uniformly distritbuted in the same range. When different dimensions exhibit significantly different distributions, adaptive ε values in DSM could be more effective. For i^{th} dimension, we can associate its ε^i with σ^i(standard deviation) by setting $\varepsilon^i = c \times \sigma^i$, where c is a scalar parameter.

2.2 Pruning Rules for DKNN Search

From Definition 1, we can simply set up the following **conditional pruning rule**: *x can be safely pruned if $\exists\ i \in \{1..D\}$ such that $|x_q^i - x^i| > \varepsilon^i$.*

Next we derive the KNN pruning rule. As we mentioned earlier, our data is organized based on vertical decomposition.Our DKNN search algorithm will access the data in a dimension-by-dimension manner. During the query processing, assume we have accessed the first t dimensions. We first derive the upper/lower bound of the distance between two points.

Proposition 1. *The upper bound of the distance between a query x_q and a data point x is defined by:*

$$||x_q, x||^2 \leq \underbrace{\sum_{i=1}^{t}(x_q^i - x^i)^2}_{\text{distance on t dimensions}} + \underbrace{\sum_{i=t+1}^{D}(\max(|x_q^i - r_{\min}^i|, |r_{\max}^i - x_q^i|))^2}_{\text{upper bound on remaining dimensions}}$$

where $r_{\min}^i$ and $r_{\max}^i$ are the minimal and maximal values on the ith dimension in the feature space respectively.

The lower bound of the distance between a query x_q and a data point x is defined by:

$$||x_q, x||^2 \geq \underbrace{\sum_{i=1}^{t}(x_q^i - x^i)^2}_{\text{distance on t dimensions}} + \underbrace{\frac{(\sum_{i=t+1}^{D} x_q^i - \sum_{i=t+1}^{D} x^i)^2}{D-t}}_{\text{Lower bound on remaining dimensions}}$$

Based on the derived upper and lower bounds, we can derive the following **KNN pruning rule**: *x can be safely pruned if its lower bound is greater than the current K^{th} smallest upper bound.*

2.3 DKNN Search Algorithm

Based on the established pruning rules, we propose a new search algorithm (Fig 1) which can maintain a small intermediate candidate set by cooperating conditional pruning and KNN pruning concurrently. As more dimensions have been accessed, the candidate set size becomes smaller and smaller. Therefore, I/O cost can be reduced significantly.

DKNN Search Algorithm
Input: x_q, Rank[ ]
Output: DKNN[ ]

```
For i=1 to D
DKNN[ ]←UpdateCandidateBound(x_q^{Rank[i]}, x^{Rank[i]});
DKNN[ ]←SortCandidate();
For j=1 to DKNN.size()
      if |DKNN[j][i]-x_q^i|>ε^i
            Remove DKNN[j−−];
      else if j>K and DKNN[j].lb>DKNN[K].ub
            Remove DKNN[j−−];
Return DKNN[ ];
```

Fig. 1. DKNN Query Processing

The order in which we access the dimensions does not change the final results. However, accessing the dimensions with higher pruning power (PP) [2] earlier will reduce the size of intermediate results, resulting in smaller computational cost and I/O cost. Normally, a larger σ^i corresponds to a larger PP on that dimension. So it is recommended to access the dimensions in the descending order of their σ^i to improve the performance.

3 Experiments

An extensive empirical performance study is conducted on Getty Image Dataset. The experimental results confirm the effectiveness of DSM and the efficiency of our DKNN search algorithm.

3.1 Experiment Setup

Getty Image Dataset contains 21,820 images[3], together with their annotations. RGB feature in 216 dimensionality is generated for each image. The average standard deviation on all dimensions is about 0.025. Precision and Mean Average Precision (MAP) are used to measure the effectiveness of DSM. Two images are considered as relevant if they share one or more keywords. PP is used to measure the efficiency of our DKNN search algorithm.

3.2 Effectiveness and Efficiency

Effectiveness: Fig 2 shows the results on precision and MAP at top-K results when we use fixed/adaptive ε for all dimensions. We can observe that DSM outperforms baseline for different values of ε. However, a too small value of

[2] *PP = No. of pruned objects/No. of total objects.*
[3] http://creative.gettyimages.com

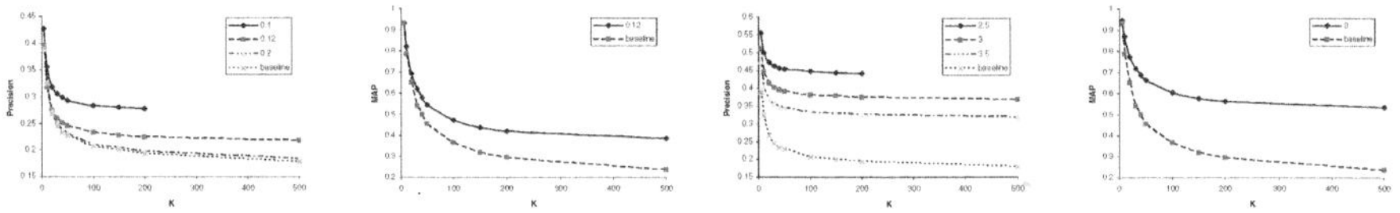

(a) precision, fixed ε (b) MAP, fixed ε (c) precision, adaptive ε (d) MAP, adaptive ε

Fig. 2. Effect of fixed/adaptive ε on precision/MAP at top-K results on Getty dataset

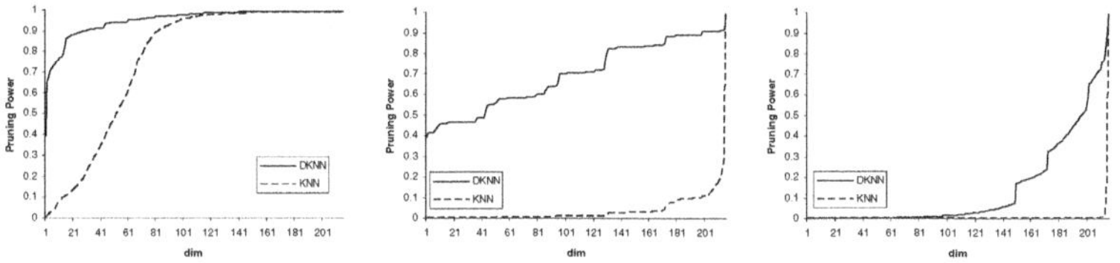

(a) descending (b) natural (c) ascending

Fig. 3. Effect of Dimension Access Order on Pruning Power with fixed ε=0.12 and K=300

the fixed ε may return insufficient number of results, while a too large value may lose the effect dimension-specific constraints. A suitable value will improve the search quality by huge gaps. Compared with MAP result in fixed ε shown in Fig 2(b), MAP is further improved by setting an adaptive ε, which is less sensitive to large variances and able to achieve better results quality than fixed ε.

Efficiency: We compare three dimension access orders (i.e.,access by the ascending/descending/natural (original)) of dimensions standard deviations. As shown in Fig 3, accessing the dimensions in the descending order of dimension standard deviations outperforms natural order greatly, which in turn outperforms accessing the dimensions in the ascending order of dimension standard deviations significantly. Furthermore, DKNN algorithm outperforms DKNN outperforms traditional KNN by an order of magnitude.

4 Conclusion

In this paper, we introduce a new type of query called Dimension-specific KNN (DKNN) query to find KNN with dimension-specific constraints. An efficient DKNN search algorithm is developed based on dimension-specific pruning and global KNN pruning rules concurrently. An extensive empirical performance study reveals that our proposals achieves significant improvements over traditional methods.

References

1. Assent, I., Wenning, A., Seidl, T.: Approximation techniques for indexing the earth mover's distance in multimedia databases. In: ICDE, p. 11 (2006)
2. Böhm, C., Berchtold, S., Keim, D.: Searching in high-dimensional spaces: Index structures for improving the performance of multimedia databases. ACM Computing Surveys 33(3), 322–373 (2001)
3. Ciaccia, P., Patella, M., Zezula, P.: M-tree: An efficient access method for similarity search in metric spaces. In: VLDB, pp. 426–435 (1997)
4. de Vries, A., Mamoulis, N., Nes, N., Kersten, M.: Efficient k-NN search on vertically decomposed data. In: SIGMOD, pp. 322–333 (2002)
5. Jagadish, H., Ooi, B., Tan, K., Yu, C., Zhang, R.: idistance: An adaptive b+-tree based indexing method for nearest neighbor searc. TODS 30(2), 364–379 (2005)
6. Lv, Q., Josephson, W., Wang, Z., Charikar, M., Li, K.: Multi-Probe LSH: Efficient Indexing for High-Dimensional Similarity Search. In: VLDB, pp. 950–961 (2007)
7. Weber, R., Schek, H., Blott, S.: A quantitative analysis and performance study for similarity search methods in high dimensional spaces. In: VLDB, pp. 194–205 (1998)

Application of Information Retrieval Techniques for Source Code Authorship Attribution

Steven Burrows, Alexandra L. Uitdenbogerd, and Andrew Turpin

School of Computer Science and Information Technology
RMIT University
GPO Box 2476V, Melbourne 3001, Australia
{steven.burrows,alexandra.uitdenbogerd,andrew.turpin}@rmit.edu.au

Abstract. Authorship attribution assigns works of contentious authorship to their rightful owners solving cases of theft, plagiarism and authorship disputes in academia and industry. In this paper we investigate the application of information retrieval techniques to attribution of authorship of C source code. In particular, we explore novel methods for converting C code into documents suitable for retrieval systems, experimenting with 1,597 student programming assignments. We investigate several possible program derivations, partition attribution results by original program length to measure effectiveness of modest and lengthy programs separately, and evaluate three different methods for interpreting document rankings as authorship attribution. The best of our methods achieves an average of 76.78% classification accuracy for a one-in-ten classification problem which is competitive against six existing baselines. The techniques that we present can be the basis of practical software to support source code authorship investigations.

Keywords: Adversarial information retrieval, authorship attribution, source code.

1 Introduction

Automatically detecting the author of a document is useful in plagiarism detection, copyright infringement, computer crime and authorship disputes. To assign authorship, a profile of each author is constructed from known sources, and then documents of unknown or uncertain authorship can be matched against these profiles manually, or with computer-based statistical analysis, machine learning or similarity calculation methods [1].

Authorship attribution techniques can be used to solve real-world natural language (plain text) problems in literature, papers, articles, essays and reports; but it can also solve problems in the domain of *structured text* such as source code. For example, will further analysis of the malicious WANK (Worms Against Nuclear Killers) and OILZ worms confirm that three authors were indeed involved [2]? Moreover, what can be learned about the author of the unnamed Internet worm of 1989 [3]? And finally, how can a lecturer know if programming assignments have indeed been written by the students themselves, obtained from

X. Zhou et al. (Eds.): DASFAA 2009, LNCS 5463, pp. 699–713, 2009.

a source code search engine such as Codase,[1] or outsourced to an anonymous programmer at Rent A Coder?[2]

In contrast to the large body of work on natural language authorship attribution (see for example Zhao and Zobel [4] for a comparison of many methods), we focus on authorship attribution for structured text; specifically C source code. Various source code authorship attribution techniques have been proposed, but they all disagree on what should comprise an author profile; use limited collections; and lack comparison to multiple baselines. We address all of these shortcomings by thoroughly investigating a wide range of features, employing a large data set, and comparing our work to the published baselines of six other research groups.

We use a collection of 1,597 C programs written by the 100 most prolific authors of C programming assignments in the School of Computer Science and Information Technology at RMIT University over an eight year period. Anonymous versions are used which are then tokenised and converted to n-gram representations which form document surrogates that are indexed and queried using standard information retrieval techniques [5,6]. We treat our experiments as being a 10-class authorship attribution problem. That is, we index a random subset of ten authors from the collection, and then attempt to attribute authorship for all documents in the subset to one of the ten authors.

In Section 2 we review methods for solving authorship attribution problems, and discuss approaches that have been applied to source code. Section 3 describes the baseline work that forms the starting point for this paper. In Section 4 we outline our methodology and present results in Section 5. We conclude in Section 6 and offer avenues for future work.

2 Previous Work

In this section we describe popular authorship attribution methods which fall within four broad categories. We then summarise our previous work, and the baselines to which we compare our new work.

2.1 Authorship Attribution Methods

Frantzeskou et al. [1] describe a taxonomy of four authorship analysis methodology categories: manual inspection, statistical analysis, machine learning and similarity measurement which we discuss in turn.

Manual inspection involves analysis by an expert to draw conclusions about coding skill or motive. Such evaluations can be used for authorship disputes in courts of law. Authorship disputes can arise because "programmers tend to feel a sense of ownership of their programs" [7] which can lead to code reproduction in successive organisations. This approach is not scalable for large problems.

[1] `http://www.codase.com`
[2] `http://www.rentacoder.com`

Statistical analysis techniques can be used to find effective feature sets for authorship analysis by eliminating features that make insignificant contributions towards identifying authorship compared to others. Discriminant analysis is one such statistical analysis technique which Ding and Samadzadeh [8] use for authorship attribution feature selection to create Java code fingerprints.

Machine learning involves finding patterns in samples of work belonging to one or more authors, and then classifying unseen works as belonging to the most likely author based upon the patterns in the training data. Many forms of machine learning algorithms have been applied to authorship attribution including support vector machines, decision trees, case-based reasoning and neural networks [9,10].

Similarity measurement involves measuring the distance between query documents and documents belonging to candidate authors by means of a similarity function. Nearest-neighbour methods [11] have been applied here to calculate the distance between feature lists of documents. One implementation is the Euclidean distance metric which computes the distance between two vectors in n-dimensional space.

2.2 Authorship Attribution Implementations

This paper continues from our previous work [12] where we investigated how n-gram representations [13] using modest numbers for n can be used to detect small repeating patterns of source code to indicate style. Scalability considerations were explored next by increasing the 10-class authorship attribution problem in increments up to 100-class to observe how severely the effectiveness of our approach degrades for larger problems. Next we further explored scalability considerations by reducing the number of submissions per author to simulate a scenario of having limited training data. Finally, we compared our work to a baseline implementation [14] and found our work to be more effective by 39% when measured with mean reciprocal rank.

We have identified works by six other research groups for source code authorship attribution which we treat as baseline implementations for our new contributions. These are the works by Ding and Samadzadeh [8] (canonical discriminant analysis), Elenbogen and Seliya [15] (C4.5 decision trees), Frantzeskou et al. [16] (nearest neighbour measurement), Krsul and Spafford [17] (discriminant analysis), Lange and Mancoridis [18] (nearest neighbour measurement) and MacDonell et al. [10] (case-based reasoning, discriminant analysis and neural networks). This list conveniently includes representative classification methods from all automated authorship attribution categories as described in Section 2.1 [1]. In addition to varying classification methods, these approaches also vary in terms of number of authors and work samples in each experiment, average program length, programming language, and level of programming experience of the authors whom contributed the work samples. We reserve a detailed comparison of the baseline systems to our own work in Section 5.3 where we explain the impact of these factors on reported classification accuracy scores.

3 Baseline

In this section, we describe the C program collection used as data in our experiments, and summarise some of our previous work [12] which forms the baseline for our new contributions.

3.1 Collection Creation

Our collection is comprised of 1,597 C programming assignments from the School of Computer Science and Information Technology at RMIT University. We refer to these assignments as COLLECTION-A. These assignments represent works from the 100 most prolific authors of C programming assignments in our school from 1999 to 2006 with fourteen to twenty-six submissions per author. No discrimination is made between undergraduate assignments, postgraduate assignments and year level — there is a mixture of assignments from all levels of ability. COLLECTION-A contains near-empty submissions of one line up to 10,789 lines of code, with the mean length being 830 lines of code. We note that the distribution of submission lengths is skewed towards shorter submissions, as we have a median of 650 lines of code.

When constructing COLLECTION-A, all submissions were made anonymous to comply with the ethical requirements of our university's Human Research Ethics Committee. This meant renaming files and removing all source code comments and output strings. We additionally deleted byte-identical copies resulting from duplicate submissions.

3.2 N-Gram Results

In our previous work, we used standard information retrieval methods to conduct 10-class authorship attribution experiments. Each C program was reduced to a "document" that contained n-grams comprised of operators and keywords only. In Section 4.1 we consider other types of features. We experimented with 1-gram to 90-gram representations [13] of these features to see if small frequent patterns are good markers of authorship.

In each 10-class experiment, documents belonging to ten random authors comprised a "collection" (about 160 documents) and each document in the collection was used as a "query" against this collection in turn. The Zettair search engine,[3] with its various ranking schemes, returned a ranked list for each query. The reciprocal rank (RR) of the highest ranked document by the same author as the query was used as the main outcome metric. For example, if the highest ranked document by the query author is 3, then RR is 0.33; or if the rank is 5, the RR is 0.2. We also used the Average Precision (AP) of the ranked list as an outcome metric, where AP is defined as "taking the set of ranks at which the relevant documents occur, calculating the precision at those depths in the

[3] `http://www.seg.rmit.edu.au/zettair`

ranking, and then averaging the set of precision values so obtained" [19]. As is typical in information retrieval, we report the mean of these two measures over many queries: mean reciprocal rank (MRR) and mean average precision (MAP). The random sampling of ten authors was repeated 100 times and we tested five different ranking schemes: Cosine [6], Pivoted Cosine [20], Dirichlet [21], Okapi BM25 [22,23], and our own scheme called Author1 [12].

The above process allowed us to compare several n-gram sizes with each of the five similarity measures and results are presented in Table 1. Note that the results differ to those reported previously [12] as byte-identical duplicates have now been removed from COLLECTION-A. We found that the Okapi similarity measure combined with 8-grams was most effective when measured in MRR (76.39%) and Okapi with 6-grams was most effective when measured in MAP (26.70%). These results are highlighted in Table 1.

We also examined the statistical significance of differences in MRR and MAP at the 95% confidence level using a permutation test (the distribution of RR scores is very skewed towards 1.00). We chose the null hypothesis to be "no difference from the most effective MRR or MAP result", and tested 2-gram up to 14-gram results for all five similarity measures. Two MRR results were found to be not significant from Okapi with 8-grams — Pivoted Cosine with 8-grams (75.89%; $p = 0.24$) and Okapi with 6-grams (75.59%; $p = 0.06$). These results are also marked on Table 1. Also, the most effective MAP result (Okapi with 6-grams) was found to be statistically significant compared to all other tested MAP results ($p < 0.05$). We carried forward Okapi with 6-grams over 8-grams for all future experiments for this reason and also because these representations take up less space.

3.3 Scalability Considerations

The previous experiment was for a 10-class problem, so we next tested larger problems up to the 100-class problem. With the problem size ten times bigger than our previous experiment, we were pleased to observe that MRR using the Okapi 6-gram system only fell from 75.59% to 64.97% (10.62%), which shows promise for dealing with even larger problems.

We also explored reducing the number of submissions per author. Beginning with our average of sixteen submissions per author, we randomly removed submissions in fixed amounts until we had only two submissions per author. In this case we found that MRR for the Okapi 6-gram system fell from 75.59% to 31.53% (44.06%). These results indicate that an information retrieval approach to attributing authorship for C code may not work well when there is a limited number of samples per author. However, we speculate that removing submissions chronologically instead of randomly may have less impact on the MRR score, as this simulates a more real-life scenario. We were not able to do this in COLLECTION-A as the school submission archive is messy in parts, and we don't have confidence in the timestamps. Experiments using reliable timestamps is an avenue for future work.

Table 1. Effect of varying the n-gram size and similarity measure for a 10-class authorship attribution problem using sixteen work samples per author on average, with authors randomly selected from COLLECTION-A. Results are averaged using all documents as queries (about 160 queries) in 100 random subsets of ten authors; a total of approximately 16,000 runs. The most effective results are highlighted — Okapi with 8-grams for MRR and Okapi with 6-grams for MAP. Two further MRR results are highlighted that were not statistically significant from Okapi with 8-grams at the 95% confidence level.

Grm	Similarity Measure MRR%					Similarity Measure MAP%				
Sze	Au1	Cos	Dir	Oka	P.Co	Au1	Cos	Dir	Oka	P.Co
1	51.53	59.41	23.36	42.66	28.49	17.05	19.85	12.22	17.72	13.31
2	65.80	68.44	28.33	67.34	53.33	20.73	22.50	12.98	23.24	18.25
4	72.00	74.10	53.43	75.52	72.79	23.91	25.10	19.33	26.00	23.70
6	73.85	74.42	59.42	**75.59**	74.70	25.71	25.82	20.44	**26.70**	25.52
8	75.49	74.94	61.17	**76.39**	**75.89**	24.96	24.65	19.58	25.61	25.00
10	73.72	73.35	60.44	74.95	74.69	22.78	22.73	17.76	23.47	23.25
12	74.17	73.45	61.39	74.95	74.55	21.57	21.42	16.95	22.01	21.75
14	72.77	72.20	62.41	73.51	73.32	19.09	18.93	15.74	19.38	19.28
16	71.63	71.12	63.55	72.20	72.25	16.81	16.71	14.73	16.98	16.97
18	70.41	70.14	64.21	70.81	70.86	14.59	14.65	13.31	14.79	14.81
20	67.96	67.92	64.48	68.27	68.33	13.36	13.41	12.66	13.53	13.54

4 Methodology

In this section, we outline the methodology for the new contributions. We first describe our approach for evaluating effective feature sets. Next we explore the effect of varying program length and the implications of particularly small queries, where a query is a program of undetermined authorship. Following that we describe an experiment to simulate a real-life authorship classification problem.

4.1 Feature Selection

Much work is available that defines good programming style. For example, Cannon et al. [24] define coding style guidelines such as commenting, white space, declarations, naming conventions and file organisation. Oman and Cook [25] collate programming style guidelines from many sources and organise them in a taxonomy which we use as a basis for categorising style features in our work. The taxonomy contains three main categories. *Typographic style* refers to all aspects of code that do not affect program execution such as commenting, naming characteristics and layout (spaces, tabs and new lines); next, *control structure style* refers to flow control tokens that affect algorithm implementation decisions such as operators, keywords and standard library functions; finally, *information structure style* refers to the organisation of program memory, input and output such as data structures, and input/output functions such as `printf` and `scanf`.

Based upon the above taxonomy, we create six classes of features for experimentation. White space, operators, literals, keywords, I/O words (from the

Table 2. Number of unique features in each feature class, and the distribution of tokens in COLLECTION-A

	W. Space	Operator	Literal	Keywd	I/O	Func	Total
Features	4	39	5	32	60	185	325
Percent	1.23	12.00	1.54	9.85	18.46	56.92	100.00
Tokens	8,109,257	1,495,730	1,409,749	384,718	143,691	76,355	11,619,500
Percent	69.79	12.87	12.13	3.31	1.24	0.66	100.00

`stdio.h` ANSI C89 header file) and function words (all standard library words not in `stdio.h`). We omit comments for de-identification reasons. We provide a summary of the number of tokens in each class and the total volume of tokens in COLLECTION-A in Table 2.

We experiment with all sixty-three ($2^6 - 1$) possible combinations of these feature classes, with each combination forming a *feature set*, using our experimental methodology from our previous work [12]. That is, we create 6-gram program representations with each feature set in turn and query each document against all works from different combinations of ten randomly selected authors.

4.2 Query Length Investigation

We next explore the effects of varying query length towards the effectiveness of source code authorship attribution; recording the query length in number of tokens against the MRR and MAP scores of each run. We still continue to run all feature class combinations described above to observe if tokens generated from some feature classes are more resilient to shorter queries than others.

4.3 Classification

Finally, we simulate a real-life 10-class authorship classification problem, attempting to correctly classify work samples as belonging to their authors using the best methods from previous sections. In the previous experiments we simply used the MRR and MAP scores as indications that a system would return a ranked list of documents with works from the correct author ranked highly in the list. In this experiment, we test three metrics for measuring the strength of style for *all* candidate authors, and we classify each work sample as belonging to the author with the highest score.

The *single best result* metric attributes authorship to the author of the top ranked document. For the *average scaled score* metric, we divide all scores returned by the ranking algorithm by the score of the top ranked document. The scores for each candidate author are then averaged and authorship is assigned to the highest average score. Finally for the *average precision* metric, we calculate the average precision for documents of each author in turn, and classify the query program as belonging to the author with the highest average precision score. For all metrics, we take care to omit the query document from all result lists. To identify the most effective metric, we average the results of 100 runs.

Table 3. Effectiveness of twelve of sixty-three tested feature sets sorted by MRR score. The six feature classes are operators, keywords, I/O tokens, function tokens, white space tokens and literal tokens respectively. We omit the remaining fifty-one results for brevity. These results demonstrate that operators, keywords and white space features together are strong markers of authorship. Note that the "feature set" column is a numeric identifier which we refer to in the text.

F. Set	Oper	Keywd	I/O	Func	W. Spc	Lit	MRR	MAP
50	Yes	Yes			Yes		82.28	41.33
58	Yes	Yes	Yes		Yes		82.20	40.36
55	Yes	Yes		Yes	Yes	Yes	81.98	39.22
51	Yes	Yes			Yes	Yes	81.74	39.74
54	Yes	Yes		Yes	Yes		81.68	41.13
62	Yes	Yes	Yes	Yes	Yes		81.61	39.90
..	...	...	...	...	...	...	.....	.....
32	Yes						74.78	26.19
..	...	...	...	...	...	...	.....	.....
16		Yes					69.40	20.33
..	...	...	...	...	...	...	.....	.....
01						Yes	65.73	23.10
08			Yes				62.07	16.79
04				Yes			56.98	9.91
02					Yes		43.26	22.15

5 Results and Analysis

In this section we provide results and analysis of our feature selection, query length investigation and classification experiments. Our classification results are then benchmarked against implementations by six other research groups.

5.1 Feature Selection Results

The top six rows of Table 3 provide a summary of the top performing feature sets (all of which have a statistically insignificant MRR from Feature Set 50 (the top row) using a permutation test at the $p < 0.05$ level). The bottom six rows show the performance of the six feature classes in isolation. As can be seen, using feature sets that contain a single feature class leads to a poor ranking of documents by the same author as the query, whereas selecting white space, operator and keyword features leads to a high ranking of similarly authored documents.

Koppel et al. [26] discuss that for text categorisation, "frequent but unstable features are especially useful", and so we would expect these to particularly effective. White space features were the most prevalent as shown in Table 2 representing 69.79% of all tokens. We believe they are especially useful as white space placement is a strong marker of programming style. For example, the following two for-loop declarations are functionally equivalent, but the use of white space can be helpful in distinguishing authorship:

```
for(i=0; i<limit; i++);          for (i = 0; i < limit; i++);
```

Our results showed that the most effective feature set without white space features had a MRR score of 4.49% less than the most effective feature set. This is a statistically significant result as shown by a permutation test at the 95% confidence interval.

Key observations were made when inspecting the volumes of each individual token. For example, the ratio of new lines to carriage returns was found to be 16:1.[4] This suggests that at least one out of sixteen submissions was not developed in our predominantly Unix-based environment which gives a useful authorship marker concerning choice of operating system for programming assignment development. The white space and literal tokens were all found in large quantities — the nine tokens that make up these categories were all within the top fifteen when totalling the volume of each token. Of the remaining token classes, the parenthesis was the most prevalent operator token, `int` was the most prevalent keyword, `NULL` was the most prevalent I/O token, and `strlen` was the most prevalent function token.

5.2 Query Length Results

In Figure 1 we summarise the results for our query length investigation. We compare the best of our feature sets (Feature Set 50) to the average of all feature sets. The goal is to show how the selection of one of our strongest feature sets can increase effectiveness particularly for the smallest queries. Our results are partitioned into program lengths of intervals of 100 tokens initially (0–99, 100–199, 200-299, and so on). Programs are then partitioned into intervals of 1,000 tokens for programs with 1,000 tokens or more given that we have less of these. The final partition is marked as 20,000 which represents all programs with at least 20,000 tokens up to the maximum in our collection (95,889 tokens).

The trends in Figure 1 show that shorter queries are markedly less effective in attributing authorship when averaged across all feature sets when compared to Feature Set 50 alone. There is a reasonable correlation between MRR/MAP and query length for Feature Set 50, however this trend drops off for queries less than 5,000 tokens for all feature sets averaged.

We conducted permutation tests on both the MRR and MAP results taking all results from Feature Set 50 and the first run for all other results. Results were found to be statistically significant ($p < 10^{-15}$). We also fitted linear models to each of the four lines and found that the gradient of the line for all feature sets compared to Feature Set 50 was 1.89 times greater for MRR and 2.86 times greater for MAP which confirms the strength of this feature set for smaller queries. Additionally, all our linear models were found to be statistically significant ($p < 10^{-13}$). The implication of these results is that we can largely retain authorship attribution effectiveness for short queries provided that an appropriate feature set is selected.

[4] New lines and carriage returns were treated as separate tokens.

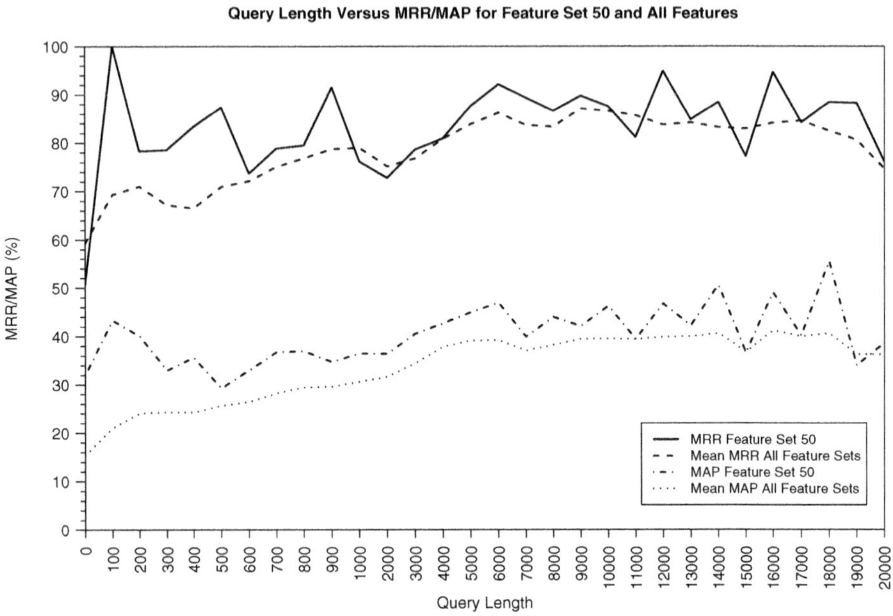

Fig. 1. Effectiveness of Feature Set 50 compared to the average of all feature sets measured in MRR and MAP. Results are partitioned based upon thirty-one query length intervals on the x-axis. These results show a downward trend for the lines representing all feature sets alone for queries less than 5,000 tokens, but results are much more consistent for Feature Set 50.

5.3 Classification Results

When using the "single best score" classification method we correctly classified work in 76.78% of cases for the 10-class problem. This is the best of our methods compared to "average scaled score" (76.47%) and "average precision" (74.68%), but this difference was not found to be statistically significant ($p > 0.05$).

In Table 4, we present our "single best score" method results for eight problem sizes ranging from 7-class to 46-class. We found that the "single best score" method was again most effective for the harder authorship attribution problems (12-class up to 46-class), however the "average scaled score" results were marginally higher for the 7-class and 8-class problems by 0.20% and 0.21% respectfully. The "single best score" results are reported here only.

We now compare our classification results to those of six other projects [8,10,15,16,17,18]. There are many variables that must be considered when comparing our results to those of other researchers. For example, choice of authorship markers, size of the test collection, level of ability of the authors whom contributed the work samples, and the choice of programming language. Therefore we don't think it is possible to select any approach as being strictly more effective than another. However, we now summarise each approach and discuss

Table 4. Comparison of our work to baseline systems of six external research groups. Codes are given in the first column indicating the first letter of the author's surname and problem size which we refer to in the text: (M07) MacDonell et al. [10] (Case-based reasoning); (F08a/b) Frantzeskou et al. [16] (Nearest neighbour); (B10) Burrows et al. (Okapi BM25); (E12) Elenbogen and Seliya [15] (C4.5 decision tree); (L20) Lange and Mancoridis [18] (Nearest neighbour); (K29) Krsul and Spafford [17] (Discriminant analysis); (F30) Frantzeskou et al. [16] (Nearest neighbour); and (D46) Ding and Samadzadeh [8] (Canonical discriminant analysis). For each baseline the remaining nine columns respectively represent the problem difficulty, range and total number of work samples, range and average lines of code of the collection, level of programming experience of the work owners ("low" for students, "high" for professionals, and "mixed" for a hybrid collection), programming language of the test collection, reported effectiveness of the approach, and our effectiveness score for the same problem size (using the "single best score" metric). To represent incomplete data, we marked non-obtainable data with a dash (—) and data obtained from personal communication with a dagger (†) or double dagger (‡) where estimates were provided.

ID	No Au	Range Work	Total Work	Range LOC	Avg LOC	Experience	Language	Effectiveness	This paper
M07	7	5–114	351	†1–1,179	†148	Mixed	C++	88.0%	78.66%
F08a	8	6–8	‡60	36–258	129	Low	Java	88.5%	78.46%
F08b	8	4–29	107	23–760	145	High	Java	100.0%	78.46%
B10	10	14–26	1,597	1–10,789	830	Low	C	76.8%	76.78%
E12	12	6–7	83	‡50–400	‡100	Low	C++†	74.7%	75.25%
L20	20	3	60	†336–80,131	11,166	High	Java	55.0%	72.47%
K29	29	—	88	—	—	Mixed	C	73.0%	70.67%
F30	30	4–29	333	20–980	172	High	Java	96.9%	70.48%
D46	46	4–10	225	—	—	Mixed	Java	67.2%	68.58%

the strengths and weaknesses of the most effective baselines to account for the differences. The major properties of each work are summarised in Table 4.

When comparing results based upon problem difficulty alone (7-class up to 46-class) using Table 4, our classification rate is well ahead of (L20) Lange and Mancoridis [18], very close to (E12) Elenbogen and Seliya [15], (K29) Krsul and Spafford [17] and (D46) Ding and Samadzadeh [8] (within 3%), and well behind (M07) MacDonell et al. [10] and (F08a/F08b/F30) Frantzeskou et al. [16]. In accounting for the strong result of (M07) MacDonell et al. [10] (9.14% higher classification rate than ours), a large portion of the collection is industry-based which we would expect to be easier to classify given the more mature programming style of the authors compared to our students. Also, the number of work samples per author varies the most compared to all other baselines. For example, the work by one author comprises 114 samples out of the 351 program collection (32.5%), and we would expect these samples to score highly as classification accuracy is 32.5% by random chance alone. Conversely, another author has only five samples which would be particularly difficult to classify correctly, but this poorer result would contribute much less to the overall reported effectiveness.

We summarised three baselines in Table 4 from Frantzeskou et al. [16] which are more effective than our equivalent result by margins of 9.85% (F08a), 21.35% (F08b) and 26.42% (F30) respectfully. The two largest margins were based on industry collections so we don't discuss these further. However, the student collection (F08a) still remained more effective by 9.85%. We observed that the results reporting various n-gram sizes and profile lengths do not agree with each other between collections so we suggest that alternative similarity measures which are less sensitive to the input parameters could be investigated as future work.

In conducting our experiments, we have witnessed varying results between runs which highlight the impact of the collection chosen on classification effectiveness. In Table 4, we reported an average result of 78.65% for our 8-class student collection problem. The eight students were randomly selected from 100 for each of 100 runs with each run attempting to classify every piece of work from eight authors. We could consider our work to have used 100 partly overlapping collections. The accuracy of these 100 results had a standard deviation of 3.92% and ranged from 64.46% up to 88.72%. This result is marginally higher than the 8-class student collection result of (F08a) Frantzeskou et al. [16] (88.5%). When considering that their approach performs poorly when an inappropriate profile length parameter is chosen for the collection in question [27], we believe our standard deviation is quite modest which suggests that our approach is more consistent and reliable.

Our collection is likely to be more difficult for authorship attribution experiments relative to the other researchers discussed above. We obtained work samples from our school's assignment submission archive from an eight-year period and students would often be completing their degrees in non-overlapping periods and have very differing assessments. Moreover, we expect some submissions to be incomplete or very small. Indeed, our smallest submission was only one line of code, and we didn't attempt to manually remove these difficult submissions.

A more precise comparison can only be performed if the above methodologies are compared on the same test collections and this remains an avenue for future work. Stein et al. [28] have cited the need to develop "publicly available large corpora" for benchmarking proposed approaches and we agree that this is important for advancing the field. Our contribution goes part way towards solving this problem as we have used more authors and more work samples than any of the benchmarked projects. However there are still privacy issues to be explored which may or may not prohibit the sharing of this student data with others.

6 Discussion and Conclusions

In this paper, we have presented a source code authorship attribution approach based on information retrieval techniques capable of solving real-world authorship attribution problems such as resolving authorship disputes, facilitating easier software project maintenance when authorship data is lacking, and supporting plagiarism detection investigations [29] where authorship evidence is needed in addition to matching content information.

The contributions built upon previous work that used a large collection of university assignments written in the C programming language [12]. The first contribution investigated effective methods for deriving surrogate documents via combining features as 6-grams for authorship attribution tasks. Several feature classes were investigated, and we found that white space, operator and keyword tokens are particularly valuable.

Next we explored the effect of varying length query documents. Smaller programs were indeed more difficult to attribute correctly when averaged across all tested feature sets, but we found that our most effective feature set containing white space, operator and keyword tokens showed greatest authorship attribution improvement for the smallest queries in particular.

Finally, using our "single best result" metric we demonstrated that we can correctly classify 76.78% of all query documents in a 10-class authorship classification experiment. When dealing with the varying collection size, data sources, programming language, underlying methodology and choice of authorship markers of competing approaches, we implemented authorship attribution experiments of equivalent difficulty. We have shown that our information retrieval approach is a competitive alternative to the six existing approaches examined [8,10,15,16,17,18] that have employed a variety of statistical analysis, machine learning and similarity measurement techniques [1].

It is hoped that our contributions will further advance the source code authorship attribution field. However, this paper represents one component of a larger research project and much remains for future work. For example, our collection contains work samples from authors of varying levels of programming ability from first year students up to final year students. We therefore anticipate differences in authorship classification effectiveness between these groups and indeed other groups such as academic staff and industry professionals. We therefore plan to explore how the evolution of programming style between these groups affects authorship classification rates.

Finally, we note that there are an increasing number of contributions towards source code authorship attribution implementations as evidenced by our work and the contributions of the authors of our baselines. Additionally, natural language authorship attribution is a particularly mature field [4,30]. We have demonstrated that with suitable choice of document surrogates, information retrieval techniques work well, so we would expect this to carry over to other domains. For example, we may consider one or more of mathematical proofs, chemical formulae, Z specifications, UML definitions, SQL, XML, HTML, LaTeX or word processing markup in future experiments.

References

1. Frantzeskou, G., Gritzalis, S., MacDonell, S.: Source code authorship analysis for supporting the cybercrime investigation process. In: Filipe, J., Belo, C., Vasiu, L. (eds.) Proceedings of the First International Conference on E-business and Telecommunication Networks, Setubal, Portugal, pp. 85–92. Kluwer Academic Publishers, Dordrecht (2004)

2. Longstaff, T.A., Schultz, E.E.: Beyond preliminary analysis of the WANK and OILZ worms: A case study of malicious code. Computers and Security 12(1), 61–77 (1993)
3. Spafford, E.H.: The internet worm: Crisis and aftermath. Communications of the ACM 32(6), 678–687 (1989)
4. Zhao, Y., Zobel, J.: Effective and scalable authorship attribution using function words. In: Lee, G.G., Yamada, A., Meng, H., Myaeng, S.-H. (eds.) AIRS 2005. LNCS, vol. 3689, pp. 174–189. Springer, Heidelberg (2005)
5. Baeza-Yates, R., Ribeiro-Neto, B.: Modern Information Retrieval, 1st edn. Addison Wesley Longman, Amsterdam (1999)
6. Witten, I., Moffat, A., Bell, T.: Managing Gigabytes: Compressing and Indexing Documents and Images, 2nd edn. Morgan Kaufmann Publishers, San Francisco (1999)
7. Glass, R.L.: Special feature: Software theft. IEEE Software 2(4), 82–85 (1985)
8. Ding, H., Samadzadeh, M.H.: Extraction of Java program fingerprints for software authorship identification. Journal of Systems and Software 72(1), 49–57 (2004)
9. Drucker, H., Wu, D., Vapnik, V.N.: Support vector machines for spam categorization. IEEE Transactions on Neural Networks 10(5), 1048–1054 (1999)
10. MacDonell, S.G., Gray, A.R., MacLennan, G., Sallis, P.J.: Software forensics for discriminating between program authors using case-based reasoning, feed-forward neural networks and multiple discriminant analysis. In: Proceedings of the Sixth International Conference on Neural Information Processing, Perth, Australia, Perth, Australia, pp. 66–71 (November 1999)
11. Indyk, P., Motwani, R.: Approximate nearest neighbors: Towards removing the curse of dimensionality. In: Vitter, J. (ed.) Proceedings of the Thirtieth Annual ACM Symposium on the Theory of Computing, ACM Special Interest Group on Algorithms and Computation Theory, Dallas, Texas, pp. 604–613. ACM Press, New York (1998)
12. Burrows, S., Tahaghoghi, S.M.M.: Source code authorship attribution using n-grams. In: Spink, A., Turpin, A., Wu, M. (eds.) Proceedings of the Twelth Australasian Document Computing Symposium, Melbourne, Australia, RMIT University, pp. 32–39 (December 2007)
13. Schleimer, S., Wilkerson, D., Aiken, A.: Winnowing: Local algorithms for document fingerprinting. In: Ives, Z., Papakonstantinou, Y., Halevy, A. (eds.) Proceedings of the 2003 ACM SIGMOD International Conference on Management of Data, ACM Special Interest Group on Management of Data, San Diego, California, pp. 76–85. ACM Press, New York (2003)
14. Jones, E.: Metrics based plagiarism monitoring. In: Meinke, J.G. (ed.) Proceedings of the Sixth Annual CCSC Northeastern Conference on The Journal of Computing in Small Colleges, Middlebury, Vermont, Consortium for Computing Sciences in Colleges, pp. 253–261 (April 2001)
15. Elenbogen, B., Seliya, N.: Detecting outsourced student programming assignments. Journal of Computing Sciences in Colleges 23(3), 50–57 (2008)
16. Frantzeskou, G., Stamatatos, E., Gritzalis, S., Katsikas, S.: Effective identification of source code authors using byte-level information. In: Anderson, K. (ed.) Proceedings of the Twenty-Eighth International Conference on Software Engineering, Shanghai, China, ACM Special Interest Group on Software Engineering, pp. 893–896 (May 2006)
17. Krsul, I., Spafford, E.H.: Authorship analysis: Identifying the author of a program. Computers and Security 16(3), 233–257 (1997)

18. Lange, R.C., Mancoridis, S.: Using code metric histograms and genetic algorithms to perform author identification for software forensics. In: Thierens, D. (ed.) Proceedings of the Ninth Annual Conference on Genetic and Evolutionary Computation, London, England, ACM Special Interest Group on Genetic and Evolutionary Computation, pp. 2082–2089. ACM Press, New York (2007)
19. Moffat, A., Zobel, J.: Rank-biased precision for measurement of retrieval effectiveness. ACM Transactions on Information Systems 27(1), 1–27 (2008)
20. Singhal, A., Buckley, C., Mitra, M.: Pivoted document length normalization. In: Frei, H.-P., Harman, D., Schaubie, P., Wilkinson, R. (eds.) Proceedings of the Nineteenth Annual International ACM SIGIR Conference on Research and Development in Information Retrieval, Zurich, Switzerland, pp. 21–29. ACM Press, New York (1996)
21. Zhai, C., Lafferty, J.: A study of smoothing methods for language models applied to information retrieval. ACM Transactions on Information Systems 22(2), 179–214 (2004)
22. Jones, K.S., Walker, S., Robertson, S.E.: A probabilistic model of information retrieval: Development and comparative experiments part 1. Information Processing and Management 36(6), 779–808 (2000)
23. Jones, K.S., Walker, S., Robertson, S.E.: A probabilistic model of information retrieval: Development and comparative experiments part 2. Information Processing and Management 36(6), 809–840 (2000)
24. Cannon, L.W., Elliott, R.A., Kirchhoff, L.W., Miller, J.H., Miller, J.M., Mitze, R.W., Schan, E.P., Whittington, N.O., Spencer, H., Keppel, D., Brader, M.: Recommended C style and coding standards. Technical report, Bell Labs, University of Toronto, University of Washington and SoftQuad Incorporated (February 1997) (accessed September 24, 2008), `http://vlsi.cornell.edu/courses/eecs314/tutorials/cstyle.pdf`
25. Oman, P.W., Cook, C.R.: A taxonomy for programming style. In: Sood, A. (ed.) Proceedings of the 1990 ACM Annual Conference on Cooperation, Association for Computing Machinery, pp. 244–250. ACM Press, New York (1990)
26. Koppel, M., Akiva, N., Dagan, I.: A corpus-independent feature set for style-based text categorization. In: Proceedings of the IJCAI 2003 Workshop on Computational Approaches to Style Analysis and Synthesis, Acapulco, Mexico (2003)
27. Frantzeskou, G., Stamatatos, E., Gritzalis, S., Katsikas, S.: Source code author identification based on n-gram author profiles. In: Maglogiannis, I., Karpouzis, K., Bramer, M. (eds.) Artificial Intelligence Applications and Innovations, vol. 204, pp. 508–515. Springer, New York (2006)
28. Stein, B., zu Eissen, S.M., Potthast, M.: Strategies for retrieving plagiarized documents. In: Kraaij, W., de Vries, A.P., Clarke, C.L.A., Fuhr, N., Kando, N. (eds.) Proceedings of the Thirtieth Annual International ACM SIGIR Conference on Research and Development in Information Retrieval, pp. 825–826. ACM Press, Amsterdam (2007)
29. Burrows, S., Tahaghoghi, S.M.M., Zobel, J.: Efficient plagiarism detection for large code repositories. Software: Practice and Experience 37(2), 151–175 (2006)
30. Zhao, Y., Zobel, J., Vines, P.: Using relative entropy for authorship attribution. In: Ng, H.T., Leong, M.-K., Kan, M.-Y., Ji, D. (eds.) AIRS 2006. LNCS, vol. 4182, pp. 92–105. Springer, Heidelberg (2006)

In the Search of NECTARs from Evolutionary Trees

Ling Chen[1] and Sourav S. Bhowmick[2]

[1] L3S, University of Hannover, Germany
[2] School of Computer Engineering, Nanyang Technological University, Singapore
lchen@l3s.de, assourav@ntu.edu.sg

Abstract. Mining trees is very useful in domains like bioinformatics, web mining, mining semi-structured data, and so on. These efforts largely assumed that the trees are static. However, in many real applications, tree data are evolutionary in nature. In this paper, we focus on mining evolution patterns from historical tree-structured data. Specifically, we propose a novel approach to discover *negatively correlated subtree patterns* (NECTARs) from a sequence of historical versions of unordered trees.The objective is to extract subtrees that are *negatively correlated* in undergoing *structural* changes. We propose an algorithm called NECTAR-Miner based on a set of *evolution metrics* to extract NECTARs. NECTARs can be useful in several applications such as maintaining mirrors of a website and maintaining XML path selectivity estimation. Extensive experiments show that the proposed algorithm has good performance and can discover NECTARs accurately.

1 Introduction

Mining tree-structured data has gained tremendous interest in recent times due to the widespread occurrence of tree patterns in applications like bioinformatics, web mining, semi-structured data mining, and so on. Existing work on mining tree-structured data can be broadly classified into three categories: *association rule mining* [3], *frequent substructure mining* [2,15], and *classification/clustering* [10,16]. While these tree pattern mining techniques have been innovative and powerful, our initial investigation revealed that majority of the existing approaches of tree mining focus only on snapshot data, while in real life tree-structured data is dynamic in nature. Consider a sequence of tree-structured data in Figure 1, where the black and gray circles respectively represent the newly inserted nodes and deleted nodes. Typically, there are two types of changes to tree-structured data: changes to *data content* (e.g., leaf nodes in an XML tree) and changes to the *structure* of tree data (internal nodes). *In this paper, we focus on the structural evolution of tree-structured data only.*

The evolutionary nature of structure of trees leads to two challenging problems in the context of data mining. The first one is to maintain the previously discovered knowledge. For instance, in frequent substructure mining, as the data source changes new frequent structures may emerge while some existing ones may not be frequent anymore. The second one is to discover novel knowledge by analyzing the evolutionary characteristics of historical tree data. Such knowledge is difficult or even impossible to be discovered from snapshot data efficiently due to the absence of evolution-related information. In this paper, we focus on the second issue. That is, *we present techniques*

X. Zhou et al. (Eds.): DASFAA 2009, LNCS 5463, pp. 714–729, 2009.

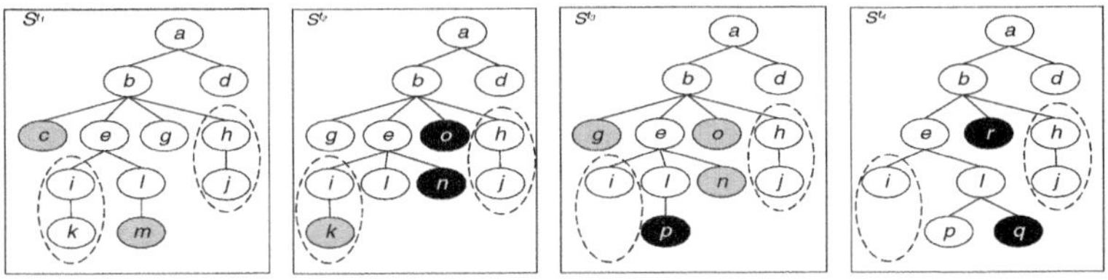

Fig. 1. Sequence of historical tree versions

to discover a specific type of novel knowledge by mining the evolutionary features of tree-structured data.

Let us elaborate informally on the types of novel knowledge one may discover by analyzing evolutionary features of trees. Consider the different versions of a tree structure in Figure 1. We may discover the following types of novel knowledge by exploiting the evolution-related information associated with the trees. Note that this list is by no means exhaustive.

- *Frequently Changing Structures (FCS):* Different parts of a tree may evolve in different ways over time. Some parts of the tree may evolve more *frequently* than other parts. Some parts may change more *significantly* in the history compared to other parts that may only change slightly. For example, the subtree rooted at node b changes in all the four versions. On the other hand, the subtree rooted at d never changes. We refer to structures that change frequently and significantly in the history as *frequently changing structures*. Here, *frequently* refers to the large number of times the corresponding parts changed, while *significantly* refers to the large percentage of nodes that have changed in the corresponding subtrees.
- *Associative Evolutionary Structures (AES):* Similar to the transactional association rules, different parts of the tree may be associated in terms of their evolutionary features over time. For example, in Figure 1 one may observe that whenever subtrees rooted at i and ℓ change frequently and significantly, the subtree rooted at h does not change. Then, a negative association rule may be extracted between the subtrees with respect to some appropriately specified thresholds. Note that a positive association rule can similarly be extracted from subtrees that change frequently and significantly together. We refer to such structures as *associative evolutionary structures*.

We have discussed frequently changing structures in [17]. In this paper, we focus on discovering a specific type of associative evolutionary structures from historical tree-structured data. Particularly, we propose a technique to discover subtrees that are *negatively correlated* in undergoing structural changes. That is, when a set of subtrees changes (significantly), another set of subtrees rarely changes (significantly). We refer to this type of patterns as NECTARS (**NE***gatively* **C***orrelated sub***T***ree p***A***tte***R***n*). While this pattern can involve subtrees which may or may not be structurally related, in this paper, we focus on discovering NECTARs from subtrees having ancestor-descendant relationships. As we shall see in Section 5, NECTARs are useful in applications such as maintaining mirrors of Web site [7] and maintaining XML path selectivity estimation.

Given a sequence of versions of a tree structure, we propose a set of *evolution metrics*, in Section 3, to quantitatively measure the *frequency* and *correlation* of *significant changes* to subtrees. Based on such metrics, we propose an algorithm called NECTAR-Miner in Section 4 to extract the NECTARs from the tree sequence. Our proposed algorithm consists of two major phases: the *GDT construction* phase and the NECTARs *discovery* phase. In the first phase, given a sequence of historical tree-structured data, the *GDT (General Delta Tree)* is constructed to efficiently represent evolutionary features of trees. In the next phase, negatively correlated subtrees are extracted by traversing the *GDT*. Our experimental results in Section 6 show that the proposed algorithm can extract all NECTARs efficiently.

2 Modeling Structural Changes to Trees

We first present different types of *structural changes* to tree-structured data and how we quantify the *degree* of change in the data. Next, we introduce the notion of *structural delta* which will be used subsequently in our discussion. We use the following notations in the sequel. Let $S = \langle N, E \rangle$ denotes a tree structure where N and E respectively represent the set of nodes and edges of the tree, $\mathcal{T} = \langle t_1, t_2, \ldots, t_n \rangle$ be a sequence of time points with some particular time granularity, S^{t_i} be the version of S at time point t_i, and s_i be a subtree of S, denoted as $s_i \prec S$.

2.1 Types of Structural Changes

An edit operation is an operation e that can be applied on a tree $S_1 = \langle N_1, E_1 \rangle$ to produce another tree $S_2 = \langle N_2, E_2 \rangle$, denoted as $S_1 \xrightarrow{e} S_2$. Given two versions of a tree structure, different types of *edit operations* have been defined in traditional change detection algorithms [5]. For instance, *atomic* edit operations such as insertion, deletion, and update can be defined based on nodes. Furthermore, complex operations on subtrees such as *move*, *copy*, and *delete a subtree*, can be defined as well. Usually, an edit operation defined on a subtree can be decomposed into a sequence of atomic edit operations defined on nodes.

In our study, we consider the atomic edit operations defined on leaf nodes that cause structural changes to an *unordered* tree. Note that our work can easily be extended to *ordered* trees as well. Hence we define only two edit operations as follows: (a) **INS**$(x(name), p)$: This operation creates a new leaf node x, with node label "name", as a child node of node p in a tree. (b) **DEL**(x): This operation is the inverse of the insertion one. It removes leaf node x from a tree.

For example, consider the first two tree versions, S^{t_1} and S^{t_2}, in Figure 1. The version S^{t_2} can be produced from S^{t_1} by applying the following structural edit operations: inserting nodes labeled n and o as a child of nodes e and b, respectively, and deleting nodes labeled c and m. Note that we do not consider the update operation because it does not incur structural changes. Certainly, depending on application the update operation can be regarded as deleting a node and inserting a node with a new label.

Given two versions of a tree (subtree), *edit distance* is usually adopted in traditional change detection algorithms to measure the distance between the two versions. Edit distance is defined based on edit script, which is a sequence of edit operations that converts one tree into another. Note that, in traditional change detection algorithms there may exist more than one valid edit script since the edit operations are decomposable. However, since we consider atomic structural edit operations, there exists only one valid edit script. Then, edit distance can be defined as follows. Given two versions of a trees S at time points t_j and t_{j+1}, denoted as S^{t_j} and $S^{t_{j+1}}$, let $E(S^{t_j} \rightarrow S^{t_{j+1}})$ be a sequence of atomic edit operations, $\langle e_1, e_2, \ldots, e_n\rangle$, which converts S^{t_j} to $S^{t_{j+1}}$, then, the edit distance between the two versions, denoted as $d(S^{t_j}, S^{t_{j+1}})$, is the number of edit operations in the edit script. That is, $d(S^{t_j}, S^{t_{j+1}}) = |E|$. For instance, let the subtree $a/b/e$ in Figure 1 be s. The following sequence of edit operations convert s^{t_1} to s^{t_2}: $E(s^{t_1} \rightarrow s^{t_2}) = \langle DEL(m), INS(n, e)\rangle$. Then, the edit distance between s^{t_1} and s^{t_2}, $d(s^{t_1}, s^{t_2})$, is 2.

2.2 Degree of Changes (DoC)

Considering that an edit distance is an absolute value which measures the difference between two tree structures, we define the *Degree of Change* of a tree in two versions by normalizing the edit distance by the size of the *consolidate tree* of the two tree versions. Given two tree versions $S^{t_j} = \langle N^{t_j}, E^{t_j}\rangle$ and $S^{t_{j+1}} = \langle N^{t_{j+1}}, E^{t_{j+1}}\rangle$, the *consolidate tree* of them, denoted as $S^{t_j} \uplus S^{t_{j+1}}$, is $\langle N, E\rangle$, where $i)$ $N = N^{t_j} \cup N^{t_{j+1}}$, $ii)$ $e = (x, y) \in E$, if and only if x is the parent of y in E^{t_j} or $E^{t_{j+1}}$. For example, the consolidate tree of $a/b/e$ in the first two versions in Figure 1 contains the nodes e, i, k, ℓ, m, and n.

Definition 1 (Degree of Change). *Given two versions of a tree S^{t_j} and $S^{t_{j+1}}$, let $d(S^{t_j}, S^{t_{j+1}})$ be the edit distance between the two versions, then the Degree of Change of S from t_j to t_{j+1}, denoted as DoC(S, t_j, t_{j+1}), is: DoC$(S, t_j, t_{j+1}) = \frac{d(S^{t_j}, S^{t_{j+1}})}{|S^{t_j} \uplus S^{t_{j+1}}|}$ where $|S^{t_j} \uplus S^{t_{j+1}}|$ is the size of the consolidate tree of S^{t_j} and $S^{t_{j+1}}$.* □

The value of *DoC* ranges from 0 to 1. If a tree does not change in two versions, then its *DoC* is zero. If a tree is totally removed or newly inserted, then the *DoC* of the tree is one. The greater the value of *DoC*, the more significantly the tree changed. For example, let the subtree a/b in Figure 1 be subtree s_1. Then the *DoC* of s_1 in the first two versions is $DoC(s_1, t_1, t_2) = 4/12 = 0.33$.

2.3 Structural Delta

When a tree evolves from one version to another version, some of its subtrees change (i.e., their *DoC* values are greater than zero), while some of its subtrees remain unchanged (i.e., their *DoC* values equal to zero). We are interested in the set of changed subtrees. Particularly, we refer to the set of changed subtrees in two successive versions as the *structural delta*.

Definition 2 (Structural Delta). *Given two versions of a tree S at time t_j and t_{j+1}, the structural delta of S from t_j to t_{j+1}, denoted as $\Delta_S(t_j, t_{j+1})$, refers to the set of changed subtrees in the two versions and $\Delta_S(t_j, t_{j+1})$=$\{s | s \prec S^{t_j}$ & $DoC(s, t_j, t_{j+1}) > 0\} \cup \{s | s \prec S^{t_{j+1}}$ & $DoC(s, t_j, t_{j+1}) > 0\}$.* □

For example, consider the first two tree versions in Figure 1. $\Delta_S(t_1, t_2) = \{a/b, a/b/c, a/b/e, a/b/e/l, a/b/e/l/m, a/b/e/n, a/b/o\}$. Note that the subscript S of Δ_S can be omitted if it is clear from the context.

3 Evolution Metrics

We now introduce the metrics defined to measure the *change frequency* and *change significance* for a set of subtrees. Particularly, we use the following notations for the subsequent definitions. Let $\Sigma = \langle S^{t_1}, S^{t_2}, \ldots, S^{t_n} \rangle$ be a sequence of versions of tree S on $\mathcal{T}$, $\langle \Delta(t_1, t_2), \ldots, \Delta(t_{n-1}, t_n) \rangle$ be the sequence of corresponding structural deltas, $\Omega = \Delta(t_1, t_2) \cup \ldots \cup \Delta(t_{n-1}, t_n)$ be the set of all changed subtrees.

3.1 Frequency of Significant Change (FoSC)

Note that DoC of a subtree measures how significantly a subtree changed in two successive versions. In order to justify whether a set of subtrees are correlated in undergoing significant changes, we need a metric to measure how frequently the set of subtrees undergoes significant changes together. Hence, we introduce the metric *Frequency of Significant Change* ($FoSC$) for a set of subtrees. Note that the metric $FoSC$ is defined with respect to a given threshold of DoC because we say a subtree undergoes significant changes only if its DoC value is no less than the DoC *threshold.*

Definition 3 (Frequency of Significant Change (FoSC)). *Let $X = \{s_1, s_2, \ldots, s_m\}$ be a set of changed subtrees, $X \subseteq \Omega$, and the threshold of DoC be α. The Frequency of Significant Change for the set X, with respect to α, denoted as $FoSC_\alpha(X)$, is:* $FoSC_\alpha(X) = \frac{\sum_{j=1}^{n-1} D_j}{(n-1)}$ *where* $D_j = \prod_{i=1}^{m} D_{j_i}$ *and*

$$D_{j_i} = \begin{cases} 1, \text{ if } DoC(s_i, t_j, t_{j+1}) \geq \alpha \\ 0, \text{ otherwise} \end{cases} \quad 1 \leq i \leq m \qquad \square$$

That is, $FoSC_\alpha$ of a set of subtrees is the fraction of subsequent versions (after the first version) where the set of subtrees undergoes significant changes together. The value of $FoSC_\alpha$ ranges in $[0, 1]$. If all subtrees in the set undergo significant changes together in each subsequent version, then the value of $FoSC_\alpha$ equals to one. If subtrees in the set never undergo significant changes together in subsequent versions, then the value of $FoSC_\alpha$ is zero. For example, reconsider the sequence of historical tree versions in Figure 1. Let $X = \{a/b, a/b/e\}$ and the threshold of DoC α be $1/4$. Then, $FoSC_{1/4}(X) = 2/3$ as both subtrees undergo significant changes in two subsequent versions S^{t_2} and S^{t_4}. Let $Y = \{a/b/e, a/b/e/i\}$. Then $FoSC_{1/4}(Y) = 1/3$ as the two subtrees undergo significant changes together only in the version S^{t_3}.

3.2 Correlation of Change (CoC)

We now define the metric to measure the correlation between subtrees in undergoing significant changes. Given a DoC threshold α, in each transition between two successive versions, a set of subtrees either undergoes significant changes together or not.

This could be considered as a binary variable. Furthermore, it is a symmetric binary variable because we are interested in both the occurrences and the nonoccurrences of significant changes to subtrees. Hence, correlation measures which are suitable for analyzing symmetric binary variables can be used, such as ϕ-coefficient, odds ratio, and the Kappa statistic etc. [11]. In our analysis, we use the ϕ-coefficient. Given the contingency table in Figure 2(a), where X (Y) represents that a set of subtrees X (Y) undergoes significant changes together and $\neg X$ $(\neg Y)$ represents subtrees in X (Y) do not undergo significant changes together, the ϕ-coefficient between variables X and Y can be computed by the following equation.

$$\phi(X,Y) = \frac{Mf_{11} - f_{1+}f_{+1}}{\sqrt{f_{1+}(M - f_{1+})f_{+1}(M - f_{+1})}} \tag{1}$$

Particularly, in our mining context, the value of M in Figure 2(a) equals to $n-1$, which is the total number of transitions between successive historical versions. Furthermore, f_{11} refers to the number of versions where all subtrees in X and Y undergo significant changes together. Hence, f_{11} equals to $FoSC_\alpha(X \cup Y) \times (n-1)$. Similarly, f_{1+} equals to $FoSC_\alpha(X) \times (n-1)$ and f_{+1} equals to $FoSC_\alpha(Y) \times (n-1)$. Thus, the Equation 1 can be transformed as the following one, which is formally defined as *Correlation of Change* (CoC).

Definition 4 (Correlation of Change (CoC)). *Let X and Y be two sets of subtrees, s.t. $X \subseteq \Omega$, $Y \subseteq \Omega$, and $X \cap Y = \emptyset$. Given a DoC threshold α, the Correlation of Change of X and Y, with respect to α, denoted as $CoC_\alpha(X,Y)$, is:*

$$CoC_\alpha(X,Y) = \frac{FoSC_\alpha(X \cup Y) - FoSC_\alpha(X) * FoSC_\alpha(Y)}{\sqrt{FoSC_\alpha(X)(1 - FoSC_\alpha(X))FoSC_\alpha(Y)(1 - FoSC_\alpha(Y))}}$$

□

According to the definition of ϕ-coefficient, if $CoC_\alpha(X,Y)$ is greater than zero, two sets of subtrees X and Y are positively correlated in undergoing significant changes. Otherwise, they are negatively correlated in undergoing significant changes. In the following discussion, the subscript α in both $FoSC_\alpha$ and CoC_α is omitted if α is understood in the context. For example, consider the sequence of historical versions of tree S in Figure 1 again. Let the threshold of DoC α be $1/3$. Let $X = \{a/b, a/b/e\}$ and $Y = \{a/b/e/i\}$. $CoC_{1/3}(X,Y) = \frac{3\times0-2\times1}{\sqrt{2\times(3-2)\times1\times(3-1)}} = -1$. Hence, $\{a/b, a/b/e\}$ and $\{a/b/e/i\}$ are negatively correlated in undergoing significant changes.

3.3 NECTAR Mining Problem

Based on the evolution metrics introduced above, various negatively correlated subtree patterns or NECTARs can be defined. For example, we can define the pattern as two sets of subtrees with a negative *CoC* value. The subtrees may or may not be structurally independent. Particularly, in this paper, we define NECTAR on subtrees with ancestor-descendant relationships.

	Y	¬Y	Σ_{row}
X	f_{11}	f_{10}	f_{1+}
¬X	f_{01}	f_{00}	f_{0+}
Σ_{col}	f_{+1}	f_{+0}	M

(a) 2X2 contingency table

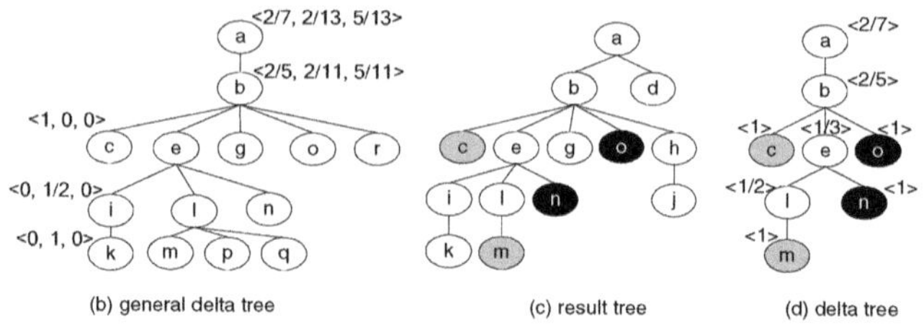

Fig. 2. 2 × 2 contingency table, general delta tree and delta tree

Let s_i be an ancestor subtree of s_{i+1} (conversely, s_{i+1} be a descendant subtree of s_i), denoted as $s_i \succ s_{i+1}$, if there is a path from the root of s_i to the root of s_{i+1}. Then, NECTARs can be defined as follows.

Definition 5 (NECTARs). *Given the DoC threshold α, the FoSC threshold η, and the CoC threshold ζ $(0 \leq \alpha, \eta \leq 1, \zeta \geq 0)$, $P = \langle X, Y \rangle$, where $X = \langle s_1, s_2, \ldots, s_{m-1} \rangle$, $Y = \langle s_m \rangle$, and $s_i \succ s_{i+1}$ $(1 \leq i < m)$, is a* NECTAR *if it satisfies the following conditions: (i) $FoSC_\alpha(X) \geq \eta$; (ii) $FoSC_\alpha(P) < \eta$; (iii) $CoC_\alpha(X, Y) \leq -\zeta$.* □

Consider the sequence of historical versions of tree S in Figure 1 again. Suppose the threshold $\alpha = 1/3$, $\eta = 2/3$, and $\zeta = 1/2$. $\langle\langle a/b, a/b/e \rangle, \langle a/b/e/i \rangle\rangle$ is a NECTAR because $FoSC(\{a/b, a/b/e\}) = 2/3 \geq \eta$, while, $FoSC$ $(\{a/b, a/b/e, a/b/e/i\}) = 0 < \eta$ and CoC $(\{a/b, a/b/e\}, \{a/b/e/i\}) = -1 < -\zeta$.

A NECTAR $\langle\langle s_1, s_2, \ldots, s_{m-1} \rangle, \langle s_m \rangle\rangle$ indicates that the sequence of subtrees $\langle s_1, s_2, \ldots, s_{m-1} \rangle$ frequently undergo significant change together. Whereas, they rarely undergo significant changes together with a descendant subtree s_m. Based on the inferred knowledge, we identified two types of subsumption relationships between NECTARs as follows.

Definition 6. (***Tail Subsumption***) *Given two* NECTARs *$P_1 = \langle X, Y_1 \rangle$ and $P_2 = \langle X, Y_2 \rangle$, where $Y_1 = \langle s_x \rangle$ and $Y_2 = \langle s_y \rangle$, we say P_1 tail-subsumes P_2 (or P_2 is tail-subsumed by P_1), denoted as $P_1 \sqsupset_t P_2$, if $s_x \succ s_y$.* □

Definition 7. (***Head Subsumption***) *Given a* NECTARs *$P_1 = \langle X_1, Y_1 \rangle$ and a subtree sequence X_2 s.t. $FoSC(X_2) \geq \eta$, where $X_1 = \langle s_1 s_2 \cdots s_m \rangle$ and $X_2 = \langle v_1 v_2 \cdots v_n \rangle$, we say P_1 is head-subsumed by X_2, denoted as $P_1 \sqsubset_h X_2$, if $\exists i (1 < i \leq m \leq n)$ such that $s_1 = v_1, \cdots, s_{i-1} = v_{i-1}$, and $s_i \prec v_i$.* □

Consider example in Figure 1 again. If both $\langle\langle a/b \rangle, \langle a/b/e \rangle\rangle$ and $\langle\langle a/b \rangle, \langle a/b/e/i \rangle\rangle$ are NECTARs, then the former tail-subsumes the latter. If the subtree sequence $S = \langle a/b, a/b/e \rangle$ satisfies the *FoSC* threshold, then all NECTARS $\langle\langle a/b, a/b/e/i \rangle, Y \rangle$ are head-subsumed by the sequence S, where Y can be any descendant subtree of $a/b/e/i$.

Often we observe that subsumed NECTARs add no further value to applications of NECTARs [7]. Hence, we exclude the subsumed NECTARs from the task of NECTAR mining. Thus, the problem of NECTAR *mining* can be formally defined as follows.

Definition 8 (NECTAR Mining). *Given a sequence of historical versions of a tree structure, the DoC threshold α, the FoSC threshold η and the CoC threshold ζ, the*

problem of **NECTAR** ***mining*** *is to find the complete set of* NECTARs *where each satisfies the specified thresholds and is neither tail-subsumed or head-subsumed.*

4 NECTAR-MINER Algorithm

Given a sequence of historical versions of a tree structure and thresholds of DoC and $FoSC$, the general algorithm NECTAR_*Miner* is shown in Figure $3(a)$. Note that, although it is optimized for nonsubsumed NECTARs, it can easily be extended to mine all NECTARs. Also, the threshold of CoC does not need to be given manually. Cohen [9] discussed the strength of the ϕ-coefficient value. It is stated that the correlation is strong if $|\phi|$ is above 0.5, moderate if $|\phi|$ is around 0.3, and weak if $|\phi|$ is around 0.1. Based on this argument, the CoC threshold can be set automatically and progressively. In our algorithm, we set the initial threshold of CoC ζ as 0.5. If no NECTAR is discovered with respect to this value and ζ is greater than 0.3, ζ is decreased successively. With a given ζ, basically, there are two phases involved in the mining procedure: the *GDT construction* phase and the NECTAR *discovery* phase. We elaborate on them in turn.

4.1 Phase I: *GDT* Construction

Given a sequence of historical versions of a tree structure, we construct a data structure, called *General Delta Tree* (GDT), that is appropriate for NECTAR mining. A GDT not only registers change information of subtrees, such as their DoC values, but also preserves the structural information of subtrees, such as the ancestor-descendant relationships.

Definition 9 (General Delta Tree). *Given a sequence of historical tree versions Σ, a sequence of consolidate trees of each two successive versions can be obtained, $\langle S_1^{'} = \langle N_1, E_1\rangle$, $S_2^{'} = \langle N_2, E_2\rangle, \cdots, S_{n-1}^{'} = \langle N_{n-1}, E_{n-1}\rangle\rangle$, where $S_i^{'} = S^{t_i} \uplus S^{t_{i+1}}$. A General Delta Tree $GDT = \langle N, E\rangle$, where $N = N_1 \cup N_2 \cup \cdots \cup N_{n-1}$ and $e = (x, y) \in E$ only if x is a parent of y in any $E_i (1 \leq i \leq n-1)$. Each node x is associated with a vector where the ith element corresponds to $DoC(s_x^{t_i}, s_x^{t_{i+1}})$.* □

Consider the sequence of historical tree versions in Figure 1. The constructed general delta tree is shown in Figure $2(b)$. Note that for clarity, we show the vector of DoC values for some of the nodes only. The node b is associated with a sequence $\langle 1/3, 2/11, 5/11\rangle$, which indicates that the DoC value of the subtree a/b in the first two versions is $1/3$, $DoC(a/b, t_2, t_3) = 2/11$, and $DoC(a/b, t_3, t_4) = 5/11$.

The algorithm of this phase is shown in Figure $3(b)$. At first, we initialize the root node of GDT. Then, for each two successive tree versions, we employ the algorithm X-Diff [12] to detect changes to tree structures[1].We adapt the algorithm slightly so that it directly takes generic tree structures as input. Given two tree versions, X-Diff generates a *result tree* which combines the nodes in both versions. For example, Figure $2(b)$ shows the example *result tree* generated by X-Diff after detecting changes between the first two tree versions in Figure 1. Here, we use gray nodes to represent the deleted nodes and black nodes to represent the inserted nodes.

[1] Note that our technique is not dependent on a specific tree differencing algorithm. Hence, it can work with any tree-diff approach.

```
Input: Σ=<S^t1, S^t2, ..., S^tn >, α, η
Output: a set of NECTARs P
Description:
    ζ = 0.5
    GDT = GDT_Constructor(Σ)
    while((| P | == 0)&&(ζ > 0.3))
        P = P ∪ Total_Pattern_Miner
            (GDT.root, α, η, ζ)
        ζ--
    return P
```

(a) Algorithm: NECTAR_Miner

```
Input: Σ=<S^t1, S^t2, ..., S^tn >
Output: a general delta tree GDT
Description:
    initialize GDT
    for ( i = 0; i < n; i++ ) do
      result_tree_i = X-Diff' (S^ti, S^ti+1);
      initialize delta_tree_i.root with result_tree_i.root
      for each child x of result_tree_i.root
          Depth_Traverse(x, delta_tree_i.root )
      end for
      delta_tree_i.root.descendants += x.descendants
      delta_tree_i.root.changed_descendants +=
                                  x.changed_descendants
      Merge_Tree(GDT.root, delta_tree_i.root, i)
    end for
    return GDT
```

(b) Algorithm: GDT_Constructor

```
Input: x, α, η, ζ
Output: total set of NECTARs P
Description:
    if(FoSCα(s_x) ≥ η)
    then
      P = <s_x>
      P.bitmap = Trans2Bit(x.DoCs, α)
      P = P ∪ Part_Pattern_Miner
            (x, P, α, η, ζ)
    end if
    for(each child y of x ) do
      Total_Pattern_Miner(y, α, η, ζ)
    end for
    return P
```

(c) Algorithm: Total_Pattern_Miner

```
Input: x, x', i
Description:
    the ith element of x.DoCs =
       x'.changed_descendants/x'.descendants
    for ( each child y' of x') do
      if (x has no matching child y)
      then
        create a child y of x corresponding to y'
      end if
      Merge_Tree(y, y', i)
    end for
```

(d) Algorithm: Merge_Tree

```
Input: nodes x, y
Description:
    Create a child x' of y corresponding to x
    x'.descendants = 1
    if(x is inserted/deleted)
    then
       x'.changed_descendants = 1
    end if
    for( each child z of x )
      Depth_Traverse(z, x' )
      x'.descendants += z.descendants
      x'.changed_descendants +=
                z.changed_descendants
    end for
    if (x'.changed_descendants == 0)
    then
       remove x' from delta_tree
    end if
```

(e) Algorithm: Depth_Traverse

```
Input: x, P, α, η, ζ
Output: NECTARs starting from s_x: P_x
Description:
    for (each child y of x) do
      if(FoSCα(P ∪ s_y) ≥ η)
      then
         P = P ∪ < s_y >
         P.bitmap = P.bitmap ∩ Trans2Bit(y.DoCs, α)
         P_x = P_x ∪ Part_Pattern_Miner
                       (y, P, α, η, ζ)
      else
         if(CoCα(P, <s_y>) ≤ - ζ)
         then
            P_x = P_x ∪ P
         end if
      end if
    end for
    return P_x
```

(f) Algorithm: Part_Pattern_Miner

Fig. 3. NECTAR-miner algorithm

For each *result tree* generated by X-Diff, we employ the algorithm *Depth_Traverse* (Figure 3(*e*)) to generate a *delta tree* which contains change information of subtrees. Basically, it traverses the *result tree* in the depth-first order. While reaching a node x in the *result tree*, we create a corresponding node in the *delta tree* and maintain two counters for the node: *descendants* and *changed_descendants*. The former records the number of descendants of x while the latter records the number of inserted/deleted descendants of x. Then, the *DoC* value of the subtree rooted at x can be computed by dividing *changed_descendants* by *descendants*. If the *DoC* value is zero, we remove the node from the *delta tree* because the subtree rooted at x is not a changed subtree. Otherwise, we associate the *DoC* value with the node. For example, after performing *Depth_Traverse* on the *result tree* in Figure 2(*c*), the generated *delta tree* is shown in Figure 2(*d*).

Then, for each generated *delta tree*, we merge it into the *GDT*. The algorithm is shown in Figure 3(d). When merging the ith *delta tree* (subtree) rooted at node $x^{'}$ with the GDT (subtree) rooted at node x, we only need to set the ith element of the DoC sequence of x with the DoC value of $s_{x^{'}}$. Then, the algorithm iteratively merges their children with the same label. For a child $y^{'}$ of $x^{'}$, if x does not have any matching child, we create a child node of x labeled as y. We associate a DoC sequence with the new child node such that the ith element of the sequence is set as the DoC value of $y^{'}$.

It can be shown that the complexity of the first phase is $O((n-1)\times(|S|^2\times deg(S)\times \log_2(deg(S))))$ where $|S|$ is the maximum size of the historical tree versions and $|\Delta S|$ is the maximum size of the *result trees*. Due to space constraints, the reader me refer to [8] for proof.

4.2 Phase II: NECTAR Discovery

The input of this phase is the GDT and the thresholds α, η and ζ of DoC, $FoSC$ and CoC. We aim to discover all non-subsumed NECTARs satisfying the thresholds from the GDT.

The complete set of NECTARS $\mathcal{P}$ can be divided into disjoint partitions, $\mathcal{P} = \mathcal{P}_1 \cup \mathcal{P}_2 \cup \ldots \cup \mathcal{P}_m$, such that each partition contains NECTARs starting from the same subtree. That is, $\forall P_x = \langle\langle s_x s_{x+1} \ldots s_{x+m}\rangle, \langle s_{x+m+1}\rangle\rangle \in \mathcal{P}, P_y = \langle\langle s_y s_{y+1} \ldots s_{y+n}\rangle, \langle s_{y+n+1}\rangle\rangle \in \mathcal{P}$, if $s_x = s_y$ then $P_x, P_y \in \mathcal{P}_i$; otherwise, $P_x \in \mathcal{P}_i, P_y \in \mathcal{P}_j$ $(i \neq j, 1 \leq i, j \leq m)$.

Therefore, the discovery of the complete set of NECTARs can be divided into subproblems of mining patterns for each partition, where each pattern begins with the same subtree. The algorithm of the second phase is shown in Figure 3(c). We mine patterns by traversing the *GDT* with the depth-first order. For each node x, which represents the subtree s_x, if the value of $FoSC(s_x)$ is no less than the threshold η, we call the function $Part\_Pattern\_Miner$ to discover NECTARs starting from s_x. Note that, the $FoSC(s_x)$ can be computed with the vector of DoC values associated with node x. For example, consider the GDT in Figure 2(a). Let $\alpha = 0.3$ and $\eta = 0.3$. The vector of DoC values of node b is $\langle 2/5, 2/11, 5/11\rangle \approx \langle 0.40, 0.18, 0.45\rangle$. Then, $FoSC(a/b) = 2/3 \approx 0.67 > \eta$. Thus, we need to find NECTARs starting from subtree a/b.

Before mining NECTARs starting from some subtree s_x, we initialize a pattern $P = \langle s_x\rangle$ and associate a bitmap with the pattern. In a bitmap, there is one bit for each element of the DoC sequence of node x. If the ith element is greater than α, then bit i is set to one. Otherwise, bit i is set to zero. For example, before mining NECTARS starting from a/b in the GDT in Figure 2(a), the bitmap associated with $P = \langle a/b\rangle$ is $\langle 101\rangle$ if α is set as 0.3.

The algorithm *Part_Pattern_Miner* is shown in Figure 3(f). For a particular node x in GDT, it generates all non-subsumed NECTARs starting from s_x. Note that, we directly mine non-subsumed NECTARs instead of performing filtering process afterwards. For this purpose, we employed the following three strategies in *Part_Pattern_Miner*. *Firstly, when mining* NECTARs *starting from a subtree* s_x, *candidate patterns are generated and examined by traversing* s_x *with the depth-first manner.* As shown in Figure 3(f), *Part_Pattern_Miner* performs a depth-first traversal on the subtree rooted at node x.

This strategy facilitates the performing of the following two strategies. *Secondly, once a sequence of subtrees satisfying the FoSC threshold is discovered, candidate patterns must be extended from this sequence.* For example, as shown in the Lines 4 to 6 of *Part_Pattern_Miner*, given the subtree rooted at node x and current pattern $\langle s_x \rangle$, for each child y of x, if $FoSC(\{s_x, s_y\})$ satisfies the threshold η, the pattern is updated as $P = \langle s_x, s_y \rangle$. Simultaneously, the bitmap of P is also updated with the DoC vector of node y. Then, *Part_Pattern_Miner* is recursively called for the updated pattern. This strategy prevents generating NECTARs which are head-subsumed. *Thirdly, once a* NECTAR *is discovered, we stop traversing nodes deeper than the current one.* For example, as shown in Lines 8 and 9 of *Part_Pattern_Miner*, if $CoC(\{s_x\}, \{s_y\}) < -\zeta$, a NECTAR $\langle\langle s_x \rangle, \langle s_y \rangle\rangle$ is discovered. *Part_Pattern_Miner* is not recursively called any more. This strategy keeps from generating NECTARs that are tail-subsumed.

For example, consider the GDT in Figure $2(a)$ again. Let $\alpha = 0.3$, $\eta = 0.3$ and $\zeta = 0.5$. $Part\_Pattern\_Miner$ mines NECTARs starting from subtree a/b as follows. The initial pattern $P = \langle a/b \rangle$ and its bitmap is $\langle 101 \rangle$. For the child node e of b, since DoC vector of e is $\langle 1/3, 1/3, 1/3 \rangle \approx \langle 0.33, 0.33, 0.33 \rangle$, then $FoSC(\{a/b, a/b/e\}) = 2/3 > \eta$. Hence, the pattern is updated as $P = \langle a/b, a/b/e \rangle$ and its updated bitmap is $\langle 101 \rangle$. Then, we further find NECTARs starting from $\langle a/b, a/b/e \rangle$. For the child node i of e, since $FoSC(\{a/b, a/b/e, a/b/e/i\}) = 0 < \eta$ and $CoC_{0.3}(\{a/b, a/b/e\}, \{a/b/e/i\}) = -1 < -\zeta$, then $\langle\langle a/b, a/b/e \rangle, \langle a/b/e/i \rangle\rangle$ is discovered as a NECTAR. For the child node l of e, since $FoSC(\{a/b, a/b/e, a/b/e/l\}) = 1/3 > \eta$, the pattern, together with its bitmap, is updated. Then, we further search NECTARs starting from $\langle a/b, a/b/e, a/b/e/l \rangle$ and discover the NECTARs $\langle\langle a/b, a/b/e, a/b/e/l \rangle, \langle a/b/e/l/p \rangle\rangle$. The algorithm terminates when all descendants of node b are visited.

Lemma 1. *No subsumed* NECTAR *will be discovered* NECTAR*_Miner.* □

The proof of the above lemma is given in [8]. The complexity of the second phase is $O(|GDT| \times dep(GDT))$ where $dep(GDT)$ is the depth of GDT [8].

Theorem 1. *The algorithm* NECTAR*_$Miner$ discovers the set of non-subsumed* NECTARs *completely.* □

The correctness of NECTAR_$Miner$ comes from the completeness of the *GDT* and the correctness of $Part\_Pattern\_Miner$. Since *GDT* not only records each changed subtree and its *DoC* in each successive versions but also preserves the ancestor-descendant relationship between subtrees, all information in relevance to NECTAR mining is maintained. The three strategies employed by $Part\_Pattern\_Miner$ ensure that each potential pattern is either examined or skipped to avoid generate subsumed NECTARs. Hence, NECTAR_$Miner$ is complete in discovering non-subsumed NECTARs.

5 Representative Applications of NECTARs

In this section, we present some representative applications of NECTARs. The reader may refer to [7,8] for details.

Maintaining autonomous Web mirrors. A mirror site is a replica of an original Website and is often located in a different place throughout the Internet so that the original site can be accessed from more than one place. Web mirroring can be generally classified as *internal mirroring* and *external mirroring*. The former refers to the situation that both the original site and mirror sites are set up and maintained by the same organization. The latter means that the original site and mirror sites are autonomous.

A mirror site should be updated frequently to ensure that it reflects the content of the original site. Although for internal mirroring this process can be performed efficiently by synchronizing mirror sites only if the original site evolves, for external mirroring the server holding the mirror site has to scan the original site frequently to detect the changes and update its local copy. Obviously, if the size of the Web site directory is large, such frequent scan and update may incur heavy workload. Hence, optimized mirror maintenance strategies are needed to improve the efficiency of external mirror maintenance.

NECTARs can be mined from historical directory hierarchies of Web sites for more efficient mirror site maintenance [7]. When maintaining the mirror of a changed target directory, its subdirectories, which have positive evolution association with it, should not be skipped from mirroring since they frequently change significantly together. On the contrary, the mirror maintenance process can be optimized by skipping the subdirectories which have negative evolution association with it since these subdirectories rarely undergo significant changes together with the target directory. Discovered patterns are then used to design optimal strategies for autonomous Web mirror maintenance.

Maintaining XML path selectivity estimation. Cost-based XML query optimization calls for accurate estimation of the selectivity of *path expressions*, which are commonly used in XML query languages to locate a subset of XML data. Recently, Bloom Histogram [13] has been proposed as a framework for XML path selectivity estimation in a dynamic context. There are three basic components in the system: the *data file*, the *dynamic summary* and the *estimator*. The data file refers to the XML file which is usually large. The dynamic summary keeps the necessary information of the data and the estimator is responsible for efficiently estimating the selectivity of path expression queries. The dynamic summary usually uses a path table to record the frequency of occurrences of paths. When an XML file evolves, the system needs to rebuild the path table *completely* before updating the estimator. We can use NECTARs to *selectively* rebuild the path table rather than building it from scratch. Specifically, some paths in the path table do not need to be updated if their frequency of occurrences change insignificantly. For example, we can mine NECTARs from historical XML tree versions to discover when the support[2]of some paths change significantly, the support of some path rarely undergoes significant changes together. Then, when updating the occurrence for a path, e.g., $//a/b$, whose support changes significantly, we check whether any child of node b forms a NECTAR with nodes on the path. Suppose a child c of b forms a NECTAR with a and b, we do not update the path $//a/b/c$ in the path table as the support of the path may not change or change insignificantly.

[2] The support of a path refers to the number of occurrences of the path in the document. The definition of DoC can be adjusted easily to measure the degree of support change.

F	Fanout of the first tree version	10
D	Depth of the first tree version	10
N	Size of the first tree version	1K
V	Number of tree versions	50
C	Change percent between tree versions	3%

(a) Parameter List

Data	Feature Parameters
D1	$V = 50$, $C = 3\%$
D2	$V = 50$, $C = 6\%$
D3	$V = 50$, $C = 12\%$
D4	$N = 1K$, $C = 3\%$
D5	$N = 1K$, $C = 6\%$

Data	Feature Parameters
D6	$N = 1K$, $C = 12\%$
D7	$N = 10K$, $V = 200$, $C = 3\%$
D8	$N = 10K$, $V = 200$, $C = 6\%$
D9	$N = 10K$, $V = 200$, $C = 12\%$

(b) Dataset List

Fig. 4. Parameters and datasets

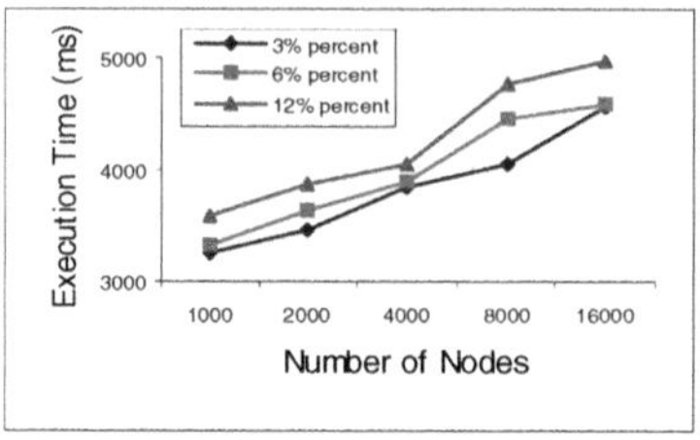

(a) Variation on tree size

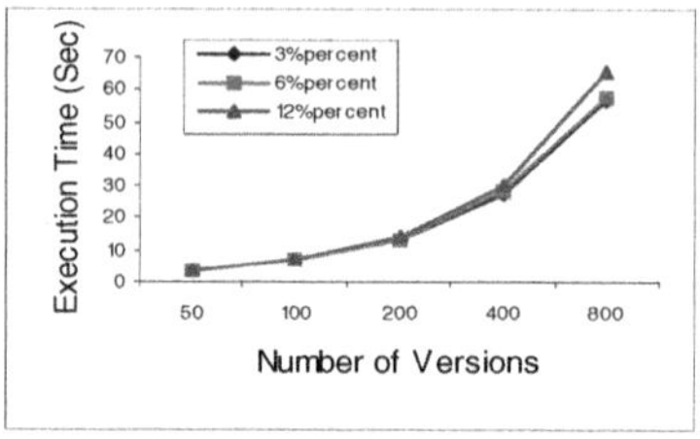

(b) Variation on version numbers

Fig. 5. Experimental results I

6 Experimental Results

The NECTAR mining algorithm is implemented in the Java programming language. All experiments are conducted on a Pentium IV 2.8GHz PC with 512 MB memory. The operating system is Windows 2000 professional.

Datasets: We implemented a synthetic data generator which generates sequential tree versions. The set of parameters used by the generator are shown in Figure 4(a). At first, the generator constructs the first historical tree version, S^{t_1}, based on parameters F, D and N. The first tree version is generated using the following recursive process. At a given node in the tree S^{t_1}, we decide how many children to generate. The number of children is sampled uniformly at random from 0 to F. Before processing children nodes, we assign random probabilities to decide whether a child node is a leaf node, as long as tree depth is less than the maximum depth D. If a child node is not a leaf node and the total number of nodes is less than N, the process continues recursively.

Once the first tree version has been created, we create as many subsequent versions as specified by the parameter V. The tree version S^{t_i} is generated based on the version $S^{t_{i-1}}$ with parameter C. For example, if the value of C is 3% and the number of nodes in the tree S^{t_i} is N_i, then the number of inserted and deleted nodes between trees S^{t_i} and $S^{t_{i-1}}$ is $3\% \times N_i$. We randomly decide the positions of inserted and deleted nodes. Basically, there are 9 datasets used in the following experiments. The key parameter values used in generating the datasets are shown in Figure 4(b).

Scaleup Evaluation: The two dominant factors that affect the complexity of the algorithm are the average size of trees and the number of historical versions. Hence, we evaluate the scalability of the algorithm by varying the two parameters.

We conduct the first experiment on the three data sets D_1, D_2 and D_3. For each data set, we vary the size of the first tree version from 1K nodes to 16K nodes. The

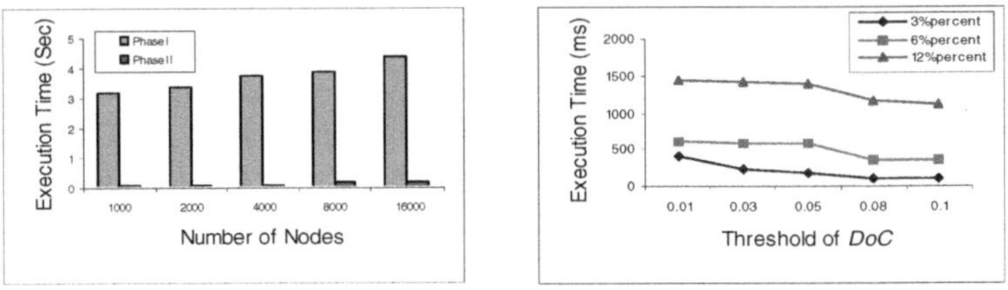

(a) Run time of phases

(b) Variation on *DoC* Threshold

Fig. 6. Experimental results II

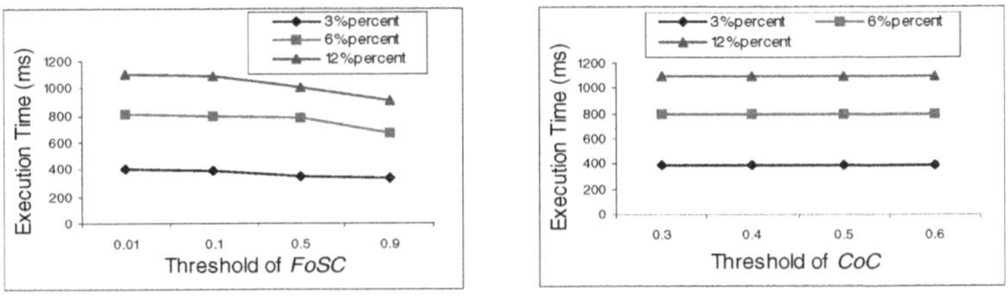

(a) Variation on *FoSC* Threshold

(b) Variation on *CoC* Threshold

Fig. 7. Experimental Results III

thresholds of DoC, $FoSC$ and CoC are set as 0.001, 0.01, and 0.3, respectively. The experimental results are shown in Figure $5(a)$. It can be observed that the scale-up feature of the mining algorithm is basically linear to the size of trees. Furthermore, the smaller the change percent, the faster the mining algorithm.

We conduct the second experiment on the three data sets D_4, D_5 and D_6. For each data set, we vary the number of versions from 50 to 800. The thresholds of DoC, $FoSC$ and CoC are set as 0.001, 0.01, and 0.3, respectively. The experimental results are shown in Figure $5(b)$. We observed that when the number of versions increases, the run time increases quickly because the sizes of subsequent trees increase as well.

The respective scaleup feature of the two phases of the mining algorithm was studied as well on D_4 with 50 versions. Figure $6(a)$ shows the results. Both phases take more time when the tree size increases. However, the first phase dominates the efficiency of the algorithm.

Efficiency Evaluation: We now study how the thresholds of DoC, FoC, and CoC affect the efficiency of the algorithm. We show the run time of the second phase as the thresholds do not interfere the performance of the first phase. Experiments are conducted on the data sets D_7, D_8 and D_9.

The results of varying the threshold of DoC are shown in Figure $6(b)$. The $FoSC$ threshold and CoC threshold are set as 0.01 and 0.3 respectively. It can be observed that the greater the DoC threshold value, the more efficient the algorithm. This is because when the DoC threshold value is greater, fewer subtrees are interesting enough so that we need to mine few negative evolution patterns. Figure $7(a)$ shows the results of the experiments on varying the threshold of $FoSC$. The thresholds of DoC and CoC are set as 0.001 and 0.3 respectively. We noticed that the greater the $FoSC$ threshold, the more efficient the algorithm. Finally, Figure $7(b)$ shows that the variation on the CoC threshold has no affection on the performance of the mining algorithm, as illustrated by the complexity analysis.

7 Related Work

Although traditional association rule mining focuses on mining positive associations between items, it was observed that negative associations were useful as well in some situations [14]. Several work have been conducted on mining negative association rules [4][1][14]. They are different from each other in the employed correlation metrics. Consequently, the developed data mining algorithms are different as well. To the best of our knowledge, there exists no work on mining negative associations from changes to tree structure. Furthermore, our rules are distinguished in the way that the trees have ancestor-descendant relationships.

In [6], we proposed a novel approach for mining structural evolution of XML data to discover FRACTURE patterns. A FRACTURE pattern is a set of subtrees that frequently change together and frequently undergo significant changes together. NECTARs are different from FRACTUREs in at least the following two aspects. First, since the former focus on the negative evolution associations while the latter capture the positive evolution associations, different evolution metrics are used. Second, the former are defined on a sequence of subtrees with ancestor-descendant relationships, while the latter are defined on subtrees which may or may not be structurally related.

8 Conclusions

This work is motivated by the fact that existing tree mining strategies focus on discovering knowledge based on statistical measures obtained from the static characteristics of tree data. They do not consider the evolutionary features of the historical tree-structured data. In this paper, we proposed techniques to discover a novel type of pattern named negatively correlated subtree pattern (NECTAR) by analyzing the associations between structural evolution patterns of subtrees over time. NECTARs can be useful in several applications such as maintaining mirrors of a website and maintaining XML path selectivity estimation. Two evolution metrics, $FoSC$ and CoC, which measure the frequency and correlation of significant changes to historical tree-structured data, respectively, are defined to evaluate the interestingness of NECTARs. A data mining algorithm NECTAR-Miner based on divide-and-conquer strategy is proposed to discover NECTARs. Experimental results showed that the proposed algorithm has good performance and can accurately identify the NECTARs.

References

1. Antonie, M.-L., Zaïane, O.R.: Mining positive and negative association rules: An approach for confined rules. In: Boulicaut, J.-F., Esposito, F., Giannotti, F., Pedreschi, D. (eds.) PKDD 2004. LNCS, vol. 3202, pp. 27–38. Springer, Heidelberg (2004)
2. Asai, T., Abe, K., et al.: Efficient Substructure Discovery from Large Semi-structured Data. In: SIAM DM (2002)
3. Braga, D., Campi, A., Ceri, S., et al.: Mining Association Rules from XML Data. In: DAWAK (2000)
4. Brin, S., Motwani, R., Silverstein, C.: Beyond Market Baskets: Generalizing Association Rules to Correlations. In: SIGMOD (1997)
5. Chawathe, S., Garcia-Molina, H.: Meaningful Change Detection in Structured Data. In: SIGMOD (1997)
6. Chen, L., Bhowmick, S.S., Chia, L.T.: FRACTURE Mining: Mining Frequently and Concurrently Mutating Structures from Historical XML Documents. Data and Knowl. Eng. 59, 320–347 (2006)
7. Chen, L., Bhowmick, S.S., Nejdl, W.: Mirror Site Maintenance Based on Evolution Associations of Web Directories. In: ACM WWW, Banff, Canada, Techical report (2007), `http://www.cais.ntu.edu.sg/~assourav/TechReports/NEAR-TR.pdf`
8. Chen, L., Bhowmick, S.S.: Finding nectars in Evolutionary Trees. Technical Report, CAIS-10-2007 (2007), `http://www.cais.ntu.edu.sg/~assourav/TechReports/NECTAR-TR.pdf`
9. Cohen, J.: Statistical Power Analysis for the Behavioral Sciences, 2nd edn. Lawrence Erlbaum, Mahwah (1988)
10. Lian, W., Cheung, D.W., Mamoulis, N., Yiu, S.M.: An Efficient and Scalable Algorithm for Clustering XML Documents by Structure. IEEE TKDE, vol. 16(1) (2004)
11. Tan, P.-N., Steinbach, M., Kumar, V.: Introduction to Data Mining. Addison Wesley, Reading (2006)
12. Wang, Y., DeWitt, D.J., et al.: X-diff: An Effective Change Detection Algorithm for XML Documents. In: ICDE (2003)
13. Wang, W., Jiang, H., et al.: Bloom Histogram: Path Selectivity Estimation for XML Data with Updates. In: VLDB (2004)
14. Wu, X., Zhang, C., Zhang, S.: Mining both Positive and Negative Association Rules. In: ICML (2002)
15. Zaki, M.J.: Efficiently Mining Frequent Trees in a Forest. In: SIGKDD (2002)
16. Zaki, M.J., Aggarwal, C.C.: XRules: An Effective Structural Classifier for XML Data. In: SIGKDD (2003)
17. Zhao, Q., Bhowmick, S.S., Mohania, M., Kambayashi, Y.: Discovering Frequently Changing Structures from Historical Structural Deltas of Unordered XML. In: ACM CIKM (2004)

Reducing Space Requirements for Disk Resident Suffix Arrays*

Alistair Moffat[1], Simon J. Puglisi[2], and Ranjan Sinha[1]

[1] Department of Computer Science and Software Engineering, The University of Melbourne, Australia
{alistair,rsinha}@csse.unimelb.edu.au
[2] School of Computer Science and Information Technology, RMIT University, Melbourne, Australia
sjp@cs.rmit.edu.au

Abstract. Suffix trees and suffix arrays are important data structures for string processing, providing efficient solutions for many applications involving pattern matching. Recent work by Sinha et al. (SIGMOD 2008) addressed the problem of arranging a suffix array on disk so that querying is fast, and showed that the combination of a small trie and a suffix array-like blocked data structure allows queries to be answered many times faster than alternative disk-based suffix trees. A drawback of their LOF-SA structure, and common to all current disk resident suffix tree/array approaches, is that the space requirement of the data structure, though on disk, is large relative to the text – for the LOF-SA, $13n$ bytes including the underlying n byte text. In this paper we explore techniques for reducing the space required by the LOF-SA. Experiments show these methods cut the data structure to nearly half its original size, without, for large strings that necessitate on-disk structures, any impact on search times.

1 Introduction

Suffix trees [17,23,24] and suffix arrays [15] offer elegant solutions to many important problems requiring string processing. The simplest and perhaps most widely applicable of these is pattern matching: given a pattern of m characters $P[0 \ldots m-1]$ and a text $T[0 \ldots n-1]$, find the set of positions at which P occurs in T. After $O(n)$ time spent preprocessing, suffix trees (and suffix arrays when suitably augmented) allow any pattern to be located in time $O(m+occ)$ (independent of the size of the text), where occ is the number of occurrences of the pattern P in T. Many more applications of these data structures are described in the literature [1,3,12,21].

Despite these theoretical advantages, the large space requirements of suffix trees, and their poor locality of memory reference, have resulted in them having only limited adoption. Two broad lines of research have attempted to address these problems. The first is the active field of *compressed full-text indexes* [18], where (in broad terms) the aim is to exploit the relationship between the suffix tree/array and the Burrows-Wheeler Transform. The constraining factor with this approach is that the index must be compressed sufficiently well that it can reside entirely in RAM, as memory access patterns

* This work was supported by the Australian Research Council.

X. Zhou et al. (Eds.): DASFAA 2009, LNCS 5463, pp. 730–744, 2009.

in the compressed index tend to be worse than in the raw index as a result of the need to "decode" each position of occurrence from the index.

The second line of research involves storing and accessing the index on secondary memory, which in most environments is vastly more abundant than RAM, but also several orders of magnitude slower to access. The majority of the work in this direction has been concerned with constructing suffix trees on disk [6,19,22,10], or constructing suffix arrays on disk [7,9], rather than actually describing how best to use the index in disk-based operations. One exception is the work of Baeza-Yates et al. [4], who describe a structure they call the SPAT array, in which the on-disk suffix array is indexed by a fixed-length prefix of symbols. The drawback of this arrangement is that a search may require multiple blocks of suffix array data to be consulted, plus multiple accesses to the underlying text string, each of which might require a disk operation.

Sinha et al. [20] contributed a structure they call the LOF-SA, in which a small in-memory trie and a suffix array-like blocked data structure on disk combine to allow queries to be answered many times faster than alternative disk-based suffix trees. A drawback of the LOF-SA (and to a greater or lesser extent, of all current disk-resident suffix tree/array approaches) is that the space requirement of the data structure, though on disk, is large relative to the text. In the case of the LOF-SA, a total of $13n$ bytes of disk storage is required, including the n bytes occupied by the underlying text string.

The key distinguishing characteristic of the LOF-SA is the storage, with each of the suffix array pointers, of a set of *fringe* characters that help cut back on the number of times the underlying text string needs to be consulted. Experiments using the LOF-SA show that the majority of short patterns can be accessed using only the trie and the sequentially-accessed LOF-SA blocks, and that the added cost of searching for a long pattern is typically just one access into the underlying text. While expensive to store compared to (say) a $5n$-byte pure suffix array index for a text of n bytes, the LOF-SA provides much faster pattern searching than the suffix array for large text strings, when the suffix array is too large for memory; and also for short text strings, when both structures fit in memory, but the LOF-SA has superior locality of reference.

In this paper we explore techniques for reducing the space requirements of the LOF-SA structure. Significant savings can be achieved by reorganizing and packing each LOF-SA item, and through the use of standard integer compression methods. Experiments on different data types show that around $7n$ bytes suffices for the resulting *compressed LOF-SA* structure, with, for large strings that necessitate on-disk structures, search times that are unaffected.

In the remainder of this section we define the key concepts of this research area. Section 2 then introduces the LOF-SA, which is our focus throughout. Section 3 examines the redundancy present in two subcomponents of the LOF-SA, with the aim of reducing overall space requirements. The efficiency of the smaller data structure relative to the original LOF-SA is evaluated experimentally in Section 4. We conclude in Section 5 by discussing future avenues for investigation.

We consider a string $T[0]T[1]\cdots T[n]$ of $n+1$ symbols. The first n symbols of T are drawn from a constant ordered alphabet Σ consisting of symbols $\sigma_j, j = 0, 1, \ldots, |\Sigma|-1$ ordered $\sigma_0 < \sigma_1 < \cdots < \sigma_{|\Sigma|-1}$. We assume each symbol requires one byte of storage, so $|\Sigma| \in 1 \ldots 256$. The final character $T[n]$ is a special "end of string" character, \$,

i	0	1	2	3	4	5	6	7	8	9	10	11	12	13	14	15	16
$T[i]$	g	t	t	a	t	g	a	t	a	t	g	t	t	t	g	t	$
SA	16	6	3	8	5	14	0	10	15	2	7	4	13	9	1	12	11
LCP	–	0	2	3	0	1	2	3	0	1	4	1	2	3	1	2	2

Fig. 1. Example suffix array and LCP array for the string $T =$ "gttatgatatgtttgt"

lexicographically smaller than all symbols in Σ. We also assume that $n < 2^{32}$, so that integers in the range $0 \ldots n$ can be stored in four bytes.

The *suffix array* of T, denoted SA_T (or just SA when the context is clear), is an array $\mathrm{SA}[0 \ldots n]$ which contains a permutation of the integers $0 \ldots n$ such that $T[\mathrm{SA}[0] \ldots n] < T[\mathrm{SA}[1] \ldots n] < \cdots < T[\mathrm{SA}[n] \ldots n]$. That is, $\mathrm{SA}[j] = i$ iff $T[i \ldots n]$ is the j th suffix of T in ascending lexicographical order. A structure often used alongside SA_T is the *longest common prefix* or *lcp array* $\mathrm{LCP} = \mathrm{LCP}[0 \ldots n]$: for every $j \in 1 \ldots n$, $\mathrm{LCP}[j]$ is the length of the longest common prefix of suffixes $\mathrm{SA}[j-1]$ and $\mathrm{SA}[j]$, with $\mathrm{LCP}[0]$ undefined. The example shown in Figure 1 illustrates these data structures. For example, in Figure 1, the longest common prefix of suffixes 0 and 10 in columns 6 and 7 is "gtt" of length 3; while that of suffixes 2 and 7 in columns 9 and 10 is "tatg" of length 4.

The LCP array can be built in $O(n)$ time and space provided SA_T and T are available [13,16]. An important property of the SA is that all the positions where a given pattern occurs in T appear in a contiguous portion of SA_T. For example consider the pattern "at", which occurs in the example string at positions 3, 6 and 8. These positions appear together (though permuted) at $\mathrm{SA}[1 \ldots 3]$. Thus pattern matching can be translated to the problem of finding an appropriate interval of the suffix array SA.

2 The LOF-SA Data Structure

The LOF-SA data structure consists of two parts [20]. The larger of the two components is an augmented suffix array, organized as a set of *LOF items*, one per symbol in the original string. Each LOF item contains three fields: an L value, which the length of the longest common prefix for this suffix; an O item, which is the pointer offset in the string corresponding to this suffix; and an F item, which is a set of *fringe* characters, drawn from the string T at locations $T[O + L \ldots O + L + f - 1]$, where f is a preset constant that indicates how many symbols will be extracted into each fringe, and O and L are the O and L values respectively for this item. For example, if $f = 2$, the next two characters past the end of the common prefix – which are, of course, the distinguishing characters for each suffix – are stored in each LOF item, and can be used to guide the searching process without a pointer deference (via the O component) being needed to the underlying string. That is, by strategically duplicating parts of the string, better locality of reference can be achieved, and some of the accesses to the underlying text either delayed or eliminated entirely. Figure 2 shows the L, O, and F items for the string shown in Figure 1.

The LOF items are stored sequentially in suffix order in *LOF blocks* of (typically) up to 4,096 items, with the requirement being that all of the LOF items in any block uniquely share a single common prefix. In the example string nodes 9 to 15 could be in

i	0	1	2	3	4	5	6	7	8	9	10	11	12	13	14	15	16
$T[i]$	g	t	t	a	t	g	a	t	a	t	g	t	t	t	g	t	$
L	–	0	2	3	0	1	2	3	0	1	4	1	2	3	1	2	2
O	16	6	3	8	5	14	0	10	15	2	7	4	13	9	1	12	11
F	–	at	ga	tt	ga	t$	ta	tg	t$	at	tt	ga	t$	tt	ta	gt	tg

Fig. 2. Example L, O, and F items with $f = 2$ for the string $T =$ "gttatgatatgtttgt" shown in Figure 1

a single LOF block, because all of the matching strings start with "t"; and nodes 11-13 start with "tg" and could be a single LOF block. But nodes 10-12 could not form a LOF block, because that arrangement would split both the set of suffixes starting with "ta" (locations 9-10) and the suffixes starting with "tg" (locations 11-13); and nor does it span all suffixes starting with just "t", which start at offset 8 and continue through to offset 16.

The first suffix in each LOF block is then inserted into the other component of the LOF-SA, which is an index of the block-identifying prefix strings, one per LOF block. Because of the way it is constructed, this index can be rapidly searched to identify the LOF block (or blocks) that correspond to any string prefix. In our implementation the index structure is organized as a trie, but other dictionary data structures can also be used. Even for large underlying strings T, the *LOF trie* occupies a relatively small amount of space, and, because it is accessed for every pattern search, can be assumed to have been buffered in to main memory by the operating system once some initial start-up period has elapsed.

To search the LOF-SA to find a pattern P, the LOF trie is used to identify the LOF block (or blocks, if P is a range query or is short enough to be exhausted within the trie structure) that must contain P. Note that only a single block can be required when P is a long pattern, because of the rule that every LOF block is identified by a unique prefix. That LOF block is then fetched into memory, and sequentially scanned. The L and F components form a rudimentary "copy and extend by f characters" data compression scheme (in a sense, an LZ78 mechanism with a simple phrase-book containing all prefixes of the previous string); and the strings indicated by each LOF item can be partially reconstructed from the sequence of LOF items prior to any accesses to the text being needed. The partially reconstructed strings can then be checked against (prefixes of) the pattern P. If any of the reconstructed strings match P, the corresponding O components from those LOF items are used to access segments of the underlying text, and a final determination made as to whether the pattern matches.

One slight complication arises if an L value in the pure suffix array SA_T is more than f larger than the previous L value, because it gives rise to a gap in the reconstructed string when the LOF block is sequentially processed. For example, in Figure 2, at item 10 there is a jump in the L value by more than $f = 2$, and so a gap in the reconstructed string would result if the gap was retained. To avoid the issue, the f fringe characters of each LOF triple are adjusted where necessary, to guarantee that all partially reconstructed strings do not contain any gaps. Returning to the example, this means that item 10 in Figure 2 would store an L of 3, and an F of "gt", rather than the 4 and

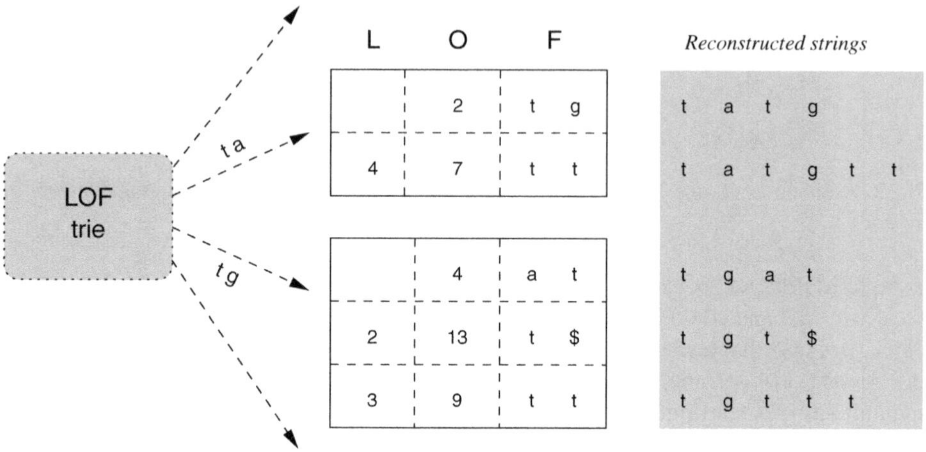

Fig. 3. Part of the LOF-SA for the string $T =$ "gttatgatatgtttgt$" in a LOF-SA with fringe length $f = 2$. Two LOF blocks are shown, presumed to be distinguished by their two-symbol common prefixes, "ta" and "tg" respectively.

"tt" that are shown in the figure. The very first LOF triple stored in each LOF block must also have its L value restricted, so that the first string of each block can be readily constructed. By adding this simple requirement, no intervening characters are ever unknown as the search proceeds, and accesses to the text are delayed for as long as possible. This arrangement is referred to as a *left justified fringe* [20].

Part of the LOF-SA for our example string is depicted in Figure 3, with a fringe of $f = 2$ characters, and an assumption that in the vicinity of the two LOF blocks that are shown, all strings in each block share two characters of common prefix. The reconstructed strings inferred from the L and F values in the LOF items are shown in the shaded box on the right.

3 Reducing Space Requirements

Before turning our attention to reducing the size of the LOF-SA we describe the two families of data used in our experiments. The DNA dataset consists of several large prefixes, ranging in size from 40 million to 400 million characters (bases) taken from the concatenated human genome downloaded from the Ensembl website (`http://www.ensembl.org`). We replaced all occurrences of "n" with random choices amongst "a", "c", "g" and "t", as is common practice. The zero-order entropy of these files was uniformly 2.0 bits per symbol, over an alphabet of $|\Sigma| = 4$. The ENG datasets contained English text derived from files of the Gutenberg Project (`http://www.gutenberg.org`), obtained from the Pizza-Chili website (`http://pizzachili.di.unipi.it/texts.html`). We further removed carriage-return, line-feed and tab symbols from these files to obtain files that typically had around 200 distinct symbols in them, and a zero-order self-entropy of around 4.5 bits per character.

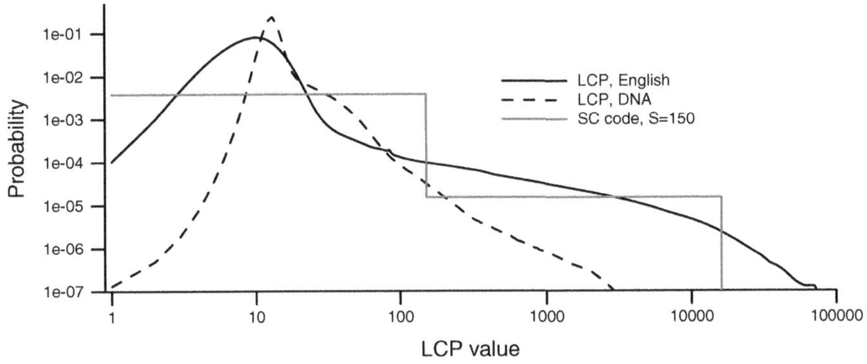

Fig. 4. Distribution of the L (longest common prefix) values for DNA100M and ENG100M. The grey lines show the corresponding inferred probabilities associated with the $S = 150$ codes that is the best match to the ENG100M L distribution.

Reordering Components. The most frequently accessed values during a LOF-SA search are the L's, since they are examined as each LOF-SA item is processed. The F values are also consulted relatively frequently. On the other hand, the O values are required only when the text is about to be checked, and our expectation is that this is only approximately once per pattern search [20].

Our first optimization of the LOF-SA structure is thus to partition each block of LOF items into two parts, retaining the L and F components of each item as a coupled pair, and putting all of their O values at the end of each block. To specify this and subsequent node arrangements, we make use of regular expression notation: each fully interleaved block in the original LOF-SA has the pattern $(L_4O_4F_4)^*$, where the subscripts indicate the number of bytes required by each component. Stripping out all of the O values and storing them together at the end of the block gives the arrangement $(L_4F_4)^*(O_4)^*$, and has the advantage that the sequential part of each search iterates over 8-byte units rather than 12-byte units. Each LOF-SA node still requires 12 bytes, but the access pattern is tighter than when every O value must be bypassed.

Compaction. At every step of the LOF block search algorithm an L value is accessed, but there is no requirement that this value be a fixed 4-byte quantity. The distribution of the L values for the DNA100M and ENG100M collections is shown in Figure 4. Small values dominate (note the logarithmic vertical scale in the graph), which suggests that *compaction* of the L values may be an option – it is clearly possible for all L values greater than 255 to be truncated to 255, and the L's to be stored in one byte each. A minor modification to the search algorithms is then needed so that in the (rare) cases in which the search depth reaches 255, false matches are identified by checking the text.

With the L values fitted into one byte, the fringe values F become the focus. One obvious way of saving space is to reduce the length of each fringe, and rather than storing the next $f = 4$ characters, store the next $f = 3$, with those three bytes packed with the L byte into a single word. This then gives rise to an $(L_1F_3)^*(O_4)^*$ arrangement, in which $9n$ bytes in total are required for an n byte text. Reducing the size of each LOF-SA node in this way means that less data is fetched from disk during each search operation.

Adding Compression. For applications other than pattern matching (in particular, applications that perform a bottom-up traversal of the suffix tree) access to the full LCP values is required, and so a lossless compression method must be employed to store the L values. The fact that the majority of the L values are small integers means that a wide variety of possible codes can be considered, including all of the standard methods such as Golomb codes, Huffman codes, and the word-aligned codes of Anh and Moffat [2]. However, all of these make the interleaving of L and F components difficult to achieve, both in terms of their alignment, and in terms of the fact that fast decoding using these methods relies on a lookahead bit-buffer.

Instead, we focus our attention on byte-aligned codes, and in particular explore the use of the (S, C)-byte code [5]. In the (S, C)-byte code each integer is represented as a sequence of *continuer* bytes, followed by a single *stopper* byte. The two types of byte are distinguished by their value when considered as 8-bit integers; and the decoded value is formed as the composition of the bits that form the balance of each byte. Because the L values are mostly small, and their probability distribution is for the most part non-increasing, there is no need to permute the L values according to frequency, and the resultant elimination of the usual symbol translation step in the decoder means that access to sequential L values can be fast. The fourth and fifth columns of the first section of Table 1 give the compression performance of the $(128, 128)$-byte code (that is, the standard byte code in which the top bit of each byte indicates whether it is a stopper or a continuer) for the two files reported, and the optimal (S, C)-byte code when applied to the L values of each of the data sets. As points of reference, the second column in Table 1 shows the zero-order entropy of the L values across the two files, and the third column shows the performance attained by a zero-order Huffman coder (`http://www.cs.mu.oz.au/~alistair/mr_coder`).

As all suffixes in a bucket share a common prefix, the length of this common prefix can be subtracted off prior to compression to obtain a smaller set of L values. Doing so results in H_0 dropping to 3.64 for DNA and to 6.88 for the English collection, but the effect is almost unnoticeable when byte codes are used. Nevertheless, the implementations we experiment with later in this paper all contain this small optimization.

Table 1. Compression performance for various methods on the L and F values for two of the test files, presented as bits per value averaged across each file. Values for the other test files were similar to the values listed

Collection	H_0	`shuff`	`byte`	(S, C)-byte
Compressing the L_4 *values*				
DNA100M	3.76	3.79	8.04	8.02
ENG100M	7.05	7.09	9.85	9.84
Compressing the F_3 *values*				
DNA100M	5.43	5.46	8.00	8.00
ENG100M	11.28	11.32	13.70	13.22
Compressing the F_4 *values*				
DNA100M	7.34	7.37	9.00	8.00
ENG100M	13.56	13.61	15.65	15.55

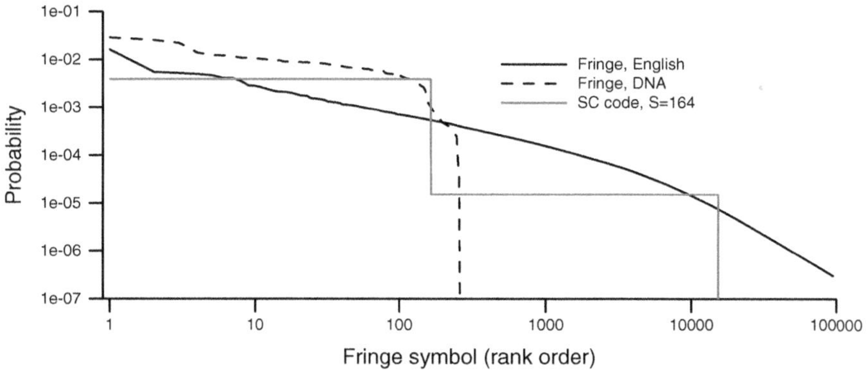

Fig. 5. Frequency sorted fringe values for DNA100M and ENG100M for a fringe length of $f = 4$. The grey line shows the corresponding inferred probabilities associated with the corresponding $S = 164$ codes that is the best match for the ENG100M file.

Compressing the Fringe. Each fringe consists of three or four consecutive symbols from the source text, so if the source text is compressible, so too is the set of fringe sequences. However, use of any kind of multi-state or conditioning encoder will add greatly to the complexity of the searching operation. Instead, we treat each fringe string as a single complex 24- or 32-bit symbol, and represent them in a zero-order sense. Taking this approach has the net effect (in terms of the product of the component symbol probabilities) of coding each fringe's first character in a zero-order model; coding each fringe's second character in a first order model, conditioned by the previous character; each fringe's third character in a second order model; and so on.

The fact that the L and F components are stored tightly interleaved means that it is beneficial to use the same coding method for both. However, unlike the L values, the natural ordering of the fringes does not make a good approximation of the frequency ordering, and if (S, C)-byte codes are to be used, the fringes need to be reordered and a mapping stored. In the terminology of Brisaboa et al. [5] a *dense code* must be generated, where each fringe is represented by its rank in the descending-frequency sorted list of 24-bit (or 32-bit) multi-byte symbols. Figure 5 illustrates the probabilities associated with the descending frequencies fringe sets for two of the test files, and also plots the implied probability distribution for the $(164, 92)$-byte code that is the best match to the measured probability distribution for file ENG100M.

Another useful attribute of the (S, C)-byte code is the ability to bypass codewords without fully reconstructing the corresponding strings, by virtue of the stopper byte always being the last one. In many situations the L value alone will be sufficient to confirm that the current LOF item cannot match the pattern P, in which case decoding the F value might be unnecessary, and bypassing it the preferred option.

Each access to a fringe value that is carried out requires a lookup in the decoder mapping, likely incurring a cache miss. Tight coupling of the search algorithm and the decoder is therefore required: rather than decompress a block in its entirety and then hand it to the search algorithm, on-the-fly decoding of LOF-SA items from within the search algorithm means that full decoding of fringe characters via the symbol mapping can be deferred until necessary. Culpepper and Moffat [8] discuss the costs associated

Table 2. Most frequent fringe strings and their probabilities

Rank	DNA100M		ENG100M	
	$f = 3$	$f = 4$	$f = 3$	$f = 4$
1	"TAA" 0.0622	"TAAA" 0.0288	"the" 0.0232	"the " 0.0163
2	"GAA" 0.0559	"GAAA" 0.0253	"of " 0.0076	"and " 0.0055
3	"CAA" 0.0471	"CAAA" 0.0220	"he " 0.0074	"that" 0.0052
4	"TAT" 0.0345	"TAAT" 0.0138	", a" 0.0067	", an" 0.0049
5	"TCA" 0.0343	"GAAG" 0.0124	". " 0.0067	"was " 0.0047
6	"TTT" 0.0318	"TTTT" 0.0121	"in " 0.0062	"ing " 0.0041
7	"TCT" 0.0318	"TATT" 0.0117	"and" 0.0060	" the" 0.0041
8	"GAG" 0.0310	"TACA" 0.0108	"to " 0.0060	"with" 0.0035
9	"CAG" 0.0292	"TAAG" 0.0107	"tha" 0.0060	"The " 0.0028
10	"TAG" 0.0280	"GAGA" 0.0107	"is " 0.0056	"had " 0.0028

Table 3. Explanation of the methods considered

Method	Description
$(L_4O_4F_4)^*$	Original LOF-SA arrangement described by Sinha et al. [20].
$(L_4F_4)^*(O_4)^*$	O components separated. No compression.
$(L_1F_3)^*(O_4)^*$	O components separated, L component restricted to 255, and fringe of three. No compression.
$(L_{4c}F_{3c})^*(O_4)^*$	O components separated, L component and three-byte F component compressed using (S, C)-byte code.
$(L_{4c}F_{4c})^*(O_4)^*$	O components separated, L component and four-byte F component compressed using (S, C)-byte code.

with large-alphabet decoder mappings, and describe a *semi-dense prelude* mechanism for reducing it. We have not adopted this approach in the experiments reported below, but plan to do so in our next round of experimentation.

Use of an (S, C)-byte code allows fringes for small alphabets like DNA to be efficiently handled. The optimal choice $S = 254$ when the fringe length is 4 on a DNA alphabet reduces the space used by the fringe to one byte per entry. This is close to what can be achieved by a Huffman coder which, for 4-base DNA sequences, typically requires 7.4 bits per symbol on average (Table 1).

Table 2 lists the ten highest frequency fringe strings that appeared in two typical data files, for $f = 3$ and $f = 4$. In the English text, There is a strong relationship between common short words and the fringe characters, which occurs because those short words typically indicate the end of a phrase of longer words which occurs more than than once in the text. In the DNA data there is no similar pattern, and the list of fringe strings is both much shorter, and also closer to being uniform in its probability distribution.

Table 3 summarizes the various LOF-SA arrangements we have considered.

4 Performance Comparison

This section reports the results of experiments using the various LOF-SA approaches described in the previous section. The purpose of the experiments was to compare the reduced-space variants of the LOF-SA to the original, and hence to the range of other disk-oriented suffix array approaches reported by Sinha et al. [20].

Hardware. All tests were conducted on a 2.8 GHz Intel Pentium IV with 1 GB of RAM, 1 MB of cache, and 250 GB local SATA disk. Note that, while the processor is hyper-threaded, the implementation made no use of parallelism, and each query was executed through to completion before the next one in the sequence was commenced. The overall elapsed time for the query sequence was measured to obtain per-query execution times, ensuring that the machine was under light load and that no other significant I/O or CPU tasks were running. The compiler was g++ (gcc version 4.1.2) executed with the `-O3` option, under the Debian GNU/Linux (sarge) operating system. Times were recorded with the standard Unix `getrusage` function.

Queries and Data. The test files have already been described. To form a set of pattern-match test queries, a sequence of random locations in each test file was used to extract 100-character strings. These strings were then used as queries. Because the strings were long, and because they were guaranteed to have at least one answer, they always required at least one access into the underlying text string to verify that an answer had indeed been found. That is, each of the test queries drove the LOF-SA to its maximally expensive searching behavior.

Methodology. In all cases the experiments were carried out using a consistent approach, to ensure that the measured times genuinely reflected long-term steady-state processing rates. A very long query stream was executed from a cold start, and throughput measured at regular intervals (of the order of one minute apart) until several consecutive intervals attained the same throughput. Only when a steady-state arrangement of the LOF-SA across memory and disk had been achieved – as evidenced by consistent query throughput rates – was measurement undertaken to record query times for any

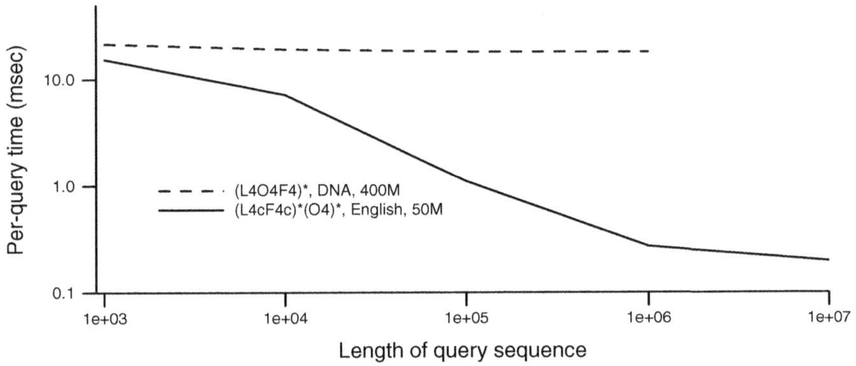

Fig. 6. Convergence of querying costs from a cold start of two different LOF-SA structures for two different files. The values listed are per-query times in milliseconds, averaged over the whole sequence of queries. Running times for the large data structure converge quickly; running times for the smaller data structure only stabilize after many queries have been processed. The top line is approximately five hours long, the lower one thirty minutes long.

Table 4. Space requirements in megabytes of different LOF-SA structures, not including the storage associated with the underlying string

Representation	DNA40M	DNA100M	DNA400M	ENG50M	ENG100M	ENG200M
$(L_4O_4F_4)^*$	457.7	1144.4	4577.7	588.4	1176.5	2353.1
$(L_1F_3)^*(O_4)^*$	305.1	762.9	3051.7	392.3	784.3	1568.7
$(L_{4c}F_{3c})^*(O_4)^*$	229.3	573.1	2292.2	342.2	679.4	1364.6
$(L_{4c}F_{4c})^*(O_4)^*$	229.3	573.1	2292.3	356.2	707.5	1420.7

particular configuration of LOF-SA and data set. At that time one or more further time intervals constituted the experimental run.

Figure 6 shows the average per-query cost of processing a lengthening stream of queries using two different LOF-SA arrangements. The $(L_4O_4F_4)^*$ arrangement for the file DNA400M represents the largest data structure of all of our experiments, totaling almost 5 GB including the string itself; and the $(L_{4c}F_{4c})^*(O_4)^*$ arrangement on file ENG50M is one of the smaller structures, taking approximately 400 MB in total. Because the operating system is unable to buffer any significant fraction of the larger structure in memory, query response times converge quickly, and what is shown is a long-term querying rate of around 20 milliseconds per query, including two disk accesses (one to fetch a LOF-SA block, and one to access the text to check the answer). On the other hand, the whole of the smaller structure slowly migrates into memory under operating system control as the query sequence is processed, and query times are still improving after more than a million queries have been processed – it takes approximately this many queries for all of the data structure and text to have been "touched" at least once.

All LOF-SA block accesses were implemented as disk operations using `fseek` and `fread`, and hence any caching of data in-memory was the result of operating system actions rather than under control of the search program. For the data points reported below that correspond to small test files, long query sequences were needed before stable per-query average running times were achieved, and on the smaller test files the query sequence used in the experiments contained 1,000,000 queries.

Caching was less of an issue for the larger files, since their data structures tended to remain on disk, and execution times stabilized over shorter query sequences (although, interestingly, over approximately the same periods of elapsed querying time, of the order of 10–20 minutes of continuous execution). These required that as few as 10,000 queries be executed from a cold start with data all on disk, before accurate timing could be started. The rate of convergence is a function of the compressed/compacted size of the LOF-SA, and the relative size of the available main memory, rather than the length of the base string.

Space Required. Recall that the principal objective in this paper was to reduce the space required by the LOF-SA structure. Table 4 shows that this goal has been achieved, and that the $12n$ space requirement of the original implementation can be reduced by approximately 50% for DNA data, and by around 40% for English text. What remains now is to document the effect this space saving has on query execution times.

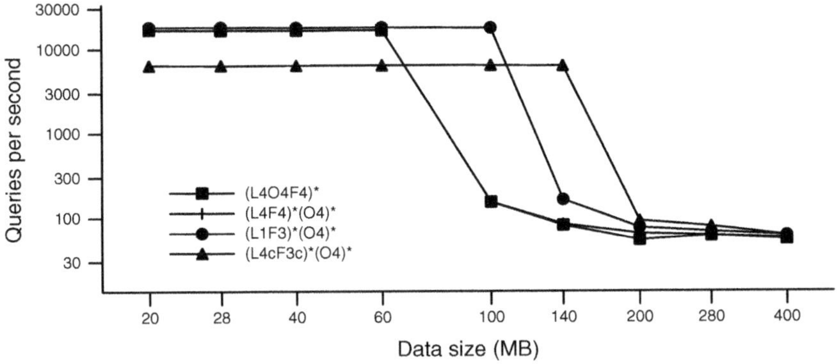

(a) DNA data

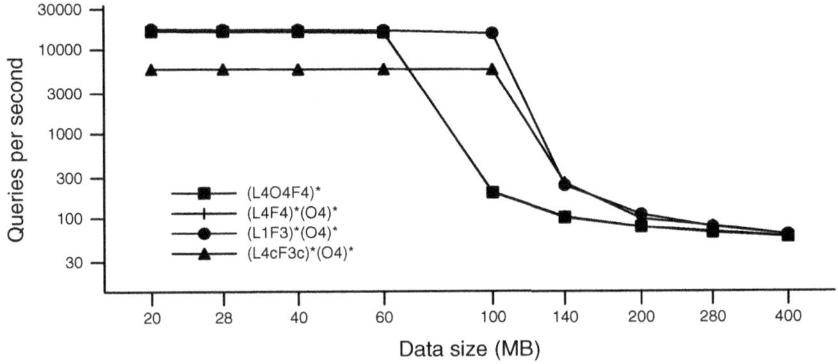

(b) English data

Fig. 7. Long-term query cost as a function of input string size, for two different types of data. Compaction and compression provide considerable benefits in throughput for mid-sized strings, by allowing more (or all) of the data structure to be buffered in memory. All values shown are throughput rates in terms of queries per second, measured in each case after a (demonstrably sufficiently) long prefix of untimed queries had been executed to ensure that query response times were stable. The $(L_4F_4)^*(O_4)^*$ arrangement gives almost identical throughput rates as the $(L_4O_4F_4)^*$. Note the one-off factor-of-100 decrease in throughput at the point at which the data structures become too large to be buffered in main memory.

Querying Time. Having established the number of queries necessary to measure long-term average times, Figure 7 shows, for three of the LOF-SA arrangements, the long-term stable cost of querying them on the 1 GB experimental machine. The range of file sizes spans the range from "small enough to become buffered in main memory" through to "too large for main memory", and the measured execution times clearly show this polarizing difference. When the LOF-SA structure fits in main memory, string searching is possible in under 0.1 milliseconds. On the other hand, when access to the data structure requires physical disk operations, each search operation requires around 20 milliseconds. Note, however, that after shifting through the transitional zone to reach that 50 queries per second zone, execution cost stabilizes again for fully-on-disk performance, and does not grow further as the text string lengthens. Note also that, when the string is beyond the transitional zone, and is sufficiently large that on-disk behavior is inevitable, that the compressed fringe arrangement is no slower than the uncompressed arrangement. In this case the cost of doing the necessary decompression is compensated for by the reduced volumes of data being transferred from disk.

The benefit of the compressed approaches we have explored in this paper is that they delay that transitional moment, and allow strings approximately twice as long as previously to be handled within a given memory limit. Figure 7 shows this effect clearly – on the intermediate file sizes (text strings of around 100 MB), the $(L_1F_3)^*(O_4)^*$ and $(L_{4c}F_{3c})^*(O_4)^*$ execute significantly more quickly than does the original $(L_4O_4F_4)^*$ arrangement, with the effect more pronounced on the DNA data than on the English text, because the fringe of the former is more compressible.

Moreover, when the text string is large and the LOF-SA operates in a disk-based mode, each search operation takes typically just two disk accesses – one to fetch a LOF block, and then in into the text to confirm the likely answer. If there are multiple answers to the query, multiple consecutive LOF blocks might be required in order to retrieve the full set of O values, but in most cases just one access to the string will be required to identify the set of matching locations.

5 Conclusions and Future Work

The experimental results of the previous section demonstrate that the space requirements of the LOF-SA can be reduced by nearly half, without adversely affecting large-string search times. To obtain these results (S, C)-byte codes were used for the L and F components, and the O component was separated out.

Further reductions in space may still be possible. For example, the first row of Table 1 shows that there is still around 4 bits per LOF item that can be saved off the L components for DNA data, and it might be that a (S, C)-nibble code is able to capture that gain, and perhaps not lose too much compression effectiveness on the F values. The word-aligned codes of Anh and Moffat [2] are also candidates for compressing the L values. Other possibilities for handling the L values are to store them as signed differences relative to their predecessor in the block, and exploit the long-range repetitions that seem an inevitable consequence of the fact that the suffixes of strings that share common prefixes are likely to be differentiated at the same points.

Another aspect for further investigation is the O values. González and Navarro [11] consider suffix array compression, and describe a scheme in which the O values are converted to signed differences, and then repeated patterns of differences are located and exploited using the Re-Pair compression technique [14]. This approach is related to the observation made in the previous paragraph about patterns in the L differences, and it may further be possible for the L and O values to be coupled, and used to bias the compression of each other in a mutually beneficial manner.

Finally, to speed query processing, we are exploring the addition of internal structure to the LOF blocks, to reduce the amount of sequential processing (and decompression) that is necessary. The ability of the (S, C)-byte code to skip past fringe values is helpful, but relies on careful organization of the searching logic. A small index at the front of each block, to indicate the next level in the hierarchical partitioning after the common prefix string, is expected to allow time savings that will further accelerate the throughput attainable when the string is short enough that in-memory operation is the steady state.

References

1. Abouelhoda, M.I., Kurtz, S., Ohlebusch, E.: Replacing suffix trees with enhanced suffix arrays. Journal of Discrete Algorithms 2(1), 53–86 (2004)
2. Anh, V.N., Moffat, A.: Inverted index compression using word-aligned binary codes. Information Retrieval 8(1), 151–166 (2005)
3. Apostolico, A.: The myriad virtues of subword trees. In: Apostolico, A., Galil, Z. (eds.) Combinatorial Algorithms on Words. NATO ASI Series F12, pp. 85–96. Springer, Berlin (1985)
4. Baeza-Yates, R.A., Barbosa, E.F., Ziviani, N.: Hierarchies of indices for text searching. Information Systems 21(6), 497–514 (1996)
5. Brisaboa, N.R., Fariña, A., Navarro, G., Esteller, M.F. (S,C)-dense coding: An optimized compression code for natural language text databases. In: Nascimento, M.A., de Moura, E.S., Oliveira, A.L. (eds.) SPIRE 2003. LNCS, vol. 2857, pp. 122–136. Springer, Heidelberg (2003)
6. Cheung, C.-F., Xu Yu, J., Lu, H.: Constructing suffix tree for gigabyte sequences with megabyte memory. IEEE Transactions on Knowledge and Data Engineering 17(1), 90–105 (2005)
7. Crauser, A., Ferragina, P.: A theoretical and experimental study on the construction of suffix arrays in external memory. Algorithmica 32, 1–35 (2002)
8. Culpepper, J.S., Moffat, A.: Enhanced byte codes with restricted prefix properties. In: Consens, M.P., Navarro, G. (eds.) SPIRE 2005. LNCS, vol. 3772, pp. 1–12. Springer, Heidelberg (2005)
9. Dementiev, R., Kärkkäinen, J., Mehnert, J., Sanders, P.: Better external memory suffix array construction. ACM Journal of Experimental Algorithmics 12(3.4), 1–24 (2008)
10. Farach-Colton, M., Ferragina, P., Muthukrishnan, S.: On the sorting-complexity of suffix tree construction. Journal of the ACM 47(6), 987–1011 (2000)
11. González, R., Navarro, G.: Compressed text indexes with fast locate. In: Ma, B., Zhang, K. (eds.) CPM 2007. LNCS, vol. 4580, pp. 216–227. Springer, Heidelberg (2007)
12. Gusfield, D.: Algorithms on strings, trees, and sequences: Computer science and computational biology. Cambridge University Press, Cambridge (1997)
13. Kasai, T., Lee, G.H., Arimura, H., Arikawa, S., Park, K.: Linear-time longest-common-prefix computation in suffix arrays and its applications. In: Amir, A., Landau, G.M. (eds.) CPM 2001. LNCS, vol. 2089, pp. 181–192. Springer, Heidelberg (2001)
14. Larsson, J., Moffat, A.: Off-line dictionary-based compression. Proceedings of the IEEE 88(11), 1722–1732 (2000)
15. Manber, U., Myers, G.W.: Suffix arrays: a new method for on-line string searches. SIAM Journal of Computing 22(5), 935–948 (1993)
16. Manzini, G.: Two space saving tricks for linear time LCP array computation. In: Hagerup, T., Katajainen, J. (eds.) SWAT 2004. LNCS, vol. 3111, pp. 372–383. Springer, Heidelberg (2004)
17. McCreight, E.M.: A space-economical suffix tree construction algroithm. Journal of the ACM 23(2), 262–272 (1976)

18. Navarro, G., Mäkinen, V.: Compressed full text indexes. ACM Computing Surveys 39(1) (2007)
19. Phoophakdee, B., Zaki, M.J.: Genome-scale disk-based suffix tree indexing. In: Chan, C.Y., Ooi, B.C., Zhou, A. (eds.) Proceedings of the 2007 ACM SIGMOD International Conference on Management of Data, pp. 833–844. ACM, New York (2007)
20. Sinha, R., Puglisi, S.J., Moffat, A., Turpin, A.: Improving suffix array locality for fast pattern matching on disk. In: Lakshmanan, L.V.S., Ng, R.T., Shasha, D. (eds.) Proceedings of the 2008 ACM SIGMOD International Conference on Management of Data, pp. 661–672. ACM, New York (2008)
21. Smyth, B.: Computing Patterns in Strings. Pearson Addison-Wesley, Essex (2003)
22. Tian, Y., Tata, S., Hankins, R.A., Patel, J.M.: Practical methods for constructing suffix trees. The VLDB Journal 14(3), 281–299 (2005)
23. Ukkonen, E.: Online construction of suffix trees. Algorithmica 14(3), 249–260 (1995)
24. Weiner, P.: Linear pattern matching algorithms. In: Proceedings of the 14th Annual Symposium on Switching and Automata Theory, pp. 1–11. IEEE Computer Society, Washington (1973)

A Right-Time Refresh for XML Data Warehouses

Damien Maurer[1,2], Wenny Rahayu[1], Laura Rusu[1], and David Taniar[3]

[1] La Trobe University, Australia
[2] University of Technology of Troyes, France
[3] Monash University, Australia
{D.Maurer,W.Rahayu,L.Rusu}@latrobe.edu.au

Abstract. The new era of business intelligence and web data warehousing has brought in new research and development issues, whereby it is imperative to have an efficient refresh process for the information in the XML warehouses. Existing *strict real-time* or *static timestamp-based* refresh mechanisms are not always efficient. In this paper, we propose to build a waiting queue to make the XML warehouse refresh processes discrete, based on (*i*) *Level of Criticality* (LoC), (*ii*) *Impact on Reports* (IoR), and (*iii*) *Frequency of Appearance* (FoA). Our experimental results have shown the effectiveness of our proposed method.

1 Introduction

Business intelligence is growingly recognized by the industry as an important tool to support decision making, through the use of data warehouses. The information is extracted from the operational databases, and then it is cleaned and transformed to be stored into a data warehouse. When new data is added or existing data is updated in the operational system, the data warehouse needs to be refreshed, so that the information contained in the warehouse will be reasonably up-to-date to support the decision making process.

In practice, it is very common that the data warehouse is refreshed on a regular basis (e.g. once every night/week/month). However, some organizations require having the most recent data available just in-time. Generally, there are two types of data warehouse refresh mechanisms:

- *Uniform Discrete Refresh*: a static time-stamp is used to determine the refresh period (e.g. daily/weekly/yearly), and
- *Continuos Refresh*: a real-time refresh mechanism is used; each update in the operational database is loaded into the data warehouse in real time.

In the former approach some important updates might not be immediately reflected in the warehouse, while in the latter approach the data is always up-to-date but the solution might be highly expensive.

In this paper, we propose a *Non-Uniform Discrete Refresh* mechanism (or a *Right-Time Refresh* mechanism) to address the lack of precision in static uniform updates but at the same time to optimize the cost of continuous real-time updates.

X. Zhou et al. (Eds.): DASFAA 2009, LNCS 5463, pp. 745–749, 2009.

2 Related Work

A substantial amount of work has been carried out during the last few years to find solutions for the issues in XML warehousing.

The work described in [5] focuses on the development a *conceptual design methodology for XML* document warehouses based on Object-Oriented concepts. The X-Warehousing approach proposed in [1] focuses on *multidimensional approach* to capture user requirements. The work in [6] proposes a framework for the *multidimensional analysis of XML documents*, called XML-OLAP. These works mainly discuss warehousing within the static XML data scenario, and do not address the distinct issue of temporality or refresh time interval in the XML warehouses.

The idea of incremental and real-time data warehousing has been around for several years, but within the context of the traditional (non-XML) warehousing. Some commercial developers such as Oracle have started to incorporate the technology to perform data propagation and synchronization in order to support real-time data integration [12]. Other works in the area deal mainly with the execution technique of the update propagation [11] or the new requirements for sophisticated technologies to solve the real-time challenges in business intelligence [10].

None of the work in incremental update has dealt with the new complex and hierarchical nature of XML web data warehousing. Some work in XML warehousing for storing multi-versions XML data has been proposed by [4, 9]. However, these works do not propose a dynamic update dedicated on optimizing the interval and the re-grouping of update queries.

To the best of our knowledge, existing works in XML warehousing have not dealt with refresh mechanisms, and this issue has been identified as critical by some of the works mentioned above. For example [1] claims the necessity of further research on the optimal update/refresh policy design, which is our main focus in this paper.

3 Proposed Right-Time Refresh Mechanism

Our proposed right-time refresh mechanism consists of three levels of refresh optimizations: *(i) Level of Criticality* (*LoC*), (*ii*) *Impact on Reports* (*IoR*), and (*iii*) *Frequency of Appearance* (*FoA*), which are used to build the waiting queue.

In most of the commercial Data Warehousing systems, update/refresh of the warehouse data is done at pre-defined intervals [2, 3, 4]. For small databases, it is possible to run updates of the system in real-time, so there is no need to determine the best update interval, but if there is a large amount of data, it could become very difficult to update regularly. This is especially true in the case of an XML data warehouse because of the specific structure of the XML documents and different types of constraints, which could appear during XML updates.

Our proposed method is to optimize the update interval by considering the following three indicators:

- A predefined *level of criticality* for each XML element/attribute,
- The *impact of the update on the reports*, and
- The *frequency of appearance* in the reports.

These three indicators are used to build a waiting queue for the updates. Each update is queued in the ETL (Extract - Transform - Load) phase, until a user-specified threshold is reached. When that happens, all the awaiting updates are pushed in the data warehouse. When a low priority update comes, it waits in the queue until a high priority update arrives, and the whole waiting queue of updates is pushed in.

Level of Criticality (**LoC**) is used to pre-define how important the attributes are. That is, if a higher importance was given to specific elements/attributes, this means the user would want to update them more frequently. The work in [2] showed that it is possible to assign specific needs for different parts of the same business entity.

This method also considers structural changes, because in XML not only the data can change, but also the structure can evolve too. We can pre-define levels of criticality for structural changes, like deletion or modification of an element/attribute.

Impact on Reports (**IoR**) shows the impact of a particular change to an element/attribute on the reports run on the data warehouse. If the change has a negligible impact, we can assume that there is no need to push this update immediately into the data warehouse. However, if the user needs a report on a specific item which has changed, the impact of this specific update is more important.

Typically an update impacts measurable variables; in this case, the impact on reports can be averaged over a number of dimensions in the warehouse. However, it is possible that the update would consist in a non-numeric change, for example a random string (a name, a description, etc), or an enumeration value (colours, Boolean values, etc).

In the case of a change in the XML structure itself, the issue is wider than the problem of updating the data only. This is because a change in the XML document structure would completely change the way how the data warehouse extracts, transforms and load data from it.

Frequency of Appearance (**FoA**) is the frequency of each XML node appearing in the warehousing reports. This is different from the *IoR* described previously, which does not use the user's feedback about the data warehouse usage, and it is just about the data itself. *FoA* considers the usage of the data, not the data itself. *FoA* works as follows: The data warehouse maintains a file that stores the usage frequency of each node that appears in the warehousing reports. Every time a warehousing report is required to be built, the data warehouse updates its *FoA* for every node involved in the query.

The waiting queue is built by combining the three abovementioned right-time refresh optimization indicators. All the incoming updates are queued without changing the order of their appearance, and they are pushed to the data warehouse when a certain total volume of update is waiting. To calculate this volume, ETL first calculates for each update a numerical value, called *Queue Value* (*QV*), by combining the values of *LoC*, *IoR*, and *FoA*. When the sum of *QV*s for all updates reaches the threshold, they are pushed to the data warehouse.

We identified two major ways of combining the values of the three indicators to calculate the *QV*s:

(*i*) using a simple aggregate – in this case each indicator has the same weight, *or*

(*ii*) using a balanced aggregate – in this case the average is calculated using a balance to attenuate the effect of very low values, and also to give a preference for a particular method. Table 1 shows an example of update using balanced aggregate.

Table 1. Example of the process to update queue using balanced average (simulated values)

Time	*LoC*	*IoR*	*FoA*	Balanced Average (1 – 1.5 – 1.2)	Queue (threshold = 100%)
Update *TV set* price	10%	24%	30%	27	27
Update *MP3 player* name	10%	18%	15%	18	45
Update *Radio set* area	24%	25%	32%	33	78
Update *DVD player* range	30%	39%	30%	42	120 (>100 : Push)
Queue pushed		*Data Warehouse Updated*			0
Update *MP3 player* price	20%	34%	14%	29	29

4 Experiments

To test our proposed method, we have built a prototype system, which generates a high amount of simulated updates and builds a queue. The prototype system uses the concept of consolidated delta to collect the updates. This is a storage technique for changes or updates in XML documents, introduced in our previous work [7,8,9].

4.1 Usefulness of the Method

To prove that our method is better than not using any optimization, we have conducted several experiments by simulating different update flows, with different measurable variables, as follows:

- The average gap between arrival dates without management (with a real time policy) and arrival dates with our method (the 'queue push' date);
- The total time to calculate the *QV*s for 1000 and 10000 updates;
- The time to retrieve a state of the data warehouse at time = $t_{end/2}$ called S_h;
- The average time of the push;
- The number of nodes created in the XML consolidated delta file by 1000 and respectively 10000 updates;

4.2 Analysis of the Results

For a high flow of updates, the gap between arrival time with and without management was negligible (0.10s and 0.12s). In this case the discrete process has negligible impact on the data warehouse reports results. With a low flow the gap is larger (4s and 3.8s). These results can be explained by the fact that when the update flow is slow, the threshold is only reached after a long time. Therefore, we believe our method is better for higher flows of update.

The time to calculate the *QV*s for 1000 updates was 1.09*s* (10.61*s* for 10000 updates). The time to retrieve the state S_h with no management was 0.56*s* and 4.89*s*, but was negligible using our method (0.012*s* for state 500/1000 and 0.089*s* for state 5000/10000). The average time before the queue is pushed is 0.14*s* (respectively 0.18*s*) for a high flow and 4.6*s* (respectively 4.4*s*) for a low flow. The slower the flow is, the longer the updates have to wait before their push.

The amount of T_n is the number of updates when there is no management, because they are pushed as they come. With our method, there is as much T_n as there is pushed. Our method allows the number of different T_n to be dramatically lowered (30 versus 1000 and 320 versus 10000).

5 Conclusion

The methodology proposed in this paper aimed to find a right time update policy for XML warehouses. This technique provides a very good basis for the update mechanism in systems which need a discrete update, but with an optimized time interval between updates. We proposed three indicators, namely *Level of Criticality*, *Impact on Reports* and *Frequency of Appearance* which are used to build a waiting queue that *does not change the order of updates, but allows a time interval to be created between the data pushes into the data warehouse.*

References

1. Boussaid, O., Messaoud, R.B., Choquet, R., Anthoard, S.: X-warehousing: An XML-based approach for warehousing complex data. In: Manolopoulos, Y., Pokorný, J., Sellis, T.K. (eds.) ADBIS 2006. LNCS, vol. 4152, pp. 39–54. Springer, Heidelberg (2006)
2. Italiano, I.C., Ferreira, J.E.: Synchronization Options for Data Warehouse Designs. IEEE Computer 39(3), 53–57 (2006)
3. Mannino, M.V., Walter, Z.D.: A framework for data warehouse refresh policies. Decision Support Systems 42(1), 121–143 (2006)
4. Marian, A., Abiteboul, S., Cobena, G., Mignet, L.: Change-Centric Management of Versions in an XML Warehouse. In: 27th International Conference on Very Large Databases (VLDB), pp. 581–590 (2001)
5. Nassis, V., Dillon, T.S., Rajagopalapillai, R., Rahayu, J.W.J.: An XML document warehouse model. In: Li Lee, M., Tan, K.-L., Wuwongse, V. (eds.) DASFAA 2006. LNCS, vol. 3882, pp. 513–529. Springer, Heidelberg (2006)
6. Park, B.-K., Han, H., Song, I.-Y.: XML-OLAP: A multidimensional analysis framework for XML warehouses. In: Tjoa, A.M., Trujillo, J. (eds.) DaWaK 2005. LNCS, vol. 3589, pp. 32–42. Springer, Heidelberg (2005)
7. Rusu, L.I., Rahayu, W., Taniar, D.: Partitioning Methods for Multi-version XML Data Warehouses. In: Distributed and Parallel Databases. Springer, Heidelberg (2008); accepted for publication July 15, 2008 (in-press) (to appear, 2009)
8. Rusu, L.I., Rahayu, J.W.J., Taniar, D.: Storage techniques for multi-versioned XML documents. In: Haritsa, J.R., Kotagiri, R., Pudi, V. (eds.) DASFAA 2008. LNCS, vol. 4947, pp. 538–545. Springer, Heidelberg (2008)
9. Rusu, L.I., Rahayu, J.W.J., Taniar, D.: Warehousing dynamic XML documents. In: Tjoa, A.M., Trujillo, J. (eds.) DaWaK 2006. LNCS, vol. 4081, pp. 175–184. Springer, Heidelberg (2006)
10. Golfarelli, M., Rizzi, S., Cella, I.: Beyond data warehousing: what's next in business intelligence? In: ACM International Workshop on Data Warehousing and OLAP, pp. 1–6 (2004)
11. Lau, E., Madden, S.: An Integrated Approach to Recovery and High Availability in an Updatable, Distributed Data Warehouse. In: 32nd International Conference on Very Large Databases (VLDB), pp. 703–714 (2006)
12. Rittman, M.: An Introduction to Real-Time Data Integration, Oracle Technology Network (2008), http://www.oracle.com/technology/pub/articles/rittman-odi.html

Demonstrating Effective Ranked XML Keyword Search with Meaningful Result Display

Zhifeng Bao[1], Bo Chen[1], Tok Wang Ling[1], and Jiaheng Lu[2]

[1] School of Computing, National University of Singapore
{baozhife,chenbo,lingtw}@comp.nus.edu.sg
[2] School of Information and DEKE, MOE, Renmin University of China,
jiahenglu@gmail.com

Abstract. In this paper, we demonstrate an effective ranked XML keyword search with meaningful result display. Our system, named ICRA, recognizes a set of object classes in XML data for result display, defines the matching semantics that meet user's search needs more precisely, captures the ID references in XML data to find more relevant results, and adopts novel ranking schemes. ICRA achieves both high result quality and high query flexibility in search and browsing. An online demo for DBLP data is available at **http://xmldb.ddns.comp.nus.edu.sg/**.

1 Introduction

The goal of XML keyword search is to find only the meaningful and relevant data fragments corresponding to interested objects that users really concern on. Most previous efforts are built on either the tree model or the digraph model. In tree model, Smallest Lowest Common Ancestor (SLCA) [5] is an effective semantics. However, it cannot capture the ID references in XML data which reflect the relevance among objects, while digraph model can. In digraph model, a widely adopted semantics is to find the *reduced subtrees* (i.e. the smallest subtrees in a graph containing all keywords). However, enumerating results by increasing the sizes of reduced subtrees is a NP-hard problem, leading to intrinsically expensive solutions. Moreover, it neither distinguishes the containment and reference edge in XML data, nor utilizes the database schema in defining matching semantics.

Besides, we observe that existing approaches in both models have two common problems. *First*, regarding to the design of matching semantics, they fail to effectively identify an appropriate information unit for result display to users. Neither SLCA (and its variants) nor reduced subtree is an appropriate choice, as neither of them is able to capture user's search target, as shown in Example 1. *Second*, the existing ranking strategies in both models are built at XML node level, which cannot meet user's search concern more precisely at object level.

Example 1. Query 1: "Suciu" is issued on the XML data in Fig. 1, intending to find papers written by "Suciu". Both SLCA and reduced subtree return the author nodes with value "Suciu", which is not informative enough to user.

X. Zhou et al. (Eds.): DASFAA 2009, LNCS 5463, pp. 750–754, 2009.

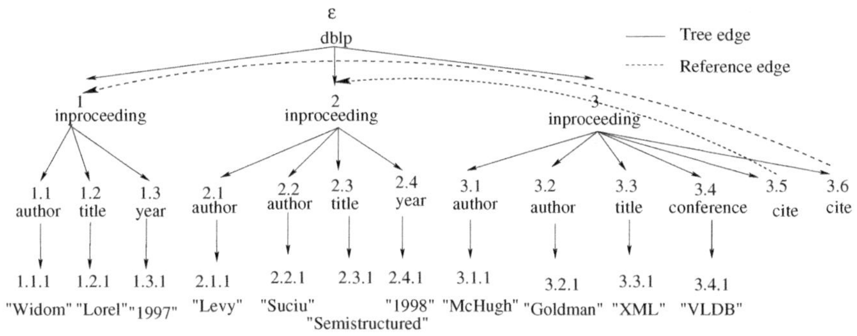

Fig. 1. Example XML document (with Dewey numbers)

Query 2: "Suciu XML" is issued on Fig. 1 to find XML papers written by Suciu. As there is no Suciu's paper containing "XML", the SLCA result is the whole subtree under the root node, which contains too much irrelevant information.□

Generally speaking, the technical challenges of this demo lie in as below:
(1) The results need to be semantically meaningful to the user to precisely meet user's search needs, and meanwhile avoid overwhelming the user with a huge number of trivial results. However, methods on graph model suffer from producing large number of trees containing the same pieces of information many times.
(2) How to define appropriate matching semantics to find more relevant results by capturing ID references in XML data while optimizing the search efficiency.
(3) How to design a general-purpose and effective ranking scheme.

To address these challenges, we present an XML keyword search system ICRA [1]. In particular, by modeling XML data as a set of *interconnected object-trees*, ICRA first automatically recognizes a set of objects of interest and the connections between them. Meanwhile, object trees as query results contain enough but non-overwhelming information to represent a real world entity, so the problem of proper result display is solved. To capture user's search concern on a single object, *ICA* is proposed; to capture user's search concern on multiple objects, *IRA pair (group)* is proposed to find a pair (group) of object trees that are related via direct or indirect citation/reference relationships and together contain all keywords. i.e. IRA helps find more relevant results. For Query 1 in Example 1, ICA returns inproceeding:2 rather than its *author* subelement, which is both informative and relevant. For Query 2, ICA cannot find any qualified single inproceeding, while IRA finds a pair of inproceedings (inproceeding:2, inproceeding:3), where inproceeding:2 written by "Suciu" is cited by inproceeding:3 containing "XML".

Compared with prior search systems, ICRA has significant features.
(1) The interconnected object-trees model guarantees meaningful result display. Compared to tree model, it can capture ID references to find more relevant results; compared to digraph model, it achieves more efficient query evaluation.
(2) It takes advantage of the schema knowledge to define the matching semantics which is of same granularity as user's search needs, and facilitate the result display and performance optimization in terms of result quality and efficiency.

(3) It designs a novel relevance oriented ranking scheme at object-tree level, which takes both the internal structure and content of the results into account.

2 Ranking Techniques

As ICA and IRA correspond to different user search needs, different ranking schemes are designed. To rank the ICA results, traditional TF*IDF similarity is extended to measure the similarity of an object tree's content w.r.t. the query. Besides considering the content, the structural information within the results are also considered: (1)**Weight of matching elements in object tree**. (2)**Pattern of keyword co-occurrence**. Intuitively, an object tree o is ranked higher if a nested element in o directly contains all query keywords, as they co-occur closely. (3)**Specificity of matching element**. Intuitively, an object tree o is ranked higher if an element nested in o exactly contains (all or some of) the keywords in Q, as o fully specifies Q. E.g. for query "Won Kim", Won Kim's publications should be ranked before Dae-Won Kim's publications. To rank the IRA results, we combine the self similarity of an IRA object o and the bonus score contributed from its IRA counterparts. Please refer to [1] for more details.

3 System Architecture

System architecture is shown in Fig. 2. During data preprocessing, the *Indexing Engine* parses the XML data, identifies the object trees, and builds the keyword inverted list storing for each keyword k a list of object trees containing k; it also captures ID references in XML data and stores them into *reference connection table*. A B+ tree is built on top of these two indices respectively. During the query processing stage, it retrieves the object trees containing the specified keywords to compute ICA results; then it computes IRA and Extended IRA (EIRA) results with the help of reference connection table. Lastly, it cleans and ranks the results.

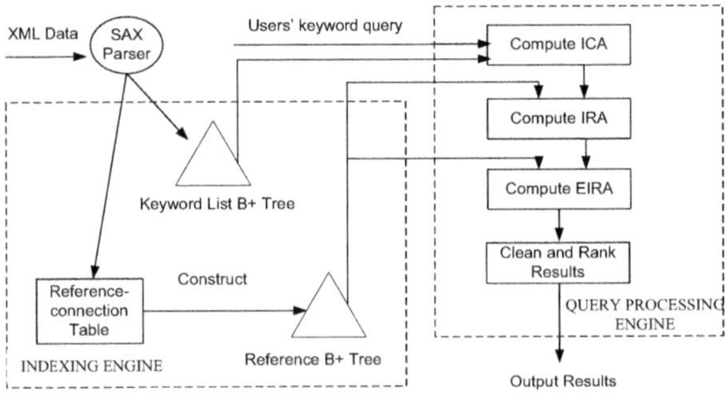

Fig. 2. System architecture

4 Overview of Online Demo Features

ICRA provides a concise interface, user can explicitly specify their search concern - *publications* (default) or *authors*. ICRA offers various query flexibility: users can issue pure keyword queries that can be any combinations of words in full or partial specification of author names, topics, conference/journal and/or year.

4.1 Search for Publications

Users can search for publications with various types of queries as below. Readers are encouraged to try more query types at ***http://xmldb.ddns.comp.nus.edu.sg/***.
Author name. - E.g. we can query "Jiawei Han" for his publications. ICRA will rank Jiawei Han's papers higher than papers co-authored by Jiawei and Han.
Multiple author names. - to search for co-authored papers.
Topic. - E.g. we can query "xml query processing".
Topic by an author. - E.g. we can query "Jim Gray transaction" for his publications related to transaction. Jim Gray's papers containing "transaction" are ranked before his papers citing or cited by "transaction" papers.
Topic of a year. - E.g. we can query "keyword search 2006".
Conference and author. - E.g. we can query "VLDB Raghu Ramakrishnan".

4.2 Search for Authors

Users can also search for authors with various types of queries as below.
Author name. - By typing an author name, ICRA returns this author followed by a ranked list of all his/her co-authors (e.g. try "Tova Milo").
Topic. - We can search for authors who have the most contributions to a research topic (e.g. try "XML keyword search").
Conference/Journal name. - We can find active authors in a particular conference or journal (e.g. try "DASFAA").
Author name and topic/year/conference/journal. - Besides the author himself/herself, we can also search for his/her co-authors in a particular topic or year or conference/journal (e.g. we can search for Ling Tok Wang and his co-authors in DASFAA 2006 with a query "Ling Tok Wang DASFAA 2006").

4.3 Browsing

Besides searching, ICRA also supports browsing from search results to improve its practical usability. E.g. users can click an author (or conference/journal) name in a result to see all publications of this author (or the proceeding/journal). Link is provided to find the references and citations of a paper. When searching for authors, we output both the number of publications containing all keywords and the number of publications that may be relevant via the reference connections.

4.4 Result Display

ICRA displays the result for ICA and IRA semantics *separately* in "AND" and "OR" part. In addition, since a same paper may appear in more than one IRA pair/group, it will annoy the user in result consumption if such paper appears many times. Therefore, ICRA only outputs one IRA object o for each IRA pair/group, and provides links to the objects that form IRA pair/group with o.

5 Effectiveness of ICRA

In the demo, we will compare the result quality of ICRA with typical academic demo systems for DBLP, such as BANKS [4], ObjectRank [3] and FacetedDBLP [2]. We will also compare ICRA with commercial systems such as Microsoft Libra and Google Scholar†. ICRA has a good overall performance in terms of both result quality and query response time, as evidenced by experiments in [1].

6 Feature Comparisons

A comparison of the features in existing demos is given from a user's perspective. BANKS produces results in form of reduced trees which is difficult for novice users to consume. Query types supported by ObjectRank and FacetedDBLP are not as flexible as ICRA. E.g. they cannot handle searching papers of co-authors, or a topic by author etc. ObjectRank doesn't support browsing. DBLP CompleteSearch doesn't employ any relevance oriented ranking functions.

References

1. Bao, Z., Chen, B., Ling, T.W.: Effective ranked XML keyword search with meaningful result display, `http://xmldb.ddns.comp.nus.edu.sg/Demo.pdf`
2. Diederich, J., Balke, W.-T., Thaden, U.: Demonstrating the semantic growbag: automatically creating topic facets for faceteddblp. In: JCDL (2007)
3. Hwang, H., Hristidis, V., Papakonstantinou, Y.: Objectrank: a system for authority-based search on databases. In: VLDB (2006)
4. Kacholia, V., Pandit, S., Chakrabarti, S., Sudarshan, S., Desai, R.: Bidirectional expansion for keyword search on graph databases. In: VLDB (2005)
5. Xu, Y., Papakonstantinou, Y.: Efficient keyword search for smallest LCAs in XML databases. In: SIGMOD (2005)

† Libra: http://libra.msra.cn/ Google Scholar: http://scholar.google.com/

In-Page Logging B-Tree for Flash Memory*

Gap-Joo Na[1], Bongki Moon[2], and Sang-Won Lee[1]

[1] Sungkyunkwan University, Suwon, Korea
{factory,wonlee}@ece.skku.edu
[2] University of Arizona, Tucson, AZ, U.S.A.
bkmoon@cs.arizona.edu

Abstract. We demonstrate the IPL B^+-tree prototype, which has been designed as a flash-aware index structure by adopting the *in-page logging (IPL)* scheme. The IPL scheme has been proposed to improve the overall write performance of flash memory database systems by avoiding costly erase operations that would be caused by small random write requests common in database workloads. The goal of this demonstration is to provide a proof-of-concept for IPL scheme as a viable and effective solution to flash memory database systems.

1 Introduction

Since NAND flash memory was invented as a sector addressable non-volatile storage medium about two decades ago, its density has increased approximately twice annually and the trend is expected to continue until year 2012 [1]. Due to its superiority such as low access latency and low energy consumption, flash-based storage devices are now considered to have tremendous potential as an alternative storage medium that can replace magnetic disk drives.

On the other hand, due to the erase-before-write limitation of flash memory, updating even a single record in a page results in invalidating the current page containing the record and writing a new version of the page into an already-erased area in flash memory. This leads to frequent write and erase operations. In order to avoid this, we have proposed the in-page logging (IPL) scheme that allows the changes made to a page to be written (or *logged*) in the database, instead of writing the page in its entirety [2]. Since flash memory comes with no mechanical component, there is no compelling reason to write log records sequentially as long as it does not cause extra erase operations. Therefore, under the in-page logging approach, a data page and its log records are co-located in the same physical location of flash memory, specifically, in the same erase unit. Since we only need to access the previous data page and its log records stored in the same erase unit, the current version of the page can be recreated efficiently

* This work was partly supported by the Korea Research Foundation Grant funded by the Korean Government (KRF-2008-0641) and MKE, Korea under ITRC IITA-2008-(C1090-0801-0046). This work was also sponsored in part by the U.S. National Science Foundation Grant IIS-0848503. The authors assume all responsibility for the contents of the paper.

X. Zhou et al. (Eds.): DASFAA 2009, LNCS 5463, pp. 755–758, 2009.

under this approach. Consequently, the IPL approach can improve the overall write performance considerably. The IPL scheme uses physiological log records primarily for improving write performance, but the log records can also be used to realize a lean recovery mechanism [2].

The goal of this work is to demonstrate that IPL is a viable and effective solution to flash memory database systems that enables them to deliver high performance promised by the desired properties of flash memory. We have designed a new index structure called IPL B^+-tree as a variant of B^+-tree for computing platforms equipped with flash memory as stable storage. We have implemented the IPL B^+-tree on a development circuit board running Linux 2.6.8.1 kernel.

In addition to providing the proof-of-concept implementation of the IPL scheme, this demonstration will showcase (1) the design of IPL B^+-tree for flash memory database systems, (2) the IPL B^+-tree engine that implements the in-page logging mechanism for B^+-tree indexes, (3) the development platform that allows the IPL storage manager to be in full control of address mapping for flash memory.

2 System Description

The traditional B^+-tree is designed for disk-based storage systems, and yields poor write performance with flash-based storage systems. This section presents the design and implementation details of the IPL B^+-tree.

2.1 Design of IPL B^+-tree

In an IPL B^+-tree, as is illustrated in Figure 1, the in-memory copy of a tree node can be associated with a small in-memory log sector. When an insertion or a deletion operation is performed on a tree node, the in-memory copy of the tree node is updated just as done by traditional B^+-tree indexes. In addition, a physiological log record is added to the in-memory log sector associated with the tree node. An in-memory log sector is allocated on demand when a tree node becomes dirty, and is released when the log records are written to a log sector in flash memory.

The log records in an in-memory log sector are written to flash memory when the in-memory log sector becomes full or when the corresponding dirty tree node is evicted from the buffer pool. When a dirty tree node is evicted, it is not necessary to write the content of the dirty tree node back to flash memory, because all of its updates are saved in the form of log records in flash memory. Thus, the previous version of the tree node remains intact in flash memory, but is augmented with the update log records.

When an in-memory log sector is to be flushed to flash memory, its content is written to a flash log sector in the erase unit which its corresponding tree node belongs to. To support this operation, each erase unit (or a physical block) of flash memory is divided into two segments – one for tree nodes and the other for log sectors, as shown at the bottom of Figure 1.

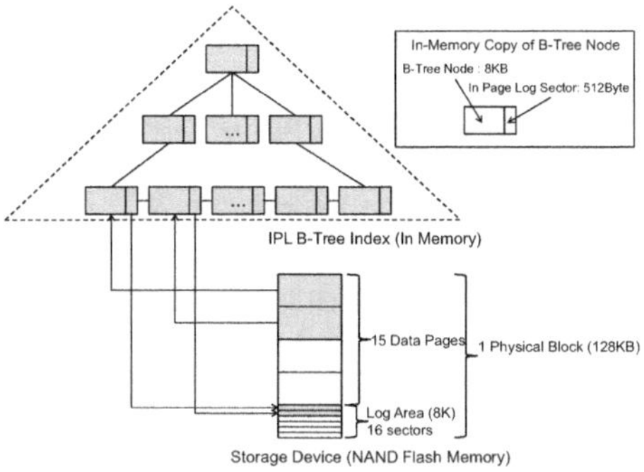

Fig. 1. IPL B^+-tree structure

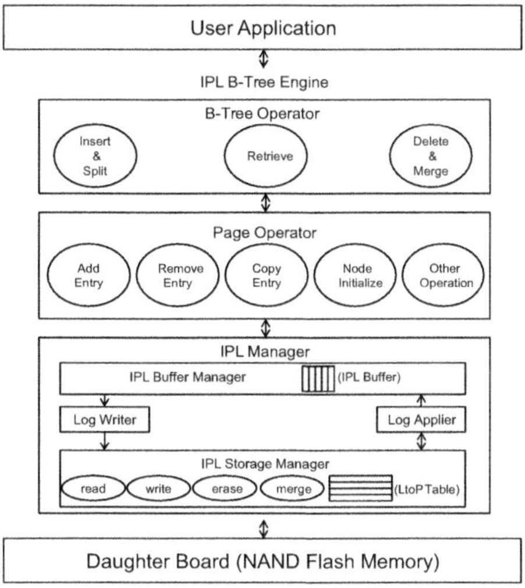

Fig. 2. IPL B^+-tree Engine

2.2 IPL B^+-tree Engine

Figure 2 shows the IPL B^+-tree engine that consists of three components: (1) B^+-tree operator module, (2) page operator module, and (3) IPL manager. The B^+-tree operator module processes traditional B^+-tree operations such as insertion, deletion and search. If a single operation involves more than a tree node, this is divided into multiple single-node requests, each of which is processed by the page operator module. For each single-node request, the page operator module adds its physiological log record to the corresponding in-memory log sector.

The most important component of the IPL B^+-tree engine is the IPL manager that provides the in-page logging mechanism for B^+-tree indexes. The IPL manager consists of four internal components: (1) IPL buffer manager, (2) IPL storage manager, (3) Log writer, and (4) Log applier. The IPL buffer manager maintains in-memory tree nodes and their corresponding in-memory log sectors in an LRU buffer pool. When an in-memory log sector becomes full and needs to be flushed to flash memory, the IPL buffer manager determines whether the log area of the corresponding erase unit in flash memory can accommodate the in-memory log sector. If it does, the in-memory log sector is written to flash memory. Otherwise, a merge request is sent to the IPL storage manager. Then, the Log applier computes the current version of tree nodes by applying the log records to the previous version of tree nodes. Since the entire tree nodes in the merged erase unit are relocated to a physically different region in flash memory, the logical-to-physical mapping is updated by the IPL storage manager, when a merge is complete.

When a tree node is to be read from flash memory, the Log applier sends a read request to the IPL storage manager, which returns an in-flash copy of the tree node along with its log records. Then, the Log applier creates the current version of the tree node by applying its log records to the tree node.

3 Demonstration Scenario

The demonstration will be set up with a host system and a target system as shown in Figure 3. The IPL B^+-tree runs on the target system that consists of an EDB9315A processor board and a flash daughter board [3].[1] The requests for B^+-tree operations are submitted from the host system, and the progress and performance of B^+-tree operations executed on the target system can be monitored on the host system.

In order to graphically illustrate the progress and performance of the B^+-tree operations to be demonstrated, we have implemented a performance monitor with GUI that runs on the Labview environment, as is shown in Figure 4. For more extensive performance comparison, a standard B^+-tree running on a comparable computing platform with a magnetic disk will also be available in the demonstration.

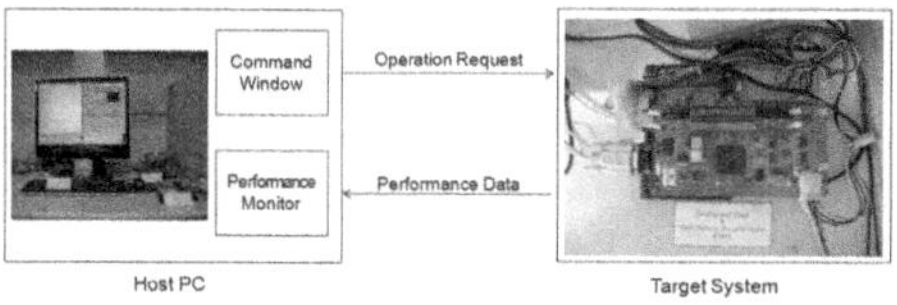

Fig. 3. Demonstration System Setup

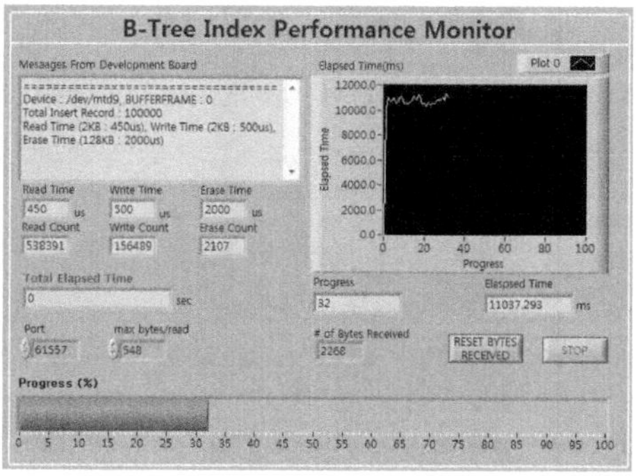

Fig. 4. Index Performance Monitor

References

1. Hwang, C.-G.: Nanotechnology Enables a New Memory Growth Model. Proceedings of the IEEE 91(11), 1765–1771 (2003)
2. Lee, S.-W., Moon, B.: Design of Flash-Based DBMS: An In-Page Logging Approach. In: Proceedings of the ACM SIGMOD, Beijing, China, pp. 55–66 (June 2007)
3. Cirrus Logic. EP9315 Data Sheet - Enhanced Universal Platform System-on-Chip Processor (March 2005) DS638PP4

[1] Most of the flash memory SSD products come with an on-device controller that runs a flash translation layer to deal with address mapping as well as command interpretation. Consequently, the logical-to-physical mapping is completely hidden from outside the flash memory storage devices. That is the reason why we have chosen an EDB9315 processor board as a hardware testbed.

Supporting Execution-Level Business Process Modeling with Semantic Technologies

Matthias Born[1], Jörg Hoffmann[1], Tomasz Kaczmarek[3], Marek Kowalkiewicz[1], Ivan Markovic[1], James Scicluna[2], Ingo Weber[1], and Xuan Zhou[1]

[1] SAP Research, Karlsruhe, Germany
{mat.born,joe.hoffmann,marek.kowalkiewicz,ivan.markovic, ingo.weber,xuan.zhou}@sap.com
[2] STI Innsbruck, Austria
james.scicluna@sti2.at
[3] Poznan University of Economics, Poland
t.kaczmarek@kie.ae.poznan.pl

Abstract. When creating execution-level process models from conceptual to-be process models, challenges are to find implementations for process activities and to use these implementations correctly. Carrying out these activities manually can be time consuming, since it involves searching in large service repositories and cycles of testing and re-designing. We present *Maestro for BPMN*, a tool that allows to annotate and automatically compose activities within business processes, and to verify the consistency of an annotated process.

1 Introduction

One of the big promises of Semantic Business Process Managament is to help bridge the gap between the business level, where the processes are modeled, and the IT level, where they are implemented. This is one of the key issues addressed in the SUPER project,[1] based on Service-Oriented Architectures, and on semantic annotations. For that purpose, we have made extensions to *Maestro for BPMN*, a modeling tool developed at SAP Research.

In our framework, business processes are modeled as a set of *tasks* together with their control flow. In order to make a process executable, a user has to associate every task in the process with an implementation in terms of services. To support this implementation activity, we propose a combination of the following steps: (1) semantically annotate the individual tasks; (2) check if the annotation and the control flow are consistent; (3) discover and compose services which implement the functionality required for each task. In our work, we developed techniques supporting the user in step (1) (published in [2]); we developed fully automated techniques for step (2) (published in [5]) and step (3) (published in [8]). The demonstration will show the Maestro tool, which implements all these techniques in a convenient GUI. Figure 1 shows a screenshot.

[1] http://www.ip-super.org

X. Zhou et al. (Eds.): DASFAA 2009, LNCS 5463, pp. 759–763, 2009.

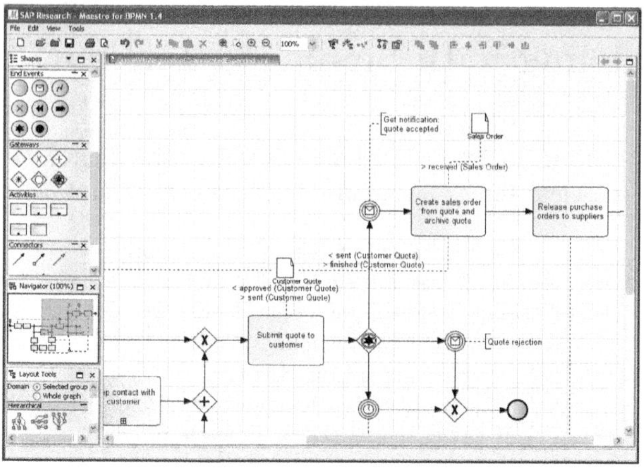

Fig. 1. A fragment of a process model represented in Maestro for BPMN

Within SUPER, business process models are represented in terms of the *sBPMN* ontology [1],[2] serving as a meta model for BPMN process models. The ontology features the concepts, relations and attributes for standard BPMN. We extended it with the ability to define a state of the process before and after execution of successive tasks. With these extensions we can derive semantic goal descriptions for tasks. This is in line with most popular approaches to Semantic Web Service description,[3] where Web services can be annotated with preconditions and postconditions.

As indicated, we developed support for the convenient creation of semantic annotations. That support is based on business object life cycles. Further, we extended Maestro with automatic consistency checking – verifying the interaction of annotations and control flow – as well as automatic discovery and composition – finding service-based implementations for individual tasks.

We define a formal framework for service composition, following A.I. methodologies [9]. We consider plug-in matches, where services do not have to match exactly, but must be able to connect in all possible situations. In particular, we take the background ontology, i.e. the domain axioms it specifies, into account. This is in contrast to many existing works that assume exact matches (of concept names). We explore restricted classes of axioms, to find solutions efficiently.

The consistency checking also deals with ontological domain axiomatizations, and it also exploits restricted classes of axioms for the sake of computational efficiency. The check determines whether any preconditions may be violated, and whether any parallel tasks are in a logical conflict.

Section 2 explains how we support semantic annotations. Sections 3 and 4 cover consistency checking and discovery/composition. Section 5 concludes.

---

[2] sBPMN is written in WSML (http://www.wsmo.org/TR/d16/d16.1/v0.21/)

[3] Followed for example in WSMO (http://www.wsmo.org).

2 Process Modeling and Semantic Annotation

From the graphical point of view, *Maestro for BPMN* implements BPMN 1.1. However, it makes use of the sBPMN ontology, by creating on-the-fly a set of instances for sBPMN classes. If a new BPMN task is created on the drawing pane, an instance of the concept *Task* is created in the in-memory working ontology. This enables supportive reasoning over the working ontology.

The tool allows a user-friendly semantic annotation of process models. The annotations link process tasks to a domain ontology. We focus on how process activities manipulate business objects in terms of their life cycles. E.g., a task "Send quote" sets the status of the object "Quote" to the state "sent". For this purpose, the domain ontology needs to specify the business objects of interest together with their life cycles. The textual descriptions of tasks (or other elements) are matched against the entities of interest in the domain ontology using linguistic methods for term similarity, synonyms, etc. Another way to restrict the matches is by employing the process structure, e.g., not suggesting the same activity twice, or comparing the process control flow to the object life cycle.

3 Consistency Checking

The consistency checking is an extended form of process verification: it deals with inconsistencies between control flow and semantic annotations. Specifically, we check two properties: Are the semantic preconditions guaranteed to be true whenever a task is activated? Are there conflicting tasks that may be executed in parallel? The basic steps taken for answering these questions are the following:

- **Compute the parallelism relation.** For every task, determine the set of tasks that are potentially executed in parallel. This is done by propagating a matrix through the process graph, where matrix entry (i, j) is true iff tasks (T_i, T_j) may be executed in parallel.
- **Detect conflicting parallel tasks.** Two parallel tasks T_i and T_j are in *precondition conflict* if pre_i (the precondition of task T_i) contradicts $post_j$, and they are in *effect conflict* if $post_i$ contradicts $post_j$.
- **Detect non-supported preconditions.** We designed a propagation algorithm that associates every edge e with the *intersection*, $I(e)$, of all logical states that may arise while e is activated (carries a token). In other words, $I(e)$ contains exactly those literals which are always true when e is activated. Hence, if task T_i has e as its incoming edge, then we can test T_i's precondition for inclusion in $I(e)$. If the precondition is not included, then we know that there exists a process execution in which T_i is activated from a workflow point of view, but is not executable according to its semantic annotation.

4 Task Discovery and Composition

Our first step in finding process task implementations is to discover a single semantic Web service for each annotated task. For each task, we check whether the domain ontology concept annotating the service matches the concept annotating the task. We follow a matching technique proposed in [7], analysing intersection of ontological elements in service descriptions and rating two descriptions as relevant whenever they overlap. For that, we use the standard reasoning task of *concept satisfiability* of conjunctions of concepts.

If a Web service cannot be found for a task, composition is performed. In the spirit of [6], we formalize the semantics of composed services based on the notion of "belief updates" from A.I. [9]. Composition is computationally hard and has two main sources of complexity: (i) combinatorial explosion of possible compositions, and (ii) worst-case exponential reasoning. We tackle (i) using heuristic search, a well known technique for dealing with combinatorial search spaces. We adapt techniques originating in A.I. Planning [4]; ours is the first heuristic of this kind that deals with ontological axiomatizations. We address (ii) by trading off expressivity of the background ontology against efficiency: we investigate *tractable classes* of ontologies, i.e., classes where the required reasoning can be performed in polynomial time. We show that a combination of binary clauses (such as subsumption relations and attribute domain/image typing) and attribute cardinality bounds is tractable. It follows from [3] that Horn clauses (corresponding to Logic Programming rules) are not tractable; we design tractable approximate update-reasoning techniques that preserve either soundness or completeness. Other features (such as QoS) are part of our ongoing work. Our experiments show that the tool can create non-trivial composed services, from repositories containing hundreds of services, within a few seconds.

5 Conclusion

We present extensions to Maestro for BPMN. The demo shows how business analysts can easily annotate process elements (in particular tasks). Maestro checks whether the annotated process model is consistent in terms of its workflow and the semantics of its tasks. Maestro automatically finds and composes Web services that can realize individual tasks. Overall, this provides significant support for the service-based implementation of process models.

References

1. Abramowicz, W., Filipowska, A., Kaczmarek, M., Kaczmarek, T.: Semantically enhanced business process modelling notation. In: SBPM Workshop (2007)
2. Born, M., Dörr, F., Weber, I.: User-friendly semantic annotation in business process modeling. In: Hf-SDDM Workshop (December 2007)
3. Eiter, T., Gottlob, G.: On the complexity of propositional knowledge base revision, updates, and counterfactuals. Artificial Intelligence 57, 227–270 (1992)
4. Hoffmann, J., Nebel, B.: The FF planning system: Fast plan generation through heuristic search. J. AI Research 14, 253–302 (2001)

5. Hoffmann, J., Weber, I., Scicluna, J., Kacmarek, T., Ankolekar, A.: Combining scalability and expressivity in the automatic composition of semantic web services. In: ICWE (2008)
6. Lutz, C., Sattler, U.: A proposal for describing services with DLs. In: DL (2002)
7. Trastour, D., Bartolini, C., Preist, C.: Semantic web support for the business-to-business e-commerce lifecycle. In: WWW, pp. 89–98 (2002)
8. Weber, I., Hoffmann, J., Mendling, J.: On the semantic consistency of executable process models. In: ECOWS (2008)
9. Winslett, M.: Reasoning about action using a possible models approach. In: AAAI, pp. 89–93 (1988)

Video Annotation System Based on Categorizing and Keyword Labelling*

Bin Cui[1], Bei Pan[1], Heng Tao Shen[2], Ying Wang[1], and Ce Zhang[1]

[1] Key Laboratory of High Confidence Software Technologies (Peking University), Ministry of Education, CHINA & School of EECS, Peking University
{bin.cui,panbei,wangying,zhangce}@pku.edu.cn
[2] School of ITEE, University of Queensland, Australia
shenht@itee.uq.edu.au

Abstract. In this work, we demonstrate an automatic video annotation system which can provide users with the representative keywords for new videos. The system explores the hierarchical concept model and multiple feature model to improve the effectiveness of annotation, which consists of two components: a SVM classifier to ascertain the category; and a multiple feature model to label the keywords. We implement the demo system using the videos downloaded from YouTube. The results show the superiority of our approach.

1 Introduction

The rapid growth and availability of videos have created new challenges to managing, exploring and retrieving videos with rich query expressions, high accuracy and efficiency. An important challenge for video retrieval is to support content based queries efficiently when keywords are not available. However, content based video retrieval is time consuming and hardly feasible currently. An alternative is to annotate the videos with meta-data serving as index in large video databases. The target of annotation research is to map the video content into high-level semantics represented by keywords.

The problem of automatic video annotation is closely related to that of content-based video retrieval, and has attracted more and more attention recently [4,3]. A basic idea is to independently train a concept model for each keyword, and detect the keywords for a given video. Since the video keywords for a certain video are generally correlated, the basic idea can be further refined with semantic fusion, such as concept ontology [3,5,1]. However, these approaches are based on the individual keyword detector, and hence the overall performance depends on the independent detection.

In this demo, we present our novel idea to design an automatic video annotation system. To build the system, we construct the training video set and hierarchical keyword sets, extract content features and generate multiple feature models to connect the features with keywords. The overall system is shown in Fig. 1. We extract multiple content features from labelled videos to generate feature models for annotation. Additionally, we construct a hierarchical correlative concept structure indicating the inherent correlation between categories and keywords. In the training stage, we use the global feature

* This research was supported by NSFC under Grant No.60603045.

X. Zhou et al. (Eds.): DASFAA 2009, LNCS 5463, pp. 764–767, 2009.

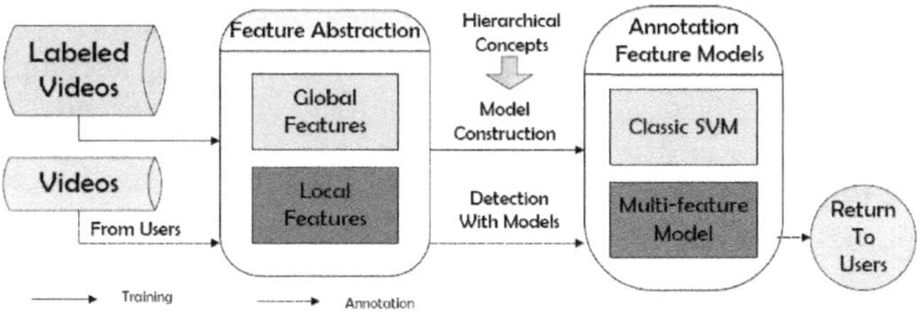

Fig. 1. The Overview of Annotation System

and SVM classifier to train a categorization model which can detect the category of videos; and use low-level content feature to train a multi-feature model for keyword labelling. When a new video comes in, we first classify the video into categories. With the identified category information, we get the keywords using the multiple feature model.

2 Video Categorizing

As we introduced previously, the enhancement of semantic relation between keywords can improve the effectiveness of annotation. Since one single video may contain different meanings at multiple semantic levels, we employ hierarchical video concept organization to represent the semantic information of keywords. A simple example is shown in Fig. 2, where the hierarchical structure consists of two levels. The first level is category, and under each category we have the relevant keywords. The seven categories shown in the figure are selected from YouTube category set, and the keyword set for each category is constructed manually in accordance with their relevant categories, which is smaller and more correlated. If we can ascertain the category of video before keyword annotation, we may reduce the search scope of keywords and improve annotation effectiveness and efficiency. We adopt the global features to train our categorization model, because the global features can better represent videos in general perspective. The features include Average Color Entropy, Co-occurrence Texture, Average shot length, Global Camera Motion coefficients, and Face vector, etc.

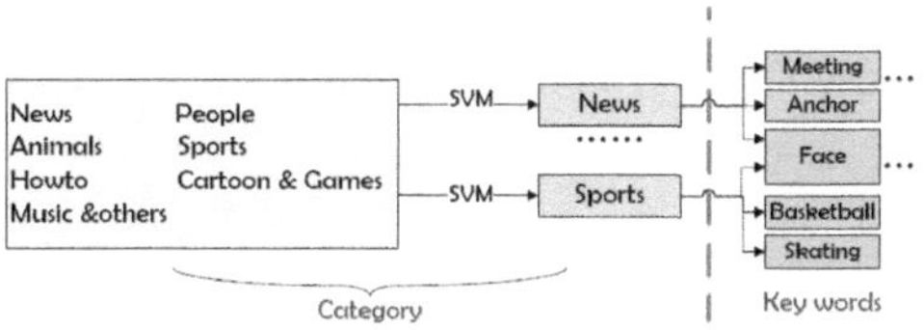

Fig. 2. Hierarchical Video Concept Organization

3 Keyword Labelling

Having gained the category information at the video level, we turn to the shot level and utilize keyframe to generate keywords. Current automatic image annotation systems, such as [2], use fixed feature set to represent concepts without considering whether individual feature is representative. Although a concept can be described with the typical features, e.g. color, texture and shape, different kinds of features may play different roles. To better represent concepts, we develop a weighted multi-feature model to accomplish the annotation of keyframes. Finally, we fuse the keywords for keyframes of a video as annotation.

3.1 Concept Representation

In this section, we focus on the representation method for keywords using various features. The local static features we use include 24-dimensional color feature, 8-dimensional texture feature and 180-dimensional shape feature. We assign keywords to related shots which is represented by keyframes, and the features of these keyframes are clustered to represent the respective keyword. The characteristics of a video are the sum of distinct features of all clusters which is relevant to the video. In the system, we design a weight tuning mechanism to evaluate the effect of different features on the annotation. Multi-features of video data provide the flexibility of allowing the system to model video based on any combination of different features.

3.2 Multi-feature Model

From the previous discussion, it is easy to conclude that different keywords require different features or their combination for better representation. We use a linear discriminative function to model the feature for a keyword. Each keyword may be attached to a number of keyframes, and we extract the three feature vectors from all the keyframes. For each keyword, we conduct the density clustering within the respective feature space. Using the feature vector space we construct the kernel representation of each cluster f_i, we calculate the weight according to the function: $w_i = \frac{1}{rad_i/\theta_i} \times dens_i \ (i \in \{color, texture, edge\})$, where rad_i refers to the radius of each feature cluster gained with the assistance of f_i; θ_i is the noise threshold; $dens_i$ indicates the density of clusters. The higher density the feature cluster generates, the more weight in the representation of feature is. If the feature has higher weight, the feature is more representative with respect to the keyword. From Fig. 3, we can see the weights for "Car", where shape is the most important and which is consistent with the human perception.

4 System Demo

To demonstrate the proposed video annotation system, we use the videos downloaded from YouTube. Our data set contains 658 video clips and 15,918 shots covering various

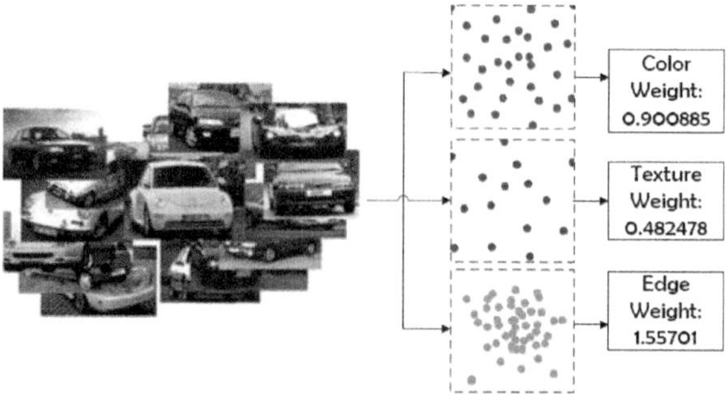

Fig. 3. An Example of Multi-Feature Model

categories discussed in Section 2. We use the same keyword set as [4]. In our annotation system, users can provide the unlabelled videos by either upload or URL. Fig. 4 shows the interface of our system, and the annotation results with a sample video.

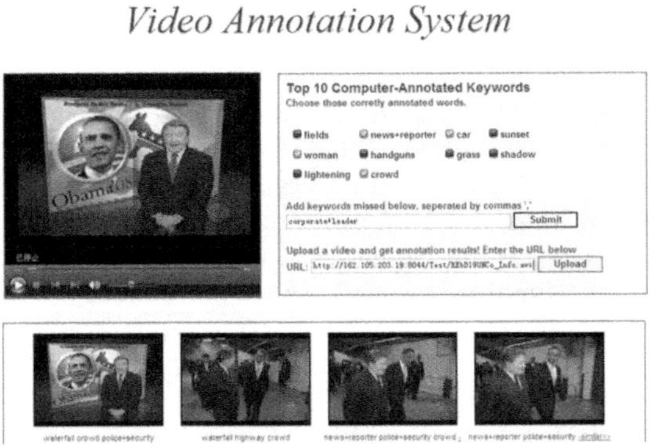

Fig. 4. The Interface of Annotation System with Sample

To evaluate the performance, we compare our method with a traditional method that uses SVM classifier to train the concept models. Overall, our approach can achieve average precision 28.4% which is about 40% better than that of SVM (17.5%). Due to the space constraint, we omit the details about comparison. Our approach has two advantages. First, the video categorization can classify the video according to the global feature, and reduce the keyword range. Second, the mutli-feature model with weighting can capture the content feature of video more effectively.

5 Conclusion and Future Work

In this demonstration, we have developed a novel video annotation system by combining the hierarchical concepts and multi-feature model. The experimental results showed that our proposed system can provide satisfactory performance. In the future, we consider to enlarge the keyword set and integrate more features to construct more representative model for semantic concepts. Expanding the system with relevance feedback is another promising topic to improve the user experience of the system.

References

1. Jiang, W., Chang, S.-F., Loui, A.: Active concept-based concept fusion with partial user labels. In: IEEE ICIP (2006)
2. Li, J., Wang, J.: Real-time computerized annotation of pictures. In: ACM MM (2006)
3. Luo, H., Fan, J.: Building concept ontology for medical video annotation. In: ACM MM (2006)
4. Snoek, C., Worring, M., Gemert, J., Geusebroek, J.-M., Smeulders, A.: The challenge problem for automated detection of 101 semantic concepts in multimedia. In: ACM Multimedia (2006)
5. Wu, Y., Tseng, B.L., Smith, J.R.: Ontology-based multi-classification learning for video concept detection. In: IEEE ICME (2004)

TRUSTER: TRajectory Data Processing on ClUSTERs

Bin Yang[1], Qiang Ma[1], Weining Qian[2], and Aoying Zhou[2]

[1] School of Computer Science, Fudan University
Shanghai, China
{byang,maqiang}@fudan.edu.cn
[2] Institute of Massive Computing, East China Normal University
Shanghai, China
{wnqian,ayzhou}@sei.ecnu.edu.cn

Abstract. With the continued advancements in location-based services involved infrastructures, large amount of time-based location data are quickly accumulated. Distributed processing techniques on such large trajectory data sets are urgently needed. We propose TRUSTER: a distributed trajectory data processing system on clusters. TRUSTER employs a distributed indexing method on large scale trajectory data sets, and it makes spatio-temporal queries execute efficiently on clusters.

1 Introduction

With the proliferation of positioning technologies and the increasing use of inexpensive location-aware sensors, large amount of spatio-temporal trajectory data have been accumulated. From the outdoor location reported by GPS devices in longitude-latitude coordinate system, to the indoor location observed by RFID readers in symbolic coordinate system, large amount of different formats of trajectory data have emerged. Facing complicated and huge amount of trajectory data, it is thought that only distributed data processing system can conquer this challenge. TRUSTER is a data processing system designed to manage large amount of trajectory data in a distributed environment.

Most existing trajectory indexing methods treat the time dimension as an additional space dimension, and use high dimensional indexing methods to index the trajectory. Suppose an object moves in a k-dimensional space, the trajectory of such object is essentially a k+1-dimensional space. Different R-Tree based indexing methods [1] can be applied on the k+1-dimensional space. Such indexing methods couple the spatial dimension and temporal dimension together, and it is difficult to partition the index and improper to use in a distributed environment.

We aim to present a distributed indexing method which could process queries over huge amount of trajectory data on clusters. TRUSTER partitions the whole spatial space into several static partitions. Trajectories are distributed into these partitions according to their spatial extent. For each partition, index is just built over temporal dimension for all trajectories in the partition. Different nodes in clusters can take charge of different spatial partitions, which makes both index creation and queries execution much more efficiently.

X. Zhou et al. (Eds.): DASFAA 2009, LNCS 5463, pp. 768–771, 2009.

2 System Design

TRUSTER is consisted of two modules: spatial partition module and query execution module. Both modules employ MapReduce [2,3] model to process trajectory data. We describe the two modules in this chapter.

2.1 Spatial Partition Module

Given a static partition method, the spatial partition module assigns each trajectory to specific partitions according to its spatial dimensions. On each partition, 1D R-Tree is created on the temporal dimension for all trajectories in the partition. Fig. 1 depicts the flow of spatial partition module.

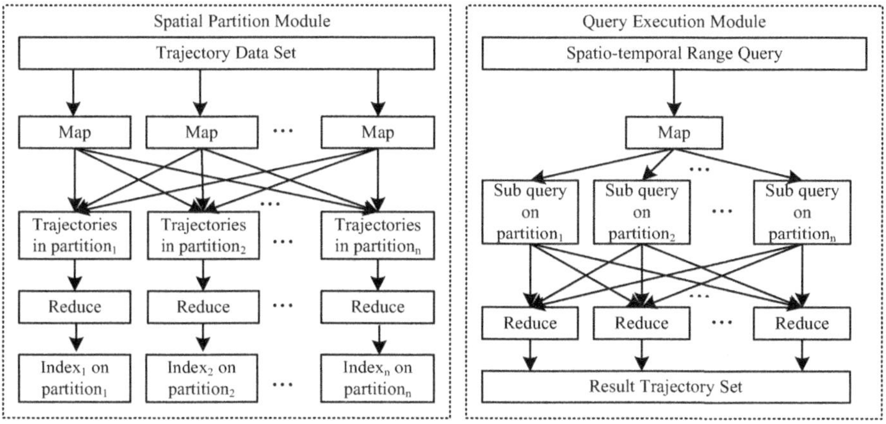

Fig. 1. Spatial Partition Module and Query Execution Module

The most commonly used line segments model for representing trajectories is applied in TRUSTER. Each line segment is indicated as $s_k(p_i,\ p_j)$, where s_k is segment identifier, p_i and p_j indicate two consecutive reported positions. Each position is in form of $(x,\ y,\ t)$, where x and y represent the coordinates of the object, and t indicates the timestamp of the report of the position. For each segment $s_k(p_i,\ p_j)$, map function transforms it into a list $(partitionID,$ $s'_k(p'_i,\ p'_j))$. If a line segment is fully covered by a partition, the line segment is assigned to the partition. For example, line segment $s_3(p_3,\ p_4)$ in Fig. 2 is transformed into $(Partition_2,\ s_3(p_3,\ p_4))$. If a line segment spans more than one partitions, such line segment is divided into several sub line segments. Each sub line segment is fully covered by one partition, and is assigned to the partition. The new end points of such sub line segments are the intersections with the original line segment and the partition boundary. The corresponding time for these end points are determined by interpolating. For example, segment $s_2(p_2,$ $p_3)$ is divided into two sub line segments and mapped to $(Partition_4,\ s_{21}(p_2,$ $p_7))$ and $(Partition_2,\ s_{22}(p_7,\ p_3))$.

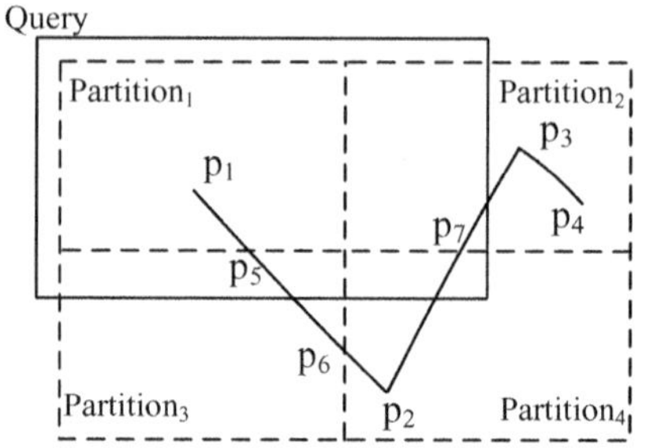

Fig. 2. Example of Trajectories, Partitions and Query

Reduce function will insert the temporal extent of all line segments and sub line segments in the same *partitionID* into 1D R-Tree, i.e. insert interval $[p'_i.t, p'_j.t]$ as the index entry for line segment $s'_k(p'_i, p'_j)$. For example, in $Partition_2$, $[p_7.t, p_3.t]$ is inserted as index entry for $s_{22}(p_7, p_3)$, $[p_3.t, p_4.t]$ is inserted as index entry for $s_3(p_3, p_4)$.

2.2 Query Execution Module

Query execution module divides each spatio-temporal range query into several temporal sub queries, and executes these sub queries on different nodes, which is described in Fig. 1. Suppose a query is represented as Q(E_s, E_t), where E_s is spatial range and E_t is temporal range. If E_s covers a partition, line segments which satisfy E_t in the partition are definitely the result of the original query. If E_s overlaps with a partition, line segments which satisfy E_t in the partition are possibly the result, and it requires further spatial refinement on such line segments.

Map function transforms Q(E_s, E_t) into a list of $Q'(partitionID, E_t, tag)$, where *tag* indicates *partitionID* is an overlapped or covered candidate partition. Using the corresponding temporal index according to the *partitionID*, all the line segments satisfied with E_t will be fetched. If *tag* indicates an overlapped partition, spatial refinement is conducted. The result of map function is a list of line segments $s'_k(p'_i, p'_j)$. For example, the query which is shown in Fig. 2 as a solid rectangle, is transformed into four sub queries $Q_1(Partition_1, E_t, COVER)$, $Q_2(Partition_2, E_t, OVERLAP)$, $Q_3(Partition_3, E_t, OVERLAP)$ and $Q_4(Partition_4, E_t, OVERLAP)$. At reduce phase, reduce function combines the sub line segments with the same end point to one line segment.

3 Demonstration Outline

A TRUSTER system is built on Hadoop [3,4] clusters. We use GSTD [5] data generator to generate different trajectory data sets. The demonstration includes three parts and the GUI of TRUSTER is shown in Fig. 3:

1. System Configuration and Monitoring: User can make configurations on clusters. Such configurations include the replication level of each data file, data

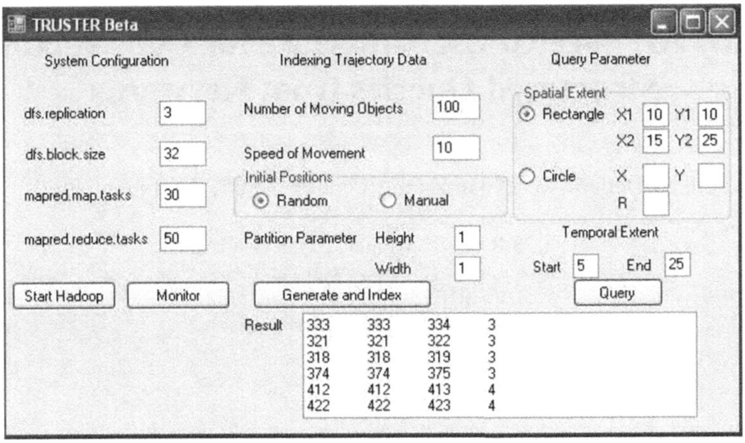

Fig. 3. User Interface of TRUSTER

block size, the number of map and reduce tasks. A system monitor is developed for TRUSTER to help users to control the clusters. Workload and the number of tasks on different nodes are monitored in real time. The monitoring information can be both shown in GUI and stored as a log file.

2. Trajectory Data Generating and Indexing: Users can make different configurations to generate trajectory data. The configurable parameter includes the number of moving objects, the speed of movement and the initial positions. Users can also configure the size of rectangular partitions. After the configuration, TRUSTER generates the trajectories and indexes them according to the given partition.
3. Query in TRUSTER. Users can search trajectories indexed in TRUSTER. The search interface accepts the spatial extent and the temporal extent separately, and returns the result trajectories.

Acknowledgments. This paper is partially supported by National Natural Science Foundation of China (NSFC) under Grant No. 60673134.

References

1. Pfoser, D., Jensen, C.S., Theodoridis, Y.: Novel approaches in query processing for moving object trajectories. In: VLDB, pp. 395–406 (2000)
2. Dean, J., Ghemawat, S.: Mapreduce: Simplified data processing on large clusters. In: OSDI, pp. 137–150 (2004)
3. Yang, B., Qian, W., Zhou, A.: Using wide table to manage web data: a survey. Frontiers of Computer Science in China 2(3), 211–223 (2008)
4. Bialecki, A., Cafarella, M., Cutting, D., OMalley, O.: Hadoop: a framework for running applications on large clusters built of commodity hardware, `http://lucene.apache.org/hadoop`
5. Theodoridis, Y., Silva, J.R.O., Nascimento, M.A.: On the generation of spatiotemporal datasets. In: Güting, R.H., Papadias, D., Lochovsky, F.H. (eds.) SSD 1999. LNCS, vol. 1651, pp. 147–164. Springer, Heidelberg (1999)

SUITS: Faceted User Interface for Constructing Structured Queries from Keywords*

Elena Demidova[1], Xuan Zhou[2], Gideon Zenz[1], Wolfgang Nejdl[1]

[1]L3S Research Center, Hanover, Germany
{demidova,zenz,nejdl}@L3S.de
[2]CSIRO ICT Centre, Australia
xuan.zhou@CSIRO.au

Abstract. Keyword search and database query are two ways for retrieving data in real world settings. In this demonstration, we show SUITS, a novel search interface over relational databases that smoothly integrates the flexibility of keyword search and the precision of database queries. SUITS allows users to start with arbitrary keyword queries, refine them incrementally by following the suggestions given by the system and finally obtain desired structured queries.

Keywords: keyword search, query construction.

1 Introduction

The digital information accessible today, such as that on the Web, possesses semantics in both structured and unstructured forms. Keyword search, which was originally developed for retrieving documents, is an intuitive and convenient interface for accessing unstructured data. However, keyword search leaves users with limited expressiveness in describing their information needs. As a consequence, users may fail to find desired information. On the contrary, database queries enable users to exploit available structures to achieve more precise queries and corresponding result sets. However, using a database system requires adequate knowledge of the database schema and proficiency in the query language, making data access a difficult task for unskilled users. In this demo, we present SUITS, a novel search interface which provides a layer of abstraction on top of relational databases to smoothly integrate the intuitiveness of keyword search and the expressiveness of database queries.

As shown in Fig. 1, the SUITS interface consists of four parts: a search field for the user to input keyword queries, a result window to present search results (at the bottom), a query window to present structured queries (on the left) and a faceted query construction panel providing query construction options (on the right). To perform search, the user first issues a keyword query, for instance "Fuzz Wright London" intended to search for the movie called "Hot Fuzz" and directed by Edgar Wright. Besides returning a ranked list of results like standard keyword search [1, 2],

* This work is partially supported by the FP7 EU Project OKKAM (contract no. ICT-215032) and TENCompetence Integrated Project (contract no. 027087).

X. Zhou et al. (Eds.): DASFAA 2009, LNCS 5463, pp. 772–775, 2009.

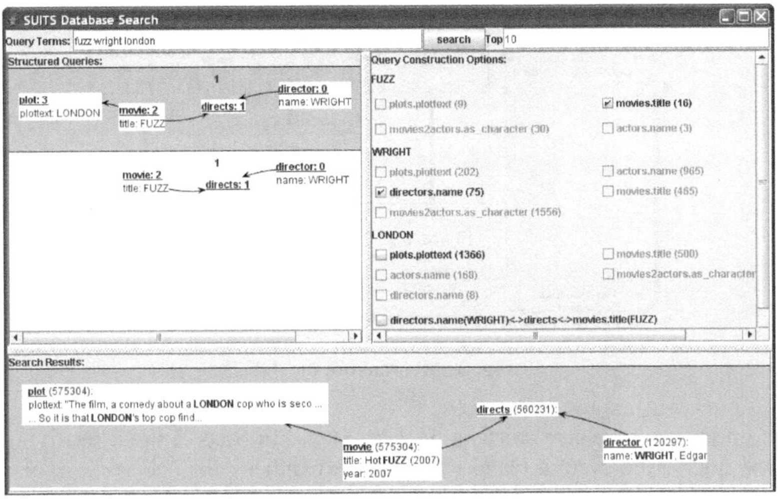

Fig. 1. The Search Interface of SUITS

SUITS suggests to the user a list of structured queries in the query window to clarify her intent. The suggested queries assign different semantics to the keywords. Some queries search for the movies with actor "Wright", while others search for the actors who acted as a character named "London". If the user identifies the structured query that represents her intent, she can click on it, such that the result window will show the results of that particular query. Otherwise, she can go to the faceted query construction panel to incrementally construct the intended query. The panel lists a number of query construction options suggested by the system. As shown in Fig. 1, the user specifies that "Fuzz" must appear in the movie title and "Wright" must be a director's name. The query window changes accordingly to zoom into the queries satisfying the chosen options. Such interaction continues iteratively until the user obtains the right query and/or satisfactory search results.

With the SUITS interface, users can start with arbitrary keywords and structure their queries progressively by following the system's suggestions. Finally, they can either select a completely structured query from the query window or a partially structured query by specifying appropriate query construction options. It depends on the degree to which users want / are able to clarify their search intents.

2 System Description

The query processing of SUITS can be split into two phases: an offline pre-computing phase and an online query phase. In the first phase, SUITS creates inverted indexes for all text columns in the database. It also generates query templates that are potentially employed by users when forming structured queries. A query template is a structural pattern used to query a database. For example, users sometimes search for the movies with a certain character, and sometimes search for the actors who have played in a certain movie. Both are commonly used query templates.

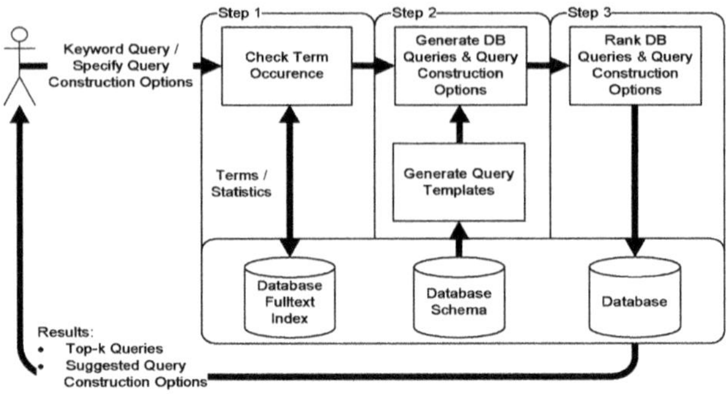

Fig. 2. Architecture of SUITS

The online query phase consists of three steps. In Step 1, upon receiving a user's keyword query, the system checks the full-text indices for occurrences of the query terms. In Step 2, it combines the term occurrences and the pre-computed query templates to create meaningful structured queries and query construction options. In Step 3, the system ranks the queries according to their likelihood of being intended by the user, and selects a set of query construction options that can more effectively clarify the user's intent. Finally, it returns the top-k queries and a set of selected options to the user. If the user chooses some of the query construction options, these options are fed back to Step 2 in order to filter out queries and query construction options that do not satisfy the user's specification.

The success of SUITS relies on the proper ranking of structured queries, effective selection of the query construction options and efficient query processing. In [3], we presented a set of techniques to accomplish these tasks. In this paper, due to the size limitations, we focus only on the query construction.

3 Query Construction

SUITS suggests appropriate query construction options to support incrementally creating the user intended structured query from keywords. Choosing a query construction option is equivalent to specifying a part of the structured query. Thus, query construction options are actually partial queries. As shown in Fig. 3, the complete queries and partial queries constructed from a set of keywords can be organized in a hierarchy. At the bottom of this hierarchy are the smallest partial queries composed of only one keyword and one attribute. In the middle are partial queries that join two or more keywords together. At the top, complete structured queries involving all keywords are located.

During the query construction, the system first presents a small set of partial queries to the user. If the user chooses any of the partial queries, she actually indicates that her intended complete query will contain that partial query. Therefore, the system can remove from the query hierarchy all the complete queries not containing that partial query. Later on, the system presents another set of partial queries to the user for selection. This process carries on until the user identifies the desired query in the query window.

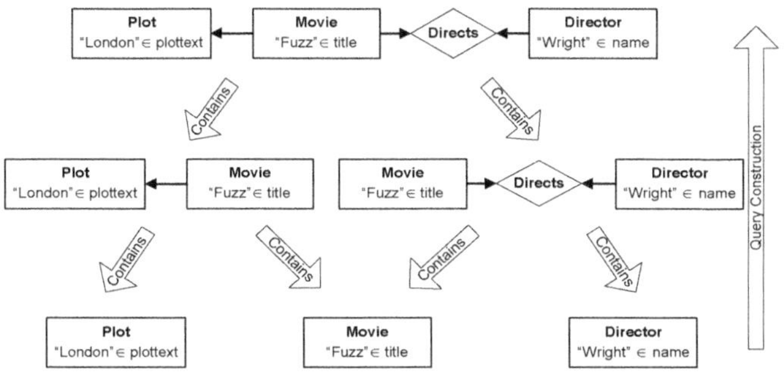

Fig. 3. Hierarchy of Partial Queries

Selecting proper partial queries is crucial in the query construction process. A good strategy allows a user to exclude as many complete queries as possible in each round, so that she can obtain the desired query quickly. In [3], we proposed a number of principles for selection and ranking of the query construction options.

4 Demonstration Overview

In this demonstration we will primarily show how SUITS works and how a user can employ it to efficiently identify desired information in a database without any knowledge of the database schema. First, we will demonstrate the complete query process. This process starts from submitting a keyword query to the system, followed by a presentation of the top-k structured queries that give different interpretations of the keywords, followed by an execution of the queries to retrieve search results. Then we show how the query construction options suggested by SUITS can guide users to quickly construct desired structured queries. We also present the result navigation component of SUITS which enables extending a search result to explore the database context. Our demonstration uses a real-world dataset, a crawl of the Internet Movie Database (IMDB).

References

1. Agrawal, S., Chaudhuri, S., Das, G.: DBXplorer: A System for Keyword-Based Search over Relational Databases. In: Proc. of the ICDE 2002 (2002)
2. Bhalotia, G., Hulgeri, A., Nakhey, C., Chakrabarti, S., Sudarshan, S.: Keyword searching and browsing in databases using BANKS. In: Proc. of the ICDE 2002 (2002)
3. Zhou, X., Zenz, G., Demidova, E., Nejdl, W.: SUITS: Constructing Structured Queries Using Keywords. Technical Report, L3S Research Center (2007)

Demonstration of a Peer-to-Peer Approach for Spatial Queries

Aleksandra Kovacevic, Aleksandar Todorov, Nicolas Liebau, Dirk Bradler, and Ralf Steinmetz

Technische Universität Darmstadt,
Merckstr. 25, 64289 Darmstadt, Germany

Abstract. This demo paper presents an prototype implementation of a decentralized and distributed approach for spatial queries. The main focus is the location-based search for all the objects or information in the particular geographical area.

1 Introduction

Services, which provide the users with data based on their geographical location, also known as Location-Based Services (LBS), are becoming increasingly popular. For the content providers and companies (such as mobile carriers) LBS is an opportunity to supply their users and/or customers with location specific data. And for the users of location-based services this means more personalized information, data relevant for the user at his current location. Nowadays, there is a variety of examples for such services, varying from vehicle tracking and navigation systems, personalized weather services, nearby traffic jam alerts, and even location-based games. The emergency telephone service, such as 112 in Europe, dispatches the calls to the nearest emergency medical service, police or firefighters according to the caller's position [3]. All of these examples make use of locality and supply the user with data based on his current position.

The present location-based services, just as most other services, require a centralized point of maintenance and management in order to provide the information and keep it up-to-date. Such solutions are often very expensive (e.g. Google) and sophisticated, and get even more expensive and harder to manage as they scale. Because of these reasons, many service providers are able to process only a limited amount of requests or process requests with a big delay in time, which might result in inaccurate, incomplete, or outdated responses, sent back to the requesting users. One could imagine a scenario, where a user is trying to find out what the current menu list in a nearby restaurant is. This involves at least two previous steps - 1. The restaurant's management has sent the updated menu card to the central server, and 2. The central server has accepted and stored the information. What if the user also wants to request the daily price list, currently free tables, currently playing song? This means, the server has to store a lot more data then just the menu list, and also

X. Zhou et al. (Eds.): DASFAA 2009, LNCS 5463, pp. 776–779, 2009.

must be able to receive frequent updates. This, as already mentioned, is a problem when addressing scalability and maintenance costs. Additionally it can take days or even weeks for the central server to process, verify, and store the new information.

The adoption of peer-to-peer systems can reduce the cost and maintenance complexity and overhead in many similar as the above described cases. Being client and server in one entity, at the same time, a peer itself is responsible for the information that it represents and thus do not need a central server for maintenance. Publishing, updating and removing of information happens directly on the peer. As a result, the service consumers get fresher information since the updates are instantly visible and the service providers have more control of their data and services, since issues, like updates and availability, are managed directly on the peer.

2 Our Approach

In our prior work [1,2] we designed Globase.KOM, a peer-to-peer overlay for location-based search is presented. The main goal of this overlay is to enable fully retrievable search for peers in a specified geographical area, exact searches and finding the closest peer. Globase.KOM is a superpeer-based overlay forming a tree enhanced with interconnections. We use the more powerful peers with good network connectivity, which tend to stay online for a long time as superpeers in Globase.KOM. Superpeers are responsible for indexing all nodes/services in one clearly defined geographical area. The normal (nonsuper) peers in the network simply offer and consume services without having additional responsibilities. The world projection is divided into disjoint, nonoverlapping zones. Each zone is assigned to a superpeer located inside the zone that keeps overlay/underlay contact addresses for all peers in that zone. Superpeers form a tree where node A is called the parent of node B when B's zone is inside A's zone (see Figure 2). A location-based search starts with contacting one superpeer, which then forwards the query message to its parent or children. Simulation results showed its scalability, efficiency in the sense of performance - costs ratio, load balance, and full retrievability.

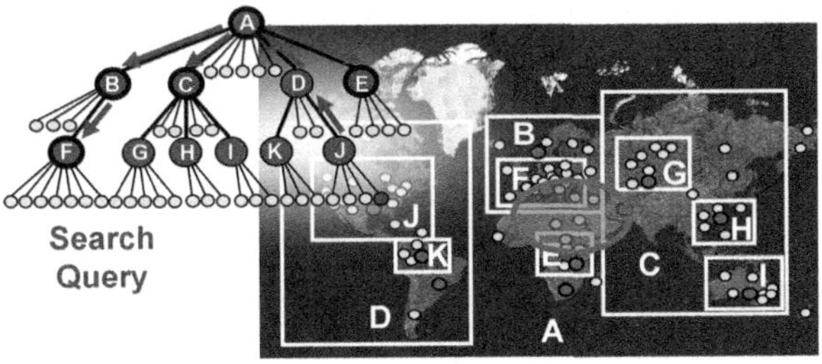

Fig. 1. Example of an Area Search in Globase.KOM

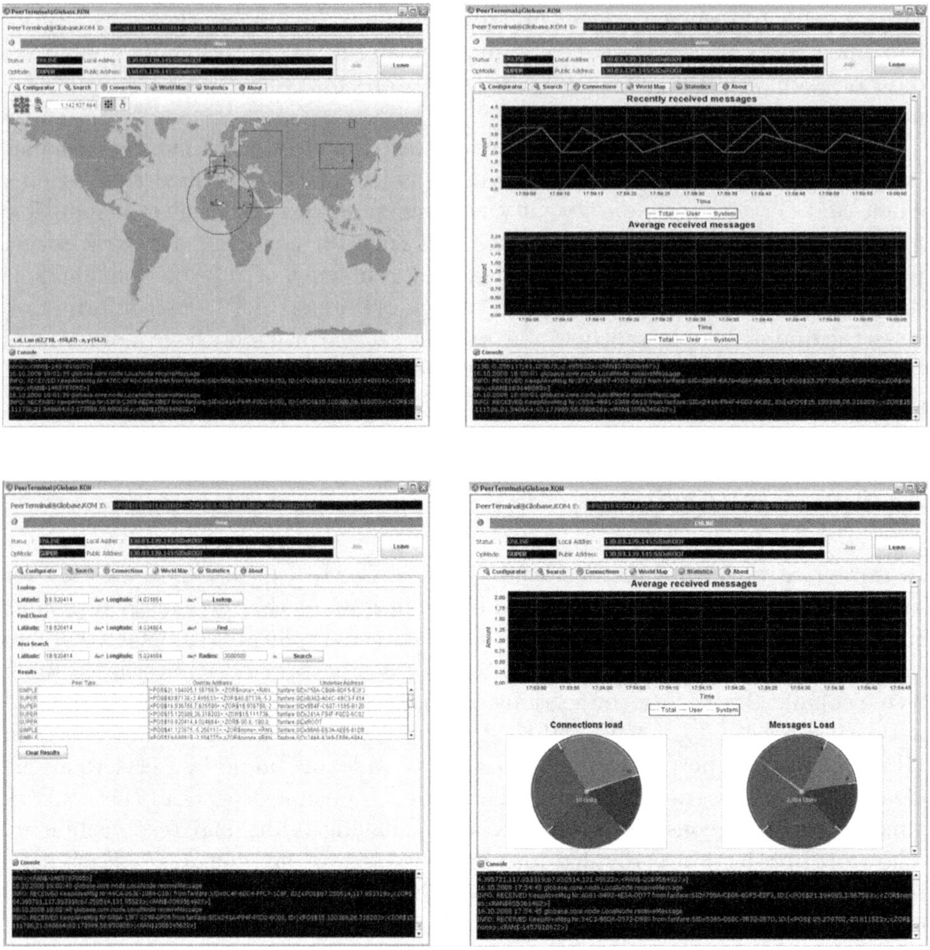

Fig. 2. Screenshots of Globase.KOM prototype

3 Prototype

Here we will descibe a prototype implementation of Globase.KOM. There are several crucial requirements that are taken into account prior implementation. *Communication* is an integral part of any network and must be reliable if the final product is to be reliable. As the performance of the prototype on a variety of parameters, an easy *configuration* is desirable. Modularity and extendibility is crucial requirement for building practical layered implementations. Every peer should be able to contact anyone and still be able to accept connections while doing so. Therefore, we used Java RMI (Remote Method Invocation) [1], built on top of the TCP/IP network in a non-blocking, asynchronous manner.

[1] http://java.sun.com/javase/technologies/core/basic/rmi/index.jsp

Graphical user interface consists of five panels: configuration, search, connections, world map, and statistics panel. In *configuration panel* it is possible to set all the parameters of the prototype through the given interface or loading the appropriate xml file. The *search panel* provides an easy way to invoke the basic operations specified in Globase.KOM by specifying the latitude and longitude of a point for exact search and find closest operation and additionally the radius of a searched area for area search. In the same panel, the overlay and underlay address and the type of the peers from the search results are listed. The *connections panel* is read-only. It simply displays the whole contact information of a peer, its routing table. *World map panel* displays all known contacts, zones, and results of performed operations on a map of the world. Diverse libraries from the OpenMap[2] project were used for this purpose. In *statistic panel* displays the amount of received messages per second in average and recently, separated into user and system messages. There are two barometers indicating the current load of the peer based on the two load criteria considered in the prototype - amount of peers in the zone of responsibility and amount of received messages per second. For this panel, the JFreeChart [3] libraries were used.

4 Demo Description

Demonstration of this prototype will consist of 100 instances runned on 2 or 3 machines. It will show three operations of Globase.KOM - search for the peer on the exact geographical location, finding the geographically closest peer, and searching for all peers in free chosed geographical area. The location of the peers will be generated based on the bitmap presenting the world map of all Skype users which we captured in our measurements in July 2008. Current statistics, routing tables, and effects of different setting of parameter will be presented.

References

1. Kovacevic, A., Liebau, N., Steinmetz, R.: Globase. KOM - A P2P Overlay for Fully Retrievable Location-based Search. In: Proceedings of the Seventh IEEE International Conference on Peer-to-Peer Computing (September 2007)
2. Kovacevic, A., Heckmann, O., Liebau, N., Steinmetz, R.: Location Awareness - Improving Distributed Multimedia Communication. Special Issue of the Proceedings of IEEE on Advances in Distributed Multimedia Communications 96(1) (January 2008)
3. Graffi, K., Kovacevic, A., Wulffert, K., Steinmetz, R.: ECHoP2P: Emergency Call Handling over Peer-to-Peer Overlays. In: International Workshop on Peer-to-Peer Network Virtual Environments (P2P-NVE) (December 2007)

---

[2] http://openmap.bbn.com/
[3] http://www.jfree.org/jfreechart/

Techniques for Efficiently Searching in Spatial, Temporal, Spatio-temporal, and Multimedia Databases

Hans-Peter Kriegel, Peer Kröger, and Matthias Renz

Institute for Informatics
Ludwig-Maximilians-Universität München, Germany
Oettingenstr. 67, 80538 Munich, Germany
{kriegel,kroegerp,renz}@dbs.ifi.lmu.de
http://www.dbs.ifi.lmu.de

Abstract. This tutorial provides a comprehensive and comparative overview of general techniques to efficiently support similarity queries in spatial, temporal, spatio-temporal, and multimedia databases. In particular, it identifies the most generic query types and discusses general algorithmic methods to answer such queries efficiently. In addition, the tutorial sketches important applications of the introduced methods, and presents sample implementations of the general approaches within each of the aforementioned database types. The intended audience of this tutorial ranges from novice researchers to advanced experts as well as practitioners from any application domain dealing with spatial, temporal, spatio-temporal, and/or multimedia data.

1 Introduction

The management and analysis of spatial, temporal, spatio-temporal, and multimedia data is a hot topic in database research because such data types occur in many applications. Querying databases of such a content is very important for these applications. In recent years, a vast amount of research has been done to explore efficient solutions for answering similarity queries on these data types. However, the existing research mainly focuses on one data type only although many proposed approaches are conceptually rather similar. As a consequence, it is a complex task to keep track with current research results not only because of the large amount of existing approaches, but also because of very different vocabularies used to express similar concepts. This tutorial aims at providing a cooperate and comprehensive view of the state-of-the-art research in similarity search for spatial, temporal, spatio-temporal, and multimedia data by identifying the general algorithmic approaches common to all solutions. This will build the bridges between the various approaches proposed for the different data types and illustrate relationships among them.

2 Searching in Spatial, Temporal, Spatio-temporal and Multimedia Databases

Real-world applications dealing with spatial, temporal, spatio-temporal, and multimedia data require efficient methods for similarity search. Example applications are shape

X. Zhou et al. (Eds.): DASFAA 2009, LNCS 5463, pp. 780–783, 2009.

based similarity search and docking queries in Geographic and CAD databases, time-series matching, proximity queries on moving objects and image and video retrieval in multimedia databases. There are a bundle of problems emerging from the aforementioned applications that challenge the problem of designing efficient solutions for answering these basic query types.

In this tutorial, we provide a detailed introduction to basic algorithmic approaches to tackle the problem of answering similarity queries efficiently. In addition, we discuss the relationships between the approaches. Last but not least, we will present sample implementations of these basic approaches for each of the four data types mentioned above.

2.1 General Concepts of Query Processing in Complex Structured Data

The first part of the tutorial details the general algorithmic approaches for efficient similarity query processing including indexing and multi-step query processing. We start with the general concept of feature based similarity search which is commonly used to efficiently support similarity query processing in non-standard databases [1,7,9]. This section gives a general overview of state-of-the-art approaches for similarity query processing using indexing methods. Thereby, diverse approaches for different similarity query types are discussed, including distance range queries (DRQ), k-nearest neighbor queries (kNNQ), and reverse k-nearest neighbor queries (RkNNQ). Furthermore, this section addresses the concept of multi-step query processing. Here, we show the optimal interaction between access methods and multi-step query processing methods [18,12] and give a survey of this topic showing the relationships among diverse existing approaches. Finally, the first part of the tutorial briefly sketches other types of queries that are specialized for a given data type, e.g. intersection queries for spatial objects.

2.2 Querying Spatial, Temporal, Spatio-temporal and Multi-media Data

The second part of the tutorial gives an overview of the sample implementations of the previously presented basic approaches for each of the four data types mentioned above. It starts with sample solutions for query processing in spatial data implementing the general algorithmic approaches presented previously. The presented methods address queries on two- or three dimensional spatially extended objects, e.g. geographic data [16,5], CAD data [11,4] and protein data.

Next, the tutorial addresses query processing in time series data which is the most important data type among temporal data. Here, we sketch the general problem of indexing time series known under the term "curse of dimensionality" and discuss diverse solutions for this problem, including dimensionality reduction and GEMINI framework [7]. In addition to matching based similarity query methods [8] we also discuss threshold based similarity search methods for time series data [3].

The sample implementations concerning spatio-temporal data mainly focuses on proximity queries in traffic networks, i.e. queries on objects moving within a spatial network. This section introduces techniques enabling efficient processing of proximity queries in large network graphs. In particular, we discuss solutions that are adequate for densely populated [17] and sparsely populated traffic networks [19,13]. The later case requires graph embedding techniques that allows the application of multi-step query processing concepts.

Finally, this tutorial outlines sample solutions for query processing in multimedia data implementing the general algorithmic approaches presented in the first part of this tutorial, e.g. [10,2]. In addition to similarity filter techniques that are specialized to multimedia data we also discuss how the multi-step query processing techniques can cope with uncertainty in multimedia data. Here we give an overview of sample solutions, e.g. [6,14,15].

References

1. Agrawal, R., Faloutsos, C., Swami, A.: Efficient similarity search in sequence databases. In: Lomet, D.B. (ed.) FODO 1993. LNCS, vol. 730. Springer, Heidelberg (1993)
2. Assent, I., Wichterich, M., Meisen, T., Seidl, T.: Efficient similarity search using the earth mover's distance for large multimedia databases. In: Proceedings of the 24th International Conference on Data Engineering (ICDE), Cancún, Mexico (2008)
3. Aßfalg, J., Kriegel, H.-P., Kröger, P., Kunath, P., Pryakhin, A., Renz, M.: Similarity search on time series based on threshold queries. In: Ioannidis, Y., Scholl, M.H., Schmidt, J.W., Matthes, F., Hatzopoulos, M., Böhm, K., Kemper, A., Grust, T., Böhm, C. (eds.) EDBT 2006. LNCS, vol. 3896, pp. 276–294. Springer, Heidelberg (2006)
4. Aßfalg, J., Kriegel, H.-P., Kröger, P., Pötke, M.: Accurate and efficient similarity search on 3D objects using point sampling, redundancy, and proportionality. In: Bauzer Medeiros, C., Egenhofer, M.J., Bertino, E. (eds.) SSTD 2005. LNCS, vol. 3633, pp. 200–217. Springer, Heidelberg (2005)
5. Brinkhoff, T., Horn, H., Kriegel, H.-P., Schneider, R.: A storage and access architecture for efficient query processing in spatial database systems. In: Abel, D.J., Ooi, B.-C. (eds.) SSD 1993. LNCS, vol. 692. Springer, Heidelberg (1993)
6. Cheng, R., Singh, S., Prabhakar, S., Shah, R., Vitter, J., Xia, Y.: Efficient join processing over uncertain data. In: Proceedings of the 15th International Conference on Information and Knowledge Management (CIKM), Arlington, VA (2006)
7. Faloutsos, C., Ranganathan, M., Manolopoulos, Y.: Fast subsequence matching in time series database. In: Proceedings of the ACM International Conference on Management of Data (SIGMOD), Minneapolis, MN (1994)
8. Keogh, E.: Exact indexing of dynamic time warping. In: Proceedings of the 28th International Conference on Very Large Data Bases (VLDB), Hong Kong, China (2002)
9. F., Sidiropoulos, N., Faloutsos, C., Siegel, E., Korn, Z.P.: Fast nearest neighbor search in medical image databases. In: Proceedings of the 22nd International Conference on Very Large Data Bases (VLDB), Bombay, India (1996)
10. Korn, F., Sidiropoulos, N., Faloutsos, C., Siegel, E., Protopapas, Z.: Fast and effective retrieval of medical tumor shapes. In: IEEE Transactions on Knowledge and Data Engineering (1998)
11. Kriegel, H.-P., Brecheisen, S., Kröger, P., Pfeifle, M., Schubert, M.: Using sets of feature vectors for similarity search on voxelized cad objects. In: Proceedings of the ACM International Conference on Management of Data (SIGMOD), San Diego, CA (2003)
12. Kriegel, H.-P., Kröger, P., Kunath, P., Renz, M.: Generalizing the optimality of multi-step *k*-nearest neighbor query processing. In: Papadias, D., Zhang, D., Kollios, G. (eds.) SSTD 2007. LNCS, vol. 4605, pp. 75–92. Springer, Heidelberg (2007)

13. Kriegel, H.-P., Kröger, P., Renz, M., Schmidt, T.: Hierarchical graph embedding for efficient query processing in very large traffic networks. In: Ludäscher, B., Mamoulis, N. (eds.) SSDBM 2008. LNCS, vol. 5069, pp. 150–167. Springer, Heidelberg (2008)
14. Kriegel, H.-P., Kunath, P., Pfeifle, M., Renz, M.: Probabilistic similarity join on uncertain data. In: Li Lee, M., Tan, K.-L., Wuwongse, V. (eds.) DASFAA 2006. LNCS, vol. 3882, pp. 295–309. Springer, Heidelberg (2006)
15. Kriegel, H.-P., Kunath, P., Renz, M.: Probabilistic nearest-neighbor query on uncertain objects. In: Kotagiri, R., Radha Krishna, P., Mohania, M., Nantajeewarawat, E. (eds.) DASFAA 2007. LNCS, vol. 4443, pp. 337–348. Springer, Heidelberg (2007)
16. Orenstein, J.A., Manola, F.A.: Probe spatial data modelling and query processing in an image database application. IEEE Trans. on Software Engineering 14(5), 611–629 (1988)
17. Papadias, D., Zhang, J., Mamoulis, N., Tao, Y.: Query processing in spatial network databases. In: Proceedings of the 29th International Conference on Very Large Data Bases (VLDB), Berlin, Germany (2003)
18. Seidl, T., Kriegel, H.-P.: Optimal multi-step k-nearest neighbor search. In: Proceedings of the ACM International Conference on Management of Data (SIGMOD), Seattle, WA (1998)
19. Shahabi, C., Kolahdouzan, M.R., Sharifzadeh, M.: A road network embedding technique for k-nearest neighbor search in moving object databases. Geoinformatica 7(3), 255–273 (2003)

Knowledge Discovery over the Deep Web, Semantic Web and XML

Aparna Varde[1], Fabian Suchanek[2], Richi Nayak[3], and Pierre Senellart[4]

[1] Department of Computer Science, Montclair State University, Montclair, NJ, USA
vardea@mail.montclair.edu
[2] Databases and Information Systems, Max Planck Institute for Informatics, Saarbrucken, Germany
suchanek@mpi-inf.mpg.de
[3] Faculty of Information Technology, Queensland University of Technology, Brisbane, Australia
r.nayak@qut.edu.au
[4] Department of Computer Science and Networking, Telecom Paristech, Paris, France
pierre@senellart.com

Abstract. In this tutorial we provide an insight into Web Mining, i.e., discovering knowledge from the World Wide Web, especially with reference to the latest developments in Web technology. The topics covered are: the Deep Web, also known as the Hidden Web or Invisible Web; the Semantic Web including standards such as RDFS and OWL; the eXtensible Markup Language XML, a widespread communication medium for the Web; and domain-specific markup languages defined within the context of XML We explain how each of these developments support knowledge discovery from data stored over the Web, thereby assisting several real-world applications.

Keywords: Information Retrieval, Standards, Web Mining.

1 Introduction

This tutorial focuses on knowledge discovery from the Web with particular emphasis on the Deep Web, Semantic Web and XML including domain-specific markup languages. The vast amount of data stored and exchanged over the World Wide Web is a huge source of knowledge that can be useful in various potential applications.

Among the recent advances in Web technology, we have the Deep Web over which stored information is not obvious but needs to be inferred, for example from queries through forms. The Semantic Web encompasses standards such as RDFS and OWL which often serve as the basis for defining ontology with reference to context.

XML, the eXtensible Markup Language has become a widely accepted means of communication with its descriptive tag sets that can be extended to add semantics to the data stored on the Web. This also facilitates the development of domain-specific markup languages that can be accepted as the lingua franca for communication in their respective fields.

All these developments provide great potential for mining the Web, i.e., discovering knowledge from the stored data. The data available on the Web is typically in a

X. Zhou et al. (Eds.): DASFAA 2009, LNCS 5463, pp. 784–788, 2009.

semi-structured format which presents additional challenges in knowledge discovery as opposed to data stored in traditional relational databases. In this tutorial we address various aspects of knowledge discovery from the Web with respect to these developments. We give an overview of the Deep Web, Semantic Web, XML and domain-specific markup languages in terms of their fundamental concepts and explain how each of these enable knowledge discovery. Suitable examples are provided at relevant points in the tutorial. Interesting real-world applications are also described. The tutorial is thus divided into four parts as described in the following four sections.

2 The Deep Web

A large part of the information present in the World Wide Web is *hidden* to current-day search engines, because it is not accessible through hyperlinks but lies in databases queried through forms. This *Deep* Web (or *Hidden* Web, or *Invisible* Web) has been estimated to contain 500 times as much data as the *Surface Web*. If such precise measures are debatable, this order of magnitude has been confirmed by recent work, and it is unquestionable that with information of the best quality (e.g., *Yellow Pages* services, U.S. *Census Bureau,* library catalogs, bibliography), the hidden Web is not only an invaluable source of information, but is also, due to its semi-structured, template nature, a rich source for knowledge discovery.

Access to content of the Deep Web requires filling in and submitting (HTML) forms, in order to retrieve some response pages, typically structured as lists or table records. Two approaches coexist for benefiting of the data hidden behind forms. The first one, the most straightforward, which has been advocated and experimented with by Google is an *extensional* one: response pages generated by the deep Web service are just stored as regular Web pages, that can be queried and retrieved as usual. The second approach, exemplified by the METAQUERIER system, is *intensional*: the goal is not to store response pages, but to understand the structure of both forms and response pages, and thus to know the semantics of this service, that can then be called as needed, depending on a user query. In either case, some schema matching and text mining techniques are used to associate form fields with concepts, in order to generate corresponding response pages. In the intensional case, understanding the structure of a response page means discovering the template this page was created from, either by unsupervised techniques such as ROADRUNNER, or by (semi-)supervised techniques.

This part of the tutorial presents different approaches for accessing the Deep Web, including but not limited to our own work, and shows how relevant data and information can be discovered and extracted from it.

3 The Semantic Web

The Semantic Web project envisions that people will publish semantic information in a computer-processable formalism that allows the information to be globally interlinked. For this purpose, the World Wide Web Consortium (W3C) has developed the knowledge representation formalisms RDFS and OWL. These formalisms are based on XML, but go beyond it by specifying semantic relationships between entities and even logical constraints on them. A collection of world knowledge in these formalisms is commonly called an *ontology*.

In this section of the tutorial, we first explain the vision and the applications of the Semantic Web project. We then give an introduction to semantic knowledge representations and ontologies in general. We also explain the knowledge representation formalisms RDFS and OWL, their syntax and semantics. We show where the Semantic Web has already taken off: Several large-scale ontologies are available online and are interlinked in the spirit of the Semantic Web. We explain how this information was gathered from different sources and how it can be queried using the SPARQL query language. Furthermore, we emphasize how this enhances knowledge discovery.

4 XML, the eXtensible Markup Language

The eXtensible Markup Language (XML) has become a standard language for data representation on the Web. With the continuous growth in XML based Web data sources, the ability to manage collections of XML documents and discover knowledge from them for decision support is increasingly important.

Mining of XML documents significantly differs from structured data mining and text mining. XML allows the representation of semi-structured and hierarchal data containing not only the values of individual items but also the relationships between data items. Element tags and their nesting therein dictate the structure of an XML document. Due to the inherent flexibility of XML, in both structure and semantics, discovering knowledge from XML data is faced with new challenges as well as benefits. Mining of structure along with content provides new insights and means into the knowledge discovery.

Recognizing the increasing interest in XML mining, this portion of the tutorial aims to discuss challenges that occur while mining the XML based Web data along with their solutions. We also provide issues and directions for research and development work in the future.

5 Domain-Specific Markup Languages

A significant expansion in the area of the Web and XML is the development of domain-specific markup languages. Such languages typically encompass the syntax of XML and capture the semantics of the given domains. Storing and exchanging data in this format, i.e., using XML based markups greatly boosts knowledge discovery. In this section of the tutorial we provide an overview of domain-specific markup languages with some real-world examples. We consider state-of-the-art markups, e.g., MML, the medical markup language, MatML, the Materials Markup Language and a few more.

We briefly outline the steps involved in the development of markup languages, the desired features of the languages, the use of XML constraints in preserving domain semantics and additional requirements, and the retrieval of information from the markups using XQuery, XPath and others in the XML family. We explain how data storage using such markup languages can assist data mining with classical techniques such as association rules. We also stress on the fact that in addition to a having adequate schemas for the markups, relevant ontological developments using standards in the literature can further assist the discovery of knowledge from data in the respective domains. A summary of our own research as well as related work by others in the area is discussed.

In conclusion, we summarize the most important points in the tutorial and briefly touch upon the potential for future work in the area of Web Mining.

References

1. Chang, C., Kayed, M., Girgis, M.R., Shaalan, K.F.: A survey of Web information extraction systems. IEEE Transactions on Knowledge and Data Engineering 18(10), 1411–1428 (2006)
2. Crescenzi, V., Mecca, G., Merialdo, P.: Roadrunner: Towards automatic data extraction from large Web sites. In: VLDB, Rome, Italy (September 2001)
3. He, B., Patel, M., Zhang, Z., Chang, K.C.: Accessing the deep Web: A survey. Communications of the ACM 50(2), 94–101 (2007)
4. Madhavan, J., Halevy, A.Y., Cohen, S., Dong, X., Jeffery, S.R., Ko, D., Yu, C.: Structured data meets the Web: A few observations. IEEE Data Engineering Bullerin 29(4), 19–26 (2006)
5. Senellart, P., Mittal, A., Muschick, D., Gilleron, R.: andTommasi, M., Automatic Wrapper Induction from Hidden-Web Sources with Domain Knowledge. In: WIDM, Napa, USA, pp. 9–16 (October 2008)
6. Auer, S., Bizer, C., Kobilarov, G., Lehmann, J., Cyganiak, R., Ives, Z.G.: DBpedia: A nucleus for a web of open data. In: Aberer, K., Choi, K.-S., Noy, N., Allemang, D., Lee, K.-I., Nixon, L., Golbeck, J., Mika, P., Maynard, D., Mizoguchi, R., Schreiber, G., Cudré-Mauroux, P. (eds.) ASWC 2007 and ISWC 2007. LNCS, vol. 4825, pp. 722–735. Springer, Heidelberg (2007)
7. Lenat, D., Guha, R.V.: Building Large Knowledge Based Systems: Representation and Inference in the Cyc Project. Addison-Wesley, Reading (1989)
8. Staab, S., Studer, R. (eds.): Handbook on Ontologies, 2nd edn. Springer, Heidelberg (2008)
9. Suchanek, F.M., Kasneci, G., Weikum, G.: YAGO: A Core of Semantic Knowledge. In: WWW 2007 (2007)
10. Word Wide Web Consortium. OWL Web Ontology Language (W3C Recommendation 2004-02-10), `http://www.w3.org/TR/owl-features/`
11. Li, H., Shan, F., Lee, S.Y.: Online mining of frequent query trees over XML data streams. In: 15th international conference on World Wide Web, Edinburgh, Scotland, pp. 959–960. ACM Press, New York (2008)
12. Kutty, S., Nayak, R.: Frequent Pattern Mining on XML documents. In: Song, M., Wu, Y.-F. (eds.) Handbook of Research on Text and Web Mining Technologies, pp. 227–248. Idea Group Inc., USA (2008)
13. Nayak, R.: Fast and Effective Clustering of XML Data Utilizing their Structural Information. Knowledge and Information Systems (KAIS) 14(2), 197–215 (2008)
14. Rusu, L.I., Rahayu, W., Taniar, D.: Mining Association Rules from XML Documents. In: Vakali, A., Pallis, G. (eds.) Web Data Management Practices (2007)
15. Wan, J.: Mining Association rules from XML data mining query. Research and practice in Information Technology 32, 169–174 (2004)

16. Boag, S., Fernandez, M., Florescu, D., Robie, J., Simeon, J.: XQuery 1.0: An XML Query Language. In: W3C Working Draft (November 2003)
17. Clark, J., DeRose, S.: XML Path Language (XPath) Version 1.0. W3C Recommendation (November 1999)
18. Davidson, S., Fan, W., Hara, C., Qin, J.: Propagating XML Constraints to Relations. In: International Conference on Data Engineering (March 2003)
19. Guo, J., Araki, K., Tanaka, K., Sato, J., Suzuki, M., Takada, A., Suzuki, T., Nakashima, Y., Yoshihara, H.: The Latest MML (Medical Markup Language) —XML based Standard for Medical Data Exchange / Storage. Journal of Medical Systems 27(4), 357–366 (2003)
20. Varde, A., Rundensteiner, E., Fahrenholz, S.: XML Based Markup Languages for Specific Domains. In: Web Based Support Systems. Springer, Heidelberg (2008)

(Please note that the references have been cited as they are used in the respective sections, i.e., references 1 through 5 are for Section 2, references 6 through 10 are for Section 3, references 11 through 15 are for Section 4 and references 16 through 20 are for Section 5.)

Top-k Algorithms and Applications

Gautam Das

Department of Computer Science and Engineering
University of Texas at Arlington
Arlington, TX, USA 75063
gdas@uta.edu

Abstract. In recent years, there has been a great deal of interest in developing effective techniques for ad-hoc search and retrieval in relational databases, document and multimedia databases, scientific information systems, and so on. A popular paradigm for tackling this problem is top-*k* querying, i.e., the ranking of the results and returning the *k* results with the highest scores. Numerous variants of the top-k retrieval problem and several algorithms have been introduced in recent years. In this tutorial we shall discuss the top-*k* problem in detail, especially the fundamental algorithms such as FA and TA, important variants such as algorithms operating under restricted sorted/random access, deterministic and probabilistic approximations, as well as distributed and streaming top-*k* computations. A significant portion of the tutorial will be focused on applications of these top-*k* algorithms, especially in the context of the Web services and online databases, multimedia, documents and relational databases.

Keywords: relational databases, top-*k* algorithms, search engines, keyword queries.

1 Introduction (5 minutes)

This portion of the lecture will introduce the top-*k* problem, give a precise problem definition, then give a high-level overview of the motivation for the problem, different problem versions, and the various important applications of top-*k* querying and retrieval. The problem will be motivated via examples of ranking/scoring functions for structured data, documents/IR repositories, *k*NN queries in Euclidean/metric space settings, ranking functions on the Web, and so on.

2 Fundamental Top-k Algorithms (20 minutes)

Basic Top-k Algorithm: This portion will be devoted to describing the simple algorithm based on a linear scan of the data repository. The advantages of this algorithm are that it is simple to implement, requires no preprocessing, maintains a bounded buffer of size k, and is usually quite fast unless the repository is large.

FA and TA Algorithms for Top-k Queries: This portion will introduce the concept of monotonic ranking functions, followed by the two fundamental top-*k* algorithms:

X. Zhou et al. (Eds.): DASFAA 2009, LNCS 5463, pp. 789–792, 2009.

Fagin's Algorithm (FA) and Threshold Algorithm (TA). The history as well as details of both algorithms will be discussed, followed by a theoretical analysis of the algorithms (e.g., instance optimality in the case of TA). Implementation issues of TA will be discussed, including a discussion of its practicality.

3 Algorithms for Important Problem Variants (30 minutes)

This portion of the seminar will be dedicated to discussing most of the important variants of the top-k problem. In particular, the following variants and their algorithms will be discussed

Threshold Algorithm with no (or limited) Sorted/Random Access: There are several applications (e.g., merging of information for multiple web services) where the standard access methods for top-*k* algorithms, i.e., sorted access and random access, are either unavailable, or are extremely expensive. Several important algorithmic variants have been developed for these problem variants, e.g., TA-SORTED, NRA, TA-ADAPT, STREAM-COMBINE and so on, and this portion of the seminar will be dedicated to discussing these variants.

Adaptive Algorithms: Another set of algorithms that are assuming increasing importance are top-*k* algorithms that adapt the rate at which they consume data from each sorted stream. Algorithms such as QUICK COMBINE shall be discussed in this context.

Approximation Algorithms: An important class of top-*k* algorithms is approximation algorithms. Such algorithms have the ability of stopping early and producing a set of results that are close to true top-*k* results. Measures of approximation will be discussed, e.g., the usual IR measures of precision and recall, as well as distance measures between ranked lists such as Kendall Tau and Spearman's footrule measures. Several variations of top-*k* algorithms that stop early yet approximate the true top-*k* results shall be discussed.

Probabilistic Algorithms: Instead of deterministic approximation guarantees, an important class of top-*k* algorithms are probabilistic in nature, i.e., they stop early by aggressively pruning the set of top-*k* candidates, but in the process are able to give probabilistic guarantees for the objects that were rejected from consideration. We shall discuss the work by Theobald et al (VLDB 2004) as an example of such an algorithm.

Algorithms using Views: Like traditional query processing, top-*k* query processing can also benefit from the existence of materialized results of previously executed top-*k* queries. There have been several efforts in this context, e.g., the PREFER algorithm (SIGMOD01) as well as the recent LPTA algorithm (linear programming based TA, VLDB06), and we shall briefly discuss these systems here.

Distributed Top-k Computations: Top-*k* querying over distributed databases is an important problem especially in the context of information retrieval of document databases distributed over P2P networks. We shall discuss the KLEE algorithm that has been recently proposed to implement top-*k* computations "in network" efficiently.

Continuous Monitoring of Top-k Queries over Data Streams: Continuous monitoring of top-k queries when the underlying data repository is changing is an important practical problem, e.g., in publish/subscribe systems. We shall discuss recent work on this problem (SIGMOD 06).

4 Applications (30 minutes)

This portion of the seminar will focus on discussing practical applications of top-*k* algorithms, and the issues that arise in their efficient implementation.

Multimedia Databases: The original applications for the FA and TA algorithms were envisioned for multimedia database retrieval. We shall discuss the efforts to develop practical systems for such applications, e.g., IBM's middleware solution.

Ranking in Information Retrieval/Document databases: Classical Information Retrieval relies on merging/intersection of inverted lists for ranking and retrieval of the top documents for a query. We shall discuss the parallels between such inverted list-based approaches and the approaches based on sorted/random access operations.

Top-k Queries in Graph Databases: Many data repositories exist in graph form. For example, the web is a graph, as are XML and semi-structured databases. Even relational databases may be represented as graphs if we view foreign-key joins as connections (i.e., edges) between tuples (i.e., nodes). There has been a series of work on keyword querying of such graph-structured databases, and the ranking and retrieval of the top-*k* sub-trees that satisfy these keyword queries. We shall discuss several of these systems, including DBXPLORER, DISCOVER, and in particular focus on how list merging techniques are efficiently used for top-*k* retrieval.

Top-k Algorithms in Relational Database Applications: Ranking of tuples in traditional relational databases has been assuming increasing importance. The need for automatic ranking and top-*k* retrieval arises in numerous applications, e.g., searching of online product catalogs such as homes, cars, etc. We shall describe implementations of several list merge-based algorithms for database ranking, e.g., the CIDR03 (tf-idf based) and VLDB04 (Probabilistic IR-based) papers. We shall describe how random (resp. sorted) access are simulated via index lookup (resp. index scans) in database applications.

A parallel effort focuses on how to implement ranking/top-*k* operators inside database query processing engines. This research involves issues such as the challenges of implementing TA inside database engines, developing costing functions for such operators, and so on. We shall discuss related work done by Ihab Ilyas et al, as well as by Gerhard Weikum (CIDR 2005 survey).

5 Conclusions and Challenges (5 minutes)

We shall conclude by emphasizing that top-*k* query processing is an important component of information systems, especially in ad-hoc search and retrieval applications. While there has been a flurry of activity in this area in recent years, there is tremendous amount of work still left to be done, and we hope that this seminar invigorates the enthusiasm of the audience in this direction.

6 Description of Target Audience

The anticipated audience will consist of database, IR, data mining, and algorithms researchers, web services developers, information systems designers, and database engine designers and developers.

7 Biographical Sketch

Gautam Das is an Associate Professor and Head of the *Database Exploration Laboratory* (DBXLAB) at the CSE department of the *University of Texas at Arlington*. Prior to joining UTA in Fall 2004, Dr. Das has held positions at Microsoft Research, Compaq Corporation and the University of Memphis. He graduated with a B.Tech in computer science from IIT Kanpur, India, and with a PhD in computer science from the University of Wisconsin, Madison. Dr. Das's research interests span data mining, information retrieval, databases, algorithms and computational geometry. He is currently interested in ranking, top-k query processing, and sampling problems in databases, as well as data management problems in P2P and sensor networks, social networks, blogs and web communities. His research has resulted in over 85 papers, many of which have appeared in premier database, data mining and algorithms conferences and journals. Dr. Das has served as Program Co-Chair of COMAD 2008, CIT 2004 and SIGMOD-DMKD 2004, Best Paper Awards Chair of KDD 2006, as well as in program committees of numerous conferences. He has served as a Guest Editor for the ACM TKDD special issue devoted to the best papers of KDD 2006. Dr. Das's research has been supported by grants from National Science Foundation, Office of Naval Research, Microsoft Research, Nokia Research, Cadence Design Systems and Apollo Data Technologies.

Information Services: Myth or Silver Bullet?

Dimitrios Georgakopoulos

CISRO ICT Centre, Australia
dimitrios.georgakopoulos@csiro.au

PANELISTS
Elisa Bertino, Purdue University, USA
bertino@cs.purdue.edu
Alistair Barros, SAP Australia, Australia
alistair.barros@sap.com
Ryszard Kowalczyk, Swinburne University, Australia
RKowalczyk@groupwise.swin.edu.au

In the Web we are drowning in data, but we are starving for useful information. Information services aim to deal with this problem by providing domain- and application-specific organization, search, and presentation of data in the web. Although information services provide for the management of data, their main objectives are to:

- promote the *sharing* of information and related knowledge in the Web
- provide *customized/tailored information search and delivery* that addresses the needs of specific user communities

There many examples of information services that are currently available in the Web. The panel discussed sample information services for various science-related disciplines, including Biochemistry (e.g., GenBank, http://www.ncbi.nlm.nih.gov/Genbank/ GenbankOverview.html; fMRIDC, http://www.fmridc.org/f/fmridc; Flytrap, http://www.fly-trap.org; and the Protein Data Bank,http://www.pdb.org/pdb/home/home.do), and Astronomy (NASA's SkyView, http://skyview.gsfc.nasa.gov/; Astronomical Image Library version 2.0, http://www.astronomy.ca/images/; OMNIweb, http://omniweb.gsfc.nasa.gov/ow.html; EU's ESO Science Archive Facility, http://archive.eso.org/cms/; and the Sloan Digital Sky Server, http://cas.sdss.org/astro/en/).

Common aspect of such information services include:

- Images are used to facilitate information search
- Queries and results often involve a mixture of data, images, and live or synthesized video
- Separate interfaces for experts and general web users are often provided
- Data schema employs some form of semantic annotation or references to an external ontology
- Emphasis on information retrieval, few use SQL
- Interface based on web and web service standards

X. Zhou et al. (Eds.): DASFAA 2009, LNCS 5463, pp. 793–794, 2009.

The panel discussed issues that included the following:

- The current and future role of database management systems in the development of information services, and in particular in the service schema and data semantics, query language, query interface, and internal functionality.
- The extend service science, SOA, and web service technologies are ready to support information services, in areas such as service discovery, composition, service query and web service interfaces.
- Trust and security in information services.
- Personalization/customization of information services for diverse user communities ranging from subject matter experts (e.g., astronomers, biochemists, etc.) to casual web users.

Author Index

GPSR Compliance

The European Union's (EU) General Product Safety Regulation (GPSR) is a set of rules that requires consumer products to be safe and our obligations to ensure this.

If you have any concerns about our products, you can contact us on ProductSafety@springernature.com

In case Publisher is established outside the EU, the EU authorized representative is:

Springer Nature Customer Service Center GmbH
Europaplatz 3
69115 Heidelberg, Germany

Batch number: 10844661

Printed by Printforce, the Netherlands